PATTY'S INDUSTRIAL HYGIENE AND TOXICOLOGY

Fourth Edition

Volume I, Parts A and B
GENERAL PRINCIPLES

Volume II, Parts A, B, C, D, E, and F
TOXICOLOGY

Third Edition

Volume III, Parts A and B
THEORY AND RATIONALE
OF INDUSTRIAL HYGIENE
PRACTICE

PATTY'S INDUSTRIAL HYGIENE AND TOXICOLOGY

Fourth Edition
Volume II, Part A
Toxicology

GEORGE D. CLAYTON
FLORENCE E. CLAYTON
Editors

CONTRIBUTORS

D. M. Aviado	T. H. Gardiner	R. J. Papciak
R. B. Balodis	C. R. Green	J. O. Pierce
E. M. Beliles	R. K. Hinderer	C. O. Schulz
R. P. Beliles	J. L. Holtshouser	E. J. Sowinski
M. J. Brabec	S. R. Howe	D. E. Stevenson
F. L. Cavender	C. J. Kirwin, Jr.	M. A. Thomson
R. A. Davis	T. D. Kuhn	J. M. Waechter, Jr.
J. Doull	H.-W. Leung	R. J. Weir
C. Farr	D. A. Morgott	J. A. Zapp, Jr.
J. B. Galvin		

A Wiley-Interscience Publication
JOHN WILEY & SONS, INC.
New York / Chichester / Brisbane / Toronto / Singapore

Library of Congress Cataloging in Publication Data:
Patty's industrial hygiene and toxicology.

   "A Wiley-Interscience publication."
   Rev. ed. of: Industrial hygiene and toxicology /
F. A. Patty.
   Includes bibliographical references and index.
   Contents: v. 1. General principles (2 v.)—v. 2
Toxicology pt. A.
   1. Industrial hygiene.   2. Industrial toxicology.
I. Clayton, George D.   II. Clayton, Florence E.
III. Allan, R. E. (Ralph E.)   IV. Patty, F. A. (Frank
Arthur), 1897–1981 Industrial hygiene and toxicology.
RC967.P37 1991   613.6'2   90-13080
ISBN 0-471-54724-7 (v. 2, pt. A)

Printed in the United States of America

10 9 8 7 6 5 4 3 2 1

Volume II of this series is dedicated to Paul D. Halley, past president of the American Industrial Hygiene Association, former Director of Environmental Health Services for AMOCO, (Industrial Hygiene, Safety and Toxicology), and Editor-in-Chief for eight years of the *AIHA Journal*, who, after retirement from the latter position, initiated contacts for the fourth revision of this volume, but was forced to relinquish its editorship because of illness.

Contributors

Domingo M. Aviado, M.D., Atmospheric Health Sciences, Short Hills, New Jersey

Rasma B. Balodis, J.D., R. T. Vanderbilt Company, Inc., Norwalk, Connecticut

Eloise M. Beliles, 3002 Phyllmar Place, Oakton, Virginia

Robert P. Beliles, Ph.D., D.A.B.T., 3002 Phyllmar Place, Oakton, Virginia

Michael J. Brabec, Ph.D., D.A.B.T., Eastern Michigan University, Department of Chemistry, Ypsilanti, Michigan

Finis L. Cavender, Ph.D., D.A.B.T., C.I.H., Abilene Christian University, Abilene, Texas

Richard A. Davis, Ph.D., D.A.B.T., American Cyanamid Co., Agricultural Research Division, Princeton, New Jersey

John Doull, M.D., Ph.D., Medical Center, University of Kansas, Kansas City, Kansas

Craig Farr, Ph.D., D.A.B.T., ATOCHEM, No. American, Inc., King of Prussia, Pennsylvania

Jennifer B. Galvin, Ph.D., D.A.B.T., Phillips Petroleum Company, Bartlesville, Oklahoma

Thomas H. Gardiner, Ph.D., 11502 Glenora Rd., Houston, Texas

Charles R. Green, Ph.D., Witco Corporation, Houston, Texas

Robert K. Hinderer, Ph.D., The B. F. Goodrich Company, Brecksville, Ohio

Joseph L. Holtshouser, C.I.H., C.S.P. Goodyear Tire & Rubber Company, Akron, Ohio

Susan R. Howe, Chemical Manufacturers Association, Washington, D.C.

Carroll J. Kirwin, Jr., D.A.B.T., Phillips Petroleum Company, Bartlesville, Oklahoma

T. Dee Kuhn, Chemical Manufacturers Association, Washington, D.C.

Hon-Wing Leung, Ph.D., D.A.B.T., C.I.H., Union Carbide Corporation, Danbury, Connecticut

David A. Morgott, Ph.D., D.A.B.T., C.I.H., Corporate Health & Environment Laboratories, Eastman Kodak Company, Rochester, New York

Raymond J. Papciak, Texaco Chemical Company, Houston, Texas

James O. Pierce, Sc.D., Professor, University of Southern California, Safety Sciences Department, Los Angeles, California

Carl O. Schulz, Ph.D., D.A.B.T., Department of Environmental Health Sciences, School of Public Health, University of South Carolina, Columbia, South Carolina

Edward J. Sowinski, Ph.D., C.I.H., D.A.B.T., EHMS, Inc., Hudson, Ohio

Donald E. Stevenson, Ph.D., Shell Oil Company, Houston, Texas

Mark A. Thomson, Ph.D., Uniroyal Chemical Co., Inc., Middlebury, Connecticut

John M. Waechter, Jr., Ph.D., D.A.B.T., The Dow Chemical Co., Health and Environmental Sciences, Midland, Michigan

Robert J. Weir, Ph.D., 1606 Simmons Court, McLean, Virginia

John A. Zapp, Jr., Ph.D. (deceased), (formerly Director, Haskell Laboratory for Toxicology and Industrial Medicine, E. I. du Pont de Nemours and Company, Wilmington, Delaware)

Preface

More than a decade has elapsed since the third revision of Volume II on Toxicology, during which time more new products have been developed, research has increased as laws regulating emissions have proliferated affecting all segments of society, and accelerated public awareness has dramatically affected the use of certain substances.

Another development during this period has been the sophistication of data and record keeping through computerization, resulting in a wealth of data, not previously easily accessible to researchers, now available for their evaluation.

Toxicology is defined as the scientific study of poisons, their actions, their detection, and the treatment of conditions produced by them. Theophrastus von Hohenheim, a Swiss physician (1493–1541), better known as "Paracelsus" stated that *all* substances are poisons, there is none that is not a poison, and it is the right dose that differentiates a poison from a remedy. It is the role of the toxicologist to specify the safe use of substances. The input of industrial toxicologists in developing threshold limit values has been preeminent during the past 50 years.

Some individuals advocate zero concentrations for all chemicals. They do not recognize that the largest source of chemicals dispersed to the atmosphere eminates from nature itself. Wind and moving objects can, and do at times, create copious amounts of dust. For example, the recent volcanic eruptions discharged into the atmosphere more contaminants than all of industry over a select period of time. For eons of years this has periodically occurred. During these naturally occurring phenomena, mankind and animals have not only continued to exist, but have multiplied and developed. In the twentieth century, during the increase of industrial activity, pollutants increased simultaneously. However, the life span of the human race also increased. It is therefore obvious that people can survive in less than a chemical-free environment.

It is the industrial toxicologist who studies to what degree individuals can be exposed without harmful effects, taking into consideration the admonition of Paracelsus that all substances are poisons—that it is the *right dose* that differentiates a poison from a remedy. This concept is the foundation for threshold limit values.

In previous revisions of Volume II, the chapters have followed consecutively those of Volume I. In this fourth revision, the decision was made to begin Volume II with Chapter 1. This revision, as was the case with previous revisions, has been expanded, with the addition of three books. We would have preferred to present the chapters by classes of chemicals, but, as in the past, this ideal arrangement was not possible. In the best interest of the authors, generally those whose manuscripts were completed first are contained in Part A, followed by Part B, and so on.

The authors of Volume IIA are outstanding scientists in their respective fields, and we are indebted to them for condensing and interpreting the huge volumes of data on various chemicals in a manner that is concise and useful to the reader. It is interesting to note that one of the authors, Dr. Robert Weir, has been a contributor to the Patty series since 1963.

With the widespread universal concern regarding the effect of chemicals on health, including those in the home, in office buildings, in industry, and by the general public, the interest in this series should be greater than previously. The authors have made sincere efforts to make these data "user-friendly" to all segments of society.

GEORGE D. CLAYTON
FLORENCE E. CLAYTON

San Luis Rey, California
August 1993

Contents

USEFUL EQUIVALENTS AND CONVERSION FACTORS

1 kilometer = 0.6214 mile
1 meter = 3.281 feet
1 centimeter = 0.3937 inch
1 micrometer = 1/25,4000 inch = 40 microinches
    = 10,000 Angstrom units
1 foot = 30.48 centimeters
1 inch = 25.40 millimeters
1 square kilometer = 0.3861 square mile (U.S.)
1 square foot = 0.0929 square meter
1 square inch = 6.452 square centimeters
1 square mile (U.S.) = 2,589,998 square meters
    = 640 acres
1 acre = 43,560 square feet = 4047 square
    meters
1 cubic meter = 35.315 cubic feet
1 cubic centimeter = 0.0610 cubic inch
1 cubic foot = 28.32 liters = 0.0283 cubic meter
    = 7.481 gallons (U.S.)
1 cubic inch = 16.39 cubic centimeters
1 U.S. gallon = 3.7853 liters = 231 cubic inches
    = 0.13368 cubic foot
1 liter = 0.9081 quart (dry), 1.057 quarts (U.S.,
    liquid)
1 cubic foot of water = 62.43 pounds (4°C)
1 U.S. gallon of water = 8.345 pounds (4°C)
1 kilogram = 2.205 pounds

1 gram = 15.43 grains
1 pound = 453.59 grams
1 ounce (avoir.) = 28.35 grams
1 gram mole of a perfect gas ≎ 24.45 liters (at
    25°C and 760 mm Hg barometric pressure)
1 atmosphere = 14.7 pounds per square inch
1 foot of water pressure = 0.4335 pound per
    square inch
1 inch of mercury pressure = 0.4912 pound per
    square inch
1 dyne per square centimeter = 0.0021 pound
    per square foot
1 gram-calorie = 0.00397 Btu
1 Btu = 778 foot-pounds
1 Btu per minute = 12.96 foot-pounds per
    second
1 hp = 0.707 Btu per second = 550 foot-pounds
    per second
1 centimeter per second = 1.97 feet per minute
    = 0.0224 mile per hour
1 footcandle = 1 lumen incident per square foot
    = 10.764 lumens incident per square meter
1 grain per cubic foot = 2.29 grams per cubic
    meter
1 milligram per cubic meter = 0.000437 grain per
    cubic foot

To convert degrees Celsius to degrees Fahrenheit: °C (9/5) + 32 = °F
To convert degrees Fahrenheit to degrees Celsius: (5/9) (°F − 32) = °C
For solutes in water: 1 mg/liter ≎ 1 ppm (by weight)
Atmospheric contamination: 1 mg/liter ≎ 1 oz/1000 cu ft (approx)
For gases or vapors in air at 25°C and 760 mm Hg pressure:
    To convert mg/liter to ppm (by volume): mg/liter (24,450/mol. wt.) = ppm
    To convert ppm to mg/liter: ppm (mol. wt./24,450) = mg/liter

CONVERSION TABLE FOR GASES AND VAPORS[a]
(Milligrams per liter to parts per million, and vice versa; 25°C and 760 mm Hg barometric pressure)

Molecular Weight	1 mg/liter ppm	1 ppm mg/liter	Molecular Weight	1 mg/liter ppm	1 ppm mg/liter	Molecular Weight	1 mg/liter ppm	1 ppm mg/liter
1	24,450	0.0000409	39	627	0.001595	77	318	0.00315
2	12,230	0.0000818	40	611	0.001636	78	313	0.00319
3	8,150	0.0001227	41	596	0.001677	79	309	0.00323
4	6,113	0.0001636	42	582	0.001718	80	306	0.00327
5	4,890	0.0002045	43	569	0.001759	81	302	0.00331
6	4,075	0.0002454	44	556	0.001800	82	298	0.00335
7	3,493	0.0002863	45	543	0.001840	83	295	0.00339
8	3,056	0.000327	46	532	0.001881	84	291	0.00344
9	2,717	0.000368	47	520	0.001922	85	288	0.00348
10	2,445	0.000409	48	509	0.001963	86	284	0.00352
11	2,223	0.000450	49	499	0.002004	87	281	0.00356
12	2,038	0.000491	50	489	0.002045	88	278	0.00360
13	1,881	0.000532	51	479	0.002086	89	275	0.00364
14	1,746	0.000573	52	470	0.002127	90	272	0.00368
15	1,630	0.000614	53	461	0.002168	91	269	0.00372
16	1,528	0.000654	54	453	0.002209	92	266	0.00376
17	1,438	0.000695	55	445	0.002250	93	263	0.00380
18	1,358	0.000736	56	437	0.002290	94	260	0.00384
19	1,287	0.000777	57	429	0.002331	95	257	0.00389
20	1,223	0.000818	58	422	0.002372	96	255	0.00393
21	1,164	0.000859	59	414	0.002413	97	252	0.00397
22	1,111	0.000900	60	408	0.002554	98	249.5	0.00401
23	1,063	0.000941	61	401	0.002495	99	247.0	0.00405
24	1,019	0.000982	62	394	0.00254	100	244.5	0.00409
25	978	0.001022	63	388	0.00258	101	242.1	0.00413
26	940	0.001063	64	382	0.00262	102	239.7	0.00417
27	906	0.001104	65	376	0.00266	103	237.4	0.00421
28	873	0.001145	66	370	0.00270	104	235.1	0.00425
29	843	0.001186	67	365	0.00274	105	232.9	0.00429
30	815	0.001227	68	360	0.00278	106	230.7	0.00434
31	789	0.001268	69	354	0.00282	107	228.5	0.00438
32	764	0.001309	70	349	0.00286	108	226.4	0.00442
33	741	0.001350	71	344	0.00290	109	224.3	0.00446
34	719	0.001391	72	340	0.00294	110	222.3	0.00450
35	699	0.001432	73	335	0.00299	111	220.3	0.00454
36	679	0.001472	74	330	0.00303	112	218.3	0.00458
37	661	0.001513	75	326	0.00307	113	216.4	0.00462
38	643	0.001554	76	322	0.00311	114	214.5	0.00466

CONVERSION TABLE FOR GASES AND VAPORS (*Continued*)
(*Milligrams per liter to parts per million, and vice versa; 25°C and 760 mm Hg barometric pressure*)

Molecular Weight	1 mg/liter ppm	1 ppm mg/liter	Molecular Weight	1 mg/liter ppm	1 ppm mg/liter	Molecular Weight	1 mg/liter ppm	1 ppm mg/liter
115	212.6	0.00470	153	159.8	0.00626	191	128.0	0.00781
116	210.8	0.00474	154	158.8	0.00630	192	127.3	0.00785
117	209.0	0.00479	155	157.7	0.00634	193	126.7	0.00789
118	207.2	0.00483	156	156.7	0.00638	194	126.0	0.00793
119	205.5	0.00487	157	155.7	0.00642	195	125.4	0.00798
120	203.8	0.00491	158	154.7	0.00646	196	124.7	0.00802
121	202.1	0.00495	159	153.7	0.00650	197	124.1	0.00806
122	200.4	0.00499	160	152.8	0.00654	198	123.5	0.00810
123	198.8	0.00503	161	151.9	0.00658	199	122.9	0.00814
124	197.2	0.00507	162	150.9	0.00663	200	122.3	0.00818
125	195.6	0.00511	163	150.0	0.00667	201	121.6	0.00822
126	194.0	0.00515	164	149.1	0.00671	202	121.0	0.00826
127	192.5	0.00519	165	148.2	0.00675	203	120.4	0.00830
128	191.0	0.00524	166	147.3	0.00679	204	119.9	0.00834
129	189.5	0.00528	167	146.4	0.00683	205	119.3	0.00838
130	188.1	0.00532	168	145.5	0.00687	206	118.7	0.00843
131	186.6	0.00536	169	144.7	0.00691	207	118.1	0.00847
132	185.2	0.00540	170	143.8	0.00695	208	117.5	0.00851
133	183.8	0.00544	171	143.0	0.00699	209	117.0	0.00855
134	182.5	0.00548	172	142.2	0.00703	210	116.4	0.00859
135	181.1	0.00552	173	141.3	0.00708	211	115.9	0.00863
136	179.8	0.00556	174	140.5	0.00712	212	115.3	0.00867
137	178.5	0.00560	175	139.7	0.00716	213	114.8	0.00871
138	177.2	0.00564	176	138.9	0.00720	214	114.3	0.00875
139	175.9	0.00569	177	138.1	0.00724	215	113.7	0.00879
140	174.6	0.00573	178	137.4	0.00728	216	113.2	0.00883
141	173.4	0.00577	179	136.6	0.00732	217	112.7	0.00888
142	172.2	0.00581	180	135.8	0.00736	218	112.2	0.00892
143	171.0	0.00585	181	135.1	0.00740	219	111.6	0.00896
144	169.8	0.00589	182	134.3	0.00744	220	111.1	0.00900
145	168.6	0.00593	183	133.6	0.00748	221	110.6	0.00904
146	167.5	0.00597	184	132.9	0.00753	222	110.1	0.00908
147	166.3	0.00601	185	132.2	0.00757	223	109.6	0.00912
148	165.2	0.00605	186	131.5	0.00761	224	109.2	0.00916
149	164.1	0.00609	187	130.7	0.00765	225	108.7	0.00920
150	163.0	0.00613	188	130.1	0.00769	226	108.2	0.00924
151	161.9	0.00618	189	129.4	0.00773	227	107.7	0.00928
152	160.9	0.00622	190	128.7	0.00777	228	107.2	0.00933

CONVERSION TABLE FOR GASES AND VAPORS (*Continued*)

(*Milligrams per liter to parts per million, and vice versa; 25°C and 760 mm Hg barometric pressure*)

Molec-ular Weight	1 mg/liter ppm	1 ppm mg/liter	Molec-ular Weight	1 mg/liter ppm	1 ppm mg/liter	Molec-ular Weight	1 mg/liter ppm	1 ppm mg/liter
229	106.8	0.00937	253	96.6	0.01035	277	88.3	0.01133
230	106.3	0.00941	254	96.3	0.01039	278	87.9	0.01137
231	105.8	0.00945	255	95.9	0.01043	279	87.6	0.01141
232	105.4	0.00949	256	95.5	0.01047	280	87.3	0.01145
233	104.9	0.00953	257	95.1	0.01051	281	87.0	0.01149
234	104.5	0.00957	258	94.8	0.01055	282	86.7	0.01153
235	104.0	0.00961	259	94.4	0.01059	283	86.4	0.01157
236	103.6	0.00965	260	94.0	0.01063	284	86.1	0.01162
237	103.2	0.00969	261	93.7	0.01067	285	85.8	0.01166
238	102.7	0.00973	262	93.3	0.01072	286	85.5	0.01170
239	102.3	0.00978	263	93.0	0.01076	287	85.2	0.01174
240	101.9	0.00982	264	92.6	0.01080	288	84.9	0.01178
241	101.5	0.00986	265	92.3	0.01084	289	84.6	0.01182
242	101.0	0.00990	266	91.9	0.01088	290	84.3	0.01186
243	100.6	0.00994	267	91.6	0.01092	291	84.0	0.01190
244	100.2	0.00998	268	91.2	0.01096	292	83.7	0.01194
245	99.8	0.01002	269	90.9	0.01100	293	83.4	0.01198
246	99.4	0.01006	270	90.6	0.01104	294	83.2	0.01202
247	99.0	0.01010	271	90.2	0.01108	295	82.9	0.01207
248	98.6	0.01014	272	89.9	0.01112	296	82.6	0.01211
249	98.2	0.01018	273	89.6	0.01117	297	82.3	0.01215
250	97.8	0.01022	274	89.2	0.01121	298	82.0	0.01219
251	97.4	0.01027	275	88.9	0.01125	299	81.8	0.01223
252	97.0	0.01031	276	88.6	0.01129	300	81.5	0.01227

[a] A. C. Fieldner, S. H. Katz, and S. P. Kinney, "Gas Masks for Gases Met in Fighting Fires," U.S. Bureau of Mines, Technical Paper No. 248, 1921.

PATTY'S INDUSTRIAL HYGIENE AND TOXICOLOGY

Fourth Edition

Volume II, Part A
TOXICOLOGY

CHAPTER ONE

Industrial Toxicology: Retrospect and Prospect

**John A. Zapp, Jr., Ph.D. (deceased), and
John Doull, M.D., Ph.D.**

In the third revision of Volume II of *Industrial Hygiene and Toxicology* produced in 1981, this chapter was written by Dr. John Zapp, who died in 1987. In planning the fourth revision of this volume, we invited Dr. John Doull to update the chapter. After giving considerable thought to it, Dr. Doull advised us that he recommended leaving much of the chapter as it was written by Dr. Zapp, because it has an elegant history of toxicology. We agreed with Dr. Doull. He has updated where necessary, and has added a section on the role of toxicology in setting exposure limits, including a current description of both the threshold and nonthreshold approach, which is basic in the regulatory process.

THE EDITORS

Industrial toxicology is a comparatively recent discipline, but its roots are shadowed in the mists of time. The beginnings of toxicology, the knowledge or science of poisons, are prehistoric. Earliest humans found themselves in environments that were at the same time helpful and hostile to their survival. They found their food among the plants, trees, animals, and fishes in their immediate surroundings, their clothing in the skins of animals, and their shelter mainly in caves. Their earliest tools and weapons were of wood and stone.

Patty's Industrial Hygiene and Toxicology, Fourth Edition, Volume 2, Part A, Edited by George D. Clayton and Florence E. Clayton.
ISBN 0-471-54724-7 © 1993 John Wiley & Sons, Inc.

1   THE BEGINNINGS OF TOXICOLOGY

It was in the very early period of prehistory that humans must have become aware of the phenomenon of toxicity. Some fruits, berries, and vegetation could be eaten with safety and to their benefit, whereas others caused illness or even death. The bite of the asp or adder could be fatal, whereas the bite of many other snakes was not. Humans learned from experience to classify things into categories of safe and harmful. Personal survival depended on recognition and avoidance, so far as possible, of the dangerous categories.

In a unique difference from other animals, humans learned to construct tools and weapons that facilitated their survival. Stone and wood gave way in time to bronze and then to iron as materials for the construction of these tools and weapons. The invention of the bow and arrow was a giant step forward in weaponry, for it gave humans a chance to kill animals or other people from a safe distance. And humans soon used their knowledge of the poisonous materials they found in their natural environment to enhance the lethality of their weapons.

One of the earliest examples of the deliberate use of poisons in weaponry was the smearing of arrowheads and spearpoints with poisons to improve their lethal effectiveness. In the Old Testament we find at Job 6:4, "The arrows of the Almighty find their mark in me, and their poison soaks into my spirit" (The New English Bible version). The Book of Job is generally dated about 400 B.C.

L. G. Stevenson (1) cites the Presidential Address of one F. H. Edgeworth before the Bristol Medico-Chirurgical Society in 1916, to the effect that Odysseus is credited in Homer's *Odyssey* with obtaining a man-killing poison from Anchialos, king of the Taphians, to smear on his bronze-tipped arrows. This particular passage does not occur in modern translations of the *Odyssey* and, according to Edgeworth, was probably expurgated from the text when Greece came under the domination of Athens, at which time the use of poisons on weapons was considered barbaric and not worthy of such a hero as Odysseus.

Because the earliest literature reference to Homer is dated at 660 B.C., well before the Pan-Athenian period, an early origin of the use of poisoned arrows can be assumed. Indeed, the word "toxic" derives from the early Greek use of poisoned arrows.

The Greek word for the bow was *toxon* and for a drug was *pharmakon*. An arrow poison was, therefore called *toxikon pharmakon*, or drug pertaining to the bow. Many Latin words are derived from the Greek, but the Romans took only the first of the two Greek works as their equivalent of "poison," that is, *toxicum*. Other Latin words for poison were *venenum* and *virus*. In the transition to English, *toxicum* became "toxin," and the knowledge or science of toxins become "toxicology."

There were practicing toxicologists in Greece and Rome. Stevenson (1) refers to a book by Sir T. C. Albutt (2) according to which the professional toxicologists of Greece and Rome were purveyors of poisons, and dealt in three kinds: those that acted quickly, those that caused a lingering illness, and those that had to be given repeatedly to produce a cumulative effect. These poisons were of vegetable or animal origin, with the exception of arsenic. Although the toxicity of lead was described by Hippocrates, and of mercury by Pliny the Elder, these metals were apparently not deliberately employed as poisons before the Renaissance.

There is little doubt that the customers of the early toxicologists were interested in assassination or suicide. Poisons offered a safer means, for the assassin, of disposing of an enemy than the more visible alternatives that posed the risk of premature discovery and possibly effective retaliation. As a means of suicide, poison often seemed more acceptable than other available means of self-destruction. Although poisons have continued to be used for both homicide and suicide, their popularity for these purposes has decreased as the popularity of firearms has increased.

The use of poisons as adjuncts to other weapons such as the spear or arrow appears to have ceased in western Europe long before the discovery of firearms. It has persisted to this day in primitive civilizations such as those of the African pygmies and certain tribes of South American Indians. The use of poison on a large scale as a primary weapon of war occurred during World War I, when poison gases were employed by both sides. In the interval between World War I and World War II, the potential of chemical and biological agents as a means of coercion was thoroughly studied by most of the powers, and both sides were prepared to use them, if necessary, in World War II. Although their use in future wars has apparently been renounced, it should not be forgotten that the chemical and biologic toxins remain viable means of coercion that could be utilized under appropriate circumstances in future conflicts. It would not be prudent to forget this in thinking in terms of national defense.

The early and sinister uses of poisons did result in contributions to toxicology. Furthermore, the knowledge obtained did not require extrapolation to the human species, for humans were the subject in early experimentation.

As mentioned above, the professional toxicologists of Greece and Rome had recognized and dealt with poisons that produced acute effects, those that produced lingering effects, and those that produced cumulative effects. We recognize these categories today. The "dose–effects" relationship was also recognized. In Plato's well-known description of the execution of Socrates (3), Socrates is required to drink a cup of hemlock, an extract of a parsley-like plant bearing a high concentration of the alkaloid coniine. When Socrates asks whether it is permissible to pour out a libation first to any god, the jailor replies, "We only prepare, Socrates, just as much as we deem enough."

The ancients also had some concept of the development of tolerance to poisons. There have come down through the ages the poison damsel stories. In one of these, related by Stevenson (1), a king of India sent a beautiful damsel to Alexander the Great because he guessed, rightly, that Alexander was about to invade his kingdom. The damsel had been reared among poisonous snakes and had become so saturated with their venom that all of her secretions were deadly. It is said that Aristotle dissuaded Alexander from doing what seemed natural under the circumstances until Aristotle performed a certain test. The test consisted in painting a circle on the floor around the girl with an extract of dittany, believed to be a powerful snake poison. When the circle was completed, the girl is said to have collapsed and died.

The poison damsel stories continued to appear from time to time, and even Nathaniel Hawthorne wrote a short story about one entitled "Rappaccini's Daughter."

Kings and other important personages, fearing assassinations, tried sometimes to protect themselves from this hazard by attempting to build up an immunity to specific poisons by taking gradually increasing doses until able to tolerate lethal doses, sometimes—it is said—with disastrous results to the queen. Other kings took the precaution of having a slave taste their food before they ate. When slaves became too scarce or expensive, they substituted a dog as the official taster, and found that it worked about as well. Perhaps we have here the birth of experimental toxicology in which a nonhuman species was deliberately used to predict human toxicity.

2 FROM AGGRESSION TO PREVENTION

2.1 The Middle Ages

Little of importance to the science of toxicology developed during the Middle Ages. Such research as was done was largely empirical, and involved the search for such things as the Philosopher's Stone, the Universal Solvent, the Elixir of Life, and the Universal Remedy. The search for the Universal Remedy is rumored to have been abandoned in the twelfth century when the alchemists learned how to make a 60% solution of ethyl alcohol through improved techniques of distillation, and found that it had some remarkable restorative properties.

2.2 The Sixteenth to the Nineteenth Centuries

Although modern science is generally held to have had its beginnings in the seventeenth century with the work of Galileo, Descartes, and Francis Bacon, there was a precursor in the sixteenth century of some importance to toxicology. This was the physician–alchemist Phillipus Aureolus Theophrastus Bombastus von Hohenheim, known as Paracelsus. Born in 1490, the son of a physician, Paracelsus studied medicine with his father, and alchemy at various universities. He was not impressed with the way that either medicine or alchemy was being taught or practiced, and decided that more could be learned from the study of Nature than by studying books by ancient authorities.

Through travel and observation, Paracelsus learned more than his contemporaries about the natural history of diseases, to the cure of which he applied his knowledge of both medicine and alchemy. He advocated that the natural substances then used as remedies be purified and concentrated by alchemical methods to enhance their potency and efficacy. He also attempted to find specific therapeutic agents for specific diseases, and became highly successful as a practicing physician; in 1526 he was appointed Town Physician to the city of Basel, Switzerland, and a lecturer in the university. Being of an egotistical and quarrelsome disposition, Paracelsus quickly antagonized the medical and academic establishment.

Syphilis in the sixteenth century was a more lethal disease than it was to become later, and the medical profession had no interest in it or cure for it. Paracelsus introduced and advocated the use of mercury for the treatment of syphilis, and it worked. The establishment, however, was outraged and denounced Paracelsus for using a poison to treat a disease. Paracelsus loved an argument and responded to this and other accusations with a series of "Defenses," of which the Third Defense (4) contained this statement with respect to his advocacy of the use of mercury or any other poison for therapeutic purposes: "What is it that is not poison? All things are poison and none without poison. Only the dose determines that a thing is not poison." Paracelsus lectured and wrote in German, which was also contrary to prevailing academic tradition. When his works were eventually translated into Latin, the last sentence of the above quotation was usually rendered, "Dosis sola facit venenum" or "The dose alone makes a poison." This principle is the keystone of industrial hygiene, and is a basic concept in toxicology.

Mercury soon became and remained the therapy of choice for syphilis for the next 300 years until Ehrlich discovered on his 606th trial an arsphenamine, Salvarsan, which was superior. Antimony was widely used as a therapeutic agent from the seventeenth to the nineteenth century, with the medical profession sharply divided as to whether it was more poison than remedy or more remedy than poison.

The period from the seventeenth to the nineteenth century witnessed little decline in the use of human subjects for the initial evaluation of remedies. In 1604, a book said to have been written by a monk named Basile Valentine, but more probably by an anonymous alchemist, was published under the title *The Triumphant Chariot of Antimony*. The book states that the author had observed that some pigs fed food containing antimony had become fat. He therefore gave antimony to some monks who had lost considerable weight through fasting, to see if it would help them to regain weight faster. Unfortunately, they all died. Up to this time, the accepted name for the element had been stibium (from which we retain the symbol, Sb), but it was renamed antimony from the words *anti—moine* meaning "monk's bane." The Oxford English Dictionary agrees that this might be the popular etymology of the word. I [John Zapp] am indebted to H. W. Haggard (5) for this anecdote.

2.3 The Nineteenth and Twentieth Centuries

Experimental toxicology as we know it followed the rise of organic chemistry, the beginning of which is usually dated around 1800. The rise was very rapid, and it is estimated that by 1880, some 12,000 compounds had been synthesized, and of these some turned out to be very toxic, in some cases proving fatal to the chemists who prepared them. Two of the war gases employed on a large scale in World War I, that is, phosgene, $COCl_2$, and mustard gas, bis(β-chloroethyl) sulfide, had been prepared in 1812 and 1822, respectively.

The early organic chemists were not deliberately looking for poisons, but for dyes, solvents, or pharmaceuticals, for example. Toxicity was an unwanted side effect, but if it was there it was there, and had to be recognized. The sheer number of new organic compounds being synthesized in the laboratory, along with a growing public disapproval of the practice of letting toxicity be discovered by its effects on humans, led to a more extensive use of convenient and available animals such as dogs, cats, or rabbits as surrogates for humans, much as some of the ancient kings used dogs instead of slaves to test their food before they dined.

Loomis (6) credits M. J. B. Orfila (7) with being the father of modern toxicology. A Spaniard by birth, Orfila studied medicine in Paris. According to Loomis:

> He is said to be the father of modern toxicology because his interests centered on the harmful effects of chemicals as well as therapy of chemical effects, and because he introduced quantitative methodology into the study of the action of chemicals on animals. He was the author of the first book devoted entirely to studies of the harmful effects of chemicals (Orfila, 1815). He was the first to point out the valuable use of chemical analyses for proof that existing symptomatology was related to the presence of the chemical in the body. He criticized and demonstrated the inefficiency of many of the antidotes that were recommended for therapy in those days. Many of his concepts regarding the treatment of poisoning by chemicals remain valid today, for he recognized the value of such procedures as artificial respiration, and he understood some of the principles involved in the elimination of the drug or chemical from the body. Like many of his immediate followers, he was concerned primarily with naturally occurring substances for which considerable folklore existed with respect to the harmfulness of such compounds.

A reading of some of the earlier nineteenth century reports indicates a lack of recognition of and concern with either intraspecies or interspecies variation. Sometimes it is not possible to determine from the report what species of animal was tested. Some reports were based on dosage of only one animal, it being assumed that all others would react similarly. In reports of inhalation toxicity, a lethal concentration might be identified without designating the length of the exposure time.

The initial experience of biologic variability came more from study of the action of drugs than from study of the action of chemicals as such. The increased interest in the action of drugs resulted from the availability of so many new organic compounds that could be explored for possible therapeutic activity.

In the second half of the nineteenth century the phenomenon of biologic variability was recognized by pharmacologists, as was also the necessity for establishing the margin of safety between a therapeutically effective dose and a toxic dose of a drug. Clinical trials of new drugs, with adequate controls, began to be accepted as good science. The traditional wisdom and beliefs about therapeutic practice were reexamined by the pharmacologists, but as Clark stated in the introduction to his 1937 monograph on general pharmacology (8),

> The energy of pharmacologists during the second half of the nineteenth century was largely expended on this task which was both wearisome and thankless. Neither the clinicians or the drug manufacturers were grateful to the pharmacologist who hampered their freer flights of fancy by captious criticism.

The thanklessness of the pharmacologist's task becomes understandable in light of the economic and social theory prevailing during the nineteenth century. Adam Smith's book, *An Inquiry into the Wealth of Nations*, published in 1776, had argued that a laissez-faire policy was the best way to assure growth of capital, increasing production and increasing prosperity, and this doctrine became the accepted belief of the Industrial Revolution. It did work well up to a point, but the increased prosperity went to the capitalists rather than to the workers.

C. T. Thackrah, M.D. (9), a pioneer in British industrial medicine writing in 1831, stated, "Most persons who reflect on the subject will be inclined to admit that our employments are to a considerable degree injurious to health. . ." and "Evils are suffered to exist, even when the means of correction are known and easily applied. Thoughtlessness or apathy is the only obstacle to success."

But laissez-faire was defended even by prominent clergymen of the time. In 1850, Archbishop Whately, in England, wrote, "More harm than good is likely to be done by almost any interference in men's money transactions, whether letting and leasing, or buying and selling of any kind" (10).

And the Reverend Horace Bushnell advised the businessman to conduct his business according to the laws of trade, "and never let his operations be mixed up with charities."

By the closing years of the nineteenth century, however, strict laissez-faire economics was being seriously questioned, and the Church was beginning to preach the Social Gospel. At the same time Science was being looked to as the great hope for bringing about Utopia, and scientific research was encouraged.

It was also in the latter part of the nineteenth century that a change in attitude occurred toward occupational disabilities. As the Industrial Revolution brought more and more men, women, and even children into work environments that were productive of occupational disabilities, and as the disabled were discharged and thrown upon public charity in strict adherence to laissez-faire policy, it began to appear reasonable to legislators in various industrialized countries that the industry responsible for occupational disabilities should bear their cost rather than the general public.

Germany led the way in 1883 with the passage of a Workingmen's Insurance Law, which set up an insurance fund into which both employers and employees paid up to 6 percent of employee earnings. For this the workers obtained free medical care, as well as some compensation during periods of disability.

Because insurance premiums depend on the payout, and because both workers and employers were paying the premiums, each now had an incentive for minimizing the payout, and each group developed an interest in applying available remedies that would reduce disabilities.

England passed a similar Workman's Compensation Law in 1897, but in the United States the first State Workman's Compensation Law was not enacted until 1910, and the most recent in 1948. The National Safety Council was established in 1911, and in 1914, the U.S. Public Health Service set up its Division of Industrial Hygiene.

In the year 1914, the first American book on occupational diseases was published

(11). It was written by W. G. Thompson, M.D., and shows a rather modern outlook as shown by the following:

> It is quite true that many of the processes of manufacture will always involve risks to health, as many trades involve risk to limb and life. One cannot handle white lead without risk of disease, just as one cannot use dynamite without risk of injury. Yet, in each case, the workman has the right of warning against the hazard, the right of such protection as modern scientific knowledge affords, and should have the right of compensation when disabled as a result of the lack of such warning and protection.

The Occupational Safety and Health Act of 1970 (OSHA) affirmed these rights, but responsible industry had moved in that direction long before OSHA. It was about the time that Dr. Thompson's book appeared that some of the larger chemical companies began hiring corporate medical directors and physicians to provide on-site medical services directed not only to therapy, but also to prevention of occupational disabilities. No law, at that time, required that they do so.

There had been no organic chemicals industry in the United States prior to World War I. It was born just after World War I because we had felt, during that war, the effects of deprivation of such useful things as aniline dyes (used for printing our stamps and currency, among other things) and pharmaceuticals (including even aspirin), which we had been importing from Germany. There was a natural desire to free ourselves from future dependence on such items, much as we have felt about oil imports since 1973. Fortunately, manpower and facilities used during the war for manufacture of munitions were available after 1918, and several companies decided to use both the manufacturing facilities and manpower to get into the organic chemicals business. And because neither employers nor workers had any previous experience in making and handling organic chemicals, the side effects of unanticipated toxicity began to be encountered. That toxicity was not wanted; it was counterproductive; it was a problem among other problems. Problems had to be managed if the industry was to survive.

To manage a problem, it must be anticipated, the causes must be identified and analyzed, and practical means of overcoming the problems must be available. As a means to this end, industrial preventive medicine, industrial toxicology, and industrial hygiene became valuable tools.

By the mid 1930s, three of the large chemical companies in the United States had established in-house laboratories of industrial toxicology. The companies were Du Pont, Dow, and Union Carbide. The purpose of these laboratories was to provide management with sufficient information about the toxicity of new chemicals to enable management to make prudent business decisions. Could the engineers design safe plants and processes for the manufacture and handling of the chemical? Could the final product be used safely for its intended use? If not, management could expect only trouble when the toxicity became manifest. No prudent management welcomes trouble.

By 1938, there were enough government-affiliated people engaged in the practice of industrial hygiene on the federal, state, and local levels to make possible the formation of the American Conference of Governmental Industrial Hygienists (ACGIH). In 1939, the American Industrial Hygiene Association (AIHA) was founded. These societies sought to bring collective knowledge and skills together in order to achieve a sound basis for all to carry out their responsibilities for recognizing, evaluating, and controlling those hazards of the workplace that cause occupational illness and disability, or even discomfort and reduction in efficiency. Above all, they believed in the possibility of controlling hazards through reduction of exposure to an acceptable level.

The *Journal of Industrial Hygiene* began publication in 1918, and the *American Industrial Hygiene Association Journal* began publication in 1939. Papers on industrial toxicology were accepted by both these journals.

The experience of the early industrial toxicologists and industrial hygienists was consistent with Paracelsus' principle that the dose alone makes a poison. Both in the toxicology laboratory and in the workplace it was observed that certain levels of exposure produced toxic effects but certain lower levels of exposure did not. Prevention of injury in the workplace could be accomplished, therefore, if exposure of workmen could be kept below a level that produced toxic effects. Industrial hygiene has operated by means of this method of control.

Following World War II, there was a marked expansion of the organic chemicals industry, particularly in the fields of organic pesticides, elastomers, and other synthetic polymers for use in textile fibers or plastic films. Food technology changed to meet increasing demands for food that kept well, was convenient to prepare, was attractively packaged, and looked, felt, and tasted good. Questions soon arose, however, about the safety of the new pesticides and food additives, and in particular about possible toxic effects from long-term low-level exposure.

By 1950, Congress was considering the necessity for amending the Food, Drug and Cosmetic Act of 1938 to meet the changed conditions, and the Food and Drug Administration (FDA) had begun asking manufacturers to conduct "lifetime" exposure of at least one species, usually the rat, to establish "proof of safety" before marketing a new food additive or pesticide.

The Food Additives Amendment of 1958 required that any new intentional or unintentional food additive have FDA approval, usually in the form of a regulation, published in the *Federal Register* as a response to the manufacturer's petition for the proposed use, before the material could be marketed. Pesticide chemicals that leave a residue in or on raw agricultural commodities were soon brought under the same system of control. The FDA would establish a tolerance level for food additives or pesticides at either a presumed no-effect level or the lowest level required to produce the desired effect, whichever was the lower of the two levels. The Delaney clause, added to the Food Additives Amendment of 1958 on the floor of the House, prohibited the FDA from setting any tolerance other than zero, however, for any substance found to induce cancer when ingested by man or other animal. Initially, substances generally recognized as safe (GRAS) or substances covered by prior FDA approval (prior sanctions) would not be considered food additives under the amendment, but subsequently, prior sanctions were canceled and many substances from the original GRAS list were brought under the formal regulation process.

As time went on, the protocols acceptable to the FDA for toxicity tests offered as proof of safety for a proposed product became increasingly complex and expensive. The Occupational Safety and Health Act of 1970 (OSHA) was brought about in part by a concern for toxic effects on workers caused by long-term exposure to chemicals in the work environment. At the same time, the National Institute for Occupational Safety and Health (NIOSH) was established in the Department of Health, Education and Welfare, to advise OSHA on the health aspects of chemical hazards of the workplace, and to recommend standards for the control of such hazards. The Toxic Substances Control Act by 1976 seeks to "regulate commerce and protect human health and the environment by requiring testing and necessary use restriction on certain substances, and for other purposes" (from the preamble to the act). The Environmental Protection Agency (EPA) took over from the FDA responsibility for establishing tolerances for pesticides and administers the Toxic Substances Control Act.

We have seen that during the nineteenth and twentieth centuries, industrial toxicology has come of age, and is no longer an option affordable by a few concerned industries resolved to protect their employees and the public by anticipating the possible toxic side effects of otherwise marketable commodities. Now, with the current laws and regulations, no industry can ignore toxicity as a key factor in business planning. Many in-house toxicology laboratories have been, and are still being, established since the 1930s. Commercial toxicology laboratories have multiplied to serve those who have no in-house laboratories, and some universities have set up institutes to carry out toxicity testing on a contract basis.

Whereas in the beginning toxicology served the purposes of aggression, the coin has now been reversed and in our time the largest number of toxicologists are serving the purposes of prevention, based on supplying accurate information and advice on the toxic potential of the multitude of chemical substances in the human environment.

3   THE DEVELOPMENT OF INDUSTRIAL TOXICOLOGIC TESTING

Knowledge of the toxic effects of chemicals on humans developed along two lines of investigation. The first was direct observation of effects produced in humans. Even up to the nineteenth century substances thought to have therapeutic value were administered directly to humans with often disastrous results. Haggard (5) tells of King Mithridates of Pontus, who in the second century B.C. had the ambition to discover a universal antidote for poisons. He is said to have sent men to search throughout the known world for all kinds of alleged poisons, and he administered these to slaves, observed the effects, and tried out various antidotes. He came up with a recipe for an antidote, found after his death. It contained 37 to 63 ingredients, all now believed to be worthless.

Direct observation of toxic effects on humans is still attempted through the discipline of epidemiology, but this involves only the observation of populations, not the deliberate administration of chemicals to humans to observe their toxic effects.

The second method of obtaining knowledge of the toxic effect of chemicals on humans is indirect. The substances are administered to animals, and the results are extrapolated to humans. This is the only acceptable method today, and is the method employed by industrial toxicology.

The question that industrial toxicologists were asking toward the end of the nineteenth century was "How much of this substance is safe?" rather than "How much is toxic?" Their thinking, however, was directed mainly to the question of acute toxicity. With the recognition that there was such a thing as intraspecies variability, experiments were directed toward estimating the minimum lethal dose, which was the largest dose that would kill only the most susceptible member(s) of the species. Something a little less than this might then be the largest dose that would *not* kill *any* members of the species. To approximate the minimum lethal dose, one obviously had to expose a number of animals to the same dose and to use several different dose levels.

Consideration of the efficient use of laboratory facilities and cost control led to a shift away from the use of dogs, cats, and rabbits for acute toxicity experiments and to the use of rats, mice, and guinea pigs as the preferred test species. The albino rat emerged as the preferred species for most studies, at least in their preliminary stages. It is worth mentioning that much progress has been made in the commercial breeding of experimental animals for use in toxicity testing. The rats, mice, guinea pigs, and rabbits now available for purchase are relatively free from the diseases that commonly afflict these species, so that any pathological changes induced by the test chemical are more readily discernible than was the case previously.

It was only in the twentieth century that a clear understanding of the quantitative aspects of biologic variation developed. For any individual the effects of a given drug or poison would increase with increasing dose and decrease with decreasing dose. This had been known for a long time. But by 1937, Clark (8) could note:

> One of the most familiar facts in medical practice is that no two persons respond in exactly the same manner to drugs. The use of biological methods for standardizing drugs has necessitated the measurement of the extent and distribution of individual variation in response to drugs, and in consequence a large [amount of] literature has accumulated.

What this large amount of literature had demonstrated was that for a population divided into subgroups receiving graded doses, a plot of a selected response, such as percent mortality, against dose (or most often the logarithm of the dose) usually assumed an S, or sigmoid, or ogival shape characteristic of the cumulative form of the normal curve of error. The greatest increase in mortality for a given increment of dose occurs around the 50% mortality point; hence the lethal dose for 50% of a population, the LD_{50}, can be calculated with greater precision and with fewer animals than any other percentage point. For these reasons, acute toxicity is usually expressed in terms of the LD_{50}, although for many purposes it might be more desirable to estimate, say, the LD_{01} or LD_{99}. If the data fit the theoretical curve fairly well, it is possible to calculate statistical confidence limits for the values obtained.

The sigmoid curve of response versus dose can be converted into a straight line if percent response is expressed as the corresponding multiples of the standard deviation. A probability graph paper soon became available on which percent response could still be plotted on the ordinate, but with the percents spaced so as to correspond to the appropriate multiples of the standard deviation. These techniques had been worked out by the middle 1930s.

There are other mathematical models besides the normal curve of error that provide reasonable approximation to observed dose–response data, but the LD_{50} remains the point that can be determined most efficiently, and it is therefore the usual index of acute toxicity. The slope of the dosage–mortality curve provides an index of potency.

If animals are exposed by the inhalation route, there are two variables that determine lethality. These are the concentration of the toxin in the air, and the time over which the animals are exposed. For a given exposure time, the most efficient measure of acute toxicity is the concentration causing 50% mortality, or LC_{50}. But for each LC_{50} the exposure time must be specified.

World War I stimulated a great many studies of acute inhalation toxicity for chemical warfare purposes. The number of compounds examined during World War I as possible chemical warfare agents is estimated to have been between 3000 and 4000, and of these, 54 were used in the field at one time or another. The German chemist Fritz Haber proposed that for a given effect such as death, the inverse relationship between concentration and exposure time could be expressed by the simple equation $Ct = k$, where C is concentration and t is exposure time. This relationship holds fairly well over a relatively narrow range of concentrations and times. The information gained through the testing for military purposes had, of course, post-war value for industrial toxicology.

Some of the World War I chemical warfare agents were selected for their irritancy to skin or eyes, rather than for systemic toxicity, and both the techniques developed for their study, as well as the information gained, were useful to postwar industrial toxicology.

Although chronic, or cumulative, toxicity had been recognized for centuries, it received much less attention than acute toxicity until recent times, possibly because acute toxic effects were more likely to occur and be recognized than were chronic effects. Chronic toxicity could, however, be investigated by any relevant route of exposure provided the dosages used were small enough to permit the animals to survive their repeated assaults over a sufficiently long period of time to permit the chronic damage to appear. The most perplexing question was, "How long should a prolonged exposure be to gain all the necessary information?" Opinions differed, but the majority of toxicologists seemed to feel that 90 days of repeated exposure would be sufficient to elicit all the important manifestations of chronic toxicity in the rat or mouse provided the daily doses were sufficiently high but still consistent with survival.

In 1938, as a consequence of the elixir of sulfanilamide tragedy, in which a number of persons died as a result of taking a solution of sulfanilamide in diethylene glycol for therapeutic purposes, the U.S. Food and Drug Administration undertook a comprehensive investigation of the toxicity of the glycols. This investigation culminated in a "lifetime" feeding study with diethylene glycol in rats. In 1945, Nelson et al. (12) reported the results at a meeting of the Federation of American Societies for Experimental Biology. A surprising result of the study was the finding that some of the rats fed a diet containing 4 percent diethylene glycol had developed bladder stones, and that some of those with bladder stones had also developed fibropapillomatous tumors of the bladder. Because neither bladder stones nor tumors had been found in tests of shorter duration, it became obvious that, for some lesions, 90 days was not a sufficient time of exposure. By 1950, the FDA had begun recommending lifetime studies, for which they considered 2 years in the rat to be proper, as part of proof of safety of proposed new intentional and unintentional food additives and pesticides. As a guide to the perplexed, members of the FDA staff prepared an article entitled "Procedures for the Appraisal of the Toxicity of Chemicals in Foods, Drugs, and Cosmetics," which was published in the September, 1949, issue of *Food Drug Cosmetic Law Journal* (13). It contained a section on how to do long-term chronic toxicity studies, and recommended a period of 2 years for the rat, plus 1 year for a nonrodent species such as the dog.

Although not an official regulation, the article advised everyone of the FDA's expectations with respect to data submitted to it as proof of safety of the proposed new food additive or pesticide. A revision of the article appeared in 1955 (14), and a third revision was published in 1959 as a monograph put out by the Association of Food and Drug Officials of the United States (15).

During the same period, the Food Protection Committee of the National Academy of Science/National Research Council was publishing and revising "Principles and Procedures for Evaluating the Safety of Food Additives" (16) which were, in general, consistent with the FDA staff's guidelines. One common thread ran through both sets of recommendations. With each revision, the complexity of the tests increased and so did the cost.

The FDA's recommended protocol in 1959 (15) for a "lifetime" test with rats called for four groups of a minimum of 25 males and 25 females each. There would be a control group, a low-dose group (a no-effect level, it was hoped), a high-dose group (chosen to be an effect level), and a mid-dose group. All animals would be necropsied for gross pathology. Selected organs would be weighed, and selected organs would be preserved for histopathology. During the course of the experiment, food consumption and weight gains would be measured, blood and urine would be monitored for deviations from normality, and any behavioral changes would be noted. A three-generation reproduction study would be carried out at all dose levels. Also, a similar experiment would be carried out with four groups of six to eight dogs each for an exposure period of 2 years, to determine whether a nonrodent species responded differently from the rat. Dog reproduction studies were not required. The lifetime of the rat was considered to be 2 years for the purposes of the test. So it was in 1959.

In the ensuing two decades, the complexity and cost of the tests again increased. On August 22, 1978, the EPA published in the *Federal Register* guidelines for registering pesticides in the United States (17). Though the guidelines were directed to pesticides, EPA stated that it believed them to be consistent with those of other federal regulatory agencies and the Interagency Regulatory Liaison Group (IRLG). They provided, therefore, a glimpse of what all federal regulatory agencies would be requiring in the future. Nor did the EPA guidelines differ much from the procedures recommended in 1977 by the National Academy of Sciences/National Research Council in their publication "Principles and Procedures for Evaluating the Toxicity of Household Substances" (18), which was prepared at the request of the Consumer Products Safety Commission.

Considering only the "lifetime" feeding study with rats, we find that both sets of guidelines call for 50 animals of each sex for the control and for each of the three dosage level groups, which is double the number recommended in 1959. Under certain circumstances it might be advisable to start the testing in utero, by feeding the test substance to weanling rats, breeding these, and then starting over with their offspring (17, 18).

If there is any suspicion that the test compound is oncogenic, the EPA guidelines would require a second chronic toxicity test with the mouse or hamster, or another species if appropriate. The mouse would be the normal choice, the duration of the test would be 18 to 24 months, and the test groups would consist of at least 50 animals of each sex per group.

It is obvious that chronic toxicity testing has come a long way in only three decades, and the burden of carrying out these tests has fallen largely upon industry, whether the tests be carried out in-house or by contract. It is not so obvious that the cost of the testing is borne ultimately by the consumer.

As of 1979, it may be said that techniques were available for carrying out, with experimental animals, tests that will answer almost any question that might be asked regarding the various aspects of toxicity to the test species. The techniques encompass acute toxicity, short-term or subchronic toxicity, chronic toxicity, oncogenicity, mutagenicity, teratogenicity, neurotoxicity, skin irritation and sensitization, and eye irritation, among others. The complete battery of tests on one compound could consume 500 to 1000 animals, could cost in excess of $500,000, and would require 3 to 5 years for completion—if all went well.

It follows that with such complexity and cost and time, it is not possible to test completely all compounds of interest with the available personnel and facilities, or even the potentially available, within the foreseeable future.

If an industry is attempting to develop a pesticide or a chemical warfare agent, toxicity to the target species is an asset. But in most cases toxicity in an industrial product is not the desired effect but rather an unwanted side effect which, if unrecognized, could create major problems for the manufacturer.

A prudent management should therefore seek answers to the following toxicity-related questions before investing capital in the production and marketing of a new product:

1. Can it be manufactured safely?
2. Can it be used safely for its intended purpose?
3. Can it be disposed of safely into the environment during manufacture and after its intended use?

If the answer to any of these questions is "No," it would not be prudent to invest capital in the project, no matter how useful and profitable the proposed product might appear to be if there were no side effects, for toxicity would be a tangible and significant product defect.

A prudent management must therefore take toxicity into consideration from a purely business point of view. But because the cost of toxicity testing becomes a part of the cost of product development, and eventually of cost of product to the consumer, management would like to keep that cost as low as possible. On the other hand, if frugality leads to inadequate testing, the business decision process is jeopardized, and a wrong decision could be very costly.

But though all industry has its own need for valid toxicity information, not all industry has had the benefit of competent toxicologic advice. In its absence, persons not trained or competent in toxicology have decided what tests, if any, are required, and have directed contract laboratories to carry out just these tests. Usually they have not been able to judge the competence of the testing laboratory, nor have they been able to interpret properly the reports submitted. Hence errors of omission and commission have occurred, and problems have arisen that have affected not only the industry but the public as well. And when problems arise that affect the public, or are even perceived as a potential threat by the public, the public demands that the government do something about it.

The government responds to public demands. Each of the laws and regulations requiring toxicity information from industry has been a response to a perceived public need for government restriction of industry's freedom to make its own decisions. Under these laws, government regulatory agencies, not industry, decide whether a material can be made, used, and disposed of safely. The Occupational Safety and Health Administration (OSHA) decides whether or under what conditions a substance can be made safely; the Food and Drug Administration (FDA) decides whether or how much of a substance can be permitted in food or drugs; the Environmental Protection Agency (EPA) decides whether or how much of a substance can be used as a pesticide, and under what conditions it can be made and used without harm to health or the environment. The Department of Transportation decides what special precautions are necessary in moving a material from one place to another.

The advantage to society of having an impartial review by experts of the basis for an industry decision that a proposed new product can be made, used, and disposed of safely is obvious, but the problem is whether the government is likely to be the source of the required impartial expertise.

In the first place, industry decides from evidence, in its judgment sufficient to reach a decision, whether a new product can or cannot be made, used, and disposed of safely. And here industry can make two kinds of error, both important for business decisions. It can decide erroneously that the product is unsafe and should therefore not be made. This would be a Type I error. Or, industry can decide erroneously that the product is safe when in fact it is not safe. This would be a Type II error. The penalty to industry for a Type I error would be loss of a potentially profitable business, and to the public, loss of a potentially useful product. The penalty to industry for a Type II error could be financial loss, as well as loss of credibility and public confidence, and to the public, harm to health and/or the environment. Industry therefore must seek to avoid both kinds of error.

The government regulatory agencies, on the other hand, have little incentive for avoiding Type I errors, but are greatly concerned to avoid Type II errors, for these could lead to, in addition to harm to health or the environment, public criticism, political disadvantage, and possible loss of jobs. To protect themselves against Type II errors, regulatory agencies usually demand a greater degree of proof of safety than industry might think sufficient to reach a reasonable judgment.

Even so, at present consumer advocates and environmental activists are pressing the agencies to demand evidence that proposed new products are *absolutely* safe. Such evidence cannot be obtained and will never be obtainable, as Aristotle pointed out centuries ago. As one government toxicologist put it, "The one thing I worry about is finding myself someday on the witness stand without a piece of paper in my hand that would justify my decision to approve this product." It is therefore understandable that regulatory agencies tend to require protocols that minimize their chance of committing Type II errors. Nor are the regulatory agencies inhibited by increasing the cost of, and time required for, expanded toxicity testing, because the initial exposure must be borne by industry, although the bulk of it will eventually be passed on to the public, which may not as yet realize that this is happening.

But may not the regulatory agencies take the position that the avoidance of Type II errors is their mandated job, and that they have no concern with Type I errors? An affirmative answer to this question might be proper if the number of suspect materials were small, and their loss of no importance. But on the contrary, thousands of substances are viewed as suspect because the *possibility* of one or more harmful effects has not been conclusively eliminated.

Under these circumstances, undue emphasis on the avoidance of Type II errors does carry penalties for society. If, for example, the FDA delays approval of a proposed new drug for 5 years in their effort to avoid making a Type II error, and if it is then found that this new drug can save 1000 lives a year, the penalty for a 5-year delay in approval is 5000 unnecessary premature deaths. A statement prepared and approved by the Committee on Chemistry and Federal Policy of the National Academy of Sciences' section on chemistry deals with broader aspects of the problem, and is worth reading in its entirety, but the following excerpts (19) are relevant here:

In his annual report to the members of the National Academy of Sciences this year, president Philip Handler stressed that "the falloff in industrial productivity, the decline of innovative new industrial starts has become a matter of widespread concern. It is clearly a threat to the vitality of the U.S. economy and, hence, to much else that we cherish. . . ."

Potentially a very significant factor is the ever-growing impact of federal regulatory actions, which have already diverted resources from basic to defensive research. Every high-technology company has experienced the necessity to develop new testing methods, new criteria for safety, and new ways to meet environmental requirements, all of which have added substantially to the time, effort, and cost of developing new products. The end result for whatever reasons has been a startling decline in innovative research and products. More broadly, 204 new publicly financed small technical companies came into being in 1969; by 1975 this number had dwindled virtually to zero. . . .

The question is not whether we need federal regulation of industrial products and processes, but whether we can put such regulation on a sounder scientific basis.

Perhaps the answer to these problems will come about eventually through risk/ benefit analysis, which is achieving some popularity. But it must be remembered that a particular risk may be avoidable only at the cost of creating other risks of as great or greater importance, and the benefit to be obtained from a particular course of action may cancel other benefits (20).

The question of how much toxicity testing is optimal is certainly a contemporary problem area worthy of serious consideration not only by scientists, but also by government policy makers and by the public.

4   THE ROLE OF TOXICOLOGY IN SETTING EXPOSURE LIMITS

Toxicology is defined (21) as the study of the adverse effects of chemicals on biologic systems, but toxicology is clearly more than just the identification and characterization of adverse effects. Toxicology, like medicine, is both a science and an art, and it is the art or predictive aspect of toxicology as much as or more than the scientific aspect that is responsible for the key role of toxicology in public health.

Toxicologic predictions are traditionally based on information describing the chemical and its adverse effects, the dose and/or exposure conditions, and the target or exposure subjects. These same three end points, the type of adverse effect, the exposure scenario, and the susceptibility of the exposed population are also the basic input for assessing the risk of exposure to chemicals in our environment and are used for setting exposure limits for chemicals in the workplace as well as in our food, water, air, soil, and drugs.

Because the quality and validity of such predictions are highest when they are based on data obtained from the target species, our first goal in setting exposure limits for chemicals in the workplace is to find epidemiologic or industrial toxicology studies carried out under conditions that are identical to those for which the recommendation is intended. However, even with these ideal conditions (good human data on the adverse effects of the chemical obtained with the correct exposure scenario), our recommendations may not be predictive for every individual in the target population because they are formulated for the majority of an average or typical population.

When this ideal information is not available, our next choice is to seek human data obtained with the desired chemical that can be interpolated (within the range of the existing data) or extrapolated (beyond the range of the data) to formulate the prediction. In using this type of data, we recognize that variation in the dose, the exposure route, rate and/or duration, or population differences in sex, age, occupation, dietary, or personal habits may prejudice or even invalidate the prediction.

In those cases where we do not have information on human exposure to the desired chemical, we are forced to use some type of surrogate data as the basis for the prediction. For example, we might use human data on a chemical that has toxicologic and chemical properties that are similar to the target chemical or even data from a class of chemicals that have appropriate structure–activity relationships. In principle, it is only when we do not have adequate human data on the target chemical or on a reasonable surrogate that animal studies are used for setting exposure limits. However, in practice, animal studies are almost always used as a part of this process because they are useful in validating existing human data and they provide the basic input for both the threshold and nonthreshold approaches to setting health-based exposure limits for chemicals.

4.1 The Threshold Approach to Setting Exposure Limits

The first step in what is traditionally referred to in toxicology as hazard evaluation is to identify all the adverse effects that can be produced by either acute or chronic exposure to the chemical, and the second step is to establish dose–response relationships for each of these adverse effects. Ideally, these studies would also provide information on the effects of administration by different routes, at different rates and durations of exposure, and information in other test species. This information together with data on the chemical and physical properties of the chemical, kinetic data in various species, gene–toxicology studies, teratology, reproduction and other types of end-organ damage, plus the mechanistic information constitutes the toxicologic data base for the chemical.

The next step is to determine whether this information is relevant to the target species and the exposure scenario. In those cases where the information is appropriate and where there is a threshold or no-effect level (NOEL), then the final step is to divide the NOEL by an appropriate safety factor to assess risk and establish the exposure limit.

A modification of this approach has been developed by the Environmental Protection Agency (EPA) in which the slope of the dose–response curve and its confidence limits are used to calculate benchmark or reference doses for low incidence responses (22). Although this approach is more responsive to the number of animals at each test dosage level and avoids the use of a single threshold value, it is basically an extension of the traditional threshold approach.

The goal of the threshold or nonlinear approach is not to demonstrate the safety of the chemical under all conditions but to predict the dose or exposure situations at which specific adverse effects are not likely to occur in the target population. The advantage of this approach is that it is simple and easy to understand and that it has been widely used for more than 50 years to assess and regulate the hazards of chemicals in our environment. However, this approach cannot be used with nonthreshold effects and it requires the exercise of judgment or a weight of evidence evaluation in three areas: the selection of the test data to be used, the decision about the relevance of the data, and the selection of the safety factor.

One of the problems in making such species to species comparisons is that the test and the target species have different physical and functional characteristics such as body weight, surface area, caloric intake, life-span, and a host of other biologic characteristics. Traditionally most such comparisons have been based on body weight, but for some chemicals, body surface area appears to be a better basis for comparison, and an interagency regulatory body (FCCSET) has recently recommended (23) using body weight to the ¾ power as the standard for all dose normalization. Another approach is to use pharmacokinetic data from the test and target species to improve the comparison, and methods have been developed to use physiology-based pharmacokinetic factors in the risk assessment process (24).

The ideal situation, of course, would be to make such comparisons on the basis of the receptor dose rather than the administered dose, and the use of biologic dosimeters is a logical first step in this direction. However, even in those cases where we are dealing with the same receptor, such as in setting limits for children and adults, there may be qualitative and quantitative differences in the response of the receptor to the same dose of drugs or toxic agents. Another approach that toxicologists have traditionally utilized to increase the validity of our trans-species predictions is to conduct the test in a variety of species and to use the most sensitive species or the average as the predictor. This works well when the most sensitive species handles and responds to the chemical in the same way as the target species, but without this knowledge about both the mechanism of action and the kinetics in the test and target species, we tend to justify our selection of test data from a sensitive species as being conservative rather than because it is the most predictive species.

The final step in using the threshold approach for extrapolating the results of animals studies to humans is the selection of the safety factor to be applied to the animal data. This is a judgmental decision that is intended to encompass both the interspecies and intraspecies variation and to reflect the confidence of the evaluator in both the quality and the relevance of the test data being used for the prediction. When this procedure was initially established by Lehman and Fitzhugh (25) at the Food and Drug Administration (FDA), a factor of 100 was used as a single value to encompass all these uncertainties, but over the years, it has become customary to divide this into two factors of 10 and to assign these to the intraspecies and the interspecies variation.

In retrospect, it is evident that these values have served society well and consequently they have been incorporated into the guidelines of other agencies in the

United States and throughout the world; moreover, they are used by various committees, such as those of the Committee on Toxicology of the National Research Council/National Academy of Sciences, the Threshold Limit Value Committee of the American Conference of Governmental Industrial Hygienists, and many other standard-setting bodies.

In developing guidelines for the benchmark dose approach, EPA recommended (26) the use of the term "uncertainty factor" rather than "safety factor" to avoid the implication of absolute safety, and the use of additional "modifying factors" when using a LOAEL (lowest observed adverse effect level) rather than the NOAEL (no observed adverse effect level) or when using subchronic test data to predict chronic effects or to protect sensitive subpopulation groups. Although the intent of these changes is to enhance the precision of the trans-species prediction, the net result is a potential cascade of multiplicative uncertainty factors that are defined largely by guidelines rather than by actual data and good toxicologic principles.

This is a critical issue for the credibility of toxicologic predictions and their role in providing a scientific basis for responsible public health decisions. To improve this situation, we need to generate data to validate these default assumptions and to increase our use of actual data as the basis for all dose to dose, route to route, duration to duration, and species to species extrapolations. We also need to document clearly and justify each step of the process as well as to identify and distinguish those elements of the safety factor that are arbitrarily included to increase the margin of safety from those elements that are based on the actual toxicologic findings in the test and target species. Improving the precision and validity of the safety factor and increasing the understanding of how it was derived and used in setting each exposure level will benefit all users of this approach and the disciplines of industrial hygiene, toxicology, occupational medicine, and epidemiology.

4.2 The Nonthreshold Approach to Setting Exposure Limits

Although it can be reasonably argued that there is a threshold for all biologic effects including cancer, the existence of a threshold for this or other toxic effects may be virtually impossible to distinguish from the zero or no-effect level or from the background. In such cases, an alternative approach must be used to characterize the dose–response relationship; various susceptibility models (probit and logit) were used initially for this purpose. These were replaced by the multistage model of Armitage and Doll, which was more consistent with the age-related incidence of human cancer. The linearized multistage model is currently the most popular of the quantitative risk assessment (QRA) models but other biologically driven models are under study, such as the Moolgavkar two-stage model. All these QRA models are focused on the high-dose to low-dose extrapolation, and thus carcinogen classification systems based on these models focus on potency rather than on the relevance of the animal data for predicting effects in humans (species to species extrapolation).

In an effort to address this problem, the Threshold Limit Value Committee of the American Conference of Governmental Industrial Hygienists recently adopted new guidelines for the classification of workplace carcinogens that include consideration of the mechanism of action, exposure route, and other factors, as well as the dose extrapolation (27). Efforts are also being made to modify other carcinogen classification systems (IARC, EPA, etc.) so that they can accommodate animal carcinogens that act through mechanisms that do not appear to be directly relevant to humans (such as peroxisome proliferation and α-2-globulin conjugation in the male rat kidney).

Carcinogenesis is a complex process believed to occur as the result of multiple events, which may be modulated by a variety of somatic and environmental stressors. Because the epidemiology or experimental frequency of cancer is affected by the heterogeneity of environmental stressors, by individual exposure experiences, by cellular genomes and phenotypes, by age-dependent physiological stressors, and by the efficiencies of defense and repair, the mathematical modeling of cancer development kinetics is at best problematic (28). The challenge for the toxicologist is to define useful general principles without reliance on simplistic models, to evaluate judiciously key chemical and biologic interactions without under- or overestimation of their importance, and to base estimates of risk on all available information without dogmatic adherence to one theoretical approach (29). Because the nonthreshold approach to setting exposure limits is a more recent development than the threshold approach, it lacks the validation that comes from long experience. However, the problem of reconciling conflicting data from epidemiologic and toxicologic studies, the need for better procedures for handling mixtures of chemicals and aggregating adverse effects, and the lack of data on which to base such limits is a continuing problem for both the threshold and nonthreshold approaches to setting exposure limits for environmental chemicals (30).

The process of establishing exposure limits for workers, military personnel, and other population groups exposed to chemicals is the major practical application of what is currently referred to as risk assessment. Groups such as the U.S. Food and Drug Administration, the TLV committee of the American Conference of Governmental Industrial Hygienists, and the Committee on Toxicology of the National Research Council have been engaged in this activity for nearly 50 years. The basic elements of this process are the case by case approach, the priority of human over animal data, the use of the threshold approach, and the reliance on good science and expert judgment rather than rigid protocols to achieve an optimal balance between benefits and risks. The process was created, developed, and validated by toxicologists, and it is a part of the heritage of toxicology.

5 PROSPECTS

Toxicology has come of age in our time in that its importance is recognized by industry, the government, and the public. Industry needs a knowledge of toxicity as an important input to prudent business decisions. Government is firmly established as the regulator that tells industry what it must and must not do to preserve health and the environment. The public fears toxicity and demands the right to know the hazards to which it may be subjected and adequate protection against those hazards.

There appears to be little prospect that the problem of toxicity will be forgotten by any of the involved groups in the foreseeable future. We might well advise our sons and daughters to consider the profession of toxicology as a possible vocation that promises reasonable job security for the competent.

Advances in the science of toxicology will undoubtedly occur, for like all science it is dynamic, and as Lord Macaulay wrote in 1837, ". . . it is a philosophy which never rests, which has never attained, which is never perfect. Its law is progress." Macaulay, of course, was speaking of science in general, but his words are applicable to toxicology.

We can anticipate more efficient methods for arriving at the truth, the realities of the situation rather than opinion and speculation. We can hope that they will supply valid answers at lower cost and in less time than do the currently available methods. We can hope that the public and the media will become better informed and able to distinguish between fact and fiction concerning problems of toxicology. And we can hope that government will view and deal with problems of toxicity in proper perspective against the broad spectrum of other problems.

REFERENCES

1. L. G. Stevenson, *The Meaning of Poison*, University of Kansas Press, Lawrence, 1959.
2. T. C. Albutt, *Greek Medicine in Rome*, London, 1921.
3. Plato, "Phaedo," in *Plato, Selections*, R. Demos, Ed., Scribner's Sons, New York, 1927.
4. Paracelsus, "Epistola Dedicatora St. Veit Karnten: Sieben Schutz-, Schirm- und Trutzreden. Dritte Defension (August 24, 1538)."
5. H. W. Haggard, *Devils, Drugs and Doctors*, Harper, New York, 1929.
6. T. A. Loomis, *Essentials of Toxicology*, 3rd ed., Lea and Febiger, Philadelphia, 1978.
7. M. J. B. Orfila, "Traite des poisons tires mineral, vegetal, et animal on toxicologie generale sous le rapports de la pathologie et de la medecine legale," Crochard, Paris, 1815.
8. A. J. Clark, "General Pharmacology," Ergänzungswerk, Vol. IV, *Handbuch der Experimentellen Pharmakologie* (Begr. A. Heffter), Springer, Berlin, 1937.
9. C. T. Thackrah, *The Effects of the Principal Arts, Trades, and Professions*, London, 1831.
10. M. W. Childs and D. Cater, *Ethics in a Business Society*, Harper, New York, 1954.
11. W. G. Thompson, *The Occupational Diseases*, Appleton, New York, 1914.
12. A. A. Nelson, O. G. Fitzhugh, and H. O. Calvery, "Diethylene Glycol," *Fed. Proc.*, **4**, 149 (1945).
13. A. J. Lehman and FDA Staff, "Procedures for the Appraisal of the Toxicity of Chemicals in Foods, Drugs, and Cosmetics," *Food Drug Cosmet. Law J.* (Sept. 1949).
14. A. J. Lehman and FDA Staff, "Procedures for the Appraisal of the Toxicity of Chemicals in Foods, Drugs, and Cosmetics," *Food Drug Cosmet. Law J.* (Oct. 1955).
15. FDA Staff, Division of Pharmacology, *Appraisal of the Safety of Chemicals in Foods,*

Drugs, and Cosmetics," Association of Food & Drug Officials of the United States, Baltimore, MD, 1959.

16. National Academy of Science/National Research Council, Food Protection Committee/ Food & Nutrition Board, "Principles and Procedures for Evaluating the Safety of Food Additives," Publ. No. 750, 1959.

17. EPA Pesticide Programs, "Proposed Guidelines for Evaluating Pesticides in the U.S.; Hazard Evaluation; Humans and Domestic Animals," *Fed. Regist.*, **43**(163) (Aug. 22, 1978).

18. National Academy of Science/National Research Council Committee on Toxicol/ Assemb. Life Sciences, "Principles and Procedures for Evaluating the Toxicity of Household Substances," National Academy of Sciences, 1977.

19. NAS Section on Chemistry—Committee on Chemistry and Public Policy, "The Chemical Industry and Federal Policy," *Chem. Eng. News*, **57**(6), 32 (1979).

20. A. Wildavsky, "No Risk is the Highest Risk of All," *Am. Sci.*, **67**, 32 (1979).

21. C. D. Klaassen and D. Eaton, "Principles of Toxicology," in Casarett and Doull's *Toxicology: The Basic Science of Poisons*, 4th ed., M. O. Amdur, J. Doull, and C. D. Klaassen, Eds., Pergamon Press, New York, 1991.

22. D. G. Barnes and M. L. Dourson, "Reference Dose (RfD): Description and Use in Health Risk Assessments," *Reg. Toxicol. Pharm.*, **8**, 471 (1988).

23. W. H. Farland, personal communication.

24. National Research Council, *Pharmacokinetics in Risk Assessment*, Vol. 8, *Drinking Water and Health*, National Academy Press, Washington, DC, 1987.

25. A. J. Lehman and O. G. Fitzhugh, "100-fold margin of safety," *Assoc. Food Drug Off. U.S. Q. Bull.*, **18**, 33 (1954).

26. US Environmental Protection Agency, *Interim Methods for the Development of Inhalation Reference Concentrations*, EPA 600/8-90/066A, Environmental Criteria and Assessment Office, Research Triangle Park, NC, 1990.

27. American Conference of Governmental Industrial Hygienists, *Documentation of the Threshold Limit Values and Biological Exposure Indices*, 6th ed., American Conference of Governmental Industrial Hygienists, Cincinnati, OH, 1992.

28. G. B. Gori, "Cancer Risk Assessment: The Science that is not," *Reg. Toxicol. Pharm.*, **16**, 10 (1992).

29. B. E. Butterworth and T. L. Goldsworthy, "The Role of Cell Proliferation in Multistage Carcinogenesis," *Proc. Soc. Exp. Biol. Med.*, **198**, 683 (1991).

30. National Research Council, *Toxicity Testing: Strategies to Determine Needs and Priorities*, National Academy Press, Washington, DC, 1984.

CHAPTER TWO

Occupational Carcinogens

Robert P. Beliles, Ph.D., D.A.B.T., and
Carl O. Schulz, Ph.D., D.A.B.T.

As early as 1775, Percivall Pott identified scrotal cancer among chimney sweeps as an occupational cancer, but with a few specific exceptions it is only in the past two decades that significant changes have been made to protect the working man and woman against harmful materials in the workplace. This is especially true of occupational carcinogens. The passage of the Occupational Safety and Health Act in 1970 led to the establishment of the Occupational Safety and Health Administration (OSHA) to regulate worker health and safety and the National Institute for Occupational Safety and Health (NIOSH), a research agency to aid OSHA in rulemaking.

Also in 1970, the Environmental Protection Agency (EPA) was established by Executive Order. The Toxic Substances Control Act (TSCA) of 1977 gave EPA broad regulatory authority over chemicals in commerce. Designed to complement the existing legislation, TSCA gave EPA the authority, among other things, to regulate exposure to chemicals and to require industry to submit health and safety data, as well as to test chemicals for health and environmental effects. TSCA distinguishes between new (premanufacturing) and existing chemicals. Prior to marketing, the manufacturer must notify the agency. The EPA has 90 days to review the chemical for potential health and environmental concerns and notify the manufacturer of its findings. TSCA also established the Interagency Testing Committee (ITC) to help EPA prioritize existing chemicals with regard to testing needs. OSHA is one of the eight statutory members of this committee. TSCA directs the ITC to pay particular attention to chemicals that are carcinogens, mu-

Patty's Industrial Hygiene and Toxicology, Fourth Edition, Volume 2, Part A, Edited by George D. Clayton and Florence E. Clayton.
ISBN 0-471-54724-7 © 1993 John Wiley & Sons, Inc.

tagens, and/or teratogens. Industry must submit health, safety, and exposure data on any chemical the ITC recommends to EPA for testing. Also, industry is required to notify EPA of any chemical that presents an imminent substantial hazard to human health or to the environment. Upon review of these data the EPA Administrator has the authority to refer the regulation of the chemical to the appropriate agency. Chemicals that are exclusively of occupational concern would be referred to OSHA for rule making.

The OSHA "worker's right-to-know" standard (Hazard Communication-CFR 1910.1200) requires that the employer inform workers of the hazards of chemicals used in their workplaces. This standard may encourage the worker to use safer work practices. It should also result in the reduction of workplace hazards, including cancer, because employers may fear legal action on the part of the harmed workers. Norton Nelson, in 1981, suggested that all occupational illnesses are conceptually preventable. He reaffirmed this theory in 1991 (1), but such prevention has not been achieved, even with cancer.

The current regulation of carcinogens in the workplace and, to some extent, the regulation of cancer by other agencies is based on three events: the adoption of the OSHA Cancer Policy (2), the Supreme Court's ruling on OSHA's benzene standard, and the publication by the National Academy of Sciences (NAS) of guidelines for risk assessment (3).

OSHA adopted in 1980 a generic cancer policy that included a lengthy discussion of the basis for determining if human or animal data suggest that a particular agent poses a risk of cancer to workers exposed in their employment. The OSHA Cancer Policy reaffirmed OSHA's determination to base the permissible exposure limits on the most technologically feasible controls. Proposals have been made to alter part of the policy, but no final regulations have been published.

Because OSHA attempted to rely on its newly adopted cancer policy, the Supreme Court in June of 1980 overturned OSHA's benzene standard on the grounds that the agency had not shown a significant health risk. However, the Court pointed out in its decision on benzene, and subsequently in its decision on the cotton dust standard, that OSHA had done acceptable risk assessments in the development of the cotton dust standard, as well as in the coke oven emission standard. Most importantly, but often overlooked in discussions of the decision, the Court indicated that only a risk as high as one in 1000 was sufficient cause for regulation. On the other hand, the Court suggested that only a risk as low as one in 1,000,000 might be acceptable.

Additionally, the Supreme Court indicated that OSHA might use the opinion of expert bodies as a basis for its standards. Thus findings of the American Conference of Governmental Industrial Hygienists (ACGIH) or those named in the Hazard Communication Standard, that is, the International Agency for Cancer Research (IARC) and the National Toxicology Program (NTP), and other organizations remain important in the determination of whether an agent has a potential to cause cancer in the workplace. OSHA used the opinion of the ACGIH in the review of standards in 1989, but the consideration of ACGIH's position on short-term exposure limits in the case of ethylene oxide delayed the adoption of that standard by OSHA.

NAS examined the risk assessment process. Its paradigm for risk assessment divides the process into four parts: hazard evaluation, dose–response relationship, exposure assessment, and risk characterization—in part, the relationship between exposure concentration, number of persons exposed, and dose–response determination. The uncertainties in the assessment of risk are also to be considered. In addition, the NAS indicated that risk assessment should be separated from risk management. This was at odds with the OSHA Cancer Policy, which combined risk assessment, at least the hazard evaluation, with the management of risk.

These events led to the development of a number of mathematical methods for the extrapolation of animals doses, generally at high levels, to the risk that humans might encounter at low levels. Because most chronic toxicity studies have 50 animals per sex per dose group, the minimum increase in cancer that could be detected (note that this increase is not statistically significant) would be 2 percent, or 1 percent if the sexes were combined. Such an increase in cancer risk is unacceptable in humans. The ACGIH, one of the expert groups that recommends exposure limits for worker health, claims not to use any of these mathematical extrapolation methods in the adoption of their threshold limit values (TLV). Instead, they rely on "experience" and safety factors.

In retrospect, because of the Hazard Communication Standard, the requirement in the Toxic Substances Control Act to report adverse effects, liability, and ethical considerations, it is doubtful that industrial hygienists, occupational physicians, or toxicologists, as well as managers, would wish to wait for the government or other expert body to perform a risk assessment before attempting to eliminate or minimize exposure to potential occupational carcinogens.

1 HISTORICAL OVERVIEW

A review of the regulation of carcinogens in the workplace by OSHA standards adopted through 1980 provides insight into the basis for the current state of worker protection from cancer (4). The regulation of occupational carcinogens has importance in the legal as well as in the scientific arena. A group of cancer-causing agents listed in the appendix of the ACGIH's TLVs in 1968 was OSHA's first attempt to regulate carcinogens, but the promulgation of standards for these chemicals was delayed by the claim that, because the chemicals were merely listed in an appendix, they were not part of the national consensus standards that OSHA was permitted to adopt in the first several years after it was established. Thus the standards for these chemicals (often called the "14 carcinogens") in this first formal governmental attempt to reduce cancer in the workplace did not become official until 1974. By that time there were already standards in place to cover other carcinogens (e.g., asbestos).

The standards for these 14 carcinogens [bis(Chloromethyl) ether, chloromethyl methyl ether, 2-naphthylamine, 1-naphthylamine, 4-aminobiphenyl, 2-acetylaminofluorene, 4-nitrobiphenyl, benzidine, 3,3'-dichlorobenzidine, β-propiolactone, 4-(dimethylamino)azobenzene, N-nitrosodimethylamine, 4,4'methylenebis(2-chlorobenzenamine), and ethylenimine] provide for certain workplace practices, including the use of air-supplied respirators to reduce human exposure to zero. These standards require medical surveillance, through which the physician is to determine if the workers are at risk because of reduced immunologic competence, pregnancy, smoking, and intake of cytotoxic drugs or steroids. It is not clear what was to be done with a determination of increased risk. There was no provision for worker removal, as in the OSHA Lead Standard, but these standards mark an attempt to identify a subpopulation that might be at increased risk.

Table 2.1 indicates the scientific bases for standards for 21 occupational carcinogens that were adopted by OSHA between 1970 and 1980. These include the "14 carcinogens" listed above (4). Review of Table 2.1 shows that some of the standards were based solely on animal data, whereas the results of in vitro mutagenic assays were considered as supporting evidence. For some compounds consideration of metabolism was also an important aspect of the classification. It is noteworthy that bladder cancer in humans was a frequent end point leading to regulation and that at least one agent, β-propiolactone, was regulated solely on the basis of injection site tumors. Additionally, for six of the 21 carcinogens regulated in that period, there was evidence of carcinogenic potential in dogs or nonhuman primates.

The federal standard on MBOCA [4,4'-methylenebis(2-chlorobenzenamine)] was set aside after a court battle, although some states maintained their standards for this agent used in the manufacture of several types of plastic. EPA notified OSHA of the risk of MBOCA, as obligated under the TSCA. In 1990, the ACGIH proposed to classify MBOCA as a confirmed human carcinogen (A1) without a TLV. OSHA has only a permissible exposure limit (PEL) of 0.02 ppm, and that was not adopted until 1989 (5).

The experimental pesticide 2-acetylaminofluorene (2-AAF) is not found in many workplaces. It is used as a positive control in rodent bioassays for cancer and is not included among the TLVs.

The Asbestos Standard was the first complete standard; it defined the agents regulated and the workplaces covered. The standard, which served as a model for all complete standards that followed, specified compliance requirements covering engineering controls, work practices, the conditions for use of respirators, special clothing and special changing rooms with lockers, laundry of clothing by the employer, air monitoring, the use of signs and labels, housekeeping, and waste disposal procedures. The Asbestos Standard is also precedent setting in that it requires medical surveillance and record keeping by the employer. This standard reduced exposure for as many as 2.5 million workers. The NIOSH/OSHA recommendation in 1976 and 1979 to lower the standard to 0.1 fibers/m^3 has not been acted upon.

The Vinyl Chloride Standard started as an emergency temporary standard reducing worker exposure from 500 to 50 ppm. The final standard of 1974 established a 1-ppm PEL, a 15-min 5-ppm limit, and an action level of 0.5 ppm. These limits were largely based on the lowest levels of detection.

The Coke Oven Emission Standard was proposed in 1975 and was under legal

Table 2.1. Bases of OSHA Cancer Standards 1970–1980

Compound	Basis
Asbestos	Human exposure causes asbestosis, lung cancer, and pleural mesothelioma
2-Acetylaminofluorene	Cancer in rats, hamster, dogs, mice, rabbits, and fowl
4-Aminobiphenyl (4-ABP)	Bladder cancer in humans and dogs, at other sites in mice and rabbits
Benzidine	Bladder cancer in humans and dogs
Bis(Chloromethyl) ether (BCME)	Human case reports of lung cancer; lung cancer in rats and mice by inhalation; positive effects by other routes
Chloromethyl methyl ether	Positive in animals; epidemiologic studies suggestive, but possibly mixed with BCME
3,3'-Dichlorobenzidine	Suggestive epidemiology because of exposure to other carcinogens, largely based on positive carcinogenic studies in hamsters, mice, and rats
4,4'-Methylenebis(2-chlorobenzenamine) (MBOCA)[a]	Bladder cancer in dogs; liver, lung, Zymbal gland, and mammary tumors in mice
4-(Dimethylamino)azobenzene	Cancer in rats, dogs, trout, and neonatal mice; similar metabolism in humans and dogs
Ethylenimine	Cancer in mice; injection site tumors in rats
2-Naphthylamine (2-NA)	Epidemiologic studies positive for bladder cancer; also bladder cancer in monkeys, dogs, and hamsters
1-Naphthylamine	Structural and metabolic similarities to 2-NA
4-Nitrobiphenyl	Bladder cancer in dogs; converted in vivo to 4-ABP
N-Nitrosodimethylamine	Rats developed liver, lung, and renal tumors after inhalation exposure. Positive effects by other routes in mice, rabbits, hamsters, guinea pigs, and fish
β-Propiolactone	Injection site tumors in mice and rats
Vinyl chloride	Reports of death due to angiosarcomas in vinyl chloride workers supported by a finding of angiosarcomas and other tumors in mice and rats
Coke oven emissions	Lung cancer in workers
1,2-Dibromo-3-chloropropane	Cancer in mice and rats; supported by in vitro mutagenic studies
Arsenic	Lung cancer in workers exposed to arsenic trioxide
Acrylonitrile	Brain tumors and tumors at other sites in rats exposed via drinking water and inhalation; lung tumors reported in workers

[a]Complete standard never adopted by OSHA.

siege until the Supreme Court noted that OSHA had stated that lowering the standard from 1 to 0.15 mg/m^3 would prevent 240 cases of human lung cancer each year. The 0.15 mg/m^3 was based on technological feasibility.

The rule making on 1,2-dibromo-3-chloropropane (DBCP) was initiated because male workers in a DBCP manufacturing plant complained that their wives were not getting pregnant. The workers' subsequent semen analyses revealed abnormal sperm. However, during the rule making, positive cancer findings in mice and rats were reported by the National Cancer Institute. Positive in vitro mutagenic findings were also cited in the final standard of 1978. Thus the manufacture of this pesticide was regulated as a carcinogen, and not a reproductive hazard, although the medical surveillance section requires radioimmune assays for follicle-stimulating hormone, luteinizing hormone, and total estrogen in female workers. The DBCP standard requires that warnings be posted as a "Cancer Hazard," but not a "Cancer and Reproductive Hazard," as does the Ethylene Oxide Standard. During the rule-making process EPA revoked registration of DBCP for use on all food products and limited its use to certified pesticide operators. The PEL for DBCP is 1 ppb.

The PEL for acrylonitrile was lowered from 20 ppm to 2 ppm, with a 15-min limit of 10 ppm and an action level of 1 ppm. This standard is also based on technological feasibility.

In 1978, OSHA published a final standard for arsenic, setting the PEL at 10 μg/m^3. The uses of inorganic arsenic compounds as pesticides or as wood preservatives were excluded from the scope of the standard because these uses are regulated by EPA. After the Supreme Court ruling on benzene, OSHA asked the Ninth Circuit Court to remand the standard to the agency for the purpose of determining the significance of risk. The standard had been challenged by industry shortly after its issuance and was before the Court at that time. The agency proposed a risk assessment based on epidemiologic studies in copper smelters and in the pesticide manufacturing industry. Hearings were held on that effort, and other assessments were reviewed. The Chemical Manufacturers Association (CMA) represented industry and challenged the risk assessment performed by OSHA. CMA initially tried to maintain that it had not been demonstrated that arsenic is a carcinogen. OSHA affirmed the risk at the previous PEL and showed that the new PEL (10 μg/m^3) reduced the risk of cancer in exposed workers. These findings were published in the *Federal Register* on January 14, 1983. The District Federal Court in San Francisco accepted this risk assessment and praised OSHA for its efforts. Thus arsenic serves as a transition to risk assessment-based standards (4).

It seems appropriate to review the benzene standard-setting efforts that led to the current activity in quantitative risk assessment. An epidemic of leukemia in Italian shoemakers was traced to benzene in the glue they used (6). Together with this report and reports from other sources, it became obvious that benzene is a leukemogen (7). OSHA, newly established in 1970, adopted the ACGIH standard of 10 ppm. In 1977, OSHA sought to lower the standard to 1 ppm and adopt other work practices that would further protect the health of the workers. Although OSHA felt that benzene was a carcinogen and that no concentration other than zero would be safe, the agency felt that technology and economic feasibility permitted lowering the standard only as low as 1 ppm. OSHA was challenged in court by both labor and industry. The AFL/CIO claimed that the standard did not protect workers to a sufficient degree, whereas the American Petroleum Institute claimed that OSHA had failed to show that 10 ppm was harmful and that the standard was too costly. In 1980, the Supreme Court set aside the standard in a decision that prepared the stage for development of the art of risk assessment. In 1986, OSHA again proposed the 1-ppm PEL. Industry did not challenge the standard, and it is enforced today. The risk assessments for benzene, based on various epidemiologic studies, suggested that between 9.5 and 174 cases of leukemia per 1000 workers would develop if exposure to 10 ppm benzene lasted for 40 to 45 years, a working lifetime (8). Assuming that the proximal leukemogen from benzene was two or more of the metabolites, and using information indicating that the amount of benzene retained and metabolized increases nonlinearly with increasing concentrations, Beliles and Totman (9) calculated that the risk in workers, based on Zymbal gland tumors in rats exposed to higher concentrations, would be 14/1000 with an upper 95 percent confidence limit of 25/1000, whereas when based on leukemia in mice, it would be 13/1000 (48/1000 UC 95 percent). Thus, at least for benzene, the animal data were consistent with epidemiologic data and did not overestimate the risk for workers. With regard to the environmental risk related to factory emissions, the EPA in 1989 issued a regulation intended to reduce benzene emissions 90 percent, thereby reducing the risk from 1/142 to 1/5000. For those living next to a coke oven, EPA estimated that there would only be one leukemia case every 20 years due to benzene (10). More recently it has been suggested that benzene may also cause multiple myeloma in exposed humans (11). In 1990, the ACGIH proposed changing their classification of benzene to confirmed human carcinogen (A1) from suspected human carcinogen (A2) which had been assigned since 1977. They also proposed lowering the TLV from 10 ppm to 0.1 ppm. They suggested the likelihood of risk from benzene-induced leukemia death in a worker exposed to 0.1 ppm would be nearly the same as that for a worker not exposed to benzene. Also the ACGIH noted that 0.1 ppm was less than the concentrations inducing genetic changes in experimental animals and humans (12).

2 SYSTEMS FOR CLASSIFYING EVIDENCE OF HUMAN CARCINOGENICITY

During the past 20 years several governmental and international organizations have attempted to develop criteria for classifying the evidence of human carcinogenic potential of chemicals. Some of these approaches serve as the basis for regulatory activity, whereas others represent statements of general principle. None of these systems can be considered entirely objective, and all of them reflect the inherent uncertainty in classifying carcinogens on the basis of less than complete information. In the Cancer Policy of 1980, OSHA (2) indicated that potential occupational carcinogens could be categorized into two classifications, based on the strength of the evidence for potential carcinogenic activity in humans. Category I occupational carcinogens are those compounds whose carcinogenicity "has been determined in humans or in two or more mammalian species of test animals or in one species if the results of that study have been replicated," and Category II is reserved for those substances for which carcinogenic activity has been reported but for which the evidence can only be described as suggestive. Included in this category are compounds that have been determined as carcinogenic in one species in unreplicated studies. In establishing this classification OSHA clearly indicated that identification of occupational carcinogens does not require positive findings in epidemiologic studies, and nonpositive findings in epidemiologic studies do not necessarily outweigh positive findings in animal studies.

The OSHA Cancer Policy was designed to indicate in part how the agency would identify and classify occupational carcinogens. Unlike the ACGIH and IARC schemes, OSHA's classification was not divided between animal evidence and human evidence. However, unlike the EPA guidelines, it did not deal with the numerical estimation of risk. The OSHA policy based the classification of an occupational carcinogen on human evidence, a single long-term study in mammals with concordance, or in some cases, a single long-term study without concordance. Concordance is defined as additional testing in the same or other species, as positive results in short-term tests, or induction of tumors at injection sites or implantation sites. The OSHA policy also provided that a substance that is metabolized by mammals to yield one or more occupational carcinogens will itself be identified as a potential occupational carcinogen. OSHA would seem to have created a new classification in the rule making for formaldehyde, because it requires warnings of "Potential Cancer Hazard" for this compound. The Hazard Communication Standard indicates that OSHA will accept the judgments of IARC and NTP (as set forth in the latest edition of the Annual Report on Carcinogens). Despite the title of "Annual Report," the foregoing has not been published yearly; the most recent report was published in 1991 (13).

IARC has developed criteria for classifying evidence of carcinogenicity. When the IARC working group evaluates available information on the carcinogenic potential of a specific chemical or an industrial process, it first reviews separately the evidence from epidemiologic studies and studies in experimental animals. For each type, the working group reaches a conclusion as to whether there is sufficient, limited, or inadequate evidence of carcinogenicity. In 1982, IARC adopted criteria for the combined evaluation of human and animal evidence relevant to possible chemical carcinogenicity in humans. Under this system chemicals that have been evaluated for their carcinogenic potential are classified into one of three categories. Chemicals for which there is sufficient evidence of carcinogenicity on the basis of epidemiologic studies are included in Group 1, that is, compounds known to be carcinogenic to humans. Group 2 consists of chemicals that are probably carcinogenic to humans and is further divided into two subgroups based on the nature of the available evidence. Group 2A includes compounds showing limited epidemiologic evidence of carcinogenicity. Group 2B includes compounds for which there is sufficient evidence of carcinogenicity in animals, but inadequate evidence from epidemiologic studies. Group 3 includes those chemicals that cannot be classified as to their carcinogenicity in humans. The classification procedure allows for the use of short-term results or information on potential mechanisms of action, based on structure–activity relationships, to upgrade the classification of a chemical from one group or subgroup to the next higher group.

The IARC classification scheme has been adopted by the NTP for identifying compounds that are "known to be carcinogens" or that are "reasonably anticipated to be carcinogens" for inclusion in the most recent NTP Annual Report on Carcinogens.

The ACGIH classifies cancer potential in two ways. Their designation of A1 indicates a confirmed human carcinogen; the A2 designation indicates a suspected human carcinogen. The A1 classification without a TLV indicates that exposure should be avoided or held to the lowest level possible. ACGIH (14) has proposed a more elaborate scheme in which some of the A2 compounds may be downgraded to "animal carcinogens," but this proposal seems overly complicated.

EPA has developed a similar system for stratifying evidence of human carcinogenicity. In the EPA system chemicals are classified in one of five groups, based on the overall weight of the evidence for carcinogenicity. Group A comprises those chemicals for which there is "sufficient evidence from epidemiologic studies to suggest a causal association between exposure to the agents and cancer." Group B includes those compounds for which there is limited evidence of human carcinogenicity. As in the IARC scheme, this group is further divided into two subgroups, B1 and B2, with the criteria for inclusion in B1 being limited evidence of carcinogenicity from epidemiologic studies. B2 is for sufficient evidence in animals. The EPA system includes a group C, which comprises compounds that are designated as possible human carcinogens on the basis of a wide range of evidence including limited long-term bioassays, short-term tests, and structure–activity relationships. Group D is for compounds that are not classifiable as to human carcinogenicity, and group E comprises compounds for which there is adequate epidemiologic and experimental evidence that they are not human carcinogens. It is worth noting that EPA's classification of a chemical as a carcinogen often does not indicate whether the hazards exist for the inhalation or oral routes, even though EPA has acknowledged that cancer hazards may be route specific. The EPA Cancer Guidelines, unlike the OSHA Cancer Policy, discuss risk assessment methodology and suggest that as a special case of the multistage model, the linearized multistage model (LMS) is generally appropriate.

Table 2.2 (15) shows the ACGIH TLVs, the OSHA PELs, and the NIOSH recommended exposure limits (REL). These are enforceable standards provided by OSHA or voluntary efforts when originating from ACGIH and NIOSH. Chemicals classified by other sources (NTP and IARC) as carcinogens are also included. Review of Table 2.2 shows consistency across groups with respect to some compounds, whereas the results are inconsistent with respect to others. Benzene, 4-aminodiphenyl, and arsenic, for example, are recognized as human carcinogens by all groups. Differences between the exposure limits with respect to benzene and arsenic are due to the methods of assessing risk or safety and the consideration of feasibility by OSHA. DBCP, an animal carcinogen, is regulated as such by OSHA, but ACGIH has no classification or air level for the compound. The classifications

Table 2.2. Classification and Standards for Occupational Carcinogens (mg/m³ as a TWA unless Otherwise Noted)

Compound	ACGIH[a] TLV (mg/m³)	OSHA[b] PEL (mg/m³)	NTP[c]	IARC[d]	NIOSH[e] REL (mg/m³)
Acetaldehyde	180 STEL 270	180 STEL 270		2B	
Acetamide	none		2	2B	
2-Acetylaminofluorene		0–wp CFR 1910.1014			Ca CFR 1910.1014
Acrylamide	0.03, A2	0.03		2B	NON-Ca 0.3
Acrylonitrile	4.3, A2	4.3 (2 ppm) C 10 ppm CFR 1910.1045	2	2A	Ca skin 1 ppm; C 10 ppm
Adriamycin			2	2A	
Aldrin	0.25	0.25		3	Ca Lowest detectable level
Allyl chloride	0.3			3	
4-Aminodiphenyl, also 4-aminobiphenyl	A1 skin	0–wp CFR 1910.1011	1	1	Ca CFR 1910.1011
Amitrole	0.2	0.2	2	2B	
Aniline and homologues	7.6 skin	8		3	
Anisidine	0.50 skin	0.5		2B	
Antimony trioxide production	A2				
Antimony trioxide, handling and use, as Sb	0.5				
Antimony and compounds, as Sb	0.5	0.5			NON-Ca 0.5 skin
Arsenic and soluble compounds, as As	0.2		1	1	Ca C 0.002
Inorganic arsenic		0.01 CFR 1910.1018			
Arsenic trioxide production	A2				
Arsine	0.16	0.2			Ca C 0.002

Asbestos	A1	CFR 1910.1001			Ca
All forms	0.02 fiber/cm^3*	CFR 1910.1101	1	1	0.1 fiber/cm^3 all forms
Amosite	0.5 fiber/cm^3	0.2 fiber/cm^3			
Chrysotile	2.0 fiber/cm^3	0.2 fiber/cm^3			
Crocidolite	0.2 fiber/cm^3	0.2 fiber/cm^3			
Other forms	2.0 fiber/cm^3	0.2 fiber/cm^3			
Tremolite	2, A1	2.0 fiber/cm^3			Ca 0.1 fiber/cm^3
Asphalt (petroleum) fumes	5	5	1	1	Ca; C 5
Benzene	0.3, A1* skin	3 STEL 15; CFR 1910.1028	1	1	Ca 0.32; C 3.2
Benzidine	A1 skin	0–wp; CFR 1910.1010	1	1	Ca; CFR 1910.1010
Benzo(a)pyrene	A2	0.2 (coal tar pitch volatiles)	2	2A	Ca 0.1
Benzo(b)fluoranthene	A2*			2B	
Benzyl chloride	5.2	5	2	2A	NON-Ca C 5
Beryllium and compounds, as Be	0.002, A2	0.002, C 0.005 (30 min peak)	2	2A	Ca 0.0005
tert-Butyl chromate, as CrO$_3$	C 0.1 skin	C 0.1 skin			Ca 0.001
1,3-Butadiene	22, A2	1000—rulemaking in progress	2	2B	Ca; Lowest feasible level
Cadmium, dust, and salts, as Cd	0.01, A2 (total dust)*; 0.002 (respirable fraction)*	0.2, C 0.6	2	2A	Ca; Lowest feasible (0.01)
Cadmium oxide fume, as Cd	0.01, A2	0.1 C 0.3			Ca Lowest feasible (0.01)
Cadmium oxide production	0.001, A2				
Carbon black	3.5	3.5		3	Ca 3.5; 0.1 in the presence of PAHs
Carbon tetrachloride	31, A2	12.6	2	2B	Ca C 12.6 (60 min)
Chlordane	0.5 skin	0.5 skin		3	
Chlordecone (kepone)			2	2B	Ca 0.001

Table 2.2. Continued

Compound	ACGIH[a] TLV (mg/m³)	OSHA[b] PEL (mg/m³)	NTP[c]	IARC[d]	NIOSH[e] REL (mg/m³)
Chlorinated camphene	0.50 skin C1	0.5 skin C1	2	2B	Ca 0.001
Chlorodiphenyl (42% Cl)	1.0 skin	1.0 skin			Ca 0.001
Chlorodiphenyl (54% Cl)	0.5 skin	0.5 skin			
Chloroform	49, A2	9.76	2	2B	Ca C 9.76 (60 min)
bis(Chloromethyl) ether	0.0047, A1	0—wp; CFR 1910.1008	1	1	Ca; CFR 1910.1008
Chloromethyl methyl ether (also methyl chloromethyl ether)	A2	0—wp; CFR 1910.1006	1		Ca; CFR 1910.1006
β-Chloroprene	36 skin	90 skin		3	Ca C 3.6
Chromite ore processing	0.05, A1			1	
Chromium (VI) compounds as Cr			1	1	
Water soluble	0.5				
Certain water insoluble	0.05, A1				
Chromic acid and chromates		C 0.1			Ca 0.001; C 0.05
Calcium chromate	0.001, A2		1	1	Ca 0.001; C 0.05 under Cr VI
Chromyl chloride	0.16				
Chrysene	A2	0.2			Ca 0.1 as coal tar pitch
Coal tar pitch volatiles as benzene solubles	0.2, A1	0.2			Ca 0.1 as cyclohexane extractable
Coke oven emissions		0.15; CFR 1910.1029			Ca 0.5–0.7
Dichlorodiphenyltrichloroethane (DDT)	1	1	2	2B	Ca 0.5
2,4-Diaminoanisole	0.34				Ca Lowest feasible
o-Dianisidine based dyes			2	2B	Ca Minimize exposure
1,2-Dibromo-3-chloropropane (DBCP)		0.001; CFR 1910.1044	2	2B	Ca; CFR 1910.1044
Dichloroacetylene	C 0.39	C 0.4		3	
p-Dichlorobenzene	60, A2*	450 STEL	2	2B	

Substance					
3,3'-Dichlorobenzidine (and its salts)	A2 skin	0–wp CFR 1910.1007	2	2B	Ca CFR 1910.1007
Dichloropropene	4.5 skin	5 skin	2	2B	
Dieldrin	0.25 skin	0.25 skin		3	Ca Lowest feasible
Di(2-ethylhexyl)phthalate, also (di-sec-octyl)phthalate	5, STEL 10	5, STEL 10	2	2B	Ca Lowest feasible
Diglycidyl ether (DGE)	0.53	0.5	2	2A	Ca C 1
Dimethyl carbamoyl chloride	A2		2	2A	Ca Lowest feasible
Dimethylaminobenzene (xylidine)	2.5, A2 skin	10	2	2B	
4-(Dimethylamino)azobenzene	none	0–wp CFR 1910.1015	2	2B	Ca CFR 1910.1015
1,1-Dimethylhydrazine	0.025, A2 skin*	1 skin	2	2B	Ca C 0.015 (120 min)
Dimethyl sulfate	0.52, A2	0.5	2	2A	NON-Ca
Dinitrotoluene	0.15, A2 Skin	1.5		2B	Ca Lowest feasible
Dioxane	90 skin	90 skin	2	2B	Ca C 3.6 (30 min)
Epichlorohydrin	0.39, A2 skin	8.0	2	2A	Ca Exposure to be minimized
Ethyl acrylate	20, A2 C 61	20, C 100	2	2B	NON-Ca
Ethyl bromide	22, A2*				
Ethyl chloride	2640	2600			Ca Handle with care
Ethylene dibromide (1,1-dibromoethane)	A2	20 ppm, C 30 ppm proposed-peak 50 ppm (5 min)	2	2B	Ca 0.038; C 3.1
Ethylene dichloride	40	4, C 8	2	2B	Ca 4; C 8
Ethylene oxide	1.8	1.8, STEL 9 CFR 1910.1047	2	2A	Ca 0.18; C 9 (10 min)
Ethylene thiourea			2	2B	Ca Minimize exposure
Ethylenimine	0.88-skin	0–wp CFR 1910.1012		3	Ca CFR 1910.1012
Formaldehyde	C 0.37, A2	1.2, C 2.5 CFR 1910.1048	2	2A	Ca 0.016 ppm C 0.1 ppm
Hematite, underground mining			1	1	

Table 2.2. *Continued*

Compound	ACGIH[a] TLV (mg/m³)	OSHA[b] PEL (mg/m³)	NTP[c]	IARC[d]	NIOSH[e] REL (mg/m³)
Heptachlor and heptachlor epoxide	0.05, A2 skin*				
Hexachlorobenzene	0.025, A2 skin*			3	
Hexachlorobutadiene	0.21, A2 skin	0.24			Ca Minimize exposure
Hexachloroethane	9.7, A2 skin*	10 skin	2	2B	
Hexamethyl phosphoramide	A2, skin		2	2B	Ca 0.04 (120 min)
Hydrazine	0.013, A2 skin*	0.1	1	3	
Isopropyl oil (product of isopropyl alcohol manufacture, strong acid method)					
Lead, inorganic dust, and fumes, as Pb	0.15	0.05 (NON-Ca) CFR 1910.1025		2B	NON-Ca <0.1
Lead arsenate as PbHAsO₄	0.15	0.01 as As CFR 1910.1018			
Lead chromate					
As Pb	0.05, A2				
As Cr	0.012, A2				
Lindane	0.5	0.5	2		
Methyl bromide	19 skin	20 skin		3	Ca Lowest feasible
Methyl chloride	103, C 207	105, C 210		3	Ca Lowest feasible
Methylene chloride	174, A2	1740, C 1000 ppm		2B	Ca Lowest feasible
4,4'-Methylenebis(2-chloroaniline)	A1 skin*	0.22 skin	2	2A	Ca 0.003
4,4'-Methylenedianiline	0.81, A2 skin		2	2B	Ca Lowest feasible
Methyl hydrazine	0.019, A2 skin*	C 0.35 skin			Ca C 0.08 (120 min)
Methyl iodide	12, A2 skin	10	2	3	Ca Lowest feasible
Naphtha (coal tar)	400			1	
1-Naphthylamine (α-napthylamine)	none	0–wp CFR 1910.1004			Ca CFR 1910.1004
2-Naphthylamine (β-napthylamine)	A1	0–wp CFR 1910.1009	1	1	Ca CFR 1910.1009

Substance					
Nickel metal	0.05, A1*	1	2	1	Ca 0.015
Insoluble cpds as Ni	0.05, A1*	1	2	1	Ca 0.015
Soluble cpds as Ni	0.05, A1*	0.1	2	1	Ca 0.015
Nickel carbonyl as Ni	Deletion Ni to apply*	0.007		1	Ca Lowest detectable (0.007)
Nickel sulfide roasting, fume and dust	Deletion Ni to apply*		1		Ca 0.015
4-nitrodiphenyl (also 4-nitrobiphenyl)	A1	0—wp CFR 1910.1003		3	Ca CFR 1910.1003
2-Nitropropane	36, A2	35		3	Ca Lowest feasible
Particulates polycyclic aromatic hydrocarbons (PPAH)-coal tar pitch	0.2, A1	0.2		1	Ca 0.1 as cyclohexane extractable
Pentachloroethane					Ca Minimize exposure
Perchloroethylene	339, STEL 1370	170		3	Ca Minimize exposure
N-Phenyl-β-napthylamine	A2				Ca Lowest feasible
Phenyl glycidyl ether (PGE)	6.1	6			Ca C 5
Phenylhydrazine	22, A2	20, STEL 45			Ca C 0.6 (120 min)
Propane sultone	A2		2	2B	Ca
β-Propiolactone	1.5, A2	0—wp CFR 1910.1013	2	2B	Ca CFR 1910.1013
Propylene imine	4.7, A2	5	2	2B	
Propylene oxide	48	50		2A	
Quartz (crystalline silica)	0.1	0.1 (respirable dust)	2		Ca 0.05
Selenium cpds as Se	0.2	0.2		2B	
Strontium chromate	0.0005, A2*				
Styrene monomer	213 skin STEL 426	215, STEL 425	2	2B	NON-Ca 215 C 425
2,3,7,8-Tetrachlorodibenzo-p-dioxin (TCDD)				2B	Ca Lowest feasible
1,1,1,2-Tetrachloroethane					Ca Minimize exposure
1,1,2,2-Tetrachloroethane				3	Ca Lowest feasible
o-Tolidine	A2 skin		2	2B	Ca 0.02 (60 min)
Toluene-2,4-diisocyanate (TDI)	0.036, STEL 0.14	0.4, STEL 0.15	2	2B	Ca 0.035; C 0.14

Table 2.2. *Continued*

Compound	ACGIH[a] TLV (mg/m³)	OSHA[b] PEL (mg/m³)	NTP[c]	IARC[d]	NIOSH[e] REL (mg/m³)
o-Toluidine	8.8, A2	22	2	2B	
p-Toluidine	8.8, A2	9	—	—	
1,1,2-Trichloroethane	55 skin	45 skin		3	Ca Minimize exposure
Trichloroethylene	269, STEL 1070	270, STEL 1080		3	Ca 135
Vinyl acetate	35, A2*				Ca CFR 1910.1017
Vinyl bromide	22, A2	20		2A	Ca
Vinyl chloride	13 (5 ppm), A1	2, C 5 ppm CFR 1910.1017	1	1	CFR 1910.1017
Vinylcyclohexene dioxide	57, A2 skin	60			
4-Vinylcyclohexane	0.4, A2*				
Vinylidene chloride	20	4		3	
Welding fumes (not otherwise classified)					Ca Lowest feasible

[a]AGGIH designation of A1 indicates a confirmed human carcinogen; the A2 designation indicates a suspected human carcinogen; C indicates a ceiling; STEL is used to denote a short-term exposure limit; the designation of "skin" indicates that skin contact should be avoided. The designation NL indicates that the material is no longer found in the workplace. *Proposed change 1991–1992.

[b]The 0–wp designation indicates that the material is one of the original "14 carcinogens."

[c]The NTP designations for carcinogens are as follows: 1, known to be carcinogenic, with evidence from human studies; 2, reasonably anticipated to be a carcinogen, with limited evidence in humans or sufficient evidence in experimental animals.

[d]IARC's classification scheme is as follows: 1, carcinogenic in humans, with sufficient epidemiologic evidence; 2A, probably carcinogenic in humans with (usually) at least human evidence; 2B, probably carcinogenic to humans, but (usually having no human evidence) sufficient evidence of carcinogenicity in experimental animals. Not all compounds classified are included in this table.

[e]NIOSH defines materials as carcinogens (Ca) with no other classification.

by IARC and EPA are consistent with the OSHA regulation. On the other hand, only IARC considers selenium compounds and inorganic lead as constituting carcinogenic hazards. ACGIH and, sometimes, NTP and IARC classify particular segments of an industrial process as having carcinogenic potential (arsenic trioxide production); or an entire industrial segment may be so classified (hematite underground mining). ACGIH can act more rapidly than the other classifying and standard-setting groups, as is illustrated by their stand on 4-vinylcyclohexane.

3   THE SCIENTIFIC BASIS OF RISK ASSESSMENT

The hazard evaluation or the classification of a material as an occupational carcinogen is based on the finding that the material induces cancer in humans or that it causes a significant increase in cancer in animals. Most often, for a material to be recognized as an occupational carcinogen, the route of exposure to the material must be through airborne concentration or impaction on the skin in at least one of the studies.

3.1   Hazard Evaluation: the Identification of Occupational Carcinogens

The methods used to identify occupational carcinogens are continuously evolving and are not without controversy. Historically, occupational carcinogens were identified and regulated almost exclusively on the basis of human experience. Thus the first official attempts to control occupational carcinogen exposure involved compounds that were identified when excess incidences of relatively rare cancers were observed in exposed workers. Notable examples were lung cancer in asbestos-exposed workers, hemangiosarcomas in rubber workers exposed to vinyl chloride monomer, and bladder cancers in workers exposed to the manufacture of azo dyes.

On the other hand, the delay in regulating 1,3-butadiene, following completion of an inhalation bioassay in mice indicating that the material was very potent at a level below the then-current PEL of 1000 ppm and caused cancer to such an extent that the study was halted after 61 weeks, seems unreasonable. Epidemiologic investigations done under such circumstances to discover the "smoking gun" can certainly be considered, in the words of Utidjian (16) as an ". . . opportunistic analysis of the results of inadvertent human experimentation in industry." In 1983, Sir Richard Doll observed that although most occupationally related cancers have been detected by clinical intuition or epidemiologic observation, they could be detected sooner if modern toxicologic techniques were employed to test the substance *before* men and women were exposed to them in the workplace (17). The reliance on proof positive as determined by sufficient numbers of workers' death certificates verifying cancer as the cause of their demise no longer seems appropriate when compared to the cost of an animal bioassay.

In recent years, scientists and regulators have sought to identify and regulate potential occupational carcinogens as early as possible in order to reduce the risk of future occupational cancer "epidemics." This has led to an increased reliance

on the consideration of the complete body of available experimental and observational evidence (i.e., a weight of the evidence approach) for the identification
of occupational carcinogens. This body of evidence comprises studies of exposed
humans, studies in whole animals, mechanistic studies, and structure–activity analysis.

3.1.1 Studies of Exposed Humans

Studies in the human population can generally be grouped into one of two categories, descriptive or analytical. Descriptive studies are useful only for generating
hypotheses regarding possible associations between exposure and disease, whereas
analytical studies can be used to test a hypothesis and verify the existence of an
association between a disease and exposure. Descriptive studies include case reports, surveillance systems, ecological studies, and cluster studies.

Case reports are descriptions of one or more cases of the "disease of interest"
and often include demographic, medical, and occupational history information that
can form the basis for an inference regarding possible etiologies. These reports are
of limited value as evidence of causal associations because they are uncontrolled,
anecdotal, and heavily reliant on recall for the identification of possible chemical
exposure. In spite of these limitations, the majority of known human carcinogens
were first suggested in case reports.

Surveillance systems are data bases rather than studies or reports. Such systems,
which include cancer registries, compile data on the incidence of certain diseases
within a specific geographic area or occupational category and provide information
on the incidence or prevalence of those diseases within the population covered by
the registry. Differences in disease incidence/prevalence between geographic areas
or occupational classifications may indicate an association between the disease and
some aspect of the geographic area or occupational environment. When the data
in a surveillance system are combined with data about demographics or occupation
from another source, the resulting study is termed an ecological study.

Ecological studies (also called correlational studies) look for correlations between disease incidence data and information on employment history. Typically,
these are record-linkage studies in which records that reflect the incidence of a
disease in a given population (such as a disease registry, hospital discharge records,
or death certificates) are linked to records reflecting exposure (such as employment
records). Such studies can indicate whether a correlation exists between disease
and specific employment categories and incidence/prevalence/mortality. As with
other descriptive studies, it is impossible to control for confounding variables.

The final type of descriptive study is referred to as a health effects survey or
"cluster study." This is a special case of ecological study in which the incidence or
severity of a specific adverse effect appears to be higher in a small geographic area
or occupational category (i.e., the workers at a single facility) within a given time
frame than in the population at large. A careful census of the incidence/prevalence
of the disease in the defined population can confirm or deny the apparent excess.
Like other descriptive studies, however, it is usually not possible to control for

potential confounding factors. Furthermore, the population covered by cluster studies are relatively small, resulting in very low statistical power.

Once a hypothetical causal association between cancer and an occupational exposure has been suggested by descriptive human studies, by studies in experimental animals, or by structure–activity relationships, the hypothesis may be tested using one or more analytical study methods. This overall category includes human experimental (clinical) studies and cohort or case-control epidemiologic studies. Human experimental studies have the advantage of controlling human exposure to the agent in question, as is the case in classical animal experiments for testing toxicity. Because it is unethical to select humans for exposure to a potentially harmful agent, the use of this study design is generally limited to randomized clinical trials of therapeutic agents and intervention studies. By controlling for confounding factors using randomization in clinical trials, the demonstration of an adverse effect may provide very convincing evidence of a causal relationship with exposure to the agent of concern. Intervention studies, in which the frequency of a specific adverse effect is compared before and after some type of corrective measure has been applied, provide data for testing the hypothesis that the application of a corrective measure reduces the frequency of the adverse outcome. The results may be conclusive if the study is carefully designed and confounding factors well controlled. Human studies are useful in developing biologic markers. Human pharmacokinetic information is invaluable in determining the biologically effective dose (BED) and in the quantitative assessment of risk (QRA).

Ethical considerations preclude the use of human experimental studies and longitudinal as opposed to historical prospective epidemiologic studies for the purpose of identifying occupational carcinogens. Thus most prospective epidemiologic studies of occupational cancer are historical in nature; that is, the studies look for an association between present or future disease and past exposure.

Cohort or prospective epidemiologic studies are those in which the incidence of, or mortality from, the disease of interest within a group (cohort) of individuals known to be exposed to the agent is compared to that within a comparison group that is as similar as possible to the cohort, except that the group was not exposed to the agent of interest. Case control or strictly retrospective studies, on the other hand, identify a set of individuals who have or had the disease of interest and a matching set of individuals within the disease. Exposure to the agent of interest is ascertained in both groups, and if such an exposure was more common within the case group than without the matched control group, an association is presumed to exist between the exposure and the disease.

Although epidemiologic studies provide the most reliable basis for identifying human carcinogens, they must always be interpreted with caution. A positive finding, that is, a statistically significant association between cancer and a specific occupational exposure, is not a sufficient basis for concluding that there is a cause and effect relationship. The actual cause of the cancer might be an unidentified factor that is common to both the exposure and the disease. In order to have definitive evidence that a causal association exists between cancer and an occupational exposure, it is necessary that the epidemiologic evidence be reproducible, demonstrate a dose–response relationship, demonstrate appropriate temporal characteristics (i.e., the exposure occurred prior to onset of the disease), not be explainable by known confounders, and be biologically plausible.

The results of "negative" (nonpositive) epidemiologic studies are even more difficult to interpret and seldom definitively rule out the carcinogenic potential of an occupational exposure. The methods of epidemiology were devised to identify the etiology of a disease when an excess incidence of the disease (epidemic) is obvious. The use of these methods to identify an excess incidence of disease associated with a given etiology is of limited utility. Epidemiologic studies may fail to detect a true association between a disease and an exposure because they lack the statistical power. Even the most powerful occupational studies are unlikely to detect less than a 50 percent increase in the background incidence of cancer. Often the expected incidence of rare cancers in occupational cohort studies is less than one, resulting in an inappropriate basis for statistical comparison between the cohort and the control group.

Classification of exposure on the basis of objective workplace or personal exposure monitoring data is unusual in historical occupational epidemiologic studies. Rather, the determination of exposure status often is based on individual recall or indirect means of ascertainment, such as a review of employment history. In the future, information obtained from OSHA inspections, NIOSH industrial hygiene surveys, voluntary industrial hygiene monitoring by industry, or TSCA submissions may be useful in epidemiologic studies.

A true association between an occupational exposure and cancer may not be detected by an epidemiologic study if there is selection bias (i.e., including unexposed individuals in the cohort or including exposed individuals in the comparison group) or if recall or ascertainment of past exposure is faulty. Finally, as mentioned above, the potential for confounding is an ever-present problem in occupational epidemiologic studies. Ideally, the cohort and the comparison population should be identical except for the exposure in question. Practically, this is impossible to achieve in occupational epidemiologic studies. Few workers are exposed to one and only one chemical in the workplace. Differences in overall life-style, including such factors as tobacco and alcohol use and risk-taking behavior, are often significant between the cohort and the comparison group, especially if the latter is representative of the general population. In some cases these differences can be identified and factored into the analysis, but in many cases they cannot. In summary, although a single epidemiologic study showing a statistically significant positive association between cancer and a given chemical exposure can be strong evidence of a causal relationship, it is generally not definitive evidence for such a relationship unless supported by other evidence.

Many epidemiologic studies suggest that certain industrial processes are associated with an increased risk of cancer, but the causative agent in the process cannot be isolated. Table 2.2 lists several such situations, including antimony trioxide production, cadmium oxide production, chromite ore processing, hematite mining, and welding fumes. In such situations one should consider that the problem is in the nature of exposure to a complex mixture. Indeed, humans are exposed daily to complex mixtures. In experimental studies the exposure may consist of the pure agent. However, some have been carried out with technical-grade materials later found to contain significant amounts (5 to 10 percent) of carcinogens.

Although human studies are often preferred for the classification of a material as a carcinogen and the quantification of risk, they have certain drawbacks. Most often the exposure is not adequately measured, and exposure over the years has declined. Often workers' exposures are not consistent, and many workers are not employed a full working lifetime, up to 45 years as defined by most risk assessors. Workers are most often exposed to more than one cancer-causing agent. Furthermore, epidemiologic studies may be limited to rather short periods that deal only with a worker's employment period, not allowing sufficient time for the cancer to develop because of the long latency between cancer-initiating exposures and the progression of clinical disease. Additional inconsistency exists in the rate of cancer in situations where the worker is exposed to the same cancer-causing agent outside the workplace. For example, a worker might be exposed regularly to formaldehyde in the course of employment and be additionally exposed at home as a result of offgassing from housing materials.

3.1.2 Whole Animal Bioassays

Long-term bioassays in experimental animals have been used for more than two decades as a means of identifying potential human carcinogens. These assays are typically conducted using inbred strains of mice and/or rats. The animals are exposed to the test material at doses up to and including a dose (maximum tolerated dose, MTD) that causes signs of gross toxicity (e.g., decreased weight gain or decreased survival) for the better part of the lifetime of the test species (18 months or more for mice and 2 years or more for rats). The detailed methodology of these studies varies, depending on the nature of the test material, its intended use, and the overall objectives of the study. If such a study is designed to test the potential carcinogenicity of the test compound in the workplace, the animals typically will be exposed by an appropriate route (inhalation and/or dermal) using an appropriate exposure schedule (e.g., 6 hr/day for 5 days/week, beginning after the animals are sexually mature). Otherwise, the protocol may involve oral exposure (feeding, gavage, or drinking water), sometimes beginning in utero.

Using the results of long-term bioassays as the primary basis for identifying human carcinogens is the subject of ongoing scientific debate. Although essentially all the occupational exposures to materials that have been identified as human carcinogens have also produced positive carcinogenic responses in animal models, many compounds that cause cancer in rodent bioassays have not been demonstrated as human carcinogens. A case in point is the whole family of chlorinated aliphatic hydrocarbons. Many of these compounds have been shown to generate a carcinogenic response, particularly liver tumors in mice, when administered orally to rodents in long-term bioassays. However, only one of them, vinyl chloride, has been shown to be a human carcinogen, despite relatively widespread occupational and environmental exposure of humans to members of this family of compounds.

A number of hypothetical and semiempirical explanations have been advanced for this apparent species difference. Most of these explanations are pharmacokinetic in nature. In other words, the carcinogenic response seen in rodents as a result of very high doses of these materials is not relevant either to lower dose exposures in rodents or to any human exposures. This is true, the argument runs, because by the oral route and at the high doses tested, the absorption, distribution, metabolism (including bioactivation), and excretion of the chlorinated aromatic hydrocarbons follow different kinetics and pathways than they do at environmentally relevant exposures. Furthermore, the rodent liver has a relatively higher metabolic rate than does the human liver, so that more rapid metabolism of the parent compound would result in higher concentration of reactive intermediates at critical targets. Finally, the liver damage caused by high doses of these compounds may overwhelm natural repair mechanisms, whereas at lower doses these mechanisms may operate to protect the target from the type of damage that can progress to neoplasia.

Although all these hypothetical explanations for the apparent interspecies differences in susceptibility to the carcinogenic effects of chlorinated aliphatic hydrocarbons are biologically plausible and there is a body of experimental evidence in support of some of them, the possibility remains that some or all of these compounds do generate a carcinogenic response in humans by some route and at some level of exposure. The absence of a clear excess of liver cancers among workers exposed to these compounds may only indicate that by other than the oral route different organs or tissues may be the targets for the carcinogenic effect. For example, whereas trichloroethylene causes liver tumors in mice exposed orally, it causes lung tumors when the mice are exposed by inhalation. If trichloroethylene results in an increased incidence of cancer at one or more sites in exposed workers, it might be difficult to detect such an effect in an epidemiologic study because of the high background incidence of all tumors and tumors at some specific sites, such as the lungs, where confounding by cigarette smoking is often a problem.

The logical conclusion is that long-term bioassays in experimental animals are inadequate as the sole basis for identifying occupational carcinogens, especially if the animals have been exposed by an irrelevant route and exposure regimen. However, when the results of such studies are considered and interpreted in relation to a larger body of observational and experimental evidence that includes studies of the mechanism of action and relative pharmacokinetics of the compound in animals and humans, they can provide the primary bases for the identification of occupational carcinogens, even in the absence of clear-cut epidemiologic evidence of a positive association.

Although the exact mechanism of action has not been elucidated for any chemical carcinogen, it is quite clear that many chemicals that cause cancer in experimental animals must be absorbed into the systemic circulation, transported to the site(s) of metabolic transformation and action, and activated to the proximal carcinogenic reactive state, which then interacts with the appropriate subcellular component to initiate the carcinogenic response. Other chemical carcinogens may not initiate the carcinogenic process at all, but instead may facilitate a carcinogenic response initiated by another xenobiotic compound or by an endogenous chemical moiety, through altered gene expression, induction of cell damage and repair, or perturbations in hormone status. There may be sex-, age-, and species-related differences in all these steps. Thus a compound that is a carcinogen in rodents may not cause cancer because, at the relatively lower levels to which humans are likely exposed, the compound is not absorbed or transported in sufficient quantities to reach a sensitive target, or it is metabolized in such a manner that it is deactivated and excreted rather than activated, or there is no receptor or gene target in humans.

3.1.3 Mechanistic Studies

Mechanistic studies focus on three different theories about the mechanism involved in the induction of cancer. These areas of study include genetic toxicity, immunosuppression, and cell proliferation. The results of such studies, considered in relation to pharmacokinetics, metabolism, and chemical structure, can broaden the basis for the assessment of risk. Because of the costs and the time involved in conducting long-term bioassays, scientists have been investigating a number of short-term experimental studies that can aid in the identification of chemical carcinogens. Some of these studies, performed in conjunction with long-term bioassays, can provide valuable mechanistic information that is useful in assessing potential human health risks. In the expanding field of risk assessment, the possibility of using human tissue is an approaching likelihood. Already the metabolic capability of the mouse lung to transform 1,3-butadiene to its reactive genotoxic metabolites has been contrasted to the inability of human lung tissue to perform the same transformation (18).

The idea that genotoxicity (alteration of DNA) can lead to cancer is currently supported by regulators. There are a number of mechanistic tests based on the idea that cancer is due to a change in DNA. These studies deal with the mutagenic potential of chemicals. Some of the tests that may reveal the potential of a chemical to cause cancer as a result of genotoxicity are listed in Table 2.3.

Although current cellular transformation studies utilize only animal tissues, it seems only a matter of time until this technique can be used with human tissue. Cellular transformation studies are unique among the mechanistic studies because changes in cellular morphology are used as an end point. If these techniques can be coupled with pharmacokinetic knowledge regarding the target tissue concentration, then the risk can be explored to a greater extent.

The role of cellular toxicity per se in the development of cancer has long been recognized as important. Recently, the importance of the role of increased cell turnover, that is, cell proliferation, has been emphasized by well-known scientists (19, 20). However, it had already been noted that after cell death caused by ionizing radiation (21) and alkylating agents (22) a rebound phenomenon that results in an increase of cellular reproduction exists. With either X-irradiation or xenobiotics a greater tendency exists for the effect on cell turnover to be manifested in tissue with a normal high cell division rate, for example, the gastrointestinal tract or the bone marrow. Accelerated cell turnover increases the likelihood that mutations

Table 2.3. Short-Term Tests For Chemical Carcinogens

In vitro Tests

Mutagenesis in bacteria (e.g., Ames test)
Mutagenesis in eukaryotic microorganisms
Mutagenesis in mammalian cells in culture
Induction of chromosome aberrations (breaks, nondisjunctions, sister chromatid exchanges) in mamallian cells in culture
Induction of unscheduled DNA synthesis (DNA repair)
Induction of hepatocellular proliferation
Inhibition of intracellular communication (metabolic cooperation) in mammalian cells in culture
Mammalian cell transformation

In vivo Tests

Mutagenesis in *Drosophila*
Induction of dominant lethal mutations in rodents
Induction of micronuclei, chromosomal aberrations or SCEs in bone marrow of mice
Induction of unscheduled DNA synthesis in rodent liver
Detection of DNA adducts using P postlabeling (any tissue, any species)
Induction of enzyme-altered foci in rodent liver
Initiation/promotion bioassays
Gene mutation and/or activation in transgenic mice

---

will occur because the DNA of the dividing cell is more sensitive to chemical alteration. An increase in cell turnover rate also increases the likelihood that the initiated (genetically altered) cell type multiplies. Often increased cellular proliferation is manifested as hyperplasia.

The role of immunosuppression is important in the induction of cancer because many patients treated with immunosuppressants do indeed develop the malady as a result of treatment with such drugs. In addition, many promoters have immunosuppression as an innate part of their apparent mechanism of action. Cancer-inducing agents that alter the immune system include ultraviolet light, ionizing radiation, benzene, trichloroethylene, cadmium, lead, nickel carbonyl, arsenicals, diethylstilbestrol, 2,3,7,8-tetrachlorobenzo-*p*-dioxin, benzo(*a*)pyrene, and many of the phorbol esters (23, 24).

Another type of evidence that is useful for identifying potential occupational carcinogens is structure–activity relationships. It has long been recognized that the presence of certain functional groups in organic molecules serve to impart genotoxic and often carcinogenic activity to the compound. Nitrosamines and epoxides are examples of such classes of compounds. Other chemical classes demonstrate carcinogenic activity associated with size and shape rather than with specific functional groups. Thus four- and five-ring polycyclic aromatic hydrocarbons in which certain positions are available for hydroxylation by aryl hydrocarbon hydroxylase are carcinogenic in experimental animals, whereas smaller and larger or sterically hindered compounds of the same general category are inactive. In another example, the ability of specific chlorinated dibenzo-*p*-dioxins and dibenzofurans to produce a carcinogenic response in experimental animals appears to be related to the affinity with which they bind to a specific receptor (the Ah receptor). Thus although only a few specific congeners have been tested in conventional long-term carcinogenesis bioassays, a number of untested congeners, that is, those in which the four lateral positions are occupied by chlorine atoms, are considered to have carcinogenic potential based on their structure; and their relative potency is determined by the relative affinity with which they bind to the Ah receptor. Other classes of chemicals, some of whose members are considered to be carcinogenic on the basis of structure alone, even in the absence of animal testing, include nitrofurans, polychlorinated biphenyls (PCBs), and aromatic azo compounds.

Mechanistic studies may also be used to indicate that a cancer seen in experimental animals, particularly rodents, is unlikely to be carcinogenic in humans. There is considerable discussion among scientists regarding the role of cell proliferation in the development of cancer, particularly for nongenotoxic chemicals (19, 20). Most importantly, if cell proliferation (particularly in response to toxicity) is involved in the induction of tumors, then only those dosages that cause toxicity will produce a proliferative response and result in tumor formation. There are several such cause-and-effect relationships that appear not to have an appropriate basis for extrapolation across species. Among those involving the urinary tract are male rat tumors related to α-2-μ-globulin accumulation, and rodent bladder tumors related to physical irritation. D-Limonene, a nongenotoxic agent, induces kidney tumors in male rats; but this observation is considered by several governmental agencies not to be a valid basis for classification of D-limonene as a human carcinogen. Likewise, the rodent bladder tumors that result from irritation by melamine and saccharin are unlikely to provide a sound basis for the classification of these materials as human carcinogens and therefore cannot function as a basis for extrapolation of risk from rodents to humans.

The B6C3F1 mouse is frequently used for chronic bioassays, and it is the standard strain used in bioassays by NTP. This strain (particularly the males) is uniquely prone to the development of liver tumors of both a spontaneous and a chemically induced nature. Liver tumors are the most frequently induced tumors in both the rat and mouse NTP bioassays, but the liver is not among the top 10 cancer sites found in men or women (25). Because cell proliferation may be the major cause of the increase in mouse liver tumors, it has been suggested that, in the case of nongenotoxic agents that produce mouse liver cancer, the highest dose at which there is no stimulation of cell proliferation in the liver might be considered a no adverse effect level (NOAEL). This level or dose would be used in the safety factor approach for assessing risk (26).

Peroxisome proliferation may be another mechanism that operates in rodents, but not in humans. Peroxisome proliferation was the apparent cause of liver tumors in both rats and mice following exposure to diethylhexyl phthalate (27). Peroxisome proliferation resulting from exposure to trichloroethylene may also be involved in the induction of liver tumors in mice. The causative metabolite is trichloroacetic acid. Because the metabolism of trichloroethylene is saturated at a lower dose in the rat than in the mouse, the lower concentration of trichloroacetic acid in the rat liver may cause the lack of liver tumor induction in that species (28).

Thyroid follicular cancer in rodents may occur largely because of inhibition of thyroid–pituitary homeostasis. Hill et al. (29) suggested that if humans develop tumors following long-term derangement in thyroid–pituitary status, they are less sensitive than rodents. In the case of nongenotoxic materials that induce thyroid cancers, it is appropriate to consider that a threshold may exist; this may be ascertained using clinical pathology measurements of thyroid status.

Consideration of the metabolism is also important with regard to the possible mechanism involved in apparent species differences. The case has been made in the example of methylene chloride that the metabolic pathway responsible for inducing tumors in rodents operates only at high levels unlikely to occur in humans. Metabolism and kinetics of absorbed materials are important considerations. OSHA recommended classification of a material as a carcinogen if it is metabolized to a known carcinogen. It is also appropriate to consider situations in which the material forms known carcinogens in certain in vitro situations. For example, azobenzene in acid forms benzidine, a known carcinogen. Another example of this is the apparent formation of vinyl chloride in groundwater contaminated with trichloroethylene under methanogenic conditions (30).

3.2 An Approach to the Classification of Occupational Carcinogens

In view of the confusion created by differing classifications among organizations dealing with chemicals, a simpler two-tiered classification scheme for occupational carcinogens may be useful. In both tiers the agent must be in the workplace. The first category, *Occupational carcinogens*, are those agents that:

A. Are metabolized to known carcinogens.
B. Produce cancer among workers as determined by epidemiologic studies or can be shown to have the potential to do so by experimental studies utilizing human tissue.
C. Produce cancer in two species of nonhuman mammals. In one species the route of exposure should be by the inhalation or dermal route, and mechanistic studies should indicate similar pathways in different species, including humans.
D. Produce cancer in one species of nonhuman mammals and, on the basis of structure or mechanistic studies, seem likely to produce cancer in other mammals, including humans.

Potential occupational carcinogens, the second category, are those that:

A. Epidemiologic studies suggest are carcinogens.
B. One or more nonhuman mammalian studies suggest have a carcinogenic potential, and mechanistic studies do not suggest that the end points are unlikely to exist in humans.

Many chemicals produce cancer in experimental animals. In most cases, epidemiologic studies have not confirmed increases at the same sites in humans. This may occur because the dose in animals is higher, the routes of exposure are different between the experimental animals and humans, the animals may have unusually susceptible tissues, or the metabolism and pharmacokinetics differ. Many classification schemes do not take into consideration the route or nature of the cancer end point produced in the animal studies. For examples, it hardly seems appropriate to consider the cancers induced by oral administration of toluene diisocyanate (TDI) in experimental animals (31) as a basis for classification of the material as an occupational carcinogen. The important route of exposure for this sensitizing agent in the workplace is inhalation and skin contact, and many workers are affected. Furthermore, there was no evidence of TDI-associated benign or malignant tumors in mice or rats exposed to 0.05 or 0.15 ppm for 2 years (32). Because the OSHA/ACGIH short-term exposure limit is 0.02 ppm, it is not unreasonable that at higher concentrations the survival in the animal studies was reduced.

4 ASSESSING RISK FROM EXPOSURE TO OCCUPATIONAL CARCINOGENS

The second element of quantitative risk assessment, dose–response characterization, is the determination of a quantitative relationship between exposure and response that will allow the estimation of the magnitude or frequency (in the case of cancer) of a response associated with any level of exposure. Typically, dose–response information is available from the same studies that were relied upon in the hazard evaluation step. However, the data base is generally much more limited. Many epidemiologic studies that provide qualitative evidence for an association between exposure to a given chemical and cancer provide little or no quantitative dose–response information.

Most occupational epidemiologic studies are historical in nature (even though prospective in mode), in that the incidence of, or mortality from, cancer in a given cohort is determined using historical exposure occurring from 5 to 30 years earlier. In most of these, no quantitative data are available that would indicate the amount of the chemical to which the workers were exposed. Often the only valid conclusion regarding exposure in these studies is that the exposures were likely to have been higher in the past than those that could be measured at the time the study was conducted. In the best available historical epidemiologic studies, there are some quantitative data on exposure during the time period of interest. Typically this information consists of the results of a few work-area samples taken at widely spaced time intervals. If the work force was relatively stable, and good records regarding job assignments and employment histories exist, it may be possible to divide the cohort into two or more exposure categories that reflect different intensities and/or durations of exposure. In these types of studies it may be possible to derive a quantitative or semiquantitative relationship between exposure and response.

In cases where epidemiologic studies provide no useful dose–response information, it may be necessary to use the results of bioassays in experimental animals as the basis for derivation of an exposure–response relationship. These data may also be incomplete or less than precise. The optimal design of a whole-animal bioassay to identify the carcinogenic potential of a chemical is not the optimal design for determining a quantitative dose–response relationship. The best design for animal bioassays includes the choice of three or more exposure levels. This should result in a wide range of response rates, allowing for better recognition of the shape of the dose–response curve. Unbalanced designs in which lower dose groups contain more animals than higher dose groups are desirable in order to define best the shape of the dose–response curve in the low-response regions. Bioassays designed solely to identify carcinogens, on the other hand, need only to provide maximum sensitivity for detecting any positive response. Thus such bioassays typically may be performed with only two dose groups, the higher dose being a maximum tolerated dose (MTD). If the MTD results in a tumor response of 90 percent or more, or if it causes significantly decreased survival (which is often the case), it can result in much uncertainty as to the true shape of the dose–response curve. Unfortunately, the only quantitative dose–response data available for many chemical carcinogens are the results of long-term bioassays designed to identify the carcinogenic potential of the compound.

In those cases where quantitative dose–response data are available from long-term bioassays in animals, it still may be necessary to select the specific dose–response relationship that is the most appropriate basis for human risk assessment. For example, exposure may result in a carcinogenic response in more than one species or strain, or in more than one target tissue, or in pre-neoplastic lesions, benign tumors, and/or malignant tumors in a single target tissue. Also there may be significant species/sex differences in dose–response relationships.

4.1 Selection of Cancer End Points

The selection of the cancer end point is one of the first steps in performing a dose–response characterization. In many cases there is no a priori basis for selecting one response over another as the foundation for human health risk assessment. Traditionally, the health regulatory agencies have chosen to select the most sensitive response in the most sensitive species/sex, in the absence of compelling scientific reasons for doing otherwise. Often benign and malignant tumors of any one type (site) are combined for total tumor count in any tissue. The basis for the combination is that the benign tumors in a particular cell type are likely to progress to malignant ones. Risk assessments using animal tumors are not generally site specific; that is, they only estimate the risk of developing tumors at any site. Multiple tumor types from several cell types within a particular organ would be consistent with accumulation or activation in that particular organ. Another choice regarding cancer site selection is to use the number of animals bearing statistically significant tumors and rare tumors. This is generally reserved for xenobiotics that produce a wide variety of tumors.

Generally, epidemiologic studies deal with the increased risk of a specific cancer, for example, lung cancer, whereas animal studies are used to predict an increase in cancer rates at all sites. Thus the predictions from animal studies are generally not site specific. In cases where epidemiologic studies suggest a type of cancer that might be increased in workers, the use of equivalent tumors in animals may be most appropriate. Alternatively, one might select tumors that occur in an inhalation or dermal study over those that occur in studies involving the oral route, when occupational cancer is the subject of the assessment.

Selection of tumor types, or sites, showing consistency across animal species is certainly an appropriate choice. In contrast, tumor sites occurring in rodents, but not likely to occur in humans for various mechanistic reasons previously discussed, serve as a reason for rejecting the use of that particular tumor type.

Normally, one is concerned about the increased incidence of cancer produced by different doses. However, shortening of the time-to-tumor is also an important consideration. In many instances this information is not available to the risk assessor. The time when the first tumor appears is often not noted. When the time at which an individual animal died as a result of cancer is available, time-to-tumor analysis should be attempted, if the time to the first tumor or the median time at which tumors were noted appears to be shortened by the treatment. In the absence of data suitable for time-to-tumor analysis, any deaths before the appearance of the first tumor should be removed from consideration.

It is wise to show a preference for studies produced since 1980, because they are likely to have been conducted under good laboratory guidelines (GLPs) and have more complete information, especially the detailed pathological evaluation including time of death (if not from the open literature, then, from the sponsor). Ideally, the detailed raw data should be reviewed.

4.2 Exposure Pharmacokinetics

The selection of the appropriate dose paradigm for risk assessment is also important. The exposure assessment process is the determination of the number of workers who are exposed to the compound of interest and the frequency, intensity, and duration of their exposure. In general, no two workers are exposed to exactly the same amounts of a potential occupational carcinogen. Thus complete exposure assessment would result in a different exposure for every individual in the work force, which is generally not feasible. In many cases the exposure of a given population of workers is most often normally distributed over a range of concentrations. Accordingly, exposure assessment for a population of workers usually involves the estimation of the median (or in some cases, mean) exposure. This may be supplemented with an estimate of a reasonable worst-case exposure in order to indicate more appropriately an upper limit of risk.

Ideally exposure assessments are based on empirical results. The most reliable measure of relevant exposure to workplace chemicals is the determination of biologic indicators. These typically involve measuring the concentration of the compound of interest or an indicator metabolite in tissues or excreta. Examples include blood lead and urinary arsenic levels. Importantly, biologic monitoring indexes do include the total exposure, not just the occupational exposure.

The next most reliable monitoring data are those derived from personal exposure monitoring, for example, sampling and analysis of air samples collected in the breathing zones of individual workers. Less reliable exposure data are the results of work-area monitoring in which air samples are collected at specific locations and analyzed to provide an estimate of the concentrations to which workers in those areas might be exposed. The latter is unfortunately the most common form of data available for exposure assessment.

In cases where limited monitoring data are available, it may be possible to estimate exposure by modeling air concentrations using dispersion models and estimated emissions. Mass balance calculations may be used to estimate emission rates of the chemical into the workplace air. Using data on ventilation rate and total air volume, it is possible to calculate vapor concentrations in various work-place locations. This type of estimation is useful only for establishing an upper limit on airborne concentrations in workplaces with rapid and even contaminant distribution.

Information on the frequency and duration of exposure is best determined from detailed employment records. In the absence of such records it is necessary to depend on the individual worker's recall, which may be faulty.

It is possible to estimate the actual risk of cancer in workers by combining epidemiology and animal studies. This process is rarely straightforward, however. If the dose–response data are derived from studies in experimental animals (as is often the case), they must be adjusted to account for differences across species. For example, the animal study may have involved a different exposure route and/ or schedule than that experienced by workers. Furthermore, there may be important interspecies differences in the relationship between administered dose and delivered dose. For most carcinogens, the frequency of the response is proportional to the concentration of the proximal carcinogen at the target tissue (often referred to as the biologically effective dose, BED). Unfortunately, the dose data from animal bioassays and the exposure information from human exposure assessment typically can only be used to calculate the administered dose of the parent compound. There are very few scientific data that indicate the quantitative relationship between administered doses and delivered doses, in either experimental animals or humans. In addition, these relationships are likely to vary from compound to compound. Traditionally, interspecies extrapolation has been performed on the basis of some measures of administered dose, such as milligrams per kilogram of body weight per day or milligrams per square meter of body surface area per day. In recent years, there has been increasing use of available pharmacokinetic data to refine interspecies extrapolation of doses.

Applied dose is the concentration in the exposure media (air, water, food), the amount applied to the skin (sometimes expressed in terms of the surface area of the skin covered), the concentration or amount instilled into the eye, or the amount gavaged (or injected), which is expressed in terms of milligrams per kilogram of body weight (mg/kg) or occasionally in terms of milligrams per body surface area (m²). The amount of the applied dose may be converted to the internal dose by measuring food, water, or air intake, or by using standard factors. For example, the standard (70-kg) man breathes 20 m³ in 24 hr or 10 m³ in 8 hr. The default position of regulators is to convert the applied dose to an internal dose, expressed in terms of milligrams per kilogram of body weight. Sometimes efforts are made to express the internal dose in terms of the amount of the chemical activated to a genotoxic metabolite, or, in cases where the specific cancer-causing metabolite is unknown, the total amount of the agent metabolized. Converting the applied dose or exposure concentration to a more accurate indication of the biologically effective dose (BED) is an important step in the risk assessment process and can reduce the uncertainties involved. In addition, pharmacokinetics can be useful in the classification of materials as carcinogens, as well as for extrapolation of risk from animal data to the human condition.

In some risk assessments the erroneous assumption is made that 100 percent of the material is absorbed. Another faulty assumption is that the experimental animals and humans absorb the same amount. Neither is likely to be true. No material is likely to be 100 percent absorbed, particularly by inhalation or ingestion. Furthermore, the percent absorbed at lower airborne concentrations is likely to be higher than the percent absorbed at higher concentrations.

Once deposited in the lungs, particulates can be absorbed, phagocytized, transported to and stored indefinitely in regional lymph nodes, or expelled up the airways by ciliary action. Depending on the concentration, the inhaled material may alter the phagocytic capability, the ciliary activity, or the passage across the respiratory membranes, as well as change the aspects of normal respiratory function. Any of these changes can alter the estimation of delivered dose. Oberdorster (33) has conceptualized a method for extrapolating from rats to humans in estimating equivalent human exposure of particulates in Figure 2.1.

The author also provides information on the comparative regional deposition. For example, in the rat the percent deposition of a 2.0-μm particulate is 23.3 in the nasopharyngeal area, 4.1 in the tracheobronchial area, and 6.4 in the pulmonary area. In contrast, the human percent deposition in the nasopharyngeal area is 31.9, in the tracheobronchial area 5.2, and in the pulmonary area 14.5. With active humans, less is deposited in the nasal area, whereas more is deposited in the pulmonary area. The particulate size effect is also quite variable between rats and men. For a 5-μm particulate 100 percent is deposited in the rat's nose, whereas in the human nose 30 percent of the particles pass through at rest, and the amount rises during mouth breathing as a result of activity.

Tumors of the nasal cavity may be more prevalent in rodents than in humans because of the greater deposition in the nasal cavity as compared to humans. However, the presence of nasal tumors in rodent inhalation studies may be predictive of human lung cancer. This may be particularly true if the material can be absorbed onto small particulates that are deposited in the lungs. In this case lung tissue may be in contact with the material for a longer time than if a vapor were inhaled. The nasal cavity may be a site of tumors because some materials or their metabolites bind to these tissues even after absorption by other routes. N'-Nitro-

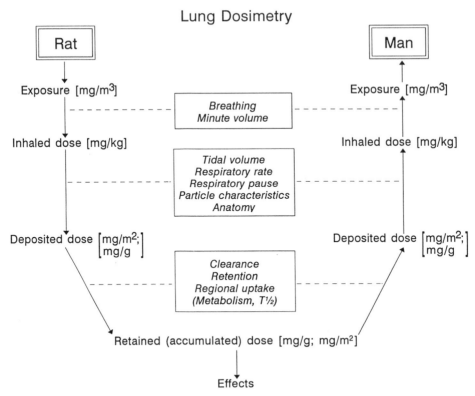

Figure 2.1. A concept for extrapolation of particle deposition from rat to humans. Adapted from Reference 33.

sonornicotine, phenacetin, 1,2-dibromoethane, 2,3,7,8-tetrachlorodibenzo-*p*-dioxin, progesterone, and testosterone are among the carcinogens that have this property, at least in rodents (34).

Gastrointestinal absorption is more important than one would expect in the occupational situation. It may occur when material is cleared from the respiratory tract by ciliary action. Additionally, hand-to-mouth contact (because of smoking, applying cosmetics, and eating) provides opportunity for oral intake.

Dermal absorption has been studied for some compounds. Unfortunately, there are few good animal models for human skin; and in vitro methods of determining percutaneous penetration using explanted human skin, animal skin, or artificial skin have not been validated. The degree of absorption often depends on the area of the skin exposed. Some simulation models have been developed for dermal absorption. However, the degree of absorption may be increased by lesions on the skin, regardless of whether they are compound induced or not. Because of the morphological differences in animal and human skin, across-species extrapolation may be difficult, if not impossible.

To estimate the delivered compound dose that results from the inhalation of its vapor at a given concentration in air, a simple one-compartment model often gives a good approximation. When an experimental animal or a human is exposed to a vapor (or gas) at a constant concentration over time, one can assume that the compound is absorbed into the blood at a constant rate (R in mg/hr). The absorption rate may vary, however, if the material changes the respiration rate or minute volume. The absorption takes place rapidly, in general, into the pulmonary capillaries; and the resulting concentration in the blood (actually, the apparent volume of distribution, V, in liters) is determined by the elimination rate ($K_e = 0.693/T_{1/2}$ in hr^{-1}). The concentration in blood (apparent volume of distribution) will reach equilibrium (steady state) at about 6 half-lives ($T_{1/2}$) and 50 percent of steady-state concentration in one $T_{1/2}$. The steady-state concentration (SSCONC) is

$$\text{SSCONC} = R/K_e \times V$$

The apparent volume of distribution is also affected by binding or storage in other organs, for example, the fat. If the $T_{1/2}$ is comparable to the length of time between daily occupational exposures, accumulation will clearly take place. If the $T_{1/2}$ is very long (weeks), then the total body-burden equilibrium will not be reached for months.

If a first-order elimination rate is operative, the decline in blood concentration will be the reverse of the increase during exposure. Thus the area under the concentration curve (AUC) equals SSCONC, or if the SSCONC is not reached during the exposure, maximum concentration times the exposure duration:

$$\text{AUC} = \text{SSCONC} \times \text{duration of exposure}$$

In the case of formaldehyde, it was reported that high concentrations altered the breathing patterns of mice, but not rats. The fact that a human breathes through the mouth as well as through the nose should be considered when using data from rodents, which breathe only through their noses. Based on their risk estimation of formaldehyde, Beliles and Parker (35) recommended that for primary direct-acting inhaled carcinogens, the airborne concentration, the duration of exposure, minute volume, and the surface area of the target organ (in this case, the nasal cavity) should be considered. These authors also corrected for the fact that humans are mouth breathers.

With regard to particulates, Oberdorster (33) has indicated that if the typical occupational exposure retention half-life is 0.7 days, then the daily deposited dose and the accumulated dose are the same. Longer half-lives of materials deposited in the lungs have been noted. For example, with Ni_3S_2 the $T_{1/2}$ is 36 days in rats and 103 days in humans.

The metabolic rate and time considerations are often overlooked in the conversion from animals to humans. Because of a slower metabolic rate in humans, the steady-steady body burden reflected by plasma concentration may be reached in animal studies, but not during human workplace exposures.

The simple one-compartment model may not be appropriate for the estimation of delivered dose for compounds that are metabolized. For those compounds that are metabolized in the liver, Michaelis–Menten kinetics may be proper. If this is biologically correct, then it is possible to have clearance in a single circulation time of the parent agent given by the oral route. If the lungs have no metabolic capability, the brain will be exposed to the highest concentrations possible by the inhalation route, provided the material can pass through the "blood-brain" barrier.

Currently, the concept of dose-to-the-target tissue is considered by many to be the correct dose paradigm. Experimental pharmacokinetists, however, have considered the maximum concentration for receptor-mediated effects or integrated concentration over time for nonreceptor effects to be more appropriate dose paradigms. They also use the plasma concentration as a surrogate for actual tissue concentration. This is probably a reasonable dose surrogate, as well as a good indicator of body burden, and would facilitate the use of biologic monitoring. Body burden itself is a reasonable paradigm of BED for chemicals that accumulate in the body.

The metabolism and kinetics of absorbed materials are important considerations. OSHA recommended classification of a material as a carcinogen if it is metabolized to a known carcinogen. Metabolism and excretion most often eliminate toxic and carcinogenic materials from the body. However, some materials may instead be activated; that is, the metabolite may be more reactive than the parent compound. This is an important consideration in any pharmacokinetic modeling. Ideally, the concentration of the reactive metabolite at the target organ is a more appropriate basis for estimating the BED than is the applied dose of the parent compound. It is necessary to consider the site of metabolism in such cases. The possibility that more than one metabolite may be carcinogenic is a necessary consideration. This is a default position often used by EPA. Because the metabolic process can be overloaded (saturated), it is possible that ever-increasing higher doses will not produce an ever-increasing BED. In addition, each species is likely to have different doses (tissue concentrations) at which the metabolism becomes saturated.

The routes of elimination are also important considerations in the pharmacokinetic analysis of a chemical. The concentration of a material in the exhaled breath or in the urine can be useful in biologic monitoring. For example, with urinary bladder carcinogens and, possibly, some renal toxicants, the urinary concentration could be a good estimator of the BED.

The ACGIH biologic exposure indexes (BEI), tabulated in Table 2.4 for selected carcinogens, provide some indication of the relationship between airborne concentrations as a time-weighted average (TWA) and the biologic exposure markers (parent compound or metabolite) in the urine, blood, or exhaled breath (37). In compounds with longer half-lives the time of sampling becomes less critical. Also, the commonality of some metabolites is represented.

The relationship has been refined more exactly in the case of arsenic. Pinto et al. (36) suggested that at low exposure concentrations (<150 μg/m^3) the following linear relationship exists between 8-hr airborne concentration (C_{air}, in μg/m^3) and urinary concentration (C_{urine}, in μg/l):

$$C_{air} = 0.3\ C_{urine}$$

Table 2.4. ACGIH Biological Exposure Indicators (BEI) for Selected Airborne Carcinogens

Determinant	Sampling Time	BEI	Notation[b]
ARSENIC AND SOLUBLE COMPOUNDS INCLUDING ARSINE[a]			
Inorganic arsenic metabolites in urine	End of work week	50 µg/g creatinine	B
BENZENE			
Total phenol in urine	End of shift	50 µg/g creatinine	B, Ns
Benzene in exhaled air	Prior to next shift		
Mixed-exhaled		0.08 ppm	Sq
End-exhaled		0.12 ppm	Sq
CADMIUM			
Cadmium in urine	Not critical	10 µg/g creatinine	B
Cadmium in blood	Not critical	10 µg/l	B
Cadmium in urine[a]	Not critical	5 µg/g creatinine	B
Cadmium in blood[a]	Not critical	5 µg/l	B
CHROMIUM (VI), Water soluble or fume			
Total chromium in urine	Increase during shift	10 µg/g creatinine	B
	End of shift at end of work week	30 µg/g creatinine	B
METHYL CHLOROFORM			
Methyl chloroform in end-exhaled air	Prior to the last shift of work week	40 ppm	
Trichloroacetic acid in urine	End of work week	10 mg/l	Ns, Sq
Total trichloroethane in urine	End of shift at end of work week	30 mg/l	Ns, Sq
Total trichloroethane in blood	End of shift at end of work week	1 mg/l	Ns
PERCHLOROETHYLENE			
Perchloroethylene in end-exhaled air	Prior to the last shift of work week	10 ppm	
Perchloroethylene in blood	Prior to the last shift of work week	1 mg/l	
Trichloroacetic acid in urine	End of work week	7 mg/l	Ns, Sq

[a]Proposed.
[b]NS, Nonspecific; B, background levels included; Sq, semiqualitative.

Even more recently, studies have suggested that at higher doses more of the arsenic may be excreted in the urine (38).

Time is another consideration, and dose should actually be expressed as a function of time. One of the factors that makes cancer a difficult assessment problem is that cancers have long latencies, leukemia (at 7 years) being an exception. Time is one of several factors that underlie the difference between risk assessments by

EPA and those applied by OSHA to occupational carcinogens. EPA estimates human risk to exist because of continuous exposure over the entire lifetime, whereas OSHA's assessment considers only the occupational lifetime. The working lifetime is assumed to be 40 hours a week, 48 weeks a year, for 45 years, or 14 percent of the entire lifetime. Fortunately for toxicologists and risk assessors, most rodent inhalation bioassays are for 6 hours a day, 5 days a week, for 2 years, or 17.5 percent of the rodent's lifetime. This difference is so slight as to not warrant further correction (39).

Pharmacokinetics (and metabolism) are useful for more appropriate estimations of biologically effective doses (BED). Converting the applied dose or exposure concentration to a more accurate indication of the BED is an important step in the risk assessment process and can reduce the uncertainties involved. Pharmacokinetics can be useful in the classification of materials as carcinogens, as well as in the extrapolation of risk from animal data to the human condition. Bolt and Filser (40) used pharmacokinetics in the risk estimation for ethylene, a compound not normally classified as a carcinogen. These investigators estimated the amount of ethylene oxide formed by the metabolism of ethylene. The carcinogenic response of ethylene oxide in rodent studies was then used as a basis for the estimation of risk from the parent compound, ethylene.

Developing appropriate pharmacokinetic models enables one to reconstruct the exposure. The duration of the target organ exposure to the reactive chemical is an important consideration. The maximum concentration in the plasma or the integrated dose (area under the concentration curve, AUC) should be considered as a possible surrogate of the BED. Theoretically, the maximum concentration is likely to be the best dose paradigm if binding (receptor) is involved in the toxicity, whereas AUC is likely to be the best dose paradigm if binding is not an important factor in the development of the carcinogenic response.

Recently pharmacokinetics has become synonymous with the computer-assisted solution of a series of differential equations to provide the concentration over time of a material in the target organ. This approach was popularized for volatile organic compounds by Anderson and various co-workers. The method, physiologically based pharmacokinetics (PB-PK), involves the consideration of partition coefficients between various media (e.g., air/blood, blood/fat) and blood flow through the target organ. Figure 2.2 represents a schematic of a PB-PK model. When this approach is used for risk assessment, it inherently assumes the concordance across species of the target organ. However, a high concentration in any tissue does not necessarily indicate that the toxicity will occur in that organ. Many organic chemicals are stored in the fat, and many metals are bone seekers. These tissues, despite their high concentrations, are not the target of most chemically induced toxicity, although they do often present long-term (even lifelong) depots for harmful body burdens to be stored, from which they can be later mobilized and re-enter the blood compartment, thereby reaching other tissues.

Bois et al. (41) reviewed three independently derived PB-PK models of benzene pharmacokinetics. These investigators indicated the parameter values (some 23 in all) leading to acceptable fits to the data spread over the entire range of plausible

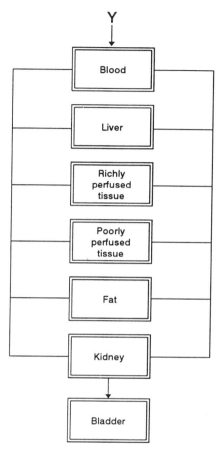

Figure 2.2. A schematic of a physiologically based pharmacokinetic model.

values. They also indicated that the use of standard values does not ensure a good prediction. In the end they recommended that, given the lack of knowledge with regard to the exact role of benzene's individual metabolites, it is more useful for regulatory purposes to consider a one-compartment model with the Michaelis–Menten terms.

Generally, one should use measured values as opposed to standard values. However, standard values should be examined to ensure the plausibility of the measured values. In the case of DEHP, the standard value for mouse food consumption was preferred to measured food consumption because of food wastage in the treated groups (42). For formaldehyde, on the other hand, measured values of minute volume in mice were better than the standard values because of compound-induced alterations in minute volume.

Simpler models with two to four compartments have long been used in the pharmaceutical industry to predict blood concentration of therapeutic agents. The apparent plasma volume is employed, along with constants related to the transfer of the agents to or from other compartments and the disappearance of the agent from the plasma, to predict the plasma concentration over time. In the drug industry, human exposure occurs early in the development of the drug, and few animal data are required to develop these models. The blood concentration over time is an often-used tool to determine appropriate dosing regimens. The effective half-time can be determined, for example, by following the decline in blood concentration after the exposure stops.

The amount of material and its metabolites excreted as a result of exposure are also useful in developing simple compartmental models. Because of the increased use of biologic monitoring, an opportunity exists to derive directly many of the parameter values used in relatively simple compartmental models. For example, Maneval et al. (43) have developed a four-compartment model that predicts the concentration over time of a nonmetabolized xenobiotic partly bound to the tissue of the kidney in the urine, plasma, and other tissues.

Differences between species in underlying physiology must also be considered. Many physiological constants can be scaled across species on the basis of allometric procedures. Davidson et al. (44) indicated that the rationale for extrapolation or scaling across species is based on the commonality of anatomic and physiologic reactions. Also, they outlined the development of the allometric equation $Y = aW^n$ relating species body weight (W) with various physiological, biochemical, pharmacological, and anatomic characteristics as the fundamental basis for extrapolation of biologic data from laboratory animals to humans. In summary, however, they recommend that the best course of action is to derive the data from the species in question, including humans.

The use of three exponents was described, W^1 for body weight, $W^{0.75}$ for metabolic rate, and $W^{0.67}$, a more conservative estimate based on surface area. In the use of modeling, especially PB-PK modeling, it has become standard practice to use approximately $W^{0.7}$ when dealing with parameter values associated with metabolic activity and W^1 for parameter values associated with body size, such as blood volume. One of the most important pharmacological expressions that can be scaled is AUC. Scaling exponents of 0.67 based on surface area (44) and 0.75 based on metabolic rate (45) have been used. Thus in extrapolation with the resulting AUC from rat data (300 g, AUC = 8 mg·hr/l) to the human (70 kg) at comparable exposure time and concentration, the following estimate, lacking human data, is appropriate:

$$\text{AUC(human)} = \text{AUC(rat)} \times (\text{body weight ratio})^{0.75}$$

$$\text{AUC(human)} = 8 \text{ mg·hr/l} \times (70000/300)^{0.75}$$

$$= 8 \times 233^{0.75}$$

$$= 8 \times 59.6$$

$$= 477 \text{ mg·hr/l}$$

The amount metabolized per day is most often scaled by allometry. Assuming that a 35-g mouse exposed to 10 ppm for 6 hr a day, 5 days a week, will metabolize 0.124 mg of the material daily, and assuming virtually complete metabolism in a day, the dose would be 3.54 mg/kg/day. Using $W^{0.67}$, a 250-g rat would be expected to metabolize 3.73 times more [$(250/35)^{0.67}$], or 0.46 mg, but the dose would be 1.85 mg/kg/day, less on a milligram per kilogram basis. The scaling factor from mouse to human (70 kg) would be 163 = [$70/0.035]^{0.67}$, 20.2 mg, or 2.89 mg/kg/day. From rats to humans, the scaling factor would be $70/0.25^{0.67} = 43.6$, and the daily metabolized dose would be 20.1 mg. The assumption is generally made that a similar exposure scenario results in the human equivalent dose, which produces the same cancer incidence in rodents and humans. Based on the allometry $W^{0.75}$ or $W^{1.0}$, the scaling factors would be 68.4 and 280, resulting in doses of 31.5 or 129 mg/day, as compared to 20.1 mg/day. Because the low-dose risk is greater when human equivalent doses are less, the use of $W^{0.67}$ is more conservative with regard to public health.

The application of scaling to risk produces the same result as scaling the dose if one is operating in the linear part of the metabolic curve. Scaling of risk should *not* be performed when metabolic differences have been taken into consideration, as in PB-PK modeling. Likewise, scaling dose with $W^{0.75}$ should not be performed when the activation occurs by other than metabolic means. For example, activation of a material by acid hydrolysis can occur in the stomach. In the case of DEHP, which is probably converted to the reactive metabolites by acid hydrolysis in the stomach, the risk from mouse-to-rat scales best in terms of body weight (42).

Pharmacokinetics need not involve solution of differential equations by computer techniques to be useful in the assessment of risk. Consideration of the absorption or retention of a material by occupationally important exposure routes is of primary importance. The International Committee on Radiation Protection has developed models on the relationship between particle size and deposition in the respiratory tract. Some industrial hygienists have often made use of devices that indicate the particle size distribution in their air-sampling strategies. Information on the particulate size distribution can be useful in estimation of the amount of dust that impacts on the respiratory surface, and this is certainly a better estimate of the biologically effective dose than the airborne concentration. *Regardless of the method used, as much information as is available should be incorporated into the calculation of the dose.*

4.3 Human Cancer Risk Estimation Methodology

Risk assessment techniques may be used to determine the likelihood of risk to an exposed worker and are an aid in establishing "safe" doses or doses with acceptable risk. Once quantitative exposure–response (dose paradigm versus cancer end-point incidence) data have been selected, the risk assessor may proceed to estimate what might be an unacceptable risk incurred by continued occupational exposure to the xenobiotic in question and the risk reduction achieved by lowering the current exposure. In cases where the assessment is based on human data, the extrapolation is usually linear. The doses are frequently expressed as concentration-years, and the risk can be converted to the occupational working lifetime.

In using human data the response or risk data are generally expressed as deaths from a particular type (site) of cancer. The EPA, when performing risk assessment based on human data, indicates that estimated risk at lower exposures is an upper plausible bound. Because exposure assessment is often the weakest link in the use of epidemiologic studies, it may be desirable to apply pharmacokinetic principles to the often limited exposure information. Pharmacokinetic information has been used to convert one type of information (urinary concentration) to another type (airborne concentration).

The oldest method of extrapolating from animals to humans involves the utilization of safety factors, or in the modern vernacular, "uncertainty factors." When this approach is used, the dose (level) that produces no observed adverse effect (NOAEL) or the lowest observed adverse effect level (LOAEL) is ascertained. In 1972, Weil (46) suggested in the case of a chronic animal bioassay producing only a LOAEL, not a NOAEL, for a carcinogenic response that a safety factor of 5000 be applied. Currently, for noncancer end points factors up to 10 are applied for use of LOAEL versus a NOAEL, when using subchronic rather than lifetime data, when extrapolating from animals to humans, and to account for more susceptible members of the human population. Additional uncertainty factors may be used if the data are incomplete, or for the scientific factors based on professional judgment (47). Thus the level established for human protection could be as little as 1/10,000ths of the lowest level producing an observed adverse effect (LOAEL). Under the conditions used by Weil (46) in suggesting his safety factor of 5000, the contemporary methodology would select a maximum uncertainty factor of 10,000. Thus the current practice is not all that extreme.

For inhalation studies, the dose is sometimes adjusted on the basis of particulate size, the mass of the toxic agent per surface area of the respiratory tract per unit time, so that it represents the human equivalent exposure (47). The uncertainty factor approach is most often used for noncancer end points, but may be applied to animal cancer end points, especially if the potential carcinogen is not genotoxic or a reasonable estimate of the comparative BED paradigm is not possible.

In determining NOAELs or LOAELs one should examine carefully the underlying data. Usually a statistical difference between the control and treated groups is considered to constitute an effect. The majority of such departures from the control may be interpreted as an indication of an adverse effect. However, it behooves toxicologists to examine the data to determine if the change is clearly compound related and is truly an adverse effect. Decreases in serum activity of enzymes released from injured cells have little biologic significance by themselves, but do indicate cell damage. Although some biologic markers of adverse effect may have greater sensitivity than others, the indicators of a toxic effect are generally displayed as a continuum of indications of effect increasing with dose. On the other hand, a statistical significance of the slope requires that nonstatistically significant changes at the lower levels be considered as biologically important. This is especially true when rare types of cancers are involved because of the small numbers of these important tumors.

The oldest conceptual cancer model is the initiator–promoter model. Although pure promoters are numerous, there are few, if any, pure initiators. Many of these initiator–promoters are known as complete carcinogens. The initiator–promoter concept has been modified slightly and has given rise to the two-stage model, sometimes referred to as a biologically based pharmacodynamic (BB-PD) model (48). The most familiar animal model is the two-stage procedure (initiation–promotion), widely used in the study of agents that cause skin tumors in mice. One of the key concepts or reasons for interest in the applicability of two-stage models is that many scientists consider cell proliferation (increased cell turnover) an indispensable "step" in the induction of cancer.

According to the theory of carcinogenesis that serves as a basis for the two-stage model, cancer develops because of mutations occurring in a small subpopulation of cells at the target site. The mutant cells, if the critical genetic loci are affected, escape from the normal control regulating cell growth. Individual cellular genetic alterations may be inherited, or they can be induced by specific chemicals. Mutations most often occur as a result of persistent DNA damage caused by chemical or physical agents at mitosis. Once the cancer cell begins to divide rapidly, similar cells accumulate, and additional mutational changes may occur. An increased rate of cell division is a characteristic of cancer cells. However, some cells of the body normally have a high cell turnover rate that may be age dependent. During development, the cell turnover rate may be higher than in the mature individual. Additionally, an increase in cell turnover rate may be induced by irritation, cell death, or a loss of genetic control of cell proliferation. Because single-strand DNA forms during cell division, dividing cells are more susceptible to chemically induced alterations of the DNA. These changes (mutations) increase the likelihood for altered cell type survival. Two critical factors in this model for considering two-stage carcinogenesis are mutation rate per cell division and cell divisions per unit time.

Noteworthy aspects of the mouse ED01 study of 2-acetylaminofluorene (2-AAF) speak to the conceptual usefulness of the two-stage model. In the liver the incidence of cancer increases linearly with time, whereas in the bladder the increase is exponential (49). The difference in the time relationship between the liver and the bladder occurs because of a difference in the pharmacokinetics related to the development of tumors at the two different sites. In the liver, 2-AAF is metabolized to the N-hydroxyl intermediate and then to a reactive sulfur-containing metabolite. Because the transformed cells or the "cells in the foci" apparently do not metabolize 2-AAF, only the first genetic event occurs in the liver. There is no compound-induced increase in cell proliferation in the liver. In the bladder N-hydroxylation occurs as in the liver, but then an N-glucuronide is formed. This glucuronide is excreted into the urine, where it is hydrolyzed, and causes DNA adduct formation. This reactive metabolite causes both initiation and proliferation. Proliferation is observed only at doses above 60 ppm, whereas DNA adduct formation occurs at dietary concentrations as low as 5 ppm. Cellular proliferation in the bladder is essential for tumor formation (19). In conclusion, the two-stage model may be used for either case, but understanding the pharmacodynamics or oncodynamics of the particular tumor-target site is a prerequisite.

Effective implementation of the two-stage model to predict cancer incidence requires more information than is now being gathered in most chronic bioassays. Serial histopathological evaluation, for example, would appear to be a necessity. The inability of current assays to detect effects of the test article on cell turnover rates and inaccuracies in tumor incidences are continuing shortcomings. It seems unlikely that sufficient information can be gathered to implement the use of this model for prediction of chemically increased cancer incidence rates. Computer simulation models could be developed to implement the two-stage model, but many of the parameter values would have to be based on the optimization (best fit). With some 23 or more parameter values in PB-PK modeling using the best-fit procedure, plus a dozen or more for the BB-PD, the uncertainty rises to absurdity.

In the case of materials that appear to be nongenotoxic, an appropriate default position would be to use hyperplasia or other indications of target organ toxicity as a biologic marker for effect. Then the "safety factor" approach previously discussed might be appropriate; however, the NOAEL should be used, not the LOAEL, to reduce the uncertainty. Again it must be emphasized that this approach will be an organ-specific assessment.

As indicated previously, the Supreme Court's decision on the OSHA benzene standard resulted in the proliferation of mathematical models for extrapolating from high doses (in animals) to low doses (generally occurring with human exposure). Among these models are the probit, Weibull, log-normal, time-to-tumor, and the multistage, with its special case, the linearized multistage (LMS) using q1*, which is the 95 percent upper confidence limit (UC 95%) on the slope of the linear term in the dose–response curve. These models are essentially curve-fitting or curve-forcing techniques and generate a maximum likelihood estimate (MLE) plus a 95 percent upper confidence limit. The federal government's position on cancer has been that a single molecule of a cancer-causing agent "may" interact with DNA, causing cancer. This is sometimes called the "one hit theory." Therefore the most conservative points of view have been adopted as a matter of policy but have not always been acted upon. This conservative point of view is also the basis for the selection of the linearized multistage model (q1*) as a default position for QRA. It should be noted that the statistical confidence limits account only for sampling variability.

It is customary to convert the doses administered to animals to a human equivalent dose (HED) and to assume that at the HEDs the magnitude of the response would be the same in humans as in animals. When adequate human dose–response data are available, human exposure over time and the cancer rate of each time-exposure group are used to extrapolate down to the current level or proposed lower level for estimating the reduction of risk associated with the reduction of exposure.

Before utilizing any of the methodologies for high-to-low dose extrapolation, it is important to consider the dose–response relationships that are likely. In Figure 2.3 are three different types of dose–risk relationships. Curve 1 has two nearly

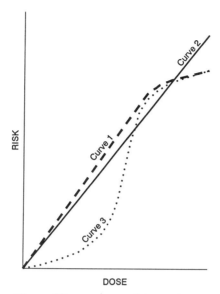

Figure 2.3. Three different types of dose–risk relationships.

linear components; curve 2 is entirely linear; and curve 3 has three linear components. It is clear that bioassays with only two treatment groups provide no information on the shape of the dose–response curve. In order to reduce the likelihood of an extrapolation error, the dose–risk relationship may be expanded. This can be accomplished by including data from studies other than the one showing the greatest sensitivity. Frequently one can adjust the dose paradigm so that other studies can be considered. Both sexes in the same study can be used, and in some cases, the incidence data from two species are compatible with a single dose–response pattern. Curve 1 may occur with the lesser slope in mid-range of the risk when the metabolism to the reactive compound is saturated or when genetic polymorphism is about 50 percent. For example, if slow acetylation occurs in about 50 percent of the workers, that group would have a greater number of adducts produced by 4-ADP than fast acetylators. Curve 2 suggests linear kinetics. Curve 3 might suggest saturable detoxification, followed by saturable activation.

When extrapolating from high exposures in animals to the lower doses to which humans might be exposed, one can express the risk in several ways. If the doses are expressed in milligrams per kilogram, then the unit risk (a slope function of the risk at 1 mg/kg) can be expressed as the maximum likelihood estimate (MLE), the 95 percent confidence limits of the MLE, or the upper bound estimate of the linear component at the unit risk (EPA's q1*). It should be remembered that with human high-to-low dose extrapolation, MLE at low doses is taken to be the upper plausible bound.

Risk can be expressed in terms of the extra or additional risk over background at a given dose. The additional risk at a dose (d) is defined as

$$P(d) - P(0)$$

Table 2.5. Multistage Analysis of Linear Data Sets A and B for Additional Risk

	Response A		Response B	
Dose	Observed	Expected	Observed	Expected
1000	30/50	0.6	30/50	0.6
300	10/50	0.2	10/50	0.2
100	NA	NA	3/50	0.064
0	0/50		0/50	
		Additional Risk		
1	MLE	6.7×10^{-4}		6.4×10^{-4}
	UC 95%	1.1×10^{-3}		1.041×10^{-3}
	q1*	1.1×10^{-3}		1.042×10^{-3}
10	MLE	6.7×10^{-3}		
	UC 95%	1.1×10^{-2}		
	q1*	1.1×10^{-3}		

that is, the probability (P) of the response at dose (*d*) minus (*P*) of the control dose of zero.

The extra risk is

$$[P(d) - P(0)]/[1 - P(0)]$$

EPA uses additional risk, but others may use extra risk (42).

Several computer programs for low-dose extrapolation will operate on personal computers. The following set of examples illustrates some of the information that can be obtained from the use of such programs and some of the differences that arise from using disparate types of data. In all examples it is assumed that the appropriate dose paradigm for the human dose has been used and that the target dose has been appropriately adjusted for all pharmacokinetic and physiological considerations. Additionally, the arithmetic capability of modern computer techniques is used to illustrate the slight differences, but the researcher should not use more than one significant figure in actual practice because of the uncertainties of the estimation of risk.

In the first example (Table 2.5), the dose response is linear in nature, and can be taken from any of the three curves illustrated in Figure 2.3. The computations were done using GLOBAL 86 (50). In Table 2.5 response A, it should be noted that the q1* does not change when the target dose is 10, or at the unit risk at a dose of 1. When an additional dose–response group is added, the impact on the statistic is minimal, mainly because of the use of whole numbers in the response. Also note that risk predicted at a dose of 1 from response A set is approximately the same as that from response B set. The 3/50 response (in response B) at the dose of 100 is not different statistically from the response (0/50) in the control. The introduction in the response B of the 100 dose causes only a very slightly nonlinear term to enter the extrapolation, accounting for the slight variation be-

Table 2.6. Multistage Analysis of Data Sets C and D

	Response C		Response D	
Dose	Observed	Expected	Observed	Expected
1000	35/50	0.72	30/50	0.63
500	25/50	0.43	20/50	0.32
250	10/50	0.26	5/50	0.15
0	5/50	0.092	0	0
		Additional Risk		
1	MLE	6.14×10^{-4}		5.49×10^{-4}
	UC 95%	1.19×10^{-3}		1.01×10^{-3}
	q1*	1.27×10^{-3}		1.01×10^{-3}
10	MLE	6.16×10^{-3}		
	UC 95%	1.18×10^{-2}		
	q1*	1.27×10^{-3}		
		Extra Risk		
10	MLE at 10	6.78×10^{-3}		
	UC 95%	1.26×10^{-2}		
	q1*	1.27×10^{-3}		

tween the q1* and the UC 95%. In this example extra risk is not different from additional risk.

The responses selected for the next example could be from curve 1 or the upper portion of curve 3. The impact of a background tumor incidence (response C) is illustrated in Table 2.6. The difference between additional and excess risk is not remarkable. The difference in the 95 percent UC and the q1* at a dose of 1 is due to nonlinear components in the dose–response curve. The MLE risk, without any background in the control (response D), is altered, but the upper confidence limit factors are stable.

In Table 2.7 all the dose–response data could be from curve 3. The first data set (two response groups) is quite linear and gives the greatest risk. The second data set (three response groups) fits a quadratic curve, and the fit to the data is very good. Note that only a dose response with no observed effect has been added. The second data set may, in fact, represent the missing information of dose–response data sets with a LOAEL. The response that the model predicts for this "no effect level" is too small to be detected in a sample of this size, but would still represent an unacceptable risk in the workplace. The third data set is of the same general nature as the previous one. However, this data set does not fit as well as the second set. Nevertheless, the MLEs in the second and third data sets are more comparable than the MLE from the first data set, whereas the upper confidence estimates in all three data sets are comparable. It is interesting to observe that the predicted risk at the lowest dose used (13) was greater than 1/1000. When the safety factor approach is applied to these data, using 100 for those sets with a

Table 2.7. The Effect of Increasing Dose Groups on the Assessment of Risk

Dose	Response 2		Response 3		Response 5	
	Observed	Expected	Observed	Expected	Observed	Expected
75	50/100	0.446	50/100	0.486	50/100	0.495
45					3/60	0.079
33	4/100	0.107	4/100	0.055	4/100	0.027
28					1/60	0.016
13			0/100	0.003	0/100	0.002
0	0/100	0	0/100	0.0	0/200	0.0

Additional Risk

1	MLE	1.05×10^{-4}	1.57×10^{-6}	7.16×10^{-6}
	UC 95%	1.20×10^{-3}	6.15×10^{-4}	1.00×10^{-3}
	(q1*)	1.12×10^{-3}	6.13×10^{-4}	1.01×10^{-3}

Safe Dose Based on Safety Factors

	0.033	0.13	0.13

NOAEL and 1000 for the data set with a LOAEL, the application of the safety factor for the LOAEL would provide the lesser risk.

Table 2.8 represents dose–response relationships that one rarely sees because of the limited number of dose groups in many studies, but that may more often exist. This series of responses is representative of curve 3, and the underlying kinetics may be saturable activation and saturable detoxification. In Table 2.8, the multistage (50) and probit model (51) are represented. The probit model (with

Table 2.8. Comparison of Predictions (Fit) and Risk by the Multistage and Probit Models

Dose	Response		
	Observed	Multistage: Expected	Probit: Expected
200	89/90	0.98	0.974
176	94/100	0.96	0.953
120	80/100	0.77	0.816
75	50/100	0.43	0.479
33	4/100	0.10	0.043
13	0/100	0.016	0.0002
0	0/100	0.0	0.0

Additional Risk

1	MLE	9.5×1^{-5}	
	UC 95% (q1*)	7.6×10^{-4}	2.05×10^{-15}
		(8.2×10^{-5})	

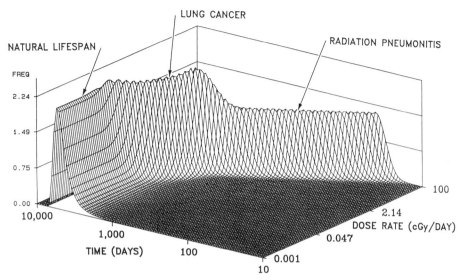

Figure 2.4. A three-dimensional representation of log-normal dose-rate (radiation cGy/day), time (years), and response (death from natural causes, bone cancer, and radiation injury) for bone burdens of injected ^{226}Ra in beagles. Supplied by O. G. Raabe, University of California at Davis.

additional background) gives by far the best fit to the data, and the risk is much less. The probit model estimates in almost all instances approach zero risk more rapidly than the more conservative multistage model.

The actual development of cancer depends on the time (both duration and chronology) that sensitive tissues are exposed, the latency of the tumor, and the presence of competing biologic effects. Competing factors are the main reason why a 100 percent tumor response is not often observed in long-term animal bioassays. Figure 2.4 illustrates the relationship between dose rate, time, and probability of death. The competing factors are death from related or natural causes versus deaths from lung cancer or pneumonitis induced by inhalation of ^{239}PuO$_2$ among dogs (52). Competing toxicities and other factors should be taken into consideration in the assessment of risk.

In the example in Table 2.9, we have used a 2-year rat study (6 hr/day, 5 days a week) and extrapolated to the occupational risk utilizing a computer program that performs the across-species calculations (53). The A response set is based on the assumption that all animals are at risk. In response B, deaths before the first tumor have been removed from consideration. For response C, rats that died after the first tumor appeared, either because of competing cancer types or other toxicity, have been censored. It should be noted that the q1* is in terms of milligrams per kilogram, whereas the MLEs and 95% UC limits have been converted into the exposure units, micrograms per cubic meter.

Because EPA does considerably more risk assessments than OSHA, one might be tempted to apply these assessments to the occupational scenario. Time is one

Table 2.9. The Effects of Censoring Data on the Assessment of Risk

Airborne Concn. (μg/m^3)	Response A		Response B		Response C	
	Observed	Expected	Observed	Expected	Observed	Expected
100	10/50	12/50	1/49	12/40	10/25	11/25
30	8/50	5/50	8/45	5/45	8/40	7/40
0	1/50	2/50	10/40	1/49	1/49	1/49
		Additional Risk				
q1*		2.4×10^{-3}		3.3×10^{-3}		5.5×10^{-3}
1 MLE		1.1×10^{-3}		1.5×10^{-3}		2.5×10^{-3}
(95% UC)		1.8×10^{-3}		2.4×10^{-3}		3.8×10^{-3}
10 MLE		1.1×10^{-2}		1.5×10^{-2}		2.5×10^{-2}
(95% UC)		1.8×10^{-2}		2.3×10^{-2}		3.7×10^{-2}

of several factors (occupational lifetime versus continuous lifetime exposures) that underlie the difference between risk assessments by EPA and those for occupational carcinogens. With regard to continuous lifetime exposure, human exposure is unlikely to be constant, even if there is a major variation between indoor (nonoccupational) exposures and outdoor exposures. Additionally, there is dermal exposure to water contaminants during bathing; and the inhalation route may become important during showering. Human exposure is neither constant nor for a full lifetime (54).

Also, EPA may use data from the oral route to predict risk of cancer via inhalation, even though inhalation may cause cancer at a different target tissue. OSHA prefers to use data from inhalation exposures. These differences are evident in the comparison of the risks estimated for arsenic. The EPA estimated the upper bound for excess cancer risk from a lifetime of continuous exposure to 1 μg/l of arsenic in the drinking water to be 5×10^{-5}, or 5 in 100,000 (55). This assessment was based on the 1968 study by Tseng et al. (56) on the prevalence of skin cancer in an endemic area of chronic exposure in Taiwan. Using the 1982 study of Enterline and Marsh (57), which showed an increase in respiratory cancer, OSHA (58) judged the best estimate of the risk of inorganic arsenic exposure for a 45-year working lifetime at 10 μg/m^3 to be eight deaths per 1000 workers. Although there is some effort (59) to extrapolate from one route (oral) to another (inhalation), too often the target organ may change. Moreover, the configuration of the internal dose paradigm (maximum concentration, area under the concentration curve) is very different. The site of metabolic activation may be in the lungs following inhalation, whereas it is frequently in the liver after oral intake. Although reservations about the applicability of the risk assessment are diminished by the use of pharmacokinetics information, validation by comparison with human data, careful attention to time factors, and consideration of competing risks, uncertainties still exist in the assessment of worker risk from exposure to occupational carcinogens. It is appropriate for the risk assessor to consider these reservations, at least in a qualitative sense. Uncertainties may result from individual or subpopulation susceptibility of workers. The basis of the susceptibility may be life-style, genetic variations, exposure from other sources, or existing disease states. EPA has based some risk assessments on the most susceptible subpopulations, for example, asthmatics and individuals living nearest to emission sources. OSHA, despite the main purpose of the OSHA act ". . . to assure so far as possible every working man and woman in the Nation safe and healthful working conditions," has not based its risk assessments on the sensitive worker (60). This may result in sensitive or susceptible workers being forced out of some workplaces.

In light of the foregoing, if the EPA assessment is based on inhalation exposure and one understands all the uncertainties that can be involved, it might be assumed that the occupational risk is 14 percent of the lifetime risk, as a "back of the envelope" calculation (39). A somewhat more complicated procedure was presented by Alavanja et al. (61). Assuming a 40-year working lifetime of 50 weeks, they indicated that the working lifetime was 12.4 percent of the total, and then suggested a further 1.5 correction to the slope ($q1^*$), based on the difference between 20 m^3 in 24 hr and 10 m^3 in 8 hr.

REFERENCES

1. N. Nelson, *Ann. Rev. Public Health*, **11**, 29 (1991).
2. Occupational Safety and Health Administration (OSHA), "Identification, Classification and Regulation of Potential Occupational Carcinogens," *Fed. Regist.*, **45**, 5002 (1980).
3. National Academy of Sciences/National Research Council (NAS/NRC), *Risk Assessment in the Federal Government: Managing the Process*, National Academy Press, Washington, D.C., 1983, p. 191.
4. R. P. Beliles, "Workplace Carcinogens: Regulatory Implications of Investigations" in *Handbook of Carcinogen Testing*, H. Milman and E. Weisburger, Eds., Noyes Publications, Park Ridge, NJ, 1985, pp. 569–586.
5. Occupational Safety and Health Administration, "Air Contaminants: Final Rule," *Fed. Regist.*, **54**, 2332 (1989).
6. M. Askoy, S. Erdem, G. Dincol, et al., *Blood*, **44**, 555 (1974).
7. B. D. Goldstein, *J. Toxicol. Environ. Health*, Suppl. 2 (1977).
8. P. Infante, *Am. J. Ind. Med.*, **11**, 559 (1987).
9. R. P. Beliles and L. C. Totman, *Reg. Toxicol. Pharmacol.*, **9**, 186 (1989).
10. L. B. Lave and F. K. Ennever, *Ann. Rev. Public Health*, **11**, 69 (1990).
11. B. D. Goldstein, *Ann. N.Y. Acad. Sci.*, **609**, 255 (1991).
12. American Conference of Governmental Industrial Hygienists (ACGIH), TLV Committee, *Appl. Occup. Environ. Hyg.*, **5**, 453 (1990).
13. National Toxicology Program (NTP), Sixth Annual Report on Carcinogens, Research Triangle Park, NC, 1991.
14. American Conference of Governmental Industrial Hygienists (ACGIH), TLV Committee, *Appl. Occup. Environ. Hyg.*, **6**, 616 (1991).

15. American Conference of Governmental Industrial Hygienists (ACGIH), *Guide to Occupational Exposure Values*, Cincinnati, OH, 1990.

16. H. M. D. Utidjian, "The Interaction Between Epidemiology and Animal Studies," in *Industrial Toxicology I: Perspectives in Basic and Applied Toxicology*, B. Ballantine, Ed., Wright, London-Boston, 1988, pp. 309–329.

17. R. Doll, "Epidemiologic Discovery of Occupational Cancers, Third Conference of Epidemiology in Occupational Health, Singapore, September 28–30, 1983," abstracted in *Scand. J. Work Environ. Health*, **10**, 121 (1984).

18. National Toxicology Program (NTP), "Toxicology and Carcinogenesis of 1,3-Butadiene in B6C3F1 Mice," NTP TR 434, NIH Publication No. 92-3165, 1991.

19. S. M. Cohen and L. B. Ellwein, *Science*, **249**, 1007 (1991).

20. B. N. Ames and L. S. Gold, *Science*, **249**, 970 (1991).

21. R. P. Beliles, J. G. Kereiakes, and A. T. Krebs, *J. Natl. Cancer Inst.*, **22**, 1045 (1959).

22. J. W. Grisham, G. J. Smith, and K. E. Beliles, Proceedings of the Fifteenth Conference on Environmental Toxicology AFAMRL-tR-84-002, 1984, pp. 158–167.

23. National Research Council (NRC), *Biologic Markers in Immunotoxicology*, National Academy Press, Washington, DC, 1992.

24. J. H. Dean, M. L. Luster, and G. Borman, in *Immunopharmacology*, P. Sirois and M. Rola-Pleszczynski, Eds., Elsevier, New York, 1982, pp. 349–397.

25. J. Huff, J. Haseman, and D. Rall, *Ann. Rev. Pharmacol. Toxicol.*, **31**, 621 (1991).

26. J. Goodman, J. M. Ward, J. A. Popp, J. E. Klaunig, and T. Fox, *Fundam. Appl. Toxicol.*, **17**, 651 (1991).

27. J. G. Conway, R. C. Cattley, J. A. Popp, and B. E. Butterworth, *Drug. Metab. Rev.*, **21**, 65 (1989).

28. I. W. F. Davidson and R. P. Beliles, *Drug Metab. Rev.*, **23**, 493 (1991).

29. R. N. Hill, L. S. Erdreich, O. E. Paynter, P. A. Roberts, S. L. Rosenthal, and C. F. Wilkinson, *Fundam. Appl. Toxicol.*, **12**, 629 (1989).

30. T. M. Vogel and P. L. McCarty, *Appl. Environ. Microbiol.*, **49**, 1080 (1985).

31. M. P. Dieter, G. A. Boorman, C. W. Jameson, H. B. Matthews, and J. E. Huff, *Toxicol. Ind. Health*, **6**, 599 (1990).

32. E. Loeser, *Toxicol. Lett.*, **15**, 71 (1983).

33. G. Oberdorster, *Health Physics*, **57** (suppl. 1), 213 (1989).

34. C. S. Barrow, *Toxicology of the Nasal Passages*, Hemisphere Publishing Corp., Washington, DC, 1986, p. 317.

35. R. P. Beliles and J. Parker, *Health Physics*, **57** (Suppl. 1), 333 (1989).

36. S. S. Pinto, M. O. Varner, and K. W. Nelson, *J. Occup. Med.*, **18**, 677 (1976).

37. American Conference of Governmental Industrial Hygienists (ACGIH), "Threshold Limit Values for Chemical Substances and Physical Agents and Biological Exposures Indices," Cincinnati, OH, 1991–1992.

38. P. E. Enterline, V. L. Henderson, and G. M. Marsh, *Am. J. Epidemiol.*, **25**, 929 (1987).

39. R. P. Beliles and J. C. Parker, *Toxicol. Ind. Health*, **3**, 371 (1987).

40. H. M. Bolt and J. G. Filser, "Kinetics and Deposition in Toxicology," in *Proceedings of the International Conference on Contributions of Toxicology to Risk Assessment and Health Regulation*, 1986.

41. F. Y. Bois, T. J. Woodruff, and R. Spear, *Toxicol. Appl. Pharmacol.*, **110**, 79 (1991).
42. R. P. Beliles, J. A. Selinas, and W. J. Kluwe, *Drug. Metab. Rev.*, **21**, 3 (1989).
43. D. C. Maneval, D. Z. Argenio, and W. Wolf, *Eur. J. Nucl. Med.*, **16**, 29 (1990).
44. I. W. F. Davidson, J. C. Parker, and R. P. Beliles, *Reg. Toxicol. Pharmacol.*, **6**, 211 (1986).
45. W. R. Campbell and J. Mordenti, *Adv. Drug Res.*, **20**, 1 (1991).
46. C. S. Weil, *Toxicol. Appl. Pharmacol.*, **21**, 454 (1972).
47. A. M. Jarabek, M. G. Menache, J. H. Overton, M. L. Dourson, and F. J. Miller, *Toxicol. Ind. Health*, **2**, 279 (1990).
48. R. B. Connonly and M. A. Andersen, *Ann. Rev. Pharmcol. Toxicol.*, **31**, 303 (1991).
49. K. G. Brown and D. G. Hoel, *Fundam. Appl. Toxicol.*, **3**, 458 (1983).
50. R. B. Howe, K. S. Crump, and C. Van Landingham, "GLOBAL 86: A Computer Program to Extrapolate Quantal Animal Toxicity Data to Low Doses," K. S. Crump and Co., Ruston, LA, 1986.
51. J. Kovar and D. Krewski, "Dores 81: A Computer Program for Low Dose Extrapolation of Quantal Response Toxicity Data," Health and Welfare Canada, 1981.
52. O. G. Raabe, *Fundam. Appl. Toxicol.*, **8**, 465 (1987).
53. K. S. Crump, R. B. Howe, C. Van Landingham, and W. G. Fuller, TOX-RISK, Version 3.1, Clement International Corp., K. S. Crump Division, Ruston, LA, 1992.
54. K. G. Brown and R. P. Beliles, "On Basing Extrapolation for a Chemical Carcinogen on Total Dose (Constant Rate × Continuous Exposure Duration): Implication and Extension," in *Risk Assessment in Setting National Priorities*, J. J. Bonin and D. E. Stevenson, Eds., Plenum Press, New York-London, 1989, pp. 59–66.
55. Environmental Protection Agency (EPA), "Arsenic, Integrated Risk Assessment," EPA, Washington, DC, 1990.
56. W. P. Tseng et al., *J. Natl. Cancer Inst.*, **40**, 453 (1968).
57. P. E. Enterline and G. M. Marsh, *Am. J. Epidemiol.*, **116**, 895 (1982).
58. Occupational Safety and Health Administration (OSHA), "Supplemental Statement of Reasons for the Final Rule," *Fed. Regist.*, **48**, 1864 (1983).
59. T. R. Gerrity and C. J. Henry, *Principles of Route-to-Route Extrapolation of Risk Assessment*, Elsevier, New York, 1990, p. 317.
60. R. P. Beliles, *Ann. Conf. Gov. Ind. Hyg.*, **3**, 71 (1982).
61. M. C. R. Alavanja, C. Brown, R. Spiratas, and M. Gomez, *Appl. Occup. Environ. Hyg.*, **5**, 510 (1990).

CHAPTER THREE

Criteria for Identifying and Classifying Toxic Properties of Chemical Substances

Edward J. Sowinski, Ph.D, D.A.B.T., C.I.H., and Finis L. Cavender, Ph.D, D.A.B.T., C.I.H.

1 INTRODUCTION

1.1 Background

Historically, a need has been perceived by industrial toxicologists and hygienists to characterize risk by developing rating schemes to address the severity of potential health effects regarding chemical toxicity. These schemes have focused primarily on acute hazards. For example, in the 1940s, members of the American Industrial Hygiene Association's Toxicology Committee determined that there was a need to express degree of toxicity in a simple manner. They collected data regarding reported poisonings where the dose was known. Using these data, ranges of toxicity were devised to characterize the severity of effects. Published in the *American Industrial Hygiene Association* (*AIHA*) *Journal* in 1949, the system of toxicity ratings serve as the basis for most classification schemes in existence today. See Table 3.1 (1).

When one utilizes various schemes for classifying chemical hazards, it is important to recognize some basic concepts upon which classification schemes are based.

Patty's Industrial Hygiene and Toxicology, Fourth Edition, Volume 2, Part A, Edited by George D. Clayton and Florence E. Clayton.
ISBN 0-471-54724-7 © 1993 John Wiley & Sons, Inc.

Table 3.1. Tabulation of Toxicity Classes (1)

Toxicity Rating	Category	Single Oral Dose	Comparable Lethal Dose for Humans[a]
1	Extremely toxic	≤ 1 mg/kg	A taste–<7 drops
2	Highly toxic	1–50 mg/kg	7 drops–1 teaspoon
3	Moderately toxic	50–500 mg/kg	1 teaspoon–1 ounce
4	Slightly toxic	0.5–5 g/kg	1 oz–1 pt or lb
5	Practically nontoxic	5–15 g/kg	1 pt–1 qt
6	Relatively harmless	≥ 15 g/kg	>1 qt

[a]For a 70-kg (150-lb) person.

Fundamentally, chemicals vary in their capacity to cause toxic effects at specific exposure levels. A specific level of exposure for one chemical may cause particular adverse health effects, whereas a similar level of exposure to another chemical may be absent of similar effects. Also, all chemicals do not possess the capacity to produce a full range of toxic effects. For example, all chemicals do not have carcinogenic potential, adverse reproductive capacity, neurological effects, or mutagenic properties. Most chemicals have a measurable degree of acute toxicity, although the degree of acute toxicity for many chemicals is at such a high level of exposure that it is insignificant from a practical standpoint. Overall, fine lines between safe and potentially harmful exposure concentrations of chemicals do not exist. Potential adverse health consequences from exposure to chemical substances occur in a continuum from low to high levels of exposure. This concept is referred to as the dose–response relationship, which was first cited by Paracelsus in the early 1500s as "All substances are poisons; there is none which is not a poison. The right dose differentiates a poison and a remedy" (2).

Unfortunately, there are wide variations in the quality and quantity of toxicity data for chemical substances. Many chemicals have no data or inadequate data. Nonetheless, a large body of experimental toxicity data does exist for a wide variety of chemical substances. These data range from acute to chronic effects and include a variety of unique effects such as carcinogenicity, reproductive effects, mutagenicity, and neurological effects, plus a wide variety of specific organ and system effects (3–6). More than 80,000 chemicals are currently cited in the EPA Toxic Substances Control Act (TSCA) Chemical Inventory (7). Generally, more toxicity information exists for those chemicals that are used in large volume or for which human exposure can be significant. This includes many commercial industrial chemicals. Additionally, pharmaceuticals, food additives, and pesticides generally have extensive toxicity data including information on beneficial levels.

Effective communication of chemical hazards requires a framework for communicating the results of animal toxicity testing, with its inherent uncertainty and differences in quality and quantity of data for various chemical substances. It is important to recognize that classification schemes do not represent fine lines between "safe" and "unsafe" chemicals. However, classification schemes are helpful in a practical sense to distinguish groups of chemicals that are more likely to possess defined hazard characteristics and to facilitate hazard communication programs.

1.2 Government Regulation of Hazardous Chemicals

The need for organized communication of chemical hazards has resulted in a variety of classification schemes for experimental toxicity data so that organized interpretations regarding significance to humans can be made. These schemes typically exist as part of hazard communication, transportation, and labeling regulations, or as a part of various consensus standards, toxicity testing protocols, or independent publications. Unfortunately, the criteria used for hazard classification purposes are not always consistent. This can result in a chemical being portrayed differently by various classification schemes, thus leading to inconsistency and confusion despite the fact that the same toxicity data have been used for classification purposes. Efforts have been initiated to "harmonize" various classification schemes worldwide (8). However, a disparity is likely to exist for some time to come. Consequently, the practicing occupational health professional needs to be aware of various health hazard classification systems in order to cope effectively with different regulations and the overall demands of hazard communication. This chapter outlines the significant chemical health hazard classification schemes that are encountered worldwide.

2 PRIMARY EXISTING CLASSIFICATION SYSTEMS

Many countries have requirements that address some part of the issue of chemical classification and labeling, but few have comprehensive systems. Chemical classification systems are generally not in one legislative package. Different agencies have responsibility and authority for different parts of overall systems. Often there are different purposes or driving forces within different pieces of legislation. Although all have a general purpose of protecting people potentially exposed to the chemicals, their specific intent may vary. For example, the United States and Canada have workplace hazard identification systems that are based on the principle that workers have the right to know this information. Hence they are driven by a necessity to communicate hazards to the ultimate users. In the European Community (EC), an added purpose is to facilitate trade within EC countries. Thus there may be less emphasis on communicability. The degree of specificity also varies significantly between various schemes. The EC system is very specific, giving exact wording for label statements on particular categories of substances. However, the scope is more narrow than comparable systems in the United States and Canada, where a performance-oriented, criteria-driven approach is used. This is not to imply that one system is inherently better than another. Rather the differences in the various schemes simply underscore how divergent results can occur when judgments are not coordinated. The following are general descriptions of the primary existing schemes in the world.

2.1 OSHA Hazard Communication Standard (9)

The Occupational Safety and Health Administration (OSHA) recognizes that hazard evaluation is a process that relies heavily on the professional judgment of the evaluator, particularly in the area of chronic hazards. The performance orientation of the hazard determination does not diminish the duty of the chemical manufacturer, importer, or employer to conduct a thorough evaluation, examining all relevant data and producing a scientifically defensible evaluation. An amplified description of the OSHA Hazard Communication Standard is presented in Chapter 6, Volume I, Part A, 4th edition of Patty's *Industrial Hygiene and Toxicology*. The following criteria are used in making hazard determinations that meet the requirements of OSHA.

2.1.1 Human Data

Where available, epidemiologic studies and case reports of adverse health effects are considered in the evaluation. Criteria have been published for use in establishing the strength of association regarding human health effects and chemical exposure (10). These include the following:

1. *The Strength of Association.* This is usually expressed as some type of risk ratio, such as the death rate of a disease in an exposed population relative to the rate in an unexposed population. The higher the risk ratio, the less likely it is that the association resulted from some confounding factor, and thus the more likely it is that the association is causal.

2. *Consistency.* A causal hypothesis is supported when positive results are seen repeatedly in several studies done independently by different investigators using different populations. If only one or two studies have been conducted, causation can nevertheless be inferred if the risk ratios are very high.

3. *Specificity.* This criterion refers to an association limited to specific workers and to particular sites and types of disease with no association between the work and other modes of illness. The relation between vinyl chloride monomer and angiosarcoma of the liver is one example. However, causality can occur in the absence of specificity.

4. *Relationship in Time.* This criterion requires that the exposure precede the development of the disease by a biologically relevant time period.

5. *Coherence of the Evidence.* This criterion is satisfied when the associations found in epidemiologic studies do not conflict with what is known of the natural history and biology of the disease.

6. *The Biologic Gradient.* To satisfy this criterion, the data should show a dose–response relationship; that is, the disease rate should be found to increase as the level and duration of exposure increase.

7. *Biologic Plausibility.* A causal hypothesis is strengthened when there are known biologic mechanisms that explain the association between the substance and the disease.

8. *Experimental Verification*. This criterion would be satisfied if removal of the substance or reduction in the level of exposure was eventually followed by a decline in the incidence of the disease.

These criteria have been used in the analysis of smoking and cancer (11).

2.1.2 Animal Data

Evidence of human health effects in exposed populations is generally not available for the majority of chemicals produced or used in the workplace. Therefore the available results of toxicologic testing in animal populations are used to predict the health effects that may be experienced by exposed workers. In particular, the definitions of certain acute health hazards utilized in classification schemes refer to specific animal testing results.

2.1.3 Adequacy and Reporting of Health Effects Data

The results of any studies that are designed and conducted according to established scientific principles, and that report statistically significant conclusions regarding the health effects of a chemical, are considered to be a sufficient basis for a hazard determination by OSHA. The results of other scientifically valid studies that tend to refute the findings of hazard should also be utilized in an overall evaluation of data.

2.2 Canada's Workplace Hazardous Materials Information System (WHMIS) (12)

The Workplace Hazardous Materials Information System (WHMIS) is a Canadian national system designed to ensure that all employers provide needed information and train employees properly in the handling of hazardous materials in the workplace. WHMIS is consensus legislation representing input from government, industry, and labor. It is intended to ensure that the hazards of materials produced or sold in, imported into, or used within Canadian workplaces are identified by suppliers and that standard classification criteria are utilized. Suppliers of chemical materials in Canada must convey hazard information in a specified manner by means of labeling on the containers of "controlled products" and by providing more detailed information in the form of material safety data sheets (MSDS). A controlled product for health purposes is defined for WHMIS under the Federal Hazardous Products Act as any material included in any of the classes outlined in the act (13). For health purposes, these classes include materials causing immediate and serious toxic effects, materials causing other toxic effects, biohazardous materials, and corrosive materials. Employers are responsible for evaluating all products produced in a workplace process using the hazard criteria identified in the Controlled Products Regulations. The criteria for classifying toxic properties of chemical substances under WHMIS are outlined in comparison with other schemes in Section 3 below.

Table 3.2. WHMIS Symbols for Acute Toxicity Classes (13)

Materials causing immediate and serious toxic effects (very toxic)	
Materials causing other toxic effects (toxic)	
Corrosive materials	

In Canada employers must ensure that supplier-provided containers of controlled products are labeled with WHMIS labels. As long as a controlled product remains in its supplier-provided container, the supplier label must remain attached to the container and be legible. For workplace processes, employers are required to furnish workplace warnings in the form of labels, tags, or appropriate markings. There is no specific format for workplace labeling; however, information on safe handling, hazard warnings, storage, and use of the controlled product must be provided. Reference must also be made to the availability of a material safety data sheet. Table 3.2 summarizes the classes of controlled products, descriptive divisions, and corresponding hazard symbols for WHMIS hazard classifications.

2.3 European Community's Classification of Dangerous Substances (14)

Within the European Community (EC), the objective of classifying chemical hazards is to identify the toxicologic, physiochemical, and ecotoxicologic properties of substances that may present a risk during normal handling and use. Individual chemicals and chemical mixtures are labeled in accord with identified hazard classes in order to warn and protect the user, the general public, and the environment. The EC label is intended to take into account all hazards in the form in which chemicals are placed on the market and not necessarily in any different form in which they may be used, that is, diluted. The most severe hazards are highlighted by symbols (Table 3.3). Other hazards are specified in standard risk phrases and safety phrases that give advice on handling precautions (14).

The data required for classification and labeling under EC provisions may be obtained from a variety of sources; for example, the results of previous tests, information required for the international transportation of chemical substances, information taken from reference publications in the scientific literature, or information derived from practical experience. It is not required under EC rules to

Table 3.3. European Community (EC) Symbols for
Acute Toxicity Classes (14)

Very toxic and toxic	
Harmful	
Corrosive	

conduct testing in order to classify chemical hazards. However, from a practical standpoint, testing is frequently conducted, especially for acute effects, by chemical suppliers in order to provide a rational basis for classifying and labeling. The criteria for classifying the toxicity of chemical substances within EC are outlined in the next section.

3 DESCRIPTION AND COMPARISON OF CRITERIA USED FOR TOXICITY CLASSIFICATION

3.1 Acute Toxicity Criteria

Acute toxicity generally refers to health effects that result from relatively short exposures. The ultimate acute effect is death. Testing to determine the potential of a chemical to cause death has been one of the most widely used parameters to characterize toxicity. By generating known concentrations of a chemical and exposing test animals to these concentrations, toxicologists have been able to establish lethality data that can be extrapolated to recommendations for "safe" levels of exposure. All the major classification schemes rely on the use of lethality data to determine acute toxicity.

Historically, oral toxicity was primarily used as a guide to acute toxicity. However, further refinement of the system extrapolated to both dermal and inhalation toxicity. Within inhalation toxicity, there is generally a differentiation between gases/vapors and dusts/mists. Classification systems allow comparisons to be made in the relative toxicity of various chemicals, thus laying the groundwork for the appropriate selection of protective measures. Knowledge of lethality data has also been the basis for the establishment of permissible exposure limits or threshold limit values. Using the lethality data, and determining an appropriate margin of safety to allow for fluctuations in exposures, occupational health professionals have

Table 3.4. Acute Oral, Dermal, and Inhalation Toxicity Criteria According to OSHA (9)

Category	LD_{50} Orally in Rat (mg/kg)	LD_{50} Skin Absorption in Rat or Rabbit (mg/kg)	LC_{50} Inhalation in Rat 4 hr (mg/l)
Highly toxic	≤ 50	≤ 200	≤ 2
Toxic	50–500	200–1000	2–20

been able to help ensure that workers can be exposed to the chemicals involved with relative confidence that they do not risk poisoning or death as a result.

3.1.1 OSHA Acute Toxicity Criteria

The following criteria, summarized in Table 3.4, apply to the classification of acute toxicity under the OSHA Hazard Communication Standard:

Highly Toxic. A chemical falling within the following categories:

a. A chemical that has a median lethal dose (LD_{50}) of 50 mg or less per kilogram of body weight when administered orally to albino rats weighing between 200 and 300 g each.

b. A chemical that has a median lethal dose (LD_{50}) of 200 mg or less per kilogram of body weight when administered by continuous contact for 24 h (or less if death occurs within 24 hr) with the bare skin of albino rabbits weighing between 2 and 3 kg each.

c. A chemical that has a median lethal concentration (LC_{50}) in air of 200 ppm by volume or less of gas or vapor, or 2 mg/l or less of mist, fume, or dust, when administered by continuous inhalation for 1 hr (or less if death occurs within 1 hr) to albino rats weighing between 200 and 300 g each.

Toxic. A chemical falling within any of the following categories:

a. A chemical that has a median lethal dose (LD_{50}) of more than 50 mg/kg but not more than 500 mg/kg of body weight when administered orally to albino rats weighing between 200 and 300 g each.

b. A chemical that has a median lethal dose (LD_{50}) of more than 200 mg/kg but not more than 1000 mg/kg of body weight when administered by continuous contact for 24 hr (or less if death occurs within 24 hr) with the bare skin of albino rabbits weighing between 2 and 3 kg each.

c. A chemical that has a median lethal concentration (LC_{50}) in air of more than 200 ppm but not more than 2000 ppm by volume of gas or vapor, or more than 2 mg/l but not more than 20 mg/l of mist, fume, or dust, when administered by continuous inhalation for 1 hr (or less if death occurs within 1 hr) to albino rats weighing between 200 and 300 g each.

Table 3.5. Acute Oral, Dermal, and Inhalation Toxicity Criteria According to EC (14)

Category	LD_{50} Orally in Rat (mg/kg)	LD_{50} Skin Absorption in Rat or Rabbit (mg/kg)	LC_{50} Inhalation in Rat 4 hr (mg/l)
Very toxic	≤25	≤50	≤0.5
Toxic	25–200	50–400	0.5–2
Harmful	200–2000	400–2000	2–20

3.1.2 European Community Acute Toxicity Criteria

Criteria for the European Community (EC) classification of acute toxicity, summarized in Table 3.5, are as follows:

> *Very Toxic.* Substances and preparations are classified in EC as very toxic and assigned the symbol "T+" in accordance with the following criteria:
>
> *a.* Very toxic if swallowed: LD_{50} oral, rat, ≤25 mg/kg.
>
> *b.* Very toxic in contact with skin: LD_{50} dermal, rat or rabbit, ≤50 mg/kg.
>
> *c.* Very toxic by inhalation: LC_{50} inhalation, rat, ≤0.5 mg/l based on a 1 to 4 hr exposure.

Toxic. Substances and preparations are classified in EC as toxic and assigned the symbol "T" in accordance with the following criteria:

> *a.* Toxic if swallowed: LD_{50} oral, rat, 25 to 200 mg/kg.
>
> *b.* Toxic in contact with skin: LD_{50} dermal, rat or rabbit, 50 to 400 mg/kg.
>
> *c.* Toxic by inhalation: LC_{50} inhalation, rat, 0.5 to 2 mg/l, 4 hr.

Harmful. Substances and preparations are classified in EC as harmful and assigned the symbol "X" in accordance with the following criteria:

> *a.* Harmful if swallowed: LD_{50} oral, rat, 200 to 2000 mg/kg.
>
> *b.* Harmful in contact with skin: LD_{50} dermal, rat or rabbit, 40 to 2000 mg/kg.
>
> *c.* Harmful by inhalation: LC_{50} inhalation, rat, 2 to 20 mg/l, 4 hr.

3.1.3 Canadian Acute Toxicity Criteria

The Canadian (WHMIS) criteria for acute toxicity, summarized in Table 3.6, follow:

> *Very Toxic Materials.* Chemicals causing immediate and serious effects under WHMIS:

Table 3.6. Acute Oral, Dermal, and Inhalation Toxicity Criteria According to WHMIS (13)

Category	LD_{50} Orally in Rat (mg/kg)	LD_{50} Skin Absorption in Rat or Rabbit (mg/kg)	LC_{50} Inhalation in Rat 4 hr (mg/l)
Very toxic	≤ 50	≤ 200	≤ 0.5
Toxic	50–500	200–1000	0.5–2.5

  a. Oral LD_{50} not exceeding 50 mg/kg of body weight when tested in accordance with OECD Test Guideline No. 401 (15).
  b. Dermal LD_{50} not exceeding 200 mg/kg of body weight when tested in accordance with OECD Test Guideline No. 402 (16).
  c. Inhalation LC_{50} not exceeding 2500 ppm by volume when tested by OECD Test Guideline No. 403 (17).
  d. Inhalation LC_{50} not exceeding 0.5 mg/l of dust, mist, or fume when tested for 4 hr in accordance with OECD Test Guideline No. 403 (17).

Toxic Materials

  a. Oral LD_{50} of more than 50 mg/kg but not exceeding 500 mg/kg of body weight when tested in accordance with OECD Test Guideline No. 401 (15).
  b. Dermal LD_{50} of more than 200 mg/kg but not exceeding 1000 mg/kg of animal body weight when tested in accordance with OECD Test Guideline No. 402 (16).
  c. Inhalation LC_{50} of more than 1500 ppm but not exceeding 2500 ppm of vapor when tested for 4 hr in accordance with OECD Test Guideline No. 403 (17).
  d. Inhalation LC_{50} of more than 0.5 but not exceeding 2.5 mg/l of dust, mist, or fume when tested according to OECD test protocol No. 403 (17).

Tables 3.4 to 3.6 provide a comparison of acute oral, acute dermal, and acute inhalation criteria for OSHA, EC, and WHMIS.

3.2   Criteria for Chemicals Corrosive to the Skin

3.2.1   OSHA Criteria for Skin Corrosive

A *corrosive* is a chemical that causes visible destruction of, or irreversible alterations in, living tissue by chemical action at the site of contact. For example, a chemical is considered to be corrosive if, when tested on the intact skin of albino rabbits by the method described by the U.S. Department of Transportation in Appendix A to 49 CFR Part 173, it destroys or changes irreversibly the structure of the tissue at the site of contact following an exposure period of 4 hr (18). This term is not associated with action on inanimate surfaces.

3.2.2 EC Criteria for Skin Corrosive

Substances and preparations are classified as corrosive by EC and assigned the symbol "C" in accordance with the following criteria:

> *Causes Severe Skin Burns.* When a chemical is applied to healthy intact skin and causes full thickness destruction as a result of up to 3 min of exposure, or if this result can be predicted.
>
> *Causes Burns.* When a chemical is applied to healthy intact animal skin which causes full thickness destruction of skin tissue as a result of up to 4 hr of exposure, or if this result can be predicted.

3.2.3 WHMIS Criteria for Skin Corrosive

A chemical substance or mixture is identified as corrosive under WHMIS if there is evidence that it causes visible necrosis of human skin tissue or if a material is an untested mixture containing a product, material, or substance that causes visible necrosis and is present at a concentration of at least 1 percent.

3.3 Skin Irritation Criteria

Criteria for scoring the degree of chemical irritancy to skin and eyes have been developed and published by Draize et al. (19). The Draize tests for skin and eye irritancy were first incorporated into the U.S. Federal Hazardous Substances Act (FHSA) in 1960, and with minor modifications are still used today. The Draize test allows chemicals to be compared based upon a numerical score following administration of the test material into the eye or onto the skin of experimental animals, usually rabbits. Criteria outlining the scoring associated with the Draize protocol for skin and eye irritants are shown in Tables 3.7 and 3.8. In recent years considerable controversy has developed regarding the use of animals in skin and eye irritancy studies. Limited progress has been made in the use of tissue cultures as opposed to whole animals. Modern practice calls for minimizing the use of experimental animals and continued research on alternative models.

3.3.1 OSHA Criteria for Irritant

A chemical that is not corrosive but that causes a reversible inflammatory effect on living tissue by chemical action at the site of contact is considered an irritant by OSHA. A chemical is a skin irritant if, when tested on the intact skin of albino rabbits by appropriate techniques, it results in a cumulative score for erythema and edema of 5 or more. A chemical is an eye irritant if so determined under the scoring procedure of Draize.

Table 3.7. Weighted Scores for Skin Irritancy (19)

Skin Reaction	Score
A. Erythema and eschar formation	
No erythema	0
Very slight erythema	1
Well-defined erythema	2
Moderate to severe erythema	3
Severe erythema (beef red) to slight eschar formations	4
B. Edema formation	
No edema	0
Very slight edema	1
Slight edema (not well defined)	2
Moderate edema (raised 1 mm)	3
Severe edema (raised >1 mm)	4

3.3.2 EC Criteria for Irritant

Noncorrosive substances and preparations are classified as irritants by EC in accordance with the following criteria:

Irritating to Skin. A substance or preparation is irritating to the skin if, when applied to healthy intact animal skin for up to 4 hr, significant inflammation is caused that is present 24 hr or more after the end of the exposure period. Inflammation is significant if the mean value of the scores is 2 or more for erythema and eschar formation or edema formation. The same shall be the case if the test has been completed using three animals if the score for either erythema formation or edema formation observed in two or more animals is equivalent to 2 or more.

Irritating to Eyes. A substance is irritating to the eyes, if, when it is applied to the eye of the animal, significant ocular lesions are caused which are present 24 hr or more after instillation of the test material.

Ocular lesions are significant if the means of the scores have any of the values: cornea opacity equal to or greater than 2 but less than 3; iris lesion equal to or greater than 1, but not greater than 1.5; redness of the conjunctivae equal to or greater than 2.5; edema of the conjunctivae (chemosis) equal to or greater than 2. The same shall be the case where the test has been completed using three animals if the lesions, on two or more animals, are equivalent to any of the above values except that for iris lesion the value should be equal to or greater than 1, but less than 2, and for redness of conjunctivae the value should be equal to 3.

3.3.3 WHMIS Criteria for Skin and Eye Irritants

A pure substance or tested mixture is considered as causing skin or eye irritation if it meets the following criteria:

Table 3.8. Weighted Scores for Ocular Lesions (19)

Lesion	Grade
A. Cornea	
No ulceration or opacity	0
Scattered or diffuse areas of opacity	1[a]
Easily discernible translucent areas, iris obscured	2
Nacreous areas, no details of iris visible	3
Complete corneal opacity, iris not discernible	4
B. Iris	
Markedly deepened folds, congestion, swelling, moderate circumcorneal injection, iris still reacting to light	1[a]
No reaction to light, hemorrhage, gross destruction	3
C. Conjunctivae	
Redness	
Vessels normal	0
Some diffuse vessels definitely injected	1
Diffuse, crimson red, individual vessels not easily discernible	2[a]
Diffuse, beefy red	3
Chemosis	
No swelling	0
Any swelling above normal	1
Obvious swelling with partial eversion of lids	2[a]
Swelling of lids—half closed	3
Swelling with lids more than half closed	4

[a]Lowest score considered positive.

a. The chemical causes an effect graded at a mean of 2 or more for erythema formation or 2 or more for edema formation when tested in accordance with OECD Test Guideline No. 404 (20).

b. The chemical causes an effect graded at a mean of 2 or more for corneal damage, 1 or more for iris damage, or 2.5 or more for conjunctival swelling or redness when tested in accordance with OECD Test Guideline No. 405 (21).

3.4   Criteria for Sensitization

3.4.1   OSHA Criteria for Skin Sensitizers

A chemical substance or preparation is considered to be a skin sensitizer if it causes a substantial proportion of exposed people or animals to develop an allergic reaction in normal tissue after repeated exposure. No specific protocol is specified for testing sensitizers by OSHA.

3.4.2   EC Criteria for Sensitizers

A chemical substance or preparation that carries a sensitizer warning may cause sensitization by skin contact if practical experience shows the substances and preparations to be capable of inducing a sensitization reaction in a substantial number of persons by skin contact, or on the basis of a positive response in experimental animals. In the case of adjuvant-type sensitization test methods, a response of at least 30 percent of the animals is considered positive. For any other test method, a response of at least 15 percent of the animals is considered positive.

3.4.3 WHMIS Criteria for Sensitization

A pure substance or tested mixture is considered to be capable of skin sensitization if in an animal assay carried out in accord with OECD test guideline No. 406 (22) it meets one of the following criteria:

   a. (1) It produces a response in 30 percent or more of the test animals, when using one of the technique incorporating the use of an adjuvant, or (2) it produces a response in 15 percent or more of the test animals, when using one of the techniques not incorporating the use of an adjuvant.
   b. Evidence shows that it causes skin sensitization in persons following exposure in the workplace.

3.5 Criteria for Chronic Effects

This section summarizes criteria for classifying carcinogens, mutagens, and teratogens. Criteria contained in OSHA, EC, and WHMIS regulations, plus criteria contained in IARC, EPA, and other guidance documents, are presented. Table 3.9 summarizes the information described below (8).

3.5.1 Criteria for Classifying Chemicals as Carcinogens

The definition of a carcinogen implies that it causes malignant tumors; however, it should be recognized that a precise distinction between benign and malignant tumors is not always possible. Nevertheless, the finding of an increase in malignant tumors is the hallmark of carcinogenicity. For a substance to be considered a suspected carcinogen to humans or a proven animal carcinogen of potential relevance to humans under expected conditions of exposure, it will normally need to be genotoxic, as well as giving a positive result in appropriately performed animal carcinogenicity bioassays. The exact relevance to humans of chemicals that are positive in an animal bioassay but negative in a battery of short-term tests for genotoxicity is not clear. These substances particularly—as with all chemicals— must be considered on a case-by-case basis for hazard classification. The very large variations in potency and latency of carcinogens are not routinely utilized in classifying substances for carcinogenic hazard identification purposes, but these aspects should be considered when making risk management decisions.

In evaluating the available data, it has to be realized that no single piece of evidence typically provides definitive evidence of carcinogenicity. It is only when all the evidence is evaluated that implications to human health can be defined. It is particularly important when considering the relevance of ambient exposures to

Table 3.9. Comparison of Approaches for Classifying Chemicals as Carcinogens, Mutagens, or Teratogens with EC, EPA, OSHA, and WHMIS (8)

EC	EPA	OSHA	WHMIS
Carcinogens			
Category 1: Known to be carcinogenic to humans *Category 2:* Regarded as carcinogenic to humans based on animal studies *Category 3:* Cause concern owing to possible carcinogenic effects	*Group A:* Human carcinogen *Group B:* Probable human carcinogen *Group C:* Possible human carcinogen	No categorization; any chemical in A or B is covered: (A) Listed in NTP Annual Report, in IARC monographs 1 or 2, or regulated by OSHA (B) Based on hazard evaluation of other relevant data	No categorization; human or animal listed under ACGIH; A1a, A1b, A2; also IARC Groups 1, 2A, 2B are covered
Mutagens			
Category 1: Known to be mutagenic to humans *Category 2:* Regarded as mutagenic to humans *Category 3:* Cause concern owing to possible mutagenic effects	Sufficient evidence Suggestive evidence Limited evidence	No categorization; based on hazard evaluation of all relevant data	Not addressed at this time
Teratogens			
Category 1: Known to be teratogenic to humans *Category 2:* Regarded as teratogenic to humans	No specific categorization	No categorization; based on hazard evaluation of all relevant data	No categorization; positive in OECD 414, 415, 416 are covered

humans that the mechanism of action and the probable shape of the dose–response curve at very low doses be considered.

The major classification systems currently employed, for example, those of the European Economic Community (EEC) (14), Environmental Protection Agency (EPA) (23), and International Agency for Research on Cancer (IARC) (24), envisage three broad categories of carcinogens. Some regulations utilize classifications made by others, for example, Occupational Safety and Health Administration (OSHA) and Workplace Hazardous Material Information System (WHMIS). An expanded classification scheme that incorporates the principles contained in the IARC and EPA schemes has been published and incorporated into an American National Standard for Chemical Labeling by the American National Standard In-

Table 3.10. Comparison of Classification Schemes for Carcinogens[a]

	Designation		
	EC	EPA	IARC
1. Proven human carcinogenic substance	1	A	1
2. Suspected human carcinogenic substance	2	B[1]	2A
3. Proven animal carcinogenic substance of potential relevance to humans	2	B[2]	2B
4. Suspected animal carcinogenic substance of potential relevance to humans	3	C	3
5. Substances nonclassifiable with regard to carcinogenicity	—	D	4
6. Negative evidence	—	E	—
7. No evidence	—	—	—

[a]*Note*: A direct comparison of categories is not always possible. This comparison is proposed as a reasonable approximation (8).

stitute (ANSI) (8, 25). A summary of classification schemes for carcinogens is presented below and summarized in Table 3.10.

3.5.1.1 International Criteria for Classifying Carcinogens. A summary of carcinogenic categories has been published (8). An overview of carcinogen classes follows.

Category 1: Proven Human Carcinogenic Substances. Evidence for inclusion in this category is provided for the positive results of formal epidemiologic studies (10).

Category 2: Suspected Human Carcinogenic Substances. One of the following types of evidence shows that human exposure to the substance may result in the development of cancer:

1. Suggestive epidemiologic data not sufficient to satisfy causality.
2. Proven evidence from animal studies carried out under conditions which are relevant to human exposure. The evidence from animal studies must be strong evidence as defined in Category 3 below. Studies of genotoxicity and of metabolic processes can assist in this understanding.

Category 3: Proven Animal Carcinogenic Substances of Potential Relevance to Humans. At least one of the following findings supports the justification for a substance being classified as a proven animal carcinogen with relevance to humans when obtained under exposure conditions which correspond to those in humans. There should be a clear statistical and biologically significant increase in the incidence of malignant tumors:

1. In more than one species or strain.

2. In multiple experiments, preferably with different routes of administration or a different dose levels.
3. To an unusual degree with regard to incidence, site or type, or age at onset.

The data would not normally be regarded as sufficient for a proven animal carcinogen classification if one of the following is true:

1. Only benign tumors are induced.
2. An excess only of tumors such as hepatic nodules in rats or mice or only pulmonary tumors in mice is induced.
3. An excess of malignant tumors is induced only in an organ that has a high spontaneous incidence.
4. An excess of malignant tumors is induced only by an irrelevant route of administration.
5. The dose level required to produce tumors in experimental animals is so high that it adversely affects the normal physiology of the experimental animals owing to its bulk or physicochemical properties.
6. No tumors are produced by exposure at or above those levels to which humans are likely to be exposed, even if tumors are found at exposures that produce chronic injury in the tissues or organ in which tumors later appear.

Category 4: Suspected Animal Chemical Carcinogenic Substances of Potential Relevance to Humans. In this case there is limited evidence from animal studies carried out under conditions that are possibly relevant to humans. Some experimental results that might lead to this categorization include the following:

1. A small increase in the incidence of malignant tumors of significance, for example, where background data suggest that this incidence could have occurred by chance, particularly if this is only toward the end of the animal's natural life-span.
2. An increased incidence of malignant tumors only in organs for which the natural incidence is high or variable.
3. An increased incidence of malignant tumors in only one species or animal strain.
4. Malignant systemic tumors are induced only by routes of exposure that are not relevant to human exposure.
5. Other information suggests that limited animal data are not relevant to humans, for example, when extensive and valid epidemiologic studies have given no evidence of a carcinogenic effect.
6. A treatment-related increase is observed only of nonmalignant tumors that are well known to progress to malignancy.

It should be noted that results from some routes of administration, for example, subcutaneous, intravenous, intramuscular, and intraperitoneal treatment, are typically not reliable indicators of carcinogenicity. The subcutaneous route particularly gives false-positive findings and much caution should be exercised in interpreting the data especially when only local tumor formation is induced.

Category 5: Substances Nonclassifiable with Regard to Carcinogenicity. This category includes substances for which some experimental evidence exists, but the evidence is limited in strength and/or is irrelevant to the human situation. Some examples follow:

1. The only positive evidence of carcinogenicity is from animal studies and occurs under one of the following conditions:
    a. At excessively high doses that result in altered physiological conditions such as tissue damage (necrosis, chronic irritation) or a change in metabolic pathway.
    b. Under exposure conditions that do not occur when the substance is handled or used by human, for example, "solid-state" sarcomas from plastic implantation.
    c. Under exposure conditions in which the substance was used in a physical form to which humans are not exposed.
2. The experimental evidence is equivocal, whereas there is valid negative evidence from epidemiologic studies.
3. The only positive evidence is from experimental studies that were improperly designed, improperly conducted, or could not be reproduced.
4. Metabolic data indicate that the metabolic products or rates of formation of those products are grossly different in the experimental animal from those of humans.

Category 6: Negative Evidence. This category is used for agent(s) that show no evidence for carcinogenicity in at least two adequate (and appropriate) animal tests in different species or in both epidemiologic and animal studies.

Category 7: No Data. In this case no relevant data are available.

It is appropriate to note that a material could be classified as having negative evidence of carcinogenicity when animal studies have been carried out under conditions that are relevant to human exposure and have positive evidence under inappropriate or exaggerated experimental conditions. The appropriate negative experimental and epidemiologic evidence should supersede data collected from inappropriate studies.

3.5.1.2 Classification of Carcinogenic Substances According to European Community (EC) Criteria (14). For the purpose of classification and labeling, chemical substances are divided into three categories by EC:

Category 1: Substances Known to be Carcinogenic to Humans. There is sufficient evidence to establish a causal association between human exposure to a substance and the development of cancer.

Category 2: Substances that Should be Regarded as if They are Carcinogenic to Humans. There is sufficient evidence to provide a strong presumption that human exposure to a substance may result in the development of cancer, generally on the basis of (1) appropriate long-term studies and (2) other relevant information.

Category 3: Substances that Cause Concern for Humans owing to Possible Carcinogenic Effects but in Respect of Which the Available Information is Not Adequate for Making a Satisfactory Assessment. There is some evidence from appropriate animal studies, but this is insufficient to place the substance in Category 2.

Placing a substance into Category 1 is done on the basis of epidemiologic data; placing into Categories 2 and 3 is based primarily on animal experiments.

For classification as a Category 2 carcinogen either positive results in two animal species should be available or there should be clear positive evidence in one species, together with supporting evidence such as genotoxicity data, metabolic or biochemical studies, induction of benign tumors, structural relationship with other known carcinogens, or data from epidemiologic studies suggesting an association.

Category 3 actually comprises two subcategories:

1. Substances that are well investigated but for which the evidence of a tumor-inducing effect is insufficient for classification in Category 2. Additional experiments would not be expected to yield further relevant information with respect to classification.
2. Substances that are insufficiently investigated. The available data are inadequate, but they raise concern for humans.

For a distinction between Categories 2 and 3, the considerations listed below are relevant. These considerations, especially in combination, would lead in most cases to classification in Category 3, even though tumors have been induced in animals:

- Carcinogenic effects only at very high dose levels exceeding the "maximal tolerated dose." The maximal tolerated dose is characterized by toxic effects which, although not yet reducing life-span, go along with physical changes such as about 10 percent retardation in weight gain.
- Appearance of tumors, especially at high dose levels, only in particular organs of certain species known to be susceptible to a high spontaneous tumor formation.
- Appearance of tumors only at the site of application in very sensitive test systems (e.g., intraperitoneal or subcutaneous application of certain locally active compounds) if the particular target is not relevant to humans.

- Lack of genotoxicity in short-term tests in vivo or in vitro.
- Existence of a secondary mechanism of action with the implication of a practical threshold above a certain dose level (e.g., hormonal effects on target organs or on mechanisms of physiological regulation, chronic stimulation of cell proliferation).
- Existence of a species-specific mechanism of tumor formation (e.g., by specific metabolic pathways) irrelevant for humans.

For a distinction between Category 3 and no classification, considerations are relevant that exclude a concern for humans:

- A substance should not be classified in any of the categories if the mechanism of experimental tumor formation is clearly identified, with good evidence that this process cannot be extrapolated to humans.
- If the only available tumor data are liver tumors in certain sensitive strains of mice, without any other supplementary evidence, the substance may not be classified in any of the categories.
- Particular attention should be paid to cases where the only available tumor data are the occurrence of neoplasms at sites and in strains where they are well known to occur spontaneously with a high incidence.

3.5.1.3 Classification of Carcinogenic Substances According to International Agency for Research on Cancer (IARC) Criteria (24)

Human Carcinogenicity Data IARC. The evidence relevant to carcinogenicity from studies in humans is classified into one of the following categories by IARC.

SUFFICIENT EVIDENCE OF CARCINOGENICITY. IARC considers that a causal relationship has been established between exposure to the agent and human cancer. That is, a positive relationship has been observed between exposure to the agent and cancer in studies in which chance, bias, and confounding could be ruled out with reasonable confidence.

LIMITED EVIDENCE OF CARCINOGENICITY. A positive association has been observed between exposure to the agent and cancer for which a causal interpretation is considered by the IARC to be credible, but chance, bias, or confounding could not be ruled out with reasonable confidence.

INADEQUATE EVIDENCE OF CARCINOGENICITY. The available studies are of insufficient quality, consistency, or statistical power to permit a conclusion regarding the presence or absence of a causal association.

EVIDENCE SUGGESTING LACK OF CARCINOGENICITY. There are several adequate studies covering the full range of doses to which human beings are known to be exposed, which are mutually consistent in not showing a positive association between exposure to the agent and any studied cancer at any observed level of exposure. A conclusion of "evidence suggesting lack of carcinogenicity" is inevitably limited to the cancer sites, circumstances and doses of exposure, and length of observation covered by the available studies. In addition, the possibility of a very small risk at the levels of exposure studied can never be excluded.

Experimental Carcinogenicity Data (IARC). The evidence relevant to carcinogenicity in experimental animals is classified into one of the following categories:

SUFFICIENT EVIDENCE OF CARCINOGENICITY. IARC considers that a causal relationship has been established between the agent and an increased incidence of malignant neoplasms or of an appropriate combination of benign and malignant neoplasms in (1) two or more species of animals or (2) in two or more independent studies in one species carried out at different times or in different laboratories or under different protocols.

In the absence of adequate data on humans, it is biologically plausible and prudent to regard agents for which there is sufficient evidence of carcinogenicity in experimental animals as if they presented a carcinogenic risk to humans.

LIMITED EVIDENCE OF CARCINOGENICITY. The data suggest a carcinogenic effect but are limited for making a definitive evaluation because (1) the evidence of carcinogenicity is restricted to a single experiment, or (2) there are unresolved questions regarding the incidence only of benign neoplasms or lesions of uncertain neoplastic potential, or of certain neoplasms that may occur spontaneously in high incidence in certain strains.

INADEQUATE EVIDENCE OF CARCINOGENICITY. The studies cannot be interpreted as showing either the presence or absence of a carcinogenic effect because of major qualitative or quantitative limitations.

EVIDENCE SUGGESTING LACK OF CARCINOGENICITY. Adequate studies involving at least two species are available which show that, within the limits of the tests used, the agent is not carcinogenic. A conclusion of evidence suggesting lack of carcinogenicity is inevitably limited to the species, tumor sites, and doses of exposure studied.

Overall Evaluation by IARC. Finally, the total body of evidence is taken into account; the agent is described according to the wording of one of the following categories, and the designated group is given. The categorization of an agent is a matter of scientific judgment, reflecting the strength of the evidence derived from studies in humans and in experimental animals and from other relevant data.

Group 1: The Agent is Carcinogenic to Humans. This category is used only when there is sufficient evidence of carcinogenicity in humans.

Group 2. This category includes agents for which, at one extreme, the degree of evidence of carcinogenicity in humans is almost sufficient as well as agents for which at the other extreme, there are no human data but for which there is experimental evidence of carcinogenicity. Agents are assigned to either 2A (probably carcinogenic) or 2B (possibly carcinogenic) on the basis of epidemiologic, experimental, and other relevant data.

Group 2A: *The Agent is Probably Carcinogenic to Humans.* This category is used when there is limited evidence of carcinogenicity in humans and sufficient evidence of carcinogenicity in experimental animals. Exceptionally, an agent may be classified into this category solely on the basis of limited evidence of carcinogenicity in humans or of sufficient evidence of carcinogenicity in experimental animals strengthened by supporting evidence from other relevant data.

Group 2B: *The Agent is Possibly Carcinogenic to Humans.* This category is generally used for agents for which there is limited evidence in humans in the absence of sufficient evidence in experimental animals. It may also be used when there is inadequate evidence of carcinogenicity in humans or when human data are nonexistent but there is sufficient evidence of carcinogenicity in experimental animals. In some instances, an agent for which there is inadequate evidence or no data in humans, but limited evidence of carcinogenicity in experimental animals together with supporting evidence from other relevant data, may be placed in this group.

Group 3: *The Agent is Not Classifiable as to its Carcinogenicity to Humans.* Agents are placed in this category when they do not fall into any other group.

Group 4: *The Agent is Probably not Carcinogenic to Humans.* This category is used for agents for which there is evidence suggesting lack of carcinogenicity in humans together with evidence suggesting lack of carcinogenicity in experimental animals. In some circumstances, agents for which there is inadequate evidence of or no data for carcinogenicity in humans but evidence suggesting lack of carcinogenicity in experimental animals, consistently and strongly supported by a broad range of other relevant data, may be classified in this group.

3.5.1.4 Classification of Carcinogens According to OSHA (9).

The OSHA Hazard Communication Standard relies on criteria and identification of carcinogens put forward from other sources. A chemical is considered to be a carcinogen if one of the following is true:

1. It has been evaluated by IARC and found to be a carcinogen within Class 2B or greater.
2. It is listed as a carcinogen in the "Annual Report on Carcinogens" published by the National Toxicology Program (NTP) (26).
3. It has been specifically regulated as a carcinogen by OSHA.

3.5.1.5 Classification of Carcinogens According to WHMIS (13). Under the provisions of Canada's Workplace Hazardous Material Information System (WHMIS), a chemical is identified as a carcinogen if it is listed in either one of the following publications:

1. "Threshold Limit Values for Chemical Substances and Physical Agents in the Work Environment" published by the American Conference of Governmental Industrial Hygienists (ACGIH) (27).
2. Group 1 or Group 2 IARC Carcinogens as published in the IARC Monographs on the Evaluation of the Carcinogenic Risk of Chemicals to Humans (24).

3.5.2 Criteria for Classifying Chemicals as Teratogens

There is a need to delineate criteria for classifying chemicals that may cause adverse effects to the overall reproductive process owing to increasing data in the literature relating to such effects (5, 6). Existing classification schemes mainly provide guidance for evaluating adverse effects on the developing conceptus, that is, the teratogenic aspect of the overall reproductive process. This limitation in scope is justified because teratogenicity is an end point that can be assessed on a reasonably valid and reproducible basis.

In this area of toxicology, many of the terms have been applied loosely and misunderstanding has been common. A number of expert groups have defined terms, but definitions are usually valid only for the particular document in question. The following two definitions represent a most common basis for approaching reproductive and teratogenic effects (8):

Reproductive Toxicity. Adverse effects of chemicals that interfere with the ability of males or females to reproduce.

Developmental Toxicity. Adverse effects of chemicals on the developing conceptus associated with exposure during pregnancy. Such effects may be manifested in the embryonic or fetal periods or postnatally.

A specific aspect of developmental toxicity is the production of structural malformations. A "teratogen" is defined as an agent that causes irreversible deleterious structural malformations in a conceptus as a consequence of exposure of the mother during pregnancy. Whether or not a specific malformation is deleterious is a matter of judgment in each particular case. This definition of teratogen excludes agents causing other manifestations of developmental toxicity, that is, death of the conceptus, growth retardation, or functional deficits. These and other types of developmental toxicity remain unclassified for the present, because morphological change, that is, teratogenicity, is the only reproducible assessment currently available (8, 28).

3.5.2.1 European Community (EC) Criteria for Teratogens. The EC has put forward the most specific guidance on classifying chemicals as developmental toxins, that is, teratogens (14).

> *Category 1: Substances Known to be Teratogenic to Humans.* Chemicals in this category must present sufficient evidence to establish a causal association between human exposure to a substance and subsequent nonheritable birth defects in the offspring. To be consistent with the word "teratogenic," the use of the term "birth defect" can be considered to relate to deleterious, structural malformations in a conceptus.

In order for a chemical to be classified in EC Category 1 as a teratogen, a clear, unequivocal relationship between human exposure during gestation and increased incidence of a specific structural malformation must be established. Evidence for inclusion in this category is provided by formal epidemiologic evidence. Several background factors must be evaluated in order to substantiate a causal relationship. These include the background level of teratogenic manifestations, the difficulty in identification of specific malformations, multiple confounding factors, and sources of bias. These factors must be addressed in the design of epidemiological studies of reproductive and developmental hazards (10).

> *Category 2: Substances That Should be Regarded as if They are Teratogenic to Humans.* Chemicals in this category must present sufficient evidence to provide a strong presumption that human exposure to the substance may result in nonheritable birth defects in the offspring, generally on the basis of appropriate animal studies or other relevant information.

Appropriate animal studies should meet at least the following criteria (8):

1. There should be a relevant route of exposure, that is, for industrial chemicals, dermal, inhalation, or oral.
2. There is exposure during pregnancy/organogenesis only.
3. Structural malformations should be considered inactive of a teratogenic event only if they occur at exposure levels that do not cause overt maternal toxicity (e.g., a reduction in maternal body weight).
4. The increase in incidence of structural malformations in comparison to that of controls must be statistically significant, and the malformations themselves must be of biological significance.
5. Good historical background data should exist on the usefulness of the species tested and data should be for the same strain and from the same laboratory. The rat and rabbit are the preferred species.
6. Adequate group sizes should be used, for example, 20 litters/group for rodents and 12 litters/group for lagomorphs should be available for analysis.

7. Dose or treatment response should be demonstrable for defects that are not rare occurrences.

8. Fetal assessment and data evaluation should be carried out using accepted statistical and laboratory methods.

In the evaluation of animal studies against the above criteria, particular consideration should be given to maternal toxicity as a confounding factor. Positive teratogenic findings seen only in the presence of maternal toxicity do not necessarily indicate specific hazard to the conceptus. By contrast, positive findings in the absence of maternal toxicity imply differential susceptibility between mother and conceptus. The occurrence of teratogenic effects in studies meeting the above criteria in one mammalian species may be regarded as sufficient for a Category 2 classification of the substance concerned, unless there are clear reasons to doubt the relevance to humans. If the above criteria are not met, then insufficient evidence exists to categorize the substance with respect to teratogenicity.

It is recognized that certain chemicals may cause adverse effects on development only at very high dose levels. In such a case, careful interpretation is necessary to determine whether or not a hazard exists. When conducting a study or evaluating results of a study, an upper dose limit, for example, 1000 mg/(kg)(day), should be considered as a dose that normally is of little practical relevance.

With regard to "other relevant information" referred to by EC in the present state of science in developmental toxicology, the relevance of structure–activity relationships, behavioral data, and in vivo and in vitro screening procedures is not sufficiently well understood to justify use of such data for classification purposes.

3.5.2.2 Canadian WHMIS Criteria for Teratogenicity and Embryo Toxicity.

A pure substance or tested mixture is identified as poisonous with regard to teratogenicity and embryo toxicity if it is shown to cause injury to the fetus or embryo in a statistically significant proportion of the test population at a concentration that has no adverse effect on the pregnant female when tested in accordance with OECD Test Guidelines for Teratogenicity (29). In this aspect of WHMIS "injury" includes death, malformation, permanent metabolic or physiological dysfunction, growth retardation, or a psychological or behavioral alteration that occurs during pregnancy, at birth, or in the postnatal period.

3.5.2.3 OSHA Criteria for Reproductive Toxins.

OSHA has not developed specific criteria for reproductive or developmental toxins. OSHA defines reproductive toxins as those chemicals that affect the reproductive capabilities including chromosomal damage (mutations) and effects on fetuses (teratogenesis) (9).

The hazard determination requirement of OSHA is performance oriented and depends on the adequacy and accuracy of the hazard determination. According to OSHA, the process relies heavily on the professional judgment of the evaluator, particularly with respect to hazards in the chronic and reproductive category. The guidance provided above in Section 3.5.2.1 on EC criteria for teratogens regarding "criteria for animal studies" and "other relevant information" may be utilized as a guide in the performance oriented approach to evaluating data advanced by OSHA.

3.5.3 Criteria for Classifying Chemicals as Mutagens

Many test systems are used today to provide indications of the mutagenicity of chemicals. Most of the test systems, however, give only results relating to part of the mutagenic spectrum. Direct evidence for the occurrence of a causal relationship linking chemical exposure and an increase in the frequency of occurrence of a heritable effect in a human population has never been shown. The absence of this sort of evidence in humans does not rule out the potential for human DNA and chromosomes to be affected in ways comparable to those that have been observed experimentally in laboratory animals, and therefore there must be concern about the possible hazards of chemicals causing heritable effects. Animal studies and supporting in vitro studies are currently used to predict genetic hazard associated with chemical exposure (30).

Predictions of relative hazard for mutagens are generally made from a weight of evidence approach. Because evidence of germ cell mutagenicity is of greatest concern, animal tests measuring this are placed in the highest possible category. Other evidence should result in the classification of a substance in a category of lower human concern.

Only results from tests conducted according to established scientific procedures should be used to evaluate mutagenicity. A description of mutagenicity assays in use and factors to be considered in their application and interpretation has been published (30).

3.5.3.1 Criteria for Mutagens. Criteria have been put forth for identifying chemical mutagens of relevance to humans (8).

> *Category 1*: *Evidence for Substances with Human Germ Cell Mutagenicity*. There is sufficient evidence to establish a causal connection between human exposure to chemicals and heritable genetic effects. No known causal relationships between chemicals and heritable genetic effects are known.
>
> *Category 2*: *Evidence for Substances with Mammalian Germ Cell Mutagenicity*. There is one of the following types of evidence for mutagenicity and chemical interaction with the genetic material in mammalian germ cells in vivo:
>> 1. Valid positive results from an in vivo mammalian germ cell study that measures mutations transmitted to offspring.
>> 2. Evidence that the chemical interacts with the genetic material of mammalian germ cells in vivo *plus* clearly positive results in at least one valid in vivo study assessing either gene mutation or chromosome aberrations in somatic cells of humans or other mammals.
>
> *Category 3*: *Evidence for Substances with Somatic Cell Mutation in Mammals without Evidence for Germ Cell Interaction*. Evidence of mutagenic activity

Table 3.11. Comparison of Classification Schemes of Mutagens

	EEC	EPA
1. Evidence for substances with human germ cell mutagenicity	Category 1	—
2. Evidence for substances with mammalian germ cell mutagenicity	Category 2	Sufficient evidence
3. Evidence for substances with somatic cell mutation in mammals without evidence for germ cell interaction	Category 3	Suggestive evidence
4. Inadequate evidence for classifying mutagenicity	—	Limited evidence
5. Negative evidence	—	—

a Note: A direct comparison of categories is not always possible. This comparison is proposed as a reasonable approximation (8).

is provided when a substance gives clearly positive results in at least one valid in vivo mutagenicity study assessing either gene mutation or chromosome aberrations in somatic cells of humans or other mammals. This category is to be used if no evidence is available to permit one to determine whether interaction occurs with the genetic material of the mammalian germ cell in vivo.

Category 4: *Substances with Inadequate Evidence for Classifying Mutagenicity*. This category is for chemicals for which there are insufficient data available to meet the criteria described above; two examples follow:

1. Positive test results only for end points without established clinical relevance and for which human genetic health hazards are not defined, such as DNA damage, sister chromatid exchange, and DNA binding in somatic cells.
2. Positive findings of gene mutations or chromosome aberrations from in vitro studies only.

Category 5: *Negative Evidence*. In a weight of evidence approach to classifying chemicals, it is important to distinguish between chemicals for which no evidence exists and those for which appropriate negative evidence is available. This category is therefore necessary to indicate that there is evidence that a chemical gives negative results in appropriate tests to measure gene mutation and chromosome aberration, or that there is appropriate evidence to indicate that a chemical does not interact with the genetic material of mammalian germ cells in vivo.

These categories for mutagenicity show concordance with the EC classification scheme and also share elements with the EPA categories of sufficient, suggestive, and limited evidence for mutagens (14, 23).

Table 3.11 summarizes the approaches for mutagenicity classification. OSHA and WHMIS contain no specific criteria for categorizing mutagens other than a hazard evaluation of all relevant data.

4 SUMMARY

This chapter presents an overview of toxicity classifications contained in several significant regulatory schemes used in various parts of the world. These include the OSHA, EC, and WHMIS classifications. Criteria contained in these classification schemes are summarized for acute and chronic toxicity. Tables summarizing toxicity criteria for classification purposes are presented.

   In addition to the regulatory schemes described in this chapter, there are other approaches to categorizing chemicals according to toxicity in use. These principally include voluntary consensus approaches. Some examples include the National Fire Protection Association (NFPA), the National Paint and Coatings Association Hazardous Materials Information System (HMIS), and the American National Standard Institute (ANSI) Chemical Labeling Guide (31, 32, 25). A detailed description of these and other systems is beyond the scope of this chapter. However, the health professional who is involved in hazard communication should be aware of these and other federal, state, and local regulations that may apply in particular circumstances. It is expected that the area of categorizing and classifying chemical hazards will be dynamic for many years to come, with efforts aimed at harmonizing the many diverse systems that currently exist.

REFERENCES

1. H. C. Hodge and J. H. Sterner, "Tabulation of Toxicity Classes," *Am. Ind. Hyg. Assoc. J.,* **10**, 93–96 (1949).

2. H. M. Pachter, "Paracelsus: Magic into Science," Collier Books, New York, 1961.

3. G. D. Clayton and F. E. Clayton, (Eds.) Patty's *Industrial Hygiene and Toxicology*, 3rd and 4th eds., Wiley, New York.

4. Casarett and J. Doull, *Toxicology, the Basic Science of Poisons*, Macmillan, New York, 1990.

5. Registry of Toxic Effects of Chemical Substances, U.S. Department of Health and Human Services, Public Health Service; National Institute for Occupational Safety and Health (on line from NIOSH), Cincinnati, OH.

6. Hazardous Substances Data Bank, National Library of Medicine (on line from NLM), Bethesda, MD.

7. EPA Toxic Substances Control Act (TSCA) Inventory of Chemical Substances (on line), EPA, Washington, DC.

8. B. Broecker, E. Sowinski, et al., "Criteria for Identifying and Classifying Carcinogens, Mutagens and Teratogens," *Regul. Toxicol. Pharmacol.*, **7**, 1–20 (1987).

9. OSHA Hazard Communication Standard, 29 CFR 1910.1200.

10. A. B. Hill, *Principles of Medical Statistics*, 9th ed., Oxford University Press, New York, 1971, pp. 309–320.

11. Advisory Committee to the Surgeon General of the Public Health Service, *Smoking and Health*, U.S. Public Health Service, 1964, pp. 182–189.

12. Canada Occupational Safety and Health Regulations; Amendment, SOR/88-68, *Canada Gazette*, Part II, January 20, 1988.

13. Control Products Regulations SOR/88-64, *Canada Gazette*, Part II, January 20, 1988.

14. "Classification and Labelling of Dangerous Substances," European Community Council Directive; Annex VI, 67/548/EEC, June 27, 1967.

15. OECD Test Guideline No. 401, "Acute Oral Toxicity," Organization Economic Community Development, Paris, May 12, 1981.

16. OECD Test Guideline No. 402, "Acute Dermal Toxicity," Organization Economic Community Development, Paris, May 12, 1981.

17. OECD Test Guideline No. 403, "Acute Inhalation Toxicity," Organization Economic Community Development, Paris, May 12, 1981.

18. U.S. Department of Transportation 49 CFR, Part 173.

19. J. H. Draize, G. Woodward, and H. O. Calvery, "Methods for the Study of Irritation and Toxicity of Substances Applied Topically to the Skin and Mucous Membranes," *J. Pharmacol. Exp. Ther.*, **82**, 377–390 (1944).

20. OECD Test Guideline No. 404 "Acute Dermal Irritation/Corrosion," Organization Economic Community Development, Paris, May 12, 1981.

21. OECD Test Guideline No. 405 "Acute Eye Irritation/Corrosion," Organization Economic Community Development, Paris, May 12, 1981.

22. OECD Test Guideline No. 406 "Skin Sensitization," Organization Economic Community Development, Paris, May 12, 1981.

23. Environmental Protection Agency "Proposed Guidelines for Carcinogenic, Mutagenic and Reproductive Risk," FR 49:227, November 23, 1984.

24. IARC Monographs on the Evaluation of Carcinogenic Risks to Humans, World Health Organization, International Agency for Research on Cancer, Lyon, France.

25. American National Standard for Hazardous Industrial Chemicals—Precautionary Labeling. ANSI Z129.1-1988, American National Standard Institute, New York.

26. National Toxicology Program (NTP) Annual Report on Carcinogens, National Technical Information Service (NTIS), Springfield, VA.

27. Threshold Limit Values for Chemical Substances and Physical Agents, American Conference of Governmental Industrial Hygienists, Cincinnati, OH.

28. T. H. Shepard, *Catalog of Teratogenic Agents*, 6th ed., Johns Hopkins University Press, Baltimore, 1989.

29. OECD Test Guideline No. 414, *Teratogenicity*, Organization Economic Community Development, Paris, May 12, 1981.

30. International Commission for Protection Against Environmental Mutagens and Carcinogens, Final Report, "Screening Strategy for Chemicals that are Potential Germ Cell Mutagens in Mammals," *Mutat. Res.*, **114**, 117–177 (1983).

31. HMIS "Hazardous Material Information System," National Paint and Coatings Association, Washington, DC.

32. NFPA "National Fire Protection Agency," Standard 704, Quincy, MA.

CHAPTER FOUR

Complex Mixtures of Tobacco Smoke and the Occupational Environment

Domingo M. Aviado, M.D.

1 INTRODUCTION

The 1990s continue to be transitional years for the U.S. Occupational Safety and Health Administration (OSHA) in finalizing workplace standards (1). Shortly after OSHA was created in 1970, this regulatory agency adopted the threshold limit values (TLV) recommended by the American Conference of Governmental Industrial Hygienists (ACGIH), and these workplace standards continue to be in force to the present time (2). The National Institute for Occupational Safety and Health (NIOSH), also created in 1970, has been comprehensively reviewing the research literature on industrial chemicals and in 1989, OSHA provisionally accepted revised permissible exposure levels (PEL) for 428 industrial substances. A Federal Court of Appeals suspended the revision and in 1992 ordered an indefinite suspension until scientific questions are satisfactorily answered.

The original TLVs and revised PELs were derived from studies of worker exposure. The purpose of workplace standards is the prevention of occupational disease. There is a conspicuous difference between the two groups in arriving at their recommended workplace standards. The ACGIH insists on toxicologic information derived from workers, whereas NIOSH conducts a broader approach, recognizing data not derived from workers. The latter has issued Criteria Docu-

Patty's Industrial Hygiene and Toxicology, Fourth Edition, Volume 2, Part A, Edited by George D. Clayton and Florence E. Clayton.
ISBN 0-471-54724-7 © 1993 John Wiley & Sons, Inc.

ments that emphasize research on a particular chemical substance. Both ACGIH and NIOSH rely on animal studies but utilize the results differently. ACGIH uses animal studies to establish margins of safety, whereas NIOSH encourages risk analysis, that is, quantitative estimates of mortality rates of workers, projected from death rates of exposed rodents.

The above differences in arriving at workplace standards demand reconciliation. The scientific question is as follows: Is nonoccupational or household exposure to chemicals relevant to workplace exposure? The question applies to recent events relating to possible health effects of worker exposure to environmental tobacco smoke (ETS).

The potential association of increased disease risk with ETS exposure to nonsmoking workers is presently being considered by OSHA (3). According to a Current Intelligence Bulletin released in 1991 by NIOSH (4), there is an association between spousal smoking and increased incidence of certain cardiopulmonary diseases, especially lung cancer, but possibly heart disease as well. The association was derived from epidemiologic studies of nonsmokers who reported exposure from cigarette smoking spouses. So far, there are no reported epidemiologic studies designed to examine the incidence of cardiopulmonary disease in nonsmoking workers sharing work facilities with smoking workers. Only respiratory tract irritation has been reported, but there are no specific studies on occupational heart disease, neoplastic disease, and respiratory tract disease.

1.1 Diseases Reportedly Associated with Spousal Smoking

The claim that household ETS exposure is associated with an increased risk of certain cardiopulmonary diseases originated from a study of Japanese nonsmoking women whose husbands were smokers (5). Additional studies conducted in the United States, Europe, and Asia have reported a higher incidence of the following diseases in nonsmoking women whose husbands were smokers, compared to those whose husbands were nonsmokers:

Heart disease	Ischemic heart disease
Neoplastic disease	Pulmonary neoplasm
	Extrapulmonary neoplasms in women, namely, nasal sinuses, brain, breast, cervix, bone marrow
Respiratory tract disease	Respiratory tract irritation
	Chronic obstructive lung disease including chronic bronchitis and pulmonary emphysema
	Bronchial asthma

The above list is derived from reviews sponsored by the Office on Smoking and Health, U.S. Department of Health and Human Services (6), National Research Council (7), Environmental Protection Agency (8), and nongovernmental groups and scientists (9–11). All publications review epidemiologic studies relating to reported household exposure and spousal or childhood diseases. There is no published review on occupational diseases in relation to ETS exposure of workers. This chapter is intended to discuss occupational disease and ETS exposure, with special reference to the three groups listed above. For completeness, coronary atherosclerosis and cardiac arrhythmia and myopathy are included in the group of occupational heart diseases.

1.2 Evaluation of Health Effects of Complex Mixtures

At the outset, it is important to emphasize that ETS is a complex mixture of more than 3700 chemicals identified in mainstream tobacco smoke. Two recent reports of the National Research Council (12, 13) have included a discussion of the toxicity of complex mixtures, such as coke oven emissions, diesel fuel emissions, and gasoline engine emissions. There are three accepted procedures for the evaluation of possible health effects of complex mixtures:

1. It is determined whether a component substance belongs to a group of similar structure and toxicologic profile, ignoring the inert components. Metallic mixtures of beryllium, chromium, or nickel have specific work standards, regardless of chemical nature.

2. Two or more components with dissimilar chemical structure and toxicologic profile are assumed to have additive effects unless proven otherwise. Synergism and antagonism between components in a complex mixture are rare and few available examples were derived from extensive epidemiologic studies with confirmatory mechanistic observations in experimental animals. There are suggested formulas to assure that components of mixtures with additive effects do not exceed work standards.

3. Products of combustion are complex mixtures containing hundreds, or even thousands, of chemical substances. In workplaces containing vehicular emissions, carbon monoxide is typically accepted as a marker for levels of other substances. The potential development of cardiopulmonary disease resulting from chronic exposure to carbon monoxide below the TLV of 50 mg/m^3 is discussed below, as it might relate to ETS exposure (Section 3.1).

There has been no reason to regulate components of vehicular emissions in workplaces. As long as carbon monoxide levels are controlled, then the levels of other constituents can be expected to be below their respective TLVs. The concentrations of particulate matter and polycyclic aromatic amines, if excessively high, have respective work standards: TLV of 10 mg/m^3 and 0.2 mg/m^3. These standards were derived from morbidity and mortality studies of workers exposed to dust, coke oven emissions, or soot. There are reports that unrealistically high levels of ETS in bars, restaurants, and kitchens are associated with respiratory tract irritation. However, there are no such reports regarding ETS exposure in the workplace. The fundamental question is whether reported nonoccupational exposure would relate to occupational disease.

1.3 Spousal Disease Versus Occupational Disease

In spite of the lack of information derived directly from workers, proponents of the claim that spousal smoking is associated with cardiopulmonary disease have extended their hypothesis to include occupational disease. Their reasons are as follows: (1) There are more than 50 chemical substances in ETS that are potentially related to cardiopulmonary disease. (2) Carbon monoxide and other cardiotoxic chemicals present in ETS can contribute to pathogenesis of heart disease in experimental animals. (3) Benzo[a]pyrene and other substances are known to cause skin tumors in mice and it has assumed that these substances cause neoplastic diseases in exposed workers. (4) Respiratory tract irritation has been reported by workers complaining of high levels of ETS in poorly ventilated enclosures.

The above list represents major reasons discussed in the literature in favor of the causal hypothesis of ETS exposure and spousal diseases. The other reasons, as well as conflicting interpretations, are discussed below under sections arranged in the following order of techniques: Section 2, on chemical analysis of ETS; Section 3, on epidemiologic and animal studies relating to heart disease; Section 4, on animal experiments using potential carcinogens present in ETS; and Section 5, on case reports and animal mechanistic studies of respiratory tract disease. Each section includes a discussion of reports of cardiopulmonary disease associated with exposure to industrial chemicals that are reported in ETS, as well as industrial chemicals not reported in ETS. This inclusion is necessary to assure a uniform assessment of occupational diseases reported from exposure to industrial emissions or ETS.

2 ENVIRONMENTAL TOBACCO SMOKE AS A COMPLEX MIXTURE

Environmental tobacco smoke (ETS) is a diluted and aged mixture of chemical substances derived from the burning end and cigarette butt: main-stream smoke (MSS) inhaled from the filtered or unfiltered tip and side-stream smoke (SSS) from the lighted end. Nonsmokers sharing a workroom with smoking workers may be exposed to ETS. Although there are more than 3700 constituents reported in cigarette MSS, approximately 100 have been reported in SSS. Of the 100-plus SSS constituents emitted from lighted cigarettes in the laboratory, 21 have actually been detected in indoor ambient air. The remaining 80-plus SSS constituents have not been detected indoors.

A key reference on constituents in ETS is the 1986 Report of the U.S. National Research Council entitled "Environmental Tobacco Smoke; Measuring Exposure and Assessing Health Effects" (7). The table it contains has been reproduced in several recent reviews, including the Surgeon General's Report (6) and a NIOSH Current Intelligence Bulletin (4). The original table containing 50 SSS constituents is reproduced in Table 4.1 with the addition of nine constituents derived from reviews since 1986 (14, 15). The chemical constituents are arranged in the order of decreasing amount of maximal SSS emissions. About half are derived from particulates and the rest are in the vapor component of SSS emissions. The first column represents line notation numbers that are consistently used in subsequent tables in this chapter. The two-letter notations represent three groups of diseases potentially associated with ETS exposure, namely:

HD = Heart disease, 21 chemicals
ND = Neoplastic disease, 34 chemicals
RT = Respiratory tract disease, 9 chemicals
? = No associated disease reported, 9 chemicals

The maximal reported SSS emission per cigarette is listed in Table 4.1 instead of range (low to high) levels reported in the original source. The original table did not include an entry of indoor concentration which has been added above. It should be noted that less than half of listed substances have been detected in ETS. More than half of substances listed above have not been reported in ETS because SSS has been so diluted by indoor air to escape detection by currently available analysis. There is also the probability that most constituents in SSS are decomposed within a very short time after entering room air.

2.1   Dilution and Uptake of ETS

Side-stream smoke is not inhaled directly by nonsmokers, but is diluted immediately by air in the workplace and continuously by air exchanges. The magnitudes of differences between concentrations of substances in MSS inhaled by the smoker and ETS exposure of nonsmokers have been summarized in a National Research Council monograph (7). The ranges reported in the literature (parts per million or parts per billion) are as follows:

	MSS	ETS
Carbon monoxide	24,900–57,400 ppm	1–18.5 ppm
Nicotine	430,000–1,080,000 ppb	0.15–7.5 ppb
Benzo[a]pyrene	5–11 ppb	0.0001–0.074 ppb

The ratios of peak values are as follows: 3100 for carbon monoxide, 144,000 for nicotine, and 148 for benzo[a]pyrene. There is no uniform dilution for all three because of varied levels in MSS relative to SSS. The unpredictable fates of vapor components (e.g., carbon monoxide) and particulates (e.g., nicotine and benzo[a]pyrene) are influenced by humidity, temperature, air movement, and adsorption by machinery and furnishings in the workplace.

   It is necessary to remember the dilution factors for three major ETS constituents (up to 144,000) because they emphasize the considerable dilution of SSS by room air. Although the concentrations of some SSS constituents may be many times higher than for MSS, the dilution of up to 144,000 silences the argument that ETS contains higher levels of chemicals than MSS.

Table 4.1. Side-Stream Smoke Constituents and Those Detected Indoors as ETS

Line No.	Associated Disease			SSS Constituent	Maximal SSS Emission (mg/cigarette)	Indoor Concn. (mg/m³)
1	HD			Carbon dioxide	440	
2	HD			Carbon monoxide	108	0.6–13
3			RT	Cigarette smoke condensate	30	0.0006–0.03
4			RT	Ammonia	22	
5	HD		RT	Nicotine	8.2	0.006–0.03
6		ND	RT	Formaldehyde	4.6	0.008–0.04
7	?			Acetic acid	2.92	
8	HD		RT	Nitrogen oxides	2.8	0.045–0.176
9			RT	Acrolein	1.7	0.03–0.300
10	HD	ND		Acetaldehyde	1.26	
11	HD			Acetone	1.0	0.36–0.710
12	HD			Methyl chloride	0.88	
13	?			Formic acid	0.78	
14	?			3-Methylpyridine	0.47	
15		ND		3-Vinylpyridine	0.45	
16		ND		Hydroquinone	0.428	
17	HD		RT	Pyridine	0.39	
18	HD	ND	RT	Phenol	0.25	0.001–0.019
19	HD	ND		Benzene	0.24	0.02–0.317
20	HD	ND		Catechol	0.14	
21	?			Lactic acid	0.121	
22	HD			Hydrocyanic acid	0.11	
23	?			d-Butyrolactone	0.11	
24		ND		Cholesterol	0.11	
25	HD			Methylamine	0.1	
26				Succinic acid	0.087	
27	HD			Dimethylamine	0.036	
28	HD			Carbonyl sulfide	0.0546	
29	?			Benzoic acid	0.027	
30		ND		Quinoline	0.02	
31	HD			Aniline	0.011	
32	?			Anatabine	0.01	
					µg/cigarette	µg/m³
33	?			Harman	5.00	
34	HD	ND		2-Toluidine	3.0	
35		ND		Nickel	2.5	
36		ND		N'-Nitrosonornicotine	1.7	
37		ND		NNK[a]	1.4	
38		ND		Fluoranthene	1.26	
39		ND		N'-Nitrosodimethylamine	1.04	0.01–0.240
40		ND		N-nitrosodiethylamine	1.00	0.01–0.200
41		ND		Benzo[a]fluorene	0.75	–0.039
42	HD	ND		Cadmium	0.72	

Table 4.1. (*Continued*)

Line No.	Associated Disease	SSS Constituent	Maximal SSS Emission (mg/cigarette)	Indoor Concn. (mg/m³)
43	?	Zinc	0.4	
44	ND	N-nitrosopyrrolidine	0.28	
45	ND	Benz[*a*]anthracene	0.28	
46	ND	4-Aminobiphenyl	0.14	
47	ND	Benzo[*e*]pyrene	0.135	0.003–0.023
48	ND	Benzo[*g,h,i*]perylene	0.098	0.006–0.010
49	HD ND	Hydrazine	0.096	
50	HD ND	Benzo[*a*]pyrene	0.09	0.0028–0.760
51	ND	2-Naphthylamine	0.07	
52	ND	*N*-Nitrosodiethanolamine	0.043	
53	ND	Dibenz[*a,h*]anthracene	0.41	6.0
54	ND	Anthanthrene	0.039	0.0041–0.0094
55	ND	Perylene	0.039	0.0007–0.0013
56	HD ND RT	Toluene	0.035	40–1040
57	ND	Pyrene		0.0041–0.0094
58	ND	Coronene		0.005–0.0012
59	ND	Polonium-210	0.4 pCi	

[a]4-(Methylnitrosamino)-1-(3′pyridyl)-1-butanone.

2.1.1 Nicotine as ETS Marker

That nicotine and its major metabolite (cotinine) are detected in blood and urine of ETS-exposed nonsmokers has been utilized by proponents of the ETS–cardiopulmonary disease hypothesis. Their reasoning is as follows: because nicotine is a major cause of heart disease seen in cigarette smokers, it follows that any nicotine derived from ETS can cause heart disease in exposed nonsmokers. However, there is disagreement concerning whether any nicotine absorbed by nonsmokers can influence the heart. The estimates of ETS exposure are as follows: a nonsmoker might absorb, at most, the nicotine equivalent of 1/20 to 1 cigarette in 1 day, which has not been reported to have a significant pharmacological action. In animal experiments, inhalation, ingestion, parenteral injection, and dermal application of nicotine have been reported to influence cardiac function, coronary circulation, and atherogenesis, but these studies used amounts of nicotine that cannot be attained by ETS exposure. Furthermore, coronary atherosclerosis has not been reproduced in experimental animals by injection of nicotine. Workers processing tobacco leaf (cigar making, leaf curing, and warehouse workers) also have not been reported to show a higher incidence of heart disease, compared to non-tobacco workers. Details of the potential health significance of absorbed nicotine are discussed under Section 3 (on occupational heart disease) and Section 5 (on occupational respiratory tract disease).

2.1.2   Carboxyhemoglobin as ETS Marker

Carboxyhemoglobin levels in nonsmokers have been utilized to estimate amounts of carbon monoxide absorbed from ETS (36). The National Research Council 1986 report (7) lists nine studies that investigated the effects of ETS exposure in nonsmokers. Seven additional studies can be added to the list and are summarized in Table 4.2.

The results in Table 4.2 were derived from volunteers in poorly ventilated rooms or experimental chambers with smoke generated by cigarettes. The three sets of mean and standard deviation listed in the table clarify some confusion caused by the absence of a statistical summary in the National Research Council table of carboxyhemoglobin levels (page 258 of Reference 7). The mean value after ETS exposure has been erroneously estimated "to be close to 3%" (35). Examination of nine cited articles indicated that the mean levels were 1.29 percent before and 1.85 percent after, with a change of $+0.56$ percent carboxyhemoglobin. Seven additional studies reported after 1986 have been added to the list. It should be noted that the average carboxyhemoglobin values are as follows: before ETS exposure 0.9 percent and after exposure 1.56 percent, with a mean change of $+0.65$ percent carboxyhemoglobin. Exposure to ETS does not appear to be associated with greater than 2.0 percent carboxyhemoglobin, a level that has not been reported to influence heart function (see Section 3.1).

2.2   Work Standards for Industrial Chemicals

The minute levels of carbon monoxide in ETS, up to 3100 times less than the concentration in MSS, poses a critical challenge to claims that ETS exposure can cause cardiopulmonary disease in nonsmokers. Proponents of the claimed association between ETS exposure and cardiopulmonary disease in general (occupational and nonoccupational) contend that three ETS constituents underlie this relationship: nicotine, carbon monoxide, and polynuclear aromatic hydrocarbons that coincidentally have workplace standards. For completeness, there are 28 reported constituents of SSS that are also used as industrial chemicals to which exposure of workers has been reportedly associated with cardiopulmonary disease (Table 4.3).

The listing in Table 4.3 of 28 SSS constituents imputed to ETS is selected from the 59 listed in Table 4.1. This selected list is a revision of my listing of suspected pulmonary carcinogens in ETS (37, 38). Also listed in Table 4.3 are corresponding threshold limit values (TLVs) for various substances, defined as the recommended standards for 8-hr daily exposure for the prevention of occupational disease (2).

2.3   Cigarette Equivalents to Attain Threshold Limit Values

The 28 SSS constituents with established workplace standards are listed in the order of increasing number of cigarette equivalents, defined as the number of cigarettes burned in a sealed enclosure of 100 m^3 to attain, but not to exceed, the corresponding TLV. The list starts with nicotine, which is the most widely investigated constituent of tobacco smoke. The maximum reported SSS collected from one burning cigarette is 8.2 mg nicotine. On the basis of TLV (0.5 mg/m^3), it would take 6.6 cigarettes to attain TLV for 100 m^3 in a sealed, unventilated enclosure (0.5×100 divided by 8.2). It is unlikely for the nicotine concentration in public places to attain the TLV level. If smoking has been at an extremely high level in poorly ventilated rooms, subjective discomforts would be expected to lead to corrective measures before nicotine levels would approach the TLV (see Section 5).

The fifth SSS constituent listed in the order of increasing cigarette equivalents is carbon monoxide: it would take 50 cigarettes burning in a 100 m^3 sealed chamber to attain the corresponding TLV of 55 mg/m^3 (12). Including nicotine and carbon monoxide, six constituents would require less than 100 cigarettes to attain the corresponding TLV. Most of the remaining 22 constituents would require more than a 1000 or even 1,000,000 cigarettes to attain the respective TLV. Such excessively high cigarette equivalents suggest that to attain TLV levels, more than 1000 cigarettes must be ignited simultaneously in an enclosed space of 100 m^3. Consideration of cigarette equivalents clearly indicates that exposure to 28 ETS constituents in workplaces rarely approximates corresponding TLVs.

3 OCCUPATIONAL HEART DISEASE

The pathogenesis of heart disease has been directed to its occurrence in the general population. There is limited research on occupational heart disease because of the overwhelming importance of nonoccupational factors, that is, dietary, hereditary, life-style, and environmental factors. Because there has been no grouping of industrial chemicals according to potential relationships with specific forms of heart disease, I have recently proposed a system that is described below (39).

Data on a possible association between workplace exposure to an ETS constituent may consist of a combination of human studies and animal experimentation. Workers such as garage and toll booth attendants may be exposed to one or more substances (such as carbon monoxide) reported both in ETS and in other sources, so the total exposure is the sum of two or more sources, that is, vehicular emissions, ambient air pollution, and ETS. The same group of workers may have varied personal habits that have been reported to be associated with heart disease, such as consumption of cholesterol and fats and xanthine beverages at the employees' cafeteria, physical inactivity on the job, and job-related stress. Outside of the workplace, there are additional potential risk factors for heart disease, such as lack of leisure time exercise, dietary cooking fat and salt content, household exposure to cooking gas, gas heaters, and household solvents. Other major risk factors reported for heart disease include the worker's familial history of heart disease, diabetes, hypertension, hyperlipidemia, and obesity. Any conclusion on a possible role of ETS in occupational heart disease necessitates controlling for such risk factors.

Although there are more than 300 potentially hazardous chemicals in the workplace, there are less than three score of industrial chemicals that have been sug-

Table 4.2. Blood Carboxyhemoglobin Levels in Subjects Exposed to ETS

Study	No. Cig./(hr) (10 m^3)	CO (ppm)	No. of Sub.	Carboxyhemoglobin Level (% Mean ± SD)		
				Control	Exposure	Change
NATIONAL RESEARCH COUNCIL 1986 REPORT						
Andersson and Dalhamn, 1973, Sweden (18)	3.1	4.5	5	2.5	2.5	0
Dahms et al., 1981, U.S. (19)		17.5	10	0.62 ± 0.08	1.06 ± 0.08	+0.03
Harke, 1970, Germany (20)	3.9	30	7	0.9 ± 0.1	2.1 ± 0.1	+1.2
Huch et al., 1980, Switzerland (21)	2.3		12	1.3	1.8	+0.5
Hugod et al., 1978, Denmark (22)	2.5	20	10	0.73 ± 0.16	1.63 ± 0.25	+0.9
Pimm et al., 1978, Canada (23)	2.4	24	20	0.6	0.85	+0.25
Polak, 1977, Belgium (24)	6.7	23	15	2.0	2.3	+0.3
Russell et al., 1973, England (25)	15.1	38	12	1.6 ± 0.6	2.6 ± 0.7	+1.03
Sappanen, 1977, Finland (26)	3.8	16	28	1.6 + 0.3	2.2 + 0.6	+0.6

	C1	C2	n			
Mean ± SD for 9 studies			119	1.29 ± 0.55	1.85 ± 0.59	+0.56
ADDITIONAL STUDIES						
Alvarez-Sala et al., 1985 and 1986, Spain (27, 28)			29	0.84 ± 0.65	1.27 ± 0.64	+0.43
			10	0.66 ± 0.32	1.13 ± 0.41	+0.49
Asano and Ohkubo, 1988, Japan (29)		12.5	6	0.1	1.0	+0.9
Davis, 1990, U.S., (30)	25	5	10	1.1 ± 0.6	1.2 ± 0.7	+0.1
Hoepfner et al., 1987, Germany (31)	2.7	25	5	0.4 ± 0.1	2.6 ± 0.2	+2.2
			9	0.9 ± 0.2	2.7 ± 0.1	+1.8
Olshansky, 1982, U.S. (32)		12	5	0.77	1.84	+1.07
				0.7 ± 0.15	3 ± 0.2	+2.3
Scherer et al., 1987, Germany (33)	10	10	10	0.18 ± 0.02	0.87 ± 0.04	+0.69
Winneke et al., 1984, Germany (34)		24		0.65 ± 0.32	2.69 ± 0.13	+2.04
		15	64	0.56	0.96	+0.4
Mean ± SD for 7 studies			163	0.64 ± 0.22	1.36 ± 0.63	+0.72
Mean ± SD for all 16 studies			282	0.91 ± 0.51	1.56 ± 0.66	+0.65

Table 4.3. Side-Stream Smoke Constituents with TLVs and Cigarette Equivalents (in 100 m³ unventilated chamber)

Line No. (Table 4.1)	Chemical Name	Max SSS (mg/cigarette)	TLV (mg/m³)	Cigarette Equivalent
5	Nicotine	8.2	0.5	6.6
6	Formaldehyde	4.6	0.37	8
9	Acrolein	1.7	0.25	15
3	Cigarette smoke condensate	30	10	33
2	Carbon monoxide	108	55	50
4	Ammonia	22	18	82
16	Hydroquinone	0.428	2	467
7	Acetic acid	2.92	25	856
13	Formic acid	0.78	9	1,154
12	Methyl chloride	0.88	10.3	1,170
42	Cadmium	0.0007	0.01	1,430
10	Acetaldehyde	1.26	180	1,430
8	Nitrogen oxides	2.8	50	1,780
1	Carbon dioxide	440	9000	2,040
17	Pyridine	0.39	16	4,100
18	Phenol	0.25	19	7,600
22	Hydrocyanic acid	0.11	11	10,000
25	Methylamine	0.1	13	13,000
19	Benzene	0.24	32	13,300
20	Catechol	0.14	23	16,500
31	Aniline	0.011	8	44,000
27	Dimethylamine	0.036	18	50,000
28	Carbonyl sulfide	0.0546	30	54,945
49	Hydrazine	0.00009	0.13	145,000
11	Acetone	1	1780	178,000
50	Benzo[a]pyrene	0.00009	0.2	222,000
34	2-Toluidine	0.003	9	300,000
56	Toluene	0.000035	375	1,000,000

gested to be associated with heart disease such as ischemic heart disease, coronary atherosclerosis, and cardiac arrhythmia and cardiomyopathy. Although heart disease is the leading cause of death in the United States, occupational exposure to chemicals is considered less prevalent and less important than major factors of age, gender, cholesterol in the diet, and familial or inherited susceptibility to cardiovascular disease.

Although it is relatively simple to establish a strong association between exposure to halogenated solvents and cardiac arrhythmias, it is more complex to obtain supportive evidence as to whether chemicals play a major role in coronary ischemic heart disease and atherosclerosis. Occupational heart diseases can be grouped into three major categories; these can be subgrouped according to the method of investigation, that is, clinical studies, pathological observations, or experimental animal studies: (1) *ischemic heart disease* (Methods A, B, and C), including mortality studies, exercise testing for angina pectoris, and coronary blood flow indicators; (2) *coronary atherosclerosis* (Methods D, E, and F), demonstrable in patients by angiography and histopathology, atherosclerosis in experimental animals, and in vitro studies of hematologic factors; and (3) *cardiac arrhythmia and myopathy* (Methods G and H), both clinically and experimentally induced. The three groups of methods and eight subgroupings (A to H) are carried over to consideration of occupational heart disease associated with exposure to chemicals in the course of manufacturing and processing of industrial products. The chemicals supposedly associated with occupational heart disease are listed in Table 4.4 under five classes, one inorganic and four organics. Each compound is identified by notations on investigative methods A to H.

Table 4.4 includes industrial chemicals reported to be associated with occupational heart diseases (39). The compounds with line numbers are also reported in SSS as indicated in Table 4.1.

Of the 32 industrial chemicals listed in Table 4.4, 21 are reported as ETS constituents. These 21 industrial chemicals, when individually used in factories below the corresponding TLV, have not been associated with heart disease or with any adverse effect on corresponding target organs, that is, mucosal surfaces, skin, blood, nervous system, lungs, kidneys, and liver. The remaining 11 industrial chemicals (no line number) are *not* reported to be present in ETS. The existing methods for establishing cardiac toxicologic profiles are listed as follows: Methods A, B, and C for ischemic heart disease; Methods D, E, and F for coronary atherosclerosis; and Methods G and H for cardiac arrhythmias and myopathy. Most industrial chemicals have been studied by one or two methods, thus contributing to the uncertainty of whether these 21 chemicals are related to heart disease. Only five chemicals have been studied by three to eight methods and have a stronger basis for claims of a relationship with occupational heart disease, namely, carbon monoxide, ethylene glycol dinitrate, nitroglycerin, carbon disulfide, and methylene chloride.

3.1 Ischemic Heart Disease

Ischemic heart disease (IHD) is represented clinically by angina pectoris, myocardial infarction, cardiac arrhythmia, cardiogenic shock, and sudden death. The epidemiologic and clinical literature on work-associated ischemic heart disease consists of the following: Method A, mortality statistics; Method B, exercise testing for anginal pain; and Method C, coronary blood flow indicators. The plan is to state how each method has been applied to the concept that ischemic heart disease is related to exposure to chemical substances in the manufacture of industrial products. Although ETS constituent levels are unlikely to attain their corresponding TLV, it is important to discuss the existing claim that the mere presence of these chemicals is sufficient to suggest an association between ETS exposure and occupational heart disease.

Table 4.4. Industrial Chemicals Suspected of Causing Heart Disease

Line No. (Table 4.1)	Industrial Chemicals	Ischemic Heart Disease	Coronary Atherosclerosis	Arrhythmias and Myopathy
INORGANICS: OXIDES AND METALS				
2	Carbon monoxide	A B C	D E F	G H
1	Carbon dioxide			G
8	Nitrogen oxides			G
	Arsenic	A		H
42	Cadmium	A		
	Cobalt	A		H
	Lead	A	F	
NITROGENOUS COMPOUNDS				
5	Nicotine	A		
31	Aniline		F	
20	Catechol			G
	Dinitrotoluene	A		
	Ethylene glycol dinitrate	A C		G
49	Hydrazine			G
22	Hydrocyanic acid	C		H
	Nitroglycerin	A C		G
17	Pyridine			G
34	2-Toluidine		F	
POLYNUCLEAR AROMATIC HYDROCARBONS				
50	Benzo[a]pyrene	A	E	
	7,12-dimethyldibenz[a,h]anthracene		E	
	3-Methylcholanthrene		E	
NONHALOGENATED SOLVENTS				
28	Carbonyl sulfide	A B C	D E	H
10	Acetaldehyde			G
11	Acetone			G
19	Benzene			G
27	Dimethylamine			G
25	Methylamine			G
18	Phenol			G H
56	Toluene			G
HALOGENATED SOLVENTS				
12	Methyl chloride			G H
	Methyl chloroform			G H
	Methylene chloride		F	G H
	Trichlorofluoromethane			G H

3.1.1 Method A: Mortality Studies

There are scant data on heart disease in workers differentiated by exposure or nonexposure to ETS in the workplace. Most published studies relate to differences in spousal smoking habits, based on the premise that mortality rates of nonsmokers might be influenced by smoking habits of their spouses. In 1984, Schievelbein and Richter reviewed the available literature and concluded that in concentrations of carbon monoxide and nicotine reportedly present in ETS, it is unlikely for ETS exposure to play any role in the development and progression of ischemic heart disease (40). The 1986 Report of the Surgeon General (6) and the National Research Council (7), after examining the available information, concluded that further studies on the potential relationship between ETS exposure and cardiovascular disease are needed in order to determine whether ETS exposure increases the risk of cardiovascular disease in general, and of ischemic heart disease in particular. Recent epidemiologic studies were reviewed by Wexler (41), who questioned the reported relationship between spousal smoking and heart disease.

Prospective (cohort) and retrospective (case control) studies have been conducted on the potential relationship between spousal ETS exposure and IHD incidence. Although some spousal studies (smoker married to nonsmoker) report a statistically significant association, most studies do not. Lee and his collaborators (42) conducted studies in England consisting of administering a questionnaire to 200 hospital patients and 200 controls for each gender and age group. Patients with ischemic heart disease and controls did not show any statistically significant difference in ETS exposure based on smoking habits of spouses. Exposure to ETS was also evaluated by an index of presence in the workplace, during travel, and at leisure. From the standpoint of worker ETS exposure, the negative results of Lee et al. are more relevant than positive results of spousal studies that do not include ETS exposure outside the home environment.

3.1.1.1 Carbon Monoxide. Heart disease mortality rates have been reported for workers exposed to high levels of carbon monoxide from vehicular emissions (tunnel workers, bus drivers, parking attendants) and industrial furnaces (steel foundry, coke oven, chemical manufacture) (43, 44). However, the results of occupational exposure to high levels of carbon monoxide do not support the argument that this substance contributes to heart disease potentially associated with ETS exposure, in which reported levels of the gas are a tiny fraction of the TLV.

3.1.1.2 Carbon Disulfide/Carbonyl Sulfide. These two compounds are linked by the fact that the former is an industrial chemical reported to be associated with heart disease among workers producing viscose rayon fibers. This compound is metabolized to carbonyl sulfide, which happens to be a reported SSS constituent. The concentration of carbonyl sulfide is so low that it is unlikely to attain the TLV (Table 4.3: 54,945 cigarettes to attain TLV). However, it is important to discuss mortality studies of rayon viscose workers, because other than carbon monoxide, carbon disulfide is the only industrial chemical for which there are extensive data on an association with ischemic heart disease. In a critical review of the toxicologic literature on carbon disulfide, Beauchamp et al. (45) reviewed data on the mortality rates of viscose rayon workers. In Finland, where there is a high incidence of ischemic heart disease, a significantly higher mortality rate has been reported among exposed workers compared to a control group. In Japan, however, where there is a notably lower incidence of ischemic heart disease, no increased mortality rate has been reported among viscose rayon workers. The excess deaths attributed to carbon disulfide became apparent if predisposing risk factors existed, such as hypertension, hyperlipidemia, and excessive intake of cholesterol and saturated fats (45, 46).

The above observations are essential to consider in attempts to interpret mortality studies on ETS exposure. Dietary intakes of cholesterol and fatty food were not considered as a confounding factor in mortality studies relating to workers exposed to the industrial chemicals listed in Table 4.4 (with Method A notation). The reported higher susceptibility of Scandinavians to heart disease is reflected by the lower TLV (15 mg/m^3) compared to the TLV in other European countries and the United States (30 mg/m^3).

3.1.1.3 Polycyclic Aromatic Hydrocarbons.

It has been suggested by proponents of the ETS heart disease hypothesis that Scandinavian roofers show excess mortality for ischemic heart disease (47). They extrapolate from roofers exposed to polycyclic aromatic hydrocarbons (PAH) to ETS-exposed workers without recognizing the difference in composition of PAH. Exposures to PAH among coke oven workers, creosote wood appliers, and asphalt road builders have not been reported to be associated with excess mortality for heart disease, but have been reported to be associated with excess mortality for lung cancer. From the standpoint of chemical composition of PAH exposures determined by nature of product, PAH exposures of roofers are irrelevant to ETS exposure (see also Method F).

3.1.1.4 Heavy Metals.

Mortality studies on work-related exposure have been reviewed by Kristensen (48). Lead and cadmium workers have been reported to show a higher mortality rate from heart disease and hypertension. In the absence of experimental animal studies, heart disease is likely to be a complication of hypertension rather than a direct effect of lead or cadmium on the heart and coronary vessels. The suggestion that heart disease may be associated with workplace exposure to arsenic or cobalt can be traced to instances of beer drinking contaminated with either of these metals, and subsequent death from cardiomyopathy.

3.1.2 Method B: Exercise Testing and Angina Pectoris

Exercise testing is essential for the diagnosis of ischemic heart disease. A positive diagnosis is based on the appearance of chest pain or classical angina pectoris after completion of standardized exercise on a treadmill or bicycle ergometer. Exercise testing has also been used to evaluate severity of arteriosclerotic heart disease based on time of onset of an ischemic pattern in the electrocardiogram as well as the appearance of cardiac arrhythmias, or the visualization of abnormal wall motion in the echocardiogram.

3.1.2.1 ETS Exposure of Anginal Patients.

All available reports on exercise testing do not relate to specific occupational groups comparing two subgroups, those with ETS exposure and those with no ETS exposure. There are two studies on anginal patients that suggested to the investigators that ETS exposure during bicycle ergometry may shorten the time period to onset of chest pain. The first study, reported in 1978, consisted of a group of 10 American male veterans (49). For various reasons, the 1978 protocol for exercise testing was evaluated by an ad hoc committee of the Environmental Protection Agency. In 1983, the committee concluded that the method used on American male veterans "did not meet a reasonable standard of scientific quality" (50). In 1987, a second study of exercise testing during ETS exposure was reported by Soviet investigators (51). The results were essentially similar to those reported from American veterans. It is my opinion that shortening onset of anginal pain during exercise testing as a result of ETS exposure has not been proven pending evaluation of the Soviet protocol. Anti-anginal drugs sold in the United States are supported by results of exercise testing in European laboratories that have been approved by the U.S. Food and Drug Administration, and so far the list does not include any Soviet laboratories.

3.1.2.2 Influence of Carbon Monoxide on Exercise Testing.

Proponents of the theory that ETS exposure aggravates angina pectoris emphasize the presence of carbon monoxide in ETS, in spite of the data that the concentration reported in ETS is 3100 times lower than in MSS (see Section 2.2). Blood carboxyhemoglobin levels of subjects exposed to ETS in public places range from 1 to 2 percent among nonsmokers (see Section 2.1.1). Slight elevations of blood carboxyhemoglobin level (to 2 and 3.9 percent) have been reported following administration of carbon monoxide in air (50 and 100 ppm, TLV and twice TLV) (52). Exercise testing of healthy subjects was reported to result in an ischemic pattern of electrocardiogram at these blood carboxyhemoglobin levels. However, as indicated in Table 4.3, this would require 50 and 100 cigarettes burning in a sealed enclosure of 100 m^3 for carbon monoxide to attain the TLV and twice the TLV, respectively.

3.1.2.3 ETS Exposure as Risk Factor for Angina Pectoris.

Proponents of the claim that ETS exposure aggravates angina pectoris have not considered the complexities of the disease separate from other manifestations or complications of ischemic heart disease (i.e., acute myocardial infarction and sudden deaths). Although prospective and retrospective studies report that cigarette smoking is one of many risk factors for acute myocardial infarction and sudden deaths, the data on angina pectoris are even more complex. The 1983 report of the Surgeon General on cardiovascular disease, referring to risk factors, concluded that "variation in the strength of association between smoking and angina pectoris may be influenced by . . . methodological considerations" (page 70 of Reference 53). More recently, it has been argued that the 30-year results of an ongoing prospective study at Framingham, Massachusetts indicate that cigarette smoking is a negative risk factor in women; that is, there is a lower incidence of angina pectoris in women smokers compared to women nonsmokers (54). The results in men have indicated either a positive or no significant relationship between cigarette smoking and angina pectoris, depending on methodological variation. Some studies relating to cardiac patients admitted to hospitals report that after an initial cardiac episode, the prognosis is not influenced by smoking (55, 56). After the initial infarction, prior smoking was not associated with the severity of subsequent complications. These observations on cigarette smoking in relation to the prognosis of myocardial infarction and the influence of angina pectoris raise additional questions. How can ETS, a dilute mixture of tobacco smoke components in air, aggravate angina pectoris or influence the prognosis of acute myocardial infarction, in light of recent inconsistencies in data derived from smokers?

3.1.3 Method C: Coronary Blood Flow Indicators

Coronary arteries visualized by angiography can show obstruction that is organic (arteriosclerosis and thrombosis) or nonorganic (vasospasm) in nature. Total coronary blood flow is measured by a tracer clearance technique. Patients with ischemic heart disease show a reduction in coronary blood flow that is limited to an infarcted area. When infarction is detected in workers previously exposed to carbon monoxide or carbon disulfide, it is not possible to isolate the potential association with chemical exposure from other potentially confounding risk factors. Carbon monoxide alone, by increasing carboxyhemoglobin, can increase coronary blood flow, but the result would be an oversupply of blood with reduced oxygen-carrying capacity.

Nitroglycerin and organic nitrates are useful vasodilators for the relief of angina pectoris. The pharmacological action of nitroglycerin is manifested in workers who are exposed daily to nitroglycerin and ethylene glycol dinitrate but after a weekend of nonexposure, develop chest pain on Monday morning. Workers suffer from vasospastic angina as a result of nitrate withdrawal during the weekend, and are relieved upon resuming nitrate work exposure. Autopsied workers did not show coronary arterial obstruction confirming the occurrence of vasospastic angina, brought about by weekend withdrawal from nitrate. Workers were acclimatized to the nitrate level in the work environment (48).

3.2 Coronary Atherosclerosis

The term coronary *atherosclerosis* used in this chapter refers to histopathological changes in arteries leading to *ischemic heart disease* (Section 3.1). Although both terms are included in *coronary heart disease*, there are differences in methodology. This section is devoted to progressive organic lesions of coronary arteries, the methods for detection, and their evolution, based on human observations and animal experimentation. The focus is on industrial exposure to carbon disulfide and carbon monoxide, because of the relatively greater amounts of data on these substances. The potential relevance of these industrial chemicals to ETS exposure is also discussed.

The demonstration of coronary atherosclerosis ideally should include histopathological evidence derived from autopsy (Method D). This has been accomplished for worker exposure to carbon disulfide and has been supported by the occurrence of hyperlipidemia in exposed workers and coronary atherosclerosis in experimental animals (Method E). On the other hand, some industrial chemicals are associated with the development of coronary atherosclerosis based on animal experiments only or on hematologic changes in workers that in animals contribute to aortic atherogenesis (Table 4.4). Some of these observations have been used to support the claim that ETS exposure is involved in coronary atherosclerosis. A distinction is made between concepts derived from human studies (Method D), animal experiments (Method E), and in vitro techniques (Method F).

3.2.1 Method D: Coronary Angiography and Histopathology

The most direct method for diagnosis of coronary atherosclerosis is by histopathological examination and coronary angiogram. Although there are isolated reports that workers exposed to carbon monoxide suffer from increased coronary atherosclerosis (antemortem or postmortem), this exposure is confounded by competing risk factors such as personal habits, familial history, and environmental pollution. Among viscose rayon workers, the occurrence of coronary atherosclerosis reported at autopsy of workers dying of heart disease led to mortality studies (Method A). Workers are also reported to suffer from hyperlipidemia that is not entirely due to carbon disulfide exposure. It is difficult to replicate earlier studies on workers using modern techniques of diagnosing coronary atherosclerosis, because exposure levels have come under strict regulation.

There are no case reports of coronary atherosclerosis in workers exposed to a single polynuclear aromatic amine because workplace exposure is to mixtures that include benzo[a]pyrene. Only research laboratory workers investigating benzo[a]pyrene are candidates for long-term exposure, and so far there has been no report of a higher incidence of heart disease. There are also no case reports of coronary atherosclerosis from prolonged exposure to the heavy metals and nitrogenous compounds listed in Table 4.4.

3.2.2 Method E: Coronary Atherosclerosis in Experimental Animals

Repeated attempts to induce coronary atherosclerosis in experimental animals by inhalation of cigarette smoke have failed. Additional feeding with a cholesterol-enriched diet has reportedly led to the development of atherosclerosis not involving coronary arteries. In baboons, after 2 to 3 years of oral feeding of cholesterol and saturated fat, and daily inhalation of cigarette smoke, arterial lesions were compared between smokers and controls. Among male baboons, the extent of carotid atherosclerosis was greater in smokers than in controls, but there were no significant differences in atherosclerosis of the aorta or coronary, iliac-femoral, and bronchial arteries. Among female baboons, there were no significant differences in atherosclerosis between smokers and controls (57).

The same general remarks apply to experimental testing of carbon monoxide in levels far exceeding those reported for ETS exposure. Rabbits, pigeons, and chickens are reported to need supplementary feeding of cholesterol to show carbon monoxide-induced aortic atherosclerosis (58).

Carbon disulfide is the only industrial chemical reported to cause atherosclerosis in animals without supplemental cholesterol feeding. Coronary and aortic atherosclerosis and myocardial lesions were detected in rats after 4 months of inhalation exposure (46). There were elevations of serum cholesterol, phospholipid, and triglycerides, indicating similarity to the human form of atherosclerosis. Other investigators have tested carbon monoxide and benzo[a]pyrene and have not observed hyperlipidemia and atherosclerosis similar to those reported for carbon disulfide. In the past, research on carbon monoxide, benzo[a]pyrene, and other polynuclear aromatic hydrocarbons has not been directed to a comparison with carbon disulfide.

Polynuclear aromatic hydrocarbons have been reported to induce aortic atherosclerosis in pigeons and chickens (59, 60). It has been speculated that these studies in birds relate to human subjects exposed to ETS (47). There are several reasons for the inapplicability of results of these bird experiments to coronary atherosclerosis: (1) 7,12-dimethylbenzo(a,h)anthracene and 3-methycholanthrene are not known to be present in ETS; (2) although benzo[a]pyrene is reportedly present in ETS, the dose administered, 50 mg/kg injection, is farfetched compared to concentration levels in SSS, which is reported to be 0.00009 mg/cigarette; (3) hepatic metabolism is essential for atherogenesis in one strain, but not in the other strain, a sequence that applies to oral or injected compounds but not to the inhalation route; and (4) the typical result is aortic atherosclerosis and rarely coronary atherosclerosis. Aortic atherosclerosis is different from coronary atherosclerosis because of myocardial extravascular support in the latter. There are intracardiac mechanisms that influence coronary circulation but are absent in other arterial beds. There are long-term animal experiments designed to study carcinogenicity of polynuclear aromatic hydrocarbons and so far, coronary atherosclerosis has not been reported in sacrificed animals.

3.2.3 Method F: In Vitro Studies of Hematologic Factors

Hematologic factors include alterations in hemoglobin oxygen transport such as carbon monoxide and methylene chloride increasing carboxyhemoglobin, and aniline and 2-toluidine leading to methemoglobinemia. The ultimate consequence is a reduced supply of oxygen and presumably atherosclerosis resulting from carbon monoxide. However, prolonged testing with methylene chloride or aniline has not been reported to produce experimental atherosclerosis, suggesting that these two industrial chemicals reduce hemoglobin oxygen transport differently from carbon monoxide.

Several techniques have been developed for the specific purpose of discovering therapeutic agents for the prevention, suppression, and reversal of atherosclerosis.

Drugs for influencing blood platelets, blood lipoprotein levels, and endothelial vulnerability evolved from application of in vitro testing of blood derived from patients with ischemic heart disease, as well as peripheral vascular diseases. The same techniques for identifying therapeutic agents have also been applied to investigating how carbon monoxide and ETS might play a role in atherosclerosis. The interpretation of results derived from one test has been extended to include the entire progression of atherosclerosis even though the test was intended to show a therapeutic, rather than toxic, effect of chemical agents.

In vitro tests have been applied to blood from ETS-exposed subjects, based on the assumption that any reported effect will contribute to coronary atherosclerosis. It should be pointed out that chemically induced platelet aggregation leads to vascular clot formation, which does not necessarily involve interaction with endothelial cells and the formation of atherosclerotic plaque. Also, in the laboratory it has not been possible to initiate aortic plaque formation by exceeding the normal level of fibrinogen. Any reported increase in fibrinogen level in the blood of ETS-exposed subjects may not be relevant to a potential relationship with coronary atherogenesis. It is conceivable that for some people, ETS exposure may be perceived as stressful, with release of catecholamines, and that catecholamines are responsible for in vitro testing results. It has not been possible to conduct a double blind testing of ETS exposure because both investigator and subject can detect ETS presence.

3.2.3.1 Platelet Aggregation.
Exposure of healthy nonsmokers to ETS is alleged to alter results of in vitro testing of platelets in platelet-rich plasma. Aggregation of platelets is tested by the following agents added in vitro: edetic acid and formaldehyde or prostaglandin. The possibility that ETS exposure increases platelet aggregation is alleged to be an important step in the evolution of coronary atherosclerosis in nonsmokers (47).

In vitro studies of platelet aggregation in blood derived from smokers have yielded inconsistent results, which bring into question the applicability of this method to ETS exposure in nonsmokers. Platelet aggregation testing using whole blood reported no statistically significant differences between nonsmokers and smokers (61). Cigarette smoking is reportedly associated with alterations in platelet factors involved in thrombus formation, but the change has been attributed to the presence of carbon monoxide levels higher than those reported in subjects exposed to ETS (62). In vitro testing does not necessarily reflect events in vivo. Although platelets may be activated in vivo, they become attached to erythrocytes or form platelet aggregates during the collection and centrifugation needed to make platelet-rich plasma. There is some evidence that activated platelets are lost from supernatant "platelet-rich plasma," which includes older or less active platelets.

3.2.3.2 Plasma Fibrinogen Levels.
Another in vitro test for a clotting factor has been added to the list of reports supporting the ETS–heart disease hypothesis. Patients with ischemic heart disease were questioned about their smoking habits, and nonsmokers were queried for ETS exposure in the workplace and household.

Control subjects were derived from the same community in Australia (63). It was reported that the collected blood samples showed higher fibrinogen concentrations among current smokers than nonsmokers. Subjects exposed to ETS had higher levels than those not exposed. The differences were not statistically significant because of the high variability of measured fibrinogen levels. According to the questionnaire responses, estimates of ETS exposure at work were reported to be higher than at home, but the estimated odds ratio for heart disease was less than one. The investigators interpreted their results to indicate inaccurate reporting of ETS exposure or the possibility that household exposure to ETS is associated more with heart disease than is workplace exposure. The potential relevance of fibrinogen levels in relation to ETS exposure is further questioned by observations that psychosocial factors may influence the plasma fibrinogen concentration in patients with ischemic heart disease (64).

3.3   Cardiac Arrhythmia and Myopathy

The third and last group of methods for establishing cardiac toxicologic profiles for industrial chemicals relates to alterations in cardiac function. The methods are intended to detect irregularities in heart beat or rhythm, to measure excitability of the intact heart, and to record electrical properties of excised atrium and papillary muscle. Cardiac output is measured by the tracer dilution technique and ventricular imaging in patients; invasive procedures are required for application of the Fick principle in patients and insertion of blood flow recorders in experimental animals. Perfusion of the excised heart offers an opportunity of measuring myocardial contractility and metabolism. Enzymatic studies and electron microscopy complete the techniques for detecting cardiomyopathy. All these procedures have been applied to determine the occurrence and mechanism for two groups of diseases: irregular heart beat, or arrhythmia, and cardiomyopathy.

3.3.1   Method G: Irregular Heart Beat

Industrial chemical poisoning can be manifested by irregularities of heart beat, or cardiac arrhythmia, in order of increasing severity, ranging from tachycardia or bradycardia, atrioventricular block, atrial or ventricular extrasystole, and atrial fibrillation to ventricular fibrillation and cardiac arrest. The benign forms (up to atrial fibrillation) are reversible by stopping chemical exposure, but ventricular fibrillation and cardiac arrest require immediate cardiopulmonary resuscitation. Poisonings characterized by cardiac arrhythmias have been reported for the following (see Table 4.4, Method G): most halogenated and nonhalogenated solvents, some nitrogenous compounds, one heavy metal (lead), and one oxide (carbon monoxide). The arrhythmia results from a direct action of the chemical on the heart, specifically by altering excitability, conduction, and refractoriness of one or more of the following: atrial muscle, atrioventricular node, conducting system, and ventricular muscle. The effects have been reported in appropriate human studies and animal experimentation. The occurrence of poisoning by industrial chemicals does not support the proposition that because the same chemicals may be reported at minute levels in ETS, then ETS also may lead to the development of heart disease in workers.

3.3.2 Method H: Experimentally Induced Cardiomyopathy

The most extreme example of unjustified application of results from animal experiments to ETS exposure of nonsmokers is as follows: In the course of attempting to determine whether long-term cigarette smoking leads to cardiomyopathy, rabbits were exposed in an infant incubator (65). It was reported that all the smoke from three burning cigarettes entered the inlet of the incubator through a mechanical device and rabbits were kept for 30 min. This description appears to me as a sealed chamber with cigarette smoke entering the inlet for 30-min periods. Several groups of rabbits were sacrificed: controls, after one 30-min exposure, twice daily exposure for 2 weeks, and twice daily exposure for 8 weeks. The heart was studied for mitochondrial oxidative processes. There was a decrease in respiration as well as in phosphorylation rate that was interpreted by the investigators as cardiomyopathy. The investigators recognized that carbon monoxide in the incubator was probably responsible for metabolic changes but they did not monitor air or blood levels.

Hugod and collaborators (66–68) exposed rabbits to one of the following mixtures: carbon monoxide 220 mg/m^3 or four times TLV; carbonyl sulfide 130 mg/m^3 or five times TLV; nitric oxide 6 mg/m^3 or one-fifth TLV. The rabbits were in airtight exposure chambers containing freely flowing air or predetermined mixtures in air for periods ranging from 1 to 7 weeks. The results of 140 rabbits sacrificed for electron microscopic examination performed blindly showed no morphological signs of myocardial damage. The four vapor constituents, in levels far exceeding ETS levels, were not associated with ultrastructural changes in rabbit heart, signifying the absence of cardiomyopathy.

The rabbit exposure studies described above were extended to include biochemical and histomorphological investigation of atherosclerosis. Exposure to each of the four gas–air mixtures was not related to intimal damage of the aorta and coronary arteries. The negative results noted for carbonyl sulfide-exposed rabbits do not support the claim that this known metabolite for carbon disulfide is responsible for coronary atherosclerosis reported by other investigators.

Cardiomyopathy has been reported following exposure to halogenated solvents, based on case reports of poisoning and experimental studies on intact and perfused heart. Cardiomyopathy from heavy metals is described in case reports of individuals drinking beer from containers that leached arsenic, cadmium, or lead (48). Cardiomyopathy from hydrocyanic acid is also based on case reports of poisoning and is readily supported by biochemical studies of heart muscle. Carbon monoxide is probably the most frequently encountered industrial and household chemical associated with death by cardiomyopathy. History of exposure to vehicular emissions or household natural gas is verifiable by blood analysis for carboxyhemoglobin. Among nonhalogenated solvents, only phenol has been reportedly related to cardiomyopathy (48).

3.4 Summary of Occupational Heart Diseases in Relation to ETS Exposure

Among more than 32 industrial chemicals potentially related to heart disease, only four substances or chemical classes have extensive supportive data: carbon monoxide (line 2 of Table 4.4), carbon disulfide/carbonyl sulfide (line 28), ethylene glycol dinitrate and organic nitrates, and methylene chloride and halogenated solvents. Other industrial chemicals (oxides, nonhalogenated solvents, nitrogenous compounds, and heavy metals), have not been adequately supported by human studies and animal experiments.

Methylene chloride is a solvent prototype for industrial chemicals that may be related to cardiac arrhythmia and myopathy in lethal or sublethal levels. Carbon disulfide is a selected prototype for industrial chemicals that may be related to ischemic heart disease or coronary atherosclerosis. There are no data indicating whether prolonged exposure to low levels of methylene chloride is associated with ischemic heart disease or whether high levels of carbon disulfide are associated with cardiac arrhythmia and myopathy. Methods to establish a cardiac toxicologic profile applied to one prototype must be applied to the other.

The cardiac toxicologic profile for carbon disulfide/carbonyl sulfide is as follows: (*a*) Mortality studies of viscose rayon workers report excess ischemic heart disease deaths, provided predisposing or other risk factors are present; (*b*) there is a high incidence of angina pectoris reported in workers exposed to carbon disulfide; (*c*) there is a reduction in coronary blood flow reported in workers developing ischemic heart disease, but there are no published results of myocardial tracer clearance studies; (*d*) coronary angiogram and postmortem histopathological studies report coronary atherosclerosis associated with carbon disulfide exposure; (*e*) coronary atherosclerosis developed in experimental animals exposed to carbon disulfide, with or without dietary cholesterol supplement. There is no information for (*f*) in vitro hematologic factors and (*g*) cardiac arrhythmia. (*h*) Experimental cardiomyopathy was reportedly not detected by electron microscopy in animals exposed to five times TLV for carbon disulfide.

The theory that ETS causes ischemic heart disease is based on inferences from the following: (*a*) epidemiologic studies of household exposures reported for nonsmoking spouses of smokers; (*b*) exercise studies of anginal patients with ETS exposure, but with questionable protocol; (*c*) coronary blood flow assumed to be insufficient because carbon monoxide present in ETS; (*d*) coronary atherosclerosis assumed to occur because aortic atherosclerosis reported in animals exposed to carbon monoxide at considerably higher levels than ETS; (*e*) coronary atherosclerosis supposedly occurring because benzo[*a*]pyrene is reportedly associated with atherosclerosis in cholesterol-fed birds; (*f*) in vitro testing for platelet aggregation and reduced fibrinogen level, suggesting atheromatous plaque formation; (*g*) cardiac arrhythmia postulated based on ventricular excitability studies of animals exposed to carbon monoxide; and (*h*) cardiomyopathy inferred from rabbit heart mitochondrial studies.

It is my opinion that the available studies do not support a judgment that ETS exposure is associated with any form of occupation-related heart disease. Although

ETS reportedly contains constituents that have been associated with occupational heart disease, the concentrations are so low that it is unlikely for any substance to attain the corresponding TLV in a work environment.

4 OCCUPATIONAL NEOPLASTIC DISEASE

Proponents of an association between ETS and neoplastic diseases rely on the estimate that there are at least 33 constituents in SSS with reported carcinogenic activity in experimental animals. These data are derived from skin-painting studies reporting topical new growths. Most purported skin carcinogens (polycyclic aromatic amines, heterocyclic compounds, and nitrosamines) have not been tested by animal inhalation, a procedure that would be required to classify occupational exposures as pulmonary carcinogens. About a dozen constituents of SSS with reported animal dermal carcinogenicity have been reported in ETS, but in concentrations so low that a nonsmoking worker must be exposed to the emission of hundreds or thousands of cigarettes to exceed threshold limit values (see Table 4.3). There are no compelling data to support the hypothesis that any SSS constituent with reported dermal carcinogenic activity (in experimental animals) is absorbed in nonsmokers exposed to ETS.

The definitions of potential carcinogens used by OSHA are as follows (69):

Category I Potential Carcinogens. A substance shall be identified, classified, and regulated as a Category I Potential Carcinogen if, upon scientific evaluation, the Secretary determines that the substance meets the definition of a potential occupational carcinogen in (1) humans, or (2) in a single mammalian species in a long-term bioassay where the results are in concordance with some other scientifically evaluated evidence of a potential carcinogenic hazard, or (3) in a single mammalian species in an adequately conducted long-term bioassay, in appropriate circumstances where the Secretary determines the requirement for concordance is not necessary. Evidence of concordance is any of the following: positive results from independent testing in the same or other species, positive results in short-term tests, or induction of tumors at injection or implantation sites.

Category II Potential Carcinogens. A substance shall be identified, classified, and regulated as a Category II Potential Carcinogen if, upon scientific evaluation, the Secretary determines that: (1) The substance meets the criteria set forth in Section 1990.112(a), but the evidence is found by the Secretary to be only "suggestive"; or (2) the substance meets the criteria set forth in Section 1990.112(a) in a single mammalian species without evidence of concordance.

It is my opinion that in an occupational setting, ETS does not meet the OSHA's criteria for designation as a "potential occupational carcinogen." My reasons are as follows: Epidemiologic data claimed to be supportive of a causative relationship between ETS and lung cancer are derived from nonoccupational exposure, mostly of nonsmoking oriental women married to smokers. Besides being generally in-

Table 4.5. Pulmonary Carcinogens in National Toxicology Program List

Group Name	No. of Substances	Animal Evidence			
Arsenic compounds	13		it		
Asbestos fibers	4	ih	it	or	
Beryllium compounds	11	ih	it		
Bis(chloromethyl)ether	2	ih		pe	
Chromium compounds	15	ih	it		pi
Nickel compounds and emissions	8	ih	it		pi
Mustard gas and nitrogen mustard	2	ih		pe	
Coke oven emissions	15	ih			
Soots, tars and mineral oils	19		it		
Polycyclic aromatic hydrocarbons	15		it		pi
Radioactive compounds (not included in NTP)		ih			

conclusive, these data are specifically questionable as a basis for claims concerning occupational exposures.

4.1 OSHA Category I Potential Carcinogens

Until candidate and priority lists are announced by OSHA, the National Toxicology Program (NTP) Sixth Annual Report on Carcinogens would be a suitable guide to acceptable data relevant to whether ETS is a potential carcinogen, specifically a cause of lung cancer (71). Among the more than 200 entries in the NTP list, 10 specifically state the potential occurrence of lung cancer in workers and, by OSHA criteria, can be properly labeled as "pulmonary carcinogens" (see Table 4.5).

The nature of "animal evidence" for the listed "pulmonary carcinogens" consists of reported occurrence of pulmonary neoplasms as a result of administration by inhalation (ih); intratracheal (it); oral (or); parenteral injection (pe); or pleural implant (pi) routes. The following substances have been tested by inhalation exposure in animals in support of "sufficient evidence" derived from human studies on workers: asbestos fibers, bis(chloromethyl) ether, chromium compounds, nickel compounds and emissions, mustard gas, coke oven emissions, and radioactive compounds. One entry (beryllium compounds) also has supportive animal inhalation studies, although available data on human carcinogenicity are rated as "limited." The following "pulmonary carcinogens" have "no supportive data" derived from inhalation studies, although experimental lung cancer has been reported from intratracheal, pleural implant, oral, parenteral, or intratracheal routes: arsenic compounds, soots, tars, and mineral oils; and polycyclic aromatic hydrocarbons.

No long-term inhalation study designed to evaluate potential effects of workplace ETS exposure has been conducted. One reason for the absence of such studies is the difficulties in simulating workplace environments in the laboratory chamber. For example, constituents of ETS are not stable and react with equipment and furniture. Lifetime exposure of mice to chambers filled with cigarette smoke (MSS) is the closest approach to ETS exposure and results have been negative (70).

It is necessary to describe briefly a selected prototype for a pulmonary carcinogen, such as bis(chloromethyl) ether. Exposure of workers to this substance is reported to increase the risk of lung cancer. The exposure occurs in chemical plant workers, ion exchange resin makers, laboratory workers, and polymer workers. Absorption of the chemical is through the lungs and skin. It is rated by NTP as a "known carcinogen" because there is sufficient evidence of carcinogenicity from studies in humans, which suggests a causal relationship between the agent and lung cancer (71).

Bis(chloromethyl) ether is associated with tumors at the site of application in mice and rats, that is, lungs after inhalation, subcutaneous tissues after injection, and skin after repeated topical application. The features of inhalation studies are as follows (72):

Rat inhalation	100 ppb for 6 hr daily for 6 weeks
Rat inhalation	100 ppb for 6 hr daily for 26 weeks
Rat inhalation	75 ppb for 6 hr daily for 2 years
Mouse inhalation	1 ppm daily for 82 days

The above results support the rating by NTP of "known carcinogen." The TLV is 0.001 ppm, which provides an acceptable factor of safety based on additional short-term inhalation studies in human and animals (2).

4.2   OSHA Category II Potential Carcinogens

So far, among the 34 SSS constituents listed as potentially related to neoplastic diseases in Table 4.1, four are in the NTP list of carcinogens, namely, benzene (line 19), nickel (line 35), aminobiphenyl (line 46), and naphthylamine (line 51). Only nickel is a "pulmonary carcinogen" according to the NTP, but the amounts detected in ETS are so minute that it would take several million ignited cigarettes to generate metallic nickel levels comparable to those used in studies of neoplastic responses in experimental animals. Benzene is associated with bone marrow neoplasms; aminobiphenyl and naphthylamine are associated with extrapulmonary neoplasms (urinary bladder). Urinary bladder neoplasm has not been reported as associated with spousal smoking.

The remaining 29 SSS constituents are classified as without sufficient evidence from epidemiologic studies to support a causal association in human neoplastic diseases. For purposes of description of animal experiments, these 29 constituents are grouped into three chemical classes: 11 polycyclic aromatic hydrocarbons (PAH) and heterocyclic compounds, 6 N-nitroso compounds, and 12 miscellaneous compounds. In Table 4.6, the following code is used for route of administration: dp = dermal painting; or = oral administration; pe = parenteral injection; it = intratracheal; ih = inhalation.

Table 4.6. Animal Carcinogens With Limited Epidemiologic Data

Line No. (Table 4.1)	Chemical Name	Route of Administration			
POLYCYCLIC AROMATIC HYDROCARBONS AND HETEROCYCLIC COMPOUNDS					
50	Benzo[a]pyrene	dp		pe	it	ih
45	Benz[a]anthracene	dp				
53	Dibenz[a,h]anthracene	dp	or			
54	Anthanthrene	dp				
41	Benzo[a]fluorene	dp				
48	Benzo[g,h,i]perylene	dp				
47	Benzo[e]pyrene	dp				
58	Coronene	dp				
38	Fluoranthene	dp				
55	Perylene	dp				
57	Pyrene	dp				
N-NITROSO COMPOUNDS					
39	N'-Nitrosodimethylamine		or	pi	it	ih
40	N-Nitrosodiethylamine		or	pi	it	ih
44	N-Nitrosopyrrolidine		or	pi		
52	N-Nitrosodiethanolamine		or	pi		
36	N'-Nitrosonornicotine		or	pi		
37	4-(Methylnitrosamino)-1-(3'-pyridyl)-1-butanone					
MISCELLANEOUS COMPOUNDS					
10	Acetaldehyde					ih
42	Cadmium					ih
20	Catechol	dp				
24	Cholesterol	dp				
6	Formaldehyde					ih
49	Hydrazine	dp				ih
18	Phenol	dp				ih
59	Polonium-210					ih
16	Hydroquinone	dp				
56	Toluene					ih
34	2-Toluidine	dp				ih
15	3-Vinylpyridine	dp				

4.2.1 Polycyclic Aromatic Hydrocarbons and Heterocyclic Compounds

The labeling of substances as pulmonary neoplastic agents is based on the following procedures applied in experimental animals and listed in the Registry of Toxic Effects of Chemical Substances (RTECS) (72):

> Oral administration in rat (lung neoplasm)
>   (53) dibenz[a,h]anthracene 4160 mg/kg for 126 weeks
> Intrathoracic implant in rat (lung neoplasm)

(50)   benzo[a]pyrene 150 mg/kg

Intratracheal injection in hamster (lung neoplasm)
   (50)   benzo[a]pyrene 120 mg/kg for 17 weeks
Intratracheal injection in mouse (lung neoplasm)
   (50)   benzo[a]pyrene 200 mg/kg for 10 weeks
Intratracheal injection in rat (lung neoplasm)
   (50)   benzo[a]pyrene 68 mg/kg for 15 weeks
Inhalation in mouse (lung neoplasm)
   (50)   benzo[a]pyrene 200 $\mu g/m^3$ 6 hr daily for 13 weeks
Inhalation in hamster (lung neoplasm)
   (50)   benzo[a]pyrene 1500 $\mu g/m^3$ 4 hr for 96 weeks

In these tests, RTECS has rated lung neoplasm results as "equivocal" when they yield uncertain but seemingly positive results (72). The positive results not rated as equivocal were derived by intrathoracic implantation or intratracheal injection. Only the inhalation study is relevant to the potential health effects of ETS exposure, and these results are designated equivocal.

Benzo[a]pyrene is a research chemical and thus only researchers would be exposed to the pure substance. However, occupational exposure is widespread because benzo[a]pyrene is present in coal tar, coke oven emissions, and creosote, all of which have corresponding work exposure standards (71). Benzo[a]pyrene and other PAHs occur in the combustion products of coal, oil, petroleum and biologic matter. Human exposure can occur indoors as a result of heating and cooking with natural gas, oil, coal, and wood. Individuals working in restaurants, airports, tarring facilities, refuse incinerators, power plants, and coke manufacturing facilities may be exposed to benzo[a]pyrene and related PAHs. These mixtures differ qualitatively and quantitatively from ETS so that it is scientifically incorrect to apply reports of carcinogenicity from such mixtures to cigarette smoke.

4.2.2   N-Nitroso Compounds

The carcinogenicity of the nine nitroso compounds is not supported by epidemiologic studies. The supporting animal experiments consist of oral studies as well as the following parenteral routes of administration (71, 72):

Oral administration in rat (liver and gastrointestinal neoplasm)
   (40)   N-Nitrosodiethylamine 19 mg/kg for 3.3 years
   (36)   N'-Nitrosonornicotine 3000 mg/kg for 30 weeks
   (39)   N-Nitrosodimethylamine 1134 mg/kg for 81 weeks
   (52)   N-Nitrosoethanolamine 8000 mg/kg for 50 weeks
   (44)   N-Nitrosopyrrolidine 685 mg/kg for 98 weeks
Oral administration in mouse (lung neoplasm)
   (40)   N-Nitrosodiethylamine 57 mg/kg for 10 days
   (39)   N-Nitrosodimethylamine 11 mg/kg for 16 weeks

(44)   N-Nitrosopyrrolidine 3235 mg/kg for 92 weeks

Intraperitoneal injection in rat (abdominal and/or intrathoracic neoplasm)
   (40)   N-Nitrosodiethylamine 75 mg/kg
   (39)   N-Nitrosodimethylamine 30 mg/kg

Intraperitoneal injection in hamster (lung and olfactory neoplasm)
   (41)   N-Nitrosopyrrolidine 1602 mg/kg for 25 weeks

Subcutaneous injection in mouse (lung and liver neoplasm)
   (40)   N-Nitrosodiethylamine 104 mg/kg for 52 weeks
   (41)   N-Nitrosopyrrolidine 396 mg/kg for 72 weeks

Subcutaneous injection in rat (olfactory neoplasm)
   (36)   Nitrosonornicotine 3160 mg/kg for 20 weeks

Subcutaneous injection in hamster (lung and olfactory neoplasm)
   (52)   N-Nitrosoethanolamine 10500 mg/kg for 42 weeks
   (36)   Nitrosonornicotine 3100 mg/kg for 25 weeks

Intratracheal injection in hamster (lung neoplasm)
   (40)   N-Nitrosodiethylamine 12 mg/kg for 15 weeks
   (39)   N-Nitrosodimethylamine 12 mg/kg for 15 weeks

Inhalation in hamster (lung neoplasm)
   (40)   N-Nitrosodiethylamine 215 $\mu g/m^3$ for 9 weeks

Inhalation in rat (liver and renal neoplasm)
   (39)   N-Nitrosodimethylamine 200 $\mu g/m^3$ for 45 weeks

Inhalation in mouse (lung neoplasm)
   (39)   N-Nitrosodimethylamine 200 $\mu g/m^3$ for 45 weeks

Most of the above results relating to the appearance of lung neoplasms have been evaluated as equivocal by RTECS. The dose administered to experimental animals is expressed in milligram per kilogram quantities, whereas nitrosamines reported in ETS have been detected indoors in nanogram per cubic meter levels. Based on these data, the biological plausibility is questionable that nitrosamines in ETS contribute to pulmonary carcinogenesis (see Chapter 11).

4.2.3 Miscellaneous Compounds

Under the third and last grouping of experimental animal carcinogens, there are 12 constituents. The suspicion of carcinogenicity derived from human observations applies to cadmium (line 42), formaldehyde (line 6), hydrazine (line 49), and 2-toluidine (line 56). However, the reported animal inhalation studies indicate development of extrapulmonary neoplasms such as nasal (formaldehyde), lymphoma (cadmium), nasal/gastrointestinal (hydrazine), and urinary bladder (2-toluidine). These four constituents of SSS, so far undetected indoors from ETS, are unlikely to contribute to associated neoplastic diseases.

   The remaining eight miscellaneous substances do not have supporting human studies and suspicion of carcinogenicity is entirely based on experimental animal observations. The substances were administered not by inhalation but by oral and parenteral routes. The dermal route has been used for the following compounds resulting in tumor initiation, tumor promotion or co-carcinogenic activity (14): catechol (line 20), cholesterol (line 24), phenol (line 18), hydroquinone (line 16), and 3-vinylpyridine (line 15).

Experimental pulmonary carcinogenesis relevant to ETS inhalation can be tested only by exposing animals in an appropriate chamber (73). Inhalation exposures of animals have assisted in identifying carcinogens in the workplace, such as bis(chloromethyl) ether (see Section 4.1). Because the relevant exposure route for ETS constituents is limited to the respiratory tract, extrapulmonary routes are not germane. The intratracheal route cannot be used to prove or disprove pulmonary carcinogenicity because of alterations in lung function relating to absorption, elimination, and pharmacokinetics. Intratracheal injection has been a useful tool for research on enzymatic, immunologic, and histochemical changes that relate to carcinogenesis. However, the procedure is not useful for bioassay testing of substances for which the only exposure route is inhalation.

4.3 Summary of Occupational Neoplastic Diseases

It is important to recognize that measurements of carboxyhemoglobin and nicotine metabolites are not indicators of exposure to any of 33 suspected carcinogens contained in SSS and sometimes imputed to ETS. There are reported indoor concentrations for the following: formaldehyde, benzene, phenol, toluene, three nitroso compounds, and nine polycyclic aromatic hydrocarbons. However, the reported levels are based on a few isolated instances of transient monitoring indoors (Table 4.1). There are no reported total daily exposure levels of nonsmokers to individual constituents because currently available technology does not permit accurate continuous detection of ETS indoors with varying levels of ventilation.

The definition of a potential occupational carcinogen (see Section 4) has not been met for ETS. The available information for 33 alleged carcinogens detectable in SSS is limited to animal experiments, mostly consisting of dermal, oral, and parenteral injections. Intratracheal injection influences physiological and biologic events in the lungs to the extent that results cannot be applied to ETS exposure. Furthermore, for SSS constituents with workplace standards, hundreds, or even thousands, of cigarettes must be ignited continuously for ETS to reach designated hazardous levels in an enclosure. The absence of relevant human or animal inhalation studies does not support the hypothesis that any of the 33 SSS constituents are pulmonary carcinogens.

Although there are five metallic trace elements reported in cigarette smoke, only nickel, cadmium, and polonium-210 have been detected in SSS and MSS, whereas arsenic and chromium were detected only in MSS. None of the five has been detected indoors in ETS. The presence of metals in cigarette smoke depends on soil conditions and proximity of tobacco farming to refineries.

There are 11 polycyclic aromatic hydrocarbons and heterocyclic compounds in SSS that, according to the NTP, have "sufficient" evidence of animal carcinogenicity, but limited or no evidence of human cancer. These 11 SSS constituents were classified by the appearance of skin tumors after skin painting in mice; most were also tested orally and by subcutaneous injections in rats and mice; and a few compounds were tested by intrathoracic implantation, intratracheal injection, and inhalation, in hamsters, mice, and rats. Benzo[a]pyrene is the most widely investigated among polycyclic aromatic hydrocarbons. The substance was tested for pulmonary carcinogenesis by the inhalation route but results were termed equivocal by RTECS criteria.

There are six nitroso compounds reported in SSS, each of which is suspected as a carcinogen on the basis of experimental animal studies. There are no reported human cases of cancer attributed to nitroso compounds. Although some of them administered orally in mice and rats have been reported to cause lung cancer, the results are relevant to the oral administration of ingested nitrosamine, rather than to their possible presence in ETS.

5   OCCUPATIONAL RESPIRATORY TRACT DISEASE

The third and last group of diseases sometimes claimed to be associated with household exposure to ETS involves the respiratory system, specifically the upper respiratory tract, the lower respiratory tract, and lung parenchyma. Almost all non-neoplastic diseases of the lung parenchyma reported in cigarette smokers have been reported to be associated with ETS household exposure, namely, chronic obstructive lung disease, including chronic bronchitis and pulmonary emphysema, and acute bronchial asthma. An association between spousal smoking and respiratory tract disease is claimed on the premise that MSS and SSS have essentially the same chemical constituents as ETS and that repeated exposure to ETS elicits similar respiratory tract effects as actual smoking of cigarettes. The fallacy of this reasoning, discussed above, is that it would take an unreasonable number of cigarettes to approach and exceed TLV in an unventilated work room (see Section 2).

5.1   Respiratory Tract Irritation

Mucosal irritation is the only association with ETS exposure that has been repeatedly reported in the workplace as well as in experimental nonventilated chambers. Respiratory tract irritation is manifested by sneezing, a running nose, coughing, and expectoration. There are complaints of conjunctival irritation such as a tearing and burning sensation of the eyelids. In questionnaire responses, there are subjective complaints of discomfort, difficulty of breathing, dizziness, and nausea. In unventilated rooms, ETS exposure has been associated with erratic breathing initiated by breath holding, an upper respiratory tract reflex response to chemical irritation.

Respiratory tract irritation has been reported by occupants of crowded and poorly ventilated bars and restaurants (6–11). The sources of reported elevated levels of carbon monoxide include tobacco use and cooking gas emission. Carbon monoxide is not associated with mucosal irritation but is the marker for excessively high concentrations of ETS. In experiments conducted in unventilated rooms, it has been reported that mucosal irritation occurs when the following ETS constituents are increased significantly above preexposure levels: acrolein, ammonia, formaldehyde, nitrogen oxides, phenol, and toluene. The levels when mucosal irritation appears are 1/100th to 1/10th of the respective TLV. There are other potential irritants reportedly in ETS in levels that have not been detected in experimental exposure rooms, such as pyridine and methyl chloride. The phenomenon of respiratory tract irritation is the only reported example of additive responses to constituents in ETS. If there is any form of synergism between two or more constituents, its occurrence can be explored in experimental animals such as mucus production, ciliary activity, and reflex responses in the respiratory tract.

5.2 Chronic Obstructive Lung Disease

The diagnosis of chronic obstructive lung disease is derived from detection of abnormal lung function combined with respiratory distress. The disease is usually progressive with disruption of gas exchange in the lung and histopathological changes of pulmonary emphysema and chronic bronchitis. It has not been possible to attribute a single case of chronic obstructive lung disease to ETS exposure because of confounding exposures of workers to industrial chemicals and atmospheric pollutants reported to cause chronic obstructive lung disease (74). There are several studies on the etiology of chronic obstructive lung disease and a study conducted by Zarkovic (75) in Yugoslavia prior to the 1992 civil war illustrates that environmental variables are important. In a survey of 12,000 adults and 10,000 school children in 10 counties of Bosnia and Herzegovina, there were regional differences reported in chronic bronchitis rates. Cor pulmonale was highest in rural areas of central Bosnia and lowest in rural regions of Herzegovina; the prevalence was not related to smoking. The high prevalence of chronic bronchitis in Sarajevo was attributed to air pollution, and in rural Bosnia to occupational dust on farms. The civil war has disrupted a followup survey to determine whether ETS exposure in household and work environments was associated with the incidence of chronic obstructive lung disease.

White and Froeb (76) reported in 1980 that nonsmoking workers exposed to ETS had lower values for expiratory flow rates than nonsmokers not exposed to ETS. Their conclusion that ETS exposure caused lung disease was criticized because of the controversial nature of lung function testing that could be influenced by subject awareness, subjects' physical training, and exposure to household dusts and atmospheric pollutants (77). Aside from possible technical problems, the experimental premise may be questionable. It is unlikely that White and Froeb were able to identify a truly representative group of subjects with work histories of 20 years or more who had never been exposed to tobacco smoke. This "no exposure" group, along with other experimental groups in the 1980 study, needed verification from an epidemiologic point of view (77).

A decade later, White et al. (78) reported on a study of workplace exposure to ETS. Instead of selecting some of the 2100 subjects originally studied, they used a new set of volunteers. There were no measurements of lung function. Instead, questionnaires were completed. The investigators' conclusion was that ETS-exposed workers reported eye irritation, cough, phlegm production, and chest colds more frequently than control subjects. Carbon monoxide levels during ETS exposure were higher during the workday but never exceeded the accepted TLV. While these reported observations are consistent with an association of ETS exposure with respiratory tract irritation, a follow-up of these subjects is necessary to determine if repeated ETS exposure is indeed associated with a parenchymal decrease, as postulated by Shephard (79).

5.3 Bronchial Asthma

The claim that ETS exposure can cause occupational asthma is not based on case reports or epidemiologic studies of workers. The concept is entirely an extrapolation from studies of maternal smoking and childhood asthma (80). The question of whether initial exposure can initiate the disease is usually examined in childhood. For adult workers, studies of whether industrial chemical exposure can provoke chronic asthma in a previously nonatopic individual are inconclusive.

Occupational asthma provoked by exposure to occupational dust and industrial chemicals has been reported consistently in atopic workers. There are case reports documenting acute asthmatic attacks provoked by exposure to more than 100 occupational inhalants (80, 81). Diagnosis of the allergen can be accomplished by provocative exposure to the offending chemical or dust. The list includes some constituents of ETS, but the levels in ETS are exceedingly low and do not approach levels known to provoke acute asthmatic attack.

It is difficult to interpret questionnaire responses relating to bronchial asthma. There is a subjective element in describing past history of bronchial asthma so that it is difficult to distinguish respiratory discomfort from real bronchospasm. Attacks that require medication are most likely to be real bronchial asthma but less severe complaints relieved without medication are questionable.

5.4 Nicotine as Prototype of Respiratory Tract Irritant

Among the dozen ETS constituents reported to cause mucosal irritation, nicotine is the most thoroughly investigated substance. The sensation of respiratory tract irritation is initiated by a "nicotine effect" on sensory nerve endings in the respiratory tract. I have cataloged the reflexes initiated by nicotinic action listed in Table 4.7 and derived from my own experimental studies (82–84).

The nicotinic receptor list (Table 4.7) is based on animal experiments and human observations. The first entry is the upper respiratory tract reflex (83). This reflex, also known as the Kratschmer reflex, represents the (I) first line of defense to inhalants, such that apnea and sneezing serve to expel the foreign substance. Nicotine and other constituents in ETS are reported to initiate the response in the lower respiratory tract; the polypnea and coughing serve to expel the irritants forming the (II) second line of defense. The bronchomotor responses, particularly

Table 4.7. Reflexes Stimulated by Nicotine

Receptor Location	Afferent Innervation	Reflex Response
I Upper respiratory tract (nose, pharynx, larynx)	Olfactory, trigeminal Nasopharyngeal, vagus	Apnea Sneezing Bradycardia
II Lower respiratory tract (trachea, bronchi, bronchioles)	Vagus	Apnea Polypnea Coughing
III Lower respiratory tract (bronchioles)	Vagus Thoracic sympathetic	Bronchoconstriction Bronchodilation
IV Pulmonary and coronary blood vessels	Vagus	Apnea, polypnea Bradycardia Hypotension
V Carotid body and aortic body	Glossopharyngeal Vagus	Polypnea Tachycardia

bronchoconstriction, are the (III) third line of defense, preventing the entrance of irritants to respiratory bronchioles. In the event of absorption, the pulmonary vascular and coronary vascular receptors serve as the (IV) fourth line of defense so that further distribution of absorbed chemical is retarded by bradycardia and hypotension. This response, also known as the Bezold–Jarish reflex, is regarded as a protective mechanism for the circulatory and respiratory systems that are challenged by the absorbed chemical substance. The (V) fifth and final line of defense consists of carotid and aortic body chemoreceptors that stimulate respiration and circulation to prolong survival and ultimate recovery.

All the above reflexes are reported to be activated by nicotine. The fourth set of receptors in pulmonary and coronary vessels are responsible for the Bezold–Jarish reflex, which is a protective mechanism for the myocardium during coronary ischemia (85). A very low concentration of nicotine is reportedly absorbed from ETS exposure, but its influence, if any, on the Bezold–Jarish reflex is not known.

The fundamental question is whether repeated respiratory tract irritation by nicotine can cause chronic obstructive lung disease, specifically chronic bronchitis and pulmonary emphysema. The histopathological examination of smokers' lungs does not necessarily implicate MSS irritants because chronic bronchitis and pulmonary emphysema are not seen in a majority of smokers. Tolerance to respiratory tract irritation associated with inhaling cigarette smoke is reported to occur in habitual smokers. My own experiments in animals indicate that cigarette smoke exposure per se does not initiate pulmonary emphysema (86). Exposure of experimental animals to SSS is not associated with pulmonary lesions (87). The production of parenchymal lung disease by repeated exposure to MSS has not been proven in experimental animals (88).

6 CONCLUSIONS

Respiratory tract irritation has been reported to be associated with constituents in ETS. Workers in crowded and smoke-filled bars, restaurants, kitchens, and other public places report episodes of mucosal irritation of the nose, mouth, throat, respiratory airways, and conjunctivae. The reported elevated levels of nicotine and other constituents in ETS are less than the respective TLVs, suggesting that any respiratory tract complaints are the summation of subthreshold effects. Any one of a dozen constituents can be used as a marker for potential irritants in ETS, but the most informative is the specific prototype, that is, nicotine. Nonventilated chambers have been used to simulate smoke-filled rooms, and the phenomenon of respiratory tract irritation seen in workplaces has been reported. The irritation is readily reversible by resumption of adequate ventilation to dispose of high levels of ETS.

The potential associations between spousal smoking and the incidence of cardiopulmonary disease cannot be extrapolated to workers. Spousal heart disease, spousal neoplastic disease, and childhood bronchial asthma are cardiopulmonary disease entities with clinical features and pathogenesis different from corresponding occupational diseases. Other than the obvious difference in age (childhood versus adult) and gender (spousal disease in women versus occupational disease in men), there are compelling differences. Epidemiologic studies relating to household exposure and spousal and childhood diseases consist almost entirely of questionnaire responses. If there is personal contact between medical personnel and subject (child or spouse), it usually occurs once for the explicit purpose of obtaining past medical history and estimating ETS exposure, but without actual measurement of ETS levels. There has been no uniformity in questionnaires used so that deficiencies of earlier epidemiologic studies are only partially corrected in later ones.

The evolution of occupational disease is an extended process lasting for several years. Medical records prior to employment and yearly examinations are necessary to detect the occurrence of cardiopulmonary disease. Diagnostic and research techniques for detection of occupational heart disease have been cataloged, together with relevant experimental animal procedures to serve as a model for reviewing whether ETS exposure of workers is associated with any form of heart disease (Section 3). Because workers not exposed to ETS and workers exposed to ETS in the same occupational group have not been medically followed for several years, there is no justification to apply the reasoning that because home exposure through spousal smoking might be associated with cardiopulmonary disease, then workplace ETS exposure can be expected to have a similar association.

Results of household exposure cannot be applied to workers. The ongoing controversy between work standards proposed by ACGIH and those proposed by NIOSH is in part the outcome of differences in opinion. Only data derived from occupational groups are used by ACGIH, whereas NIOSH relies on data derived from nonoccupational and family groups. The controversy extends to the consideration of occupational diseases purportedly associated with ETS. Confounding factors related to ETS exposure in workplaces are different from factors influencing spouses and children in studies of household exposure. Dietary factors and household pollution may influence the incidence of spousal and childhood diseases. On the other hand, workers are exposed to industrial chemicals and outdoor pollutants as well as work-related stress. Familial patterns of inherited or acquired susceptibility to cardiopulmonary disease do not apply to occupational groups.

Unlike household exposure, workplace exposure can be monitored by good industrial hygiene practice. There are adequate ETS markers for exposures to certain constituents: carbon monoxide for ischemic heart disease, benzo[a]pyrene for neoplastic diseases, and nicotine for respiratory tract irritation. These markers can continue to be monitored in a prospective study of workers differentiated according to ETS exposure. Such a study will require considerable expense and human effort. A simple recourse is to regard ETS as a complex mixture, similar to vehicular emission. As long as carbon monoxide is not allowed to accumulate in workplaces by adequate ventilation, ETS is unlikely to be associated with reports of respiratory tract irritation. ETS levels that are sufficiently high to provoke complaints of irritation can serve as an indication that corrective measures should be initiated to increase workplace ventilation.

REFERENCES

1. National Institute for Occupational Safety and Health, *NIOSH Recommendations for Occupational Safety and Health*, DHHS (NIOSH) Publication No. 92-100, 1992, 205 pp.

2. American Conference of Governmental Industrial Hygienists, *Documentation of the Threshold Limit Values and Biological Exposure Indices*, 5th ed., Cincinnati, OH, 1986, 690 pp.

3. Occupational Safety and Health Administration, "Occupational Exposure to Indoor Air Pollutants; Request for Information," *Fed. Regist.*, **56**, 47892–47897 (1991).

4. National Institute for Occupational Safety and Health, "Environmental Tobacco Smoke in the Workplace—Lung Cancer and Other Health Effects," *NIOSH Current Intelligence Bulletin 54*, DHHS (NIOSH) Publication No. 91-108, 1991, 18 pp.

5. T. Hirayama, "Non-smoking Wives of Heavy Smokers Have a Higher Risk of Lung Cancer: A Study from Japan," *Brit. Med. J.*, **282**, 183–185 (1981).

6. U.S. Department of Health and Human Services, *The Health Consequences of Involuntary Smoking—a Report of the Surgeon General*, DHHS (CDC) Publication No. 87-8398, 1986, 359 pp.

7. National Research Council, *Environmental Tobacco Smoke—Measuring Exposures and Assessing Health Effects*, National Academy Press, Washington, DC, 1986, 337 pp.

8. U.S. Environmental Protection Agency, *Respiratory Health Effects of Passive Smoking: Lung Cancer and Other Disorders*, EPA/600/6-90/006B, Draft, 1992.

9. D. J. Ecobichon and J. M. Wu, Eds., *Environmental Tobacco Smoke*, Proceedings of the International Symposium at McGill University 1989, Lexington Books, Lexington, MA, 1990, 388 pp.

10. W. O. Spitzer, V. Lawrence, R. Dales, G. Hill, et al., "Links between Passive Smoking and Disease: A Best-Evidence Synthesis," *Clin. Invest. Med.*, **13**, 17–42 (1990).

11. M. P. Eriksen, C. A. LeMaistre, and G. R. Newell, "Health Hazards of Passive Smoking," *Ann. Rev. Publ. Health*, **9**, 47–70 (1988).

12. National Research Council, *Complex Mixtures—Methods for In Vivo Toxicity Testing*. National Academy Press, Washington, DC, 1988, 227 pp.

13. National Research Council, *Human Exposure Assessment for Airborne Pollutants*, National Research Council Publication, Washington, DC, 1991, 207–255 pp.

14. D. Hoffmann and S. S. Hecht, "Advances in Tobacco Carcinogenesis," in *Handb. Exp. Pharmacol.*, **94**, 63–102 (1990).

15. M. R. Guerin, R. A. Jenkins, and B. A. Tomkins, *The Chemistry of Environmental Tobacco Smoke: Composition and Measurement*, Lewis Publishers, Ann Arbor, MI, 1992, 330 pp.

16. G. B. Gori and N. Mantel, "Mainstream and Environmental Tobacco Smoke," *Reg. Toxicol. Pharmacol.*, **14**, 88–105 (1991).

17. D. Hoffmann, J. D. Adams, and K. D. Brunnemann, "A Critical Look at *N*-Nitrosamines in Environmental Tobacco Smoke," *Toxicol. Lett.*, **35**, 1–8 (1987).

18. G. Andersson and T. Dalhamn, "Health Risks from Passive Smoking" (in Swedish), *Lakartidningen*, **70**, 2833–2836 (1973).

19. T. E. Dahms, J. F. Bolin, and R. G. Slavin, "Passive Smoking—Effects on Bronchial Asthma," *Chest*, **80**, 530–534 (1981).

20. H. P. Harke, "The Problem of 'Passive Smoking' " (in German), *Munch. Med. Wochenschr.*, 112, 2328–2334 (1970).

21. R. Huch, J. Danko, L. Spatling, and A. Huch, "Risks the Passive Smoker Runs," *Lancet*, **1980-II**, 1376.

22. C. Hugod, L. H. Hawkins, and P. Astrup, "Exposure of Passive Smokers to Tobacco Smoke Constituents," *Int. Arch. Occup. Environ. Health* **42**, 21–29 (1978).

23. P. E. Pimm, F. Silverman, and R. J. Shephard, "Physiological Effects of Acute Passive Exposure to Cigarette Smoke," *Arch. Environ. Health*, **33**, 201–213 (1978).

24. E. Polak, "Cigarette Paper—Its Role in the Pollution of Domestic Areas. Passive Smoking—a New Concept Summarized" (in French), *Bruxelles Med.*, **57**, 335–340 (1977).

25. M. A. H. Russell, P. V. Cole, and E. Brown, "Passive Smoking: Absorption by Nonsmokers of Carbon Monoxide from Room-air Polluted by Tobacco Smoke," *Postgrad. Med. J.*, **49**, 688–692 (1973).

26. A. Seppanen, "Smoking in Closed Space and Its Effect on Carboxyhemoglobin Saturation of Smoking and Nonsmoking Subjects," *Ann. Clin. Res.*, **9**, 281–283 (1977).

27. J. L. Alvarez-Sala, J. M. Tello, A. Villegas, J. Montero, E. Montero, and D. Espinos, "Carboxihemoglobina en Sangre en Fumadores Pasivos," *N. Arch. Fac. Med.* Madrid, **43**, 377– (1985).

28. J. L. Alvarez-Sala, J. M. Tello, T. G. Tenorio, and D. Espinos, "Passive Smokers" (in Spanish), *Med. Clin.*, **86**, 864– (1986).

29. M. Asano and C. Ohkubo, "A Study on Physiological Responses to Passive Smoking by Standard and Nicotine-less Cigarettes, with Special Regard to Smoking Habit in Young Healthy Males," in *Smoking and Health*, M. Aoki et al., Eds., Elsevier, 1988, 255–257.

30. J. W. Davis, "Some Acute Effects of Smoking on Endothelial Cells and Platelets," *Adv. Exp. Med. Biol.*, **273**, 107–118 (1990).

31. I. Hoepfner, G. Dettbarn, G. Scherer, G. Grimmer, and F. Adlkofer, "Hydroxy-

phenanthrenes in the Urine of Nonsmokers and Smokers," *Toxicol. Lett.*, **35**, 67–71 (1987).

32. S. J. Olshansky, "Is Smoker/Nonsmoker Segregation Effective in Reducing Passive Inhalation among Nonsmokers," *Am. J. Publ. Health*, **72**, 737–739 (1982).

33. G. Scherer, K. Estphal, A. Biber, I. Heopfner, and F. Adlkofer, "Urinary Mutagenicity After Controlled Exposure to Environmental Tobacco Smoke—ETS," *Toxicol. Lett.*, **35**, 1355–1400 (1987).

34. G. Winneke, K. Plischeke, A. Roscovanu, and H-W. Schlipkoeter, "Patterns and Determinants of Reaction to Tobacco Smoke in an Experimental Exposure Setting," *Proc. Int. Conf. Indoor Air Qual. Clim.*, **2**, 351–356 (1984).

35. K. Steenland, "Passive Smoking and the Risk of Heart Disease," *J. Am. Med. Assoc.*, **267**, 94–99 (1992).

36. D. M. Aviado, "Carbon Monoxide as an Index of Environmental Tobacco Smoke Exposure," *Eur. J. Resp. Dis.*, **133**, 47–60 (1984).

37. D. M. Aviado, "Suspected Pulmonary Carcinogens in Environmental Tobacco Smoke," *Environ. Tech. Lett.*, **9**, 539–544 (1988).

38. D. M. Aviado, "Health Effects of 50 Selected Constituents of Environmental Tobacco Smoke," in *Indoor Air Quality*, H. Kasuga, Ed., Springer-Verlag, Berlin, 1990, pp. 383–389.

39. D. M. Aviado, "Cardiovascular Toxicology of Halogenated Hydrocarbons and Other Solvents," Chapter 15 in *Cardiovascular Toxicology*, D. Acosta, Ed., Raven Press, New York, 1992, pp. 455–479.

40. H. Schievelbein and F. Richter, "The Influence of Passive Smoking on the Cardiovascular System," *Prev. Med.*, **13**, 626–644 (1984).

41. L. M. Wexler, "Environmental Tobacco Smoke and Cardiovascular Disease: A Critique of the Epidemiological Literature and Recommendations for Future Research," in *Environmental Tobacco Smoke: Proceedings of the International Symposium at McGill University 1989*, D. J. Ecobichon and J. M. Wu, Eds., Lexington Books, Lexington, MA, 1990, pp. 139–152.

42. P. M. Lee, J. Chamberlain, and M. R. Alderson, "Relationship of Passive Smoking to Risk of Lung Cancer and Other Smoking-associated Diseases," *Brit. J. Cancer*, **54**, 97–105 (1986).

43. L. H. Kuller and E. P. Radford, "Epidemiological Bases for the Current Ambient Carbon Monoxide Standards," *Environ. Health Perspect.*, **52**, 131–139 (1983).

44. F. B. Stern, W. E. Haperin, R. W. Hornung, V. L. Ringenburg, and C. S. McCammon, "Heart Disease Mortality Among Bridge and Tunnel Officers Exposed to Carbon Monoxide," *Am. J. Epidemiol.*, **128**, 1276–1288 (1988).

45. R. O. Beauchamp, Jr., J. S. Bus, J. A. Popp, C. J. Boreiko, and L. Goldberg, "A Critical Review of the Literature on Carbon Disulfide Toxicity," *Crit. Rev. Toxicol.*, **11**, 169–278 (1983).

46. T. Wronska-Nofer, S. Szendzikowski, and M. Obrebska-Parke, "Influence of Chronic Carbon Disulphide Intoxication on the Development of Experimental Atherosclerosis in Rats," *Brit. J. Ind. Med.*, **37**, 387 (1988).

47. S. A. Glantz, and W. W. Parmley, "Passive Smoking and Heart Disease. Epidemiology, Physiology, and Biochemistry," *Circulation*, **83**, 1–12 (1991).

48. T. S. Kristensen, "Cardiovascular Diseases and the Work Environment. A Critical

Review of the Epidemiologic Literature on Chemical Factors," *Scand. J. Work Environ. Health*, **15**, 245–264 (1989).

49. W. S. Aronow, "Effect of Passive Smoking on Angina Pectoris." *N. Engl. J. Med.*, **299**, 21–24 (1978).

50. U.S. Environmental Protection Agency, *Revised Evaluation of Health Effects Associated with Carbon Monoxide Exposure: An Addendum to the 1979 EPA Air Quality Criteria Document for Carbon Monoxide*, Final Report, NTIS, U.S. Department of Commerce, Springfield, IL, 1984.

51. E. Khalfen and V. Klochkov, "Effect of Passive Smoking on Physical Tolerance of Ischemic Heart Disease Patients" (in Russian), *Ter. Arkh.*, **59**, 112–115 (1987).

52. E. N. Allred, E. R. Bleecker, B. R. Chaitman, T. E. Dahms, S. O. Gottlieb, J. D. Hackney, M. Pagano, R. H. Selvester, S. M. Walden, and J. Warren, "Effects of Carbon Monoside on Myocardial Ischemia," *Environ. Health Perspect.*, **91**, 89–132 (1991).

53. U.S. Department of Health and Human Services, *The Health Consequences of Smoking. Cardiovascular Disease*, A Report of the Surgeon General, U.S. Government Printing Office, Washington, DC, 1983, p. 70.

54. C. C. Seltzer, "The Negative Association in Women Between Cigarette Smoking and Uncomplicated Angina Pectoris in the Framingham Heart Study Data," *J. Clin. Epidemiol.*, **44**, 871–876 (1991).

55. T. L. Kelly, E. Gilpin, S. Ahnve, H. Henning, and J. Ross, Jr., "Smoking Status at the Time of Acute Myocardial Infarction and Subsequent Prognosis," *Am. Heart J.*, **110**, 535–541 (1985).

56. K. Robinson, R. M. Conroy, and R. Mulcahy, "Smoking and Acute Coronary Heart Disease: A Comparative Study," *Brit. Heart J.*, **60**, 465–469 (1988).

57. W. R. Rogers, K. D. Carey, C. A. McMahan, M. M. Montiel, G. E. Mott, H. S. Wigodsky, and H. C. McGill, Jr., "Cigarette Smoking, Dietary Hyperlipidemia, and Experimental Atherosclerosis in the Baboon," *Exp. Mol. Pathol.*, **48**, 135–151 (1988).

58. K. Kjeldsen, *Smoking and Atherosclerosis*, Tulein & Koch, Copenhagen, 1969.

59. R. Albert, F. Vanderlaan, and M. Nishizumi, "Effect of Carcinogens on Chicken Atherosclerosis," *Cancer Res.*, **37**, 2232–2235 (1977).

60. A. Penn, G. Batastini, J. Soloman, F. Burns, and R. Albert, "Dose-Dependent Size Increases of Aortic Lesions Following Chronic Exposure to 7,12-Dimethylbenz(*a*)anthracene," *Cancer Res.*, **41**, 588–592 (1981).

61. P. C. Elwood, A. D. Beswick, D. S. Sharp, J. W. G. Yarnell, S. Rogers, and S. Renaud, "Whole Blood Impedence Platelet Aggregometry and Ischemic Heart Disease—the Caerphilly Collaborative Heart Disease Study," *Atherosclerosis*, **10**, 1032–1036 (1990).

62. A. Mansouri and C. A. Perry, "Alteration of Platelet Aggregation by Cigarette Smoke and Carbon Monoxide," *Thromb. Haemost.*, **48**, 286–288 (1982).

63. A. J. Dobson, H. M. Alexander, R. F. Heller, and D. M. Lloyd, "Passive Smoking and the Risk of Heart Attack or Coronary Death," *Med. J. Aust.*, **154**, 793–797 (1991).

64. T. W. Meade, J. Imeson, and Y. Stirling, "Effects of Changes in Smoking and Other Characteristics on Clotting Factors and the Risk of Ischaemic Heart Disease," *Lancet*, **1987**, 986–988.

65. A. Gvozdjakova, V. Bada, L. Sany, J. Kucharska, F. Kruty, P. Bozek, L. Trstansky, and J. Gvozdjak, "Smoke Cardiomyopathy: Disturbance of Oxidative Processes in Myocardial Mitochondria," *Cardiovasc. Res.*, **18**, 229–232 (1984).

66. C. Hugod and P. Astrup, "Exposure of Rabbits to Carbon Monoxide and Other Gas Phase Constituents of Tobacco Smoke. Influence of Coronary and Aortic Intimal Morphology," *Munch. Med. Wochenschr.*, **122**, S18–S24 (1980).

67. C. Hugod, "Myocardial Morphology in Rabbits Exposed to Various Gas-phase Constituents of Tobacco Smoke," *Atherosclerosis*, **40**, 181–190 (1981).

68. O. Kamstrup, and C. Hugod, "Exposure of Rabbits to 50 ppm Carbonyl Sulfide—A Biochemical and Histomorphological Study," *Int. Arch. Occup. Environ. Health*, **44**, 109–116 (1979).

69. Occupational Safety and Health Administration, "Identification, Classification and Regulation of Potential Occupational Carcinogens," *Code Federal Regulations*, Title 29, Part 1990 (1991).

70. C. J. Henry, and R. E. Kouri, *Chronic Exposure of Mice to Cigarette Smoke*, Field, Risch Associates, New York, 1984, 187 pp.

71. National Toxicology Program, *Sixth Annual Report on Carcinogens—Summary*, USDHHS-PHS, 1991, 461 pp.

72. National Institute for Occupational Safety and Health, *Registry of Toxic Effects of Chemical Substances: 1985–1986 Edition User's Guide*, USDHHS-PHS Publication No. 87–114, 1987.

73. D. M. Aviado and H. Salem, "Respiratory and Bronchomotor Reflexes in Toxicity Studies," in *Inhalation Toxicology—Research, Methods, Applications, and Evaluation*, H. Salem, Ed., Marcel Dekker, New York, 1987, pp. 135–151.

74. U.S. Department of Health and Human Services, *The Health Consequences of Smoking—Cancer and Chronic Lung Disease in the Workplace*, A Report of the Surgeon General, 1985, 542 pp.

75. G. Zarkovic, "Etiology of Non-specific Chronic Respiratory Illness and Cor Pulmonale in Bosnia and Herzegovina," *Int. J. Epidemiol.*, **1**, 167–176 (1972).

76. J. R. White and H. F. Froeb, "Small-Airways Dysfunction in Nonsmokers Chronically Exposed to Tobacco Smoke," *N. Engl. J. Med.*, **30**, 720–723 (1980).

77. D. M. Aviado, "Small-Airways Dysfunction in Passive Smokers," *N. Engl. J. Med.*, **303**, 393 (1992).

78. J. R. White, H. F. Froeb, and J. A. Kulik, "Respiratory Illness in Nonsmokers Chronically Exposed to Tobacco Smoke in the Work Place," *Chest*, **100**, 39–43 (1991).

79. R. J. Shephard, "Respiratory Irritation from Environmental Tobacco Smoke," *Arch. Environ. Health.*, **47**, 123–130 (1992).

80. C. A. Frazier, Ed., *Occupational Asthma*, Van Nostrand Reinhold, New York, 1980, pp. 361.

81. R. J. Davies and A. D. Blainey, "Occupational Asthma," in *Asthma*, 2nd ed., T. J. H. Clark and S. Godfrey, Eds., Chapman and Hall, London, 1983.

82. D. M. Aviado and M. Samanek, "Bronchopulmonary Effects of Tobacco and Related Substances, I to IV," *Arch. Environ. Health*, **11**, 141–176 (1965).

83. D. M. Aviado, M. Samanek, and L. Folle, "Cardiopulmonary Effects of Tobacco and Related Substances," *Arch. Environ. Health*, **12**, 705–724 (1966).

84. D. M. Aviado and F. Palacek, "Pulmonary Effects of Tobacco and Related Substances, I and II," *Arch. Environ. Health*, **15**, 187–203 (1967).

85. D. M. Aviado, "Nicotinic Receptors in Healthy and Ischemic Heart with Special Reference to the Bezold–Jarisch Reflex," *Arch. Int. Pharmacodyn*, **319**, 7–23 (1992).

86. H. Ito and D. M. Aviado, "Pulmonary Emphysema and Cigarette Smoke," *Arch Environ. Health*, **16**, 865–870 (1968).

87. F. Adlkofer, G. Scherer, R. Wenzel-Hartung, H. Brune, and C. Thomas, "Exposure of Hamsters and Rats to Sidestream Smoke of Cigarettes: Preliminary Results of a 90-Day-Inhalation Study," in *Indoor and Ambient Air Quality*, R. Perry and P. W. Kirk, Eds., Selper, London, 1988, pp. 252–276.

88. D. M. Aviado and T. Watanabe, "Functional and Biochemical Effects on the Lung Following Inhalation of Cigarette Smoke and Constituents, I and II," *Toxicol. Appl. Pharmacol.*, **30**, 201–209 (1974); **35**, 403–412 (1976).

CHAPTER FIVE

Acetone

David A. Morgott, Ph.D., D.A.B.T., C.I.H.

Acetone is widely regarded as one of the least toxic solvents in commercial use today. This opinion stems from many years of clinical study, laboratory testing, and industrial experience with the chemical. Perhaps no other organic solvent, with the exception of ethanol, has been investigated to the same extent as acetone. The long-standing interest in the biochemical, pharmacological, and toxicologic properties of acetone can be traced to three important characteristics of the chemical:

1. Acetone is a normal by-product of mammalian metabolism; levels within the body can, however, be altered by changes in nutrition or energy balance.
2. Acetone is a highly volatile organic solvent that is miscible with water; thus large amounts of the vapor can be absorbed through the lungs and quickly distributed throughout the body.
3. Acetone is manufactured and used in large amounts for a variety of commercial and industrial applications; thus the potential for exogenous human exposure is widespread.

1 PRODUCTION AND USE

Acetone production in the United States has risen slowly but erratically over the past decade. The average annual increase in production has been estimated at 0.7 percent for the past 10 years with large year-to-year fluctuations (1). Production

Patty's Industrial Hygiene and Toxicology, Fourth Edition, Volume 2, Part A, Edited by George D. Clayton and Florence E. Clayton.
ISBN 0-471-54724-7 © 1993 John Wiley & Sons, Inc.

for 1990 has been reported at 2220 million lb, which represents about 86 percent of U.S. capacity (1, 2). The United States is capable of producing about 30 percent of the worldwide capacity, which was estimated to be 6794 million lb in 1989 (3). Large-scale commercial production of acetone is generally accomplished by either of two processes. The first, and by far the most common, is by acid hydrolysis of cumene hydroperoxide, whereby acetone and phenol are formed as co-products. The second process, catalytic dehydrogenation of isopropyl alcohol, accounts for less than 5 percent of U.S. production. Other production methods, such as biofermentation, propylene oxidation, and diisopropylbenzene oxidation, together account for only a small percentage of worldwide production.

Approximately 40 percent of the acetone manufactured in the United States is used to synthesize methyl methacrylate and methacrylic acid; another 23 percent is used to prepare methyl isobutyl ketone, methyl isobutyl carbinol, and bisphenol-A (1). Use of acetone as a solvent in paint, ink, resin, and varnish formulations account for about 20 percent of production. Most of the remaining production is used as a process solvent in the manufacture of cellulose acetate yarn, smokeless gun powder, surface coatings, and various pharmaceutical and cosmetic products (3).

2 CHEMICAL AND PHYSICAL PROPERTIES

Acetone (CAS 67-64-1), otherwise known as 2-propanone or dimethyl ketone, is a clear, colorless liquid that is highly flammable and infinitely soluble with water. Reagent-grade acetone can contain up to 0.4 percent water as well as small amounts of other polar solvents. A partial list of acetone's physical and chemical properties appears in Table 5.1 (4).

At low concentrations acetone vapors have a characteristic sweet and fruity odor. The odor threshold for humans has been reported as ranging from about 10 to 680 ppm, with 100 to 140 ppm being the odor recognition threshold for most individuals (5–8). The wide range in reported values may be associated with the rapid accommodation and olfactory fatigue that can occur following exposure to the vapor (9).

3 ANALYTICAL DETERMINATION

3.1 Biological Specimens

Virtually every organ and tissue within the human body contains some acetone, which is one of three biochemicals collectively referred to as "ketone bodies" (the others are β-hydroxybutyrate and acetoacetate). Acetone is produced within the body as a result of the breakdown and utilization of stored fats and lipids as a source of energy. Consequently conditions such as strenuous physical exercise and prolonged dieting, which lead to a breakdown of fat within the body, may result

Table 5.1. Chemical and Physical Properties of Acetone (4)

Empirical formula	C_3H_6O
Molecular weight	58.08 g/mol
Boiling point	56.1°C at 760 torr
Freezing point	−94.7°C at 760 torr
Density	0.7990 g/cm³ at 20°C
	0.7844 g/cm³ at 25°C
Vapor pressure	182 torr at 20°C
	231 torr at 25°C
Refractive index	0.3587 at 20°C
	1.3560 at 25°C
Viscosity	0.303 cP at 25°C
Surface tension	23.2 dyn/cm at 20°C
Heat capacity	29.8 cal/K-mol at 25°C
Heat of vaporization	7.48 kcal/mol at 25°C
Evaporation rate	82 sec (90%)
Air flammable limits	Lower 2.55% (v/v)
	Upper 12.8% (v/v)
Flash point	Open cup −9°C
	Closed cup −17°C
Minimum ignition temperature	465°C

in higher than average amounts of acetone in the bloodstream. Excess production is found in several disease states, particularly diabetes, but also in association with fever, anorexia, and starvation and following anesthesia. As a result of its high volatility and solubility in water, measurable amounts of acetone are continuously being excreted in the breath and urine of humans.

Numerous analytical procedures have been developed over the years for measuring acetone in biologic fluids. The analytical methods for acetone in biologic samples may be roughly divided into two classes: those used clinically to measure ketone body levels in body fluids and those used investigatively to measure the acetone arising from endogenous and exogenous sources.

3.1.1 Clinical Methods for Ketone Bodies

Most of the clinical methods for total ketone bodies involve the chemical or biochemical (enzymatic) breakdown of β-hydroxybutyrate and acetoacetate into acetone, which is then isolated and quantified by any of various instrumental techniques. Older, less reliable clinical methods generally measured total acetone in the blood without differentiating the relative contribution of each ketone body. These early methods were notoriously insensitive and laborious. Distillation techniques were typically used to isolate the liberated acetone, which was then measured nephelometrically (10, 11), titrimetrically (12, 13), or colorimetrically (14, 15). The earliest clinical methods capable of distinguishing endogenous acetone from the other ketone bodies relied upon the reaction of acetone with 2,4-dinitrophenylhydrazine or salicylaldehyde to produce a colored end product (16–19). The reaction of acetone with salicylaldehyde was later modified, along with the techniques used simultaneously to measure β-hydroxybutyrate and acetoacetate, to produce the first spectrophotometric method with an acceptable degree of accuracy (20). The primary disadvantage of the spectrophotometric procedures was the excessively long analysis time. The delay between sample collection and analysis could lead to spuriously elevated acetone concentrations because of spontaneous decarboxylation of acetoacetate in the blood or serum specimen. Acetoacetate decarboxylation is catalyzed by serum amino acids and albumin in a pH- and temperature-dependent fashion (21, 22).

Head-space gas–liquid chromatography was developed as a clinical alternative to spectrophotometry approximately 20 years ago (23, 24). The gas chromatographic techniques provided advantages of sensitivity, specificity, and speed that were previously unattainable. Furthermore, sample preparation and storage techniques were optimized to minimize the spontaneous formation of acetone from acetoacetate (23, 25, 26). Further refinements in head-space methodologies focused on the perfection of ezymatic techniques for oxidizing the β-hydroxybutyrate to acetone (27, 28). The most recently developed method for the determination of acetone in the clinical laboratory involves derivatization of the sample with 2,4-dinitrophenylhydrazine (DNPH) followed by isolation and quantification of the hydrazone derivative using high-pressure liquid chromatography (29). This procedure was found suitable for use with both blood and urine specimens, and is vastly superior to an alternative procedure for urinary acetone that uses pentafluorobenzyloxyammonium chloride as the derivatizing reagent (30). Despite the availability of sensitive and specific quantitative techniques, considerable interest still exists in the semiquantitative "dipstick" methods for ketone bodies in the urine. Most of these nitroprusside-based techniques are sensitive to the presence of both acetone and acetoacetate (31).

Many authors have published methods for measuring the level of acetone in expired air. Although primarily used to monitor treatment in cases of diabetes or obesity, breath acetone levels have occasionally been used to estimate acetone blood levels in cases of acute intoxication (32–34). The utility of breath acetone analysis for this purpose arises from the close correlation that has been observed between breath acetone levels and the concentration of acetone and the other ketone bodies in blood or serum (25, 35, 36). Although spectrophotometric (37, 38) and nephelometric (39) methods have been developed for the determination of acetone in expired air, the majority of published methods have involved gas chromatography with flame ionization detection (GC/FID). The points of departure for the various chromatography methods have been with the nature of the column packing and with the manner of sample collection. The most common procedures for sample collection have been (1) rebreathing into a polyester bag (26, 35, 36); (2) single end-tidal breath collection in a glass syringe or plastic bag (40, 41, 42); and (3) direct breathing into a gas chromatograph (34). Because of the need for patient cooperation and involvement, the measurement of acetone in expired air has not enjoyed wide clinical acceptance.

3.1.2 Investigative Methods

The investigative methods described below have been designed for use in experimental research using either human subjects or laboratory animals. In contrast to the clinical methods, the investigative procedures focus specifically on the measurement of endogenously created or exogenously administered acetone in particular biologic tissues or fluids. Methods are available for the direct gas chromatographic analysis of whole blood using a packed column or deproteinized serum using a capillary column (43, 44). Head-space gas chromatographic procedures have also been devised to measure acetone levels in blood, urine, and milk samples from dairy cattle (45). Still other gas chromatographic methods have been developed for the measurement of acetone in rodent tissues and organs using both head-space analysis and absorption/desorption (purge and trap) techniques (46, 47). Spectrophotometry and liquid scintillation counting have been combined to yield an assay for both unlabeled and radio-labeled acetone in blood and urine specimens (48). Gavino et al. (49) have designed and optimized a very accurate and sophisticated procedure for determining unlabeled and radio-labeled acetone in either blood or liver perfusate. One version of their method involves the initial reduction of acetone to isopropanol using sodium borohydride followed by separation on a high-pressure liquid chromatograph (HPLC) and final quantification by gas chromatography. The primary advantage of this procedure is that the thermal decomposition of acetoacetate to acetone is obviated (26).

Investigational techniques for the determination of acetone in expired air are distinguished from the clinical methods by their complexity and sophistication. These procedures have generally been used to measure acetone levels either in normal human breath or in the breath of laboratory animals dosed with acetone. Methods have been published for measuring breath acetone levels both by gas chromatography–mass spectrometry (GC/MS) (50–52) and by GC/FID (53–55). The methods of Henderson et al. and of Conkle et al. use cryogenic trapping systems to concentrate the acetone in the breath sample (50, 51), whereas in the method of Phillips and Greenberg, acetone is adsorbed and concentrated using a porous resin (55). Brechner and Bethune used a vapor phase chromatographic method to measure acetone levels in the alveolar air of anesthetized dogs given acetone intravenously (53).

3.2 Environmental Media

The two most often cited procedures for the determination of acetone in water and air samples have been direct GC/MS of a sample concentrate and HPLC of the 2,4-dinitrophenylhydrazine derivative (56–59). Techniques for collecting and concentrating the sample include the use of solid adsorbents, cryogenic freeze-out traps, liquid impingers, and DNPH-treated resins (60–63). Alternative methods for air analysis include gas chromatography with a reduction gas detector (64), chemical-ionization mass spectrometry (65), and HPLC with fluorescence and chemiluminescence detection (66). All three of these methods have sensitivities in the parts-per-billion range.

Table 5.2. Normal Values for Acetone in Human Blood and Urine Specimens

Type of Specimen	Number of Subjects	Average Acetone Concentration (mg/l)	Standard Deviation (mg/l)	Reference
Plasma	20	4.35	1.31	29
Plasma	20	1.74	11.6	26
Plasma	31	0.41	0.17	28
Whole blood	6	0.93	0.06	49
Serum	11	2.9	0.3	36
Spot urine	20	3.02	1.25	29
Spot urine	15	0.76	0.63	75
Spot urine	3	0.31	0.09	30
Spot urine	10	0.8	0.2	36

4 BACKGROUND LEVELS

Acetone is a normal endogenous biochemical that can routinely be detected and measured in body fluids. Detectable amounts of acetone have been found in a variety of biologic specimens including whole blood (fetal through adult), cerebrospinal fluid, urine, exhaled air, and breast milk (67–70). Because endogenous acetone formation is so closely linked with the utilization of stored fats as a source of energy, levels can fluctuate depending on an individual's health, nutrition, and level of physical activity.

4.1 Blood, Urine, Exhaled Air

The background levels of acetone in an average human may be affected by a large number of physiological and analytical factors. The following variables appear most likely to influence the background levels of acetone measured in an average adult: (1) the duration and degree of fasting prior to sample collection (71); (2) the time of sample collection relative to the normal circadian cycle for ketogenesis (72); (3) the type of test specimen collected and length of time before analysis; and (4) the procedures used for separation and analysis of the specimen.

Table 5.2 lists normal values for blood acetone in five groups of nonfasting adults. It should be noted that ketogenesis is at a minimum following a meal and peaks between 10:00 and 12:00 PM in healthy adults (72). The values presented in Table 5.2 are in general agreement with the often cited upper normal limit of 10 mg/l (73). Trotter et al. have reported that the acetone concentration in plasma is 8 to 11 percent greater than the level in whole blood; whereas Gavino et al. found no difference for these two types of specimens (26, 49). Gavino et al. also reported that the average blood acetone concentration in nonfasted rats was no different from the level found for nonfasting human subjects (0.99 versus 0.93 mg/l, respectively). Other investigators have reported that the average acetone level in a small group ($n = 6$) of unfasted female Wistar rats was 1.20 mg/l (46). Recent data have shown that the decarboxylation of acetoacetate to acetone is catalyzed by plasma proteins at storage temperatures as low as $-20°C$ (74). Sample deproteinization was therefore recommended in those cases where rapid analysis was impossible. Normal values for the concentration of endogenous acetone in human urine specimens are also presented in Table 5.2. The values presented are for spot urine collections and are not necessarily reflective of the values to be obtained for 24-hr urine specimens. Importantly, Pezzagno et al. have reported that the urine acetone concentration in repetitive urine collections from resting humans did not increase appreciably when the subjects performed light physical exercise. The same study, however, showed a consistent diurnal trend, with the urine acetone concentration being higher in the late evening and early morning than during the day (75).

A large number of investigators have published normal values for the concentration of acetone in human expired air. An inspection of the values listed in Table 5.3 shows that acetone levels in expired air vary over a very broad range (0.13 to 11.27 $\mu g/l$) regardless of whether the human subjects were fed or fasted overnight. Separate studies have shown the rate of acetone exhalation to vary over a broad range in healthy human volunteers with the values typically falling between 29 and 230 $\mu g/hr$ (50, 51). Some of the variation in the reported average concentrations in expired air may be due to the sex-related difference in acetone exhalation that has been shown to exist for a large group of nonfasting healthy subjects (50). In this very well-designed and well-conducted study, acetone exhalation was measured by condensing the vapors from rebreathed oxygen in a cryogenic apparatus, followed by flash evaporation for GC/MS analysis. The significantly higher values found for normal females ($n = 64$) over normal males ($n = 42$) was not as distinct when diabetic outpatients were examined. The authors of this study also observed that exhaled acetone levels in normal subjects (1) increased and decreased in response to weight loss or gain, respectively; (2) increased rapidly and dramatically when carbohydrates were removed from the diet; and (3) remained relatively constant for the different age groups.

Using the blood-to-air partition coefficient for acetone, the concentration of acetone in expired air can be calculated from the level found in blood. Numerous in vivo determinations of the blood-to-air partition coefficient following exogenous acetone administration suggest that the acetone concentration in expired air should be about 330-fold lower than the value for blood (76–78). A comparison, however, of the average normal expired air acetone levels (Table 5.3) with the average normal plasma values (Table 5.2) indicates that the blood-to-air ratio is well over 1000. This discrepancy is apparently the result of acetoacetate decarboxylation in the plasma specimens following their collection (40). The best clinical method for minimizing the increase in plasma acetone following specimen collection is perhaps the procedure of Kimura et al. (28, 79).

4.2 Air and Water

Acetone is a commonly found contaminant in air and water samples. The background levels of acetone in these media can be the result of both natural and man-

Table 5.3. Normal Values for Acetone in the Expired Air of Adult Humans

Subject Description	Number of Subjects	Average Acetone Concentration (µg/l)	Standard Deviation (µg/l)	Range (µg/l)	Reference
Fasting males	20	1.53	0.36	0.87–4.22	120
Fasting adults	89	1.63	—	0.29–11.27	40
Fasting males and females	23	1.04	0.29	0.74–1.60	42
Fasting adults	40	1.1	0.5	—	41
Fasting adults	15	1.35	0.70	0.58–2.81	55
Nonfasting healthy adults	13	1.19	—	0.50–1.99	52
Nonfasting adults	67	1.10	0.88	0.13–4.22	34
Nonfasting adults	14	0.97	0.07	0.7–1.7	36
Nonfasting adults	20	1.16	0.58	—	26
Nonfasting adults and juveniles	35	1.18	—	0.29–3.25	40
Nonfasting adult females	13	1.06	0.54	0.40–1.80	33
Nonfasting obese adults	7	0.71	0.50	0.23–1.52	33
Nonfasting obese adults	10	1.34	0.88	0.43–2.78	33

made emissions. Natural sources of acetone include animal manure decay, volcanic eruptions, forest fires, and the growth of vegetation (80–83). Anthropogenic sources include automobile exhaust, cigarette smoke, curing of finished products, and industrial releases. (84–87). The photochemical oxidation of alkanes in the troposphere can also contribute substantially to atmospheric acetone levels (88).

Background levels of acetone in the atmosphere have been assessed from both ground level and airborne monitoring stations located throughout the world. The average acetone concentration at two rural ground level sites in the southwestern United States (2.8 ppb) was somewhat higher than the values reported for remote regions of Idaho (0.5 ppm), Alaska (1.0 ppb), the Pacific northwestern United States (0.9 ppb), and central Ontario (1.7 ppb) (89–93). Air samples from the Smoky Mountains and from rural Pennsylvania contained acetone levels ranging from 0.7 to 4.0 ppb and 0.2 to 1.8 ppb, respectively (94, 95).

The concentrations of acetone in urban areas show large unpredictable variations. The average concentration at urban ground level sites in Salvador, Brazil, Raleigh, North Carolina, and Tucson, Arizona, ranged from 12 to 24 ppb (89, 96, 97); however, acetone levels at specific urban sites often ranged as low as 1 to 4 ppb. Airborne meaurements in the upper troposphere and lower stratosphere have revealed planetary acetone concentrations of 120 ppt or less (65, 98, 99). A compilation of available outdoor data on acetone was used to calculate an average daily outdoor concentration of 6.9 ppb ($n = 17$), whereas the average indoor concentration was calculated to be 8.0 ppb ($n = 4$) (100).

Although dissolution in rainwater does not appear to be an important process for the removal of acetone vapors from the air, small amounts are routinely detected in water samples from various sources. Low levels of acetone have reportedly been found in Los Angeles rainwater (<50 µg/l), Philadelphia municipal drinking water (1.0 µg/l), and surface seawater from the Straits of Florida (22 µg/l) (101–103). Volatilization and biodegradation have been found to be the most important processes for the removal of acetone from water, whereas photolysis and reactions with hydroxyl radicals were the predominant processes for eliminating atmospheric acetone (104, 105).

5   ENDOGENOUS ACETONE ELEVATIONS

The acetone level in the body at any instant is reflective of acetoacetate production, which is in turn affected by free fatty acid utilization by the liver. Consequently normal and abnormal physiological conditions and disease states, such as those described in Table 5.4, can appreciably increase ketogenesis and thereby the body burden of acetone. For instance, infants, pregnant women, and training athletes can have ketone body levels that are elevated 2- to 20-fold above normal because of ketogenesis resulting from their higher energy requirements (106, 107). Likewise, children can have appreciably higher blood acetone levels than adults because of their higher energy expenditure and the possible ketogenic influence of growth hormone (115). The average blood acetone levels in infants 2 to 5 days old have

Table 5.4. Common Physiological Conditions
and Disease States Leading to Hyperketonemia
in Humans (106–112)

Physiological Conditions
Pregnancy
Postnatal growth
High Fat consumption
Dieting
Lactation
Physical exercise
Perinatal development
Disease States
Starvation
Alcoholism
Diabetes mellitus
Hypoglycemia
Eating disorders
Prolonged vomiting
Prolonged fasting
Acute trauma
Inborn errors of metabolism

been reported to be three-fold higher than in adolescents 10 to 15 years of age (20). The hyperketonemia that develops in certain diseases is also linked to carbohydrate deprivation and a compensatory increase in the utilization of stored fat. Numerous common clinical conditions can elevate blood acetone levels including acute trauma, starvation, prolonged vomiting, alcoholism, and others as listed in Table 5.4.

Aside from cases of accidental or intentional exposure, the most dramatic increases in blood ketone bodies occur in individuals with uncontrolled diabetes mellitus. As shown in Table 5.5, patients with severe diabetic ketoacidosis can have plasma acetone levels as high as 750 mg/l, which is up to 300 times the normal limit (35, 48). Clinical findings in cases of acute acetone intoxication suggest that acetone blood levels in excess of 1000 mg/l are necessary to cause unconsciousness in humans (116, 117). Although the acetone levels characteristic of diabetic ketoacidosis may cause persistent drowsiness and mild proteinuria, acetone is not responsible for diabetic coma or any prominent symptoms of diabetic shock (118, 119). Brega et al. have reported that occupationally exposed workers can have plasma acetone levels ranging from 6.0 to 74.9 mg/l (29). Similarly Kobayashi et al. have indicated that acetone-intoxicated patients had urine levels ranging from 31.0 to 650.9 mg/l (30).

Other clinical conditions, such as diabetic ketosis and prolonged fasting, lead to more moderate increases in the acetone level. In these situations, the elevations in blood acetone are typically accompanied by even larger increases in the other ketone bodies; however, unlike acetone, acetoacetate and β-hydroxybutyrate are

Table 5.5. Blood and Urine Acetone Levels in Diabetics and Fasting Humans

Subject Description	Sample Type	Number of Subjects	Average Acetone Concentration (mg/l)	Range (mg/l)	Reference
Controlled diabetics	Plasma	50	5.8	2.3–9.3	29
Ketotic diabetics	Plasma	10	23.2	5.8–52.3	48
Ketoacidotic diabetics	Plasma	12	290.4	98.7–720.2	48
Ketoacidotic diabetics	Plasma	27	424.0	145.2–749.2	68
Ketoacidotic diabetics	Serum	10	148.7	0.0–604.0	27
Controlled diabetics	Serum	10	3.4	1.5–10.6	36
Uncontrolled diabetics	Serum	8	45.5	1.3–212.0	36
Nonobese adults fasted 3 days	Plasma	6	46.5	29.0–81.3	48
Obese adults fasted 3 days	Plasma	6	17.4	11.6–29.0	48
Obese adults fasted 21 days	Plasma	3	81.3	58.1–98.7	48
Controlled diabetics	Spot urine	9	6.8	0.4–44.0	36
Uncontrolled diabetics	Spot urine	8	14.0	0.9–35.0	36
Diabetics	Spot urine	6	3.1	0.6–0.9	30

ionizable organic acids that can disrupt normal acid–base balance when formed in sufficient amounts. Acetone, in contrast, is nonionic and is produced together with carbonic acid during the breakdown of acetoacetate. Some authors have postulated that the formation of acetone and carbonic acid in the body represents a control mechanism whereby the potential acid–base imbalance from the strongly acidic acetoacetate can be ameliorated through the formation of the weakly acidic carbonic acid (35).

Numerous studies have documented slight to moderate increases in breath acetone in diabetics and fasting adults. As shown in Table 5.6, elevations in breath acetone are generally less severe than those described previously for plasma acetone; there are, however, several notable exceptions. Stewart and Boettner found that breath acetone levels in cases of juvenile diabetes were, on average, nearly 100 times greater than normal (41). Likewise, diabetics who tested strongly positive for glucosuria had breath levels that averaged about 340 times the normal level. Rooth and Øestensen reported that an individual who consumed a paint thinning solvent had a breath acetone level of 2200 μg/l 2 days after the event (34). Several reports have shown that a 36-hr fast can result in a 40-fold elevation in breath acetone level, and that the increase could be immediately reduced by consuming small amounts of ethanol (120–122).

6 ACUTE EXPOSURES—HUMAN

Following acute exposure to relatively large amounts of vapor or liquid, acetone toxicity is characterized by nonspecific local and systemic effects. The local effects of acetone are primarily irritative and are observed in the mucous membranes of the eyes, nose, and throat. Except in cases of prolonged liquid contact or high vapor exposures, acetone irritation is not particularly severe or debilitating compared with the effects of other primary sensory irritants. Respiratory tract irritation appears to be the most sensitive indicator of acetone overexposure in humans and serves as the basis for limiting occupational exposures. The systemic effects of acetone are associated with its anesthetic properties and its ability to disrupt normal intermediary metabolism at high concentrations. The most noticeable effect of systemic overexposure is a generalized central nervous system (CNS) depression, which can range in severity from lightheadedness to narcosis depending on the magnitude and length of the exposure. Accidental exposures to extremely large amounts of acetone vapor are necessary for the development of any overt signs of acetone-induced toxicity. The signs and symptoms of acetone overexposure are perhaps most notable for their lack of specificity; the clinical picture in cases of human acetone intoxication can be easily confused with other medical conditions such as diabetes and nonspecific mixed-solvent-induced narcosis.

6.1 Laboratory Studies

Exogenous exposure to acetone can occur by ingestion, inhalation, or dermal contact; however, the most probable route in humans is through vapor inhalation.

Table 5.6. Acetone Levels in Expired Air of Ketotic Humans

Subject Description	Number of Subjects	Average Acetone Concentration (μg/l)	Range or Standard Deviation (μg/l)	Reference
Healthy adults, 12-hr fast	6	3.1	0.6	121
Healthy adults, 36-hr fast	6	63.1	47.0	121
Alcoholics, 7-day abstinence	6	1.9	0.4	120
Controlled diabetics	10	2.3	0.5–9.5	36
Uncontrolled diabetics	7	7.9	1.6–24.5	36
Controlled diabetics, overnight fast	129	30.1	116.2	41
Juvenile diabetics, overnight fast	33	104.7	213.9	41
Adult diabetics, overnight fast	96	4.5	13.6	41
Borderline diabetics	40	0.8	0.3–1.7	34
Late-onset diabetics	20	1.7	0.2–3.3	34
Juvenile diabetics	49	4.4	0.1–25.2	34
Diabetics, normal carbohydrate diet	22	1.37	0.56	42
Diabetics, reduced carbohydrate diet	19	2.78	2.01	42

Early human studies with acetone generally involved oral or intravenous administration and focused on either the in vivo pharmacokinetics or the acute narcotic effects of the chemical. Consequently, little effort was expended on describing any subtle clinical changes that followed the treatment. More recent investigations have generally dealt with specific questions concerning the irritative and neurotoxicologic properties of acetone. Acetone has received considerable attention in two specific areas of neurotoxicology; neurobehavioral performance and neurophysiological response. The former have assessed cognitive skills, sensorimotor response, and mood or temperament, whereas the latter have assessed neuronal function in the central and peripheral nervous systems (123, 124).

6.1.1   Oral and Intravenous Routes

Albertoni reported that the consumption of 15 to 20 g of acetone per day by healthy humans for several days caused no adverse effects other than slight drowsiness (125). Likewise, Frerichs found no apparent effects in diabetics who consumed 20 g of acetone per day for 5 days (126). Koehler et al. showed that the intravenous administration of 200 ml of a 0.5 percent solution of acetone in normal saline over 2 hr to healthy and diabetic test subjects resulted in a small decrease in blood pressure and a slight transient drowsiness (127). Widmark administered 8 to 16 g of acetone in aqueous solution to humans by the oral or anal route without any apparent adverse effects (128). Haggard et al. reported no ill effects in several test subjects who drank 4.7 to 5.4 g of acetone (78). The acetone blood levels observed in the latter two studies never exceeded 280 mg/l.

6.1.2   Inhalation Route

In self-exposure trials, Kagen found that 22 mg/l (~9300 ppm) of acetone vapor could not be tolerated for longer than 5 min because of throat irritation (129). DiVincenzo et al. reported that human volunteers exposed to either 200 or 500 ppm of acetone for 2 hr did not experience any subjective symptoms of irritation other than an odor awareness at 500 ppm (130). Blood specimens collected before and after treatment did not reveal any exposure-related alterations in serum biochemistry or hematology. An inhalation study performed by Haggard et al. indicated that a single resting individual exposed to either 211 or 2110 ppm for 8 hr did not experience any loss of judgment or coordination (78). The same results were obtained when moderate exertion was required of a test subject receiving a 2110-ppm exposure or when the exposure duration was extended to 3 days at 2110 ppm (8 hr/day).

Stewart et al. examined the local and systemic effects of repetitive exposures to acetone vapors in male and female volunteers (131). The stated goals of this project were to develop a biomonitor for acetone exposures and to examine the physiological response of healthy adults to differing levels and lengths of exposure to acetone. Two series of experiments were performed. In the first series, two groups of male subjects were exposed to each of four vapor concentrations (0, 200, 1000, and 1250 ppm) for either 3.0 hr or 7.5 hr/day for 4 days/week (the first day of each week was a control exposure at 0 ppm). The subjects were exposed to progressively higher vapor levels of acetone on each succeeding week of treatment. Following the fourth week of exposure at 1250 ppm of acetone, the subjects were given a fifth week of exposure at 0 ppm and then a final week where the vapor concentration was allowed to fluctuate between 750 and 1250 ppm (average of 1000 ppm) on each of four exposure days. The second series of studies was performed in four female subjects who were exposed to 1000 ppm of acetone for either 1.0, 3.0, or 7.5 hr/day for 4 days/week.

All the subjects in the investigation were given a complete medical and physical exam at the beginning and end of the study. The medical exam included the weekly collection of blood for a complete blood count and a 23-element clinical chemistry analysis. Blood pressure, temperature, subjective responses, clinical signs and symptoms, and urinalysis were recorded daily. Electrocardiograms were recorded continuously during each exposure session. Alveolar air breath samples were collected at 0, 0.25, 0.50, 1, 2, and 3 hr following the exposure. Cardiopulmonary testing (heart rate, minute ventilation, expiratory flow rate, alveolar-capillary gas exchange, and vital capacity) was performed on male subjects shortly before ending each weekly exposure session. Metabolic, hematologic, and cardiopulmonary measurements were conducted before and after physical exercise on a separate group of male subjects exposed for 7.5 hr. A battery of neurophysiological and neurobehavioral tests was also performed at various times throughout the exposures. The neurophysiological tests included spontaneous electroencephalograms, visual evoked response using a strobe light, and a Romberg heel-to-toe equilibrium examination. Cognitive neurobehavioral testing including an arithmetic test, a coordination test, and a visual inspection test.

These investigators recorded few adverse effects from the acetone exposures. Urinalysis, complete blood counts, and clinical chemistry values were within normal ranges for both the male and female subjects. No statistically significant changes were noted in the electrocardiograms, and no trends or consistent changes were noted in the pulmonary function tests. The neurophysiological and neurobehavioral exams were unaffected in all but one case. Although male subjects exposed to 1250 ppm of acetone for 7.5 hr showed a statistically significant increase in the amplitude of the visual evoked response when compared to background values, this effect was not observed in male or female subjects exposed to 1000 ppm of acetone for similar lengths of time.

Subjective complaints of eye irritation, throat irritation, headache, and tiredness were noted in the male subjects at all of the exposure concentrations including the initial 0 ppm control level. Exposures of 1000 and 1200 ppm caused three of the four subjects to complain of eye irritation, throat irritation, and tiredness; however, none of these complaints were noted during the final week of exposures when the vapor concentration fluctuated between 750 and 1250 ppm. Inexplicably, three of the four women exposed to 1000 ppm of acetone were noted to have begun menstruating earlier than normal. The average blood and urine levels of acetone 30 min after ending the last 7.5-hr exposure to 1000 ppm were 69.3 and 63.8 mg/l, respectively. The maximum blood acetone level found was 117.6 mg/l for an individual exposed for 7.5 hr to vapor concentrations fluctuating between 750 and 1250 ppm. The authors of the study concluded that no serious or deleterious effects occurred in the subjects upon repeated exposure to acetone concentrations of 1000 ppm or less for periods up to 7.5 hr. Moreover, the increase in visual evoked response at 1250 ppm suggested a depression or synchronization of cerebral cortical activity in the subjects exposed for 7.5 hr.

Matsushita et al. described two separate studies where male university students received either a single exposure or six consecutive daily exposures to acetone vapors (132, 133). Single-day exposures were performed with groups of five subjects at concentrations of either 0, 100, 250, 500, or 1000 ppm, whereas the multiple-day exposures were conducted with groups of six subjects at concentrations of 250 and 500 ppm. In both studies, the exposures lasted 6 hr, with a 45-min lunch break separating the 3-hr morning and afternoon segments. Using a test questionnaire, up to seven different symptoms were subjectively ranked (0 to 2); the results were multiplied by the number of students to obtain the total score for each complaint. The seven symptoms were unpleasant odor, tension, headache, general weakness, lack of energy, heavy eyes, and mucous membrane irritation. For any single symptom, a maximum score of 10 or 12 could be calculated depending on whether five or six subjects took part in the exposure. Subject scores were recorded three times during both the morning and afternoon sessions (10, 30, and 90 min into each exposure period) and again on the morning following the exposure, giving a total of seven scoring interviews for each exposure day.

Using the scoring system described above, the subjects exposed to 500 and 1000 ppm for a single day responded with total scores of 4 or 5 for each of the first five interviews, whereas the groups exposed to 100 and 250 ppm had scores of 0 or 1. The primary complaints during these scoring sessions were an unpleasant odor and mucous membrane irritation. The total score for all exposure groups at the sixth interview was zero, indicating that complete sensory adaptation had occurred at all levels by the end of the exposure. On the morning following the exposure, students in the 500- and 1000-ppm exposure groups had scores of 6 to 10 for the following four symptoms: tension, general weakness, heavy eyes, and lack of energy. The average postexposure blood and urine concentrations of acetone were about 60 and 53 mg/l, respectively, for the 1000-ppm exposure. The authors also reported that exposures at 500 or 1000 ppm resulted in a temporary decrease in the phagocytic activity of neutrophils and an increase in the eosinophil and leukocyte counts in peripheral blood specimens collected at 3, 7, 24, and 32 hr postexposure. All of the above measurements returned to normal after 48 hr. The authors attributed the changes in white blood cell count to an inflammatory reaction caused by the irritating effects of the vapor.

Many of the findings from the single-day exposures were also reported for the multiple-day treatment (133). For the 500-ppm exposure, mucous membrane irritation was reported to be greatest immediately after entering the chamber in the morning and afternoon sessions. Appreciable accommodation was noted as each exposure continued, but no day-to-day adaptation was observed. Irritation to the throat was described to be much less severe than that to the eyes and nose. The complaints recorded on the day after each exposure were similar to those described for the single-day treatments. Hematologic abnormalities detected in daily blood specimens from the subjects exposed to 500 ppm were the same as those described for the single-day exposures. It should be noted that the white blood cell effects reported in this study are unique and have not been observed in other studies with human volunteers or industrially exposed workers.

The Matsushita et al. study also included a neurobehavioral examination of the reaction times for the repeatedly exposed subjects (133). The test measured reaction time to a visual stimulus and involved an additional 250-ppm 6-hr/day acetone exposure for subjects that were required to double their metabolic rate through physical exercise. The absolute values for the reaction time test showed considerable variation between individuals and were not statistically different from the values obtained in untreated controls; however, when the data were averaged for all subjects in an exposure group and expressed relative to the average preexposure value, several statistically significant changes were obtained. The response times were longer on each of the six exposure days for the 500-ppm group and on two of the six exposure days for both of the 250-ppm exposure groups (resting and exercising). A close examination of the data suggests that learning may have occurred during the control periods and contributed to the positive results. The data did not reveal any consistent dose- or time-related trends in the magnitude of the response.

Dick et al. performed a series of neurobehavioral studies on approximately 20 male and female volunteers who were exposed to either 250 ppm of acetone for 4 hr or to a combination of 125 ppm of acetone and 200 ppm of methyl ethyl ketone (MEK) for 4 hr (134, 135). Four psychomotor tests, one sensorimotor test, and one psychological test were performed on the subjects before, during, and after the exposure session. Acetone was shown to cause an effect on the responses obtained in two of these tests, a dual auditory tone discrimination compensatory tracking test and a profile of mood states (POMS) test. The dual auditory tone discrimination compensatory tracking test involved the presentation of two tasks either in series (single presentation) or simultaneously (multiple presentation). The auditory tone task required the determination of a 760-Hz tone from a series of 750-Hz signals, whereas the compensatory tracking test required the subject to use a joystick to position a horizontally moving arrow under a stationary arrow as the former swept across a video screen. The tracking error (i.e., the extent the arrows were misaligned), response time, number of correct hits, and number of false responses were recorded. Relative to preexposure control values, the 250-ppm acetone exposure caused an increase in both the response time and the percentage of incorrect responses in the auditory tone portion of the dual task when the tests were presented in series. The response measurements were not affected by the exposure when both portions of the dual task were presented simultaneously.

The POMS test involved the submission of written responses to a questionnaire that was graded on six different measures of the individual's affective state: tension–anxiety, depression–dejection, anger–hostility, vigor–activity, fatigue–inertia, and confusion–bewilderment. Male subjects taking the POMS test showed an increase in the anger–hostility portion of the test. Except for a small change in the percentage of incorrect responses in the dual auditory discrimination test, none of these effects were noted when the subjects were exposed to the acetone–MEK mixture. The authors noted that the results from their study needed careful interpretation and that additional research was needed to detect more distinct declines in human performance. Statistically different results were obtained only during the first 2-hr exposure session and during the 2-hr post-exposure session; no differences in response time were noted during the second 24-hr exposure session, indicating a lack of consistency in the response.

Suzuki performed several neurophysiological tests on two groups of eight or nine male student volunteers exposed to acetone at concentrations of 250 to 270 ppm or 500 to 750 ppm for two 3-hr sessions with a 1-hr break (136). Spontaneous and evoked changes in five separate physiological functions were recorded before, during, and at the end of the exposure, but because of wide variations in the initial base-line response, only those data collected at the end of the experiment were compared with control values. The five physiological function tests were galvanic skin response (GSR), vasoconstriction, heart beat interval, respiration interval, and cerebral activity. The following nonstatistically significant tendencies were noted for the exposed groups: (1) a decrease in the spontaneous GSR and an increase in the evoked GSR, (2) a decrease in evoked vasoconstriction activity, (3) a decrease in the mean time interval for 10 heart beats at the high exposure concentration, and (4) an increase in cerebral activity. The authors indicated that an increase in air temperature within the exposure chamber was positively correlated with several of the observed positive responses. The degree of correlation between the air temperature and the physiological response was then noted to be affected by the acetone vapor concentration. Despite the temperature interference, the authors concluded that the occupational limit for acetone should be lowered to a level below 250 ppm.

Nakaaki performed a neurobehavioral time estimation test on two male and two female student volunteers exposed to acetone vapors for 4 hr on a single day (137). The 4-hr exposure was composed of two 2-hr sessions with a 2-hr rest period separating the sessions. The study was to be conducted at a constant acetone exposure concentration; however, because of fluctuating chamber concentrations, the exposure sessions were described as either a low or high exposure. The low-level exposures were performed at acetone vapor concentrations ranging from 170 to 450 ppm, whereas the high level were between 450 and 690 ppm. At 30-min intervals, the test subjects were asked to estimate the passage of time for periods lasting from 4 to 30 sec. The authors reported that the time estimations tended to be more prolonged for both the exposed male and female subjects relative to control values. An examination of the data, however, shows that the results were quite variable with no apparent difference for either of the exposure ranges.

Seeber et al. studied the neuropsychological relationship between inherent susceptibility for a particular chronic symptom and an individual's acute subjective response to an organic solvent exposure (138, 139). The primary goal of this study was to determine whether the chronic neurological symptoms reported by an individual could be related to the acute symptoms that were reported during a solvent exposure. Groups of 16 male volunteers received a 1000-ppm acetone exposure, a combination exposure with 500 ppm acetone and 200 ppm ethyl acetate, or a filtered-room-air exposure. Urine specimens were collected every 2 hr during and after the exposure (14-hr duration) and analyzed for acetone in order to calculate urinary excretion rates. A chronic symptom questionnaire was completed by each subject one week prior to their 4-hr exposure. The responses on the questionnaire were grouped into one of six chronic symptom categories: psycho/neurovegetative lability, neurological symptoms, lack of activity, excitability, lack of concentration, and special symptoms.

Information from the chronic symptom questionnaire was subsequently compared to the results obtained from a second set of ranked-response questions that dealt with four acute symptoms: tension, tiredness, complaints, and annoyance. These ranked-response questions were answered at 2-hr intervals during and after the exposure. A correlation between any of the six chronic symptom categories and any of the four acute symptoms suggested a possible temporal relationship between the chronic symptom and the acute response (i.e., the chronic symptom of the subject could be used to predict the acute response). No statistically significant cross-correlations were observed in the subjects exposed to a 1000 ppm of acetone; however, the group receiving a combined exposure to ethyl acetate and acetone displayed a cross-correlation for psycho/neurovegetative lability and their scores for both complaints and annoyance. Further analysis revealed that three of the four acute responses correlated with the rate of acetone excretion in the urine, tiredness being the exception. The relationships, however, were not strong, and the associations with excretion rates did not show any clear time-related patterns or trends.

6.1.3 Ocular and Dermal Contact

As mentioned above, subjective complaints of slight to mild eye irritation have often been reported for workers exposed to acetone vapor. There have been several laboratory studies of the irritative effects of acetone vapor to ocular tissue; however, they have often been poorly conceived or inadequately described. Nelson et al. examined the irritative effects of acetone and other volatile solvents using male and female university students exposed for 3 to 5 min in an exposure chamber (140). Slight irritation was described by some of the subjects exposed to 300 ppm of acetone, but most of the subjects reportedly tolerated 500 ppm without severe effect. The authors of this study acknowledged that adaptation was not considered and that the results were not representative of an occupational environment. In a recent study of the irritative effects of ketonic solvents, the vapors of acetone were reported to be nonirritating to the eyes and lungs of human volunteers. The ocular exposures were performed for 15 sec using goggles placed over the eyes, and the inhalation exposure required 10 breaths through a mouthpiece connected to a vapor-generating device (141). The report contained very few details and no information was given on the range of acetone vapor concentrations examined. Based on clinical experience, Kechijian concluded that the overzealous use of acetone-containing nail polish removers could cause onychoschizia, a condition characterized by split and peeling nails (142).

There have been several reports of direct ocular contact with liquid acetone, and although the effects were more serious than with acetone vapor, the data seem to show that prompt treatment will prevent permanent corneal damage. In fact, acetone has been used with no apparent difficulty to remove polymerized cyanoacrylate adhesives from the eyelid following accidental fusion (143). The ocular damage from an accidental splash of acetone has often been limited to the corneal epithelium, which heals completely after several days. Grant reported that small amounts of acetone in the eye caused an immediate stinging sensation, but when promptly irrigated, the damage was confined to the epithelium (144). McLaughlin reported three cases of human corneal burns from liquid acetone that healed within 48 hr of irrigation and removal of the damaged corneal epithelium (145).

Mayou once speculated that the cataract and lens opacity observed in a male patient may have resulted from his occupational exposures to acetone (146). The assertion was based on the similarity between the patient's cataract and those previously observed in diabetics. Weill reported a case of severe eye injury in a worker exposed to the vapors of a methanol solution that contained 5 percent acetone (147). Permanent corneal opacity has been reported in a single incident involving an employee who was splashed by a solution of cellulose acetate in acetone (148). A tenacious film of cellulose acetate was deposited on the surface of the affected employee's cornea following evaporation of the acetone. After reviewing many reports, Grant concluded that there was no substantive evidence of cataract formation or optic nerve damage in humans systemically exposed to acetone (149).

An ultrastructural examination of acetone-treated human skin has shown that topical exposure can affect the architectural integrity of the epidermis. Lupulescu et al. obtained biopsy specimens of the epidermis from the forearms of six male volunteers who were treated with acetone for periods of 30 or 90 min (150). The biopsies were collected immediately after treatment and at 72-hr post-treatment for analysis by light and electron microscopy. Acetone contact was found to cause cellular damage in both the stratum corneum and stratum spinosum that was treatment-time related. Tissue specimens collected at 72-hr post-treatment showed a high degree of repair and restoration, with little evidence of damage still present. In a follow-up study using seven subjects, Lupulescu and Birmingham found that pretreatment of the skin with a protective gel substantially reduced the severity of the cellular and structural damage caused by a 90-minute acetone exposure (151).

6.2 Industry Studies

Raleigh and McGee performed two occupational health surveys on filter press operators and support personnel exposed to acetone vapor in a cellulose fiber facility (152). The surveys were conducted a year apart and involved the collection of breath, breathing zone, and room air samples for acetone analysis. In the first study, nine employees were monitored for seven 8-hr workdays and were asked to rate (slight, mild, or strong) any symptoms of sensory irritation following each sample collection. Medical exams performed at the end of each work shift provided an independent record of any eye, nose, or throat irritation resulting from the acetone exposure. Breathing zone samples produced a mean daily time-weighted average (TWA) exposure of 1006 ppm (range 950 to 1060 ppm) for the 7-day survey period. Individual breathing zone samples ranged as high as 5500 ppm when the filters were removed from the presses. Reports of irritation among the nine employees were recorded as follows: eye irritation in seven, throat irritation in four, headache and lightheadedness in three, and nasal irritation in two employees. The symptoms were transient and generally occurred when the vapor concentrations exceeded 1000 ppm. Of the 31 individual reports of eye irritation, 21 occurred when the acetone concentration was greater than 1500 ppm; four slight to mild responses were obtained when the instantaneous concentration was between about 750 and 1000 ppm. In every case, the effect was transient, and no evidence of eye irritation was observed by the examining physician. Medical examinations were essentially normal in all respects except for a slight redness in the nasal mucosa of one individual and slight congestion in the nose and throat of another.

Four filter press operators were involved in the second Raleigh and McGee survey, half of whom were monitored for three 8-hr work shifts and the other half for two 8-hr shifts. Air samples collected from the breathing zones of each employee revealed a TWA exposure concentration of 2070 ppm (range 155 to 6596 ppm) during the 3-hr monitoring period. In this portion of the study, two of the four employees complained of eye irritation, three of nasal irritation, and one of throat irritation. Physical examinations were negative except for one individual whose throat was slightly congested. The authors noted that reports of eye irritancy were highly variable in each survey and that the subjective responses showed only a general concentration-related increase in severity.

Israeli et al. conducted neurobehavioral testing on five employees working on an office appliance production line where acetone was used in a solvent-based glue and in a cleaning fluid (153). The workplace concentration was reported to be about 200 ppm according to the instantaneous readings from acetone-sensitive detector tubes. The subject's reaction time to a light and sound stimulus was measured before and after an 8-hr work shift. Control values were obtained on the same employees 2 days after being removed to an acetone-free work area. The highly variable results obtained on repeated testing were ignored by averaging the values for each individual. A statistically significant increase was obtained in the light stimulus response time when the mean values for each individual were averaged for the five subjects who took part in the study. Examination of the data shows that both the pre-shift and post-shift response-time measurements were increased relative to the control session, which suggests that the effect was not exposure related. Nevertheless, the authors concluded that the increase in response time was due to acetone and that occupational exposure limits ought to be modified in accordance with the results.

6.3   Medical Case Histories

This section describes instances of accidental or intentional exposures to large amounts of acetone. The 1988 yearly summary of the American Association of Poison Control Centers listed 1001 incidents of human acetone poisoning throughout the United States (154). Approximately 95 percent of these incidents involved the accidental ingestion of pure acetone (excluding nail polish removers), with about 44 percent of the cases involving children less than 6 years old. No fatalities were reported in any of these incidents, and only one major medical outcome was reported (details not provided). Minor organ damage has occasionally been reported in cases of severe acetone narcosis, but this damage may have been associated with oxygen deprivation and tissue hypoxia. The hyperglycemia that often accompanies cases of severe acetone intoxication is likely associated with acetone's well-known gluconeogenic effects (see below). The available information suggests that individuals succumbing to the acute effects of acetone recover completely and relatively quickly following the incident.

6.3.1   Occupational Setting

Ross described an instance of overt acetone intoxication in an industrial environment (155). The incident involved four individuals responsible for removing standing water from an unventilated pit (12 ft deep) within a large building. Water had been diverted to the pit following rupture of a water main. Two members of the crew entered the pit to shovel the water into buckets, which were then hauled to the surface by the others. Before taking a lunch break, both members of the pit crew experienced some sensory irritation, and one of the members reportedly felt inebriated. Upon returning to the pit, the latter worker became unconscious, whereupon another member of the crew entered to provide assistance. Nearby employees, who helped with the rescue, were also affected by the vapors. Upon arrival at a hospital, one employee was unconscious and another was confused, drowsy, and unsteady. The comatose individual was hospitalized for 4 days and discharged; the other was released immediately. Five of the six workers exposed to acetone vapor during the rescue operation reported feeling symptoms such as dizziness, eye and throat irritation, and leg weakness. The acetone vapor concentration in the pit was found to be greater than 12,000 ppm. Up to 50 ppm of trichloroethane was also detected in the pit at various times after the accident. The incident was attributed to the migration of acetone vapor into the pit from several holding tanks located nearby.

Foley reported an unusual case of acetone poisoning in an employee who attempted suicide by inhaling acetylene gas (156). The employee attached a hose from a cylinder of acetylene gas to a paper cup that he fastened to his face with a rubber band. He was found unconscious shortly after the start of his work shift and was immediately taken to an emergency room where a physical examination revealed a rapid heart beat, rapid breathing, and cyanosis. Laboratory tests showed elevations in serum glucose, creatinine, phosphorus, and lactic acid as well as hematuria, glycosuria, and ketonuria. The elevations in serum creatinine were considered to be the result of acetone interference, which was known to give spuriously elevated creatinine values with the alkaline picrate method of analysis (157, 158). The marked acidosis noted in this patient was attributed to the hypotension and hypoxia that occurred when the employee became unconscious. Cylinders of acetylene gas were discovered to contain a small amount of acetone, which functions as a carrier. The author concluded that the hyperglycemia and acetonuria observed in this patient were the result of acute acetone poisoning.

Sack documented an instance of acute intoxication in an employee who had worked in an acetone environment for many years (159). On the day in question, the employee was cleaning a kettle used to filter a solution that consisted of a synthetic fiber dissolved in acetone. The employee wore a respirator while working in the kettle; however, a poor fit resulted in a severe acetone overexposure. The individual called out for assistance and was observed to be in a very agitated and excited state. He quickly became unconscious while inside the kettle and was taken to the clinic in a coma. The employee vomited several times at the clinic and showed marked salivation and hyperactivity. After revival with a CNS stimulant, the patient was able to speak but was still very excitable. Blood acetone levels were reportedly 430 and 302 mg/l at 9 and 11 hr after the mishap, respectively. Acetone was detected in the urine together with urobilin, red and white blood cells, and some albumin. These findings, together with a rise in serum glucose and bilirubin levels, suggested that slight liver and kidney damage had occurred.

Smith and Mayers reported two cases of acute poisoning in a raincoat manufacturing operation where the seams were waterproofed with a resin dissolved in either acetone or MEK (160). Three coats of the resin were applied manually at 2-hr intervals by an individual known as a brusher. The first two coats contained the acetone-based resin, whereas the final coat was applied with the MEK-based resin. Samples of the workroom air showed that the acetone vapor concentration ranged from 330 to 495 ppm, whereas the MEK concentration varied between 398 and 561 ppm. Two female brushers were affected by the solvent vapors on different occasions. One complained of stomach distress and watery eyes one morning and later was found unconscious in a rest room. On admission to the local hospital, she was in a coma and had a strong acetone odor on her breath, but no acetonuria. She was treated with a CNS stimulant and awoke immediately, but complained of a severe headache. She showed no medical signs on the day following the incident and was discharged on the second day of hospitalization. The second case followed the first by one day and was apparently less severe. The individual fainted and convulsed at the work site but regained consciousness immediately afterward. At the hospital she was confused and had a headache, but after 1 hr of observation she was allowed to leave. The authors ascribed these incidents to the additive effects of both solvents and surmised that the total vapor concentration likely exceeded 1000 ppm in the area where the employees worked.

6.3.2 Hospital Setting

There has been a total of nine nosocomial cases of acetone poisoning reported in the literature. The incidents generally involved hospital patients who required joint immobilization for a broken hip or leg. In each case, a synthetic plaster substitute in use at the time was used to construct the patient's cast. The plaster substitute contained a large amount of acetone, which was used as a setting fluid; other components, such as vinyl acetate, nitrocellulose, and boric acid, were often included. Patient descriptions and clinical observations are presented in Table 5.7 for each of the reported cases. Males and females were both involved; their ages ranged from about 2 to 42 years. The acetone exposure typically occurred via vapor inhalation; however, in some cases, skin absorption was considered to be the portal of entry. The onset of symptoms, typified by initial lethargy and drowsiness, occurred within 1 to 12 hr of the exposure; nausea and vomiting were often seen later. Many patients lapsed into unconsciousness, and glycosuria, acetonuria, and ketosis were commonly observed. Other frequently noted clinical signs and symptoms included hematemesis, labored breathing, tachycardia, and throat irritation. In several instances, the attending physician mistakenly suspected a diabetic coma and prescribed treatment according to this diagnosis. Further summary and analysis of the signs and symptoms of acetone intoxication in a hospital setting is presented in the case described by Hift and Patel (169).

6.3.3 Commercial Products

A recent report by Sakata et al. describes a male patient who attempted suicide by ingesting approximately 100 ml of an adhesive cement that contained 39 percent cyclohexanone, 28 percent MEK, 18 percent acetone, and 15 percent polyvinyl chloride (170). The individual reportedly drank about 720 ml of sake (10 percent ethanol) 30 min prior to drinking the adhesive solution. The patient was unconscious when taken to a hospital about 2 hr after the acetone ingestion. Gastric lavage, plasma exchange, and charcoal hemoperfusion were all performed, and the patient regained consciousness about 7 hr after ingestion. A persistent hyperglycemia was observed on days 1 through 6 of hospitalization, which was followed by a second rise in serum glucose on days 9 through 16. Serum transaminase levels began to rise markedly on day 6 and peaked on days 12 and 13 of treatment. After declining for several days, the ALT and AST activities began to rise again, and prednisone treatment was instituted. High concentrations of the metabolite cyclohexanol were present in the blood and were thought to be responsible for the coma, whereas the hyperglycemia was attributed to acetone. The authors were not certain what caused the increased serum transaminase levels. Although not specifically stated, the patient apparently recovered and was released from the hospital.

Gitelson et al. described a case involving an attempted suicide by an adult male who consumed 200 ml of pure acetone (171). The patient was observed to be stuporous with shallow respiration when observed at the hospital 1 hr after the incident. He lapsed into a coma shortly after admission and his throat was red and swollen; some tissue erosion was noted on the soft palate. The intravenous administration of a CNS stimulant resulted in temporary arousal. Complete consciousness was not regained until about 12 hr after admission. Acetone and small amounts of albumin were present in the urine, but glucosuria was not detected. Slightly elevated

Table 5.7. Signs and Symptoms of Acute Acetone Intoxication Resulting from the Application of a Synthetic Plaster Cast

	Year (Reference)								
	1903 (161)	1944 (162)	1946 (163)	1947 (164)	1950 (165)	1952 (166)	1952 (167)	1956 (168)	1961 (169)
Age (yr)	12	42	10	21	32	3	10	1.5	7
Sex	Male	Male	Female	Female	Female	Male	Male	Female	Male
Onset (hr)	<24	10	10	1?	12	1	9	12	6?
Recovery (days)	—	5	3	1?	1.5	2?	4	1	3.5
Circumstances	Hip cast	Body cast	Leg cast	Hip cast	Body cast	Leg cast	Hip cast	Hip cast	Leg cast
Exposure route	Dermal?	Inhalation	Inhalation	Inhalation	Inhalation	Inhalation	Inhalation	Inhalation	Dermal
SIGNS AND SYMPTOMS									
Unconscious	Yes	Yes	No	Yes	No	Yes	No	Yes	No
Drowsiness	Yes	Yes	Yes	Yes	Yes	Yes	Yes	Yes	Yes
Vomiting	Yes	Yes	Yes	Yes	No	—	Yes	Yes	Yes
Hematemesis	—	Yes	Yes	Yes	No	—	Yes	Yes	Yes
Acetonuria	Yes	Yes	Yes	Yes	Yes	Yes	Yes	Yes	Yes
Glycosuria	—	Yes	Yes	—	Yes	—	Yes	Yes	Yes
Tachycardia	Yes	Yes	Yes	Yes	Yes	Yes	Yes	Yes	Yes
Acetone on breath	Yes	Yes	Yes	Yes	Yes	—	Yes	Yes	Yes
Labored breathing	Yes	Yes	Yes	No	—	Yes	Yes	No	Yes
Throat irritation	—	Yes	Yes	—	—	—	No	—	Yes
Dermal irritation	—	Yes	No	—	—	—	No	No	—

blood glucose levels were observed during the first 9 days of hospitalization, but liver function tests were normal. Leg pain and a marked disturbance of gait were reported when the patient became ambulatory on day 6. Lower extremity involvement was still evident when the patient was discharged on day 13; however, the patient's condition eventually returned to normal. Elevated blood glucose levels were again observed about 4 weeks after ingestion of acetone when the patient returned to the hospital complaining of polydypsia and polyuria. Glucose levels returned to normal after 2 months of dietary restriction. A glucose tolerance test performed 5 months after the incident was within the normal range. The authors noted that the hyperglycemia was more prolonged than in most cases of acetone poisoning.

Ramu et al. described a suspected case of acetone intoxication due to the ingestion of nail polish remover (116). The patient had a history of alcohol abuse and was known to have suffered from liver cirrhosis, peripheral neuropathy, cerebral atrophy, and gastrointestinal bleeding at various times prior to admission. The patient was lethargic but conscious upon examination, and no throat inflammation was observed. A neurological examination reportedly gave essentially normal results. Ketones were detected in the urine, and gastric lavage showed acetone in the stomach contents. Extremely high acetone blood levels (2500 mg/l) were also observed, but no hyperglycemia or glycosuria was reported. The patient denied drinking from a half-empty bottle of nail polish remover found in her home. It is important to note that on admission the patient was being treated with antihypertensive and antiepileptic drugs (phenobarbital, phenytoin, and furosemide) that may have affected acetone metabolism and elimination. The patient's level of alertness returned to normal within 3 days. The authors reported that the same patient was readmitted after about 1 month, displaying the same signs and symptoms as in the first instance. Very high acetone blood levels were again observed, and the patient was referred to a rehabilitation center after recovery.

Gamis and Wasserman filed a recent case report involving the accidental ingestion of acetone by a young child (117). A 2.5-year-old child consumed nearly all of a 6-oz bottle of fingernail polish remover that contained 65 percent acetone and 10 percent isopropanol. The child was unconscious when found in his home and began having a seizure while being taken to a hospital. Phenobarbital was used to control the seizure, but the patient was unconscious when examined at the hospital approximately 45 min after being discovered. Notable clinical findings from the first day included acetonuria, acetonemia, metabolic acidosis, respiratory depression, hypothermia, and hyperglycemia. The patient gradually regained consciousness by the second day; however, evidence of acetonuria, hyperglycemia, and an acid–base imbalance were still present. The patient was discharged on the fourth day after a neurological examination showed no abnormalities. Acetone blood levels at 1, 18, 48, and 72 hr after the onset of symptoms were 4450, 2650, 420, and 40 mg/l, respectively. The initial acetone blood level of 4450 mg/l was the highest ever reported in a human regardless of circumstance. The decline in blood acetone was noted to follow closely the course of recovery. A 6-month follow-up examination of the patient showed no signs of neurodevelopmental complications.

Mirchev described several cases of human liver and kidney damage following acetone intoxication (172). Four individuals (two male and two female) reportedly displayed gastrointestinal and central nervous system symptoms that were typical of a solvent overdose. Three of the examined patients ingested acetone, whereas the fourth was exposed by inhalation. All four showed evidence of liver damage based on unstated clinical measurements. Two of the patients who ingested acetone reportedly displayed signs of mild renal damage. Few details were provided in this report; information regarding the solvent composition, clinical methodologies, and the clinical course were not described in the English abstract.

7 ACUTE EXPOSURES—LABORATORY ANIMALS

The acute exposure of laboratory animals to acetone has generally focused on three toxicologic end points: lethality, sensory irritation, and narcosis. Early interest in the anesthetic effects of acetone has also inspired some investigators to study the cardiovascular and pulmonary effects of acetone exposure. The lethality and irritative effects of acetone have repeatedly been examined using both in vivo and in vitro techniques. These studies have shown that high acetone exposures are necessary to cause death at either the cellular or clinical level. Animal bioassays have revealed that acetone vapors are weakly irritating to the mucous membranes of the eyes, nose, and throat. Skin and eye injury can occur on direct liquid contact; however, the tissue damage is localized to the exposed surfaces and recovery is complete with prompt and proper medical intervention.

A large body of research has focused on the narcotic effects of acetone. Studies from the 1920s and 1930s centered on the exposure concentrations and durations necessary to cause narcosis. Particular attention was given to describing the neurological signs and the stages of CNS depression that preceded unconsciousness. More recent studies have examined the neurophysiological and neurobehavioral effects of acetone exposure. These later studies have shown considerable variability and have not produced a clear and unequivocal description of the cognitive effects caused by acetone vapor exposures.

7.1 In Vivo Studies

The acute effects of acetone have been examined in mice, rats, rabbits, guinea pigs, dogs, cats, and monkeys. Although vapor inhalation has been the preferred route of exposure for most neurotoxicology and sensory irritation studies with acetone, other exposure routes have frequently been used to collect valuable toxicity information. The ocular and dermal routes have been used to study the local effects of acetone and its activity as a contact sensitizer. The intravenous and subcutaneous routes have provided data on the pulmonary and cardiovascular effects of acetone. Although the oral route has typically been used to determine the LD_{50} of acetone, a considerable amount of lethality data has been obtained following vapor inhalation, intravenous injection, and intraperitoneal administration.

7.1.1 Inhalation Route

The hallmarks of many past inhalation studies with acetone have been the extremely high vapor concentrations or long exposure intervals needed to cause a serious adverse effect. Pozzani et al. reported an 8-hr LC_{50} value of 21,150 ppm (50.1 mg/l) and 95 percent confidence limits of 17,950 to 24,850 ppm (42.5 to 58.9 mg/l) for female Carworth-Nelson rats (173). Smyth et al. reported that five of six rats survived a 4-hr exposure to 16,000 ppm (38.0 mg/l) of acetone (174). Mashbitz et al. reported that mice exposed to acetone at concentrations ranging from 42,200 to 84,400 ppm (100 to 200 mg/l) became unconscious after about 35 min (175). Specht et al. examined the female guinea pigs that succumbed to acetone exposures ranging from 10,000 to 50,000 ppm (23.7 to 118.7 mg/l) and found the following gross abnormalities in varying degrees: pulmonary congestion and edema, splenic congestion and hemorrhage, renal congestion, and glomerular distension (176).

Animal inhalation studies have shown that the narcotic effects of acetone depend upon both the length and magnitude of the exposure. The acute data summarized in Table 5.8 suggest that vapor concentrations in excess of 8000 ppm are required to elicit any sign of acetone intoxication, regardless of the exposure duration. These data also indicate that the degree of CNS involvement can intensify with increase in either the exposure concentration or the exposure duration. The narcotic effects of acetone appear to proceed through several distinct phases that can be described as follows: drowsiness, incoordination, loss of autonomic reflexes, unconsciousness, respiratory failure, and death. Using rats, Haggard et al. demonstrated that the onset of narcosis was closely associated with the acetone levels in the blood and that many of the stages could be observed by manipulating either the exposure concentration or the exposure duration (78). The authors found that blood levels of 1000–2000 mg/l were necessary for slight incoordination, 2910–3150 mg/l for loss of the righting reflex, and 9100–9300 mg/l for respiratory failure.

Bruckner and Petersen examined the anesthetic properties of acetone in rats and mice by determining the exposure concentration necessary to cause narcosis (179). Groups of male Sprague-Dawley rats were exposed for 3 hr to acetone concentrations of 12,600, 19,000, 25,300, and 50,600 ppm. The degree of narcosis was measured at regular intervals during and after the exposure by performing five tests (wire maneuverability, visual pacing, grip strength, tail pinch, and righting reflex) that measured unconditioned performance and involuntary reflex. Each animal was scored between 0 and 8 on each of these tests, and the individual results were averaged to obtain a mean performance score. Performance scores showed a dose-related decline at all but the highest acetone exposure level. Animals exposed to 50,600 ppm of acetone died within 2 hr of initiating the exposure. The performance score for the group of rats exposed to 19,000 ppm of acetone returned to the preexposure level after 9 hr, whereas the group exposed to 25,300 ppm of acetone required 21 hr before its performance score returned to base-line levels.

Goldberg et al. examined the avoidance and escape behavior of female Carworth rats exposed to acetone vapors for 10 days at 4 hr/day (180). Groups of animals were trained to avoid (conditioned response) or escape (unconditioned response)

Table 5.8. Acute Toxicity of Acetone to Laboratory Animals Following an Inhalation Exposure

Species	Exposure Concentration		Duration (hr)	Observed Effects	Reference
	mg/l	ppm			
Mice	20	8,440	4.0–7.75	Loss of righting reflex	178
	20	8,440	7.75	Narcosis in some animals	
	48	20,256	1.5	Deep narcosis	
	48	20,256	1.0–1.2	Loss of righting reflex	
	110	46,420	0.6–1.0	Deep narcosis	
	110	46,420	1.0	Lethal	
Rats	5	2,100	8.0	None	78
	10	4,220	8.0	None	
	25	10,550	1.7–4.2	Incoordination	
	50	21,100	2.2–2.7	Loss righting reflex	
	100	42,200	1.75–1.9	Loss corneal reflex	
	100	42,200	4.5–5.5	Respiratory failure	
	200	84,400	2.5–3.0	Respiratory failure	
	200	84,400	0.35–0.83	Loss of corneal and righting reflexes	
	300	126,600	1.75–2.25	Respiratory failure	
	300	126,600	0.17–0.42	Loss of corneal and righting reflexes	
Cats	40	16,880	3.75–4.0	Loss of righting reflex	129
	114	48,108	1.5	Loss of righting reflex	
	178	75,116	1.0–1.25	Deep narcosis	
Cats	8–10	3,375–4,220	5.0	Eye and nose irritation	177
	20–50	8,440–21,100	3.0–4.0	Drowsiness and stupor	
	80–100	33,760–42,200	4.0	Narcosis with convulsions	
	125	52,750	1.5	Narcosis with convulsions	
Guinea pigs	23.7	10,000	47–48	Some lethality	176
	47.4	20,000	9.0	Narcosis	
	118.5	50,000	3.0–4.0	Lethal	

177

a shock stimulus by climbing a pole situated in a test chamber with an electric grid on the floor. Each animal was evaluated before and after daily acetone exposures of 3000, 6000, 12,000, and 16,000 ppm. The 3000-ppm exposures were without effect on all exposure days, the 6000-ppm exposure inhibited the avoidance but not the escape response, and the two highest exposures inhibited both responses. Normal responses were obtained after three days of exposure to 6000 and 12,000 ppm, which indicated adaptation and tolerance developed on repeated exposure to acetone vapors.

Geller et al. reported that acetone caused minimal effects in juvenile baboons taught to perform a complex operant discrimination task (181). A group of male baboons was exposed continuously (24 hr/day) to 500 ppm acetone for 7 days. The percentage of correct and incorrect responses was recorded along with the time necessary to respond precisely to a stimulus-induced discrimination task that resulted in a food reward when performed correctly. The acetone exposure caused no change in the number of correct responses, a highly variable change in the number of extra incorrect responses, and a consistent increase in the response time relative to control values. The authors did not measure blood or urine acetone levels; however, the uninterrupted exposure undoubtedly resulted in an extremely high acetone body burden. The same group of authors examined the operant behavior of rats exposed to much lower levels of acetone (182). Acetone exposures were performed at 150 ppm for times ranging from 30 min to 4 hr. Male Sprague-Dawley rats were trained to respond on a fixed-ratio (FR) or a fixed-interval (FI) schedule of reinforcement (food reward) following an auditory stimulus. The results were highly variable with the 30-min exposure causing no effect, the 1-hr exposure causing an increase in FR and FI values, the 2-hr exposure resulting in a decrease in both values, and the 4-hr exposure causing inconsistent changes among the animals. The authors ascribed the variable results to differences in the rates of elimination among the animals and to adaptive changes.

Garcia et al. studied the lever-pressing activity of rats operantly trained on a FR schedule of reinforcement (183). Up to five animals were exposed to acetone at concentrations ranging from 25 to 100 ppm for 3 hr. Base-line response rates were measured 2 days prior to the exposure and compared to values obtained both during and following the treatment. The results were extremely variable with the preexposure base-line values varying over a 7-fold range. Most of the animals were reportedly unaffected by the acetone exposure; however, several showed either a relative increase or decrease in their response rate for up to 5 days post-exposure.

De Ceaurriz et al. subjected mice to a behavioral despair swimming test following a 4-hr acetone exposure ranging from about 2000 to 3000 ppm (184). Male Swiss mice were placed in a container of water and the lag time between water contact and the initiation of swimming behavior was measured relative to a control group. A nominal concentration of 2000 ppm caused no change in swimming behavior, whereas concentrations ranging from about 2600 to 3000 ppm caused the swimming lag time to decrease up to 59 percent. The authors used these data to calculate an ID_{50} value, which was defined as the acetone concentration necessary to cause a 50 percent decrease in the time that the mice remained immobile in the water. The

ID_{50} value for acetone was determined to be 2800 ppm, which was higher than the values for six other ketonic solvents.

Glowa and Dews examined the effects of acetone on schedule-controlled operant behavior in mice (185). Male CD-1 mice were each serially exposed to six nomimal concentrations of acetone ranging from 100 to 56,000 ppm. The mice were trained to seek a food reward when two feeding stimuli (blinking lights and white noise) were presented in a specific manner. The animals received the food reward on a fixed-interval schedule when their noses tripped a photocell located behind the feeding hole. The response rates were measured during the last 10 min of each 40-min exposure session and compared to the total number of correct responses obtained during a control session. The response rate in mice was not affected by acetone concentrations less than 1000 ppm, whereas 56,000 ppm completely abolished the response. The response rate returned to the base-line level 30 min after terminating the 56,000 ppm exposure. The response data were used to determine an EC_{50} value, which was defined as the acetone concentration needed to decrease the response rate to half the base-line level. The EC_{50} value for acetone was calculated to be 10,964 ppm, which was higher than the value obtained with four other solvents.

Acute studies in mice have indicated that acetone is an extremely weak sensory irritant. Kane et al. developed a short-term test to predict acceptable levels of human exposure to airborne sensory irritants (186). Previous studies had shown that sensory irritation of the eyes, nose, and throat was caused by stimulation of the same unspecialized trigeminal nerve endings that caused a reflexive decline in respiration rate following exposure to irritant vapors or gases (187). Thus the maximum decrease in respiration rate, relative to background, provided an accurate physiological measure of trigeminal nerve irritation. The concentration-related decline in respiration allowed for the circulation of an RD_{50} value, which corresponded to the irritant concentration needed to produce a 50 percent decrease in the initial respiration rate. By examining a large number of chemicals, the RD_{50} value in mice was shown to be empirically related to the irritation threshold in humans, and the procedure was offered as a reliable means of estimating permissible exposure limits (188). The interspecies multiplier judged to provide an adequate margin of safety and to be suitable for calculating an acceptable 8-hr TLV–TWA was 0.03.

Two separate groups of investigators have examined acetone in the short-term irritancy bioassay described above. Kane et al. used male Swiss-Webster mice and a 10-min exposure period to examine the irritation potential of 11 solvents, including acetone (189). Of the solvents tested, acetone was the least potent with an RD_{50} value of 77,516 ppm. Acetone was also found to be one of only two chemicals that showed rapid and complete tolerance during the exposure session. Using a slightly different method, De Ceaurriz et al. studied the irritancy of 22 solvents in male Swiss OF_1 mice exposed for 5 min (190). As before, acetone was the least irritating of the solvents; however, the RD_{50} value of 23,480 ppm was appreciably lower than the value obtained by Kane et al. The discrepancy between the two values has been ascribed to either differences in mouse strain sensitivity or to variations

Table 5.9. Acute Lethality of Acetone to Sprague-Dawley Rats at Different
Stages of Maturity (198)

Age Group	Weight Range (g)	LD$_{50}$ Value ml/kg	LD$_{50}$ Value g/kg	95% Confidence Limits ml/kg	95% Confidence Limits g/kg
Newborn (24–48 hrs)	5–8	2.2	1.8	1.7–3.8	1.4–3.0
Immature (14 days)	16–50	5.6	4.5	3.9–8.0	3.1–6.4
Young adult	80–160	9.1	7.3	6.8–12.1	5.4–9.7
Older adult	300–470	8.5	6.8	7.8–9.3	6.2–7.4

in the test protocol (191). De Ceaurriz et al. failed to provide the confidence interval information and time-response profiles needed for a detailed comparison of the two studies. A comparison of nearly 40 chemical vapors and gases using the protocol of Kane et al. showed that acetone was the weakest irritant to be tested (192). The mechanism for acetone-induced irritation of the trigeminal nerve has been suggested to involve a loose physical adsorption of the solvent vapors at the nerve receptor site, rather than a tight chemical interaction with the receptor protein (193). Correlations have been established between the potency of nonreactive irritants, such as acetone, and certain physicochemical properties of the solvents such as the boiling point, molecular weight, and gas–liquid partition coefficient (194–196).

7.1.2 Oral Route

Single-dose oral lethality studies have been performed in mice, rats, and rabbits. Smyth et al. reported a 14-day oral LD$_{50}$ of 10.7 ml/kg (8.5 g/kg) and 95 percent confidence limits of 7.7 to 15.0 ml/kg (6.2 to 11.9 g/kg) for female Carworth-Wistar rats (174). Using the same test conditions and test species, Pozzani et al. found the LD$_{50}$ value to be 12.6 ml/kg (9.8 g/kg) and the 95 percent confidence limits to be 10.6 to 14.9 ml/kg (8.5 to 11.9 g/kg) (173). Clothier et al. reported an oral LD$_{50}$ value of 168 mmol/kg (9.8 g/kg) for an unspecified strain and sex of rat (197).

Kimura et al. reported oral LD$_{50}$ values for male and female Sprague-Dawley rats at different stages of maturity (198). The results, presented in Table 5.9, show that acetone was more lethal for newborn rats than for rats 14 days of age and older. Krasavage et al. reported an oral LD$_{50}$ value of 5.3 g/kg for an unstated sex and strain of rabbit and an LD$_{50}$ value between 4 and 8 g/kg for an unstated sex and strain of mouse (199). Using male ddY mice, Tanii et al. found the oral LD$_{50}$ of acetone to be 90.39 mmol/kg (5.25 g/kg) with 95 percent confidence limits of 61.68 to 132.5 mmol/kg (3.58 to 7.70 g/kg) (200). Pretreatment of the mice with an intraperitoneal dose of CCl$_4$ (6.4 ml/kg) 24 hr prior to acetone administration caused the LD$_{50}$ value to decrease slightly to 73.35 mmol/kg (4.26 g/kg). Kennedy and Graepel have reviewed much of the acute lethality data for rats and concluded that acetone was of low oral toxicity (201).

Unlike studies by the parenteral routes, the oral route has not frequently been used to study acetone-induced narcosis. Albertoni reported that a minimum oral dose of 4 g/kg of acetone would cause narcosis in dogs, whereas 8 g/kg would cause death (125). Similarly, Walton et al. reported the minimum narcotic and lethal doses of acetone to be 7 and 10 ml/kg (5.6 and 8.0 g/kg) for rabbits (202). Panson and Winek examined the effects of acetone aspiration in anesthesized Sprague-Dawley rats of each sex (203). Aspiration toxicity was determined from the degree of lung hemorrhage and the lung weight/body weight ratio in rats that succumbed to treatment. Despite problems caused by instantaneous volatilization and inhalation of acetone vapor, the authors concluded that acetone was hazardous when aspirated and that induced vomiting and gastric lavage were contraindicated in cases of human ingestion.

7.1.3 Ocular and Dermal Routes

Studies conducted in rabbits have shown that acetone can be a severe eye irritant when applied undiluted and left in contact with the cornea for an extended length of time. Ocular damage and irritancy was much less severe when the exposures involved brief contact or diluted aqueous solutions of acetone. Carpenter and Smyth developed a ranking system for eye irritants based on the total number of points scored 24 hr after treatment with a chemical (204). Points were accumulated according to the degree of corneal opacity, necrosis, and iritis observed in the treated eye, with the maximum total score being 20 points. The authors found that the application of 5 ml of undiluted acetone to the cornea of rabbits resulted in an average point score of 5 or less, whereas the application of 20 ml produced a total score of 5 points or higher. When these values were ranked together with those of more than 175 different chemicals, acetone was judged to cause a grade 5 injury (scale of 1 to 10), which was considered to be the minimum grade for a severe eye irritant. Larson et al. developed a method to measure the relative capacity of chemicals to cause ocular edema (205). The method determined the chemical concentration necessary to cause a 50 percent increase in the fluid content of the eye after 1 hr. Using a series of aqueous acetone solutions, the authors found that an acetone concentration of 3.9 M (225 mg/ml) caused the requisite amount of damage. Of the 32 solvents tested, only two were less toxic than acetone.

Turss et al. examined whether undiluted acetone could be used as a solvent to remove cyanoacrylate adhesive from the cornea of rabbits (206). Gross and histopathological examinations of ocular tissue were performed following application of acetone to the eyes of rabbits that had an artificial corneal injury repaired with the adhesive. Several minutes of acetone contact was found to destroy the corneal epithelium; however, the stroma was not permanently affected, even when acetone was applied directly to this tissue layer. All corneal injury was reversible within 4 to 6 days, and adhesive removal using acetone was found to be far less traumatic than mechanical scraping.

Kennah et al. compared the Draize irritation method for scoring ocular injury to an ocular swelling technique that entailed corneal thickness measurements (207). Ocular irritation (Draize) and swelling (corneal thickness) were determined in rabbits following treatment with aqueous acetone solutions ranging from 1 to 100 percent in strength. A maximum Draize score of 66 (severely irritating) was obtained 24 hr post-treatment with undiluted acetone. Corneal swelling was severe, increasing approximately two-fold relative to control. The diluted acetone solutions (1, 3, 10, and 30 percent) caused minimal injury in both procedures. A similar technique termed corneal pachymetry was developed by Morgan et al. to measure ocular injury in rabbits as a function of corneal thickness (208). The authors compared the irritation index, using a modified Draize scoring procedure, with the thickness measurements obtained by corneal pachymetry. When the eyes were evaluated 3 days after treatment, undiluted acetone was rated as a corrosive eye irritant by the Draize criteria and as a severe irritant using corneal thickness measurements. Bolková and Čejková examined the enzymatic activity in rabbit corneal tissue that was briefly treated with undiluted acetone (209). Alkaline and acid phosphatase activities were determined biochemically and histochemically in both the corneal epithelium and the stroma. Beginning on day 7 post-treatment, the regenerating corneal epithelial tissue showed a large increase in alkaline phosphatase activity that lasted for at least 3 weeks. The acid phosphatase activity in the corneal epithelium was minimally affected by the treatment, and the decreased enzyme activities observed in the corneal stroma were present only on the first day following treatment. The authors noted that the acetone-induced damage was generally limited to the epithelium and that no permanent damage occurred to the cornea.

The ability of acetone to dehydrate and defat unprotected skin is well known from industrial and laboratory experience. Animal studies have confirmed this observation and demonstrated that acetone is neither a skin irritant nor a contact allergen. Smyth et al. reported that the dermal LD_{50} for male New Zealand rabbits was greater than 20 ml/kg (174). In addition, the authors found that 10 ml of undiluted acetone applied to the depilated skin of five rabbits did not result in any sign of irritation within 24 hr of application. Anderson et al. reported that acetone was nonirritating when dermally applied to guinea pigs (210). Skin irritation was measured on three scales: visual erythema and edema, inflammatory cell count in histology specimens, and epidermal thickness in biopsy specimens. Similarly, Descotes reported that acetone did not cause contact hypersensitivity in male or female Balb/c mice (211). The allergic potential of acetone was determined from the amount of ear swelling that occurred 24 hr after topical challenge (the sensitizing dose of acetone was applied 9 days earlier).

Grubauer et al. examined the effects of acetone treatment on the barrier function of hairless mouse skin (212, 213). The rate of transcutaneous water loss was measured along with the amount of lipid removed when undiluted acetone was applied directly to the skin. The rate of water loss through the skin was shown to be linearly related to the amount of lipid removed by the acetone treatment. Perturbations in the barrier function of the skin were primarily related to the amount of sphingolipids and free sterols removed from the stratum corneum by the acetone treatment. If the mouse skin remained exposed to the air following the acetone treatment, the barrier function of skin returned to normal after 2 days. In contrast, when the treatment site was occluded with a water-impermeable membrane, the recovery was prevented. Nearly complete recovery was also obtained when the treatment site was occluded with a water-permeable membrane. Additional biochemical and histochemical studies allowed the authors to conclude that transcutaneous water loss served as a stimulus for lipid resynthesis following acetone-induced delipidation of the skin.

7.1.4 Secondary Parenteral Routes

The secondary parenteral routes of administration used most frequently in acute studies of acetone toxicity were intramuscular (im), intravenous (iv), intraperitoneal (ip), and subcutaneous (sc) injection. The acetone LD_{50} values obtained by the parenteral routes are not substantially different from values obtained by oral and inhalation exposure. A 5-g/kg sc dose of acetone has been reported to be the minimum lethal amount for both dogs and guinea pigs (125, 178). A 4-ml/kg (3.2 g/kg) iv dose of acetone was found to be immediately lethal to rabbits, and a dose of 2 ml/kg (1.6 g/kg) caused delayed mortality in several of the animals (202). The same authors found the maximum tolerated and minimum lethal iv doses of acetone for rats to be 5 ml/kg (4.0 g/kg) and 6 to 8 ml/kg (4.8 to 6.4 g/kg), respectively. Sanderson estimated the average lethal ip dose of acetone to be 0.5 g/kg for female Wistar rats (214), whereas Clothier et al. reported an ip LD_{50} value of 22 mmol/kg (1.3 g/kg) for an unspecified strain and sex of mouse (197). Zakhari reported an ip LD_{50} value of 3.1 g/kg (standard deviation 0.42 g/kg) for male CF-1 mice (215). DiVincenzo and Krasavage found that two of four guinea pigs given a 3.0-g/kg ip dose of acetone died and that all animals survived a 1.5-g/kg dose (216). Histopathology and serum values of ornithine carbamyl transferase showed no evidence of liver damage other than slight lipid accumulation in the rats that succumbed to treatment.

Parenteral routes of administration have been commonly used to examine the effects of acetone on respiration, circulation, and the CNS. Early interest in the physiological effects of acetone was inspired by two unrelated observations, the anesthetic potential of acetone and the high blood levels of acetone that accompanied a diabetic coma. The studies summarized in Table 5.10 describe the clinical signs that developed following the iv and im administration of acetone. Saliant and Kleitman studied the circulatory and respiratory effects of acetone in great detail (220). Acetone was administered intravenously to anesthetized cats and dogs at doses ranging from 125 to 500 mg/kg. The treatment caused an immediate 15 to 40 percent drop in blood pressure in both species. The blood pressure decline was followed by either respiratory stimulation or respiratory depression depending upon the magnitude of the dose, the concentration of acetone in the dose solution, the rate of administration, and the number of treatments administered. The authors concluded that acetone was not a potent toxicant; however, its toxicity could be increased by repetitive injections over a short period of time.

Schlomovitz and Seybold administered acetone by iv infusion to four species of laboratory animals and calculated the blood concentration that caused death (221).

Table 5.10. Acute Toxicity of Acetone to Laboratory Animals Following Parenteral Exposure

Species	Exposure Route	Approximate Dose (g/kg)	Treatment Regimen	Observed Effect	Reference
Dog	iv	0.47	Single bolus dose	Listlessness, weakness, incoordination, recovery	217
	iv	2.1	Single 15-min infusion	Excitation, listlessness, weakness, atoxia, recovery	217
	iv	0.29–0.36	Four injections over 7-hr period	Vomiting, slight weakness, listlessness, recovery	217
	iv	1.62	Single bolus dose	Prostration, incoordination, labored breathing, recovery	217
	iv	2.55	Two injections 1 hr apart	Prostration, coma, convulsions, recovery	217
	iv	0.38–0.76	Six injections over 4-hr period	Uneasiness, atoxia, weakness, convulsions, dyspnea, cauce, recovery	217
Rabbit	iv	0.80	Single 60-min infusion	CNS and respiratory depresson, recovery	202
	im	4.0	Single 40-min infusion	Paralysis at injection site, CNS and respiratory depression, recovery	202
	iv	3.2	Single 50-min infusion	Incoordination, drowsiness, slightly increased respiration, recovery	202
	im	2.4	Single 70-min infusion	No effect	202
	iv	0.5	Single 40-min infusion	No effect	218
Dog	iv	1.14, 0.38	Two injections 10 min apart	Unsteadiness, drowsiness, dyspnea, recovery	219
	iv	0.62–3.11	Seven injections at 1-hr intervals	Excitement, dyspnea, prostration, coma, death	219
	sc	0.76	Five injections over 1-hr period	Excitement, dyspnea, prostration, coma, death	219

Continuous iv administration of acetone to dogs at rates ranging from 0.045 to 1.19 ml/(min)(kg) was found to cause death in less than an hour. Respiratory depression, irregular heart beat, and a drop in blood pressure commonly preceded death. Lethal blood levels in dogs ranged from 450 to 11,900 mg/l, with the average value being about 3000 mg/l. The average lethal blood levels of acetone were 42,000, 46,000, and 72,300 mg/l for guinea pigs, cats, and rabbits, respectively. The authors concluded that respiratory paralysis was the primary cause of death when the rate of increase in blood acetone exceeded 30,000 mg/(l) (min). There was no distinct cause of death when the rate of increase was lower than 1000 mg/(l) (min).

7.2 In Vitro Studies

In vitro studies have been useful for examining the subtle physiological changes that accompany brief solvent contact. Because of the weak and nonspecific effects observed at high doses, acetone has not been extensively studied using in vitro methodologies.

7.2.1 Isolated Tissues and Organs

Many of the studies with isolated tissues and organs have focused on the decrement in cardiac rhythm or cardiovascular function following acetone treatment. Saliant and Kleitman found that the perfusion of isolated frog and turtle hearts with a 1 percent acetone solution caused no change in the amplitude or frequency of the contractions (220). A 5 percent solution of acetone caused cardiac amplitude to decrease initially in the isolated frog heart, but a rapid tolerance occurred during the 15 to 25 min perfusion. Acetone concentrations of 10 percent caused immediate cardiac arrest; however, a gradual recovery occurred when the perfusion time was extended up to 8 min. The isolated turtle heart was generally more resistant to the effects of acetone than the frog heart, and no cumulative injury was observed to occur on repeated perfusion. Sklianskaya et al. performed similar experiments with isolated frog hearts that were perfused for 10 to 12 min with acetone concentrations ranging from 2.5 to 30 ml/l (2.0 to 23.0 g/l) (222). Cardiac activity (contraction amplitude multiplied by contraction frequency) was found to decrease 15 to 20 percent with acetone concentrations ranging from 2.5 to 5.0 ml/l (2.0 to 4.0 g/l), whereas concentrations ranging from 10 to 30 ml/l (8.0 to 23.0 g/l) caused a much greater decline in cardiac activity ranging from 22 to 42 percent.

Raje et al. used isolated and perfused rabbit hearts to determine what effect acetone concentrations of 0.5 and 1.0 percent had on contractile force, contraction rate, and cardiac flow (223). A dose-related decrease in the contraction force was observed together with an increase in the contraction frequency. A 70 to 90 percent increase in cardiac flow was observed, which was attributed to dilation of the coronary vessels. Chentanez et al. used the isolated atrium portion of rat hearts to examine the effect of acetone on the contraction rate and the release of norepinephrine from cardiac cells (224). Acetone concentrations ranging from 10 to 210 mM (0.58 to 12.2 g/l) caused a dose-related increase in atrial contraction, but at concentrations of 300 and 400 mM (17.4 and 23.2 g/l) the effect began to diminish. The increase in atrial contraction was associated with the release of norepinephrine by parasympathetic nerve terminals in the atrium. The authors hypothesized that the higher acetone concentrations may have caused some damage to the muscle fibers, thereby making them insensitive to the effects of the norepinephrine release.

Several groups of investigators have examined the efficacy of acetone as a treatment vehicle for dissolving and applying polar test chemicals in percutaneous absorption experiments. Hinz et al. examined whether the in vitro treatment of skin from a hairless mouse with acetone altered the permeability characteristics for other chemicals (225). Skin sections were placed in flow-through glass diffusion cells and treated with 50 ml of acetone prior to assessing the damage with tritiated water (^3H_2O). The rate of ^3H_2O penetration was measured after both single and multiple applications of acetone. Regardless of the test conditions, the acetone pretreatment did not cause any significant change in ^3H_2O permeation relative to untreated control values. Because human skin was known to be less permeable than mouse skin, the authors concluded that acetone mediation of percutaneous absorption was not likely to occur on human skin.

King and Monteiro-Riviere examined the effect of acetone on skin viability, vascular response, and epidermal morphology using an isolated perfused porcine skin flap model (226). Application of acetone to the unoccluded skin surface did not alter glucose utilization, lactate production, or lactate dehydrogenase (LDH) activity. Likewise, arterial pressure and flow rate were unaffected. The authors concluded that acetone could be safely used as a vehicle in percutaneous absorption and cutaneous toxicity studies.

7.2.2 Cell Cultures

The use of cell cultures has been primarily aimed at devising alternative in vitro techniques for determining the acute lethality and ocular toxicity of acetone together with other common chemicals. Shopsis and Sathe developed an in vitro bioassay that could substitute for the Draize ocular irritation test typically performed in rabbits (227). The bioassay assessed the chemical concentration necessary to cause a 50 percent inhibition (UI_{50} value) in the normal uptake of tritiated uridine by Balb/c 3T3 cells grown in culture. The acetone UI_{50} value was determined to be 546 mM (31.7 g/l), which the authors felt showed a good correlation with the mild to moderate ocular irritation potential of acetone. Another in vitro bioassay, developed by Clothier et al., was proposed as a substitute for the standard LD_{50} determination in rodents (197). The assay used mouse 3T3-L1 cells grown in culture and determined the chemical concentration necessary to reduce the protein content of the cells by 50 percent (ID_{50} value). Of the 60 chemicals examined in the bioassay, only two were found to have an ID_{50} value greater than that of acetone, which had a value of 617 mM (35.8 g/l). Cultures of human keratinocytes formed the basis of an in vitro assay, developed by Hoh et al., to evaluate the skin and eye irritation potential of chemicals (228). The concentration of test chemical needed to cause a 50 percent decrease (NR_{50} value) in cell viability was determined by measuring the decrease in dye uptake by the cells. Acetone was examined in this assay along with 13 other chemicals and ranked sixth in potency with an NR_{50} value of 0.38 M (22 g/l). The authors reported that the assay performed better with severe irritants than with weaker types such as acetone.

Story et al. examined the cytotoxic effects of acetone in isolated hepatocytes from Sprague-Dawley rats (229). The incubation of hepatocytes with 10 mM (0.58 g/l) of acetone for up to 5 hr at 37°C did not affect glutamate oxaloacetate transaminase (AST) or LDH release, cell viability, urea synthesis capability, or steady-state ATP levels. Ebert et al. studied whether acetone could stimulate hemoglobin synthesis and porphyrin production in three different cell cultures (230). Incubation of murine erythroleukemia cells, tetraploid erythroleukemia cells, and T3Cl2 cells with 330 mM (19.2 g/l) acetone for 5 days caused a marked stimulation in hemoglobin synthesis. Because this response could be triggered by a relatively large variety of solvents, the authors concluded that the stimulation of hemoglobin synthesis in malignant cell cultures was not sensitive to any particular chemical moiety.

8 CHEMICAL INTERACTIONS

Acetone has the ability to induce the enzymatic activity of a specific constitutive cytochrome P450 isozyme that plays an important role in the metabolism of endogenous and exogenous substrates. A considerable amount of information is available on the physiological function and toxicologic consequence of P450 induction by acetone. Because acetone is metabolized by the same P450 isozyme that is induced following high-dose administration (see below), the auto-inductive increase in cytochrome P450 levels provides a mechanism for increasing the elimination of acetone when high body burdens develop. The pretreatment of laboratory animals with acetone can also potentiate or antagonize the acute effects of known systemic toxicants that are metabolized by the induced P450 isozyme. The potentiation observed following acetone administration has involved treatments with known nephrotoxins and hepatotoxins; the potentiating effects were generally quantitative in nature and involved an increase in the extent of damage without altering the types of tissues or organs affected. Minimally effective dose thresholds have been shown to exist for the induction of cytochrome P450 by acetone.

8.1 Cytochrome P450 Induction and Enhancement

Acetone administration has been shown to induce a specific cytochrome P450 protein that has enzymatic activity toward a wide variety of chemical substrates. Cytochrome P450 IIE1, known also as P450j, P450$_{ac}$, or P450 3a, is a constitutive isozyme in the P450 subfamily that can be induced several fold following high-dose treatments with acetone. Acetone appears to induce the P450 IIE1 activity in rat and rabbit liver by post-transcriptional stabilization of the enzyme protein against autophagosomal and lysosomal degradation (231–235). In addition to acetone, a wide variety of chemical exposures and physiological conditions have been shown to induce P450 IIE1 activity in rat and rabbit liver. These include, among others, ethanol (236), MEK (237), isoniazid (238), 95 percent oxygen (239), trichloro-ethylene (240), high fat/low carbohydrate diets (241), diabetes (242), and fasting (243). P450 IIE1 induction by ethanol, fasting, and diabetes has been related to their own inherent activity and not to secondary acetone production (244–246). Chromium(VI) and disulfiram administration along with surgical hypophysectomy have been shown partially to block P450 IIE1 activity in rat hepatic microsomes (247–249). Acetone-induced P450 IIE1 activity has been detected in the lung, liver, and kidney of rabbits and hamsters (250–252). Likewise, chicken and mouse liver, rabbit bone marrow and olfactory mucosa, and rat Kupffer cells have all been shown to possess acetone-inducible P450 IIE1 activity (253–257). The acetone-induced P450 IIE1 from rabbit and rat liver has been found to be immunochemically and catalytically similar to the constitutive P450 IIE1 from human liver (258, 259).

The induction of hepatic P450 IIE1 by acetone typically requires repeated administration either by gavage or via the drinking water. The treatment for P450 IIE1 induction has generally been a 1 to 10 percent solution of acetone in the drinking water for at least 7 days; however, enzyme induction has also been observed following four intragastric doses of acetone at concentrations ranging from 1 to 3 g/kg (5 ml/kg of a 20 to 33 percent solution). Under these conditions, a two- to sixfold elevation in hepatic P450 IIE1 has been observed 18 to 24 hr after the final treatment (233, 240, 249, 251). The minimum drinking water regimen needed to induce hepatic P450 IIE1 activity in Syrian golden hamsters was 0.5 percent acetone for 7 days (249), whereas the minimum treatment needed to induce hepatic P450 IIE1 activity in Sprague-Dawley rats was a single oral dose of 0.25 ml/kg (0.20 g/kg) of acetone (260). High acetone doses have also been shown to sharply induce cytochrome P450 IIB1 (phenobarbital-induced form) and mildly induce glutathione transferse (231, 232, 261) in rat liver.

Cytochrome P450 IIE1 has been shown to have a wide range of enzymatic activities toward a diverse group of substrates. Hepatic microsomes from acetone-treated rodents have shown a two- to ten-fold increase in aniline hydroxylase (262, 263), p-nitrophenol hydroxylase (264, 265), diethylnitrosamine N-deethylase (266, 267), dimethylnitrosamine N-demethylase (268, 269), and ethoxycoumarin O-deethylase activity (270, 271). In addition, acetone-induced microsomes from rat and rabbit liver sharply increased the in vitro hydroxylation of benzene and phenol (272, 273). The induction of P450 IIE1 by acetone has been shown to alter the in vivo metabolism of ethanol and ethyl carbamate in rats (274, 275). Likewise, hepatic microsomes from acetone-treated rats have increased the in vitro metabolism of n-propanol and n-butanol (276), azoxymethane and methylazoxymethanol (277), acetaminophen (278, 279), acetonitrile (280), diethyl ether (281), glycerol (282), and methyl t-butyl ether (283).

The in vitro addition of acetone to hepatic microsomal preparations from rats, mice, rabbits, and dogs was capable of enhancing aniline hydroxylation by two- to fourfold (284–286). The mechanism for acetone-induced catalytic enhancement has not been clearly elucidated; however, the increase in activity was not associated with P450 IIE1 induction. Acetone enhancement of acetanilide and N-butylaniline hydroxylation have also been observed (278), together with the O-dealkylation of p-anisidine and p-phenetidine (287). The in vitro enhancement of aniline hydroxylation has been ascribed by some to be a nonspecific solvent effect involving membrane alterations because acetone shares this trait with several common solvents, including ethanol, isopropanol, n-butanol, and tetrahydrofuran (288, 289). Most researchers, however, have related acetone enhancement of aniline hydroxylase activity to a specific configurational or conformational change in cytochrome P450 following binding at a specific site on the protein (290–293).

8.2 Antagonism

Acetone pretreatments have been shown to antagonize the adverse effects of several xenobiotics in animals. Jenney and Pfeiffer have shown acetone to be effective in preventing the seizures and convulsions that accompany treatment with hydrazides (294). The treatment of Harlan mice with a single 168-mg/kg ip dose of semicarbazide resulted in the rapid onset of convulsions, seizures, and death in 91 to 98 percent of the animals. The oral administration of a 4.0 g/kg dose of acetone prior to ip treatment with semicarbazide prevented seizures and death in all mice. Acetone was the only ketone examined that was capable of modifying the effects of semicarbazide, which indicated that the mechanism was not due to the reactivity of semicarbazide toward carbonyl groups. Additional studies by Kohli et al. revealed that acetone could prevent the seizures caused by either electroshock treatment or the administration of isoniazid (295). An oral acetone dose of 490 mg/kg completely protected rats of both sexes from the clonic convulsions and death caused by a single 300-mg/kg ip dose of isoniazid. Other ketonic solvents, such as cyclohexanone and MEK, were not able to provide the same degree of protection as acetone. Further studies with convulsive electroshock treatment showed that acetone could also protect against physically induced convulsions. The acetone ED_{50} value for preventing electroshock-induced convulsions was 220 mg/kg with 95 percent confidence limits of 134 to 360 mg/kg.

Acetone has been shown to decrease the incidence and severity of liver necrosis in male Long-Evans rats treated with acetaminophen. Price and Jollow showed that rats pretreated with 2.5 ml/kg of acetone exhibited appreciably less liver necrosis at each of the acetaminophen doses examined (296). The dose of acetaminophen required to produced liver necrosis in 50 percent of the animals (TD_{50}) increased from 590 to 910 mg/kg following acetone pretreatment. The authors provided metabolic and pharmacokinetic data to suggest that the mechanism involved a decline in the rate of reactive mercapturate formation. A similar antagonistic effect was observed in streptozotocin-induced diabetic rats; however, the mechanism was dissimilar and could not be related to secondary acetone production (297). These in vivo studies with acetaminophen contrast with the in vitro studies of Moldéus and Gergely, who reported that acetone potentiated the effects of acetaminophen (298). Hepatocyte cultures from Sprague-Dawley rats pretreated with phenobarbital rapidly lost viability when 100 mmol acetone and 5 mmol acetominophen were added to the incubation mixture. In contrast to the in vivo study, there was a dramatic increase in the rate of metabolite production in the mercapturate pathway. The changes in cell viability and metabolite formation were not observed with hepatocytes from male NMRI mice. No reasonable explanation was offered for the apparent ability of acetone to potentiate acetaminophen toxicity in vitro and antagonize the hepatotoxicity in vivo.

Lo et al. recently showed that acetone could reduce the nephrotoxicity of N-(3,5-dichlorophenyl)succinimide (NDPS) in rats (299). Male Fischer rats were orally administered 10 mmol/kg (0.6 g/kg) of acetone 16 hr before an ip dose of NDPS. The acetone pretreatment mildly reduced the NDPS-induced increases in BUN and kidney weight; however, the changes in urine volume and renal morphology were unaffected. Acetone doses less than 10 mmol/kg were ineffective and did not antagonize NDPS-induced renal damage.

8.3 Potentiation

Acetone pretreatment has repeatedly been shown to potentiate halogenated solvent hepatotoxicity and nephrotoxicity. Acetone shares this ability with several other polar solvents, including ethanol, isopropanol, and 1,3-butanediol (300–302). The ability of these solvents to potentiate halocarbon-induced organ damage has been related to the induction of microsomal enzymes that metabolize halogenated chemicals to reactive intermediates (303–306). Past research has generally established that the mechanism responsible for the potentiation of halocarbon toxicity in rats involves the induction of cytochrome P450 IIE1 by the polar solvents (307–309).

8.3.1 Halogenated Hydrocarbons

Traiger and Plaa first demonstrated that the oral pretreatment of male Sprague-Dawley rats with 1.0 ml/kg (0.8 g/kg) of acetone could result in the development of severe hepatocellular damage following a single ip injection of carbon tetrachloride (CCl_4) (300, 310). Large increases in serum ALT activity, liver glucose-6-phosphatase activity, and triglyceride content were observed 24 hr after treatment with the haloalkane. No liver damage was observed in the absence of an acetone pretreatment. Further studies with pyrazole, an inhibitor of alcohol dehydrogenase, and the cytochrome P450 inhibitor aminotriazole established that the metabolism of isopropanol to acetone was only partially responsible for the potentiating effects of the alcohol (311). The authors concluded that unmetabolized isopropanol possessed some ability to potentiate CCl_4 hepatoxicity. Further studies by the same group of authors demonstrated that the hepatotoxicity of other halogenated solvents could be potentiated by acetone pretreatment of mice (301). The oral administration of 2.5 ml/kg (2.0 g/kg) of acetone to male Swiss-Webster mice 18 hr before an ip injection of chloroform or 1,1,2-trichloroethane exacerbated the hepatocellular toxicity of these solvents. Little or no potentiation was observed when 1,1,1-trichloroethane or perchloroethylene were administered following acetone pretreatment. The authors concluded that acetone pretreatment was not capable of substantially increasing the degree of toxicity of weak hepatotoxicants.

In the only well-documented instance of potentiation of halocarbon toxicity in humans by polar solvents, Folland et al. documented four cases of renal or hepatic damage in workers exposed to CCl_4 in an isopropanol bottling plant (312). The mishap occurred when CCl_4 was substituted for acetone in an equipment cleaning operation. Rags were soaked with CCl_4 from open buckets and used for about 2 hr by two employees. Within 48 hr of exposure, 18 of the 43 employees in the plant became ill with the following signs and symptoms: nausea, vomiting, headache, weakness, and abdominal pain. Four severely affected employees showed clinical signs of liver or kidney damage (elevated serum BUN, bilirubin, or AST) and required hospitalization and supportive treatment before recovery occurred. Follow-up monitoring at the work site revealed an average airborne isopropanol concentration of 410 ppm in the vicinity of the cleaning operation. The mean alveolar air concentrations of isopropanol and acetone for employees working near the site were 100 ppm and 19 ppm, respectively. The authors concluded that both these solvents may have had a role in the potentiation of CCl_4 toxicity. Deng et al. cited a recent episode of CCl_4 intoxication in a printing plant where isopropanol was also in use; however, the authors failed to provide sufficient evidence to support their claim of CCl_4 potentiation by isopropanol (313).

Hewitt et al. demonstrated that ketonic solvents other than acetone were capable of potentiating haloalkane toxicity (314). Male Sprague-Dawley rats were given equimolar oral dosages of either acetone, methyl n-butyl ketone, or 2,5-hexanedione 18 hr prior to a single ip injection of chloroform ($CHCl_3$). Acetone pretreatment was found to cause a large increase in plasma ACT (32-fold) and OCT (132-fold) activities without any concomitant rise in either the liver-to-body-weight ratio or total bilirubin concentration. Marked centrilobular necrosis with degenerated hepatocytes and balloon cells was observed on histological analysis of liver specimens. Pretreatment with acetone also caused an increase in the kidney-to-body weight ratio and an elevation in the uptake of p-aminohippurate (PAH) and tetraethylammonium ion (TEA) in excised renal cortical slices. The authors concluded that methyl n-butyl ketone and 2,5-hexanedione were more effective than acetone in potentiating $CHCl_3$-induced hepatotoxicity and nephrotoxicity.

Hewitt et al. examined the relationship between carbon chain length and the capacity to potentiate $CHCl_3$ hepatoxicity in a homologous series of ketones (315). Five ketonic solvents, acetone, 2-butanone (MEK), 2-pentanone (MPK), 2-hexanone (MnBK), and 2-heptanone (MnAK), were administered to male Sprague-Dawley rats 18 hr before an ip dose of $CHCl_3$. A positive correlation was observed between the number of carbon atoms in the solvent and the increase in plasma ALT and OCT activities. Although the correlation was not strong, the data suggested that acetone was the weakest potentiator of $CHCl_3$ hepatotoxicity. Brown and Hewitt investigated the dose–response and structure–activity relationships for ketone solvent-induced potentiation of $CHCl_3$ hepatotoxicity and nephrotoxicity (316). A challenge dose of $CHCl_3$ was administered 18 hr after oral pretreatment of male Fisher 344 rats with 1.0–15.0 mmol/kg of the following five ketonic solvents: acetone, 2-butanone, 2-pentanone, 2-hexanone, and 2-heptanone. Acetone doses of 5.0 mmol/kg (290 mg/kg) or greater caused statistically significant increases in plasma creatinine and ALT activity; however, the degree of damage showed no apparent dose–response relationship. Acetone pretreatment also caused a decrease in PAH uptake by kidney slices and an increase in the severity of renal tubular necrosis when the dosages exceeded 1.0 mmol/kg (58 mg/kg). The results indicated that pretreatment with 1.0 mmol/kg of acetone was not effective in potentiating $CHCl_3$ nephrotoxicity or hepatotoxicity. In contrast to the previously described study, the authors concluded that there was no correlation between the degree of $CHCl_3$ potentiation and length of the carbon chain in the ketonic solvent.

Hewitt et al. also studied the relationship between the pretreatment time delay and the severity of $CHCl_3$-induced hepatobiliary dysfunction (317). Male Sprague-Dawley rats were given a 15 mmol/kg (870 mg/kg) oral dose of acetone 10 to 90 hr before an oral dose of $CHCl_3$. The administration of acetone 18 to 24 hr prior to $CHCl_3$ treatment was found to cause the greatest increase in plasma bilirubin and the greatest decrease in bile flow rate. There was no potentiation of the hepatobiliary damage when the time interval between pretreatment and $CHCl_3$ administration was greater than 24 hr. The potentiation of hepatobiliary dysfunction by acetone was noted to be less severe than that caused by either MnBK or MEK. Hewitt et al. also studied the activation mechanism responsible for the potentiation of $CHCl_3$ hepatotoxicity (318). The studies were designed to determine if increased mixed-function oxidase (MFO)-mediated metabolism or decreased glutathione-mediated detoxification was responsible for the potentiating effects of acetone. Male Sprague-Dawley rats were administered an oral dose of $CHCl_3$ at various times following an oral dose of acetone. Liver microsomal fractions from acetone-treated animals contained slightly higher cytochrome-c reductase and ethoxycoumarin O-deethylase activities; however, cytochrome P450 levels, aminopyrine N-demethylase activity, and aniline hydroxylase activity were unaffected. The increases in microsomal metabolism paralleled the relative potency for liver damage, with acetone showing the least effect on microsomal function. Acetone pretreatment did significantly affect hepatic glutathione levels relative to controls. The authors suggested that the potentiating effects of acetone were mostly attributable to increased rates of $CHCl_3$ metabolism by the monooxygenase system; however, a specific mechanism was not proposed.

Plaa et al. reported that a threshold or minimally effective dosage (MED) of acetone was necessary for the potentiation of CCl_4 liver toxicity (319). Dose-related increases in serum ALT and OCT were observed when male Sprague-Dawley rats were orally administered graded doses of acetone ranging from 0.025 to 2.5 ml/kg (20 mg/kg to 2.0 g/kg) before an ip dose of CCl_4. The MED of acetone was determined to be 0.25 ml/kg (200 mg/kg) and the noneffective dose (NED) was 0.10 ml/kg (80 mg/kg). Oral pretreatment with acetone twice a day for 3 days at the NED did not lead to any hepatocellular potentiation; however, when a similar treatment schedule was used with the MED, the degree of potentiation was greater than that observed after a single dose at the same level. The authors determined that the pretreatment dosing schedule could decidedly influence the potentiating effects of acetone. The effect of the pretreatment dosing schedule on the degree of potentiation was further examined by administering a total acetone dose of 1.5

ml/kg (1.2 g/kg) at four different rates: 1.5 ml/kg orally in a single bolus dose, 0.25 ml/kg every 12 hr, 0.125 ml/kg every 6 hr, or a continuous iv infusion at 0.0208 ml/(kg)(hr). As shown in Table 5.11, a single oral dose of acetone elicited the greatest effect on CCL_4 hepatotoxicity, with the graded dosages producing far less potentiation and the iv infusion causing no potentiation. Pharmacokinetic studies under the same dosing conditions revealed that acetone potentiation was correlated with the peak blood level of acetone and not with the area under the blood acetone concentration-versus-time curve. The data indicated that acetone blood levels in excess of 300 mg/l were necessary to induce potentiation. The authors concluded that a definite threshold existed in the rat for the potentiating effects of acetone on CCl_4-induced liver damage, regardless of the exposure regimen.

Charbonneau et al. conducted a detailed examination of the dose–response relationship for acetone potentiation under various treatment conditions (320). Using male Sprague-Dawley rats, the authors assessed what influence the exposure regimen, dosing vehicle, and route of administration had on the acetone dosage necessary for CCl_4 potentiation. Preliminary experiments established that the 4-hr inhalation NED and MED for acetone potentiation of CCl_4 hepatotoxicity were 1000 and 2500 ppm, respectively. In addition, the inhalation NED and MED were shown to be independent of the number of exposure sessions because plasma ALT and bilirubin measurements were similar for animals receiving 1 versus 10 daily exposures. A large range of acetone exposures were performed both orally and by inhalation using multiples of the MED. The results indicated that when corn oil was used as a dosing vehicle the absorption rate was slower and the peak blood levels of acetone were lower than the values obtained when water was used as the treatment vehicle. As shown in Table 5.12, acetone blood levels below approximately 100 mg/l were not associated with any potentiation of CCl_4 hepatotoxicity regardless of the vehicle or exposure route; however, the degree of potentiation at comparable blood levels appeared to be greater when acetone was administered orally rather than by inhalation. This finding was ascribed to the liver first-pass effect that occurred following oral, but not inhalation, exposure. The relationship between acetone blood levels and the degree of potentiation was also observed when the exposure concentration was held constant (5000 ppm) and the exposure duration increased in 1-hr increments. A detailed pharmacokinetic analysis of the acetone concentration-versus-time curves revealed that the degree of potentiation was very closely related to the area under that portion of the blood concentration-versus-time curve that was above the threshold blood level. These findings explained why a divided dose of acetone was not as effective as a single large dose in potentiating CCl_4 hepatotoxicity.

The ability of acetone to potentiate the hepatotoxicity of weak hepatotoxins has been investigated by several groups of investigators. MacDonald et al. found that the oral pretreatment of male Sprague-Dawley rats with acetone had a biphasic effect on the hepatocellular damage caused by the ip administration of 1,1,2-trichloroethane (321). Acetone was administered at dosages ranging from 0.1 to 3.5 ml/kg (0.08 to 2.8 g/kg) and the potentiating effects were characterized by an increase in plasma ALT, a decrease in liver glutathione content, and the histological

Table 5.11. Effects of Different Acetone Pretreatment Regimens in Carbon Tetrachloride Hepatotoxicity in Male Sprague-Dawley Rats (319)

Acetone Dose (ml/kg)	Treatment Interval (hr)	Number of Treatments	Plasma ALT Activity (units/ml)		Plasma OCT Activity (units/ml)		Plasma Bilirubin Concn. (mg/dl)	
			Mean	Std. Dev.	Mean	Std. Dev.	Mean	Std. Dev.
1.5	—	1	2588	347	1626	241	0.40	0.09
0.25	12	6	247	48	143	79	0.16	0.01
0.125	6	12	144	28	58	16	0.12	0.02
0.0208[a]	72	continuous	65	9	14	5	0.03	0.01

[a]IV infusion rate in mg/(kg)(hr).

Table 5.12. Acetone Blood Levels and Clinical Chemistry Values Associated with Acetone-Induced Potentiation of Carbon Tetrachloride Hepatotoxicity in Male Sprague-Dawley Rats (320)

Exposure Route	Acetone Exposure Level (ppm or ml/kg)	Peak Acetone Blood Concn. (mg/l)		Plasma Bilirubin Concn. (mg/dl)		Plasma GPT Activity (units/ml)	
		Mean	Std. Dev.	Mean	Std. Dev.	Mean	Std. Dev.
Inhalation	1000	91	3	0.33	0.01	44	1
	2500	312	9	0.32	0.01	206	91
	5000	727	68	0.57	0.03	748	93
	10,000	2114	266	0.46	0.07	1061	443
	15,000	3263	220	1.02	0.19	2144	231
	Air	0	0	0.27	0.02	33	3
Oral							
Corn oil	0.10	45	3	0.32	0.04	70	9
	0.25	130	17	0.32	0.04	216	40
	0.50	266	41	0.36	0.18	404	103
	1.00	580	32	0.63	0.07	1134	369
	1.50	730	37	0.94	0.24	1459	355
	Corn oil	0	0	0.25	0.01	60	10
Water	0.10	74	4	0.33	0.01	46	3
	0.25	156	6	0.33	0.02	103	3
	0.50	385	24	0.44	0.02	242	43
	1.00	697	29	0.56	0.03	808	174
	1.50	1010	97	0.63	0.10	1088	241
	Water	0	0	0.32	0.01	54	9

appearance of centrilobular coagulative necrosis. Unlike the dose-related increase seen with CCl_4, acetone doses greater than 0.5 ml/kg did not cause a further increase in 1,1,2-trichloroethane hepatotoxicity, but appeared to protect the animals against the liver damage caused by a large dose of the latter halocarbon. A similar biphasic response was observed by Hewitt and Plaa when investigating the potentiation of 1,1-dichloroethylene toxicity by acetone (322). Male Sprague-Dawley rats were given an oral dose of 1,1-dichloroethylene after treatment with acetone at five doses ranging from 1 to 30 mmol/kg (0.06 to 1.74 g/kg). A mild increase in plasma ALT and OCT activities and total bilirubin levels was observed when acetone doses of 5 or 10 mmol/kg (0.3 and 0.6 g/kg) were administered; however, little if any potentiation was observed when the two highest acetone doses were given. In further studies, Hewitt et al. examined the degree of acetone-induced potentiation after treatment with two weakly hepatotoxic halocarbons (323). Pretreatment of male Sprague-Dawley rats with a 15 mmol/kg (0.9 g/kg) oral dose of acetone before the oral administration of bromodichloromethane and dibromochloromethane resulted in a significant potentiation of liver but not kidney toxicity. Plasma ALT and OCT activity and the liver-to-body-weight ratios were each increased with halocarbon doses greater than 0.25 ml/kg. A comparison with $CHCl_3$ hepatotoxicity revealed that the degree of potentiation was greater with bromodichloromethane.

Brondeau et al. also observed a biphasic interaction in rats and mice exposed to acetone and 1,2-dichlorobenzene (DCB) vapors (324). The animals were exposed to three vapor concentrations of acetone before being exposed to DCB vapor concentrations of about 270 ppm (mice) or 380 ppm (rats). Depending upon the exposure concentrations employed, a 4-hr inhalation pretreatment of male Sprague-Dawley rats or OF1 mice with acetone was found to potentiate or antagonize 1,2-dichlorobenzene (DCB) hepatotoxicity. The studies were performed in rats preexposed to acetone vapors at concentrations of 4785, 10,670, or 14,790 ppm and in mice exposed at 6747, 8910, or 14,345 ppm. Liver damage and metabolic capacity were assessed in rats by measuring serum glutamate dehydrogenase (GLDH) activity, liver cytochrome P450 content, and liver glutathione S-transferase (GST) activity. As shown in Table 5.13, the lowest acetone concentration caused a 12-fold increase in serum GLDH activity relative to control animals that received DCB alone. In contrast, the GLDH activity at the highest acetone concentration was not elevated relative to untreated controls. Liver cytochrome P450 levels and GST activity were uniformly elevated by about 50 to 100 percent in rats preexposed to acetone. Parallel pretreatment of rats and mice with MEK, MiBK, or cyclohexanone revealed that the potentiating capacity of acetone was distinctly unlike the other ketones. The authors attributed the unusual effects of acetone pretreatment to the induction of a detoxification pathway at high acetone concentrations or to the saturation of a metabolic pathway that acetone and DCB shared in common.

Charbonneau et al. investigated the ability of acetone to potentiate the hepatotoxicity of a mixture of CCl_4 and trichloroethylene (TCE) (325). Dose-response relationships were examined in male Sprague-Dawley rats administered an oral dose of acetone in a corn oil vehicle at levels of 0.25, 0.75, and 1.5 ml/kg (0.2, 0.6,

Table 5.13. Effects of Acetone Preexposures on Liver Function and Liver Metabolism in 1,2-Dichlorobenzene-Treated Sprague Dawley Rats (324)

Preexposure Concentration (ppm)	Serum GLDH Activity (units/l)		Liver GST Activity [nmol/(mg)(min)]		Liver Cyt. P450 Concn. (nmol/mg)	
	Mean	Std. Error	Mean	Std. Error	Mean	Std. Error
4785	1046	476*	96	2.6*	0.77	0.02*
10,670	19.4	6.4*	86	2.1*	0.64	0.02*
14,790	3.6	0.6*	88	5.7*	0.73	0.01*
control	87.9	26.4	66	1.6	0.43	0.01

*Significantly different from the control administered DCB alone ($p < .05$).

1.2 g/kg). The pretreatment was followed with the ip administration of either TCE, CCl_4, or a TCE–CCl_4 combination. Regardless of the acetone dose, the pretreatment had no effect on either ALT or bilirubin levels when the rats were administered TCE. The animals administered CCl_4 following acetone pretreatment showed large increases in either plasma ALT activity or bilirubin levels, depending on the dose of CCL_4 administered. Acetone pretreatment caused a marked dose-related potentiation of the liver damage caused by the TCE–CCl_4 combinations. The authors concluded that the potentiating effects of acetone were most severe when the 0.75 ml/kg dose was administered and that synergistic effects could occur with the administration of halocarbon combinations. In follow-up studies, Charbonneau et al. examined the dose–response relationship for the potentiation by acetone of the liver toxicity induced by a TCE–CCl_4 mixture (326). The acetone MED for potentiating the liver damage caused by TCE, CCl_4, or a TCE–CCl_4 mixture was determined by pretreating male Sprague-Dawley rats at each of five oral dose levels ranging from 0.05 to 0.25 ml/kg (0.04 to 0.2 g/kg). Regardless of the dose level, acetone pretreatment did not elicit any hepatotoxicity following TCE exposure. Using plasma ALT as the most sensitive indicator of liver damage, the acetone MED for potentiation of CCl_4 hepatotoxicity was 0.25 ml/kg, and the MED for potentiation of the TCE–CCl_4 mixture was at least 0.05 ml/kg. Because the animals pretreated with a corn oil vehicle also showed a large increase in hepatotoxicity following TCE–CCl_4 treatment, it was not possible to determine an NED for acetone. The relative contributions of TCE and CCl_4 to the decrease in the MED was investigated by altering the dosage of each component used in the mixture. Three combinations of TCE and CCl_4 were given intraperitoneally following an oral acetone dosage of 0.75 ml/kg. The plasma ALT levels were increased uniformly for all three mixtures when the animals were pretreated with either acetone or the corn oil vehicle. In contrast, the mixtures containing a relatively higher concentration of TCE were found to cause a greater increase in plasma bilirubin levels.

Charbonneau et al. have continued their investigation into the effects of acetone pretreatment on the hepatotoxicity of halocarbon mixtures and have recently reported on the results obtained with 28 different binary mixtures (327). Male Sprague-Dawley rats received a 0.75 ml/kg (0.6 g/kg) dose of acetone in corn oil prior to oral treatment with all possible binary combinations of CCl_4, $CHCl_3$, TCE, 1,1-dichloroethylene, tetrachloroethylene, 1,1,1-trichloroethane, 1,1,2-trichloroethane, and 1,1,2,2-tetrachloroethane. The halocarbons were administered individually following pretreatment with either acetone or corn oil in order to determine whether the observed effects of the mixtures on plasma ALT activity were less than additive (infra-additive), additive, or greater than additive (supra-additive). Following the corn oil pretreatment, 26 of the 28 mixtures showed an additive effect and two mixtures were supraadditive. In contrast, with acetone pretreatment 10 mixtures showed infra-additive effects, 17 showed additivity, and one showed supra-additivity. The only supra-additive effect that followed acetone pretreatment was with a TCE–CCl_4 combination. The mechanism responsible for the apparent protection afforded by the acetone pretreatment on the hepatotoxicity of the solvent mixtures was not investigated.

Table 5.14. Toxic Interactions Between Acetone and Other Solvents When Co-Administered by the Inhalation and Oral Routes to Female Carsworth Nelson Rats (173)

Paired Solvent	Inhaln. LC_{50} Value (mg/l)		Oral LD_{50} Value (ml/kg)	
	Predicted	Observed	Predicted	Observed
Hexane	100.7	123.4	—	—
Dioxane	61.2	61.6	8.28	7.13
Ethyl acetate	53.0	68.1	11.7	12.9
Carbon tetrachloride	50.8	61.2	4.78	3.73
Acetonitrile	39.7	14.6	9.99	2.75
Toluene	30.4	37.3	10.59	7.96
Propylene oxide	14.0	20.5	1.04	2.38
Epichlorohydrin	2.46	1.84	0.32	0.26

8.3.2 Other Chemicals

Acetone has been shown to affect the toxicity of chemicals other than the halogenated hydrocarbons. Although some attention has been given to the effects of acetone on barbiturate-induced narcosis (328, 329), thiobenzamide hepatotoxicity (330), and cisplatin nephrotoxicity (331), the vast majority of research has focused on the chemical interactions observed following ethanol administration (332–334). The toxicity of acetonitrile, styrene, N-nitrosodimethylamine, and 2,5-hexanedione has, however, been closely examined following acetone pretreatment of rats or mice.

The ability of acetone to affect acetonitrile (ACN) toxicity was first discovered in a series of experiments that examined the interactions that occurred when groups of organic solvents were co-administered. Pozzani et al. administered various binary solvent mixtures to female Carworth-Nelson rats and determined the 4-hr inhalation LC_{50} and oral LD_{50} values (173). The results obtained with individual solvents were used to identify potential additive responses with nine different solvents that were mixed in various combinations and ratios to obtain the solvent pairs. Table 5.14 shows the experimentally observed results and those predicted from a strictly additive relationship for the eight solvent pairs that included acetone. The acetone–ACN mixture was the only mixture that displayed LC_{50} and LD_{50} values that were significantly less than the predicted result, which strongly suggested a synergistic relationship between the two solvents. Smyth et al. expanded upon these findings and examined the oral LD_{50} values for 26 different solvent pairs that included acetone as one of the components (335). The solvents were mixed in equal volumes and administered undiluted to female albino rats. As before, the observed LD_{50} results were compared to the values obtained when the individual solvents were assumed to behave in an additive manner. The only acetone-containing mixture that showed a potential synergistic effect was the acetone–ACN pair, which gave an observed LD_{50} value that was nearly fourfold lower than the predicted result.

The mechanistic basis for the synergistic relationship between acetone and ACN was investigated by Freeman and Hayes (336). Using female Sprague-Dawley rats, the authors monitored the development of toxic signs when acetone and ACN were co-administered. The appearance of severe signs (tremors and convulsions) of toxicity was delayed from 3 hr when pure ACN was administered to 32 to 36 hr when 1.96 g/kg acetone was co-administered. Compared to ACN alone, the oral administration of a 25 percent solution of an ACN–acetone mixture delayed the time when peak cyanide concentrations appeared in the blood, with the values increasing from 9 to 15 hr when ACN was given to 39 to 48 hr when the solvents were co-administered. In addition, the amount of ACN metabolized to cyanide was increased by the acetone co-treatment. Interestingly, the authors found that the administration of a second oral dose of acetone to the rats just prior to the development of toxicity was capable of providing complete protection against ACN toxicity. The authors concluded that the acetone co-treatment caused a delayed potentiation of ACN toxicity by initially inhibiting cyanide formation and later stimulating the degree of ACN metabolism to cyanide. Subsequent in vitro studies by these same authors confirmed that acetone initially inhibited, then induced a specific liver microsomal cytochrome P450 isozyme responsible for the metabolism of ACN to cyanide (337). The ability of acetone to potentiate ACN toxicity was recently demonstrated by Boggild et al., who described a human fatality that followed the co-ingestion of acetone and ACN (338). The case involved a young woman who ingested a mixture of acetone and ACN with suicidal intent. The patient's physical and clinical appearance was unremarkable for about 18 hr and then rapidly deteriorated with convulsions, acidosis, and cardiovascular depression. The patient died 30 hr after ingestion of the mixture, and the authors attributed the delayed development of ACN toxicity to the acetone that was co-consumed.

Elovaara et al. reported that acetone pretreatment could potentiate the lung damage caused by inhaled styrene (339). Acetone was administered to a group of male Wistar rats for 1 week in their drinking water at a concentration of 1 percent, and the rats were then exposed to 2100 mg/m^3 (488 ppm) of styrene for 24 hr. Compared to a group of rats that were exposed only to styrene, the acetone-pretreated animals did not show any greater effects on liver glutathione levels or relative lung and liver organ weights; however, the pretreated rats did show a larger decrease in lung glutathione content, a greater urinary excretion of glutathione reaction products (i.e., thioether compounds), and a larger decrease in lung P450-dependent MFO activities. The authors concluded that acetone pretreatment had potentiated the lung toxicity of styrene. Further studies by Elovaara et al. established that acetone pretreatment could induce the in vivo metabolism of styrene to mandelic and phenylglyoxylic acids in rats and that the isoenzyme responsible for the metabolic activation of styrene was cytochrome P450 IIE1 (340). These findings were in agreement with a human inhalation study that did not show any pharmacokinetic interactions when 70 ppm of styrene and 523 ppm of acetone were co-administered for 2 hr (341).

The effects of acetone pretreatment on N-nitrosodimethylamine (NDMA) toxicity have been examined in two separate studies. Lorr et al. reported that NDMA hepatoxicity could be potentiated by a single oral dose of acetone (342). Male

Sprague-Dawley rats given a 2.5 ml/kg (2.0 g/kg) dose of acetone prior to NDMA showed a twofold elevation in plasma ALT and an increase in the severity of the centrilobular necrosis observed histologically. The authors concluded that the acetone pretreatment induced the cytochrome P450-dependent NDMA-demethylase activity in the liver and that this reaction was responsible for the formation of a toxic intermediate from NDMA. Haag and Sipes examined the mutagenicity of NDMA in the Ames assay using the liver microsomal fraction from mice treated ip with 3 ml/kg (2.4 g/kg) acetone (343). When compared to untreated controls, the microsomes from pretreated C57B1/6J mice were capable of dramatically increasing the number of revertant colonies found following NDMA addition. The metabolic activation of NDMA was again ascribed to the induction of a high-affinity dimethylnitrosamine N-demethylase in the microsomes of acetone-treated mice.

A synergistic relationship has been demonstrated in rats that were co-treated with acetone and 2,5-hexanedione (HD) for 6 weeks. Ladefoged et al. maintained groups of male Wistar rats on drinking water that contained either HD, acetone, or a combination of the two chemicals at concentrations of 0.5 percent (344). Caudal motor nerve conduction velocities were measured weekly during the last 4 weeks of treatment, and the balance time on a rotating rod was recorded weekly for all 6 weeks of treatment. The body weight gain and water consumption was significantly decreased in the animals co-treated with acetone and HD. HD and the HD–acetone mixture also caused a significant reduction in the rotarod balance time and the nerve conduction velocities. The difference in balance time between HD-treated rats and those given HD and acetone was statistically significant during the last three treatment weeks. The difference in conduction velocities for these two groups was significant during the fourth and sixth weeks of treatment. The authors concluded that acetone co-treatment appeared to potentiate the neurophysiological and neurobehavioral effects of HD; however, the authors were not certain whether acetone caused a purely neurotoxic interaction or interfered with the metabolism and elimination of HD. Previous studies by Ladefoged and Perbellini had shown that body clearance of HD was retarded in New Zealand white rabbits that were administered an iv dose of HD and acetone (345).

9 METABOLISM

The rates and routes of acetone metabolism have been extensively examined in both humans and laboratory animals. Although the scientific research in this area is somewhat confounded by several questionable reports, a critical examination of the literature provides a clear picture of the metabolic fate of acetone.

Acetone is one of three "ketone bodies" that arise from the production of acetyl coenzyme A within the liver. The enzyme systems catalyzing the formation of acetone reside primarily within the mitochondria of hepatocytes and to a much smaller extent in renal cells. Because acetone is nonionic and miscible with water, the chemical is capable of passively diffusing across cell membranes and distributing throughout the body fluids. The other two ketone bodies, acetoacetate and β-hydroxybutyrate, are organic acids that can cause metabolic acidosis when produced in large amounts. Acetone, in turn, is derived from both the spontaneous and catalytic breakdown of acetoacetate. Endogenous and exogenous acetone is eliminated from the body either by excretion into the urine and exhaled air or by enzymatic metabolism. Under normal circumstances, metabolism is the predominant route of elimination; however, metabolic saturation can result in disproportionate changes in acetone excretion. Under saturation conditions, the elimination half-life can greatly increase, and the secondary excretion pathways can become the primary route of elimination for acetone.

The first step in the enzymatic biotransformation of endogenous acetone is the cytochrome P450-dependent oxidation of acetone to acetol by acetone monooxygenase. The acetol is then biotransformed by either of two pathways, an extrahepatic propanediol pathway or an intrahepatic methylglyoxal pathway. The oxidation of acetol to methylglyoxal is also cytochrome P450-dependent and is catalyzed by acetol monooxygenase. Recent in vivo and in vitro studies have conclusively shown that the acetone and acetol monooxygenase activity in rat and rabbit liver is specifically associated with cytochrome P450 IIE1 (346, 347). Consequently, the induction of P450 IIE1 activity seen after the administration of large acetone doses provides a mechanism whereby acetone can reflexively increase its elimination when high body burdens are encountered. The oxidation of acetone by acetone monooxygenase can be saturated at physiologically and environmentally attainable blood levels; therefore the inductive effects of acetone provide a mechanism for amplifying metabolism when high acetone body burdens develop. Recent studies indicate that a third metabolic pathway leading to acetate and formate can also be recruited to assist in the elimination of acetone when the acetone monooxygenase pathway is overloaded.

Once formed or taken up in the body, acetone can be eliminated by either metabolic breakdown or direct excretion. Under normal conditions, the primary route of elimination from the body is through metabolic breakdown by either an intrahepatic or extrahepatic pathway. Excretion of unmetabolized acetone by direct exchange at the lung or kidney is an important route of removal when the metabolic capacity of the body has been overwhelmed. The metabolic pathways for acetone have been extensively studied in both humans and laboratory animals. These species possess metabolic enzymes for breaking down acetone into substances that can be incorporated into the carbon pool of the body. The monooxygenase enzymes that initiate the conversion of acetone to D-lactate are specific for acetone and are located primarily within the liver. The formation of D-lactate from acetone provides the body with a mechanism for recovering a portion of the energy that is lost when acetone is formed from acetoacetate. Acetone is generally considered to be gluconeogenic, because glucose has consistently been shown to be an end product of acetone metabolism. In addition, basic amino acids, proteins, glycogen, cholesterol, and other biochemicals have been shown to contain carbon atoms derived from the breakdown of acetone. Metabolism studies with acetone have shown it to be an important biochemical constituent of the human body that can be efficiently

molecules of acetyl-CoA by the enzyme acetoacetyl-CoA thiolase to yield aceto-acetyl-CoA (349). A synthase-mediated reaction then catalyzes the addition of a third acetyl-CoA molecule to yield β-hydroxymethyl-β-methylglutaryl-CoA (HMG CoA). This later reaction is considered to be the rate limiting step governing the mitochondrial production of acetone. HMG CoA is next cleaved by a lyase to produce acetoacetate and a molecule of the starting material, acetyl-CoA. The acetoacetate can be reversibly converted to β-hydroxybutyrate by the enzyme β-hydroxybutyrate dehydrogenase. These two ketone bodies can diffuse out of the cell and be circulated to peripheral tissues where they can serve as a source of energy by being reconverted into acetyl-CoA units. In addition, excess production of these anionic compounds can result in metabolic ketoacidosis when the buffering capacity of the body is exceeded. The terminal step in this reaction sequence leads to the formation of acetone as a result of both the spontaneous and catalytic decarboxylation of acetoacetate.

There has been considerable confusion regarding the last step in the endogenous formation of acetone. Many researchers have regarded endogenous acetone to be formed spontaneously from acetoacetate (348, 350); however, Gavino et al. have shown that proteins can catalyze the decarboxylation of acetoacetate to acetone to a large degree (351). Other researchers have argued that a specific enzyme, ace-toacetate decarboxylase, catalyzes this reaction in mammals and that enzymatic activity can be detected in the albumin fraction of human blood as well as in rat liver and kidney (352, 353). The enzyme was shown to have a low substrate affinity with a Michaelis constant, K_m, between 0.1 and 0.4 M. The level of enzymatic activity has been reported to be higher in the placenta and fetus of pregnant rats fasted for 2 days (354). Some researchers have pointed out that acetoacetate de-carboxylase may be an important regulator of metabolic pH in those circumstances where excess acetoacetate is being produced (354, 355). Kimura et al. recently showed that acetoacetate decarboxylase activity was associated with albumin and a wide variety of amino acids (74). the authors argued that the K_m (0.7 M) for the reaction was far too high for a true enzyme and that blood proteins nonspecifically catalyzed the carboxylation of acetoacetate to acetone.

Unusual elevations in blood acetone have been observed in laboratory animals treated with certain halogenated hydrocarbons. This condition was first reported by Filser et al. after exposing male Wistar rats to the vapors of the following four halogenated ethylenes: cis-1,2-dichloroethylene (1100 ppm), trans-1,2-dichloro-ethylene (900 ppm), tetrachloroethylene (1200 ppm), and vinylidene fluoride (500 ppm) (356). The animals were exposed in a closed recirculating inhalation chamber for a period of 25 hr at the initial vapor concentrations noted above. After a lag period of about 5 hr, there was a linear increase in the acetone concentration within the chamber. The rate of acetone production and exhalation was found to be relatively similar for all four compounds, ranging from 1.2 to 2.5 mmol/(hr)(kg). In contrast, control values for nontreated animals were 0.05 mmol/(hr)(kg) or less. Because the test compounds contained only two carbons atoms, the exhaled acetone was not considered to be a metabolite of the halocarbons. Furthermore, the exhaled acetone was not deemed to be a product of in vivo lipid peroxidation because independent studies in Sprague-Dawley rats had previously shown that acetone was not produced and exhaled following the administration of agents known to cause lipid peroxidation in vivo (357).

Further studies by Filser and Bolt suggested that a metabolite, rather than the parent halocarbon, was likely responsible for the high levels of acetone in the exhaled air (358). Five halogenated solvents were examined at an initial vapor concentration that caused metabolic saturation and ensured maximum zero-order production of the active metabolites. The fastest acetone expiration rate was observed with cis-1,2-dichloroethylene (163 mmol/kg in 25 hr) and the slowest with perchloroethylene (47 mmol/kg in 25 hr). Treatment of the animals with phenobarbital and other P450 inducers prior to a vinylidene fluoride exposure caused the acetone formation to increase by about 165 percent. Pretreatment with pyrazole, a potent inhibitor of vinylidene fluoride metabolism, caused a slight decrease in acetone production. Interestingly, pyrazole, a known inducer of cytochrome P450 IIE1, was capable by itself of stimulating acetone production. Because the amount of acetone formed was far greater than the amount of halocarbon vapor absorbed, the authors concluded that acetone was not a halocarbon metabolite and that the increased production depended upon the initial metabolism of the halocarbon to a reactive metabolite.

The metabolic basis for halocarbon-induced acetonemia was investigated in a series of inhalation experiments performed by Filser et al. (359). A large number of halogenated solvents were compared according to their ability to cause acetonemia in rats. Male Wistar rats were exposed to 14 different solvent vapors for 50 hr, during which time the acetone production rate was measured. Halogenated ethylenes generally resulted in greater acetone production than did the halogenated ethanes or methanes. The only solvents not resulting in any appreciable exhalation of acetone were 1,1,1-trichloroethane and the control solvent n-hexane. The results obtained with these two compounds were attributed to their low toxicities and to their relatively nonreactive metabolites. Fluoroacetate and chloroacetate were also examined for their acetone-producing ability because they were considered to be the putative reactive metabolites for several of the halogenated ethylenes, including vinylidene fluoride. Both compounds caused increases in acetone production and excretion; however, the rates of formation were not as great as with vinylidene fluoride. The authors hypothesized that the halocarbons were metabolized to dihaloacetic acids that could cause acetonemia by interfering with the enzymatic synthesis of acetyl-CoA in the Krebs cycle. Bolt et al. showed that metabolic activation was a necessary prerequisite for the elevated production and excretion of acetone following 1,3-butadiene administration (360). When male Sprague-Dawley rats were exposed to 1,3-butadiene, the fall in chamber concentration paralleled the rise in concentrations of acetone and the reactive epoxide, 1,2-epoxybutene-3.

The studies of Simon et al. were aimed at describing the biochemical mechanism responsible for halocarbon-induced acetonemia in rats (361). Using vinylidene fluoride and male Sprague-Dawley rats, the authors demonstrated that appreciable quantities of acetone, acetoacetate, and β-hydroxybutyrate were excreted into the urine during the exposure. These data demonstrated that the synthesis of all three ketone bodies could be affected by the halocarbon exposure and that their progenitor, acetyl-CoA, was likely involved in the process. These studies went on to show that vinyl chloride vapors could reduce cytosolic, but not mitochondrial, coenzyme A activity through the covalent interaction of a reactive metabolite. The results of these experiments suggested that the mechanism responsible for ketone production following halocarbon exposure involved a complex compensatory increase in the rate of fatty acid transport to the mitochondria that ultimately resulted in the excessive production of all three ketones. Attempts by Buchter et al. to use acetone production as a biomonitor for tetrachloroethylene exposure in humans were not successful (362). Although acetone exhalation in the solvent-exposed workers tended to be higher than in the nonexposed controls, high individual variations prevented any meaningful application of the results.

Disulfiram, a therapeutic agent used to treat chronic alcoholics and a potent inhibitor of cytochrome P450 IIE1 activity, has been reported to cause acetonemia in both humans and rats (248). DeMaster and Nagasawa showed that unfasted male Sprague-Dawley rats given daily oral doses of disulfiram had elevated acetone blood levels (363). Blood acetone levels were increased about fivefold (60 mg/l) following a single dose and about 25-fold (310 mg/l) after three doses. Similarly, the authors showed that five alcoholics receiving disulfiram therapy for at least a month showed an average increase of 15-fold (18 mg/l) for acetone in the expired air. These results were later confirmed by Stowell et al. with nonalcoholic volunteers who were given disulfiram on each of 2 days (364). All of the subjects showed a rapid and marked rise in blood acetone that lasted for 3 to 5 days. Disulfiram-induced acetonemia has not been observed in subjects treated subcutaneously with the drug (365). Studies suggest that disulfiram-induced acetonemia may be due to a metabolite rather than the parent chemical. Filser and Bolt observed elevated blood levels of acetone when diethyldithiocarbamate (dithiocarb) was used to inhibit hepatic metabolism (358). This chemical was known to be a major in vivo metabolite of disulfiram, and a single ip injection was observed to cause an immediate increase in acetone exhalation.

In a recent follow-up study, DeMaster and Stevens reported that a single dose of either disulfiram or cyanamide could increase blood acetone levels in rats (366). The blood from male Sprague-Dawley rats given an oral dose of disulfiram or an ip dose of cyanamide showed acetone increases of 16-fold and 10-fold, respectively. Because the increase in blood acetone was not accompanied by any changes in the remaining ketone bodies, the authors attributed the disulfiram-induced acetonemia to the enzymatic inhibition of acetone metabolism by either acetone monooxygenase or aldehyde dehydrogenase. In contrast, the acetonemia observed following cyanamide administration was accompanied by an increase in blood acetoacetate and β-hydroxybutyrate levels; therefore, this chemical was considered to increase the rate of acetone formation by some unknown mechanism.

9.2 Catabolism

9.2.1 Endogenous and Exogenous Acetone

Early studies of acetone metabolism focused primarily on the fate of the carbon atoms and on their eventual incorporation into different types of macromolecules.

Price and Rittenberg orally administered 1.2 mg/kg of carbon-14 labeled acetone ([1,3-^{14}C]acetone) to a female Wistar rat and found 47 percent of the radioactivity in the expired air as radio-labeled carbon dioxide ($^{14}CO_2$) within about 14 hr (367). Further studies in a rat given a daily oral dose of acetone for seven days [7.1 mg/ (kg)(day)] showed that about 7 percent of the administered dose could be recovered in the expired air as [1,3-^{14}C]acetone, with an additional 76 percent appearing as $^{14}CO_2$. Radioactivity could be detected in the following tissue macromolecules: cholesterol, heme, glycogen, fatty acids, urea, tyrosine, leucine, arginine, aspartic acid, and glutamic acid. The authors concluded that the administered acetone was being metabolized to acetate and possibly another two-carbon fragment that were being utilized in the citric acid cycle or as a precursor for amino acid synthesis. Similar results were obtained by Zabin and Bloch in feeding studies with Sprague-Dawley rats (368). Animals given 0.5 mmol/100 g/day (~0.03%) of [1,3-^{14}C]acetone in their diet showed the presence of radio-labeled fatty acids and cholesterol in their livers. Because the percentage of radioactivity in these molecules was nearly identical to the amount obtained when radio-labeled acetate was administered, the authors concluded that acetone was being metabolized to a two-carbon acetyl fragment.

Sakami provided evidence for the biotransformation of acetone to a one-carbon formate metabolite in the rat liver (369). Male rats (strain not specified) were given sc injections of 0.50 to 0.62 mmol/kg (29 to 36 mg/kg) of [1,3-^{14}C]acetone over a 10-hr period. An average of 53 percent of the total dose could be recovered as $^{14}CO_2$ in the expired air 14 hr after treatment. Radio-labeled glycogen, serine, choline, and methionine were isolated from the liver and gastrointestinal (GI) tract following sacrifice. Degradation of the labeled products revealed that the radio-activity in choline and methionine was confined to the carbon atom in the terminal methyl group. Isotopic analysis of the glucose derived from the liver glycogen revealed that the carbon atoms in positions 1 and 6 were labeled to a much higher degree than those in positions 2 and 5. Because this was the labeling pattern observed when radio-labeled formate was administered to rats, the authors concluded that both formate and acetate were possible metabolites of acetone.

Sakami and Lafaye were the first to suggest that pyruvate could be formed as an intermediate during the biotransformation of acetone (370). Male albino rats were orally administered 2.9 mmol/kg (168 mg/kg) of carbonyl-labeled acetone ([2-^{14}C]acetone). A total of 27 percent of the dose was recovered in the expired air as $^{14}CO_2$. Determination of the isotopic distribution in the glucose isolated from liver glycogen revealed that positions 2 and 5 were labeled to a higher degree than positions 1 and 6. This labeling pattern suggested that the glucose derived from acetone metabolism could be the result of oxidation to pyruvate and that two different metabolic pathways could account for the differences in glucose labeling. Sakami and Rudney subsequently proposed a separate carbon cleavage and carbon oxidation pathway for the metabolism of acetone (371).

Rudney demonstrated that acetone could be metabolized to 1,2-propanediol (372). Albino rats of both sex were given a 3.0 mmol/kg (174 mg/kg) sc injection of [2-^{14}C]acetone and the isotopic labeling pattern was determined for the glucose and 1,2-propanediol phosphate isolated from the liver. Carbon atom 2 of 1,2-propanediol was found to be labeled, and carbon atoms 3 and 4 were preferentially labeled in the glucose isolated from the liver glycogen. Additional in vitro studies with [1-^{14}C]1,2-propanediol showed that this compound was capable of being metabolized to both lactate and formate by a rat liver preparation. Previous in vivo studies had shown that [1-^{14}C]1,2-propanediol was capable of labeling serine, choline, and glycogen in much the same manner as [1,3-^{14}C]acetone (373). These studies indicated that 1,2-propanediol was a metabolite of acetone in the rat and that pyruvate was not a likely metabolite based on the carbon labeling pattern of glucose.

Mourkides et al. examined the pyruvate and acetate pathways in greater detail using male Wistar rats that were given a single ip injection of [2-^{14}C]acetone at dose levels ranging from about 10 to 460 mg/kg (374). When the dose was less than 25 mg/kg approximately 25 to 45 percent of the dose could be recovered as $^{14}CO_2$ within 3 hr of treatment; however, at dose levels greater than about 260 mg/kg, the recovery of $^{14}CO_2$ declined to about 2 to 6 percent of the dose, indicting less efficient metabolism at the higher doses. Isolation and purification of alanine, glutamic acid, and aspartic acid from the liver and carcass of the treated animals revealed a carbon labeling pattern that was consistent with the metabolism of acetone to pyruvate and acetate by separate pathways.

Bergman et al. described the distribution of radioactivity in pregnant and non-pregnant guinea pigs administered a single dose of radio-labeled acetone (375). Female guinea pigs (strain not stated) were fasted for 3 days and administered [2-^{14}C]acetone by intracardiac injection at dose levels ranging from about 1 to 210 mg/kg. The cumulative percentage of the dose recovered as $^{14}CO_2$ within 6 hr decreased from about 53 percent at the lowest dose to about 4 percent at the highest. Although the authors ascribed the decline in $^{14}CO_2$ elimination to differences in total body burden, the elimination profiles clearly indicate that the rates of elimination were faster in the low-dose animals and that the metabolism of acetone to $^{14}CO_2$ was saturated at the higher doses. The animals with pregnancy ketosis contained a smaller percentage of the administered radioactivity in their liver lipids and expired $^{14}CO_2$ than the nonpregnant animals. A comparison of the recovery data obtained in guinea pigs treated with the high and low doses of acetone revealed that recovery from tissues and excreta was unrelated to the acetone body burden and that dose-dependent metabolism was responsible.

Lindsay and Brown examined the incorporation of radio-labeled acetone into plasma lactate and glucose from normal and ketotic sheep (376). Normal male and ketotic female (2-day fast during pregnancy) Merino sheep were infused intravenously with small amounts (1.7 to 4.6 mg/min) of [1,3-^{14}C]acetone in order to determine the gluconeogenic utilization of endogenous acetone. By comparing the amount of radioactivity in expired $^{14}CO_2$, plasma lactate, and plasma glucose with the amount found in blood acetone, the rate of incorporation of endogenous acetone into these end products was estimated. The metabolism of endogenous acetone accounted for less than 0.2 percent of the $^{14}CO_2$ being excreted and about 1 to 3 percent of the plasma lactate, regardless of whether normal or ketotic animals were being examined. The percentage of plasma glucose from acetone metabolism averaged about 2.2 percent in ketotic sheep and about 0.5 percent in normal animals. The authors estimated the rate of acetoacetate decarboxylation to be less than 1 mg/min in normal sheep and 2.4 to 6.5 mg/min in ketotic sheep. Luick et al. performed similar glucose incorporation studies in cows that were either spontaneously ketotic or were fasted for 5 days prior to use (377). Ketotic and normal cows were given a tracer amount (exact dose unknown) of [2-^{14}C]acetone by bolus iv injection. The percent incorporation of endogenous acetone into $^{14}CO_2$ was found to be 0.5 to 1.0 percent in ketotic cows and 9 to 16 percent in normal animals. The degree and rate of label incorporation into plasma glucose was found to be several times greater in normal than in ketotic cows. Five constituents were isolated from milk: lactose, casein, fat, albumin, and citrate. The degree of radioactivity incorporation into these milk constituents was 12 to 29 percent for normal cows, and 2.4 to 3.5 percent for ketotic animals. These data indicated that acetone was strongly gluconeogenic in normal cows with relatively large amounts of milk lactose being synthesized from endogenous acetone. The results from this study contrast sharply with those of the previous study in sheep.

Black et al. investigated the metabolic mechanism responsible for the incorporation of acetone into the glucose and amino acids of lactating cows (378). Normal and spontaneously ketotic cows (unspecified type) were given a single bolus iv dose of [2-^{14}C]acetone. The casein and lactose isolated from milk specimens were digested to obtain their constituent amino acids and hexoses. The glucose and galactose from lactose were labeled to the same degree, indicating that the galactose was derived from labeled glucose. The labeling intensity in the amino acids from casein increased in the following order: glycine < serine < aspartic acid < glutamic acid < alanine. The results indicated that acetone was gluconeogenic in the cow and that 40 to 70 percent of the endogenous acetone was metabolized in the citric acid cycle through a common precursor, oxaloacetate. The authors concluded that the utilization of acetone for glucose synthesis was not enhanced in ketotic cows and that the glucose from acetone constituted a small and insignificant portion of the total production.

Coleman found that the synthesis of gluconeogenic precursors from acetone could be induced by fasting or prior acetone administration (379). The studies were performed in two genetic mouse variants (obese and diabetic) along with normal male C57BL mice. All mice were fasted for 1 to 6 days, and then administered 0.55 mmol/kg (32 mg/kg) of [2-^{14}C]acetone by an unspecified route. The two genetic variants were found to convert 20 to 40 percent of the dose into $^{14}CO_2$ within 3 hr, whereas the corresponding range for the normal mice was 15 to 25 percent. Approximately 10 to 20 percent of the acetone was excreted unmetabolized in the expired air, and an average of 1.5 percent was excreted in the urine. The rate of synthesis of liver glycogen from the administered acetone was found to be slightly greater for the two mutant strains than for the normal mice [1.55 versus 0.98 mmol/ (g)(hr)]. Additional in vitro studies using whole liver homogenates from fasted and fed mice revealed that an appreciable amount of lactate was formed upon incubation of the cytosolic fraction with [2-^{14}C]acetone and NADPH. Fasting of the animals of each genotype for 3 to 6 days was found to induce the rate of lactate synthesis by two- to threefold, with the induction being slightly greater in the mutant than the normal mice. Liver homogenates from normal and mutant mice administered 1.0 to 2.5 percent acetone in their drinking water for up to 3 days showed a fivefold increase in the rate of lactate formation. The degree of induction was found to be dose- and time-related in all types of mice, with significantly greater induction observed in the genetic variants than in the normal mice. The authors hypothesized that acetone was capable of inducing its own metabolism to lactate and that the enzyme being affected was likely a cytochrome P450-dependent monooxygenase.

Casazza et al. made the first deliberate attempt at describing the alternate pathways for acetone catabolism in mammals (380). Fasted male Wistar rats were given 1 percent acetone in their drinking water for 3 days and then administered 5 mmol/kg (290 mg/kg) of [2-^{14}C]acetone by the ip route. Blood specimens collected 1 hr after treatment were found to contain ^{14}C-labeled 1,2-propanediol, 2,3-butanediol, and D-lactate. Acetol was formed when the liver microsomal fractions from the treated rats were incubated with acetone. Likewise, methylglyoxal and D-lactate were formed when the microsomes were incubated with acetol under similar conditions. Incubations of either acetone, acetol, or methylglyoxal with intact rat hepatocytes instead of liver homogenates resulted in the formation of D-lactate and glucose. The formation of 1,2-propanediol from acetol was demonstrated following iv infusion and ip injection of acetol. The hepatocytes isolated from rats given acetone in their drinking water were found to be capable of metabolizing 1,2-propanediol to L-lactate and glucose. The authors proposed two alternate pathways for the biotransformation of acetone. Both pathways proceeded through a common oxidized intermediate, acetol, which was then either oxidized to methylglyoxal by the enzyme acetol monooxygenase or, alternatively, reduced to the levorotatory (L) isomer of 1,2-propanediol by an unknown extrahepatic mechanism. Thus the methylglyoxal pathway led to the formation of pyruvate through D-lactate, and the 1,2-propanediol pathway led the formation of pyruvate through L-lactate. The use of metabolic inhibitors suggested that the methylglyoxal pathway was slightly more active than the 1,2-propanediol pathway. The metabolic intermediates in the 1,2-propanediol pathway have been described in detail by Ruddick (381).

Argilés compiled much of the available metabolic information on acetone metabolism and described the alternate pathways that appear to operate in mammals (382). The metabolic scheme presented in Figure 5.3 shows that the acetone formed from the enzymatic and spontaneous breakdown of acetoacetate can be converted to acetol via the microsomal enzyme acetone monooxygenase. The metabolism of acetol can proceed via two separate pathways, the intrahepatic methylglyoxal pathway or the extrahepatic 1,2-propanediol pathway. Liver microsomal enzymes such as acetone monooxygenase often have a limited capacity to metabolize large amounts of substrate. This fact together with the pivotal position of acetol in the reaction sequence suggests that acetone oxidation by cytochrome P450 IIE1 is the rate-limiting factor that controls the overall elimination of acetone from the body. Consequently, acetone monooxygenase is the enzyme most likely responsible for

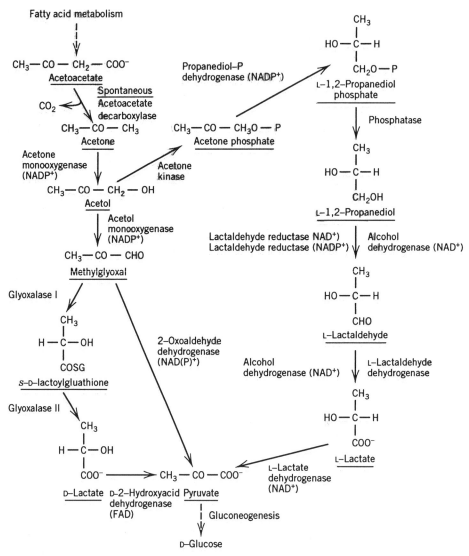

Figure 5.3. Mammalian metabolism of acetone (382). Copyright© 1986, reprinted with permission.

the metabolic saturation and zero-order kinetics observed with high acetone body burdens.

Perinado et al. provided supportive evidence for the biotransformation of acetone by two separate pathways in the rat (383). Pregnant and virgin Wistar rats were administered a 100 mg/kg iv dose of unlabeled acetone. Peak plasma acetone levels for virgin rats were two to threefold higher than for fasted pregnant rats after correcting for the difference in distribution volumes. Fasted pregnant rats were found to eliminate acetone at a much faster rate than either the fed pregnant rats or the virgin rats. Variable amounts of acetol, 1,2-propanediol, and methylglyoxal were detected in plasma and liver specimens from the pregnant dams, term fetuses, and virgin rats. The metabolite levels were generally higher in virgin than in pregnant animals, and the levels of acetol and methylglyoxal were many fold higher in fed virgin versus fasted virgin rats.

Kosugi et al. provided strong evidence for the biotransformation of acetone by a third pathway involving an acetate intermediate (384). Wistar and Sprague-Dawley rats of both sex were intravenously infused over a 3-hr period with 323 to 1113 mg/kg of [2-^{14}C]acetone. The animals were given an initial diabetes-inducing streptotozin injection or were pretreated for 3 to 4 days with 1 percent acetone in their drinking water and then fasted for 1 to 2 days. The resulting blood acetone concentrations ranged from about 4.1 to 9.7 mM (273 to 563 mg/l), and the incorporation of radioactivity into whole body glucose was generally 1 percent or less. Digestion of the glucose revealed that carbon atoms 3 and 4 were preferentially labeled, which was different from the pattern observed with [2-^{14}C]pyruvate. Similar results were obtained with liver perfusion and hepatocyte incubation studies. The authors concluded that a third pathway involving acetate or acetyl-CoA was involved in the metabolism of acetone at high dose levels. Gavino et al. provided unequivocal evidence for the formation of acetate during acetone metabolism (385). The liver perfusates of male Sprague-Dawley rats perfused in situ with 4.5 mM (260 mg/l) of [2-^{14}C]acetone were shown to contain a large percentage of [1-^{14}C]acetate, with small amounts of labeled lactate and 1,2-propanediol also present.

Follow-up studies by Kosugi et al. provided an explanation for the acetate and formate metabolites reported by earlier investigators and established that a third metabolic pathway was operable when large doses of acetone were administered (386). Fasted, fed, and diabetic (streptozotocin-induced) Wistar rats were given a 3-hr iv infusion of [2-^{14}C]acetone at a low dose level between 0.17 and 0.23 mg/kg or at a higher dose level between 313 and 383 mg/kg. Several of the fasted animals were administered 1 percent acetone in their drinking water for 5 days prior to [2-^{14}C]acetone treatment. Acetone blood levels in fed and fasted rats administered the high dose of acetone ranged from 4.0 to 5.4 mM (232 to 314 mg/l), whereas the blood levels did not exceed 0.3 mM (17 mg/l) in the rats given the tracer dose. When a tracer dose of acetone was administered, glucose carbon atoms 1, 2, 4, and 5 were predominantly labeled, which suggested that the metabolism of acetone proceeded through methylglyoxal and lactate. In contrast, when the high dose was administered, glucose carbons 3 and 4 were primarily labeled, which indicated that acetate was formed as an intermediate metabolite. Studies with fed, fasted, or acetone-pretreated animals indicated that the metabolism of acetone to acetate depended upon the acetone blood level and not on any condition that would raise endogenous acetone production. The authors concluded that the pathway leading to acetate formation was active only at high acetone doses and that the lactate pathway predominated at more moderate levels.

9.2.2 Alcohols and Other Compounds

Acetonemia and acetonuria have repeatedly been shown to be a clinical consequence of acute isopropanol (IPA) intoxication in humans (387–389). Most cases of IPA intoxication have involved the direct ingestion of a 70 percent solution by chronic alcoholics seeking an ethanol substitute. Occasional case reports of suicidal or accidental intoxication with IPA have also been documented in the literature. Fatalities due to IPA ingestion are rare and typically involve individuals who are in poor health or in a chronically debilitated state (390). Because IPA is rapidly and extensively metabolized to acetone, elevations in blood, urine, and breath acetone invariably accompany occupational exposures to IPA (391, 392).

The acetone blood levels that follow a severe IPA overdose are generally very high, and the time course for elimination from the blood is often substantially longer than for the parent alcohol (393, 394). Consequently, clinicians can often detect acetone in the blood and urine of intoxicated patients long after the IPA has disappeared from the body (395, 396). The relatively slow elimination of acetone from the body is caused by the saturation of the metabolic step that controls the clearance from tissues and organs. Once metabolic saturation has occurred, the elimination rate of acetone declines and blood half-life increases in response to the excess body burden.

Several case reports of human IPA intoxication have included data that clearly demonstrate the metabolic saturation that accompanies high acetone body burdens. Daniel et al. described the time course for IPA and acetone elimination from the blood of two patients who consumed a large amount of an IPA-containing solution (393). The blood levels of acetone were greater than those of IPA at all times, and the elimination of IPA was virtually complete when the acetone concentrations were still at a maximum. The tendency for the acetone blood levels to remain between 1.0 and 3.0 g/l for an extended length of time and the slow nonlinear return of the blood levels to normal base-line levels indicated metabolic saturation. Natowicz et al. reported similar findings in a patient who was admitted in a coma following the ingestion of rubbing alcohol (394). In this instance, blood levels peaked at about 1.3 g/l and slowly declined to normal over a 4-day period. Despite the apparent zero-order decline in the acetone concentration, a first-order half-life of 22.4 hr was calculated for acetone. Pappas et al. reported that the elimination half-life of acetone ranged from 7.6 to 26.2 hr in six human cases of isopropanol intoxication; the corresponding peak blood levels of acetone ranged from about 1.4 to 5.8 g/l (397). These data and a visual inspection of the time-course plots suggest that the metabolism of acetone was saturated. Similar zero-order declines in plasma acetone have been observed in rats and dogs administered a 1.0 g/kg dose of IPA by the ip and iv routes, respectively (398). Likewise, rats exposed by inhalation to 4000 or 8000 ppm IPA for 8 hr had peak acetone blood levels (1.1 to 1.5 g/l) that were disproportionately related to the exposure (399).

The rate and extent of IPA metabolism to acetone has been examined in both laboratory animals and in human volunteers. Ashkar and Miller reported that 80 to 90 percent of the IPA ingested by humans could be metabolized to acetone

(400). Pappas et al. agreed with these estimates based on the calculation of rate constants for the formation of acetone from IPA in acutely intoxicated patients (397). Ellis conducted 90-min liver perfusion experiments with rabbits and determined that an average of about 52 percent of the IPA dose was converted to acetone when the IPA concentration was 1.0 mg/ml (401). The percent conversion decreased to an average of 33 percent when the IPA perfusion concentration was raised to 3.0 mg/ml. These data indicate that the metabolism of IPA to acetone was itself saturated and that the percent conversion to acetone would be higher with smaller doses of IPA. Similar pharmacokinetic results were obtained in a study performed by Nordmann et al., in which fasted rats were treated with either a 1 g/kg ip dose or a 6 g/kg oral dose of IPA (402).

Several reports have documented unusually high IPA blood levels in human postmortem tissue specimens. Lewis et al. reported blood levels of IPA ranging up to 0.44 g/l in autopsy specimens from humans dying of liver disease, cardiovascular disease, or diabetes (403). Additional studies with normal and diabetic (alloxan-induced) male Sprague-Dawley rats given a single oral dose of acetone at levels of 1, 2, or 4 g/kg showed a dose-related increase in the IPA blood levels. Davis et al. corroborated these findings and reported that high levels of IPA were present in postmortem blood, liver, kidney, and brain specimens from individuals whose deaths were unrelated to an IPA exposure (404). The highest IPA levels were generally found in the liver specimens (70 to 590 mg/l) and the lowest in the brain (20 to 120 mg/l). In vitro incubations of yeast alcohol dehydrogenase with acetone and NADH for up to 130 min resulted in a buildup of IPA in the reaction medium, which suggested that this enzyme was responsible for the appearance of IPA in the tissue specimens. These findings were consistent with those of Leibman, who reported that the hepatic cytosol from rabbits was capable of slowly reducing acetone to IPA (405).

Acetone has been shown to be a metabolite of several chemicals other than IPA. Dayal et al. reported that acetone was an in vitro metabolite of 2-nitropropane in mouse hepatic microsomal incubations (406). Acetone has also been shown to be a metabolite of t-butanol both in vivo and in vitro. Baker et al. found acetone in the urine and expired air specimens of Long-Evans and Sprague-Dawley rats given a single ip dose of radio-labeled t-butanol (407). The recovery of labeled acetone was reported to be highly variable, ranging from 0.5 to 9.5 percent of the dose. The enzymatic mechanism responsible for the formation of acetone from t-butanol was found to involve an intermediate alkoxy radical that spontaneously decomposed to acetone and a methyl radical (408). A similar liver microsomal oxidizing system involving hydroxyl radical production has been shown to be effective in metabolizing IPA to acetone (409).

10 PHARMACOKINETICS

The pharmacokinetics of acetone have been studied in numerous species using various routes of administration. These data have shown that acetone possesses several unusual characteristics with regard to both uptake from the lung and elimination via metabolism. Owing to its high volatility, inhalation represents the predominant route of uptake into the body. Once absorbed, acetone is rapidly and evenly distributed throughout the nonadipose tissues and is not preferentially stored in any tissue of the body.

Under normal conditions acetone is efficiently and effectively metabolized to a variety of products that can be used as building blocks for the synthesis of complex proteins and carbohydrates. Once the elimination capacity of the body has been overwhelmed by either excessive endogenous production or exogenous uptake, a number of compensatory processes are called upon to handle the surplus. The excretion of acetone in the urine and exhaled air increase nonlinearly as the metabolic pathways are overloaded. In addition, alternate routes of metabolism appear to be recruited to assist in the elimination process. Sustained high blood levels of acetone can also result in the induction of enzymes responsible for its own metabolism and the metabolism of the other compounds. This compensatory response to saturation is likely responsible for the ability of acetone to potentiate the toxicity of chemicals that require metabolic activation to a reactive intermediate.

10.1 Human Studies

Pharmacokinetic descriptions of the overall absorption and elimination of acetone have been performed following oral, inhalation, and dermal exposures with humans; however, exogenous exposures to acetone typically occur by the pulmonary route. A considerable amount of research has been conducted on the pulmonary uptake and elimination of acetone vapors by human subjects. These studies have demonstrated that acetone is capable of being absorbed into tissues lining the respiratory tract upon inspiration and then partially desorbed upon exhalation. The high blood-to-air partition coefficient for acetone suggests that a large percentage of inhaled acetone will be absorbed into the body; however, the occurrence of a peculiar wash-in/wash-out effect effectively reduces the uptake into the blood. The miscibility of acetone in the fluid layers lining the lung appears to be responsible for the wash-in/wash-out phenomenon. This unusual characteristic provides a likely explanation for the wide range of pulmonary uptake values reported for acetone.

Acetone elimination displays Michaelis–Menten saturation kinetics in rats and humans, with the K_m in rats being about 160 ppm. At low acetone concentrations (i.e., less than the K_m), the elimination kinetics are first-order and are governed by the rate of metabolism in the liver. At higher levels significantly exceeding the K_m, the elimination of acetone becomes zero-order and the respiratory and urinary excretion pathways may become the predominant routes of elimination. The respiratory and urinary routes of excretion of absorbed acetone constitute minor pathways of elimination under normal first-order conditions where metabolic saturation is not a factor. The complex interaction of acetone vapors in the respiratory system together with the unusual Michaelis–Menten kinetics exhibited during metabolic elimination complicate the calculation or prediction of an acetone body burden from exposure information. Traditional linear assumptions regarding the applied dose to internal dose relationship are not valid for acetone except at lower exposure concentrations.

10.1.1 Laboratory Setting

The blood-to-air partition coefficient is an important factor affecting the overall pulmonary uptake and excretion of acetone vapors. A large number of studies have focused on the measurement (410) or calculation (411) of acetone partition coefficients in various biological fluids and tissues. The earliest of these studies was performed by Widmark, who used both in vivo and in vitro techniques to measure the blood-to-air partitioning of acetone (76). The in vitro experiments were repeated several times using defibrinated calf serum and blood, and the results were compared to the water-to-air partition coefficient for acetone. The average water-, serum-, and blood-to-air partition values were reported to be 406, 392, and 314, respectively. Widmark proceeded to measure the value in vivo using a human subject who consumed 9.5 g of acetone in 500 ml of water (130 mg/kg). The average blood-to-air partitioning of acetone was calculated to be 334 by this method.

Using a similar approach, Briggs and Shaffer measured the partitioning of acetone into water or cow serum following incubation at a wide range of temperatures (77). When the temperature was maintained between 37 and 38°C, the average water-to-air and serum-to-air partition coefficients were reported to be 334 and 337, respectively. The in vitro results were compared with data obtained from anesthetized dogs given an iv (1.46 g/kg) or an ip (0.8 g/kg) dose of acetone. The serum-to-air partition coefficients ranged from 307 to 369, and the average values for the iv and ip treatments were 336 and 330, respectively. The authors also measured the in vivo blood-to-air partitioning of acetone in fasting subjects and in patients with diabetic ketoacidosis and obtained an average value of 355. Haggard et al. determined the in vivo blood-to-air and urine-to-blood partition coefficients in four subjects that consumed a 50 mg/kg dose of acetone (78). The average blood-to-air value was reported to be 330 (range of 322 to 339) and the urine-to-blood value 1.34 (range of 1.33 to 1.35). Widmark examined the urinary excretion of acetone in two human subjects who consumed a dose ranging from 110 to 219 mg/kg (412). The concentration of acetone in the urine was found to be governed by simple diffusion processes. Widmark concluded that the amount of acetone excreted in the urine was primarily governed by the rate of urine formation and the urine-to-blood partition coefficient, which was determined to be 1.235. Wigaeus et al. determined that the in vitro blood-to-air partition coefficient for acetone was 275 at 37°C; however, in vivo measurements of acetone in post-exposure specimens of blood and air suggested a value between 340 and 400 (413).

Young and Wagner reported that both the type of animal blood and the hematocrit could affect the blood-to-air partitioning of acetone (414). The blood-to-air partition coefficient for both humans and dogs declined linearly as the hematocrit was increased from 0 percent (plasma) to 100 percent (packed red blood cells). The values for dog blood were consistently greater than the coefficients obtained with human blood specimens. The average results in human blood ranged from a high of 315 with plasma to a low of 210 with packed red blood cells. In contrast, the partitioning in dog blood ranged from 373 with plasma to 193 with packed cells. The blood-to-air partition coefficients in human and dog blood possessing a normal hematocrit of 45 percent were calculated to be 268 and 292, respectively. Sato and Nakajima used a more modern and precise vial equilibration procedure to measure the water-to-air, oil-to-air, and blood-to-air distributions of acetone and 16 other solvents (415). The average in vitro water-, oil-, and blood-to-air values for acetone were 395, 86, and 245, respectively. Acetone was shown to have the highest partition coefficients of the 17 solvents tested.

Fiserova-Bergerova and Diaz used a slightly modified method to measure the distribution of acetone between air and organ tissues from five autopsied humans (416). Separate measurements on control samples were used to correct for the amount of endogenous acetone in each tissue. The tissue-to-air partition coefficients were determined for the following specimens: muscle, kidney, lung, cerebral white matter, cerebral gray matter, fat, whole blood, packed erythrocytes, and plasma. A liver-to-air partition coefficient was not determined because of the metabolic consumption that occurred during incubation. The results obtained with acetone and six other hydrophilic solvents are shown in Table 5.15. The largest tissue-to-air partitioning was observed for whole blood and the smallest for fat. The published values for 28 other solvents were used to arrive at a regression equation to predict the liver-to-air partition coefficients.

Landahl and Hermann investigated the retention of acetone vapors in the nose and lungs of two human volunteers (417). The nasal retention of acetone vapors was measured at concentrations of 300 and 3000 mg/m^3 (126 and 1264 ppm) by drawing air through the nose and out the mouth while the subjects held their breath. The lung measurements involved the inspiration of a controlled volume of acetone vapor at 800 and 11,000 mg/m^3 (337 and 4635 ppm) followed by a short breath-holding period and then a forced expiration of a fixed volume of air. The results showed that the retention of acetone vapors in the nose varied from 18 to 40 percent when the flow rate through the nose was set at 18 l/min. In addition, the nasal retention was found to be independent of the acetone vapor concentration. The lung retention of acetone vapor ranged from 53 to 61 percent of the inhaled concentration and was independent of either the vapor concentration, the length of time the breath was held, or the volume of air exhaled.

Dalhamn et al. obtained higher retention values when they examined the absorption of the acetone vapors contained in cigarette smoke (418, 419). Using male and female subjects, they determined the percentage of the acetone contained in the smoke from two nonfilter cigarettes (0.56 mg per 35-ml puff) that was retained by the mouth or lung. The authors found that the average retention of acetone was 56 percent for the mouth and 86 percent for the lung; however, these studies did not take into consideration the reversible absorption of acetone into the fluid lining the inside surface of the nose and lung. Nomiyama and Nomiyama examined the respiratory retention, alveolar uptake, and alveolar excretion of acetone in male and female subjects (five of each sex) exposed to about 130 ppm for 4 hr (420). The three measures of vapor movement into and out of the lung were related

Table 5.15. A Comparison of Human Tissue-to-Air Partition Coefficients for Seven Hydrophilic Solvents (416)

Compound	Muscle	Kidney	Lung	Brain White	Brain Gray	Fat	Whole Blood	Packed RBC's	Plasma	Liver (predicted)
Methanol	1309	1355	1733	1091	1413	231	1626	1522	1871	948
Ethanol	850	940	1172	790	1044	215	1332	1246	1455	778
2-Propanol	502	503	590	441	560	180	719	605	812	422
1-Propanol	678	686	817	584	720	296	955	799	969	564
Acetone	151	146	160	121	148	86	196	170	217	113
2-Butanol	343	371	400	337	387	388	541	519	591	329
2-Butanone	103	107	103	96	111	162	125	106	133	79

by the following equation:

$$\text{retention (R)} = \text{alveolar uptake (U)} - \text{alveolar excretion (E)}$$

where

$$R = \frac{\text{concn inhaled} - \text{concn exhaled}}{\text{concn inhaled}}$$

Lung retention and alveolar excretion were both calculated from the amount of acetone vapor in the inspired and expired air of the subjects during and after exposure. The retention of acetone declined during the first 2 hr of the exposure and then reached an apparent steady state with a mean value of 11 to 18 percent. A statistically significant sex difference was observed for body retention and alveolar uptake, with men displaying higher percentages than women. The authors acknowledged that the values reported in their study were lower than those of other investigators, a fact they attributed to the unusual breath holding techniques and longer exposure durations employed. In a follow-up report, the authors published additional kinetic data for the terminal respiratory excretion rates in the same set of subjects (421). The respiratory excretion data were described by a two-compartment model for the men and a one-compartment model for the women. The compartmental models were used to calculate the amount of acetone excreted by the lungs during a 4-hr post-exposure period. The ratio of respiratory excretion to body retention was then computed to determine what fraction of the retained acetone vapor was eliminated by the respiratory route. The authors reported that the rate of respiratory excretion of acetone was very slow and that less was expired by the women than the men; however, they also reported that up to 76 percent more acetone was exhaled than was absorbed, a fact consistent with the normal formation and respiratory removal of metabolically produced acetone.

Cander and Forster suggested that the acetone absorbed in the upper respiratory tract during inspiration could be re-entrained into the alveolar air upon expiration (422). Two subjects inspired 20,000 ppm of acetone in a single breath and forcibly exhaled after holding their breath for a period of time ranging from 1.5 to 15 sec. The acetone concentration in the air at the beginning and end of the breath-holding period provided a measurement of the acetone removed by the capillary blood flowing to the lung. The amount of residual acetone vapor in the alveolar air was compared to the theoretical value expected when the volume of alveolar air, the volume of lung parenchymal tissue, the solubility of acetone in both parenchymal tissue and blood, and the pulmonary blood flow rate were all taken into consideration. A comparison of the theoretical and actual values for residual alveolar acetone revealed that the rate of acetone removal by the lung was about eightfold slower than predicted. The authors concluded that a substantial portion of the acetone adsorbed into the pulmonary tissue upon inspiration was removed during expiration, thus contaminating the expired alveolar air sample used for analysis.

Schrikker et al. provided additional data to support the fact that highly soluble vapors, such as acetone, were partially desorbed from epithelial tissues of the respiratory tract upon exhalation (423). Measurements were performed continuously as the subjects breathed normally while at rest or while exercising with a work load of 50, 100, or 150 W. Data were collected from four male subjects that inspired 0.01 percent (100 ppm) of acetone during 12 wash-in breaths followed by 10 wash-out breaths with room air. There was an appreciable difference in the amount of acetone expired at the beginning and end of any single breath during the wash-in cycle. The excretion of acetone from venous blood accounted for about 50 to 60 percent of the acetone found in the wash-out air samples, and this amount increased only slightly when the work load was elevated. At the start of the wash-out cycle, there was more acetone in the air at the beginning of the breath than at the end; however, as the wash-out cycle progressed, the difference in acetone concentrations at the beginning and end of the breath became smaller and eventually reversed such that by the sixth or seventh breath there was more acetone expired at the end of the breath than at the beginning. This highly unusual behavior was ascribed to the initial absorption of acetone into the nonperfused tissues that lined the upper airways (i.e., the nose, pharynx, and bronchi) during the wash-in period. This zone of high acetone was gradually displaced into the deeper and more distal regions of the respiratory tract; then, as breathing continued, the high concentration zone moved back up toward the mouth and the vapors were re-entrained in the relatively acetone-free expired air of the wash-out cycle.

Further studies by Schrikker et al. showed that the pulmonary excretion of acetone was affected to a large degree by the exchange of vapor between the inspired air and the tissues lining the respiratory tract (424). Seven male subjects were exposed to 0.1 percent (1000 ppm) of acetone for 4 to 5 min at work loads of 0, 25, 50, and 75 W. Resting individuals (0 W) were also examined when they held their breath for 1 sec at the end of each inspiration. A minimum excretion value of about 0.5 was obtained for acetone when the individuals were at rest; however, during exercise, there was a significant increase in the excretion of acetone. The authors concluded that vapor exchange in the tissues and fluid lining the respiratory tract was an important process that contributed to the overall excretion of acetone by the lung. Likewise, they pointed out that the extent of dissolution and re-entrainment of acetone vapors was affected by the volume of dead space air in the lung, which changed in accordance with the ventilation rate.

Jakubowski and Wieczorek investigated the effects of a work load on the pulmonary uptake of acetone and four other solvent vapors by human subjects (425). Five male subjects were exposed on four different occasions to about 200 mg/m^3 (84 ppm) of acetone vapor for 2 hr. Each session was separated by a 3-day interval and was conducted while the subjects were either at rest or exercising on a bicycle ergometer at a rate of 25, 50, or 75 W. Inspired and expired air samples were used to calculate the percentage of acetone retained in the body, the acetone intake rate, and the acetone uptake rate. The results displayed in Table 5.16 show that the percentage of acetone retained by the body remained relatively constant as the pulmonary ventilation increased with higher work loads. The rate of acetone uptake by the body was therefore determined to be directly proportional to the ventilation

Table 5.16. Influence of a Work Load on the Pulmonary Disposition of Acetone Vapors in Human Volunteers[a] (425)

Avg. Inspired Air Concn. (mg/m³)	Work Load (W)	Fractional Lung Retention		Ventilation Rate (m³/hr)		Rate of Intake (mg/hr)		Rate of Uptake (mg/hr)	
		Mean	Std. Dev.	Mean	Std. Dev.	Mean	Std. Dev.	Mean	Std. Dev.
197	rest	0.43	0.04	0.39	0.13	78	24	34	11
194	25	0.42	0.06	1.08	0.11	210	20	88	14
195	50	0.40	0.10	1.51	0.15	293	32	118	34
197	75	0.44	0.02	1.85	0.21	364	45	159	22

[a]Exposed for 2 hr to about 85 ppm.

rate. Because the concentration of acetone in an expired air sample can be affected by a number of processes other than diffusion into the blood (i.e., dead-space absorption, re-entrainment from tissues, and alveolar excretion), the rates reported in this study represent the sum of all these processes.

In an early series of pharmacokinetics experiments, Widmark measured the absorption and excretion of acetone following oral administration (128). The author noted that fasting affected the absorption rate when a 137 mg/kg oral dose of acetone was administered. On an empty stomach, the blood levels of acetone rose rapidly and a peak blood level of 310 mg/g was observed at 10 min post-treatment. In contrast, the absorption was much slower after a meal, with a peak blood level of 190 mg/g occurring 42 min after ingestion. Koehler et al. examined the disappearance of acetone from the blood of diabetic and healthy subjects administered 10 g (~140 mg/kg) by iv infusion over a 2-hr period (127). The average peak blood level of acetone at the end of the infusion was found to be 230 mg/l in the healthy subjects and 195 mg/l in the diabetic patients. The elimination of acetone from the blood was noted to be extremely slow for both groups, with the rate of decline being somewhat faster in the diabetic patients. When a separate group of healthy subjects was similarly treated, the blood acetone levels remained elevated at concentrations ranging from 135 to 145 mg/l throughout the 4-hr post-treatment period. The authors suggested that the initially slow disappearance of acetone from the blood was due to a derangement of acetone metabolism that occurred when the blood concentrations were excessively high.

Haggard et al. conducted a comparison of acetone pharmacokinetics in both resting and exercising humans exposed by either the oral or the respiratory route (78). The tissue distribution of acetone was estimated in subjects given oral doses ranging from 40 to 60 mg/kg. After correcting for the cumulative amount of acetone eliminated by excretion (urine and expired air) and metabolism, the remainder was used to compute an average tissue-to-blood distribution factor of 0.82. The percentage of absorbed acetone eliminated by excretion was found to decrease as the blood levels declined, ranging from a high of 36 percent to a low of 7 percent. The percentage of acetone metabolized was calculated to increase from 64 to 94 percent as the blood levels fell from 73 to 2 mg/l. Periodic estimations of the metabolic rate indicated a decrease from about 2.1 to 1.1 mg/(kg)(hr) over a 24-hr period. The authors noted that the metabolic rate was related to the amount of acetone in the blood (i.e., the rate was apparently first-order). The metabolic rates were found to increase from about 2.7 to 6.0 mg/(kg)(hr) when examined before and then during exercise by an individual that was given a 70 mg/kg dose of acetone. This rate change was attributed to an increase in the basal metabolic rate of the individual because the relative increase matched the estimated rise in energy expenditure. No attempt was made to describe the kinetic data from these experiments using classical compartmental modeling techniques.

Haggard et al. also examined the kinetics of acetone in resting and exercising male volunteers exposed for 8 hr to 1, 3, or 5 mg/l (422, 1266, and 2110 ppm) of acetone (78). The end-exposure blood concentrations of acetone were found to be 30, 99, and 165 mg/l for the individual subjects exposed to 422, 1266, and 2110 ppm of acetone, respectively. Increases in the acetone blood concentrations were noted to be approximately proportional to the rise in the exposure concentrations. Acetone blood levels were observed to increase from 30 to 62 mg/l when a single subject exposed to 422 ppm was required to perform moderate exercise. A relatively small increase in blood acetone (165 to 330 mg/l) was also observed in an individual who performed moderate exercise while exposed to 2110 ppm of acetone. These increases in blood acetone were regarded as being small considering the large increase in ventilation rate that accompanied the exercise and the expected pro-portionality between the ventilation rate and the pulmonary uptake of acetone. The authors concluded that a ninefold increase in the metabolic rate was necessary to account for the less than expected rise in the acetone blood levels. The accu-mulation of acetone in the body was examined in a single subject repeatedly exposed to 2110 ppm for 8 hr/day on each of three consecutive days. The end-exposure blood levels of acetone were 162, 180, and 182 mg/l for the first, second, and third day, respectively. The blood level obtained 24 hr after the start of the last exposure was 91 mg/l. These data indicated that acetone accumulated in the body on repeated high level exposures and that there was a limit to the amount of acetone that accumulated in the body as demonstrated by the apparent plateau in the blood levels. The authors estimated that exposures below 1266 ppm in a resting individual or 422 ppm in a moderately active person would not cause any accumulation of acetone in the body with repeated 8-hr exposures.

Wigaeus et al. performed inhalation studies that were used to describe the relative uptake and excretion of acetone in resting and exercising humans (413). Eight healthy male subjects were divided into three exposure groups that were defined according to the exercise regimen; the 2-hr acetone exposure for each group was also varied between 300 and 552 ppm (see Table 5.17). Physiological variables such as heart rate, oxygen uptake, pulmonary ventilation rate, and pul-monary air volume were recorded along with venous and arterial blood levels. As shown in Table 5.17, the relative uptake of acetone was found to remain constant at an unexpectedly low value of about 44 percent of the exposure concentration, regardless of the activity regimen. The blood acetone concentration increased con-tinuously during exposure and did not reach an apparent steady-state. The half-life ($t_{1/2}$) for acetone excretion by the lung was calculated to be 4.3 hr and pulmonary excretion was shown to account for about 15 to 26 percent of the amount absorbed. Only about 1 percent of the absorbed acetone was excreted unchanged in the urine. The average $t_{1/2}$ for acetone elimination from venous and arterial blood was 3.9 and 6.1 hr, respectively. After considering the volume of air exchanged in the lungs as function of the exercise regimen, the authors determined that the amount of acetone taken up by the lungs (i.e., the absolute uptake) was greater in the ex-ercising subjects because of their higher ventilation rates. The authors reasoned that the lower than expected values for the relative uptake of acetone may have been due to the nonspecific absorption of acetone vapors into the mucosal surfaces of the nose and throat.

Pezzagno et al. examined the pharmacokinetics of acetone in humans and derived linear mathematical equations to describe the relationship between inhaled acetone

Table 5.17. Uptake and Elimination of Acetone Vapor in Groups of Human Subjects Receiving a Two-Hour Inhalation Exposure (413)

Group[a]	Activity Level	Exposure Concentration		Respiratory Uptake		Respiratory Excretion		Urinary Excretion		Total Excretion	
		ppm	mg/m³	mg	%	mg	%	mg	%	mg	%
1	Rest	552	1309	594	44	88	15	6.8	1	95	16
2	Rest/50 W	300	712	705	44	134	19	7.2	1	141	20
3	Rest/50, 100, 150 W	311	737	1170	44	304	26	12.2	1	316	27

[a]Group 1 was exposed to 522 ppm and remained at rest during the entire exposure period. Group 2 received a 300-ppm exposure, and the subjects were required to exercise at a work load of 50 W for the last 1.5 hr of the exposure period. Group 3 individuals were exposed at 311 ppm and increased their work load every 30 min over the 2-hr exposure period.

and the concentrations in blood and urine (426, 427). Resting and exercising subjects were exposed by inhalation to intentionally varying vapor concentrations of acetone that ranged from 56 to 500 mg/m³ (21 to 211 ppm) for periods of 4 hr or less. The average relative uptake of acetone was found to be 54 percent for the subjects at rest and 53 percent for the individuals performing light physical exercise. The amount of acetone taken up by the lung was thereby shown to depend strongly on the rate of pulmonary ventilation. A strong linear correlation was observed between the inhaled concentration of acetone and the end-exposure concentrations in the blood (r = .86 to .99) and alveolar air (r = .91) for all of the treatment groups. A good linear relationship was also observed between the inhaled acetone concentration the acetone level in the urine; however, the relationship showed some variability for 2-hr versus 4-hr exposures and for the resting versus the exercising subjects. Tada et al. also examined the relationship between inhaled acetone and the concentration in urine and expired air (428). Human subjects were exposed on four separate occasions to concentrations ranging from 200 to 600 ppm of acetone for 2 to 4 hr on three consecutive days. The authors reported that the breath levels of acetone at 30 min post-exposure and the urinary acetone levels at 8 hr post-exposure were generally proportional to the exposure concentration. Urine acetone levels were at a maximum 2 to 4 hr after exposure termination and returned to normal within 16 to 18 hr.

DiVincenzo et al. exposed exercising male volunteers to vapor concentrations of 100 or 500 ppm of acetone for 2 hr and measured the relative uptake, organ excretion, and blood elimination (130). Four exposure groups were defined based on the exposure concentration and level of exercise. Using expired air measurements, the respiratory uptake was determined to be 75 to 80 percent for both the 100 and 500 ppm exposures. The blood levels of acetone increased steadily during the exposure sessions, and a steady-state concentration was not attained. Peak end-exposure blood levels of acetone were barely above background, measuring 2 and 10 mg/l for the 100 and 500 ppm exposures, respectively. Although the authors concluded that the rate of acetone elimination from the blood was independent of the blood concentration (i.e., apparently zero-order), a first-order $t_{1/2}$ of about 3 hr was calculated. The amount of acetone recovered in the urine during the 24-hr collection period differed by less than threefold for the two exposure concentrations, a fact inconsistent with the authors' conclusion that acetone elimination occurred by a zero-order process. The authors ascribed the discrepancy in the urine data to the collection of casual specimens rather than a total 24-hr void.

Five alternate physiological-pharmacokinetic models have been developed for use with acetone (429–433). The models provide a method for predicting the blood and tissue levels of acetone in humans following an inhalation exposure. The models constructed thus far have limited utility because they are applicable only to exposure levels that result in first-order elimination kinetics.

Reichard et al. measured the rates of acetone production in obese and nonobese humans during starvation-induced ketonemia and examined the contribution to glucose formation under these conditions (434). As described in Table 5.18, three groups of human volunteers were given a small iv dose of [2-¹⁴C]acetone (1.01 to

Table 5.18. Elimination of the Acetone Produced in Fasting Obese and Nonobese Human Volunteers Given A Small Tracer Dose of [2-^{14}C]Acetone and Monitored for Six Hours (434)

Group	Avg. Body Weight (kg)	Duration of Fast (days)	Avg. Blood Concentration		Avg. Urine Concentration		Respiratory Excretion (%)	Urinary Excretion (%)	Metabolism (%)	Glucose Incorporation (%)	Acetone Turnover Rate [μmol/(m²)(min)]
			mM	mg/l	mM	mg/l					
1	72.3	3	0.76	44.1	1.2	69.7	14.7	1.4	83.9	4.2	51
2	108.7	3	0.29	16.8	0.4	2.3	5.3	0.6	94.1	3.1	31
3	124.1	21	1.37	79.6	2.6	151.0	25.2	1.3	73.5	11.0	61

2.72 μmol) following a prolonged fast. The elimination of endogenous acetone by excretion and metabolism were measured along with the overall average concentrations in the plasma and urine. Radioactivity from the administered acetone was detected in plasma glucose, lipids, and proteins, but not in plasma free fatty acids, acetoacetate, or β-hydroxybutyrate. A linear first-order decline in the plasma radioactivity was observed while acetone blood levels remained constant. These data allowed the average acetone turnover rate to be calculated. The authors concluded that there was a direct relationship between the rate of endogenous acetone production and the plasma levels that were observed.

Owen et al. examined the pharmacokinetics of endogenous acetone in patients with moderate to severe diabetic ketoacidosis (435). A group of adult patients suffering from a variety of serious clinical ailments in addition to being diabetic were administered a small tracer dose (0.75 to 1.56 μmol) of 2[^{14}C]-acetone by bolus iv injection. The initial mean plasma acetone concentration in the patients was 4.96 mM (288 mg/l) with a range of 1.55 to 8.91 mM (90 to 517 mg/l). The blood values for each patient remained relatively constant throughout the observation period. The rate of acetone excretion in the breath and urine were measured and used to calculate the turnover rate in each individual. Acetone turnover rates ranged from 68 to 581 μmol/min (values normalized to the standard human body surface area, 1.73 m^2) and were unrelated to the acetone plasma concentrations. Distinct nonlinearities in the rates of acetone production and elimination were described. When the plasma acetone concentration was below 5 mM (290 mg/l), there was a direct linear relationship between the rate of formation and the amount in the plasma; however, at higher plasma levels there was a marked decrease in acetone production. The excretion of acetone in the expired air accounted for about 20 percent of the production rate at plasma levels below 5 mM, but then increased to about 80 percent of the production rate when the plasma concentration was above this level. Corresponding differences were described for the amount of acetone metabolized. These data indicated that acetone metabolism was capacity limited and that deviations from normal first-order kinetics could be expected when acetone levels in the blood were in the vicinity of 300 mg/l.

Reichard et al. extended the studies of Owen et al. in ketoacidotic diabetics (436). In these experiments, the patients were administered 5.7 to 6.7 μmol of [2-C^{14}]acetone by constant iv infusion rate rather than by bolus injection. The radiolabeled acetone was infused over a 4-hr period in order to attain steady-state levels of radioactivity in the plasma. The initial average concentration of acetone in the plasma was measured at 3.26 mM (189 mg/l) and ranged from 0.50 to 6.02 mM (29 to 349 mg/l) for the individual subjects. A minimum of 0.5 to 4.1 percent of the plasma glucose from the treated patients was derived from endogenously produced acetone. The acetone turnover rate was calculated from the rate of [2-^{14}C]acetone infusion required to maintain a steady state. In contrast to the previous study, the authors reported that the acetone turnover rate appeared to be linearly related to the plasma concentration up to a level of 7.61 mM (442 mg/l).

Several reports have documented the cutaneous absorption of acetone in humans. Parmeggiani and Sassi placed volunteers in a static chamber with their heads remaining outside the enclosure (437). Acetone-saturated pads were placed on the skin of the subjects for 30 min in order to saturate the air with acetone vapor. After 90 min of exposure, acetone concentrations in the blood and urine were reported to be substantially elevated above preexposure levels. Fukabori et al. used human volunteers and applied acetone to 12.5 cm^2 of skin on four consecutive days (438). The acetone concentration ranges in the blood, urine, and expired air were 5 to 12 mg/l, 8 to 14 mg/l, and 5 to 12 ppm, respectively. These values increased by three- to fivefold when the exposure duration was increased from 2 to 4 hr/day. The authors reported that skin penetration was rapid and that pulmonary excretion predominated over urinary excretion.

10.1.2 Industrial Setting

Pharmacokinetic studies in acetone-exposed workers have focused on the relationship between the exposure concentration and the amount of acetone excreted in the air and urine. These studies have generally sought to describe the mathematical relationships that would allow biologic exposure indexes to be established for acetone-exposed employees. At the range of exposure concentrations occurring in an occupational environment, however, the whole body elimination of acetone occurs according to Michaelis–Menten kinetics. The mathematical discontinuities caused by this type of elimination process invalidate the use of linear regression techniques and complicate the development of a reliable biomonitor that can be applied at a full range of exposure concentrations.

Baumann and Angerer reported that four workers exposed to 71.1 mg/m^3 (30 ppm) of acetone for 2 hr retained about 80 percent of the inhaled acetone (439). The amount of acetone absorbed over an 8-hr work shift was calculated to be about 200 mg. The concentration of acetone in the urine increased from about 0.75 mg/l at the beginning of the work shift to about 2.0 mg/l by the end of the shift. Similarly, the acetone in venous blood specimens increased from 1.0 mg/l at the start of the shift to 3.3 mg/l by the end. The urine and blood acetone levels returned to normal within 24 hr. The authors reportedly found elevated levels of formic acid in the urine of the exposed workers.

Brugnone et al. measured the alveolar air and venous blood levels of acetone in 20 shoe factory employees exposed to small amounts of acetone vapor (440, 441). Measurement of the instantaneous concentration of acetone in the workroom air over a 4.5-hr monitoring period ranged from 1.1 to 49.9 mg/m^3 (0.5 to 21.1 ppm). The unweighted average value for the four values obtained during the monitoring period was 27.2 mg/m^3 (11.5 ppm). Calculation of the relative uptake at 1, 2.75, and 4.5 hr gave results of 81.4, 81.2, and 71.2 percent, respectively. The acetone concentration in venous blood was linearly related to the alveolar air levels, but not with the acetone concentration in the ambient air. The acetone blood levels ranged between 0.47 and 3.0 mg/l, which were within physiologically normal limits. The alveolar air acetone concentration was well correlated with the ambient air levels of acetone; however, the slope of the regression line showed considerable variability at different times during the day. The authors noted, but did not explain, the apparent decrease in pulmonary uptake as the workday progressed.

Pezzagno et al. examined the relationship between the inhaled air concentration of acetone and the concentration excreted in the urine by studying occupationally exposed volunteers (426, 427). The urinary excretion of acetone was determined in 104 workers from three different factories (paint, plastics, and artificial fibers), and the values were compared to the 4-hr TWA exposure to acetone, which ranged as high as 1500 ppm. A close linear relationship was observed for the 4-hr urine acetone level and the 4-hr TWA exposure concentrations ($r = .91$). The authors suggested that urinary acetone levels could be used to monitor acetone exposures in the workplace biologically and that a urine acetone concentration of 55 mg/l would equate to a 4-hr TLV–TWA exposure concentration of 1000 ppm. This value required a correction for background urine acetone levels, which averaged about 0.8 mg/l but were highly variable. Kawai et al. examined the relationship between urinary acetone levels and the 8-hr TWA exposure concentration for a group of 28 workers in a plastics plant (442). Urine acetone levels ranged from about 0.5 to 23 mg/l for employees whose personal 8-hr TWA exposure concentrations ranged from less than 1 ppm to about 45 ppm. A good correlation was observed between the TWA exposure level and the end-exposure concentration of acetone in the urine ($r = .90$). The correlation was not substantially altered after correcting for the background level of acetone in the urine. The authors concluded that urine acetone measurements could be used as an indicator of occupational exposure.

10.2 Laboratory Animal Studies

Numerous pharmacokinetic studies have been performed in laboratory animals treated with acetone by various routes. These studies have been performed in mice, rats, rabbits, and dogs, and the findings have generally agreed with those from humans. Animal studies have provided additional evidence that acetone elimination from the body is retarded once metabolic saturation has occurred and that uptake by the lung is subject to a wash-in/wash-out effect.

Widmark conducted several elegantly designed and interpreted pharmacokinetic studies in rabbits given a 700 mg/kg sc injection of acetone (128). Acetone blood levels reportedly declined in a logarithmic (i.e., first-order) manner following the absorption. Peak blood levels of acetone ranged from about 675 to 728 mg/g and typically occurred 50 to 90 min following treatment. A single exponential equation was found to describe accurately the loss of acetone from the blood, which was virtually complete by 24 hr post-treatment. The average volume of distribution (V_d) for acetone was 1.8 liters and the average blood elimination $t_{1/2}$ was 7.5 hr. When an acetone dose of approximately 350 mg/kg was injected iv into the jugular vein, an average $t_{1/2}$ of 7.4 hr was obtained. When rabbits were given 350 mg/kg of acetone by oral, intraperitoneal, or rectal administration, the post-exposure blood levels could be described by an exponential equation; however, specific rate constants were not provided. These data were used to show that the rate of acetone absorption and the coinciding peak blood levels were affected by the route of administration. DiVincenzo et al. determined the blood $t_{1/2}$ for acetone in groups of male Beagle dogs exposed for 2 hr to concentrations of 100, 500, or 1000 ppm (130). The amount of acetone eliminated from the blood and excreted into the expired air was approximately proportional to the exposure concentration. The rate of acetone elimination in the breath and blood declined in a first-order manner throughout the measurement period. The $t_{1/2}$ for acetone in the blood was calculated to be about 3 hr, regardless of the exposure concentration.

In a series of studies, Haggard et al. examined the pharmacokinetics of acetone in rats administered acetone by the intraperitoneal, oral, and pulmonary routes (78). Rats of an unknown sex and strain were initially exposed by inhalation to acetone vapor concentrations ranging from 5 to 300 mg/l (2100 to 126,000 ppm) for up to 8 hr. The peak blood levels obtained following an 8-hr exposure to 5, 25, or 50 mg/l (2100, 10,500, or 21,000 ppm) were about 250, 2000, and 4400 mg/l, respectively; the values were noted to rise disproportionately relative to the rise in exposure concentration. Appreciable day-to-day accumulation of acetone was observed in rats exposed for nine consecutive days (8 hr/day) to 5.35 mg/l (2250 ppm) of acetone. When continuous 24-hr exposures were performed at the same exposure concentration, the blood levels were found to plateau at about 1000 mg/l after 3 days of treatment.

In further studies, an acetone dose of 3 g/kg was administered orally to groups of rats. Following treatment, the blood levels of acetone peaked at about 2300 mg/l and tended to remain elevated for about 46 hr. The rate of acetone elimination from the blood could not be fully described by a single exponential equation. When the acetone blood levels ranged between 500 and 2300 mg/l, the rate of elimination was slower than the rate observed at lower blood concentrations. Rats given an ip dose of acetone ranging from 25 to 2000 mg/kg displayed an average blood-to-tissue distribution ratio of 0.82. The rate of acetone metabolism was calculated at regular intervals by subtracting the rate of excretion from the total rate of elimination. The rate of acetone metabolism tended to plateau and exhibit evidence of saturation at blood concentrations exceeding about 400 to 500 mg/l. The metabolic rate failed to increase substantially above 12 mg/(kg)(hr) once the blood level reached about 800 to 900 mg/l. At blood levels above 1000 mg/l, approximately 75 percent of the absorbed acetone was eliminated by excretion pathways, whereas only about 40 percent was eliminated by excretion at blood levels of 150 percent mg/l. The rate of acetone metabolism in the fasted rats was found to be about 20 to 30 percent greater than in fed animals. These studies demonstrated that large doses (concentrations) were capable of causing a disproportionate increase in the acetone body burden and that the pharmacokinetics of acetone displayed typical Michaelis–Menten behavior.

Hallier et al. studied the pharmacokinetics of acetone vapor in rats using gas uptake from a closed-recirculating exposure chamber (443). Male Wistar rats were exposed at either of four initial chamber concentrations ranging as high as 62,000 ppm, and the rate of acetone loss from the exposure chamber was monitored for up to about 30 hr. After an absorption and distribution phase lasting about 8 hr, acetone disappeared from the chamber at a rate equal to its rate of metabolism within the animal. Acetone elimination exhibited apparent first-order kinetics when the chamber concentration was 100 ppm or less. Graphical analysis of rate data indicated that acetone elimination could be described by a Michaelis–Menten equation. The Michaelis constant (K_m) and maximum velocity (V_{max}) for the process were calculated to be 160 ppm and 230 μmol/(kg)(hr), respectively. Acetone blood-to-air and urine-to-air partition coefficients averaged 210 and 330, respectively. Analysis of the urine specimens indicated that about 4.7 percent of the metabolized acetone was excreted as formate and that the excretion was linear with time.

Wigaeus et al. exposed mice to [2-^{14}C]acetone and determined the tissue distribution and respiratory excretion of acetone and its metabolites (444). Male white mice of an unstated strain received both single and repetitive exposures to 1200 mg/m^3 (500 ppm) of acetone. Steady-state concentrations of unmetabolized acetone and total radioactivity were found in most tissues and organs after 6 hr of exposure. Exposures lasting longer than 6 hr caused further accumulation in the liver and brown adipose tissue. After 24 hr of continuous exposure, the highest concentration of radioactivity was observed in the liver ($\sim$0.3 mg/g) and the lowest in the adipose tissue ($\sim$0.06 mg/g). Only about 10 percent of the radioactivity found in the liver was composed of unmetabolized acetone. Except for the adipose tissue, repetitive 6 hr/day exposures to acetone caused little or no additional accumulation of radioactivity in the tissues analyzed. The accumulation of radioactivity in the adipose tissue was noted to be relatively small compared to values obtained with lipophilic solvents, and the increase was attributed to the incorporation of the radioactive carbon from the metabolized acetone into lipids and lipid-soluble biomolecules.

The rate of elimination of total radioactivity following a single 6-hr exposure to acetone was greatest for the blood, kidneys, lungs, muscle, and brain with $t_{1/2}$ values of 2 to 3 hr and slowest for the subcutaneous adipose tissue where the $t_{1/2}$ was slightly greater than 5 hr. The acetone concentration in all tissues was found to be at normal levels within 24 hr of exposure termination. Calculation of the distribution ratios between the various tissues and blood revealed that unmetabolized acetone preferentially remained in the blood (i.e., the ratios were all less than unity). In vivo and in vitro determinations of the blood-to-air partition coefficient for acetone yielded values of 70 and 248, respectively. The authors attributed this large discrepancy to the possibility that a portion of the inhaled acetone was reversibly dissolved into the saliva and mucous membranes that lined the upper respiratory tract. The authors concluded that acetone was evenly distributed throughout the body and that acetone did not accumulate in the body after repeated exposures to 500 ppm.

Plaa et al. examined the elimination pharmacokinetics of both acetone and isopropanol in male Sprague-Dawley rats administered a single oral dose of 0.10, 0.25, 1.00, or 2.50 ml/kg (319). As shown in Table 5.19, the maximum blood concentration, the volume of distribution, and the blood half-life all showed changes indicative of metabolic saturation and Michaelis–Menten kinetics for both acetone and IPA. Values for the area under the blood concentration versus time curve (AUC) and for total body clearance failed to show any clear evidence of dose

Table 5.19. Pharamacokinetic Parameters for Acetone Elimination Following Oral Treatment of Male Sprague-Dawley Rats with Either Acetone or Isopropanol (319)

Dose (ml/kg)	Maximum Plasma Concn. (mg/l)	Area Under the Curve [mg/(hr)(L)]	Total Body Clearance (ml/hr)	Volume of Distribution (ml)	Elmination Half-life (hr)
		Acetone Administration			
0.25	311	3,049	64	227	2.4
1.00	871	9,143	86	608	4.9
2.50	1292	26,195	75	779	7.2
		Isopropanol Administration			
0.25	207	1,850	62	361	4.0
1.00	634	6,978	84	372	3.1
2.50	917	21,120	75	779	7.2

dependency, which may have been due to the linear one-compartment model used to describe the elimination kinetics of each compound. A separate determination of systemic bioavailability (F) following iv infusion and oral gavage revealed that acetone was completely absorbed from the GI tract with F equaling 1.02. These data indicated that acetone and IPA elimination were saturated in rats when the blood levels rose above 300 to 400 mg/l.

Hetenyi and Ferrarotto examined the rate and extent of acetone turnover in male Sprague-Dawley rats that were fasted for 3 days and then administered a small tracer amount (0.17 to 0.20 mg/kg) of uniformly radio-labeled acetone ([U-^{14}C]acetone) by intra-arterial injection (445). The metabolic clearance and turnover rates of acetone were calculated to be 2.25 ml/(min)(kg) [1.78 g/(min)(kg)] and 0.74 μmol/(min)(kg) [43 μg/(min)(kg)], respectively. In a separate study, Filser and Bolt reported that the endogenous rate of acetone production in untreated rats was about 100-fold lower, with a value of 5 μmol/(hr)(kg) [290 μg/(hr)(kg)] (446).

Egle determined the uptake of acetone vapors from surgically isolated regions of the canine respiratory tract (447). Anesthetized mongrel dogs of both sexes were exposed to acetone vapor concentrations ranging from 0.36 to 0.80 mg/l (151 to 337 ppm) for an unstated length of time. The respiratory uptake was determined after either a one-way or two-way pass of the vapors through the upper and lower respiratory tracts. The uptake of acetone vapors by the entire respiratory tract averaged about 52 percent at ventilation rates of 4 to 18 breaths/min; however, the uptake dropped to about 42 percent when the ventilation rate exceeded 20 breaths/min. Similar results were obtained for the lower respiratory tract. The uptake of acetone from the upper respiratory tract averaged about 57 percent for both the one-way and two-way passes; however, the values tended to decline when the respiration rate was increased. With the respiration rate remaining constant, the uptake of acetone increased noticeably from 52.1 to 58.7 percent when the inspired concentration was increased from 0.46 to 1.75 mg/l (194 to 737 ppm). These findings were consistent with the wash-in/wash-out effect observed in previous studies with humans. Aharonson et al. observed a rise in acetone uptake when the pulmonary ventilation rate was increased in anesthetized mongrel dogs (448). The uptake of an unstated concentration of acetone vapors by the nose and throat more than doubled when the air flow rate was increased from 2 to 5 l/min. The increase in uptake was thought to be associated with an increase in the perfusion of blood to the tissues, an increase in the transfer of vapors into the tissues, or an increase in the effective area for vapor uptake.

Morris et al. studied the regional deposition of acetone vapors in the upper respiratory tract (URT) of various animal species (449). Anesthetized male Fischer 344 rats and Hartley guinea pigs with a tracheostomy were exposed for less than 20 min to acetone vapor concentrations ranging from 810 to 1800 mg/m^3 (340 to 756 ppm). The acetone vapors were drawn through the URT at flow rates ranging from 50 to 300 ml/min for the rats and 70 to 575 ml/min for the guinea pigs. The deposition of acetone vapors in the URT was found to reach an apparent steady state in both species rapidly. Furthermore, the ratio of the deposited to nondeposited fraction was shown to be inversely related to the inspiratory flow rate. The deposited fraction did not depend on the inspired acetone concentration, which indicated that URT metabolism was not an important factor in the overall uptake. The URT deposition in the Fischer rat was found to be about twofold higher than in the guinea pig. The species difference was not associated with differences in the blood-to-air partition coefficient, which was found to average 260 and 270 in the rat and guinea pig, respectively. The authors concluded that differences in the anatomy or physiology of the URT may have been responsible for the higher URT deposition in the Fischer rat. Additional studies on URT deposition in anesthetized Sprague-Dawley rats revealed an even greater uptake of acetone (450). In this case, the acetone vapor concentrations ranged from 1000 to 1300 mg/m^3 (420–546 ppm) and the nominal flow rates through the URT were between 70 and 500 ml/ min. The absolute deposition of acetone vapors at flow rates of 70, 150, 300, and 500 ml/min were 0.66, 0.96, 1.18, and 1.76 μmol/min, respectively, which were about two to three times greater than with either the Fischer rat or the Hartley guinea pig.

The uptake of acetone vapors has also been examined in male Syrian golden hamsters (451) and B6C3F1 mice (452). As in previous studies, the deposition of acetone vapors in the isolated URT was inversely related to flow rate and independent of the vapor concentration, and tissue metabolism did not contribute to the overall uptake of acetone. At a ventilation rate 2.5- to 3-fold higher than the resting rate, the uptake efficiency in various species declined in the following fashion: Sprague-Dawley rat (21%); B6C3F1 mouse (14%); Fischer 344 rat (12%); Hartley guinea pig (7%); Syrian hamster (5%). The inefficient uptake of acetone vapors by the Syrian hamster was attributed to the slow perfusion of blood through the nasal tissue, whereas other anatomic or physiological factors were thought to be responsible for the differences observed among the remaining species.

11 SUBCHRONIC TOXICITY

The subchronic toxicity of acetone has been examined in rodents exposed by the oral, dermal, and pulmonary routes. These studies have typically involved the treatment of animals at a rate of 5 days/week for periods lasting from 2 to 13 weeks. Because subchronic studies have often been designed to determine the maximally tolerated dose (MTD) of acetone, extremely large doses (concentrations) of acetone have been involved. A comparison of the results from oral and inhalation studies has resulted in some conflicting results, particularly with regard to effects on the hematopoietic system. The organs most sensitive to large acetone exposures appear to be the liver and kidney, which show increases in relative or absolute organ weights without any corresponding effect on tissue morphology. The organ weight changes observed in these studies may be related to the auto-inductive effects of acetone and the resulting increase in protein turnover.

11.1 Inhalation Studies

Bruckner and Peterson reported that repeated inhalation of high vapor concentrations of acetone were well tolerated by male Sprague-Dawley rats (453). Animals exposed 3 hr/day to 19,000 ppm of acetone for up to 8 weeks (5 days/week) were sacrificed and examined after either 4 or 8 weeks of exposure or 2 weeks after the last exposure. Blood and tissue (lung, liver, brain, kidney, and heart) specimens were examined biochemically and histologically for evidence of tissue damage. Serum specimens were analyzed for lactate dehydrogenase (LDH) activity, AST activity, and BUN concentration and the liver was assayed for triglyceride content. No statistically significant increases were observed in these parameters for any of the treatment groups. Likewise, tissue histopathology did not reveal any treatment-related evidence of organ damage. A slight, statistically significant decrease was reported for the absolute weights of the brain and kidney (relative organ weights were not calculated) of animals in the interim and terminal treatment groups. No organ weight changes were observed in the 2-week recovery group.

Goldberg et al. recorded mild neurobehavioral changes in rats repeatedly exposed to high vapor concentrations of acetone (454). Female Carworth rats exposed for 2 weeks (5 days/week) at 4 hr/day to 3000, 6000, 12,000, or 16,000 ppm of acetone were examined for their response to avoidance and escape stimuli before and after each exposure. Repeated daily exposures to 6000 ppm of acetone produced an inhibition of avoidance behavior, but did not produce any signs of motor imbalance. Acetone concentrations of 12,000 or 16,000 ppm produced ataxia in several animals after a single exposure; however, a rapid tolerance developed and ataxia was not seen on subsequent days. Kurnayeva et al. studied the combined effect of vapors and noise on the immune, cardiovascular, endocrine, and nervous system of Wistar rats exposed for 1.5 to 2.0 months (455). The animals were exposed for 4 hr/day, 5 days/week, to 2000 mg/m³ (843 ppm) of acetone and 85 db of noise. The combined exposure was reportedly without effect on the various measures of physiological function; however, results were only presented for the immune system.

11.2 Oral Studies

Spencer et al. compared the neurotoxicity of various ketonic solvents and reported that repeated acetone treatments did not cause neuropathological lesions in rats (456). Sprague-Dawley rats (sex not stated) were initially administered acetone in their drinking water for eight weeks at 0.5 percent; the concentration was then increased to 1.0 percent for an additional 4 weeks. Tissue specimens were collected from the spinal cord and spinocerebellar area as well as from the plantar, tibial, and sciatic nerves. Histological examination of nerve tissue from acetone-treated animals was found to be indistinguishable from the controls. The authors concluded that acetone did not cause central-peripheral distal axonopathy in rats. Ladefoged et al. examined the performance of male Wistar rats on a rotarod following a 6-week drinking water exposure to 5 percent acetone (344). Balance time on the rotarod was measured once a week for a maximum of 30 sec during the treatment period. The authors reported that the acetone exposure had no effect on the balance times observed throughout the study.

Hètu and Joly examined liver microsomal MFO activities in female Sprague-Dawley rats given 1 percent acetone in their drinking water for 2 weeks (457). The treatment resulted in an average acetone dose of 1.2 g/(kg)(day) and did not cause any change in body weight, liver weight, or liver microsomal protein content. Livers removed at the end of the treatment period showed elevated levels of cytochrome P450 as well as increases in p-nitrophenol hydroxylase, aniline hydroxylase, and 7-ethoxycoumarin O-deethylase activities. The authors noted that the increases in hepatic enzyme activity were nearly abolished in animals that were allowed a 24-hr recovery. Sinclair et al. found that male C57BL/6 mice given a single ip injection of iron-dextran 3 days prior to the administration of 1 percent acetone and 5-aminolevulinic acid in their drinking water exhibited high levels of uroporphyrin in their liver and urine after 35 days of treatment (458). These increases were associated with a large decrease in liver uroporphyrinogen decarboxylase activity.

Skutches et al. showed that acetone treatment could affect insulin-stimulated glucose oxidation in rat adipose tissue (459). Male Sprague-Dawley rats were given acetone in their drinking water for seven days at concentrations of 0.5, 1.0, 2.0, or 3.0 percent. The average plasma acetone concentrations at the end of the treatment period were proportional to the drinking water concentration and ranged from 0.52 mM (30 mg/l) at 0.5 percent acetone to 15.9 mM (923 mg/l) at 3.0 percent acetone. The in vitro rate of glucose oxidation by epididymal adipose tissue from the acetone-treated rats was significantly lower than the control rate when insulin was added to the incubation mixture. The effect was noted in the adipose tissue of rats that were treated with 3.0 percent acetone for 7 days and in the tissue of rats that were allowed 4 days of recovery after 5 days of treatment at 3 percent. The authors concluded that the insulin resistance observed in diabetic patients may be due in part to the acetone or acetol levels observed in these individuals.

Acetone has been administered by oral gavage to groups of 30 male and female Sprague-Dawley rats for 90 consecutive days (460, 461). The treatment was performed at three dose levels (100, 500, and 2500 mg/kg) using water as a dosing vehicle control. No toxicologically significant differences in food consumption or body weight gain were noted at any dose. Animals of both sexes treated at 2500 mg/kg revealed increases in several hematological parameters (hemoglobin concentration, hematocrit, and erythrocyte indexes) and an increase in the serum activity of three enzymes (ALT, AST, and alkaline phosphatase). Male rats also showed an increase in serum cholesterol at the two highest doses and a decrease in serum glucose at the 2500 mg/kg dose. Female rats showed increases in the absolute liver and kidney weight at the two highest doses; however, parallel increases in the organ-to-body weight ratios were only observed at the highest dose level. Male rats administered 2500 mg/kg displayed an increase in the organ-to-body weight ratios for the liver and kidney, but the absolute weights of these organs were not affected. Histopathological examination of tissue sections from the liver did not reveal any structural abnormalities. The renal tubular cells of both male and female rats showed some morphological abnormalities at the high dose; however, the changes were observed in both the control and test animals. The authors reported that the acetone treatment appeared to accentuate the renal nephropathy that accompanies aging.

Dietz et al. conducted a similar subchronic study using both rats (Fischer 344) and mice (B6C3F$_1$) of both sexes (462, 463). Acetone was administered in the drinking water of both species for either 14 days or 13 weeks. The 14-day study was conducted at five or six concentrations with five male and five female animals per species per dose, whereas the 13-week study used five concentrations and 10 animals per sex per species per dose. The drinking water concentrations and the calculated time-weighted average doses to the mice and rats used in these studies are listed in Table 5.20. No mortality was seen in the rats or mice used in either the 14-day or 13-week study. Overt clinical signs of toxicity were observed only in the rats treated at the 10 percent dose level in the 14-day study and were described as an emaciated appearance from decreased weight gain. Acetone-induced increases in relative kidney weights were observed at the 2 percent dose level in female rats and the 5 percent dose level in male rats. The kidney weight changes were reportedly associated with a nephropathy that occurred spontaneously in untreated control rats. The authors suggested that large doses of acetone accelerated the occurrence and severity of this condition in the treated rats. Male and female rats also exhibited an increase in the relative liver weight at the 2 percent dose level, and the male rats exhibited an increase in the relative testicular weight at the 5 percent dose level. The liver weight changes were not associated with treatment-related morphological alterations. No ophthalmic lesions were observed in either the rats or mice used in this study. Hematologic effects consistent with macrocytic anemia were noted in male rats dosed at the 0.5 percent level and above. Minimal to mild splenic pigmentation (hemosiderosis) was observed exclusively in male rats dosed at the 2 percent and 5 percent levels and was probably related to the hematologic effects. Anemia was not apparent in female rats although some mild hematologic

Table 5.20. Time-Weighted-Average Dose for Male and Female Fischer 344 Rats and B6C3F$_1$ Mice Exposed To Acetone in Their Drinking Water (463)

Concentration %	14-Day Study TWA Dose [mg/(kg)(day)]				13-Week Study TWA Dose [mg/(kg)(day)]			
	Rats		Mice		Rats		Mice	
	Male	Female	Male	Female	Male	Female	Male	Female
0	0	0	0	0	0	0	0	0
0.125	—	—	—	—	—	—	380	—
0.25	—	—	—	—	200	200	611	892
0.5	714	751	965	1569	400	600	1353	2007
1	1616	1485	1579	3023	900	1200	2258	4156
2	2559	2328	3896	5481	1700	1600	4858	5945
5	4312	4350	6348	8804	3400	3100	—	11,298
10	6942	8560	10,314	12,725	—	—	—	—

changes were noted in female rats dosed with 5 percent acetone. The most notable findings in the mice were increased liver and decreased spleen weights, which were confined exclusively to female mice given 5 percent acetone. In contrast to the rats, mice showed no compound-related effects in the hematologic indexes. Centrilobular hepatocellular hypertrophy was noted histologically in several of the female mice dosed at the 5 percent level. The authors concluded that the MTD for acetone in the drinking water was 2 percent for male rats and male mice and 5 percent for female mice. No toxic effects were identified for female rats.

11.3 Subcutaneous and Dermal Studies

Misumi and Nagano performed a neurophysiological study with mice treated subcutaneously with acetone (464). Male Donryu mice treated for 15 weeks (5 days/week) with a 400 mg/(kg)(day) dose of acetone showed no evidence of neurological dysfunction relative to control animals. The acetone treatment did not cause any difficulty in walking or dullness in movement, and there were no significant changes in motor or sensory nerve conduction velocities. The authors concluded that acetone was not neurotoxic to the peripheral nervous system.

Considerable controversy has surrounded the reported appearance of cataracts in guinea pigs, but not rabbits, treated dermally with acetone. Rengstorff et al. reported that guinea pigs treated either cutaneously or subcutaneously with acetone (3 days/week) for 3 weeks developed cataracts (465). Male and female and guinea pigs (strain not stated) were treated dermally with 0.5 ml of acetone, subcutaneously with 0.05 ml of 50% acetone solution; or subcutaneously with 0.05 ml of a 5 percent solution of acetone. The treated animals were examined using an ophthalmoscope and slit-lamp for evidence of ocular damage at 2, 3, 4, 5, and 6 months posttreatment. By the third month, bilateral cataracts were observed in the lenses of the animals treated both dermally and subcutaneously. When 1 percent acetone was applied dermally to guinea pigs or two rabbits twice daily (5 days/week) for either 4 or 8 weeks, several of the guinea pigs, but neither of the rabbits, developed cataracts during the 6-month examination period. None of the control animals showed histological evidence of cataract formation. The authors noted that morphological appearance of the lesion in the acetone-treated animals bore some resemblance to the cataracts observed in diabetics and that their above-average levels of acetone could be the cause.

Rengstorff et al. further examined the susceptibility of New Zealand White rabbits and albino guinea pigs to the cataractogenic effects of dermal acetone treatments (466, 467). Male and female rabbits treated with 1 ml of acetone 3 days/week for 3 weeks showed no lens abnormalities during the 6-month observation period. Male and female guinea pigs were treated with 0.5 ml of acetone 5 days/week for 6 weeks and examined with an ophthalmoscope and slip-lamp at regular intervals for 1 year post-treatment. By 3 months post-treatment, cataracts were observed in 30 percent of the test animals and none of the control animals. The ascorbate levels in aqueous humor specimens from the test animals with cataracts were appreciably below those found in the control animals throughout the 1-year observation period. The authors concluded that the development of cataracts in guinea pigs was species specific and related to ascorbate synthesis.

12   DEVELOPMENTAL AND REPRODUCTIVE TOXICITY

Repeated exposures of male and female rodents to acetone has resulted in mild reproductive or developmental effects. No increase in fetal lethality or teratogenicity was observed in pregnant mice and rats exposed to acetone vapors. A recent 1-year drinking water study in rats showed that acetone did not affect any of a variety of reproductive indexes, including the number of matings, number of pregnancies, and testicular morphology. Because of its favorable properties as a treatment vehicle, acetone has been tested in wide array of in vitro test systems designed to detect any adverse effects on a developing embryo. These assays have generally used chicken eggs and rodent embryos as test systems for evaluating the adverse effects on growth or structural development (468, 469). Other in vitro systems have included frog eggs and embryos (470, 471), mallard embryos (472), and hydra (473). Many of the studies presented below have evaluated the use of acetone as a treatment vehicle or carrier for administering water insoluble compounds.

12.1   In Vivo Studies

Acetone tested positive in the Chernoff and Kavlock screening assay for reproductive toxicants (474). Acetone was administered orally at a dose of 3500 mg/(kg)(day) to a group of 50 female CD-1 albino mice on days 6 through 15 of gestation. The weight and survival of the pups in each litter were then determined postnatally on days 1 and 3. Fetal weight and survival and maternal mortality were evaluated against a ranking system that scaled the severity of each adverse effect. The changes observed due to acetone administration yielded a total score of 19 out of a possible 22 points, which indicated that acetone warranted high priority for additional developmental testing.

The potential developmental effects of acetone vapor have been examined in virgin and pregnant Sprague-Dawley rats and Swiss CD-1 mice (475). Groups of 32 positively mated rats were exposed by inhalation for 6 hr/day to 440, 2200, or 11,000 ppm of acetone on days 6 through 19 of gestation. The mice were exposed similarly at concentrations of 0, 440, 2200, or 6600 ppm of acetone on days 6 through 17 of gestation. The mice did not receive an 11,000 ppm exposure because of the severe narcosis observed at the end of the first exposure session. For comparative purposes, three groups of 10 virgin female rats and mice were also exposed to acetone vapors for 14 days and 12 days, respectively. Maternal uterine and fetal body weights were recorded upon sacrifice and live fetuses were examined for gross, visceral, skeletal, and craniofacial defects. No deaths were observed in the treated rats. The mean pregnancy rate was greater than or equal to 93 percent for all treatment groups. Blood specimens were collected from pregnant rats on three of the exposure days and analyzed for the level of all three ketone bodies. Table

Table 5.21. Acetone Plasma Levels Resulting from the Daily Inhalation Exposure of Pregnant Sprague-Dawley Rats on Days 6 Through 19 of Gestation (475)

Exposure Concentration (ppm)	Acetone Level 30 min Post-Exposure (mg/l)			Acetone Level 17 hr Post-Exposure (mg/l)		
	Day 7	Day 14	Day 19	Day 7	Day 14	Day 19
0	<0.06	<0.6	1.7	<0.6	<0.6	<0.6
440	38.3	43.0	52.9	<4.1	<0.6	<1.7
2,200	296.2	273.0	296.2	<0.6	0.6	23.2
11,000	2149.0	2090.9	1974.7	174.2	174.2	348.5

5.21 shows the acetone blood levels in the specimens collected from pregnant rats at 30 min and 17 hr post-exposure. The 30-min acetone blood levels rose in a disproportionate fashion relative to the change in exposure concentration. Additionally, acetone blood levels had not returned to normal by 17 hr post-exposure when the exposure concentration was 11,000 ppm. Blood levels of acetoacetate and β-hydroxybutyrate were not increased following the acetone treatment.

Clinical signs of toxicity in rats were confined to the 11,000 ppm treatment group and included a statistically significant reduction in maternal body weight gain starting on day 14 of gestation and a decrease in the uterine and extragestational weight gain of the dams. The fetal weights were found to be significantly lower for the 11,000 ppm group relative to the control group. Mean body weights of treated virgin female rats were also reduced, but not significantly. No effects were seen in the mean liver or kidney weights of pregnant dams, the organ-to-body weight ratios, the number of implantations, the mean percentage of live pups per litter, the mean percentage of resorptions per litter, or the fetal sex ratio. The incidence of rat fetal malformations was not significantly increased by gestational exposure to acetone vapors, although the percentage of litters with at least one pup exhibiting a malformation was reportedly greater for the 11,000-ppm group (11.5 percent) than for the control group (3.8 percent).

No treatment-related effects were seen on maternal or virgin body weight or maternal uterine weight in the treated mice. The mean pregnancy rate in groups of treated mice was greater than or equal to 85 percent. A statistically significant reduction in fetal weight and a slight but statistically significant increase in the percent incidence of late resorptions were seen in mice treated with 6600 ppm of acetone. The increased incidence of late resorptions in mice did not affect the mean number of live fetuses per litter. The incidence of fetal malformations was not altered at any exposure level; however, a treatment-related increase in the liver-to-body weight ratio was observed for the pregnant dams. The authors concluded that 2200 ppm of acetone was the no-observable-effect level (NOEL) for developmental toxicity in both rats and mice. Minimal maternal toxicity was seen at levels of 11,000 ppm of acetone in rats and 6600 ppm in mice.

Dietz et al. detected some mild adverse spermatogenic effects in male rats, but not mice, that consumed 5 percent acetone [3.1 g/(kg)(day)] in their drinking water for 13 weeks (462, 463). Fischer 344 rats showed a relative decrease in testicular weight along with depressed sperm motility, increased epididymal weight, and an increased incidence of abnormal sperm. No histopathological changes were observed in the testes of the test animals. The authors noted that the effects were subtle in nature and only occurred at very high dose levels.

Larsen et al. examined the reproductive effects of acetone in male Wistar rats administered 0.5 percent acetone in their drinking water for 6 weeks (476). On the fifth week of treatment, the rats were allowed to mate with untreated females and the number of matings were recorded together with the number of pregnancies and the number of fetuses per pregnancy. The absolute weight of the testes was measured along with the diameter of the seminiferous tubules in the treated and control rats. Semiquantitative histopathological scoring was used to detect any effects on vacuole formation, chromatic margination, epithelial disruption, multinucleated giant-cell formation, intracellular debris, or atrophy of the testes. None of these measures of reproductive and testicular toxicity were affected by the acetone treatment relative to the control animals.

12.2 In Vitro Studies

Kitchin and Ebron developed an in vitro rodent whole-embryo culture system as a screening assay for teratogenic agents (477). Various solvents were evaluated as potential delivery systems for water-insoluble chemicals using embryos from Sprague-Dawley rats. Acetone was added to the culture medium at concentrations of 0.1, 0.5, 2.5, and 10.0 percent. After a 48-hr incubation, acetone concentrations of 0.5 percent and greater were found to cause structurally abnormal embryos and a high rate of embryo lethality. Equivalent amounts of ethanol, dimethyl sulfoxide, and Tween produced a similar degree of dysmorphogenesis and teratogenicity. Picard and Van Maele-Fabry used these data to quantify acetone embryo lethality as an ELC_{50} value (478). The acetone concentration that caused 50 percent mortality in the embryos was calculated to be 0.20 M (11.6 g/l).

Schmid evaluated the effects of acetone and 38 other chemicals in a similar postimplantation rat embryo culture (479). Acetone did not cause any structural malformations when incubated with embryo cultures from Wistar rats at concentrations of 0.1, 0.3, 0.6, or 1.0 percent; however, evidence of embryo toxicity was observed at acetone concentrations of 0.6 percent or greater. An acetone level of 3 percent was found to inhibit the growth and differentiation of all the embryos tested. Guntakatta et al. developed a mouse embryo limb-bud cell culture system to estimate teratogenic potential of chemicals (480). Embryos from Swiss-Webster mice were incubated with acetone concentrations ranging from 10 to 100 mg/ml for periods up to 96 hr. Acetone did not cause any increase in [^{3}H]thymidine or $^{35}SO_4^{-2}$ incorporation into the proteoglycans that were biosynthesized.

DiPaolo et al. added polycyclic hydrocarbons dissolved in acetone to cell cultures derived from Syrian hamster embryos and then morphologically evaluated the clones that formed from the cultured cells (481). Cloning efficiency and induced transformants were determined when acetone was added at concentrations of up to 0.02 percent. No transformants and a high cloning efficiency were found in the acetone-treated control cultures. Acetone was also examined for use as a solvent in a similar assay that involved the ip injection of the test chemical into pregnant hamsters 3 days before isolating and preparing the embryo cell cultures (482). Treatment of animals with 3.0 ml/kg of acetone did not result in morphological evidence of cell transformation in any of the subcultures.

Acetone did not decrease the number of fertile eggs, the number of viable embryos, or the percentage of live embryos when used as a vehicle to treat developing chick embryos (483). Acetone caused less mortality than dimethyl sulfoxide when injected into the eggs at a level of about 0.8 ml/kg. Further studies with chicken embryos showed that the use of acetone as a treatment vehicle did not alter the liver weights or the liver RNA and DNA content of viable embryos (484). McLaughlin et al. injected acetone and 13 other chemicals into the yolk sacs of fresh fertile chicken eggs before a 20-day incubation (485, 486). Toxicity was judged by the percentage of hatched eggs and an examination of the embryonic development in eggs that did not hatch. The percentage of eggs that hatched was 80 and 50 percent at acetone levels of 39 and 78 mg/kg, respectively. According to the authors, acetone showed a relatively low order of toxicity and no teratogenic effects. Walker used a very similar yolk sac injection procedure and found that the treatment of chicken eggs with 0.05 to 0.10 ml of acetone caused a 20 to 37 percent survival depending on when the eggs were treated (487). The embryo mortality was associated with the formation of a protein coagulum immediately following injection of acetone. Acetone volumes less than 0.05 ml did not cause any increase in mortality. Korhonen et al. found that the injection of 5 ml of acetone into the eggs of developing chick embryos did not significantly increase the number of early or late deaths or the percentage of malformations in the embryos that failed to survive (488, 489). Swartz treated chicken embryos with 0.05 ml of acetone for either 5 or 12 consecutive days and did not observe any increase in mortality relative to untreated controls (490).

13 CHRONIC TOXICITY

Information on the potential long-term or chronic effects of acetone is generally limited to the results obtained from mouse dermal application studies, employee health surveys, and in vitro mutagenicity assays. In many instances, the available information has been derived from studies where acetone was used as a control to evaluate the effects of an unrelated chemical. When compared to unexposed control populations, acetone exposures did not cause any increase in tumorigenicity in mouse dermal studies or have any effect on mortality or clinical chemistry values in the occupational surveys. The carcinogenicity of acetone has not been formally examined in a 2-year rodent bioassay; however, in vitro tests for mutagenicity, chromosome damage, and DNA interaction indicate that acetone is not genotoxic except under severe conditions.

13.1 Human Studies

Ott et al. examined the mortality rates and clinical laboratory results for 948 employees exposed to a median TWA acetone exposure of 380, 770, or 1070 ppm over a span of 23 years in a cellulose fiber production plant (491–493). These workers were compared to the U.S. general population and used as a reference group for a larger methylene chloride epidemiology study. No statistical differences were found for either men or women between the acetone-exposed groups and the control group. The observed deaths for all causes, cardiovascular disease, and total malignant neoplasms were below expectation by 55, 61, and 43 percent, respectively. Likewise, there was no indication that occupational acetone exposures up to 1070 ppm had an adverse affect on selected hematologic and clinical chemistry determinations.

Oglesby et al. summarized 18 years of industrial experience with employees in a cellulose acetate production facility where medical information from thousands of workers and spanning over 21 million man-hours of acetone exposure were compiled and reviewed (494). Using medical department visits, lost-time records, and comprehensive medical examinations, the authors did not detect an increased incidence of illness relative to appropriate control populations. Mild transient symptoms of irritation were recorded when the average exposure concentrations exceeded 2500 ppm. The authors concluded that, based on years of employee experience, acetone concentrations up to 1500 ppm would be entirely without injurious or objectional effects for repeated continuous exposure of up to 8 hr. Rösgen and Mamier examined 45 men and 39 women in a container manufacturing establishment where acetone was used as a solvent (495). Blood specimens from the employees were used to determine the hemoglobin concentration, coagulation time, sedimentation rate, red and white blood cell counts, and white blood cell differential counts. The clinical measurements revealed a below-normal hemoglobin concentration in several of the women, which was attributed to their poor nutritional status. The authors concluded that the acetone exposures did not cause any hematologic abnormalities.

Parmeggiani and Sassi examined workers at three separate manufacturing sites where acetone was used in combination with other solvents (437, 496). The first work site involved a group of six employees in a plant manufacturing acetate rayon. Employees exposed to acetone at concentrations ranging from 309 to 918 ppm for up to 3 hr over a 7- to 15-year period reportedly complained of drowsiness, eye and throat irritation, dizziness, inebriation, and headache. The employees also displayed muscular weakness, vertigo, and chronic inflammation of the stomach, duodenum, and air passages. A physical examination showed signs of pharyngeal, conjunctival, and lung irritation in five of the six employees. Very similar signs and symptoms were obtained in a separate employee exposed to only about 25 ppm of acetone. The second work site employed four workers in a bottle lacquering operation where acetone concentrations ranged from 84 to 147 ppm. Workers at this site experienced nausea, abdominal pain, headache, vertigo, loss of appetite, vomiting, and other debilitating symptoms. The third site was a plant producing polyvinyl chloride fibers and involved 11 employees. Acetone vapor concentrations ranging from 13 to 86 ppm reportedly caused irritation to eyes, nose, throat, and bronchi, as well as some severe CNS disturbances. High concentrations of carbon disulfide, a potent CNS toxicant, were also found in this plant. No apparent relation was observed between the acetone exposure concentration and the amount of sensory irritation or CNS involvement experienced by the subjects used in these studies. In addition, the results from this study contrast with others that show acetone to be a mild sensory irritant. There is a high probability that the results from the Parmeggiani and Sassi investigation were distorted by confounding variables.

Grampella et al. examined the possibility that long-term occupational exposures to acetone could cause systemic organ damage (497). A group of 60 volunteers employed for at least 5 years in an acetate fiber manufacturing facility were divided into two equal groups according to their level of exposure. The low-exposure group had TWA acetone exposures (time not stated) that ranged from 1303 to 1550 mg/m^3 (549 to 653 ppm) and an average urine acetone value of 62 mg/l (measurements recorded on a spot urine specimen collected midway through the work shift). In contrast, the high exposure group received personal TWA exposures ranging from 2251 to 2488 mg/m^3 (948 to 1048 ppm) and had an average midshift urine acetone level of 93 mg/l. The two groups of test subjects were compared to a single group of 60 controls that had never been exposed to acetone. Blood specimens were collected from all of the subjects and submitted to the following hematologic and clinical analyses: glucose, glutamic-pyruvic transaminase (ALT), glutamic-oxaloacetic transaminase (AST), λ-glutamyl transpeptidase, protein electrophoresis, blood urea nitrogen, creatinine, platelet count, and red and white blood cell counts. After taking into consideration risk factors, past medical histories, and age, no statistically significant difference was noted between the different test groups for any of the clinical measurements.

13.2 Laboratory Animal Studies

The only long-term acetone feeding study found in the open literature was performed by Sollmann (498) and involved only three female albino rats given 2.5 percent acetone in their drinking water [average daily dose of 1.8 ml/(kg)(day)] for 18 weeks. None of the animals died during the study, but the authors noted a sharp decrease in fluid consumption, a 23 percent decrease in food consumption, and a 34 percent drop in the expected weight gain over the course of the study. These rough measures of toxicity were the only end-points monitored during the study. Additional information on the effects of long-term exposure to acetone can be obtained from dermal application studies performed in mice. Because of its utility as a vehicle for the dissolution and application of water-insoluble chemicals in dermal carcinogenicity studies, data are available from the vehicle control animals used in these studies.

Van Duuren et al. conducted a skin carcinogenicity study with a flame retardant that was applied to the shaved dorsal skin of mice (499). Vehicle-control groups of 29 female ICR/Ha Swiss mice were treated topically with 0.1 ml of acetone or

0.1 ml of an acetone–water mixture (9:1) three times a week for up to 424 days. Histopathological analysis of all major organs revealed a total of 14 lung tumors, one liver tumor, one forestomach tumor, and no skin tumors in the acetone and acetone–water treatment groups. Lung papillary tumors were seen in 37 percent of the untreated mice, whereas 24 percent of the acetone and acetone–water treated mice showed similar tumors. The incidence of forestomach tumors in acetone or acetone–water treated mice was comparable to untreated mice (3 percent). Except for one undifferentiated malignant liver tumor, which was not cited as a remarkable finding by the authors, the incidence of systemic tumors in acetone and acetone–water treated mice was not different from the background incidence of tumors in untreated mice.

In a separate study, Van Duuren et al. reported that acetone was not a tumor promoting agent on the skin of 20 female ICR/Ha Swiss mice treated with 0.1 ml of acetone given three times per week over 441 days (500). The initiator, 7,12-dimethylbenz[a]anthracene was applied 2 weeks before beginning the acetone treatment. Weiss et al. reported that acetone was capable of inhibiting tumor formation when used in a mouse two-stage initiation-promotion experiment (501, 502). A 56 percent decrease in the total number of tumors per mouse was observed when 7,12-dimethylbenz[a]anthracene-initiated CD-1 mice were treated with 0.1 to 0.4 ml of acetone 3 to 5 min prior to treatment with a promoter (phorbol 12-myristate 13-acetate) twice a week for 10 weeks.

Ward et al. provided a historical review of the organ pathology in the acetone-treated mice used in skin painting studies (503). Two groups of 30 female SENCAR mice were combined and examined together because there were no observed differences in the incidence of neoplastic and non-neoplastic lesions. The first group received 0.2 ml of acetone twice per week for 92 weeks whereas the second group was treated with 0.2 ml of acetone once per week for 88 weeks. The major organs and tissues from all of the animals were examined both macroscopically and microscopically following necropsy. Fifty percent of the animals survived past 96 weeks of age with 15 of the 60 mice dying because of neoplastic lesions and 27 because of non-neoplastic lesions. The two major contributing causes of death were identified as glomerulonephritis and histiocytic sarcoma. The authors concluded that there was no evidence that the tumors or lesions were associated with the acetone treatment. Iverson reported extremely low malignant skin tumor rates and yields in hairless *hr/hr* Oslo mice treated with 0.1 ml of acetone once or twice a week for 20 weeks (504).

Peristianis et al. conducted a 2-year dermal carcinogenicity study with three epoxy resins dissolved in acetone and applied to the shaved dorsal skin of mice (505). A group of 99 male and 11 female CF1 mice were treated with 0.2 ml of acetone twice a week for up to 103 weeks. The percentages of male and female mice that survived until the end of the study were 26 and 22 percent, respectively. The animals not surviving the entire treatment regimen were necropsied. The most common causes of death or illness were identified to be systemic neoplasia, renal failure, urethral obstructions, and suppurative lesions. One skin tumor was found in a male mouse from the acetone-treated control group, and about 15 percent of the male and female mice were found to have irritant lesions of the skin. The observed incidence of each tumor type in the acetone control group was not statistically different from the control value.

Zakova et al. applied 0.2 ml of acetone to the shaved dorsal skin of male and female CF1 mice once a week for 2 years as a solvent control group in an epoxy resin carcinogenicity study (506). Acetone had no effect on the survival of the 300 animals tested and resulted in dermal inflammatory reactions (focal acanthosis, dermal fibrosis) in only 6 percent of the animals. DePass et al. included an acetone control group in a dermal oncogenicity study of four alkoxysilane compounds (507). The shaved dorsal skin in a control group of 40 male C3H/HeJ mice was treated with 25 µl (20 mg) of acetone three times a week for life. The mean survival time in days for the acetone-treated animals was reported to be 504 days. No carcinomas and two sarcomas (a fibrosarcoma and a lymphosarcoma) were found on or below the skin in the control group. The authors reported that the historical incidence of sarcomas of various types in acetone-treated mice from their laboratory was 5/724.

14 GENOTOXICITY

The genotoxicity of acetone has been tested in a wide variety of prokaryotic and eurkaryotic test systems. Studies in the *Salmonella* assay have shown acetone to be nonmutagenic and to be an acceptable vehicle for dissolving and delivering water-insoluble chemicals to the tester stains. Acetone has been shown to be preferable to dimethyl sulfoxide as a solvent vehicle in the Ames assay when testing the mutagenic activity of several chlorinated and polynuclear aromatic hydrocarbons (508, 509). Likewise, acetone was been used as a delivery vehicle in yeast cell mutagenicity assays and has not produced genotoxic effects in mammalian cell systems in culture.

14.1 Bacterial Cell Mutagenicity Assays

Acetone has been tested in several histidine-requiring strains of *Salmonella typhimurium* for its ability to cause a reverse mutation to histidine prototrophy (Ames test). These tests have been conducted using both the standard plate incorporation procedure and the more sensitive preincubation technique. Arimoto et al. showed that acetone, unlike many other solvents, failed to enhance the mutagenicity of tryptophan pyrolysate mutagens following preincubation with strain TA 98 and rat liver S9 (510). Maron et al. conducted genotoxicity studies with strains TA98, TA100, TA1535, and TA1537 and showed that acetone concentrations of 500 mg/plate affected the survival of each strain when the solvent was pre-incubated for 30 min with the bacteria and a rat liver S9 fraction (511). The authors recommended that the acetone concentration should not exceed 200 mg/plate when it is used as a delivery vehicle.

Acetone was found to be negative in *Salmonella* strains TA97, TA98, TA100, and TA1535 at levels up to 1 mg/plate (512). The preincubation-type assays were carried out both in the presence and absence of liver S9 fractions from Aroclor® 1254-induced Sprague-Dawley rats and Syrian Golden hamsters. Ishidate et al. reported that acetone was negative for strains TA92, TA94, TA98, TA100, TA1535, and TA1537 at a maximum concentration of 10 mg/plate (513). The assays were conducted using a liver S9 fraction from Fischer rats treated with polychlorinated biphenyls for 5 days. De Flora et al. reported that acetone was negative for strains TA97, TA98, TA100, TA1535, and TA1537 when tested in the presence and absence of a liver S9 fraction from Aroclor® 1254-induced male Sprague-Dawley rats (514). The results were obtained using the plate incorporation procedure at a maximum plate concentration that did not exceed the toxicity limit (value not stated).

De Flora et al. also examined the mutagenic activity of acetone in three strains of *Escherichia coli* that lacked specific DNA repair mechanisms (514). The minimum inhibitory concentration of acetone causing growth inhibition provided a measure of chemical interaction with bacterial DNA. The results showed acetone to be low in potency, with the minimum inhibitory concentration being greater than 40 mg/plate for strains WP2, WP67, and CM871. Kubinski et al. reported that acetone gave negative results in a similar DNA-cell-binding assay for the detection of chemical mutagens (515). The incubation of acetone at concentrations up to 500 ppm with viable cells and DNA isolated from *E. coli* did not alter the binding to macromolecules within the bacterial cells. These results were obtained in the presence and absence of either lysosomal enzymes or an S9 liver fraction from Aroclor®-induced Sprague-Dawley rats.

DeMarini et al. examined acetone and several other solvents for use as a carrier in the microscreen prophage-induction assay (516). No increase in plaque-forming units were observed when the lambda prophage $WP2_s(\lambda)$ was incubated with up to 10 percent acetone before plating onto *E. coli* indicator strain TH-008. Likewise, Rossman et al. found no evidence of genotoxicity or mutagenicity when 10 percent acetone was incubated with Aroclor® 1254-induced rat liver S9 and the lambda prophage $WP2_s(\lambda)$ prior to plating with *E. coli* strain SR714 (517). Acetone concentrations up to 100 mM were not mutagenic (i.e., β-galactosidase was not activated) in the SOS chromotest using *E. coli* PQ37 as the indicator organism (518).

14.2 Yeast Cell Mutagenicity Assays

Acetone genotoxicity has been examined in several different yeast strains both with and without metabolic activation by a liver homogenate fraction. Abbondandolo et al. reported that acetone was not mutagenic (i.e., red to white phenotypic change) for strain P_1 of *Schizosaccharomyces pombe* when tested at a concentration of 5 percent both with and without activation by a liver S10 fraction from male Swiss albino mice (519). Preliminary toxicity studies showed that acetone was toxic to this strain of yeast at a concentration of 10 percent. The authors concluded that acetone was compatible with this strain of yeast and could be used as a solvent vehicle for compounds having limited water solubility.

Zimmermann et al. examined the ability of acetone to induce either chromosomal malsegregation, mitotic recombination, or point mutations in *Saccharomyces cerevisiae* strain D61.M (520, 521). The authors reported that acetone-induced chromosomal malsegregation, characterized as aneuploidy, but did not induce mitotic recombinations or point mutations. An overnight incubation of the cells in ice-cold water with 7 to 8 percent acetone was needed to observe the aneuploidy consistently. The authors concluded that acetone acted directly on tubulin during the growth phase causing the formation of labile microtubules that functioned poorly during mitosis. Whittaker et al. reported further on acetone-induced chromosomal malsegregation in *Saccharomyces cerevisiae* (522). Results from repeated trials in two separate laboratories revealed that acetone caused weak and ambiguous results. Aneuploidy was only observed in one laboratory when cold incubation and a concentration of about 50 mg/ml were used in the assay.

14.3 Chinese Hamster Cell Mutagenicity Assays

Abbondandolo et al. reported that V79 Chinese hamster cells were able to survive a 24-hr incubation with a 5 percent, but not a 10 percent, solution of acetone (519). The cells were treated with acetone both as a cultured monolayer and as single cells in a buffered medium. Subsequent mutagenicity studies were not performed by the authors using this cell line. Tates and Kriek showed that 1 mg/ml of acetone did not cause chromosomal aberrations or sister chromatid exchanges in Chinese hamster ovary cells when tested in the presence or absence of a liver S9 fraction from Aroclor® 1254-induced male Sprague-Dawley rats (523). Latt et al. reported that an acetone concentration of 87 mM (5 mg/ml) failed to cause an increase in sister chromatid exchanges in V79 Chinese hamster diploid lung cells when tested with metabolic activation (524). Acetone did not cause chromosomal aberrations or sister chromatid exchanges in Chinese hamster ovary cells when tested at concentrations up to 5 mg/ml (525). The assays were performed with and without metabolic activation by a liver S9 fraction from Aroclor®-induced male Sprague-Dawley rats. Hatch et al. reported that acetone vapors did not enhance the transformation of Syrian hamster embryo cells by SA7 adenovirus (526). Basler performed a micronucleus assay with the bone marrow of Chinese hamsters treated with an 865 mg/kg ip dose of acetone and found no increase in the number of micronuclei in the polychromatic erythrocytes (527).

Chen et al. reported that 25 to 250 μl of acetone caused appreciable inhibition of gap junction-mediated metabolic cooperation between V79 Chinese hamster lung fibroblasts (528). The metabolic cooperation assay examined the ability of acetone to alter the in vitro transfer of 6-thioguanine across gap junctions in normal and hypoxanthine phosphoribosyl transferase deficient cells. The authors concluded that the toxicity of acetone may be related to disturbances in membrane integrity. Ishidate et al. found that an acetone concentration of 40 mg/ml caused chromosomal aberrations in Chinese hamster fibroblasts when the test was conducted without metabolic activation (513). The structural aberrations included chromosome gaps, breaks, and fragmentations. Positive results were not obtained when a liver S9 fraction from polychlorinated biphenyl-induced Fischer rats was used in the assay.

14.4 Other Mammalian Cell Mutagenicity Assays

Acetone produced negative results in the following cytogenetic assays using cell lines derived from a species other than the hamster. Norppa reported that acetone concentrations ranging from 10.5 to 20.9 mM (0.6 to 1.2 mg/ml) did not cause chromosomal aberrations or sister chromatid exchanges in cultured human lymphocytes (529). Freeman et al. examined the ability of acetone to transform Fischer rat embryo cells that were infected with Rauscher leukemia virus (530). The assay revealed that acetone concentrations up to 0.1 mg/ml did not cause the cells to lose their normal contact inhibition. Amacher et al. reported that acetone did not cause any point mutations at the thymidine kinase locus in a L5178Y mouse lymphoma cell assay (531). In addition, acetone caused little cytotoxicity to the cells when tested at a level of 1 percent. Similarly, McGregor et al. found that an acetone concentration of 10 µl/ml was not cytotoxic or mutagenic to L5178Y mouse lymphoma cells when the liver S9 concentration (Aroclor™ 1254-induced male Fischer rats) was varied over a 20-fold range (532). Sina et al. found that an acetone concentration of 1 percent did not cause cytotoxicity or single-strand breaks in an alkaline elution/rat hepatocyte assay and was suitable for use as a solvent vehicle (533). Lake et al. reported that acetone concentrations up to 10 percent inhibited DNA synthesis but did not cause unscheduled DNA synthesis in human epithelial cell cultures (534). Kabarity reported that human leukocytes treated with high concentrations of acetone could cause a nonspecific effect on cell division (535). Incubation of the cells with 25 or 50 mM (1.4 or 2.9 g/l) acetone caused the formation of abnormal multipolar anaphase and star metaphase cells to be produced.

REFERENCES

1. *U.S. Chemical Industry Statistical Handbook*, Chemical Manufacturers Association, Washington, D.C., 1991, p. 40.
2. Anonymous, "Chemical Profile on Acetone," *Chem. Mark, Rep.*, **238**, 38 (1990).
3. W. L. Howard, "Acetone," in Kirk-Othmer *Encyclopedia of Chemical Technology*, 4th ed., Vol. 1, J. I. Kroschwitz and M. Howe-Grant, Eds., Wiley, New York, 1991. p. 176.
4. J. A. Riddick, W. B. Bunger, and T. K. Sakano, *Organic Solvents. Physical Properties and Methods of Purification*, 4th ed., Vol. II, Wiley, New York, 1986.
5. G. Leonardos, D. Kendall, and N. Barnard, "Odor Threshold Determinations of 53 Odorant Chemicals," *J. Air Poll. Control Assoc.*, **19**, 91–95 (1969).
6. J. H. Ruth, "Odor Thresholds and Irritant Levels of Several Chemical Substances: A Review," *Am. Ind Hyg. Assoc. J.*, **47**, 142–151 (1986).
7. J. E. Amoore, and E. Hautala, "Odor as an Aid to Chemical Safety: Odor Thresholds Compared with Threshold Limit Values and Volatilities for 214 Industrial Chemicals in Air and Water Dilution," *J. Appl. Toxicol.*, **3**, 272–290 (1983).

8. T. M. Hellman, and F. H. Small, "Characterization of the Odor Properties of 101 Petrochemicals Using Sensory Methods," *J. Air Pollut. Control Assoc.*, **24**, 979–982 (1974).

9. L. E. Kane, B. S. Dombroske, and Y. Alarie, "Evaluation of Sensory Irritation from Some Common Industrial Solvents," *Am. Ind. Hyg. Assoc. J.*, **41**, 451–455 (1980).

10. R. A. Shipley and C. N. H. Long, "Studies on the Ketogenic Activity of the Anterior Pituitary. II. A Method for the Assay of the Ketogenic Activity," *Biochem. J.*, **32**, 2246–2256 (1938).

11. S. C. Werch, "A Microdiffusion Method for the Estimation of Acetone," *J. Lab. Clin. Med.*, **25**, 414–420 (1940).

12. R. H. Barnes and A. N. Wick, "A Method for the Determination of Blood Ketone Bodies," *J. Biol. Chem.*, **131**, 413–423 (1939).

13. T. E. Weichselbaum and M. Somogyi, "A Method for the Determination of Small Amounts of Ketone Bodies," *J. Biol. Chem.*, **140**, 5–20 (1941).

14. R. H. Barnes and A. N. Wick, "A Method for the Determination of Blood Ketone Bodies," *J. Biol. Chem.*, **131**, 413–423 (1939).

15. J. B. Lyon and W. L. Bloom, "The Use of Furfural for the Determination of Acetone Bodies in Biological Fluids," *Can. J. Biochem.*, **36**, 1047–1056 (1958).

16. P. A. Mayes and W. Robson, "The Determination of Ketone Bodies," *Biochem. J.*, **67**, 11–15 (1957).

17. L. A. Greenberg and D. Lester, "A Micromethod for the Determination of Acetone and Ketone Bodies," *J. Biol. Chem.*, **154**, 177–190 (1944).

18. C. Thin and A. Robertson, "The Estimation of Acetone Bodies," *Biochem. J.*, **51**, 218–223 (1952).

19. J. Procos, "Modification of the Spectrophotometric Determination of Ketone Bodies in Blood Enabling the Total Recovery of β-hydroxybutyric Acid," *Clin. Chem.*, **7**, 97–106 (1961).

20. V. H. Peden, V. H., "Determination of Individual Serum 'Ketone Bodies,' with Normal Values in Infants and Children," *J. Lab. Clin. Med.*, **63**, 332–343 (1964).

21. E. M. P. Widmark, "The Kinetics of the Ketonic Decomposition of Aceto-acetic Acid," *Acta Med. Scand.*, **53**, 393–421 (1920).

22. G. J. Van Stekelenburg and G. Koorevaar, "Evidence for the Existence of Mammalian Acetoacetate Decarboxylase: With Special Reference to Human Blood Serum," *Clin. Chem. Acta*, **39**, 191–199 (1972).

23. C. J. P. Eriksson, "Micro Method for Determination of Ketone Bodies by Head-space Gas Chromatography," *Anal. Biochem.*, **47**, 235–243 (1972).

24. G. J. Van Stekelenburg and J. W. De Bruyn, "A Simple Gas Chromatographic Determination of Acetone and β-Hydroxybutyric Acid in Blood Serum by Means of Head Space Gas Sampling," *Clin. Chem. Acta*, **28**, 233–237 (1970).

25. P. Hradecký, P. Jagoš, and J. Janák, "Gas Chromatographic Head-space Analysis of Clinically Interesting Ketone Bodies," *J. Chromatogr.*, **146**, 327–332 (1978).

26. M. D. Trotter, M. J. Sulway, and E. Trotter, "The Rapid Determination of Acetone in Breath and Plasma," *Clin. Chem. Acta*, **35**, 137–143 (1971).

27. L. Siegel, N. I. Robin, and L. J. McDonald, "New Approach to Determination of Total Ketone Bodies in Serum," *Clin. Chem.*, **23**, 46–49 (1977).

28. M. Kimura, K. Kobayashi, A. Matsuoka, K. Hayashi, and Y. Kimura, "Head-space

Gas-chromatographic Determination of 3-Hydroxybutyrate in Plasma after Enzymic Reactions, and the Relationship Among the Three Ketone Bodies," *Clin. Chem.*, **31**, 596–598 (1985).

29. A. Brega, P. Villa, G. Quadrini, A. Quadri, and C. Lucarelli, "High-performance Liquid Chromatographic Determination of Acetone in Blood and Urine in the Clinical Diagnostic Laboratory," *J. Chromatogr.*, **553**, 249–254 (1991).

30. K. Kobayashi, M. Okada, Y. Yasuda, and S. Kawai, "A Gas Chromatographic Method for the Determination of Acetone and Acetoacetic Acid in Urine," *Clin. Chem. Acta*, **133**, 223–226 (1983).

31. S. K. Kundu and A. M. Judilla, "Novel Solid-phase Assay of Ketone Bodies in Urine," *Clin. Chem.*, **37**, 1565–1569 (1991).

32. G. Rooth, "Insulin Action Measured by Acetone Disappearance," *Acta Med. Scand.*, **182**, 271–272 (1967).

33. G. Rooth and G. Tibbling, "Free Fatty Acids, Glycerol and Alveolar Acetone in Obese Women during Phenformin Treatment," *Acta Med. Scand.*, **184**, 263–267 (1968).

34. G. Rooth and S. Östenson, "Acetone in Alveolar Air, and the Control of Diabetes," *Lancet, 1966– **II**, 1102–1105.

35. M. J. Sulway and J. M. Mullins, "Acetone in Diabetic Ketoacidosis," *Lancet, 1970– **I**, 736–740.

36. S. Levey, O. J. Balchum, V. Medrano, and R. Jung, "Studies of Metabolic Products in Expired Air. II. Acetone," *J. Lab. Clin. Med.*, **63**, 574–584 (1964).

37. R. D. Stewart, E. A. Boettner, and B. T. Stubbs, "Rapid Infra-red Determination of Acetone in the Blood and the Exhaled Air of Diabetic Patients," *Nature*, **191**, 1008 (1961).

38. D. Glaubitt and J.-G. Rausch-Stroomann, "Eine Neue Methode zur Aceton-bestimmung in der Atemluft," *Clin. Chem. Acta*, **4**, 165–169 (1959).

39. R. S. Hubbard, "Determination of Acetone in Expired Air," *J. Biol. Chem.*, **43**, 57–65 (1920).

40. O. B. Crofford, R. E. Mallard, R. E. Winton, N. L. Rogers, J. C. Jackson, and U. Keller, "Acetone in Breath and Blood," *Trans. Am. Clin. Climatol. Assoc.*, **88**, 128–139 (1977).

41. R. D. Stewart and E. A. Boettner, "Expired-air Acetone in Diabetes Mellitus," *New Engl. J. Med.*, **270**, 1035–1038 (1964).

42. C. N. Tassopoulos, D. Barnett, and T. R. Fraser, "Breath-acetone and Blood-sugar Measurements in Diabetes," *Lancet, 1969– **II**, 1282–1286.

43. N. C. Jain, "Direct Blood-injection Method for Gas Chromatographic Determination of Alcohols and Other Volatile Compounds," *Clin. Chem.*, **17**, 82–85 (1971).

44. N. B. Smith, "Determination of Volatile Alcohols and Acetone in Serum by Non-polar Capillary Gas Chromatography after Direct Sample Injection," *Clin. Chem.*, **30**, 1672–1674 (1984).

45. H. E. K. Winterbach and P. J. Apps, "A Gas-chromatographic Headspace Method for the Determination of Acetone in Bovine Milk, Blood and Urine," *Onderstepoort J. Vet. Res.*, **58**, 75–79 (1991).

46. F. J. López-Soriano and J. M. Argilés, "Simultaneous Determination of Ketone Bodies in Biological Samples by Gas Chromatographic Headspace Analysis," *J. Chromatogr. Sci.*, **23**, 120–123 (1985).

47. S. Holm and E. Lundgren, "A Purge-and-trap Method for the Analysis of Acetone in Biological Tissues," *Anal. Biochem.*, **136**, 157–160 (1984).

48. A. C. Haff and G. A. Reichard, "A Method for Estimation of Acetone Radioactivity and Concentration in Blood and Urine," *Biochem. Med.*, **18**, 308–314 (1977).

49. V. C. Gavino, B. Vinet, F. David, N. Garneau, and H. Brunengraber, "Determination of the Concentration and Specific Activity of Acetone in Biological Fluids," *Anal. Biochem.*, **152**, 256–261 (1986).

50. M. J. Henderson, B. A. Karger, and G. A. Wrenshall, "Acetone in Breath. A Study of Acetone Exhalation in Diabetic and Nondiabetic Human Subjects," *Diabetes*, **1**, 188–200 (1952).

51. J. P. Conkle, B. J. Camp, and B. E. Welch, "Trace Composition of Human Respiratory Gas," *Arch. Environ. Health*, **30**, 290–295 (1975).

52. B. O. Jansson and B. T. Larsson, "Analysis of Organic Compounds in Human Breath by Gas Chromatography–Mass Spectrometry," *J. Lab. Clin. Med.*, **74**, 961–966 (1969).

53. V. L. Brechner and R. W. Bethune, "Determination of Acetone Concentration in Arterial Blood by Vapor Phase Chromatography of Alveolar Gas," *Diabetes*, **14**, 663–665 (1965).

54. S. P. Eriksen and A. B. Kulkarni, "Methanol in Normal Human Breath," *Science*, **141**, 639–640 (1963).

55. M. Phillips and J. Greenberg, "Detection of Endogenous Acetone in Normal Human Breath," *J. Chromatogr.*, **422**, 235–238 (1987).

56. P. Ciccioli, G. Bertoni, F. Brancaleoni, R. Fratarcangeli, and F. Bruner, "Evaluation of Organic Pollutants in the Open Air and Atmospheres in Industrial Sites Using Graphatized Carbon Black Traps and Gas Chromatographic–Mass Spectrophotometric Analysis with Specific Detectors," *J. Chromatogr.*, **126**, 757–770 (1976).

57. P. Clair, M. Tua, and H. Simian, "Capillary Columns in Series for GC Analysis of Volatile Organic Pollutants in Atmospheric and Alveolar Air," *J. High Resolut. Chromatog.*, **14**, 383–387 (1991).

58. M. C. Facchini, G. Chiavari, and S. Fuzzi, "An Improved HPLC Method for Carbonyl Compound Speciation in the Atmospheric Liquid Phase," *Chemosphere*, **15**, 667–674 (1986).

59. K. Fung and D. Grosjean, "Determination of Nanogram Amounts of Carbonyls as 2,4-dinitrophenylhydrazones by High-performance Liquid Chromatography," *Anal. Chem.*, **53**, 168–171 (1981).

60. L. H. Keith, A. W. Garrison, F. R. Allen, M. H. Carter, T. L. Floyd, J. D. Pope, and A. D. Thurston, "Identification of Organic Compounds in Drinking Water from Thirteen U.S. Cities," in *Identification and Analysis of Organic Pollutants in Water*, L. H. Keith, Ed., Ann Arbor Press, Ann Arbor, MI, 1979, pp. 329–373.

61. E. Robinson, R. A. Rasmussen, H. H. Westberg, and M. W. Holdren, "Nonurban Methane Low Molecular Weight Hydrocarbon Concentrations Related to Air Mass Dentification," *J. Geophys. Res.*, **78**, 5345–5351 (1973).

62. J-O. Levin and L. Carleborg, "Evaluation of Solid Sorbents for Sampling Ketones in Workroom Air," *Ann. Occup. Hyg.*, **31**, 31–38 (1987).

63. Y. Tsuchiya, "Volatile Organic Compounds in Indoor Air," *Chemosphere*, **17**, 79–82 (1988).

64. D. O'Hara and H. Singh, "Sensitive Gas Chromatographic Detection of Acetaldehyde

and Acetone Using a Reduction Gas Detector," *Atmos. Environ.*, **22**, 2613–2615 (1988).

65. G. Hauck and F. Arnold, "Improved Positive-ion Composition Measurements in the Upper Troposphere and Lower Stratosphere and the Detection of Acetone," *Nature*, **311**, 547–550 (1984).

66. L. Nondek, R. E. Milofsky, and J. W. Birks, "Determination of Carbonyl Compounds in Air by HPLC Using On-line Analyzed Microcartridges, Fluorescence and Chemiluminescence Detection," *Chromatographia*, **32**, 33–39 (1991).

67. B. J. Dowty, J. L. Laseter, and J. Storer, "The Transplacental Migration and Accumulation in Blood of Volatile Organic Compounds," *Pediat. Res.*, **10**, 696–701 (1976).

68. M. J. Sulway, M. D. Trotter, E. Trotter, and J. M. Malins, "Acetone in Uncontrolled Diabetes," *Postgrad. Med. J.*, **47** (Suppl.), 383–387 (1971).

69. A. Zlatkis, W. Bertsch, H. A. Lichtenstein, A. Tishbee, F. Shunbo, H. M. Liebich, A. M. Coscia, and N. Fleischer, "Profile of Volatile Metabolites in Urine by Gas Chromatography Mass Chromatography," *Anal. Chem.*, **45**, 763–767 (1973).

70. E. D. Pellizzari, T. D. Hartwell, B. S. H. Harris, R. D. Waddell, D. A. Whitaker, and M. D. Erickson, "Purgeable Organic Compounds in Mother's Milk," *Bull. Environ. Contam. Toxicol.*, **28**, 322–328 (1982).

71. H. Göschke and T. Lauffenburger, "Aceton in der Atemluft und Ketone im Venenblut bei Vollständigem Fasten Normal und Übergewichtiger Personen," *Res. Exp. Med.*, **165**, 233–244 (1975).

72. K. E. Wildenhoff, "Diurnal Variations in the Concentrations of Blood Acetoacetate, 3-Hydroxybutyrate and Glucose in Normal Persons," *Acta Med. Scand.*, **191**, 303–306 (1972).

73. N. W. Teitz, in *Clinical Guide to Laboratory Tests*, Saunders, Philadelphia, PA, 1983, pp. 2–3.

74. M. Kimura, M. Shimosawa, K. Kobayashi, T. Sakoguchi, A. Igaki, M. Hashimoto, A. Matsuoka, and Y. Kimura, "Acetoacetate Decarboxylase Activity of Plasma Components," *Jap. J. Clin. Chem.*, **15**, 37–43 (1986).

75. G. Pezzagno, M. Imbriani, S. Ghittori, and E. Capodaglio, "Urinary Concentration, Environmental Concentration, and Respiratory Uptake of Some Solvents: Effect of the Work Load," *Am. Ind. Hyg. Assoc. J.*, **49**, 546–552 (1988).

76. E. M. P. Widmark, "XXXII. Studies in the Acetone Concentration in Blood, Urine, and Alveolar Air. III. The Elimination of Acetone through the Lungs," *Biochem. J.*, **14**, 379–394 (1920).

77. A. P. Briggs and P. A. Shaffer, "The Excretion of Acetone from the Lungs," *J. Biol. Chem.*, **48**, 413–428 (1921).

78. H. W. Haggard, L. A. Greenberg, and J. M. Turner, "The Physiological Principles Governing the Action of Acetone together with the Determination of Toxicity," *J. Ind. Hyg. Toxicol.*, **26**, 133–151 (1944).

79. M. Kimura, N. Ogasahara, K. Kobayashi, A. Hitoi, A. Matsuoka, and Y. Kimura, "Improvement in the Head-space Gas-chromatographic Method for Determination of Three Ketone Bodies in Plasma," *Clin. Chem.*, **36**, 160–161 (1990).

80. M. S. Smith, A. J. Francis, and J. M. Duxbury, "Collection and Analysis of Organic Gases from Natural Ecosystems: Application to Poultry Manure," *Environ. Sci. Technol.*, **11**, 51–55 (1977).

81. R. E. Stoiber, R. E. Leggett, T. F. Jenkins, R. P. Murrmann, and W. I. Rose, "Organic Compounds in Volcanic Gas from Santiaguito Volcano, Guatemala," *Am. Geolog. Soc. Bull.*, **82**, 2299–2302 (1971).

82. F. Lipari, J. M. Dasch, and W. F. Scruggs, "Aldehyde Emissions from Woodburning Fireplaces," *Environ. Sci. Technol.*, **18**, 326–330 (1984).

83. V. A. Isidorov, I. G. Zenkevich, and B. V. Ioffe, "Volatile Organic Compounds in the Atmosphere of Forests," *Atmos. Environ.*, **19**, 1–8 (1985).

84. J. E. Sigsby, S. Tejada, W. Ray, J. M. Lang, and J. W. Duncan, "Volatile Organic Compound Emissions from 46 In-use Passenger Cars," *Environ. Sci. Technol.*, **21**, 468–477 (1987).

85. C. E. Higgins, W. H. Griest, and G. Olerich, "Application of Tenax Trapping to Analysis of Gas Phase Organic Compounds in Ultra-low Tar Cigarette Smoke," *J. Assoc. Off. Anal. Chem.*, **66**, 1074–1083 (1983).

86. A. Colombo, M. De Bortoli, E. Pecchio, H. Schauenburg, H. Schlitt, and H. Vissers, "Chamber Testing of Organic Emission from Building and Furnishing Materials," *Sci. Total Environ.*, **91**, 237–249 (1990).

87. P. H. Howard, G. W. Sage, W. F. Jarvis, and D. A. Gray, "Acetone," in *Handbook of Environmental Fate and Exposure Data for Organic Chemicals*, Lewis Publishers, New York, 1990, pp. 9–18.

88. A. P. Altshuller, "Chemical Reactions and Transport of Alkanes and Their Products in the Troposphere," *J. Atmos. Chem.*, **12**, 19–61 (1991).

89. R. J. Snider and G. A. Dawson, "Tropospheric Light Alcohols, Carbonyls, and Acetonitrile: Concentrations in the Southwestern United States and Henry's Law Data," *J. Geophys. Res.*, **90**, 3797–3805 (1985).

90. R. A. Duce, V. A. Mohnen, P. R. Zimmerman, D. Grosjean, W. Cautreels, R. Chatfield, R. Jaenicke, J. A. Ogren, E. D. Pelliari, and G. T. Wallace, "Organic Material in the Global Troposphere," *Rev. Geophys. Space Phys.*, **21**, 921–952 (1983).

91. L. A. Cavanagh, C. F. Schadt, and E. Robinson, "Atmospheric Hydrocarbon and Carbon Monoxide Measurements at Point Barrow, Alaska," *Environ. Sci. Technol.*, **3**, 251–257 (1969).

92. E. Robinson, R. A. Rasmussen, H. H. Westberg, and M. W. Holdren, "Nonurban Methane Low Molecular Weight Hydrocarbon Concentrations Related to Air Mass Dentification," *J. Geophys. Res.*, **78**, 5345–5351 (1973).

93. P. B. Shepson, D. R. Hastie, H. I. Schiff, M. Polizzi, J. W. Bottenheim, K. Anlauf, G. I. Mackay, and D. R. Karecki, "Atmospheric Concentrations and Temporal Cariations of C_1–C_3 Carbonyl Compounds at Two Rural Sites in Central Ontario," *Atmos. Environ.*, **25A**, 2001–2015 (1991).

94. R. R. Arnts and S. A. Meeks, "Biogenic Hydrocarbon Contribution to the Ambient Air of Selected Areas," *Atmos. Environ.*, **15**, 1643–1651 (1981).

95. R. B. Chatfield, E. P. Gardner, and J. G. Calvert, "Sources and Sinks of Acetone in the Troposphere: Behavior of Reactive Hydrocarbons and a Stable Product," *J. Geophys. Res.*, **92**, 4208–4216 (1987).

96. D. Grosjean, A. H. Miguel, and T. M. Tavares, "Urban Air Pollution in Brazil: Acetaldehyde and other Carbonyls," *Atmos. Environ.*, **24B**, 101–106 (1990).

97. R. B. Zweidinger, J. E. Sigsby, S. B. Tajada, F. D. Stump, D. L. Dropkin, W. D.

Ray, and J. W. Duncan, "Detailed Hydrocarbon and Aldehyde Mobile Source Emissions from Roadway Studies," *Environ. Sci. Technol.*, **22**, 956–962 (1988).

98. F. Arnold, G. Knop, and H. Ziereis, "Acetone Measurements in the Upper Troposphere and Lower Stratosphere—Implications for Hydroxyl Radical Abundances," *Nature*, **321**, 505–507 (1986).

99. G. Knop and F. Arnold, "Stratospheric Trace Gas Detection Using a New Balloonborne ACIMS Method: Acetonitrile, Acetone, and Nitric Acid," *Geophys. Res. Lett.*, **14**, 1262–1265 (1987).

100. J. J. Shah and H. B. Singh, "Distribution of Volatile Organic Chemicals in Outdoor and Indoor Air. A National VOCs Database," *Environ. Sci. Technol.*, **22**, 1381–1388 (1988).

101. D. Grosjean and B. Wright, "Carbonyls in Urban Fog, Ice Fog, Cloudwater and Rainwater," *Atmos. Environ.*, **17**, 2093–2096 (1983).

102. L. H. Keith, A. W. Garrison, F. R. Allen, M. H. Carter, T. L. Floyd, J. D. Pope, and A. D. Thurston, "Identification of Organic Compounds in Drinking Water from Thirteen U.S. cities," in *Identification and Analysis of Organic Pollutants in Water*, L. H. Keith, Ed., Ann Arbor Press, Ann Arbor, MI, 1979, pp. 329–373.

103. J. F. Corwin, "Volatile Oxygen-containing Organic Compounds in Sea Water: Determination," *Bull. Marine Sci.*, **19**, 504–509 (1969).

104. R. E. Rathbun, D. W. Stephans, D. J. Shultz, and D. Y. Tai, "Fate of Acetone in Water," *Chemosphere*, **11**, 1097–1114 (1982).

105. H. Meyrahn, J. Pauly, W. Schneider, and P. Warneck, "Quantum Yields for the Photodissociation of Acetone in Air and an Estimate for the Life Time of Acetone in the Lower Troposphere," *J. Atmos. Chem.*, **4**, 277–291 (1986).

106. P. Paterson, J. Sheath, P. Taft, and C. Wood, "Maternal and Foetal Ketone Concentration in Plasma and Urine," *Lancet*, *1967*-**II**, 862–865.

107. J. H. Koeslag, T. D. Noakes, and A. W. Sloan, "Post-exercise Ketosis," *J. Physiol.*, **301**, 79–90 (1980).

108. R. E. Kimura and J. B. Warshaw, "Metabolic Adaptations of the Fetus and Newborn," *J. Pediatr. Gastroenterol. Nutr.*, **2** (Suppl. 1), S12–S15 (1983).

109. S. B. Lewis, J. D. Wallin, J. P. Kane, and J. E. Gerich, "Effect of Diet Composition on Metabolic Adaptations to Hypocaloric Nutrition: Comparison of High Carbohydrate and High Fat Isocaloric Diets," *Am. J. Clin. Nutr.*, **30**, 160–170 (1977).

110. L. J. Levy, J. Duga, M. Girgis, and E. E. Gordon, "Ketoacidosis Associated with Alcoholism in Nondiabetic Subjects," *Ann. Int. Med.*, **78**, 213–219 (1973).

111. R. Smith, D. J. Fuller, J. H. Wedge, D. H. Williamson, and K. G. G. M. Alberti, "Initial Effect of Injury on Ketone Bodies and Other Blood Metabolites," *Lancet*, *1975*-**I**, 1–3.

112. G. Rooth and S. Carlström, "Therapeutic Fasting," *Acta Med. Scand.*, **187**, 455–463 (1970).

113. E. H. Williamson and E. Whitelaw, "Physiological Aspects of the Regulation of Ketogenesis," *Biochem. Soc. Symp.*, **43**, 137–161 (1978).

114. A. Walther and G. Neumann, "Über den Verlauf Azetonausscheidung in der Ausatmungsluft bei Radsportlern vor, während und nach einer spezifischen Laborbelastung," *Acta Biol. Med. Ger.*, **22**, 117–121 (1969).

115. P. Felig, E. B. Marliss, and G. F. Cahill, "Metabolic Response to Human Growth Hormone during Prolonged Starvation," *J. Clin. Invest.*, **50**, 411–421 (1971).

116. A. Ramu, J. Rosenbaum, and T. F. Blaschke, "Disposition of Acetone Following Acute Acetone Intoxication," *West. J. Med.*, **129**, 429–432 (1978).

117. A. S. Gamis and Wasserman, "Acute Acetone Intoxication in a Pediatric Patient," *Pediat. Emerg. Care*, **4**, 24–26 (1988).

118. M. J. Sulway, M. D. Trotter, E. Trotter, and J. M. Malins, "Acetone in Uncontrolled Diabetes," *Postgrad. Med. J.* **47** (Suppl.), 383–387 (971).

119. P. Fisher, "The Role of the Ketone Bodies in the Etiology of Diabetic Coma," *Am. J. Med. Sci.*, **221**, 384–387 (1951).

120. A. W. Jones, "Breath-acetone Concentrations in Fasting Healthy Men: Response of Infrared Breath-alcohol Analyzers," *J. Anal. Toxicol.*, **11**, 67–69 (1987).

121. A. W. Jones, "Breath Acetone Concentrations in Fasting Male Volunteers: Further Studies and Effect of Alcohol Administration," *J. Anal. Toxicol.*, **12**, 75–79 (1988).

122. J. Neiman, A. W. Jones, H. Numminen, and M. Hillbom, "Combined Effect of a Small Dose of Ethanol and 36 hr Fasting on Blood-glucose Response, Breath-acetone Profiles and Platelet Function in Healthy Men," *Alcohol Alcoholism*, **22**, 265–270 (1987).

123. F. Gamberale, "Use of Behavioral Performance Tests in the Assessment of Solvent Toxicity," *J. Work Environ. Health*, **11** (Suppl. 1), 65–74 (1985).

124. A. M. Seppäläinen, "Neurophysiological Aspects of the Toxicity of Organic Solvents," *Scand. J. Work Environ. Health*, **11** (Suppl. 1), 61–64 (1985).

125. P. Albertoni, "Die Wirkung und die Verwandlungen Einiger Stoffe im Organismus in Beziehung zur Pathogenese der Acetonämie und des Diabetes," *Arch. Exp. Pathol. Pharmakol.*, **18**, 219–241 (as reported by Browning, 1953) (1884).

126. F. T. Frerichs, "Über den plötzlichen Tod und über das Coma bei Diabetes (diabetische Intoxication)," *Ztschr. Klin. Med.*, **6**, 3–53 (as reported by Fisher, 1951) (1883).

127. A. E. Koehler, E. Windsor, and E. Hill, "Acetone and Acetoacetic Acid Studies in Man," *J. Biol. Chem.*, **140**, 811–825 (1941).

128. E. M. P. Widmark, "Studies in the Concentration of Indifferent Narcotics in Blood and Tissues," *Acta Med. Scand.*, **52**, 87–164 (1919).

129. E. Kagan, "Experimentelle Studien über den Einfluss technisch und hygienisch wichtiger Gase und Dämfpe auf den Organismus. XXXVI. Aceton," *Arch. Hyg. Berl.*, **94**, 41–53 (as reported by Browning, 1953) (1924).

130. G. E. DiVincenzo, F. J. Yanno, and B. D. Astill, "Exposure of Man and Dog to Low Concentrations of Acetone Vapor," *Am. Ind. Hyg. Assoc. J.*, **34**, 329–336 (1973).

131. R. D. Stewart, C. L. Hake, A. Wu, S. A. Graff, H. V. Forster, W. H. Keeler, A. J. Lebrun, P. E. Newton, and R. J. Soto, "Acetone: Development of a Biologic Standard for the Industrial Worker by Breath Analysis," U.S. Dept. of Commerce, National Technical Information Service PB82-172917, 1975.

132. T. Matsushita, A. Yoshimune, T. Inoue, S. Yamada, and H. Suzuki, "Experimental Studies for Determining the MAC Value of Acetone. 1. Biologic Reactions in the "One-day Exposure" to Acetone," *Jap. J. Ind. Health*, **11**, 477–485 (1969).

133. T. Matsushita, E. Goshima, H. Miyakaki, K. Maeda, Y. Takeuchi, and T. Inoue, "Experimental Studies for Determining the MAC Value of Acetone. 2. Biologic Reactions in the Six-day Exposure to Acetone," *Jap. J. Ind. Health*, **11**, 507–515 (1969).

134. R. B. Dick, W. D. Brown, J. V. Setzer, B. J. Taylor, and R. Shukla, "Effects of Short Duration Exposures to Acetone and Methyl Ethyl Ketone," *Toxicol. Lett.*, **43**, 31–49 (1988).

135. R. B. Dick, J. V. Setzer, B. J. Taylor, and R. Shukla, "Neurobehavioural Effects of Short Duration Exposures to Acetone and Methyl Ethyl Ketone," *Brit. J. Ind. Med.*, **46**, 111–121 (1989).

136. H. Suzuki, "An Experimental Study on Physiological Functions of the Autonomic Nervous System of Man Exposed to Acetone Gas," *Jap. J. Ind. Health*, **15**, 147–164 (1973).

137. K. Nakaaki, "An Experimental Study on the Effect of Exposure to Organic Solvent Vapor in Human Subjects," *J. Sci. Labour*, **50**, 89–96 (1974).

138. A. Seeber and E. Kiesswetter, "Exposure to Mixtures of Organic Solvents: Subjective Symptoms as Valid Adverse Effects?," in *Proceedings of the 4th International Conference on the Combined Effects of Environmental Factors*, L. D. Fechter, Ed., 1991, pp. 71–74.

139. A. Seeber, E. Kiesswetter, and M. Blaszkewicz, "Correlations between Subjective Disturbances Due to Acute Exposure to Organic Solvents and Internal Dose," *Neurotoxicology*, **13**, 265–270 (1992).

140. K. W. Nelson, J. F. Ege, Jr., M. Ross, L. W. Woodman, and L. Silverman, "Sensory Response to Certain Industrial Solvent Vapors," *J. Ind. Hyg. Toxicol.*, **25**, 282–285 (1943).

141. R. B. Douglas and J. E. Coe, "The Relative Sensitivity of the Human Eye and Lung to Irritant Gases," *Ann. Occup. Hyg.*, **31**, 265–267 (1987).

142. P. Kechijian, "Nail Polish Removers: Are They Harmful?" *Sem. Dermatol.*, **10**, 26–28 (1991).

143. A. M. Mindlin, "Acetone Used as a Solvent in Accidental Tarsorrhaphy," *Am. J. Opthalmol.*, **83**, 136–137 (1977).

144. W. M. Grant, "Acetone," in: *Toxicology of the Eye*, 1st ed., Charles C Thomas, Springfield, IL, 1962, pp. 9–10.

145. R. S. McLaughlin, "Chemical Burns of the Human Cornea," *Am. J. Opthalmol.*, **29**, 1355–1362 (1946).

146. M. S. Mayou, "Cataract in an Acetone Worker," *Proc. R. Soc. Med.*, **25**, 475 (1932).

147. M. G. Weill, "Severe Ocular Disturbance from Inhalation of Impure Acetone," *Bull. Soc. Fr. Opthalmol.*, **47**, 412–414 (1934) (as reported by Grant, 1986).

148. J. J. Gomer, "Corneal Opacities from Pure Acetone," *VonGraefe's Arch. Opthalmol.*, **150**, 622–624 (1950) (as reported by Grant, 1986).

149. W. M. Grant, "Acetone," in *Toxicology of the Eye*, 3rd ed., Charles C Thomas, Springfield, IL, 1986, pp. 41–42.

150. A. P. Lupulescu, D. J. Birmingham, and H. Pinkus, "Electron Microscopic Study of Human Epidermis after Acetone and Kerosene Administration," *J. Invest. Dermatol.*, **60**, 33–45 (1973).

151. A. P. Lupulescu and D. J. Birmingham, "Effect of Protective Agent Against Lipid-solvent-induced Damages," *Arch. Environ. Health*, **31**, 29–32 (1976).

152. R. L. Raleigh and W. A. McGee, "Effects of Short, High-concentration Exposures to Acetone as Determined by Observation in the Work Area," *J. Occup. Med.*, **14**, 607–610 (1972).

153. R. Israeli, Y. Zoref, Z. Tessler, and J. Braver, "Reaktionszeit als Mittel zur Aceton-TLV-(MAK)-Wertbestimmung," *Zbl. Arbeitsmed.*, **27**, 197–199 (1977).

154. T. L. Litovitz, B. F. Schmitz, and K. C. Holm, "1988 Annual Report of the American Association of Poison Control Centers National Data Collection System," *Am. J. Emerg. Med.*, **7**, 495–545 (1989).

155. D. S. Ross, "Acute Acetone Intoxication Involving Eight Male Workers," *Ann. Occup. Hyg.*, **16**, 73–75 (1973).

156. R. J. Foley, "Inhaled Industrial Acetylene. A Diabetic Ketoacidosis Mimic," *J. Am. Med. Assoc.*, **254**, 1066–1067 (1985).

157. S. K. Gerard and H. Khayam-Bashi, "Characterization of Creatinine Error in Ketotic Patients. A Prospective Comparison of Alkaline Picrate Methods with an Enzymatic Method," *Am. J. Clin. Pathol.*, **84**, 659–664 (1985).

158. P. J. Watkins, "The Effect of Ketone Bodies on the Determination of Creatinine," *Clin. Chem. Acta*, **18**, 191–196 (1967).

159. G. Sack, "Ein Fall von gewerblicher Azetonvergiftung," *Arch. Gewerbepathol. Gewerbehyg.*, **10**, 80–86 (1940).

160. A. R. Smith and M. R. Mayers, "Study of Poisoning and Fire Hazards of Butanone and Acetone," *Ind. Bull. NYS Dept. Labor*, **23**, 174–176 (1944).

161. T. Cossmann, "Acetonvergiftung nach Anlegung eines Zelluloid-Mullverbandes," *Münch. Med. Wochenschr.*, **50**, 1556–1557 (1903) (as reported by Browning, 1953).

162. G. F. Strong, "Acute Acetone Poisoning," *Can. Med. Assoc. J.*, **51**, 359–362 (1944).

163. C. C. Chatterton and R. B. Elliott, "Acute Acetone Poisoning from Leg Casts of a Synthetic Plaster Substitute," *J. Am. Med. Assoc.*, **130**, 1222–1223 (1946).

164. L. J. Fitzpatrick and D'D.C. Claire, "Acute Acetone Poisoning (Resulting from Synthetic Plaster Substitute in Spica Cast)," *Curr. Res. Anesth.*, **26**, 86–87 (1947).

165. R. B. Pomerantz, "Acute Acetone Poisoning from Castex," *Am. J. Surg.*, **80**, 117–118 (1950).

166. Anon., "Acetone Poisoning and Immobilizing Casts," *Brit. Med. J.*, **2**, 1058 (1952).

167. L. C. Harris and R. H. Jackson, "Acute Acetone Poisoning Caused by Setting Fluid for Immobilizing Casts," *Brit. Med. J.*, **2**, 1024–1026 (1952).

168. P. K. Renshaw and R. M. Mitchell, "Acetone Poisoning Following the Application of a Lightweight Cast," *Brit. Med. J.*, **1**, 615 (1956).

169. W. Hift and P. L. Patel, "Acute Acetone Poisoning due to a Synthetic Plaster Cast," *S. Afr. Med. J.*, **35**, 246–250 (1961).

170. M. Sakata, J. Kikuchi, and M. Haga, "Disposition of Acetone, Methyl Ethyl Ketone and Cyclohexanone in Acute Poisoning," *Clin. Toxicol.*, **27**, 67–77 (1989).

171. S. Gitelson, A. Werczberger, and J. B. Herman, "Coma and Hyperglycemia Following Drinking of Acetone," *Diabetes*, **15**, 810–811 (1966).

172. H. Mirchev, "Hepatorenal Lesions in Acute Acetone Poisoning," *Vutr. Bolesti*, **17**, 89–92 (1978).

173. U. C. Pozzani, C. S. Weil, and C. P. Carpenter, "The Toxicological Basis of Threshold Limit Values: 5. The Experimental Inhalation of Vapor Mixtures by Rats, with Notes upon the Relationship between Single Dose Inhalation and Single Dose Oral Data," *Am. Ind. Hyg. Assoc. J.*, **20**, 364–369 (1959).

174. H. F. Smyth, C. S. Weil, J. S. West, and C. P. Carpenter, "An Exploration of Joint

Toxic Action: Twenty-seven Industrial Chemicals Intubated in Rats in all Possible Pairs," *Toxicol. Appl. Pharmacol.*, **14**, 340–347 (1969).

175. L. M. Mashbitz, R. M. Sklianskaya, and F. I. Urieva, "The Relative Toxicity of Acetone, Methylalcohol and Their Mixtures: II. Their Action on White Mice," *J. Ind. Hyg. Toxicol.*, **18**, 117–122 (1936).

176. H. Specht, J. W. Miller, and P. J. Valaer, "Acute Response of Guinea Pigs to the Inhalation of Dimethyl Ketone (Acetone) Vapor in Air," *Public Health Rep.*, **52**, 944–954 (1939).

177. F. Flury and W. Wirth, "Zur Toxikologie der Lösungsmittel (Verschiedene Ester, Aceton, Methylalkohol)," *Arch. Gewerbepath. Gewerbhyg.*, **5**, 1 (1934) (as reported by Browning, 1953).

178. E. Browning, "Ketones," in *Toxicity of Industrial Organic Solvents*, Chemical Publishing Co., New York, 1953, pp. 320–339.

179. J. V. Bruckner and R. G. Petersen, "Evaluation of Toluene and Acetone Inhalant Abuse. I. Pharmacology and Pharmacodynamics," *Toxicol. Appl. Pharmacol.*, **61**, 27–38 (1981).

180. M. E. Goldberg, H. E. Johnson, U. C. Pozzanni, and H. F. Smyth, Jr., "Effect of Repeated Inhalation of Vapors of Industrial Solvents on Animal Behavior. I. Evaluation of Nine Solvent Vapors on Pole-climb Performance in Rats," *Am. Ind. Hyg. Assoc. J.*, **25**, 369–375 (1964).

181. I. Geller, E. Gause, H. Kaplan, and R. J. Hartman, "Effects of Acetone, Methyl Ethyl Ketone and Methyl Isobutyl Ketone on a Match-to-sample Task in the Baboon," *Pharmacol. Biochem. Behav.*, **11**, 401–406 (1979).

182. I. Geller, R. J. Hartmann, S. R. Randle, and E. M. Gause, "Effects of Acetone and Toluene Vapors on Multiple Schedule Performance of Rats," *Pharmacol. Biochem. Behav.*, **11**, 395–399 (1979).

183. C. R. Garcia, I. Geller, and H. L. Kaplan, "Effects of Ketones on Lever-pressing Behavior of Rats," *Proc. West. Pharmacol. Soc.*, **21**, 433–438 (1978).

184. J. De Ceaurriz, J. C. Micillino, B. Marignac, P. Bonnet, J. Muller, and J. P. Guenier, "Quantitative Evaluation of Sensory Irritating and Neurobehavioral Properties of Aliphatic Ketones in Mice," *Food Chem. Toxicol.*, **22**, 545–549 (1984).

185. J. R. Glowa and P. B. Dews, "Behavioral Toxicology of Volatile Organic Compounds. IV. Comparison of the Rate-decreasing Effects of Acetone, Ethyl Acetate, Methyl Ethyl Ketone, Toluene, and Carbon Disulfide on Schedule-controlled Behavior of Mice," *J. Am. Coll. Toxicol.*, **6**, 461–469 (1987).

186. L. E. Kane, C. S. Barrow, and Y. Alarie, "A Short-term Test to Predict Acceptable Levels of Exposure to Airborne Sensory Irritants," *Am. Ind. Hyg. Assoc. J.*, **40**, 207–229 (1979).

187. Y. Alarie, "Sensory Irritation by Airborne Chemicals," *CRC Crit. Rev. Toxicol.*, **2**, 299–363 (1973).

188. Y. Alarie, "Bioassay for Evaluating the Potency of Airborne Sensory Irritants and Predicting Acceptable Levels of Exposure in Man," *Food Cosmet. Toxicol.*, **19**, 623–626 (1981).

189. L. E. Kane, B. S. Dombroske, and Y. Alarie, "Evaluation of Sensory Irritation from Some Common Industrial Solvents," *Am. Ind. Hyg. Assoc. J.*, **41**, 451–455 (1980).

190. J. C. De Ceaurriz, J. C. Micillino, P. Bonnet, and J. P. Guenier, "Sensory Irritation Caused by Various Industrial Airborne Chemicals," *Toxicol. Lett.*, **9**, 137–143 (1981).

191. P. M. J. Bos, A. Zwart, P. G. J. Reuzel, and P. C. Bragt, "Evaluation of the Sensory Irritation Test for the Assessment of Occupational Health Risk," *Crit. Rev. Toxicol.*, **21**, 423–450 (1992).

192. Y. Alarie and J. I. Luo, "Sensory Irritation by Airborne Chemicals: A Basis to Establish Acceptable Levels of Exposure," in: *Toxicology of the Nasal Passages*, C. S. Barrow, Ed., Hemisphere Publishing Corporation, Washington, DC, 1986, pp. 91–100.

193. G. D. Nielsen, "Mechanisms of Activation of the Sensory Irritant Receptor by Airborne Chemicals," *Crit. Rev. Toxicol.*, **21**, 183–208 (1991).

194. J. Muller and G. Greff, "Recherche de Relations Entre Toxicite de Molecules d'Interet Industriel et Proprietes Physio-chimiques: Test d'irritation des Voies Aeriennes Superieures Applique a Quatre Familles Chimiques," *Food Chem. Toxicol.*, **22**, 661–664 (1984).

195. D. W. Roberts, "QSAR for Upper-respiratory Tract Irritants," *Chem.-Biol. Interact.*, **57**, 325–345 (1986).

196. M. H. Abraham, G. S. Whiting, Y. Alarie, J. J. Morris, P. J. Taylor, R. M. Doherty, R. W. Taft, and G. D. Nielsen, "Hydrogen Bonding 12. A New QSAR for Upper Respiratory Tract Irritation by Airborne Chemicals in Mice," *Quant. Struct.-Act. Relat.*, **9**, 6–10 (1990).

197. R. H. Clothier, L. M. Hulme, M. Smith, and M. Balls, "Comparison of the *in vitro* Cytotoxicities and Acute *in vivo* Toxicities of 59 Chemicals," *Mol. Toxicol.*, **1**, 571–577 (1987).

198. E. T. Kimura, D. M. Ebert, and P. W. Dodge, "Acute Toxicity and Limits of Solvent Residue for Sixteen Organic Solvents," *Toxicol. Appl. Pharmacol.*, **19**, 699–704 (1971).

199. W. J. Krasavage, J. L. O'Donoghue, and G. D. DiVincenzo, "Ketones," in: Patty's Industrial Hygiene and Toxicology, G. D. Clayton and F. E. Clayton, Eds., 3rd ed., Vol. IIC, Wiley, New York, 1982, pp. 4720–4727.

200. H. Tanii, H. Tsuji, and K. Hashimoto, "Structure–toxicity Relationship of Monoketones," *Toxicol. Lett.*, **30**, 13–17 (1986).

201. G. L. Kennedy and G. J. Graepel, "Acute Toxicity in the Rat Following Either Oral or Inhalation Exposure," *Toxicol. Lett.*, **56**, 317–326 (1991).

202. D. C. Walton, E. F. Kehr, and A. S. Lovenhart, "A Comparison of the Pharmacological Action of Diacetone Alcohol and Acetone," *J. Pharmacol. Exp. Therap.*, **33**, 175–183 (1928).

203. R. D. Panson and C. L. Winek, "Aspiration Toxicity of Ketones," *Clin. Toxicol.*, **17**, 271–317 (1980).

204. C. P. Carpenter and H. F. Smyth, Jr., "Chemical Burns of the Rabbit Cornea," *Am. J. Opthalmol.*, **29**, 1363–1372 (1946).

205. P. S. Larson, J. K. Finnegan, and H. B. Haag, "Observations on the Effect of Chemical Configuration on the Edema-producing Potency of Acids, Aldehydes, Ketones and Alcohols," *J. Pharmacol. Exp. Therap.*, **116**, 119–122 (1956).

206. U. Turss, R. Turss, and M. F. Refojo, "Removal of Isobutyl Cyanoacrylate Adhesive from the Cornea with Acetone," *Am. J. Opthalmol.*, **70**, 725–728 (1970).

207. H. E. Kennah, S. Hignet, P. E. Laux, J. D. Dorko, and C. S. Barrow, "An Objective

Procedure for Quantitating Eye Irritation Based upon Changes of Corneal Thickness," *Fund. Appl. Toxicol.*, **12**, 258–268 (1989).

208. R. L. Morgan, S. S. Sorenson, and T. R. Castles, "Prediction of Ocular Irritation by Corneal Pachymetry," *Food Chem. Toxicol.*, **8**, 609–613 (1987).

209. A. Bolková and J. Čejková, "Changes in Alkaline and Acid Phosphatases of the Rabbit Cornea Following Experimental Exposure to Ethanol and Acetone: A Biochemical and Histochemical Study," *Graefe's Arch. Clin. Exp. Opthalmol.*, **220**, 96–99 (1983).

210. C. Anderson, K. Sundberg, and O. Groth, "Animal Model for Assessment of Skin Irritancy," *Contact Derm.*, **15**, 143–151 (1986).

211. J. Descotes, "Identification of Contact Allergens: The Mouse Ear Sensitization Assay," *J. Toxicol. Cut. Ocular Toxicol.*, **7**, 263–272 (1988).

212. G. Grubauer, K. R. Feingold, R. M. Harris, and P. M. Elias, "Lipid Content and Lipid Type as Determinants of the Epidermal Permeability Barrier," *J. Lipid Res.*, **30**, 89–96 (1989).

213. G. Grubauer, P. M. Elias, and K. R. Feingold, "Transepidermal Water Loss: The Signal for Recovery of Barrier Structure and Function," *J. Lipid Res.*, **30**, 323–333 (1989).

214. D. M. Sanderson, "A Note on Glycerol Formal as a Solvent in Toxicity Testing," *J. Pharm. Pharmacol.*, **11**, 150–156 (1959).

215. S. Zakhari, "Acute Oral, Intraperitoneal, and Inhalational Toxicity of Methyl Isobutyl Ketone in the Mouse," Chapter 11 in *Isopropanol and Ketones in the Environment*, L. Goldberg, Ed., CRC Press, Cleveland, OH, 1977, pp. 101–104.

216. G. D. DiVincenzo and W. J. Krasavage, "Serum Ornithine Carbamyl Transferase as a Liver Response Test for Exposure to Organic Solvents," *Am. Ind. Hyg. Assoc. J.*, **35**, 21–29 (1975).

217. A. R. J. Dungan, "Experimental Observations on the Acetone Bodies," *J. Metab. Res.*, **4**, 229–295 (1924).

218. R. Schneider and H. Droller, "The Relative Importance of Ketosis and Acidosis in the Production of Diabetic Coma," *Quart. J. Exp. Physiol.*, **28**, 323–333 (1938).

219. F. M. Allen and M. B. Wishart, "Experimental Studies in Diabetes. Series V. Acidosis. 9. Administration of Acetone Bodies and Related Acids," *J. Metab. Res.*, **4**, 613–648 (1923).

220. W. Saliant and N. Kleitman, "Pharmacological Studies on Acetone," *J. Pharmacol. Exp. Ther.*, **19**, 293–306 (1922).

221. B. H. Schlomovitz and E. G. Seybold, "The Toxicity of the 'Acetone Bodies'," *Am. J. Physiol.*, **70**, 130–139 (1924).

222. R. M. Sklianskaya, F. E. Urieva, and L. M. Mashbitz, "The Relative Toxicity of Acetone, Methylalcohol and Their Mixtures: I," *J. Ind. Hyg. Toxicol.*, **18**, 106–116 (1936).

223. R. R. Raje, A. Schwartz, and M. Schwartz, "*In-vitro* Toxicity of Acetone Using Coronary Perfusion in Isolated Rabbit Heart," *Drug Chem. Toxicol.*, **3**, 333–342 (1980).

224. T. Chentanez, K. Tantrarungroj, and C. Sadavongvivad, "Correlation between Positive Chronotropic Effect and Norepinephrine Release Induced by Acetone in the Rat Right Atrium," *Toxicol. Lett.*, **37**, 183–193 (1987).

225. R. S. Hinz, C. D. Hodson, C. R. Lorence, and R. H. Guy, "*In vitro* Percutaneous

Penetration: Evaluation of the Utility of Hairless Mouse Skin," *J. Invest. Dermatol.*, **93**, 87–91 (1989).

226. J. R. King and N. A. Monteiro-Riviere, "Effects of Organic Solvent Vehicles on the Viability and Morphology of Isolated Perfused Porcine Skin," *Toxicology*, **69**, 11–26 (1991).

227. C. Shopsis and S. Sathe, "Uridine Uptake Inhibition as a Cytotoxicity Test: Correlations with the Draize Test," *Toxicology*, **29**, 195–206 (1984).

228. A. Hoh, K. Meier, and R. M. Dreher, "Multilayered Keratinocyte Culture Used for *in vitro* Toxicology," *Mol. Toxicol.*, **1**, 537–546 (1987).

229. D. L. Story, S. J. Gee, C. A. Tyson, and D. H. Gould, "Response of Isolated Hepatocytes to Organic and Inorganic Cytotoxins," *J. Toxicol. Environ. Health*, **11**, 483–501 (1983).

230. P. S. Ebert, H. L. Bonkowsky, and I. Wars, "Stimulation of Hemoglobin Synthesis in Murine Erthyroleukemia Cells by Low Molecular Weight Ketones, Aldehydes, Acids, Alcohols and Ethers," *Chem.-Biol. Interact.*, **36**, 61–69 (1981).

231. I. Johansson , G. Ekström, B. Scholte, D. Puzycki, H. Jörnvall, H., and M. Ingelman-Sundberg, "Ethanol-, Fasting-, and Acetone-inducible Cytochromes P-450 in Rat Liver: Regulation and Characteristics of Enzymes Belonging to the IIB and IIE Gene Subfamilies," *Biochemistry*, **27**, 1925–1934 (1988).

232. M. J. J. Ronis, I. Johansson, K. Hultenby, J. Lagercrantz, H. Glaumann, and M. Ingelman-Sundberg, "Acetone-regulated Synthesis and Degradation of Cytochrome P4502E2 and Cytochrome P4502B1 in Rat Liver," *Eur. J. Biochem.*, **198**, 383–389 (1991).

233. M. J. Ronis and M. Ingelman-Sundberg, "Acetone-dependent Regulation of Cytochrome P-450J IIE1 and P-450B IIB1 in Rat Liver," *Xenobiotica*, **19**, 1161–1166 (1989).

234. B-J. Song, R. L. Veech, S. S. Park, H. V. Gelboin, and F. J. Gonzalez, "Induction of Rat Hepatic *N*-Nitrosodimethylamine Demethylase by Acetone is Due to Protein Stabilization," *J. Biol. Chem.*, **264**, 3568–3572 (1989).

235. C. M. Hunt, P. S. Guzelian, D. T. Molowa, and S. A. Wright, "Regulation of Rat Hepatic Cytochrome P450IIE1 in Primary Monolayer Hepatocyte Culture," *Xenobiotica*, **21**, 1621–1631 (1991).

236. E. T. Morgan, D. R. Koop, and J. M. Coon, "Catalytic Activity of Cytochrome P-450 Isozyme 3a Isolated from Liver Microsomes of Ethanol-treated Rabbits. Oxidation of Alcohols," *J. Biol. Chem.*, **257**, 13951–13957 (1982).

237. S. Imaoka and Y. Funae, "Induction of Cytochrome P450 Isozymes in Rat Liver by Methyl *n*-Alkyl Ketones and *n*-Alkylbenzenes," *Biochem. Pharmacol.*, **42** (Suppl.), S143–S150 (1991).

238. D. E. Ryan, L. Ramanathan, S. Iida, P. E. Thomas, M. Haniu, J. E. Shively, C. S. Lieber, and W. Levin, "Characterization of a Major Form of Rat Hepatic Microsomal Cytochrome P-450 Induced by Isoniazid," *J. Biol. Chem.*, **260**, 6385–6393 (1985).

239. N. Tindberg and M. Ingelman-Sundberg, "Cytochrome P-450 and Oxygen Toxicity. Oxygen-dependent Induction of Ethanol-inducible Cytochrome P-450 (IIE1) in Rat Liver and Lung," *Biochemistry*, **28**, 4499–4504 (1989).

240. D. R. Koop, B. L. Crump, G. D. Nordblom, and M. J. Coon, "Immunochemical Evidence for Induction of the Alcohol-oxidizing Cytochrome P-450 of Rabbit Liver Microsomes by Diverse Agents: Ethanol, Imidazole, Trichloroethylene, Acetone, Pyrazole, and Isoniazid," *Proc. Natl. Acad. Sci.*, **82**, 4065–4069 (1985).

241. J-S. H. Yoo, S. M. Ning, C. B. Pantuck, E. J. Pantuck, and C. S. Yang, "Regulation of Hepatic Microsomal Cytochrome P450IIE1 Level by Dietary Lipids and Carbohydrates in Rats," *J. Nutr.*, **121**, 959–965 (1991).

242. M. R. Past and D. E. Cook, "Effect of Diabetes on Rat Liver Cytochrome P-450. Evidence for a Unique Diabetes-dependent Rat Liver Cytochrome P-450," *Biochem. Pharmacol.*, **31**, 3329–3334 (1982).

243. J. Hong, J. Pan, F. J. Gonzalez, H. V. Oelboin, and C. S. Yang, "The Induction of a Specific Form of Cytochrome P-450 (P450j) by Fasting," *Biochem. Biophys. Res. Commun.*, **142**, 1077–1083 (1987).

244. C. S. Lieber, L. M. Lasker, L. M. DeCarli, J. Saeli, and T. Wojtowicz, "Role of Acetone, Dietary Fat and Total Energy Intake in Induction of Hepatic Microsomal Ethanol Oxidizing System," *J. Pharmacol. Exp. Ther.*, **247**, 791–795 (1988).

245. J-Y. Hong, J. Pan, Z. Dong, S. M. Ning, and C. S. Yang, "Regulation of N-Nitrosodimethylamine Demethylase in Rat Liver and Kidney," *Cancer Res.*, **47**, 5948–5953 (1987).

246. Z. Dong, J. Hong, Q. Ma, D. Li, J. Bullock, F. J. Gonzalez, S. S. Park, H. V. Gelboin, and C. S. Yang, "Mechanism of Induction of Cytochrome P-450ac (P-450j) in Chemically-induced and Spontaneous Diabetic Rats," *Arch. Biochem. Biophys.*, **263**, 29–35 (1988).

247. A. Mikalsen, J. Alexander, R. A. Andersen, and M. Ingelman-Sundberg, "Effect of *in vivo* Chromate, Acetone and Combined Treatment on Rat Liver *in vitro* Microsomal Chromium(VI) Reductive Activity and on Cytochrome P450 Expression," *Pharmacol. Toxicol.*, **68**, 456–463 (1991).

248. J. F. Brady, F. Xiao, M-H. Wang, Y. Li, S. M. Ning, J. M. Gapac, and C. S. Yang, "Effects of Disulfuram on Hepatic P450IIE1, other Microsomal Enzymes, and Hepatotoxicity in Rats," *Toxicol. Appl. Pharmacol.*, **108**, 366–373 (1991).

249. J-Y. Hong, S. M. Ning, B-L. Ma, M-J. Lee, J. Pan, and C. S. Yang, "Roles of Pituitary Hormones in the Regulation of Hepatic Cytochrome P450IIE1 in Rats and Mice," *Arch. Biochem. Biophys.*, **281**, 132–138 (1990).

250. T. D. Porter, S. C. Khani, and M. J. Coon, "Induction and Tissue-specific Expression of Rabbit Cytochrome P450IIE1 and IIE2 Genes," *Mol. Pharmacol.*, **36**, 61–65 (1989).

251. S. Menicagli, P. Puccini, V. Longo, and P. G. Gervasi, "Effect of Acetone Administration on Renal, Pulmonary and Hepatic Monooxygenase Activities in Hamster," *Toxicology*, **64**, 141–153 (1990).

252. T-H. Ueng, J-N. Tsia, J-M. Ju, Y-F. Ueng, M. Iwasaki, and F. P. Guengerich, "Effects of Acetone Administration on Cytochrome P-450-dependent Monooxygenases in Hamster Liver, Kidney, and Lung," *Arch. Toxicol.*, **65**, 45–51 (1991).

253. J. F. Sinclair, S. Wood, L. Lambrecht, N. Gorman, L. Mende-Mueller, L. Smith, J. Hunt, and P. Sinclair, "Isolation of Four Forms of Acetone-induced Cytochrome P-450 in Chicken Liver by h.p.l.c. and their Enzymic Characterization," *Biochem. J.*, **269**, 85–91 (1990).

254. P. G. Forkert, T. E. Massey, A. B. Jones, S. S. Park, H. V. Gelboin, and L. M. Anderson, "Distribution of Cytochrome CYP2E1 in Murine Liver after Ethanol and Acetone Administration," *Carcinogenesis*, **12**, 2259–2268 (1991).

255. X. Ding and M. J. Coon, "Induction of Cytochrome P-450 Isozyme 3a (P-450IIE1) in Rabbit Olfactory Mucosa by Ethanol and Acetone," *Drug Metab. Disp.*, **18**, 742–745 (1990).

256. G. G. Schnier, C. L. Laethem, and D. R. Koop, "Identification and Induction of Cytochromes P450, P450IIE1 and P450IA1 in Rabbit Bone Marrow," *J. Pharmacol. Exp. Ther.*, **251**, 790–796 (1989).

257. D. R. Koop, A. Chernosky, and E. P. Brass, "Identification and Induction of Cytochrome P450 2E1 in Rat Kupffer Cells," *J. Pharmacol. Exp. Therap.*, **258**, 1072–1076 (1991).

258. R. C. Robinson, R. G. L. Shorr, A. Varrichio, S. S. Park, H. V. Gelboin, H. Miller, and F. K. Friedman, "Human Liver Cytochrome P-450 Related to a Rat Acetone-inducible, Nitrosamine-metabolizing Cytochrome P-450: Identification and Isolation," *Pharmacology*, **39**, 137–144 (1989).

259. J. Raucy, P. Fernandes, M. Black, S. L. Yang, and D. R. Koop, "Identification of a Human Liver Cytochrome P-450 Exhibiting Catalytic and Immunochemical Similarities to Cytochrome P-450 3a of Rabbit Liver," *Biochem. Pharmacol.*, **36**, 2921–2926 (1987).

260. K. W. Miller and C. S. Yang, "Studies on the Mechanisms of Induction of *N*-Nitrosodimethylamine Demethylase by Fasting, Acetone, and Ethanol," *Arch. Biochem. Biophys.*, **229**, 483–491 (1984).

261. H. Sippel, K. E. Penttilä, and K. O. Lindros, "Regioselective Induction of Liver Glutathione Transferase by Ethanol and Acetone," *Pharmacol. Toxicol.*, **68**, 391–393 (1991).

262. H. Clark and G. Powis, "Effect of Acetone Administration *in vivo* upon Hepatic Microsomal Drug Metabolizing Activity in the Rat," *Biochem. Pharmacol.*, **23**, 1015–1019 (1974).

263. C. J. Patten, S. M. Ning, A. Y. Lu, and C. S. Yang, "Acetone-inducible Cytochrome P-450: Purification. Catalytic Activity, and Interaction with Cytochrome b$_5$," *Arch. Biochem. Biophys.*, **251**, 629–638 (1986).

264. L. E. Rikans, "Effects of Ethanol on Microsomal Drug Metabolism in Aging Female Rats. I. Induction," *Mech. Ageing Develop.*, **48**, 267–280 (1989).

265. G. Bànhegyi, T. Garzo, F. Antoni, and J. Mandl, "Accumulation of Phenols and Catechols in Isolated Hepatocytes in Starvation or After Pretreatment with Acetone," *Biochem. Pharmacol.*, **21**, 4157–4162 (1988).

266. R. Puccini, R. Fiorio, V. Longo, and P. G. Gervasi, "High Affinity Diethylnitrosamine-deethylase in Liver Microsomes from Acetone-induced Rats," *Carcinogenesis*, **10**, 1629–1634 (1989).

267. R. Puccini, R. Fiorio, V. Longo, and P. G. Gervasi, "Effects of Acetone Administration on Drug-Metabolizing Enzymes in Mice: Presence of a High-Affinity Diethylnitrosamine De-ethylase," *Toxicol. Lett.*, **54**, 143–150 (1990).

268. I. G. Sipes, M. L. Slocumb, and G. Holtzman, "Stimulation of Microsomal Dimethylnitrosamine-*N*-demethylase by Pretreatment of Mice with Acetone," *Chem. Biol. Interact.*, **21**, 155–166 (1978).

269. M. F. Argus, B. J. Neuburger, S. C. Myers, and J. C. Arcos, "Induction of Dimethylnitrosamine-demethylase by Polar Solvents," *Proc. Soc. Exp. Biol. Med.*, **163**, 52–55 (1980).

270. Y. Y. Tu, R. Peng, Z-F. Chang, and C. S. Yang, "Induction of a High Affinity Nitrosamine Demethylase in Rat Liver Microsomes by Acetone and Isopropanol," *Chem.-Biol. Interact.*, **44**, 247–260 (1983).

271. C. Hètu and J-G. Joly, "Effects of Chronic Acetone Administration of Ethanol-inducible Monooxygenase Activities in the Rat," *Biochem. Pharmacol.*, **37**, 421–426 (1988).

272. I. Johansson and M. Ingelman-Sundberg, "Benzene Metabolism by Ethanol-, Acetone-, and Benzene-inducible Cytochrome P-450 (IIE1) in Rat and Rabbit Liver Microsomes," *Cancer Res.*, **48**, 5387–5390 (1988).

273. D. R. Koop, C. L. Laethem, and G. G. Schnier, "Identification of Ethanol-inducible P450 Isozyme 3a (P450IIE1) as Benzene and Phenol Hydroxylase," *Toxicol. Appl. Pharmacol.*, **98**, 278–288 (1989).

274. J. Cunningham, M. Sharkawi, and G. L. Plaa, "Pharmacological and Metabolic Interactions between Ethanol and Methyl *n*-Butyl Ketone, Methyl Isobutyl Ketone, Methyl Ethyl Ketone, or Acetone in Mice," *Fund. Appl. Toxicol.*, **13**, 102–109 (1989).

275. R. A. Kurata, H. E. Hurst, F. W. Benz, R. A. Kemper, and W. J. Waddell, "Studies on Inhibition and Induction of Metabolism of Ethyl Carbamate by Acetone and Related Compounds," *Drug Metab. Disp.*, **19**, 388–393 (1991).

276. E. Albano, A. Tomasi, J. O. Persson, Y. Terelius, L. Goria-Gatti, M. Ingelman-Sundberg, and M. U. Dianzani, "Role of Ethanol-inducible Cytochrome P450 (P450IIE1) in Catalyzing the Free Radical Activation of Aliphatic Alcohols," *Biochem. Pharmacol.*, **41**, 1895–1902 (1991).

277. O. S. Sohn, H. Ishizaki, C. S. Yang, and E. S. Fiala, "Metabolism of Azoxymethane, Methylazoxymethanol, and *N*-Nitrosodimethylamine by Cytochrome P450IIE1," *Carcinogenesis*, **12**, 127–131 (1991).

278. C. A. Lee, K. E. Thummel, T. F. Kalhorn, S. D. Nelson, and J. T. Slattery, "Activation of Acetaminophen-reactive Metabolite Formation by Methylxanthines and Known Cytochrome P-450 Activators," *Drug Metab. Disp.*, **19**, 966–971 (1991).

279. J. Liu, C. Sato and F. Marumo, "Characterization of the Acetaminophen-glutathione Conjugation Reaction by Liver Microsomes: Species Difference in the Effects of Acetone," *Toxicol. Lett.*, **56**, 269–274 (1991).

280. J. J. Freeman and E. P. Hayes, "Microsomal Metabolism of Acetonitrile to Cyanide. Effects of Acetone and Other Compounds," *Biochem. Pharmacol.*, **37**, 1153–1159 (1988).

281. J. F. Brady, M. J. Lee, H. Ishizaki, and C. S. Yang, "Diethyl Ether as a Substitute for Acetone/ethanol-inducible Cytochrome P-450 and as an Inducer for Cytochrome(s) P-450," *Mol. Pharmacol.*, **33**, 148–154 (1988).

282. D. K. Winters and A. I. Cederbaum, "Oxidation of Glycerol to Formaldehyde by Rat Liver Microsomes. Effects of Cytochrome P-450 Inducing Agents," *Biochem. Pharmacol.*, **39**, 697–705 (1990).

283. J. F. Brady, F. Xiao, S. M. Ning, and C. S. Yang, "Metabolism of Methyl *Tertiary*-Butyl Ether by Rat Hepatic Microsomes," *Arch. Toxicol.*, **64**, 157–160 (1990).

284. M. W. Anders, "Acetone Enhancement of Microsomal Aniline Parahydroxylase Activity," *Arch. Biochem. Biophys.*, **126**, 269–275 (1968).

285. H. Vainio and O. Hänninen, "Enhancement of Aniline *p*-Hydroxylation by Acetone in Rat Liver Microsomes," *Xenobiotica*, **2**, 259–267 (1972).

286. M. W. Anders, "Effect of Phenobarbital and 3-Methylcholanthrene Administration on the *In Vitro* Enhancement of Microsomal Aromatic Hydroxylation," *Arch. Biochem. Biophys.*, **153**, 502–507 (1972).

287. M. Kitada, T. Kamataki, and H. Kitagawa, "NADH-synergism of NADPH-dependent *O*-Dealkylation of Type II Compounds, *p*-Anisidine and *p*-Phenetidine, in Rat Liver," *Arch. Biochem. Biophys.*, **178**, 151–157 (1977).

288. T. Wolfe, H. Wanders, and F. P. Guengerich, "Organic Solvents as Modifiers of Aldrin Epoxidase in Reconstituted Monooxygenase Systems and in Microsomes," *Biochem. Pharmacol.*, **38**, 4217–4223 (1989).

289. J. C. Kawalek and A. W. Andrews, "The Effect of Solvents on Drug Metabolism *in vitro*," *Drug Metab. Disp.*, **8**, 380–384 (1980).

290. M. W. Anders and J. E. Gander, "Acetone Enhancement of Cumene Hydroperoxide Supported Microsomal Aniline Hydroxylation," *Life Sci.*, **25**, 1085–1090 (1979).

291. M. Kitada, M. Ando, S. Ohmori, S. Kabuto, T. Kamataki, and H. Kitagawa, "Effect of Acetone on Aniline Hydroxylation by a Reconstituted System," *Biochem. Pharmacol.*, **21**, 3151–3155 (1983).

292. W. R. Bidlack and G. L. Lowery, "Multiple Drug Metabolism: *p*-Nitroanisole Reversal of Acetone Enhanced Aniline Hydroxylation," *Biochem. Pharmacol.*, **31**, 311–317 (1982).

293. G. Powis and A. R. Boobis, "The Effect of Pretreating Rats with 3-Methylcholanthrene upon the Enhancement of Microsomal Aniline Hydroxylation by Acetone and Other Agents," *Biochem. Pharmacol.*, **24**, 424–426 (1975).

294. E. H. Jenney and C. C. Pfeiffer, "The Convulsant Effect of Hydrazides and the Antidotal Effect of Anticonvulsants and Metabolites," *J. Pharmacol. Exp. Therap.*, **122**, 110–123 (1958).

295. R. P. Kohli, K. Kishor, P. R. Dua, and R. C. Saxena, "Anticonvulsant Activity of Some Carbonyl Containing Compounds," *Ind. J. Med. Res.*, **55**, 1221–1225 (1967).

296. V. F. Price and D. J. Jollow, "Mechanism of Ketone-induced Protection from Acetaminophen Hepatotoxicity in the Rat," *Drug Metab. Disp.*, **11**, 451–457 (1983).

297. V. F. Price and D. J. Jollow, "Increased Resistance of Diabetic Rats to Acetaminophen-induced Hepatotoxicity," *J. Pharmacol. Exp. Ther.*, **220**, 504–513 (1982).

298. P. Moldèus and V. Gergely, "Effect of Acetone on the Activation of Acetaminophen," *Toxicol. Appl. Pharmacol.*, **53**, 8–13 (1980).

299. H-H. Lo, V. J. Teets, D. J. Yang, P. I. Brown, and G. O. Rankin, "Acetone Effects on *N*-(3,5-Dichlorophenyl)succinimide-induced Nephrotoxicity," *Toxicol. Lett.*, **38**, 161–168 (1987).

300. G. J. Traiger and G. L. Plaa, "Differences in the Potentiation of Carbon Tetrachloride in Rats by Ethanol and Isopropanol Pretreatment," *Toxicol. Appl. Pharmacol.*, **20**, 105–112 (1971).

301. G. J. Traiger and G. L. Plaa, "Chlorinated Hydrocarbon Toxicity. Potentiation by Isopropyl Alcohol and Acetone," *Arch. Environ. Health*, **28**, 276–278 (1974).

302. W. R. Hewitt, H. Miyajima, M. G. Côté, A. Hewitt, D. J. Cianflone, and G. L. Plaa, "Dose-response Relationships in 1,3-Butanediol-induced Potentiation of Carbon Tetrachloride Toxicity," *Toxicol. Appl. Pharmacol.*, **64**, 529–540 (1982).

303. T-H. Ueng, L. Moore, R. G. Elves, and A. P. Alvares, "Isopropanol Enhancement of Cytochrome P-450-dependent Monooxygenase Activities and Its Effects on Carbon Tetrachloride Intoxication," *Toxicol. Appl. Pharmacol.*, **71**, 204–214 (1983).

304. I. G. Sipes, B. Stripp, G. Krishna, H. M. Maling, and J. R. Gillette, "Enhanced Hepatic Microsomal Activity by Pretreatment of Rats with Acetone or Isopropanol." *Proc. Soc. Exp. Biol. Med.*, **142**, 237–240 (1973).

305. H. M. Mailing, B. Stripp, I. G. Sipes, B. Highman, Saul and M. A. Williams, "Enhanced Hepatotoxicity of Carbon Tetrachloride, Thioacetamide, and Dimethylnitro-

samine by Pretreatment of Rats with Ethanol and Some Comparisons with Potentiation by Isopropanol," *Toxicol. Appl. Pharmacol.*, **33**, 291–308 (1975).

306. Kobush, G. L. Plaa, and P. Du Souich, "Effects of Acetone and Methyl *n*-Butyl Ketone on Hepatic Mixed-function Oxidase," *Biochem. Pharmacol.*, **38**, 3461–3467 (1989).

307. A. Sato and T. Nakajima, "Enhanced Metabolism of Volatile Hydrocarbons in Rat Liver Following Food Deprivation, Restricted Carbohydrate Intake, and Administration of Ethanol, Phenobarbital, Polychlorinated Biphenyl and 3-Methylcholanthrene: A Comparative Study," *Xenobiotica*, **15**, 67–75 (1985).

308. J. F. Brady, D. Li, H. Ishizaki, M. Lee, S. M. Ning, F. Xiao, and C. S. Yang, "Induction of Cytochromes P450IIE1 and P450IIB1 by Secondary Ketones and the Role of P450IIE1 in Chloroform Metabolism," *Toxicol. Appl. Pharmacol.*, **100**, 342–349 (1989).

309. I. Johansson and M. Ingelman-Sundberg, "Carbon Tetrachloride-Induced Lipid Peroxidation Dependent on an Ethanol-inducible Form of Rabbit Liver Microsomal Cytochrome P450," *FEBS Lett.*, **183**, 265–269 (1985).

310. G. J. Traiger and G. L. Plaa, "Effect of Aminotriazole on Isopropanol- and Acetone-induced Potentiation of CCl₄ Hepatotoxicity," *Can. J. Physiol. Pharmacol.*, **51**, 291–296 (1973).

311. G. J. Traiger and G. L. Plaa, "Relationship of Alcohol Metabolism to the Potentiation of CCl₄ Hepatotoxicity Induced by Aliphatic Alcohols," *J. Pharmacol. Exp. Ther.*, **183**, 481–488 (1972).

312. D. S. Folland, W. Schaffner, H. E. Ginn, O. B. Crofford, and D. R. McMurray, "Carbon Tetrachloride Toxicity Potentiated by Isopropyl Alcohol. Investigation of an Industrial Outbreak," *J. Am. Med. Assoc.*, **236**, 1853–1856 (1976).

313. J-F. Deng, T-F. Wang, T-S. Shih, and F-L. Lan, "Outbreak of Carbon Tetrachloride Poisoning in a Color Printing Factory Related to the Use of Isopropyl Alcohol and an Air Conditioning System in Taiwan," *Am. J. Ind. Med.*, **12**, 11–19 (1987).

314. W. R. Hewitt, H. Miyajima, M. G. Côté, and G. L. Plaa, "Acute Alteration of Chloroform-induced Hepato- and Nephrotoxicity by *n*-Hexane, Methyl *n*-Butyl Ketone, and 2,5-Hexanedione," *Toxicol. Appl. Pharmacol.*, **53**, 230–248 (1980).

315. W. R. Hewitt, E. M. Brown, and G. L. Plaa, "Relationship between the Carbon Skeleton Length of Ketonic Solvents and Potentiation of Chloroform-induced Hepatotoxicity in Rats," *Toxicol. Lett.*, **16**, 297–304 (1983).

316. E. M. Brown and W. R. Hewitt, "Dose-response Relationships in Ketone-induced Potentiation of Chloroform Hepato- and Nephrotoxicity," *Toxicol. Appl. Pharmacol.*, **76**, 437–453 (1984).

317. L. A. Hewitt, P. Ayotte, and G. L. Plaa, "Modifications in Rat Hepatobiliary Function Following Treatment with Acetone, 2-Butanone, 2-Hexanone, Mirex, or Chlordecone and Subsequently Exposed to Chloroform," *Toxicol. Appl. Pharmacol.*, **83**, 465–473 (1986).

318. L. A. Hewitt, C. Valiquette, and G. L. Plaa, "The Role of Biotransformation-detoxification in Acetone-, 2-Butanone-, and 2-Hexanone-potentiated Chloroform-induced Hepatotoxicity," *Can. J. Physiol. Pharmacol.*, **65**, 2313–2318 (1987).

319. G. L. Plaa, W. R. Hewitt, P. du Souich, G. Caillè, and S. Lock, "Isopropanol and Acetone Potentiation of Carbon Tetrachloride-induced Hepatotoxicity: Single versus Repetitive Pretreatment in Rats," *J. Toxicol. Environ. Health*, **9**, 235–250 (1982).

320. M. Charbonneau, J. Brodeur, P. du Souich, and G. L. Plaa, "Correlation between

Acetone-potentiated CCl$_4$-induced Liver Injury and Blood Concentrations after Inhalation or Oral Administration," *Toxicol. Appl. Pharmacol.*, **84**, 286–294 (1986).

321. J. R. MacDonald, A. J. Gandolfi, and I. G. Sipes, "Acetone Potentiation of 1,1,2-Trichloroethane Hepatotoxicity," *Toxicol. Lett.*, **13**, 57–69 (1982).

322. W. R. Hewitt and G. L. Plaa, "Dose-dependent Modification of 1,1-Dichloroethylene Toxicity by Acetone," *Toxicol. Lett.*, **16**, 145–152 (1983).

323. W. R. Hewitt, E. M. Brown, and G. L. Plaa, "Acetone-induced Potentiation of Trihalomethane Toxicity in Male Rats," *Toxicol. Lett.*, **16**, 285–296 (1983).

324. M. T. Brondeau, M. Ban, P. Bonnet, J. P. Guenier, and J. De Ceaurriz, "Acetone Compared to Other Ketones in Modifying the Hepatotoxicity of Inhaled 1,2-Dichlorobenzene in Rats and Mice," *Toxicol. Lett.*, **49**, 69–78 (1989).

325. M. Charbonneau, S. Oleskevich, J. Brodeur, and G. L. Plaa, "Acetone Potentiation of Rat Liver Injury Induced by Trichloroethylene–Carbon Tetrachloride Mixtures," *Fund. Appl. Toxicol.*, **6**, 654–661 (1986).

326. M. Charbonneau, F. Perreault, E. Greselin, J. Brodeur, and G. L. Plaa, "Assessment of the Minimal Effective Dose of Acetone for Potentiation of the Hepatotoxicity induced by Trichloroethylene–Carbon Tetrachloride Mixtures," *Fund. Appl. Toxicol.*, **10**, 431–438 (1988).

327. M. Charbonneau, E. Greselin, J. Brodeur, and G. L. Plaa, "Influence of Acetone on the Severity of the Liver Injury Induced by Haloalkane Mixtures," *Can. J. Physiol. Pharmacol.*, **69**, 1901–1907 (1991).

328. V. Postolache, T. Safta, B. Cuparencu, and L. Steiner, "Die Einwirkung des Acetons auf die Barbiturat-Narkose," *Arch. Toxikol.*, **25**, 333–337 (1969).

329. B. D. Astill and G. D. DiVincenzo, "Some Factors Involved in Establishing and Using Biological Threshold Limit Values (BTLV's)," in *Proceedings of the 3rd Annual Conference on Environmental Toxicology*, Paper No. 7, Aerospace Medical Research Laboratory, Wright-Patterson Air Force Base, OH, 1972, pp. 115–135.

330. E. Chieli, P. Puccini, V. Longo, and P. G. Gervasi, "Possible Role of the Acetone-inducible Cytochrome P-450IIE1 in the Metabolism and Hepatotoxicity of Thiobenzamide," *Arch. Toxikol.*, **64**, 122–127 (1990).

331. L. A. Scott, E. Maden, and M. A. Valentovic, "Influence of Streptozotocin (STZ)-induced Diabetes, Dextrose Diuresis and Acetone on Cisplatin Nephrotoxicity in Fisher 344 (F344) Rats," *Toxicology*, **60**, 109–125 (1990).

332. H. Hoensch, "Ethanol as Enzyme Inducer and Inhibitor," *Pharmacol. Ther.*, **33**, 121–128 (1987).

333. H. J. Zimmerman, "Effects of Alcohol on Other Hepatotoxins," *Alcoholism: Clin. Exp. Res.*, **10**, 3–15 (1986).

334. M. J. Coon and D. R. Koop, "Alcohol-inducible Cytochrome P-450 (P-450$_{ALC}$)," *Arch. Toxikol.*, **60**, 16–21 (1987).

335. H. F. Smyth, C. S. Weil, J. S. West, and C. P. Carpenter, "An Exploration of Joint Toxic Action: Twenty-seven Industrial Chemicals Intubated in Rats in all Possible Pairs," *Toxicol. Appl. Pharmacol.*, **14**, 340–347 (1969).

336. J. J. Freeman and E. P. Hayes, "Acetone Potentiation of Acute Acetonitrile Toxicity in Rats," *J. Toxicol. Environ. Health*, **15**, 609–621 (1985).

337. J. J. Freeman and E. P. Hayes, "Acetone Potentiation of Acute Acetonitrile Toxicity in Rats," *J. Toxicol. Environ. Health*, **15**, 609–621 (1985).

338. M. D. Boggild, R. W. Peck, and C. V. R. Tomson, "Acetonitrile Ingestion: Delayed Onset of Cyanide Poisoning due to Concurrent Ingestion of Acetone," *Postgrad. Med. J.*, **66**, 40–41 (1990).

339. H. Elovaara, H. Vainio, and A. Aitio, "Pulmonary Toxicity of Inhaled Styrene in Acetone-, Phenobarbital- and 3-Methylcholanthrene-treated Rats," *Arch. Toxicol.*, **64**, 365–369 (1990).

340. H. Elovaara, K. Engstrröm, T. Nakajima, S. S. Park, H. V. Gelboin, and H. Vainio, "Metabolism of Inhaled Styrene in Acetone-, Phenobarbital- and 3-Methylcholanthrene-pretreated Rats: Stimulation and Stereochemical Effects by Induction of Cytochrome P450IIE1, P450IIB, and P450IA," *Xenobiotica*, **21**, 651–661 (1991).

341. E. Wigaeus, A. Löf, and M. B. Nordqvist, "Uptake, Distribution, Metabolism, and Elimination of Styrene in Man. A Comparison between Single Exposure and Co-exposure with Acetone," *Br. J. Ind. Med.*, **41**, 539–546 (1984).

342. N. A. Lorr, K. W. Miller, H. R. Chung, and C. S. Yang, "Potentiation of the Hepatotoxicity of *N*-Nitrosodimethylamine by Fasting, Diabetes, Acetone, and Isopropanol," *Toxicol. Appl. Pharmacol.*, **73**, 423–431 (1984).

343. S. M. Haag and I. G. Sipes, "Differential Effects of Acetone or Aroclor 1254 Pretreatment in the Microsomal Activation of Dimethylnitrosamine to a Mutagen," *Mutat. Res.*, **74**, 431–438 (1980).

344. O. Ladefoged, H. Hass, and L. Simonsen, "Neurophysiological and Behavioral Effects of Combined Exposure to 2,5-Hexanedione and Acetone or Ethanol in Rats," *Pharmacol. Toxicol.*, **65**, 372–375 (1989).

345. O. Ladefoged and L. Perbellini, "Acetone-induced Changes in the Toxicokinetics of 2,5-Hexanedione in Rabbits," *Scand. J. Work Environ. Health*, **12**, 627–629 (1986).

346. I. Johansson, E. Eliasson, C. Norstend, and M. Ingelman-Sundberg, "Hydroxylation of Acetone by Ethanol- and Acetone-inducible Cytochrome P-450 in Liver Microsomes and Reconstituted Membranes," *Fed. Europ. Biochem. Soc. Lett.*, **196**, 59–64 (1986).

347. D. R. Koop and J. P. Casazza, "Identification of Ethanol-inducible P-450 Isozyme 3a as the Acetone and Acetol Monooxygenase of Rabbit Microsomes," *J. Biol. Chem.*, **260**, 13607–13612 (1985).

348. F. N. LeBaron, "Lipid Metabolism I. Utilization and Storage of Energy in Lipid Form," in *Textbook of Biochemistry with Clinical Correlations*, T. M. Devlin, Ed., Wiley, New York, 1982, pp. 467–479.

349. D. E. Vance, "Metabolism of Fatty Acids and Triacylglycerols," Chapter 13 in *Biochemistry*, G. Zubay, Ed., Addison-Wesley, Reading, MA, 1984, pp. 471–503.

350. O. Wieland, "Ketogenesis and Its Regulation," *Adv. Metab. Disorders*, **3**, 1–47 (1967).

351. V. C. Gavino, J. Somma, L. Philbert, F. David, M. Garneau, J. Bélair, and H. Brunengraber, "Production of Acetone and Conversion of Acetone to Acetate in the Perfused Rat Liver," *J. Biol. Chem.*, **262**, 6735–6740 (1987).

352. G. J. Van Stekelenburg and G. Koorevaar, "Evidence for the Existence of Mammalian Acetoacetate Decarboxylase: With Special Reference to Human Blood Serum," *Clin. Chem. Acta*, **39**, 191–199 (1972).

353. G. Koorevaar and G. J. Van Stekelenburg, "Mammalian Acetoacetate Decarboxylase Activity. Its Distribution in Subfractions of Human Albumin and Occurrence in Various Tissues of the Rat," *Clin. Chem. Acta*, **71**, 173–183 (1976).

354. F. J. Lopez-Soriano and Argilés, "Rat Acetoacetate Decarboxylase: Its Role in the Disposal of 4C-Ketone Bodies by the Fetus," *Horm. Metab. Res.*, **18**, 446–449 (1986).

355. F. J. Lopez-Soriano, M. Alemany, and J. M. Argilés, "Rat Acetoacetic Acid Decarboxylase Inhibition by Acetone," *Int. J. Biochem.*, **17**, 1271–1273 (1985).

356. J. G. Filser, H. M. Bolt, K. Kimmich, and F. A. Bencsàth, "Exhalation of Acetone by Rats on Exposure to *trans*-1,2-Dichloroethylene and Related Compounds," *Toxicol. Lett.*, **2**, 247–252 (1978).

357. C. J. Dillard and A. L. Tappel, "Volatile Hydrocarbon and Carbonyl Products of Lipid Peroxidation: A Comparison of Pentane, Ethane, Hexanal, and Acetone as *in vivo* Indices," *Lipids*, **14**, 989–995 (1979).

358. J. G. Filser and H. M. Bolt, "Characteristics of Haloethylene-induced Acetonemia in Rats," *Arch. Toxicol.*, **45**, 109–116 (1980).

359. J. G. Filser, P. Jung, and H. M. Bolt, "Increased Acetone Exhalation Induced by Metabolites of Halogenated C_1 and C_2 Compounds," *Arch. Toxicol.*, **49**, 107–116 (1982).

360. H. M. Bolt, G. Schmiedel, J. G. Filser, H. P. Rolzhäuser, K. Lieser, D. Wistuba, and V. Schurig, "Biological Activation of 1,3-butadiene to Vinyl Oxirane by Rat Liver Microsomes and Expiration of the Reactive Metabolite by Exposed Rats," *J. Cancer Res. Clin. Oncol.*, **106**, 112–116 (1983).

361. P. Simon, H. M. Bolt, and J. G. Filser, "Covalent Interaction of Reactive Metabolites with Cytosolic Coenzyme A as Mechanism of Haloethylene-induced Acetonemia," *Biochem. Pharmacol.*, **34**, 1981–1986 (1985).

362. A. V. Buchter, R. Korff, H. W. Georgens, H. Peter, and H. M. Bolt, "Studie zur Perchlorethylen-(Tetrachlorethen-) Exposition unter Berücksichtigung von Expositions-Spitzen und Aceton-Bildung," *Zbl. Arbeitsmed.*, **34**, 130–136 (1984).

363. E. G. DeMaster and H. T. Nagasawa, "Disulfuram-induced Acetonemia in the Rat and Man," *Res. Comm. Chem. Pathol. Pharmacol.*, **18**, 361–364 (1977).

364. A. Stowell, J. Johnsen, A. Ripel, and J. Morland, "Disulfiram-induced Acetonemia," *Lancet*, **1983-I**, 882–883.

365. J. Johnsen, A. Stowell, T. Stensrud, A. Ripel, and J. Morland, "A Double-blind Placebo Controlled Study of Healthy Volunteers Given a Subcutaneous Disulfiram Implantation," *Pharmacol. Toxicol.*, **66**, 227–230 (1990).

366. E. G. DeMaster and J. M. Stevens, "Acute Effects of the Aldehyde Dehydrogenase Inhibitors, Disulfiram, Paragyline and Cyanamide, on Circulating Ketone Body Levels in the Rat," *Biochem. Pharmacol.*, **37**, 229–234 (1988).

367. T. D. Price and D. Rittenberg, "The Metabolism of Acetone. I. Gross Aspects of Catabolism and Excretion," *J. Biol. Chem.*, **185**, 449–459 (1950).

368. I. Zabin and K. Bloch, "The Utilization of Isovaleric Acid for the Synthesis of Cholesterol," *J. Biol. Chem.*, **185**, 131–138 (1950).

369. W. Sakami, "Formation of Formate and Labile Methyl Groups from Acetone in the Intact Rat," *J. Biol. Chem.*, **187**, 369–377 (1950).

370. W. Sakami and J. M. Lafaye, "The Metabolism of Acetone in the Intact Rat," *J. Biol. Chem.*, **193**, 199–203 (1951).

371. W. Sakami and H. Rudney, "The Metabolism of Acetone and Acetoacetate in the Mammalian Organism," in *The Major Metabolic Fuels, Brookhaven Symposia in Biology*, Vol. 5, Brookhaven National Laboratory, Upton, NY, 1952, pp. 176–190.

372. H. Rudney, "Propanediol Phosphate as a Possible Intermediate in the Metabolism of Acetone," *J. Biol. Chem.*, **210**, 361–371 (1954).

373. H. Rudney, "The Metabolism of 1,2-Propanediol," *Arch. Biochem. Biophys.*, **29**, 231–232 (1950).

374. G. A. Mourkides, D. C. Hobbs, and R. E. Koeppe, "The Metabolism of Acetone-2-C^{14} by Intact Rats," *J. Biol. Chem.*, **234**, 27–30 (1959).

375. E. N. Bergman, A. F. Sellers, and F. A. Spurrell, "Metabolism of C^{14}-labeled Acetone, Acetate and Palmitate in Fasted Pregnant and Nonpregnant Guinea Pigs," *Am. J. Physiol.*, **198**, 1087–1093 (1960).

376. D. B. Lindsay and R. E. Brown, "Acetone Metabolism in Sheep," *Biochem. J.*, **100**, 589–592 (1966).

377. J. R. Luick, A. L. Black, M. G. Simesen, M. Kametaka, and D. S. Kronfeld, "Acetone Metabolism in Normal and Ketotic Cows," *J. Dairy Sci.*, **50**, 544–549 (1967).

378. A. L. Black, J. R. Luick, S. L. Lee, and K. Knox, "Glucogenic Pathway for Acetone Metabolism in the Lactating Cow," *Am. J. Physiol.*, **222**, 1575–1580 (1972).

379. D. L. Coleman, "Acetone Metabolism in Mice: Increased Activity in Mice Heterozygous for Obesity Genes," *Proc. Natl. Acad. Sci.*, **77**, 290–293 (1980).

380. J. P. Casazza, M. E. Felver, and R. L. Veech, "The Metabolism of Acetone in Rat," *J. Biol. Chem.*, **259**, 231–236 (1984).

381. J. A. Ruddick, "Toxicology, Metabolism, and Biochemistry of 1,2-Propanediol," *Toxicol. Appl. Pharmacol.*, **21**, 102–111 (1973).

382. J. M. Argilés, "Has Acetone a Role in the Conversion of Fat to Carbohydrate in Mammals?," *Trends Biochem. Sci.*, **11**, 61–63 (1986).

383. J. Peinado, F. J. Lopez-Soriano, and J. M. Argilés, "The Metabolism of Acetone in the Pregnant Rat," *Biosci. Rep.*, **6**, 983–989 (1986).

384. K. Kosugi, R. F. Scofield, V. Chandramouli, K. Kumaran, W. C. Schumann, and B. R. Landau, "Pathways of Acetone's Metabolism in the Rat," *J. Biol. Chem.*, **261**, 3952–3957 (1986).

385. V. C. Gavino, J. Somma, L. Philbert, F. David, M. Garneau, J. Bélair, and H. Brunengraber, "Production of Acetone and Conversion of Acetone to Acetate in the Perfused Rat Liver," *J. Biol. Chem.*, **262**, 6735–6740 (1987).

386. K. Kosugi, V. Chandramouli, K. Kumaran, W. C. Schumann, and B. R. Landau, "Determinants in the Pathways Followed by the Carbons of Acetone in Their Conversion to Glucose," *J. Biol. Chem.*, **261**, 13179–13181 (1986).

387. S. K. Agarwal, "Non-acidotic Acetonemia: A Syndrome due to Isopropyl Alcohol Intoxication," *J. Am. Med. Soc. N.J.*, **76**, 914–916 (1979).

388. P. C. Hawley and J. M. Falko, "'Pseudo' Renal Failure after Isopropyl Alcohol Intoxication," *S. Med. J.*, **75**, 630–631 (1982).

389. S. J. Rosansky, "Isopropyl Alcohol Poisoning Treated with Hemodialysis: Kinetics of Isopropyl Alcohol and Acetone Removal," *J. Toxicol. Clin. Toxicol.*, **19**, 265–271 (1982).

390. L. Adelson, "Fatal Intoxication with Isopropyl Alcohol (Rubbing Alcohol)," *Am. J. Clin. Pathol.*, **38**, 144–151 (1962).

391. F. Brugnone, L. Perbellini, P. Apostoli, M. Bellomi, and D. Caretta, "Isopropanol Exposure: Environmental and Biological Monitoring in a Printing Works," *Brit. J. Ind. Med.*, **40**, 160–168 (1983).

392. G. Triebig, M. Fritz, K. H. Schaller, F. Helbing, E. M. Bünte, G. Kufner, and D. Weltle, "Arteitsmedizinische Untersuchungen bei beruflich Iso-Propanol-exponierten Frauen," *Arbeitsmed. Sozialmed. Praventivmed.*, **24**, 27–31 (1989).

393. D. R. Daniel, B. H. McAnalley, and J. C. Garriott, "Isopropyl Alcohol Metabolism after Acute Intoxication in Humans," *J. Anal. Toxicol.*, **5**, 110–112 (1981).

394. M. Natowicz, J. Donahue, L. Gorman, M. Kane, J. McKissick, and L. Shaw, "Pharmacokinetic Analysis of a Case of Isopropanol Intoxication," *Clin. Chem.*, **31**, 326–328 (1985).

395. P. G. Lacouture, D. D. Heldreth, M. Shannon, and F. H. Lovejoy, "The Generation of Acetonemia/Acetonuria Following Ingestion of a Subtoxic Dose of Isopropyl Alcohol," *Am. J. Emerg. Med.*, **7**, 38–40 (1989).

396. A. W. Freireich, T. J. Cinque, G. Xanthaky, and D. Landau, "Hemodialysis for Isopropanol Poisoning," *New Engl. J. Med.*, **277**, 699–700 (1967).

397. A. A. Pappas, B. H. Ackerman, K. M. Olsen, and E. H. Taylor, "Isopropanol Ingestion: A Report of Six Episodes with Isopropanol and Acetone Serum Concentration Time Data," *Clin. Toxicol.*, **29**, 11–21 (1991).

398. U. Abshagen and N. Rietbrock, "Kinetik der Elimination von 2-Propanol und seines Metaboliten bei Hund und Ratte," *Arch. Pharmakol.*, **264**, 110–118 (1969).

399. S. Laham and M. Potvin, "Studies on Inhalation Toxicity of 2-Propanol," *Drug. Chem. Toxicol.*, **3**, 343–360 (1980).

400. F. S. Ashkar and R. Miller, "Hospital Ketosis in the Alcoholic Diabetic: A Syndrome due to Isopropyl Alcohol Intoxication," *S. Med. J.*, **64**, 1409–1411 (1971).

401. F. W. Ellis, "The Role of the Liver in the Metabolic Disposition of Isopropyl Alcohol," *J. Pharmacol. Exp. Ther.*, **105**, 427–436 (1952).

402. R. Nordmann, C. Ribiere, H. Rouach, F. Beauge, Y. Giusicelli, and J. Nordmann, "Metabolic Pathways Involved in the Oxidation of Isopropanol into Acetone by the Intact Rat," *Life Sci.*, **13**, 919–932 (1973).

403. G. D. Lewis, A. K. Laufman, B. H. McAnalley, and J. C. Garriott, "Metabolism of Acetone to Isopropyl Alcohol in Rats and Humans," *J. Forensic Sci.*, **29**, 541–549 (1984).

404. P. L. Davis, L. A. Dal Cortivo, and J. Maturo, "Endogenous Isopropanol: Forensic and Biochemical Implications," *J. Anal. Toxicol.*, **8**, 209–212 (1984).

405. K. C. Leibman, "Reduction of Ketones in Liver Cytosol," *Xenobiotica*, **1**, 97–104 (1971).

406. R. Dayal, B. Goodwin, I. Linhart, K. Mynett, and A. Gescher, "Oxidative Denitrification of 2-Nitropropane and Propane-2-nitronate by Mouse Liver Microsomes: Lack of Correlation with Hepatocytotoxic Potential," *Chem.-Biol. Int.*, **79**, 103–114 (1991).

407. R. C. Baker, S. M. Sorensen, and R. A. Deitrich, "The *in vivo* Metabolism of Tertiary Butanol by Adult Rats," *Alcoholism: Clin. Exp. Res.*, **6**, 247–251 (1982).

408. A. I. Cederbaum, A. Qureshi, and G. Cohen, "Production of Formaldehyde and Acetone by Hydroxyl-radical Generating Systems During the Metabolism of Tertiary Butyl Alcohol," *Biochem. Pharmacol.*, **32**, 3517–3524 (1983).

409. A. I. Cederbaum, A. Qureshi, and P. Messenger, "Oxidation of Isopropanol by Rat Liver Microsomes. Possible Role of Hydroxyl Radicals," *Biochem. Pharmacol.*, **30**, 825–831 (1981).

410. G. Pezzagno, S. Ghittori, M. Imbriani, and E. Capodaglio, "La misura dei coefficienti di solbilità degli aeriformi nel sangue," *G. Ital. Med. Lav.*, **5**, 49–63 (1983).

411. S. Paterson and D. Mackay, "Correlation of Tissue, Blood and Air Partition Coefficients of Volatile Organic Chemicals," *Brit. J. Ind. Med.*, **46**, 321–328 (1989).

412. E. M. P. Widmark, XXXII. "Studies in the Acetone Concentration in Blood, Urine, and Alveolar Air. II. The Passage of Acetone and Aceto-acetic Acid into the Urine," *Biochem. J.*, **14**, 364–378 (1920).

413. E. Wigaeus, S. Holm, and I. Astrand, "Exposure to Acetone. Uptake and Elimination in Man," *Scand. J. Work Environ. Health*, **7**, 84–94 (1981).

414. I. H. Young and P. D. Wagner, "Effect of Intrapulmonary Hematocrit Maldistribution on O_2, CO_2, and Inert Gas Exchange," *J. Appl. Physiol.*, **46**, 240–248 (1979).

415. A. Sato and T. Nakajima, "Partition Coefficients of Some Aromatic Hydrocarbons and Ketones in Water, Blood and Oil," *Brit. J. Ind. Med.*, **36**, 231–234 (1979).

416. V. Fiserova-Bergerova and M. L. Diaz, "Determination and Prediction of Tissue-gas Partition Coefficients," *Int. Arch. Occup. Environ. Health*, **58**, 75–87 (1986).

417. H. D. Landahl and R. G. Hermann, "Retention of Vapors and Gases in the Human Nose and Lung," *Arch. Ind. Hyg.*, **1**, 36–45 (1950).

418. T. Dalhamn, M-L. Edfors, and R. Rylander, "Mouth Absorption of Various Compounds in Cigarette Smoke," *Arch. Environ. Health*, **16**, 831–835 (1968).

419. T. Dalhamn, M-L. Edfors, and R. Rylander, "Retention of Cigarette Smoke Components in Human Lungs," *Arch. Environ. Health*, **17**, 746–748 (1968).

420. K. Nomiyama and H. Nomiyama, "Respiratory Retention, Uptake and Excretion of Organic Solvents in Man. Benzene, Toluene, *n*-Hexane, Trichloroethylene, Acetone, Ethyl Acetate and Ethyl Alcohol," *Int. Arch. Arbeitsmed.*, **32**, 75–83 (1974).

421. K. Nomiyama and H. Nomiyama, "Respiratory Elimination of Organic Solvents in Man. Benzene, Toluene, *n*-Hexane, Trichloroethylene, Acetone, Ethyl Acetate and Ethyl Alcohol," *Int. Arch. Arbeitsmed.*, **32**, 85–91 (1974).

422. L. Cander and R. E. Forster, "Determination of Pulmonary Parenchymal Volume and Pulmonary Capillary Blood Flow in Man," *J. Appl. Physiol.*, **59**, 541–551 (1959).

423. A. C. M. Schrikker, W. R. de Vries, A. Zwart, and S. C. M. Luijendijk, "Uptake of Highly Soluble Gases in the Epithelium of the Conducting Airways," *Pflügers Archiv Eur. J. Physiol.*, **405**, 389–394 (1985).

424. A. C. M. Schrikker, W. R. de Vries, A. Zwart, and S. C. M. Luijendijk, "The Excretion of Highly Soluble Gases by the Lung in Man," *Pflügers Archiv Eur. J. Physiol.*, **415**, 214–219 (1989).

425. M. Jakubowski and H. Wieczorek, "The Effects of Physical Effort on Pulmonary Uptake of Selected Organic Compound Vapours," *Pol. J. Occup. Med.*, **1**, 62–71 (1988).

426. G. Pezzagno, M. Imbriani, S. Ghittori, E. Capodaglio, and J. Huang, "Urinary Elimination of Acetone in Experimental and Occupational Exposure," *Scand. J. Work Environ. Health*, **12**, 603–608 (1986).

427. G. Pezzagno, M. Imbriani, S. Ghittori, and E. Capodaglio, "Urinary Concentration, Environmental Concentration, and Respiratory Uptake of Some Solvents: Effect of the Work Load," *Am. Ind. Hyg. Assoc. J.*, **49**, 546–552 (1988).

428. O. Tada, K. Nakaaki, and S. Fukabori, "An Experimental Study on Acetone and

Methyl Ethyl Ketone Concentrations in Urine and Expired Air after Exposure to Those Vapors," *J. Sci. Labour*, **48**, 305–336 (1972).

429. V. Fiserova-Bergerova, "Toxicokinetics of Organic Solvents," *Scand. J. Work Environ. Health*, **11**, 7–21 (1985).

430. W. D. Brown, J. V. Setzer, R. B. Dick, F. C. Phipps, and L. K. Lowry, "Body Burden Profiles of Single and Mixed Solvent Exposures," *J. Occup. Med.*, **29**, 877–883 (1987).

431. J. Brodeur, S. Laparé, K. Krishnan, R. Tardif, and R. Goyal, "Le problème de l'ajustement des Valeurs Limites d'Exposition (VLE) Pour des Horaires de Travail Non-conventionnels: Utilité de la Modélisation Pharmacocinétique à Base Physiologique," *Trav. Santé*, **6**, S11–S16 (1991).

432. G. Johanson and P. H. Näslund, "Spreadsheet Programming—A New Approach in Physiologically Based Modeling of Solvent Toxicokinetics," *Toxicol. Lett.*, **41**, 115–127 (1988).

433. G. Johanson, "Modelling of Respiratory Exchange of Polar Solvents," *Ann. Occup. Hyg.*, **35**, 323–339 (1991).

434. G. A. Reichard, A. C. Haff, C. L. Skutches, C. P. Holroyde, and O. E. Owen, "Plasma Acetone Metabolism in the Fasting Human," *J. Clin. Invest.*, **63**, 619–626 (1979).

435. O. E. Owen, V. E. Trapp, C. L. Skutches, M. A. Mozzoli, R. D. Hoeldtke, G. Boden, and G. A. Reichard, Jr., "Acetone Metabolism during Diabetic Ketoacidosis," *Diabetes*, **31**, 242–248 (1982).

436. G. A. Reichard, C. L. Skutches, C. P. Holroyde, and O. E. Owen, "Acetone Metabolism in Humans during Diabetic Ketoacidosis," *Diabetes*, **35**, 668–674 (1986).

437. L. Parmeggiani and C. Sassi, "Occupational Poisoning with Acetone—Clinical Disturbances, Investigations in Workrooms and Physiopathological Research," *Med. Lav.*, **45**, 431–468 (1954).

438. S. Fukabori, K. NaKaaki, and O. Tada, "On the Cutaneous Absorption of Acetone," *J. Sci. Labour*, **55**, 525–532 (1979).

439. K. Baumann and J. Angerer, "Untersuchungen zur Frage der beruflichen Lösungsmittelbelastung mit Aceton," *Krebsgefaehrdung Arbeitsplatz Arbeitsmed.*, **19**, 403–408 (1979).

440. F. Brugnone, L. Perbellini, L. Grigolini, and P. Apostoli, "Solvent Exposure in a Shoe Upper Factory. I. *n*-Hexane and Acetone Concentration in Alveolar and Environmental Air and in Blood," *Int. Arch. Occup. Environ. Health*, **42**, 51–62 (1978).

441. F. Brugnone, L. Perbellini, E. Gaffuri, and P. Apostoli, "Biomonitoring of Industrial Solvent Exposures in Workers' Alveolar Air," *Int. Arch. Occup. Environ. Health*, **47**, 245–261 (1980).

442. T. Kawai, T. Yasugi, Y. Uchida, O. Iwami, and M. Ireda, "Urinary Excretion of Unmetabolized Acetone as an Indicator of Occupational Exposure to Acetone," *Int. Arch. Occup. Environ. Health*, **62**, 165–169 (1990).

443. E. Hallier, J. G. Filser, and H. M. Bolt, "Inhalation Pharmacokinetics Based on Gas Uptake Studies. II. Pharmacokinetics of Acetone in Rats," *Arch. Toxicol.*, **47**, 293–304 (1981).

444. E. Wigaeus, A. Löf, and M. B. Nordqvist, "Distribution and Elimination of 2-[^{14}C]-acetone in Mice after Inhalation Exposure," *Scand. J. Work Environ. Health*, **8**, 121–128 (1982).

445. G. Hetenyi and C. Ferrarotto, "Gluconeogenesis from Acetone in Starved Rat," *Biochem. J.*, **231**, 151–155 (1985).

446. J. G. Filser and H. M. Bolt, "Inhalation Pharmacokinetics Based on Gas Uptake Studies. IV. The Endogenous Production of Volatile Compounds," *Arch. Toxicol.*, **52**, 123–133 (1983).

447. J. L. Egle, Jr., "Retention of Inhaled Acetone and Ammonia in the Dog," *Am. Ind. Hyg. Assoc. J.*, **34**, 533–539 (1973).

448. E. F. Aharonson, H. Menkes, G. Gurtner, D. L. Swift, and D. F. Proctor, "Effect of Respiratory Airflow Rate on Removal of Soluble Vapors by the Nose," *J. Appl. Physiol.*, **37**, 654–657 (1974).

449. J. B. Morris, R. J. Clay, and D. G. Cavanagh, "Species Differences in Upper Respiratory Tract Deposition of Acetone and Ethanol Vapors," *Fundam. Appl. Toxicol.*, **7**, 671–680 (1986).

450. J. B. Morris and D. G. Cavanagh, "Deposition of Ethanol and Acetone Vapors in the Upper Respiratory Tract of the Rat," *Fundam. Appl. Toxicol.*, **6**, 78–88 (1986).

451. J. B. Morris and D. G. Cavanagh, "Metabolism and Deposition of Propanol and Acetone Vapors in the Upper Respiratory Tract of the Hamster," *Fundam. Appl. Toxicol.*, **9**, 34–40 (1987).

452. J. B. Morris, "Deposition of Acetone in the Upper Respiratory Tract of the B6C3F1 Mouse," *Toxicol. Lett.*, **56**, 187–196 (1991).

453. J. V. Bruckner and R. G. Petersen, "Evaluation of Toluene and Acetone Inhalant Abuse. II. Model Development and Toxicology," *Toxicol. Appl. Pharmacol.*, **61**, 302–312 (1981).

454. M. E. Goldberg, H. E. Johnson, U. C. Pozzanni, and H. F. Smyth, Jr., "Effect of Repeated Inhalation of Vapors of Industrial Solvents on Animal Behavior. I. Evaluation of Nine Solvent Vapors on Pole-climb Performance in Rats," *Am. Ind. Hyg. Assoc. J.*, **25**, 369–375 (1964).

455. V. P. Kurnayeva, N. L. Burykina, M. L. Zeltser, A. D. Dasayeva, L. I. Kolbeneva, K. A. Veselovskaya, Yu. M. Demin, and Yu. A. Loshchilov, "Experimental Study of Combined Effect of Solvents and Noise," *J. Hyg. Epidemiol. Microbiol. Immunol.*, **30**, 49–56 (1986).

456. P. S. Spencer, M. C. Bischoff, and H. H. Schaumburg, "On the Specific Molecular Configuration of Neurotoxic Aliphatic Hexacarbon Compounds Causing Central-peripheral Distil Axonopathy, *Toxicol. Appl. Pharmacol.*, **44**, 17–28 (1978).

457. C. Hètu and J-G. Joly, "Effects of Chronic Acetone Administration of Ethanol-inducible Monooxygenase Activities in the Rat," *Biochem. Pharmacol.*, **37**, 421–426 (1988).

458. P. R. Sinclair, W. J. Bement, R. W. Lambrecht, J. M. Jacobs, and J. F. Sinclair, "Uroporphyrin Caused by Acetone and 5-Aminolevulinic Acid in Iron-loaded Mice," *Biochem. Pharmacol.*, **23**, 4341–4344 (1989).

459. C. L. Skutches, O. E. Owen, and G. A. Reichard, "Acetone and Acetol Inhibition of Insulin-stimulated Glucose Oxidation in Adipose Tissue and Isolated Adipocytes," *Diabetes*, **39**, 450–455 (1990).

460. B. Sonawane, C. de Rosa, R. Rubenstein, D. Mayhew, S. V. Becker, and D. Dietz, "Estimation of Reference Dose (RfD) for Oral Exposure to Acetone," *J. Am. Coll. Toxicol.*, **5**, 605 (abstract) (1986).

461. D. A. Mayhew and L. D. Morrow, "Ninety Day Gavage Study in Albino Rats Using

Acetone," U.S. Environmental Protection Agency, EPA Contract No. 68-01-7075, 1988.

462. D. D. Dietz, J. R. Leininger, E. J. Rauckman, M. B. Thompson, R. E. Chapin, R. L. Morrissey, and B. S. Levine, "Toxicity Studies of Acetone Administered in the Drinking Water of Rodents," *Fundam. Appl. Toxicol.*, **17**, 347–360 (1991).

463. D. Dietz, "Toxicity Studies of Acetone in F344/N Rats and B6C3F$_1$ Mice (Drinking Water Studies)," NTP TOX 3, USDHHS, National Toxicology Program, Research Triangle Park, NC, NIH 91-3122, 1991, pp. 1–38.

464. J. Misumi and M. Nagano, "Neurophysiological Studies on the Relation between the Structural Properties and Neurotoxicity of Aliphatic Hydrocarbon Compounds in Rats," *Brit. J. Ind. Med.*, **41**, 526–532 (1984).

465. R. H. Rengstorff, J. P. Petrali, and V. M. Sim, "Cataracts Induced in Guinea Pigs by Acetone, Cyclohexanone, and Dimethyl Sulfoxide," *Am. J. Optom.*, **49**, 308–319 (1972).

466. R. H. Rengstorff, J. P. Petrali, and V. M. Sim, "Attempt to Induce Cataracts in Rabbits by Cutaneous Application of Acetone," *Am. J. Optom. Physiol. Optics*, **53**, 41–42 (1975).

467. R. H. Rengstorff and H. I. Khafagy, "Cutaneous Acetone Depresses Aqueous Humor Ascorbate in Guinea Pigs," *Arch. Toxicol.*, **58**, 64–66 (1985).

468. S. Ameenuddin and M. L. Sunde, "Sensitivity of Chick Embryo to Various Solvents Used in Egg Injection Studies," *Proc. Soc. Exp. Biol. Med.*, **175**, 176–178 (1984).

469. K. T. Kitchin and M. T. Ebron, "Combined Use of a Water-insoluble Chemical Delivery System and a Metabolic Activation System in Whole Embryo Culture," *J. Toxicol. Environ. Health*, **13**, 499–509 (1984).

470. J. R. Rayburn, D. J. DeYoung, J. A. Bantle, D. J. Fort, and R. McNew," Altered Developmental Toxicity Caused by Three Carrier Solvents," *J. Appl. Toxicol.*, **11**, 253–260 (1991).

471. D. Marchal-Segault and F. Ramade, "The Effects of Lindane, an Insecticide, on Hatching and Postembryonic Development of *Xenopus laevis* (Daudin) Anurian Amphibian," *Environ. Res.*, **24**, 250–258 (1981).

472. D. J. Hoffman and W. C. Eastin, "Effects of Industrial Effluents, Heavy Metals, and Organic Solvents on Mallard Embryo Development," *Toxicol. Lett.*, **9**, 35–40 (1981).

473. E. M. Johnson, L. M. Newman, B. E. G. Gabel, T. F. Boerner, and L. A. Dansky, "An Analysis of the Hydra Assay's Applicability and Reliability as a Developmental Toxicity Prescreen," *J. Am. Coll. Toxicol.*, **7**, 111–126 (1988).

474. M. Pereira, P. Barnwell, and W. Bailes, "Screening of Priority Chemicals for Reproductive Hazards: Benzethonium Chloride, 3-Ethoxy-1-propanol, and Acetone," Environmental Health Research and Testing Inc., Project No. ETOX-85-1002, 1987.

475. T. J. Mast, J. J. Evanoff, R. L. Rommereim, K. H. Stoney, R. J. Weigel, and R. B. Westerberg, "Inhalation Developmental Toxicology Studies: Teratology Study of Acetone in Mice and Rats," prepared for the National Institute of Environmental Health Sciences, National Toxicology Program, Contract DE-AC06-76RLO 1830, Pacific Northwest Laboratory, Battelle Memorial Institute, 1988.

476. J. J. Larsen, M. Lykkegaard, and O. Ladefoged, "Infertility in Rats Induced by 2,5-Hexanedione in Combination with Acetone," *Pharmacol. Toxicol.*, **69**, 43–46 (1991).

477. K. T. Kitchin and M. T. Ebron, "Further Development of Rodent Whole Embryo

Culture: Solvent Toxicity and Water Insoluble Compound Delivery System," *Toxicology*, **30**, 45–57 (1984).

478. J. J. Picard and G. Van Maele-Fabry, "A New Method to Express Embryotoxic Data Obtaine *in Vitro* on Whole Murine Embryos," *Teratology*, **35**, 429–437 (1987).

479. B. P. Schmid, "Xenobiotic Influences on Embryonic Differentiation, Growth and Morphology *in Vitro*," *Xenobiotica*, **15**, 719–726 (1985).

480. E. J. Guntakatta, E. J. Matthews, and J. O. Rundell, "Development of a Mouse Embryo Limb Bud Cell Culture System for the Estimation of Chemical Teratogenic Potential," *Teratogen. Carcinogen. Mutagen.*, **4**, 349–364 (1984).

481. J. A. DiPaolo, P. Donovan, and R. Nelson, "Quantitative Studies of *in Vitro* Transformation by Chemical Carcinogens," *J. Natl. Cancer Inst.*, **42**, 867–876 (1969).

482. J. M. Quarles, M. W. Sega, C. K. Schenley, and W. Lijinsky, "Transformation of Hamster Fetal Cells by Nitrosated Pesticides in a Transplantation Assay," *Cancer Res.*, **39**, 4525–4533 (1979).

483. J. R. Strange, P. M. Allred, and W. E. Kerr, "Teratogenic and Toxicological Effects of 2,4,5-Trichlorophenoxyacetic Acid in Developing Chick Embryos," *Bull. Environ. Contamin. Toxicol.*, **15**, 682–688 (1976).

484. P. M. Allred and J. R. Strange, "The Effects of 2,4,5-Trichlorophenoxyacetic Acid and 2,3,7,8-Tetrachlorodibenzo-*p*-dioxin on Developing Chicken Embryos," *Arch. Environ. Contam. Toxicol.*, **6**, 483–489 (1977).

485. J. McLaughlin, J-P. Marliac, M. J. Verrett, M. L. Mutchler, and O. G. Fitzhugh, "The Injection of Chemicals into the Yolk Sac of Fertile Eggs Prior to Incubation as a Toxicity Test," *Toxicol. Appl. Pharmacol.*, **5**, 760–771 (1963).

486. J. McLaughlin, J-P. Marliac, M. J. Verrett, M. K. Mutchler, and O. G. Fitzhugh, "Toxicity of Fourteen Volatile Chemicals as Measured by the Chick Embryo Method," *Am. Ind. Hyg. J.*, **25**, 282–284 (1964).

487. N. E. Walker, "Distribution of Chemicals Injected into Fertile Eggs and Its Effect upon Apparent Toxicity," *Toxicol. Appl. Pharmacol.*, **10**, 290–299 (1967).

488. A. Korhonen, K. Hemminki, and H. Vainio, "Embryotoxicity of Industrial Chemicals on the Chicken Embryo: Thiourea Derivatives," *Acta Pharmacol. Toxicol.*, **51**, 38–44 (1982).

489. A. Korhonen, K. Hemminki, and H. Vainio, "Embryotoxic Effects of Acrolein, Methacrylates, Guanidines and Resorcinol on Three Day Chicken Embryos," *Acta Pharmacol. Toxicol*, **52**, 95–99 (1983).

490. W. J. Swartz, "Long- and Short-term Effects of Carbaryl Exposure in Chick Embryos," *Environ. Res.*, **26**, 463–471 (1981).

491. M. G. Ott, L. K. Skory, B. B. Holder, J. M. Bronson, and P. R. Williams, "Health Evaluation of Employees Occupationally Exposed to Methylene Chloride. General Study Design and Environmental Considerations," *Scand. J. Work Environ. Health*, **9**(Suppl. 1), 1–7 (1983).

492. M. G. Ott, L. K. Skory, B. B. Holder, J. M. Bronson, and P. R. Williams, "Health Evaluation of Employees Occupationally Exposed to Methylene Chloride. Mortality," *Scand. J. Work Environ. Health*, **9**(Suppl. 1), 8–16 (1983).

493. M. G. Ott, L. K. Skory, B. B. Holder, J. M. Bronson, and P. R. Williams, "Health Evaluation of Employees Occupationally Exposed to Methylene Chloride. Clinical Laboratory Evaluation," *Scand. J. Work Environ. Health*, **9**(Suppl 1), 17–25 (1983).

494. F. L. Oglesby, J. L. Williams and D. W. Fassett, "Eighteen-year Experience with Acetone," presented at the Annual Meeting of the American Industrial Hygiene Association, Detroit, MI, 1949.

495. Rösgen and Mamier, "Sind Azetongase blutschädigend?," *Öffentl. Gesundheitsdienst*, **10**, 83–86 (1944).

496. E. C. Vigliani and N. Zurlo, "Experiences of the Clinica del Lavoro with Several Maximum Workplace Concentrations (MAC) for Industrial Toxic Substances," *Arch. Gewerbepath. Gewerbehyg.*, **13**, 528–534 (1955).

497. C. Grampella, G. Catenacci, L. Garavaglia, and S. Tringali, "Health Surveillance in Workers Exposed to Acetone," in *Proceedings of the VII International Symposium on Occupational Health in the Production of Artificial Organic Fibres*, Wolfheze, Holland, 1987, pp. 137–141.

498. T. Sollmann, "Studies of Chronic Intoxications on Albino Rats. II. Alcohols (Ethyl, Methyl and Wood) and Acetone, *J. Pharmacol. Exp. Ther.*, **16**, 291–309 (1921).

499. B. L. Van Duuren, G. Loewngart, I. Seidman, A. C. Smith, and S. Melchione, "Mouse Skin Carcinogenicity Tests of the Flame Retardants Tris(2,3-dibromopropyl)phosphate, Tetrakis(hydroxymethyl)phosphonium Chloride, and Polyvinyl Bromide," *Cancer Res.*, **38**, 3236–3240 (1978).

500. B. L. Van Duuren, A. Sivak, C. Katz, and S. Melchionne, "Cigarette Smoke Carcinogenesis: Importance of Tumor Promoters," *J. Natl. Cancer Inst.*, **47**, 235–240 (1971).

501. H. S. Weiss, J. F. O'Connell, A. G. Hakim, and W. T. Jacoby, "Inhibitory Effect of Toluene on Tumor Promotion in Mouse Skin," *Proc. Soc. Exp. Biol. Med.*, **181**, 199–204 (1986).

502. H. S. Weiss, K. S. Pacer, and C. A. Rhodes, "Tumor Inhibiting Effect of Pre-promotion with Acetone," *Fed. Am. Soc. Exp. Biol. J.*, **2** A1153 (abstract) (1988).

503. J. M. Ward, R. Quander, D. Devor, M. L. Wenk, and E. F. Spangler, "Pathology of Aging Female SENCAR Mice Used as Controls in Skin Two-stage Carcinogenesis Studies," *Environ. Health Perspect.*, **68**, 81–89 (1986).

504. O. H. Iversen, "The Skin Tumorigenic and Carcinogenic Effects of Different Doses, Numbers of Dose Fractions and Concentrations of 7,12-Dimethylbenz[*a*]anthracene in Acetone Applied on Hairless Mouse Epidermis. Possible Implications for Human Carcinogenesis," *Carcinogenesis*, **12**, 493–502 (1991).

505. G. C. Peristianis, S. M. A. Doak, P. N. Cole, and R. W. Hend, "Two-year Carcinogenicity Study on Three Aromatic Epoxy Resins Applied Cutaneously to CF1 Mice," *Food Chem. Toxicol.*, **26**, 611–624 (1988).

506. N. Zakova, F. Zak, E. Froelich, and R. Hess, "Evaluation of Skin Carcinogenicity of Technical 2,2-Bis(*p*-glycidyloxyphenyl)propane in CF1 Mice," *Food Chem. Toxicol.*, **23**, 1081–1089 (1985).

507. L. R. DePass, B. Ballantyne, E. H. Fowler, and C. S. Weil, "Dermal Oncogenicity Studies on Two Methoxysilanes and Two Ethoxysilanes in Male C3H Mice," *Fund. Appl. Pharmacol.*, **12**, 579–583 (1989).

508. E. R. Nestmann, G. R. Douglas, D. J. Kowbel, and T. R. Harrington, "Solvent Interactions with Test Compounds and Recommendations for Testing to Avoid Artifacts," *Environ. Mutagen.*, **7**, 163–170 (1985).

509. D. Anderson and D. B. MacGregor, "The Effect of Solvents Upon the Yield of Revertants in the *Salmonella*/activation Mutagenicity Assay," *Carcinogenesis*, **1**, 363–366 (1980).

510. S. Arimoto, Y. Nakano, Y. Ohara, K. Tanaka, and H. Hayatsu, "A Solvent Effect on the Mutagenicity of Tryptophan-pyrolysate Mutagens in the *Salmonella*/mammalian Microsome Assay," *Mutat. Res.*, **102**, 105–112 (1982).

511. D. Maron, J. Katzenellenbogen, and B. N. Ames, "Compatibility of Organic Solvents with the *Salmonella*/microsome test," *Mutat. Res.*, **88**, 343–350 (1981).

512. "NTP, Cellular and Genetic Toxicology," in *National Toxicology Program Fiscal Year 1987 Annual Plan*, Department of Health and Human Services, Research Triangle Park, NC, NTP-87-001, 1987, pp. 43–96.

513. M. Ishidate, T. Sofuni, K. Yoshikawa, M. Hayashi, T. Nohmi, M. Sawada, and A. Matsuoka, "Primary Mutagenicity Screening of Food Additives Currently Used in Japan," *Food Chem. Toxicol.*, **22**, 623–636 (1984).

514. S. De Flora, P. Zanacchi, A. Camoirano, C. Bennicelli, and G. S. Badolati, "Genotoxic Activity and Potency of 135 Compounds in the Ames Reversion Test and in a Bacterial DNA-repair Test," *Mutat. Res.*, **133**, 161–198 (1984).

515. H. Kubinski, G. E. Gutzke, and Z. O. Kubinski, "DNA-cell-binding (DCB) Assay for Suspected Carcinogens and Mutagens," *Mutat. Res.*, **89**, 95–136 (1981).

516. D. M. DeMarini, B. K. Lawrence, H. G. Brooks, and V. S. Houk, "Compatibility of Organic Solvents with the Microscreen Prophage-induction Assay: Solvent–Mutagen Interactions," *Mutat. Res.*, **263**, 107–113 (1991).

517. T. G. Rossman, M. Molina, L. Meyer, P. Boone, C. B. Klein, Z. Wang, F. Li, W. C. Lin, and P. L. Kinney, "Performance of 133 Compounds in the Lambda Prophage Induction Endpoint of the Microscreen Assay and a Comparison with *S. typhimurium* Mutagenicity and Rodent Carcinogenicity Assays," *Mutat. Res.*, **260**, 349–367 (1991).

518. W. Von der Hude, C. Behm, R. Gürtler, and A. Basler, "Evaluation of the SOS Chromotest," *Mutat. Res.*, **203**, 81–94 (1988).

519. A. Abbondandolo, S. Bonatti, C. Corsi, G. Corti, R. Fiorio, C. Leporini, A. Mazzaccaro, R. Nieri, R. Barale, and N. Loprieno, "The Use of Organic Solvents in Mutagenicity Testing," *Mutat. Res.*, **79**, 141–150 (1980).

520. F. K. Zimmermann, V. W. Mayer, I. Scheel, and M. A. Resnick, "Acetone, Methyl Ethyl Ketone, Ethyl Acetate, Acetonitrile and Other Polar Aprotic Solvents are Strong Inducers of Aneuploidy in *Saccharomyces cerevisiae*," *Mutat. Res.*, **149**, 339–351 (1985).

521. F. K. Zimmermann, "Mutagenicity Screening with Fungal Systems," *Ann. New York Acad. Sci.*, **407**, 186–196 (1983).

522. W. G. Whittaker, F. K. Zimmermann, B. Dicus, W. W. Piegorsch, S. Fogel, and M. A. Resnick, "Detection of Induced Mitotic Chromosome Loss in *Saccharomyces cerevisiae*—An Interlaboratory Study," *Mutat. Res.*, **224**, 31–78 (1989).

523. A. D. Tates and E. Kriek, "Induction of Chromosomal Aberrations and Sister Chromatid Exchanges in Chinese Hamster Cells *in Vitro* by Some Proximate and Ultimate Carcinogenic Arylamide Derivatives," *Mutat. Res.*, **88**, 397–410 (1981).

524. S. A. Latt, J. Allen, S. E. Bloom, A. Carrano, E. Falke, D. Kram, E. Schneider, R. Schrack, R. Tice, B. Whitfield, and S. Wolfe, "Sister-chromatid Exchanges: A Report of the Gene-tox Program," *Mutat. Res.*, **87**, 17–62 (1981).

525. "NTP, Toxicology Research and Applied Studies Overview," in *National Toxicology Program Fiscal Year 1989 Annual Plan*, Chpt. VIII, Department of Health and Human Services, Research Triangle Park, NC, NTP-89-168, 1989, pp. 41–134.

526. G. G. Hatch, P. D. Mamay, M. L. Ayer, B. C. Casto, and S. Nesnow, "Chemical Enhancement of Viral Transformation in Syrian Hamster Embryo Cells by Gaseous and Volatile Chlorinated Methanes and Ethanes," *Cancer Res.*, **43**, 1945–1950 (1983).

527. A. Basler, "Aneuploidy-inducing Chemicals in Yeast Evaluated by the Micronucleus Test," *Mutat. Res.*, **174**, 11–13 (1986).

528. T-H. Chen, T. J. Kavanagh, C. C. Chang, and J. E. Trosko, "Inhibition of Metabolic Cooperation in Chinese Hamster Cells by Various Organic Solvents and Simple Compounds," *Cell Biol. Toxicol.*, **1**, 155–171 (1984).

529. H. Norppa, "The *in Vitro* Induction of Sister Chromatid Exchanges and Chromosome Aberrations in Human Lymphocytes by Styrene Derivatives," *Carcinogenesis*, **2**, 237–242 (1981).

530. A. E. Freeman, E. K. Weisburger, J. H. Weisburger, J. H. Wolford, J. M. Maryak, and R. J. Huebner, "Transformation of Cell Cultures as an Indication of the Carcinogenic Potential of Chemicals," *J. Natl. Cancer Inst.*, **51**, 799–808 (1973).

531. D. E. Amacher, S. C. Pailler, G. N. Turner, V. A. Ray, and D. S. Salsburg, "Point Mutations at the Thymidine Kinase Locus in L5178Y Mouse Lymphoma Cells. II. Test Validation and Interpretation," *Mutat. Res.*, **72**, 447–474 (1980).

532. D. B. McGregor, I. Edwards, C. G. Riach, P. Cattanach, R. Martin, A. Mitchell, and W. J. Capary, "Studies of an S9-based Metabolic Activation System Used in the Mouse Lymphoma L5178Y Cell Mutation Assay," *Mutagenesis*, **3**, 485–490 (1988).

533. J. F. Sina, C. L. Bean, G. R. Dysart, V. I. Taylor, and M. O. Bradley, "Evaluation of the Alkaline Elution/Rat Hepatocyte Assay as a Predictor of Carcinogenic/ Mutagenic Potential," *Mutat. Res.*, **113**, 357–391 (1983).

534. R. S. Lake, M. L. Kropko, M. R. Pezzulti, R. H. Shoemaker, and H. J. Igel, "Chemical Induction of Unscheduled DNA Synthesis in Human Skin Epithelial Cell Cultures," *Cancer Res.*, **38**, 2091–2098 (1978).

535. A. Kabarity, "Wirkung von Aceton auf die Chromosomen-Verteilung bei menschlichen Leukocyten," *Virchows Arch. Abt. B Zellpath.*, **2**, 163–170 (1969).

CHAPTER SIX

Aldehydes and Acetals

Michael J. Brabec, Ph.D., D.A.B.T.

1 GENERAL CONSIDERATIONS

The aldehydes are a major class of industrial chemicals. Formaldehyde is the preeminent aldehyde, ranking twenty-fifth in U.S. production (6.43 billion lb) of all industrial chemicals in 1991, and tenth among the organics (1). Aldehydes are intermediates in the synthesis of chemicals important in the construction, agricultural, garment, pharmaceutical, and rubber industries. Formaldehyde is used directly to preserve biologic tissue in embalming, tanning, taxidermy, and histology. Aldehydes of higher molecular weight contribute characteristic flavors and odors to perfumes and essential oils.

The major ultimate consumer of aldehydes is the construction industry, which uses formaldehyde resin-bonded wood products and adhesives. The fluctuating activity of this industry makes the output of formaldehyde difficult to predict but the modest average annual expansion of 1 percent for the past decade may be expected to continue (1). Similar growth rates may be expected for aldehydes in general. Recent production figures and uses for selected aldehydes are listed in Table 6.1.

1.1 Physical and Chemical Properties of Aldehydes

The physical and chemical properties of various classes of aldehydes are listed in Tables 6.2 to 6.5. Most of the aldehydes below 110 Da molecular weight are liquids at room temperature, although formaldehyde and ketene are gases. The aromatic

Patty's Industrial Hygiene and Toxicology, Fourth Edition, Volume 2, Part A, Edited by George D. Clayton and Florence E. Clayton.
ISBN 0-471-54724-7   © 1993 John Wiley & Sons, Inc.

Table 6.1. United States Production and Use of Aldehydes

Compound	Total Production (million lb)	Major Use	Ref
Formaldehyde (37% by wt.)	6430	Urea–formaldehyde resins	1
Acrolein	900	Acrylics, methionine	8
Acetaldehyde	680	Resins, esters, acetic acid	179
Proprionaldehyde	280	Intermediate	179
Butyraldehyde	3000	Intermediate	179
Salicyaldehyde	6	Medicinal	179
Anisaldehyde	4	Flavoring	179
Furfural	350	Specialty solvent, intermediate	152

and heterocyclic aldehydes, particularly those with alkyl side groups, are solids at conventional temperatures. Solubility of aldehydes in water varies from insoluble to very soluble. The flash points of short-chain aliphatic aldehydes are low, warning of their flammability.

1.2   Chemical Reactions of Aldehydes

Any advanced organic chemistry text presents a comprehensive discussion of the reactions and reaction mechanisms of aldehydes (2). Some reactions commonly encountered are presented below.

1.2.1   Reduction to Alcohols and Oxidation to Acids

The importance of aldehydes as intermediates in the chemical industry is due, in part, to the ease with which aldehydes undergo oxidation and reduction. The biologic metabolism of aldehydes generally proceeds by oxidation to acids, catalyzed by NAD-requiring and NADP-requiring aldehyde dehydrogenases (3, 4). Self-oxidation–reduction can occur with low-molecular-weight aldehydes via the Cannizzaro reaction:

$$2H_2CO \xrightarrow{\text{OH}^-} HCO_2^- + H_3COH \tag{1}$$

1.2.2   Polymerization Reactions

The tendency of aldehydes to condense spontaneously to polymers, which may crystallize and precipitate, complicates aldehyde chemistry. For example, formaldehyde forms a series of paraformaldehyde polymers, and acetaldehyde condenses to paraldehyde. These crystalline solids, when heated, release the gaseous form of the monomer. Low-molecular-weight alcohols are often included in aldehyde solutions to suppress the polymerization reactions. The reactions of aldehydes with phenolic and amine groups are the basis for the formation of the formaldehyde

Table 6.2. Physical and Chemical Properties of Saturated Aliphatic Aldehydes

Compound	Formula	Mol. Wt.	Specific Gravity	M.P. (°C)	B.P. (°C)	Vapor Pressure (mm Hg) (°C)	Vapor Density (Air = 1)	Flash Point (°F)	Solubility in H$_2$O (g/100 ml)	Conversion Factors 1 mg/l (ppm)	1 ppm (mg/m³)
Formaldehyde (gas)	HCHO	30.0	0.815 (20°C)	−92	−19.5	10 (−88.0)	1.075	572 (ignit. temp.) 7.0–7.3% (exp. limits by vol.)	Very sol.	815	1.2
Formaldehyde (37% in H$_2$O with 0–15% methanol)			1.075–1.081		~98			122 (with 15% methanol)	Very sol.		
Sodium formaldehyde bisulfite	CH$_2$O HSO$_3$Na H$_2$O	152.1		Dec.					Very sol.		
Sodium formaldehyde sulfoxalate	CH$_2$O HSO$_2$Na 2H$_2$O	154.1		65	Dec.				Sol.		
Paraformaldehyde	(CH$_2$O)$_x$	(30)$_x$		25 (solid)	Dec.			158	Slightly sol. dec.		
Acetaldehyde	CH$_3$CHO	44.0	0.788 (16/4)	−123.5	21	740 (20)	1.52	−36 4.1–55% (exp. limits by vol.)	Sol.	556	1.8
Acetaldehyde sodium bisulfite	CH$_3$CHO NaHSO$_3$	148.1									
Fluoroacetaldehyde	FCH$_2$CHO	62.0			64–65		2.14			394	2.5
Hydroxyacetaldehyde (glycolaldehyde)		60.0	1.366 (100)	95–97			2.1		Very sol.	408	2.6

Table 6.2. (Continued)

Compound	Formula	Mol. Wt.	Specific Gravity	M.P. (°C)	B.P. (°C)	Vapor Pressure (mm Hg) (°C)	Vapor Density (Air = 1)	Flash Point (°F)	Solubility in H_2O (g/100 ml)	Conversion Factors 1 mg/l (ppm)	Conversion Factors 1 ppm (mg/m³)
Chloroacetaldehyde (40% in water)		78.5	1.190	−16.3	90–100		2.7	190	Very sol.	309	3.2
Trichloroacetaldehyde (chloral)	CCl₃CHO	147.4	1.512	−57.5	98	35 (20)	5.1	167	Sol. (forms hydrate)	166	6.0
Paraldehyde	(C₂H₄O)₃	132.2	0.9943 (20/4)	12.6	124 (752 mm Hg)			111.2	10.5		
Propionaldehyde (propanal)	CH₃CH₂CHO	58.1	0.807	−81	48.8	687 (45) ~300 (25)	2.0	15–19	20	422	2.3
n-Butyraldehyde	CH₃(CH₂)₂CHO	72.1	0.817	−99	76	92 (20)	2.48	15	4.0	340	2.9
Isobutyraldehyde	(CH₃)₂CHCHO	72.1	0.7938	−66	62	~170 (20)	2.48	−11	11	340	2.9
3-Hydroxybutanal (aldol)	CH₃CHOHCH₂CHO	88.1	1.098	<0	85 (dec.)	21 (20)	3.0	181 (c.c.)	Very sol.	278	3.6
4-Fluorobutanal	F(CH₂)₃CHO	90.0				50 (49)	2.4			322	3.1
n-Valeraldehyde (pentanal)	CH₃(CH₂)₃CHO	86.1	0.819	−91.5	103	~50 (25)	3.0		Sl. sol.	284	3.5
Isovaleraldehyde	(CH₃)₂CHCH₂CHO	86.1	0.7845	−51	92.5	~50 (25)	3.0			284	3.5
2-Methylbutyraldehyde	CH₃CH₂CH(CH₃)CHO	86.1				~50 (25)	3.0			284	3.5
n-Hexaldehyde (caproaldehyde, hexanal)	CH₃(CH₂)₄CHO	100.2	0.8335	−56	128	10 (20)	3.5	90	Insol.	245	4.1
2-Ethylbutyraldehyde (α-ethylbutyraldehyde)	(C₂H₅)₂CHCHO	100.2	0.818	−89	116 (80–135)	14 (20)	3.5	70	0.3	245	4.1
n-Heptylaldehyde (heptanal, enanthaldehyde)	CH₃(CH₂)₅CHO	114.2	0.850	−45	154	3 (25)	3.9		Sl. sol.	215	4.6
2-Ethylhexylaldehyde (2-Ethylcaproaldehyde, 2-ethylhexanal)	CH₃(CH₂)₄(C₂H₅)CHO	128.2		<−100	163 (155–185)	2 (22)	4.4	125	0.04	191	5.2

Table 6.3. Physical and Chemical Properties of Unsaturated Aliphatic Aldehydes

Compound	Formula	Mol. Wt.	Specific Gravity	M.P. (°C)	B.P. (°C)	Vapor Pressure (mm Hg) (°C)	Vapor Density (Air = 1)	Flash Point (°F)	Solubility in H₂O (g/100 ml)	Conversion Factors 1 mg/l (ppm)	Conversion Factors 1 ppm (mg/m³)
Ketene	CH_2=C=O	42.04		−150	−56		1.45		Dec.	582	1.7
Acrolein (2-propenal, acrylic aldehyde)	CH_2=CHCHO	56.1	0.8389	−87	52. 7	214 (20)	1.94	<0	22	437	2.3
Methacrylaldehyde (methacrolein)	CH_2=C(CH_3)CHO	70.1	0.837		68		2.4	5	6.4	349	2.8
2-Ethyl-3-propylacrolein	C_3H_7CH=C(C_2H_5)CHO	126.19	0.8518		175	1 (20)	4.4	155	0.07	194	5.2
Crotonaldehyde (β-Methylacrolein)	CH_3CH=CHCHO	70.1	0.852	−69	103 (99–104)	19 (20)	2.4	128	18	349	2.8
Methyl-β-ethylacrolein	C_2H_5CH=C(CH_3)CHO	98.14	0.854		137		3.1		Insol.	250	4.0
Mucochloric acid	CHOCCl=CClCOOH	168.97		125–127					Sl. sol.		
Citral (gerenial)	$(CH_3)_2$C=CH$(CH_2)_2$C=C—CHO (with CH_3H branch)	152.23	0.8898		229 (dec.)	5 (91–95)			Insol. (hot)	161	6.2
Citronellal (rhodinal)	$(CH_3)_2$C=CH$(CH_2)_2$CH(CH_3)CH_2CHO	154.25	0.856		204	11 (88)			Sl. sol.	159	6.3

287

Table 6.4. Physical and Chemical Properties of Aliphatic Dialdehydes

Formula	Mol. Wt.	Specific Gravity	M.P. (°C)	B.P. (°C)	Vapor Pressure (mm Hg)(°C)	Vapor Density (Air = 1)	Solubility in H_2O (g/100 ml)	Conversion Factors 1 mg/l (ppm)	1 ppm (mg/m³)
Glyoxal									
O=CH—CH=O	58.0	1.14	15	51	220 (20, approx.)	2.0	Very sol.	422	2.4
Succinaldehyde (Butanedial)									
O=CH(CH₂)₂CH=O	86.1	1.064		169	1 (25, approx.)	3.0	Sol.	284	3.5
Glutaraldehyde									
O=CH(CH₂)₃CH=O	100.1			187		3.4		245	4.1
β-Methylglutaraldehyde									
CH₃CH(CH₂CH=O)₂	114.1					4.0		215	4.7
Adipaldehyde (Hexanedial)									
CHO(CH₂)₄CHO	114.1			9 (92–94)		4.0	Sl. sol.	215	4.7

288

Table 6.5. Physical and Chemical Properties of Acetals

Compound	Formula	Mol. Wt.	Specific Gravity	M.P. (°C)	B.P. (°C)	Vapor Pressure (mm Hg) (°C)	Vapor Density (Air = 1)	Flash Point (°F)	Solubility in H₂O (g/100 ml)	Conversion Factors 1 mg/l (ppm)	1 ppm (mg/m³)
Methylal (formal, dimethoxy-methane)	$CH_2(OCH_3)_2$	76.1	0.8630	−104.8	43	~400 (25)	2.6	0	33	322	3.1
Diethoxymethane (diethyl formal, ethylal)	$CH_2(OC_2H_5)_2$	104.2	0.824	−67	89	~60 (25)	3.6		7.0	235	4.2
Dichloroethyl formal (bis-2-chloroethyl formal)	$CH_2(OCH_2CH_2Cl)_2$	173.0	1.234		218.1	0.1 (20)	6.0	230	0.78	141	7.1
1,1-Dimethoxyethane (dimethyl acetal)	$CH_3CH(OCH_3)_2$	90.1	0.850	−58	82–83	61 (20)	3.1	40	∞	272	3.7
Acetal (1,1-diethoxyethane, diethyl acetal)	$CH_3CH(OC_2H_5)_2$	118.2	0.825		107–112	20 (19.6)	4.1	−5 (c.c.) 1.6–10.4% exp. limits	5.5	207	4.8
Chloroacetal (chloroacetaldehyde diethyl acetal)	$CH_2ClCH(OC_2H_5)_2$	152.6	1.026		157	20 (62)	5.3		Sl. sol.	160	6.2
Acetaldehyde dibutyl acetal	$CH_3CH(OC_4H_9)_2$	174					6.0			141	7.1
Crotonaldehyde acetal		144.2					5.0			170	5.9
Chloral hydrate	$CCl_3CH(OH)_2$	165.4	1.9081	61–63	98 (dec.)				Sol.		
Ketoacetal (4,4-dimethoxy-2-butanone)	$CH_3COCH_2C(OCH_3)_2$	132.1					4.5		Sol.	185	5.4

resins. Bakelite, a copolymer of phenol and formaldehyde, is probably the most familiar of such resins.

1.2.3 Addition Reactions

The addition reactions are often used in the determination of aldehydes:

1. With sodium bisulfite:

$$RCHO + NaHSO_3 \rightarrow RHCOHSO_3 \tag{2}$$

The sulfite addition product crystallizes in excess sodium bisulfite solution, which is useful for the analysis of aldehydes in the presence of other organic vapors (5).

2. With alcohols to form hemiacetals and acetals:

$$RCHO + R'OH \rightarrow RCHOHOR' + R'OH \rightarrow RCH(OR')_2 \tag{3}$$

Polyhydroxy aldehydes, as the sugars, can undergo internal cyclization and hemiacetal formation:

$$CH_2OH(CHOH)_4CHO \tag{4}$$
(α-D(+)-Glucose)

β-D(+)-Glucopyranose

1.2.4 Condensation Reactions

1. With ammonia or primary amines to form imines (Schiff's bases). The reaction of formaldehyde with ammonia produces hexamethylenetetramine, which is used as a source of formaldehyde in subsequent reactions.
2. With hydrazines to form hydrazones.
3. With hydroxylamine to form oximes.
4. With sulfhydryls to form thiohemiacetals:

$$RSH + R'CHO \rightarrow RSHCOHR' \tag{5}$$

Typically, these are acid-catalyzed reactions. The products are usually solids and suitable for the isolation and characterization of the aldehyde precursors. Because sulfhydryl, amine, and amine derivatives are common reactive groups in biologic systems, this class of reactions may in part be responsible for the biologic potency exhibited by the aldehydes and acetals.

1.3 Synthesis of Aldehydes

Aldehydes are generally produced by the reduction of acids, the oxidation of alcohols, or the pyrolysis of natural products (e.g., furfural from oat bran) (4). Formaldehyde is produced commercially by the oxidation of natural gas (methane) in the presence of a metal catalyst and heat (6). Oxidation of ethylene over palladium and copper chloride catalysts is the major synthesis route for acetaldehyde (7). Other aliphatic aldehydes are synthesized by the hydroformylation of alkenes:

$$RCH{=}CH_2 + CO + H_2 \xrightarrow[Pd]{Me} RCH_2CH_2CHO + \begin{matrix} CH_3 \\ | \\ RCHCHO \end{matrix} \qquad (6)$$

Aromatic aldehydes are purified from essential oils and synthesized from products distilled from coal tar. Crotonaldehyde is synthesized by the aldol condensation of acetaldehyde. Acrolein, a major unsaturated aldehyde, is synthesized by the catalytic oxidation of propylene (8).

The peroxidative scission of double bonds to produce two aldehydes is of minor application in commercial and laboratory usage, but has biologic significance. Such a reaction occurs during the peroxidation of unsaturated lipids and the production of these reactive products near sensitive biological sites may have traumatic consequences for the cell (9).

1.4 Methods of Determination of Aldehydes

Standard methods for the determination of formaldehyde and acrolein are described in detail in the NIOSH criteria document for formaldehyde (5) and the *OSHA Analytic Methods Manual* (10). The analysis is based on the formation of a bisulfite adduct that is subsequently hydrolyzed in the presence of chromotropic acid to produce a chromophore that absorbs at 580 nm. The sensitivity limit is about 0.16 ppm in a 25-liter air sample. The method is reasonably specific for formaldehyde, with slight interference from other aldehydes. The sample must be analyzed shortly after collection, or precautions taken to minimize the self-polymerization of the captured sample. Several monographs contain protocols for analysis of aldehydes and acetals (11–13). The use of innovative spectrophotometric methods to identify and quantify aldehydes in the atmosphere, including laser spectroscopy, gas diffusion filters, and long-path-length cells, has been described (14).

1.5 Physiological Effects of Aldehydes

The interest in the toxicity of aldehydes has focused on specific compounds, particularly formaldehyde (4–6, 15–17), acetaldehyde (18), and acrolein (19). The basis for the biological monitoring of aldehyde exposure has been described (20). The aldehydes of the essential oils, used in the food and cosmetic industries, were included in a comprehensive series of monographs describing the toxicity and sensitizing properties of individual compounds (6, 21). The possibility that specific aldehydes might be mutagens (and, by implication, possibly carcinogens) has been investigated, and the data reviewed (6, 15–18).

At this time, little evidence exists to suggest that mutations would result from occupational levels of exposure to aldehydes, although some aldehydes are clearly mutagenic in some test systems (22–25). Ward et al. (26) examined a group of 11 hospital autopsy service workers exposed to formalin, and a set of matched controls, for alterations of sperm parameters. Formaldehyde exposures were episodic, but reported to be between 0.61 and 1.32 ppm (TWA). No statistical differences between groups were noted for sperm count, abnormal sperm morphology, and fluorescent-body frequency. A later report described the assay of the urines of the formalin exposed and control group for mutagenic compounds by the Ames assay (27). No increase in mutagenicity was seen, although a large proportion of the exposed groups' urine demonstrated the presence of toxic compounds (not, however, believed to be metabolites of formaldehyde). A similar survey of pathology staff did not discover any change in the frequency of sister chromatid exchange and other chromosomal aberrations related to formaldehyde exposure (28). The carcinogenicity of aldehydes is addressed further below.

Four acute effects of aldehydes are noted.

1.5.1 Primary Irritation of Eyes, Skin, and the Respiratory Tract

The low-molecular-weight aldehydes, the halogenated aliphatic aldehydes, and the unsaturated aldehydes are particularly irritating. The mucous membranes of the nasal and oral passages as well as the upper respiratory tract are affected, producing a burning sensation, an increased ventilation rate, bronchial constriction, choking, and coughing. The eyes tear, and a burning sensation is noted on the skin of the face. During low exposures, the initial discomfort may abate after 5 to 10 min but will recur if exposure is resumed after the interruption.

Furfural, the acetals, and the aromatic aldehydes in general are much less irritating than formaldehyde or acrolein. If the exposure is brief, recovery is rapid and no residual effects are noted. However, the halogenated aldehydes capable of being metabolized to α-halogenated acids are quite toxic (29).

1.5.2 Sensitization

Sensitization refers to the elicitation of an allergic reaction upon repeated exposure to a chemical agent. The aldehydes, particularly formaldehyde, tend to be offenders in this regard.

Sensitization reactions to formaldehyde are well-known (for a review, see Reference 6). The exposure to formaldehyde need not be intense. For example, complaints of sensitization have arisen when exposure was incurred while wearing crease-resistant (permanent press) fabric impregnated with a formaldehyde–melamine resin (30). In another instance, three subjects wearing arm casts impregnated with melamine–formaldehyde resin were sensitized (31).

A few of the more complex odorous aldehydes can elicit a mild sensitization reaction in human patch tests (20). Cinnamaldehyde has also been implicated as a chemical sensitizing agent. Cinnamaldehyde imparts a pleasant flavor to cosmetics, topical medicinals, and dentrifices. Contact dermatitis was reported among six patients exposed to an antiseptic ointment. Three of the six patients reacted to oil of cinnamon (32). Similar reports have been made concerning mouthwash (33) and toothpaste (34) containing cinnamaldehyde. Glutaldehyde is commonly used as a cold sterilizing agent of medical devices (e.g., endoscopy instruments) and has been associated with both irritating and sensitizing reactions (35–38).

Respiratory sensitization has been the subject of scattered reports. In a plant producing formaldehyde–phenol resin-impregnated filters, in which excursions past the recommended threshold values probably occurred, workers suffered a decrement in pulmonary function tests. Some workers developed asthma-like symptoms that persisted for several days after exposure had stopped (39). However, other chemicals were present in the environment, and it would be impossible to ascribe all the symptoms to formaldehyde alone. In another study, hospital aides sterilizing kidney filtration equipment with formalin developed bronchitis and some of the workers evinced symptoms similar to those of asthma (40).

The widespread use of formaldehyde–urea insulation and other formaldehyde-based resins in residential construction has raised concerns that respiratory sensitization may be triggered in the occupants. A group of asthmatic subjects were exposed to formaldehyde concentrations between 0.008 and 0.85 mg/m^3 for 90 min (considered to be within the range of indoor domestic environments). No significant changes in respiratory performance parameters could be demonstrated (41). The authors concluded that these levels of formaldehyde exposures were insignificant in the elicitation of pulmonary symptoms.

1.5.3 Anesthesia

Materials that have unquestionable anesthetic properties are chloral hydrate, paraldehyde, dimethoxymethane, and acetal diethyl acetal. Paraldehyde and acetal diethyl acetal yield acetaldehyde upon hydrolysis. Acetaldehyde has been shown to produce both narcosis and an alcohol-type dependency in rats (42). Although when administered experimentally, other aliphatic aldehydes will give anesthetic-like symptoms in large parenteral or oral doses, in industrial exposures this action is overshadowed by the primary irritant action, which prevents voluntary inhalation of any significant quantities. The small quantities that can be tolerated by inhalation are usually so rapidly metabolized that no anesthetic symptoms occur. Some nausea, vomiting, headache, and weakness were reported by Wilkinson in chemists exposed to high concentrations of isovaleraldehyde (2-methylbutylaldehyde) (43). None of these, however, could be classed as definite anesthetic reactions. Disturbances in memory, mood, equilibrium, and sleep in histology technicians correlated with exposure to formaldehyde, although xylene and toluene vapors were also present. The neurobehavioral changes were accompanied by irritation of the eyes and respiratory tract (44).

1.5.4 Organ Pathology

The principal pathology experimentally produced in animals exposed to aldehyde vapors is damage to the respiratory tract and pulmonary edema. Multiple hemorrhages and alveolar exudate may be present, although these effects are usually much less dramatic than those produced with gases such as phosgene, and are usually confined to the upper respiratory tract. The effects produced with the very reactive aldehydes, such as ketene, acrolein, crotonaldehyde, and chloracetaldehyde, however, are much more pronounced and include deep lung damage. High dosages of materials such as methylal and furfural have been reported to cause various changes in the liver, kidneys, and central nervous system, but there appears to have been no confirmation of this type of action in human industrial exposures. In general, the aldehydes are remarkably free of actions that lead to definite cumulative organ damage to tissues other than those that may be associated with primary irritation or sensitization. However, the questions of mutagenicity, carcinogenicity, and teratogenicity hang over many reactive organic compounds of widespread commercial use, and aldehydes are no exception. Formaldehyde, acetaldehyde, crotonaldehyde, chloral hydrate, malonaldehyde (22), and acrolein (23) have been reported to be mutagenic in various bacterial, fungal, fruit fly, and plant systems.

1.5.5 Carcinogenicity

The results of animal tests of aldehydes for carcinogenicity indicate that any tumorigenicity revealed depends heavily upon the specific chemical, the route and extent of exposure, and the species. Repeated, high-dose injections of acetaldehyde (45), acrolein (46), and formaldehyde (47), delivered subcutaneously or subdermally, have produced papillomas and sarcomas at the sites of injection in rodents.

Inhalation studies with acrolein (48) and furfural (49) did not reveal tumors in hamsters. An initial study of the simultaneous administration of furfural and benzo[a]pyrene indicated an increase in the number of tumors over that produced by exposure to benzo[a]pyrene alone (50). A subsequent study by the same authors failed to verify any co-carcinogenic effect of furfural (51). Chloroacetaldehyde diethyl acetal was tested, as were three other chemical analogues of bis(chloromethyl) ether, in repeated topical application, long-term experiments with mice and did not demonstrate initiating, promoting, or carcinogenic activity, although bis(chloromethyl) ether was an extremely potent carcinogen in such a system (52). A brief communication has reported that a high dose of 3,4,5-trimethoxycinnamaldehyde produced tumors in four of six rats (two rats succumbed to the acute effects of the compound) (53). Because the compound can be purified from lignin, the authors raised the possibility that unsaturated aldehydes in wood might be responsible for the unusually high incidence of nasal cavity tumors among woodworkers, although no data were given to support the latter statement. Acetaldehyde exposure at high doses caused tumors in the respiratory passages of rats (54) and hamsters (55). Relevant to the discussion below, considerable tissue damage was associated with exposures to acetaldehyde at these doses and evidence of DNA/ protein cross-linking was found (56).

The potential of formaldehyde to cause cancer has been studied extensively (for reviews, see References 6, 15–17, 57, and 58). Formaldehyde is mutagenic in bacterial assays and forms covalent bonds with DNA and DNA-associated protein (59), suggesting its potential as a mutagen and carcinogen. Chronic bioassays clearly indicate that formaldehyde can cause tumors in the respiratory passages of rats (60, 61). Nasal tumors were reported in rats exposed for 13 weeks to 20 ppm formaldehyde (60). Fifty percent of the rats exposed to 14.3 ppm for 24 months developed squamous cell tumors (62). Mice exposed under similar conditions appear to be more resistant, for only 3.3 percent developed tumors. Horton et al. (48) exposed mice to inhalation of formaldehyde at levels of 0.05 mg/l for 29 weeks followed by 0.15 mg/l for 35 weeks. Although the mice displayed atypical meta- and hyperplasia of the epithelium of the major airway, pulmonary neoplasms were not found at autopsy. Hamsters also appear to be resistant (63).

In rodent studies, the induction of tumors was associated with histopathological changes in the nasal tissue, and it was suspected that formaldehyde-induced cell proliferation was essential to the expression of carcinogenicity (64). This led to the study of the location of damage in the respiratory passages produced by formaldehyde exposure, and attempts to correlate the differences in cancer susceptibility noted between various species with molecular dosimetry (15, 59, 65).

A number of studies have surveyed cancer patients and workers to determine whether exposure to formaldehyde was associated with increased cancer risk. Olsen et al. (66) found the increased rate of carcinoma of the nasal cavity and sinuses among Danish workers exposed to formaldehyde was insignificant when the elevated rates of tumors were adjusted for simultaneous exposure to wood dust, paint lacquer, and glue. In a case-control study of Dutch workers, a relative risk of 1.9 to 2.5 was associated with occupations where formaldehyde exposure was thought to be possible, but because no dose history could be constructed, the authors felt that the evidence was inconclusive (67). Several National Cancer Institute studies (67–70) implicate formaldehyde exposure in the increased incidence of tumors of the respiratory passages (for a review, see Reference 71). A recent review panel concluded that the weight of the evidence and the presence of confounding factors do not support a relationship between formaldehyde exposure in humans and any malignancy, or, if present, the excess risk is very small (72).

Risk assessment has been conducted for oral (food and water) formaldehyde exposure, taking cognizance that formaldehyde is a metabolite formed at low concentrations in the body and is ingested as a natural ingredient of some foods. Formaldehyde levels as high as 100 ppb in drinking water were considered to present negligible risk, based on formaldehyde intake from other dietary sources (73). The use of formaldehyde as a bacteriostatic agent in cheese manufacturing at up to several parts per million in the final product was argued to be well below the body's capacity to remove formaldehyde by metabolism (16).

At present the only study implicating other aldehydes as carcinogens in humans is an epidemiologic survey of workers in an acetal production plant in which the incidence of tumors of the nasal and oral cavities and bronchial airways was reported to be far above that of age-matched workers in other chemical plants (74). Furthermore, the tumors appeared after relatively short exposure periods. Acetaldehyde, butyraldehyde, crotonaldehyde, n-butanol, ethylhexanol, and acetal were detected in the plant. Further work has not been published that addresses the questions of whether one of the implicated chemicals, or a combination thereof, could be responsible for the reported increase in tumors.

Generally the aldehydes, with the possible outstanding exception of formaldehyde, cannot be regarded as potent carcinogens. The intolerable irritant properties of the compounds preclude substantial worker exposure under normal working conditions. The extreme reactivity of formaldehyde, acrolein, chloroacetaldehyde, and so on that produces reactions at epithelial surfaces tends to prevent their passage into the body. The rapid metabolic conversion to innocuous materials also may limit those critical reactions necessary to initiate tumorigenesis. However, the tissue irritation even at low exposures may promote tumor formation initiated by another compound. Therefore caution is warranted, and certainly further epidemiologic studies must be performed to define whether or not a hazard may exist.

1.6 Metabolism of Aldehydes

Aldehydes are readily metabolized by three principal routes: (1) oxidation to acids; (2) reduction to alcohols; and (3) conjugation with sulfhydryls, such as glutathione. At least for formaldehyde, and perhaps acetaldehyde, conjugation with glutathione within the cell occurs rapidly, and the S-acylglutathione is the direct substrate for subsequent oxidation steps (75, 76). The predominant metabolic route of other aldehydes depends upon species of animal, the specific aldehyde involved, and the conditions of exposure.

1.6.1 Oxidation of Aldehydes

The aldehyde dehydrogenases are a rather large group of enzymes that have been well characterized. Formaldehyde dehydrogenases have been purified from a number of sources, including human (75, 76), and are remarkably similar. The enzyme is a dimer of 75,000 to 80,000 Da molecular weight, with a V_m of around 3 μmol/(min)(mg). The reaction requires GSH and NAD(P)$^+$:

$$H_2CO + GSH + NAD(P)^+ \leftrightarrow GSCHO + NAD(P)H + H^+ \qquad (7)$$

The physiological role of the enzyme may relate to the detoxification of endogenously generated formaldehyde.

It is relevant to the expression of carcinogenicity of inhaled formaldehyde to note that formaldehyde dehydrogenase activity has been demonstrated in rat nasal mucosal tissue (77). An earlier report that methylglyoxal is also a substrate of formaldehyde dehydrogenase (75) has been challenged (78) on the basis that the demonstrated activity of the enzyme toward methylglyoxal could be explained by the contamination of the substrate with formaldehyde.

Aldehyde dehydrogenase activity has been found in horse liver (76), erythrocytes (79), human liver (80), kidney (81), heart (4), and placenta (82). At least seven forms exist in the mammalian hepatocyte that differ in size (ranging from 110,000 to 250,000 Da mol. wt.); subcellular location, kinetic behavior, inducibility, inhibitor sensitivity, and substrate specificity (83–86). The high affinity mitochondrial aldehyde dehydrogenase is considered to be the principal route of oxidation of acetaldehyde. The enzyme requires NAD, and is inhibited by chloral hydrate (87). Interestingly, this enzyme is largely inactive in as many as 50 percent of individuals that are of Oriental descent (86). The inability to remove effectively the toxic acetaldehyde formed from ethanol oxidation was suggested by the authors as one possible explanation for the incidence of acute alcohol intoxication in Orientals. At least two isozymes of aldehyde dehydrogenase are induced by phenobarbital and tetrachlorodibenzodioxin (TCDD) (83). Both forms are sensitive to inhibition by micromolar concentrations of disulfiram, although the TCDD-induced enzyme prefers NADP as coenzyme and aromatic aldehydes as substrates. Disulfiram [Antabuse, bis(N,N-diethylthiocarbamoyl) disulfide], is a drug used in rehabilitation of alcoholics. The accumulation of acetaldehyde due to disulfiram inhibition of aldehyde dehydrogenase is thought to exacerbate the toxic action of ethanol and, it is hoped, reinforce the resolve of the patient to avoid further alcohol consumption (88). Another chemical alcohol deterrent, cyanamide, inhibits aldehyde dehydrogenase in vivo, but not in vitro (86). Investigations indicate that the inhibition mechanism requires cyanamide and catalase (89).

Two unsaturated aldehydes, acrolein and citral, are also potent inhibitors of aldehyde dehydrogenase activity in vitro. Acrolein is an irreversible inhibitor (90) whereas citral displays mixed type inhibition (91). However, acrolein is metabolized by subcellular fractions if mercapatoethanol is included in the incubation mixture, suggesting that a thio-acrolein conjugate is the proximate substrate (90, 92). Despite the inhibition of aldehyde dehydrogenase observed in vitro, citral is rapidly metabolized by rats; at least seven oxidized metabolites and a glucuronide conjugate were isolated from rat urine (93). Citral is also reduced by alcohol dehydrogenase (89). The reduced product may then enter an alternate oxidation pathway, or it may be speculated that citral also forms a sulfhydryl conjugate that is subsequently metabolized.

Less well understood is the role of aldehyde oxidase and xanthine oxidase in the oxidation of aldehydes. These metalloenzymes require molecular oxygen to act on aldehydes in vitro, and their importance in vivo is as yet unclear (94, 95).

The existence of a formidable array of dehydrogenase isozymes and oxidases results in the ability of an organism to oxidize a range of substrates, including acetaldehyde, propionaldehyde, n-butyraldehyde, n-valeraldehyde, and benzaldehyde (88). Indeed, the oxidation of unsaturated aldehydes may be an important defense of the cell against toxic products formed by lipid peroxidation (96). The acid oxidation products of aldehydes may be either excreted or condensed with Coenzyme A to produce the acyl-CoA derivative. At least three enzymes (acyl-CoA synthetases) exist, each exhibiting specificity based on the acid's chain length (97).

1.6.2 Reduction of Aldehydes

Reduction to alcohols represents another major route of aldehyde metabolism. Aldehyde reductases exist as monomers of 33,000 to 36,000 Da molecular weight (3, 80). Typically, these enzymes require NADP as a coenzyme, are located in both mitochondrial and cytoplasmic compartments of the cell, and may have biogenic amines as the endogenous substrate. Reduction may also occur by the alcohol dehydrogenases (3, 91). Citral is rapidly reduced by hepatic cytosolic fractions containing alcohol dehydrogenase (91). However, the formation of aldehydes by alcohol dehydrogenase is favored at pH conditions perhaps too alkaline to be a major route in vivo.

1.6.3 Formation of Thiohemiacetals

Aldehydes tend to react with cell sulfhydryl groups. The reaction of aldehydes with glutathione produces thiohemiacetals. To what extent this reaction occurs spontaneously, versus enzymatically, is unclear, but glutathione condensation products have been isolated from the urine of animals fed either acrolein (98) or chloroethanol (99). A metabolic precursor of dicloroacetaldehyde, 1,2-dichloroethylene, is reported to depress glutathione levels in hepatic mitochondria (100). Depletion of glutathione both in hepatocyte cell cultures (101) and in whole rodents (102) decreases formaldehyde metabolism. Direct formation of S-(hydroxymethyl)glutathione has been demonstrated by NMR spectroscopy in $E. coli$ (103). Although the subsequent metabolism of the adduct is enzyme catalyzed, the formation of the adduct proceeds rapidly in heat-killed cells, indicating that it is likely a direct chemical reaction. Other formaldehyde adducts of glutathione have been demonstrated to form in vitro under nonphysiological conditions and are thought not to be important physiologically (104). Cinnamaldehyde demonstrated cytotoxicity in a mouse leukemia cell line L1210 (105). The toxicity was attributed to blockage of protein synthesis by the adduct formation between sulfhydryl-containing amino acids and cinnamaldehyde.

Low-molecular-weight thiols, such as cysteine, N-acetyl-L-cysteine, methionine, glutathione, and synthetic thiols such as 2-mercaptoethanesulfonate, protect isolated hepatocytes from acrolein and allyl alcohol toxicity (106). Acrolein is believed to be the toxic product of allyl alcohol. Similarly, the depletion of glutathione by buthionine sulfoximine enhanced the embryo toxicity of acrolein in rat embryos placed in culture (107), and the concurrent addition of glutathione prevented the embryo toxicity elicited by acrolein (108). Conversely, the decrease of glutathione levels in a cell by aldehydes may potentiate the toxicity of other agents. A depletion of glutathione would presumably make an organism less tolerant of other oxidant materials of both exogenous and endogenous origin.

The ability of sulfhydryl agents to protect cells in vitro has prompted several investigations into their ability to reverse or prevent the toxic effects of aldehydes. Sprince et al. (109) reported that the anesthetic and lethal effects of acetaldehyde could be blocked by administration of a mixture of 2 mmol/kg L-ascorbic acid, 1 mmol/kg L-cysteine, and 0.3 mmol/kg thiamine. Similarly, Guerri et al. (110) injected rats and mice with lethal doses of acetaldehyde and formaldehyde and found that a subsequent injection with either 2,3-dimethylmercaptopropanol, L-cysteine, or mercaptoethanol would reduce mortality. The authors suggest a possible therapeutic application of such agents could be made in cases of acute exposure.

Attempts have been made to exploit the cellular toxicity of aldehydes in drug therapy. Aldehyde dehydrogenase levels were found to be very low in three tumor cell lines. The growth of the cells was sensitive to 2,3-dihydroxybutyraldehyde. When the cells were transplanted into animals, administration of 2,3-dihydroxybutyraldehyde slowed and stopped the further development of the tumor cells (111).

In summary, the detoxification of aldehydes can be seen to proceed basically via oxidation to yield readily metabolized acids, by reduction to alcohols, and by reaction with sulfhydryl groups, particularly glutathione. The thiohemiacetal may be the proper substrate for metabolizing enzymes. Under conditions that either deplete glutathione levels or result in an inhibition of aldehyde dehydrogenase (e.g., Antabuse treatment), the acute and chronic effects of aldehyde toxicity might be more fully expressed. Such situations should be monitored carefully.

2 SPECIFIC COMPOUNDS

2.1 Saturated Aliphatic Aldehydes

The physical and chemical properties of saturated aliphatic aldehydes are given in Table 6.2. Physiological responses of animals are presented in Table 6.6. The discussion that follows focuses primarily on effects on humans.

2.1.1 Formaldehyde

Formaldehyde is a gaseous aldehyde that may be encountered either directly, dissolved in aqueous solution (often with aliphatic alcohols present) or in combination with other agents, particularly in the preparation, processing, and use of formaldehyde resins. Exposure is also associated with the inclusion of formaldehyde as a preservative and/or disinfectant in cosmetics, foodstuffs (e.g., cheese) (16), cleaning fluids, dyes and inks, medicinals, and dentrifices (6). The use of methanol as a substitute for gasoline and diesel fuel could result in widespread, low-level exposure to formaldehyde in exhausts. Formaldehyde has recently been extensively reviewed (4, 6, 15, 17).

Formaldehyde is detectable by most people at levels below 1 ppm. It produces a mild sensory irritation of the eyes, nose, and throat at 2 to 5 ppm, becomes unpleasant at 5 to 10 ppm, and is intolerable at levels in excess of 25 ppm. Although the intensity of symptoms diminishes during exposures to levels of about 5 ppm, the tolerance is lost after a 1- to 2-hr interruption of exposure. Tissue damage likely occurs at levels of 25 to 50 ppm. Recovery tends to be rapid and complete, however. Human deaths have resulted from massive inhalation of formaldehyde (53).

The irritant properties of formaldehyde affect the mucous membrane surfaces

Table 6.6. Toxicity and Irritant Properties of Saturated Aliphatic Aldehydes

LD$_{50}$			Skin LD$_{50}$		Inhalation Toxicity				Irritant Effect			Ref.
Species	Route	g/kg	Species	ml or g/kg	Species	ppm	Time (hr)	Mortality	Species	Skin	Eye	
Formaldehyde												
Rat	Oral	0.2–0.8			Rat	250	4	LC$_{50}$	Rabbit		Severe	123
Guinea pig	Oral	0.26			Rat	815	0.5	LC$_{50}$	Man	Moderate		158
Rat	SC	0.42			Cat	650	8	~LC$_{50}$				155, 157
					Cat	200	3.5	All survived				158
Sodium Formaldehyde Bisulfite												
Mouse	Oral	3.2–6.4	Guinea pig	>20								158
Mouse	IP	1.6–3.2										
Sodium Formaldehyde Sulfoxylate												
Mouse	IP	>0.5										159
Rat	Oral	>1.0										160
Rat	IV	>2.0										
Paraformaldehyde												
Rat	Oral	>1.6										160
Acetaldehyde												
Rat	Oral	1.93			Rat	Sat. vap	2 min	LC$_{100}$	Man		Moderate	123
Rat	SC	0.64			Rat	16,000	4	0/6				123,
					Rat	20,000	30 min	LC$_{50}$				161,
					Cat	13,600	0.25	1/1				162
						4,100	3–5	0/1				

Acetaldehyde Sodium Bisulfite

Species	Route	Dose	Species	Dose	Species	Conc.	Time	Resp.	Species			Ref.
Rat	Oral	>3.2							Guinea pig	Slight		158
Guinea pig	Oral	1.6–3.2	Guinea pig	>20								

Fluoroacetaldehyde

| Mouse | IP | 0.006 | | | | | | | | | | 163 |

Glycolaldehyde (Hydroxyacetaldehyde)

| Rabbit | SC | 4.0 | | | | | | | | | | 160 |

Chloroacetaldehyde

| Rat | Oral | 0.05–0.4 | Guinea pig | 0.1-1.0 | Mice | *a* | 2.6 min | 50% | Rabbit | Severe | Severe | 158 |
| | | | | | | | | | Guinea pig | Severe | | |

Trichloroacetaldehyde (Chloral)

| Rat | Oral | 0.05–0.4 | Guinea pig | 1.0–10 | | | | | Guinea pig | Severe | | 158 |

n-Hexaldehyde (Hexanal)

| Rat | Oral | 4.9 | | | Rat | Concd. vap. | 1 | 0/6 | Rabbit | Slight | Slight | 165 |
| | | | | | Rat | 2,000 | 4 | 1/6 | | | | |

2-Ethylbutyraldehyde

| Rat | Oral | 3.98 | | | Rat | Concd. vap | 5 min | 0/6 | | | | 158, 165 |
| | | | | | Rat | 8,000 | 4 | 5/6 | | | | |

n-Heptaldehyde

| Mouse | Oral | 25 | | | | | | | | | | 159 |
| Mouse | IP | >0.5 | | | | | | | | | | 167 |

Table 6.6. (Continued)

Species	LD₅₀ Route	g/kg	Skin LD₅₀ Species	ml or g/kg	Inhalation Species	ppm	Time (hr)	Mortality	Irritant Species	Skin	Eye	Ref.
	Route	g/kg	Species	ml or g/kg	Species	ppm	Time (hr)	Mortality	Species	Skin	Eye	
2-Ethylhexylaldehyde (α-Ethylcaproaldehyde)												
Rat	Oral	3.73	Rabbit	5.04	Rat	25,000	13 min	3/3	Rabbit	Moderate	Slight	158
			Guinea pig	>20	Rat	4,000	4	1/6	Rabbit		Moderate	161
					Rat	2,000	23 min	3/3	Guinea pig	Moderate		
					Rat	145	6	0/3				
Isobutyraldehyde												
Rat	Oral	1.6–3.7	Rabbit	7.1	Rat	8,000	4	1/6	Rabbit	Slight	Severe	158
Rat	IP	1.6–3.2	Guinea pig	>20					Guinea pig	Moderate		165
β-Hydroxybutyraldehyde (Aldol, Acetaldol)												
Rat	Oral	2.2	Rabbit	10	Rat	4,000	4	2/6	Rabbit	Slight	Slight	166
			Rabbit	0.14	Rat	Satd. vap.	0.5	No deaths				
4-Fluorobutanal												
Mouse	IP	0.002										163
n-Valeraldehyde												
Mouse	Oral	6.4–12.8	Guinea pig	>20	Rat	48,000	1.2	3/3	Guinea pig	Severe		158
Rat	Oral	3.2–6.4			Rat	1,400	6	0/3	Rabbit		Severe	

Species	Route	Dose	Species	Dose	Species	Conc	Time	Ratio	Species	Irritation	Species	Irritation	Ref
Isovaleraldehyde													
Rat	Oral	>3.2	Guinea pig	>10					Guinea pig	Moderate			158
2-Methylbutyraldehyde													
Mouse	Oral	3.2–6.4	Guinea pig	>20	Rat	67,000	0.3	3/3	Guinea pig	Moderate			158
Rabbit	Oral	6.4–12.8			Rat	3,800	6.0	0/3	Rabbit		Severe		
Paraldehyde													
Rat	Oral	1.65	Rabbit	5.0					Rabbit		Severe		160
Rabbit	Oral	5.0	Guinea pig	10–20									
Dog	Oral	3.0–4.0											
Propionaldehyde													
Rat	Oral	0.8–1.6			Rat	26,000	0.5	LC$_{50}$	Rabbit		Severe		123, 158
Rat	SC	0.8							Guinea pig	Severe			161
Ethoxypropionaldehyde													
Rat	Oral	0.9	Rabbit	1.0	Rat	500	4	6/6	Rabbit	Slight		Severe	164
α, β-Dichloropropionaldehyde													
Rat	Oral	0.16	Rabbit	0.078	Rat	Concd. vap.	2 min	6/6	Rabbit	Severe		Severe	161
					Rat	16	4	4/6					
n-Butyraldehyde													
Rat	Oral	5.9	Guinea pig	>20	Rat	60,000	0.5	LC$_{50}$	Guinea pig	Moderate			123
Rat	IP	0.8							Rabbit		Severe		158
Rat	SC	10.0											161

^{a}45% equilibrium of chamber air and air bubbled through a 30% aqueous solution.

303

of the upper respiratory tract, the eyes, and the exposed surfaces of the skin. Numerous studies of exposures in the 2 to 30 ppm range describe the respiratory symptoms suffered. The subjects complained of a prickling irritation of the throat, wheezing, headache, and excessive thirst, accompanied by tearing and stinging of the eyes (5). A study in a filter manufacturing plant reported bronchitis-type symptoms with a loss of vital capacity and the presence of a productive cough in those workers involved in a phenol–formaldehyde resin-impregnation step (39). The symptoms were most severe early in the shift, with some remission noted after 15 min. Similar effects were noted in the staff of a hospital using formalin solutions to sterilize kidney dialysis units (40). In two of the staff, a single challenge of formaldehyde vapor was sufficient to trigger respiratory distress, implying that sensitization to the chemical had occurred. The symptoms also were listed in the self-described case of a medical resident, in whom respiratory sensitization to formaldehyde apparently had occurred (113).

Formaldehyde as a gas or in solution will act as a primary irritant on skin, causing an erythmatous or eczematous dermatitis reaction of exposed areas. Sensitization can result. The widespread use of formaldehyde resins to improve the durability and crease resistance of fabrics has resulted in the exposure of garment workers, sales personnel, and the general public to the resins, and to the formaldehyde released from the fabrics. Industrial procedures have improved, and the American Textile Manufacturing Institute reported a decline from 534 ppm in 1975 to 345 ppm in 1982 (114). However, the formaldehyde content of various fabrics can still be rather high and is quite variable, ranging up to 12,000 ppm (115). Contact dermatitis is occasionally the result, especially on areas of the body where the garments chaff the wearer, or where body heat and evaporation are concentrated, such as the neck or inside the thighs. Generally the person will show a positive patch test to formalin, although not necessarily to the offending fabric. The sensitizing properties of formaldehyde have been thoroughly reviewed (6).

Humans are rather tolerant of formaldehyde by ingestion, the chief sites of damage being the laryngeal, esophageal, and gastric membranes at high doses. Aspiration of formaldehyde solutions into the lungs can quickly produce respiratory distress. Corrosive gastritis has been reported after ingestion of formaldehyde solutions. Further systemic damage due to formaldehyde exposure, by either inhalation or ingestion, is generally not observed.

On the presumption that visual disturbances might be a more sensitive parameter of exposure than respiratory function, an intensive survey of workers in two wood products plants and a garment factory was carried out. The on-site concentrations of formaldehyde were measured and found to be less than the 3 ppm federal standard, and generally less than 1 ppm. No impairment of visual function was found (114).

Formaldehyde reacts with hydrochloric acid to form the potent animal carcinogen bis(chloromethyl) ether. Concern has been expressed that, should this reaction proceed even slightly in the vapor phase, parts per billion of bis(chloromethyl) ether could arise in the industrial environment. Studies have indicated that the reaction is unlikely if the recommended limits of formaldehyde and hydrochloric acid are observed (115, 116).

The present (1992) federal standard for exposure to formaldehyde is 3 ppm (5 ppm ceiling). The American Conference of Governmental Industrial Hygienists (AGCIH) recommends 1 ppm. Based on its potential human carcinogenicity (see preceding discussion, Section 1.5.4), NIOSH recommends the lowest achievable exposure (119).

2.1.2   Paraformaldehyde

Paraformaldehyde is a solid polymer of formaldehyde, which releases formaldehyde upon heating or hydration in acid or alkaline solutions (4). Contact with the solid material produces the same symptoms as with formaldehyde, although onset may be slower. It has been suggested that inhalation of paraformaldehyde particles may facilitate deeper penetration into the lung than would be suffered by inhalation of formaldehyde vapors (120).

2.1.3   Acetaldehyde

Because of the explosive hazards of acetaldehyde, it is usually handled in industry under closed systems and exposures are not apt to be continuous or at high levels. Humans can readily detect levels of 50 ppm, and some can detect acetaldehyde at levels below 25 ppm. At 50 ppm a majority of volunteers exposed for 15 min showed some signs of eye irritation, and at 200 ppm all subjects had red eyes and transient conjunctivitis. Eye irritation and, to a lesser extent, nose and throat irritation, are the only signs noted during usual industrial exposures.

Although industrial exposure to acetaldehyde has not been intensively examined, the possible etiologic role of acetaldehyde in ethanol toxicity has received considerable attention. As discussed earlier, ethanol is metabolized almost exclusively to acetaldehyde, which is in turn oxidized by aldehyde dehydrogenase (see Section 1.6.1). Acetaldehyde inhibits a number of sulfhydryl-dependent enzymes, including aldehyde dehydrogenase. Typically, the inhibition is reversible by sulfhydryl agents.

Acetaldehyde, as a component of cigarette smoke, is reported to be ciliotoxic and to depress pulmonary macrophage numbers (121, 122). Acetaldehyde will produce chromosomal aberrations in plant and animal test systems (22) and is weakly mutagenic in the Ames assay. Acetaldehyde will cause cross-linking of protein and DNA in nasal tissue of rats exposed to 1000 ppm (56). Hamsters and rats exposed for a year to 1000 ppm developed laryngeal and nasal tumors, respectively (55). Tumor occurrence was limited to tissue that suffered extreme irritation and cell proliferation. Because acetaldehyde is readily metabolized, the extrapolation of tumor incidence at high exposure levels in animals, to the human experience, is problematic.

The recommended threshold limit (1992) for acetaldehyde is 200 ppm, although AGCIH has recommended a limit of 100 ppm.

2.2 Bisulfite and Hydrosulfite Addition Products of Aldehydes

Sodium formaldehyde bisulfite and sodium formaldehyde sulfoxylate are solids and are used as reducing agents. These addition products possess few of the irritating qualities of the aldehyde. Skin sensitization to these materials is rare, and they can be handled without unusual precautionary measures. They have a relatively low oral and parenteral toxicity in experimental animals. The same applies to acetaldehyde sodium bisulfite.

2.3 Higher Aliphatic Aldehydes

Aliphatic aldehydes larger than three carbon atoms have received a certain amount of preliminary toxicologic study, as indicated in Table 6.6. In recent years there has been increasing industrial experience with these materials as a result of their availability from petrochemical reactions. The C-4 and C-5 homologues are intermediates in chemical synthesis and accelerators in the rubber industry. Propionaldehyde, n-butyraldehyde, isobutyraldehyde, and isovaleraldehyde are components of the exhaust of internal combustion engines (4). These aldehydes are characterized by lower toxicity, particularly by oral administration, than the lower-molecular-weight homologues. They appear to be relatively well tolerated by inhalation, although local reactions on skin and eyes may still be quite pronounced. Sensitization may occur but is not as troublesome as in the case of formaldehyde.

Skog (123) noted anesthesia in rats at high levels of inhalation for propionaldehyde and butyraldehyde. The survivors recovered promptly. Autopsies showed principally evidence of bronchial and alveolar inflammation. Although subcutaneous injection of rats with large doses of butyraldehyde caused hemoglobinuria, this was not seen on inhalation.

Salem and Cullumbine (124) studied the acute toxicity of aldehydes in mice, guinea pigs, and rabbits. All animals exposed to high levels by inhalation developed fatal pulmonary edema. The toxicity of the aldehyde decreased as the chain length increased. Branched-chain aldehydes are similar to straight-chain aldehydes in potency and type of effect.

Gage (125) reported that rats tolerated inhalation of 90 ppm of propionaldehyde, for 20 days, 6 hr/day, with no obvious pathology, although 1300 ppm for 6 days produced hepatic damage. Isobutyraldehyde at 1000 ppm, 6 hr/day, produced slight nasal irritation, whereas butyraldehyde was innocuous under the same conditions. When mice were supplied with butyraldehyde in drinking water for a total of 50 days (total dose of 15 g/kg), abnormal spermatogonial cells of all stages were observed, suggesting that at high levels, butyraldehyde may be a mutagen in mice (126).

Wilkinson (43) reported an instance in which several chemists engaged in distilling isovaleraldehyde (2-methylbutryaldehyde) developed some signs of chest discomfort, nausea, vomiting, and headaches. Exposure levels were not measured although the odor was very pronounced, and it is possible that fairly high levels were present. All recovered in a few days without any particular after effects.

Discussions of homologues up to C-14 are included in an extensive monograph on flavors and essences (21). Generally, these compounds are of low toxicity and human subjects do not show sensitization to patch tests of the individual aldehydes (21).

The precautions for the higher aldehydes are essentially those for most other reactive organic compounds, and should include adequate ventilation in areas where high exposures are expected, fire and explosion precautions, and proper instruction of employees in the use of respiratory, eye, and skin protection. Exposure standards do not exist for the higher aliphatic aldehydes.

2.4 Halogenated and Other Substituted Aldehydes

Halogenation tends to increase the local irritant action and general toxicity of the aldehyde. The toxicity of chloroacetaldehyde has recently been reviewed (99). Chloroacetaldehyde reacts with reduced glutathione to form S-carboxymethylglutathione and rapidly depletes the glutathione stores in the cell (127). Inhalation of the material is acutely toxic in mice, demonstrating a lethal time of 2.6 min under conditions in which the chamber reached 45 percent equilibrium with an incoming mixture of air bubbled through a 30 percent solution of chloroacetaldehyde (99). When rats were injected daily with a 0.3 or 0.6 LD_{50} dose for 30 days, 25 and 67 percent of the animals succumbed, respectively. The organ pathology at death revealed some hematologic disturbances, but the most obvious effects were bronchitis and pneumonitis (90). The recommended threshold limit of chloroacetaldehyde is 1 ppm (1992).

The monofluorinated aldehydes can be oxidized to highly toxic fluoroacetate. Trifluoroacetaldehyde and trichloroacetaldehyde, on the other hand, though displaying potent local irritant action and systemic toxicity, are not oxidized to a product capable of condensing with oxaloacetate and therefore are less toxic than either of the monohaloacetaldehydes (126). The presence of a hydroxyl group on the aldehyde does not seem to alter toxicity markedly (e.g., acetaldehyde versus hydroxyacetaldehyde and benzaldehyde versus hydroxybenzaldehyde).

2.5 Unsaturated Aliphatic Aldehydes

Data on the unsaturated aliphatic aldehydes are presented in Tables 6.3 and 6.7. Ketene, acrolein, and crotonaldehyde are important members of this group. Each is an acute irritant. Ketene is a gas, whereas acrolein and crotonaldehyde are liquids with appreciable solubility in water. The presence of the double bond considerably enhances the toxicity, as can be seen by the comparisons presented in Table 6.8.

2.5.1 Acrolein

On acute exposures, acrolein produces severe eye and respiratory irritation (19). During experimental exposures, animals succumbed to pulmonary edema and survivors were susceptible to bronchiopneumonia (123). Low, chronic exposures may present the development of tolerance.

Table 6.7. Toxicity and Irritant Properties of Unsaturated Aliphatic Aldehydes

LD$_{50}$			Skin LD$_{50}$		Inhalation Toxicity				Irritant Effect			Ref.
Species	Route	g/kg	Species	ml or g/kg	Species	ppm	Time (hr)	Mortality	Species	Skin	Eye	
					Ketene							
					Mouse	70	4	20/20				168
					Rat	250	150 min	4/4				
					Rat	120	10 days	0/4				
					Cat	370	8–12	1/2				
					Cat	230	15 days	0/1				
					Rat	53	100 min	2/2				
					Rat	1	7 × 14	0/2				
					Cat	23	4 × 2	0/1				
					Cat	1	7 × 55	0/2				
					Monkey	1	7 × 55	0/1				
					Acrolein							
Rat	Oral	0.046			Cat	690–1150	2	3/3				125
					Cat	18–92	3–4	0/2				169
					Cat	11	3–10	0/2				
					Rat	130	30 min	LC$_{50}$	Rabbit	Severe		172
					Rat	26	60 min	LC$_{50}$				
					Rat	8.3	240 min	LC$_{50}$				
					Human	150	10 min	Fatal				169
					Human	5	1 min	Intolerable				
					Human	0.25	5 min	Moderate irritation				

Methacrylaldehyde (Methacrolein)

Rat	Oral	0.14	Rabbit	0.43	Rat	250	4	5/6	Rabbit	Slight	Severe	166

2-Ethyl-3-propylacrolein

Rat	Oral	3.0	Guinea pig	>20	Rat	Conc. vap.	8	0/6	Rabbit	Slight	Slight	170

Crotonaldehyde (β-Methylacrolein)

Rat	Oral	0.22	Rabbit	0.38	Rat	1,400	30 min	LC₅₀	Rabbit		Severe	123
Rat	IP	0.07	Guinea pig	0.5–1.0	Rat	600	30 min	LC₅₀				134
Rat	SC	0.14	Rabbit	0.15–0.2								

Mucochloric Acid

Rat	Oral	0.05–0.1	Guinea pig	>5					Guinea pig	Severe	Severe	158
Rat	IP	0.01–0.025							Rabbit	Severe	Severe	165

Methyl-β-ethylacrolein (2-Methyl-2-penten-1-al)

Rat	Oral	4.3	Rabbit	4.5	Rat	2,000	4	3/6	Rabbit	Slight	Severe	171

trans-2-Hexenal

Rat	Oal	0.8–1.1										
Rat	IP	0.2										
Mouse	Oral	1.5–1.7										
Mouse	IP	0.2										

Table 6.8. Effect of Unsaturation on the Inhalation Toxicity of Aldehydes

Compound	Formula	LC_{50} (ppm) in Rats	Time of Exposure (min)
Acetaldehyde	CH_3CHO	20,000	30
Ketene	$CH_2{=}CHO$	130	30
Propionaldehyde	CH_3CH_2CHO	26,000	30
Acrolein	$CH_2{=}CHCHO$	130	30
Isobutyraldehyde	$(CH_3)_2CHCHO$	>8,000	4 hr
Methacrolein	$CH_2{=}C(CH_3)CHO$	250	4 hr
n-Butyraldehyde	$CH_3(CH_2)_2CHO$	60,000	30
Crotonaldehyde	$CH_3CH{=}CHCHO$	1,400	30

Subcellular effects of acrolein include an inhibition of mammalian DNA poly-merase (19) and an interference with several mitochondrial functions (128). The carcinogenic and mutagenic effects of acrolein have been reviewed (19). In long-term studies with hamsters exposed to 4 ppm acrolein, no evidence for an increased incidence of any type of tumor was found, although considerable alteration of the epithelial structure in the respiratory tract was produced (52). Acrolein was embryo lethal in chick embryo assays (129, 130) and embryo toxic in cultured rat embryos (131, 132). However, the latter two studies yielded conflicting conclusions about the potential teratogenicity of acrolein. Acrolein was negative in a two-generation continuous breeding study conducted on rats at concentrations [oral, as low as 3 mg/(kg)(day)] that produced other toxicologic effects (133). Humans are sensitive to levels as low as 0.25 ppm, and most find levels above 1 ppm unbearable. The recommended (1992) exposure limit is 0.1 ppm, with 15-min excursions to 0.3 ppm permitted.

2.5.2 Crotonaldehyde

Crotonaldehyde has effects very similar to those of acrolein. Skog (123) measured a 30-min LC_{50} of 1400 ppm in rats and concluded that crotonaldehyde was 10-fold less toxic than either ketene or acrolein. Rinehart (134) contested this, based on a measurement of a LC_{50} of 600 ppm in rats. Rinehart observed in his system that a considerable loss of material occurred between the point of generation of the vapors and the breathing zone of the animals that might explain the higher value obtained by other workers. Skog also reported that only high doses of crotonal-dehyde delivered subcutaneously would produce a hyperexcitable condition in the exposed rats, whereas Rinehart found that an inhalation exposure of 1000 ppm would elicit such symptoms. Auerbach et al. (22) state that crotonaldehyde is spermatocidal in mice. Humans can detect crotonaldehyde at 15 ppm. The exposure limit is 2 ppm.

2.5.3 Ketene

Ketene produces acute irritation of epithelial tissues in rodents, cats, and monkeys (135, 136). The pulmonary action tends to be in the deep lung, producing alveolar edema. The exposure value (1992) is 0.5 ppm, with excursions limited to 1.5 ppm.

2.5.4 Other Unsaturated Aldehydes

Complex unsaturated aldehydes are found in fruit essences and are used to provide the pleasant odors and flavors associated with cosmetic products and flavoring. These compounds tend to be easily metabolized and nontoxic but may be slightly irritating. Citronellal, for example, irritates abraded rabbit skin (137). Hypersensitivity reactions to cinnamaldehyde may occur in individual instances (33, 138, 139).

The precautions for handling reactive unsaturated aldehydes such as ketene, acrolein, methacrolein, and crotonaldehyde should be the same as those for other highly reactive eye and pulmonary irritants. Their safe handling is described elsewhere (142).

2.6 Aliphatic Dialdehydes

A number of dialdehydes have become available commercially (see Table 6.4) and although not all their properties are completely known, some toxicologic data can be summarized (Table 6.9). These materials have many of the same properties as the monoaldehydes but because of their bifunctional nature may provide different types of useful cross-linking reactions. They tend to polymerize readily and are sometime available only in a water solution in the presence of polymerization inhibitors.

Glyoxal probably has the most widespread use of these compounds. Henson (143) states that glyoxal vapors do not irritate the skin or mucous membranes. Glyoxal vapors are irritating to the eyes, although less so than formaldehyde. A 30 percent solution in water produces severe irritation of guinea pig skin and is toxic to rats (LD_{50} = 0.2 to 0.4 g/kg orally and <100 mg/kg intraperitoneally). Male rats exposed to 6000 mg/l glyoxal in drinking water displayed a decreased weight gain, but the authors concluded few other systemic effects were present (144). Water solutions of succinaldehyde, glutaraldehyde, and 3-methylglutaraldehyde also are relatively strong irritants to the skin and eyes. Their low vapor pressures, however, reduce the likelihood that inhalation would be a substantial route of exposure.

Although the extent of precautions needed under various types of industrial uses is not known with certainty, it would appear reasonable to handle these dialdehydes with the same general precautions as used for formaldehyde and other low-molecular-weight monoaldehydes.

3 ACETALS

The acetals are produced by reactions of aldehydes with alcohols (see Section 1.2.3.2). They may be utilized as solvents, chemical intermediates, and plasticizers, and may be used to generate aldehydes in the presence of acid. These materials have some of the properties of ethers and are stable under neutral or slightly alkaline

Table 6.9. Toxicity and Irritant Properties of Aliphatic Dialdehydes

LD$_{50}$			Skin LD$_{50}$		Inhalation Toxicity				Irritant Effect			Ref.
Species	Route	g/kg	Species	ml or g/kg	Species	ppm	Time (hr)	Mortality	Species	Skin	Eye	
Glyoxal (30% in H$_2$O)												
Rat	Oral	2.02	Guinea pig	5–10					Guinea pig	Severe		158
Rat	Oral	0.2–0.4	Rabbit	6.6								165
Guinea pig	Oral	0.76										
Rat	IP	<0.1										
Succinaldehyde (25% in H$_2$O)												
Rat	Oral	0.3	Rabbit	1.0	Rat	Conc. vap. (~15 mg/l)	6	0/3	Rabbit		Severe	158
									Guinea pig	Severe		
Hexa-2,4-dienal												
Rat	Oral	0.7	Rabbit	0.27	Rat	2000	4	1/6	Rabbit	Severe	Severe	165

Glutaraldehyde

Rat	Oral	0.82	Rabbit	0.64					Rabbit	Moderate	Severe	158
Glutaraldehyde Disodium Bisulfite												
Rat	Oral	1.6–3.2	Guinea pig	>20					Guinea pig	Moderate		158
Mouse	Oral	1.6–3.2										
Rat	IP	0.4–0.8										
3-Methylglutaraldehyde												
Rat	Oral	0.78	Rabbit	0.3	Rat	Conc. vap.	8	0/6	Rabbit	Severe	Severe	158
Rat	Oral	0.1–0.2	Guinea pig	>20	Rat	Conc. vap.	6	0/3	Guinea pig	Moderate	Severe	165
Rat	IP	0.005–0.010	Guinea pig (4% in H$_2$O)									
3-Methylglutaraldehyde Disodium Bisulfite												
Rat	Oral	1.6–3.2	Guinea pig	>20					Guinea pig	Moderate		158
Mouse	Oral	0.8–1.6										
Rat	IP	0.2–0.4										
α-Hydroxyadipaldehyde												
Rat	Oral	17	Rabbit	>20	Rat	Conc. vap.	8	0/6	Rabbit	Slight	Slight	165

conditions, but hydrolyze readily in the presence of acids to generate aldehydes (Table 6.5). This reaction makes them capable of hardening natural adhesives, such as glue and casein.

The hazards of the use of acetals in industry are not known with certainty. A summary of some of the results of the experimental study of these compounds is presented in Table 6.10. The physiological properties of the simple unsubstituted acetals are characterized by an ether-like anesthetic action and by a relatively low degree of primary irritation compared to the parent aldehyde.

A number of the unsubstituted acetals have been studied for their anesthetic properties, although at present they are not used for this purpose. Apparently, anesthesia can be produced in humans, but the onset is slower than with ether and the effects more transitory. The experiments of Weaver et al. (145) were concerned principally with the effects of methylal (dimethoxymethane) inhalation on guinea pigs and mice. With extremely high levels, 153,000 ppm, anesthesia occurred in 20 min and death occurred in about 2 hr. At these levels, definite evidence of irritation was noted in the guinea pig, including squinting, lacrimation, sneezing, and nasal discharge. Other pronounced signs of eye and respiratory tract irritation were also noted at lower levels. Histopathological studies on guinea pigs exposed to very high levels of methylal indicated moderate to severe fatty degeneration of the liver and kidney and extensive bronchiopneumonia.

The effect of repeated inhalations of methylal was studied in mice. The LC_{50} in mice for a 7-hr exposure was found to be about 18,000 ppm. Most of the deaths occurred during the course of exposure. A group of 50 mice received 15 7-hr exposures at concentrations of ~11,000 ppm. Only minor irritation was noted at this level, although lack of coordination appeared after about 4 hr of exposure. Recovery was usually complete 1 hr after removal from the chamber. Six deaths occurred in the 50 animals during the experiment. Mice exposed to 14,000 ppm showed more evidence of irritation and a greater degree of anesthesia. About 30 percent of the group of mice succumbed during a 17-day exposure. Histopathological studies indicated occasional pulmonary edema and slight fatty changes in the liver. No changes were found in the optic nerves or retinas of mice that could be attributed to methylal. The authors estimated that 1000 ppm might be safe for an 8-hr exposure for humans. No studies of human populations exist to verify this estimate. Safe handling should respect the flammability of methylal and maintain exposure levels below 1000 ppm. Skin contact should be avoided.

Ethylal produces only minor symptoms of weakness in rats with no signs of anesthesia. The halogenated compound dichloroethyl formal is toxic by the oral route in rats and by skin contact in guinea pig. It is also toxic by inhalation, with 100 percent fatalities at 120 ppm (no exposure periods given). The halogenated materials must obviously be handled with considerable care.

Dimethyl acetal toxicity in animal experiments is somewhat similar to methylal, without its pathological effects. The influence of unsaturation on an acetal is indicated by the high degree of intraperitoneal toxicity in the mouse for crotonaldehyde acetal (Table 6.10).

Acetal also appears to have anesthetic properties (137). Hydrolysis in the stomach would give rise to either a hemiacetal or to acetaldehyde and ethyl alcohol.

Ketoacetal is interesting in that the presence of the keto group in the beta position to the acetal grouping did not appear to enhance the toxicity. Not enough compounds have been studied to predict the effect of unsaturation on aldehyde groups adjacent to acetal groups, but from the fragmentary data available, the same tendencies should be evident as discussed for aldehydes.

In view of the lack of specific information, it would seem well to regard substituted acetals as potentially capable of being hydrolyzed to the component alcohols and aldehydes and to take precautions to avoid excessive skin contact or inhalation.

3.1 Aromatic and Heterocyclic Aldehydes

The physical and biologic properties of a number of aromatic and heterocyclic aldehydes are summarized in Tables 6.11 and 6.12. A number of these aldehydes occur naturally as components of essential oils or plant products. They are widely used in perfumes and as flavoring agents.

The literature references to the aromatic aldehydes do not give many details of the type of toxicity found. It may therefore be of interest to mention the few that have been reported.

p-Acetamidobenzaldehyde. The only symptoms noted were those of moderate weakness in rats receiving up to 3200 mg/kg orally.

p-(Dimethylamino)benzaldehyde. Weakness, ataxia, unconsciousness, and tremors were noted in mice receiving up to 1600 mg/kg orally or up to 400 mg/kg intraperitoneally. Repeated intraperitoneal injection in mice at levels of 100 to 200 mg/kg caused weakness and ataxia, but there was no reduction of hemoglobin during such treatment. No difference was found between the pure and technical-grade samples.

p-Nitrobenzaldehyde. In doses of 50 and 400 mg/kg orally in the rat, prostration and cyanosis were symptoms. 2,4-Dihydroxybenzaldehyde: Rats receiving 50 to 3200 mg/kg orally showed weakness, tremors, and violent convulsions.

2-hydroxy-5-Bromobenzaldehyde. Oral and intraperitoneal administration in mice and rats caused weakness, ataxia, and unconsciousness.

p-Tolualdehyde. Irreversible inhibition of liver and lung microsomal enzymes was observed following treatment of rats with *p*-xylene. The inhibition was attributed to *p*-tolualdehyde, because the inhibition could be partially prevented by pretreatment with pyrazole, an inhibitor of alcohol dehydrogenase (145). However, rats dosed orally daily at levels from 50 to 500 mg/kg for 13 weeks displayed no untoward effects except a decrease in the relative pituitary weight of females in the highest dose range (147). The authors concluded that the "no observable effect level," chronic exposure, was 250 mg/kg.

Furfural is extensively used in industry for the solvent refining of lubricating

Table 6.10. Toxicity and Irritant Properties of Acetals

LD50 Species	Route	g/kg	Skin LD50 Species	ml or g/kg	Inhalation Toxicity Species	ppm	Time (hr)	Mortality	Irritant Effect Species	Skin	Eye	Ref.
Methylal (Formal, Dimethoxymethane)												
Guinea pig	SC	>5			Guinea pig	150,000	2	Fatal	Guinea pig		Moderate	144
Rabbit	Oral	5.7			Mouse	18,000	7	LC50				
Ethylal (Diethoxymethane, Diethylformal)												
Rat	Oral	>3.2	Guinea pig	>10					Guinea pig		Slight	158
Rabbit	Oral	2.6										
Dichloroethylformal												
Rat	Oral	0.065	Guinea pig	0.17	Rat	120	4	6/6	Rabbit	Slight	Slight	164
					Rat	60	4	0/6				
Dimethylacetal (1,1-Dimethoxyethane)												
Rat	Oral	6.5	Rabbit	20	Rat	16,000	4	3/6	Rabbit	Slight	Moderate	166
Rabbit	Oral	4.5			Rat	3,000	4	LC50				

Acetal (Diethylacetal or 1,1-Diethoxyethane)										
Rat	Oral	4.6	Rabbit	10	Conc. vapor 4,000	5 min.	0/6	Rabbit	Slight	Slight
Rat	IP	0.9				4	2/6			
										166
										170
Dibutylacetal (1,1-Dibutoxyethane)										
Rat	Oral	3.25	Rat		Conc. vapor	8	0/6	Rabbit	Moderate	Slight
										165
Glyoxal Tetrabutyl Acetal										
Rat	Oral	8.9	Rabbit	2.24	Conc. vapor	8	0/6	Rabbit	Slight	Slight
										166
Chloroacetal (Chloroacetaldehyde Diethyl Acetal)										
Rat	Oral	0.05–0.4	Guinea pig	<10 (loss of weight)				Guinea pig	Severe	
										158
Keto Acetal (4,4-Dimethoxy-2-butanone)										
Mouse	Oral	1.6–3.2	Guinea pig	>5				Guinea pig	Slight	
Mouse	IP	3.6–6.4								
										158
Chloral Hydrate (Trichloroacetaldehyde Monohydrate)										
Rat	Oral	0.8								
Rabbit	Oral	1.3 (LD_{100})								
Cat	Oral	0.5 (LD_{25})								
Dog	Oral	1.1 (LD_{80})								
										160
										174

Table 6.11. Physical and Chemical Properties of Aromatic and Heterocyclic Aldehydes

Compound	Formula	Mol. Wt.	Specific gravity	M.P. (°C)	B.P. (°C)	Vapor Pressure (mm Hg) (°C)	Vapor Density (Air = 1)	Flash Point (°F)	Solubility in H_2O (g/100 ml)	1 mg/l (ppm)	1 ppm (mg/m³)
Benzaldehyde	C_6H_5CHO	106.1	1.046	−26	178	1 (26) 40 (90)	3.66	165	Sl. sol.	231	4.3
p-Hydroxybenzaldehyde	HOC_6H_4CHO	122.1		116	Subl.	1 (121) 10 (170)	4.2		1.38	200	5.0
o-Hydroxybenzaldehyde (salicylaldehyde)	HOC_6H_4CHO	122.1	1.153	−7	197	1 (33) 10 (74)	4.2		Sl. sol.	200	5.0
p-Tolualdehyde (p-methylbenzaldehyde)	$CH_3C_6H_4CHO$	120.1	1.020		204		4.1		Sl. sol.	204	4.0
p-Methoxybenzaldehyde (anisaldehyde)	$CH_3OC_6H_4CHO$	136.1	1.119	0	248	1 (73) 10 (118)	4.7		Sl. sol.	180	5.6
p-Aminobenzaldehyde	$NH_2C_6H_4CHO$	121.1		71.5			4.2		Sol.	202	5.0
3,4-Dihydroxybenzaldehyde (protocatechucaldehyde)	$(HO)_2C_6H_3CHO$	138.1		153			4.8		5	177	5.6
3-Methoxy-4-hydroxybenzaldehyde (vanillin)	$(CH_3O)C_6H_3(OH)CHO$	152.1	1.056	82	285	1 (107) 10 (154)	5.2		1	161	6.2
3,4-Dimethoxybenzaldehyde (Veratric aldehyde)	$(CH_3O)_2C_6H_3CHO$	166.2		42	281	10 (155)	5.7		Sl. sol.	147	6.8
Piperonal	$3,4\text{-}OCH_2OC_6H_3CHO$	150.1		36	263	1 (87) 10 (132)	5.2		Sl. sol.	163	6.1
Furfural (2-Furfuraldehyde)	$OCH=CHCH=CCHO$	96.1	1.1563	−37	162	1 (19) 15 (60)	3.31	155	8.3	255	3.9
Phenylacetaldehyde (α-toluicaldehyde)	$C_6H_5CH_2CHO$	120.2	1.023	33	195	10 (78)	4.1		Sl. sol.	204	4.9
Cinnamaldehyde (phenylacrolein)	$C_6H_5CH=CHCHO$	132.2	1.048	−8	246	1 (76) 10 (120)	4.6		Sl. sol.	185	5.4

Table 6.12. Toxicity and Irritant Properties of Aromatic and Heterocyclic Aldehydes

Compound	LD$_{50}$			Skin LD$_{50}$		Irritant Effect			Ref.
	Species	Route	g/kg	Species	ml or g/kg	Species	Skin	Eye	
Benzaldehyde	Rabbit	SC	5.0						160
Salicylaldehyde (2-hydroxybenzaldehyde)	Rat	SC	0.9						
p-Aminobenzaldehyde	Mouse	IP	0.9						158
p-Acetamidobenzaldehyde	Rat	Oral	>3.2	Guinea pig	>1.0	Guinea pig	Slight		
p-(Dimethylamino)benzaldehyde	Mouse	Oral	0.8–1.6						160
	Mouse	IP	0.2–0.4						
p-n-Propylbenzaldehyde	Rat	Oral	4.2						160
	Mouse	Oral	1.8						
p-(n-Propoxy)benzaldehyde	Rat	Oral	1.6	Rabbit	9.0				160
	Mouse	Oral	1.8						
m-Nitrobenzaldehyde	Rat	Oral	0.05–0.4	Guinea pig	>1.0	Guinea pig	Moderate		175
	Mouse	IP	>0.5						
2,4-Dihydroxybenzaldehyde	Rat	Oral	0.4–3.2	Guinea pig	>1.0	Guinea pig	Slight		158
	Mouse	IP	>0.5						
2,5-Dimethoxybenzaldehyde									176
2-Hydroxy-5-chlorobenzaldehyde	Rat	Oral	0.8–1.6	Guinea pig	10–20	Guinea pig	Moderate		158
	Rat	IP	0.1–0.2						
2-Hydroxy-5-bromobenzaldehyde	Rat	Oral	0.8–1.6	Guinea pig	>20	Guinea pig	Slight		158
	Rat	IP	0.2–0.4						
p-Hexoxybenzaldehyde	Mouse	IP	>0.5						
Piperonal	Mouse	IP	>0.5						177
Cinnamaldehyde	Mouse	IP	0.2						
Furfural	Rat	Oral	0.05–0.1	Guinea pig	<10	Guinea pig	Slight		178
	Rat	IP	0.02–0.05						158
	Mouse	Oral	0.5						
	Dog	Oral	0.65						152

319

oils, resins, and other organic materials. It is also used in connection with rubber manufacturing, is a constituent of some insecticidal preparations and rubber cements, and is widely used as a reagent. The toxicity of furfural has been summarized in the AGCIH Biological Response Indices (20). Furfural can be derived from pentosans present in straws and bran by hydrolysis and dehydration with sulfuric acid. Although the vapor is quite irritating, it has relatively low volatility (1 mm Hg at 19°C) so that the inhalation of toxicologically significant quantities at room temperature is unlikely. It is a relatively strong skin irritant and may be capable of producing dermatitis in humans. The hazard of handling furfural resins is similar to that of handling other aldehyde resins.

In animals, furfural has a relatively high degree of toxicity. The oral LD_{50} in the rat is 149 mg/kg (148). The symptoms upon acute exposure are weakness, ataxia, and unconsciousness. No signs of central nervous system stimulation were seen in dogs, cats, or rabbits. The inhalation exposure of cats to very high levels of furfural (2800 ppm) for half an hour resulted in death due to pulmonary edema (149).

Korenman and Resnik (150) state that levels of 1.9 to 14 ppm caused eye and throat irritation and headache. Balance measurements of excretions versus amount of inspired material indicate that, in humans, cutaneous absorption routes can be important (151). The dermal LD_{50} of furfural in rabbits is between 500 and 1000 mg/kg (occluded exposure) (152).

Furfural is metabolized in rats, dogs, and rabbits to furoic acid, furoyl glycine, and furanacrylic acid, and is then excreted (153, 154). In humans, the chief excretory product in the urine is furoyl glycine, with slight amounts of furanacrylic acid, leading to the suggestion that furoic acid may be followed as a biologic indicator of exposure (20). Little is known about the metabolism of furfural derivatives, although 5-nitrofurfural is excreted in rats as 5-nitro-2-furoic acid.

Although no studies on the health of workmen exposed over long periods of time have been published, Dunlop and Peters (155) state that no injury due to furfural exposure, other than an occasional allergic skin manifestation, has been reported. In hamsters, furfural, in conjunction with benzo[a]pyrene, was tested as a co-carcinogen, with negative results (51). However, furfural was positive in a mouse carcinogenesis skin test model, indicating activity as an initiator (156) and raising concern about the risk of furfural as a risk factor in the diet (133). The current exposure limit (5 ppm) is based on the primary irritant effects at this concentration.

The precautions in handling furfural should include adequate ventilation and provision of skin and eye protection. In case of contact with skin and eyes, the chemical should be removed promptly by copious washing with water. Although no unusual medical examinations are indicated, individuals with demonstrated tendencies toward contact dermatitis should avoid contact with furfural.

REFERENCES

1. *Chem. Eng. News*, **70**, 34 (1992).
2. R. Brettle, "Aldehydes," in *Comprehensive Organic Chemistry—The Synthesis and*

Reactions of Organic Compounds, J. F. Stoddart, Ed., Pergamon Press, New York, 1979, pp. 943–1015.

3. R. E. McMahon, "Aldehydes, Alcohols and Ketones," in *Metabolic Basis of Detoxification*, W. D. Jakoby, I. R. Bend, J. Caldwell, Eds., Academic Press, New York, 1983, pp. 91–104.

4. *Formaldehyde and Other Aldehydes*, National Academy Press, Washington, DC, 1981.

5. "Criteria for a Recommended Standard . . . Occupational Exposure to Formaldehyde," DHEW (NIOSH) Publication No. 77-126, 1976.

6. S. E. Feinman, "Chemistry of Formaldehyde," in *Formaldehyde Sensitivity and Toxicology*, Susan E. Feinman, Ed., CRC Press, Boca Raton, FL, 1988, pp. 3–15.

7. H. J. Hageman, "Acetaldehyde," in *Encyclopedia of Chemical Technology*, Vol. 1, 4th ed., Mary Howe-Grant, Ed., Wiley-Interscience, New York, 1991, pp. 94–109.

8. W. G. Etzkorn, J. J. Kurland, and W. D. Neilsen, "Acrolein," in *Encyclopedia of Chemical Technology*, Vol. 1, 4th ed., Mary Howe-Grant, Ed., Wiley-Interscience, New York, 1991, pp. 232–251.

9. B. Das and D. F. Church, *Chem. Res. Toxicol.*, **4**, 341 (1991).

10. *OSHA Analytic Methods Manual*, Vol. 1, OSHA Analytic Laboratory, Salt Lake City, 1985.

11. F. P. Snell and C. L. Hilton, Eds., *Encyclopedia of Industrial Chemical Analysis*, Vol. 4, Interscience, New York, 1967.

12. G. R. Umbreit, "Carbonyl Compounds," in F. D. Snell and L. S. Ettre, Eds., *Encyclopedia of Industrial Chemical Analysis*, Vol. 8, Wiley-Interscience, New York, 1969.

13. E. Sawicki and C. R. Sawicki, *Aldehydes–Photometric Analysis*, Vols. 1–4, Academic Press, New York, 1976.

14. P. L. Haust, "Spectrophotometric Methods for Air Pollution Measurements," in *Advances in Environmental Science and Technology*, J. N. Pitts and R. L. Metcalf, Eds., Vol. 2, Wiley-Interscience, New York, 1971, pp. 91–213.

15. H. D. Heck, M. Casanova, and T. B. Starr, *Crit. Rev. Toxicol.*, **26**(6), 397 (1990).

16. P. Restani and C. L. Galli, *Crit. Rev. Toxicol.*, **21**(5), 315 (1991).

17. J. E. Gibson, Ed., *Formaldehyde Toxicity*, Hemisphere Publishing Corp., Washington, DC, 1983.

18. V. J. Feron, H. P. Til, F. de Vrijer, R. A. Woutersen, F. R. Cassee, and P. J. van Bladeren, *Mutat. Res.*, **259**(3–4), 363 (1991).

19. R. O. Beauchamp, D. A. Andjelkovich, A. D. Kligerman, K. T. Morgan, and H. D. Heck, *Crit. Rev. Toxicol.*, **14**(4), 309 (1985).

20. *Documentation of the Biological Exposure Indices*, 6th ed., American Conference of Government and Industrial Hygienists, Inc., Cincinnati, OH, 1991.

21. P. L. J. Opdyke, *Food Cosmet. Toxicol.*, **11**, 95 (1973).

22. C. Auerbach, M. Moutschen-Dahman, and J. Moutschen, *Mutat. Res.*, **39**, 317 (1977).

23. C. Izard and C. Libermann, *Mutat. Res.*, **47**, 115 (1978).

24. T. Neudecker, D. Lutz, E. Eder, and D. Henschler, *Mutat. Res.*, **91**(1), 27 (1981).

25. O. Sterner, R. E. Carter, and L. M. Nilsson, *Mutat. Res.*, **188**(3), 169 (1987).

26. J. B. Ward, J. A. Hokanson, E. R. Smith, L. W. Chang, M. A. Pereira, E. B. Whorton, and M. S. Legator, *Mutat. Res.*, **130**(6), 417 (1984).

27. T. H. Connor, J. B. Ward, and M. S. Legator, *Int. Arch. Occup. Environ. Health*, **56**(3), 225 (1985).

28. E. J. Thomson, S. Shackleton, and J. M. Harrington, *Mutat. Res.*, **141**(2), 89 (1984).

29. J. L. Cicmanec, L. W. Condie, G. R. Olson, and S.-R. Wang, *Fundam. Appl. Toxicol.*, **17**(2), 376 (1991).

30. S. E. O'Quinn and C. B. Kennedy, *J. Am. Med. Assoc.*, **194**, 123 (1965).

31. W. S. Logan and H. O. Perry, *Arch. Derm.*, **106**, 717 (1972).

32. C. D. Calnan, *Contact Dermatitis*, **2**(3), 167 (1976).

33. C. G. Mathias, R. R. Chappler, and H. I. Maibach, *Arch. Dermatol.*, **116**(1), 74 (1980).

34. C. G. Mathias, H. I. Maibach, and M. A. Conant, *Arch. Dermatol.*, **116**(10), 1172 (1980).

35. D. Norback, *Scand. J. Work Environ. Health*, **14**(6), 366 (1988).

36. J. R. Fowler, *J. Occup. Med.*, **31**(10), 852 (1989).

37. O. J. Corrado, J. Osman, and R. J. Davies, *Hum. Toxicol.*, **5**(5), 325 (1986).

38. P. Wiggins, S. A. McCurdy, and W. Zeidenberg, *J. Occup. Med.*, **31**(10), 854 (1989).

39. J. B. Schoenberg and C. A. Mitchell, *Arch. Environ. Health*, **30**, 574 (1975).

40. D. J. Hendrick and D. J. Lane, *Brit. J. Ind. Med.*, **34**, 11 (1977).

41. H. Harving, J. Korsgaard, O. F. Pedersen, L. Molhave, and R. Dahl, *Lung*, **108**(1), 15 (1990).

42. A. Ortiz, P. J. Griffiths, and J. M. Littleton, *J. Pharm. Pharmacol.*, **26**, 249 (1974).

43. J. F. Wilkinson, *J. Hyg.*, **40**, 555 (1940).

44. K. H. Kilburn, B. C. Seidman, and R. Warshaw, *Arch. Environ. Health*, **40**(4), 229 (1985).

45. M. Umeda, *Gann*, **47**, 153 (1956).

46. M. H. Salaman and F. J. C. Roe, *Brit. J. Cancer*, **10**, 70 (1956).

47. F. Watanabe, *Gann*, **45**, 45 (1954).

48. A. W. Horton, R. Tye, and K. L. Stemmer, *J. Natl. Cancer Inst.*, **30**, 31 (1963).

49. H. P. Til, V. J. Feron, and A. P. Degroot, *Food Cosmet. Toxicol.*, **10**, 291 (1972).

50. W. Straks and V. J. Feron, *Eur. J. Cancer*, **9**, 359 (1972).

51. V. J. Feron and A. Kruysse, *Toxicology*, **11**, 127 (1978).

52. B. L. Van Duuren, A. Sivak, B. M. Goldschmidt, C. Katz, and S. Melchionne, *J. Natl. Can. Inst.*, **45**, 481 (1969).

53. R. Schoental and S. Gibbard, *Brit. J. Cancer*, **26**, 504 (1972).

54. R. A. Woutersen, L. M. Appelman, V. J. Feron, and C. A. Van Der Heijden, *Toxicology*, **31**, 123 (1984).

55. V. J. Feron, A. Kruysse, and R. A. Woutersen, *Eur. J. Cancer Clin. Oncol.*, **18**, 13 (1982).

56. C.-W. Lam, M. Casanova, and H. D. Heck, *Fundam. Appl. Toxicol.*, **6**(3), 541 (1986).

57. *Formaldehyde.* IARC Monograph Evaluation of Carcinogenic Risk Chem Human, **29**, 345 (1982).

58. R. A. Squire and L. L. Cameron, *Regul. Toxicol. Pharmacol.*, **4**(2), 107 (1984).

59. M. Casanova, K. T. Morgan, W. H. Steinhagen, J. I. Everitt, J. A. Popp, and H. D. Heck, *Fundam. Appl. Toxicol.*, **17**(2), 409 (1988).

60. V. J. Feron, J. P. Bruyntjes, R. A. Woutersen, H. R. Immel, and L. M. Appelman, *Cancer Lett.*, **39**(1), 101 (1988).

61. J. A. Swenberg, W. D. Kerns, R. I. Mitchell, E. J. Gralland, and K. L. Pavkov, *Cancer Res.*, **40**(9), 3398 (1980).

62. W. D. Kerns, K. L. Pavkov, D. J. Donofrio, E. J. Gralla, and J. A. Swenberg, *Cancer Res.*, **43**, 4382 (1983).

63. G. M. Rusch, J. J. Clary, W. E. Rinehart, and H. F. Bolte, *Toxicol. Appl. Pharmacol.*, **68**(3), 329 (1983).

64. J. W. Wilmer, R. A. Woutersen, L. M. Appelman, W. R. Leeman, and V. J. Feron, *Toxicol. Lett.*, **47**(3), 287 (1989).

65. Banbury Report 19: "Risk Quantitation and Regulatory Policy," D. G. Hoel, R. A. Merrill, and F. P. Perera, Eds., Cold Spring Harbor Laboratory, 1985.

66. J. H. Olsen, S. P. Jensen, M. Hink, K. Faurbo, N. O. Breum, and O. M. Jensen, *Int. J. Cancer*, **34**(5), 639 (1984).

67. R. B. Hayes, J. W. Raatgever, A de Bruyn, and M. Gerin, *Int. J. Cancer*, **37**(4), 487 (1986).

68. G. C. Roush, J. Walrath, L. T. Stayner, S. A. Kaplan, J. T. Flannery, and A. Blair, *J. Natl. Cancer Inst.*, **79**(6), 1221 (1987).

69. A. Blair, P. A. Stewart, M. O'Berg, W. Gaffey, J. Walrath, J. Ward, R. Bales, S. Kaplan, and D. Cubit, *J. Natl. Cancer Inst.*, **76**, 1071 (1986).

70. A. Blair, P. A. Stewart, R. N. Hoover, J. F. Fraumeni, J. Walrath, M. O'Berg, and W. Gaffey, *J. Natl. Cancer Inst.*, **78**, 191 (1987).

71. S. E. Feinman, "Carcinogenicity," in *Formaldehyde Sensitivity and Toxicity*, S. E. Feinman, Ed., CRC Press, Boca Raton, FL, 1988, pp. 179–196.

72. "Epidemiology of Chronic Occupational Exposure to Formaldehyde: Report of the Ad Hoc Panel on Health Aspects of Formaldehyde," Universities Associated for Research and Education in Pathology, Inc. *Toxicol. Ind. Health*, **4**(1), 77 (1988).

73. B. A. Owen, C. S. Dudney, E. L. Tan, and C. E. Easterly, *Regul. Toxicol. Pharmacol.*, **11**(3), 220 (1990).

74. G. Bittersohl, "Epidemiological Research on Cancer Risk by Aldol and Aliphatic Aldehydes," in *Environmental Quality and Safety, Global Aspects of Chemistry, Toxicology and Technology as Applied to the Environment*, Vol. 4, F. Coulson and F. Korte, Eds., Academic Press, New York, 1975, pp. 235–238.

75. L. Uotila and M. Koivusalo, *J. Biol. Chem.*, **249**, 7653 (1974).

76. J. H. Eckfeldt and Y. Yonetani, "Isozymes of Aldehyde Dehydrogenase from Horse Liver," in *Methods in Enzymology*, W. A. Wood, Ed., Vol. 89, Academic Press, New York, 1982, pp. 474–479.

77. M. Casanova-Schmitz, R. M. David, and H. D. Heck, *Biochem. Pharmacol.*, **33**(7), 1137–1142 (1984).

78. T. Pourmotabbed and D. J. Creighton, *J. Biol. Chem.*, **261**(30), 14240 (1986).

79. K. Inoue, U. Ohbora, and K. Yamasama, *J. Neurochem.*, **19**, 273 (1972).

80. J. P. Wartburg and B. Wermuth, "Aldehyde Reductase from Human Tissue," in *Methods in Enzymology*, W. A. Wood, Ed., Vol. 89, Academic Press, New York, 1982, pp. 506–513.

81. T. G. Flynn, J. Shires, and D. J. Walton, *J. Biol. Chem.*, **260**(8), 2933 (1975).

82. M. Kouri, T. Loivula, and M. Koivusalo, *Acta Pharmacol. Toxicol.*, **40**, 460 (1977).

83. T. Loivula and M. Koivusalo, *Biochim. Biophys. Acta*, **410**, 1 (1975).

84. R. Lindahl and S. Evces. *J. Biol. Chem.*, **259**(19), 11986 (1984).

85. R. Lindahl and S. Evces, *J. Biol. Chem.*, **259**(19), 11991 (1984).

86. M. Ikawa, C. C. Impraim, G. Wang, and A. Yoshida, *J. Biol. Chem.*, **258**(10), 6282 (1983).

87. R. I. Feldman and H. Weiner, *J. Biol. Chem.*, **247**(1), 260 (1972).

88. A. H. Blair and F. H. Bodley, *Can. J. Biochem.*, **47**, 265 (1965).

89. J. Prunonosa, M. L. Sagrista, and J. Bozal, *Drug Metab. Disp.*, **19**(4), 787, (1991).

90. D. Y. Mitchell and D. R. Petersen, *Drug Metab. Disp.*, **16**(1), 37 (1988).

91. C. S. Boyer and D. R. Petersen, *Drug Metab. Disp.*, **19**(1), 81 (1991).

92. L. E. Rikans, *Drug Metab. Disp.*, **15**(3), 356 (1987).

93. J. J. Diliberto, P. Srinivas, D. Overstreet, G. Usha, L. T. Burka, and L. S. Birnbaum, *Drug Metab. Disp.*, **18**(6), 866 (1990).

94. S. Yoshihara and K. Tatsumi, *Drug Metab. Disp.*, **18**(6), 876 (1990).

95. A. J. Sherratt and L. A. Damani, *Drug Metab. Disp.*, **17**(1), 20 (1989).

96. D. Y. Mitchell and D. R. Petersen, *Toxicol. Appl. Pharmacol.*, **87**(3), 169 (1985).

97. D. E. Vance, "Catabolism of Fatty Acid," G. Zubay, *Biochemistry*, 2nd ed., Macmillan, New York, 1988, pp. 598–614.

98. C. M. Kaye, *Biochem. J.*, **134**, 1093 (1973).

99. M. Robinson, R. J. Bull, G. R. Olson, and J. Stober, *Cancer Lett.*, **48**(3), 197 (1989).

100. R. J. Jaeger, *Res. Commun. Chem. Path. Pharmacol.*, **18**, 83 (1977).

101. D. P. Jones, H. Thor, B. Andersson, and S. Orrenius, *J. Biol. Chem.*, **53**(17), 6031 (1978).

102. M. Casanova and H. D. Heck, *Toxicol. Appl. Pharmacol.*, **89**(1), 105 (1987).

103. R. P. Mason and J. K. M. Sanders, *Biochemistry*, **25**, 4504 (1986).

104. S. Naylor, R. P. Mason, J. K. M. Sanders, D. H. Williams, and G. Moneti, *Biochem. J.*, **249**, 573 (1988).

105. K. H. Moon and M. Y. Pack, *Drug Chem. Toxicol.*, **6**(6), 521 (1983).

106. J. Ohno, K. Ormstad, D. Ross, and S. Orrenius, *Toxicol. Appl. Pharmacol.*, **78**(2), 169 (1985).

107. V. L. Slott and B. F. Hales, *Biochem. Pharmacol.*, **36**(12), 2019 (1987).

108. V. L. Slott and B. F. Hales, *Biochem. Pharmacol.*, **36**(13), 2187 (1987).

109. H. Sprince, C. M. Parker, G. G. Smith, and L. J. Gonzales, *Agents, Actions*, **5**, 164 (1975).

110. C. Guerri, W. Godfey, and S. Grisolia, *Physiol. Chem. Phys.*, **8**, 543 (1976).

111. A. Perin, A. Sessa, and E. Ciaranfi, *Cancer Res.*, **38**, 2180 (1978).

112. L. A. Levison, *J. Am. Med. Assoc.*, **42**, 1492 (1904).

113. J. A. Porter, *Lancet*, **1975-II**, 603 (1975).

114. H. Poole, *Formaldehyde Survey*, American Textile Manufacturers Institute, Charlotte, NC, 1984.

115. G. Houding, *Acta Derm. Venerol.*, **41**, 194 (1961).

116. L. G. Wayne, R. J. Bryan, and K. Ziedman, *Irritant Effects of Industrial Chemicals: Formaldehyde*, DHEW (NIOSH) Publ. No. 77-117, 1976.

117. J. C. Tou and G. J. Kallos, *Am. Ind. Hyg. Assoc. J.*, **35**, 419 (1974).

118. G. J. Kallos and R. A. Solomen, *Am. Ind. Hyg. Assoc. J.*, **34**, 469 (1974).

119. "Formaldehyde," Hygienic Guide Series, *Am. Ind. Hyg. Assoc. J.*, **26**, 189 (1965).

120. E. J. Kerfoot and T. F. Mooney, *Am. Ind. Hyg. Assoc. J.*, **36**, 533 (1975).

121. T. Dalhamn and R. Rylander, *Acta Pharmacol. Toxicol.*, **25**, 369 (1967).

122. R. Rylander, *Am. Rev. Respir. Dis.*, **108**, 1279 (1973).

123. E. Skog, *Acta Pharmacol. Toxicol.*, **6**, 299 (1950).

124. H. Salem and H. Cullumbine, *Toxicol. Appl. Pharmacol.*, **2**, 183 (1960).

125. J. C. Gage, *Brit. J. Ind. Med.*, **27**, 1 (1970).

126. J. Moutschen-Dahmen, M. Moutschen-Dahmen, N. Houbrechts, and A. Colizzi, *Bull. Soc. Roy. Sci. Leige*, **45**, 58 (1976).

127. M. K. Johnson, *Biochem. Pharmacol.*, **16**, 185 (1967).

128. F. D. Hayes, R. D. Short, and J. E. Gibson, *Toxicol. Appl. Pharmacol.*, **26**, 93 (1973).

129. A. Kdorhonen, K. Hemminki, and H. Vainio, *Acta Pharmacol. Toxicol.*, **52**(2), 95 (1983).

130. G. Chhibber and S. H. Gilani, *Environ. Res.*, **39**(1), 44 (1986).

131. V. L. Slott and B. F. Hales, *Teratology*, **34**(2), 155 (1986).

132. B. P. Schmid, E. Goulding, K. Kitchin, and M. K. Sanyal, *Toxicology*, **22**(3), 235 (1981).

133. R. A. Parent, H .E. Caravello, and A. M. Hoberman, *Fundam. Appl. Toxicol.*, **19**(2), 228 (1992).

134. W. E. Rinehart, *Am. Ind. Hyg. Assoc. J.*, **28**, 561 (1967).

135. G. R. Cameron and A. Neuberger, *J. Pathol. Bacteriol.*, **45**, 653 (1937).

136. J. F. Treon, H. E. Sigmon, K. V. Kitzmiller, F. F. Heyroth, W. J. Younker, and J. Cholak, *J. Ind. Hyg. Toxicol.*, **31**, 209 (1949).

137. P. L. J. Opdyke, *Food Cosmet. Toxicol.*, Suppl 13, 685 (1975).

138. T. E. Drake and H. I. Maibach, *Arch. Dermatol.*, **112**(2), 202 (1976).

139. P. J. Danneman, K. A. Booman, J. Dorsky, K. A. Hohrman, A. S. Rothenstein, R. I. Sedlak, R. J. Steltenkamp, and G. R. Thompson, *Food Chem. Toxicol.*, **21**(6), 721 (1983).

140. J. A. Hoskins, *J. Appl. Toxicol.*, **4**(6), 283 (1984).

141. G. Witz, *Free Rad. Biol. Med.*, **7**(3), 333 (1989).

142. N. I. Sax, *Dangerous Properties of Industrial Materials*, 7th. ed., Van Nostrand Reinhold, New York, 1989.

143. E. V. Henson, *J. Occup. Med.*, **1**, 457 (1959).

144. H. Ueno, T. Segawa. T. Hasegawa, K. Nakamuro, H. Maeda, Y. Hiramatsu, S. Okada, and Y. Sayato, *Fundam. Appl. Toxicol.*, **16**(4), 763 (1991).

145. F. L. Weaver, A. R. Hough, B. Highman, and L. T. Fairhall, *Brit. J. Ind. Med.*, **8**, 279 (1951).

146. J. M. Patel, C. Harper, and R. T. Drew, *Drug Metab. Disp.*, **6**, 368 (1978).

147. P. G. Branton, I. F. Gaunt, P. Grasso, A. B. Lansdown, and S. D. Gangolli, *Food Cosmet. Toxicol.*, **10**, 637 (1972).

148. W. J. McKillip and E. Sherman, "Furan Derivatives," in Kirk-Othmer, *Encyclopedia of Chemical Technology*, Vol. 11, 3rd ed., Wiley, New York, 1980, pp. 499–527.

149. *API Toxicological Review, Furfural*, American Petroleum Institute, Dept. Safety, New York, 1948.

150. J. Korenman and J. B. Resnick, *Arch. Hyg.*, **104**, 344 (1930).

151. J. Flek and V. Sedivec, *Int. Arch. Occup. Environ. Health*, **41**, 159 (1978).

152. L. A. Wood and M. H. Seevers, unpublished reports to The Quaker Oats Company, through W. J. McKillip and E. Sherman, "Furan Derivatives," in Kirk-Othmer, *Encyclopedia of Chemical Technology*, Vol. 11, 3rd ed., Wiley, New York, 1980, pp. 499–527.

153. R. T. Williams, *Detoxification Mechanisms*, 2nd ed., Wiley, New York, 1959.

154. A. A. Nomeir, D. M. Silveira, M. F. McComish, and M. Chadwick, *Drug Metab. Disp.*, **20**(2), 198 (1992).

155. A. P. Dunlop and F. N. Peters, *Furans*, Reinhold, New York, 1953.

156. Y. Miyakawa, Y. Nishi, K. Kato, H. Sata, M. Takahasi, and Y. Hayashi, *Carcinogenesis*, **12**(7), 1169 (1991).

157. C. P. Carpenter, H. F. Smyth, and U. C. Pozzani, *J. Ind. Hyg. Toxicol.*, **31**, 343 (1949).

158. D. W. Fassett, "Aldehydes and Acetals," in *Industrial Hygiene and Toxicology*, D. W. Fassett and D. D. Irish, Eds., Vol. 2, 2nd ed., Wiley-Interscience, New York, 1962.

159. *Summary Tables of Biological Tests*, **8**, 743 (1956).

160. *Handbook of Toxicology*, Vol. I, W. Spector, Ed., Saunders, Philadelphia, 1956.

161. H. F. Smyth, C. P. Carpenter, and C. S. Weil, *A.M.A. Arch. Ind. Health Occup. Med.*, **4**, 119 (1951).

162. N. Iwanoff, *Arch. Hyg.*, **73**, 307 (1910–1911); through *Industrial Hygiene and Toxicology*, D. W. Fassett and D. D. Irish, Eds., Vol. 2, 2nd ed., Wiley-Interscience, New York, 1962.

163. F. L. M. Pattison, *Toxic Aliphatic Fluorine Compounds*, Elsevier, Princeton, NJ, 1959.

164. H. F. Smyth and C. P. Carpenter, *J. Ind. Hyg. Toxicol.*, **30**, 63 (1948).

165. H. F. Smyth, C. P. Carpenter, C. S. Weil, and U. C. Pozzani, *A.M.A. Arch. Ind. Health Occup. Med.*, **10**, 61 (1954).

166. H. F. Smyth, C. P. Carpenter, and C. S. Weil, *J. Ind. Hyg. Toxicol.*, **31**, 60 (1949).

167. E. Boyland, *Biochem. J.*, **34**, 1196 (1940).

168. H. A. Wooster, C. C. Lushbaugh, and C. E. Redemann, *J. Ind. Hyg. Toxicol.*, **29**, 56 (1947).

169. H. F. Smyth and C. P. Carpenter, *J. Ind. Hyg. Toxicol.*, **26**, 269 (1944).

170. I. F. Gaunt, J. Colley, M. Wright, M. Creasy, and P. Grasso, *Food Cosmet. Toxicol.*, **9**, 755 (1971).

171. H. F. Smyth, in *Glycols*, ACS Monograph No. 114, G. Curme and F. Johnston, Eds., Reinhold, New York, 1952.

172. B. Ballantyne, D. E. Dodd, I. M. Pritts, D. J. Nachreiner, and E. H. Fowler, *Hum. Toxicol.*, **8**(3), 229 (1989).

173. P. Knoefel, *J. Pharmacol.*, **50**, 88 (1934).

174. J. Adams, *J. Pharmacol. Exp. Therap.*, **78**, 340 (1943).

175. J. H. Draize, E. Alvarez, and M. F. Whitesell, *J. Pharmacol. Exp. Therap.*, **93**, 26 (1968).

176. *Summary Tables of Biological Tests*, **6** (1954).

177. Sloan-Kettering Institute Screening Data, New York, through D. W. Fassett, "Aldehydes and Acetals," in *Industrial Hygiene and Toxicology*, D. W. Fassett and D. D. Irish, Eds., Vol. 2, 2nd ed., Wiley-Interscience, New York, 1962.

178. *Summary Tables of Biological Tests*, **7** (1955).

179. D. J. Miller, "Aldehydes," in *Encyclopedia of Chemical Technology*, Vol. 1, 4th ed., M. Howe-Grant, Ed., Wiley-Interscience, New York, 1991, pp. 926–937.

CHAPTER SEVEN

Epoxy Compounds

Thomas H. Gardiner, Ph.D., John M. Waechter, Jr., Ph.D., D.A.B.T., and Donald E. Stevenson, Ph.D.

1 GENERAL CONSIDERATIONS

1.1 Chemistry

An epoxy compound is defined as any compound containing one or more oxirane rings. An oxirane ring (epoxide) consists of an oxygen atom linked to two adjacent (vicinal) carbon atoms as follows:

$$\text{>C}\underset{\diagdown\;\diagup}{\overset{O}{-}}\text{C<}$$

The term alpha-epoxide is sometimes used for this structure to distinguish it from rings containing more carbon atoms. The alpha does not indicate where in a carbon chain the oxirane ring occurs.

The oxirane ring is highly strained and is thus the most reactive ring of the oxacyclic carbon compounds. The strain is sufficient to force the four carbon atoms nearest the oxygen atom in 1,2-epoxycyclohexane into a common plane, whereas in cyclohexane the carbon atoms are in a zigzag arrangement or boat structure (1). As a result of this strain, epoxy compounds are attacked by almost all nucleophilic substances to open the ring and form addition compounds. For example,

$$RNH_2 + \overset{O}{\overset{\diagup\diagdown}{CH_2 - CH - R_1}} \longrightarrow R - NHCH_2 - \overset{OH}{\overset{|}{CHR_1}}$$

Patty's Industrial Hygiene and Toxicology, Fourth Edition, Volume 2, Part A, Edited by George D. Clayton and Florence E. Clayton.
ISBN 0-471-54724-7 © 1993 John Wiley & Sons, Inc.

Among agents attacking epoxy compounds are halogen acids, thiosulfate, carboxylic acids, hydrogen cyanide, water, amines, aldehydes, and alcohols.

1.2 Sources and Uses

The epoxides are prepared from parent unsaturated hydrocarbons, and most often by low-temperature dehydrohalogenation of their halohydrins with bases. Another method of choice is the direct epoxidation of unsaturated hydrocarbons either with preformed peroxy acids or, where possible, with hydrogen peroxide and an acid (usually a carboxylic acid), the peroxy acid being formed in situ.

A third commonly used method of synthesis is the reaction of epichlorohydrin with an active hydrogen compound to form a halohydrin, which is then followed by a basic dehydrohalogenation to generate another epoxide group (2):

$$RH + H_2C \overset{}{\underset{O}{-}} CH-CH_2-Cl \longrightarrow RCH_2\overset{OH}{\overset{|}{C}}HCH_2Cl \overset{OH}{\longrightarrow} RCH_2CH \overset{O}{-} CH_2$$

Most of the epoxy compounds encountered in industrial practice today are the uncured epoxy resins. They are marketed in a variety of physical forms from low-viscosity liquids to tack-free solids and require admixture with curing agents to form the desired hard and infusible cross-linked polymers. These are in demand because of their toughness, high adhesive properties (polarity), low shrinkage in molds, and chemical inertness.

It is the uncured resins that are of main interest to toxicologists, for a well-cured resin should have few or no unreacted epoxide groups remaining in it. The toxicology of the curing agents is not treated in this chapter. They are most frequently bi- or trifunctional amines, di- or tricarboxylic acids and their anhydrides, polyols, and compounds containing mixed functional groups, such as aminols and amino acids, as well as other resins containing such groups (2).

Two monomeric epoxy compounds find use as medical sterilants but the main use of the monomers is as reactive diluents in resin mixtures. Monomerics are also used as intermediates in chemical synthesis; for example, reaction with water produces glycols, and with amines, alkanolamines. If less than equivalence of either of these reagents is used, polymeric glycol or aminol ethers result. These addition reactions must be well controlled to prevent polymer formation or, at the limit, explosions.

Resins have found application as protective coatings, adhesives for most substrates (metals included), caulking compounds, flooring and special road paving, potting and encapsulation resins, low-pressure molding mixtures, and binding agents for fiber glass products. Uncured, they are used as plasticizers and stabilizers for vinyl resins.

Epoxy resin coating formulations can generally be limited to one of three forms: solution coatings, high-solids formulations, and epoxy powder coatings. Solid epoxies are used in coating applications as solid solutions or heat converted coating.

Solution coatings are often room-temperature applications and typically there is little potential for vapor exposure. Skin contact is not uncommon during application of coatings of this type. Heat-converted coatings are usually applied and cured by mechanical means and exposure to vapors or contact with skin is minimal.

Solid resins are used for other applications such as electrical molding powders and decorative or industrial powder coatings. For applications of this kind, exposure to vapors and dust can occur and is greatest during formulating and grinding.

1.3 Methods of Determination

1.3.1 Determination in the Atmosphere

In the National Institute for Occupational Safety and Health (NIOSH) standard method (3, 4), the air sample is adsorbed on charcoal and the collected epoxide displaced for gas–liquid chromatography (GLC) determination with carbon disulfide. The GLC peak for the epoxide is compared with standard samples for quantification.

Similar standard methods for the determination in air by GLC of ethylene oxide, propylene oxide, n-butyl glycidyl ether, isopropyl glycidyl ether, phenyl glycidyl ether, epichlorohydrin, and glycidol have been described (5).

Other methods include hydration, followed by oxidation of trapped epoxide to aldehydes and colorimetric determination of the latter with sodium chromotropate (6), or reaction of trapped epoxide with hydrogen halide and determination of chloride or hydrogen ion consumption as a result of chlorohydrin formation. This method fails where chloride or acid is present in the air sample. Airborne dusts are trapped and are estimated, as are condensed phases.

1.3.2 Analysis of Condensed Phases

The most generally used method for the determination of the epoxide content of a substance involves the addition of hydrogen chloride to the epoxide to form the chlorohydrin:

$$\underset{O}{\overset{\diagdown}{C}} - \overset{\diagup}{C} + H^+ + Cl^- \longrightarrow \underset{OH}{\overset{\diagdown}{C}} - \underset{Cl}{\overset{\diagup}{C}}$$

followed by determination of chloride or hydrogen ion; because of limited solvent properties, water, alcohol, or ether is not generally used. Most epoxide groups can be determined by reaction with pyridinium chloride in pyridine under a 20-min reflux (83 ml concentrated HCl in 1 liter pyridine is the reagent). The excess acid is back-titrated with 1 N NaOH in methanol to phenolphthalein end point.

This method cannot be used with epoxides where the ring contains a tertiary carbon or with styrene oxide, because the epoxide isomerizes to a ketone without chloride incorporation. For these the addition of thiosulfate (7) or chlorohydrin formation using pyridine chloride in chloroform (8) may be employed.

Refractory epoxides such as dieldrin and dicyclopentadiene dioxide are determinable with anhydrous hydrogen bromide in dioxane (9).

1.4 Physiological Properties

1.4.1 Summary

Considering that epoxides can react with nucleophiles, particularly basic nitrogens, one might expect the epoxides to have toxic effects on physiological structures and processes. The magnitude and nature of physiological disruption depend on the reactivity of the particular epoxide, its molecular weight, and its solubility, both of which control its access to molecular targets and the number of epoxide groups present in a given molecular weight. The toxicity ranges from the highly active, low-molecular-weight mono- and diepoxides to the inert cured resin systems possessing only a few epoxy groups per molecule. Effects most commonly observed in animals have been dermatitis (either irritative or secondary to induction of sensitization), eye irritation, pulmonary irritation, and gastric irritation, which are typically found in the tissues that are the first to come into contact with the epoxy compound.

Systemic toxicity in experimental animals has included loss of body weight and moderate liver or kidney damage. Disruption of hematopoesis, primarily leukopenia, has also been demonstrated in laboratory animals with some compounds, but similar changes have not been observed as a result of occupational exposure. Although most compounds are mutagenic to bacteria, not all have produced tumors in animals. A few of the epoxides have been shown to be teratogenic or embryo toxic in animals when administered by routes not relevant to occupational exposure.

1.4.2 Irritation of Surface Tissues and Sensitization

1.4.2.1 Skin and Eye Effects. Irritation of the skin and respiratory tract are the most commonly encountered toxic manifestations of contact with epoxy compounds. In general, it appears that epoxy compounds of higher molecular weight (e.g., epoxy novolac resins and diglycidyl ether of bisphenol A) produce less dermal irritation than those of lower molecular weight. Remarkably, some glycidyloxy compounds are only slightly irritating to the eyes, even though they are significant skin irritants (e.g., cresyl glycidyl ether and *t*-butyl glycidyl ether); however, most are severe eye irritants as liquids or vapors. In some instances, liquid epoxy compounds splashed directly into the eye may cause pain and, in severe cases, corneal damage.

Skin irritation is usually manifested by more or less sharply localized lesions that develop rapidly on contact, more frequently on the arms and hands. Signs and symptoms usually include redness, swelling, and intense itching. In severe cases, secondary infections may occur. Workers show marked differences in sensitivity. Devices made from epoxy resins have produced severe dermatitis when not properly cured and when in prolonged contact with the skin (10). Skin irritation also has been reported from exposure to epoxy vapors (11).

1.4.2.2 Skin Sensitization. Most of the epoxy compounds have the ability to produce delayed contact skin sensitization, although there are notable exceptions, such
as the advanced bisphenol A/epichlorohydrin resins. The higher molecular weight
of these resins may be responsible for the absence of dermal sensitization (12, 13),
although the response of some lower-molecular-weight aliphatic glycidyloxy compounds was equivocal.

The human data available on skin sensitization of epoxy compounds does not
assist in determining the structural requirements necessary to produce sensitization,
but does provide some practical guidance for industrial hygiene purposes. Specifically, of the alkyl glycidyl ethers, only the C_8–C_{10} alkyl glycidyl ether appears to
be a human sensitizer. Despite equivocal results in tests for delayed contact sensitization in guinea pigs, *n*-butyl glycidyl ether and cresyl glycidyl ether do produce
dermal sensitization in some humans.

Skin sensitization reactions can be elicited from much less agent than is required
for an irritative response. Because this condition is difficult to treat, sensitized
individuals may require transfer to other working areas. Particular attention should
be paid to vapors and fine airborne dusts.

1.4.2.3 Acute Pulmonary Effects. Animals exposed to vapors of gaseous or volatile
epoxy compounds, primarily ethylene oxide, propylene oxide, and epichlorohydrin,
have shown pulmonary irritation. Sequelae of this effect may include pulmonary
edema, cardiovascular collapse, and pneumonia. However, this route of exposure
is unlikely for a large proportion of epoxy compounds owing to their low volatility.

For the glycidyloxy compounds for which LC_{50}s have been determined, it appears
that none of these compounds can be considered highly acutely toxic by the inhalation route. However, although there are no data to demonstrate clearly pulmonary sensitization to epoxy compounds, the potential for this reaction exists.

1.4.3 Systemic Effects

1.4.3.1 Acute Toxicity. In general, the acute toxicity of epoxy compounds as observed in laboratory animals can be considered as moderate to very low following
oral and dermal routes of exposure. Liver and kidney toxicity are most frequently
observed in experimental animals, but have not been reported in humans, perhaps
owing to a lack of occupational exposure to extreme amounts. Lung irritation
following inhalation or gastrointestinal irritation following gavage has also been
observed in animals.

Oral and dermal LD_{50} values generally range from 1000 to 5000 mg/kg in rodents,
and there are not marked differences in acute toxicity among the structurally diverse
categories of epoxy compounds. Some epoxide compounds with LD_{50} values less
than 1000 mg/kg are butadiene dioxide, ethylene oxide, propylene oxide, allyl
glycidyl ether, and resorcinol diglycidyl ether. Nevertheless, it is generally difficult
to achieve acutely toxic levels of epoxy compounds by dermal exposure. Usually
the irritating properties of epoxy liquids or vapors limit significant exposure to
produce systemic toxicity.

It is worth noting that acute toxicity of some epoxy compounds decreases with increasing molecular size or chain length. For example, C_8-C_{10} alkyl glycidyl ethers are more acutely toxic orally than $C_{12}-C_{14}$ or $C_{16}-C_{18}$ alkyl glycidyl ethers. Also, greater unsaturation of the epoxy molecule may translate into greater acute toxicity, as evidenced by the greater acute oral toxicity of allyl glycidyl ether compared to the alkyl glycidyl ethers.

1.4.3.2 Subchronic Toxicity. For those epoxy compounds for which repeat exposures have been conducted, generally the liver, kidneys, respiratory epithelium or nasal mucosa (when inhalation was the route of exposure), and stomach (when given by gavage) appear to be the major target organs. Effects on the liver and kidney have been relatively nonspecific or adaptive, as indicated by an increased organ weight without accompanying histopathology. Exceptions include ethylene oxide, which caused renal tubular degeneration and necrosis in mice; vinylcyclohexene dioxide, which produced kidney tubule cell necrosis; *n*-butyl glycidyl ether, which produced liver necrosis; and phenyl glycidyl ether, which produced atrophic liver and kidney effects in rats. Depression of hematopoesis has also been observed in laboratory animals for butadiene dioxide, polyglycidyl ether of substituted glycerin, and vinylcyclohexene dioxide. Respiratory epithelium and nasal mucosa effects have been responses typical of irritation, such as flattening or destruction of epithelial cells.

There are some data to suggest that the testes may be a target organ for certain epoxide compounds, such as glycidol, ethylene oxide, and vinylcyclohexene dioxide; however, most of the studies reporting testicular effects had deficiencies that place the validity of the results in question. Deficiencies included lack of controls [C_8-C_{10}-glycidyl ether, *n*-butyl glycidyl ether (BGE)], use of sexually immature rats (BGE), exposure by a nonoccupational route [allyl glycidyl ether (AGE), C_8-C_{10}-glycidyl ether], effects observed at lethal or near-lethal doses only (AGE), and results that were not reproducible in longer-term subchronic studies [C_8-C_{10}-glycidyl ether, AGE, BGE, and phenyl glycidyl ether (PGE)]. Furthermore, two of the compounds reported to produce testicular atrophy (AGE and PGE) were negative in additional studies conducted specifically to examine mammalian reproduction. A one-generation reproduction study in rats on the diglycidyl ether of bisphenol A also indicated that this material did not produce adverse effects on either male or female reproduction.

There is evidence in rats from a National Toxicology Program study that glycidol produces neurotoxicity (14), and this finding suggests that glycidyl esters, if metabolized to glycidol, could have this effect. However, glycidyl ethers have shown no evidence of neurotoxic effects in numerous acute or repeated dosing subchronic studies in rodents. Propylene oxide appeared neurotoxic in rats at lethal doses, but this effect was absent at nonlethal doses. Ethylene oxide was neurotoxic in laboratory animals at nonlethal doses.

1.4.4 Carcinogenicity

Some studies (15, 16) have shown that epoxides are precursors of reactive intermediates that alkylate DNA. In the cell, the epoxide ring may open with the production of a carbonium ion:

$$R-\overset{\displaystyle \overset{O}{\diagup}\diagdown}{CH-CH_2} \quad \xrightarrow{\text{Nucleophile}} \quad R-\overset{\displaystyle \overset{O^-}{|}}{CH}-\overset{+}{CH_2}$$

Then in the presence of DNA (R—NH), the following reaction may proceed:

$$R-\overset{\displaystyle \overset{O^-}{|}}{CH}-\overset{+}{CH_2} \; + \; R^1-NH \;\longrightarrow\; R-\overset{\displaystyle \overset{OH}{|}}{CH}-CH_2-N-R^1$$

resulting in the alkylation of the DNA. In vitro, the reaction may be quantitatively observed by the accompanying chemiluminescence. This reaction is analogous to the carbonium ion attack of DNA by nitrogen mustards.

In the case of diepoxides and also monoepoxides that contain another reactive functional group (e.g., epichlorohydrin and glycidaldehyde) cross-linking of the DNA strands may occur, which in turn may result in genotoxic effects. Epoxy compounds containing an olefinic structure may also be more reactive because the double bond may be converted to an epoxide group in vivo by the microsomal enzyme aryl hydrocarbon hydroxylase (microsomal monoxygenase), AHH (17).

Epoxy compounds are metabolized by microsomal epoxide hydrase (EH) (hydrolase, hydratase). This enzyme hydrolyzes epoxides to glycols:

$$R-\overset{\displaystyle \overset{O}{\diagup}\diagdown}{CH-CH_2} \; + \; H_2O \;\longrightarrow\; R-\overset{\displaystyle \overset{OH}{|}}{CH}-\overset{\displaystyle \overset{OH}{|}}{CH_2}$$

which are not thought to be carcinogenic or mutagenic, barring other carbonium ion generators in the molecule.

Epoxide hydrase activity is distributed throughout the body but it is organ, species, and even strain variant (18). The liver, testis, lung, and kidney contain considerable amounts of the enzyme; the skin and gut, however, are relatively poorly supplied. In this regard it should be noted that mouse tissues have a much lower level of EH activity than does human tissue; in fact, at least two strains of mice, C57BL/6N and DBA/2N, have *no* EH activity in their skins (19). Therefore it may be questioned if positive dermal mouse tests are indicative of potential human epoxide carcinogenicity.

Epoxide hydrase is inducible by Arochlor® 12, epoxides, and phenobarbital, but not by 3-methylcholanthrene (20). It is therefore under separate control from AHH. It is noncompetitively inhibited by styrene oxide and particularly by 3,3,3-trichloropropene oxide (21). The inhibition and induction are also species variant (22).

Epoxy compounds may also be metabolized by the cytoplasmic enzyme glutathione-S-transferase, which converts epoxides to 2-alkylmercapturic acids. This enzyme, because it is in the aqueous phase, may play a minor role in the detoxification of large lipophilic epoxides but is active against low-molecular-weight epoxides (23, 24).

A further mechanism for detoxification of epoxides is their reaction with proteins and extranuclear RNA. This results in necrosis of the affected cells, with reduction of the concentration of the epoxide available for binding to DNA.

A number of epoxide compounds have been found to be carcinogenic in rodents, although there has been no clear epidemiologic evidence for cancer in the workplace. In rats and/or mice, many epoxy compounds produce a carcinogenic response in the tissues of first contact. These compounds include allyl glycidyl ether, phenyl glycidyl ether, neopentyl glycol diglycidyl ether, resorcinol diglycidyl ether, butylene oxide, propylene oxide, styrene oxide, bis-2,3-epoxycyclopentyl ether, epichlorohydrin, and glycidaldehyde. A few other epoxide compounds, such as ethylene oxide, butadiene dioxide, vinylcyclohexene dioxide, and glycidol have produced tumors at sites other than the "portal of entry." Larger-molecular-weight glycidyloxy compounds such as 1,4-butanediol diglycidyl ether, castor oil glycidyl ether, the diglycidyl ether of bisphenol A, and advanced bisphenol A/epichlorohydrin epoxy resins have been negative in dermal bioassays, that is, in studies where the exposure was by an occupationally relevant route. The previous edition of this book (25) tabulated the results from several carcinogenesis studies on 86 epoxy compounds, many of limited or no industrial importance. Because the scientific literature on many of the industrially important epoxy compounds has increased greatly, and is expanding, a similar table has not been included in this edition for reasons of space.

1.4.5   Genetic Toxicity

Generally, in vitro genetic toxicity testing of the epoxide compounds important in industry has resulted in positive (genotoxic) responses. The largest amount of testing of epoxy compounds for genotoxic effects has been carried out using the Ames test. The Ames test is widely used because of its simplicity, sensitivity, and economy. This involves treating enteric bacteria, usually *Salmonella typhimurium*, premutated to require histidine for proliferation in histidine-free culture, with the test compound which, if mutagenic, causes reversion to the histidine-independent strain. Thus each reverted mutant develops a colony and a quantitative evaluation of the mutagenic effect results from colony counts. Most of the epoxy compounds tested in this bacterial assay have been found to have mutagenic activity. This result is not surprising because many of these compounds have been tested in strains TA1535 and TA100 of *S. typhimurium* or in other gene mutation assays that are specifically sensitive to base-pair substitution. Many other in vitro assays examining both gene mutation and chromosomal effects have been employed to test the epoxy compounds including assays in *E. coli*, yeast, Chinese hamster ovary cells (CHO/HPGRT), mouse lymphoma cells, and cultured human lymphocytes; the results have usually been positive (i.e., producing gene mutations or chromosomal effects).

Fewer epoxy compounds have been tested in the in vivo assays for genotoxic effects, although some have been extensively studied. Some of the lower-molecular-weight epoxy compounds have shown genotoxic effects in various in vivo assays. For example, butadiene dioxide was positive in the *Drosophilia* sex-linked recessive lethal assay and produced chromosomal aberrations and an increased incidence of sister chromatid exchange in bone marrow cells of rats exposed by inhalation. Diepoxybutane was positive in the *Drosophilia* sex-linked recessive lethal assay. Ethylene oxide was positive in the mouse micronucleus assay and mouse dominant lethal assay. Glycidol was positive in the *Drosophilia* sex-linked recessive lethal assay and mouse micronucleus assay and produced chromosomal aberrations in the bone marrow of mice dosed orally or intraperitoneally. Glycidaldehyde was positive in the *Drosophilia* sex-linked recessive lethal assay.

In contrast, epichlorohydrin, also a low-molecular-weight epoxy compound, was negative in both the mouse micronucleus test following intraperitoneal administration and the mouse dominant lethal assay following oral or intraperitoneal administration, although it was positive in many of the in vitro assays. In addition, propylene oxide, although positive in all of the in vitro assays tried, was negative in the mouse micronucleus assay, mouse dominant lethal assay, and *Drosophilia* sex-linked recessive lethal assay, and failed to cause chromosomal changes in monkey bone marrow cells following exposures to 300 ppm. Other compounds showing positive or equivocal effects in vitro but negative effects in vivo are styrene oxide, the diglycidyl ether of bisphenol A, and many of the glycidyloxy compounds used in epoxy resin formulations.

1.4.6 Developmental Toxicity

There has been no evidence of teratogenicity for glycidol, glycidyl ethers, or olefin oxides, except ethylene oxide (EO), when tested by oral or inhalation exposure in conventional developmental toxicity studies; fetal toxicity has been observed at maternally toxic doses for glycidol orally in mice, propylene oxide by inhalation in rats, and 1,2-epoxybutane by inhalation in rabbits. No evidence for fetal toxicity, in some instances even at maternally toxic doses, has been observed for diglycidyl ether of bisphenol A orally in rats or rabbits, phenyl glycidyl ether by inhalation in rats, 1,2-epoxybutane by inhalation in rats, or propylene oxide or styrene oxide by inhalation in rabbits.

When glycidol was tested in rats by intra-amniotic injection, a nonrelevant route of potential human exposure, teratogenicity was observed. Additionally, repeated intravenous infusion of ethylene oxide was teratogenic in mice. Inhalation of extremely high levels of EO (600 to 1200 ppm compared to an American Conference of Governmental Industrial Hygienists (ACGIH) threshold limit value of 1 ppm) in mice at the time of fertilization or early zygote development has led to fetal deaths or malformations in some survivors. However, no teratological effects have been demonstrated by inhalation exposures up to 150 ppm in rats or rabbits.

2 SPECIFIC COMPOUNDS

2.1 Olefin Oxides

2.1.1 Introduction

Direct catalytic oxidation of olefinic bonds produces olefin oxides as one of many oxidation products. Ethylene is approximately 60 percent converted to the oxide at temperatures in the range of 100 to 150°C. The other olefins are less efficiently converted. Peroxidation with peracids is sometimes used to produce olefin oxides; this synthesis route is especially well suited to cyclic olefins. Thus the cycloaliphatic epoxides are often products of peracid oxidation of the corresponding cyclic olefin.

An indirect but more general and more specific synthesis path consists of adding hypochlorous acid to olefins to form the chlorohydrins. Subsequent treatment with strong bases results in dehydrochlorination and the formation of the epoxide. Olefin oxides, which are herein classified solely on the basis of their chemical structure, differ widely in chemical reactivity, physiological response, industrial and commercial use, and exposure patterns. Specific compounds that have commercial or industrial significance are described individually below.

2.1.2 Ethylene Oxide (CASRN 75-21-8)*

Ethylene oxide is also described as a 1,2-epoxyethane, oxirane, and dimethylene oxide. Abbreviations in common use are EO and EtO.

$$\begin{array}{cc} H & H \\ HC & - & CH \\ & \diagdown \; O \; \diagup & \end{array}$$

2.1.2.1 Source, Uses, and Industrial Exposure. Ethylene oxide is currently produced by the direct oxidation of ethylene with oxygen or air over a catalyst. Industrial production for 1990 in the United States has been estimated at 6.2 billion lb, with worldwide production estimated at over 16.5 billion lb. About 60 percent of the ethylene oxide produced in the United States is converted into monoethylene glycol for use in antifreeze, polyester resins, industrial chemicals, and solvents. About 15 percent is used to produce various nonionic surfactants. Most of the remainder is used to produce polyethylene glycols, glycol ethers, ethanolamines, and mixed polyglycols. A small amount of the total production is used as a sterilant, for medical and hospital supplies and as a fumigant for furs and certain foods, such as spices. These uses are limited because of the possibility of significant operator exposures and undesirable residues.

During the past few years, permissible exposure levels (PELs) to ethylene oxide have been reduced significantly in all occupational settings. The Occupational Safety and Health Administration (OSHA) PEL is 1 ppm, and modern manufacturing plants are generally well below this figure. Exposures resulting from the sterilizer use of ethylene oxide have historically been much higher (in the range of 20 to 50 ppm time-weighted average (TWA), with higher short-term peaks) but have now been very significantly reduced and are typically less than 1 ppm. Exposures are sometimes described as "high" or "low." This leads to some confusion, because experimentalists often regard exposures in the 100-ppm range as "low." High dose rates are used in order to establish effects that may not be easily measurable at lower concentrations, whereas such exposures would now not be regarded as tolerable in the context of human exposure.

*CASRN = Chemical Abstracts Service Registry Number.

Ethylene oxide is formed in the body from exogenous or endogenous sources of ethylene. The latter source may be formed as a result of lipid peroxidation, for ethane and pentane are also formed from this source (26). It has been estimated that about 2 percent of inhaled ethylene is converted to ethylene oxide in humans (27).

Several areas of research related to ethylene oxide are particularly active; among these are the use of "biomarkers" in human populations, including hemoglobin adducts and lymphocytic chromosomal changes. Other active areas are the study of mutagenicity in mice and the pharmacokinetics and metabolic disposition of ethylene oxide. Although this summary provides an overview, it is not intended to be complete, particularly in areas where important information is emerging at a rapid pace.

2.1.2.2 Physical and Chemical Properties.
The physical and chemical properties of ethylene oxide are summarized below. It is a highly flammable gas at room temperature and normal pressure, condensing to a liquid at 10°C. Polymerization may occur violently if initiated by acids, bases, or heat. Ethylene oxide is highly reactive toward molecules with active hydrogen atoms and toward nucleophilic agents such as amines and alcohols. For instance, it readily forms ethylene glycol with water, ethylene chlorohydrin with HCl, and hydroxyethylamines with primary or secondary amines. It also reacts with sulfhydryl compounds such as cysteine and glutathione. The major residue that occurs in products such as sterilized foodstuffs and medical equipment is 2-chloroethanol, as a result of reaction with chloride ions. In spite of its generally reactive nature, ethylene oxide is more slowly removed from the atmosphere or water at a neutral pH.

Physical and Chemical Properties of Ethylene Oxide

Molecular formula	C_2H_4O
Molecular weight	44.05
Specific gravity	
0/4°C	0.8966
20/20°C	0.8711
Freezing point	−112.5°C
Boiling point (760 mm Hg)	10.4°C
Vapor density (40°C)	1.49
Refractive index (4°C)	1.3614
Solubility	Miscible with water, acetone, methanol, ether, benzene, carbon tetrachloride
Flash point	−18°C (open cup)
Flammability limits	3 to 100%
Color	Colorless
Odor	Ethereal—characteristic sweet olefinic odor
Odor threshold	260 ppm for perception and 500 to 700 ppm for recognition

1 ppm = 1.80 mg/m^3 at 25°C, 760 mm Hg; 1 mg/l = 555 ppm at 25°C, 760 mm Hg

The high chemical reactivity and exothermic nature of the reactions of ethylene oxide present several problems in storage, handling, and use. Although liquid ethylene oxide is relatively stable, the vapor in concentrations ranging from 3 to 100 percent is highly flammable and subject to explosive decomposition. Ignition may result from many common sources of heat and the resulting pressure rise may cause the violent rupture of containing equipment. Ethylene oxide vapor is heavier than air and will travel rapidly as a layer to the lowest points. It reacts with materials containing a labile hydrogen. Polymerization is catalyzed by acids, alkalies, some carbonates, oxides of iron and aluminum, and boron. No acetylide-forming metals such as copper or copper alloys should be allowed to come into contact with ethylene oxide. When boron trifluoride is used as a catalyst for ethylene oxide, very toxic organofluorine compounds may be produced (28).

2.1.2.3 Determination in the Atmosphere. Because of the high odor threshold for ethylene oxide, sensory recognition does not provide an adequate indication of hazardous exposures. The current method used by OSHA (29) for monitoring potential exposures in the workplace involves adsorption on activated charcoal that has been treated with hydrobromic acid. The ethylene oxide reacts to form bromo-ethanol, which is desorbed with dimethylformamide. The sample is derivatized to a heptafluorobutyrate ester by reaction with heptafluorobutylimidazole and analyzed by gas chromatography, using an electron capture detector.

An alternative method involves adsorption on activated charcoal, desorption with cold carbon disulfide, and analysis by gas chromatography. Care must be taken to keep the sample tube and the desorbed sample cold to avoid loss of ethylene oxide.

Ambient air monitoring may be done directly, using gas chromatography with flame ionization or photoionization detection. Grab samples may be taken for this type of analysis, but inert sampling bags (such as Tedlar) must be used to avoid loss of ethylene oxide.

Colorimetric detector tubes may also be used to give an approximation of ethylene oxide levels but should not be relied on for quantitative data.

2.1.2.4 Physiological Response. Ethylene oxide must be handled with the full appropriate precautions. As a reactive substance, ethylene oxide is highly irritating to exposed tissues. Dilute solutions may cause irritation or necrosis of the eyes and irritation, blistering, and necrosis of the skin. Excessive exposure to the vapor may also cause irritation of the eyes, skin, and respiratory tract, including the lungs. Systemic effects may include depression of the central nervous system, nausea, vomiting, convulsive seizures, and profound limb weakness. Tissue damage may result in secondary infections such as pneumonia. Other toxicologic end points are discussed in more detail in sections below. The aim of this section is to provide an overview of this and also provide sources for additional information (30, 31).

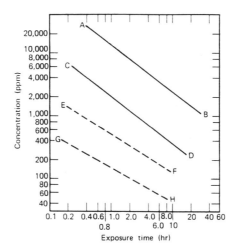

Figure 7.1. Inhalation toxicity of ethylene oxide in air. Severe injury or death likely in areas ABCD. EF probable maximum without injury for one exposure. GH probable maximum without injury for repeated daily exposures.

Acute Toxicity. The oral LD_{50} of ethylene oxide administered in water or corn oil to rats and mice was in the range of 250 to 350 mg/kg. The dose–response curve for mortality was quite steep (31).

The most likely route of exposure is by inhalation. A recent study in rats provided a 1-hr LC_{50} of just over 5000 ppm (W. M. Snellings, personal communication). The 4-hr LC_{50} has been estimated to be approximately 1460 ppm for rats, 835 ppm for mice, and 960 ppm for dogs. Again, the dose–response curve for mortality appears to be steep. Figure 7.1 summarizes estimates of acute inhalational toxicity for varying exposure times. The area bounded by AB and CD represents intensities of exposure that may be expected to cause more or less severe injury or death. The broken line EF should given an indication of the maximum single exposures that are without observable acute clinical injury to humans. This is based on an interpretation of animal data. The line GH represents a similar estimate, but for repeated daily exposures.

The first signs of inhalation toxicity in animals relate to respiratory and eye irritation. These include scratching of the nose, nasal discharge, lacrimation, and salivation. Respiration may become labored or gasping as the lungs become congested and edematous. Secondary lung infections and pneumonia can ensue, leading to delayed deaths. There may also be signs of central nervous system effects, including ataxia, prostration, convulsions, and vomiting. Corneal opacities have been seen in some species, including guinea pigs and monkeys.

Skin Effects. Liquid ethylene oxide does not adversely affect skin unless in sufficient quantities to cause "frostbite" by evaporative cooling; however, liquid ethylene oxide can cause severe chemical burns and blistering if left in contact with human skin, such as inside a leather shoe or underneath clothing. Even dilute aqueous solutions can produce irritation, edema, or skin burns in rabbits. Similar effects may be seen in humans, if contact is prolonged, for example, by the continued wearing of contaminated clothing. The vapor may also be irritant, particularly if the skin is wet or oily. The data are mixed concerning skin sensitization in humans. Guinea pig tests have been negative. Certainly occupational dermatitis can occur in people handling materials containing high residues of ethylene oxide. Ethylene oxide will "off-gas" from materials that have been fumigated. The reduction in the levels of residues is a function of time and ventilation, with the residual ethylene oxide dissipating over a period of a few hours up to 2 or 3 days, depending on the initial level and type of material. Thus the incidence of occupational dermatitis would reflect this decay period in specific contact situations.

Eye Irritation. Liquid ethylene oxide can cause severe irritation and corneal injury. Immediate flushing with water or saline may reduce the severity of these effects. High concentrations of the vapor are irritating to the eyes of both animals and humans. Cataracts have been reported in men after working on a leaking sterilizer (30). There is also an indication that cataracts may occur in exposed monkeys. Deschamps et al. (32, 33) examined a population of 55 persons exposed to ethylene oxide and found that the incidence of lens opacities was similar to controls. A significant percentage of people over 45 in both exposed and nonexposed groups demonstrated ocular opacities. However, in terms of cataracts, there was an increased incidence in the exposed population. Eye protection (chemical goggles) should be worn when handling ethylene oxide or its solutions.

Subchronic Toxicity. Subchronic studies have been conducted in rats, mice, rabbits, guinea pigs, and rhesus monkeys. The concentrations studied range from 50 ppm to as high as 830 ppm (31). These species generally tolerate 50 ppm for 7 hr/day, 5 days/week. 830 ppm is lethal in these species after repeated exposures. These are some species differences in susceptibility, but at doses in the order of 110 to 330 ppm there is increasing evidence of lung damage, including hemorrhages, edema, collapse, and pneumonia. At lower levels of exposure there appears to be minimal irritation or evidence of frank tissue damage or increased cell turnover.

In addition to pulmonary effects there is also evidence of nervous system involvement, which is discussed more fully below. Figure 7.1 provides information on short-term exposures.

Many of the clinical signs of exposure to high concentrations of ethylene oxide suggest observable pharmacological effects on the nervous system. In several species (rat, mouse, rabbit, monkey, dog) repeated high exposures have a characteristic peripheral effect on the lumbosacral nerves. This results in some paralysis, muscle atrophy, and decrease in pain perception and reflexes in the hind limbs. Exposures sufficient to cause these effects lie above 200 ppm, with 100 ppm being regarded as a no-effect level. It is possible that more subtle neurobehavioral effects will be detected in animals at somewhat lower exposures.

Chronic Toxicity/Carcinogenicity. Ethylene oxide has been tested for carcinogenic potential by several routes of exposure. Dermal carcinogenesis has not been demonstrated, although subcutaneous injection has produced tumors in mice. Intragastric administration was associated with stomach tumors in rats.

There have been two inhalational chronic bioassays in rats (30) and one in mice (34). The rat studies showed an increased incidence of mononuclear cell leukemias, peritoneal mesotheliomas, and brain tumors. The study in mice showed a statistical excess of tumors of the lung and Harderian gland. An increased incidence of tumors of the uterus and mammary gland and malignant lymphoma also occurred. These bioassays have been subjected to several intensive analyses (30, 35). Given the similarity in tissue concentrations between organs and the range of tumors found, there is some ambiguity as to what are the "target organs." In the rat studies the incidence of tumors rose towards the end of the study.

Preneoplastic foci have been demonstrated in the livers of rats. These foci have not been demonstrated to lead to hepatocarcinogenesis in rats exposed to ethylene oxide (36).

Reproductive and Developmental Toxicity. There is ample evidence that ethylene oxide reaches the testes of exposed animals and it must be assumed that this is also true for the ovaries. Testicular tubular degeneration and atrophy have been reported in rats and guinea pigs exposed to ethylene oxide. Maximum alkylation of sperm occurs at the mid to late spermatid stage. In rats, abnormal sperm heads are also seen. Some of these testicular effects are prevented in rats by the administration of methylcobalamin (37, 38). Sega et al. (39) reported that in mice, the alkylation of sperm DNA was greater when a given dose was given in a short time frame than when the same total dose (concentration × time) was given over a longer period. This is consistent with the corresponding mutagenicity data. There is a very active area of research and interested readers are advised to review the current literature.

In a classical rat reproductive toxicity study, a reduction in the fertility index and litter sizes was seen at 100 ppm, whereas 33 ppm was considered a no observed effect exposure. No teratological effects were demonstrated by inhalational exposures up to 100 to 150 ppm in rats and rabbits, although effects have been seen after the repeated intravenous infusion of up to 150 mg/kg in mice (see Reference 30 for references). There are several studies indicating that exposure of female mice to 600 to 1200 ppm ethylene oxide at the time of fertilization or early zygote development leads to midgestational or late fetal deaths and malformations in some of the survivors (40). Rutledge and Generoso (41) showed that these effects were most obvious when exposure was 1 or 6 hr after mating and were marginal by 9 hr. Generoso et al. (42) have suggested that altered gene expression is responsible rather than a mutation per se. The no-effect concentration for these effects is 600 ppm.

There is an eastern European report of a range of gynecological and pregnancy disorders resulting from the exposure of women to ethylene oxide. This study is difficult to evaluate without further data (30). Other studies have not confirmed these effects but warn that potential reproductive effects from ethylene oxide should not be disregarded (43). (Although ethylene oxide does not appear to be a highly active reproductive toxin as judged by classical reproductive toxicologic techniques, the potential for mutagenic events to affect offspring adversely should be considered; see also the section below on genetic toxicity).

Metabolism and Pharmacokinetics. Ethylene oxide is very soluble in water and also in blood. Uptake of the vapor via the lungs into the body is rapid, and it has been calculated that almost all inhaled vapor is absorbed in mice (31). There are at least two types of event that may occur after absorption, namely, interaction with a range of macromolecules and metabolic conversion into metabolites. The half-life of ethylene oxide in the body is minutes. In humans the half-life has been estimated to be an average of 42 min (27). Fost et al. (44) have suggested that a polymorphism may occur in humans in relation to their ability to metabolize ethylene oxide, possibly related to the presence of strong and weak conjugating enzyme systems.

There are two possible pathways for the metabolism of ethylene oxide in animals, hydrolysis and conjugation with glutathione. Another potential metabolite, namely, 2-chloroethanol, which is formed by the reaction with chloride ions, has not been demonstrated in animals, although it is found in materials of plant origin such as foodstuffs fumigated with ethylene oxide. Golberg (30) suggested that the alcohol is an intermediate in the formation of other metabolites, but there is not general agreement on this (45). The major product of hydrolysis is ethylene glycol or 1,2-ethanediol. This is excreted via the urine, although some may be further metabolized to hydroxyacetic acid, oxoacetic acid, ethanioic acid, formic acid, and carbon dioxide.

Ethylene oxide reacts with glutathione (possibly via the formation of chloroethanol) through the action of glutathione transferase to form hydroxyethyl cysteine derivatives which are excreted via the urine. The major portion of absorbed ethylene oxide appears to be excreted via the urine. A small proportion (10 percent in rats) may be exhaled as carbon dioxide or as metabolites in feces (4 percent in rats). High or repeated exposures may deplete tissue glutathione levels (46). The general pattern of metabolism appears to be similar in all the species investigated, with rates of decline following exposure being of a similar magnitude. It remains unclear whether differences in metabolism are responsible for the species differences in response, for example, tumor formation.

Ethylene oxide can be measured directly in tissues. Concentrations in tissues can also be estimated from the amounts of adducts formed with a variety of macromolecules. Rate constants for the clearance of ethylene oxide have been estimated for various species by Osterman-Golkar and Bergmark (47), who concluded that there was approximately a twofold range. Brugnone et al. (48) estimated that in exposed workers during work, the ratio of the blood concentrations ($\mu g/l$) to the environmental concentration (mg/m^3) was, on average, 3. Several groups have utilized adduct formation to study the tissue dose of ethylene oxide at specific target organ sites (45).

Ethylene oxide readily reacts with macromolecules such as proteins, RNA, and DNA. In the case of the former, hemoglobin has been investigated extensively. It has been found that the nucleophilic amino acids cysteine, histidine, and valine become ethoxylated and can be used as biomarkers of exposure. Erythrocytes have a measured lifetime in humans and animals and, because they lack suitable repair enzymes, the altered amino acids in hemoglobin act as a cumulative indicator of exposure with a predictable lifetime. These properties have been utilized successfully, particularly in Europe, as a way of monitoring occupational exposure. An approximation has been established between atmospheric exposures and alkylated hemoglobin bases ("adducts"). Lars Ehrenberg of Sweden has been a leading pioneer of this approach (49).

There is need for caution in interpreting the results of these biomonitoring methods. It should be noted that erythrocytes with high levels of adducts may have a shorter half-life than unexposed cells. Erythrocytes from different species have very different survival curves. The data summarized by Rhomberg et al. (50) indicated that the relationship between hemoglobin adduct formation and exposure in parts per million × hours is rather similar for mice, rats, rabbits, and humans when these parameters are plotted on a log–log scale.

The kinetics of the interaction of ethylene oxide with hemoglobin differs from that with DNA, because the latter is more protected and also has repair enzymes. However, there is some parallelism of results, with the alkylated levels of DNA being lower than those of hemoglobin. The rate constant for protein binding is three to four times greater than that for DNA. Fost et al. (44) reported that in humans, although inter-individual differences in binding with the lymphocyte DNA were not significant when compared with the ability to form conjugates with methyl halides, there were significant differences in the formation of hemoglobin adducts.

It must be emphasized that these biomonitoring techniques should be regarded as indicators of exposure rather than as a direct prediction of a toxicologic effect at this stage of the science. This is confirmed by the fact that tissue concentration does not appear to be related to the incidence of adverse responses in specific tissues. Clearly, the response is multifactorial and the other factors require elucidation.

Studies employing such biomonitoring techniques in animals and humans have demonstrated that there is probably a low level of endogenous production of ethylene oxide from ethylene in the body. The latter is one end product of lipid peroxidation. It has also established that environmental exposure from ethylene oxide in tobacco smoke occurs because the level of adducts is somewhat higher in smokers than nonsmokers.

The major reaction product of ethylene oxide with DNA is 7-(2-hydroxethyl) guanine, or 7-HEG. This represents about 90 percent of the alkylated sites in DNA and is detectable in all tissues of treated rats. The concentrations in the brain tend to be highest and the testes have the lowest levels, but there is no great difference between tissues in which tumors form and those where this has not been demonstrated. The initial rate of disappearance of these adducts suggests a half-life of about 7 days. 7-HEG is removed by normal DNA repair processes (45, 51).

Genetic Toxicity. Ethylene oxide has been shown to be mutagenic in a wide range of test systems, ranging from bacteria, fungi, and yeasts to mammalian cells including human lymphocytes (51a). It is often utilized as a positive control substance in such tests. Ethylene oxide is regarded as a direct mutagen; that is, metabolic activation is not required. There are two particular aspects of the mutagenicity of ethylene oxide that require detailed evaluation, namely, ability to induce mutations in germ cells and the demonstration of cytogenetic and other mutational end poinds in human lymphocytes.

Any mutation arising in a germ cell has the potential of affecting a fetus resulting from the fusion of the cell with the corresponding germ cell arising from another individual. These gene mutations may be either dominant or recessive. In the case of the former, a dominant lethal mutation leads to the death of the offspring during pregnancy. Ethylene oxide has been shown to produce such dominant lethal mutations in mice via the DNA of sperm.

It has been shown that there is a gradient of susceptibility during the spermatogenic germ-cell stages, with late spermatids being particularly susceptible. Protamine, which is protein unique to sperm cells, is also alkylated. Chromosome aberrations may be seen in the sperm of treated animals, as reviewed by Dellarco (52). Among these are heritable translocations in which the genetic codes of DNA become rearranged, leading to potential adverse consequences for the fetus and progeny. No specific similar consequence has yet been identified for humans as a result of exposure to ethylene oxide. Rhomberg et al. (50) have examined the available mouse data and have concluded that the dose response is nonlinear. They estimated that for human males the risk of fathering a translocation carrier when exposed to an average of 0.5 ppm for 60 ppm·hr was less than 1 in 10,000 (ppm concentration was multiplied by 120 hr exposures); exposure to 100 ppm for 12,000 ppm·hr might result in a more than 1 in 10 risk. Because sperm have a limited lifespan, the time sequence of exposure and fertilization is a key factor; that is, there is a relatively narrow window for an effect to occur.

There are many reports of ethylene oxide producing mutagenic effects in the blood cells of animals and humans. For instance, sister chromatid exchanges (SCEs) can be seen in the chromosomes of peripheral lymphocytes. Micronuclei may also be seen in bone marrow or red blood cells. Mutations have also been demonstrated in the HPGRT locus, which controls a specific enzyme. In general, these effects have been seen in populations with regular exposures in the range of 10 ppm or more. The background rates are such that controlled studies must be conducted on exposed and control populations to measure any effect and that little weight can usually be placed on the results of single samples.

Wiencke et al. (52) suggested that there may be a bimodal distribution of SCEs in the unexposed population. Tates et al. (52a) concluded that in terms of sensitivity for detection, hemoglobin adducts were more readily detected than SCEs > chromosomal aberrations > micronuclei > HPGRT mutations. This work has been extended by Mayer et al. (53), who showed in a relatively small population of hospital workers exposed to ethylene oxide at or below the current OSHA limitations that although control smokers had more hemoglobin adducts and SCEs than exposed nonsmokers, there was an exposure-related increase when non-smokers and smokers were analyzed separately (see also Reference 54). Other end points such as chromosomal aberrations, micronuclei, DNA repair index, and single-strand breakage did not show these trends. This information also confirms that at low exposure levels, extraneous sources must also be considered.

De Jong et al. (55) explored the use of cytogenetic monitoring at low levels of exposure to several compounds, including ethylene oxide, and concluded that the methods were not sufficiently sensitive for the typical current occupational exposures.

Effects in Humans. No case histories of deaths in humans exposed to high concentrations of ethylene oxide have been located.

Clinical observations suggest that continued overexposure (possibly several hundred parts per million) in humans may lead to nystagmus, ataxia, incoordination, weakness, and slurred speech. Sensorimotor neuropathy may occur in both the upper and lower limbs. Full or partial recovery may occur over a several month period. The available evidence does not suggest that any neurotoxic hazard will exist at the current exposure limits, although there is some evidence of effects within the older occupational exposure limits (56, 57).

There is no evidence of cardiovascular effects such as electrocardiogram changes following exposure to ethylene oxide (30). Renal effects have not been reported in humans, although high exposures in mice have led to tubular degeneration and necrosis.

The human evidence relating to carcinogenicity is equivocal. There is one epidemiologic report that suggested ethylene oxide exposure may be related to leukemias. Three cases versus 0.2 expected (acute myelogenous and "associated with macroglobulinemia") were found in a population exposed to methyl formate and ethylene oxide (58). Several other studies have failed to confirm this conclusion (59–61). The current assessment is that the epidemiologic study of high-exposure populations has not revealed a consistent cancer response in humans and, given the reductions in exposure that have occurred, it seems unlikely that studies on current or future workers will show any response (62).

A recent epidemiology study conducted by Hagar et al. (63) utilized hemoglobin adducts to validate environmental exposure assessments related to sterilizer usage. They indicated that in the sterilizer plants examined, exposures about 30 years ago would be 75 ppm but for the past 10 years were usually well below 1 ppm (<0.2 ppm). More than 2000 workers were included and although there had been few deaths, the mortality and cancer ratios were fewer than expected, with no cases of leukemia.

The U.S. Environmental Protection Agency (EPA) and some other regulatory agencies derive a potency estimate of carcinogens based on an analysis of animal studies. A single number cannot realistically characterize the carcinogenic risk posed by ethylene oxide because there are many assumptions and data choices built into such a single number. The reader is referred to the extensive discussion by Golberg (30) and Austin and Sielken (35) on the range of potency values that can be derived from the animal data. The U.S. EPA has classified ethylene oxide as a B1 carcinogen and the International Agency for Cancer Research (IARC) has classified ethylene oxide as a 2A carcinogen, both of which imply that there is adequate evidence in animals and limited evidence in humans.

2.1.2.5 Hygienic Standards of Permissible Exposure. The Occupational Safety and Health Administration (OSHA) has established a permissible exposure limit (PEL) for ethylene oxide of 1 ppm in air, as an 8-hr time-weighted average (TWA). An excursion limit of 5 ppm, as a 15-min TWA, has also been established. An action level of 0.5 ppm, as an 8-hr TWA, has been set which, if met or exceeded for 30 or more days per year, triggers certain requirements such as ongoing workplace monitoring, medical surveillance, and annual employee training. For further information on the OSHA requirements, refer to the Ethylene Oxide Standard, 29 CFR 1910.1047.

The American Conference of Governmental Industrial Hygienists (ACGIH) has established a threshold limit value (TLV) for ethylene oxide of 1 ppm, as an 8-hr TWA, and has assigned an A2 designation (suspected human carcinogen).

The National Institute for Occupational Safety and Health (NIOSH) has classified ethylene oxide as a potential human carcinogen and recommends maintaining workplace exposure below 0.1 ppm, as an 8-hr TWA.

2.1.2.6 Odor and Warning Properties. Because of the high odor threshold for ethylene oxide, sensory recognition does not provide an adequate indication of hazardous exposures.

2.1.3 Propylene Oxide (CASRN 75-56-9)

This compound, also called 1,2-epoxypropane, propene oxide, and methyl oxirane, has the formula

$$H_3C - \underset{\diagdown O \diagup}{CH - CH_2}$$

2.1.3.1 Sources, Uses, and Industrial Exposure. Propylene oxide is synthesized commercially from propylene through the intermediate propylene chlorohydrin. However, it also can be made by direct oxidation of propylene with either air or oxygen, but this method produces other materials as well. It is used largely for the production of propylene glycol and its derivatives. Substantial quantities are used also in the preparation of hydroxypropylcelluloses and sugars, surface-active agents, and isopropanolamine, and as a fumigant, herbicide, preservative, and solvent (64).

Propylene oxide is highly reactive chemically, being intermediate between ethylene oxide and butylene oxide. It is also extremely flammable. The liquid is relatively stable but may react violently with materials having a labile hydrogen, particularly in the presence of catalysts such as acids, alkalies, and certain salts. It polymerizes exothermically. No acetylide-forming metals such as copper or copper alloys should be in contact with propylene oxide. The same general handling procedures as described for ethylene oxide (65) should be employed.

2.1.3.2 Physical and Chemical Properties. Propylene oxide is a colorless liquid with the following properties:

Molecular weight	58.03
Specific gravity	
20/20°C	0.8304
25/25°C	0.826
Freezing point	-112°C
Boiling point (760 mm Hg)	34.2°C
Vapor density (air = 1)	2.0
Refractive index (25°C)	1.363
Solubility	40.5% by wt. in water at 20°C, 59% by wt. in water at 25°C; miscible with acetone, benzene, carbon tetrachloride, ether, and methanol
Flash point	-30°C
Flammability limits	2.1 to 38.5% by vol. in air

1 ppm = 2.376 mg/m^3 at 25°C, 760 mm Hg; 1 mg/l = 421 ppm at 25°C, 760 mm Hg

2.1.3.3 Determination in the Atmosphere. An OSHA standard method for the analysis of propylene oxide by absorption on charcoal, displacement with CS_2, and determination by GLC has been described (66).

2.1.3.4 Physiological Response. Propylene oxide is only moderately toxic acutely, but even dilute solutions may cause irritation and necrosis of the skin, and the vapors are irritating to the eyes and respiratory system. Non-neoplastic effects were minimal (with the exception of severe contact irritation) following repeated exposures; however, the compound caused cancer in laboratory animals, although tumor formation appeared to be confined to tissues which were the site of initial contact with propylene oxide. It did not produce reproductive or teratogenic effects in animals. It was mutagenic in several in vitro assays but produced equivocal results in in vivo genotoxicity assays.

Acute Toxicity. Oral LD$_{50}$ values by gavage of a 5 percent aqueous solution were 1.14 g/kg in rats and 0.69 g/kg in guinea pigs (67). When gavaged as a 10 percent olive oil solution, rats survived 0.3 g/kg and died at 1.0 g/kg (68). The dermal LD$_{50}$ in rabbits was 1.3 g/kg (69). Four-hour inhalation LC$_{50}$s of propylene oxide were 4126 mg/m^3 (1740 ppm) in mice and 9486 mg/m^3 (4000 ppm) in rats (67). At 7200 ppm in rats, deaths were 0/10, 2/10, and 5/10 after exposures of 0.25, 0.5, and 1.0 hr, respectively. There was no evidence of systemic toxicity at necropsy in rats exposed by inhalation of 4000 ppm for 0.5 hr, 2000 ppm for 2 hr, or 1000 ppm for 7 hr. Sensitivity to the acute lethal effects of propylene oxide appeared to be in the order dog > rat > guinea pig (68).

Skin Effects. Undiluted propylene oxide has low irritation potential if the skin is uncovered to allow for evaporation. However, when the skin is covered, as when contaminated clothing or shoes are worn, even dilute (10 percent) water solutions can cause severe irritation.

Eye Irritation. Propylene oxide vapor is irritating to the eyes of animals, and liquid caused severe eye irritation in rabbits (69).

Subchronic Toxicity. When inhaled repeatedly, propylene oxide vapor appears to be about one-third as toxic as ethylene oxide. The primary toxic effect is lung irritation. When vapor exposures were conducted for 5 days/week for approximately 6 to 7 months, at a vapor concentration of 102 ppm, 9/40 rats died, but there were 0/16 deaths in guinea pigs, and there were no pathological findings at necropsy in either species. At 195 ppm, 7/40 rats died, there were no deaths in 16 guinea pigs, four rabbits, or two monkeys, and only female guinea pig lungs increased in weight. At 457 ppm, no adverse effects were seen in rabbits and monkeys; growth depression, lung damage, and eye irritation were seen in rats and guinea pigs; and deaths occurred in rats (68). In a more recent National Toxicology Program (NTP) inhalation study, however, no treatment-related gross or microscopic pathology was observed in mice or rats exposed to up to 500 ppm propylene oxide vapor for 6 hr/day, 5 days/week, for 13 weeks (71).

Repeated gavage doses of 0.2 g/kg as a 10% solution in olive oil given five times a week until 18 doses had been given failed to produce significant toxicity in rats; a dose of 0.3 g/kg caused loss of body weight, gastric irritation, and slight liver injury (68).

Propylene oxide has been reported to cause central-peripheral distal axonopathy in rats exposed five times a week for 7 weeks to a vapor concentration of 1500 ppm, a nearly lethal level (72). However, in a separate study at lower, nonlethal concentrations, there was no evidence of neurotoxicity associated with inhalation exposure in rats at concentrations as high as 300 ppm for 6 hr/day for 24 weeks (73).

Chronic Toxicity/Carcinogenicity. In an NTP inhalation bioassay in F344 rats and B6C3F1 mice, propylene oxide vapor was administered for 6 hr/day, 5 days/week, for 103 weeks at concentrations of 0, 200, or 400 ppm. The NTP concluded there was some evidence of carcinogenicity in rats, as indicated by increased incidences of papillary adenomas of the nasal turbinates in high-dose males and females, and clear evidence of carcinogenicity in mice, as indicated by increased incidences of hemangiomas or hemangiosarcomas of the nasal turbinates at 400 ppm. Suppurative inflammation, hyperplasia, and squamous metaplasia in rats and inflammation in mice were also observed in the respiratory epithelium of the nasal turbinates (71). These studies have been extensively reviewed (74, 75).

In a 28-month inhalation bioassay in Wistar rats (76), rats were exposed at concentrations of 0, 30, 100, or 300 ppm. Nasal and respiratory tract tumors were observed at the high dose, and there was a dose-related increase in mammary tumors in females. Hyperplasia of the respiratory epithelium and degeneration of the olfactory epithelium were also observed in the high-dose group.

In an oral carcinogenicity study in Sprague–Dawley rats, propylene oxide was administered twice weekly by gavage in salad oil for 109.5 weeks at doses of 0, 15, or 60 mg/kg. Treatment resulted in a dose-dependent increase in the incidence of squamous cell carcinomas of the forestomach. Tumor incidence at other sites was not increased compared to controls. Propylene oxide-treated rats also developed papillomas, hyperplasia, or hyperkeratosis of the forestomach (77).

A dose-related increase in local sarcomas was observed in female NMRI mice given subcutaneous injections of 0.1, 0.3, 1.0, or 2.5 mg/mouse once weekly for 95 weeks (77).

IARC has determined there is sufficient evidence for carcinogenicity of propylene oxide in experimental animals (78).

Developmental Toxicity. Sprague–Dawley rats and New Zealand White rabbits were exposed by inhalation to 500 ppm propylene oxide in a developmental toxicity study (79). There was no evidence of fetal toxicity or developmental defects in rabbits. In rats, however, fetal toxicity was evident by decreased fetal body weight and crown–rump length. An increase in wavy ribs and reduced ossification in fetal vertebrae and ribs were also noted in rats. Reduction in number of corpora lutea, implantation sites, and live fetuses occurred in rats exposed during the pregestational period, suggesting an effect on fertility and survival of offspring. Because maternal toxicity was also evident in the dams, clear interpretation of the rat developmental effects in this study is difficult. The data have been extensively reviewed (80, 81).

In another subsequent inhalation developmental toxicity study in F344 rats, exposures were for 6 hr/day on gestational days 6 to 15 at concentrations of 0, 100, 300, and 500 ppm. There was no evidence of fetal toxicity or teratogenicity at any dose, with the exception of increased frequency of seventh cervical ribs in fetuses at the maternally toxic dose of 500 ppm. The no observed adverse effect level (NOAEL) was considered to be 300 ppm (82).

Reproductive Toxicity. Inhalation exposure to propylene oxide at levels up to 300 ppm over two generations did not produce any adverse effects on reproductive function, including a lack of effect on fertility and survival of offspring (83).

Genetic Toxicity. Mutagenicity data have been reviewed extensively (78, 84). Propylene oxide was mutagenic in vitro to *Salmonella* strains TA100 and TA1535 in both the presence and absence of a metabolizing system in the Ames assay (85, 86), and was positive in the *E. coli* WP2 strain (87), indicating it is a base substitution mutagen. It also induced forward mutations in the yeast *Schizosaccharomyces pombe* (88).

In in vitro mammalian cell assays, propylene oxide induced chromosomal aberrations in an epithelial-like cell line derived from rat liver (89) and in human lymphocytes in cultures established from peripheral blood (86, 90). It also was positive in the CHO/HGPRT mutation assay (91), induced chromosomal aberrations and sister chromatid exchanges in Chinese hamster ovary cells (92), and was mutagenic in the mouse lymphoma forward mutation assay (93).

Sex-linked recessive lethal mutation was observed in *Drosophila* exposed to 645 ppm propylene oxide vapor for 24 hr, with spermatocytes and mature sperm being sensitive stages (94).

Evidence for propylene oxide mutagenicity following in vivo exposure is not conclusive. There was no evidence for either chromosome aberrations or SCEs in lymphocytes from monkeys exposed for 2 years to propylene oxide vapor at concentrations as high as 300 ppm (95). Orally administered propylene oxide at a dose as high as 500 mg/kg was negative in a mouse micronucleus assay, and gave no evidence of mutagenic action on sperm at a dose of 250 mg/kg/day for 14 days in a mouse dominant lethal test. However, when it was injected intraperitoneally at a dose of 300 mg/kg, propylene oxide was positive in the mouse micronucleus assay; lower doses were without effect (86). In male rats or mice exposed to 300 ppm propylene oxide vapors for 7 hr/day on five consecutive days, there were no sperm head abnormalities in mice, and no dominant lethal mutation was observed in rats following mating with virgin females weekly for 6 weeks (94).

In comparing the genotoxic effects of propylene oxide with those of ethylene oxide, the latter was found to be 5 to 10 times more effective than propylene oxide with respect to gene conversion, reverse mutation, and sister chromatid conversion in yeast, but the two compounds were approximately equal in abilities to induce point mutations and SCEs in human lymphocytes (96).

Other Effects. Numerous studies have demonstrated the expected formation of DNA adducts with propylene oxide following in vitro and in vivo exposure (97–101). However, correlation of DNA binding to specific nucleic acid bases with mutagenic events remains unclear. Determination of hemoglobin adducts as a means to monitor exposure to propylene oxide has been utilized effectively for occupational levels as low as 1 ppm (102–105).

Effects in Humans. Contact dermatitis (106, 107) and corneal burns (108) have been reported in humans exposed to propylene oxide. No significant excess of mortality could be found due to any cause in a cohort of 602 workers exposed occupationally to propylene oxide and ethylene oxide over the period 1928 to 1980 (109). However, in 43 male workers from the cohort, there was an increase in chromosome aberrations in peripheral blood lymphocytes of workers exposed for more than 20 years (110). Additionally, a reduced capacity for unscheduled DNA synthesis in lymphocytes from individuals exposed to propylene oxide and ethylene oxide has been reported, suggesting a reduced ability to repair DNA damage.

Exposures were for 1 to 20 years to time-weighted average air concentrations of 0.6 to 12 ppm with short-term exposures to air levels as high as 1000 ppm (111).

2.1.3.5 Hygienic Standards of Permissible Exposure.
The ACGIH TLV and OSHA TWA for 1991 are 20 ppm. The concentration "immediately dangerous to life and health" (IDLH) is 2000 ppm. Propylene oxide is considered a carcinogen by NIOSH.

2.1.3.6 Odor and Warning Properties.
The odor threshold is 45 ppm, which is above a level suitable for prolonged exposure.

2.1.4 Butylene Oxides (CASRN 106-88-7, 3266-23-7)

The structures of 1,2-butylene oxide (CASRN 106-88-7) and 2,3-butylene oxide (CASRN 3266-23-7) are as follows:

$$H_2C\!-\!CH\!-\!CH_2\!-\!CH_3 \qquad CH_3\!-\!CH\!-\!CH\!-\!CH_3$$

2.1.4.1 Source, Uses, and Industrial Exposures.
Butylene oxide is available commercially as the mixed isomers or as the single 1,2 isomer. The oxides are prepared commercially from butylene through the intermediate butylene chlorohydrin. The butylene oxides are used for the production of the corresponding butylene glycols and their derivatives, such as polybutylene glycols, mixed polyglycols, and glycol ethers and esters. They are also used to make butanolamines, surface-active agents, and gasoline additives, and as acid scavengers and stabilizers for chlorinated solvents.

The butylene oxides are highly flammable and highly reactive chemically, but are less reactive than ethylene or propylene oxide. The liquids are relatively stable but they may react violently with materials having a labile hydrogen, particularly in the presence of catalysts such as acids, alkalies, and certain salts. They are capable of polymerizing exothemically. The same general precautions should be taken when handling the butylene oxides as when handling ethylene oxide (112).

2.1.4.2 Physical and Chemical Properties.
1,2-Butylene oxide (1,2-epoxybutane) and butylene oxide(s), a mixture of 1,2- (80 to 90 percent) and 2,3-epoxybutane (10 to 20 percent), are water-white liquids with the following properties:

	1,2-Butylene Oxide	Butylene Oxides
Molecular formula	C_3H_5O	C_3H_5O
Molecular weight	72.1	72.1
Specific gravity (25°C)	0.826	0.824
Freezing point (°C)	Below -60	Below -50
Boiling point (°C)(760 mm Hg)	62.0 to 64.5	59 to 63
Refractive index (25°C)	1.381	1.378
Density of satd. air (air = 1)	~0.977	~1.36
Vapor density	~177	~183

	1,2-Butylene Oxide	Butylene Oxides
Solubility at 25°C		
g/100 g H$_2$O	~8.24	~9
Other common solvents	Miscible with common aliphatic and aromatic solvents	Miscible with aliphatic and aromatic solvents
Flash pt (°F/closed cup)	−15	5
Flammability limits		1.5 to 18.3% by volume in air

1 ppm = 2.94 mg/m^3 at 25°C, 760 mm Hg; 1 mg/l = 340 ppm at 25°C, 760 mm Hg

2.1.4.3 Determination in the Atmosphere. The methods available for the determination of butylene oxide in air for industrial hygiene purposes are the same as those described for propylene oxide and are subject to the same limitations.

2.1.4.4 Physiological Response. Toxicologic data are available mostly for the 1,2 isomer, which is the primary component of the commercial butylene oxides. They tend to be generally less reactive than ethylene oxide and propylene oxide. Butylene oxides are moderately acutely toxic and are substantial irritants, which react with "portal of entry" tissues, such as nasal epithelium and lung when inhaled. The isomers are direct-acting alkylating agents and have been shown to be genotoxic in several in vitro bacterial and mammalian cell assays and in vivo in *Drosophila*. There is limited evidence for carcinogenicity in rodents; however, IARC determined that 1,2-epoxybutane is not classifiable as to its carcinogenicity to humans (Group 3). Because fetal toxicity was observed only in rabbits at maternally toxic doses of 1,2-epoxybutane vapor, NIOSH concluded that the fetus would be protected at maternally safe exposure levels. 1,2-Epoxybutane is extensively metabolized and rapidly eliminated.

Acute Toxicity. The LD$_{50}$ of 1,2-epoxybutane in rats was 1.17 g/kg orally and 1.76 g/kg dermally (69). The 4-hr inhalation LC$_{50}$ was reported to be about 1000 ppm; clinical signs of eye and respiratory tract irritation were observed (113).

Skin Effects. 1,2-Epoxybutane was not a skin sensitizer in guinea pigs (114), but elicited marked skin irritation when applied occluded in rabbits (69).

Eye Irritation. Marked irritation with corneal injury resulted in rabbits when liquid chemical was instilled in the eye (69).

Subchronic Toxicity. In rats inhaling 400 to 1600 ppm 1,2-epoxybutane for 6 hr/day, 5 days/week for 2 weeks, there was inflammation of the nasal mucosa and lungs, bone marrow hyperplasia, and elevated white blood cell count in the high-dose animals. Surviving mice exposed to the same levels had similar effects on the respiratory tract (115). In an NTP study (113) at the same doses, compound-related lesions included pulmonary hemorrhage and rhinitis in rats at 1600 ppm and nephrosis in mice at 800 and 1600 ppm.

In a 13-week subchronic inhalation study (115) at doses of 75 to 600 ppm, inflammatory and degenerative changes in the nasal mucosa were observed in mice and rats and myeloid hyperplasia in bone marrow was observed in male rats. In a comparable NTP study (113) at doses of 50 to 800 ppm, inflammation of nasal turbinates was seen in rats and renal tubular necrosis was seen in mice at 800 ppm. Rhinitis was observed in mice at levels as low as 100 ppm.

Chronic Toxicity/Carcinogenicity. In a dermal carcinogenicity study, 10 percent 1,2-epoxybutane in acetone was applied to the backs of mice three times a week for 77 weeks and the animals were killed in week 85. No visible skin reactions or tumors were observed in test or 100 percent acetone control mice (116).

Chronic 2-year inhalation exposures to 1,2-epoxybutane vapor were conducted by NTP in mice and rats (113). Exposure concentrations were 0, 200, or 400 ppm in rats and 0, 50, or 100 ppm in mice. Inflammatory, degenerative, and proliferative lesions occurred in the nasal cavity of both rats and mice. No exposure-related neoplastic lesions were seen in mice. However, in high-dose rats, nasal papillary adenomas occurred in males (7/50) and females (2/50) compared with none in controls, and alveolar/bronchiolar adenomas or carcinomas (combined) occurred with increased incidence in exposed males (5/49) relative to controls (0/50) (75).

Developmental Toxicity. Inhalation teratology studies were conducted in rats and rabbits at vapor concentrations of 0, 250, or 1000 ppm 1,2-epoxybutane (117). In rats, although maternal mortality and depressed body weight gain were observed at the high dose, fetal growth and viability were not affected, and there was no evidence of teratogenicity. In rabbits, litter size was reduced and fetal mortality was increased in the high-dose group; however, maternal mortality in this group was high (14/24).

Genetic Toxicity. The genetic toxicity of 1,2-epoxybutane has been reviewed (118). It was positive for gene mutation in the Ames assay in *Salmonella* strains TA100 and TA1535 with or without metabolic activation, but was negative in strains TA98 and TA1537, indicative of a direct-acting mutagen that induces base-pair substitutions. Forward mutations were also induced in vitro in the mouse lymphoma assay with and without activation. Both chromosomal aberrations and SCEs were induced in cultured Chinese hamster ovary cells. When fed to male *Drosophila*, 1,2-epoxybutane caused significant increases in the number of sex-linked recessive lethal mutations and reciprocal translocations in the germ cells (113). It produced no effect in the hepatocyte rat primary culture/DNA repair test (119).

Pharmacokinetics/Metabolism. 1,2-Epoxybutane is extensively metabolized and rapidly eliminated following either inhalation exposure or gavage in male rats.

Acute exposures of males to vapor concentrations of 2000, 1000, or 400 ppm caused a dose-related depletion of nonprotein sulfhydryl groups in liver and kidney tissue. Steady-state uptake rates of 1,2-epoxybutane were determined to be 0.0433 mg/kg/min at 50 ppm and 0.720 mg/kg/min at 1000 ppm. These rates correspond to an estimated uptake of 15.6 and 252 mg/kg during a 6-hr exposure. It appears that the physical and biologic processes involved in absorption, metabolism, and elimination of 1,2-epoxybutane are essentially linear throughout the exposure range of 50 to 1000 ppm (120).

Effects in Humans. No data were available in humans; however, excessive exposure to vapors would likely result in disagreeable respiratory irritation.

2.1.4.5 Hygienic Standards of Permissible Exposure. The American Industrial Hygiene Association (AIHA) has established a TWA workplace environmental exposure limit (WEEL) of 2 ppm.

2.1.4.6 Odor and Warning Properties. The odor of the mixed straight-chain isomers of butylene oxide may be described as sweetish, somewhat like butyric acid, and disagreeable. It is doubtful that the odor can be relied on to prevent excessive exposure.

2.1.5   Butadiene Dioxide (CASRN 1464-53-5)

The structure of this compound, also known as diepoxybutane, is as follows:

2.1.5.1 Source, Uses, and Industrial Exposures. Diepoxybutane is prepared by chlorination of butadiene followed by epoxidation with peracetic acid, with subsequent hydrolysis of the epoxide group and final reepoxidation with caustic. It is suggested as a chemical intermediate, cross-linking agent, and in the preparation of erythritol and pharmaceuticals.

2.1.5.2 Physical and Chemical Properties. Diepoxybutane is a water-white, low-viscosity liquid with the following properties (121):

Molecular weight	86.09
Specific gravity (25/4°C)	0.962
Boiling point (760 mm Hg)	138°C
Vapor pressure (20°C)	3.9 mm Hg
Refractive index (20°C)	1.435
Solubility	Miscible in water in all proportions

1 ppm = 3.52 mg/m^3 at 25°C, 760 mm Hg; 1 mg/l = 284 ppm at 25°C, 760 mm Hg

2.1.5.3 Determination in the Atmosphere. No OSHA standard method for the determination of diepoxybutane was found. The pyridinium chloride–chloroform method for epoxy groups (122) and GLC is suggested.

2.1.5.4 Physiological Response. Diepoxybutane is a severe pulmonary irritant. It is classed as highly toxic on inhalation, and moderately toxic following ingestion or skin absorption. It causes severe eye and skin irritation. The *meso* form appears to have half the acute fatal toxicity to mice of the *dl* form. The compound produced skin tumors and sarcomas in mice and depression of the hematopoietic system. It is a direct alkylating agent and is genotoxic in several assays.

Acute Toxicity. The single-dose LD_{50} value for oral in the rat was 0.078 g/kg and for dermal in the rabbit was 0.089 g/kg. Inhalation of concentrated vapors killed all of six rats in 15 min; the 4-hr LC_{50} in rats was 90 ppm (123).

Skin Effects. The undiluted material is a severe primary skin irritant resulting in necrosis in rabbits (123).

Eye Irritation. Diepoxybutane is a severe eye irritant, causing corneal necrosis in rabbit eyes (123).

Chronic Toxicity/Carcinogenicity. Both the *dl* and *meso* isomers of diepoxybutane produced significant numbers of skin tumors in mouse dermal bioassays. Skin applications (10 mg/animal in 0.1 ml acetone), repeated three times weekly for one year, also resulted in sebaceous gland suppression, hyperkeratosis, and hyperplasia (124). Repeated once-weekly subcutaneous injections of the *dl* isomer in mice (0.1 or 1.1 mg/animal) and rats (1 mg/animal) resulted in local fibrosarcomas in 9/50 rats and 5/50 and 5/30 mice, respectively, compared to none in controls (125, 126). In mice given 12 thrice-weekly intraperitoneal injections of the *l* isomer dissolved in water or in tricaprylin, the incidence of lung tumors was increased significantly at doses in water of 27, 108, and 192 mg/kg (127). Intragastric administration in rats once weekly for a year was nontoxic, presumably because of rapid acid hydrolysis of the epoxide in the stomach (126).

Genetic Toxicity. Diepoxybutane was mutagenic in *Salmonella* strains TA98, TA100 (127, 128), TA1535 (129), and TA1530 (130) with or without metabolic activation, and *E. coli* (131, 132). In mammalian cell assays, in vitro, diepoxybutane was mutagenic in the mouse lymphoma cell forward mutation assay (133), induced chromosome aberrations in Chinese hamster ovary cells (134, 135), induced SCEs in Chinese hamster V79 cells (136, 137), in Chinese hamster ovary cells (138), and in human lymphocytes (139, 140, 141), and caused chromosome damage in cultured rat liver cells (142). However, the compound was negative in the rat hepatocyte DNA-repair test (143) and did not induce unscheduled DNA synthesis in rat or mouse hepatocytes in vitro (144).

In in vivo studies, diepoxybutane was found to increase the frequency of structural chromosomal abnormalities and SCEs in bone marrow cells of mice and Chinese hamsters, when inhaled from an aerosol during a 2-hr head-only exposure or administered as a single intraperitoneal injection; inhalation doses, calculated from blood concentrations, were 102 mg/kg in mice and 335 mg/kg in hamsters; intraperitoneal doses were 32 and 147 mg/kg (145). Diepoxybutane also induced SCEs in bone marrow, alveolar macrophages, and regenerating liver cells of mice following intraperitoneal injection (146). A number of studies have shown the compound to be mutagenic in *Drosophila* (147–152).

Effects In Humans. No documented reports of human exposure were available. However, diepoxybutane would be expected to be severely irritating acutely. The compound has been used to detect Fanconi anemia by virtue of its cytogenetic action (153).

2.1.5.5 Hygienic Standards of Permissible Exposure. No standards of permissible occupational exposure were found.

2.1.5.6 Odor and Warning Properties. No information on odor, sensory threshold limits, or warning properties was found.

2.1.6 Epoxidized Glycerides (Linseed, CASRN 8016-11-3, Soya, CASRN 8013-07-8, and Tall Oils)

2.1.6.1 Source, Uses, and Industrial Exposure. These epoxidized oils are made by epoxidizing the unsaturated bonds of unsaturated carboxylic acid–glycerin esters (triglycerides) with peracids (peracetic acid or its equivalent, hydrogen peroxide in acetic acid). Because the molecular weights of these esters approach 900 before epoxidation and because the unsaturated esters are diluted with inert palmitates and stearates, there are sufficient epoxy groups to bind them into polymers, but too few to constitute much of a handling hazard even though there may be a di- or triepoxide content. The epoxidized glycerides are primarily reactive diluents and find use in coatings of food cans. They are recommended for PVC homo- and copolymer stabilization and plasticization for rigid, flexible, extruded, calendered, and molded compounds. Applications are found in intravenous tubing, blood bags, food wrap film, cap liners and seals, meat trays, upholstery, pipe, and construction materials.

2.1.6.2 Physical and Chemical Properties. The properties of epoxidized linseed and soya bean oil are tabulated below; there are no data on epoxidized tall oil.

	Epoxidized Linseed Oil	Epoxidized Soya Bean Oil
Oxirane oxygen (%)	9.3	7.0 to 7.2
Specific gravity (25/25°C)	1.03	0.992 to 0.996
Color, Gardner	1	1

	Epoxidized Linseed Oil	Epoxidized Soya Bean Oil
Viscosity, Stokes (25°C)	3.3	3.2 to 4.2
Fire Point (°F)	650	600
Flash Point (°F)	590	590
Odor	Very low	Mild
Vapor pressure (mm Hg)(25°C)	0.1	0.1
Freezing Point (°C)	0	0

2.1.6.3 Determination in the Atmosphere. No OSHA standard method for the determination of these materials was found. These products are nonvolatile until heated to about 500°F, when acrolein is released from glyceride decomposition. The pyridinium chloride–chloroform method for epoxy groups (122) is suggested.

2.1.6.4 Physiological Response. Although the unepoxidized oils are practically nontoxic, it is conceivable that sensitive or sensitized individuals could develop symptoms upon repeated dermal contact to the epoxidized oils. Good hygienic practice demands minimal dermal contact.

Genetic Toxicity. Epoxidized soya bean oil has been tested for mutagenicity in the Ames assay (strains TA98 and TA100) (154) and the CHO/HGPRT gene mutation assay (155), and was negative in both in vitro assays. Additionally, stomach and intestine homogenates of rats dosed orally with epoxidized soya bean oil were tested in the Ames assay (TA98 and TA100) at 6 hr after dosing, and were found negative (156). Epoxidized soya bean and linseed oils were also tested for cytotoxicity in the HeLa cell MIT-24 in vitro test system; both materials demonstrated low toxicity (157).

2.1.6.5 Hygienic Standards of Permissible Exposure. No standards have been established for these materials.

2.1.6.6 Odor and Warning Properties. Not distinctive.

2.1.7   Vinylcyclohexene Monoxide (CASRN 106-86-5)

The structure of vinylcyclohexene monoxide is as follows:

2.1.7.1 Source, Uses, and Industrial Exposure. Vinylcyclohexene monoxide is a chemical intermediate; it can be copolymerized with other epoxides to yield polyglycols having unsaturation available for further reaction.

2.1.7.2 Physical and Chemical Properties. Vinylcyclohexene monoxide is a color-less, mobile liquid combining readily with water, alcohols, phenols, and other agents containing active hydrogens. It has the following properties:

Molecular weight	124.18
Specific gravity (20/20°C)	0.9598
Freezing point	Sets to a glass below -100°C
Boiling point (760 mm Hg)	169°C
Vapor density	3.75
Vapor pressure (20°C)	2.0 mm Hg
Refractive index (20°C)	1.4700
Solubility	0.5% in water
Concentration in "saturated" air	0.263%

1 ppm $\simeq$ 5.07 mg/m^3 at 25°C, 760 mm Hg; 1 mg/l $\simeq$ 197 ppm at 25°C, 760 mm Hg

2.1.7.3 Determination in the Atmosphere. No OSHA standard methods for the determination of vinylcyclohexene monoxide were found. The pyridinium chloride–chloroform method for epoxy groups (122) and GLC are suggested but the HCl–dioxane method for epoxides (122) might also be used.

2.1.7.4 Physiological Response. The LD$_{50}$ in rats given the material by mouth was 2 ml/kg. The percutaneous LD$_{50}$ for rabbits was 2.83 g/kg. Rats exposed to saturated vapors for 2 hr survived, but three of six died after 4 hr exposure. The compound produced moderate skin irritation in rabbits, although eye irritation was not marked. Inhalation exposure caused pulmonary irritation in rats (25).

2.1.7.5 Hygienic Standards of Permissible Exposure. No standards for permissible occupational exposure were found.

2.1.7.6 Odor and Warning Properties. No information on odor or warning properties were found. Irritation to the respiratory tract might serve as a warning property for overexposure.

2.1.8 Vinylcyclohexene Dioxide (CASRN 106-87-6)

This compound, also called 1-epoxyethyl-3,4-epoxycyclohexane (VCHD), has the following structure:

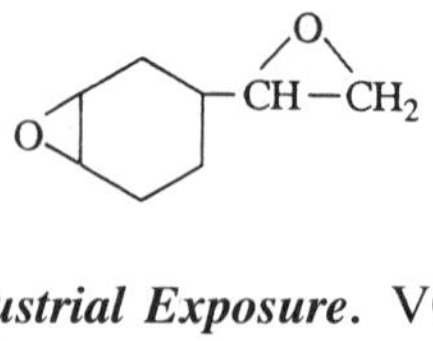

2.1.8.1 Sources, Uses, and Industrial Exposure. VCHD is used as a chemical intermediate and as a monomer for preparation of polyglycols containing unreacted epoxy groups.

2.1.8.2 Physical and Chemical Properties. VCHD is a clear, water-soluble liquid with a faintly olefinic odor and the following properties (158):

Molecular weight	140
Specific gravity (20/20°C)	1.0986
Freezing point	Sets to glass at −55°C
Boiling point (760 mm Hg)	227°C
Vapor density	4.07
Vapor pressure (20°C)	0.1 mm Hg
Refractive index (20°C)	1.4787
Solubility	18.3% in water at 20°C

1 ppm = 5.73 mg/m^3 at 25°C, 760 mm Hg; 1 mg/l = 1.74 ppm at 25°C, 760 mm Hg

2.1.8.3 Determination in the Atmosphere. No OSHA standard methods for the determination of vinylcyclohexene dioxide were found. The pyridinium chloride–chloroform method for epoxy groups (122) or GLC is suggested but the HCl–dioxane method for epoxides (122) might also be used.

2.1.8.4 Physiological Response. VCHD is moderately acutely toxic and is severely irritating. With repeated dermal exposure, it produces marked skin lesions, which progress to skin cancer in rodents; additionally, ovarian and lung cancer is seen in mice. Systemically, VCHD causes reproductive organ toxicity in mice and forestomach and kidney lesions in rodents when given orally. It is mutagenic in the Ames assay, and there is evidence that it is immunosuppressive in mice. VCHD is a contact allergen in humans and has the potential to induce respiratory sensitivity.

Acute Toxicity. Limited unpublished data indicate the compound has an oral LD$_{50}$ of 2.13 g/kg and an inhalation 4-hr LC$_{50}$ of 800 ppm in rats.

Skin Effects. The undiluted chemical causes severe irritation in the rabbit Draize test.

Eye Irritation. The compound would be expected to be a severe eye irritant.

Subchronic Toxicity. In 14-day dermal toxicity studies, rats receiving 139 mg/rat (males) or 112 mg/rat (females) had congestion and/or hypoplasia of the bone marrow; most had acute nephrosis; skin lesions included necrosis and ulceration, epidermal hyperplasia, and hyperkeratosis. Similar skin effects were seen in mice receiving 5 mg/mouse (159).

In subsequent 13-week NTP studies (160), VCHD toxicity was evaluated by both the oral and dermal routes of exposure. In the dermal studies, groups of 10 rats of each sex received 0.3 ml of VCHD in acetone daily 5 days/week at concentrations of 3.75 to 60 mg/ml, and mice received 0.1 ml at concentrations of

0.625 to 10 mg/ml. At the two highest dose levels tested, skin lesions consisted of hyperkeratosis of the epidermis and sebaceous gland hyperplasia, and follicular atrophy of the ovary was seen in mice. In oral studies, mice and rats (10 of each sex) were given 62.5 to 1000 mg/kg VCHD in corn oil by gavage daily 5 days/week. In rats and mice, systemic effects included hyperplasia and hyperkeratosis of the forestomach. In rats, there was also kidney tubule cell necrosis, and in mice there were also reproductive organ effects, which included follicular atrophy in ovaries and degeneration of germinal epithelium in testes.

Chronic Toxicity/Carcinogenicity. NTP conducted dermal cancer bioassays in F344 rats and B6C3F1 mice by administering VCHD in acetone 5 days/week for 105 weeks to rats at 0, 15, or 30 mg/animal or for up to 103 weeks to mice at 0, 2.5, 5, or 10 mg/animal. There was an increased incidence of squamous cell papillomas and carcinomas of the skin in both species, and mid- and high-dose female mice had benign and malignant ovarian tumors. The incidence of lung tumors in female mice was also marginally increased (160).

Genetic Toxicity. There are numerous studies that demonstrate the mutagenicity of VCHD in the Ames assay in *Salmonella* strains TA100 and TA1535 in the presence or absence of S9 metabolic activation (127, 161–165).

Metabolism/Pharmacokinetics. VCHD is absorbed by rodents exposed dermally, orally, or by inhalation (69). In dermal absorption studies conducted by NTP in rats and mice, 30 percent of a dose of radio-labeled VCHD was absorbed over a 24-hr period in both species. By 24 hr, 70 to 80 percent of the absorbed dose was eliminated from the body, mostly in the urine. Of the radioactivity remaining in the body, no tissue contained more than 1 percent of the applied dose, and tissue/blood ratios ranged from 0.3 to 1.5 in rats and 0.8 to 2.8 in mice (159).

Other Effects. The NTP has carried out immunotoxicity studies in B6C3F1 mice. Immune function tests indicated that VCHD was immunosuppressive at 10 mg/mouse and, to a lesser extent, at 5 mg/mouse when applied dermally daily for 5 days and at 2.5 and 5 mg/mouse when applied dermally for 14 days (159).

Effects in Humans. Allergic contact dermatitis has been reported in an individual exposed occupationally to VCHD (166). Immunologic responses have also been observed in plant workers exposed dermally and via inhalation to VCHD (167), indicating sensitization can occur.

2.1.8.5 Hygienic Standards of Permissible Exposure. The ACGIH TLV and OSHA TWA for 1991 are 10 ppm (skin) with an A2 carcinogen classification.

2.1.8.6 Odor and Warning Properties. Although no odor threshold data were found, the irritating properties of VCHD may serve to warn of overexposure.

2.1.9 Styrene Oxide (CASRN 96-09-3)

The structure of this compound, styrene 7,8-oxide, hereafter referred to as styrene oxide, is shown below:

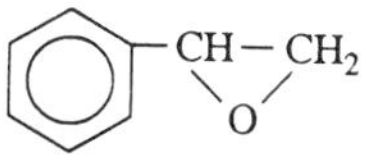

2.1.9.1 Source, Uses, and Industrial Exposure. Styrene oxide is made commercially from styrene through the intermediate chlorohydrin. It exists largely as an intermediate in the production of styrene glycol and its derivatives. Limited amounts may be used as a liquid diluent in the epoxy resin industry.

Styrene oxide has a flash point of about 80°C (175°F) as determined by the open cup procedure. On this basis, it presents a hazard of flammability similar to that encountered with such well-known chemical products as *o*-cresol, *o*-dichlorobenzene, naphthalene, phenol, and dimethylaniline. A definite hazard exists whenever styrene oxide is heated to temperatures at and above the flash point.

Styrene oxide will polymerize exothermally or react vigorously with compounds having a labile hydrogen, including water, in the presence of catalysts such as acids, bases, and certain salts. Experience with the chemical reactivity of styrene oxide has shown that styrene oxide is not as hazardous as ethylene oxide; however, precautions should be taken to prevent excessive pressure under storage or reaction conditions and to relieve such pressure should it occur.

2.1.9.2 Physical and Chemical Properties. Styrene oxide is a colorless liquid with the following properties:

Molecular weight	120.1
Specific gravity (25/25°C)	1.054
Freezing point	36.7°C
Vapor density	4.30
Boiling point (760 mm Hg)	194.2°C
Vapor pressure (20°C)	0.3 mm Hg
Refractive index (25°C)	1.533
Solubility	0.28% in water at 25°C; miscible with methanol, ether, carbon tetrachloride, benzene, and acetone
Flash point (TOC)	175°F
Flammability limits	Precautions necessary at elevated temperature

1 ppm $\simeq$ 4.91 mg/m^3 at 25°C, 760 mm Hg; 1 mg/l $\simeq$ 203.6 ppm at 25°C, 760 mm Hg

2.1.9.3 Determination in the Atmosphere. No OSHA standard methods for the determination of styrene oxide were found. A method for measuring styrene oxide in the blood using gas chromatographic/mass spectrometry has been described (168); a similar approach may be applicable for the determination of styrene oxide in air. The pyridinium chloride–chloroform method for epoxy groups (122) and GLC have been suggested but the HCl–dioxane method for epoxides (122) might also be used.

2.1.9.4 Physiological Response. The acute oral and dermal toxicity of styrene oxide is low. Acute inhalation toxicity studies have shown lethality in laboratory animals at concentrations close to saturation. Styrene oxide may produce severe eye irritation, moderate skin irritation, and skin sensitization. Styrene oxide has been shown to be genotoxic in several in vitro assays, although only weakly mutagenic or negative in in vivo assays. Several investigators have shown that styrene oxide produced malignant tumors in the forestomach of rodents when administered by gavage without producing tumors at other sites. Dermal application of styrene oxide has not resulted in an increased incidence of tumors in mice although it has been reported to induce malignant lymphoma in mice after administration by an unspecified route.

Acute Toxicity (169). The LD_{50} for guinea pigs and rats is about 2.0 g/kg. Single 24-hr skin exposures in rabbits gave an LD_{50} value of 2.83 g/kg. Rats exposed to air saturated with vapors (theoretically calculated to be 395 ppm) survived a 2-hr exposure, but three of six died following a 4-hr exposure. Forty-four of 106 rats exposed to 300 ppm styrene oxide for a single 7-hr inhalation exposure died within a 3-day period post-exposure (170).

Eye Irritation. Undiluted styrene oxide may cause relatively severe irritation and pain to the eyes, but it is not apt to cause serious burns with permanent loss of vision. Solutions as dilute as 1 percent may have some irritating action.

Skin Irritation. Tests with laboratory animals and human subjects indicate that styrene oxide is capable of causing moderate skin irritation and skin sensitization. These effects may result from single or repeated contact with undiluted material and with solutions as dilute as 1 percent. Experience indicates that persons who have become hypersensitive may react rather severely to contact with the vapor as well as with the liquid material. There is some evidence that styrene oxide is absorbed slowly through the skin. This absorption could be significant only from exposures that produced extensive and serious injury to the skin.

Chronic Toxicity/Carcinogenicity. Several chronic toxicity/carcinogenicity studies have been conducted on styrene oxide; five studies involved administration of the material at high doses by gavage; in two others styrene oxide was applied to the skin. Maltoni et al. (171) and Conti et al. (172) administered styrene oxide to Sprague–Dawley rats by gavage at dose levels of 50 or 250 mg/kg/day in olive oil, 4 to 5 days/week, for 52 weeks. The surviving animals were then held until spontaneous death for up to 52 additional weeks. A dose-related increase in the incidence of forestomach neoplasias, papillomas, and precursor lesions (acanthosis and dysplasia) was observed without any other indications of an oncogenic response in any other organ or tissue.

In another study (173), styrene oxide was given in olive oil to pregnant rats at a dose of 200 mg/kg on the 17th day of pregnancy and the offspring dosed with styrene oxide by gavage for 96 weekly doses at a dose level of 100 to 150 mg/kg. A statistically significant increase in forestomach tumors together with papillomas and other early changes indicative of chronic irritation (hyperkeratosis, hyperplasia, and dysplasia) were observed in the treated animals. The incidence of tumors in the styrene oxide groups was not increased relative to controls for any other tissues.

Lijinski (174) reported a study in which styrene oxide was administered in corn oil by gavage three times a week for up to 104 weeks to rats and mice at dose levels of 275 or 550 mg/kg/day and 375 or 750 mg/kg/day, respectively. Consistent with the results of previous gavage studies, a high incidence of squamous cell carcinomas, papillomas, and other non-neoplastic lesions indicative of irritation were found in the forestomach of styrene oxide-treated animals. As in other studies, the incidence of tumors in the styrene oxide groups was not increased relative to controls for any other tissues.

Long-term dermal studies in which styrene oxide was applied three times per week to the skin as a 5 or 10 percent solution in acetone or benzene showed no indication of an oncogenic response in mice (69, 124), but it has been reported to induce malignant lymphoma in mice after administration by an unspecified route (175).

Developmental Toxicity. Injection of styrene oxide into the air space of fertilized chicken eggs resulted in embryo death ($LD_{50} = 1.5$ μmol/egg) and a 7 percent incidence of malformations as compared to none in controls (176). Fetal malformations were not increased in the offspring of female rats exposed to 100 ppm styrene oxide for 7 hours/day, 5 days/week for 3 weeks prior to mating and then exposed to 100 ppm styrene oxide for 7 hr/day daily during days 0 to 18 of gestation even though this exposure regimen resulted in severe maternal toxicity (170). Exposure of pregnant rabbits to 15 or 50 ppm styrene oxide for 7 hr/day on days 1 through 24 of gestation resulted in maternal toxicity but no fetotoxicity or teratogenicity (177, 170).

Reproductive Toxicity. A statistically significant decrease in the fraction of mated female rats that were found pregnant following exposure to styrene oxide (100 ppm) has been reported (80). Because corpora lutea but no implantation sites were found in the nonpregnant animals it was suggested by the authors that styrene oxide may cause pre-implantation loss of fertilized ova.

Genetic Toxicity. Styrene oxide and styrene have been tested in several in vitro and in vivo assays for genotoxicity; the results have been reviewed (178–180). In the in vitro assays, styrene oxide shows evidence of producing both gene mutations and chromosomal effects that appear to be modulated in part by the presence of exogenous metabolic systems. In vivo, styrene oxide was weakly genotoxic or showed no genotoxicity depending on the species and test system, possibly due to metabolic inactivation prior to its distribution from the site of administration to the target cells. Cantoreggi and Lutz (181) have recently reported that despite a relatively long half-life of styrene oxide in animals, the chemical reactivity of styrene oxide appears to be too low to result in a detectable production of DNA adducts in vivo. A comparison of these results to the DNA binding of other carcinogens suggests it is unlikely that the tumorigenic effect of styrene oxide results from a purely genotoxic mechanism.

Pharmacokinetics and Metabolism. The metabolism of styrene to styrene 7,8-oxide and the subsequent metabolism of styrene oxide has been widely described and summarized (182, 183). Styrene oxide is metabolized by two different pathways to mandelic and hippuric acid; these can be conjugated with glucuronic acid or mandelic acid can be dehydrogenated to phenylglyoxylic acid. Styrene oxide is also directly conjugated with glutathione and subsequently metabolized to mercapturic acids. The metabolites of styrene oxide are excreted primarily in the urine.

2.1.9.5 Hygienic Standards of Permissible Exposure. No current recommendations for exposure limits to styrene oxide could be found; however, current recommended threshold limits established by various nations for styrene range from 20 to 50 ppm (180). It is recommended that every precaution be taken when handling or using styrene oxide to prevent contact with the person.

2.1.9.6 Odor and Warning Properties. No information on odor or warning properties were found. Irritation to the respiratory tract might serve as a warning property for overexposure.

2.2 Glycidyl Ethers

2.2.1 Introduction

Compounds containing the epoxide group react readily with a number of other compounds. Reactions of epoxides and such materials can lead to linear polymer formation if the average functionality is approximately two equivalents per molecule. Higher functionalities produce cross-linked systems. Epoxy materials with two or more epoxide rings per molecule are convertible into thermoset plastics by such reactions and are useful as epoxy resins. Materials that will convert epoxides into these cross-linked polymers are called curing agents.

Epoxy resins tend to be highly viscous liquids or solids. The use of solvents to reduce viscosity or reduce the solid to a solution is not practical because, on polymerization, the solvent would be trapped in the cured resin network, altering the physical properties of the polymer. The use of low-viscosity epoxy compounds, that is, reactive diluents, can reduce and often eliminate this difficulty; these materials not only reduce viscosity but also enter into reaction with the curing agent and become an integral part of the polymer. Monofunctional epoxy compounds are usually more effective reactive diluents for reducing the viscosity, but tend to compromise the high performance physical properties afforded by epoxy resins. For some applications, however, reactive diluents can be used and still provide acceptable physical properties.

2.2.2 Aliphatic Monoglycidyl Ethers

2.2.2.1 Allyl Glycidyl Ether (CASRN 106-92-3). This compound, 1-allyloxy-2,3-epoxypropane, has the structure

$$CH_2=CH-CH_2-O-CH_2-\overset{\displaystyle O}{\overset{\diagup\diagdown}{CH}}-CH_2$$

Source, Uses, and Industrial Exposure. Allyl glycidyl ether (AGE) is manufactured through the condensation of allyl alcohol and epichlorohydrin with subsequent dehydrochlorination with caustic to form the epoxy ring. It is a commercial chemical of primary interest as a resin intermediate, and is also used as a stabilizer of chlorinated compounds, vinyl resins, and rubber.

Physical and Chemical Properties. AGE is a colorless liquid of characteristic, but not unpleasant, odor with the following properties (185):

Molecular weight	114.14
Specific gravity (20/4°C)	0.9698
Freezing point	Forms glass at $-100°C$
Boiling point (760 mm Hg)	153.9°C
Vapor density (25°C)	3.32
Vapor pressure (25°C)	4.7 mm Hg
Refractive index (20°C)	1.4348
Solubility	14.1% in water; miscible with acetone, toluene, octane
Flash point (Tag open cup)	135°F
Concentration in "saturated" air (25°C)	0.62%

1 ppm = 4.66 mg/m^3 at 25°C, 760 mm Hg; 1 mg/l = 214 ppm at 25°C, 760 mm Hg

Determination in the Atmosphere. No OSHA standard methods for the determination of AGE were found. Gas–liquid chromatography or the HCL–dioxane method for epoxy groups (122) is suggested.

Physiological Response. AGE is a CNS depressant and also causes respiratory tract irritation when inhaled. It is moderate to low in acute oral and dermal toxicity, with the liver and kidney as main target organs. Severe lung damage results from repeated inhalation at lethal doses in rats. AGE is appreciably irritating to the skin and severely irritating to the eyes and causes occupational dermatitis and skin sensitization in humans. It appears to be a weak carcinogen in mice and is mutagenic in vitro and in vivo.

ACUTE TOXICITY (186, 187). Oral administration of AGE to rats and mice resulted in labored breathing and central nervous system (CNS) depression, preceded by incoordination, ataxia, and reduced motor activity. Piloerection, diarrhea, and coma immediately preceded death. Oral LD_{50} values were 390 mg/kg in mice and 830 to 1600 mg/kg in rats. Necropsy of rats surviving a dose of 500 mg/kg AGE revealed irritation of the nonglandular stomach, including hyperkeratosis, erosion, and ulceration. Occasionally the liver showed focal areas of necrosis, and adhesions of the stomach wall to adjacent tissues were noted. The dermal LD_{50} in rabbits was 2550 mg/kg.

The LC_{50} following 7-hr inhalation exposures in rats to atmospheric concentrations of 100 to 2600 ppm AGE was calculated to be 308 ppm. Rats exposed to 300 ppm and higher had dyspnea, irritation of the nasal turbinates and lungs, nasal discharge, and discoloration, and gross pathological effects were noted in the liver and kidneys. No visible lesions were found at necropsy in rats exposed to 100 or 250 ppm. LC_{50}s of 670 ppm (8 hr) in rats and 270 ppm (4 hr) in mice have also been reported. In these latter studies, the most common finding waَ lung irritation, and pneumonitis was confirmed by microscopic examination. Discoloration of liver and kidneys was also noted frequently at necropsy, but tissue damage was not always confirmed microscopically. Occasionally, hepatic focal inflammatory cells and moderate congestion of the central zones were observed. Although intramuscular injection has been reported to affect the hematopoietic system at lethal doses (188), subchronic and chronic inhalation studies have not demonstrated any effect on hematopoiesis even at concentrations that result in increased mortality (189).

SKIN EFFECTS. Undiluted AGE was slightly irritating to intact rabbit skin, but moderate to severe irritation was observed on abraded skin (190).

EYE IRRITATION. Undiluted compound was severely irritating to the rabbit eye in the Draize test (190). Corneal opacity was seen in some rats after a single 8-hr whole-body inhalation exposure (concentration not specified) (186) and following a 7-hr exposure to air concentrations of 300 to 500 ppm; no corneal effects were seen at 250 ppm or less (187).

SUBCHRONIC TOXICITY (186, 187). Rats were exposed, in groups of 10, to 260, 400, 600, or 900 ppm AGE vapor for 7 hr/day, 5 days/week for 10 weeks. At 600 and 900 ppm, seven and eight animals in each group, respectively, died between the 7th and 21st exposure; rats in these groups had bronchopneumonic consolidation, severe emphysema, bronchiectasis, and inflammation of the lungs. Necrotic spleens were found in two of the rats exposed at 900 ppm. At 400 ppm, one rat died after the 18th exposure; the kidney/body weight ratio was significantly increased in animals exposed at this concentration, and necropsy revealed a decrease in peritoneal fat, severe emphysema, mottled liver, and enlarged and congested adrenal glands. At 260 ppm, slight eye irritation and respiratory distress persisting throughout the exposure period were observed. Decreased body weight gain was seen at all concentrations.

CHRONIC TOXICITY/CARCINOGENICITY. In a 2-year inhalation carcinogenicity study in Osborne Mendel rats and B6C3F1 mice (50 of each sex at each exposure level), animals were exposed to concentrations of 0, 5, or 10 ppm AGE, 6 hr/day, 5 days/ week (189). Although occasional respiratory epithelial tumors were observed, the NTP concluded the data provided only equivocal evidence of carcinogenicity in male rats and female mice. No evidence was obtained to support a carcinogenic effect in female rats. Some evidence was provided for a carcinogenic response in male mice, which included three adenomas of the respiratory epithelium, dysplasia in four mice, and focal basal cell hyperplasia of the respiratory epithelium in the nasal passages of seven mice.

REPRODUCTIVE TOXICITY. Testicular degeneration was reported in rats after intramuscular injection of AGE; however, the results of the study were not statistically significant (188). Male rats were dosed with 400 mg/kg on days 1, 2, 8, and 9, and sacrificed on day 12. Focal necrosis of the testis was observed in one of the three surviving rats. Inhalation exposure of rats to 300 ppm AGE vapor 7 hr/day, 5 days/week for a total of 50 exposures also resulted in testicular atrophy in 5 of 10 rats, and 1 of 10 had small testes (191). In an 8-week inhalation study of reproductive effects in rats and mice of both sexes, the NTP exposed rats to 0 to 200 ppm and mice to 0 to 30 ppm AGE (189). Although the mating performance of exposed male rats was markedly reduced, there was no effect on sperm morphology, motility, or number. Exposed female rats and male and female mice showed no effect on reproductive performance.

GENETIC TOXICITY. AGE has been shown to be positive in the Ames test in strains TA100 and TA1535 with and without activation and negative in strains TA1537 and TA98 (127, 191, 189). AGE induced SCEs and chromosomal aberrations in Chinese hamster ovary cells in both the presence and absence of metabolic activation (189). In the same study, it was reported that AGE induced a significant increase in sex-linked recessive lethal mutations in *Drosophila*, but did not induce reciprocal translocations.

OTHER EFFECTS. Oronasal exposure of mice to 1.9 to 8.6 ppm AGE for 15 min produced a concentration-dependent expiratory bradypnea, indicative of irritation of the nasal mucosa (192). The RD_{50} (airborne concentration producing a 50% decrease in respiratory rate) was 5.7 ppm. When mice were exposed via tracheal cannulation to 105 to 185 ppm AGE for 120 min, there was a concentration-dependent decrease in the respiratory rate due to pulmonary toxicity. The RD_{50} via tracheal cannulation was 134 ppm.

EFFECTS IN HUMANS. Twenty-three cases of occupational dermatitis were reported in one study in which four of the workers developed sensitivity reactions to AGE; in one instance there was eye irritation from exposure to AGE vapor (186). Of 20 patients tested for allergenic properties of AGE, two had allergic reactions (193).

Hygienic Standards of Permissible Exposure. The ACGIH and OSHA TLV for 1991 are 5 ppm (skin) with a short-term exposure limit (STEL) of 10 ppm. The NIOSH ceiling limit is 9.6 ppm. The concentration immediately dangerous to life and health (IDLH) is 270 ppm.

Odor and Warning Properties. The odor threshold is 10 ppm.

2.2.2.2 Isopropyl Glycidyl Ether (CASRN 4016-14-2). This compound has the following structure:

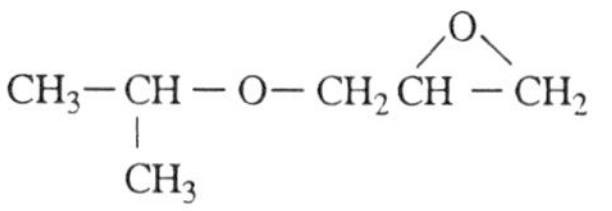

$$CH_3-CH-O-CH_2\overset{\displaystyle O}{\overset{\displaystyle /\,\backslash}{CH}}-CH_2$$
$$\underset{\displaystyle CH_3}{|}$$

Source, Uses, and Industrial Exposure. Isopropyl glycidyl ether is manufactured through the condensation of isopropyl alcohol and epichlorohydrin with subsequent dehydrochlorination with caustic to form the epoxy ring. It is used as a reactive diluent for epoxy resins, stabilizer for organic compounds, and intermediate for synthesis of ethers and esters.

Physical and Chemical Properties. Isopropyl glycidyl ether is a mobile, colorless liquid with the following properties:

Molecular weight	116.16
Specific gravity (20/4°C)	0.9186
Boiling point (760 mm Hg)	137°C
Vapor density	4.15
Vapor pressure (25°C)	9.4 mm Hg
Solubility	18.8% in water, soluble in ketones and alcohols
Concentration in "saturated" air	1.237%

1 ppm = 3.74 mg/m^3 at 25°C, 760 mm Hg; 1 mg/l = 211 ppm at 25°C, 760 mm Hg

Determination in the Atmosphere. An OSHA standard method for the analysis of isopropyl glycidyl ether by absorption on charcoal, displacement with CS_2, and determination by GLC has been described (66).

Physiological Response. Isopropyl glycidyl ether was slightly toxic following acute oral or inhalation exposure, and practically nontoxic via dermal exposure. It is moderately irritating to the eyes and skin and capable of causing skin sensitization. Slight systemic toxicity occurred following repeated inhalation exposure in rats.

ACUTE TOXICITY. Oral LD_{50} values were 1.3 and 4.2 g/kg, respectively, in mice and rats. The dermal LD_{50} in rabbits was 9.65 g/kg. Inhalation LC_{50} values were 1500 ppm in the mouse after 4-hr exposure and 1100 ppm in the rat after 8-hr exposure. Signs of toxicity following oral and inhalation exposures were mainly depressed motor activity, incoordination, and respiratory depression; irritation of the lungs was observed at necropsy following inhalation exposure (186).

SKIN EFFECTS. Isopropyl glycidyl ether was moderately irritating to rabbit skin, with a Draize score (166) of 4.3/8.0 (186).

EYE IRRITATION. The compound was moderately irritating to rabbit eyes, with a Draize score of 40/110 (186).

SUBCHRONIC TOXICITY. Rats exposed to 400 ppm for 50 7-hr periods had a slight retardation of weight gain and elevated hemoglobin but no other evidence of cumulative toxicity. Occasional animals had patchy bronchopneumonia at autopsy (186).

GENETIC TOXICITY. Isopropyl glycidyl ether was positive in the Ames assay in *Salmonella* strains TA100 and TA1535 both with and without S9 activation (191). It also induced SOS repair in *E. coli* PQ37 (132). In an in vitro mammalian assay, it induced sister chromatid exchanges in Chinese hamster V79 cells (137). It has been shown to be a direct alkylating agent (195).

EFFECTS IN HUMANS. No untoward effects, other than slight skin irritation on repeated contact, have been reported.

Hygienic Standards of Permissible Exposure. The ACGIH TLV and OSHA TWA for 1991 are 50 ppm; the STEL is 75 ppm.

Odor and Warning Properties. No information on odor or warning properties was found.

2.2.2.3 n-Butyl Glycidyl Ether (CASRN 2426-08-6). The structure of this compound, 1-(*n*-butoxy)-2,3-epoxypropane, and its analogue, *t*-butyl glycidyl ether, is as follows:

$$C_4H_9OCH_2 - \overset{\displaystyle O}{\overset{\displaystyle \diagup \diagdown}{CH - CH_2}}$$

Source, Uses, and Industrial Exposure. n-Butyl glycidyl ether (BGE) is made by the condensation of n-butyl alcohol and epichlorohydrin with subsequent dehydrochlorination with caustic to form the epoxy ring. Uses include the roles of viscosity-reducing agent for easier handling of conventional epoxy resins, acid acceptor for stabilizing chlorinated solvents, and chemical intermediate. Some curing agents may produce hazardous polymerizations in large quantities.

Physical and Chemical Properties. BGE is a colorless liquid with a slightly irritative odor and the following properties:

Molecular weight	130.21
Specific gravity (25/4°C)	0.9087
Boiling point (760 mm Hg)	164°C
Vapor pressure (25°C)	3.2 mm Hg
Concentration in "saturated" air (25°C)	0.42%
Density of vapor (25°C)	3.78
Flash point	147°F
Solubility	2% in water at 20°C

1 ppm = 5.32 mg/m^3 at 25°C, 760 mm Hg; 1 mg/l = 188 ppm at 25°C, 760 mm Hg

Determination in the Atmosphere. An OSHA standard method for the analysis of BGE by absorption on charcoal, displacement with CS_2, and determination by GLC has been described (66).

Physiological Response. BGE is a CNS depressant and causes irritation of the respiratory tract when inhaled. It is low in acute toxicity, but is appreciably irritating to the skin and eyes and causes occupational dermatitis and skin sensitization in humans. It appears to be mutagenic in vitro, but mixed results have been obtained in vivo.

ACUTE TOXICITY. The oral LD_{50} of BGE was reported to be 1.53 g/kg for mice and 2.26 g/kg for rats. Clinical signs prior to death were incoordination, ataxia, agitation, and excitement. The acute oral toxicity of t-BGE was comparable (186). The dermal LD_{50} of BGE in rabbits has ranged from 0.788 g/kg (196) to 4.93 g/kg (186). In 4-hr inhalation studies, air saturated with BGE vapors did not kill any mice (186), but one of six rats died at an air concentration of 4000 ppm (69). The LC_{50} for an 8-hr exposure in rats was 1030 ppm (186). Intraperitoneal injection caused essentially the same pattern of signs as oral gavage, with LD_{50} values of 1.14 and 0.70 g/kg for rats and mice, respectively.

SKIN EFFECTS. BGE was found to be a mild skin irritant in rabbits, with a Draize score of 2.8 out of 8 possible (186). Repeated dermal application in humans has resulted in appreciable skin irritation and sensitization (196).

Eye Irritation. BGE appeared to be a slight to moderate eye irritant in rabbits, with a Draize score of 23/110; there were corneal effects in 3/3 rabbits (197). In another study, BGE was reported to cause slight pain, slight conjunctival irritation, and slight corneal injury, all healing within 48 hr (198).

Subchronic Toxicity. Male rats inhaled BGE vapors for 50 7-hr exposures to air concentrations of 38, 75, 150, or 300 ppm (199). There were no signs of treatment-related toxicity at the two lowest doses. At 150 ppm, 1 of 10 rats died, and survivors were significantly retarded in growth. At 300 ppm there was 50 percent mortality with additional signs of toxicity in the survivors, such as emaciation, unkempt fur, liver necrosis, and a significant increase in kidney/body and lung/body weight ratios. Testicular atrophy was noted in four out of five of the surviving rats at the lethal level of 300 ppm and in one rat exposed to 75 ppm; however, the rats used in this study were juveniles (57 to 97 g) in which the testes were probably immature at the start of the study, and there was a high incidence of bronchopneumonia, which may have also contributed to their general poor health. Thus it is difficult to identify this effect clearly as treatment related.

In a more recent 28-day inhalation study, rats were exposed to BGE for 6 hr/day, 5 days/week at levels of 0.1, 0.5, or 1.0 mg/l air (equivalent to 18, 94, and 188 ppm) (200). No testicular toxicity was observed even though systemic toxicity was clearly evident. The exposures resulted in decreased body weights in the high dose group, changes in fasting glucose in a high-dose reversibility group, elevated aspartate transferase levels in serum of high dose males, and slightly increased hemoglobulin in high-dose males, which was reversible. Histopathological examination revealed a degeneration of the olfactory mucosa and metaplasia of the ciliated respiratory epithelium. These latter changes were evident at the middle dose level but not in the low-dose group.

In a repeated intramuscular injection study (6 days over a 10-day period) in rats (188), BGE did not suppress bone marrow or significantly alter leukocyte counts.

In a 90-day vapor inhalation study with the structural analogue t-BGE, male and female rats, mice, and rabbits were exposed at dose levels of 0, 25, 75, or 225 ppm for 6 hr/day, 5 days/week for 13 weeks. There were no deaths, and all animals appeared normal and healthy during the course of the study. Microscopic examination revealed that the only tissue affected was the nasal mucosa. Hyperplasia and/or flattening of the nasal respiratory epithelium and inflammation of the nasal mucosa were seen in the 225-ppm group. Decreased body weight gain and concomitant decreases in organ weights were observed in all three species at this highest dose. Minimal effects, primarily in the nasal respiratory epithelium, were observed in most rats and mice exposed to 75 ppm, and no adverse effects were found in animals exposed to 25 ppm (201).

Genetic Toxicity. BGE has been found positive in a number of in vitro genetic toxicity assays, including the Ames assay (191, 202–205) and unscheduled DNA synthesis assay in cultured human blood lymphocytes (203, 204, 206) and W138 cells (205), with and without metabolic activation. In vivo, mixed results have been obtained. BGE was negative in the mouse micronucleus assay when it was dosed by gavage (204) and negative in the host-mediated assay in mice when injected intraperitoneally (203). However, positive results were seen in a micronucleus assay in mice exposed intraperitoneally to >225 mg/kg/day for 2 days and >675 mg/kg for 1 day (207).

A dominant lethal test was conducted in mice in which males were treated topically three times a week for 8 weeks with doses of 375, 750, or 1500 mg/kg BGE and then mated with untreated females (208). Although the results suggested a dominant lethal effect, the results were tentative, because the control fetal death rate was as high as the BGE fetal death rate at the top dose in the second experiment of the study. There was no gross pathology or histopathology in liver, lungs, or testes. A separate dominant lethal test at similar doses was negative (204).

METABOLISM/PHARMACOKINETICS. When ^{14}C-BGE was administered orally to male rats and rabbits (20 mg/kg), it was rapidly absorbed and metabolized. Most of the compound, 87 percent in the rat and 78 percent in the rabbit, was eliminated in the 0- to 24-hr urine. In both species, a major route of biotransformation was via the hydrolytic opening of the epoxide ring followed by oxidation of the resulting diol to 3-butoxy-2-hydroxypropionic acid and subsequent oxidative decarboxylation to yield butoxyacetic acid (209). However, 23 percent of the dose administered to the rats was excreted in the urine as 3-butoxy-2-acetylaminopropionic acid, a metabolite that was not found in rabbits.

EFFECTS IN HUMANS. Appreciable skin irritation and sensitization have been reported in humans (193, 210, 211).

Hygienic Standards of Permissible Exposure. The ACGIH and OSHA TLV for 1991 is 25 ppm. The NIOSH ceiling limit is 5.6 ppm. The concentration "immediately dangerous to life and health" (IDLH) is 3500 ppm.

Odor and Warning Properties. Although an odor threshold has not been measured, the odor is not unpleasant but leads to irritation.

2.2.2.4 2-Ethylhexyl Glycidyl Ether (CASRN 2461-15-6). This compound (Heloxy MK116) has the structure

$$CH_3-(CH_3)_3-CH-CH_2-O-CH_2-CH-CH_2$$
$$| \qquad\qquad\qquad \diagdown O \diagup$$
$$CH_2-CH_3$$

Source, Uses, and Industrial Exposure. 2-Ethylhexyl glycidyl ether is made by condensation of 2-ethylhexanol with epichlorohydrin followed by dehydrochlorination with caustic to form the epoxy ring. In constitutes a reactive diluent and may be used as a relatively nonvolatile chloride-scavenging agent and stabilizer for vinyl resins and rubber.

Physical and Chemical Properties. The properties are as follows:

Color, Gardner-Holdt	1 to 2
Specific gravity, 25°C	0.91
Viscosity, 25°C	3 cm/sec
Epoxide equivalent weight	230

Determination in The Atmosphere. No OSHA standard methods for the determination of 2-ethylhexyl glycidyl ether were found. The pyridinium chloride–chloroform method for epoxy groups (122) and GLC are suggested but the HCl–dioxane method for epoxides (122) might also be used.

Physiological Response. The compound has low acute toxicity (rat oral LD_{50} of 7.8 g/kg) and is slightly irritating to the skin and eyes. It was found to be mutagenic in the Ames assay to *Salmonella* strains TA 100 and TA 1535 with metabolic activation (191, 212).

Hygienic Standards of Permissible Exposure. No standards of permissible occupational exposure were found.

Odor and Warning Properties. No information on odor or warning properties was found.

2.2.2.5 Alkyl Glycidyl Ethers (CASRN 68609-96-1, 68609-97-2, 6881-84-5). The general formula of the alkyl glycidyl ethers is

$$CH_3 \, (CH_2)_n - O - CH_2 - \overset{\displaystyle O}{\overset{\diagup \diagdown}{CH}} - CH_2$$

Sources, Uses, and Industrial Exposure (213). Three fractions of straight-chain alcohols derived from reduction of fats are converted to their respective glycidyl ethers by reaction with epichlorohydrin followed by dehydrohalogenation. The glycidyl derivatives are the C_8–C_{10} fraction (CASRN 68609-96-1), the C_{12}–C_{14} fraction (CASRN 68609-97-2), and the C_{10}–C_{16} fraction (CASRN 6881-84-5). The products are used mainly as reactive diluents for epoxy resins. The manufacturers claim low volatility hazard, improved substrate wettability (better adhesion), reduction of surface tension, and a lower tendency to promote resin crystallization. Although vaporization hazard is low, hot mixing should still be done with adequate ventilation and adequate precautions should be taken to prevent skin contact.

Physical and Chemical Properties. The properties of two of the epoxides are summarized as follows:

	C_8–C_{10} (Epoxide 7)	C_{12}–C_{14} (Epoxide 8)
Molecular weight	229	286
Epoxide equivalent weight	230	291

	C_8-C_{10} (Epoxide 7)	$C_{12}-C_{14}$ (Epoxide 8)
Specific gravity (4/25°C)	0.9	0.89
Melting point (°F)	10	35
Boiling point (°F)(100 mm Hg)	283	420
Vapor pressure at 70°F (mm Hg)	0.08	0.06
Flash point, ASTM D-1393-59 (°F)	245	310
Viscosity at 77°F (cm/sec)	10	10

These materials, because of their high epoxide equivalent weight, are claimed not to produce explosively hazardous polymerizations no matter what the quantity of curing agent added. This is in contrast to more reactive diluents.

Determination in the Atmosphere. Alkyl glycidyl ethers are essentially nonvolatile at room temperature. No OSHA standard methods for their determination were found.

Physiological Response. The alkyl glycidyl ethers are very low in acute toxicity, are moderate skin and mild eye irritants, and are skin sensitizers in guinea pigs. Only the C_8-C_{10} fraction appears to be a human skin sensitizer. No evidence of significant systemic toxicity has been found in repeated dermal exposure studies. Although mutagenic in some strains of *Salmonella* in the Ames assay, the alkyl glycidyl ethers are negative in other in vitro and in vivo genetic toxicity assays.

ACUTE TOXICITY. Acute toxicity studies of alkyl glycidyl ethers by oral, dermal, and inhalation routes have demonstrated very low acute toxicity, with oral LD_{50}s for C_8-C_{10}, $C_{12}-C_{14}$, and $C_{16}-C_{18}$ alkyl glycidyl ethers of 10.4, 19.2, and >31.6 ml/kg, respectively (214). Thus toxicity becomes less with increasing carbon chain length. In an acute dermal toxicity study, undiluted $C_{12}-C_{14}$ alkyl glycidyl ether was applied to the skin of 10 rabbits at 1 ml/kg. Exposure produced moderate erythema without edema at all test sites; no other effects were observed (215). In a similar dermal study at doses as high as 4.5 ml/kg, no treatment-related toxic effects were noted in major organs at necropsy (216).

SKIN EFFECTS. Alkyl glycidyl ethers are considered to be moderate skin irritants with Draize scores in rabbit irritation studies of 3 or 4 out of a possible 8 (217, 218). All three alkyl glycidyl ethers have been reported to be significant skin sensitizers in guinea pig assays.

EYE IRRITATION. Alkyl glycidyl ethers produced only mild eye irritation in rabbit Draize tests (219–221). There was slight conjunctivitis, which cleared within 1 day, and no corneal involvement.

SUBCHRONIC TOXICITY. In a 20-day dermal toxicity test in rabbits, C_8-C_{10} alkyl glycidyl ether was dosed at 2 ml/kg (5 percent solution in dimethyl phthalate). At necropsy, body weight and hematologic values were within the normal range, and there was no histological evidence of toxicity in any major organ (222). In a 90-day dermal toxicity study in rabbits with a mixture of C_{16}–C_{18} and C_8–C_{10} alkyl glycidyl ethers dosed at 2 ml/kg (5 percent solution in mineral oil), there was also no evidence of systemic toxicity (223).

GENETIC TOXICITY. The C_8–C_{10}, C_{12}, and C_{14} alkyl glycidyl ethers were found to be weakly mutagenic in Ames tests with strains TA1535 and TA100 only in the presence of metabolic activation (205). Negative results have been obtained for C_{12}–C_{14} alkyl glycidyl ether in the mouse lymphoma assay and unscheduled DNA synthesis assay using a human cell line (205). Negative results were also obtained for this compound in a host-mediated assay in which mice were dosed orally once a day for 4 days and urine was collected and tested with TA1535 for mutagenicity and in a dominant lethal assay in which compound was applied dermally at a dose of 2 g/kg, three times a week, for 8 weeks (203).

EFFECTS IN HUMANS. Although alkyl glycidyl ethers appear to be skin sensitizers in guinea pig assays, only the C_8–C_{10} fraction appears to be a human skin sensitizer under the conditions of the repeated insult patch test (224).

Hygienic Standards of Permissible Exposure. None have been established.

Odor and Warning Properties. The alkyl glycidyl ethers have a mild fatty alcohol citrus-like odor. There is no room temperature vapor hazard.

2.2.3 Aromatic Monoglycidyl Ethers

2.2.3.1 Phenyl Glycidyl Ethers (CASRN 122-60-1, 3101-60-8). This compound (1,2-epoxy-3-phenoxypropane) (PGE) (CASRN 122-60-1) has the structure

and *p*-(tert-butyl)phenyl glycidyl ether (TBPGE) (CASRN 3101-60-8) has the structure

Source, Uses, and Industrial Exposure. PGE is synthesized by condensation of phenol with epichlorohydrin, with subsequent dehydrochlorination with caustic to form the epoxy ring. As an acid acceptor it is very effective as a stabilizer of halogenated compounds. Its high solvency for halogenated materials offers many possibilities as an intermediate.

Physical and Chemical Properties. PGE is a relatively high boiling, colorless liquid with the following properties (225):

Molecular weight	150.17
Specific gravity (20/4°C)	1.1092
Melting point	3.5°C
Boiling point (760 mm Hg)	245°C
Vapor density	4.37
Vapor pressure (20°C)	0.01 mm Hg
Refractive index	1.5314
Solubility	0.24% in water; 12.9% in octane; completely soluble in acetone and toluene

Determination in the Atmosphere. An OSHA standard method for the analysis of PGE by absorption on charcoal, displacement with CS_2, and determination by GLC has been described (66).

Physiological Response. The acute oral and dermal toxicity of PGE is low, with the liver and kidney as target organs for systemic effects. PGE is an appreciable skin and eye irritant and skin sensitizer. These acute effects were comparable for TBPGE. Systemic toxicity of PGE occurs with repeated inhalation exposure, and PGE is classified by IARC as possibly carcinogenic to humans, because nasal cancer has occurred in a 2-year rat bioassay. PGE was not a developmental or reproductive toxin in rats. The compound was mutagenic in in vitro but not in in vivo assays. TBPGE was not mutagenic in vitro.

ACUTE TOXICITY. Oral LD_{50}s for PGE were 2.5 to 6.4 g/kg in the rat (69, 123, 186, 226) and 1.4 g/kg in the mouse (186). Incoordination, ataxia, and a decrease in motor activity were followed by coma and death in rats dosed orally. Extensive congestion of the liver and kidney were also observed at necropsy. Dermal LD_{50}s in the rabbit were 2.28 to 2.99 g/kg (69, 123, 186). Although irritation of the lungs occurred, no deaths were produced in mice exposed for 4 hr or rats exposed for 8 hr to saturated vapors at room temperature. The acute toxicity of TBPGE is comparable (227–229).

SKIN EFFECTS. PGE was moderately to severely irritating to rabbit skin, particularly upon prolonged or repeated treatment (69, 123, 186). PGE was positive in skin sensitization tests in guinea pigs after topical and intradermal administration (69, 230, 231). Skin effects of TBPGE were similar (227, 232, 233).

EYE IRRITATION. PGE has produced irritation ranging from mild to severe in the rabbit eye (69, 123, 186, 226, 234).

SUBCHRONIC TOXICITY. When rats inhaled PGE at a concentration of 29 ppm for 4 hr/day, 5 days/week for 2 weeks, the animals exhibited weight loss, atrophic changes in the liver, kidneys, spleen, thymus, and testes, depletion of hepatic glycogen, and chronic catarrhal tracheitis (235). In a subsequent study (236) rats and dogs were exposed to concentrations of 1, 5, and 12 ppm for 6 hr/day, 6 days/ week for 3 months. In rats at the high dose there was hyperkeratosis and inflammatory atrophy of hair follicles; this change was not seen in dogs. There were no compound-related changes in the histopathology of major organs or in blood, urine, or biochemical indexes at any exposure concentration in either rats or dogs. Six intramuscular injections of 400 mg/kg PGE in rats failed to produce any effects on hematopoiesis (237).

CHRONIC TOXICITY/CARCINOGENICITY. An inhalation carcinogenicity study in rats at concentrations of 0, 1, and 12 ppm for 2 years resulted in nasal carcinomas, 11 percent in males and 4.4 percent in females at 12 ppm, and dose-related squamous cell metaplasia, which corresponded to the increased incidence of nasal tumors (238). The International Agency for Research on Cancer (IARC) classified PGE in Group 2B (possibly carcinogenic to humans) (239).

DEVELOPMENTAL/REPRODUCTIVE TOXICITY. (240) Male rats were exposed to PGE at concentrations of 0, 2, 6, or 11 ppm for 6 hr/day for 19 consecutive days and were mated with untreated females for 6 consecutive weeks. Offspring from those matings were mated with the treatment group. There were no significant effects on reproduction. Although one of eight males from each of the three treatment groups showed testicular atrophy upon histological examination, there did not appear to be any compromise in their functional capacity to reproduce. In the same study, pregnant rats were exposed to PGE by inhalation on the 4th and 15th days of gestation at concentrations of 1, 5, and 12 ppm for 6 hr/day. No clinical signs of toxicity were observed, and there was no evidence of developmental toxicity.

GENETIC TOXICITY. PGE was mutagenic in *Salmonella* strains TA100 and TA1535 with and without metabolic activation, suggesting it is a direct-acting mutagen causing base substitutions (241). Mutagenic activity was also observed in *Klebsiella* (242) and *E. coli* (195, 243), and a positive result was obtained in a DNA-repair test using different repair deficient strains of *E. coli*, suggesting that PGE may cause DNA damage that can be repaired by recombination (240). PGE has been reported to alkylate nucleic acid bases in vitro (244); however, PGE did not bind to DNA in *E. coli* with or without metabolic activation. In mammalian cells, PGE did not induce chromosomal aberrations in cultured Chinese hamster cells (241, 245), but did induce transformation of hamster embryo cells in culture (241). TBPGE was not mutagenic in vitro in bacteria or yeast (246).

PGE given orally at a dose of 2.5 g/kg was active in two out of five mice in the host-mediated assay (241); however, DNA synthesis in mouse testes was not inhibited by oral administration of 0.5 g/kg PGE (240). PGE was also negative in the mouse micronucleus assay at an oral dose of 1 g/kg (245). There was also no increase in the incidence of chromosomal aberrations in bone marrow cells from male rats exposed to PGE by inhalation at concentrations of 0, 1, 5 or 12 ppm for

6 hr/day for 19 consecutive days (240). PGE was also negative in a rat dominant lethal assay (240).

METABOLISM/PHARMACOKINETICS. When PGE was administered orally to rats and rabbits, urinary metabolites were 2-hydroxy-3-phenoxypropionic acid and *N*-acetyl-*S*-(2-hydroxy-3-phenoxypropyl)-L-cysteine (247). When administered dermally, the absorption rates were 4.2 mg/cm^2/hr for rats and 13.6 mg/cm^2/hr for rabbits (226).

EFFECTS IN HUMANS. Skin sensitization occurs in humans, and cross-sensitization with other glycidyloxy compounds may also occur (186, 193); however, no systemic effects have been reported.

Hygienic Standards of Permissible Exposure. The ACGIH and OSHA TLV for 1991 is 1 ppm. The NIOSH considers PGE a potential carcinogen with a ceiling limit of 1 ppm.

Odor and Warning Properties. No information on odor or warning properties was found.

2.2.3.2 Cresyl Glycidyl Ether (Mixed Isomers CASRN 26447-14-3). The structure of this glycidyl ether (mixed isomers) is

The isomers of CGE include o-CGE (CASRN 2210-79-9), m-CGE (CASRN 2186-25-6), and *p*-CGE (CASRN 2186-24-5).

Source, Uses, and Industrial Exposure. CGE may be made from cresols by reaction with allyl chloride followed by epoxidation or with epichlorohydrin followed by dehydrochlorination. Suggested uses are as a reactive diluent for viscosity reduction of liquid epoxy resin systems, to increase the level of filler loading of such systems, and to reduce the tendency of the resins to crystallize.

Physical and Chemical Properties. The properties of CGE are summarized as follows:

Viscosity (25°C)	5–25 cm/sec
Color, Gardner-Holdt	4 max.
Epoxide equivalent weight	190–195
Specific gravity (25°C)	1.07–1.09
Flash point (open cup)	250°F min.

The substance is incompatible with strong oxidizing agents, strong acids, or bases.

Determination in the Atmosphere. No OSHA standard methods for the determination of CGEs were found. The pyridinium chloride–chloroform method for epoxy groups (122) and GLC are suggested but the HCl–dioxane method for epoxides (122) might also be used.

Physiological Response. CGE is of low acute toxicity, but is a moderate to severe skin irritant and a potent skin sensitizer. CGE is a weak genotoxin in vitro and in vivo in mutagenicity assays, but there was no evidence of chromosomal aberrations in exposed workers.

ACUTE TOXICITY. The oral LD_{50} of CGE in the rat was 5.8 g/kg (248). Signs of toxicity included dyspnea and lacrimation, and congested liver was observed at necropsy. The acute dermal LD_{50} in rats was >2.15 g/kg, and no toxic effects were observed (249). In a 4-hr inhalation study in rats, the LC_{50} was 1220 ppm (250).

SKIN EFFECTS. CGE is a moderate skin irritant in rabbits, with a Draize score of 5.2 out of a possible 8 (251). Undiluted o-CGE produced severe skin irritation at 24 hr, which progressed to necrosis by 14 days (252). CGE is a potent skin sensitizer in guinea pigs (253, 254).

EYE IRRITATION. Remarkably, CGE appears to be only a slight irritant to the rabbit eye, and healing is rapid (255).

GENETIC TOXICITY. CGE is considered a weak genotoxin. o-CGE was a direct-acting mutagen in *Salmonella* strains TA1535 and TA100 in the Ames assay (191). p-CGE was also mutagenic in strains TA1535 and TA100 (256). No mutagenic activity was observed in strain TA98 (257), indicating that CGE exerts its mutagenic potential by causing base-pair mutations. In an unscheduled DNA synthesis assay, o-CGE produced significant increases at 10 and 100 ppm; it was cytotoxic at 1000 ppm (203). In a host-mediated assay in mice in which o-CGE was dosed at 125 mg/kg/day for 4 days, a positive result was obtained in the urine from ICR strain mice when β-glucuronidase was added; results were negative in $B_6D_2F_1$ mice whether or not the enzyme was present (203). In a dominant lethal assay, in which CGE was administered topically to $B_6D_2F_1$ mice at a dose of 1.5 g/kg, three times a week, for 8 weeks, there was suggestive evidence of a dominant lethal effect, but it may have been related to the systemic toxicity of CGE rather than a genotoxic effect (203). In a host-mediated micronucleus test in mice, o-CGE was found not to be genotoxic (203).

METABOLISM/PHARMACOKINETICS. o-CGE was converted rapidly to the corresponding diol compounds when incubated with guinea pig liver homogenate in vitro (258).

EFFECTS IN HUMANS. CGE is a potent skin sensitizer in humans (193). No biologically significant increase in the frequency of chromosomal aberrations was found in peripheral blood lymphocytes of exposed workers (259).

Hygienic Standards of Permissible Exposure. A TWA value of 10 ppm was established by Sweden and Denmark in 1987.

Odor and Warning Properties. No information on odor or warning properties was found.

2.2.4 Aliphatic Diglycidyl Ethers

2.2.4.1 Diglycidyl Ether (CASRN 2238-07-5). The structure of diglycidyl ether (DGE), or di(2,3-epoxypropyl) ether, $C_6H_{10}O_3$ is

$$CH_2-CH-CH_2-O-CH_2-CH-CH_2$$

Source, Uses, and Industrial Exposure. Diglycidyl ether is significant only because it is a possible trace component of epoxy compounds derived from epichlorohydrin.

Physical and Chemical Properties. The properties are as follows:

Molecular weight	130
Specific gravity (20/4°C)	1.1195
Boiling point (760 mm Hg)	260°C
Vapor pressure (25°C)	0.09 mm Hg
Vapor concentration in saturated air (25°C)	0.0121%
Vapor density (air = 1)	3.78 at 25°C

Determination in Atmosphere. No OSHA standard methods for the determination of diglycidyl ether were found. The pyridinium chloride–chloroform method for epoxy groups (122) and GLC are suggested but the HCl–dioxane method for epoxides (122) might also be used.

Physiological Response. Diglycidyl ether is a severe irritant to skin, eyes, and respiratory tract. It is a strong skin sensitizer. It exhibits radiomimetic effects following acute and chronic exposure as evidenced by depression of bone marrow and other rapidly growing cells. It is genotoxic in lower organisms, and has been shown to be carcinogenic in mice following repeated skin application. At high levels of exposure it may cause adverse kidney and liver effects.

ACUTE TOXICITY (25). The acute oral LD_{50} ranged from 170 to 192 mg/kg in mice and from 450 to 510 mg/kg in rats. At gross necropsy, pulmonary hemorrhage, hyperemia and irritation of the enteric tract, liver and renal changes, and "inflammation" of the adrenal gland were found.

The intravenous LD_{50} in rabbits was 141 mg/kg. Autopsy showed severe congestion of the lungs, congestion of patchy ischemia of the liver, slight ischemia of the kidneys, and ascites. Single injections of 50, 100, and 200 mg/kg have an effect on the peripheral blood cell counts and cell morphology. These consisted of leukopenia due to a decrease of polynuclear cells, the duration of which increased with the dose. In animals given 100 mg/kg, an increase was observed in nucleated red blood cells which appeared 7 days after injection. In surviving animals there was eventually a full recovery.

The percutaneous LD_{50} ranged from 1000 to 1500 mg/kg in either rats or rabbits. The application induced severe skin irritation in both species and also caused systemic effects similar to those following intravenous injection, that is, weight loss and leukopenia on the third day after application, which after a few days changed into a leukocytosis. In rabbits, a decrease of the hemoglobin concentration also occurred.

The LC_{50} values by inhalation are listed as follows:

Species	Time of Exposure (hr)	LC_{50} (ppm)
Rabbit	24	13.3
Mouse	4	86
Rat	4	200
Mouse	8	30
Rat	8	68

The immediate effects of vapor were few and due to irritation of the mucosal membranes; however, after removal from the chamber within 24 hr the rabbits showed cloudy corneas as well as irritation of the mucous membranes and the skin. High vapor exposure (113 ppm) in rats also caused dyspnea. Necropsy showed lung congestion, granular, discolored livers, enlarged kidneys, and prominent adrenals. At single 24-hr vapor exposures of rabbits (to 3, 6, 12, and 24 ppm) the morphology of the blood cells was not affected at the 3 and 6 ppm level; at 12 ppm a possible thrombocytosis was noted; and at 24 ppm there was a marked leukocytosis pre-terminally. At 3 ppm there was still clear evidence of mucous membrane irritation, which became very severe at the highest level.

SKIN EFFECTS (25). Single and repeated application to the intact or scarified skin of rabbits induced severe irritation and chemical burns. Repeated application led to necrosis of subcutaneous tissues. Diglycidyl ether was a strong skin sensitizer in guinea pigs.

EYE IRRITATION (25). In the eyes, instillation of liquid diglycidyl ether induced severe irritation and corneal necrosis. A 15 percent solution in propylene glycol was rated as severely irritating.

SUBCHRONIC TOXICITY (25). Rats received daily percutaneous applications (5 days/week, 4 weeks) of 15, 30, 60, 125, 250, and 500 mg/kg. At the two highest dose levels there was a high mortality and at 125 mg/kg and higher there were pronounced symptoms of systemic toxicity, necrosis of the skin, and corneal opacity. The symptoms of systemic toxicity were weight loss and leukopenia due to a decrease in polynuclear cells. Autopsy of these animals showed general bone marrow aplasia, atrophy of the thymus, focal lymphoid necrosis, necrosis of the proximal convoluted tubules of the kidneys, focal necrosis of the testes, and hemorrhage of the adrenal medulla. At 30 and 60 mg/kg there were only weight loss and a decrease in polynuclear cells in the blood but no leukopenia. The bone marrow showed no changes. The thymus only was decreased in weight in the 60 mg/kg animals; the testes were all normal. At the 15 mg/kg level no effect was observed.

Dogs receiving a few repeated intravenous injections at weekly intervals of 25 mg/kg developed leukopenia. Severe inflammation in the muscles occurred when the injections were given intramuscularly. Several dogs died with secondary pulmonary infections. Also a weekly intravenous dose of 12.5 mg/kg induced leukopenia. In surviving dogs, killed a month after the last injection, the bone marrow appeared normal.

Rats were exposed to DGE by inhalation of vapors with concentrations of 0.0, 0.3, or 3.0 ppm 4 hr/day, 5 days/week. The rats exposed to 3.0 ppm received only 19 exposures; the exposures to 0.3 ppm amounted to a maximum of 60. The animals at 3.0 ppm showed mortality due to pulmonary infection and showed evidence of systemic intoxication: diminished weight gain, blood changes appearing after the seventh exposure, and bone marrow changes at autopsy. In 10 animals surviving for 1 year after the final exposure to 3.0 ppm the signs of systemic intoxication had disappeared and the bone marrow was normal but several rats showed peribronchiolitis and one a fatty dystrophy of the liver. Thirty animals were exposed to 0.3 ppm (estimated to be ±0.2 ppm actual value) of which 10 were sacrificed after 20 exposures. These animals appeared normal. Ten other animals were sacrificed after 60 exposures and 10 others kept for 1 year after the final exposure and then sacrificed. There was no effect on the peripheral blood or the bone marrow in these two groups. The group sacrificed immediately after the final exposure showed "poorly defined focal degeneration of the germinal epithelium." Thus it appears that any changes occurring at low levels of exposure are reversible. However, a clear no-effect level has not been demonstrated.

CHRONIC TOXICITY/CARCINOGENICITY (25). In mice, diglycidyl ether has been shown to produce epithelioma following repeated skin application. A total dose of 100 mg produced these tumors in 4/20 animals; a dose of 33 mg produced only 1/20.

GENETIC TOXICITY (25). Diglycidyl ether has been shown to have a mutagenic effect in bacteriophage T2 and *Neurospora* and to induce chromosome aberrations in *Vicia faba* and several other plant cells. It has a mutagenic action in bacterial systems such as *Salmonella typhimurium*.

EFFECTS IN HUMANS. No reports have appeared on adverse effects of industrial handling.

Hygienic Standards of Permissible Exposure. The 1991 ACGIH TLV is 0.1 ppm (0.53 mg/m^3 ceiling value).

Odor and Warning Properties (25). The odor threshold is not established, but the odor is recognizable at approximately 5 ppm. It is irritating to eyes at 10 ppm.

2.2.4.2 1,4-Butanediol Diglycidyl Ether (CASRN 2425-79-8). This compound has the structure

$$(-CH_2-CH_2-O-CH_2-CH \overset{O}{\overset{\diagup\diagdown}{-}} CH_2)_2$$

Source, Uses, and Industrial Exposure. The compound results from condensation of butanediol with epichlorohydrin followed by dehydrochlorination with caustic. Its primary use is as a reactive diluent in epoxy resin systems to reduce viscosity and allow higher filler loading.

Physical and Chemical Properties. 1,4-Butanediol diglycidyl ether is a liquid with the following properties:

Molecular weight	202.34
Boiling point (°C)	155 to 160
Specific gravity	1.1

Determination in the Atmosphere. No OSHA standard methods for the determination of 1,4-BDGE were found. The pyridinium chloride–chloroform method for epoxy groups (122) and GLC are suggested but the HCl–dioxane method for epoxides (122) might also be used.

Physiological Response. The acute toxicity of 1,4-BDGE by the oral and dermal routes is low; however, it is a severe skin and eye irritant and skin sensitizer. 1,4-BDGE was positive in in vitro bacterial and mammalian cell mutagenicity assays and in an in vivo mutagenicity assay, but was negative in a 2-year dermal cancer bioassay in mice.

ACUTE TOXICITY. The oral LD_{50} of 1,4-BDGE was 1.41 to 1.88 g/kg in rats (260, 261) and 3.61 g/kg in hamsters (262). The dermal LD_{50} was >2.15 g/kg in rats; no evidence of systemic toxicity was apparent (263).

SKIN EFFECTS. A single dermal application to rabbits resulted in marked skin irritation with a Draize score of 4.3 out of a possible 8.0 (260); five daily consecutive applications produced extreme irritation with a maximum score of 8.0 (261). 1,4-BDGE is a severe sensitizer in guinea pigs (266, 267).

EYE IRRITATION. Draize scores in rabbits have been 44 to 80 out of a possible 110, indicating severe irritation (268).

CHRONIC TOXICITY/CARCINOGENICITY. A 2-year dermal cancer bioassay was conducted in CF1 mice at doses of 0, 0.05, and 0.2 percent in acetone (269). Treatment did not adversely affect survival, did not increase the incidence of skin tumors, and did not result in significant skin irritation. There were no statistically significant increases in incidence of any systemic tumors, except for lymphatic tumors in females. Because there was a high background incidence of this tumor type in CF1 mice used in the testing laboratory (270), there was no clear evidence of 1,4-BDGE induced carcinogenicity.

GENETIC TOXICITY. 1,4-BDGE was mutagenic in several *Salmonella* strains in the Ames assay (191) and was positive in a mouse lymphoma assay (271), in both instances with and without activation. When 1,4-BDGE was gavaged in Chinese hamsters at daily doses of 0.6 to 3.0 g/kg for 2 days, there was a significant increase in the percentage of bone marrow cells with nuclear anomalies at 24 hr after the second dose (272).

EFFECTS IN HUMANS. 1,4-BDGE is a skin sensitizer in exposed workers (273).

Hygienic Standards of Permissible Exposure. No standards of permissible occupational exposure were found.

Odor and Warning Properties. No information on odor or warning properties was found.

2.2.4.3 Neopentyl Glycol Diglycidyl Ether (CASRN 17557-23-2). The structure is

$$(CH_3)_2 = C = (CH_2 O - CH_2 - \overset{\displaystyle O}{\overset{\diagup \diagdown}{CH}} - CH_2)_2$$

The ether is made by condensing neopentyl glycol with epichlorohydrin followed by dehydrochlorination with caustic. This may be used as a bifunctional reactive diluent for viscosity reduction of resins with minimum change in cured resin properties and to allow increased filler incorporation. Vapor exposure should not be hazardous unless the compound is heated without adequate ventilation.

Physical and Chemical Properties. The properties are as follows:

Specific gravity (25/4°C)	1.07
Flash point (open cup)	190°F min.
Viscosity (25°C)	10–16 cm/sec
Color, Gardner-Holdt	3 max.
Epoxide equivalent weight	135 to 155

Hazardous polymerization may occur with aliphatic amines in masses greater than 1 lb.

Determination in Atmosphere. No OSHA standard methods for the determination of NPGDGE were found. The pyridinium chloride–chloroform method for epoxy groups (122) and GLC are suggested but the HCl–dioxane method for epoxides (122) might also be used.

Physiological Response. Acute toxicity of NPGDGE is low by both the oral and dermal routes of exposure. The compound is slightly irritating to the rabbit eye but is a severe skin irritant and skin sensitizer. Systemic toxicity was not apparent following repeated dermal exposure. NPGDGE appears to be a weak genotoxin and weak carcinogen in mice.

ACUTE TOXICITY. In rats, the oral LD_{50} was 4.5 g/kg, and the dermal LD_{50} was >2.15 g/kg (274).

SKIN EFFECTS. NPGDGE was a moderate skin irritant in rabbits, with a Draize irritation score of 2.3 out of a possible 8.0 (275). In a 5-day repeated dermal exposure study, skin irritation became severe with necrosis (276). NPGDGE was a potent skin sensitizer in guinea pigs (277, 267).

EYE IRRITATION. NPGDGE was only slightly irritating to the washed rabbit eye (278).

SUBCHRONIC TOXICITY. In two separate 5-day repeated dermal application studies in rabbits, in which 0.5 ml of NPGDGE was applied occluded for 24 hr, daily, there was no evidence of systemic toxicity at necropsy carried out at 5 days after the last exposure (277).

CHRONIC TOXICITY/CARCINOGENICITY. A 2-year dermal carcinogenicity study of NPGDGE in C3H mice was conducted at dose levels of 0.94, 1.87, and 3.75 mg/mouse week (280). Skin tumor incidences were 10/50, 6/50, and 0/50 at 3.75, 1.87, and 0.94 mg/mouse week; no tumors occurred in the 300 acetone-treated controls. Tumor potency was calculated to be 1/700th that of benzo[a]pyrene, which was tested concurrently.

GENETIC TOXICITY (203). NPGDGE was positive in the Ames test with strain TA1535 without activation, but was negative with activation; it was negative in strain TA98 with or without activation. In a host-mediated assay in mice, minimal positive results were obtained in strain TA1535 with the addition of β-glucuronidase, but not without activation. Positive results were obtained with induction of DNA repair in cultured human leukocytes. Negative results were seen in a micronucleus test and in a dominant lethal test in mice with dermal application of 1.5 g/kg three times a week for 8 weeks to the males.

Hygienic Standards of Permissible Exposure. No standards of permissible occupational exposure were found.

Odor and Warning Properties. No information was found on odor or warning properties.

2.2.5   Aromatic Diglycidyl Ethers

2.2.5.1 Resorcinol Diglycidyl Ether (CASRN 101-90-6). This resin has the formula

$$O-CH_2-CH-CH_2$$
$$O-CH_2-CH-CH_2$$

Source, Uses, and Industrial Exposures. Resorcinol diglycidyl ether (RDGE) is made by the reaction of epichlorohydrin and resorcinol in the presence of caustic. It is suggested for use as an epoxy resin, as a stabilizer of organic chemicals, as a curing agent for "Thiokol" rubber, and for the solubilizing of protein adhesives.

Physical and Chemical Properties. Resorcinol diglycidyl ether is a colorless solid with a slight phenolic odor and following properties:

Molecular weight	222
Specific gravity	1.2183
Melting point	32 to 33°C
Boiling point	
0.05 mm Hg	150 to 160°C
12 mm Hg	208 to 210°C
Vapor density	7.95
Refractive index (20°C)	1.5409
Epoxy number	110

Determination in the Atmosphere. Inhalation exposure to vapors from these resins is unlikely because these materials are essentially nonvolatile. Although dust conditions have not been encountered, sampling could be carried out by entrapment of airborne particles by standard filter paper or impinger techniques. Subsequent analysis may be carried out through the HCl–dioxane method (122).

Physiological Response. RDGE is low in acute oral, dermal, and inhalation toxicity but moderately to severely irritating on skin contact. RDGE is severely irritating and damaging to the eye.

ACUTE TOXICITY. The single-dose oral LD_{50} values for rats, mice, and rabbits, respectively, were 2.57, 0.98, and 1.24 g/kg (190). Intraperitoneally the LD_{50} values were 0.178 and 0.243 for rats and mice, respectively (190).

The percutaneous LD_{50} in the rabbit was 2.0 ml/kg when RDGE (60 percent in xylene) was applied to the skin but not occluded (290). When RDGE remained in continuous contact with the skin, the percutaneous LD_{50} was 0.64 ml/kg (291).

Air "saturated" with RDGE vapors did not produce death to mice or rats after 8 hr of exposure (190). Rats exposed to a concentrated aerosol of 44.8 mg RDGE

(60 percent in xylene) per liter of air for 4 hr died within 5 days post-exposure (290).

SKIN EFFECTS. Single applications of RDGE to the skin of rabbits resulted in moderate irritation with a Draize score of 5 out of a possible 8 (190). Repeated applications were severely irritating with a leathery appearance of the skin and a Draize score of 7 out of 8 (190). Topical application of 0.01 ml of a 10 percent solution of RDGE in acetone to the skin of five rabbits produced a definite erythema and edema; 0.5 ml of RDGE (60 percent in xylene) applied topically to rabbits for 24 hr produced severe irritation which progressed to necrosis (290).

EYE IRRITATION. The score by the Draize method was 45/110; the resin was severely irritating (190). In another study, instillation of 0.5 ml RDGE (60 percent in xylene) in the rabbit eye resulted in severe inflammation and corneal necrosis (290).

SUBCHRONIC TOXICITY. In an early investigation, there was no evidence of toxicity as measured by body weight and organ weights in rats exposed for 50 7-hr exposures to air saturated with RDGE vapors (190). However, seven repeated skin applications of 1 ml total dose caused mortality in rabbits (190).

Monkeys receiving 100 to 200 mg/kg intravenously once monthly showed a progressively increasing depression of the white blood cell count (292).

Fourteen-day and 13-week repeated dose gavage toxicity studies in rats and mice have been conducted by NTP. The test material was 81 percent pure and was gavaged in a corn oil vehicle. Dose levels were sufficient to result in increased mortality for both rats and mice in the 14-day study, mean body weight was depressed in nearly all groups, and gross pathological examination of both species revealed lesions of the stomachs and renal medulla (reddening) with papillary growths in the stomachs of many of the dosed rats. In the 13-week study, partial mortality (rats, 1/20; mice, 16/20) occurred at the top dose levels of 200 mg/kg/ day and 400 mg/kg/day for rats and mice, respectively. Mean body weight was depressed in male rats dosed with 100 mg/kg and above and females dosed at 200 mg/kg; mice of the 400 mg//kg/day group had depressed body weights. Compound-related observations in the nonglandular stomach of both rats and mice included inflammation, ulceration, squamous cell papilloma, hyperkeratosis, and basal cell hyperplasia at dose levels of 12.5 mg/kg and above in rats and 25 mg/kg/day and above in mice. Some histopathological changes in the liver occurred in both rats and mice including necrosis and fatty metamorphosis at the top dose levels only (293).

CHRONIC TOXICITY/CARCINOGENICITY. Earlier studies have reported that mice receiving repeated applications to the skin for 1 year developed cancers, and rats similarly receiving repeated subcutaneous injections developed sarcomas (294). In what appeared to be a lifetime dermal study, 5 percent RDGE in methyl ethyl ketone was applied to the skin of C3H mice (number unspecified). The test material was an 81 percent pure commercial product, with the balance comprising other impurities. Approximately 50 mg of the solution was applied to each mouse twice a week. After 36 weeks a benign papilloma appeared on one mouse that survived for an additional 15 weeks. At week 48 a subdermal growth appeared on another mouse which was subsequently diagnosed as a squamous cell cancer. Apparently there was no concurrent control group included in this study (295). In a more recent lifetime dermal study, 1 percent RDGE in benzene was applied to the skin of 30 female Swiss-Mellerton mice at a dose level of 100 mg dosing solution/mouse three times per week. No benign or malignant skin tumors were observed even though moderate to severe crusting and/or scarring and hair loss occurred at the site of application. The median survival time was 70 weeks for the treated group, 71 weeks for the negative control group (benzene-treated), and 63 weeks for the untreated control group (125).

A 2-year oral carcinogenicity study on RDGE has been conducted using rats and mice by NTP (293). The test material was an 81 percent pure commercial product, dosed via gavage in corn oil five times a week for 2 years. The dose levels for mice were 0, 50, or 100 mg/kg/day; the dose levels for rats were 0, 12, 25, or 50 mg/kg/day. At the 50 mg/kg/day dose level there was a significant decrease in body weights and a significant increase in mortality for rats. At the 25 mg/kg dose level there was a transient decrease in body weight in rats (at weeks 80 to 100) and the survival was significantly decreased. There were no effects on body weight at the 12 mg/kg/day dose level, but male rats had a significantly lower survival rate (46 percent) than controls (78 percent) at week 104. Histologically, RDGE produced hyperkeratosis, hyperplasia, and neoplasia of the squamous epithelium of the nonglandular stomach in all treated groups. Squamous cell carcinomas were observed in the nonglandular stomach; the respective incidence in the 0, 12, and 25 mg/kg/day group males was 0/100, 39/50, and 38/50; and the incidence in females was 0/99, 27/50, and 34/50.

Genetic Toxicity. RDGE was mutagenic in *S. typhimurium* strains TA100 and TA1535 with or without metabolic activation (245, 191). Seiler (245) also showed RDGE to produce chromosomal aberrations in Chinese hamster ovary cells in vitro at a concentration of 8 and 25 μg/ml, but reported that RDGE was negative in the mouse micronucleus test following oral gavage doses of 300 or 600 mg/kg.

Pharmacokinetics and Metabolism. Intraperitoneal administration of RDGE results in an approximately 10-fold lower LD_{50} than observed following oral administration (190), suggesting RDGE is poorly absorbed from the gastrointestinal tract. Following oral administration of RDGE in an aqueous 10 percent DMSO solution it was shown to be metabolized in part to the bis-diol derivative (296). RDGE has also been shown to conjugate with glutathione via gluthathione-*S*-transferase in vitro (297).

Effects Observed in Humans. Severe burns and skin sensitization on local contact have been observed (25). Leukopenia and the appearance of atypical monocytes in the peripheral blood have been reported in humans exposed to RDGE (298).

Hygienic Standards of Permissible Exposure. No occupational exposure standards were found.

Odor and Warning Properties. The phenolic odor is easily perceptible.

2.2.5.2 Diglycidyl Ether of Bisphenol A (CASRN 1675-54-3, 25068-38-6, 25085-99-8, 25036-25-3).

These compounds include the ethers of bisphenol A [2,2-bis(4-hydroxyphenyl)propane] and condensation products of their further reaction or advancement with bisphenol A (some trademarks are EPON® resin series, D.E.R.® series, Epotuff® series, Araldite® series, EPI-Rez® series, and the ERL Bakelite® epoxy series. The general structure is

Source, Uses, and Industrial Exposure. The synthesis of the basic epoxy resin molecule involves the reaction of epichlorohydrin with bisphenol A, the latter requiring two basic intermediates for synthesis, acetone and phenol. Theoretically, the production of the diglycidyl ether of bisphenol A requires 2 mol of epichlorohydrin for each mole of the phenol.

Epoxy resins of higher molecular weight are obtained by reducing the epichlorohydrin/bisphenol A ratio. This reaction involves consumption of the initial epoxy groups in the epichlorohydrin and of some of the groups formed by dehydrohalogenation.

The properties of epoxy resins make them ideally suited for sealing, encapsulating, making castings and pottings, and formulating lightweight foams. Castings may be used for patterns, molds, and finished products. These resins are used as binders in the preparation of laminates of paper, polyester cloth, fiber glass cloth, and wood sheets. The epoxy resins have outstanding adhesive properties.

The amines were the first materials to gain general acceptance as curing agents for epoxy resins. Polyfunctional primary aliphatic amines give fast cures and provide overall properties satisfactory for a wide variety of applications. These materials are usually considerably more active physiologically than the epoxy resins and more volatile, and skin and eye irritation may occur. Other curing agents, including the acid anhydrides and organic acids, have given fewer problems in handling. A number of diluents are also physiologically more active than the resins themselves.

Some of these are the epoxy esters, ethers, and aliphatic compounds of low molecular weight.

Resin modifiers include phenolic substances, aniline formaldehyde resins, furfural, isocyanates, and silicone resins, all of which may contribute to the handling problems.

Physical and Chemical Properties. The resins are usually mixtures and may contain homologues of higher weight, isomers, branched-chain homologues, and occasionally, monoglycidyl ethers.

The general formula may be written as indicated for the molecular formula above, where n is the number of repeating units in the resin chain.

	$n = 0$	$n = 2$	$n = 9$
Molecular weight (approx.)	350	900	2900
Specific gravity	1.168	1.204	1.146
Melting point (°C)	8 to 12	64 to 76	127 to 133
Epoxy equivalent	190 to 210	450 to 525	1650 to 2050

The lower-molecular-weight resins are liquids, and as the molecular weight increases, they become increasingly viscous and finally solids. Hazardous polymerizations may occur with aliphatic amines in masses greater than one pound.

Determination in the Atmosphere. Inhalation exposure to vapors from these resins is unlikely because these materials are essentially nonvolatile. Dust from the solid resins may be trapped according to standard filter paper and liquid entrapment methods and determined by the HCl–dioxane method (122).

Physiological Response. The acute systemic toxicity of pure DGEBPA and the low-molecular-weight DGEBPA-based resins is low by either dermal or oral administration. A single dermal dose of these materials produces only slight irritation to the skin of rabbits, but repeated dermal application may produce greater irritation. Dermal exposure with liquid DGEBPA-based resins has produced skin sensitization in guinea pigs. Thus contact with the skin presents the greatest potential problem, and it has been well-documented that the lower-molecular-weight epoxy resins can cause dermatitis and skin sensitization in certain individuals upon prolonged or repeated exposure. Manufacturers of epoxy resins recommend that precautionary action be taken to avoid all skin contact with uncured resins as well as other components of epoxy resin systems. Owing to their sticky nature, resins on the skin are difficult to remove, and there often is a tendency to use solvents to remove them. This is NOT recommended, because the solvents may facilitate penetration of the resin through the skin. These materials produce minimal irritation to the eye. Inhalation exposure to these resins is unlikely owing to the very low vapor pressure of these materials. Oral subchronic studies on these materials have shown that the systemic toxicity of these materials is very low. In addition, four separate studies have indicated that DGEBPA or DGEBPA-based resins are not teratogenic. No adverse effects on reproductive parameters or reproductive organs were noted in either male or female rats gavaged with a low-molecular-weight DGEBPA-based resin. DGEBPA and various DGEBPA-based epoxy resins have been tested in several in vitro and in vivo assays for genotoxicity. Although some of the in vitro mutagenicity assays for these materials have given positive results, the results of the same and other in vitro tests have been negative, and all of the in vivo tests for mutagenicity have been negative. Metabolism studies have shown that DGEBPA is only slowly absorbed through the skin, and is detoxified by metabolism to the bis-diol of DGEBPA by the enzyme epoxide hydrolase. There is some evidence to suggest that very little parent compound would be systemically available following the oral administration of DGEBPA. A number of carcinogenicity studies involving the topical application of pure DGEBPA, as well as EPON® Resin 828 and other commercial DGEBPA-based resins, have been carried out in experimental animals. Viewing the studies as a whole, the weight of evidence does not show that DGEBPA or DGEBPA-based epoxy resins are carcinogenic.

The acute systemic toxicity of solid, that is, the advanced bisphenol A/epichlorohydrin epoxy resins (advanced DGEBPA-based) is low by either dermal or oral routes. Inhalation of these materials is unlikely because of their low volatility. Instillation into the eyes of rabbits produced only slight irritation. Dermal contact may result in slight irritation, especially with repeated or prolonged exposure. However, unlike DGEBPA-based resins of lower molecular weights, these materials do not appear to cause delayed contact hypersensitivity on the skin. Subchronic oral administration of an advanced DGEBPA-based epoxy resin did not result in demonstrable toxicity. Mutagenicity testing has produced mixed results. Chronic dermal administration did not induce tumor formation in mice.

ACUTE TOXICITY. Single-dose oral toxicity of DGEBPA is very low; early limit tests resulted in estimations of the LD_{50} at values greater than 2000 or 4000 mg/kg (299–301). Hine et al. (190) reported "exact" oral LD_{50} values of 11,400 mg/kg in rats, 15,600 mg/kg in mice, and 19,800 mg/kg in rabbits for a commercial DGEBPA-based epoxy resin. Weil et al. (69) reported an oral LD_{50} of 19.6 ml/kg (~19,600 mg/kg) for rats with a commercial DGEBPA-based epoxy resin. More recent studies with pure DGEBPA or commercial DGEBPA-based resins have produced results consistent with those previously reported; the single-dose oral LD_{50} value was reported as >1000 mg/kg in the rat and >500 mg/kg in the mouse (302–304). Lockwood and Taylor (305) found the single-dose oral LD_{50} value for rats to be >2000 mg/kg. The pharmacological effects observed even in lethal doses were not remarkable. Moderate antemortem depression occurs, and loss of body weight and diarrhea are often observed in surviving animals. Intraperitoneal toxicity is greater than oral toxicity by 5- to 10-fold (2400 and 4000 mg/kg in rats and mice, respectively) (190).

The potential for absorption through the skin in acutely toxic amounts is low; the single-dose dermal LD_{50} value in rabbits has been reported to be 20 ml/kg for a DGEBPA-based commercial resin (69). Lockwood and Taylor (305) reported 100 percent survival with no adverse effects for rabbits treated with a single dermal dose of DGEBPA-based commercial resin at a dose level of 2000 mg/kg. In other species, studies show dermal LD_{50} values of pure DGEBPA to be >800 and >1600 mg/kg for mice and rats, respectively (302). The acute dermal toxicity of a commercial DGEBPA-based resin was similar with single-dose dermal LD_{50} values of >800 and >1200 mg/kg for mice and rats, respectively (303, 304, 306).

The inhalation toxicity of DGEBPA or DGEBPA-based resins has not been studied because vapor exposures are unlikely owing to the low vapor pressure of the material. Nolan et al. (307) reported difficulty in generating an atmosphere at a respirable temperature (~22°C) that contained sufficient DGEBPA to conduct a rodent inhalation study even when using large surface areas and high temperatures initially to generate an atmosphere prior to cooling to respirable temperatures.

SKIN EFFECTS. Single prolonged (24 hr) application to the skin of rabbits showed DGEBPA-based resins to be only slightly irritating at most, even when occluded or if skin was abraded (190, 299, 300, 305, 308–311). Repeated applications were reported to be more irritating (190, 312). Application of the same liquid resin for 4 hr/day for 20 days resulted in Draize scores of 0 to 8, indicating that prolonged and repeated skin contact with liquid resins may cause severe irritation.

When representative solid resins (mol. wt. = 900 and 2900) were applied for 4 hr/day for 20 days to rabbits, Draize scores of 0 were seen, which indicates that the solid resins are less likely to cause primary irritation even with repeated or prolonged skin contact. However, it is still strongly recommended that skin contact with either solid or liquid resins be avoided (190).

SKIN SENSITIZATION. Several studies have reported lower-molecular-weight DGEBPA-based resins to produce skin sensitization in guinea pigs (303, 304, 306, 313–318). However, higher-molecular-weight resins have not produced this effect (319, 320). For additional information see the section below on Effects in Humans.

EYE IRRITATION. DGEBPA-based resins have been reported to cause only minimal eye irritation (190, 299, 300, 303, 305, 311, 321). Eye irritation in an industrial setting usually results from inadvertent transfer of resin from the hands when rubbing the eyes, and may occur even if the hands are protected with gloves. The use of goggles is an effective way to prevent accidental eye exposure.

Solid resins of higher molecular weight (~900 and above) are capable of causing moderate eye irritation by the Draize test, and scores as high as 41/110 have been reported (322). The greater degree of eye irritation potential with the solid resins is due to their ability to form dust. The small particles of resin can thus cause mechanical abrasion of the eye and surrounding tissue. Again, use of goggles is effective in preventing exposure to resin dusts.

SUBCHRONIC TOXICITY—ORAL ADMINISTRATION. Incorporation of 1 percent undiluted uncured liquid resin of molecular weight 450 (EPON® resin 834) or as 75 and 50 percent in dioctyl phthalate into the diet of rats for 26 days resulted in body weight loss and decreased food consumption, but no gross or histopathological lesions (323). Incorporation of a semisolid resin of molecular weight about 600 (EPON® resin 836) as 1 percent indiluted, uncured resin, or as 50 percent in dioctyl phthalate in the diet of rats for 26 days gave the same results (323). A lower-molecular-weight resin (mol. wt. = 350, EPON® resin 828) was cured separately with three different curing agents (boron fluoride, metaphenylenediamine, and EPON® curing agent E); 1, 5, and 10 percent of each cured resin was added to the diet of rats for 6 weeks (323). There was no evidence of behavioral change during treatment, and no diarrhea or other abnormalities were seen. Rats in the 10 percent groups ate more than controls, but actual "food" intake was about equivalent to controls. There were no statistically significant differences from control for growth rate or for liver/body and kidney/body weight ratios. The average amount of cured resin consumed by each rat was 2.4 g. In one subchronic dietary study, rats were fed DGEBPA in their diets for 3 months at concentrations up to 3 percent (324). Rats at the highest level rejected the diets and failed to gain weight; these rats showed effects upon gross and histopathological examination that were consistent with under-nutrition. There was no evidence of systemic toxicity at any levels. In another subchronic study, DGEBPA was fed to rats at dietary concentrations of 0.2, 1, or 5 percent for 26 weeks (190). All rats at the highest dose died by the end of 20 weeks, but gross and histopathological examination did not reveal evidence of systemic toxicity at any dose (190). A 28-day study with a low-molecular-weight DGEBPA-based resin (Araldite GY250) in which rats were gavaged daily with the resin in an aqueous solution of 0.5 percent carboxymethylcellulose/0.1% Tween 80 at doses of 0, 50, 200, or 1000 mg/kg/day did not alter any of the following parameters relative to controls: body weight, food consumption, water consumption, food conversion, mortality, clinical observations, eyes or hearing, hematology, blood chemistries, organ weights, gross pathology or histopathology of the spleen, heart, liver, kidney, or adrenal gland (325).

The effects of incorporating solid resins into the diet of rats has also been studied. A resin of molecular weight 950 (EPON® resin 1001) was cured separately with two different curing agents (diethylenetriamine and Versamide 115), and 1, 5, and 10 percent of each cured resin was added to the diet of rats for 6 weeks (323). There were no adverse effects noted in any group with respect to behavior, toxic effects, growth rate, or liver/body and kidney/body weight ratio. There was increased food consumption in the 10 percent groups. No gross or microscopic lesions were found. A resin of molecular weight 3000 (EPON® resin 1007) was added at 2 percent to the diet of rats for 12 weeks as the uncured resin, cured with diethylenetriamine, and cured with urea–formaldehyde resin (323). An unspecified number of deaths occurred in each group, including controls, which were described as being unrelated to treatment. Growth rate of treated groups was not statistically different from control, and no other toxic effects were described.

DERMAL ADMINISTRATION. A DGEBPA-based resin (Epidian® 5) was applied at 6.8 mg/day for 17 days to pregnant and nonpregnant rats (326, 327). Local effects (erythema, edema, and erosions) were seen on the skin of the treated sites with all groups. Nonpregnant rats had reduced body weight, hyperemia of livers and kidneys, and elevated kidney and liver enzymes. There was a decrease in the brain acetylcholinesterase activity of nonpregnant (but not pregnant) rats treated with Epidian® 5 reported in one study (326), but in the other study by these same investigators neither pregnant nor nonpregnant rats had reduced acetylcholinesterase activity (327). In pregnant rats, there was a slight body weight gain, atrophy of adipose tissue, reduced liver and kidney enzymes, hyperemia of liver and kidney, elevation of other enzymes, and slight changes in amniotic fluid compared to controls. In other related studies by these investigators, pregnant guinea pigs treated dermally with Epidian® 5, (328, 329) appeared more susceptible to the actions of the resin than the nonpregnant ones.

INTRAMUSCULAR ADMINISTRATION. An undiluted liquid resin of molecular weight 350 (EPON® 828) was injected intramuscularly in rats at 800 mg/kg once a day for 12 days (237). The only significant effect was a reduction in the rate of body weight gain.

CHRONIC TOXICITY AND CARCINOGENICITY. Older studies in which cured and uncured, solid and liquid DGEBPA-based epoxy resins were orally administered at concentrations up to 10 percent in the diet produced no adverse effects in rats with exposures up to 26 weeks (190). Neither did the oral administration in the diet or intraperitoneal or subcutaneous injection of solid and liquid epoxy resins into mice susceptible to lung tumors cause any statistically significant increase in incidence of lung tumors compared to negative controls (330). The subcutaneous implantation of 1 × 1 cm disks of cured epoxy resin in mice for 575 days did not result in tumors (331). Other, older studies where EPON® resin 828 was repeatedly injected subcutaneously into rats caused 4 of 30 animals to develop fibrosarcoma at the injection site (190), but this result is not felt to be surprising because other materials not considered to be carcinogenic also produce this same type of tumor under the same conditions (e.g., nylon, table salt, glucose).

Information more relevant to the potential occupational exposure to epoxy resins is provided by the numerous studies involving long-term skin application of DGEBPA-based resins. In an early study, these materials were applied to the skin of mice as acetone solutions (0.3 or 5 percent), three times weekly for 2 years. A solvent control as well as a positive control group were also included. There was no increase in the incidence of grossly detectable skin tumors in any of the resin-treated groups (190).

In another skin painting study in mice, "one brushful" of undiluted resin was applied to the skin of C3H mice three times weekly for up to 23 months. A skin papilloma was detected in a single mouse after 16 months of treatment, at which time 32 of the 40 mice started on the study were still alive (69). When the study was repeated, twice for 24 months and once for 27 months, no skin tumors were found (332).

Holland et al. (333) investigated the carcinogenic potential of a modified commercial EPON® 828 type resin. The test material was applied as a 50 percent solution in acetone to the skin of C3H and C57BL/6 mice of both sexes, three times weekly for 2 years at a dose of 15 or 75 mg/kg per week. No skin tumors were found in the C3H mice, but a weak carcinogenic response was noted in the C57BL/6 strain. However, it was subsequently found that the resin sample used in this test contained atypically high levels of several active contaminants, including epichlorohydrin (1500 ppm), phenyl glycidyl ether (830 ppm), and diglycidyl ether (3400 ppm), as well as about 10 percent of a presumed diluent identified only as an epoxidized polyglycol. In view of the presence of the contaminants, and particularly the high level of the epoxidized polyglycol, the weak carcinogenic response noted in the one mouse strain cannot be clearly ascribed to the resin.

In a subsequent study by the group at Oak Ridge National Laboratory (280), EPON® resin 828 and two comparable commercial resins from other manufacturers were evaluated in the C3H mouse following the protocol used in the earlier study. None of the three resins elicited skin or systemic tumors in the test animals.

Of potentially greater interest is the fact that a 1:1 mixture of DGEBPA based resins and bis(2,3-epoxycyclopentyl) ether used as an epoxy diluent produced a clear increase in the incidence of skin tumors despite the fact that neither compound gave positive results when applied individually (280). A recent study has provided evidence that mechanism of this effect may be related to the ability of bis(2,3-epoxycyclopentyl) ether to inhibit microsomal epoxide hydrolase activity, which results in DNA binding by a metabolite of DGEBPA (glycidaldehyde) formed via a different metabolic pathway (334). These data add further emphasis to recommendations for the proper handling of epoxy resins (particularly with respect to diluted resins), and they should be considered in establishing industrial hygienic practices, recommending protective equipment, and advising remedial action in case of overexposure.

The carcinogenic potential and chronic dermal toxicity of three commercially available DGEBPA-based resins was more recently investigated by Agee et al. (335). The test materials were dissolved in acetone and 50 μl applied topically, twice a week for 94 weeks, to the backs of C3H/HeJ male mice, 50 per treatment group. The three DGEBPA-based resins tested were 42 percent DGEBPA (Code C-618), 76 percent DGEBPA (Code C-621), and 27 percent DGEBPA (Code C-660) (336) and were tested in acetone at concentrations of 50 percent, 25 percent, and undiluted, respectively. Thus the actual concentrations of DGEBPA applied were 21 percent (C-618), 19 percent (C-621), and 27 percent (C-660). Two groups of 50 mice each were treated twice weekly with 50 μl of acetone or 0.025 percent benzo[a]pyrene in acetone to serve as negative and positive control groups, respectively. An additional group of 50 mice received no treatment as a negative control group. The skin from all animals was examined by light microscopy for non-neoplastic and neoplastic lesions, and histopathological examination of internal organs was conducted on half of the mice from each group. Forty-eight of the mice in the positive control group developed skin tumors with an average latent period of 32.4 weeks, whereas no skin neoplasms were observed in either of the negative control groups or in the groups treated with resins in acetone at a final concentration of 19 or 27 percent DGEBPA (i.e., C-621 and C-660). Three mice of the 50 treated with C-618 in acetone (21% DGEBPA) had microscopically detected skin papillomas, but no malignant neoplasms of the skin were present in any of the animals in this treatment group. The incidence of hepatocellular carcinoma observed in the treated and control groups was within the range of those detected in historical control animals from the same laboratory and below values reported by the animal supplier for this strain of mouse. Thus under the conditions of this study, dermal application of these DGEBPA-based resins did not produce a carcinogenic response in mice.

Zakova et al. (337) evaluated the carcinogenic potential of Araldite GY250 a DGEBPA-based epoxy resin in CF1 mice. Groups of 50 male and 50 female mice were treated for 2 years by repeated epidermal application of a 1 or 10 percent (v/v) solution in acetone. Controls were treated with acetone alone. The treatment had no effect on survival, and no excess incidence of skin or systemic neoplasia occurred.

Comprehensive studies on the carcinogenic potential of the EPON® 828-type resins have also been carried out recently at the Shell Toxicology Laboratory in the United Kingdom. Groups of CF1 mice of each sex were exposed to pure DGEBPA, EPON® resin 828, or another comparable commercial resin. The test materials were applied as a 1 or 10 percent solution in acetone, 0.2 ml twice weekly, for 2 years. Solvent (acetone) and positive (β-propiolactone) groups were also included (270).

The animals treated with β-propiolactone showed a high incidence of skin tumors in comparison to the solvent control groups, demonstrating the susceptibility of the CF1 mouse to a chemical known to produce skin cancer. In the EPON® resin 828-treated mice, the incidence of cutaneous tumors of the treated site or of the skin at all sites was not statistically significantly different from controls. In the two other treatment groups, some skin tumors were observed, but statistical analysis of this tumor data revealed the incidence was not significantly different from controls. This skin tumor data was compared to the incidence of skin tumors in control CF1 mice from two other chronic studies conducted in the Shell laboratory. Based on the low incidence of skin tumors in these "historical" control mice, the authors suggested that DGEBPA and one of the commercial resins, but not EPON® resin 828, exhibited a low order of carcinogenic potential to the skin of CF1 mice. However, the historical control data that were used for comparison by the authors were very limited, with only 100 males and 200 females on test for 2 years. The study of Zakova et al. (337) with a similar DGEBPA-based epoxy resin (Araldite GY 250), conducted by another laboratory at the same time using the same protocol and with CF1 mice supplied by the Shell laboratory, did not result in an excess of either skin or systemic tumors (see above). When the results of these two studies were combined (including additional historical control data from Zakova et al.), Peristianis and co-workers concluded that there was no evidence of carcinogenic activity of these resins in mouse skin (270).

With regard to systemic neoplasia in the mice treated with acetone solutions of EPON® resin 828, there was a statistically significant linear trend in the dose response for renal tumors in male mice when the data were analyzed by the method of Peto et al. (338). The renal tumor incidences were 6, 0, and 12 percent in the control, low-dose, and high-dose groups, respectively. This finding is not considered to be related to treatment because the authors state that renal neoplasms in male CF1 mice are common in the testing laboratory, and the absence of this tumor in the 1 percent EPON® resin 828 was rather unusual. Furthermore, if DGEBPA was truly a causative agent for renal neoplasia, one would have expected kidney tumor incidence to have shown a statistically significant increase in the test statistic for trend in dose response in all of the groups treated with DGEBPA containing resins. Yet there was not a statistically significant trend in dose response for renal neoplasia in male or female mice treated with EPIKOTE® resin 828, pure DGEBPA (270), or Araldite GY 250 (337), indicating the evidence for linking DGEBPA to renal neoplasia in mice is tenuous.

Peristianis and co-workers (270) also reported a statistically significant increase in the dose–response trend for lymphoreticular/hematopoietic tumors in female mice treated with pure DGEBPA or EPIKOTE® resin 828 when the data were analyzed by the method of Peto et al. (338). However, it is likely that this finding is not treatment-related, because the CF1 mice raised in this testing laboratory have a relatively high background incidence of these lesions. It was considered likely that the mice were susceptible to the development of lymphoreticular/hematopoietic tumors as a result of the presence of virus and/or a genetic tendency to viral infection. There was not a statistically significant increase in the dose–response trend for lymphatic tumors in female or male CF1 mice treated with other DGEBPA-based resins, either EPON® resin 828 (270) nor Araldite GY 250 (337), suggesting that DGEBPA was not the causative agent for these lesions.

In summary, a number of carcinogenicity studies involving the topical application of pure DGEBPA, as well as EPON® resin 828 and other commercial DGEBPA-based resins have been carried out in experimental animals. Viewing the studies as a whole, the weight of evidence does not show that DGEBPA- or DGEBPA-based epoxy resins are carcinogenic.

DEVELOPMENTAL TOXICITY. EPON® 828 was not teratogenic in rats or a chick embryo assay but was embryo toxic at doses of 10 percent of the oral LD_{50} (339). The reference cited for this statement is an abstract and does not provide any information regarding specific chemicals tested. In a dermal teratology probe study, rabbits were administered doses of 0, 100, 300, or 500 mg/kg/day on days 6 to 18 (340). No embryo toxicity was observed at any dose. The full teratology study, conducted at dermal doses of 0, 30, 100, or 300 mg/kg in the rabbit, showed no evidence of embryo/fetal toxicity or teratogenicity (312). Gavage teratology studies using both rats and rabbits with a low-molecular-weight DGEBPA-based epoxy resin (Araldite GY250 or TK10490) have also been conducted (341, 342). Dose levels of 0, 60, 180, and 540 mg/kg/day were used for rats and dose levels of 0, 20, 60, and 180 mg/kg/day were used for rabbits. The test material in both studies was suspended in an aqueous solution of 0.5 percent carboxymethylcellulose and 0.1 percent Tween 80 (polysorbate 80). Treatment at the top dose levels resulted in signs of material toxicity in both studies, but there were no adverse effects on mean litter size, pre- and post-implantation losses, or any evidence of a teratogenic or embryotoxic effect at any dose level.

REPRODUCTIVE TOXICITY. A one-generation reproduction study in rats has been conducted in which a DGEBPA-based epoxy resin (Araldite GY250 or TK 10490) was administered by gavage at dose levels of 0, 20, 60, 180, and 540 mg/kg/day (343). The vehicle used was an aqueous solution of 0.5 percent carboxymethyl-cellulose, 0.1 percent Tween 80. Oral administration of this resin to males for 10 weeks and females for 2 weeks prior to mating produced a lower mean body weight in males at 540 mg/kg/day, but did not affect mating performance, gestation period, or the ability of females to rear their offspring successfully to weaning (343). No treatment-related macroscopic changes, differences in mean organ weights, or histological changes to the reproductive and alimentary tracts (top dose only) in either sex of the F_0 generation were observed.

GENETIC TOXICITY. Both liquid and solid DGEBPA-based resins have been tested in several in vitro and in vivo assays for genotoxicity; the results of these tests have been reviewed (344). In vitro mutagenicity assays for these materials have given mixed results, whereas all the in vivo tests for mutagenicity have been negative.

PHARMACOKINETICS AND METABOLISM. DGEBPA was very slowly absorbed through the skin of mice (345). Following a single oral administration of ^{14}C-DGEBPA to mice the dose was relatively rapidly excreted as metabolites in the urine and feces and the profile of fecal and urinary metabolites was independent of the route of exposure (345). The major metabolite was the bis-diol of DGEBPA formed by hydrolysis of epoxides by epoxide hydrolase (346). The bis-diol was excreted in both free and conjugated forms, and was also further metabolized to various carboxylic acids (346). DGEBPA did not appear to be metabolized to phenyl glycidyl ether by mice. Metabolic pathways for DGEBPA in the rabbit appear similar to those described for the mouse (347).

Route-dependent differences in the plasma ^{14}C concentration–time profiles, tissue/plasma ^{14}C ratios, and urinary excretion following the intravenous or oral administration of ^{14}C-DGEBPA to rats have been observed (307). These data suggest that very little DGEBPA is absorbed unchanged following oral administration (307). The primary route of excretion in the rat was the feces after either intravenous or oral administration, although the plasma data suggest that only 13 percent of the orally administered radioactivity was absorbed so that some of the fecal radioactivity following oral administration may represent unabsorbed material (307). Consistent with the observations of Climie et al. (346), no unchanged DGEBPA was excreted in the urine or bile following oral or intravenous administration.

Bentley et al. (334) have investigated the hydrolysis of the epoxide functionalities of DGEBPA by the microsomal and cytosolic fractions of mouse liver and skin. These investigators reported that DGEBPA was rapidly hydrolyzed by the epoxide hydrolase of both tissues, with skin microsomal activity toward DGEBPA about 10 times greater than that found in the cytosol of skin.

Bentley et al. (334) also reported that the dermal administration of ^{14}C-DGEBPA (radio-labeled in the glycidyl side chain) resulted in radioactivity being covalently associated with the protein, DNA, and RNA purified from the skin at the site of application. Most of this radioactivity appeared to be a result of the metabolism of the glycidyl side chain to glyceraldehyde, a normal endogenous product of intermediary metabolism. Glyceraldehyde was subsequently metabolized to single carbon units which entered the one-carbon pool and were then incorporated into tissue macromolecules via normal catabolic pathways. These findings are consistent with those of Climie et al. (346), who could find no evidence for the in vivo metabolism of DGEBPA to glycidaldehyde in mice. However, Bentley et al. (334) did report the formation of small amounts of a DNA adduct that was tentatively identified as the reaction product of glycidylaldehyde and deoxyguanosine when doses of 0.8 or 2 mg/mouse of ^{14}C-DGEBPA were applied to the skin of mice (334). However, Bentley et al. based their identification on the liquid chromatography retention volume of the unknown peak as compared to a known standard and did not conclusively identify this "adduct" using other analytical techniques such as mass spectrometry. No adducts were found at the dose level of 0.4 mg DGEBPA/mouse, the lowest dose level used.

EFFECTS IN HUMANS. Occupational dermatitis from epoxy resins has been described by a number of workers including Hine et al. (190), Pluss (348), and Grandjean (349). The usual lesion is typical of contact dermatitis, the early manifestations being redness and edema, with weeping followed by crusting and scaling. Following initial contact there is usually an erythematous, discrete area confined to the point of contact. Because this frequently involves the face, it is likely that it is caused by the vapors of the hardener or active diluent, though contact with contaminated gloves or droplets may also play a part, as do occasionally the vapors of the liquid-type epoxy resin. The initial lesion usually persists for 48 hr to 10 days, after which the erythema fades and gives way to a macular rash followed by scaling.

Sensitization may follow the initial contact, resulting in the development of a papular, vesicular eczema. This is accompanied by considerable itching, and extension beyond the point of original contact. Only occasionally are areas other than the backs of the hands, the forearms, the face, and the neck involved. Recommended treatment consists of bland ointments and soaps. The worker is withdrawn from further contact, and the lesion usually subsides in 10 to 14 days. However, it may recur on further contact. If the worker is not withdrawn from contact, the dermatitis usually persists for longer periods, but usually does not become more intense. The lesions may assume a brownish color, and scaling is frequently noted.

Fregert and Thorgeirsson (350) reported that of 34 individuals previously experiencing skin sensitization as a result of occupational exposure to epoxy resins, all demonstrated a positive skin sensitization response to a DGEBPA-based epoxy resin of average molecular weight (MW) 340 following patch testing. Twenty-three of these individuals patch tested with resins of average MW 624 and MW 908 did not experience sensitization (350), and seven of these individuals patch tested with a resin of average MW 1192 did not react. However, eight patients tested with commercial mixtures of epoxy resins with average MW 1280 and MW 1850 reacted to these mixtures, which contained the MW 340 oligomer as determined by gel permeation chromatography (350). The authors concluded that the MW 340 oligomer is the component responsible for contact allergy to epoxy resins in humans. These results in humans are consistent with the absence of skin sensitization in guinea pigs for higher molecular weight resins (267, 351).

Because of the persistence of dermatitis and the possibility of sensitization, preventive measures are especially important when handling these compounds. The most important measure for combating dermatitis is good personal hygiene. This requires strong administrative controls, supervisory instruction of personnel, good work habits, and provision of adequate facilities for removing the material periodically, together with a designated cleanup period prior to "break" and quitting times. In addition, protective devices offer considerable aid in minimizing personal contact. Gloves should be worn, and contamination of the skin should be scrupulously avoided. Protective clothing and personal protective creams are of help. When possible, only persons with no history of allergic conditions or eczematous eruptions should be selected for epoxy resins applications. Procedures should be carefully supervised to ensure that mixing of volatile agents is carried out in properly ventilated areas. Bench and floor areas should be protected with disposable paper.

Remedial measures following skin contact should include thorough cleansing with soap and water, followed by a waterless hand cleanser when absolutely necessary. The use of solvents may promote epidermal penetration of materials that would otherwise not penetrate the skin. Accidental eye contamination is unlikely, but treatment consists of the usual measures. The source of contact should be identified when dermatitis develops, and the improper work condition corrected.

Reference has been made to occupational asthma in workers exposed to "fumes" of epoxy resins (352, 353). However, a closer examination of the literature revealed that the occupational asthma attributed to an "epoxy resin system" was due to fumes of phthalic anhydride, trimellitic anhydride, or triethylenetetramine, and that sensitized workers did not respond to the epoxy resin when exposed to that material alone (354). It should be further noted that it is unsure whether the epoxy resin contained unreacted DGEBPA.

Hygienic Standards of Permissible Exposure. No standards of permissible exposure to DGEBPA have been established.

Odor and Warning Properties. There are no odor and warning properties; the resins are sticky when handled.

2.2.5.3 Diglycidyl Ethers of Brominated Bisphenol A (CASRN 26265-087-7, 40039-93-8).

These resins are the condensation product of 2,2-bis(3,5-dibromo-4 hydroxyphenyl)propane and epichlorohydrin. Some of the trademarks for the polymeric brominated resins are D.E.R.® epoxy resin 500 series, EPON® resin 1000 series, Araldite LT 8000 series, and Epi-Rez 5163. The monomeric resins have the formula:

$$CH_2-CHCH_2-O-\text{(ring)}-C(CH_3)_2-\text{(ring)}-O-CH_2CH-CH_2$$

Sources, Uses, and Industrial Exposure. Bromine has been incorporated into DGEBPA-type resins to increase the ignition resistance of such resins by the use of tetrabrominated bisphenol A for the synthesis of the glycidyl ethers. Because bromine substitution reduces the thermal stability of the resins, the products are usually used with or condensed with bisphenol A resin. They therefore contain less than the theoretical amount of bromine indicated by the chemical formula (18 to 48 percent Br). The specific gravity of the resin is raised (1.78 for resins containing the most bromine).

Determination in the Atmosphere. Like the DGEBPA-based resins, inhalation exposure to vapors from these resins is unlikely because these materials are essentially nonvolatile. In the event of combustion, bromine and hydrogen bromide vapors may be released from these resins.

Physiological Response. The acute systemic toxicity of resins based on the diglycidyl ether of tetrabromobisphenol A (TBBAER) is very low. Only slight skin and eye irritation was observed in rabbits treated with TBBAER. Unlike several other epoxy resin materials, TBBAER did not cause delayed contact skin sensitization in guinea pigs. In vitro genotoxicity tests with TBBAER have produced genotoxic effects in a limited number of assays. TBBAER did not induce chromosomal aberrations in the bone marrow of rats given repeated dermal doses of 1000 mg/kg.

ACUTE TOXICITY. The acute toxicity of TBBAER has been observed to be very low. The acute oral and dermal LD_{50}s have been reported to be >12,000 and 6000 mg/kg, respectively (355, 356). In other studies, TBBAER had an oral LD_{50} of >2000 mg/kg in both the rat and rabbit; no signs of systemic intoxication were observed (357, 358). A 1-hr exposure of four female rats to an atmosphere of dust generated at room temperature (calculated to contain 0.33 mg TBBAER/l air) did not produce any apparent adverse effects with no visible lesions upon pathological examination 14 days post-exposure (357).

SKIN EFFECTS. TBBAER was found to be only slightly irritating to the skin of rabbits (357, 359).

In studies to determine the skin sensitization potential of TBBAER, eight female Hartley strain guinea pigs received a total of 10 intradermal injections of TBBAER. Fourteen days after the final injection, the animals were challenged with an intradermal dose of TBBAER. None of the animals exhibited any reaction to the challenge dose at 24 hr post-challenge (360). In addition, TBBAER was not found to be a skin sensitizer in guinea pigs when tested topically according to a modified Buehler test (361).

EYE IRRITATION. TBBAER was found to be only slightly irritating to the eyes of rabbits (357, 359).

GENETIC TOXICITY. TBBAER was mutagenic in *Salmonella typhimurium* strain TA100, but not in strains TA98, TA1535, TA1537, or TA1538. In the presence of metabolic activation, the mutagenic response in TA100 was eliminated at all concentrations tested (362).

TBBAER was tested in the presence and absence of metabolic activation for its potential to induce chromosomal aberrations in cultured Chinese hamster ovary cells. A dose-related increase in the percent of cells with chromosomal aberrations was observed, with a greater response found in the presence of metabolic activation (363). TBBAER was negative when tested in BALB/C-3T3 cells to determine its potential to induce morphological transformation (364).

Five daily dermal doses of 1000 mg/kg of TBBAER did not induce chromosomal aberrations in the bone marrow of rats (365).

Hygienic Standards of Permissible Exposure. No standards of permissible exposure to TBBAER were found.

Odor and Warning Properties. TBBAER are solids and they have no odor or warning properties.

2.2.6 Polyglycidyl Ethers

2.2.6.1 Polyglycidyl Ethers of Cresolic and Phenolic Novolacs (CASRN 9003-36-5, 28064-14-4, 40216-08-8, 92183-42-1, phenolics; CASRN 29690-82-2, 37382-79-9, 64425-89-4, 68609-31-4, cresolics). These ethers and the condensation products of their further reaction Ciba-Geigy ECN® 1200 series and Dow D.E.N.® series 400 have the general formula:

Source, Uses, and Industrial Exposure. The term novolac refers to the reaction products of phenolic compounds and formaldehyde. Glycidation of the phenolic hydroxyl groups by epichlorohydrin under similar conditions to those used for forming the diglycidyl ether of bisphenol A produces the corresponding epoxy novolac resins. The average "*n*" in the above formula varies from 0.2 to 1.8, depending on the ultimate use for the resin. The epoxy novolac resins used are cured as the diglycidyl ether of bisphenol A is. These resins differ from bisphenol A epoxy resins in that whereas a bisphenol A polymeric chain can contain a maximum of two epoxide groups regardless of chain length, cresolic and phenolic novolacs can contain an epoxide group for each phenol or cresol molecule incorporated in the chain.

Because of their high functionality, epoxy novolac resins, when cured, produce tightly cross-linked systems with improved high-temperature performance, chemical resistance, and adhesion over the resins based on the diglycidyl ether of bisphenol A. The thermal stability of epoxy novolac resins has made them useful for structural and electrical laminates and as coatings and castings for elevated temperature service. Owing to the chemical resistance of these resins they are used for lining storage tanks, pumps, and other process equipment as well as for corrosion resistant coatings.

Physical and Chemical Properties. The properties of this type of resins are as follows:

Specific gravity (25°C)	122
Water solubility	Insoluble
Flash point (ASTM-D-1310-67)	490°F
Color, Gardner, max.	2
Viscosity (52°C)	20,000 to 50,000 cm/sec
Epoxide equivalent weight	178*

*Fairly constant for all resins unless modified by partial esterification.

Hazardous polymerizations can occur with aliphatic amines in quantities greater than 1 lb.

Determination in the Atmosphere. Like the diglycidyl ether of bisphenol A these resins are essentially nonvolatile. Dusts and vapors from high-temperature reactions may be trapped by filter or liquid entrapment methods and determined by the HCl–dioxane method (122).

Physiological Response. Toxicity studies on the phenolic and cresolic novolac resins are limited to acute data and mutagenicity testing. Both the phenolic and cresolic novolac resins appear to have a low order of acute oral and dermal toxicity potential. Both are minor irritants to the skin upon single contact. The phenolic resins have been studied for irritation potential following repeated application, and the results indicate appreciable irritation can develop following repeated or prolonged skin contact. Both resins can cause minor transient eye irritation. The phenolic novolac resins were shown to be weak skin sensitizers in humans. There are no sensitization data available for the cresolic novolac resins. Cresolic novolac resin has been reported to have caused gene mutations in *S. typhimurium* strain TA1535.

PHENOLIC NOVOLAC RESINS (EPNRs)
(CASRN 9003-36-5, 28064-14-4, 40216-08-8, 92183-42-1)

ACUTE TOXICITY. The EPNRs have been shown to have low acute toxicity potential by the oral route. Acute oral toxicity testing in rats shows that the oral LD_{50} was >4000 mg/kg for a liquid resin with molecular weight of 427 (366), whereas a solid EPNR advanced with bisphenol A of unknown molecular weight had an oral LD_{50} >2000 mg/kg (367).

Ciba-Geigy (368) reported LD_{50} of >10,000 mg/kg in rats. The acute dermal toxicity was also low, with LD_{50} values of >4 ml/kg and 3000 mg/kg in New Zealand white rabbits (368).

In an acute inhalation study, rats were exposed 4 hr to a 1.7 mg/l dust concentration, which was the highest level attainable in the test system. All 10 animals survived the 14-day observation period. There were no behavioral reactions, adverse body weight effects or gross pathological findings (369).

SKIN EFFECTS. EPNRs were practically nonirritating to rabbits' skin. A solution of partially hydrolyzed EPNR (85% in methyl ethyl ketone) did not produce any effects in rabbits when applied dermally at a dose of 2000 mg/kg (370). The liquid resin was slightly to moderately irritating to the skin of rabbits (366), but essentially no skin irritation resulted from contact with the solid resin (367) or with partially hydrolyzed EPNR (370). A study by Ciba-Geigy (368) showed pale red erythema and slight edema after repeated application of EPNR. Continued observations showed circumscribed elevations of the skin through day 7 and finally a more pronounced lesion containing yellowish exudate (pus) on day 14. These studies showed that the phenolic novolac resins were minimal primary irritants, but that repeated application could cause appreciable irritation. The studies also showed that the resins were not absorbed in sufficient quantities to cause any deaths or other signs of toxicity.

It was initially reported that EPNR did not produce skin sensitization in guinea pigs using a modified Buehler assay (371) but subsequent testing showed delayed contact skin sensitization in 3 of 20 guinea pigs tested (372). When patch tested in humans, the resins were irritating to the skin but were not found to have sensitizing potential (see below).

EYE IRRITATION. Eye irritation studies in rabbits indicated only minor transient irritation with no corneal injury produced by either liquid or solid EPNRs (369, 373) or a partially hydrolyzed EPNR (370).

EFFECTS IN HUMANS. A study using a solid epoxy novolac resin in 50 male and female volunteers showed that the material was moderately to severely irritating when applied as a 10 percent solution in sesame oil. A 1 percent solution produced slight to moderate irritation. The incidence of irritation in the volunteers was 7 to 9 percent. None of the volunteers responded to a 5 percent challenge application made 2 weeks following the ninth application, suggesting the absence of skin sensitization potential (374).

CRESOLIC NOVOLAC RESINS
(CASRN 19690-82-2, 37382-79-9, 64425-89-4, 68609-31-4)

The cresolic novolac resins were found to be practically nontoxic by the oral route. The acute oral LD_{50} in rats was greater than 10,000 mg/kg (375). The acute dermal LD_{50} in rabbits was greater than 3.98 g/kg, indicating very low toxicity potential by the dermal exposure route. The cresolic novolac resins were minimally to mildly irritating to rabbit skin and eyes (376, 377). Eye irritation response involved only the conjunctivas, and no corneal injury was seen. There are no data available on sensitization potential. A cresolic novolac resin was positive in strains TA1535 and TA100 when tested for mutagenic potential in the Ames assay using *Salmonella typhimurium* strains TA98, 100, 1535, and 1537 at concentrations of 0.9, 3.4, 10.1, 30.4, and 91.2 μg/ml (378).

Hygienic Standards of Permissible Exposure. No standards of permissible occupational exposure were found.

Odor and Warning Properties. There are no warning properties; the resins are sticky when handled.

2.2.6.2 Polyglycidyl Ether of Substituted Glycerin (CASRN 25038-04-4). The structure of the compound is

$$CH_2-CH-CH_2-O-CH_2-CH-CH_2-O-CH-CH_2-CH_2$$

with epoxide groups at each end and an OR substituent below the central CH.

OR

Source, Uses, and Industrial Exposures. The polyglycidyl ether of substituted glycerine is obtained as a reaction mixture of epichlorohydrin and glycerin. It is used in conjunction with other epoxy resins in the manufacture of adhesives. Because the compound is generally used in conjunction with a curing agent, frequently active amines, the hazardous properties of these substances must also be considered.

Physical and Chemical Properties. The polyglycidyl ether of substituted glycerin is the reaction product of epichlorhydrin and glycerin. Although the exact molecular arrangement is not known, it is of uniform composition and has the following general properties:

Molecular weight	~300
Specific gravity (20/4°C)	1.023
Refractive index (25°C)	1.478
Solubility	Miscible with water, soluble in ketones
Epoxy equivalent*	140 to 165

*Grams of resin containing 1 g-equivalent of epoxide.

Determination in the Atmosphere. No OSHA standard method was found for this material. The HCl–dioxane method (122) is suggested.

Physiological Response. Systemic toxicity is low, the compound being practically nontoxic by ingestion and percutaneous absorption. No vapor hazard exists based on results of animal studies and extremely low vapor pressure. Repeated skin contact causes irritation and occasionally sensitization. Slight effects on hematopoesis have been reported (25).

ACUTE TOXICITY. The oral LD_{50} was 5 g/kg in rats and 1.87 g/kg in mice. The dermal LD_{50} in rabbits was 14.4 g/kg, and the intraperitoneal LD_{50} was 0.38 g/kg in rats and 0.30 g/kg in mice. No effects other than slight eye irritation were observed in mice and rats exposed for 8 hr to air saturated with the polyglycidyl ether of substituted glycerine.

SKIN EFFECTS. A score of 0 was obtained by the Draize test for a single application. The compound is classified as nonirritating by this method. Applications of larger amounts or repeated applications give rise to severe skin injury on rabbits.

EYE IRRITATION. A score of 82/110 was obtained by the Draize test. The compound is classified as severely irritating.

SUBCHRONIC TOXICITY. Groups of 10 male rats were exposed for 7 hr/day, 5 days a week, for a total of 50 exposures to air saturated with the vapors of the polyglycidyl ether of substituted glycerine. Aside from a slight excrustation of the eyelids in some animals, none or the rats showed any signs of toxicity or irritation attributable to the exposure.

Rabbits received 20 dermal applications of 0.2 g total dose of the resin, which was allowed to remain for 1 or 7 hr. Scoring according to the Draize method gave high scores of 8 in both cases with a final mean of 7.6 to 7.8, indicating the compounds to be highly irritating on repeated applications. In a second experiment, rabbits receiving 1 g/kg cutaneously on five successive days developed severe subcutaneous hemorrhage and skin necrosis; two of four died. The uncured resin was fed in the diet of rats for 26 weeks at a level of (0.04, 0.2, and 1 percent). No

Table 7.1. Effects on the Hematopoietic System of Exposure to the Polyglycidyl Ether of Substituted Glycerin (161)

Route	Species	Dose (g/kg)	No. of Doses	Response[a]
Respiratory	Rat	Saturated vapors	50 (8 hr each)	No effect noted
Intramuscular	Rat	0.10	6	No effect
		0.20	6	Depression of WBC count and bone marrow nucleated cell count
Intramuscular	Dog	0.2	2 (1 week apart)	Marked depression of WBC count and relative neutropenia, followed by leukocytosis; ulceration and abscess of injection site
Intravenous	Dog	0.2	1	Progressive decline in WBC to count of 500 WBC; death from overwhelming infection
Intravenous	Rabbit	0.1	2	Decrease in total WBC
Percutaneous	Rat	1.0	20	No effect
		2.0	20	No effect
		4.0	20	Depression of bone marrow nucleated cell count (only)

[a]WBC = white blood cells.

mortalities occurred at any of the levels but at the highest level there was retardation of weight gain. No significant pathology occurred.

Chronic Toxicity/Carcinogenicity. Carcinomas were produced on the skin of mice painted with the material from once to thrice weekly over a period of a year. No tumors were produced on rabbit ears in a similar test. Sarcomas were produced in rats by subcutaneous injection. When fed to strain A mice at a concentration of 0.2 percent in the diet there was no increase in the incidence of spontaneous pulmonary adenomas as compared to controls (330).

Genetic Toxicity. Assay in the Ames test showed negative response with *Salmonella typhimurium* strains TA1531 and TA1533, both with and without activation, and positive response in strains TA98 and TA100, with and without activation, at 1000 μg but negative at 50 μg (25).

Other Effects. The slight effects of repeated exposure on the hematopoietic system are summarized in Table 7.1.

Effects in Humans. Primary irritation and occasionally sensitization occur in humans. In three of eight persons receiving the material intravenously, changes in the blood occurred. These consisted of decreases in the white cell count, the total number of lymphocytes and monocytes, and the platelet count. No impairment in liver or kidney function was observed. The effects were reversible (25).

Hygienic Standards of Permissible Exposure. No standards of permissible occupational exposure were found.

Odor and Warning Properties. The compound produces skin irritation shortly after contact. There is no characteristic odor and no irritation from the vapors.

2.2.6.3 Castor Oil Glycidyl Ether (CASRN 74398-71-3). This compound is also called the 1,2,3-propanetriyl ether of 12-(oxiranylmethoxy)-9-octadecanoic acid (COGE).

Physical and Chemical Properties. The molecular formula of COGE is $C_{66}H_{116}O_{12}$.

Determination in the Atmosphere. No OSHA standard method was found for this material. The HCl–dioxane method (122) is suggested.

Physiological Response. The acute toxicity of COGE is extremely low by both the oral and dermal routes of exposure, and the compound is not significantly irritating to the skin and eyes. No systemic toxicity or carcinogenicity were apparent following repeated dermal exposure for up to nearly 2 years in mice.

ACUTE TOXICITY (379). The oral LD_{50} of COGE in rats was >5 g/kg, and the dermal LD_{50} in rabbits was >2 g/kg. There was no evidence of systemic toxicity by either route of exposure.

SKIN EFFECTS. There was only very slight edema, with an irritation score of 0.7 out of a possible 8.0, in the Draize test in rabbits (379).

EYE IRRITATION. COGE was not irritating to the rabbit eye (379).

SUBCHRONIC TOXICITY. COGE was tested in a 90-day skin painting study in C3H mice in which solutions of 12.5 or 25 percent COGE in acetone, or 100 percent COGE were applied twice weekly (380). No treatment-related effects were observed with regard to clinical signs, body weight, blood or urine parameters, or histopathology of the skin or major organs.

CHRONIC TOXICITY/CARCINOGENICITY. There was no evidence of skin or systemic carcinogenicity in a cancer bioassay in mice treated topically with 50 μl of 50 percent COGE in acetone twice weekly for 94 weeks (380).

Hygienic Standards of Permissible Exposure. No standards of permissible occupational exposure were found.

Odor and Warning Properties. No information on odor or warning properties of COGE was found.

2.3 Glycidyl Esters

2.3.1 Phthalic Acid Diglycidyl Ester (CASRN 7195-45-1)

The structure of phthalic acid diglycidyl ester [diglycidyl ester of phthalic acid; bis(oxiranylmethyl) ester of 1,2-benzenedicarboxylic acid] is as follows:

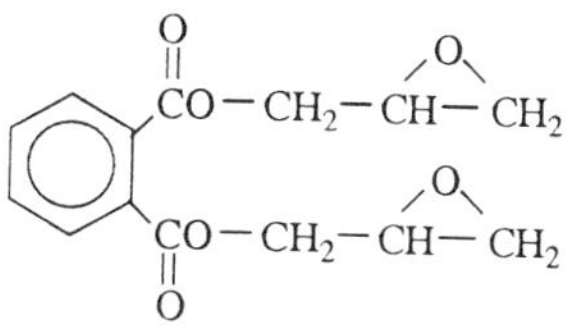

2.3.1.1 Physical and Chemical Properties

Molecular weight	278
Boiling point (760 mm Hg)	167
Refractive index (20°C)	1.523

1 ppm $\simeq$ 11.37 mg/m^3 at 25°C, 760 mm Hg; 1 mg/l $\simeq$ 87.9 ppm at 25°C, 760 mm Hg

2.3.1.2 Determination in the Atmosphere. No OSHA standard methods for DGP were found.

2.3.1.3 Physiological Response. The acute oral and dermal toxicity of DGP was low. DGP was moderately irritating to the skin of rabbits following a single topical application. Severe injury was produced in rabbits following the instillation of DGP into the eye. DGP has been shown to be mutagenic in bacteria and has been reported to cause skin sensitization in humans exposed occupationally.

Acute Toxicity. An acute oral toxicity test conducted with DGP in male rats showed the single dose oral LD$_{50}$ to be between 500 and 5000 mg/kg (381). DGP administered topically to male and female rabbits showed the dermal LD$_{50}$ to be greater than 2000 mg/kg (381).

Skin Effects. Undiluted DGP (0.5 ml) applied topically to rabbit skin resulted in a well-defined to moderate erythema and edema; the primary irritation score was 2.3/8 (381).

Eye Irritation. Instillation of 0.1 ml of DGP into the eyes of rabbits produced conjunctivitis, redness, chemosis, and lesions to the cornea and iris (381).

Genetic Toxicity. The mutagenic potential of DGP was examined in a series of in vitro microbial assays using *Salmonella typhimurium* and *Saccharomyces cerevisiae*. DGP was tested with and without metabolic activation at dose levels from 0.001 to 5.0 µl per plate. Results demonstrated that DGP was mutagenic to *S. typhimurium* strains TA1535 and TA100 in the presence and absence of metabolic activation (382).

Effects Observed in Humans. Contact allergy tests were conducted in an aircraft factory using resin composite materials. Five of six workers patch tested with 1 percent solution o-DGP showed a positive response (383).

2.3.1.4 Hygienic Standards of Permissible Exposure. No standards of permissible exposure to DGP have been established.

2.3.1.5 Odor and Warning Properties. No information was found on odor or warning properties of DGP.

2.3.2 Hexahydrophthalic Acid Diglycidyl Ester (CASRN 5493-45-8)

The structure of hexahydrophthalic acid diglycidyl ester (diglycidyl-1,2-cyclohexanedicarboxylate) is as follows:

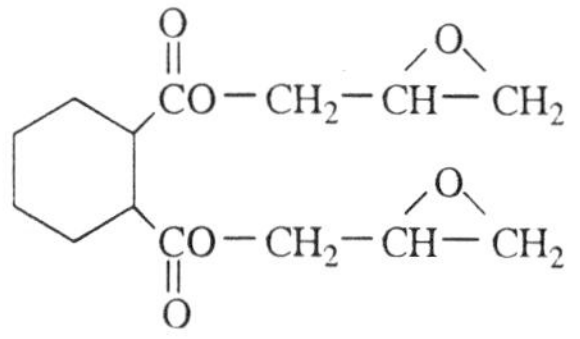

2.3.2.1 Source, Uses, and Industrial Exposures. Hexahydrophthalic acid diglycidyl ester (HADGE) is a stable, clear, amber liquid with a pH of approximately 10 and is insoluble in water. It is used in epoxy-based industrial coatings and paints.

2.3.2.2 Physical and Chemical Properties. Some of the physical and chemical properties of HADGE are as follows:

Molecular weight	284
Specific gravity (25°C)	1.22 g/ml
Boiling point (°C at 2 mm Hg)	200 to 205
Flash point (closed cup)	374°F
Appearance	Clear liquid
Vapor pressure (25°C)	<0.2 mm HG

1 ppm $\simeq$ 11.62 mg/m^3 at 25°C, 760 mm Hg; 1 mg/l $\simeq$ 86.1 ppm at 25°C, 760 mm Hg

2.3.2.3 Determination in the Atmosphere. No OSHA standard methods for DGP were found.

2.3.2.4 Physiological Response. Acute toxicity studies of HADGE by oral, dermal, and inhalation routes have demonstrated very low acute toxicity. HADGE produced no irritation to the skin of rabbits after prolonged (24 hr) exposure but more severe irritation upon repeated exposure. It has also been reported that HADGE produced delayed contact skin sensitization in guinea pigs. HADGE produced slight to moderate discomfort, moderate conjunctival redness and swelling, and slight reddening of the iris, and in the eye of one rabbit it produced corneal injury which persisted 21 days post-exposure. HADGE has been shown to be mutagenic when tested in vitro in bacteria and mouse lymphoma cells. In vivo, HADGE has been reported to produce an increase in SCEs in the bone marrow cells from Chinese hamsters treated orally and also an increased number of micronuclei in the mouse bone marrow micronucleus test.

Acute Toxicity. The acute oral LD_{50} in rats was reported to be 1030 mg/kg in the rat (344). Other studies have reported the acute oral LD_{50} in rats to be between 500 and 2000 mg/kg (384). The single-dose dermal LD_{50} has been reported to be greater than 2000 mg/kg (384) and greater than 4600 mg/kg (344). Nose-only exposures of rats to a liquid aerosol of HADGE (1.52 mg/l with a mass median aerodynamic diameter of 3.39 μm for 4 hr) revealed no treatment related effects other than transient signs of stress; all animals survived until sacrifice at 14 days post-exposure (385).

Skin Effects. Topical application of HADGE to the skin of a rabbit resulted in no irritation within 24 hr but moderate to marked erythema, moderate edema, and very slight exfoliation and necrosis on the abdominal skin of rabbits was observed after three applications (384). HADGE produced delayed contact skin sensitization in guinea pigs (384, 386).

Eye Irritation. HADGE produced slight to moderate discomfort, moderate conjunctival redness and swelling, slight reddening of the iris, and in the eye of one rabbit, corneal injury that persisted 21 days post-exposure (384).

Genetic Toxicity. In vitro, HADGE was tested for gene mutation effects in *S. typhimurium* at concentrations up to 231 μg/ml and was positive in strains TA100 and TA1535 (387). HADGE has also been reported as positive in the mouse lymphoma assay (388). Negative results were found in the BALB/3T3 fibroblast transformation assay at concentrations up to 30 μg/ml (388). In vivo, HADGE administered orally in a single dose of 1250, 2500, or 5000 mg/kg to Chinese hamsters produced a statistically significant increase in SCEs in bone marrow cells at the two higher dose levels (389). Mice given a single oral dose of 500 or 1000 mg/kg had a statistically significant increase in the frequency of micronucleated polychromatic erythrocytes as compared to negative controls (390).

Effects in Humans. Allergic contact dermatitis was observed in five workers operating machinery using a cutting fluid containing HADGE. Positive patch test reactions were obtained in all five workers to HADGE and were negative in 25 controls (391).

2.3.2.5 Hygienic Standards of Permissible Exposure. No standards of permissible occupational exposure to HADGE were found.

2.3.2.6 Odor and Warning Properties. No information was found on odor or warning properties of HADGE.

2.3.3   Glycidyl Ester of Neodecanoic Acid (CASRN 26761-45-5)

The structure of the glycidyl ester of neodecanoic acid (2,3-epoxypropyl ester of neodecanoic acid; Cardura E10) is as follows:

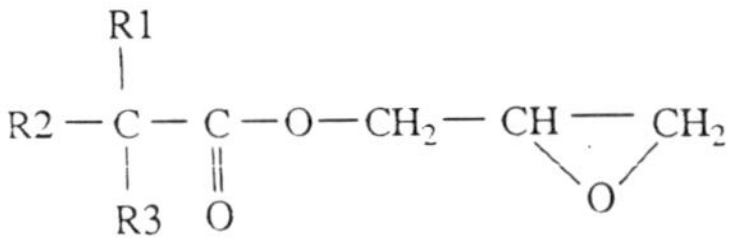

$$\text{R2}-\underset{\underset{\text{R3}}{|}}{\overset{\overset{\text{R1}}{|}}{C}}-\underset{\overset{||}{O}}{C}-O-CH_2-CH\diagdown CH_2$$

2.3.3.1 Source, Uses, and Industrial Exposure. The glycidyl ester of neodecanoic acid (GENA) is produced by the reaction of neodecanoic acid with epichlorohydrin in the presence of base:

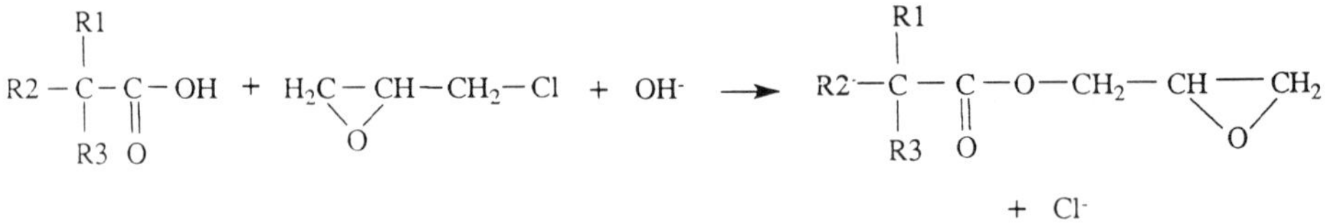

$$\text{R2}-\underset{\underset{\text{R3}}{|}}{\overset{\overset{\text{R1}}{|}}{C}}-\underset{\overset{||}{O}}{C}-OH \ + \ H_2C-CH-CH_2-Cl \ + \ OH^- \ \longrightarrow \ \text{R2}-\underset{\underset{\text{R3}}{|}}{\overset{\overset{\text{R1}}{|}}{C}}-\underset{\overset{||}{O}}{C}-O-CH_2-CH\diagdown CH_2$$

$$+ \ Cl^-$$

Because the neodecanoic acid reacts with epichlorohydrin to give a mixture of branched aliphatic 10-carbon carboxylic acids, the R_1, R_2, and R_3 in the structure above represent a mixture of aliphatic carbon chains, one of which will always be a methyl group; the total number of carbons in the other two R groups will be seven. This material may be used as an epoxy resin reactive diluent, as a component of epoxy coatings systems, and as an ingredient in alkyl resins to modify film properties. Improved chemical resistance, weatherability, film hardness, and acid number control are the properties enhanced in epoxy resin applications by the use of this material.

2.3.3.2 Physical and Chemical Properties. The tertiary configuration of the acid radical confers extreme stability to the ester linkage. It reacts readily with active hydrogen compounds and polymerizes readily in the presence of strong Lewis acids.
   Properties typical of this glycidyl ester are as follows:

Epoxide equivalent weight	240 to 250
Color	Water white

Melting point	$< -60°C$
Boiling point	$260°C$
Flash point (Plensky Martens closed cup)	$265°C$
Miscibility with water (water in the ester)	0.7% (wt.)

2.3.3.3 Determination in the Atmosphere. No OSHA standard methods for GENA were found. Because of the high boiling point and low volatility, it is not likely that detectable amounts could develop in the atmosphere.

2.3.3.4 Physiological Response. The acute oral and dermal toxicity of GENA is low. GENA was only mildly irritating to the skin of rabbits but produced delayed contact skin sensitization in guinea pigs. GENA was practically nonirritating to the eyes of rabbits. Rats fed high doses of GENA for 5 weeks showed decreased body weights, decreased hematocrit, decreased plasma alkaline phosphatase, increased plasma urea, protein, and sodium, increased urinary ketones, and increased relative liver and kidney weights with degenerative histopathological changes in the kidneys. In vitro, GENA produced a weak increase in mutation frequency in bacteria but not yeast. In other in vitro studies, GENA did not produce an increase in transformation of cultured baby hamster kidney cells but did induce a small increase in the frequency of chromosomal aberrations in cultured rat hepatocytes. GENA did not cause DNA single-strand damage in the liver of rats given single oral doses of 5 ml/kg.

Acute Toxicity. The single-dose oral LD_{50} in rats was greater than 9600 mg/kg and the single-dose dermal LD_{50} in rats was greater than 3800 mg/kg (392). In rabbits, the acute (24 hr) dermal LD_{50} was >10 ml/kg. The 4-hr acute inhalation LC_{50} in rats for GENA was greater than 240 mg/m^3 (393).

Skin Effects. The compound is moderately irritating to rat and rabbit skin at 24 hr post-application (392). When tested by the Magnusson and Kligman maximization test in the guinea pig, GENA was a severe skin sensitizer (392).

Eye Irritation. GENA was found to be practically nonirritating in the nonwashed eyes of rabbits (392).

Subchronic Toxicity. Rats were fed dietary concentrations of 0, 100, 500, 1000, and 10,000 mg GENA mg/kg/day for 5 weeks. There were no intermediate deaths and no effect on general health or behavior in any dose group. At termination, rats receiving 10,000 mg/kg/day had decreased body weight and feed intake, decreased erythrocyte count and hematocrit, increased plasma urea, protein, and sodium, decreased plasma alkaline phosphatase, increased urinary ketones, and increased relative liver and kidney weights. Degenerative, occlusive, and regenerative changes were seen in the proximal renal tubules of male and to a much lesser extent female rats. Similar changes were seen in the 5000 mg/kg dose group, except that the erythrocyte count and hematocrit were normal. No treatment-related effects were observed at a dose level of 1000 mg/kg/day or below (394).

Genetic Toxicity. The mutagenic activity of glycidyl ester of neodecanoic acid was investigated in cultures of *Salmonella typhimurium* TA98, TA100, TA1535, and TA1538, *Escherichia coli* WP2 and WP2 *uvr A*, and in cultures of *Saccharomyces cerevisiae* JDI, both with and without the incorporation of rat liver microsomal enzymes (191, 395, 396). The results indicated that glycidyl ester of neodecanoic acid induced an increase in mutation frequency, only after metabolic activation, in *Salmonella* strains TA100, TA1535, and TA1538. The mutation frequency was within the spontaneous frequency range in the absence of S9 fractions for these three strains. For TA98, there was no increase in mutation frequency either with or without S9. There was no increase in mutation frequency in either of the *E. coli* strains with or without activation. In general, no significant effects were detected below 100 μg of material per plate, suggesting that the mutagenic activity in bacteria was relatively weak. The bacterial results indicate that mutagenicity was expressed by both base-pair substitution and frameshift mechanisms (395). Canter et al. (191) have reported that GENA was weakly mutagenic in several strains of *Salmonella* with and without metabolic activation. There were no significant increases in gene conversion of *Saccharomyces cerevisiae* JD1 either with or without presence of S9 (396).

In other studies, monolayer slides of cultures of rat liver (RLI) cells were exposed to culture medium containing glycidyl ester of neodecanoic acid and after 24 hr incubation the slides were processed for metaphase chromosome analysis (396). In the RLI cells, glycidyl ester of neodecanoic acid induced a low frequency of chromatid aberrations at concentrations just below the cytotoxic dose (50 μg/ml).

When suspension cultures of baby hamster kidney cells were exposed to GENA at concentrations of approximately 44, 88, and 350 μg/ml, no increased frequency of transformed cells was observed (397).

In an in vivo study, a single oral dose of 5 ml/kg of GENA did not cause significant DNA single-strand damage in rat liver cells when measured at 6 hr post-dosing (398).

Effects in Humans. Contact dermatitis has been observed in workers exposed to GENA (399, 400).

2.3.3.5 Hygienic Standards of Permissible Exposure. No standards for permissible occupational exposure were found.

2.3.3.6 Odor and Warning Properties. The ester has a low musty odor.

2.4   Miscellaneous Epoxy Compounds

2.4.1   Epichlorohydrin (CASRN 106-89-8)

This compound, 1-chloro-2,3-epoxypropane, γ-chloropropylene oxide, *a*-epichlorohydrin (ECH), or glycidyl chloride has the formula:

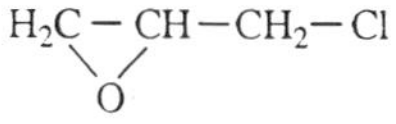

$$H_2C - CH - CH_2 - Cl$$
$$\diagdown \diagup$$
$$O$$

2.4.1.1 Source, Uses, and Industrial Exposure. ECH is available in large-scale commercial quantities through the discovery and development of processes for its production from propylene. It is employed as a raw material for the manufacture of a number of glycerol and glycidol derivatives, in the manufacture of epoxy resins, as a stabilizer in chlorine-containing materials, and as an intermediate in preparation of condensates with polyfunctional substances.

2.4.1.2 Physical and Chemical Properties. ECH is a colorless, mobile liquid which is flammable and reactive. ECH has a sweet, pungent or chloroform-like odor and the following properties:

Molecular weight	92.53
Specific gravity (20/4°C)	1.1807
Freezing point	$-57.2°C$
Boiling point (760 mm Hg)	116.1°C
Vapor density	3.21
Vapor pressure (20°C)	13 mm Hg
Refractive index (20°C)	1.43805
Solubility	6.48% in water; miscible with ethers, alcohols, CCl_4, and benzene
Flash point (Tag open cup)	105°F
Flash point (Tag closed cup)	87°F, 31°C
Concentration in "saturated air" (20°C)	1.7%
Flammability limits	3.8% vol. to 21% vol. in air

1 ppm $= 3.78$ mg/m^3 at 25°C, 760 mm Hg; 1 mg/l $= 265$ ppm at 25°C, 760 mm Hg

2.4.1.3 Determination in the Atmosphere. Standard methods for air sampling and analysis of the ECH by GLC with flame ionization detection have been described (66, 401).

2.4.1.4 Physiological Response. ECH has been found to be moderately toxic to laboratory animals following a single oral, percutaneous, intravenous, or subcutaneous dose. By inhalation the 1-hr LC_{50} has been determined to be 3617 ppm and 2165 ppm in male and female rats, respectively (402). Because the theoretical saturated vapor concentration is 17,105 ppm, excessive vapor concentrations are readily attainable and may cause unconsciousness and death. Skin contact may cause severe irritation and sensitization. ECH vapors have been reported to be irritating to the mucous membranes of the eye and respiratory tract. The subchronic inhalation toxicity observed in laboratory animals was destruction of the nasal turbinate epithelium with slight nonprogressive kidney effects and slight nondegenerative liver effects. Repeated exposure to 25 ppm or 50 ppm of ECH vapor or repeated oral administration of 15 mg ECH mg/kg/day also produced reversible sterility in male rats. The available data provide good evidence of gene and chromosomal mutagenicity of ECH in several experimental systems; however, cytogenetic studies of workers exposed to ECH have yielded equivocal evidence for a clastogenic effect on lymphocytes. Chronic inhalation exposure to 100 ppm ECH for 30 days produced tumors in the nasal epithelium of rats. Some indication of the potential of ECH to produce nasal tumors was also observed in rats inhaling 30 ppm for their entire lifetime. The oral administration of ECH by gavage or in drinking water resulted in hyperplasia, papillomas, and carcinomas in the nonglandular stomach of rats. ECH alone did not induce tumors when applied to the skin of mice but appeared to act as an initiator in groups treated subsequently with phorbol myristate acetate as a promotor. Tumor formation appears to be confined to tissues that are the site of initial ECH contact. Epidemiology studies have not provided definitive evidence for an association between occupational ECH exposure and cancer or any other adverse health effects.

Acute Toxicity. The doses or exposures of ECH producing lethality have been defined following administration by the oral, dermal, intravenous, subcutaneous, or inhalation routes of exposure. ECH has been found to be moderately toxic to laboratory animals following administration of a single oral, percutaneous, intravenous, or subcutaneous dose. Acute inhalation studies in laboratory animals have shown that lethal vapor concentrations are readily attainable at room temperature. A summary of some of the key acute toxicity data appears in Table 7.2.

Skin Effects. Undiluted epichlorohydrin (0.5 ml) was intensely irritating and necrotic to the depilated skin of laboratory rabbits when allowed to remain in contact with the skin for 24 hr (403). Smaller volumes of undiluted ECH (0.1 to 0.2 ml) applied to rabbit skin for 2 hr produced less severe irritation over a smaller area (403). A much smaller volume (0.01 ml) of undiluted ECH applied for 2 hr to rabbit skin produced only a trace of redness and capillary injection (69). Weaker solutions (0.2 ml of a 0.3 percent solution of ECH in cottonseed oil) applied to rabbit skin for 24 hr under an occlusive dressing produced no irritation whereas a 5 percent solution resulted in a marked irritation of the skin (408). Repeated applications may lead to widespread necrosis. Skin sensitization tests in guinea pigs have produced mixed or equivocal results although from the information available it appears that ECH has the potential to produce delayed contact skin sensitization (69, 314, 408, 409).

Table 7.2. Summary of Acute Toxicity Data on Epichlorohydrin[a]

Route	Species	Dose	Parameter of Toxicity
Oral	Rats	0.09 g/kg	LD_{50}
Oral	Guinea pigs	0.178 g/kg	LD_{50}
Oral	Mice	0.238 g/kg	LD_{50}
Intravenous	Rats	0.154 g/kg	LD_{50}
Intravenous	Mice	0.178 g/kg	LD_{50}
Percutaneous	Rabbits	0.88 ml/kg	LD_{50}
Percutaneous	Rats (3 applications)	0.5 ml/kg	LD_{50}
Inhalation	Mice	2370 ppm	$0/30$[b]
Inhalation	Mice	8300 ppm	$20/20$[b]
Inhalation	Rats	250 ppm (8 hr)	LC_{50}
Inhalation	Rats	500 ppm (4 hr)	LC_{50}
Inhalation	Guinea pigs	561 ppm (4 hr)	LC_{50}
Inhalation	Rabbits	445 ppm (4 hr)	LC_{50}
Inhalation	Rats (males)	3617 ppm (1 hr)	LC_{50}
Inhalation	Rats (females)	2165 ppm (1 hr)	LC_{50}

[a]Data from References 402 and 404–407.
[b]Number of deaths over the number of animals exposed.

Eye Irritation. Undiluted material or concentrated solutions (80 percent ECH in cottonseed oil) were markedly irritating to the eye and produced corneal damage on local contact (69, 409). Less concentrated solutions (20 percent ECH in cottonseed oil) produced conjunctival irritation and edema, whereas 5 percent in cottonseed oil produced no eye irritation (408). The eye irritation potential of ECH appeared to be increased in rabbits if the ECH was in a solvent that is more readily soluble in water than is ECH itself (410). ECH vapors may also give rise to eye irritation.

Subchronic Toxicity. In well documented studies, repeated exposure of rats, mice, or rabbits to concentrations of ECH vapor up to 120 ppm (studies ranged from 4 to 13 weeks) resulted in substantial degenerative changes in the nasal turbinate mucosa (411–413). Eye irritation, slight nonprogressive kidney effects (rats), and slight nondegenerative liver effects (decreased glycogen in rats and mice) and minimal changes in the content of the epididymides (rats) were also noted with transient infertility observed in male rats (but not rabbits) (412, 413). The no observed effect level for all subchronic toxicologic effects was 5 ppm. Other previously reported and poorly documented studies have confirmed nasal irritation and some liver and kidney effects at higher concentrations and also reported some effects not repeated in later studies (414, 415). A single 4-hr inhalation exposure of rats to 100 ppm ECH or exposure for 4 consecutive days for 4 hr/day to 100 ppm did not result in marked changes in several end points used to assess hepatotoxicity and kidney toxicity (416).

Chronic Toxicity/Carcinogenicity. Tumorigenic effects in laboratory animals treated with ECH have been reported, although only in tissues that were the first tissues in direct contact with ECH following either inhalation or oral administration, and only at doses that caused significant chronic irritation. Specifically, Laskin et al. (417) reported that tumors in the nasal cavity (primarily squamous cell carcinomas) developed in 15 of 140 rats exposed for 30 days to 100 ppm ECH and observed for their lifetime, as compared to no nasal tumors in 150 concurrent controls. Severe inflammation of the nasal turbinates that preceded tumor development was also produced by ECH in this study. In addition, lifetime exposure of 100 rats to 30 ppm resulted in one squamous cell carcinoma of the nasal cavity and one nasal papilloma; no tumors were produced by lifetime exposure at 10 ppm. Rats administered ECH by gavage for 2 years at dose levels of 2 or 10 mg/kg/day, or by drinking water for 81 weeks [29, 52, or 89 mg/kg/day], developed hyperplasia, papillomas, and carcinomas of the nonglandular stomach at all dose levels (418, 419). The dose levels in the gavage study were high enough to produce 7.7, 22.4, and 44.9 percent reductions in body weight for the lowest to highest dose levels, respectively.

The tumorigenic effects produced at the "portals of entry," that is, the nasal turbinates following inhalation or the nonglandular stomach following oral administration, are consistent with the chemical reactivity of ECH, a bifunctional alkylating agent that may covalently bind to many cell constituents to produce toxicity. This alkylating ability of ECH is likely responsible for the genotoxic effects observed in various in vitro genotoxicity test systems; these have been reviewed (420).

Chronic percutaneous application of ECH alone did not result in skin tumors in either mice or rats (69, 421). However, 9 of 30 mice developed either skin papillomas or, in one case, carcinoma when treated with a single application of 2 mg of ECH followed 2 weeks later by three applications per week with the active component of croton oil, phorbol myristate acetate (PMA) (421). This result compared to 3/30 mice developing papillomas in the PMA-only treated group suggests ECH may act as a tumor initiator in mice.

Developmental Toxicity. No embryo toxic, fetotoxic, or teratogenic effects were noted in the offspring of pregnant rats and rabbits inhaling 0, 2.5, or 25 ppm ECH during gestation (422). A teratology study in which rats and mice were administered daily oral doses of 0, 80, or 120 mg/kg ECH by gavage did not result in an increase in the frequency of fetal malformations in either species (423). Some evidence of fetotoxicity was noted in mice at the higher dose levels of 80 and 120 mg/kg; however, this may have been a reflection of maternal toxicity because the highest dose level was lethal to 3/32 dams (423).

Reproductive Toxicity. ECH has been found to have an effect on fertility in male rats. Rats receiving an oral dose of 15 mg/kg/day for 7 days in a 12-day period became infertile within 1 week. This effect was reversible within 1 week after treatment was discontinued. No histological changes were found in the testes, epididymis, prostate, and seminal vesicles. Libido and ejaculation capacity were not adversely affected (424). Repeated exposure (5 days/week, 6 hr/day for 10 weeks) to 25 or 50 ppm of ECH vapor resulted in reduced fertility in male rats (422). Males exposed to 25 ppm were able to impregnate females but marked pre-implantation losses were observed. The effects were reversible as early as 2 weeks post-exposure and 5 ppm did not affect fertility. No adverse effects on fertility in female rats were observed following ECH administration by either oral or inhalation routes.

Genetic Toxicity. The genotoxic potential of ECH has been extensively investigated using both in vitro and in vivo assays, the results of these have been summarized (420, 425, 426). As expected for a bifunctional aklylating agent, ECH had genotoxic activity in several in vitro assays including those conducted in bacteria, fungi, and mammalian cells in culture. In vivo, ECH appears to have the ability to express its genotoxic potential at the tissue of first contact but there are conflicting results in other tissues. For example, following intraperitoneal administration, metaphase analysis of the bone marrow cells indicated a positive effect but three micronucleus assays were negative although ECH was also administered intraperitoneally in these studies. Negative results were obtained in dominant lethal tests using mice by both the oral and intraperitoneal routes of administration (427, 428).

Pharmacokinetics and Metabolism. Studies in rats and mice have demonstrated that ECH is readily metabolized and rapidly excreted either as urinary metabolites (approximately 50 percent of the absorbed dose) or as CO_2 in the expired air (429–433). The majority of the absorbed dose was eliminated within the first 12 to 24 hr post-dosing. Hydrolysis of the epoxide ring and glutathione conjugation are two metabolic routes whereby ECH is metabolized and eliminated from the body, although the exact details of the metabolic fate of ECH have not been fully elucidated.

Effects in Humans: Single Exposures. A few days after direct skin contact with epichlorohydrin, erythema, edema, and papules associated with a burning, itching sensation were reported (434). In more severe cases blisters developed. Recovery was complete. It should be mentioned that ECH can penetrate leather shoes or rubber gloves and so may cause chemical burns. A single case was reported of chronic asthmatic bronchitis which developed after a single, though severe over-exposure. At 2 years post-exposure the exposed individual was reported to have a pronounced fatty liver, attributed by the authors to the prior ECH exposure, although a causal relationship is questionable (435).

Repeated Exposures and Epidemiology. Occasional cases of skin sensitization have been reported (436–441). A number of studies have investigated the potential relationships between repeated exposure to ECH and health effects in workers including specific organ toxicity, reproductive effects, genotoxicity, and carcino-genicity; these have been reviewed (426, 450). In general, none of these studies has provided definitive evidence for an association between occupational ECH exposure and an increased incidence of organ injury or disease, decrease in fertility, or other biologic effects, such as alterations in cytogenetics. Nevertheless, the evidence for the ability of ECH to produce toxicity in laboratory animals dictates that appropriate handling precautions and hygienic standards be followed during the manufacture and use of ECH.

2.4.1.5 Hygienic Standards of Permissible Exposure. The ACGIH TLV–TWA value for 1991 is 2 ppm (skin); a lower TLV of 0.1 ppm is proposed.

2.4.1.6 Odor and Warning Properties. The odor is generally perceived as a slightly irritating chloroform-like odor. Sensory perception studies have indicated that the mean threshold for odor recognition is approximately 10 ppm, and that at 25 ppm it is recognized by the majority of persons. Marked nose and eye irritation occur only at levels exceeding 100 ppm. It is concluded that local irritation of the eyes is not severe enough to force workers to evacuate potentially harmful areas. Eye irritation may be accepted as indicating an undesirable atmospheric contamination.

2.4.1.7 Handling Precautions. These recommendations for handling precautions are taken from the recently published document entitled *Epichlorohydrin: A Safety and Handling Guide* (442). Skin, eye, and respiratory contact should be avoided. Protective equipment, which should be routinely used even where exposure is not expected, includes chemical goggles, safety showers, and eye wash stations in the immediate working area, and respiratory protection equipment available for use during escape. Additional personal protective equipment should be used when the potential for exposure is high, but should not be considered a substitute for proper handling and engineering controls. Protective clothing (suits, boots, and gloves) made of neoprene or butyl rubber are preferred. Protective clothing made of PVC and nitrile are penetrated more readily. For short-term exposure or low concentrations, a NIOSH-approved, fitted, full-facepiece cartridge-type respirator with organic vapor cartridge or organic vapor/acid gas combination cartridge should be used. Where there is potential for longer-term exposure, or exposure to higher concentrations can be expected, supplied air or a positive pressure breathing apparatus should be used. Leather articles should not be worn as they cannot be decontaminated and must be cut up and burned.

2.4.2 Glycidol (CASRN 556-52-5)

This compound has the structure

$$CH_2\!\!-\!\!CH-CH_2OH$$
$$\diagdown\,O\,\diagup$$

2.4.2.1 Source, Uses, and Industrial Exposure. Glycidol is made through dehydrochlorination of glycerol monochlorohydrin with caustic. It is commercially available. It is suggested for use in preparation of glycerol and glycidyl ethers, esters, and amines, in the pharmaceutical industry, and in sanitation chemicals.

2.4.2.2 Physical and Chemical Properties. Glycidol is a colorless, slightly viscous liquid with the following properties:

Molecular weight	74.05
Specific gravity (20/4°C)	1.115
Boiling point (760 mm Hg)	160°C
Vapor density	2.15
Vapor pressure (25°C)	0.9 mm Hg
Solubility	Completely water soluble
Concentration in "saturated" air	0.118%

1 ppm = 3.03 mg/m^3 at 25°C, 760 mm Hg; 1 mg/l = 330 ppm at 25°C, 760 mm Hg

2.4.2.3 Determination in the Atmosphere. An OSHA standard method for the analysis of glycidol by absorption on charcoal, displacement with CS_2, and determination by GLC has been described (66).

2.4.2.4 Physiological Response

Acute Toxicity. The oral LD_{50} of a 10 percent solution of glycidol in propylene glycol is 850 mg/kg for the rat and 450 mg/kg for the mouse (186). Symptoms of intoxication were initial depression of the CNS, followed by hypersensitivity to sound with muscular tremors and facial muscular fibrillation; some rats showed terminal convulsions. All animals had dyspnea and showed lacrimation. Neat glycidol had an oral LD_{50} in rats of 640 mg/kg in females and 760 mg/kg in males; the LD_{50} following intraperitoneal injection in rats was 210 mg/kg in females and 350 mg/kg in males (443). The dermal LD_{50} in rabbits after a 7-hr exposure was 1980 mg/kg. Minimal signs of systemic toxicity were observed, but skin irritation was present. By inhalation, the 4-hr LC_{50} was 580 ppm in the rat and 450 ppm in the mouse. Signs of severe irritation included dyspnea, lacrimation, salivation, and nasal discharge. Signs of CNS stimulation were also apparent. Death usually resulted from pulmonary edema; emphysema was also detected. There was some discoloration of the kidneys and liver. An 8-hr vapor exposure caused corneal opacity (186).

Skin Effects. Glycidol is rated a moderate skin irritant, with a Draize score of 4.5/8.0, for a single application on occluded rabbit skin. It was severely irritating, with necrosis, when applied repeatedly over 4 days. Although data are not available, glycidol would be expected to be a skin sensitizer (186).

Eye Irritation. Glycidol is severely irritating to the rabbit eye, with a Draize score of 68/110. However, there were no permanent corneal defects after a single instillation (186).

Subchronic Toxicity. Rats exposed to 400 ppm glycidol, 7 hr/day for 50 days showed no signs of systemic toxicity. Only slight irritation of mucous membranes was noted. Repeated intramuscular injections failed to affect hematopoiesis in the rat (186). Following gavage of 300 mg/kg glycidol in rats for 16 days, there was edema and degeneration of the epididymal stroma, atrophy of the testis, and granulomatous inflammation of the epididymis in males. Focal demyelination in the medulla and thalamus of the brain occurred in all female mice that received 300 mg/kg glycidol for 16 days. In a subsequent 13-week gavage study, sperm count and motility were reduced in male rats at doses of 100 or 200 mg/kg. Necrosis of the cerebellum, demyelination in the medulla of the brain, tubular degeneration and necrosis of the kidney, and testicular atrophy and degeneration occurred in rats given a 400-mg/kg dose. In a 13-week mouse gavage study, there was demyelination of the brain at doses of 150 or 300 mg/kg, testicular atrophy at 19 mg/kg, and renal tubular cell degeneration in males that received 300 mg/kg (14).

Chronic Toxicity/Carcinogenicity. In a dermal carcinogenicity study, 20 mice were topically administered 100 mg of a 5 percent solution of glycidol in acetone, three times weekly, for 520 days. No tumors of any type resulted (116). Another lifetime skin carcinogenicity study in mice was negative (330). However, an oral gavage cancer study in mice and rats conducted by NTP was positive with multiple tumors (14). NTP concluded there was clear evidence for carcinogenic activity in both species. Non-neoplastic lesions observed in the cancer study included hyperkeratosis and epithelial dysplasia of the forestomach, fibrosis of the spleen in rats, and cysts of the preputial gland and kidney in male mice.

Developmental Toxicity. The teratogenic potential of glycidol was assessed in rats by intra-amniotic injections on day 13 of gestation. There was a 50 percent resorption rate with 10, 100, and 1000 μg/fetus. Of the surviving fetuses treated with 1000 μg glycidol, 44 percent were malformed, with limb defects being the most frequent malformation (444). By contrast, in a mouse developmental toxicity study, glycidol gavaged at 200 mg/kg showed no evidence of teratogenicity; there was a significant increase in the number of stunted fetuses at this dose, but all of these were present in a single litter, and this dose killed 5 of 30 dams (423).

Genetic Toxicity. (14) Glycidol was mutagenic in several in vitro tests. Mutagenic activity was observed with and without metabolic activation in *Salmonella* strains TA97, TA98, TA100, TA1535, and TA1537 in the Ames assay. It was also positive without activation in the mouse lymphoma assay; it was not tested with activation. In the Chinese hamster ovary cell assay, glycidol induced both SCEs and chromosomal aberrations in the presence and absence of metabolic activation. Glycidol was also genotoxic in in vivo assays. It induced sex-linked recessive lethal mutations and reciprocal translocations in the germ cells of male *Drosophila* exposed by feeding. It was also positive in the micronucleus test following intraperitoneal injection in male B6C3F1 mice, and induced chromosome aberrations in bone marrow cells from rats dosed orally or intraperitoneally (443).

Effects in Humans. No effects have been reported. Effective hygiene practices should be taken to prevent the irritative and potential systemic effects of glycidol.

2.4.2.5 Hygienic Standards of Permissible Exposure. The 1991–1992 ACGIH TLV–TWA value for glycidol is 25 ppm.

2.4.2.6 Odor and Warning Properties. Eye and respiratory irritation should help prevent excessive overexposure.

2.4.3 Glycidaldehyde (CASRN 765-34-4)

This compound has the stucture

$$CH_2-CH-\overset{H}{C}=O$$
$$\diagdown O \diagup$$

2.4.3.1 Source, Uses, and Industrial Exposure. Glycidaldehyde is prepared from the hydrogen peroxide epoxidation of acrolein. It is suggested as a bifunctional chemical intermediate and as a cross-linking agent for textile treatment, leather tanning, and protein insolubilization.

2.4.3.1 Physical and Chemical Properties. Glycidaldehyde is a mobile, colorless liquid with a pungent odor, having the following properties:

Molecular weight	72.1
Specific gravity	1.1403
Freezing point	−61.8°C
Boiling point	
760 mm Hg	112 to 113
100 mm Hg	57 to 58
Vapor density	2.58
Refractive index (20°C)	1.4200
Solubility	Completely soluble in all common solvents; insoluble in petroleum ether
Flash point (Tag open cup)	88°F

1 ppm = 2.94 mg/m^3 at 25°C, 760 mm Hg; 1 mg/l = 339 ppm at 25°C, 760 mm Hg

2.4.3.3 Determination in the Atmosphere. No OSHA standard methods for the determination of glycidaldehyde in air were found. The standard methods described for glycidol (adsorption on charcoal, displacement with CS_2, and determination by gas–liquid chromatography) would likely be suitable (66).

2.4.3.4 Physiological Response. Glycidaldehyde is moderately toxic following oral and dermal exposure, and is extremely irritating to the skin. It is a moderate eye and respiratory tract irritant. It is mutagenic in vitro in nonmammalian assays, and is a skin carcinogen in rodents.

Acute Toxicity. Following ingestion in rats, transitory excitement occurs succeeded by depression and labored breathing. The LD_{50} in rats is 0.23 g/kg. Respiratory exposure causes marked pulmonary tract irritation and tearing. The LC_{50} in rats exposed 4 hr is 251 ppm. Percutaneous absorption gives an approximately lethal dose of 0.2 g/kg in rabbits. At a dermal dose of 0.04 g/kg, severe local injury to the skin occurred, but none of the three rabbits died.

Eye Irritation. Glycidaldehyde is classified as moderately irritating to the rabbit eye, with a Draize score of 3.6/8.0.

Carcinogenicity/Chronic Toxicity. Glycidaldehyde is a skin carcinogen in mice following topical application (445) and subcutaneous injection (126) and in rats by subcutaneous injection (116, 126). It also actively alkylates proteins and DNA (446). IARC classified glycidaldehyde as an animal carcinogen in 1987.

Genetic Toxicity. Glycidaldehyde was mutagenic in bacteria, yeast *Klebsiella*, and in the recessive lethal test in *Drosophila* (447). It was negative in the mouse lymphoma HGPRT test (447).

Effects in Humans. Glycidaldehyde will produce marked skin irritation and sensitization, but effective hygiene practices should prevent exposure.

2.4.3.5 Hygienic Standards of Permissible Exposure. No standards for permissible occupational exposure were found.

2.4.3.6 Odor and Warning Properties. There is a pronounced aldehyde-like odor at low levels. Voluntary exposure to serious lung-irritating levels is unlikely.

2.4.4 Bis(2,3-epoxycyclopentyl) Ether (CASRN 23860-90-5)

This compound has the structure

2.4.4.1 Source, Uses, and Industrial Exposure. This product was developed as a high-performance epoxy resin and epoxy resin diluent.

2.4.4.2 Physical and Chemical Properties

Molecular weight	192
Viscosity (25°C)	35 cm/sec

Epoxy equivalent weight, g/equiv.   96
Density   1.17 g/cm³

2.4.4.3 Determination in Atmosphere. No OSHA standard methods for the analysis of this material were found. A method of analysis by GLC has been described (448).

2.4.4.4 Physiological Response. No acute toxicity data were available. The compound was positive in in vitro genetic toxicity assays, but mixed results have been reported in cancer bioassays in mice.

Chronic Toxicity/Carcinogenicity. Dermal application in C3H/JAX mice for 18–20 months at 30 percent in acetone resulted in no mice with tumors. All mice had died by 20 months. When applied in C3H/HeJ mice for up to 24 months at 50 percent in acetone for the first 15 months and 25 percent in acetone for the rest of the study, there were again no skin tumors Weil (449). In another test in C3H mice at 50 percent and 10 percent in acetone, there were 5/80 with skin tumors at the 50 percent level and 0/40 with skin tumors at 10 percent (280). When a 1:1 mixture of this compound and ERL 2774 was applied at 50 percent and 10 percent in acetone to the same two strains of mice, there were 51/80 and 19/40, respectively, C57BL/6 mice with skin tumors. The sample of ERL 2774 tested was shown to have 10 percent of an epoxidized polyglycol (mol. wt. 7500) and up to 5000 ppm phenyl glycidyl ether present. Nothing is known of the toxicity potential of the epoxidized polyglycol, but phenyl glycidyl ether is known to cause nasal carcinoma in rats following chronic inhalation (238).

Genetic Toxicity. The compound has been shown to be positive in the Ames test for bacterial mutagenesis, positive in a cell transformation assay in Chinese hamster ovary cells, and negative for induction of DNA repair in cultured human leukocytes (25).

2.4.4.5 Hygienic Standards of Permissible Exposure. No standards of permissible occupational exposure were found.

2.4.4.6 Odor and Warning Properties. No information on odor or warning properties was found.

2.4.5   3,4-Epoxycyclohexylmethyl-3,4-Epoxycyclohexanecarboxylate (CASRN 2386-87-0)

The structure of this compound is

2.4.5.1 Source, Uses, and Industrial Exposure. This compound can be made by treatment of the corresponding bis-unsaturated ester with peracetic acid. It is used as a reactive diluent, for filament winding, as an acid scavenger, or as a plasticizer. Vapor hazard is slight at ambient temperature because of its moderately high boiling point. Skin contact is to be avoided.

2.4.5.2 Physical and Chemical Properties. The properties are as follows:

Viscosity (25°C)	350–450 cm/sec
Specific gravity (25/25°C)	1.175
Color, Gardner, 1933 max.	1
Epoxy equivalent weight, g/g-mol oxirane equiv.	131 to 143
Boiling point (760 mm Hg)	354°C
Vapor pressure (20°C)	0.1 mm Hg
Freezing point	−20°C
Solubility in water, 25°C	0.03% wt.
Solubility of water in, 25°C	2.8% wt.

Hazardous polymerizations with amines may occur with masses of 1 lb.

2.4.5.3 Determination in the Atmosphere. No OSHA standard methods for the determination of this material in air were found. The HCl–dioxane method may be used on trapped condensed phases (122) or GLC.

2.4.5.4 Physiological Response (25). The oral LD_{50} in rats is 4.49 ml/kg. Saturated vapors at ambient temperature for 8 hr did not kill rats. The compound is slightly irritating to rabbit skin and has an LD_{50} dermally of 20 g/kg. It produces moderate eye irritation in rabbits. Lifetime mouse skin painting studies were negative for carcinogenesis.

2.4.5.5 Hygienic Standards of Permissible Exposure. No standards for permissible occupational exposure were found.

2.4.5.6 Odor and Warning Properties. No information on odor or warning properties was found.

REFERENCES

1. B. Ottar, *Acta Chem. Scand.*, **1,** 283 (1947), through Hine et al., Chapter 32, "Epoxy Compounds," Patty's Industrial Hygiene and Toxicology, 3rd ed., Vol. IIa, G. D. Clayton and F. E. Clayton, Eds., J Wiley, New York, 1981.

2. C. A. May and Y. Tanaka, *Epoxy Resins*, Dekker, 1973, p. 16, through Hine et al., Chapter 32, "Epoxy Compounds," Patty's Industrial Hygiene and Toxicology, 3rd ed., Vol. IIa, G. D. Clayton and F. E. Clayton, Eds., J Wiley, New York, 1981.

3. "NIOSH Special Occupational Hazard Review with Control Recommendations for the Use of Ethylene oxide as a Sterilant in Medical Facilities," DHEW No. 77-200 (1977),

through Hine et al., Chapter 32, "Epoxy Compounds," Patty's Industrial Hygiene and Toxicology, 3rd ed., Vol. IIa, G. D. Clayton and F. E. Clayton, Eds., J Wiley, New York, 1981.

4. "NIOSH Stock No. PB-262-404," National Technical Information Service, Springfield, VA, 1977, through Hine et al., Chapter 32, "Epoxy Compounds," Patty's Industrial Hygiene and Toxicology, 3rd ed., Vol. IIa, G. D. Clayton and F. E. Clayton, Eds., J Wiley, New York, 1981.

5. G. R. Schultz, Ed., OSHA Analytical Methods Manual (Organic), 1989.

6. N. E. Bolton and N. H. Ketcham, Arch. Environ. Health., 8, 711 (1964).

7. D. D. Sully, Analyst, 85, 895 (1960), through Hine et al., Chapter 32, "Epoxy Compounds," Patty's Industrial Hygiene and Toxicology, 3rd ed., Vol. IIa, G. D. Clayton and F. E. Clayton, Eds., J Wiley, New York, 1981.

8. R. E. Burge et al., Analytical Chemistry of Polymers, Interscience, 1959, p. 123, through Hine et al., Chapter 32, "Epoxy Compounds," Patty's Industrial Hygiene and Toxicology, 3rd ed., Vol. IIa, G. D. Clayton and F. E. Clayton, Eds., J Wiley, New York, 1981.

9. H. Jahn, J. Pract. Chem., 37, 113 (1968), through Hine et al., Chapter 32, "Epoxy Compounds," Patty's Industrial Hygiene and Toxicology, 3rd ed., Vol. IIa, G. D. Clayton and F. E. Clayton, Eds., J Wiley, New York, 1981.

10. L. B. Bourne et al., Brit. J. Ind. Med., 16, 81 (1959), through Hine et al., Chapter 32, "Epoxy Compounds," Patty's Industrial Hygiene and Toxicology, 3rd ed., Vol. IIa, G. D. Clayton and F. E. Clayton, Eds., J Wiley, New York, 1981.

11. A. Welker, Dermatol. Wochenschr. dd, 871 (1955), through Hine et al., Chapter 32, "Epoxy Compounds," Patty's Industrial Hygiene and Toxicology, 3rd ed., Vol. IIa, G. D. Clayton and F. E. Clayton, Eds., J Wiley, New York, 1981.

12. A. Thorgeirsson, S. Fregert, and O. Ramnas, Acta Dermatol., 57, 253, (1977).

13. D. C. Mensik and D. D. Lockwood, unpublished report of The Dow Chemical Company, 1987.

14. R. Irwin, National Toxicology Program Technical Report 374, NIH Publication No. 90-2829, 1990.

15. P. B. Hulbert, Nature, 256, 146 (1975), through Hine et al., Chapter 32, "Epoxy Compounds," Patty's Industrial Hygiene and Toxicology, 3rd ed., Vol. IIa, G. D. Clayton and F. E. Clayton, Eds., J Wiley, New York, 1981.

16. J. P. Hamman, Biochem. Biophys. Res. Commun., 70, 675 (1976), through Hine et al., Chapter 32, "Epoxy Compounds," Patty's Industrial Hygiene and Toxicology, 3rd ed., Vol. IIa, G. D. Clayton and F. E. Clayton, Eds., J Wiley, New York, 1981.

17. E. W. Maynert et al., J. Biol. Chem., 245, 5234 (1970); K. C. Leibman, J. Pharmacol. Exp. Ther., 173, 242 (1970), through Hine et al., Chapter 32, "Epoxy Compounds," Patty's Industrial Hygiene and Toxicology, 3rd ed., Vol. IIa, G. D. Clayton and F. E. Clayton, Eds., J Wiley, New York, 1981.

18. F. Oesch et al., Biochem. Pharmacol., 26, 603 (1977), through Hine et al., Chapter 32, "Epoxy Compounds," Patty's Industrial Hygiene and Toxicology, 3rd ed., Vol. IIa, G. D. Clayton and F. E. Clayton, Eds., J Wiley, New York, 1981.

19. D. W. Nebert et al., Mol. Pharmacol., 8, 374 (1972), through Hine et al., Chapter 32, "Epoxy Compounds," Patty's Industrial Hygiene and Toxicology, 3rd ed., Vol. IIa, G. D. Clayton and F. E. Clayton, Eds., J Wiley, New York, 1981.

20. F. Oesch et al., *Chimia*, **29**, 66 (1975), through Hine et al., Chapter 32, "Epoxy Compounds," Patty's Industrial Hygiene and Toxicology, 3rd ed., Vol. IIa, G. D. Clayton and F. E. Clayton, Eds., J Wiley, New York, 1981.

21. A. C. C. Craven, *Pest. Biochem. Physiol.*, **6**, 132 (1976), through Hine et al., Chapter 32, "Epoxy Compounds," Patty's Industrial Hygiene and Toxicology, 3rd ed., Vol. IIa, G. D. Clayton and F. E. Clayton, Eds., J Wiley, New York, 1981.

22. C. S. Yang et al., *Biochem. Pharmacol.*, **24**, 646 (1975), through Hine et al., Chapter 32, "Epoxy Compounds," Patty's Industrial Hygiene and Toxicology, 3rd ed., Vol. IIa, G. D. Clayton and F. E. Clayton, Eds., J Wiley, New York, 1981.

23. F. Gesch et al., *Arch. Toxicol. (Berlin)*, **39**, 97 (1977), through Hine et al., Chapter 32, "Epoxy Compounds," Patty's Industrial Hygiene and Toxicology, 3rd ed., Vol. IIa, G. D. Clayton and F. E. Clayton, Eds., J Wiley, New York, 1981.

24. T. Hayakawa et al., *Arch. Biochem. Biophys.*, **170**, 438 (1975), through Hine et al., Chapter 32, "Epoxy Compounds," Patty's Industrial Hygiene and Toxicology, 3rd ed., Vol. IIa, G. D. Clayton and F. E. Clayton, Eds., J Wiley, New York, 1981.

25. C. H. Hine, V. K. Rowe, E. R. White, K. I. Darmer, and G. T. Youngblood, through Hine et al., Chapter 32, "Epoxy Compounds," Patty's Industrial Hygiene and Toxicology, 3rd ed., Vol. IIa, G. D. Clayton and F. E. Clayton, Eds., J Wiley, New York, 1981.

26. M. Tornqvist, B. Gustafsson, A. Kautianinen, M. Harms-Ringdahl, F. Granath, and L. Ehrenberg, *Carcinogenesis*, **10**, 39–41 (1989).

27. J. G. Filser, B. Denk, M. Tornqvist, W. Kessler, and L. Ehrenberg, *Arch. Toxicol.*, **66**, 157–163 (1992).

28. C. T. Bedford, D. Blair, and D. E. Stevenson, *Nature*, **267**, 335 (1977).

29. G. R. Schultz, Ed., *OSHA Analytical Methods Manual (Organic) Part I*, Method 49 and Method 50, 1989.

30. L. Golberg, *Hazard Assessment of Ethylene Oxide*, CRC Press, Boca Raton, FL, 1986.

31. World Health Organization (WHO), International Programme on Chemical Safety: Environment Health Criteria 55, Ethylene Oxide, Geneva, 1985.

32. D. Deschamps, M. Leport, and S. Cordier, *J. Fr. Ophthalmol.*, **13**, 189–197 (1990).

33. D. Deschamps, M. Leport, A. M. Laurent, S. Cordier, B. Festy, and F. Conso, *Brit. J. Indust. Med.*, **47**, 308–313 (1990).

34. National Toxicology Program, "Toxicology and Carcinogenesis Studies of Ethylene Oxide in B6C3F1 Mice," NTP TR 326, NIH Publication No. 88-2582, U.S. Department of Health and Human Services: National Institutes of Health, 1987.

35. S. G. Austin and R. L. Sielken, Jr., *J. Occup. Med.*, **30**, 236–245 (1988).

36. B. Denk, J. G. Filser, D. Oesterle, et al., *J. Cancer Res. Clin. Oncol.*, **114**, 35–38 (1988).

37. K. Mori, M. Kaido, K. Fujishiro, et al., *Arch. Toxicol.*, **65**, 369–401 (1991).

38. K. Mori, M. Kaido, K. Fujishiro, et al., *Brit. J. Ind. Med.*, **48**, 270–274 (1991).

39. G. A. Sega, P. A. Brimer, and E. E. Generoso, *Mutat. Res.*, **249**, 339–349 (1991).

40. M. Katoh, N. L. A. Cacheiro, C. V. Cornett, K. T. Cain, J. C. Rutledge, and W. M. Generoso, *Mutat. Res.*, **210**, 337–344 (1989).

41. J. C. Rutledge and W. M. Generoso, *Teratology*, **39**, 563–572 (1989).

42. W. M. Generoso, A. G. Shourbaji, W. W. Piegorsch, and J. B. Bishop, *Mutat. Res.*, **250**, 439–446 (1991).

43. E. I. M. Florack and G. A. Zielhaus, *Int. Arch. Occup. Environ. Health.*, **62**, 273–277 (1990).

44. U. Fost, E. Hallier, H. Ottenwalder, H. M. Bolt, and H. Peter, *Human Exp. Toxicol.*, **10**, 25–31 (1991).

45. V. E. Walker, T. R. Fennell, J. A. Boucheron, N. Fedtke, F. Ciroussel, and J. A. Swenberg, *Mutat. Res.*, **233**, 151–164 (1990).

46. T. Katoh, K. Higashi, N. Inoue, et al., *Toxicology*, **58**, 1–9 (1989).

47. S. Osterman-Golkar and E. Bergmark, *Scand. J. Environ. Health*, **58**, 105–112 (1986).

48. F. Brugnone, L. Perbellini, G. Faccini, F. Pasini, G. B. Bartolucci, and E. DeRosa, *Int. Arch. Occup. Environ. Health*, **58**, 105–112 (1986).

49. H. Bartsch, K. Hemminki, and I. K. O'Neill, "Methods for Detecting DNA Damaging Agents in Humans," IARC Scientific Publication No. 89, Lyon, 1988.

50. L. Rhomberg, V. L. Dellarco, C. Siegel-Scott, K. L. Dearfield, and D. Jacobson-Kram, *Environ. Mol. Mutagen.*, **16**, 104–125 (1990).

51. D. Segerbaeck, *Carcinogenesis*, **11**, 307–312 (1990).

51a. V. L. Dellarco, W. M. Generoso, G. A. Saga, J. R. Fowle, III, and D. Jacobson-Kram, *Environ. Mol. Mutagen.*, **16**, 85–103 (1990).

52. J. K. Wiencke, M. R. Wrensch, R. Miike, and N. L. Petrakis, *Cancer Res.*, **51**, 5266–5269 (1991).

52a. A. D. Tates, T. Grummt, M. Tornqvist, P. B. Farmer, F. J. Van Dam, H. Van Mossel, H. M. Schoemaker, S. Osterman-Golkar, Ch. Uebel, Y. S. Tang, A. Ha Zwinderman, A. T. Natarajan, and L. Ehrenberg, *Mutat. Res.*, **250**, 483–497 (1991).

53. J. Mayer, D. Warburton, A. M. Jeffrey, R. Pero, S. Walles, L. Andrews, M. Toor, L. Latriano, L. Wazneh, D. Tang, W-Y Tsai, M. Kuroda, and R. Perera, *Mutat. Res.*, **248**, 163–176 (1991).

54. P. A. Schulte, M. Boeniger, J. T. Walker, S. E. Schober, M. A. Pereira, D. K. Gulati, J. P. Wojciechowski, A. Garza, R. Froelich, G. Strauss, W. E. Halperin, R. Herrick, and J. Griffith, *Mutat. Res.*, **278**, 237–251 (1992).

55. G. de Jong, N. J. Van Sittert, and A. T. Natarajan, *Mutat. Res.*, **204**, 451–464 (1988).

56. H. A. Crystall, H. H. Schaumberg, E. Cooper, et al., *Neurology*, **38**, 567–569 (1988).

57. W. J. Estrin, R. M. Bowler, A. Lash, et al. *J. Toxicol. Clin. Toxicol.*, **28**, 1–20 (1990).

58. C. Hogstedt, O. Rohlen, B. S. Berndtsson, O. Axelson, and L. Ehrenberg. *Brit. J. Ind. Med.*, **36**, 766–780 (1979).

59. M. J. Gardner, D. Coggon, B. Pannett, and E. C. Harris, *Brit. J. Ind. Med.*, **46**, 860–865 (1989).

60. K. Steenland, L. Stayner, A. Greife, W. Halperin, R. Hayes, R. Hornung, and S. Nowlin, *N. Engl. J. Med.*, **324**, 1402–1407 (1991).

61. H. L. Greenberg, M. G. Ott, and R. E. Shore, *Brit. J. Ind. Med.*, **47**, 221–230 (1990).

62. M. J. Teta, L. O. Benson, and J. N. Vitale, "Mortality Study of Ethylene Oxide Workers in Chemical Manufacturing: A Ten-year Update," submitted for publication, 1992.

63. L. Hagmar, H. Welinder, K. Linden, R. Attewell, S. Osterman-Golkar, and M. Tornqvist, *Int. Arch. Occup. Environ. Health*, **63**, 271–278 (1991).

64. G. O. Curme, Jr., and F. Johnston, Eds., *Glycols*, ACS Monograph Series No. 114, Reinhold, New York, 1952.

65. Manufacturing Chemists' Association, Chemical Safety Data Sheet SD-38, 1951, (revised 1971).

66. G. R. Schultz, Ed., *OSHA Analytical Methods Manual (Organic) Part I*, Method 07, 1989.

67. K. H. Jacobsen, E. B. Hackley, and L. Feinsilver, *A.M.A. Arch. Ind. Health*, **13**, 237 (1956).

68. V. K. Rowe, R. L. Hollingsworth, et al., *A.M.A. Arch. Ind. Health*, **13**, 228 (1956).

69. C. S. Weil, N. Condra, C. Haun, and J. A. Striegel, *Am. Ind. Hyg. Assoc. J.*, **24**, 305 (1963).

70. C. P. Carpenter and H. F. Smyth, Jr., *Am. J. Ophthalmol.*, **29**, 1963 (1946).

71. National Toxicology Program, TR 267, 1985.

72. A. Ohnishi, T. Yamamoto, Y. Murai, et al., *Arch. Environ. Health*, **43**, 353 (1988).

73. J. T. Young, J. L. Mattsson, R. R. Albee, and D. J. Schuetz, unpublished report of The Dow Chemical Company, 1985.

74. R. A. Renne, W. E. Giddens, G. A. Boorman, R. Kovatch, J. E. Haseman, and W. J. Clarke, *J. Natl. Cancer Inst.*, **77**, 573 (1986).

75. J. K. Dunnick, S. L. Eustis, W. W. Piegorsch, and R. A. Miller, *Toxicology*, **50**, 69 (1988).

76. C. F. Kuper, P. G. J. Reuzel, and V. J. Feron, *Food Chem. Toxicol.*, **26**, 159 (1988).

77. H. Dunkelbert, *Brit. J. Cancer*, **46**, 924 (1982).

78. International Agency for Research on Cancer, *IARC Monogr.*, **36**, 227 (1985).

79. P. L. Hackett, M. G. Brown, and R. L. Buschbom, et al., NIOSH, U.S. Department of Health and Human Services, 1982.

80. B. D. Hardin, R. W. Niemeier, M. R. Sikov, et al., *Scand. J. Work Environ. Health*, **9**, 94 (1983).

81. C. A. Kimmel, J. B. LaBorde, and B. D. Hardin, *Toxicology Newborn* (1984).

82. S. B. Harris, J. L. Schardein, C. E. Ulrich, et al., *Fundam. Appl. Toxicol.*, **13**, 323 (1989).

83. W. C. Hayes, H. D. Kirk, T. S. Gushow, et al., *Fundam. Appl. Toxicol.*, **10**, 82 (1988).

84. W. Meylan, L. Papa, C. T. DeRosa, and J. F. Stara, *Toxicol. Ind. Health*, **2**, 219 (1986).

85. E. H. Pfeiffer and H. Dunkelberg, *Food Cosmet. Toxicol.*, **18**, 115 (1980).

86. J. D. Bootman, C. Lodge, and H. E. Whalley, *Mutat. Res.*, **67**, 101 (1979).

87. R. E. McMahon, J. C. Cline, and C. Z. Thompson, *Cancer Res.*, **39**, 682 (1979).

88. L. Migliore, A. M. Rossi, and N. Loprieno, *Mutat. Res.*, **102**, 425 (1982).

89. B. J. Dean and G. Hodson-Walker, *Mutat. Res.*, **64**, 329 (1979).

90. J. D. Tucker, J. Xu, J. Stewart, et al., *Teratog. Carcinog. Mutagen.*, **6**, 15 (1986).

91. P. O. Zamora, J. M. Benson, A. P. Li, et al., *Environ. Mutagen.*, **5**, 795 (1983).

92. D. K. Gulati, K. Witt, B. Anderson, et al., *Environ. Mol. Mutagen.*, **13**, 133 (1989).

93. D. McGregor, A. G. Brown, P. Cattanach, et al., *Environ. Mol. Mutagen.*, **17**, 122 (1991).

94. B. D. Hardin, R. L. Schuler, P. M. McGinnis, et al., *Mutat. Res.*, **177**, 337 (1983).

95. D. W. Lynch, T. R. Lewis, W. J. Moorman, et al., *Toxicol. Appl. Pharmacol.*, **76**, 85 (1984).

96. E. Agurell, H. Cederberg, L. Ehrenbert, et al., *Mutat. Res. Fundam. Mol. Mech. Mutagen.*, **250**, 229 (1991).

97. K. Hemminki, J. Paasivirta, T. Kurkirinne, and L. Virkki, *Chem. Biol. Interact.*, **30**, 259 (1980).

98. Z. Djuric, B. H. Hooberman, L. Rosman, and J. E. Sinsheimer, *Environ. Mutagen.*, **8**, 369 (1986).

99. J. J. Solomon, F. Mukai, J. Fedyk, et al., *Chem. Biol. Interact.*, **67**, 275 (1988).

100. J. J. Solomon and A. Segal, *Environ. Health Perspect.*, **81**, 19 (1989).

101. K. Svensson, K. Olofsson, and S. Osterman-Golkar, *Chem. Biol. Interact.*, **78**, 55 (1991).

102. P. B. Farmer, S. M. Gorf, and E. Bailey, *Biomed Mass Spectrom.*, **9**, 69 (1982).

103. S. Osterman-Golkar, E. Bailey, P. B. Farmer, et al., *Scand. J. Work Environ. Health*, **10**, 99 (1984).

104. E. Bailey, P. B. Farmer, and D. E. Shuker, *Arch Toxicol.*, **60**, 187 (1987).

105. A. Kautianinen and M. Tornqvist, *Int. Arch. Occup. Environ Health*, **63**, 27 (1991).

106. R. Jolanki, Riitta, Estlander, et al., *ASTM Spec. Tech. Publ.*, (1988).

107. O. Jensen, *Contact Dermatitis*, **7**, 148 (1981).

108. R. L. McLaughlin, *Am. J. Ophthalmol.*, **29**, 1355 (1946).

109. A. M. Theiss, R. Frentzel-Beyme, R. Link, and W. G. Stocker, *Occupational Safety & Health Series*, 1981.

110. A. M. Theiss, H. Schwegler, I. Fleig, and W. G. Stocker, *J. Occup. Med.*, **23**, 343 (1981).

111. R. W. Pero, T. Bryngelsson, B. Widegren, et al., *Mutat. Res.*, **104**, 193 (1982).

112. Manufacturing Chemists' Association, Chemical Safety Data Sheet SD-38, 1951 (revised 1971).

113. National Toxicology Program, 1988.

114. The Dow Chemical Company, unpublished report, 1984.

115. R. R. Miller, J. F. Quast, J. A. Ayres, and M. J. McKenna, *Fundam. Appl. Toxicol.*, **1**, 319 (1981).

116. B. L. Van Duuren, L. Langseth, B. M. Goldschmidt, and L. Orris, *J. Natl. Cancer Inst.*, **39**, 1217 (1967).

117. M. R. Sikov, W. C. Cannon, D. B. Carr, R. A. Miller, L. F. Montgomery, and D. W. Phelps, NIOSH Technical Report 81-124, 1981.

118. *IARC Monogr.*, **47**, 222–223 (1989).

119. G. M. Williams, H. Mori, and C. A. McQueen, *Mutat. Res.* **221**, 263–286 (1989).

120. R. H. Reitz, T. R. Fox, and E. A. Hermann, unpublished report, The Dow Chemical Company, 1983.

121. B. Phillips, Union Carbide Chemicals Co., New York, 1992.

122. J. L. Jungnickel et al., *Organic Analysis*, Vol. 1, Wiley-Interscience, New York, 1963, p. 127.

123. H. F. Smyth, C. P. Carpenter, C. S. Weil, and U. C. Pozzani, *A.M.A. Arch. Indust. Hyg. Occup. Med.*, **10**, 61 (1954).

124. B. L. Van Duuren, N. Nelson, L. Orris, E. D. Palmes, and F. L. Schmitt, *J. Natl. Cancer Inst.*, **31**, 41 (1963).

125. B. L. Van Duuren, L. Orris, and N. Nelson, *J. Nat. Cancer Inst.*, **35**, 707 (1965).

126. B. L. Van Duuren, L. Langseth, L. Orris, G. Teebor, N. Nelson, and M. Kuschner, *J. Natl. Cancer Inst.*, **37**, 825 (1966).

127. M. B. Shimkin, J. H. Weisburger, E. K. Weisburger, N. Gubareff, and V. Suntzeff, *J. Natl. Cancer Inst.*, **36**, 915 (1966).

127. M. J. Wade, J. W. Moyer, and C. H. Hine, *Mutat. Res.*, **66**, 367 (1979).

128. P. G. Gervasi, L. Citti, M. Del Monte, V. Longo, and D. Benetti, *Mutat. Res.*, **156**, 77 (1985).

129. V. C. Dunkel, E. Zeiger, D. Brusick, E. McCoy, D. McGregor, K. Mortelmans, H. S. Rosenkranz, and V. F. Simmon, *Environ. Mutagen.*, **6**, 1 (1984).

130. C. DeMeester, M. Mercier, and F. Poncelet, *Mutat. Res.*, **97**, 204 (1982).

131. D. Ichinotsubo, H. F. Mower, J. Setliff, and M. Mandel, *Mutat. Res.*, **46**, 53 (1977).

132. W. von der Hude, A. Seelbach, and A. Basler, *Mutat. Res.*, **231**, 205 (1990).

133. D. B. McGregor, A. Brown, P. Cattanach, I. Edwards, D. McBride, and W. J. Caspary, *Environ. Molc. Mutag.*, **11**, 91 (1988).

134. F. Abka'i, E. Wachter, H. Tittelbach, and E. Gebhart, *Cell Biol. Toxicol.*, **3**, 285 (1987).

135. F. Darroudi, A. T. Natarajan, and P. H. M. Lohman, *Mutat. Res.*, **212**, 103 (1989).

136. Y. Nishi, M. M. Hasegawa, M. Taketomi, Y. Ohkawa, and N. Inui, *Cancer Res.*, **44**, 3270 (1984).

137. W. von der Hude, S. Carstensen, and G. Obe, *Mutat. Res.*, **249**, 55 (1991).

138. M. Sasiadek, H. Jarventaus, and M. Sorsa, *Mutat. Res.*, **263**, 47 (1991).

139. M. Sasiadek, H. Norppa, and M. Sorsa, *Mutat. Res.*, **261**, 117 (1991).

140. J. K. Wiencke, D. C. Christiani, and K. T. Kelsey, *Mutat. Res.*, **248**, 17 (1991).

141. K. T. Kelsey, D. C. Christiani, and J. K. Wiencke, *Mutat. Res.*, **248**, 27 (1991).

142. B. J. Dean and G. Hodson-Walker, *Mutat. Res.*, **64**, 329 (1979).

143. G. M. Williams, H. Mori, and C. A. McQueen, *Mutat. Res.*, **221**, 263 (1989).

144. G. T. Arce, D. R. Vincent, M. J. Cunningham, W. N. Choy, and A. M. Sarrif, *Environ. Health Perspect.*, **86**, 75 (1990).

145. R. A. Walk, J. Jenderny, G. Rohrborn, and U. Hackenberg, *Mutat. Res.*, **182**, 333 (1987).

146. M. K. Conner, J. E. Luo, and O. Gutierrez de Gotera, *Mutat. Res.*, **108**, 251 (1983).

147. O. G. Fahmy and M. J. Fahmy, *Cancer Res.*, **30**, 195 (1970).

148. R. E. Denell, M. C. Lim, and C. Auerbach, *Mutat. Res.*, **49**, 219 (1978).

149. M. J. Bird, *Genetics*, **50**, 480 (1952).

150. M. L. Goldberg, R. A. Colvin, and A. F. Mellin, *Genetics*, **123**, 145 (1989).

151. J. T. Reardon, C. A. Liljestrand-Golden, R. L. Dusenbery, and P. D. Smith, *Genetics*, **115**, 323 (1987).

152. O. A. Olsen and M. M. Green, *Mutat. Res.*, **92**, 107 (1982).

153. A. D. Auerbach, B. Adler, and R. S. K. Chaganti, *Pediatrics*, **67**, 128 (1981).

154. D. P. Ward, unpublished report of Monsanto Proprietary, 1986.

155. A. P. Li, unpublished report of Monsanto Proprietary, 1987a.

156. D. P. Ward, unpublished report of Monsanto Proprietary, 1987b.

157. B. Ekwall, C. Nordensten, and L. Albanus, *Toxicology*, **24**, 199 (1982).

158. Industrial Medical and Toxicology Deparatment, Union Carbide Corp., New York, 1958.

159. National Toxicology Program, 1989.

160. R. S. Chhabra, M. R. Elwell, and A. Peters, *Fundam. Appl. Toxicol.*, **14**, 745 (1990).

161. M. P. Murray and J. E. Cummins, *Environ. Mutagen.*, **1**, 307 (1979).

162. M. A. El-Tantawy and B. D. Hammock, *Mutat. Res.*, **79** (1980).

163. G. Gervasi, L. Citti, and G. Turchi, *Mutat. Res.*, **74**, 202 (1980).

164. C. E. Voogd, J. J. Van Der Stel, and A. A. Jacobs, *Mutat. Res.*, **89** (1981).

165. S. W. Frantz and J. E. Sinsheimer, *Mutat. Res.*, **90 (1981).**

166. C. J. Dannaker, *J. Occup. Med.*, **30**, 641 (1988).

167. R. Patterson, K. E. Harris, W. Stopford, et al., *Int. Arch. Allergy Appl. Immunol.*, **85**, 467 (1988).

168. P. W. Langvardt and R. J. Nolan, *J. Chromatogr.*, **567**, 93 (1991).

169. Industrial Medicine and Toxicology Department, Union Carbide Corp., New York, 1958, through Hine et al., Chapter 32, "Epoxy Compounds," Patty's Industrial Hygiene and Toxicology, 3rd ed., Vol. IIa, G. D. Clayton and F. E. Clayton, Eds., J Wiley, New York, 1981.

170. M. R. Sikov, W. C. Cannon, D. B. Carr, R. A. Miller, R. W. Niemeier, and B. D. Hardin, *J. Appl. Toxicol.*, **6**, 155 (1986).

171. C. Maltoni, A. Ciliberti, and D. Carrietti, *Ann. N.Y. Acad. Sci.*, **381**, 216 (1982).

172. B. Conti, C. Maltoni, G. Perino, and A. Ciliberti, *Ann. N.Y. Acad. Sci.*, **534**, 203 (1988).

173. V. Ponomarkov and L. Tomatis, *Cancer Lett.*, **24**, 95 (1984).

174. W. Lijinski, *J. Natl. Cancer Inst.*, **77**, 471 (1986).

175. P. Kotin et al., *Radiat. Res. Suppl.*, **3**, 193 (1963) through Hine et al., Chapter 32, "Epoxy Compounds," Patty's Industrial Hygiene and Toxicology, 3rd ed., Volume IIa, G. D. Clayton and F. E. Clayton, Eds., J Wiley, New York, 1981.

176. H. Vaino, K. Hemminki, and E. Elovaara, *Toxicology*, **8**, 319 (1977).

177. B. D. Hardin, G. P. Bond, M. R. Sikov, F. D. Andrew, R. P. Beliles, and R. W. Niemeier, *Scand. J. Work Environ. Health* **7**, Suppl. 4, 66 (1981).

178. R. J. Preston, *S.I.R.C. Review* **1**(1), 23 (1990), Styrene Information and Research Center, Washington, DC.

179. R. J. Preston, *S.I.R.C. Review* **1**(2), 24 (1990), Styrene Information and Research Center, Washington, DC.

180. R. Barale, *Mutat. Res.*, **257**, 107 (1991).

181. S. Cantoreggi and W. K. Lutz, *Carcinogenesis*, **13**, 193 (1992).

182. "ICPS (1987) Environmental Health Criteria 26, Styrene," United Nations Environment Programme, International Labour Organization and World Health Organization, Geneva.

183. J. A. Bond, *CRC Critic. Rev. Toxicol.*, **19**, 227 (1989).

184. *Organic Chemicals*, Tech. Publ. SC: 52-10, Shell Chemical Corp., New York, 1952, p. 68.

185. C. H. Hine, J. K. Kodama, J. S. Wellington, M. K. Dunlap, and H. H. Anderson, *AMA Arch. Ind. Health*, **14**, 250 (1956).

186. J. W. Henck, D. D. Lockwood, and H. O. Yakel, unpublished report of The Dow Chemical Company, 1978.

187. J. K. Kodama, R. J. Guzman, M. K. Dunlap, G. S. Loquvam, R. Lima, and C. H. Hine, *J. Arch. Environ. Health*, **2**, 50 (1961).

188. National Toxicology Program, NTP TR 376, 1990.

189. K. J. Olson, unpublished report of The Dow Chemical Company, 1957.

190. C. H. Hine, J. K. Kodama, H. H. Anderson, D. W. Simonson, and J. S. Wellington, *AMA Arch. Ind. Health*, **17**, 129 (1958).

191. D. A. Canter, E. Zeiger, S. Haworth, T. Lawlor, K. Mortelmans, and W. Speck, *Mutat. Res.*, **172**, 105 (1986).

192. F. Gagnaire, D. Zissu, P. Bonnet, and J. DeCeaurriz, *Toxicol. Lett.*, **39**, 139 (1987).

193. S. Fregert and H. Rorsman, *Acta Allerg.*, **19**, 296 (1964).

194. J. H. Draize, G. Woodward, et al., *J. Pharmacol. Exp. Ther.*, **82**, 377 (1944).

195. K. K. Hemminki, K. Falck, and H. Vainio, *Arch. Toxicol.*, **46**, 277 (1980).

196. D. D. Lockwood and H. W. Taylor, unpublished report of The Dow Chemical Company, 1982.

197. Procter and Gamble Company, unpublished report, 1973.

198. K. J. Olson, unpublished report of The Dow Chemical Company, 1957.

199. M. Andersen, P. Kiel, H. Larsen, unpublished data, 1957.

200. Ciba-Geigy Limited, unpublished data, 1985.

201. J. F. Quast and R. R. Miller, *Toxicologist*, **5**, 1 (1985).

202. T. H. Connor, T. C. Pullin, J. Meyne, A. F. Frost, and M. S. Legator, *Environ. Mutat.*, **2**, 284 (1980).

203. T. G. Pullin, unpublished report to The Dow Chemical Company, 1977.

204. Reichhold Chemicals, Inc., studies submitted under Section 8(d) of TSCA (40-7840108), 1978.

205. E. D. Thompson, W. J. Coppinger, C. E. Piper, N. McCarroll, T. J. Oberly, and D. Robinson, *Mutat. Res.*, **90**, 213 (1981).

206. A. F. Frost and M. S. Legator, *Mutat. Res.*, **102**, 193 (1982).

207. T. H. Connor, J. B. Ward, J. Meyne, T. G. Pullin, and M. S. Legator, *Environ. Mutat.*, **2**, 521 (1980).

208. E. B. Whorton, T. G. Pullin, A. F. Frost, A. Onofre, M. S. Legator, and D. S. Folse, *Mutat. Res.*, **124**, 225 (1983).

209. L. V. Eadsforth, D. H. Hutson, C. J. Logan, and D. J. Morrison, *Xenobiotica*, **15**, 579 (1985).

210. M. A. Wolf and V. K. Rowe, unpublished data, Dow Chemical Company, 1958.

211. R. Jolanki, L. Kanerva, T. Estlander, K. Tarvainen, H. Keskinen, and M. Eckerman, *Contact Dermatitis*, **23**, 172 (1990).

212. E. Zeiger, B. Anderson, S. Haworth, T. Lawlor, and K. Mortelmans, *Environ. Mol. Mutagen.*, **11**, 1 (1988).

213. Procter & Gamble Tech. Inf. Sheets DUP 2660-128A, DUP 2660-129.

214. EPA/OTS Document 878212388 (V-1146-153, BTS 171), 1982.

215. Procter and Gamble Miami Valley Laboratories, 1979.

216. Springborn Institute for Bioresearch, Inc., 1980.

217. Industrial Bio-Test Laboratories, Inc., 1973.

218. International Bio-Research, Inc., 1975.

219. EPA/OTS Document 878212424 (Y152-116, BTS 1393), 1982.

220. EPA/OTS Document 878212429 (Y154-103, BTS 1463), 1982.

221. EPA/OTS Document 878212408 (V200043), 1982.

222. EPA/OTS Document 878212396 (V1439-143), 1982.

223. EPA/OTS Document 878212410 (V1927-43, BTS 345S and 368), 1982.

224. Hill Top Research Institute, Inc., 1964.

225. Shell Chemical Company, Tec. Publ SC:52-10, New York, 1952, p. 56.

226. T. Czajkowska and J. Stetkiewica, *Med. Pharm.*, **23**, 363 (1972).

227. Ciba-Geigy Limited, unpublished report, 1978.

228. Ciba-Geigy Limited, unpublished report, 1975.

229. Ciba-Geigy Limited, unpublished report, 1975.

230. E. Zschunke and P. Behrbohm, *Dermatol. Wochenschr.*, **151**, 480 (1965).

231. J. E. Betso, R. E. Carreon, P. W. Langvardt, and E. Martin, unpublished report of the Dow Chemical Toxicology Research Laboratory, 1986.

232. Ciba-Geigy Limited, unpublished report, 1978.

233. Ciba-Geigy Limited, unpublished report, 1978.

234. Procter and Gamble, unpublished report, 1973.

235. J. B. Terrill and K. P. Lee, *Toxicol. Appl. Pharmacol.*, **42**, 263 (1977).

236. K. P. Lee, J. B. Terrill, and N. W. Henry, *J. Toxicol. Environ. Health*, **3**, 859 (1977).

237. J. K. Kodama, R. J. Guzman, M. K. Dunlap, G. S. Loquvam, R. Lima, and C. H. Hine, *Arch. Environ. Health*, **2**, 50 (1961).

238. K. P. Lee, P. W. Schneider, and H. J. Trochimowica, *Am. J. Pathol.*, **3**, 140 (1983).

239. International Agency for Research on Cancer (IARC), Final Draft Working Papers, IARC, Lyon, 1988.

240. J. B. Terrill, K. P. Lee, R. Culik, and G. L. Kennedy, *Toxicol. Appl. Pharmacol.*, **64**, 204 (1982).

241. E. J. Greene, M. A. Friedman, J. A. Sherrod, and A. J. Salerno, *Mutat. Res.*, **67**, 9 (1979).

242. C. E. Voogd, J. J. Stel Van, and J. A. Jacobs, *Mutat. Res.*, **89**, 269 (1981).

243. H. Ohtani and H. Nishioka, *Sci. Eng. Rev.*, **21**, 247 (1981).

244. K. Hemminki and H. Vainio, *Dev. Toxicol. Environ. Sci.*, **8**, 241 (1980).

245. J. P. Seiler, *Mutat. Res.*, **135**, 159 (1984).

246. Unpublished report of Ciba-Geigy Limited, 1978.

247. S. P. James, A. E. Pheasant, and E. Solheim, *Xenobiotica*, **8**, 219 (1976).

248. Ciba-Geigy Limited, unpublished report, 1983.

249. Ciba-Geigy Limited, unpublished report, 1983.

250. Ciba-Geigy Limited, unpublished report, 1978.

251. Ciba-Geigy Limited, unpublished report, 1975.

252. J. R. Albert, H. C. Wimberly, and N. L. Wilburn, unpublished report of Shell Oil Company, 1983.

253. M. N. Pinkerton and R. L. Schwebel, unpublished report of the Dow Chemical Company, 1977.

254. L. Ullman, C. Kroling, P. Lucini, and T. Janiak, unpublished report of the RCC Research and Consulting Company, 1991.

255. Ciba-Geigy Limited, unpublished report, 1983.

256. S. H. Neau, B. J. Hooberman, S. W. Frantz, and J. E. Sinsheimer, *Mutat. Res.*, **93**, 297 (1982).

257. Ciba-Geigy Limited, unpublished report, 1978.

258. K. Sollner and Irrgana, *K. Arzneim. Forsch.*, **15**, 1355 (1965).

259. G. De Jong, N. J. Van Sittert, and A. T. Natarajan, *Mutat. Res.*, **204**, 451 (1988).

260. Ciba-Geigy, Limited, unpublished report, 1981.

261. Ciba-Geigy, Limited, unpublished report, 1983.

262. Ciba-Geigy, Limited, unpublished report, 1983.

263. Ciba-Geigy, Limited, unpublished report, 1972.

264. Ciba-Geigy, Limited, unpublished report, 1981.

265. Ciba-Geigy, Limited, unpublished report, 1982.

266. A. Thorgeirsson, *Acta Dermatol.*, **58**, 219, 1978.

267. S. Clemmensen, *Drug Chem. Toxicol.*, **7**, 527 1977.

268. Ciba-Geigy, Limited, unpublished report, 1981.

269. E. Thorpe, P. E. Cox, R. W. Hend, S. M. A. Doak, P. F. Hunt, A. N. Crabtree, and D. E. Wiggins, unpublished report of Shell Research Limited, 1980.

270. G. C. Peristianis, M. A. Doak, P. N. Cole, and R. W. Hend, *Food Chem. Toxicol.*, **26**, 611 (1988).

271. Ciba-Geigy, Limited, unpublished report, 1983.

272. EPA/OTS Document 878210058, 1987.

273. R. Jolanki, L. Kanerva, T. Estlander, K. Tarvainen, H. Keskinen, and M. Eckerman, *Contact Dermatitis*, **16**, 87 (1987).

274. Ciba-Geigy, Limited, unpublished report, 1972.

275. Ciba-Geigy, Limited, unpublished report, 1975.

276. Ciba-Geigy, Limited, unpublished report, 1982.

277. Ciba-Geigy, Limited, unpublished report, 1976.

278. Ciba-Geigy, Limited, unpublished report, 1972.

279. Unpublished report of Ciba-Geigy, Limited, 1982.

280. J. M. Holland, L. C. Gipson, M. J. Whitaker, B. M. Eisenhower, and T. J. Stephens, Oak Ridge National Laboratory Report No. ORNL 5762, 1981.

290. M. L. Westrick and P. Gross, Industrial Hygiene Foundation of America, Inc., 1960.

291. M. L. Westrick and P. Gross, Industrial Hygiene Foundation of America, Inc., 1960.

292. Wilmington Chemical Corporation, Tech Data Sheet No. 100-774.

293. National Toxicology Program, National Institute of Health, Technical Report Series No. 257, NIH Publication 87-2513, 1987.

294. C. J. McCammon et al., *Proc. Am. Assoc. Cancer Res.*, **2**, 229, (1957).

295. Kettering Laboratory, University of Cincinnati, 1958 through Hine et al., Chapter 32, "Epoxy Compounds," Patty's *Industrial Hygiene and Toxicology, 3rd ed., Vol. IIa*, G. D. Clayton and F. E. Clayton, Ed., Wiley, New York, 1981.

296. J. P. Seiler, *Chem. Biol. Interact.*, **51**, 347, (1984).

297. E. Boyland and K. Williams, *Biochemistry*, **94**, 190, (1965).

298. Imperial Chemical Industries, Limited, unpublished report, 1959.

299. V. K. Rowe, unpublished report of The Dow Chemical Company, 1948.

300. M. A. Wolf, unpublished report of The Dow Chemical Company, 1956.

301. V. K. Rowe, unpublished report of The Dow Chemical Company, 1958.

302. R. W. Hend, D. G. Clark, and A. D. Coombs, unpublished report of Shell Chemical Company, 1977a.

303. D. G. Clark and S. L. Cassidy, unpublished report of Shell Chemical Company, 1978.

304. R. W. Hend, D. G. Clark, and A. D. Coombs, unpublished report of Shell Chemical Company, 1977.

305. D. D. Lockwood and H. W. Taylor, unpublished report of The Dow Chemical Company (1982).

306. R. W. Hend, D. G. Clark, and A. D. Coombs, unpublished report of Shell Chemical Company, 1977.

307. R. J. Nolan, S. Unger, and L. S. Chatterton, unpublished report of the Dow Chemical Company, 1981.

308. M. A. Wolf, unpublished report of The Dow Chemical Company, 1957.

309. K. Olson, unpublished report of The Dow Chemical Company, 1958.

310. K. Olson, unpublished report of The Dow Chemical Company, 1958.

311. V. K. Rowe, unpublished report of The Dow Chemical Company, 1958.

312. W. J. Breslin, H. D. Kirk, and K. A. Johnson, *Fundam. Appl. Tox.*, **10**, 736, (1988).

313. S. Fregert and B. Lundin, "Sensitization Capacity of Epoxy Resin Compounds," report to The Swedish Plastics Federation and Swedish Environmental Fund, 1977.

314. A. Thorgeirsson and S. Fregert, *Acta Dermatovener (Stockholm)*, **57**, 253 (1977).

315. H. P. Til, Report of Central Institute for Nutrition and Food Research (Civo, Switzerland) to The Dow Chemical Company, Europe, 1977.

316. D. D. Lockwood, unpublished report of The Dow Chemical Company, 1978.

317. M. N. Pinkerton, unpublished report of The Dow Chemical Company, 1979.

318. M. N. Pinkerton, unpublished report of The Dow Chemical Company, 1979.

319. A. Thorgeirsson and S. Fregert, *Acta Dermatovener (Stockholm)*, **58**, 17 (1978).

320. D. C. Mensik and D. D. Lockwood, unpublished report of The Dow Chemical Company, 1987.

321. K. Olson, unpublished report of The Dow Chemical Company, 1958.

322. C. S. Weil, 1978, personal communication, through Hine et al., Chapter 32, "Epoxy Compounds," Patty's Industrial Hygiene and Toxicology, 3rd ed., Vol. IIa, G. D. Clayton and F. E. Clayton, Eds., J Wiley, New York, 1981.

323. Shell Chemical Company, Houston, Texas, through Hine et al., Chapter 32, "Epoxy Compounds," Patty's Industrial Hygiene and Toxicology, 3rd ed., Vol. IIa, G. D. Clayton and F. E. Clayton, Eds., J Wiley, New York, 1981.

324. M. A. Wolf, unpublished report of The Dow Chemical Company, 1958.

325. W. Basler, W. Gfeller, F. Zak, and V. Skorpil, unpublished report of Ciba-Geigy Ltd., Basel, Switzerland, 1984.

326. W. Dobryszycka, M. Warwas, A. Woyton, J. Woyton, and J. Szacki, *Arch. Toxicol.*, **33**, 73 (1974).

327. J. Woyton, J. Szacki, A. Woyton, M. Warwas, and W. Dobryszycka, *Toxicol. Appl. Pharmacol.*, **32**, 5 (1975).

328. W. Dobryszyka et al., *Arch. Immunol. Ther. Exp.*, **22**, 135 (1974), through W. Dobryszyka et al., *Arch. Toxicol.*, **33**, 73 (1974).

329. J. Woyton et al., *Arch. Immunol. Ther. Exp.*, **22**, 129 (1974), through W. Dobryszyka et al., *Arch. Toxicol.*, **33**, 73 (1974).

330. C. H. Hine et al., *Cancer Res.* **18**, 20 (1958).

331. W. L. Kydd and L. M. Sreenby, *J. Natl. Cancer Inst.*, **25**, 749 (1960).

332. C. S. Weil, personal communication, through Hine et al., Chapter 32, "Epoxy Compounds," Patty's Industrial Hygiene and Toxicology, 3rd ed., Vol. IIa, G. D. Clayton and F. E. Clayton, Eds., J Wiley, New York, 1981.

333. J. M. Holland, D. G. Gosslee, and N. J. Williams, *Cancer Res.*, **39**, 1718 (1979).

334. P. Bentley, F. Bieri, H. Kuster, S. Muakkassah-Kelly, P. Sagelsdorff, W. Staubli, and F. Waechter, *Carcinogenesis*, **10**, 321, (1989).

335. J. Agee, W. Barkley, K. LaDow, R. Linneman, S. Spalding, and K. Stemmer, Report of Kettering Laboratory, Department of Environmental Health, University of Cincinnati Medical Center, Cincinnati, OH, to Celanese Corporation, 1987.

336. W. T. Solomon, Zeon Chemicals, Louisville KY, personal communication, 1990.

337. N. Zakova, F. Zak, E. Froehlich, and R. Hess, *Food Chem. Toxicol.*, **23**, 1081 (1985).

338. R. Peto, M. C. Pike, N. E. Day, R. G. Gray, P. N. Lee, S. Parish, J. Peto, S. Richards, and J. Wahrendorf *Long-term and Short-term Screening Assays for Carcinogens; A Critical Appraisal*, IARC Monographs on the Evaluation of the Carcinogenic Risk of Chemicals to Humans Suppl. 2, 1980, p. 311.

339. S. M. Roche and C. H. Hine, *Toxicol. Appl. Pharmacol.*, **12**, 327 (1968).

340. W. J. Breslin, H. D. Kirk, E. L. Wolfe, and K. A. Johnson, unpublished report of The Dow Chemical Company, 1986.

341. J. A. Smith, R. E. Masters, and I. S. Dawe, Report to Ciba-Geigy Limited, Huntingdon Research Centre Ltd., 1988.

342. J. A. Smith, D. A. John, and I. S. Dawe, report to Ciba-Geigy Limited, Huntingdon Research Centre Ltd., 1988.

343. J. A. Smith, A. Bryson, J. M. Offer, C. A. Parker, and A. Anderson, report to Ciba-Geigy Limited, Huntingdon Research Centre Ltd., 1989.

344. T. H. Gardiner, J. M. Waechter, Jr., A. Wiedow, and W. T. Solomon, *Reg. Pharmacol. Toxicol.*, **15**, S1 (1992).

345. I. J. G. Climie, D. H. Hutson, and G. Stoydin, *Xenobiotica*, **11**, 391 (1981).

346. I. J. G. Climie, D. H. Hutson, and G. Stoydin, *Xenobiotica*, **11**, 401 (1981).

347. P. C. Coveney, unpublished report of Shell Chemical Company (SBGR.83.073), 1983.

348. J. Pluss, Z. *Unfallmed.*, **47**, 83 (1954).

349. E. Grandjean, *Brit J. Ind. Med.*, **14**, 1 (1957).

350. S. Fregert and A. Thorgeirsson, *Contact Dermatitis* **3**, 301 (1977).

351. D. C. Mensick and D. D. Lockwood, unpublished report of the Dow Chemical Company, 1987.

352. R. S. Cotran, V. Kumar, and S. L. Robbins, *Robbins Pathologic Basis of Disease*, 4th ed., W. B. Saunders, Philadelphia, 1989, p. 774.

353. J. C. Hogg, Bronchial Asthma, Chapter 10 of *The Lung: Structure, Function & Disease*, W. M. Thurlbeck & M. P. Abell, Eds., Williams & Wilkins, Baltimore, 1978.

354. I. W. Fawcett, A. J. Newman Taylor, and J. Pepys, *Clin. Allergy*, **7**, 1 (1977).

355. Unpublished report of Ciba-Geigy, Limited, 1969.

356. Unpublished report of Ciba-Geigy, Limited, 1969.

357. D. D. Lockwood and V. Borrego, unpublished report of The Dow Chemical Company, 1981.

358. W. L. Wilborn, C. M. Parker, and D. R. Patterson, unpublished report of Shell Chemical Company, 1982.

359. V. A. Jud and C. M. Parker, unpublished report of Chemical Company, 1982.

360. Unpublished report of Ciba-Geigy, Limited, 1969c.

361. A. S. Lam and C. M. Parker, unpublished report of Shell Chemical Company, 1982.

362. D. M. Glueck, unpublished report of Shell Chemical Company, 1982.

363. V. L. Sawin and W. M. Smith, unpublished report of Shell Chemical Company, 1983.

364. J. O. Rundell, B. S. Tang and V. L. Sawin, unpublished report of Shell Chemical Company, 1984.

365. W. M. Smith, B. S. Tang and V. L. Sawin, unpublished report of Shell Chemical Company, 1984.

366. M. A. Wolf, Unpublished report of The Dow Chemical Company, 1959.

367. D. D. Lockwood and V. Borrego, unpublished report of The Dow Chemical Company, 1979.

368. Unpublished report of Ciba-Geigy, Limited, 1974.

369. Unpublished report of Ciba-Geigy, Limited, 1974.

370. D. D. Lockwood and H. W. Taylor, unpublished report of The Dow Chemical Company, 1982.

371. J. R. Jones, unpublished report of The Dow Chemical Company, 1990.

372. J. R. Jones, unpublished report of The Dow Chemical Company, 1991.

373. Ciba-Geigy, Limited, unpublished report, 1969.

374. M. A. Wolf, unpublished report of The Dow Chemical Company, 1958.

375. Ciba-Geigy, Limited, unpublished report, 1971.

376. Ciba-Geigy, Limited, unpublished report, 1972.

377. Ciba-Geigy, Limited, unpublished report, 1972.

378. Ciba-Geigy, Limited, unpublished report, 1979.

379. Bionetics Research Laboratories, 1970.

380. Kettering Laboratory, University of Cincinnati, 1987.

381. Bionetics Research Laboratories, unpublished report to Ciba-Geigy, 1970.

382. Litton Bionetics, Inc., unpublished report to Ciba-Geigy, 1977.

383. D. Burrows, S. Fregert, H. Campbell, and L. Trulsson, *Contact Dermatitis*, **11**, 80, (1984).

384. J. W. Lacher and J. W. Crissman, unpublished report of The Dow Chemical Company, 1990.

385. M. J. Beekman and J. W. Crissman, unpublished report of The Dow Chemical Company, 1992.

386. Ciba-Geigy, Limited, unpublished report, 1986.

387. Ciba-Geigy, Limited, unpublished report, 1986.

388. Ciba-Geigy, Limited, unpublished report, 1985.

389. Ciba-Geigy, Limited, unpublished report, 1984.

390. B. B. Gollapudi and Y. E. Samson, unpublished report of The Dow Chemical Company, 1992.

391. J. S. C. English, I. Foulds, I. R. White, and R. J. G. Rycroft, *Contact Dermatitis*, **15**, 66, (1986).

392. D. G. Clark and A. D. Coombs, unpublished report of Shell Chemical Company, 1977.

393. D. Blair, Unpublished report of Shell Chemical Company, 1983.

394. R. G. Pickering, Unpublished report of Shell Chemical Company, 1981.

395. B. J. Dean, T. M. Brooks, G. Hodson-Walker, and G. Pook, unpublished report of Shell Chemical Company, 1979.

396. B. J. Dean, T. M. Brooks, and G. Hodson-Walker, unpublished report of Shell Chemical Company, 1979.

397. A. L. Meyer, unpublished report of Shell Chemical Company, 1980.

398. M. F. Wooder and C. L. Creedy, unpublished report of Shell Chemical Company, 1980.

399. I. Dahlquist and S. Fregert, *Contact Dermatitis*, **5**, 121 (1979).

400. C. R. Lovell, R. J. G. Rycroft, and J. Matood, *Contact Dermatitis*, **11**, 190 (1984).

401. P. M. Eller, Ed., *NIOSH Manual of Analytical Methods*, 3rd ed., Vol. 1, Method 1010, 1984.

402. F. K. Dietz, M. Grandjean, J. T. Young, unpublished report of The Dow Chemical Company, 1985.

403. S. Pallade, M. Dorobantu, G. Rotaru, and E. Gabrielescu, *Arch. Mal. Prof. Med. Trav. Secur. Soc.*, **28**, 505 (1967)

404. C. D. Leake, *Univ. Calif. Publ. Pharmacol.*, **2**, 69 (1941).

405. H. F. Smyth, Jr., and C. P. Carpenter, *J. Ind. Hyg. Toxicol.*, **30**, 63 (1948)

406. C. P. Carpenter, H. F. Smyth, Jr., and V. C. Pozzani, *J. Ind. Hyg. Toxicol.*, **31**, 343 (1949).

407. Shell Chemical Company, Ind. Hyg. Bull SC: 57–86, New York, 1957.

408. W. H. Lawrence, M. Malik, J. E. Turnee, and J. Aution, *J. Pharm. Sci.*, **61**, 1712 (1972).

409. K. S. Rao, J. E. Betso, and K. J. Olson, *Drug Chem. Toxicol.*, **4**, 331 (1981).

410. A. Kandel, unpublished report of The Dow Chemical Company, 1953.

411. J. C. Gage, *Brit. J. Ind. Med.*, **16**, 11 (1959).

412. J. F. Quast, J. W. Henck, B. J. Postma, D. J. Schuetz, and M. J. McKenna, unpublished report of The Dow Chemical Company, 1979.

413. J. A. John, J. F. Quast, F. J. Murray, L. G. Calhoun, and R. E. Staples, *Toxicol. Appl. Pharmacol.*, **68**, 415 (1983).

414. A. P. Fomin, *Gig. Sanit.*, **31**, 7 (1966)

415. R. Grigorowa, G. M. Muller, R. Rothe, and R. Gohlke, *Int. Arch. Arbeitsmed.*, **33**, 297 (1974).

416. B. L. Robinson, J. W. Allis, J. E. Andrews, A. McDonald, and J. E. Simmons, *The Toxicologist*, **9**, (Abstr. #909) 227 (1989).

417. S. Laskin, A. R. Sellakumar, M. Kuschner, N. Nelson, S. LaMendola, G. M. Rusch, G. V. Katz, N. C. Dulak, and R. E. Albert, *J. Natl. Cancer Inst.*, **65**, 751 (1980)

418. Y. Konishi, A. Kawabata, A. Denda, T. Ikeda, H. Katada, H. Maruyama, and R. Higashiguchi, *Gann*, **71**, 922 (1980).

419. P. W. Wester, C. A. Van der Heijden, A. Bisschop, and G. J. Van Esch, *Toxicology*, **36**, 325 (1985).

420. Dynamac Corporation, *Health Assessment Document for Epichlorohydrin, Draft Report #1*, submitted to the US-EPA, Contract No. 68-03-3111, 1982.

421. B. L. Van Duuren, B. M. Goldschmidt, C. Katz, I. Seidman, and I. S. Paul, *J. Natl. Cancer Inst.*, **53**, 695 (1974).

422. J. A. John, T. S. Gushow, J. A. Ayres, T. R. Hanley, J. F. Quast, and K. S. Rao, *Fundam. Appl. Toxicol.*, **3**, 437 (1983).

423. T. A. Marks, F. S. Gerling, and R. E. Staples, *J. Toxicol. Environ. Health*, **9**, 87 (1982).

424. J. D. Hahn, *Nature*, **226**, 87 (1970).

425. R. J. Sram, L. Tomatis, J. Clemmesen, and B. A. Bridges, *Mutat. Res.*, **87**, 299 (1981).

426. Health and Safety Executive UK, *Toxicity Review 24; Ammonia, 1-Chloro-2,3-epoxypropane (Epichlorohydrin), Carcinogenicity of Cadmium and Its Compounds*, 1991.

427. S. S. Epstein, E. Arnold, L. Andrea, W. Bass, and Y. Bishop, *Toxicol. Appl. Pharmacol.*, **23**, 288 (1972).

428. R. J. Sram, M. Cerna, and M. Kucerova, *Biol. Zentralbl.*, **95**, 451 (1976).

429. W. W. Weigel, H. B. Plotnick, and W. L. Conner, *Res. Commun. Chem. Pathol. Pharmacol.*, **20**, 275 (1978).

430. F. A. Smith, P. W. Langvardt, and J. D. Young, *Toxicol. Appl. Pharmacol.*, **48**, A116 (1979).

431. G. Fakhouri and A. R. Jones, *Aust. J. Pharm. Sci.*, **8**, 11 (1979).

432. A. M. Rossi, L. Migliore, D. Lascialfari, I. Sbrana, N. Loprieno, M. Tortoreto, F. Bidoli, and C. Pantarotto, *Mutat. Res.*, **118**, 213 (1983).

433. R. Gingell, H. R. Mitschke, I. Dzidic, P. W. Beatty, V. L. Swain, and A. C. Page, *Drug Metab. Disp.*, **13**, 333 (1985).

434. H. Ippen and V. Mathias, *Berufsdermatosen*, **18**, 144 (1970).

435. C. Schultz, *Dtsch. Med. Wochenschr.*, **89**, 1342 (1964).

436. E. Epstein, *Contact Dermatitis Newsl.*, **16**, 475 (1974).

437. D. Lambert, M. Lacroix, G. Ducombs, F. Journet, and J. L. Chapuis, *Ann. Dermatol. Venerol.*, **105**, 521 (1978).

438. M. H. Beck and C. M. King, *Contact Dermatitis*, **9**, 315 (1983).

439. E. P. Prens, G. De Jong, T. Van Joost, *Contact Dermatitis*, **15**, 85 (1986).

440. T. Van Joost, *Contact Derm.*, **19**, 278 (1988).

441. T. Van Joost, I. D. Roesyanto, and I. Satyawan, *Contact Dermatitis*, **22**, 125 (1990).

442. Society of the Plastics Industry, Inc., *Epichlorohydrin: A Safety and Handling Guide*, 1992.

443. E. D. Thompson and D. P. Gibson, *Food Chem. Toxicol.*, **22**, 665 (1984).

444. V. L. Slott and B. F. Hales, *Teratology*, **32**, 65 (1985).

445. IARC Monograph, Vol. 11, International Agency for Research on Cancer, Lyon, 1976.

446. S. Steiner and W. P. Watson, *Carcinogenesis*, **13**, 119 (1992).

447. A. G. A. C. Knaap, C. E. Voogd, and P. G. N. Kramers, *Mutate Res.*, **101**, 199 (1982).

448. J. M. Holland, D. G. Goss'ee, L. C. Gipson, and M. J. Whitaker, Oak Ridge Natl. Lab. Rep. ORNL-5375, 1978.

449. C. S. Weil, unpublished report of the Carnegie-Mellon Institute of Research, through Hine et al., Chapter 32 "Epoxy Compounds," Patty's Industrial Hygiene and Toxicology, 3rd ed., Vol. IIa, G. D. Clayton and F. E. Clayton, Eds., J Wiley, New York, 1981.

450. G. W. Olsen, S. E. Lacy, S. R. Chamberlin, D. L. Albert, T. G. Arceneaux, L. F. Bullard, B. A. Stafford, and J. M. Boswell, *Am. J. Industrial Med.*, (in press, 1993).

CHAPTER EIGHT

Ethers

Carroll J. Kirwin, Jr., D.A.B.T., and Jennifer B. Galvin, Ph.D., D.A.B.T.

1 GENERAL

1.1 Sources

Naturally occurring ethers may be constituents of essential oils and may be extracted from these sources. Although some ethers may appear naturally, they may be prepared synthetically from other chemicals or other ethers (1).

Symmetrical ethers are produced by the catalytic dehydration of their corresponding alcohols, for example, diethyl ether from ethanol (2). They are also obtained as by-products from the formation of their corresponding esters or alcohols. Ethers may also be made by special synthesis procedures (3, 4). Some ethers are obtained through the destructive distillation of selected hardwoods.

1.2 Uses

Ethers have a wide variety of industrial uses. Their commercial value is recognized in the following industries: rubber, plastics, paints and coatings, refrigeration, medicine, dentistry, petroleum, chemical, perfume, cosmetics, toiletries, and food.

The more volatile ethers have been used as liquid refrigerants, general anesthetics, commercial solvents, primers for gasoline engines, fuel additives (5), and rocket propellants. Other ethers have been used as alkylating agents in chemical

Patty's Industrial Hygiene and Toxicology, Fourth Edition, Volume 2, Part A, Edited by George D. Clayton and Florence E. Clayton.
ISBN 0-471-54724-7   © 1993 John Wiley & Sons, Inc.

syntheses of organic chemicals and in the manufacture of polymers. They are used to denature alcohol (6). Halogenated ethers are used in the preparation of ion-exchange resin (7), which is a modified polystyrene resin that is chloromethylated and then treated with a tertiary amine or with a polyamine. Ethers have wide use as commercial solvents and extractants for esters, gums, hydrocarbons, alkaloids, oils, resins, dyes, plastics, lacquers, and paints. They are used as dewaxing extractants for lubricating oils. Ethers have had limited use as cleaning and spotting agents. They are used as chemical intermediates in the manufacture of textile aids, such as dyes and resins. In the pharmaceutical industry ethers are used as solvents, suspending agents, flavorings for oral drugs, and dental products. They are used to increase viscosity, as penetrants and wetting agents, and as antioxidants and stabilizers. Ethers are used in foods as flavorings and in perfumes as fragrances. They are used as solvents for elastomers and for regenerating rubber. They have use as anti-skinning agents in surface coatings and as weathering agents for paints and plastics. Ethers are also used in soaps. Ethers appear in heat transfer agents. Several industries use specific ethers for thickening, dispersing, suspending, binding, and film forming.

1.3 Physical and Chemical Properties

A summary of physical and chemical properties of ethers is presented in Table 8.1 (8–15).

1.4 Hygienic Standards of Permissible Exposure

Very few ethers have mandatory human exposure controls established by the Occupational Safety and Health Administration (OSHA). The atmospheric concentrations that are permissible for a normal 8-hr workday and a 40-hr work week are listed in Table 8.2 (16, 17). Those limits recommended by the American Conference of Governmental Industrial Hygienists (ACGIH) (17) are listed for comparative uses. Two carcinogenic materials are listed, bis(chloromethyl) ether and chloromethyl methyl ether, which OSHA presumes will be controlled with no permissible human exposure. Solids or liquids that contain less than 0.1 percent of these two chemicals by weight or volume are not included in this exposure restriction (16).

Air sampling techniques and chemical analytic methods that are recommended for use in industrial hygienic control of exposure of ethers are presented in Table 8.3 (18). These methods are provided by the National Institute for Occupational Safety and Health (NIOSH) in an effort to standardize air monitoring and chemical assay procedures.

Odor detection and odor recognition levels that are available for ethers are presented in Table 8.4 (19). As may be seen from the values in this table, odor detection for any individual ether may be different by orders of magnitude because methods for making these determinations are totally subjective and scales used in these evaluations are not standardized. Additionally, the olfactory receptors may accommodate very rapidly to odors, resulting in inconsistent human response. This latter situation is the major reason why ether odors in an industrial atmosphere cannot be used as a reliable monitor for estimating either the presence or absence of ether concentration in the air.

Fritz (22) has discussed general isolation and separation techniques for ethers. Detection and estimation assays for ethers are also presented.

1.5 Hazards and Photochemical Degradation

For a number of years it was believed that the OSHA-controlled carcinogen bis(chloromethyl) ether could form spontaneously whenever formaldehyde and hydrogen chloride coexist in ordinary humid air (23). When these two latter chemicals are combined in humid air at 100 ppm each, bis(chloromethyl) ether is not observed at a detection limit of 0.1 ppb (24). This information has confused the concern over human exposure by this mechanism since Frankel et al. found 300 ppb under similar conditions (25). Sellakumar et al. premixed up to 1500 ppm of formaldehyde with up to 1000 ppm HCl and detected less than 1 ppb of bis(chloromethyl) ether (26).

Ethers exposed to oxygen in air may form peroxides by the following general equation (4):

$$R_2CHOR' + O_2 \rightarrow R_2C - OR'$$
$$|$$
$$O - OH$$

Peroxides formed in solvent ethers may oxidize the product being solubilized or extracted. Some of the peroxides formed may present an explosion hazard (27). Shaking or abrupt disturbances may trigger an explosion. Peroxides can be detected in ethers by mixing in a colorless, glass-stoppered cylinder 10 ml ether with 1 ml freshly prepared 10 percent potassium iodide solution. After the solution is shaken occasionally over a 1-hr period, while being protected from light, color will be seen against a white background if peroxides are present (28). Peroxides formed in ethers may be destroyed by the addition of sodium sulfite or 5 percent aqueous ferrous solution. Keeping ethers from becoming anhydrous plus the addition of antioxidants will help reduce this explosion hazard (28a).

Photochemical rearrangement of aromatic ethers was evaluated by irradiation under nitrogen in 95 percent ethanol using a 125-W medium-pressure mercury arc. It was found that aromatic ethers undergo consistent rearrangement reactions to form varying phenol derivatives (29). Another report by Suzuki and Ono (30) states that isopropyl ether is photoxidized at wavelengths of 240 to 310 nm and higher.

1.6 General Toxicity

The data presented in Table 8.5 (31–39) are arranged according to the chemical structure of the compounds. An effort has been made to place the chemicals within each group in an order that represents an increase in chain length. Even though

Table 8.1. Physical and Chemical Properties Data for Ethers

Material	Molecular Formula	Mol. Wt.	B.P. (°C, 760 mm Hg)	M.P. (°C)	Specific Gravity (20/20°C)
Dimethyl ether	$(CH_3)_2O$	46.07	-23.65	-138.5	—
Diethyl ether	$(CH_3CH_2)_2O$	74.12	34.6	-116.3 (stable crystals)	0.7146
Methyl propyl ether	$CH_3O(CH_2)_2CH_3$	74.1	38.8	—	0.738 (20/4°C)
Dipropyl ether	$(CH_3CH_2CH_2)_2O$	102.18	88	-123	0.736
Diisopropyl ether	$[(CH_3)_2CH]_2O$	102.17	68.3	-60	0.7258 (20/4°C)
Ethyl butyl ether	$C_2H_5OC_4H_9$	102.2	91.5	—	0.749 (20/4°C)
Dibutyl ether	$(C_4H_9)_2O$	130.22	142.4	-95.2	0.7694
Di-(2-ethylhexyl) ether	$[CH_3(CH_2)_3CH(C_2H_5)CH_2]_2O$	242.5	—	—	—
Dihexyl ether	$[CH_3(CH_2)_5]_2O$	186.3	226	-43	0.7942
Methyl t-butyl ether	$CH_3OC(CH_3)_3$	88.15	55.2	108.9	0.7404
T-Amyl methyl ether	$CH_3OC(CH_3)_2CH_2CH_3$	102.18	85	—	0.770
Divinyl ether	$(CH_2=CH)_2O$	70.09	28.3	—	0.774
Ethyl vinyl ether	$C_2H_5OCH=CH_2$	72.1	35.5	-115	0.755
Butyl vinyl ether	$C_4H_9OCH=CH_2$	100.2	93.8	-113	0.780
2-Methoxyethyl vinyl ether	$CH_3OC_2H_4OCH=CH_2$	102.1	108.8	-83	0.897
2-Butoxyethyl vinyl ether	$C_4H_9OC_2H_4OCH=CH_2$	125.2	—	—	—
2-Ethylhexyl vinyl ether	$C_8H_{17}OCH=CH_2$	156.3	177.5	-100	0.810
2,6,8-Trimethylnonyl vinyl ether	$C_{12}H_{25}OCH=CH_2$	212.4	—	—	—
Methyl isopropenyl ether	$CH_3OC(CH_3)=CH_2$	88.0	35.8	—	—
Diallyl ether	$(CH_2=CHCH_2)_2O$	98.1	94.3	—	0.808 (20/4°C)
Chloromethyl methyl ether	$ClCH_2OCH_3$	80.5	59.5 (759 mm Hg)	-103.5	1.074 (20/4°C)
Bis(chloromethyl) ether	$(ClCH_2)_2O$	115.0	104	—	—
Dichloroethyl ether	$ClCH_2CH_2OCH_2CH_2Cl$	143.02	178.5	-51.7	1.222 (20/4°C)
Dichloroisopropyl ether	$[CH_3(CH_2Cl)CH]_2O$	171.07	187.3	$-96.8 - -101.8$	1.1122
Anisole	$C_6H_5OCH_3$	108.1	154	-37.4	0.989 (25/4°C)
Phenetole	$C_6H_5OC_2H_5$	122.1	170	-29.5	0.960 (25/4°C)
Guaiacol	$CH_3OC_6H_4OH$	124.1	205	32	1.097 (25/25°C)
Hydroquinone monomethyl ether	$CH_3OC_6H_4OH$	124.16	243	53	—
Hydroquinone dimethyl ether	$CH_3OC_6H_4OCH_3$	138.16	210–212	55–57	1.053 (55/55°C)
Hydroquinone monobenzyl ether	$C_6H_5CH_2OC_6H_4OH$	200.23	—	122.5	—
Eugenol	OH	164.15	253.2	-9.2	1.064 (25°C)

$CH_2CH=CH_2$

Vapor Density (Air = 1)	Vapor Pressure (mm Hg) (20°C)	% in "Saturated" Air (25°C, 760 mm Hg)	Density of "Saturated" Air (25°C, 760 mm Hg)	Refractive Index, n_D^{20}	Solubility	Flash Point (°F)	Explosive Limits (%) (v/v)
1.617	3982	—	—	1.3441 ($n_D^{42.5}$)	Water, alcohol	−41	3.4–26.7
2.55	438.9	68	2.1	1.3526	Water, miscible in alcohols	−40 (O.C.)	1.9–36.5
—	520 (28°)	—	—	1.3602	Water, alcohol	—	Flammable liquid
—	—	—	—	1.3800	Water, alcohol	40	—
3.5	119.4	21	1.5 (air = 1)	1.3682	Water, alcohol	15 (T.O.C.) −18 (C.C.)	1.4–7.9
—	44.4	—	—	1.3875	—	—	Flammable liquid
4.48	4.8	0.9	1.1 (air = 1)	1.4010	Water, miscible in alcohols	100 (C.C.)	—
—	—	—	—	—	—	—	—
—	0.07	—	—	1.4204	Water	170 (O.C.)	—
3.0	245 (25°)	—	—	1.3694	Water, organic solvents	82	1.65–8.4
—	—	—	—	1.3896	Organic solvents	−11	—
2.4	430	56	1.8 (air = 1)	1.3989	Water, miscible in alcohols	<−22 (C.C.)	1.7–27.0
—	428	—	—	1.3763	Water	—	—
—	42	—	—	1.3997	Water	30 (O.C.)	—
—	18	—	—	—	Water	—	—
—	—	—	—	—	—	—	—
—	0.90	—	—	1.4232	Water	—	—
—	—	—	—	—	—	—	—
—	—	—	—	—	—	—	—
—	—	—	—	—	Water	20 (O.C.)	—
—	—	—	—	—	Decomposes in water and hot alcohol	—	—
—	—	—	—	1.435	Miscible in ethanol, ether	—	—
4.93	0.73	0.18	1.007 (air = 1)	1.457	Water, miscible in alcohol	185 (O.C.) 131 (C.C.)	—
5.9	0.71–0.85	0.12	1.05 (air = 1)	1.4451 (n_D^{25})	Water miscible in organic solvents	185 (O.C.)	—
3.7	3.1 (25°C)	0.41	1.1 (air = 1)	1.514 (25°C)	Alcohol, ether	—	—
4.2	1.7 (25°C)	0.224	1.01 (air = 1)	1.505 (25°C)	Alcohol, ether	160 (O.C.)	—
4.27	5 (79°C) 1 (52.4°C)	0.0135	1.005 (air = 1)	—	Miscible alcohol, ether	180 (approx)	—
—	—	—	—	—	Water	—	—
—	—	—	—	—	Water, acetone	—	—
—	—	—	—	—	Water	—	—
—	1 (78.4°C) 0.03 (25°C)	0.004	—	1.541	Water, miscible alcohol	—	—

Table 8.1. (Continued)

Material	Molecular Formula	Mol. Wt.	B.P. (°C, 760 mm Hg)	M.P. (°C)	Specific Gravity (20/20°C)
Isoeugenol	OH / OCH$_3$ / CH=CHCH$_3$	164.15	266	—	1.091 (15°C)
Butylated hydroxyanisole	CH$_3$OC$_6$H$_3$(OH)C(CH$_3$)$_3$	180	264–270 (733 mm Hg)	48–55	—
Vanillin	(CH$_3$O)C$_6$H$_3$(OH)CHO	152.1	285	82–83.5	1.056
Ethyl vanillin	(C$_2$H$_5$O)C$_6$H$_3$(OH)CHO	166.2	—	77–78	—
Phenyl ether	(C$_6$H$_5$)$_2$O	170.2	257	27	1.070 (20/4°C)
Dowtherm A		166 (av.)	257.4	12	1.06 (25/25°C)
Bis(phenoxyphenyl) ether Mixed isomers	C$_{24}$H$_{15}$O$_3$	354.4	443	none	—
p-Isomer	C$_{24}$H$_{15}$O$_3$	354.4	444	110	—
Chlorinated phenyl ethers Monochloro-	C$_{12}$H$_9$ClO	204.5	153 (8 mm Hg)	—	1.19 (25/25°C)
Dichloro-	C$_{12}$H$_8$Cl$_2$O	238.9	168.2 (8 mm Hg)	—	1.21 (25/25°C)
Hexachloro-	C$_{12}$H$_4$Cl$_6$O	376.9	230–260 (8 mm Hg)	—	1.60 (20/60°C)
Decabromodiphenyl ether	C$_{12}$Br$_{10}$O	960	—	290	3.0

the number of chemicals in any one group is limited, it is possible to make general, comparative statements using Table 8.5. This corresponds with the acute toxicity data available in the *NIOSH Registry of Toxic Effects of Chemical Substances* (40).

The oral toxicity and concentrated vapor data indicate that as the chain length increases in the symmetrical ethers the toxicity is reduced. The inverse is true for skin penetration toxicity. Skin irritation rises with extending chain length to a maximum of moderate irritation potential for the dihexyl derivative and would probably exhibit a slow drop in irritancy with continued increase in chain length, if the data were available. The eye irritancy potential does not appear to change with variable compound structure.

Toxicity of aliphatic ethers has been shown elsewhere (41) to have a mathematical (parabolic) relationship to molecular weight and lipophilicity. Additionally (42), structure–activity relationships for aliphatic ethers were found to be parabolic functions of the octanol/water partition coefficients.

The data presented in Table 8.5 on the unsaturated group show that increasing

Vapor Density (Air = 1)	Vapor Pressure (mm Hg) (20°C)	% in "Saturated" Air (25°C, 760 mm Hg)	Density of "Saturated" Air (25°C, 760 mm Hg)	Refractive Index, n_D^{20}	Solubility	Flash Point (°F)	Explosive Limits (%) (v/v)
—	16 (142°C) 0.02 (25°C)	0.0026	—	1.573	Water, miscible alcohol	—	—
—	—	—	—	—	Fats, acetone	—	—
—	1 (10°C) 0.0022 (25°C)	0.00029	—	—	Water, alcohol, ether	—	—
—	—	—	—	—	Water, alcohol, ether	—	—
5.86	0.0213 (25°C, 760 mm Hg)	0.0028	1.0014 (air = 1)	1.579 (25°C)	Miscible in alcohol, ether	205 (O.C.)	—
—	144 psig (750°F) 0.08 mm Hg (25°C)	—	—	—	Alcohol, ether	255 (O.C.)	—
—	—	—	—	—	Alcohol, miscible benzene	505	—
—	—	—	—	—	Miscible benzene	—	—
—	0.007 (25°C, extrapolated	0.0009	—	1.5868 (n_D^{25})	Methanol miscible ether	258.8 (O.C.)	—
—	0.0006 (25°C, extrapolated	0.00008	—	1.5980 (n_D^{25})	Methanol, miscible ether	298.4 (O.C.)	—
—	Very low	Low	—	1.621 (n_D^{25})	Water, methanol	—	—
—	<1 (250°C)	—	—	—	Organic solvents	—	—

the chain length decreases oral toxicity, but the addition of a second double bond appears to render a compound more toxic. Skin irritation potential is increased to a moderate level with di- or polyunsaturation.

With the vinyl ethers, the oral toxicity decreases through C_4, then increases with chain length.

Not enough information is available to be definitive, but evaluation of the data of Table 8.5 suggests that it is possible that skin penetration toxicity for the vinyl ethers would increase with chain length through C_4. Branching of the chain may tend to decrease toxicity from percutaneous absorption. Skin irritation potential appears to increase with chain length. Eye changes, exhibited as corneal damage, rise to a maximum level of moderate at a carbon chain length of eight. Beyond this length there is a reduction in this type of irritancy. The allyl group appears to be the most toxic of the alkene ethers.

Table 8.2 Threshold Limit Values of Ethers

| Material | OSHA (16) | | ACGIH (17) | | | |
| | | | TWA | | STEL[a] | |
	ppm	mg/m³	ppm	mg/m³	ppm	mg/m³
Diethyl ether	400	1200	400	1210	500	1520
Diisopropyl ether	500	2100	250	1040	310	1300
Phenyl ether	1	7	1	7	2	14
Diphenyl ether–diphenyl mixture	1	7	1	7	2	14
Chloromethyl methyl ether	Controlled carcinogen no permissible limit[c]		A2[b]	A2[b]	—	—
Bis(chloromethyl) ether	Controlled carcinogen no permissible limit[c]		0.001(A1)[b]	0.0047(A1)[b]	—	—
Dichloroethyl ether	15[d,e]	90[d,e]	5[e]	29[e]	10[e]	58[e]
Chlorinated diphenyl oxides	—	0.5	—	0.5	—	—

[a]STEL = short-term exposure limit.
[b]A1 = Confirmed human carcinogen; A2 = suspected human carcinogen.
[c] = Does not apply to solid or liquid mixtures containing less than 0.1% by weight or volume.
[d] = Ceiling value—at no time exceed value given.
[e] = Skin—refers to potential overall exposure by the cutaneous route by airborne or direct contact with the substance.

Table 8.3. Air Sampling and Analytical Procedures for Ethers (18)

Material	NIOSH Method No.	Air Sampling Technique	Flow Rate of Air Sample (l/min)	Desorbant	Analytic Technique
Diethyl ether	1610	Charcoal tube	0.2 or less	Ethyl acetate	Gas chromatography with flame ionization detector
Diisopropyl ether	S368	Charcoal tube	0.2 or less	Carbon disulfide	Gas chromatography with flame ionization detector
Methyl t-butyl ether	1615	Charcoal tube	0.2 or less	Carbon disulfide	Gas chromatography with flame ionization detector
bis(Chloromethyl) ether	213	Chromosorb 101	0.02 or less	Helium	Gas chromatography/Mass spectroscopy
Chloromethyl methyl ether	220	Impinger	0.5	—	Gas chromatography with electron capture detector
Dichloroethyl ether	1004	Charcoal tube	1 or less	Carbon disulfide	Gas chromatography with flame ionization detector
Phenyl ether	S72	Charcoal tube	0.2 or less	Carbon disulfide	Gas chromatography with flame ionization detector
Phenyl ether–biphenyl mixture	S73	Silica gel	0.2 or less	Benzene	Gas chromatography with flame ionization detector
Chlorinated diphenyl oxide	5025	Filter	0.5	Isooctane	Gas chromatography with electrolytic conductivity detection

453

Table 8.4. Odor Detection/Recognition of Ethers

Material	Type of Threshold	Media	Threshold Units (ppm)
Diethyl ether (19)	Detection	Water	5.83
	Detection	Water	0.001
	Detection	Air	0.0019
	Detection	Air	0.700
Diisopropyl ether (20)	Detection	Air	0.170
Dibutyl ether (20)	Detection	Air	0.070
Methyl *t*-butyl ether (21)	Recognition	Air	7.000
	Detection	Air	0.700
Dichloroethyl ether (20)	Detection	Air	15.00
Dichloroisopropyl ether (19)	Detection	Water	0.020
Diphenyl ether (19)	Recognition	Air	0.100
	Detection	Air	0.001
	Detection	Air	0.0099
	Detection	Air	0.0008
Vanillin (19)	Recognition	Water	4.000
	Detection	Water	0.200
	Detection	Air	1.10×10^{-8} ppb
	Detection	Air	2.00×20^{-4} ppb
Ethyl vanillin (19)	Recognition	Water	2.00
	Detection	Water	0.100

With the polyoxy ether group it may be seen that as the chain length between the oxygens increases or as the overall chain length increases the oral toxicity, percutaneous toxicity and inhalation toxicity are reduced.

As would be expected, the more volatile ethers have the highest inhalation toxicity potential, as is exhibited by the concentrated vapor data. If the Ct concept (concentration times time) is used to evaluate inhalation toxicity from the metered vapor data, the haloethers and cyclic ethers appear to be the most toxic.

Bicyclic ethers appear to be more toxic than those having a single ring structure. Fishbein (7) has reviewed the carcinogenic chloroethers and has noted in comparing carcinogenicity that bifunctional alpha chloroethers are more reactive than their monofunctional analogues. As the chain length increases, activity decreases, and as chlorine occurs more distant from the ether oxygen, carcinogenic activity also decreases. It was also noted that in a general way, the more carcinogenically active compounds are the most labile; as stability increases, carcinogenicity decreases. Tabulations of carcinogenic ethers are found in Tables 8.6 and 8.7.

Relative comparisons, such as those given above, will become more reliable for describing acute industrial health hazards when more data on ethers appear in the published literature.

Table 8.5. Summary of Acute Toxicity Testing of Ethers (31, 32)

Material and Structural	Oral LD$_{50}$ for Rats (ml/kg)	Skin Penetration LD$_{50}$ for Rabbits (ml/kg)[a]	Concentrated Vapor Inhalation by Rats, Maximum for No death[b]	Inhalation of Metered Vapor Concentration by Rats — Concentration (ppm)	Time (hr)	Mortality	Irritation to Uncovered Rabbit Skin[c]	Corneal Injury in Rabbits[c]
Saturated symmetrical								
Diethyl ether (36) CH$_3$CH$_2$OCH$_2$CH$_3$	3.56 (3.23–3.92)[d]	>20	5 min	32,000[e]	4	3/6	1	2
Dibutyl ether (35) CH$_3$(CH$_2$)$_3$O(CH$_2$)$_3$CH$_3$	7.40 (6.41–8.53)[d]	10.08 (4.41–23.04)	30 min	4,000	4	2/6	4	1
Dihexyl ether (35) CH$_3$(CH$_2$)$_5$O(CH$_2$)$_5$CH$_3$	30.9 (27.8–34.4)[d]	6.9 (3.6–10.0)	8 hr	—	—	—	5	1
Di(2-ethylhexyl) ether (35) [CH$_3$(CH$_2$)$_3$CH(C$_2$H$_5$)CH$_2$]$_2$O	33.9 (22.0–52.0)[d]	—	4 hr	—	—	—	3	1
Saturated asymmetrical								
Methyl t-butyl ether (36a) CH$_3$OC(CH$_3$)$_3$	4.0	>10		85[g]	4		2	Mild eye irritancy
Ethyl butyl ether (34) CH$_3$CH$_2$O(CH$_2$)$_3$CH$_3$	1.87 (1.34–2.60)[d]	—	5 min	1,000	4	0/6	2	2
Vinyl								
Ethyl vinyl ether (37) CH$_3$CH$_2$OCH=CH$_2$	8.16 (7.03–9.48)	>20	—	64,000	4	0/6	1	1
Butyl vinyl ether (35) (CH$_3$CH$_2$)$_3$OCH=CH$_2$	10.30 (8.40–12.63)[d]	4.24 (3.02-5.95)	5 min	8,000[e] / 16,000	4 / 4	0/6 / 6/6	3	2
Isobutyl vinyl ether (36) CH$_3$CH(CH$_3$)CH$_2$OCH=CH$_2$	17.0 (12.2–23.8)	20	10 min[f]	16,000	4	3/6	1	2
2-Ethylhexyl vinyl ether (35) CH$_3$(CH$_2$)$_3$CH(CH$_2$CH$_3$)CH$_2$OCH=CH$_2$	1.35 (1.17–1.56)[d]	3.56 (2.20–5.76)	4 hr	—	—	—	3	1
2,6,8-Trimethylnonyl vinyl ether (35) CH$_3$CH(CH$_3$)CH$_2$CH(CH$_3$)-(CH$_2$)$_3$CH(CH$_3$)CH$_2$OCH=CH$_2$	1.22 (0.98–1.52)[d]	5.0	4 hr	—	—	—	5	1
Unsaturated								
Methyl 2-propenyl ether (37) CH$_3$OCH$_2$CH=CH$_2$	1.87 (1.26–2.76)	>20	—	64,000	4	4/6	1	1
Methyl 1,3-butadienyl ether CH$_3$OCH=CHCH=CH$_2$	2.14 (1.54–2.99)	—	15 min[f]	—	—	—	4	2
Ethyl 1-propenyl ether (36, 37, 39) CH$_3$CH$_2$OCH=CHCH$_3$	19.0 (13.4–27.0)	>10	5 min	8,400	4	0/6	2	2
	4.66 (2.89–7.52)	3.97 (2.43–6.48)	9 min	4,000	4	0/6	1	2
1-Propenyl 2-buten-1-yl ether (36) CH$_3$CH=CHOCH$_2$CH=CHCH$_3$	8.0 (5.10–12.5)	>10	1 hr	5,000[e]	4	2/6	4	2
Allyl								
Allyl vinyl ether (36) CH$_2$=CHCH$_2$OCH=CH$_2$	0.55 (0.30–1.02)[d]	—	5 min[f]	8,000	4	2/6	1	2
Diallyl ether (33) CH$_2$=CHCH$_2$OCH$_2$CH=CH$_2$	0.32[d]	0.6	—	—	—	—	2	4

455

Table 8.5. (Continued)

Material and Structural	Oral LD$_{50}$ for Rats (ml/kg)	Skin Penetration LD$_{50}$ for Rabbits (ml/kg)[a]	Concentrated Vapor Inhalation by Rats, Maximum for No death[b]	Inhalation of Metered Vapor Concentration by Rats			Irritation to Uncovered Rabbit Skin[c]	Corneal Injury in Rabbits[c]
				Concentration (ppm)	Time (hr)	Mortality		
2-Methoxyethyl vinyl ether (35) CH$_3$O(CH$_2$)$_2$OCH=CH$_2$	3.90 (3.49–4.36)[d]	7.13 (4.41–11.52)	1 hr	8,000	4	4/6	1	2
Polyoxy								
1-Butoxy-2-ethoxyethane (35) CH$_3$(CH$_2$)$_3$O(CH$_2$)$_2$OCH$_2$CH$_3$	2.83 (1.93–4.15)[d]	2.12 (1.26–3.56)	2 hr	—	—	—	2	4
2-Butoxyethyl vinyl ether (35) CH$_3$(CH$_2$)$_3$O(CH$_2$)$_2$OCH=CH$_2$	3.10 (2.77–3.46)[d]	3.00 (2.13–4.21)	4 hr	2,000	8	2/6	2	5
1,2-Dibutoxyethane (35) CH$_3$(CH$_2$)$_3$O(CH$_2$)$_2$O(CH$_2$)$_3$CH$_3$	3.25 (2.82–3.74)[d]	3.56 (2.20–5.76)	8 hr	—	—	—	4	2
2,2'-Dibutoxyethyl ether (35) CH$_3$(CH$_2$)$_3$O(CH$_2$)$_2$O-(CH$_2$)$_2$O(CH$_2$)$_3$CH$_3$	3.90 (3.35–4.50)[d]	4.04 (3.41–7.46)	8 hr	—	—	—	3	2
1,3-Dimethoxybutane (36) CH$_3$O(CH$_2$)$_2$CH(OCH$_3$)CH$_3$	3.73 (2.68–5.21)	10.0 (4.6–21.9)	1 hr	8,000	4	4/6	2	2
Halo								
2-Chloro-1,1,2-trifluoroethyl methyl ether (34) CH$_3$OCF$_2$CH(Cl)F	5.13 (3.91–6.74)[d]	0.2	5 min	—	—	—	1	7
2-Chloroethyl vinyl ether (33) (Cl)CH$_2$CH$_2$OCH=CH$_2$	0.25[d]	3.2	—	500	4	1/6	1	2
Dichloroethyl ether (32) (Cl)CH$_3$CH$_2$OCH$_2$CH$_2$(Cl)	0.075[d]	(0.3)[e]	—	1,000	3/4	3/6	—	4
Dichloroisopropyl ether (34) (Cl)CH$_2$CH(CH$_3$)OCH(CH$_3$)CH$_2$(Cl)	0.24 (0.22–0.27)[d]	3.00 (1.78–5.04)	—	1,000	4	1/6	1	2
Cyclic								
α-Methylbenzyl ether (37) C$_6$H$_5$CH$_2$OCH$_3$	9.80 (6.38–15.06)[d]	—	8 hr	—	—	—	4	1
Cyclopentyl ether (36) C$_5$H$_9$OC$_5$H$_9$	0.47 (0.34–0.66)	1.41	2 hr	250	4	5/6	4	2
Bicyclo[2.2.1]hept-2,5-ylene bis allyl ether (37)	3.73 (2.43–5.74)	5.95 (3.57–9.09)	8 hr	—	—	—	3	1

[a]Figure in parentheses was determined by poultices on guinea pigs.
[b]Maximum time for no death in rats when exposed to concentrated vapor obtained by passing air through compound at room temperature.
[c]Grade of primary irritation hazard to rabbit skin as explained by Smyth et al. (33); grade of hazard to rabbit eye as explained by Carpenter and Smyth (38).
[d]As g/kg in a suitable vehicle. [e]Twice concentration shown killed all rats. [f]Inhalation time shown killed all six rats. [g]LC$_{50}$, mg/l.

Table 8.6. Carcinogenicity of Halogenated Ethers (7)

Compound	Mouse Skin (Mice with Papillomas/Total Mice)[a]		Subcutaneous Injection in Mice (Sarcomas at Injection Site/Group Size)	Subcutaneous Injection in Rats (Sarcomas at Injection Site/Group Size)
	Carcinogen	Initiating Agent		
Chloromethyl methyl ether $ClCH_2OCH_3$	0	12/40 (5)	10/30	1/20
Bis(chloromethyl) ether $ClCH_2OCH_2Cl$	13/20 (12)	5/20 (2)	—	7/20
Bis(α-chloroethyl) ether $CH_3CHClOClCHCH_3$	—	7/20 (0)	4/30	—
Dichloromethyl methyl ether Cl_2CHOCH_3	0/20 (0)	3/20 (1)	—	—
Bis (β-chloroethyl) ether $ClCH_2CH_2OCH_2CH_2Cl$	—	3/20 (0)	2/30	—
Octachloro-di-n-propyl ether $CCl_3CHClCH_2OCH_2CHClCCl_3$	0/20 (0)	2/20 (1)	—	—

[a]Number of mice with carcinomas given in parentheses.

457

Table 8.7. Comparison of Carcinogenic and Mutagenic (in Microbial Systems) Activity (7)

Compound	Mutagenic Activity[a]	Carcinogenic Activity
Chloromethyl methyl ether	+	+
Bis(chloromethyl) ether	+	+
Bis(α-chloroethyl) ether	+	+
Dichloromethyl methyl ether	+	+
Bis(β-chloroethyl) ether	−	±
Octachloro-di-*n*-propyl ether	Not tested	+

[a]*E. coli* and *S. typhimurium.*

2 ALKANE ETHERS

2.1 Dimethyl Ether (Methyl Ether, Methoxymethane)

2.1.1 Determination in the Atmosphere

There is no validated NIOSH method for dimethyl ether; however, dimethyl ether can be absorbed from the atmosphere with activated charcoal, titania gel, or silica gel. Use of silica gel has the advantage of ease of elution with ethanol and subsequent analysis by gas chromatography. Air samples may also be analyzed directly by infrared analysis with a gas cell of long path length. The physical and chemical properties data are presented in Table 8.1.

2.1.2 Physiological Response

2.1.2.1 Acute. An extensive evaluation of dimethyl ether was conducted by Daly and Kennedy (47). The inhalation LC_{50} in mice was 490,000 ppm for a 15-min exposure and 380,000 ppm for a 30-min exposure. The rat 4-hr exposure LC_{50} was 164,000 ppm. The major toxic signs were sedation and narcosis. Dimethyl ether was found to be free of genetic interference in five separate genetic/mutagenic assays. At 200,000 ppm dimethyl ether causes weak cardiac sensitization in dogs. At 10,000 ppm rats show slight evidence of sedation and at 50,000 ppm rats are asleep most of the time for exposures longer than 30 min. Over a range of 750 to 2,000 ppm dimethyl ether exhibits a two-phase half-life cycle in the blood of rats of 10 and 90 min. No tissue storage was detected.

In 1924, Brown (43) studied the anesthetic effect of this gas on the cat. A mixture of 85 percent dimethyl ether–15 percent air caused profound anesthesia with gradual cessation of respiration. Approximately 20 min was necessary for complete recovery after 50 min of anesthesia. Brown also reported that dimethyl ether concentrations of 50 percent were inhaled by humans in the laboratory, with the observation that the gas is most unpleasant to inhale, being distinctly suffocating, even when taken with a high percentage of oxygen. From this report it appears that the acute toxicity of dimethyl ether is very low.

As with other alkyl ethers, its principal physiological effect is that of anesthesia.

Dimethyl ether has been proposed by Brody et al. (44) for biologic and medical use as an indicator for cardiac output and lung tissue volume studies. It has a lower fat solubility and lower anesthetic potency than other ethers used for biologic purposes. Its higher water solubility plus the fact that it is a gas at room temperature prevent it from developing a liquid ether phase in blood. It remains relatively nonpolar over a wide pH range (pH 4.0 to 10.0).

2.1.2.2 Subchronic. Rats were exposed by inhalation to dimethyl ether up to 20,000 ppm for 6 hr/day, 5 days/week for 13 weeks (46). A slight decrease in lymphocyte and increase in neutrophils was observed at 20,000 ppm. An increase in serum glutamic–pyruvic transaminase (SGPT) and a reduction in total serum protein were also observed at 20,000 ppm. These changes were not considered biologically significant. This conclusion was supported by the absence of chemically related significant gross toxic signs or abnormal histopathology.

A series of subchronic inhalation studies in rats and hamsters lasting from 2 to 4 weeks from 100 to 50,000 ppm, 6 hr/day, 5 days/week, found only mild, transitory changes, which consisted of increased leukocytes, a decrease in red blood cells, and increased liver and kidney weights. All changes were completely recoverable (47). More extensive exposures lasting up to 30 weeks resulted in increased neutrophils and decreased lymphocytes plus increased SGPT levels at 20,000 ppm. These changes were not supported by histopathology changes.

Collins et al. (45) exposed rats to dimethyl ether 6 hr/day, 5 days/week, for 30 weeks. Atmospheric concentrations of 0.02, 0.2, and 2 percent dimethyl ether were evaluated. No adverse observations were noted for body weight, food consumption, hematology, urinalysis, or histopathology. Elevated SGPT and serum glutamic–oxaloacetic transaminase (SGOT) levels were elevated, suggesting the possible onset of hepatotoxicity. Additionally, the liver weight to body weight ratio of the high dose group was reduced by a statistically significant level. However, these observations were not supported by adverse histopathological alterations.

These functional changes may be similar to those described for diethyl ether, which are transitory, recoverable shifts having questionable biologic significance.

2.1.2.3 Developmental Toxicity. Dimethyl ether was evaluated up to 40,000 ppm in rats for developmental toxicity. At 20,000 ppm and higher, reduced fetal weights and increased skeletal variations were noted. However, neither terata nor increased resorptions were observed (47).

2.1.2.4 Chronic. A lifetime study in rats did not produce cancer or clear, statistically significant evidence of chronic toxicity at 25,000 ppm of dimethyl ether (47).

2.2 Diethyl Ether (Ethoxyethane, Ether, Ethyl Ether, Diethyl Oxide, Ethyl Oxide, Sulfuric Ether)

2.2.1 Determination in the Atmosphere

NIOSH has proposed an air sampling technique for diethyl ether. The airflow rate, trapping medium, elution medium, and analytic procedure for this chemical are all presented in Table 8.3 (18).

Table 8.8. Atmospheric Concentration of Diethyl Ether and Reported Effects

Concentration (ppm)	Effect
0.7–1920	Reported range of human odor detection
200	Nasal irritation
19,000 (1.9%)	Lower flammability limit, lowest anesthesia limit
42,000 (4.2%)	LC_{50} in mice
50,000 (5.0%)	Maintenance of surgical anesthesia
64,000 (6.4%)	Lethal concentration for rats
100,000–150,000 (10–15%)	Induction of human anesthesia

Additionally, it can be detected at fairly low concentrations in the atmosphere by infrared analysis with a gas cell of long path length. A sensitive explosimeter would also be of value for concentrations known to be below explosive limits. Physical and chemical data are presented in Table 8.1.

2.2.2 Physiological Response

Without question, the primary physiological response to diethyl ether is that of general anesthesia. Absorption through the lungs occurs quickly at the onset of exposure, and elimination by this route, post-exposure, is also rapid. Absorption through the skin is of no consequence in humans. It may cause a slight irritating effect on the skin, or cause it to become dry and cracked after repeated applications. Exposure of the eyes or mucous membranes to the liquid is irritating and should be avoided. Odor detection/recognition is presented in Table 8.4.

2.2.2.1 Human Experience

Diethyl ether has been used extensively as a general anesthetic with medical safety. Few reports of death appear in the literature and Table 8.8 displays the wide range of concentrations between nasal irritation and more serious health hazards.

Concentrations of diethyl ether anesthetic ranging from 10 to 15 percent (v/v) in air are needed for human anesthesia induction (48). Maintenance of anesthesia is achieved at approximately 5 percent (v/v). The lowest anesthetic limit is 1.9 percent (v/v) (48). Above 10 percent Flury and Klimmer caution that fatalities may result (49).

Cases of human death in industry due to acute inhalation of diethyl ether are rare. Browning (50) cites one such case where diethyl ether was used in perfumery manufacture. The subject developed acute mania and died in uremic convulsions. Doubtless there are a few other such cases. Cases of narcosis are more frequent, largely owing to the rarity of permanent aftereffects. Early symptoms of acute overexposure range from excitement to drowsiness, vomiting, paleness of the face, lowering of the pulse and body temperature, irregular respiration, muscular relaxations, and excessive salivation. Temporary aftereffects of acute exposure are vomiting, salivation, irritation of the respiratory passages, headaches, and depression or excitation (49). Kidney injury has been listed as a result of acute intoxication but this has been open to question. Clinically, albumin may appear in the urine, and polycythemia in the blood (50). Nephritis may develop in rare cases. Diethyl ether used for general anesthesia has produced abnormal liver functions in patients (48). This has been characterized by elevated SGOT and alkaline phosphatase plus a depression of serum albumin (51). These changes are of short duration and return to normal.

Because of its irritating effect on mucous membranes, humans generally refrain from the oral consumption of diethyl ether. However, repeated drinking of ether has taken place (50), resulting in the development of the "ether habit" and general debility. It has been used as an expectorant and in other medicinal preparations. It has been employed for intravenous anesthesia as a 5 percent solution, and has been introduced into the common bile duct in concentrated form to dissolve biliary calculi. Sewall (52) reports a fatality from the intravenous injection of diethyl ether in the treatment of vascular diseases. Nelson et al. (53) reported that human subjects found diethyl ether irritating to the nose, but not to the eyes or throat, at a vapor concentration of 100 ppm. This may be in part due to the peroxide content. Odor detection is reported in Table 8.4.

Standard textbooks in pharmacology may be consulted for additional information on blood levels and anesthetic effects of diethyl ether in humans.

2.2.2.2 Animal Studies

Inhalation. Kärber (54) reported that the lethal concentration for mice was 133.4 mg/l, or 4.4 percent (v/v), during a continuous exposure for 97 min. Similar results for the mouse were obtained by Molitor (55), who calculated the LC_{50} for a continuous 3-hr exposure to be 127.4 mg/l or 4.2 percent (v/v). The lethal concentrations for the rat, dog, and monkey are reported to be 6.4, 10.6, and 7.16 to 19.25 percent (v/v) (56), respectively. Swann et al. (57) showed that 3.2 percent (v/v) of diethyl ether produced excitation and anesthesia in mice. At 6.4 percent (v/v) deep anesthesia was produced and respiratory arrest occurred at 12.8 percent (v/v). Upon their removal from the atmosphere spontaneous respiration occurred in the animals and normal breathing resumed. It would appear that the smaller animal species succumb to a lower concentration of diethyl ether in air than do the larger animal species. However, Robbins (58) found a lethal concentration for dogs of 6.7 to 8 percent (v/v). He states that differences in induction mixtures and in the duration of anesthesia lead to the differences in results for lethal concentrations. This is further illustrated by more recent work (59) in which the concentration necessary to produce respiratory arrest in mice was found to be 18 percent (v/v). Rapid induction and short duration of anesthesia were used in this experiment. Additional acute exposure data are presented in Table 8.5. The effect of learning performance after exposure to diethyl ether has not been defined satisfactorily (60).

Schwetz and Becker (61) evaluated the age-related susceptibility of diethyl ether to adult and neonatal rats. The median time to death for neonates is 5 to 6.5 times greater than for adult rats and the mean concentration of diethyl ether in blood is

2.5 to 3 times greater in neonatal rats than adult rats. Diethyl ether overdosage in animals may be successfully reversed by resuscitating with compressed air (62).

Oral. The oral LD_{50} (63) in 14-day-old, young adult, and adult rats, respectively, is 2.2, 2.4, and 1.7 ml/kg.

Effect on Skin and Eyes. Diethyl ether has no deleterious effect on skin if contact is of short duration. Repeated exposure of the skin causes cracking and drying due to the extraction of oils. Absorption through the skin is not great enough to cause a deleterious effect, as seen in Table 8.5. Fisher (64) had identified ether as a contact allergen, but the reliability of this claim is uncertain owing to the high concentration (60 percent) of ether recommended for patch-test challenge. Grant (65) has summarized the effects of diethyl ether to the eye and should be consulted for detailed information. Slight, reversible injury to the eye occurs with both liquid (Table 8.5) and vapor contact of diethyl ether.

Absorption, Excretion, and Metabolism. Diethyl ether is instantaneously absorbed from the inhaled air into the bloodstream, from which it passes rapidly into the brain. Dybing and Shovlund (66) showed that it is also taken up rapidly by fatty tissue in rats. The concentration in muscle tissue remains much lower than the concentration in the brain and fatty tissue. Haggard (67) showed in dogs and rabbits that 87 percent of an absorbed dose is expired in the breath and 1 to 2 percent is excreted in the urine. The concentration in the urine does not exceed that in the blood flowing through the kidneys. Upon discontinuance of exposure the diethyl ether deposited in the fatty tissues remains at a fairly high level until the concentration in the blood has become relatively low. In the rat (66), the concentration of diethyl ether in the fat was 0.12 mg/g, whereas in the blood it was 0.03 mg/g, 24 hr after an exposure to 300 mg/l (100,000 ppm) for 1 hr. Chenoweth et al. (68), using infrared and mass spectrometric methods of analysis, also found high levels of ether in the fatty depots of the dog after 2.5 hr of anesthesia. Diethyl ether was still detectable in the fat 24 hr after discontinuance of anesthesia, whereas the concentration in the blood had reached a similar low level approximately 17 hr earlier. Measurements of these concentrations in 11 organs and tissues and three body fluids revealed that the adrenals contained an unexplained high concentration of diethyl ether during anesthesia.

Diethyl ether is biodegraded to $^{14}CO_2$ in amounts of 1 to 5 percent in rats, as demonstrated in radio-labeling experiments (69).

Mechanism of Action. Krantz and Carr (70) have summarized the pertinent theories that pertain to the mechanism of action of general anesthetics. Enzymatic and other systemic changes have not yet been determined. Possible mechanisms of action on organ systems have been summarized by Booth and McDonald (71).

Table 8.9. Physiological Properties of Other Saturated Alkyl Ethers

| Material and Structural Formula | Anesthetic (ml/kg) | | Odor | Remarks |
	Surgical Anesthesia	Respiratory Arrest		
Methyl propyl ether (75, 76) CH₃OCH₂CH₂CH₃	Dog, 0.8	2.1	Ethereal	Caused no deleterious effect in dog, monkey, or rat
Methyl isopropyl ether (95) CH₃OCH(CH₃)₂	Dog, 1.12	2.62	Ethereal	Caused no deleterious effect in dog, monkey, or rat

2.2.3 Hygienic Standards of Permissible Exposure

Table 8.2 contains the permissible exposure limit of diethyl ether for industrial control, as set by OSHA (16). It also contains recommended limits suggested by ACGIH in 1992 (17) for normal workday limits.

2.3 Methyl Propyl Ether

Toxicity data on methyl propyl ether are presented in Table 8.9 (75, 76). Physical and chemical property data are presented in Table 8.1.

2.4 Propyl Ether (Dipropyl Ether, Dipropyl Oxide, 1,1′-Oxybispropane, Di-*n*-propyl ether)

Physical and chemical properties data of propyl ether are presented in Table 8.1.

The acute inhalation LC_{50} in mice for propyl ether (72) is reported to be 163 mg/m³ (39 ppm). The accuracy of this value is questionable based on the inhalation toxicity of similar ethers. Additionally, Nielsen et al. exposed mice to synthesis-grade propyl ether (90 percent) for 30 min up to 15,000 ppm with no deaths occurring (73). The respiratory rate was altered, indicating a sensory irritation response, which was completely recoverable. In a second report, mice were exposed up to 12,950 ppm for 30 min to pure (99.5 percent) propyl ether (74). The dose calculated to reduce respiratory rate by 50 percent was 11,200 ppm. The anesthetic dose, as defined by depression of the righting reflex, for a 30-min exposure was 11,600 ppm. All adverse responses were recoverable after cessation of exposure.

2.5 Diisopropyl Ether (Isopropyl Ether, 2-Isopropoxypropane, 2,2′-Oxybispropane)

2.5.1 Determination in the Atmosphere

NIOSH has proposed an air sampling technique for diisopropyl ether (18). The airflow rate, trapping medium, elution medium, and analytic procedure for this chemical are all presented in Table 8.3. These recommended techniques have been proposed by NIOSH in an attempt to standardize the analytical identification of diisopropyl ether in air. Physical and chemical properties data are presented in Table 8.1.

2.5.2 Physiological Response

The primary physiological response to diisopropyl ether is that of general anesthesia. Even though it is somewhat more toxic than diethyl ether, the vapor concentration necessary for this anesthetic effect is much higher than the concentration that causes irritation and unpleasant odors. This compound is only slightly irritating to the skin and is capable of producing only minor injury to the eyes. It is not absorbed through the skin in harmful amounts. The odor detection limit is reported to be 0.17 ppm (20) (see Table 8.4).

2.5.2.1 Human Experience

Silverman et al. (77) reported that 35 percent of humans exposed to a vapor concentration of 300 ppm objected to the unpleasant odor of this solvent. At 800 ppm for 5 min most subjects reported irritation of the eyes and nose, and the most sensitive reported respiratory discomfort (78).

The chief hazards from this chemical seem to be flammability and explosions due to peroxide formation rather than toxicity.

2.5.2.2 Animal Tests

Inhalation. Machle et al. (79) exposed animals (monkey, rabbit, and guinea pig) to a vapor concentration of 6.0 percent of isopropyl ether in air. All died owing to respiratory failure. At a vapor concentration of 3.0 percent all animals survived a 1-hr exposure but all showed signs of anesthesia, with the monkey being most susceptible. At a vapor concentration of 0.3 percent for 2 hr there was no visible indication of anesthetic action. Spector (56) reported the lethal vapor concentration for rats to be 1.6 percent for a 4-hr exposure.

Machle et al. (79) repeatedly exposed the animals referred to above for various periods of time. The animals exposed to a vapor concentration of 3.0 percent were given 10 exposures. Recoveries from the anesthesia were prompt and seemed not to be additive. However, weight loss and blood changes were apparent during and for several weeks after exposure. Animals exposed to a vapor concentration of 1.0 percent for 1 hr daily, although exhibiting signs of intoxication and depression, revealed no significant weight or blood changes during or after 20 exposures. At 0.3 percent for 2 hr and 0.1 percent for 3 hr daily, there was no deleterious effect noted during or after 20 exposures.

Oral. The minimum lethal dose for rabbits (79) was found to be between 7 and 9 ml/kg (5 to 6.5 g/kg). A rapid, intense intoxication was produced. Death was due to respiratory failure caused by depressant action. Kimura et al. (63) determined the acute oral LD_{50} for diisopropyl ether in 14-day-old, young adult, and adult rats, respectively, as 6.4, 16.5, and 16.0 ml/kg.

Effect on Skin, Eyes, and Mucous Membranes. Machle et al. (79) tested the effect of diisopropyl ether on the skin of rabbits. Single exposures to the liquid for 1 hr produced no deleterious effect. Skin absorption in harmful amounts seems to not occur. However, repeated exposure for 10 days caused dermatitis. Isopropyl ether is irritating and capable of producing minor injury to the rabbit eye (80). Vapors of diisopropyl ether do not produce a sense of irritation to the eyes or mucous membranes of the nose and throat until concentrations exceeding 300 ppm are reached (77).

2.5.3 Hygienic Standards of Permissible Exposure

Table 8.2 contains the permissible exposure limit of diisopropyl ether for industrial control, as set by OSHA (16). It also contains recommended limits suggested by ACGIH (17) in 1992 for normal workday limits.

2.6 Methyl *t*-Butyl Ether (MTBE, 2-Methoxy-2-methylpropane, *t*-Butyl methyl ether)

2.6.1 Determination in the Atmosphere

The NIOSH Method No. 1615 (18) for capturing MTBE from the atmosphere stipulates a charcoal tube as the trapping medium (see Table 8.3). If desorption from charcoal is not vigorous then tenaciously held MTBE on the entrapping charcoal will result in an underestimate of the atmospheric concentration. The analytic method includes gas chromatography with a flame ionization detector (FID). The sensitivity of this method has a range from 0.06 mg to over 10 mg per sample. Physical and chemical properties are presented in Table 8.1.

2.6.2 Physiological Response

MTBE has a distinct chemical odor, described as "terpene-like" (83). The odor detection level in air is reported to be 0.7 ppm and the odor recognition level is 7 ppm (21) (see Table 8.4). MTBE is slightly irritating to eyes and mucous membranes and minimally irritating to intact skin (84). It has a clear anesthetic effect, which is completely recoverable, and is not classified as nerve toxicity.

2.6.2.1 Human Response

Infusion of MTBE into the gallbladder of humans to dissolve gallstones has been reported by Ponchon et al. (85). Illness manifested by this treatment includes confusion, somnolence, coma, and anuria, all of which were reversible in a few hours to 18 days. Dialysis treatment was performed to help the renal failure recovery.

2.6.2.2 Animal

Acute Toxicity. The acute oral LD_{50} of MTBE in rats has been reported to be 2962.8 and 3865.9 mg/kg (86). The rat acute inhalation LC_{50} after 4 hr of exposure has been reported to range from 23,630 to 39,000 ppm. Another study reported an inhalation LC_{50} value of 18,000 ppm in rats after 5.6 min (86).

Laboratory animals have been used to evaluate the capacity for MTBE to dissolve gallstones. Human gallstones were dissolved in MTBE in vitro (87). Additionally, human gallstones were surgically implanted into dogs. The gallbladder was catheterized and MTBE was instilled and aspirated out at varying time intervals. MTBE has been shown to be effective in dissolving gallstones high in cholesterol content but ineffective in dissolving pigment stones (88).

Subchronic Toxicity. Robinson et al. reported dosing rats orally, by gavage, for 14 consecutive days up to 1428 mg/kg (89). Profound anesthesia was observed in this pilot study, which was followed by a 90-day oral study. In the 90-day study rats were dosed daily up to 1200 mg/kg by gavage. Recoverable anesthetic effects were noted at 1200 mg/kg. Lung weights were reduced. Male rat hydrocarbon nephropathy was reported. Hemoglobin and hematocrit values were elevated at dose levels above 900 mg/kg. Liver weights were also elevated but no corresponding changes in blood chemistry or adverse histopathology accompanied the liver weight changes.

Inhalation Toxicity. Data on MTBE were presented in a symposium by Klan et al. (90). Rats were exposed to MTBE by inhalation up to 8000 ppm, 6 hr/day, 5 days/week for 13 weeks. They exhibited stress-related elevated weight changes in liver, adrenal glands, and kidneys plus increased serum corticosterone. The no observable effect level was reported to be 800 ppm. Transient sedation and anesthetic effects were reported at 4000 ppm and above. Rats were also exposed by inhalation to MTBE up to 1000 ppm, 6 hr/day, 5 days/week for 13 weeks (91). A dose-related anesthetic effect was the major finding. Reduced lung weights and elevated hemoglobin at 1000 ppm were noted.

Mutagenicity. MTBE was found to be negative for mutagenic activity in the *Drosophila melanogaster* sex-linked recessive lethal assay (90). It was also reported to be nonclastogenic in a rat in vivo bone marrow cytogenetics assay (90). MTBE exhibited a positive response in the L5178Y mouse lymphoma cell gene mutation assay using a metabolic activation test system (86).

Neurotoxicity. Rats given a single exposure of 4000 or 8000 ppm MTBE by the inhalation route for 6 hr were evaluated for nervous system effects (90). In both a functional observation battery and motor activity evaluation a transient central nervous system depression, characterized as sedation, was reported. No learning or memory changes were reported.

Gill et al. (90) also reported on the neurotoxicity of MTBE in rats after 13 weeks of inhalation exposure up to 8000 ppm, 6 hr/day, 5 days/week. At 800 ppm no changes in the functional observation battery or motor activity were noted. Reduced activity and unsteady gait were observed as transient changes above 800 ppm. No treatment-related histopathology changes were reported in the nervous system.

Reproductive, Developmental Toxicity. Biles et al. reported on a two-litter, one-generation reproduction study on MTBE. Male rats were exposed 6 hr/day, 5 days/week for 12 weeks up to 3400 ppm of MTBE (92). The only statistically significant change was a decreased viability in the second f_1 litter. Neeper-Bradley et al. exposed rats of both sexes to MTBE by the inhalation route 6 hr/day, 5 days/week for 10 weeks. Exposures were conducted daily during mating, gestation, and lactation. No exposure occurred for a brief period during parturition (90). No treatment-related reproductive effects were reported. Signs of maternal toxicity were observed above 400 ppm.

Conaway et al. reported exposing rats and mice by inhalation to at least 2500 ppm during the period of pregnancy when organs develop (93). Changes observed were not statistically significant. MTBE was not found to be maternally toxic, embryo toxic, or teratogenic. In another study mice and rabbits were exposed by inhalation to MTBE up to 8000 ppm (86). Cleft palate and skeletal variation were noted in mice at or above 4000 ppm. No morphology changes were noted in rabbits. The no observable effect levels for mice and rabbits were considered to be 1000 ppm and 8000 ppm, respectively.

Metabolism. Savolainen et al. reported exposing rats to MTBE 6 hr/day, 5 days/week for 15 weeks (94). Blood levels of *t*-butanol correlated with MTBE exposures through 6 weeks of exposure, then decreased. Brain concentrations of both compounds followed a similar trend. *t*-Butanol was concluded to be a metabolite of MTBE. In another report MTBE was found to produce *t*-butanol and formaldehyde in equimolar amounts by rat liver microsomes (90). Klan et al. evaluated the uptake, distribution, metabolism, and excretion of MTBE by four routes of administration in the rat, namely, oral, dermal, inhalation, and intravenous (90). They reported several useful findings:

1. MTBE is rapidly absorbed into the blood after both oral and inhalation exposure.
2. MTBE is not readily absorbed through the skin.
3. MTBE is rapidly metabolized and excreted with a half-life of approximately 30 min.
4. The major metabolic by-products are *t*-butyl alcohol, 2-methyl-1,2-propanediol, and α-hydroxybutyric acid.
5. Neither MTBE nor its metabolites tend to accumulate in any particular tissue.
6. Some evidence indicates that MTBE can saturate metabolic enzyme systems after both high oral and high inhalation exposure.
7. The metabolic response to MTBE was similar in both sexes of the rat.

8. After inhalation exposure, MTBE is readily absorbed into the bloodstream, resulting in an appreciable body burden similar to that seen after oral exposure. This indicates that toxicologic data derived from inhalation studies can be directly correlated with oral (drinking water) exposures.

9. The plasma concentration of MTBE resulting from an oral dose of 400 mg/kg appeared to be equivalent to the plasma concentration obtained in the 6-hr inhalation exposure to 400 ppm MTBE.

2.6.3 Hygienic Standards for Permissible Exposure

There is no established permissible exposure limit (PEL) from OSHA or ACGIH on MTBE. A suggested PEL in the range of 50 to 100 ppm is based on the testing results described above.

2.7 t-Amyl Methyl Ether (TAME, 2-Methoxy-2-methylbutane, Methyl t-Pentyl Ether)

The acute oral LD_{50} for TAME in rats is 2.15 g/kg of body weight (90). Rats dosed orally, by gavage, for 29 consecutive days up to 1.0 g/kg of body weight produced two deaths in 10 rats in the high dose group. Food consumption and body weights were reduced. Adrenal gland and kidney weights were elevated, without the presence of histopathological abnormalities. The no observable effect level was claimed to be 0.125 g/(kg)(day), orally. TAME was negative in the *Salmonella* microsomal plate test and the in vivo mouse micronucleus test.

The evaluation of neurotoxicity for TAME was reported by Ferguson et al. (90). A functional observation battery evaluated neuromuscular function, reflex response, and sensory perception during inhalation exposure up to 4000 ppm 6 hr/day, 5 days/week for 4 weeks. At the high dose level 7 of 28 rats died from compound exposure. At 2000 and 4000 ppm transient central nervous system depression, that is, sedation, ataxia, and reduced activity, were noted. Changes in body temperature, hind limb splay, rotorod performance, tail pinch, and righting reflex were observed at and above 2000 ppm. No histopathological changes were reported. Physical and chemical properties data are presented in Table 8.1.

2.8 Dibutyl Ether (n-Butyl Ether, 1-Butoxybutane)

2.8.1 Determination in the Atmosphere

Dibutyl ether can be determined qualitatively and quantitatively by silica gel absorption, elution with ethanol, and gas chromatography of the eluate. It can also be determined by a sensitive explosimeter or by infrared analysis in a gas cell of long path length. Physical and chemical properties data are presented in Table 8.1.

2.8.2 Physiological Response

Dibutyl ether, although not highly toxic orally, seems to have a greater toxicity by inhalation than the lower ethers of the series. It is also more irritating to the skin.

However, little work has been done from which definite conclusions can be drawn. The odor detection limit for dibutyl ether is 70 ppb (20) (see Table 8.4).

2.8.2.1 Human Experience. Silverman et al. (77) showed that humans exposed to 100 ppm of dibutyl ether estimated it satisfactory for 8-hr exposures. At 200 ppm for 15 min the vapors produced a sensation of irritation to the eyes and nose although at 300 ppm the odor was not objectionable to the majority of subjects.

2.8.2.2 Animal Toxicity. Dibutyl ether has been reported in the range-finding studies of Smyth et al. (35). These data are presented in Table 8.5. Dibutyl ether was reported by Kosyan (81) to have an LD_{50} of 567 mg/kg. It was also reported to have a distinctly expressed local irritating effect on the skin. Dibutyl ether was evaluated for oral toxicity in combinations of two chemicals with 26 other industrial chemicals by Smyth et al. (82). In this work it was shown to have an overall effect of reducing the oral toxicity when combined with other industrial chemicals. Its median oral LD_{50} was 13.6 ml/kg with other chemicals as compared to 7.40 ml/kg alone.

2.9   Other Alkane Ethers

Physiological data on additional industrially important saturated, symmetrical and asymmetrical alkyl ethers are presented in Tables 8.5 and 8.9. Data will be found for dihexyl ether (35), di(2-ethylhexyl) ether (35), methyl isopropyl ether (95), and ethyl butyl ether (34). Physical and chemical properties data are presented in Table 8.1.

3   ALKENE ETHERS

Physiological data on several symmetrical and asymmetrical alkene ethers are presented in Table 8.5.

3.1   Divinyl Ether (Vinyl Ether, Divinyl Oxide, Ethenyloxyethene)

3.1.1   Determination in the Atmosphere

Grab sampling with subsequent infrared analysis with a gas cell of long path length should provide a highly specific method for analysis of divinyl ether. Absorption on silica gel or charcoal, elution with ethanol, and analysis of the eluate by gas chromatography also should be of value. A sensitive explosimeter could be used only when the vapor concentrations are known to be below the lower explosive limit. Physical and chemical properties data are presented in Table 8.1.

3.1.2   Physiological Response (72)

Divinyl ether has been studied extensively in animals and humans from the standpoint of anesthesia. The lethal concentration for mice in air is 51,233 ppm (97).

Table 8.10. Physiological Properties of Some Alkene Ethers

Material	Species	Anesthetic	
		Surgical Anesthesia	Respiratory Arrest
Ethyl vinyl ether (59, 99) $CH_3CH_2OCH=CH_2$	Mouse	6% (v/v)	16% (v/v)
	Rat	—	16% (v/v) for 4 hr killed 2, 3, or 4 of 6 (100)
	Dog	0.56 ml/kg	1.66 ml/kg
Isopropyl vinyl ether (104) $(CH_3)_2CHOCH=CH_2$	Dog	0.50 ml/kg	3.08 ml/kg

Analgesia in humans is obtained by inhalation of 0.2 percent (v/v) and unconsciousness by 2 to 4 percent (v/v) (48). Surgical anesthesia is induced by continuous breathing of 4 percent (v/v). Respiratory arrest occurs with the respiration of 10 to 12 percent (v/v) (69). It is considered more toxic than diethyl ether; however, onset and recovery time for surgical anesthesia are smoother and faster with divinyl ether than with diethyl ether. Cardiac irregularities may occur at the level of respiratory arrest. It is less irritating to mucous membranes and produces less salivation than diethyl ether. Prolonged anesthesia or frequent short administrations can cause liver damage. Cavaliere et al. (98) indicated that divinyl ether causes a sharp increase in bile products in the dog. Additionally, central lobular necrosis of the liver has been reported (48).

Further details of the physiological response to divinyl ether may be obtained from textbooks on pharmacology or anesthesiology.

3.2 Ethyl Vinyl Ether (Vinyl Ethyl Ether, Ethoxy Vinyl Ether, Vinamar)

3.2.1 Physical and Chemical Properties

Physical and chemical properties are presented in Table 8.1.

3.2.2 Physiological Response

Acute toxicity of ethyl vinyl ether from range-finding studies by Smyth et al. (37) is presented in Table 8.5. Additional acute toxicity data appear in Table 8.10 (59, 99, 100). McLaughlin (101) reported transitory injury of the human cornea in one instance, with return to normal within 48 hr.

3.3 Allyl Ethers

Table 8.5 exhibits acute toxicity data for allyl vinyl ether (36) and diallyl ether (33), which has an odor similar to horseradish. The *Merck Index* (97) reports the oral LD_{50} for diallyl ether in rats to be 250 mg/kg. It is also identified as an irritant that can be absorbed through the skin. No systemic toxicity data on this chemical have been found. In the same reference allyl ethyl ether is identified as irritating to eyes, skin, and mucous membranes. Nielson et al. found diallyl ether to be an upper respiratory irritant, capable of depressing the respiratory rate of mice by 50 percent at 5 ppm (102). This latter chemical has also been reported as having minimal effectiveness as a fumigant for the eggs and larvae of the fruit fly (103).

3.4   Other Alkene Ethers

Physical and chemical properties data are presented in Table 8.1. Table 8.5 contains acute toxicity data for butyl vinyl ether, isobutyl vinyl ether, 2-ethylhexyl vinyl ether, 2,6,8-trimethylnonyl vinyl ether, methyl 2-propenyl ether, methyl 1,3-butadienyl ether, ethyl-1-propenyl ether, and 1-propenyl 2-buten-1-yl ether.

Table 8.10 contains acute toxicity data for isopropyl vinyl ether (104).

4   POLYOXY ETHERS

Physical and chemical data are presented in Table 8.1.

Physiological data of polyoxy ethers are presented in Table 8.5. Acute exposure values are presented for 1-butoxy-2-ethoxyethane, 2-butoxyethyl vinyl ether, 1,2-dibutoxyethane, 2,2'-dibutoxyethyl ether, 2-methoxyethyl vinyl ether, and 1,3-dimethoxybutane.

5   HALOGENATED ALKYL ETHERS

The halogenated ethers are more potent anesthetics than the alkane and alkene ethers, and also have higher acute toxicity. Three references that describe the anesthetic potential of halogenated ethers, which are too numerous to report here, are those by Poznak and Artusio (106), Terrell et al. (107), and Speers et al. (108). Some industrially important halogenated ethers have been shown to be carcinogenic in animals and humans. See Section 1.6 for discussion of this characteristic of these alkylating agents.

5.1   Chloromethyl Methyl Ether (CMME, CME, Dimethyl Chloroether, Methyl Chloromethyl Ether)

5.1.1   Determination in the Atmosphere

Using gas–liquid chromatography (109) can help detect chloromethyl methyl ether in air at the part per billion level. Also, see Table 8.3. Physical and chemical properties data are presented in Table 8.1.

5.1.2   Physiological Response

This material is considered a carcinogenic risk to humans. Oat cell lung cancer has been statistically identified in workers exposed to chloromethyl methyl ether. White males working in three buildings that used CME exhibited statistically significant increases in death due to respiratory tract cancer (110). The authors indicated the death rate was dose related. Unfortunately, actual atmospheric concentrations of CME were not available. The authors used an arbitrary scale of exposure intensity to arrive at a dose response conclusion. Previously it was noted that the epidemic nature of CME cancer cases peaked 15 to 19 years after the start of exposure and has declined since chemical exposure ceased due to engineering improvements (111).

The IARC monograph (112) and the more recent report by Fishbein (7) should be consulted for detailed evidence of animal and human carcinogenic potential. Tables 8.6 and 8.7 present limited carcinogenic data on this compound. More recently, VanDuuren reviewed both animal and human evidence of the cancer-inducing potential of CME (113).

5.1.3 Hygienic Standards of Permissible Exposure

Table 8.2 contains the permissible exposure limit of chloromethyl methyl ether for industrial control, as set by OSHA (16).

5.2 Bis(Chloromethyl) Ether [BCME, Chloromethyl Ether, Dichloromethyl Ether, 1,1'-Dichlorodimethyl Ether, Chloro(Chloromethoxy)methane]

5.2.1 Determination in the Atmosphere

Gas–liquid chromatography (109) can help detect bis(chloromethyl) ether in air at the part per billion level. Also, see Table 8.3 and Section 1.5. Physical and chemical properties are presented in Table 8.1.

5.2.2 Physiological Response

Because technical-grade CME is contaminated with BCME workers become exposed to both materials. The carcinogenic risk of BCME has become an important industrial health consideration. Earlier work is reviewed in an IARC monograph (112). More recent data are presented by Fishbein (7). VanDuuren provides the most current review of both animal and human evidence of the cancer-inducing potential of BCME (113). These references should be consulted for detailed evidence of carcinogenic potential for BCME. Tables 8.6 and 8.7 present limited carcinogenic data on this compound.

5.2.3 Hygienic Standards of Permissible Exposure

Table 8.2 contains the permissible exposure limit of bis(chloromethyl) ether for industrial control as set by OSHA (16).

5.3 Dichloroethyl Ether [sym-Dichloroethyl Ether, β,β'-Dichloroethyl Ether, 2,2'-Dichlorodiethyl Ether, 1-Chloro-2-(β-chloroethoxy)ethane, Bis(2-chloroethyl) Ether]

5.3.1 Determination in the Atmosphere

NIOSH has proposed an air sampling technique for dichloroethyl ether. The airflow rate, trapping medium, elution medium, and analytic procedure for this chemical are all presented in Table 8.3 (18). Physical and chemical properties are presented in Table 8.1.

5.3.2 Physiological Response

Dichloroethyl ether vaporizes to an irritant gas, which can cause a delayed response of several days resulting in lung lesions. However, the vapor concentration necessary to cause damage in the respiratory tract is easily detected by its odor, and is intolerable in most cases. The compound is injurious to the eyes, in both liquid and concentrated vapor forms. It has no deleterious effect on the intact skin. However, it is rapidly absorbed in lethal amounts by the skin; therefore skin contact should be avoided. Oral consumption also should be avoided. The odor detection limit is reported to be 15 ppm (20) (see Table 8.4).

5.3.2.1 Human Experience

Schrenk et al. (114) exposed human volunteers to dichloroethyl ether. Brief exposures to concentrations above 500 ppm were very irritating to the eyes and nasal passages and were considered intolerable. They also caused coughing, retching, and nausea. At 100 to 260 ppm the irritating effects were still present to some extent but were not considered intolerable. The nauseous odor was still detectable at 35 ppm but there was no irritation. Schwartz et al. (115) reported an investigation of dermatitis in textile employees exposed to a new resin fabric finish containing this compound. McLaughlin (101) treated cases of eye injury; the eyes were minimally irritated and healed promptly within 48 hr.

5.3.2.2 Animal

Inhalation. Schrenk et al. (114) reported concentrations of 0.055 to 0.1% (v/v) in air decreased mobility and induced retching and loss of consciousness. These signs had a slower onset at 0.026 percent, and at 0.01 percent the onset was very slow and fewer signs were noted. At the high levels congestion of the lungs, liver, brain, and kidneys were noted. In Table 8.5 (32) acute inhalation toxicity data are tabulated. Repeated exposures of rats and guinea pigs to an average concentration of 69 ppm of dichloroethyl ether vapor have been carried out (116). There were 93 7-hr exposures, five per week, over a period of 130 days. No adverse effect was observed as judged by gross appearance, behavior, mortality, hematologic values, and gross and microscopic examination of the tissues at necropsy. However, all the animals showed a varying degree of growth depression, which was significant in males of both species. Organ weights also showed significant variance from controls; however, these variances were not consistent among species or sex. It was concluded that repeated exposure to 69 ppm of dichloroethyl ether produces no serious injury and that the abnormalities observed reflected only mild physiological stress.

Rats (100) exposed to 250 ppm for 4 hr had a death rate of 2/6, 3/6, or 4/6 within 14 days. It is considered a definite inhalation hazard.

Oral. Data for the acute oral toxicity of dichloroethyl ether are presented in Table 8.5 (32). Smyth (117) reports the following oral LD_{50} values, giving unpublished data: rat, 105 mg/kg; mouse, 136 mg/kg; and rabbit, 126 mg/kg. Further work (116) using the rat has given an LD_{50} value of 150 mg/kg with 19/20 confidence limits of 110 to 210 mg/kg. Innes et al. (118) reported that a dose of 100 mg/kg given orally from 7 to 28 days of age resulted in elevated tumor incidence in mice.

Effects on Animal Skin and Eyes. Irritancy data describing effects on skin and eyes appear in Table 8.5. Allen (119), in his review of the safety hazards of dichloroethyl ether, stated, "The pure liquid has no irritative effect or indeed any other effect on the skin, a factor greatly in its favor." This has essentially been confirmed by studies with rabbits (116), which showed that a 10 percent solution of the compound has very little or no effect on the intact or abraded skin after repeated applications. However, in the same studies, when the pure material was used, acutely toxic amounts rapidly penetrated the skin and caused death within a day. When graded doses of a 10 percent solution of dichloroethyl ether in propylene glycol were used, the LD_{50} for 24-hr skin absorption in rabbits was calculated to be 90 mg/kg, with 19/20 confidence limits of 55 to 145 mg/kg. Percutaneous toxicity data also appear in Table 8.5. Although dichloroethyl ether is not a primary skin irritant, contact with the skin should be avoided because of its rapid absorption and deleterious consequences.

VanDuuren et al. (120) demonstrated a weak carcinogenicity potential with dichloroethyl ether in a skin painting study in mice. Additionally, when the ether was injected subcutaneously at a total dose of 1 mg in mice, slight carcinogenicity was noted as sarcomas in 2 of 30 mice. When the compound was evaluated as an initiator of carcinogenicity with acetone, no mice developed papillomas (Tables 8.6 and 8.7). Norpoth et al. found no product-related tumors after subcutaneous injection once a week for 2 weeks (123).

Besides the eye irritancy data in Table 8.5, studies (116) with rabbits have revealed that both the pure material and a 10 percent solution in propylene glycol cause moderate pain, conjunctival irritation, and corneal injury, which generally heals within 24 hr. Immediate irrigation of the eye with flowing water greatly reduces the deleterious effects. Carpenter and Smyth (38) reported moderate eye irritancy after contact with 0.005 ml of the undiluted product. Schrenk et al. (114) reported concentrations of 0.055 to 0.1 percent (v/v) in air are immediately irritating to guinea pig eyes and mucous membranes. From these results it is evident that exposure of human eyes to dichloroethyl ether could cause moderate damage from which recovery should be satisfactory.

Chronic. Weisburger et al. reported dosing rats orally, by gavage, for 78 weeks with dichloroethyl ether (121). The compound was determined to be non-cancer forming.

Mode of Action. From the work of Schrenk et al. (114) it is evident that the physiological action of dichloroethyl ether is that of a primary respiratory irritant. Although the compound is capable of causing narcosis at high concentrations, the delayed deaths at lower concentrations have been due to severe irritation of the respiratory tree, with a resultant respiratory collapse. Allen (119) makes the interesting comparison of the chemical structure of dichloroethyl ether to that of mustard gas, dichloroethyl sulfide. The replacement of the S atom by the O atom in dichloroethyl ether abolishes the blistering action of mustard gas on the skin. However, many of the respiratory irritant effects are retained. Lingg et al. (122) determined two urine metabolites in rats, thiodiglycolic acid and 2-chloroethanol-β-*d*-glucuronide. Norpoth et al. found thiodiglycolic acid and hydroxyethyl mercapturic acid as metabolites in rats (123).

5.3.3 Hygienic Standards of Permissible Exposure

Table 8.2 contains the permissible exposure limits of dichloroethyl ether for industrial control, as set by OSHA (16). It also contains limits recommended by ACGIH (17) in 1992 that are normal workday limits.

5.4 Dichloroisopropyl Ether (Bis(β-chloroisopropyl) Ether, 2,2'-Dichlorodiisopropyl Ether, β,β'-Dichlorodiisopropyl Ether)

5.4.1 Physical and Chemical Properties

Physical and chemical properties data are presented in Table 8.1.

5.4.2 Physiological Response

As can be seen from the data in Table 8.5 (34) dichloroisopropyl ether is less toxic and less irritating than dichloroethyl ether. However, it is still fairly toxic, particularly to the liver and kidneys. The odor detection level in water for this compound is presented in Table 8.4.

5.4.2.1 Oral Tests. In addition to the data in Table 8.5 other studies indicate similar results (116). None of the five animals died when fed at 0.2 g/kg, but four of six died at 0.4 g/kg. All five rats fed 0.8 g/kg died. Smyth (117) found the LD_{50} for guinea pigs to be 0.45 g/kg.

Rats were fed 22 doses of dichloroisopropyl ether in olive oil by stomach tube during a period of 31 days (116). Even the lowest dose administered, 0.01 g/kg, caused a decrease in growth rate when compared to the controls. However, there were no differences in organ weights or hematologic values at this level. At the highest dosage level 0.20 g/kg, both the liver and kidney weights (per unit of body weight) were increased when compared to the controls. Spleen weights also increased but there was no indication of effects on the blood. Intermediate dosage levels showed only a decrease in growth rate. It is evident from this work that repeated, low oral doses of dichloroisopropyl ether, although not rapidly lethal, have deleterious effects on rats.

5.4.2.2 Inhalation. Table 8.5 has acute inhalation toxicity data on dichloroisopropyl ether. In other studies (116), rats exposed to an atmosphere believed to be essentially saturated with dichloroisopropyl ether exhibited signs of immediate eye irritation and incoordination: the maximum exposure time causing no death was 1 hr. When rats were exposed to 700 ppm, deaths occurred after 6 hr of exposure. Necropsy revealed slight lung irritation and moderate to severe liver damage. All 10 rats survived a 6-hr exposure to 350 ppm but two of five died after an 8-hr exposure. These animals exhibited moderate lung congestion and some liver necrosis. One of four animals died after an 8-hr exposure to 175 ppm. It appears that the length of exposure to dichloroisopropyl ether vapors is a highly significant factor in its capacity to cause death.

5.4.2.3 Effects to Skin and Eyes. Dichloroisopropyl ether evidently has little effect on rabbit skin. After 20 applications on the ear there was no response. Moderate scaliness was produced after 20 applications to the rabbit abdomen by the poultice technique (116). Data in Table 8.5 indicate that dichloroisopropyl ether is minimally irritating to skin and eyes. It is absorbed through the skin and at high doses can be lethal.

5.4.2.4 Chronic. Mitsumori et al. exposed mice to dichloroisopropyl ether (124). The compound was blended with olive oil and added to solid food up to 10,000 ppm. Exposures were conducted for 2 years. Reduction in food consumption in the high-dose group corresponded with reduced body weight growth. Additionally, signs of anemia were seen at 2000 ppm and higher. Hemoglobin concentration and red blood cell counts were reduced. This was accompanied with splenic hemosiderin deposition and polychromatic erythrocyte increase. The anemia observed was considered to be erythrocyte destruction produced by dichloroisopropyl ether. No biologically significant formation of tumors was identified in this 2-year study.

5.4.2.5 Metabolism. Male rats were given a single administration of dichloroisopropyl ether orally, by stomach tube (125). The excretion half-life was determined to be 19 hr. Major metabolites in urine were identified as 2-(2-chloro-1-methylethoxy)propanoic acid plus N-acetyl-S-(2-hydroxypropyl-L-cysteine). Dichloroisopropyl ether tends to be excreted 20 percent as CO_2, 47.5 percent in urine, and 4 percent in feces.

5.5   Other Haloethers

Table 8.5 contains physiological response data for the haloethers 2-chloro-1,1,2-trifluoroethyl methyl ether and 2-chloroethyl vinyl ether. Tables 8.6, 8.7, and 8.11 contain physiological response data for the haloethers dichloromethyl methyl ether (120, 126), bis (α-chloroethyl) ether (120), and octachlorodi-*n*-propyl ether (120, 126, 127). Table 8.11 contains, in addition, physiological response data for the haloethers 2-fluoroethyl methyl ether (128), 2,2,2-trifluoroethyl methyl ether (125), 2,2-dichloro-1,1-difluoroethyl methyl ether (57, 130–132), 2,2-difluoroethyl ethyl ether (129), 2,2,2-trifluoroethyl ethyl ether (129), 2-fluoro-1′,2′,2′,2′-tetrachlorodiethyl ether (128), bis(2,2,2-trifluoroethyl) ether (133, 134), 2,2,2-trifluoroethyl vinyl ether (69, 135, 136), bis-1,2-(chloromethoxy)ethane (137), 5-fluoroamyl methyl ether (128), bis-1,4-(chloromethoxy)butane (137), perfluoroisobutenyl ethyl ether (138), 6-fluorohexyl methyl ether (128), 4,4′-difluorodibutyl ether (128), bis-1,6-(chloromethoxy)hexane (137), 4-fluoro-4′-chlorodibutyl ether (128), and 2-fluoro-2′-*n*-butoxydiethyl ether (128).

The compounds in Table 8.11 are arranged predominantly by increasing chain length and by increases in halogen substituents. Increased fluorination of diethyl ether progressively diminishes its anesthetic potency so that perfluoroethyl ether is devoid of anesthetic properties. On the other hand, a partially fluorinated compound, bis(2,2,2-trifluoroethyl) ether, in addition to some anesthetic action, is a powerful convulsant (134). This chemical is included in Table 8.11. Buckle and Saunders (139) believe that the toxicity of these compounds can be explained by β-oxidation. Pattison et al. (128) propose that the ether link is ruptured in vivo. This latter concept is reinforced by work of Johnston et al. (140), who found that animals biotransform trifluoroethyl vinyl ether to its alcohol, trifluoroethanol.

6   AROMATIC ETHERS

6.1   Anisole (Phenyl Methyl Ether, Methoxybenzene)

6.1.1   Determination in the Atmosphere

Collection on charcoal followed by analysis employing ultraviolet spectrophotometry is recommended. When in isooctane, anisole has strong absorption bands at 271 and 277.5 nm, and 0.05 mg/100 ml can be measured readily. Vapor-phase chromatography and mass spectrometry also can be used. Physical and chemical data properties are presented in Table 8.1.

6.1.2   Physiological Response

6.1.2.1 Acute Toxicity. Acute toxicity data for anisole administered subcutaneously and intraperitoneally are presented in Table 8.12 (141).

6.1.2.2 Metabolism. Coombs and Hele (142) in 1926 reported that administering anisole to the dog caused an increase in the excretion of ethereal sulfate. Williams

Table 8.11. Physiological Response of Miscellaneous Haloethers

Material and Structural Formula	Remarks
Dichloromethyl methyl ether $CHCl_2OCH_3$	See Table 8.6. Data originally from VanDuuren et al. (120) exhibiting weak carcinogenicity in mice. When evaluated as an initiator of carcinogenesis with phorbol ester, 3 mice out of 20 developed papillomas and 1 developed carcinoma (126)
2-Fluoroethyl methyl ether $CH_2FCH_2OCH_3$	Intraperitoneal LD_{50} in mice is 15 mg/kg (128)
2,2,2-Trifluoroethyl methyl ether $CF_3CH_2OCH_3$	Using 10 min for the exposure time Robbins (129) found the anesthetic dose (AD_{50}) to be 8% (v/v); the fatal dose (FD_{50}) to be 16% (v/v) using 20 mice. The time necessary for induction of anesthesia of mice exposed to the LC_{50} is 20 sec. The time necessary for recovery of pain sensation after exposure to the LC_{50} for 10 min is 2 min. The time necessary to return to normal gait after 10 min exposure to the LC_{50} is 4 min
2,2-Dichloro-1,1-difluoroethyl methyl ether $CHCl_2CF_2OCH_3$ (methoxyfluorane)	Artusio et al. (130) evaluated the anesthetic effects in 100 patients. Induction was smooth; no delirium, hypotension, or ventricular arrhythmia was noted. EEG changes, associated with anesthesia were noted. Nausea, vomiting, and the need for analgesics were reduced. Desmond (131) exposed 11 patients to this anesthetic in operations lasting 3–4 hr at a concentration of 0.16%. Renal dysfunction lasting for days postoperatively was characterized by a large volume of low osmolality urine produced. Other reports of dramatic renal failure with use of this anesthetic are discussed. Swann et al. (57) found that 6.4% (v/v) produced respiratory arrest in 2 of 4 mice in 5.25 min. Rats exposed (132) to a 0.5% (v/v) atmosphere for 2, 4, or 6 hr showed decreased food intake, diuresis, and reduced urinary osmolality
2,2-Difluoroethyl ethyl ether $CHF_2CH_2OCH_2CH_3$	Using 10 min for the exposure time Robbins (129) found the anesthetic dose (AD_{50}) to be 4% (v/v) and the fatal dose (FD_{50}) to be 9% (v/v) using 24 mice. The time necessary for induction of anesthesia of mice exposed to the LC_{50} is 1 min. The time necessary for recovery of pain sensation after exposure to the LC_{50} for 10 min is 2 min. The time necessary to return to normal gait after 10 min exposure to the LC_{50} is 5–20 min

2,2,2-Trifluoroethyl ethyl ether
$CF_3CH_2OCH_2CH_3$

Using 10 min for the exposure time Robbins (129) found the anesthetic dose (AD_{50}) to be 4% (v/v) and the fatal dose (FD_{50}) to be 8% (v/v) using 16 mice. The time necessary for induction of anesthesia of mice exposed to the LC_{50} is 30 sec. The time necessary for recovery of pain sensation after exposure to the LC_{50} for 10 min is 20 sec. The time necessary to return to normal gait after 10 min exposure to the LC_{50} is 1 min.

1-Chloroethyl ether
$CH_3CHClOCHClCH_3$

See Table 8.6. Data originally from VanDuuren et al. (120) exhibiting moderate carcinogenicity in mice. When injected subcutaneously at a dose of 300 mg in mice slight carcinogenicity was noted as sarcomas in 4 of 30 mice. When evaluated as an initiator of carcinogenesis with phorbol myristate acetate 7 out of 9 mice developed papillomas

2-Fluoro-1′,2′,2′,2′-tetrachlorodiethyl ether
$CH_2FCH_2OCHClCCl_3$

Pattison et al. (128) reported the intraperitoneal LD_{50} in mice to be 48 mg/kg

Bis(2,2,2-trifluoroethyl)ether
$CF_3CH_2OCH_2CF_3$

The clonic convulsion dose (CD_{50}) reported by Truitt and Ebersberger (133) as intravenous in 30 mice is 26 mg/kg. The intravenous LD_{50} is 46 mg/kg. The intraperitoneal clonic convulsion dose (CD_{50}) in 88 rats is 600 mg/kg. The intraperitoneal LD_{50} is 1260 mg/kg in rats. The mouse clonic convulsion dose (CD_{50}) in an inhalation jar is 0.26 ml as volume of product. The tonic convulsion dose (TD_{50}) as volume in an inhalation jar in mice is 0.9 ml and 0.8 ml observed 30 sec and 10 min after onset of convulsion, respectively. The mouse inhalation LD_{50} is 2.75 ml and 0.68 ml of product in a jar for the same time intervals. Krantz et al. (134) found the convulsive threshold in rats to be between 29 ppm and 41 ppm for a 90-sec exposure time

2,2,2-Trifluoroethyl vinyl ether
$CF_3CH_2OCH{=}CH_2$ (Fluoromar)

Krantz et al. (135) reported it to be an excellent analgesic. It does not produce liver damage or sensitize the myocardium to exogenous catecholamines. It produces minimal cardiovascular and respiratory depression. An anonymous report (136) states that animals metabolize this compound to the more toxic trifluoroethanol, which appears only as a trace in humans

Bis-1,2-(chloromethyoxy)ethane
$ClCH_2OCH_2CH_2OCH_2Cl$

VanDuuren et al. (137) demonstrated that 1 mg/0.1 ml of solvent applied to mouse skin 3 times a week produced a statistically elevated incidence of squamous carcinoma. Subcutaneous injections once weekly of 0.3 mg/0.05 ml vehicle produced a statistically elevated incidence of sarcomas. Intraperitoneal injections once weekly of 0.3 mg/0.05 ml vehicle produced a statistically elevated incidence of sarcomas

Table 8.11. (Continued)

Material and Structural Formula	Remarks
5-Fluoroamyl methyl ether $CH_2F(CH_2)_4OCH_3$	Pattison et al. (128) reported that the intraperitoneal LD_{50} for mice is 90 mg/kg
Octachloro-di-n-propyl ether $CCl_3CH(Cl)CH_2OCH_2CHClCCl_3$	VanDuuren et al. (120, 126) reported that skin application in 20 mice as an initiator evaluation with phorbol ester produced 3 papillomas and 1 carcinoma. Evaluated alone it was noncarcinogenic. VanDuuren (127) reported mouse skin tests were negative for carcinogenicity
Bis-1,4-(chloromethoxy)butane $CH_2ClO(CH_2)_4OCH_2Cl$	VanDuuren et al. (137) demonstrated that 1.0 mg/0.1 ml of vehicle applied to mouse skin 3 times weekly did not produce a statistically elevated incidence of cancer. Subcutaneous injections once weekly of 0.3 ml/0.05 ml of vehicle produced no evidence of cancer. Intraperitoneal injections once weekly at a dose of 0.1 mg/0.05 ml of vehicle produced no evidence of cancer
Perfluoroisobutenyl ethyl ether $(CF_3)_2C=CFOCF=C(CF_3)_2$	Schwartzman (138) presented data identifying the olfactory threshold as a level of 2.8 mg/l. Its LD_{50} in mice is 164 mg/kg. A dose of 6 mg/kg in chronic studies in rats and rabbits produced toxic action. A dose of 0.15 mg/kg produced disturbances and minor reversible morphological changes of the nervous system and internal organs. A dose of 0.015 mg/kg produced no changes in 7 months
6-Fluorohexyl methyl ether $CH_2F(CH_2)_5OCH_3$	Intraperitoneal LD_{50} in mice is 4.0 mg/kg (128)
4,4'-Difluorodibutyl ether $CH_2F(CH_2)_3O(CH_2)_3CH_2F$	Intraperitoneal LD_{50} in mice is 0.82 mg/kg (128)
Bis-1,6-(chloromethyloxy)hexane $CH_2ClO(CH_2)_6OCH_2Cl$	VanDuuren et al. (137) demonstrated that 1.0 mg/0.1 ml of vehicle applied to mouse skin 3 times a week produced no evidence of cancer. Subcutaneous injections once weekly of 0.3 mg/0.05 ml of vehicle produced a single sarcoma in 50 mice. Intraperitoneal injections once weekly at a level of 0.3 mg/0.05 ml of vehicle produced no evidence of cancer
4-Fluoro-4'-chlorodibutyl ether $CH_2F(CH_2)_3O(CH_2)_3CH_2Cl$	Intraperitoneal LD_{50} in mice is 1.32 mg/kg (128)
2-Fluoro-2'-n-butoxydiethyl ether $CH_2FCH_2OCH_2CH_2O(CH_2)_3CH_3$	Intraperitoneal LD_{50} in mice is 43 mg/kg (128)

(143) interpreted these observations to indicate cleavage of the ether linkage with subsequent conjugation of the formed phenol with sulfuric acid. Later, however, he revised this conclusion (144), probably as a result of the work of Bray et al. (146). Bray's studies with the rabbit suggest that the increase in ethereal sulfate observed by Coombs and Hele is probably due to *p*-hydroxylation and conjugation with sulfuric acid. These authors showed the major metabolite of anisole to be *p*-hydroxyphenol methyl ether, which was excreted unconjugated (2 percent) and conjugated with glucuronic acid (48 percent) and sulfuric acid (29 percent).

Aryl-substituted anisoles such as *m*- and *p*-nitro, *p*-chloro-, *p*-methoxy-, and *p*-cyanoanisoles are reported by Bray et al. (147) to be demethylated to form the phenols, which are excreted largely at the glucuronates and ethereal sulfates. Axelrod (148, 149) has found an ether-splitting enzyme system in the microsomes of rat liver cells that would account for the observation of Bray et al. (147).

6.2 Phenetole (Phenyl Ethyl Ether, Ethoxybenzene)

6.2.1 Determination in the Atmosphere

Collection on charcoal followed by analysis employing ultraviolet spectrophotometry is recommended. When in isooctane, phenetole has strong absorption bands at 271 and 277.5 nm, and 0.05 mg/100 ml can be measured readily. Vapor-phase chromatography and mass spectrometry also can be used. Physical and chemical data properties are presented in Table 8.1.

6.2.2 Physiological Response

6.2.2.1 Human Experience

No report of adverse effects from phenetole has been found in the literature.

6.2.2.2 Animal

Acute Toxicity. Physiological response data for acute exposures of phenetole in animals are presented in Table 8.12 (116, 141).

Metabolism. According to Williams (144), phenetole, like anisole, is hydroxylated in the *para* position and excreted as the glucuronide and ethereal sulfate.

6.3 Guaiacol (*o*-Methoxyphenol, 1-Hydroxy-2-methoxybenzene, Methylcatechol)

6.3.1 Determination in the Atmosphere

Guaiacol vapor can be trapped by passing the samples of contaminated air through methanol or other suitable organic solvent or through 0.1 to 1 *N* sodium hydroxide. Guaiacol, as the sodium salt, absorbs strongly in the ultraviolet region at a wave length of 289 nm and as little as 0.01 mg/100 ml of solution can be determined

Table 8.12. Physiological Response of Acute Exposure to Cyclic Ethers

Substance	Oral LD$_{50}$	Intraperitoneal LD$_{50}$	Subcutaneous LD$_{50}$	Intravenous LD$_{50}$	Dermal LD$_{50}$
Anisole	—	—	—	—	—
Phenetole	Guinea pig: 3.0–10 g/kg (116)	Rat: 100–900 mg/kg (141)	Rat: 3500–4000 mg/kg (141) Rat: 3.5–4.0 g/kg (141)[a]	—	—
Guaiacol	Rat: 1.5 g/kg (116) Rabbit: 4 g (150) Cat: 60 drops (150) Mouse: 621 mg/kg (151)	—	Pigeon: 0.2 g (150) Rabbit: 2.5 g (150) Guinea pig: 0.9 g/kg (150) Rat: 0.9 g/kg (150)	Mice: 170 mg/kg (145)	Human: 2 g/man produces chills, temperature drop, collapse, death due to respiratory failure (150) Rabbit: 4.6 g/kg (145)
Hydroquinine monomethyl ether	Rat: 1.6 g/kg (156)	Mouse: 725 mg/kg (156) Rat: 430 mg/kg (156) Rabbit: 720–970 mg/kg (156)	—		
Hydroquinone dimethyl ether	Rat: 8.5 g/kg (156)	—	—	—	—
Hydroquinone monobenzyl ether (Na salt)	Rat: >3.2 g/kg (157)	—	—	—	—
Eugenol	Rat: 2680 mg/kg (183) Rat: 1930 mg/kg (184) Mouse: 3000 mg/kg (183) Guinea pig: 2130 mg/kg (183)	Mouse: 630 mg/kg (185)[b]	—	—	—
Isoeugenol	Rat: 1.56 g/kg (193) Guinea pig: 1.41 g/kg (193)	Mouse: 600 mg/kg (185)[b] Mouse: 365 mg/kg (193) Mouse: 540 mg/kg (193)	—	—	—
Methyl eugenol	Rat: 1179 mg/kg (195)	Mouse: >640 mg/kg (185)[b] Mouse: 540 mg/kg (196)	—	Mouse: 112 mg/kg (196)	Rabbit: >2025 mg/kg (195)
Methyl isoeugenol		Mouse: 640 mg/kg (185)[b] Mouse: 570 mg/kg (196)		Mouse: 181 mg/kg (196)	
Butylated hydroxyanisole	Rat: 4–5 g/kg in corn oil (153, 155, 156) Rat: 2.5 g/kg in propylene glycol (153, 155, 156)	—	—	—	—

					Fish LC$_{50}$
Vanillin (245)	Rabbit: 3.0 g/kg[a] Rat: 1.58 g/kg Rat: 2.0 g/kg Rat: 2.8 g/kg Guinea pig: 1.40 g/kg	Rat: 1.16 g/kg Mouse: 475 mg/kg Mouse: 0.78 g/kg Guinea pig: 1.19 g/kg	Rat: 1.8 g/kg[c] Rat: 1.5 g/kg	Dog: 1.32 g/kg slow infusion	—
Ethyl vanillin (247)	Rat: >2000 mg/kg Rabbit: 3000 mg/kg	Mouse: 750 mg/kg Guinea pig: 1140 mg/kg	Rat: 1800 mg/kg	Dog: 760 mg/kg	—
Phenyl ether (116)	Rat: 4.0 g/kg[d] Rat: 2.0 g/kg[e] Rat: 3.99 g/kg Guinea pig: 4.0 g/kg[d] Guinea pig: 1.0 g/kg[e]	—			—

	Eye Irritation	Skin Irritation	Sensitization	Fish LC$_{50}$
Phenetole	—	Rabbit: slight (116)	—	—
Guaiacol	Rabbit: Severe, necrosis (116) Rabbit: 10% in propylene glycol—mild (116)	Rabbit: Several exposures produced severe irritation, burning, loss of sensation, dermatitis with vesication (150)	—	—
Hydroquinone monomethyl ether	Rabbit: moderate corneal damage (116)	Guinea pig: 40% solution in olive oil and acetone—slight or moderate	Guinea pig: negative (157)	
Hydroquinone dimethyl ether	—	Guinea pig: 40% solution in olive oil and acetone—slight or moderate	Guinea pig: negative (157)	—
Methyl eugenol	Rabbit: Slight (195)	Rabbit: Slight (195)	—	Rainbow trout: 6 ppm, 96 hr (195) Bluegill sunfish: 8.1 ppm, 96 hr (195)

[a]Minimum lethal dose
[b]Dosed simultaneously with hexobarbital or zoxazolamine.
[c]Lethal dose.
[d]Total mortality.
[e]Total survival.

when using a 10-cm cell. Vapor-phase or paper chromatography and mass spectrometry may be useful analytical tools under specific circumstances. Physical and chemical properties are presented in Table 8.1.

6.3.2 Physiological Response

Guaiacol appears to be about one-third as toxic as phenol and to have pharmacological properties similar to phenol. It causes muscular weakness, cardiovascular collapse, and paralysis of the vasomotor centers. Medical experience indicates that toxic quantities can be absorbed through the skin quite readily. The material is not especially irritating to the skin but prolonged contact may cause injury particularly if the skin is abraded. It is severely injurious to the eyes, however. It does not seem likely that the inhalation of vapors would constitute a serious industrial hazard, unless, of course, material were handled hot or fumes were generated. An extensive literature review on guaiacol has been reported by Opdyke and Letizia and should be consulted for details (145).

6.3.2.1 Human Experience

Industrial experience to date has not indicated that guaiacol presents any unusual handling hazard. However, medical experience indicates that the material may be more hazardous to human beings than to lower animals. Solis-Cohen and Githens (150) report that a 9-year-old girl died after swallowing 5 ml and that doses above 2 g are hazardous when applied to the skin. When given to humans, the material causes irritation, burning pain with vomiting, and diarrhea that may be bloody. A tolerance develops when repeated small doses are given. Goodman and Gilman (152) report that its medical use as an expectorant has shown that large doses can cause cardiovascular collapse and that clinical doses sometimes cause gastrointestinal irritation. It would seem from this experience and from the pharmacological action of guaiacol that it is more like phenol in its effects on human subjects. Guaiacol was tested at 2 percent petrolatum on a human panel as a 48-hr skin irritancy test under closed patches (145). No irritation was noted. No skin allergic sensitization reactions were reported in a 25-person panel when guaiacol was tested at 2 percent in petrolatum (145).

6.3.2.2 Animal

Acute Toxicity. Physiological responses from single exposures of guaiacol by various routes of administration are presented in Table 8.12 (116, 150). The inhalation LC_{50} in mice is 7.57 mg/l (145).

Metabolism. Wong and Sourkes (153) demonstrated that a single dose of 50 mg/kg of guaiacol, as an intraperitoneal injection, in rats was *o*-demethylated, forming catechol.

6.3.3 Hygienic Standard for Industrial Exposure

There is no hygienic standard for guaiacol. It would be considered wise, however, to avoid vapor concentrations irritating to the eyes or respiratory passages. The prevention of topical contact and swallowing would seem quite important in industrial operations.

6.4 Hydroquinone Ethers

6.4.1 General Determination in the Atmosphere

Trapping the dust of hydroquinone ethers with filter paper devices and their vapors with organic solvents and subsequent analysis by chromatography, spectroscopy, or other method suitable for the detection of the aromatic ring should be adequate for an accurate atmospheric analysis. Physical and chemical properties are presented in Table 8.1.

6.4.2 General Physiological Response

Repeated exposure to these chemicals may result in depigmentation of skin. Excessive skin contact may cause dermatitis.

6.5 Hydroquinone Monomethyl Ether (4-Hydroxyanisole, p-Methoxyphenol)

6.5.1 Physiological Response

An extensive review of the literature was completed on human and animal toxicity experience (154). This reference should be consulted. Hydroquinone monomethyl ether produces dermal allergic sensitization and skin depigmentation in guinea pigs. Human evaluations for these two biologic effects are equivocal.

Chivers (155) described the occurrence of leukoderma in two patients who were industrially exposed to hydroquinone monomethyl ether. During industrial handling of this material skin contact should be avoided in order to preclude the onset of skin depigmentation.

The physiological responses of acute toxicity evaluations are presented in Table 8.12 (116, 156, 157). The gross toxic signs of acute intraperitoneal poisoning were those of paralysis and anoxia at lower levels and narcosis at high levels. Hodge et al. (156) reported that rats maintained on a diet of hydroquinone monomethyl ether for 7 weeks exhibited slight growth depression in males at 0.1 percent and a moderate depression at 2 percent. At the 5 percent level maximum growth depression was noted. Because no other deleterious effects were noted at these levels, the authors concluded that the flavor may have reduced the palatability for the ration and thus reduced food intake. Support for this theory is gained by a comparison with the rabbit-feeding studies wherein this compound, at a level of 10 percent in the diet, caused little or no growth depression. There were no deleterious effects noted in dogs fed this compound in amounts up to 6 g/day, and 12 g/day caused only minor body weight losses. Hodge et al. (156) also showed that application of this material at a 10 percent concentration in a suntan lotion base produced erythema and a scarification. Rabbits failed to gain any weight over a 17-day period of application. The use of suntan lotion base obscures the conclusions that can be drawn from this study. Further rabbit studies with hydroquinone monomethyl ether (116) have shown that the undiluted material can cause considerable necrosis if exposure is not limited to 1 day. Prolonged contact can cause a severe burn as shown in the rabbit. There is also an indication in this study that the material was absorbed in toxic amounts when in solution, especially through abraded skin. Fassett (157) found this material was not a skin sensitizer in guinea pigs.

Riley (158) reported depigmentation of the skin of black guinea pigs within 5 to 10 days of the start of treatment using a 20 percent concentration of hydroquinone monomethyl ether (4-hydroxyanisole) in lanolin. This was a comparative evaluation which showed that 3-hydroxyanisole had a weak depigmenting effect and that guaiacol (2-hydroxyanisole) had no depigmenting effect. Riley (159) conducted in vitro experiments using primary cultures of normal guinea pig melanocytes to elicit the mechanism of depigmentation. He proposed that a tyrosinase oxidation product from hydroquinone monomethyl ether caused cell damage by initiating lipid peroxidation. Mahler and Cordes (160) had previously found the conversion of quinone to melanin to depend on copper-containing enzymes. Hydroquinones interfere with that enzymatic conversion. Riley proposed that the melanocyte toxicity produced by hydroquinone monomethyl ether may result from semiquinone radicals formed by tyrosinase metabolism (161).

Woods and Smith (162) reported investigations in which a 20 percent concentration of hydroquinone monomethyl ether in lanolin was applied three times a week for a period of 45 days in the cheek pouches of golden Syrian hamsters. Periodic biopsies were taken and histopathological changes at the epithelial connective tissue junction were reported to be comparable with disturbances observed in developing experimental hamster cheek pouch carcinomas. Experiments conducted to confirm these observations as benign or malignant cancer changes have not been found.

6.6 Hydroquinone Monoethyl Ether (p-Ethoxyphenol)

6.6.1 Physiological Response

Brun (163) demonstrated the depigmentation effects of hydroquinone monoethyl ether by repeated applications of this chemical over a 5-week period to the skin of pigmented guinea pigs at a concentration of 15.5 percent in vaseline. Frenk (164) demonstrated that this chemical is capable of depigmenting both the fur and skin of black guinea pigs. A subsequent investigation by Frenk and Ott (165), using pigmented guinea pigs, pigmented mice, and Syrian golden hamsters, showed that a 10 percent concentration of hydroquinone monoethyl ether in white petrolatum was capable of depigmenting skin and fur after prolonged application. Microscopically it was noted that the dopa-positive epidermal melanocytes were greatly reduced after weeks of treatment.

6.7 Hydroquinone Dimethyl Ether (1,4-Dimethyoxybenzene)

6.7.1 Acute Exposure

Physiological responses after acute exposures to hydroquinone dimethyl ether are presented in Table 8.12 (156, 157).

6.7.2 Chronic Exposure

Hodge et al. (156) reported maintaining rats for 7 weeks on a diet containing hydroquinone dimethyl ether. The females on a diet of 0.5 percent produced questionable effects on growth with more definite depression when the concentration was 2 percent or greater. As in the case of the monomethyl ether, no other deleterious effects were noted at these levels and the authors concluded that the flavor may have reduced the palatability of the ration and thus reduced food intake. Again, support for this theory was gained by a comparison with the rabbit-feeding studies wherein this compound at a level of 10 percent in the diet caused little or no growth depression. Once again there were no deleterious effects noted in dogs fed this ether in amounts up to 6 g/day, and 12 g/day caused only minor body weight losses. Hodge et al. (156) studied the effects of hydroquinone dimethyl ether on rabbit skin. It was applied at a 10 percent concentration in suntan lotion base. As in the case of the monomethyl ether, some erythema and scarification developed, and the rabbits failed to gain weight during a 17-day application period. As before, however, the use of suntan lotion base clouds the conclusions that can be drawn from this study. Fassett (157) found that this material was not a sensitizer in guinea pigs and that 40 percent solutions in olive oil and acetone caused only slight or moderate irritation to guinea pig skins on a single 24-hr contact.

6.8 Hydroquinone Monobenzyl Ether (p-Benzyloxyphenol)

This material is of interest because of its ability to inhibit melanin formation in the skin and to cause depigmentation on local contact. This quality was first noted because of its use as an antioxidant in the rubber used for making gloves. Its use in rubber also caused considerable contact dermatitis. Therefore skin contact must be avoided.

6.8.1 Human Experience

Denton et al. (166) and Blank and Miller (167) refer to many cases of dermatitis and leukoderma that were evidently caused by skin contact with rubber that contained hydroquinone monobenzyl ether as an antioxidant. Lerner and Fitzpatrick (168) report a 13 percent incidence of contact dermatitis cases where this compound has been used for depigmentation purposes. They have reviewed the use of this compound for such treatment and felt it is useful under correct therapeutic use. Kelly et al. (169) found that cancer patients tolerated ingestion of up to 10 g/day of this compound without toxic symptoms. Nausea and vomiting occurred above this dose. Its use for depigmentation is therefore limited by its ability to cause contact dermatitis rather than its effect when absorbed internally. Catona and Lanzer reported a confetti-like depigmentation in three clinical cases (170).

Studies of this ether's effects on the eyes have not been located.

6.8.2 Animal Studies

The acute oral LD_{50} of the sodium salt of this compound is presented in Table 8.12 (157). Subacute exposures by Denton et al. (166) report that no toxic changes were observed in guinea pigs ingesting this compound at the level of 160 mg/kg daily for 2 months for a total dose of 4.37 g. The only change noted was a depigmentation of the hair. Peck and Sobotka (171) fed the same species approximately 12 g over a period of 5 months with neither deleterious effects nor depigmentation of skin or hair.

Lerner and Fitzpatrick (168) reported that no toxic changes were observed in mice that were injected intraperitoneally with 600 mg/kg of hydroquinone monobenzyl ether daily for 3 weeks.

Brulos et al. (172) incorporated a 5 percent concentration of hydroquinone monobenzyl ether in ethanol and applied it to the skin of guinea pigs, and determined that this material was nonsensitizing.

6.8.3 Metabolism

Fassett (157) has predicted that hydroquinone monobenzyl ether, by its low chronic toxicity, may readily be metabolized in the same manner as phenol.

6.8.4 Mode of Action

Peck and Sobotka (171), Schwartz et al. (173), and Lea (174) have studied the mechanism of the action of hydroquinone monobenzyl ether with regard to depigmentation of the skin. Mahler and Cordes (160) identified the conversion of quinone to melanin as being dependent on copper-containing enzymes. Hydroquinones interfere with that enzymatic conversion.

Recently, Bleehen and Riley (175) suggested that compounds that have nonpolar side chains diffuse into the melanosomes of the pigment cell where they are oxidized by tyrosinase to form a semiquinone free radical. This then diffuses out into the cytoplasm of the cell and initiates a chain reaction of lipid peroxidation, involving the lipids of the cell membrane. This produces widespread cellular damage leading to the death of the melanocyte. There is evidence that these cells are destroyed by the substituted phenols as a result of a specific lethal synthesis.

6.9 Eugenol (1-Allyl-3-methoxy-4-hydroxybenzene, Allyl Guaiacol)

6.9.1 Determination in the Atmosphere

A method for the collection and determination of eugenol in the air has not been found. It could be expected, however, that the material could be trapped by passing air through 1 N caustic solution or a suitable organic solution, such as liquid trapping media. The method of Meyer and Meyer (176) using ultraviolet spectrophotometry may be used to yield results accurate to concentrations as small as 0.005 mg/ml. Additionally, the method recommended by Korany et al. (177) may be modified to produce a reaction of sodium cobaltinitrite with eugenol. Measurements of absorbance at 450 nm for eugenol can then be determined. Other spectrographic and chromatographic methods (178) should also be applicable. Physical and chemical properties are presented in Table 8.1.

6.9.2 Physiological Response

6.9.2.1 Human Eugenol has been repeatedly identified as a sensitizing agent (179–181) when used in dentistry. However, Opdyke (182) has summarized the results of irritation and sensitization evaluations by several investigators. He reports that eugenol is either nonirritating or only mildly irritating and that in repeated sensitization evaluations, it is shown to be negative.

6.9.2.2 Animal

Acute Tests. The acute toxicity data available on eugenol are presented in Table 8.12 (183–185). In other studies, dogs exhibited vomiting after single doses of 250 or 500 mg/kg. Death occurred at the high level. Pulmonary edema was noted in some dogs exposed intravenously to eugenol. Intravenous injection of eugenol diluted to 1:20 transiently decreased the systemic arterial blood pressure and myocardial contractile force, impaired motor activity, and increased salivary flow. The stomachs of rats and guinea pigs given oral doses of eugenol showed desquamation of the epithelium, with punctate hemorrhages in the pyloric and glandular regions of the stomach. Additional evaluations to mucous membranes showed that application of eugenol to the ventral surface of the tongue of dogs caused erythema and occasionally ulcers with a diffuse inflammatory infiltration. The intratracheal instillation LD_{50} of eugenol in rats and hamsters is 11 mg/kg and 17 mg/kg, respectively (186). The inhalation LC_{50} in rats is greater than 2.58 mg/l (est. 240 mg/kg) (187).

Subacute Tests. Opdyke summarized the gross findings in several species of animals after subacute exposure to eugenol (182). Rats given four daily oral doses of approximately 900 mg/kg showed minor liver damage. No liver damage was observed in rats fed eugenol at 1 percent in the diet for about 4 months. Feeding of eugenol at 0.1 or 1 percent in the diet in groups of 10 male and 10 female rats for 19 weeks exhibited no effect on growth, hematology, or organ weights and histology. No adverse effect was observed in a group of 15 male and 15 female rats fed eugenol at 79.3 mg/kg of body weight per day for 12 weeks.

In a group of 20 male rats given an initial oral dose of 1.4 g eugenol/kg, which was gradually increased to 4.0 g/kg, eight rats survived 34 days, and 15 rats lived long enough to receive the maximum dose. Slight enlargement of the liver and the adrenal glands was observed and histological examination of the forestomach revealed moderately severe hyperplasia and hyperkeratosis of the stratified squamous epithelium, with focal ulceration. A small degree of osteoporosis was also seen.

Chronic Tests. Rats and mice were fed a diet of eugenol up to 6000 ppm in a 2-year study. Only mice exhibited statistically significant liver tumors (188). Young questioned the product related interpretation of the low-dose male mouse tumor incidence (189). He considered the housing placement in the test facility as a possible cause of male mouse liver tumors. The IARC considers eugenol to be in Group 3, not classifiable as to its carcinogenicity to humans (190).

Neurology. Ozeki (191) evaluated eugenol on the nerve and muscle in crayfish. Using concentrations up to 1000 ppm in water, he reported that eugenol is an effective reversible anesthetic. Conversely, Kozam (192) used the frog's sciatic nerve and exposed it to eugenol at concentrations as low as 0.05 percent (500 ppm) and found there was no reversal in the anesthetic effect after 3 hr. His conclusion was that eugenol is neurotoxic and its anesthetic activity is nonreversible at this concentration.

6.10   Isoeugenol (1-Propenyl-3-methoxy-4-hydroxybenzene)

6.10.1   Determination in the Atmosphere

The analytical procedures that are recommended for the determination of eugenol in the atmosphere are also recommended for isoeugenol. Physical and chemical properties are presented in Table 8.1.

6.10.2   Physiological Response

6.10.2.1 Human Tests

Irritation and Sensitization. Opdyke (193) summarized the irritation and sensitization potential of isoeugenol. Isoeugenol tested at 8 percent in petrolatum produced mild irritancy after a 48-hr closed-patch test in human subjects. Isoeugenol in concentrations of 2 and 5 percent in vaseline or ointment produced erythema in 1 of 30 normal human subjects and 3 of 35 normal human subjects, respectively. At a concentration of 0.1 percent in 99 percent ethanol, erythema resulted in 1 of 54 human subjects. A maximization test was carried out on 25 human volunteers to evaluate sensitization. The material was applied at a concentration of 8 percent in petrolatum and produced no sensitization reaction.

6.10.2.2 Animal Tests

Acute. Acute toxicity responses of isoeugenol are presented in Table 8.12 (185, 193). Opdyke (193) provided a summary of the toxicity of isoeugenol. Acute oral administration of isoeugenol in rats caused the animals to become comatose, with a scrawny appearance that persisted for several days, and death occurred within the first week after exposure. Isoeugenol injected intraperitoneally into mice promptly produced a distinct hypothermia, the maximum lowering of rectal temperature for a dose of 0.1 g/kg being 4°C in 30 min. Isoeugenol altered performance in rats by producing depression and hindquarter paralysis. It does not produce catatonic effects and it does not alter spontaneous motor activity in mice at 100 mg/kg.

Subacute. Opdyke (193) also summarized the subacute toxicity of isoeugenol. When it was given orally to rats at 520 mg/kg daily for 4 days, one out of six animals died and no macroscopic liver lesions were observed. Rats receiving 10,000 ppm of isoeugenol in the diet for 16 weeks showed no effect on growth or hematology and no microscopic changes in tissues. A guinea pig sensitization study was conducted at a concentration of 0.1 percent suspension of isoeugenol. Intradermal injections daily for 10 days produced a moderate skin reaction when a challenge dose of 1 percent isoeugenol in peanut oil was applied.

Percutaneous Absorption. Isoeugenol was not absorbed within 2 hr by the intact, shaved abdomen of the mouse (193).

Anesthetic Action. Sell and Carlini (194) reported that intraperitoneal injections of 200 mg/kg of isoeugenol in the mouse had a sleep induction time of 1 min and a total sleep time of 7.6 min. No deaths occurred during the subsequent 24 hr.

6.11   Methyleugenol

6.11.1   Physiological Response

Results of acute exposures to methyleugenol are presented in Table 8.12 (185, 195, 196). The anesthetic action of methyleugenol in mice, rats, and rabbits was evaluated by Sell and Carlini (194). An intraperitoneal injection of 200 mg/kg of methyl eugenol was administered to mice. The induction time to sleep was 2.1 min and the total sleeping time was 23.2 min. No death occurred during the subsequent 24-hr period. Rats given a 200 mg/kg intraperitoneal dose of methyleugenol had an induction to sleep time of 2.7 min and a total sleep time of 49.3 min. No deaths occurred at this dose. Continuous intravenous infusion of methyleugenol in rabbits produced a loss of righting reflex in 0.7 min. After discontinuance of intravenous infusion, recovery occurred in 11.55 min. During this experiment loss of the corneal reflex occurred after 4.22 min and recovery after cessation of the intravenous infusion occurred at 2.73 min. Further investigation in rats and mice receiving daily intraperitoneal injections of 100 ml/kg of methyleugenol for 26 and 42 days, respectively, produced no changes in mice, but the body weight gain of rats was significantly reduced. However, withholding the exposure resulted in weight increase in these rats.

Additional pharmacological evaluation of methyleugenol was conducted by Engelbrecht et al. (196). Mice were used to evaluate the potential for methyleugenol to induce the inability to regain posture after a single exposure. Loss of the righting reflex (LRA) as the end point was determined when the animals were unable to regain their proper posture for 30 sec or longer. The loss of righting ability (LRA_{50}) after intraperitoneal exposure was 88 mg/kg. The intravenous dose (LRA_{50}) for methyleugenol was 26 mg/kg. A rotating rod was used to evaluate the ability of animals to stay on this rod while it was rotating for a period of at least 1 min. The intraperitoneal neurotoxic dose (NT_{50}) for methyleugenol was 72 mg/kg.

6.12 Methylisoeugenol

6.12.1 Physiological Response

The physiological responses, after acute exposures to animals for methylisoeugenol, are presented in Table 8.12 (185, 196). Engelbrecht et al. (196) evaluated the loss of righting ability median dose (LRA_{50}) was 174 ml/kg. The intravenous dose was 33 ml/kg. The neurotoxic dose (NT_{50}) in mice after intraperitoneal injections was 122 mg/kg. Sell and Carlini (194) evaluated the anesthetic action of methylisoeugenol by administering intraperitoneal injections of 200 mg/kg in mice. The time to induction of sleep and sleep time were not evaluated. Eight out of 10 mice died at this dose level.

6.13 Butylated Hydroxyanisole (BHA)

6.13.1 Determination in the Atmosphere

Although there have been few occasions to determine air concentrations during the use of BHA, this can be done by collecting it in dilute caustic solution or in organic solvents and then employing colorimetric methods suitable for phenolic materials, ultraviolet spectrophotometry, or polarographic methods for its analysis (175). Physical and chemical properties are presented in Table 8.1.

6.13.2 Physiological Response

6.13.2.1 Human Experience

Fisherman and Cohen (197) described human sensitization to BHA plus BHT (butylated hydroxytoluene) in patients fed 125 mg of each chemical. After challenge, acute symptoms consisted of exacerbation of vasomotor rhinitis, headaches, asthma, flushing, suffusion of the conjunctivas, occasional dull retrosternal pain, marked diaphoresis, and somnolence. Cloninger and Novey (198) were able to confirm only the somnolence response. Fisher (199) summarized reports of sensitization to BHA by allergic contact dermatitis as being of low incidence. Patch test reports by Roed-Petersen and Hjorth (200) and Turner (201) also report that BHA produces allergic contact dermatitis. No industrially related problems have been found in the technical literature on BHA. A thorough summary of the biologic effects in humans has been reported. Although BHA is not an irritant, it can cause skin reactions owing to its weak allergic sensitization potential (202).

6.13.2.2 Animal

Reproduction and Teratology. Six female rhesus monkeys were fed equal amounts of BHA and BHT at a total dose of 100 mg/kg (50 mg/kg each) for 2 years. During this time no abnormal findings were noted for hematology, serum cholesterol, sodium, potassium, total protein, serum glutamic pyruvic transaminase, or serum glutamic oxaloacetic transaminase. The monkeys maintained normal menstrual cycles and their food consumption and body weights were unaffected. During the course of the second year, five of the six experimental animals gave birth to normal infants (203). Clegg (204) evaluated BHA in rats and mice at dose levels up to 1000 and 500 mg/kg, respectively, for teratological effects. No abnormal findings attributable to the exposures were noted. Hansen and Meyer (205) evaluated the teratological effects of BHA in rabbits at dose levels ranging from 50 to 400 mg/kg. No abnormal findings were reported. A review on BHA included unpublished reports of maternal toxicity in the absence of teratology effects (202).

Neurology. Stokes et al. (206) showed that intraperitoneal exposure to BHA at levels ranging from 5 to 500 mg/kg produced neurological changes in mice by increasing serotonin utilization in the central nervous system. Stokes and Scudder (207) exposed mice prenatally and postnatally to 0.5 percent (w/w) BHA in their diet. The treated animals exhibited increased exploration, decreased sleeping, decreased grooming of self, slow learning, and a decreased orientation reflex. The authors related these behavioral changes to altered levels of neurotransmitters. Vorhees et al. exposed rats to BHA in the diet prior to, during, and after pregnancy to evaluate developmental neurobehavioral effects (208). Newborn toxicity was observed as early death. Additionally, early growth retardation recovered prior to weaning. Neurobehavioral changes included delayed onset of the startle reflex in the fetus. However, the reflex did develop normally. No long-term behavioral abnormalities were reported.

Liver Function. Creaven et al. (209) reported no effect in liver or enzyme activity in rats fed 0.1 and 0.25 percent BHA in their diet for 12 days. At a level of 0.5 percent liver weight was not affected but enzyme activity was increased. Rat feeding studies by Gaunt et al. (210) produced liver enlargement and increased urinary ascorbic acid after six daily doses of 500 mg/kg of BHA. Adrenal weights were also elevated. It was concluded that ascorbic acid excretion in the urine may be a way of differentiating hyperfunctional liver enlargement from toxic liver enlargement. Gilbert and Golberg (211) reported increases in relative liver weight and urinary ascorbate in rats after 1 week of treatment with 500 mg/kg of BHA. After 12 weeks of this treatment liver weight increase is persistent. Allen and Engblom (212) reported on ultrastructural and biochemical liver changes in monkeys given 500 mg/kg of BHA for 28 days.

Metabolism. Absorption of BHA from the digestive tract is by passive diffusion (213). Branen et al. (214) reported changes in lipid metabolism in monkeys exposed to 50 and 500 mg/kg of BHA for 4 weeks. Total liver cholesterol was lowered at both levels and total plasma cholesterol was lowered at the high dose level. The metabolism of BHA in the rabbit has been studied by Dacre and Denz (215) and in the rat by Astill et al. (216). The metabolic pattern is similar in these species, the principal excretion product in the urine being a glucuronide ester. The excretion is relatively rapid and complete even at high dose levels. Studies have shown a similar effect in humans (217). Wilder et al. (218) showed that in the dog BHA is conjugated and excreted both as a glucuronide and as the ethereal sulfate.

Chronic Toxicity. A number of long-term feeding studies have been carried out. Wilder and Kraybill (219, 220) found that rats could tolerate up to 0.12 percent in their diet (1200 ppm) for a period of 21 months. No histopathological or carcinogenic effects were found to be associated with the ingestion, nor were there any effects on reproduction. Attempts to feed higher levels resulted in refusal to eat. There was no evidence of storage of BHA in fatty tissues of these rats. Brown et al. (221) also found no evidence of toxic effects in rats fed up to 0.1 percent (1000 ppm) in the diet for 2 years; at 0.5 percent (5000 ppm) slight initial growth changes and slight liver weight increases were present without histopathological changes.

Graham et al. (222) fed rats a diet consisting of about 75 percent bread. The bread contained among other chemical additives BHA and other antioxidants in an amount 50 times greater than the normal use level. No toxic effects were discernible after a year's feeding. The daily dosage level of BHA was from 3.3 to 7.0 mg/kg of body weight.

Hodge and Fassett (223) fed dogs up to 100 mg/kg for a period of more than a year. No pathological changes resulted from the treatment and there was no evidence of storage of BHA in brain, liver, kidney, or fat. Dogs fed an antioxidant mixture containing BHA and hydroquinone over long periods also showed no toxic effects (220).

Wilder et al. (218) maintained groups consisting of four cocker spaniel pups each for 15 months on diets that contributed a daily dosage of 0, 5, 50, and 250 mg/kg of BHA. Those animals receiving the highest dosage level ate poorly, probably because of objectionable taste, grew poorly, and exhibited signs of toxicity characterized by the presence of sugar in the urine and definite liver injury. When judged by general appearance, behavior, growth, hematologic studies, urine examination, and gross and microscopic examination of tissues, all the other groups of animals were normal. Hodge et al. (224) reported exposing mice to a single subcutaneous injection of BHA diluted in trioctanoin (50 mg/ml). A single subcutaneous injection of 0.02 ml was administered to each animal. The mean survival time in this study was 508 days for males and 408 days for females. In a separate evaluation BHA was used in a 5 percent solution in acetone for skin application studies to mice. The low dose in the evaluation was 0.05 percent solution. A volume of 0.20 ml of the acetone solution was applied once each week to the animals. The mean survival time for males was 519 days and 323 days for females exposed to BHA. In both evaluations BHA showed no carcinogenic potential.

Carcinogenicity. Hodge et al. (224) summarized the chronic exposure studies conducted on BHA up to 1966, indicating that there is little likelihood that BHA is a carcinogen. Subsequent reports have described the cancer-inhibiting character of BHA (225–232). Reports of carcinoma in the forestomach of rats (233) and papillomas in the forestomach of hamsters (233) fed dietary BHA up to 2 percent for 2 years were published from Japan. Additionally, rats fed 2 percent BHA in the diet for 24 months developed epithelial papillomas of the forestomach that recovered if they proliferated upward from basal cells (235). If proliferation from basal cells was downward no reversibility was observed. The induction of mice forestomach carcinomas was equivocal (236). Williams et al. concluded that BHA produced forestomach papillomas in rats fed 12,000 ppm of BHA in the diet for 110 weeks (237). These investigators concluded that BHA is not an initiator of cancer development but it is a promotor of cancer development through epigenetic mechanisms due to high dose exposure.

6.14 Vanillin (3-Methoxy-4-hydroxybenzaldehyde, Vanillic Aldehyde, Methylprotocatechuic Aldehyde)

6.14.1 Determination in the Atmosphere

Vanillin can be trapped from the air by scrubbing with methanol. Methanol solutions can be analyzed by ultraviolet spectrophotometry. Strong absorption bands occur from vanillin at 308 and 278 nm. Concentrations of about 0.01 mg/100 ml of methanol can be measured accurately. There are methods described in the literature for the detection of vanillin and ethyl vanillin in foods (238, 239) and for the differentiation of these chemicals from each other and from related materials (240–244). Physical and chemical properties are presented in Table 8.1.

6.14.2 Physiological Response

6.14.2.1 Human Experience

Opdyke (245) prepared a summary of human exposure to vanillin which described both irritation potential and sensitization potential. In closed-patch tests on human skin vanillin caused no primary irritation when tested at concentrations of 20 percent on 29 normal subjects, at 2 percent on 30 normal subjects, and at 0.4 percent on 35 subjects with dermatoses. Maximization tests were conducted on groups of 25 volunteers. The material was tested at concentrations of 2 and 5 percent in petrolatum and produced no sensitization reactions.

Vanillin applied undiluted for 48 hr in the standard occluded aluminum-patch test used by the North American Contact Dermatitis Research Group did not produce any irritation or sensitization in a 62-year-old subject with a perfume dermatitis.

Positive reactions to vanillin were reported in 8 out of 142 patients who were already sensitized to balsam of Peru. In studies of sensitization to balsam of Peru and its components, vanillin, pure or 10 percent in vaseline, produced positive patch-test reactions in 21 out of 164 patients sensitive to the balsam. Vanillin was considered to be a secondary allergen because sensitivity was found only in patients sensitive to vanilla, isoeugenol, and coniferyl benzoate. Cross-sensitization to other benzaldehydes was particularly uncommon. Vanillin was found not to be responsible for most cases of sensitivity to natural vanilla.

Vanillin, which appears on the published list of 400 Canadian food additives and is used in artificial flavorings, is known to cause reactions in people previously sensitized to balsam of Peru, benzoin, rosin, benzoic acid, orange peel, cinnamon, and clove. Odor detection is reported in Table 8.4.

6.14.2.2 Animal

Acute Toxicity. Physiological responses for acute exposure to vanillin are presented in Table 8.12 (245). Robinson et al. found vanillin to be a moderate skin allergic sensitizing agent (246).

Subacute and Chronic Toxicity. Opdyke (245) summarized a series of evaluations that describe the subacute and chronic toxicity data that have been generated on vanillin. Intragastric administration of 300 mg of vanillin/kg to rats, twice weekly, for 14 weeks produced no adverse effects. Groups of 16 rats were fed diets containing vanillin at levels to provide 20 mg/kg body weight per day for 18 weeks without any adverse effects, but 64 mg/(kg)(day) for 10 weeks caused growth depression and damage to the myocardium, liver, kidney, lung, spleen, and stomach. When 10 male and 10 female rats were fed diets containing 0.3, 1.0, or 5.0 percent vanillin for 13 weeks there was growth depression and enlargement of liver, kidney, and spleen at the highest level, mild changes at 1.0 percent, and none at 0.3 percent. In another study matched groups of 10 male and 10 female rats, 4 to 6 weeks old, were maintained for 91 days on diets containing up to 50,000 ppm vanillin, equivalent to about 2500 mg/(kg)(day). Records of appearance, behavior, growth, mortality, terminal body and organ weights, terminal hematologic examination, and histological studies revealed no adverse effects when the diet contained 3000 ppm vanillin, equivalent to as much as 150 mg/(kg)(day). Mild adverse effects followed ingestion of a 10,000 ppm diet, and at 50,000 ppm growth was depressed and the liver, kidneys, and spleen were enlarged.

Fed to rats at dietary levels of 10,000 ppm for 16 weeks, 1000 ppm for 27 to 28 weeks, 20,000 or 50,000 ppm for 1 year, or 5,000, 10,000, or 20,000 ppm for 2 years, vanillin had no effects on growth or hematology and produced no macroscopic or microscopic changes in the tissues.

Rats fed for 5 weeks on a diet containing a mixture of the maximum permissible amounts of 15 compounds, including vanillin at 0.5 g/kg of the diet, showed symptoms of intoxication, including decreases in adrenal vitamin C and in liver protein.

Vanillin injected intraperitoneally into strain A mice in total doses of 3.6 to 18.0 g/kg over a period of 24 weeks produced no excesses of lung tumors and was not considered to be carcinogenic.

Metabolism. Wong and Sourkes (153) demonstrated that vanillin administered intraperitoneally to rats gave rise to a number of urinary products; chief among these was vanillic acid in both free and conjugated forms. Other metabolites were conjugated vanillin, conjugated vanillyl alcohol, and catechol.

Opdyke (245) summarized other metabolic information that has appeared in the literature. In addition to the metabolites already described, he reported that when vanillin was fed to rats at doses of 100 mg/kg most metabolites were excreted in the urine within 24 hr, chiefly as glucuronide and/or sulfate conjugates, although the acids formed were also excreted free and as their glycine conjugates. In 48 hr 94 percent of the dose was accounted for, 7 percent as vanillin, 19 percent as vanillyl alcohol, 47 percent as vanillic acid, 10 percent as vanilloylglycine, 8 percent as catechol, 2 percent as 4-methylcatechol, 0.5 percent as guaiacol, and 0.6 percent as 4-methylguaiacol.

6.15 Ethyl Vanillin (3-Ethoxy-4-hydroxybenzaldehyde, Vanillal, Ethylprotocatechuic Aldehyde, Burbonal)

6.15.1 Determination in the Atmosphere

Ethyl vanillin may be trapped from the air by a scrubbing procedure as described for vanillin. Analytical technique recommended for vanillin also apply to ethyl vanillin. Physical and chemical properties are presented in Table 8.1.

6.15.2 Physiological Response

Human Experience

Opdyke (247) has summarized the irritation and sensitization information available on ethyl vanillin. He has noted that when tested as 2 percent in petrolatum, ethyl vanillin produced mild irritation after a 48-hr closed-patch test in 25 human subjects. A maximization test was carried out on 25 volunteers. The material was tested at a concentration of 2 percent in petrolatum and produced no sensitization reactions. Odor detection is reported in Table 8.4.

Animal

Acute Toxicity. Opdyke (247) has summarized the acute toxicity data available on ethyl vanillin. This information is presented in Table 8.12.

Subacute and Chronic Toxicity. Opdyke (247) has summarized available studies on ethyl vanillin. When 20 mg/(kg)(day) was fed to rats for 18 weeks no effects were produced, whereas rats fed 64 mg/(kg)(day) for 10 weeks showed a reduction in growth rate and mild cardiac, renal, hepatic, lung, spleen, and stomach injuries.

Neither 20,000 and 50,000 ppm of ethyl vanillin fed to male rats in the diet for 1 year, nor 5000, 10,000, and 20,000 ppm fed to male and female rats in the diet for 2 years produced any effects.

6.16   Phenyl Ether (Diphenyl Oxide, Phenoxybenzene, Diphenyl Ether)

6.16.1   Determination in the Atmosphere

Phenyl ether can be trapped from the air by scrubbing with methanol or isooctane. Such a solution can be analyzed by ultraviolet spectrophotometry. Experience has shown that concentrations as low as 0.05 mg of phenyl ether per 100 ml of methanol can be measured accurately at wavelengths of either 271 or 278 nm when using a 10-cm cell. When phenyl ether reacts with piperonal chloride in trifluoroacetic acid it produces a red color with a maximum absorption at a wavelength of 537 nm (see Table 8.3). Physical and chemical properties are presented in Table 8.1.

6.16.2   Physiological Response (247a)

Ethyl vanillin appears to be low in oral toxicity, it is not appreciably irritating to the skin, and its vapors do not present a toxicologic problem, but may be a nuisance because of its disagreeable odor. Odor detection is reported in Table 8.4.

6.16.2.1 Human Experience

Experience in manufacturing, handling, and selling this material over many years has indicated that the material does not present an appreciable hazard to health as ordinarily used.

6.16.2.2 Animal

Acute Oral. Physiological response to acute oral exposure of phenyl ether is presented in Table 8.12 (116). It is reported that injury to the liver, spleen, kidney, thyroid, and intestinal tract in surviving animals has occurred.

Repeated Oral. Repeated feeding studies with phenyl ether have been conducted that produce organic injury, but sufficient details were not found in published literature to allow a discussion of the results.

Skin Irritation. Skin irritation tests conducted on rabbits indicate that the undiluted material is somewhat irritating if exposures are prolonged or repeated. Effects from such an exposure are characterized by erythema and exfoliation, which clears promptly on cessation of exposure. When the material is diluted, as in perfume compositions, it does not appear to present any hazard of skin irritation.

Inhalation. The vapor toxicity of diphenyl ether has not been determined directly. However, some experiments on materials that consisted largely of this compound have indicated that vapor concentrations that can occur at ordinary room temperatures present no hazard of systemic injury. It should be noted that such concentrations are quite low because of the lower vapor pressure of the material and furthermore that such concentrations have an odor that may be disagreeable.

Metabolism. Phenyl ether, when injected into the rabbit, was excreted in the urine as *p*-hydroxyphenyl phenyl ether and none was excreted as the unchanged ether. Confirmation of these observations showed that in the rabbit no cleavage of the ether linkage occurs. The major metabolite was the *p*-hydroxylated ether, which is excreted both unconjugated (15 percent) and conjugated with glucuronic acid (63 percent) and sulfuric acid (12 percent). Another metabolite was also isolated from the rabbit urine and fairly well identified as di(*p*-hydroxyphenyl) ether.

6.16.3 Hygienic Standards of Permissible Exposure

In order to avoid complaints of disagreeable odor and, in some cases, nausea, it probably is necessary to control vapor concentrations to less than 1 ppm (see Table 8.2).

6.17 Phenyl Ether–Biphenyl Eutectic Mixture (Dowtherm A)

6.17.1 Determination in the Atmosphere

Dowtherm A can be trapped from air by scrubbing with methanol or isooctane. Ultraviolet spectrophotometry appears to offer the simplest method of analysis of such samples. In methanol, the biphenyl absorbs strongly at 247 nm and the phenyl ether at 271 and 278 nm. It is recommended, however, that because the band at 271 nm is more specific, it be employed if possible. Standards should be prepared with Dowtherm A. A sensitivity of 0.05 mg/100 ml can be expected. Also, see Table 8.3. Physical and chemical properties are presented in Table 8.1.

6.17.2 Physiological Response (247a)

The data from oral administration reveal a comparatively low systemic toxicity, sufficiently low that health problems would not be expected in the normal handling and use of Dowtherm A. Neither would a particular problem be anticipated in the food industry, where there might conceivably be accidental contamination of food. The acute toxicity is low so that the probability of harm from ingestion is remote, and the low solubility in aqueous media plus the outstanding odor and taste of Dowtherm A would cause contamination to be readily detected and excessive ingestion unlikely.

The hazard to health from inhalation appears to be low, for no adverse effects were detected in animals exposed to vapor concentrations that were quite painful, and approaching saturation. These data, together with considerable experience in use, indicate that vapor concentrations that are not painful or particularly disagreeable present no health hazards.

Skin irritation of more or less mild degree may be expected only from prolonged and repeated contacts. No adverse effect should result from occasional short contacts. The liquid may case pain and transient irritation of the eyes.

6.17.2.1 Human Experience

Dowtherm A has been used rather widely in diverse fields for a considerable number of years without causing serious difficulty due to toxicity. There have been occasional cases of mild skin irritation, as well as complaints in regard to the unpleasantness of the odor.

6.17.2.2 Animal

Oral Tests. Rats and guinea pigs were fed by intubation single doses of Dowtherm A as manufactured. Guinea pigs were fed Dowtherm A that had been in commercial use in a boiler for 5 years. The doses of the unused material permitting survival and causing the death of all rats were 2.0 and 4.4 g/kg, respectively; for guinea pigs these doses were 0.3 and 4.4 g/kg. When the used material was fed the corresponding dosages were 1.0 and 4.4 g/kg, indicating no change in acute oral toxicity as a result of prolonged commercial use.

The LD_{50} for Dowtherm A when administered as a 25 percent solution in olive oil is 5.66 ± 1.28 g/kg for the rat. Degenerative changes in the liver and kidneys in surviving animals were observed.

Small groups of rats were given repeated doses by stomach tube of either unused Dowtherm A or the material that had been in use for 5 years. In each case doses of 0.5 to 1.0 g/kg were given 5 days a week until 132 doses were administered. Those rats receiving 0.5 g/kg daily showed only a very slight depression of growth and slight weight increases of the liver and kidneys; there were no histopathological changes. Those rats receiving 1.0 g/kg daily showed a moderate depression of growth, with slight to moderate histopathological changes in the kidneys and slight changes in the liver and spleen. The kidneys showed degeneration of tubules and hyaline cast formation. A striking observation in those rats receiving 1.0 g/kg was the gross liver enlargement, up to 188 percent of normal, with no morphological abnormalities. There were no changes in the blood cells nor was there any evidence of irritation in the gastrointestinal tract. The results of repeated feeding experiments lasting for 4 weeks indicated that doses of 1.5 g/kg and larger resulted in deaths, and that doses of the order of 0.1 g/kg had no detectable adverse effect. Large doses causing death after two to four doses produced only slight to moderate histopathological change in the kidneys and minor changes in the liver.

Rats were fed daily doses of 50 and 100 mg in the diet [equivalent to 0.25 and 0.5 g/(kg)(day)] and moderate degenerative changes to the livers and kidneys were observed in about 2 months. These changes were neither increased nor extended in rats fed for periods up to 13 months.

Skin Contact. When applied daily to the ear of the rabbit, undiluted Dowtherm A produced only a slight irritation manifested by hyperemia, edema, exfoliation, hair loss, and enlargement of hair follicles. When applied daily to the rabbit's abdominal skin with a cloth bandage, undiluted Dowtherm A produced hyperemia and blistering in 3 days. There have been a very few persons who displayed an idiosyncrasy toward Dowtherm A. These cases are so rare that they are of no practical importance.

Eye Contact. Eye contact with liquid Dowtherm A causes pain and conjunctival irritation but no necrosis in the rabbit.

Inhalation. Concentrations of vapor sufficiently high to cause toxic effects from a single exposure of up to 7 hr duration are not attainable.

A group of 12 rats, eight guinea pigs, and one monkey was given repeated 8-hr exposures to an atmospheric concentration of 0.182 mg/l of Dowtherm A, partly as a vapor and partly as a fine mist. Most of the rats and guinea pigs became ill and many of them died between the 22nd and 35th exposures. The remaining animals were autopsied after the 37th exposure. Most of the effects observed were those of emaciation and starvation because the rats and guinea pigs refused to eat even when not exposed. The study of the organ weights of the animals indicated that there possibly may have been a slight increase in the weight of the liver on a g/100 g body weight basis. The results of blood studies made on the exposed rats showed that no significant changes occurred. It is noteworthy that during the first few exposures the monkey vomited in the chamber and showed other signs of nausea. However, after the feeding time was changed from immediately before exposure to after the exposure, much of the nausea was avoided. At the end of the experiment, the animal appeared to be in good condition. Hematologic studies and blood pressure measurements made during the course of the experiment revealed no evidence of adverse effect.

In a subsequent experiment, 15 rats, nine guinea pigs, four rabbits, and two monkeys were exposed 7 hr/day, 5 days/week, for a period of 6 months to a vapor concentration of 7 to 10 ppm (0.05 to 0.07 mg/l). These animals showed no adverse effect as judged by studies of mortality, growth, organ weights, and hematologic and histopathological observations. The vapors in these concentrations are extremely nauseating and even painful to the eyes and upper respiratory passages of human subjects.

Metabolism. Both biphenyl and phenyl ether are converted in the body largely to the 4-hydroxy compounds, and are excreted in the urine conjugated with glucuronic and sulfuric acids (see section above on phenyl ether metabolism).

6.17.3   Hygienic Standards of Permissible Exposure

In order to avoid complaints of disagreeable odor and, in some cases, nausea, it probably is necessary to control vapor concentration to less than 1.0 ppm and perhaps to less than 0.1 ppm for some persons. The exact level depends on the individuals concerned (see Table 8.2). Used phenyl ether–biphenyl mixtures may contain low concentrations of benzene.

6.17.4   Odor and Warning Properties

Experience with human subjects has indicated that concentrations of vapors of Dowtherm A ranging from 0.1 to 1 ppm are readily detected by the olfactory organs and are unpleasant, sometimes producing nausea. It would appear, therefore, that warning properties of Dowtherm A are excellent and will serve to prevent excessive exposure. Odor detection is reported in Table 8.4.

6.18 Bis(Phenoxyphenyl) Ether

6.18.1 Determination in the Atmosphere

The material probably can be removed from the air by scrubbing through a nonpolar solvent. Analysis of the resultant solution can probably be accomplished by ultraviolet spectrophotometry. Physical and chemical properties are presented in Table 8.1.

6.18.2 Physiological Response (247a)

Bis(phenoxyphenyl) ethers seem to be very low in toxicity by any of the routes likely to be encountered in industrial handling.

6.18.2.1 Human Experience

No cases of adverse effects due to the bis(phenoxyphenyl) ethers have been found nor would any be expected from ordinary industrial handling and use. A repeated-insult test was conducted with the undiluted mixed isomers of bis(phenoxyphenyl) ether on 50 human subjects of mixed racial background. It resulted in neither primary irritation, fatiguing reactions, nor allergenic responses in any of the subjects. Another evaluation studied the crystalline *para* isomer on 300 human subjects. There was no primary irritation or evidence of allergenicity.

6.18.2.2 Animal

Single-Dose Oral. The mixed isomers as a 20 percent solution in corn oil were fed to rats by intubation. Doses of 4 g/kg caused no deaths but did cause diarrhea and slight liver and kidney injury apparent 24 hr after feeding.

Repeated-Dose Oral. Matched groups of five male and five female young adult rats each were maintained for 31 days on diets containing 0, 0.1, 0.3, 1, 3, and 10 percent of the mixed isomers of bis(phenoxyphenyl) ether. The well-being of these animals was judged by gross appearance and behavior, food consumption, growth, mortality, hematologic determinations, final body and organ weights, and gross and microscopic examination of the tissues. By these criteria the females on the diet containing 0.1 percent of the test material were unaffected, whereas the males exhibited abnormal livers characterized by cloudy swelling in the central lobular region and slight necrosis. At the 0.3 percent level the females showed only slightly depressed growth and the males exhibited liver injury. As dosage increased, evidence of adverse effects became more prevalent in both sexes with poor growth, and injury to the liver, kidney, spleen, and testes. No hematologic changes occurred even at the highest dosage level.

Eye Contact. When a drop of the undiluted mixed isomer was placed in the rabbit eye it caused only very slight pain, which disappeared very quickly.

Skin Contact—Rabbits. The mixed isomers failed to cause any apparent irritation even when confined to the skin of rabbits for 14 days. Furthermore, there was no evidence observed to indicate that toxic amounts were absorbed. Two 48-hr exposures to the crystalline *para* isomer, 14 days apart, also failed to cause any irritation.

6.19 Chlorinated Phenyl Ethers (Monochlorodiphenyl Oxide to Hexachlorodiphenyl Oxide)

6.19.1 Determination in the Atmosphere

The chlorinated diphenyl ethers can be determined by direct combustion technique followed by absorption and titration of the liberated halogen. They also may be scrubbed from the air by appropriate solvents and determined spectrophotometrically using standard ultraviolet methods. Also, see Table 8.3. Physical and chemical properties are presented in Table 8.1.

6.19.2 Physiological Response (274a)

The chlorinated phenyl ethers are a class of synthetic organic compounds that present potential industrial handling hazards. In general, the toxicity of these materials increases with the degree of chlorination. The mono-, di-, and trichlorophenyl ethers do not present serious handling hazards, whereas the more highly chlorinated commercial products do. Toxic effects, which may result from exposure to the more highly chlorinated derivatives, may appear as either an "acneiform" dermatitis, a systemic intoxication, or both. The more common manifestation of excessive exposure to these materials is the skin condition, which develops at the site of exposure, resembles common acne in appearance, and may be accompanied by intensive itching. This condition generally is observed among men handling the chlorinated phenyl ether. However, toxicologic work on laboratory animals has indicated that the effect is cumulative, and that liver damage will result if the material is taken into the body repeatedly. Repeated inhalation of vapors or fumes from the higher chlorinated materials would be most likely to cause such an effect. Special handling precautions should be used, particularly when the higher chlorinated products are handled routinely or when heated.

A mixture of butylated monochlorodiphenyl oxide, which is predominantly monobutyl monochlorodiphenyl oxide, appears to be less toxic and less irritating than the nonbutylated forms. It also does not appear to be acneigenic or sensitizing. Repeated oral exposures have minimal, reversible effects and it is nonteratogenic. Biologic effects are presented in Table 8.13.

6.19.2.1 Human Experience

Experience with human subjects has not been extensive but is sufficient to show conclusively that exposure to even small amounts of hexachlorophenyl ether may result in appreciable acneiform dermatitis. No cases of systemic toxicity have been located.

Table 8.13. Biologic Effects of Butylated Monochlorodiphenyl Oxide Mixtures and Related Compounds

Compound	Species	Test	Results
		Acute Exposure	
Butylated monochlorodiphenyl oxide mixture (248a)	Rat	Oral LD$_{50}$	>10 g/kg
	Rat	Inhalation at 25°C, 50°C	No observable effect
	Rabbit	Eye Irritation	No observable effect
	Rabbit	Skin Irritation	No observable effect
	Rabbit	Ear acneigenicity	No observable effect
	Guinea pig	Skin sensitization	No observable effect
	Fathead minnow	96-hr static LC$_{50}$	15.4 mg/l
		Flow-through threshold	1.81 mg/l
	Daphnia magna	48-hr static LC$_{50}$	0.24 mg/l
	Rat (248b)	Pharmacokinetic profile, gavage	**Radio-labeled** compound is 90% absorbed from the gut. The elimination is biphasic and **75%** excreted in feces within 3 days. Monobutylated component half-life is 65 hr. Dibutylated component half-life is 71 hr
	Monkey (248b)	Pharmacokinetic profile, gavage	Radio-labeled compound was 90% recovered. Major route of excretion was urine. Animal was constipated

Monochlorodiphenyl oxide	Fathead minnow (248a)	96-hr static LC$_{50}$	1.75 mg/l
	Daphnia magna (248a)	Flow-through threshold	0.090 mg/l
		48-hr static LC$_{50}$	0.39 mg/l
		Subacute Exposure	
Butylated monochlorodiphenyl oxide mixture	Rabbit (248c)	Skin irritation	Very slight effect
	Rat (248c)	Diet feeding	At levels of 5 to 90 mg/(kg)(day) for 90 or 156 days, no treatment effect noted for demeanor, hematology, urinalysis, clinical chemistry, porphyrin excretion, or ophthalmology. Body weight reduction noted. Reversible liver and kidney changes noted at 45 and 90 mg/(kg)(day). Test material stored in fat
	Rat (248d)	Teratology	At levels of 500–1000 mg/(kg)(day), no teratogenic response was noted
	Rabbit (248d)	Teratology	At levels of 1 to 10 mg/(kg)(day), no teratogenic response was noted

6.19.2.2 Animal

Single-Dose Oral. The various chlorinated phenyl ethers were fed to guinea pigs by intubation. The results are summarized in Table 8.14 and show both the marked delayed effect and also the high toxicity of those materials containing four or more equivalents of chlorine.

Repeated-Dose Oral. Rabbits were fed by intubation five times a week for 4 weeks unless death intervened. The results obtained with the various chlorinated products are summarized in Table 8.15. In one instance noted, specially purified pentachlorophenyl ether was studied and the results indicated that it was appreciably less toxic than the regular reaction product. Presumably this reflects the presence of some higher chlorinated products in the regular commercial-grade material. This probably accounts for the high toxicity of the tetrachloro compound also.

The organic injury in all cases was centered in the liver and was characterized by congestion and varying degrees of fatty degeneration. The other organs did not appear to be injured.

Topical Contact. The effects of the various chlorinated diphenyl ethers were studied on rabbits. All the materials examined produced an irritation when applied to the skin. The more severe exposures (except in the case of hexachlorophenyl ether) produced a necrosis and sloughing. This acute reaction was followed by a hyperplastic reaction of the epithelium. Milder exposures of a more chronic nature, which failed to produce necrosis, did, however, produce the hyperplastic reaction in the epithelium. Progressive chlorination of the phenyl ethers modified the type of reaction produced. The ability to cause the necrotic type of reaction increased with chlorine content, apparently reaching the maximum at four equivalents of chlorine, then declining. Hexachlorophenyl ether produced a marked irritation but never definite necrosis. The ability to induce the epithelial hyperplasia increased throughout the series, the hexachlorophenyl ether being the most active. There was not a marked difference between the tetra-, penta-, and hexachloro derivatives, but there seemed to be a sharp increase in the ability to induce hyperplasia as the degree of chlorination increased from three to four or more equivalents of chlorine.

Hexachlorophenyl ether does not appear to be absorbed rapidly through the skin, but when exposures were prolonged and repeated there was definite evidence of toxic effects, weight loss, and liver injury. It should be noted, however, that the methods employed in these skin application studies were not such as to exclude the possibility of some ingestion.

6.19.3 Hygienic Standards for Industrial Exposure

No hygienic standards for the chlorinated phenyl ethers have been proposed. In the case of those materials having four equivalents of chlorine or more it may be recommended that special handling procedures be employed to prevent all possible exposure, particularly repeated exposure. The OSHA threshold limit for chlorinated diphenyl oxide is 0.5 mg/m³ (see Table 8.2).

Table 8.14. Chlorinated Phenyl Ethers Survey of Single-Dose Oral Feeding Studies on Guinea Pigs

| | After 4 Days | | After 30 days | |
| | Lethal Dose | Survival Dose | Lethal Dose | Survival Dose |
Material	(g/kg)	(g/kg)	(g/kg)	(g/kg)
1X	0.7	0.2	0.60	0.1
2X	1.3	0.4	1.00	0.05
3X	2.2	0.4	1.20	0.2
4X	3.0	0.4	0.05	0.0005
5X	3.4	1.8	0.10	0.005
6X	3.6	0.4	0.05	0.005

6.20 Decabromodiphenyl Oxide [Bis(pentabromophenyl) Ether, DBDPO, Decabromodiphenyl Ether]

6.20.1 Determination in the Atmosphere

Decabromodiphenyl oxide is a solid white powder at room temperature. Physical and chemical properties are presented in Table 8.1.

6.20.2 Physiological Response

Decabromodiphenyl ether is a flame-retardant material used in plastics, synthetic fiber, coatings, and adhesives. Being a solid at ambient temperature, it is considered to be a nuisance particulate.

Table 8.15. Chlorinated Phenyl Ethers Results of Repeated Oral Feeding of Rabbits

Material	Dose (g/kg)	No. of Doses	No. of Days	Effect
1X	0.1	19	29	None
2X	0.1	19	29	Mild liver injury
3X	0.1	5	12	Death
	0.05	20	29	Slight liver injury
	0.01	20	29	No effect
4X	0.05	4	10	Death
	0.005	20	29	Severe liver injury
5X	0.05	8	21	Death
Pentachlorophenyl ether (highly purified)	0.1	20	29	Moderate liver injury; no growth
	0.01	20	29	Slight liver injury
6X	0.001	20	29	No effect
	0.005	8	10	Death
	0.001	20	28	Severe liver injury
	0.0001	20	28	No effect

6.20.2.1 Human Experience

Repeated application of a 5 percent suspension of DBDPO in petrolatum, three times a week for 3 weeks, on the skin of 50 human subjects did not result in allergic sensitization responses during the induction phase or the challenge phase 2 weeks after the last application (249).

6.20.2.2 Animal

Acute Toxicity. The acute oral LD_{50} in rats for DBDPO was determined to be greater than 2 g/kg of body weight (249).

Effect on Skin and Eyes. DBDPO applied to the shaved, intact skin of rabbits as a one-time exposure or repeated exposures produced no response (249). Single or repeated exposures to abraded skin produced a slight redness and swelling at the application site. Dry, solid DBDPO instilled into the eyes of rabbits produced transient irritation to conjunctival tissue which disappeared within 24 hr.

Decabromodiphenyl oxide applied to the inner ear surface of rabbits for 1 month as a 10 percent chloroform solution did not show signs of bromoacne, even though redness and tissue sloughing were evident (249).

Subchronic Toxicity. Norris et al. fed male rats a diet containing up to 1 percent DBDPO [800 mg/(kg)(day)] for 30 days (249). Enlarged livers were observed grossly at the high dose level. Concomitant histopathology findings included centrilobular cytoplasmic enlargement and vacuole formation. Thyroid hyperplasia at the 1.0 and 0.1 percent dose levels was also reported. In other studies DBDPO was fed to rats and mice up to 100,000 ppm of the diet for 14 days and 50,000 ppm for 90 days. No abnormalities were reported (250).

Chronic. Rats and mice were fed up to 50,000 ppm in the diet for 2 years (250). The only biologically relevant finding for tumor formation was neoplastic nodules of the liver in male and female rats. Findings in mice were not statistically significant.

Metabolism. Rats were dosed by an unspecified route with 1 mg/kg of DBDPO suspended in corn oil (249). The excretion half-life was less than 24 hr. In 24 hr over 90 percent of the radio-labeled dose was detected in feces. Over 99 percent of the exposure dose was recovered in 48 hr. In a second study El Dareer et al. dosed rats orally and intravenously with radio-labeled DBDPO (251). After oral consumption 99 percent of the radio-labeled material was recovered in the feces and gut contents after 72 hr. Three unidentified metabolites constituted 1.5 to 28 percent of the material recovered. Liver weights were elevated. At 72 hr after the intravenous administration 74 percent of the dose was detected in feces. Of this amount 63 percent was metabolites and 37 percent was unchanged DBDPO.

Reproduction, Developmental Toxicity. Norris et al. reported no ill effects in a one-generation reproduction study in rats up to an oral dose of 3 mg/(kg)(day) of

DBDPO in the diet (249). Additionally, they reported on a rat developmental toxicity study. Rats were dosed orally by stomach tube from days 6 to 15 of pregnancy up to 1000 mg/(kg)(day). No teratogenic response was observed.

6.20.3 Hygienic Standards for Industrial Exposure

No permissible exposure limit has been established for DBDPO. It is a solid at ambient temperature and is considered a nuisance particulate. The International Agency for Research on Cancer (IARC) has classified DBDPO as Group 3, not classifiable as to its carcinogenicity to humans (252).

7 CYCLIC ETHERS

7.1 Cellulose Ethers

7.1.1 General Physical and Chemical Properties (247a)

The cellulose ethers—methylcellulose, hydroxypropyl methylcellulose, carboxy-methylcellulose sodium salt, ethylcellulose, and hydroxyethylcellulose—are all formed by reacting alkali-cellulose of predetermined average molecular weight with various materials to form the respective ether.

Each anhydroglucose unit of the cellulose polymer has three free hydroxyl groups that can be etherified. The degree to which this is effected and the nature of the substituent group influence markedly the physical properties, particularly solubility, of the product. The molecular weight of the alkali-cellulose markedly affects the viscosity of the final product. All the ethers are odorless, tasteless, and very stable chemically.

7.1.2 Determination in the Atmosphere

All the cellulose ethers may be trapped from the air by filtration through a membrane filter. The water-soluble ethers can also be trapped in cold water. Ethylcellulose can be trapped in organic solvents.

The determination of methylcellulose and ethylcellulose can be accomplished in various media for methoxyl and ethoxyl groups. ASTM methods D1347-56 for methylcellulose and D914-50 for ethylcellulose should be consulted. The colorimetric methods of Samsel and DeLap (253) and of Kanzaki and Berger (254) for methylcellulose and of Samsel and Aldrich (255) for ethylcellulose also are very useful.

The method of Morgan (256) is generally the basis for determining the hydroxyalkyl ethers of cellulose. It is based on the hydrolysis of the ether with hydriodic acid, which yields the alkyl iodide and the corresponding olefin. Measurement of the olefin formed can be accomplished by absorbing in Wijs solution with subsequent determination of the iodine number (116).

Hydroxypropyl methylcellulose may be determined by employing the method of Samsel and McHard (257) to determine the methoxyl content, and the method of Lemieux and Purves (258) as modified (259) for determining hydroxypropyl content.

Carboxymethylcellulose can be determined by the anthrone colorimetric method described by Black (260). It can also be estimated by hydrolyzing and determining the resulting glycolic acid by the method of Calkins (261). ASTM method No. D1439-58T may also be adaptable.

7.1.3 Physiological Response (247a)

The cellulose ethers are all very low in toxicity when administered by normal routes. They are not irritating to the skin or other delicate membranes of the body. When swallowed they are not absorbed to any appreciable degree and appear unchanged in the feces. No inhalation studies have been conducted, but exposure of humans to the dust in manufacturing operations over many years has not led to any known adverse effects. Parenteral administration of the water-soluble ethers has led to serious adverse effects in animals and it would seem unwise to administer them by such routes to human subjects.

7.2 Methylcellulose (Cellulose Methyl Ether, Methocel MC)

7.2.1 Determination in the Atmosphere

See Section 7.1.2 on cellulose ethers.

7.2.2 Physiological Response (247a)

As a result of extensive toxicologic studies conducted on both animals and human subjects, it may be concluded that methylcellulose is quite innocuous when swallowed or when in contact with the skin or other delicate membranes of the body. On the other hand, the evidence is equally as strong that it should not be administered parenterally.

7.2.2.1 Repeated-Dose Oral. The material was given to rats at the rate of about 0.44 g/day, part in their food and part in their drinking water, for a period of 8 months without any evidence of adverse effect being observed. Rats fed up to 6.2 g/(kg)(day) of methylcellulose for 6 months showed no adverse effect. Maintaining rats through three generations on a diet containing 5 percent methylcellulose did not produce any adverse effects, including reproductive function. Methylcellulose did not appear to be absorbed from the intestinal tract or to be hydrolyzed to methanol and cellulose. Human subjects fed single doses of 5 or 10 g of methylcellulose essentially voided all of it in the feces.

7.2.2.2 Parenteral Administration. The intravenous administration of 400-cP (centipoise) methylcellulose solution was effective in restoring blood volume associated with shock. It was considered to have certain advantages over some other macromolecular substances, but also certain limitations. When methylcellulose was given intravenously to the dog and rabbit it caused hematologic reactions, which were designated as "hematologic macromolecular syndrome." It was noted that aside from the effect upon the circulating blood, the inability of the body to degrade the substance led to its retention and accumulation in the liver, spleen, lymph nodes, kidney, and vascular walls.

Repeated administration of low viscosity (15-cP) methylcellulose intravenously to dogs produced serious effects characterized by a progressive decrease in the volume of urine and the presence in the urine of albumin, casts, and both red and white blood cells. Nonprotein nitrogen levels of the blood rose sharply but blood pressure was not increased. Generalized necrotizing vascular disease was frequently observed and death was generally due to renal failure.

Groups of rats were injected intraperitoneally with methylcellulose (400-cP) twice weekly for 15 weeks. In intact animals observations were made of grossly enlarged spleens, changes in the bone marrow and cellular elements of the blood, ascites, and infiltration of the spleen, liver, and kidneys with "storage-cell" macrophages. When the methylcellulose was given to previously splenectomized rats similar histological lesions resulted but none of the hematologic changes were observed.

When methylcellulose was injected subcutaneously in the rat, it failed to cause a neoplastic response, whereas a member of other water-soluble polymers tested in a similar manner did cause such a response.

7.3 Hydroxypropyl Methylcellulose (Propylene Glycol Ether of Methylcellulose, Methocel HG)

7.3.1 Determination in the Atmosphere

See Section 7.1.2 on cellulose ethers.

7.3.2 Physiological Response (247a)

The hydroxypropyl methylcelluloses have not been studied as extensively as has methylcellulose. However, the data available and the similarity in chemical nature strongly suggest that they can be expected to behave physiologically as methylcellulose; that is, they are essentially inert upon direct contact, essentially innocuous when swallowed, and unsuitable for parenteral administration.

7.3.2.1 Repeated-Dose Oral. High gel point methylcellulose, Methocel HG, now designated commercially as Methocel 65 HG, was fed to groups of 50 male and 50 female rats each for 2 years on diets containing 1, 5, and 20 percent of the test material. The only effect observed was growth retardation in the males fed the 20 percent diet. At oral doses of 0.1, 0.3, 1.0, and 3.0 g/(kg)(day) to groups of two dogs each for 1 year no adverse effects were noted. A dog fed 25 g/(kg)(day) for 30 days suffered no apparent ill effects. Another dog fed 50 g/(kg)(day) exhibited some diarrhea, slight weight loss, and slight depression in red blood cell count but no evidence of histological injury was found.

Methocel 65 HG was fed to 25 human subjects. Each individual received three graduated doses ranging from 0.6 to 8.9 g each at least 1 week apart. Essentially all of the ingested material was eliminated in the feces within 96 hr following single doses of 3.0 to 8.9 g. Recovery of the smaller doses was not attempted. It was concluded that the high gel methylcellulose was similar to methylcellulose in physiological properties.

Methocel 2602 (now designated as Methocel 60 HG) was fed to rats as a part of their diet for 121 days. It was found that dietary levels of 3 percent or less caused no adverse effects, that 10 percent caused slight retardation in the growth of male rats, and that 30 percent in the diet caused marked growth depression and excessive mortality in both males and females. No specific organic injury occurred even at the highest dose and the adverse effects were considered to be due to poor nutrition caused by the excessive bulk of such high level diets.

Further dietary feeding studies on rats have been carried out with other hydroxypropyl methylcelluloses, namely, Methocel 70 HG and Methocel 90 HG. The results obtained are almost identical with those described for Methocel 60 HG.

7.3.2.2 Parenteral Administration. Methocel HG administered intraperitoneally in aqueous solution to rats and mice had an LD_{50} of about 5 g/kg for both species.

7.4 Carboxymethylcellulose (Cellulose Glycolic Acid, Sodium Salt; CMC)

7.4.1 Determination in the Atmosphere

See Section 7.1.2 on cellulose ethers.

7.4.2 Physiological Response (247a)

Carboxymethylcellulose is very low in oral toxicity when fed in either single or repeated doses. It is not irritating to the skin or delicate membranes of the body. In rats and humans it passes through the gut unchanged. When given parenterally it causes undesirable effects characterized by deposition in various organs and in the walls of the blood vessels. It has been reported to cause sarcomas when injected subcutaneously in massive repeated doses.

7.4.2.1 Single-Dose Oral. Carboxymethylcellulose and its sodium and aluminum salts cannot be fed in single doses large enough to cause any apparent illness in rats, guinea pigs, and rabbits. The sodium salt has a single-dose oral LD_{50} of 27 g/kg in rats and 16 g/kg in guinea pigs. The fate of the sodium salt was studied in the rat and approximately 90 percent of a single oral dose was recovered in the feces.

7.4.2.2 Repeated-Dose Oral. The first published information regarding the physiological properties of carboxymethylcellulose appeared in 1941. It was concluded that the material is harmless when swallowed. Groups of rats maintained for 201 to 250 days on diets containing 5 percent of either the free acid, the sodium salt, or the aluminum salt did not exhibit any adverse effects.

Groups of rats fed doses of 1.0 g/(kg)(day) for 25 months did not show any detectable adverse effects as judged by appearance, growth, urine examinations, hematologic studies, fertility studies, and histological examination of numerous tissues. No neoplasms were found in any of the experimental animals. Also, groups of guinea pigs fed 1.0 g/(kg)(day) for 1 year and groups of dogs 1.0 g/(kg)(day) for 6 months did not exhibit any adverse effects as judged by appearance, growth, and histological examination of numerous organs.

Rats were unaffected by diets containing up to 14 percent of carboxymethylcellulose and the material passes through the digestive tracts of rats and humans unchanged. It was observed that, in three humans fed 20 to 30 g/day, a depression of protein digestion and an increase in fat digestion occurred. In vegetarian dogs only about one-half of that fed could be recovered in the feces, and it is postulated that this may be the result of degradation by the bacterial flora of the intestines.

7.4.2.3 Topical Contact. Carboxymethylcellulose was nonirritating to the skin of the rabbits even upon prolonged and repeated contact. A study employing the method of Schwartz et al. (262) on 200 human subjects concluded that the material was neither a primary irritant nor a skin sensitizer. It was also nonirritating when in contact with the membranes of the external genitalia and the vagina.

7.4.2.4 Parenteral Administration. Carboxymethylcellulose given intravenously to dogs as single doses caused only mild transitory shifts in the cellular elements of the blood and repeated doses caused a decrease in hemoglobin and an increase in sedimentation rate. The material was stored in the Kupffer cells, the reticulum cells of the spleen, the endothelial cells of the glomeruli, and on the walls and large branches of the aorta. In general it appeared that carboxymethylcellulose was tolerated more readily than methylcellulose under similar conditions.

Sarcomas were produced at the site of repeated subcutaneous injection of aqueous solutions of carboxymethylcellulose. It should be noted, however, that massive doses were given and therefore there must have been considerable local trauma.

7.5 Ethylcellulose (Cellulose Ethyl Ether, Ethocel)

7.5.1 Determination in the Atmosphere

See Section 7.1.2 on cellulose ethers.

7.5.2 Physiological Response (247a)

Ethylcellulose is very inert physiologically.

7.5.2.1 Repeated Oral Administration. A group of 80 rats was maintained for 8 months on a diet containing 1.2 percent of ethylcellulose, which amounted to an average dose of 182 mg/day. No evidence was found of adverse effects in any of the rats as judged by appearance, behavior, growth, and gross and microscopic examination of the tissues.

7.5.2.2 Parenteral Administration. A dose of 500 mg of powdered ethylcellulose introduced subcutaneously into 25 rats did not induce a greater than normal incidence of tumors.

7.6   Hydroxyethylcellulose (Natrosol, Cellosize)

7.6.1   Determination in the Atmosphere

See Section 7.1.2 on cellulose ethers.

7.6.2   Physiological Response (247a)

Hydroxyethylcellulose, like the other water-soluble cellulose ethers, is very low in oral toxicity but causes undesirable effects when given parenterally.

7.6.2.1 Repeated-Dose Oral. Groups of rats maintained for 2 years on diets containing 5, 1, and 0.2 percent hydroxyethylcellulose did not exhibit any adverse effects. Criteria employed included growth, food intake, life-span, frequency of extraneous infections, body measurements, kidney and liver weights, litters, hematologic examinations, occurrence of neoplasms, and histological examination of numerous organs.

7.6.2.2 Parenteral Administration. Up to 55 intravenous injections of hydroxyethylcellulose were given to dogs without causing injury other than that typical of the other water-soluble cellulose ethers. Only transitory changes in the blood picture and the deposition of the material on the intima of the blood vessels were noted.

7.7   Crown Ethers (Ethylene Oxide Polymeric Ethers)

7.7.1   General Physical and Chemical Properties

Very little information exists in the literature on the physical and chemical properties of the crown ethers. The name "crown" was first used in 1967 (263) and derives from the stereochemical structure, which for 15-crown-5 may be drawn as follows:

During distillation 18-crown-6 may convert to *p*-dioxane and present an explosion hazard due to clogging of traps (265–267). Available properties are presented in Table 8.16.

Table 8.16. Physical and Chemical Properties of the Crown Ethers

Compound	Molecular Formula	Molecular Weight	B.P. (°C)	M.P. (°C)	Specific Gravity	Vapor Pressure (mm Hg)	Octanol/Water Partition Coefficient
6-Crown-2		88	—	—	—	—	2.2
12-Crown-4		176	238 (760 mm Hg) 69 (0.5 mm Hg)	—	1.104	0.03 at room temp.	0.92
15-Crown-5		220	78 (0.05 mm Hg)	—	—	—	0.33
18-Crown-6		264	125–160 (0.2 mm Hg)	36.5	—	—	0.21
Dicyclohexyl-18-crown-6		372	—	61–70	—	—	—

7.7.2 Determination in the Atmosphere

These compounds may be liquid or solid at room temperature. No satisfactory procedure has been developed for analyzing these materials in the atmosphere (268).

7.7.3 Physiological Response

7.7.3.1 Acute Exposure. The oral LD_{50} for the following crown ethers has been determined to be as follows:

Ether	Mice (269) Oral LD_{50} (g/kg)	Time to Death	Rats (273) Oral LD_{50} (ml/kg)
6-Crown-2	6.0		
12-Crown-4	3.15	<15 min	2.83
15-Crown-5	1.02	1 hr	1.41
18-Crown-6	0.705	>1 day	1.39

Oral administration of 18-crown-6 to healthy dogs produced transitory signs of tremulous movement and paralysis of the hind limbs 2 to 12 hr after administration (270). Leong (271) reported that a single oral dose of 12-crown-4 at a level of approximately 100 mg/kg in rats produced central nervous system (CNS) effects and testicular atrophy. The dosages of higher aliphatic crowns, which were capable of producing CNS effects, were 1 to 10 times higher. This author also reported that skin absorption of 12-crown-4 in rabbits could produce CNS effects, but larger macrocyclic crowns only produced slight skin redness.

Pedersen (263) reported the approximate lethal dose of 300 mg/kg for dicyclohexyl-18-crown-6 by ingestion in rats. Death occurred within 11 min. No deaths occurred at 200 mg/kg. Skin absorption in experimental animals was fatal at a dose level of 130 mg/kg for dicyclohexyl-18-crown-6. This material was irritating to the skin and eyes. It was reported that permanent eye damage may result if the eye is not washed with water after exposure.

7.7.3.2 Repeated Exposure. Leong et al. (272) exposed rats to 0.5 and 1.0 ppm of 12-crown-4 vapors 7 hr/day, 5 days/week, for 3 weeks. Both levels produced marked testicular atrophy that was associated with degeneration of the germinal epithelium. This effect persisted up to 4 months post-exposure. Atrophy of the prostate and seminal vesicles was also noted. Exposure to 1 ppm of 12-crown-4 vapors produced prominent degradation of conditioned behavioral performances, depression of food and water intake, retardation of growth, and body tremors. These effects were reversible.

7.7.3.3 Neurotoxicity. Gad et al. exposed three species to 18-crown-6 with a repeated dose regimen that involved ever-increasing dose levels in each animal over the exposure time period (273). Rabbits were exposed intravenously 5 days/week at dose levels increasing from 6 mg/kg through 12.5 mg/kg for 3 weeks. Rats were dosed intraperitoneally to ever-increasing dose levels of 20 mg/kg through 80 mg/kg for more than 1 month. Mice were intraperitoneally dosed from 20 mg/kg through 160 mg/kg. Signs of nervous system effects included tremors, hyperactivity, loss of limb strength, muscle twitching, and decreased awareness of light stimuli. Exposure levels were increased in each species because of accommodation to each dose level in a few hours to 3 days. No clinical or histopathological changes were observed. These authors claim to have observed increasing behavior changes, which are completely reversible, with increasing molecular size of crown ethers.

7.7.3.4 Mode of Action. It is believed that the crown ethers complex with electrolytes, thus changing the permeability of Na^+ and K^+ across cell membranes (272, 274–277). Investigations thus far indicate the nerve conduction changes are fully reversible.

7.7.4 Hygienic Standards for Industrial Exposure

No hygienic standards for occupational exposure have been established for the crown ethers. However, the significant biologic changes noted for 12-crown-4 at 0.5 ppm indicate that the time-weighted average for this chemical should be lower than 0.05 ppm.

REFERENCES

1. P. Z. Bedoukian, *Perfumery Synthetics and Isolates*, Van Nostrand, New York, 1951.
2. A. I. Vogel, *J. Chem. Soc.*, 616 (1948).
3. H. R. Fleck, *Chem. Prod.*, **19**, 144 (1956).
4. C. R. Noller, *Chemistry of Organic Compounds*, 3rd ed., Saunders, Philadelphia, 1966, Chapter 9.
5. *Chem. Market. Rep.*, **214**(4), 5, 40 (1978).
6. *Code of Federal Regulations*, Title 27, Chap I, Part 212, 1977.
7. L. Fishbein, *EPA Report 560-5-77-005*, NTIS, Springfield, VA, May, 1977.
8. R. C. Weast, Ed., *Handbook of Chemistry and Physics*, 57th ed., CRC Press, Cleveland, 1976.
9. N. I. Sax, *Dangerous Properties of Industrial Materials*, 4th ed., Van Nostrand Reinhold, New York, 1975.
10. *Toxic and Hazardous Industrial Chemicals Safety Manual*, International Technical Information Institute, Tokyo, 1976.
11. *Laboratory Waste Disposal Manual*, Manufacturing Chemists Association, Washington, DC, 1973.
12. R. H. Dreisbach, *Handbook of Poisoning: Diagnosis and Treatment*, 9th ed., Lange, Los Altos, CA, 1977.
13. T. Flint and H. D. Cain, *Emergency Treatment and Management*, 5th ed., Saunders, Philadelphia, 1975.

14. *The Merck Index*, 9th ed., Rahway, NJ, 1976.

15. *Handbook of Organic Industrial Solvents*, 3rd ed., American Mutual Insurance Alliance, Chicago, 1966.

16. *Code of Federal Regulations*, Title 29, Chapter XVII, Subpart Z, 1910.1000, 1988.

17. American Conference of Governmental Industrial Hygienists, *TLVs: Threshold Limit Values for Chemical Substances and Physical Agents and Biological Exposure Indices*, Cincinnati, 1991–1992.

18. *NIOSH Manual of Analytical Methods*, 3rd ed., 1984.

19. W. H. Stahl, *ASTM Data Series DS48*, American Society for Testing and Materials, Philadelphia, 1973.

20. J. H. Ruth, *Am. Ind. Hyg. J.*, **47**, A-142 (1986).

21. A. D. Astle, TRC Environmental Consultants, Inc., Wethersfield, CT, unpublished data dated May 20, 1981.

22. J. S. Fritz, in *The Chemistry of the Ether Linkage*, S. Patal, Ed., Interscience, New York, 1967, Chapter 15.

23. *Chem. Eng. News*, 13 (Jan. 8, 1973).

24. G. J. Kallos and R. A. Solomon, *Am. Ind. Hyg. Assoc. J.*, **34**, 469 (1973).

25. L. S. Frankel et al., *Environ. Sci. Technol.*, **8**, 356 (1974).

26. A. R. Sellakumar et al., *Toxicol. Appl. Pharmacol.*, **81**, 401 (1985).

27. A. H. Owen-Jones, *Firemen*, **33**, 15 (1966).

28. *U. S. Pharmacopoeia XIX*, Mack, Easton, PA, 1975, pp. 185, 616, 734; (*a*) American Chemical Society, *Specifications for Reagent Chemicals*, 5th ed., 1974, p. 253.

29. D. P. Kelly, J. T. Pinhey, and R. D. G. Rigby, *Aust. J. Chem.*, **22**, 977 (1969).

30. H. Suzuki and H. Ono, *Kogyo Kagaku Zasshi*, **71**, 1931 (1968); through *Chem. Abstr.*, **70**, 48013d (1969).

31. H. F. Smyth, Jr. and C. P. Carpenter, *J. Hyg. Toxicol.*, **26**, 269 (1944).

32. H. F. Smyth, Jr. and C. P. Carpenter, *J. Hyg. Toxicol.*, **30**, 63 (1948).

33. H. F. Smyth, Jr., C. P. Carpenter, and C. S. Weil, *J. Hyg. Toxicol.*, **31**, 60 (1949).

34. H. F. Smyth, Jr., C. P. Carpenter, and C. S. Weil, *Arch. Ind. Hyg. Occup. Med.*, **4**, 119 (1951).

35. H. F. Smyth, Jr., C. P. Carpenter, C. S. Weil, and U. C. Pozzani, *Arch. Ind. Hyg. Occup. Med.*, **10**, 61 (1954).

36. H. F. Smyth, Jr., C. P. Carpenter, C. S. Weil, U. C. Pozzani, and J. A. Striegel, *Am. Ind. Hyg. Assoc. J.*, **23**, 95 (1962); (*a*) R. W. Reynolds et al., American Chemical Society, Meeting Sept. 8–23, 1974, Atlantic City, NJ.

37. H. F. Smyth, Jr., C. P. Carpenter, C. S. Weil, U. C. Pozzani, J. A. Striegel, and J. S. Nycum, *Am. Ind. Hyg. Assoc. J.*, **30**, 470 (1969).

38. C. P. Carpenter and H. F. Smyth, Jr., *Am. J. Ophthalmol.*, **29**, 1363 (1946).

39. C. P. Carpenter, C. S. Weil, and H. F. Smyth, Jr., *Toxicol. Appl. Pharmacol.*, **28**, 313 (1974).

40. E. J. Fairchild, Ed., R. J. Lewis, Sr., and R. L. Tatken, *Registry of Toxic Effects of Chemical Substances*, DHEW Publ. No. (NIOSH) 78-104-B, Superintendent of Documents, Washington, DC, 1977.

41. R. Jeppsson, *Acta Pharmacol. Toxicol.*, **37**, 56 (1975).

42. W. R. Glave and C. Hansch, *J. Pharm. Sci.*, **61**, 589 (1972).

43. W. E. Brown, *J. Pharmacol. Exp. Ther.*, **23**, 497 (1924).

44. A. W. Brody et al., *J. Appl. Physiol.*, **31**, 125 (1971).

45. C. J. Collins, L. M. Cobb, and D. A. Purser, *Toxicology*, **11**, 65 (1978).

46. R. G. J. Reuzel, J. P. Bruyntjes, and R. B. Beems, *Aerosol Rep.*, **20**, 23 (1981).

47. J. J. Daly, Jr. and G. L. Kennedy, Jr., *Chem. Times Trends*, **10**, 40 (1987).

48. H. L. Price and R. D. Dripps, in *The Pharmacological Basis of Therapeutics*, 4th ed., L. S. Goodman and A. Gilman, Eds., Macmillan, New York, 1970.

49. F. Flury and O. Klimmer, in *Toxicology and Hygiene of Industrial Solvents*, K. B. Lehmann and F. Flury, Eds. (transl. by E. King and H. F. Smith, Jr.), Williams and Wilkins, Baltimore, 1943.

50. E. Browning, *Toxicity of Industrial Organic Solvents*, Chemical Publishing, New York, 1953.

51. B. Dawson et al., *Mayo Clin. Proc.*, **41**, 599 (1966).

52. S. Sewall, *Bull. Hosp. Joint Dis.*, **8**, 33 (1947).

53. K. W. Nelson, J. F. Ege, Jr., M. Ross, L. E. Woodman, and L. Silverman, *J. Ind. Hyg. Toxicol.*, **25**, 282 (1943).

54. Kärber, *Arch, Exp. Pathol. Pharmakol.*, **142**, 1 (1929).

55. H. Molitor, *J. Pharmacol. Exp. Ther.*, **57**, 274 (1936).

56. W. S. Spector, Ed., *Handbook of Toxicology*, Vol. 1, Saunders, Philadelphia, 1956.

57. H. E. Swann, B. K. Kwon, G. K. Hogan, and W. M. Snellings, *Am. Ind. Hyg. Assoc. J.*, **35**, 511 (1974).

58. B. H. Robbins, *J. Pharmacol. Exp. Ther.*, **53**, 251 (1935).

59. E. T. Mörch, J. B. Aycrigg, and M. S. Berger, *J. Pharmacol. Exp. Ther.*, **117**, 184 (1956).

60. R. E. Wimer and C. Huston, *Behavior. Biol.*, **10**, 385 (1974).

61. B. A. Schwetz and B. A. Becker, *Toxicol. Appl. Pharmacol.*, **18**, 703 (1971).

62. L. F. Sancillo, P. Kraus, C. Myers, and L. Wagner, *J. Pharm. Sci.*, **57**, 1248 (1968).

63. E. T. Kimura, D. M. Ebert, and P. W. Dodge, *Toxicol. Appl. Pharmacol.*, **19**, 699 (1971).

64. A. A. Fisher, *Contact Dermatitis*, 2nd ed., Lea and Febiger, Philadelphia, 1975, p. 378.

65. W. M. Grant, *Toxicology of the Eye*, 2nd ed., Charles C Thomas, Springfield, IL, 1974, p. 464.

66. O. Dybing and K. Shovlund, *Acta Pharmacol. Toxicol.*, **13**, 252 (1957).

67. H. W. Haggard, *J. Biol. Chem.*, **59**, 737 (1924).

68. M. B. Chenoweth, D. N. Robertson, D. S. Erley, and R. Golhke, *Anesthesiology*, **23**, 101 (1962).

69. J. C. Krantz, Jr., and C. J. Carr, *The Pharmocologic Principles of Medical Practice*, 7th ed., Williams and Wilkins, Baltimore, 1969, pp. 97, 112.

70. *Ibid.*, p. 93.

71. N. H. Booth and L. E. McDonald, Eds., *Veterinary Pharmacology and Therapeutics*, Iowa State University Press, Ames, IA, 1982, pp. 191–193.

72. *NIOSH Registry of Toxic Effects of Chemical Substances*, Microfiche Ed., DHHS (NIOSH) Publ. No. 91-101-2, Oct. 1991.

73. G. D. Nielsen et al., *Acta Pharmacol. Toxicol.*, **56**, 158 (1985).

74. G. D. Nielsen et al., *Acta Pharmacol. Toxicol.*, **56**, 165 (1985).

75. L. Scheflan and M. B. Jacobs, *The Handbook of Solvents*, Van Nostrand, New York, 1953.

76. J. C. Krantz, Jr., W. E. Evans, Jr., C. J. Carr, and D. V. Kibler, *J. Pharmacol. Exp. Ther.*, **86**, 138 (1946).

77. L. Silverman, H. F. Schulte, and M. W. First, *J. Ind. Hyg. Toxicol.*, **28**, 262 (1946).

78. Shell Chemical Corp., *Safety Data Sheet, Isopropyl Ether, No. SC: 57-104*, 1959.

79. W. Machle, E. W. Scott, and J. Treon, *J. Ind. Hyg. Tox.*, **21**, 72 (1939).

80. Union Carbide Chemicals Co., *Technical Information, Isopropyl Ether, No. F-40003*, 1955.

81. S. A. Kosyan, *Tr. Erevan. Gos. Inst. Usoversh. Vrachei*, **3**, 617 (1967); through *Chem. Abstr.*, **71**, 37213p (1969).

82. H. F. Smyth, Jr., C. S. Weil, J. S. West, and C. P. Carpenter, *Toxicol. Appl. Pharmacol.*, **14**, 340 (1969).

83. P. Garrett, M. Moreau, and J. D. Lowry, Maine Department of Environmental Protection, undated report, titled "Methyl Tertiary Butyl Ether as a Ground Water Contaminant."

84. Florida Dept. of Environmental Regulation, Tallahassee, "Toxicant Profile-Gasoline and Selected Gasoline Constituents," November, 1987.

85. J. Ponchon et al., *Lancet*, **1988**, 276 (1988).

86. A. Sivak and J. Murphy, International Symposium on Alcohol Fuels, Florence, Italy, August, 1991.

87. M. J. Allen et al., *Gastroenter.*, **88**, 122 (1985).

88. I. A. D. Bouchier, *Gut*, **29**, 137 (1988).

89. M. Robinson, R. H. Bruner, and G. R. Olson, *J. Am. Coll. Toxicol.*, **9**, 525 (1990).

90. International Symposium on the Health Effects of Gasoline, Miami, FL, November 5–8, 1991.

91. Inveresk Research Int'l, Project No. 413038, 1980.

92. R. W. Biles, R. E. Schroeder, and C. E. Holdsworth, *Toxicol. Ind. Health*, **3**, 519 (1987).

93. C. C. Conaway, R. E. Schroeder, and N. K. Snyder, *J. Toxicol. Envir. Health*, **16**, 797 (1985).

94. H. Savolainen, P. Pfaffli, and E. Elovaara, *Arch. Toxicol.*, **57**, 285 (1985).

95. J. C. Krantz, Jr., C. J. Carr, W. E. Evans, Jr., and R. Musser, *J. Pharmacol. Exp. Ther.*, **87**, 132 (1946).

96. T. Sollman, *A Manual of Pharmacology*, 8th ed., Saunders, Philadelphia, 1957, p. 915.

97. *The Merck Index*, 9th ed., Rahway, NJ, 1976.

98. R. Cavaliere, B. Giovanella, and G. Moricca, *Med. Sper.*, **31**, 253 (1957); *Chem. Abstr.*, **52**, 12228 (1958).

99. Union Carbide Chemicals Co., *Technical Information, Vinyl Ethers, No. F-6701-A*, 1948.

100. C. P. Carpenter, H. F. Smyth, Jr., and U. C. Pozzani, *J. Ind. Hyg. Toxicol.*, **31**, 343 (1949).

101. R. S. McLaughlin, *Am. J. Ophthalmol.*, **29**, 1355 (1946).

102. G. D. Nielson, J. C. Bakbo, and E. Hoist, *Acta Pharmacol. Toxicol.*, **54**, 292 (1984).

103. A. K. Burditt, Jr., F. G. Hinman, and J. W. Balock, *J. Econ. Entomol.*, **56**, 261 (1963).

104. J. C. Krantz, Jr., C. J. Carr, G. Lu, and M. J. Fassel, *J. Pharmacol. Exp. Ther.*, **105**, 1 (1952).

105. G. D. Nielsen et al., *Acta Pharmacol. Toxicol.*, **54**, 292 (1984).

106. A. V. Poznak and J. F. Artusio, Jr., *Toxicol. Appl. Pharmacol.*, **2**, 374 (1960).

107. R. C. Terrell et al., *J. Med. Chem.*, **14**, 517 (1971).

108. L. Speers et al., *J. Med. Chem.*, **14**, 593 (1971).

109. R. A. Solomon and G. J. Kallos, *Anal. Chem.*, **47**, 955 (1975).

110. K. V. Maher and L. R. DeFonso, *J. Natl. Cancer Inst.*, **78**, 839 (1987).

111. W. Weiss, *J. Natl. Cancer Inst.*, **69**, 1265 (1982).

112. *IARC Monographs*, International Agency for Research in Cancer, Lyon, Vol. 4, 1974 (from WHO, Geneva, Switzerland).

113. B. L. VanDuuren, *Environ. Res.*, **49**, 143 (1989).

114. H. H. Schrenk, F. A. Patty, and W. P. Yant, *Publ. Health Rep.*, **48**, 1389 (1933).

115. L. Schwartz et al., *J. Am. Med. Assoc.*, **115**, 906 (1940).

116. Biochemical Research Laboratory, Dow Chemical Co., unpublished data.

117. H. F. Smyth, Jr., in *Handbook of Toxicology*, W. S. Spector, Ed., Vol. I, Saunders, Philadelphia, 1956.

118. J. R. M. Innes et al., *J. Natl. Cancer Inst.*, **42**, 1101 (1969).

119. H. Allen, *Chem. Prod.*, **19**, 482 (1956).

120. B. L. VanDuuren et al., *J. Natl. Cancer Inst.*, **48**, 1431 (1972).

121. E. K. Weisburger et al., *J. Natl. Cancer Inst.*, **67**, 75 (1981).

122. R. D. Lingg et al., *Soc. Toxicol.*, 17th Ann. Meet., March 12–16, 1978, Abstr. 69.

123. K. Norpoth et al., *J. Cancer Res. Clin. Oncol.*, **112**, 125 (1986).

124. K. Mitsumori et al., *J. Pesticide Sci.*, **4**, 323 (1979).

125. R. D. Lingg et al., *Arch. Envir. Contam. Toxicol.*, **11**, 173 (1982).

126. B. L. VanDuuren et al., *J. Natl. Cancer Inst.*, **43**, 481 (1969).

127. B. L. VanDuuren, *Ann. N.Y. Acad. Sci.*, **163**, 633 (1969).

128. F. L. M. Pattison, W. C. Howell, and R. G. Woolford, *Can. J. Chem.*, **35**, 141 (1957).

129. B. H. Robbins, *J. Pharmacol. Exp. Ther.*, **86**, 197 (1946).

130. J. F. Artusio et al., *Anesthesiology*, **21**, 512 (1960).

131. J. W. Desmond, *Can. Anaesth. Soc. J.*, **21**, 294 (1974).

132. L. E. Arthaud and T. A. Loomis, *Toxicol. Appl. Pharmacol.*, **45**, 845 (1978).

133. E. B. Truitt, Jr. and E. M. Ebersberger, *Arch. Int. Pharmacodyn.*, **129**, 223 (1960).

134. J. C. Krantz, Jr., E. B. Truitt, Jr., A. S. C. Ling, and L. Speers, *J. Pharmacol. Exp. Ther.*, **121**, 362 (1957).

135. J. C. Krantz, Jr., C. J. Carr, G. Lu, and F. K. Bell, *J. Pharmacol. Exp. Ther.*, **108**, 488 (1953).

136. Anonymous, *Can. Med. Assoc. J.*, **109**, 168 (1973).

137. B. L. VanDuuren, B. M. Goldschmidt, and I. Seidman, *Cancer Res.*, **35**, 2553 (1975).

138. I. E. Schwartzman, *Gig. Sanit.*, **5**, 20 (1964).

139. F. J. Buckle and B. C. Saunders, *J. Chem. Soc.*, **1949**, 2774.

140. R. R. Johnston, T. H. Cromwell, and E. I. Eger, Jr., *Anesthesiology*, **38**, 313 (1973).

141. W. S. Spector, Ed., *Handbook of Toxicology*, Vol. 1, Saunders, Philadelphia, 1956.

142. H. I. Coombs and T. S. Hele, *Biochem. J.*, **20**, 606 (1926).

143. R. T. Williams, *Detoxication Mechanisms*, Chapman and Hall, London, 1947.

144. R. T. Williams, *Detoxication Mechanisms*, 2nd ed., Wiley, New York, 1959.

145. D. L. J. Opdyke and C. Letizia, *Food Chem. Toxicol.*, **20**, 697 (1982).

146. H. G. Bray, S. P. James, W. V. Thorpe, and M. R. Wasdell, *Biochem. J.*, **54**, 547 (1953).

147. H. G. Bray, V. M. Craddoch, and W. V. Thorpe, *Biochem. J.*, **60**, 225 (1955).

148. J. Axelrod, *J. Pharmacol. Exp. Ther.*, **115**, 259 (1955).

149. J. Axelrod, *Biochem. J.*, **63**, 634 (1956).

150. S. Solis-Cohen and T. S. Githens, *Pharmacotherapeutics*, Appleton, New York, 1928.

151. *Drugs of the Future*, **5**, 539 (1988).

152. L. S. Goodman and A. Gilman, *The Pharmacological Basis of Therapeutics*, 2nd ed., Macmillan, New York, 1955.

153. K. P. Wong and T. L. Sourkes, *Can. J. Biochem.*, **44**, 635 (1966).

154. Cosmetic Ingredient Review, *J. Am. College Toxicol.*, **4**, 31 (1985).

155. C. P. Chivers, *Brit. J. Ind. Med.*, **29**, 105 (1972).

156. H. C. Hodge, J. H. Sterner, E. A. Maynard, and J. Thomas, *J. Ind. Hyg. Toxicol.*, **31**, 79 (1949).

157. D. W. Fassett, Laboratory of Industrial Medicine, Eastman Kodak Co., unpublished data.

158. P. A. Riley, *J. Pathol.*, **97**, 185 (1969).

159. P. A. Riley, *J. Pathol.*, **101**, 163 (1970).

160. H. K. Mahler and E. H. Cordes, *Biological Chemistry*, Harper and Row, NY, p. 570, 1966.

161. P. A. Riley, *Phil. Trans. R. Soc. London*, **B311**, 229 (1985).

162. D. A. Woods, and C. J. Smith, *Exp. Mol. Pathol.*, **10**, 107 (1969).

163. R. Brun, *Dermatologica*, **134**, 125 (1967).

164. E. Frenk, *Ann. Dermatol. Syphilol.*, **97**, 287 (1970).

165. E. Frenk and F. Ott, *J. Invest. Dermatol.*, **56**, 287 (1971).

166. C. R. Denton, A. B. Lerner, and T. B. Fitzpatrick, *J. Invest. Dermatol.*, **18**, 119 (1952).

167. I. H. Blank and O. G. Miller, *J. Am. Med. Assoc.*, **149**, 1371 (1952).

168. A. B. Lerner and T. B. Fitzpatrick, *J. Am. Med. Assoc.*, **152**, 577 (1953).

169. K. H. Kelly, H. R. Bierman, and M. B. Shimkin, *Proc. Soc. Exp. Biol. Med.*, **79**, 589 (1952).

170. A. Catona and D. Lanzer, *Med. J. Austral.*, **146**, 320 (1987).

171. S. M. Peck and H. Sobotka, *J. Invest. Dermatol.*, **4**, 325 (1941).

172. M. F. Brulos, J. P. Guillot, M. C. Martini, and J. Cotte, *J. Soc. Cosmet. Chem.*, **28**, 367 (1977).

173. L. Schwartz, E. A. Oliver and L. H. Warren, *Publ. Health Rep., U.S.*, **55**, 1111 (1940).

174. A. J. Lea, *Nature*, **167**, 906 (1951).

175. Editorial, *Trans. St. John's Hosp. Dermatol. Soc.*, **56**, 182 (1970).

176. F. Meyer and E. Meyer, *Arch. Pharm.*, **290**, 109 (1957); *Chem. Abstr.*, **51**, 11660 (1957).

177. M. A. Korany, N. Abdel-Salam, and M. Abdel-Salam, *J. Assoc. Offic. Anal. Chem.*, **61**, 169 (1978).

178. K. Hayashi and Y. Hashimoto, *Pharm. Bull. Tokyo*, **4**, 496 (1956); *Chem. Abstr.*, **51**, 13320 (1957).

179. R. C. Paulsom, *J. Am. Dent. Assoc.*, **89**, 895 (1974).

180. G. Koch et al., *Odontol. Revy,* **24**, 109 (1973).

181. G. Koch, B. Magnusson, and G. Nyquist, *Odontol. Revy*, **22**, 275 (1971).

182. D. L. J. Opdyke, *Food Cosmet. Toxicol.*, **13**, 545 (1975).

183. E. C. Hagan et al., *Toxicol. Appl. Pharmacol.*, **7**, 18 (1965).

184. H. A. Sober, F. Hollander, and E. K. Sober, *Proc. Soc. Exp. Biol. Med.*, **73**, 148 (1950).

185. K. Fujii et al., *Toxicol. Appl. Pharmacol.*, **16**, 482 (1970).

186. E. J. LoVoie et al., *Arch. Toxicol.*, **59**, 78 (1986).

187. G. C. Clark, *Arch. Toxicol.*, **62**, 381 (1988).

188. National Toxicology Program, NIH Publication No. 84-1779 (1983).

189. S. S. Young, *Fundam. Appl. Toxicol.*, **8**, 1 (1987).

190. International Agency for Research on Cancer, Monograph No. 36, 1985.

191. M. Ozeki, *Comp. Biochem. Physiol.*, **50C**, 183 (1975).

192. G. Kozam, *Oral Surg.*, **44**, 799 (1977).

193. D. L. J. Opdyke, *Food Cosmet. Tox., Suppl.*, **13**, 815 (1975).

194. A. B. Sell and E. A. Carlini, *Pharmacol.*, **14**, 367 (1976).

195. M. Beroza et al., *Toxicol. Appl. Pharmacol.*, **31**, 421 (1975).

196. J. A. Engelbrecht, J. P. Long, D. E. Nichols, and C. F. Barfknecht, *Arch. Int. Pharmacodyn.*, **199**, 226 (1972).

197. E. W. Fisherman and G. Cohen, *Ann. Allergy*, **31**, 126 (1973).

198. P. Cloninger and H. S. Novey, *Ann. Allergy*, **32**, 131 (1974).

199. A. A. Fisher, *Cutis*, **17**, 21 (1976).

200. J. Roed-Petersen and N. Hjorth, *Brit. J. Dermatol.*, **94**, 233 (1976).

201. T. W. Turner, *Contact Dermatol.*, **3**, 282 (1977).

202. Cosmetic Ingredient Review, *J. Am. College Toxicol.*, **3**, 83 (1984).

203. J. R. Allen, *Arch. Environ. Health*, **31**, 47 (1976).

204. D. J. Clegg, *Food Cosmet. Toxicol.*, **3**, 387 (1965).

205. E. Hansen and O. Meyer, *Toxicology*, **10**, 195 (1978).

206. J. D. Stokes, C. L. Scudder, and B. M. Boulos, *Fed. Proc.*, **35**, 429 (1976).

207. J. D. Stokes and C. L. Scudder, *Dev. Psychobiol.*, **7**, 343 (1974).

208. C. V. Vorhees et al., *Neurobehav. Toxicol. Teratol.*, **3**, 321 (1981).

209. P. F. Creaven, W. H. Davies, and R. T. Williams, *J. Pharm. Pharmacol.*, **18**, 485 (1966).

210. I. F. Gaunt, G. Feuer, F. A. Fairweather, and D. Gilbert, *Food Cosmet. Toxicol.*, **3**, 433 (1965).

211. D. Gilbert and L. Golberg, *Biochem. J.*, **97**, 28P (1965).

212. J. R. Allen and J. F. Engblom, *Food Cosmet. Toxicol.*, **10**, 769 (1972).

213. P. Fritsch, G. de S. Blanquat, and R. Derache, *Food Cosmet. Toxicol.*, **13**, 359 (1975).

214. A. L. Branen, T. Richardson, M. C. Goel, and J. R. Allen, *Food Cosmet. Toxicol.*, **11**, 797 (1973).

215. J. C. Dacre and F. A. Denz, *Biochem. J.*, **64**, 777 (1956).

216. B. D. Astill, D. W. Fassett, and R. L. Roudabush, *Biochem. J.*, **75**, 643 (1960).

217. B. D. Astill, J. Mills, and D. W. Fassett, Laboratory of Industrial Medicine, Eastman Kodak Co., unpublished data, 1960.

218. O. H. M. Wilder, P. C. Ostby, and B. R. Gregory, *J. Agric. Food Chem.*, **8**, 504 (1960).

219. O. H. M. Wilder and H. R. Kraybill, *Fed. Proc.*, **8**, 165 (1949).

220. O. H. M. Wilder and H. R. Kraybill, *Report of American Meat Institute Foundation*, University of Chicago, Chicago, 1948.

221. W. D. Brown, A. R. Johnson, and M. W. O'Halloran, *Aust. J. Exp. Biol. Med. Sci.*, **37**, 533 (1959).

222. W. D. Graham, H. Teed, and H. C. Grice, *J. Pharm. Pharmacol.*, **6**, 534 (1954).

223. H. D. Hodge and D. W. Fassett, University of Rochester and Eastman Kodak Co., unpublished data.

224. H. C. Hodge et al., *Toxicol. Appl. Pharmacol.*, **9**, 583 (1966).

225. R. J. Shamberger, S. Tytko, and C. E. Willis, *Cleveland Clin. Quart.*, **39**, 119 (1972).

226. L. W. Wattenberg, *J. Natl. Cancer Inst.*, **48**, 1425 (1972).

227. L. W. Wattenberg, *Proc. Am. Assoc., Cancer Res.*, **14**, 7 (1973).

228. L. W. Wattenberg, *J. Natl. Cancer Inst.*, **50**, 1541 (1973).

229. D. L. Berry et al., *Res. Commun. Chem. Path. Pharmacol.*, **20**, 101 (1978).

230. J. L. Speier, L. K. T. Lam, and L. W. Wattenberg, *J. Natl. Cancer Inst.*, **60**, 605 (1978).

231. L. W. Wattenberg, *Adv. Cancer Res.*, **26**, 197 (1978).

232. L. W. Wattenberg, *J. Natl. Cancer Inst.*, **60**, 11 (1978).

233. N. Ito et al., *J. Natl. Cancer Inst.*, **70**, 343 (1983).

234. N. Ito et al., *Gann*, **74**, 459 (1983).

235. T. Masui et al., *Jap. J. Cancer Res. (Gann)*, **77**, 854 (1986).

236. T. Masui et al., *Jap. J. Cancer Res. (Gann)*, **77**, 1083 (1986).

237. G. M. Williams, C. X. Wang, and M. J. Iatropoulos, *Food Chem. Toxicol.*, **28**, 799 (1990).

238. H. Böhme and O. Winkler, *Z. Lebensm. Untersuch. Forsch.*, **99**, 22 (1954); *Chem. Abstr.*, **48**, 13111 (1954).

239. K. G. Bergner and H. Sperlich, *Deut. Lebensm. Rundsch.*, **6**, 134 (1951): *Chem. Abstr.*, **45**, 8156 (1951).

240. W. R. Galley, *Chem. Anal.*, **39**, 59 (1950).

241. A. Bevenue and K. T. Williams, *Chem. Anal.*, **41**, 5 (1952).

242. H. W. Chenoweth, *Ind. Eng. Chem. Anal. Ed.*, **12**, 98 (1940).

243. H. Nechamkin, *Ind. Eng. Chem. Anal. Ed.*, **15**, 268 (1943).

244. F. Meyer and E. Meyer, *Arch. Pharm.*, **290**, 109 (1957).

245. D. L. J. Opdyke, *Food Cosmet. Toxicol.*, **15**, 633 (1977).

246. M. K. Robinson et al., *Toxicol.*, **61**, 91 (1990).

247. D. L. J. Opdyke, *Food Cosmet. Toxicol.*, **13**, 103 (1975); (*a*) C. L. Hake and V. K. Rowe, in F. A. Patty, *Industrial Hygiene and Toxicology*, 2nd ed., Wiley, New York, 1963.

248. (*a*) D. R. Branson, *ASTM STP-634* ASTM, 1977, p. 44; (*b*) R. J. Nolan et al., presented at the *17th Ann. Meet. S.O.T., March 12–16, 1978, San Francisco*, Abstr. No. 48; (*c*) D. R. Branson et al., presented at the *17th Ann. Meet. S.O.T., March 12–16, 1978, San Francisco*, Abstr. No. 60; (*d*) J. A. John et al., presented at the *33rd ASTM, October, 1977*.

249. J. M. Norris et al., *Environ. Health Perspect.*, **11**, 153 (1975).

250. National Toxicology Program, NTP TR309, NIH Publ. No. 85-2565 (August, 1985).

251. S. M. El Dareer et al., *J. Toxicol. Environ. Health*, **22**, 405 (1987).

252. IARC Monograph, **48**, 73 (1990).

253. E. P. Samsel and R. A. DeLap, *Anal. Chem.*, **23**, 1795 (1951).

254. G. Kanzaki and E. Y. Berger, *Anal. Chem.*, **31**, 1383 (1959).

255. E. P. Samsel and J. C. Aldrich, *Anal. Chem.*, **29**, 574 (1957).

256. P. W. Morgan, *Ind. Eng. Chem. Anal. Ed.*, **18**, 500 (1946).

257. E. P. Samsel and J. A. McHard, *Ind. Eng. Chem. Anal. Ed.*, **14**, 750 (1942).

258. R. U. Lemieux and C. B. Purves, *Can. J. Res.*, **25-B**, 485 (1947).

259. The Dow Chemical Co., *Anal. Method No. 200*, Nov. 17, 1955.

260. H. C. Black, Jr., *Anal. Chem.*, **23**, 1792 (1951).

261. V. P. Calkins, *Ind. Eng. Chem. Anal. Ed.*, **15**, 762 (1943).

262. L. Schwartz et al., *J. Am. Med. Assoc.*, **115**, 906 (1940).

263. C. J. Pedersen, *J. Am. Chem. Soc.*, **89**, 7017 (1967).

264. G. W. Gokel and H. D. Durst, *Synthesis*, 168 (1976).

265. P. E. Stott, *Chem. Eng. News*, **54**(37), 5 (1976).

266. T. H. Gouw, *Chem. Eng. News*, **54**(44), 5 (1976).

267. P. E. Stott, *Chem. Eng. News*, **54**(51), 5 (1976).

268. B. K. J. Leong, T. O. T. Ts'o, and M. B. Chenoweth, *Toxicol. Appl. Pharmacol.*, **27**, 342 (1974).

269. R. R. Hendrixson et al., *Toxicol. Appl. Pharmacol.*, **44**, 263 (1978).

270. K. Takayama et al., *Chem. Pharm. Bull.*, **25**, 3125 (1977).

271. B. K. J. Leong, *Chem. Eng. News*, **53**, 5 (1975).

272. B. K. J. Leong et al., *Toxicol. Appl. Pharmacol.*, **27**, 342 (1974).

273. S. C. Gad et al., *Drug Chem. Toxicol.*, **1**, 339 (1978).

274. H. K. Frensdorff, *J. Am. Chem. Soc.*, **93**, 600 (1971).

275. D. C. Tosteson, *Fed. Proc.*, **27**, 1269 (1968).

276. P. Arhem, B. Frankenhaeuser, and H. Kristbjarnarson, *Acta Physiol. Scand.*, **114**, 593 (1982).

277. E. W. Bethge et al., *Gen. Physiol. Biophys.*, **10**, 225 (1991).

CHAPTER NINE

Organic Peroxides

Jennifer B. Galvin, Ph.D., D.A.B.T., and Craig Farr, Ph.D., D.A.B.T.

1 INTRODUCTION

Peroxides as a class are highly reactive molecules owing to the presence of an oxygen–oxygen linkage in the molecular structure. Under conditions causing activation, for example, heating, the oxygen–oxygen bond is cleaved to form free radicals that are highly reactive. These highly reactive radicals can be used to initiate polymerization or curing. Consequently, organic peroxides are used as initiators for free-radical polymerization, curing agents for thermoset resins, and cross-linking agents for elastomers and polyethylene (1). An example of a peroxide chemical structure is

$$R\!-\!\underset{\substack{\| \\ O}}{C}\!-\!O\!-\!O\!-\!\underset{\substack{\| \\ O}}{C}\!-\!R'$$

The very properties that make organic peroxides valuable to industry require that these materials be handled and stored with caution. If free radicals are formed during storage in concentrated form, an accelerated decomposition may result, leading to the release of considerable heat and energy. It has been determined that decompositions of commercially available peroxides are generally low-order deflagrations rather than detonations (2). Consequently, proper storage is one of the most important factors to observe when working with organic peroxides.

Patty's Industrial Hygiene and Toxicology, Fourth Edition, Volume 2, Part A, Edited by George D. Clayton and Florence E. Clayton.
ISBN 0-471-54724-7   © 1993 John Wiley & Sons, Inc.

There have been several investigations into the types of physical hazards rep-resented by organic peroxides (2–5). These compounds may possess the combi-nation of thermal instability, sensitivity to shock, and/or friction, as well as flam-mability. Organic peroxides tend to be unstable, with the instability increasing with greater concentrations. Because of their instability, many peroxides are stored/handled in inert vehicles (6). It has been shown by Tamura (3) that the ignition sensitivity and the violence of deflagration for each organic peroxide may have a tendency to increase with increasing active oxygen content among the same type of organic peroxide, with a few exceptions. The ignition sensitivity and the violence of deflagration for each type of organic peroxide may decrease in the following order, given the same active oxygen content: diacyl peroxides > peroxyesters > dialkyl peroxides > hydroperoxides (3).

Because of the obvious immediacy of the physical hazards, basically only acute health testing has been performed on organic peroxides. Exposures should be well controlled, primarily owing to the decomposition (break-down) and/or deflagration hazard of the organic peroxide. The health data presented in Tables 9.1–9.9 were collected and furnished by the Organic Peroxide Producers Safety Division of the Society of the Plastics Industry. This represents an effort by industry to evaluate their products and provide that information to the public. Most of the information in the tables have been previously published (7) in an industry bulletin.

1.1 Detection Methods for Organic Peroxides

To determine workplace exposure to employees two methods are necessary, the first being a collection method capable of capturing the organic peroxide by some method in the breathing zone of the worker. This is referred to as the air sampling method. The second is a method of detection for analyzing that material trapped from the breathing zone of the worker. This is the analytical method. These two methods used in concert can determine the workplace air concentration to which the employee is exposed.

1.2 Air Sampling Methods

Presently only two organic peroxides have methods established for collecting air samples in the breathing zone of an employee. Research is currently underway to determine appropriate air sampling methods for some of the other organic per-oxides. The known methods are outlined in the body of this chapter.

Many organic peroxides are sold in inert carriers or vehicles such as odorless mineral spirits or dimethyl phthalate. Depending on the use and formulation of the organic peroxide, the air concentration of the carrier should also be investigated. This would require an air monitoring method specific for that material and is not covered in this chapter.

1.3 Analytic Methods

The analytic method should also be specific for each organic peroxide. NIOSH has fully validated a high-performance liquid chromatography/ultraviolet light method for benzoyl peroxide (8). Very few of the other organic peroxides have fully validated analytic methods.

One of the first conventional methods to determine concentrations of organic peroxides was the titration of iodine from sodium iodide. However, it was not specific for organic peroxides. The polarographic method came into use with the visible-recording polarograph because hydroperoxides could be distinguished from other peroxides. This method could identify the functional groups and also quantify mixtures. Di-*t*-butyl peroxide is an exception because it is not reduced polarographically (9).

Many analytical methods that have not been subjected to review by consensus standard organizations or regulatory agencies are in use. Some examples include gas chromatography (dialkyl peroxides such as di-*t*-butyl peroxide), high-pressure liquid chromatography (peroxyketals), and iodometric titration (peroxyesters, diacyl peroxides, hydroperoxides, and peroxydicarbonates). Some degree of selectivity in iodometric titrations may be obtained by variation of the reducing agent employed and the reaction conditions (10). The future will look to highly sophisticated technology such as an electrochemical detection system coupled to a high-performance liquid chromatograph (11) and cool on-column injection methods. These two methods provide the potential for both enhanced sensitivity and selectivity in peroxide determinations.

2 SPECIFIC COMPOUNDS

This section summarizes the accumulated information on the toxicology and safe handling of industrial peroxides. Because of the generous help of the Organic Peroxide Producers Safety Division of the Society of the Plastics Industry in acquiring information from various companies, this section represents information on a large number of the organic peroxides in industrial use today.

2.1 Peroxydicarbonates

This group of peroxides is of the general chemical structure found below. Most commercially available peroxydicarbonates are symmetrical in chemical structure.

$$
\begin{array}{ccc}
\text{O} & & \text{O} \\
\parallel & & \parallel \\
\text{ROC}\!-\!\text{O}\!-\!\text{O}\!-\!\text{C}\!-\!\text{OR}
\end{array}
$$

The most striking characteristic of this group is that most of them can be readily absorbed through the skin. This becomes apparent after comparing the acute dermal and oral lethality data (LD_{50}) for this group as a whole. Where data are available for both routes of exposure, 50 percent of the compounds are more acutely toxic when the material is applied to the skin than when administered orally. Further review of these data, however, shows that most of these materials are moderate to severe skin irritants, and can disrupt or destroy the natural protective function of the skin, thus permitting greater systemic absorption. Therefore systemic toxicity following dermal exposure to these compounds may result from their inherent irritancy rather than from any systemic actions per se.

Dermal exposure to these materials should not be permitted in the workplace because of the irritancy of these peroxides and not necessarily because of their systemic toxicity. These compounds would not be classified as toxic by ingestion or skin absorption based on Occupational Safety and Health Administration (OSHA) criteria, that is, toxic-ingestion $LD_{50} > 50 \leq 500$ mg/kg, and toxic-skin absorption $LD_{50} > 200 \leq 1000$ mg/kg.

These compounds are generally eye irritants; however, there are some that are not. Table 9.1 should be referred to for the irritant properties of a specific material. A limited number of peroxydicarbonates have been tested for mutagenic activity using the *Salmonella typhimurium* assay. Those tested were negative.

Presently there are insufficient data to recommend permissible exposure limits for any of the peroxydicarbonates.

2.1.1 Diisopropyl Peroxydicarbonate (IPP, Diisopropyl Perdicarbonate, Isopropyl Perdicarbonate)

$$[CH_3 - \overset{\overset{\displaystyle H}{|}}{\underset{\underset{\displaystyle CH_3}{|}}{C}} - O - \overset{\overset{\displaystyle O}{||}}{C} - O-]_2$$

Mol. Wt. 206.22
CASRN 105-64-6*

*CASRN = Chemical Abstracts Service Registry Number.

2.1.1.1 Determination in the Atmosphere. Presently there is no method for monitoring the workplace air for the concentration of IPP; however, it liberates iodine from acidified solutions of potassium iodide, providing the basis for an analytic method.

2.1.1.2 Physical/Chemical Properties and Uses. Diisopropyl peroxydicarbonate is a colorless, coarse granular crystalline solid (12). It has a melting point of 8–10°C (6). This peroxide is produced by careful reaction of isopropyl chloroformate with aqueous sodium peroxide at low temperature. IPP is commercially available in greater than 99 percent purity with an active oxygen content of 7.8 percent. It has attained commercial importance as an efficient polymerization catalyst at low reaction temperature. It is particularly useful for the polymerization of unsaturated esters such as diethylene glycol bis(allyl carbonate) and has found application as an initiator for the polymerization of ethylene and vinyl chloride (13).

Diisopropyl peroxydicarbonate is one of the few commercially available organic peroxides to be classified as a high-order deflagration hazard. It has a low decomposition temperature and its volatile decomposition products can be ignited in air. Comparably, however, IPP is much less easily ignited than black powder or dry benzoyl peroxide. IPP in the frozen state is not sensitive to friction and is less sensitive to impact. The National Fire Protection Association (NFPA) (12) recommends storing this material in a freezer with a temperature alarm in a detached, noncombustible building.

2.1.1.3 Physiological Response. Animal toxicity test data are found in Table 9.1. However, it should be emphasized that rats inhaling vapor at 9–13 ppm on a regular basis over a 3-week period produced manifestations of an irritant gas. IPP, however, has a low vapor pressure, which should reduce the risk in the workplace unless an operation or process causes this compound to be vaporized or aerosolized. No ill effects have been observed among laboratory or plant personnel, except for isolated individual cases of dermatitis and sensitivity to the characteristic odor of IPP (13).

2.1.2 Di-n-Propyl Peroxydicarbonate

$$[CH_3-CH_2-CH_2-O-\overset{\overset{\displaystyle O}{\displaystyle \|}}{C}-O-]_2$$

Mol. Wt. 206.22
CASRN 16066-38-9

2.1.2.1 Determination in the Atmosphere. No air collection method or analytic method was located for this compound.

2.1.2.2 Physical/Chemical Properties and Uses. No information was located for this compound.

2.1.2.3 Physiological Response. The data elements provided in Table 9.1 were the only toxicity information located on this chemical.

2.1.3 Di(sec-butyl) Peroxydicarbonate (Di-sec-butyl Ester Peroxydicarbonic Acid, sec-Butyl Peroxydicarbonate)

$$[CH_3-CH_2-\overset{\overset{\displaystyle CH_3}{\displaystyle |}}{\underset{\underset{\displaystyle H}{\displaystyle |}}{C}}-O-\overset{\overset{\displaystyle O}{\displaystyle \|}}{C}-O-]_2$$

Mol. Wt. 234.28
CASRN 19910-65-7

Table 9.1. Toxic Properties of Peroxydicarbonates (7)*

Peroxydicarbonate	CASRN	Oral LD$_{50}$ (mg/kg)	Dermal LD$_{50}$ (mg/kg)	Primary Skin Irritation[a]	Eye Irritation[a]	LC$_{50}$, ppm (mg/l)[b]	Salmonella typhimurium Assay
Diisopropyl peroxydicarbonate							
100%	105-64-6	2140		Mod. irritant	Ext. irritating		
30% in toluene		3720	2025	Ext. irritating	Mod. irritating		
45% in Soltrol 130		8500	1.7[c]	V. severe irritant	Irritant	Irritant gas	
45% in cyclohexane–benzene		6500	4.0[c]	V. severe irritant	Irritant		
Di-n-propyl peroxydicarbonate							
100%	16066-38-9	3400	>6800	Ext. irritating	Ext. irritating	>22.7 (>0.19) 1 hr	
85% in methylcyclohexane		4600	3500	Ext. irritating	Ext. irritating	>1433 (>12) 1 hr	
Di-sec-butyl peroxydicarbonate							
100%	19910-65-7	7600				172 ppm—no adverse effects (1 hr)	
75% in odorless mineral spirits		>4640	Not toxic at 2000[d]		Irritant		
75% in Soltrol 130		9300	1200				
Di-(2-ethylhexyl) peroxydicarbonate	16111-62-9						
97% min			Not toxic at 2000[d]		Irritant		
75% in Soltrol 130		1020					
40% in Soltrol 130		20,800					
40% in dimethyl phthalate		3690					

Compound	CAS No.	LD50[b]		Skin irritation	Eye irritation	Inhalation	Mutagenicity
Di-(4-t-butylcyclohexyl) per-oxydicarbonate	15520-11-3	>5000		Not an irritant	Not an irritant		Negative
Di-n-butyl peroxydicarbonate 50% in aromatic-free mineral spirits	16215-49-9	10[c]		V. severe irritant	Slight irritant		
Di-(3-methylbutyl) peroxydicarbonate 20% in white spirits	4113-14-8					1.7 ppm—no toxic signs; slight nose irritation	
Dicetyl peroxydicarbonate 75% wet	26322-14-5	5000		Not an irritant	Not an irritant		Negative
Di-(2-phenoxyethyl) peroxydi-carbonate	41935-39-1	>20,000	>20,000		Mild irritant		Negative
Di-(2-chloroethyl) peroxydicar-bonate	6410-72-6	4000					
Di-(3-chloropropyl) peroxydi-carbonate	34037-78-0	400					
		1500					
Di-(4-chlorobutyl) peroxydi-carbonate	14245-74-0	5200					
Di-(2-butoxyethyl) peroxydi-carbonate	6410-72-6	4000					

[a]V. = Very; ext. = extremely; mod. = moderately.
[b]All LC50 tests lasted 4 hr unless noted otherwise.
[c]LD50 reported in ml/kg.
[d]According to the Federal Hazardous Substances Act.
*All studies used rats.

2.1.3.1 Determination in the Atmosphere. No air collection method or analytic method was located for this compound.

2.1.3.2 Physical/Chemical Properties and Uses. No information was located for this compound.

2.1.3.3 Physiological Response. The data elements provided in Table 9.1 were the only toxicity information on this chemical.

2.1.4 Di(2-ethylhexyl) Peroxydicarbonate (Peroxydicarbonic Acid, Di(2-ethylhexyl) Ester)

$$[CH_3-(CH_2)_3-\underset{\underset{H}{|}}{\overset{\overset{C_2H_5}{|}}{C}}-CH_2-O-\overset{\overset{O}{||}}{C}-O-]_2$$

Mol. Wt. 346.52
CASRN 16111-62-9

2.1.4.1 Determination in the Atmosphere. No air collection method or analytic method was located for this compound.

2.1.4.2 Physical/Chemical Properties and Uses. No information was located for this compound.

2.1.4.3 Physiological Response. The data elements provided in Table 9.1 were the only toxicity information located on this chemical.

2.1.5 Di-(4-t-butylcyclohexyl) Peroxydicarbonate

$$\left[CH_3-\underset{\underset{CH_3}{|}}{\overset{\overset{CH_3}{|}}{C}}-\bigcirc-O-\overset{\overset{O}{||}}{C}-O-\right]_2$$

Mol. Wt. 398.52
CASRN 15520-11-3

2.1.5.1 Determination in the Atmosphere. No air collection method or analytic method was located for this compound.

2.1.5.2 Physical/Chemical Properties and Uses. No information was located for this compound.

2.1.5.3 Physiological Response. The data elements provided in Table 9.1 were the only available toxicity information on this chemical.

2.1.6 Di-n-Butyl Peroxydicarbonate (Dibutyl Ester Peroxydicarbonic Acid, Butyl Peroxydicarbonate, n-Butyl Peroxydicarbonate)

$$[CH_3-(CH_2)_3-O-\overset{\displaystyle O}{\overset{\displaystyle \|}{C}}-O-]_2$$

Mol. Wt. 234.28
CASRN 16215-49-9

2.1.6.1 Determination in the Atmosphere. No air collection method or analytic method was located for this compound.

2.1.6.2 Physical/Chemical Properties and Uses. It is forbidden by DOT to ship this peroxide at greater than 52 percent in solution (14).

2.1.6.3 Physiological Response. The data elements provided in Table 9.1 were the only toxicity information located on this chemical.

2.1.7 Di-(3-methylbutyl) Peroxydicarbonate (Diisoamyl Peroxydicarbonate)

$$[CH_3-\overset{\displaystyle CH_3}{\underset{\displaystyle H}{\overset{\displaystyle |}{\underset{\displaystyle |}{C}}}}-CH_2-CH_2-O-\overset{\displaystyle O}{\overset{\displaystyle \|}{C}}-O-]_2$$

Mol. Wt. 238.28
CASRN 4113-14-8

2.1.7.1 Determination in the Atmosphere. No air collection method or analytic method was located for this compound.

2.1.7.2 Physical/Chemical Properties and Uses. No information was located for this compound.

2.1.7.3 Physiological Response. The data elements provided in Table 9.1 were the only acute toxicity information located on this chemical.

A subacute inhalation study by Gage (15) used the peroxide, 20 percent by weight in white spirits, at 1 to 7 ppm to expose unidentified animals for two 6-hr exposures. There were no toxic signs observed other than slight nose irritation attributed to the white spirits. Necropsy revealed all organs were normal.

In the same report, Gage (15) exposed four male rats to a mist of this peroxide at a concentration of 44 mg/m^3 for eight 5-hr exposures. This exposure produced nose and eye irritation, respiratory difficulty, and weight loss in the animals. The necropsy revealed thickened alveolar walls in the lungs with peribronchiolar leukocytic infiltration. When the exposure level was increased to 140 mg/m³ pneumonia was produced after three 4-hr exposures.

2.1.8   Dicetyl Peroxydicarbonate (Dihexadecyl Peroxydicarbonate)

$$[CH_3-(CH_2)_{15}-O-\overset{\overset{\displaystyle O}{\|}}{C}-O-]_2$$

Mol. Wt. 571.00
CASRN 26322-14-5

2.1.8.1 Determination in the Atmosphere. No air collection method or analytic method was located for this compound.

2.1.8.2 Physical/Chemical Properties and Uses. No information was located for this compound.

2.1.8.3 Physiological Response. The data elements provided in Table 9.1 were the only toxicity information located on this chemical.

2.1.9   Di-(2-phenoxyethyl) Peroxydicarbonate

$$\left[\bigcirc\!\!-O-(CH_2)_2-O-\overset{\overset{\displaystyle O}{\|}}{C}-O-\right]_2$$

Mol. Wt. 362.32
CASRN 41935-39-1

2.1.9.1 Determination in the Atmosphere. No air collection method or analytic method was located for this compound.

2.1.9.2 Physical/Chemical Properties and Uses. No information was located for this compound.

2.1.9.3 Physiological Response. The data elements provided in Table 9.1 were the only toxicity information located on this chemical.

2.1.10   Di-(2-chloroethyl) Peroxydicarbonate

$$Cl-(CH_2)_2-O-\overset{\overset{\displaystyle O}{\|}}{C}-O-O-\overset{\overset{\displaystyle O}{\|}}{C}-O-(CH_2)_2-Cl$$

Mol. Wt. 247.03
CASRN 6410-72-6/34037-78-0

2.1.10.1 Determination in the Atmosphere. No air collection method or analytic method was located for this compound.

2.1.10.2 Physical/Chemical Properties and Uses. Di-(2-chloroethyl) peroxydicarbonate is an excellent catalyst for the polymerization of vinyl compounds and particularly for the suspension copolymerization of vinylidene chloride and vinyl chloride. It imparts superior properties with respect to odor and thermal stability as compared with those produced from conventional catalysts.

This peroxide is nonexplosive and has a half-life of 5.7 hr at 55°C. It also has a very weak odor associated with its use (16).

2.1.10.3 Physiological Response. The data elements provided in Table 9.1 were the only toxicity information located on this chemical.

2.1.11   Di-(3-chloropropyl) Peroxydicarbonate

$$Cl-(CH_2)_3-O-\overset{\overset{\displaystyle O}{\|}}{C}-O-O-\overset{\overset{\displaystyle O}{\|}}{C}-O-(CH_2)_3-Cl$$

Mol. Wt. 275.09
CASRN Undetermined

2.1.11.1 Determination in the Atmosphere. No air collection method or analytic method was located for this compound.

2.1.11.2 Physical/Chemical Properties and Uses. This peroxide is an excellent catalyst for the polymerization of vinyl compounds and particularly for the suspension copolymerization of vinylidene chloride and vinyl chloride. It imparts superior properties with respect to odor and thermal stability as compared with those produced from conventional catalysts.

This peroxide is nonexplosive and has a half-life of 4.0 hr at 55°C. It also has a very weak odor associated with its use (16).

2.1.11.3 Physiological Response. The data elements provided in Table 9.1 were the only toxicity information located on this chemical.

2.1.12   Di-(4-chlorobutyl) Peroxydicarbonate

$$Cl-(CH_2)_4-O-\overset{\overset{\displaystyle O}{\|}}{C}-O-O-\overset{\overset{\displaystyle O}{\|}}{C}-O-(CH_2)_4-Cl$$

Mol. Wt. 303.14
CASRN 14245-74-0

2.1.12.1 Determination in the Atmosphere. No air collection method or analytic method was located for this compound.

2.1.12.2 Physical/Chemical Properties and Uses. This peroxide is an excellent catalyst for the polymerization of vinyl compounds and particularly for the suspension copolymerization of vinylidene chloride and vinyl chloride. It imparts superior properties with respect to odor and thermal stability as compared with those produced from conventional catalysts.

This peroxide is nonexplosive and has a half-life of 4.3 hr at 55°C. It also has a very weak odor associated with its use (16).

2.1.12.3 Physiological Response. The data elements provided in Table 9.1 were the only toxicity information located on this chemical.

2.1.13 Di-(2-butoxyethyl) Peroxydicarbonate (Butoxyethyl Peroxydicarbonate)

$$[CH_3-(CH_2)_3-O-(CH_2)_2-O-\overset{\displaystyle O}{\overset{\displaystyle \|}{C}}-O-]_2$$

Mol. Wt. 322.35
CASRN 6410-72-6

2.1.13.1 Determination in the Atmosphere. No air collection method or analytic method was located for this compound.

2.1.13.2 Physical/Chemical Properties and Uses. Di-(2-butoxyethyl) peroxydicarbonate has a strong odor. It has a short half-life, even at low temperatures. This catalyst also has the problem of safe handling because there is the danger of decomposition accompanied by explosion (16).

2.1.13.3 Physiological Response. The data elements provided in Table 9.1 were the only toxicity information located on this chemical.

2.2 Diacyl Peroxides

This group of peroxides is of the general chemical structure:

$$R-\overset{\displaystyle O}{\overset{\displaystyle \|}{C}}-O-O-\overset{\displaystyle O}{\overset{\displaystyle \|}{C}}-R$$

Most commercially available diacyl peroxides are symmetrical in chemical structure. As a class, these peroxides are practically nontoxic by ingestion (oral LD_{50} >5000 mg/kg) or by skin absorption (dermal LD_{50} >8000 mg/kg). However, the information concerning skin absorption is quite limited (Table 9.2). None of the diacyl peroxides were considered mutagenic under the conditions of the *Salmonella typhimurium* assay.

2.2.1 Dibenzoyl Peroxide (Benzoyl Peroxide, BPO)

$$C_6H_5-\overset{\overset{\displaystyle O}{\|}}{C}-O-O-\overset{\overset{\displaystyle O}{\|}}{C}-C_6H_5$$

Mol. Wt. 242.2
CASRN 94-36-0

2.2.1.1 Determination in the Atmosphere. Benzoyl peroxide is one of the few organic peroxides for which an OSHA permissible exposure limit exists. The 8-hr time-weighted average is 5 mg/m^3. The air sampling method uses a mixed cellulose ester membrane (0.8 μm) and is analyzed using high-pressure liquid chromatography (8).

2.2.1.2 Physical/Chemical Properties and Uses. This material is a tasteless, white, granular, crystalline solid with a slight almond-like odor. The compound is formed by the reaction of sodium peroxide with benzoyl chloride in water. Dry benzoyl peroxide is a friction- and shock-sensitive material that must be handled with care. Consequently this material is generally sold in a wetted or diluted (phlegmatized) form. Handling procedures must avoid spark generation, heat, flame, or contamination. This peroxide should not be added to hot reaction mixtures, for this could result in explosive decomposition.

Benzoyl peroxide is used in industry for bleaching flour and edible oils and as an additive in the self-curing of plastics, such as acrylic dentures (17). Benzoyl peroxide is widely used as an initiator for the polymerization of vinyl monomers and styrene-unsaturated polyester resin compositions. It has been used in the vulcanization of various natural and synthetic rubbers including saturated and unsaturated hydrocarbon, polyester, and silicone rubber stocks. It has also been used as a source of free radicals in organic synthesis involving addition and deletion reactions, for example, the chlorination and the addition of hydrogen halides to unsaturated compounds. It has been described as a therapeutic aid in the treatment of burns, ulcerations, and various infected cutaneous and mucous membrane lesions. Recently topical preparations containing this organic peroxide have been most effectively used to treat acne. This successful use of benzoyl peroxide has been attributed to its antibacterial action.

2.2.1.3 Physiological Response. Human exposure up to 12.2 mg/m^3 have been reported to cause pronounced irritation of the nose and throat (18). Systemic toxicity in humans has not been reported. Skin sensitization has been reported by acne cream users (19) and in volunteers patch tested on the upper lateral arm (20). Applied experimentally to the eyes of animals, it produces superficial opacities in the cornea and inflammation of the conjunctiva, but according to another report, no injury results from a single application (21).

Animal experiments have shown that BPO is well tolerated by rats in oral dosages of 950 mg/kg (22). Other acute toxicity data are found in Table 9.2. Benzoyl peroxide has a large volume of experimental animal data surrounding it. However, most of these data concern BPO's ability to promote cancer but not initiate it in multiple strains and species, mutagenicity, and short-term studies (23–27). The relevance of these types of studies in relation to human risk has been questioned.

When repeatedly applied to the skin of mice, BPO was not carcinogenic (24). However, benzoyl peroxide is a tumor promoter in mice and hamsters but has shown no complete carcinogenic or tumor-initiating activity (28). There has been one controversial Japanese report (29) that was interpreted as BPO being a complete carcinogen. However, when the data were critically evaluated it was found consistent with BPO acting as a skin tumor promoter and not as a carcinogen. Additionally, there are other animal and in vitro studies which continue to support the lack of carcinogenic or mutagenic properties for BPO (26, 30–34). The International Agency for Research on Cancer (IARC) has evaluated the carcinogenicity of benzoyl peroxide. They classified it as Group 3. This means there is limited or inadequate evidence of carcinogenicity for animals and inadequate or absent information for humans (35a).

Benzoyl peroxide is one of the few organic peroxides that has been tested for embryo toxicity. The test method was to add solutions of the organic peroxide diluted in acetone directly onto the inner shell membrane in the air chamber of the egg, focusing it exactly on the embryo visible under the membrane (36). This organic peroxide showed 13 to 33 percent malformed embryos over a range of concentrations from 0.05 to 1.7 μmol/egg, which showed no dose response. However, it is unlikely these data can be accurately extrapolated to humans owing to the direct delivery of the dose to the embryo.

2.2.2 Di-(2,4-dichlorobenzoyl) Peroxide (2,4-Dichlorobenzoyl Peroxide, Bis(2,4-Dichlorobenzoyl) Peroxide

$$Cl_2H_3C_6 \overset{\displaystyle O}{\overset{\displaystyle \|}{-C}} -O-O- \overset{\displaystyle O}{\overset{\displaystyle \|}{C}} -C_6H_3Cl_2$$

Mol. Wt. 380.0
CASRN 133-14-2

2.2.2.1 Determination in the Atmosphere. No air collection method or analytic method was located for this compound.

2.2.2.2 Physical/Chemical Properties and Uses. The Department of Transportation (DOT) forbids it to be shipped in more than a 77 percent wetted solid in water owing to its hazardous physical properties (14).

2.2.2.3 Physiological Response. The data in Table 9.2 detail the acute toxicity of this compound. This compound appears to be of low-order toxicity based on these limited data. Mice received a single intraperitoneal injection with 500 mg/kg of di-(2,4-dichlorobenzoyl) peroxide in butyl succinate, which was reported to be the maximum tolerated dose (37a). The maximum tolerated dose is defined as that dose that slightly suppresses body weight gain. However, a later report found the LD_{50} dose to be 225 mg/kg when given intraperitoneally (38).

2.2.3 Di(p-chlorobenzoyl)Peroxide (p-Chlorobenzoyl Peroxide, Bis(p-chlorobenzoyl)Peroxide)

$$\underset{\substack{\| \\ ClH_4C_6-C-O-O-C-C_6H_4Cl}}{\overset{\substack{O \qquad O}}{}}$$

Mol. Wt. 311.12
CASRN 94-17-7

2.2.3.1 Determination in the Atmosphere. No air collection method or analytic method was located for this compound.

2.2.3.2 Physical/Chemical Properties and Uses. The Department of Transportation (DOT) forbids this material to be shipped in more than a 77 percent wetted solid in water because of its hazardous physical properties (14).

2.2.3.3 Physiological Response. This compound is structurally very similar to the preceding one, which has two chlorines each in the *para* position on the benzene rings. When comparing toxicity following an intraperitoneal injection, this compound was considerably less toxic, with an LD_{50} of 500 mg/kg in butyl succinate. The only other data available for this organic peroxide show that it had no effects on tumor cells when injected intraperitoneally at 125 mg/kg in the chest wall of mice. (37b). The data are presented in Table 9.2.

2.2.4 Di-2-methylbenzoyl Peroxide [o-Toluoyl Peroxide, Bis-(2-methylbenzoyl) Peroxide, Di-(o-methylbenzoyl) Peroxide]

$$\left[\overset{CH_3}{\underset{}{\bigcirc}} - \overset{\overset{O}{\|}}{C} - O - \right]_2$$

Mol. Wt. 270.30
CASRN 3034-79-5

2.2.4.1 Determination in the Atmosphere. No air collection method or analytic method was located for this compound.

2.2.4.2 Physical/Chemical Properties and Uses. No information was located for this compound.

2.2.4.3 Physiological Response. The data elements provided in Table 9.2 were the only toxicity information located on this chemical.

Table 9.2. Toxic Properties of Diacyl Peroxides (7)*

Diacyl Peroxide	CASRN	Oral LD$_{50}$ mg/kg	Dermal LD$_{50}$ mg/kg	Primary Skin Irritation	Eye Irritation	LC$_{50}$ (mg/l)[a]	Salmonella typhimurium Assay (Unless Specified)
Dibenzoyl peroxide 78% wet	94-36-0	Not toxic at 5000[b]		Not an irritant	Not irritating (5 min wash) Strongly irritating, but not corrosive (24 hr wash)	>24.3	Negative Negative[c]
Di-(2,4-dichlorobenzoyl) peroxide 50% in silicone fluid	133-14-2	>12,918	>8000				
Di-p-chlorobenzoyl peroxide	94-17-7	500 (ip)			Not an irritant		Negative[c]
Di-(2-methylbenzoyl) peroxide 78% wet	3034-79-5	>5000		Severe irritant	Irritant (unwashed)		Negative
Didecanoyl peroxide	762-12-9	>5000		Mod. irritating	Slight irritation		Negative
Dilauroyl peroxide	105-74-8	>5000[b]		Not an irritant	Not an irritant RTECS-moderate[f]	Toxic at 200	Negative Negative[d]

Substance	CAS No.					
Diacetyl peroxide	110-22-5					
Dipropionyl peroxide 22.7% in white spirits	3248-28-0			Severe	Negative	Saturated—1.5 hr—all animals died; 100 ppm—nose and eye irritation, respiratory difficulty, 1 death
Di-n-octanoyl peroxide	762-16-3		Severe irritation	Slight irritation	Negative	
50% in Shellsol T		>5000				
Di-(3,5,5-trimethylhexanoyl) peroxide	3851-87-4					
75% in isododecane		12.7[e]	Very severe irritation	Irritant	Negative	

[a] All LC$_{50}$ tests lasted 4 hr unless noted otherwise.
[b] According to the Federal Hazardous Substances Act.
[c] Tumor cell growth assay.
[d] Mouse lymphoma forward mutation assay, with/without metabolic activation.
[e] LD$_{50}$ reported in ml/kg.
[f] Registry of Toxic Effects of Chemical Substances classified it as moderate irritant.
* All studies used rats.

2.2.5   Didecanoyl Peroxide [*Decanoyl Peroxide, Bis(1-oxodecyl) Peroxide*]

$$[CH_3-(CH_2)_8-\overset{\overset{\displaystyle O}{\displaystyle \|}}{C}-O-]_2$$

Mol. Wt. 342.58
CASRN 762-12-9

2.2.5.1 Determination in the Atmosphere. No air collection method or analytic method was located for this compound.

2.2.5.2 Physical/Chemical Properties and Uses. The deflagration hazard was measured for this organic peroxide (3). The didecanoyl peroxide in powder form showed no pressure rise using 5 g of igniter at 98 percent purity when using the revised time-pressure test. This measurement determined that didecanoyl peroxide is not a deflagration hazard.

2.2.5.3 Physiological Response. The data elements in Table 9.2 were the only toxicity information located on this chemical.

2.2.6   Dilauroyl Peroxide [*Dodecanoyl Peroxide, Lauroyl Peroxide, Laurydol, Bis(1-oxododecyl) Peroxide, Didodecanoyl Peroxide*]

$$[CH_3-(CH_2)_{10}-\overset{\overset{\displaystyle O}{\displaystyle \|}}{C}-O-]_2$$

Mol. Wt. 398.70
CASRN 105-74-8

2.2.6.1 Determination in the Atmosphere. A simple colorimetric method has been used to determine microgram amounts of dilauroyl peroxide in solution (39).

2.2.6.2 Physical/Chemical Properties and Uses. Dilauroyl peroxide is a tasteless coarse white powder (40) with a faint pungent, soapy odor. It is produced by reaction of lauroyl chloride with sodium peroxide. Its major use is as an initiator for vinyl chloride. It is used as a polymerization agent in the plastics industry and as a curing agent for rubber. It has also been used as a burn-out agent for acetate yarns. The pharmaceutical industry uses it in topical creams in combination with antibiotics for acne treatment (6).

Dilauroyl peroxide is not a deflagration hazard. However, it has been judged an intermediate fire hazard by Noller (2). When all the physical tests available for this peroxide are evaluated collectively, it actually ranks as a low physical hazard (1).

2.2.6.3 Physiological Response. The International Agency for Research on Cancer (IARC) has evaluated the carcinogenicity of lauroyl peroxide. They classified it as a Group 3 material, which means there is limited or inadequate evidence for animals and inadequate or absent information for humans (35b). The carcinogenicity of this peroxide has primarily been studied using skin applications. After a single topical application of 10, 20, or 40 mg of lauroyl peroxide, the epidermal thickness increased markedly. This hyperplasia was characterized by a sustained production of dark basal keratinocytes (41). This peroxide is inactive as a tumor initiator or as a complete carcinogen. However, it is as effective as benzoyl peroxide as a skin tumor promoter.

Dilauroyl peroxide is one of the few organic peroxides that has been tested for embryo toxicity. The test method was to add solutions of the organic peroxide diluted in acetone directly onto the inner shell membrane in the air chamber of the egg, focusing it exactly on the embryo visible under the membrane (36). This organic peroxide showed a maximum of 25 percent malformed embryos over a range of concentrations from 0.25 to 0.50 µmol/egg, which showed no dose response. However, it is unlikely these data can be accurately extrapolated to humans owing to the direct delivery of the dose to the embryo.

Studies in male rats indicate that the following parameters should be used to detect poisoning from lauroyl peroxide: hemoglobin, methemoglobin, erythrocytes, reticulocytes, and blood peroxidase activity (42).

2.2.7 Diacetyl Peroxide (Acetyl Peroxide)

$$CH_3-\overset{\overset{\displaystyle O}{\|}}{C}-O-O-\overset{\overset{\displaystyle O}{\|}}{C}-CH_3$$

Mol. Wt. 118.10
CASRN 110-22-5

2.2.7.1 Determination in the Atmosphere. No air collection method or analytic method was located for this compound.

2.2.7.2 Physical/Chemical Properties and Uses. Diacetyl peroxide is a colorless crystal with a strong pungent odor. It must be kept in solution because of its shock sensitivity and high explosion risk. DOT forbids transport as a solid at greater than 27 percent in solution (14).

2.2.7.3 Physiological Response. Human exposure has produced eye, skin, and mucous membrane irritation after inhalation or ingestion (airborne concentrations unknown) (43). Two drops of this material (30 percent in dimethyl phthalate) applied to the eyes of rabbits caused severe corneal damage (21).

2.2.8   Di-(3-carboxypropionyl) Peroxide (Succinoyl Peroxide, Disuccinic Acid Peroxide, Disuccinoyl Peroxide, Peroxydisuccinic Acid, Succinyl Peroxide, Succinic Acid Peroxide)

$$
\underset{\text{HOC}}{\overset{\text{O}}{\|}}\text{CH}_2\text{CH}_2\underset{\text{C}}{\overset{\text{O}}{\|}}\text{—O—O—}\underset{\text{C}}{\overset{\text{O}}{\|}}\text{—CH}_2\text{CH}_2\underset{\text{C}}{\overset{\text{O}}{\|}}\text{OH}
$$

Mol. Wt. 234.18
CASRN 123-23-9

2.2.8.1 Determination in the Atmosphere. No air collection method or analytic method was located for this compound.

2.2.8.2 Physical/Chemical Properties and Uses. The compound is produced from succinic anhydride and hydrogen peroxide (44). The odorless, fine white crystalline powder has a tart taste (45).

This material must be stabilized by using a dehydrating agent such as disodium sulfate or magnesium sulfate. It is used for disinfecting, sterilizing, as a polymerization catalyst in deodorants, and for use in antiseptic agents (40). This compound mixed with a dehydrating agent does not have the friction sensitivity of a dry powder (46). It has also been reported to be stable at ambient temperature for 14 months.

2.2.8.3 Physiological Response. The only available toxicity information is for intraperitoneal (ip) injection. One reference (37c) states that the maximum dose given ip once daily to three animals for 5 to 7 successive days causing no adverse toxic effect was 15.6 mg/kg (in butyl succinate). Another reference cites the ip LD_{50} to be 10 to 12 μm/mouse in CBA mice (47). Morphological evaluations following the ip injections produced pyknonecrotic changes in the spleen and mesenteric ganglia (48). The DOT (49) reports that contact may cause burns to skin and eyes, but other reports call it only a skin irritant (40). This material has been found lacking mutagenic effectiveness in the *Drosophila melanogaster* (50). Even though not mutagenic, other reports have shown disuccinyl peroxide to induce inactivating DNA alterations (51). There are no acute toxicity data presented in Table 9.2.

2.2.9   Diproprionyl Peroxide

$$
[\text{CH}_3\text{—CH}_2\text{—}\underset{\text{C}}{\overset{\text{O}}{\|}}\text{—O—}]_2
$$

Mol. Wt. 146.14
CASRN 3248-28-0

2.2.9.1 Determination in the Atmosphere. No air collection method or analytic method was located for this compound.

2.2.9.2 Physical/Chemical Properties and Uses. No information was located for this compound.

2.2.9.3 Physiological Response. The only available toxicity information on this compound was a series of inhalation studies that used very limited exposures and numbers of animals. Table 9.2 outlines the most severe reactions. Exposures of two male and two female rats to 30 ppm four times for 5 hr each exposure caused nose irritation only. As the number of exposures increased to 19 5-hr exposures and the concentration was dropped to 10 ppm, more severe effects were observed, that is, lethargy and retarded weight gain. Necropsy found all organs normal. When the concentration was decreased to 7 ppm no toxic signs were observed or found at necropsy when four male and four female rats were exposed to 14 5-hr exposures (7).

2.2.10 Di-n-Octanoyl Peroxide

$$[CH_3-(CH_2)_6-\overset{\overset{\displaystyle O}{\displaystyle \|}}{C}-O-]_2$$

Mol. Wt. 286.40
CASRN 762-16-3

2.2.10.1 Determination in the Atmosphere. No air collection method or analytic method was located for this compound.

2.2.10.2 Physical/Chemical Properties and Uses. No information was located for this compound.

2.2.10.3 Physiological Response. The data elements provided in Table 9.2 were the only toxicity information located for this chemical.

2.2.11 Di-(3,5,5-trimethylhexanoyl) Peroxide (Diisononanoyl Peroxide)

$$[CH_3-\overset{\overset{\displaystyle CH_3}{\displaystyle |}}{\underset{\underset{\displaystyle CH_3}{\displaystyle |}}{C}}-CH_2-\overset{\overset{\displaystyle H}{\displaystyle |}}{\underset{\underset{\displaystyle CH_3}{\displaystyle |}}{C}}-CH_2-\overset{\overset{\displaystyle O}{\displaystyle \|}}{C}-O-]_2$$

Mol. Wt. 314.45
CASRN 3851-87-4

2.2.11.1 Determination in the Atmosphere. No air collection method or analytic method was located for this compound.

2.2.11.2 Physical/Chemical Properties and Uses. This material is not a deflagration hazard; the revised time-pressure test showed no pressure rise with 5 g of igniter. The peroxide tested was 75 percent pure (3).

2.2.11.3 Physiological Response. The data elements provided in Table 9.2 were the only toxicity information located on this chemical.

2.3 Peroxyesters

This group of peroxides is of the general chemical structure:

$$R\text{—}O\text{—}O\text{—}\overset{\overset{\displaystyle O}{\|}}{C}\text{—}R'$$

This class of organic peroxides has no outstanding toxicologic features. Some tend to be absorbed through the skin; however, Table 9.3 should be reviewed for the particular organic peroxide of use or interest because this is not a general trend for this group. Generally, they are not eye irritants, but some tend to be skin irritants. Only a few compounds in this group have been tested for mutagenicity. All those tested were negative in the *Salmonella typhimurium* assay, with the exception of one that was slightly positive using Chinese hamster ovary cells, in the *S. typhimurium* assay and in the mouse lymphoma forward mutation assay.

2.3.1 t-Butyl Peroxyacetate (t-Butyl Peracetate)

$$CH_3\text{—}\overset{\overset{\displaystyle O}{\|}}{C}\text{—}O\text{—}O\text{—}\overset{\overset{\displaystyle CH_3}{|}}{\underset{\underset{\displaystyle CH_3}{|}}{C}}\text{—}CH_3$$

Mol. Wt. 132.18
CASRN 107-71-1

2.3.1.1 Determination in the Atmosphere. A nonvalidated method used for air sampling employed two tubes in series of porous aromatic polymer (SKC, Catalog 226-30-02, 1989) at flow rates of 200 cm³/min. Immediately following sampling these tubes were desorbed with carbon disulfide and analyzed using gas chromatography (52).

2.3.1.2 Physical/Chemical Properties and Uses. *t*-Butyl peroxyacetate is a clear, water-like liquid. At 74 to 76 percent purity it has an active oxygen content of 8.9 to 9.2 percent. It has a half-life in benzene of 10.0 hr at 217°F (103°C).

This material decomposes to *t*-butoxy and acetoxy radicals. The *t*-butoxy radicals may be used to initiate the polymerization of vinyl monomers such as ethylene and styrene and in the cross-linking of styrene–unsaturated polyester compositions, silicone rubber, and other unsaturated resins (53b).

2.3.1.3 Physiological Response. *t*-Butyl peroxyacetate is the one member of the peroxyesters that is a primary eye irritant.

An 8-hr inhalation study at concentrations of 3, 15, 29, 305, 400, 460, 507, and 622 ppm (50% in dimethyl phthalate) was performed. No toxic effects were observed up to 29 ppm. At 305 ppm respiration difficulties appeared during the second half of the exposure period. At the four highest exposure levels animals appeared restless briefly, then lay down, showing lacrimation, nasal discharge, and slight narcosis. Labored respiration accompanied by mouth breathing occurred after 1 to 2 hr. At all levels, deaths, if any, occurred within 24 hr after the start of the exposure (54c).

The inhalation of this peroxide at 50 percent in dimethyl phthalate produced lethal concentrations for 50 percent of the animals at 450 ppm after an 8-hr exposure. The OSHA criteria for labeling a chemical toxic by inhalation is an LC_{50} >200 ≤2000 ppm. However, the time of the exposure specified by OSHA is 1 hr (55). The regulation does not clarify how the difference in time of exposure is to be rectified. This product would probably not be labeled as toxic by inhalation, however, a later study performed for 4 hr determined the LC_{50} to be 6.1 mg/l or approximately 1128 ppm (56a). Care should be taken when evaluating these data, because the second study used 75 percent peroxide in odorless mineral spirits, whereas the first used 50 percent in dimethyl phthalate. Both inhalation studies revealed that male rats were more sensitive to the peroxide than were female rats. It is unknown if this is true in the human population. Special care should be taken to protect employees from any process that might aerosolize this material. A no observed effect level was documented at 29 ppm (~156.7 mg/m³). See Table 9.3 for additional toxicity information.

2.3.2 t-Butyl Peroxypivalate (t-Butyl perpivalate, t-Butyl Trimethylperoxyacetate)

$$(CH_3)_3-C-O-O-\overset{\displaystyle O}{\overset{\displaystyle \|}{C}}-C-(CH_3)_3$$

Mol. Wt. 174.27
CASRN 927-07-1

2.3.2.1 Determination in the Atmosphere. No air collection method or analytic method was located for this compound.

2.3.2.2 Physical/Chemical Properties and Uses. *t*-Butyl peroxypivalate has 6.43 to 6.97 percent active oxygen and is commercially available at 70 to 76 percent purity (3). This organic peroxide is a highly efficient low-temperature initiator for the polymerization of numerous commercially important monomers including ethylene, vinyl chloride, vinyl acetate, styrene, acrylonitrile, and methyl methacrylate. In benzene it has a half-life of 10.0 hr at 49°C (120°F) (53c).

Noller (2) classified *t*-butyl peroxypivalate, 75 and 50 percent in mineral spirits, as a deflagration hazard after testing rate and/or violence of decomposition and ease of ignition and/or decomposition. Tamura (3), on the other hand, found no

Table 9.3. Toxic Properties of Peroxyesters (7)*

Peroxyester	CASRN	Oral LD$_{50}$ (mg/kg)	Dermal LD$_{50}$ (mg/kg)	Primary Skin Irritation	Eye Irritation	LC$_{50}$ (mg/l)[a]	Salmonella typhimurium Assay (Unless Specified)[b]
t-Butyl peroxyacetate 75% in OMS[c]	107-71-1	2562	4757	Slight irritant	Irritant	6.1	
70% in benzene (mice)		1900					
50% in Shellsol T						450 (8 hr)	
t-Butyl peroxypivalate 75% in OMS[c]	927-07-1	4169–4640	2500	Mod. to severe	Not an irritant	7.79	
t-Amyl peroxypivalate 75% in OMS[c]	29240-17-3	4270	>2000	Not an irritant	Not an irritant	>9.5	
t-Butyl peroxybenzoate	614-45-9	3639–4838	3817	Not an irritant or sensitizer	Slight irritation	>.26	Slightly positive/negative +CHO± +ML
Mice		914–2500					
t-Butyl peroxy-2-ethylhexanoate	3006-82-4	>10,000	16,818	Not an irritant	Not an irritant	42.2	
t-Amyl peroxy-2-ethylhexanoate	686-31-7	>5000	>2000	Slightly irritating	Not an irritant		
t-Butyl peroxy-3,5,5-trimethylhexanoate	13122-18-4	17.4[d]		Mod. to severe	Not an irritant	>0.8	Negative
t-Butyl peroxyneodecanoate 50% Shellsol T	26748-41-4	>12,918	>8000	Mod. to severe	Not an irritant	50.0	
t-Butyl peroxy-2-ethylhexylcarbonate	34443-12-4	>5,000	>2000	Mildly irritating	Not an irritant		
t-Butyl peroxycrotonate	23474-91-1	4100		Moderate irritant			

Compound	CAS No.	LD50	LC50	Skin irritation	Eye irritation	Inhalation/Other	Mutagenicity
t-Amyl peroxybenzoate							
Cumyl peroxyneodecanoate 90%	4511-39-1 26748-47-0						Negative
75% in OMS[c]		5126	>7940 <19,800	Not an irritant		20.2 >20.4 slight dyspnea, eye squint and wt. loss	
2,5-Dimethyl-2,5-di-(2-ethylhexanoylperoxy)-hexane	13052-09-0	>12,918	>8000	Not an irritant	Not an irritant	>800	
Di-t-butyl diperoxyazelate 75% in OMS[c]	16580-06-6	>5000		Severe irritant	Not an irritant		
1,1,3,3-Tetramethylbutyl Peroxyphenoxyacetate 30% in Shellsol T (OMS[c])	59382-51-3	>12.0[d]	>2000	Severe irritant	Not an irritant	>24 ppm	
t-Butylperoxyisopropyl-carbonate	2372-21-6	5.0[d]	>10,000	Not an irritant	Conjunctivitis		
t-Butyl monoperoxy-maleate	1931-62-0	16 (ip)					
Di-t-butyl diperoxy phthalate	15042-77-0	128 (ip)					
Cumyl peroxyneoheptanoate 75% in OMS[c]	130097-36-8 104852-44-0	~5000		Mildly irritating			

[a] All LC50 tests lasted 4 hr unless noted otherwise.
[b] ±CHO indicates a positive or negative finding in Chinese hamster ovary cells. The following +/− indicates with/without metabolic activation. ±ML indicates a positive or negative finding in the mouse lymphoma forward mutation assay.
[c] Odorless mineral spirits.
[d] LD50 reported in ml/kg.
*All studies used rats unless otherwise specified.

pressure rise with 5 g of igniter in the revised time-pressure test to evaluate deflagration.

2.3.2.3 Physiological Response. All the toxicity data located are presented in Table 9.3. However, additional subchronic data noted that no adverse effects were found after rats inhaled 50 ppm for 20 6-hr exposures. After a single 5-hr exposure of 200 ppm rats gave the appearance of nasal irritation, respiratory difficulty, and lethargy. Weight loss was observed, but at necropsy all organs appeared normal (15).

2.3.3   t-*Amyl Peroxypivalate*

$$CH_3-CH_2-\overset{\overset{\displaystyle CH_3}{|}}{\underset{\underset{\displaystyle CH_3}{|}}{C}}-O-O-\overset{\overset{\displaystyle O}{||}}{C}-\overset{\overset{\displaystyle CH_3}{|}}{\underset{\underset{\displaystyle CH_3}{|}}{C}}-CH_3$$

Mol. Wt. 188.3
CASRN 29240-17-3

2.3.3.1 Determination in the Atmosphere. No air collection method or analytic method was located for this compound.

2.3.3.2 Physical/Chemical Properties and Uses. *t*-Amyl peroxypivalate is a colorless, to slightly yellow liquid. It has an active oxygen content of 6.29 to 6.46 percent at 74 to 76 percent purity. In benzene at 0.2 mol/l it has a half-life of 10.0 hr at 58°C (137°F). It is soluble in most organic solvents and is used as a low-temperature initiator for the polymerization of ethylene, vinyl chloride, vinyl acetate, styrene, acrylonitrile, and methyl methacrylate (53d).

2.3.3.3 Physiological Response. All located toxicity data are provided in Table 9.3.
    A 4-hr inhalation study was performed and the only information provided was that during the last 2 hr of the exposure the animals experienced labored respiration. They recovered completely within a couple of hours after termination of exposure. No mortality occurred at 9.5 g/m³ of air. Necropsy did not reveal any abnormalities that were considered treatment related (54e).

2.3.4   t-*Butyl Peroxybenzoate (t-Butyl Perbenzoate, Perbenzoic Acid, t-Butyl Ester)*

$$(CH_3)_3C-O-O-\overset{\overset{\displaystyle O}{||}}{C}-\bigcirc$$

Mol. Wt. 194.25
CASRN 614-45-9

2.3.4.1 Determination in the Atmosphere. No air collection method or analytic method was located for this compound.

2.3.4.2 Physical/Chemical Properties and Uses. *t*-Butyl peroxybenzoate is produced by reacting *t*-butyl hydroperoxide with benzoyl chloride in the presence of a base. It is a colorless to slight yellow liquid with a mild aromatic odor (45). It has a half-life in benzene of 10 hr at 224°F (107°C). It is a liquid peroxyester catalyst of low volatility and high purity, effective as a medium temperature initiator for the polymerization and/or cross-linking of methyl methacrylate, acrylonitrile, isoprene, styrene, and butadiene. Low-density polyethylene of excellent mechanical properties is obtained when using this peroxide as an initiator. This peroxide is also used in various organic synthesis involving coupling reactions of olefins and paraffinic compounds as well as phenolic derivatives (53a).

The physical properties of *t*-butyl perbenzoate are that of an intermediate fire hazard (1); however, other sources refer to it as a fire hazard (2) with 8.16 percent active oxygen (3). The liquid at 99 percent purity showed a slow pressure rise with only 1 g of igniter; therefore, it is not a deflagration hazard (3). It also has low shock or impact sensitivity (1).

2.3.4.3 Physiological Response. Table 9.3 reports the acute toxicity data available on this compound. Mutagenicity tests performed by Hazleton Laboratories showed this organic peroxide to be non-mutagenic in the *Salmonella typhimurium* assay. In addition, these tests showed a slightly positive response in the Chinese hamster ovary cell and the mouse lymphoma forward mutation assays. The extrapolation of these tests to the human experience has become controversial in recent years.

Other studies have determined the possibility that *t*-butyl peroxybenzoate can be metabolized to free radicals by human carcinoma skin keratinocytes. Free radicals are suggested to be involved in the cascade of events occurring during tumor promotion (57).

Acute inhalation of this material is reported to produce no adverse effects at the maximum attainable vapor concentration of 0.26 mg/l (56b). Repeated daily inhalations of 0.006 mg/l for 4 hr (number of days not specified) brought about a decrease in body weights and urinary output. An increase in adrenal ascorbic acid and a decrease in ^{131}I uptake denote adrenal and thyroid hypofunction. No morphological alterations were found in the liver or blood. Spermatogenesis, the content of nucleic acids, and testicular depolymerase activity were not affected (58).

t-Butyl peroxybenzoate is one of the few organic peroxides that has been tested for embryo toxicity. The test method was to add solutions of the organic peroxide dissolved in acetone directly onto the inner shell membrane in the air chamber of the egg, focusing it exactly on the embryo visible under the membrane (36). This organic peroxide showed a maximum of 13 percent malformed embryos over a range of concentrations from 0.13 to 1.0 μmol/egg which showed no dose response. However, it is unlikely these data can be accurately extrapolated to humans owing to the direct delivery of the dose to the embryo.

The National Toxicology Program released a report in March, 1991 that *t*-butyl perbenzoate induced negligible systemic toxicity after dosing rats orally with 500 mg/kg.

2.3.5   t-*Butyl Peroxy-2-ethylhexanoate (t-Butyl Peroctoate)*

$$(CH_3)_3C-O-O-\overset{\overset{\displaystyle O}{\|}}{C}-\overset{\overset{\displaystyle CH_2CH_3}{|}}{\underset{\underset{\displaystyle H}{|}}{C}}-(CH_2)_3CH_3$$

Mol. Wt. 216.3
CASRN 3006-82-4

2.3.5.1 *Determination in the Atmosphere.* No air collection method or analytic method was located for this compound.

2.3.5.2 *Physical/Chemical Properties and Uses.* *t*-Butyl peroxy-2-ethylhexanoate is a colorless liquid. It has a half-life in benzene of 10.0 hr at 162°F (72°C). It is used as a medium temperature initiator for the polymerization of vinyl monomers and the curing of styrene-unsaturated polyester resins. It is regarded as an intermediate fire hazard; however, it has low impact sensitivity (53f).

2.3.5.3 *Physiological Response.* All located acute toxicity data are presented in Table 9.3. The acute inhalation study in rats used four dose levels, 103.4, 46.5, 20.8, and 9.2 mg/l. During the 4-hr exposures, nasal discharge and slight dyspnea were observed in all groups of rats. Redness of ears and paws developed shortly after exposure in all rats exposed to 103.4 and 46.5 mg/l. Deaths occurred 1 or 2 days after exposure. Necropsy revealed red patches in the lungs (56c).

2.3.6   t-*Amyl Peroxy-2-ethylhexanoate (t-Amyl Peroctoate, Peroxyhexanoic Acid, 2-Ethyl-t-pentyl ester, 2-Ethylperoxyhexanoic Acid, t-Pentyl Ester)*

$$CH_3-CH_2-(CH_3)_2-C-O-O-\overset{\overset{\displaystyle O}{\|}}{C}-\overset{\overset{\displaystyle CH_2CH_3}{|}}{C}-(CH_2)_3CH_3$$

Mol. Wt. 230.39
CASRN 686-31-7

2.3.6.1 *Determination in the Atmosphere.* No air collection method or analytic method was located for this compound.

2.3.6.2 *Physical/Chemical Properties and Uses.* No information was located for this compound.

2.3.6.3 Physiological Response. There is very little information available for *t*-amyl peroxy-2-ethylhexanoate. It has been tested in vivo (whole animal) and in cell culture for its antimalarial potency, but it was inactive in the whole animal (59).

All located acute toxicity data are presented in Table 9.3.

2.3.7 t-Butyl Peroxy-3,5,5-trimethylhexanoate (t-Butyl Peroxyisononanoate, Peroxyhexanoic Acid, 3,5,5-Trimethyl-t-butyl Ester)

$$CH_3-\overset{\overset{\displaystyle CH_3}{|}}{\underset{\underset{\displaystyle CH_3}{|}}{C}}-O-O-\overset{\overset{\displaystyle O}{||}}{C}-CH_2-\overset{\overset{\displaystyle H}{|}}{\underset{\underset{\displaystyle CH_3}{|}}{C}}-CH_2-\overset{\overset{\displaystyle CH_3}{|}}{\underset{\underset{\displaystyle CH_3}{|}}{C}}-CH_3$$

Mol. Wt. 230.39
CASRN 13122-18-4

2.3.7.1 Determination in the Atmosphere. No air collection method or analytic method was located for this compound.

2.3.7.2 Physical/Chemical Properties and Uses. No information was located for this compound.

2.3.7.3 Physiological Response. The data elements provided in Table 9.3 were the only toxicity information located on this chemical. The acute inhalation study used a finely dispersed aerosol. The particle size was measured using a Cascade impactor; the diameter of the droplets ranged from 1.2 to 4.7 μm, fully in the respirable range. The rats were somewhat restless during the first half hour of exposure to 0.8 mg/l; however, gradually all the animals fell asleep for the remainder of the 4-hr exposure. No mortality occurred; therefore, the LC_{50} is considered greater than 0.8 mg/l (54f).

2.3.8 t-Butyl Peroxyneodecanoate (t-Butyl Perneodecanoate, Peroxyneodecanoic Acid, t-Butyl Ester)

$$CH_3-\overset{\overset{\displaystyle CH_3}{|}}{\underset{\underset{\displaystyle CH_3}{|}}{C}}-O-O-\overset{\overset{\displaystyle O}{||}}{C}-\overset{\overset{\displaystyle R}{|}}{\underset{\underset{\displaystyle R_2}{|}}{C}}-R_1$$

Mol. Wt. 244.42
CASRN 26748-41-4

Each R must be at least a methyl group, and the total number of carbon atoms for all three R groups must be 8.

2.3.8.1 Determination in the Atmosphere. No air collection method or analytic method was located for this compound.

2.3.8.2 Physical/Chemical Properties and Uses. *t*-Butyl peroxyneodecanoate is a colorless liquid with 4.85 to 4.98 percent active oxygen at 75 percent purity. This peroxide is normally used dissolved 75 percent in an organic solvent. It has a half-life in benzene of 10 hr at 122°F (50°C). It is useful as a low-temperature initiator for radical catalyzed polymerization of vinyl monomers. It is considered a combustible oxidizing liquid and should be handled with care (53g).

2.3.8.3 Physiological Response. All toxicity data located are presented in Table 9.3.

2.3.9 t-Butyl Peroxy-2-Ethylhexylcarbonate

$$(CH_3)_3-C-O-O-\overset{\overset{\displaystyle O}{\|}}{C}-O-CH_2-\overset{\overset{\displaystyle H}{|}}{\underset{\underset{\displaystyle C_2H_5}{|}}{C}}-(CH_2)_3-CH_3$$

Mol. Wt. 246.34
CASRN 34443-12-4

2.3.9.1 Determination in the Atmosphere. No air collection method or analytic method was located for this compound.

2.3.9.2 Physical/Chemical Properties and Uses. No information was located for this compound.

2.3.9.3 Physiological Response. The data elements provided in Table 9.3 were the only toxicity information located on this chemical.

2.3.10 t-Butyl Peroxycrotonate (t-Butyl percrotonate)

$$(CH_3)_3-C-O-O-\overset{\overset{\displaystyle O}{\|}}{C}-CH=\overset{}{\underset{\underset{\displaystyle H}{|}}{C}}-CH_3$$

Mol. Wt. 158.2
CASRN 23474-91-1

2.3.10.1 Determination in the Atmosphere. No air collection method or analytic method was located for this compound.

2.3.10.2 Physical/Chemical Properties and Uses. No information was located for this compound.

2.3.10.3 Physiological Response. The data elements provided in Table 9.3 were the only toxicity information located on this chemical.

2.3.11   t-Amyl Peroxybenzoate (t-Amyl Perbenzoate)

Mol. Wt. 208.25
CASRN 4511-39-1

2.3.11.1 Determination in the Atmosphere. No air collection method or analytic method was located for this compound.

2.3.11.2 Physical/Chemical Properties and Uses. No information was located for this compound.

2.3.11.3 Physiological Response. The data elements provided in Table 9.3 were the only toxicity information located on this chemical.

2.3.12   Cumyl Peroxyneodecanoate

Mol. Wt. 306.43
CASRN 26748-47-0

2.3.12.1 Determination in the Atmosphere. No air collection method or analytic method was located for this compound.

2.3.12.2 Physical/Chemical Properties and Uses. No information was located for this compound.

2.3.12.3 Physiological Response. The data elements provided in Table 9.3 were the only toxicity information located on this chemical. The acute inhalation study in rats was only conducted for 1 hr. The "metered" concentration of 20.4 mg/l resulted in slight dyspnea, eye squint, and slight body weight loss measured 1 day after exposure (56f).

2.3.13 2,5-Dimethyl-2,5-di-(2-ethylhexanoylperoxy)hexane (2-Ethylperoxyhexanoic Acid, 1,1,4,4-Tetramethyltetramethylene Ester, 2-Ethylhexaneperoxoic Acid, 1,1,4,4-Tetramethyl-1,4-butanediyl Ester)

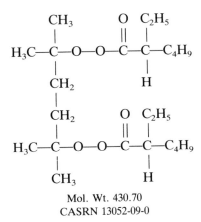

Mol. Wt. 430.70
CASRN 13052-09-0

2.3.13.1 Determination in the Atmosphere. No air collection method or analytic method was located for this compound.

2.3.13.2 Physical/Chemical Properties and Uses. No information was located for this compound.

2.3.13.3 Physiological Response. The data elements provided in Table 9.3 were the only toxicity information located on this chemical.

2.3.14 Di-t-butyl Diperoxyazelate (Di-t-butyl Diperazelate)

$$(CH_3)_3-C-O-O-\overset{\displaystyle O}{\overset{\displaystyle \|}{C}}-(CH_2)_7-\overset{\displaystyle O}{\overset{\displaystyle \|}{C}}-O-O-C-(CH_3)_3$$

Mol. Wt. 332.4
CASRN 16580-06-6

2.3.14.1 Determination in the Atmosphere. No air collection method or analytic method was located for this compound.

2.3.14.2 Physical/Chemical Properties and Uses. No information was located for this compound.

2.3.14.3 Physiological Response. The data elements provided in Table 9.3 were the only toxicity information located on this chemical.

2.3.15 1,1,3,3-Tetramethylbutyl Peroxyphenoxyacetate (2,4,4-Trimethylpentyl 2-Peroxyphenoxyacetate)

Mol. Wt. 280.35
CASRN 59382-51-3

2.3.15.1 Determination in the Atmosphere. No air collection method or analytic method was located for this compound.

2.3.15.2 Physical/Chemical Properties and Uses. No information was located for this compound.

2.3.15.3 Physiological Response. The data elements provided in Table 9.3 were the only toxicity information located on this chemical. The acute inhalation study exposed rats for 4 hr to an aerosol of 30 percent solution of the substance in Shellsol T at a concentration of 24 ppm. During the first half hour of the exposure period the rats were slightly restless. This symptom gradually disappeared and after 1 hr all rats were asleep. No mortalities occurred; therefore the LC_{50} is greater than 24 ppm (54g).

2.3.16 t-Butyl Peroxyisopropylcarbonate (Peroxycarbonic Acid, OO-t-Butyl O-Isopropyl ester, OO-t-Butyl O-Isopropyl Monoperoxycarbonate)

Mol. Wt. 176.24
CASRN 2372-21-6

2.3.16.1 Determination in the Atmosphere. No air collection method or analytic method was located for this compound.

2.3.16.2 Physical/Chemical Properties and Uses. This liquid is not a deflagration hazard with 8.9 percent active oxygen at 98 percent purity. It gives a slow pressure rise using 4 g of igniter in the revised time-pressure test (3).

2.3.16.3 Physiological Response. The data elements provided in Table 9.3 were the only toxicity information located on this chemical.

2.3.17 t-Butyl Monoperoxymaleate (Peroxymaleic Acid, OO-t-Butyl Ester, t-Butyl Peroxymaleic Acid)

$$(CH_3)_3C-O-O-\overset{\overset{\displaystyle O}{\|}}{C}-\underset{\underset{\displaystyle H}{|}}{C}=\underset{\underset{\displaystyle H}{|}}{C}-\overset{\overset{\displaystyle O}{\|}}{C}-OH$$

Mol. Wt. 188.18
CASRN 1931-62-0

2.3.17.1 Determination in the Atmosphere. No air collection method or analytic method was located for this compound.

2.3.17.2 Physical/Chemical Properties and Uses. This powder with 8.4 percent active oxygen and 98.8 percent purity showed small pressure rises in the time-pressure assay when using only 1 g of igniter (3). This indicates that this material poses a deflagration hazard.

2.3.17.3 Physiological Response. The data elements provided in Table 9.3 were the only toxicity information located on this chemical.

2.3.18 Di-t-Butyl Diperoxyphthalate

Mol. Wt. 310.00
CASRN 15042-77-0

2.3.18.1 Determination in the Atmosphere. No air collection method or analytic method was located for this compound.

2.3.18.2 Physical/Chemical Properties and Uses. This peroxide powder with 8.45 percent active oxygen and 82 percent purity was mixed 20 percent in water and showed small pressure rises in the time-pressure assay when using 1 g of igniter (3). This indicates that this material poses a deflagration hazard.

2.3.18.3 Physiological Response. The data elements provided in Table 9.3 were the only toxicity information located on this chemical.

2.3.19   Cumyl Peroxyneoheptanoate (Cumyl Perneoheptanoate)

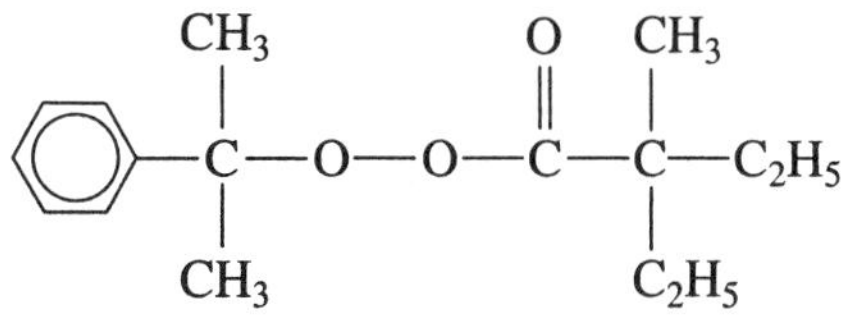

(a major isomer)
Mol. Wt. 264.35
CASRN 130097-36-8 or 104852-44-0

2.3.19.1 Determination in the Atmosphere. No air collection method or analytic method was located for this compound.

2.3.19.2 Physical/Chemical Properties and Uses. No information was located for this compound.

2.3.19.3 Physiological Response. The data elements provided in Table 9.3 were the only toxicity information located on this chemical.

2.4   Ketone Peroxides

This group of peroxides is of the general structure:

$$(RR')C(OOH)_2$$

The most striking toxicologic feature of this group is that they all have irritating properties toward the eyes and skin, ranging from moderately irritating to corrosive. Eye and face protection should be worn at all times when working with these compounds. Impervious gloves and, if warranted, arm protection, should be worn where dermal exposure could occur. Clothes should be removed promptly if splashes to the body occur to prevent skin irritation.

2.4.1   Methyl Ethyl Ketone Peroxide (2-Butanone Peroxide, MEK Peroxide)

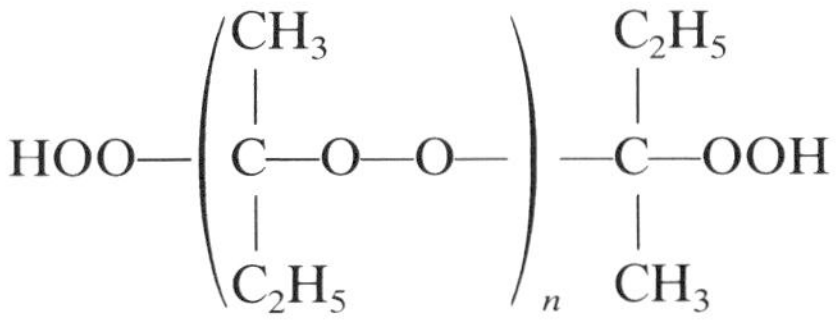

Monomer Mol. Wt. 122.12
Dimer Mol. Wt. 210.22
Trimer Mol. Wt. 298.34
CASRN 1338-23-4

2.4.1.1 Determination in the Atmosphere. MEK peroxide is one of the few organic peroxides for which an OSHA permissible exposure limit exists. This limit is a ceiling of 0.2 ppm (1.5 mg/m³)(18). The air sampling method uses a standard size XAD-4 tube and is analyzed using high-pressure liquid chromatography (60).

2.4.1.2 Physical/Chemical Properties and Uses. There are many synonyms for this compound; 2-butanone peroxide and MEK peroxide (MEKP) are the most common. This compound is a colorless liquid with a fragrant, mint-like, moderately sharp odor. Commercially available forms are 9.0 percent maximum active oxygen and are not considered a deflagration hazard. MEK peroxide is used as a curing agent with polyester resins for adhesives, plastics, lacquers, and fiber glass resin kits for boat and automobile body repair. The peroxide is usually one component used in a two-component mixture referred to as "epoxies." However, this is a misnomer because they do not contain epoxy groups (61).

2.4.1.3 Physiological Response. This compound is the most studied toxicologically of this group. It has a workplace exposure limit of 0.2 ppm (1.5 mg/m³), which is a ceiling limit and should not be exceeded at any time during the work shift (18). It is a respiratory tract irritant and a severe eye and skin irritant. The maximum nonirritating strength for skin was 1.5 percent, for eyes 0.6 percent (9). Skin irritation was delayed, with erythema and edema appearing within 2 to 3 days. Eyes washed within 4 sec after exposure resulted in no adverse effects. However, it has been noted that single MEK peroxide exposures to the eye can exacerbate preexisting corneal and limbal disease (62).

Acute intoxication with this organic peroxide is accompanied by a drop in blood pressure. The cause of this effect is unknown, but may involve lipid peroxidation of cell membranes (63).

Subchronic oral exposures three times a week for 7 weeks at one-fifth the LD_{50} caused all five rats to die. Necropsy revealed mild liver damage with glycogen depletion (64). These data suggest accumulation of the material in the body. This material is currently being assessed for carcinogenicity by the National Toxicology Program (NTP) in skin painting studies. Short term in vitro mutagenic studies have given mixed results (65).

An MEK peroxide technical product (50 percent MEKP in phosphoric acid ester and phthalic acid ester) was tested for embryo toxicity in 3-day-old chicken embryos. A dose of 0.29 μmol/egg caused early deaths in 50 percent of treated eggs and malformations in 40 percent of the survivors (36). The authors classified MEK peroxide as a moderately potent embryo toxin. However, the relevance of these data in relation to humans is unknown. The test method applied solutions directly to the inner membrane, focusing it on the embryo. It is unlikely these data can be accurately extrapolated to humans owing to the direct delivery of the dose.

Human exposure attests to the corrosive nature of this compound, causing chemical burns of the gastrointestinal tract with residual scarring and stricture of the esophagus (64). Death 2 to 3 days later was due to hepatic failure after ingestion of this material. An oral dose of 50 to 100 ml is toxic and potentially lethal to adults (66).

2.4.2 Methyl Isobutyl Ketone Peroxides (4-Methyl-2-pentanone Peroxide, Isobutyl Methyl Ketone Peroxide)

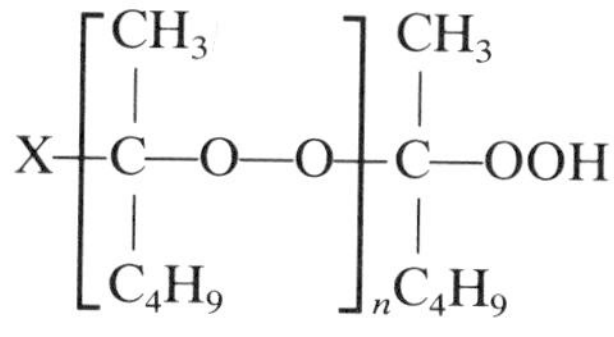

$$X \left[\begin{array}{c} CH_3 \\ | \\ C-O-O \\ | \\ C_4H_9 \end{array} \right]_n \begin{array}{c} CH_3 \\ | \\ C-OOH \\ | \\ C_4H_9 \end{array}$$

$n = 0 \text{ to } 4$
X = OH, OOH
Mol. Wt. various
CASRN 37206-20-5

2.4.2.1 Determination in the Atmosphere. No air collection method or analytic method was located for this compound.

2.4.2.2 Physical/Chemical Properties and Uses. Methyl isobutyl ketone peroxide (MIKP) is obtained from the reaction of methyl isobutyl ketone with hydrogen peroxide. During the production of this peroxide, a mixture of MIKP is obtained (67).

2.4.2.3 Physiological Response. The data elements provided in Table 9.4 were the only toxicity information located on this chemical.

The acute inhalation study used 60 percent methyl isobutyl ketone peroxides and 40 percent diisobutyl phthlate at 2.56, 1.77, 1.55, 1.38, and 1.30 mg/l (g/m^3). The particle size was determined using a Cascade impactor. Eighty percent of the mist consisted of droplets with a diameter of 1.7 to 3.3 μm. The maximum droplet size was 6.7 μm. During the first half hour of exposure the rats were restless. During the whole exposure period the animals kept their eyes closed. After the exposure period most of the rats of the higher dose groups showed mouth breathing. The exposed animals lost weight during the first week post-exposure, but regained the weight the following week. Based on these data the LC$_{50}$ of this material was determined to be 1.5 mg/l (g/m^3), with 1.4 and 1.61 mg/l (g/m^3) representing the 95 percent confidence limits (54h).

2.4.3 Acetyl Acetone Peroxide (2,4-Pentanedione Peroxide, 3,5-Dimethyl-3,5-dihydroxy-1,2-dioxolone)

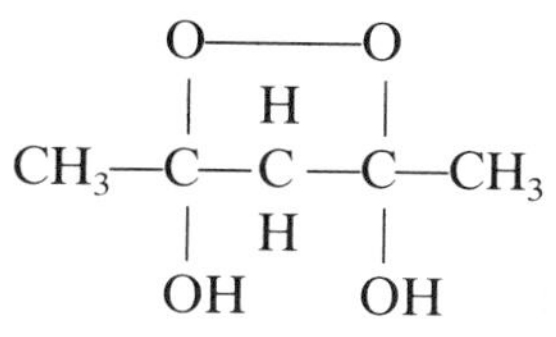

$$\begin{array}{ccc} O\!\!-\!\!-\!\!-\!\!O \\ | \quad H \quad | \\ CH_3\!-\!C\!-\!C\!-\!C\!-\!CH_3 \\ | \quad H \quad | \\ OH \quad\quad OH \end{array}$$

Mol. Wt. 134.13
CASRN 37187-22-7

Table 9.4. Toxic Properties of Ketone Peroxides (7)*

Ketone Peroxide	CASRN	Oral LD$_{50}$ (mg/kg)	Dermal LD$_{50}$ (mg/kg)	Primary Skin Irritation	Eye Irritation	LC$_{50}$ (mg/l)[a]	Salmonella typhimurium Assay
Methyl ethyl ketone peroxide	1338-23-4						Positive
OPPSD composite		>500<5000		Moderate irritant	Corrosive	>200	Negative
Lupersol DDM		681					
Lucidol & Cadet		484		Irritant	Irritant	200 ppm	
Noury (40% in DMP[b])		1017	4000		Corrosive	17	
Noury & Lucidol						33	
Methyl isobutyl ketone peroxides	37206-20-5	1.77[d]		Very severe irritant	Severe irritant	1.5	Negative
Acetyl acetone peroxide	37187-22-7	2870		Not an irritant	Severe irritant	Not a hazard at 13.1 mg/l for 1 hr	
Diacetone alcohol peroxide	54693-46-8	2.68[d]		Very severe irritant	Severe irritant	0.54	Negative
1,1-Dihydroperoxycyclohexane 21% in DMP[b]	2699-11-9	1.08[d]		Very severe irritant	Very severe irritant		Slightly positive
Cyclohexanone peroxide	12262-58-7				Severe irritant		
Di-(1-hydroxycyclohexyl) peroxide	2407-94-5						
100% (mice)		3080		Irritating	Irritating		
60% in DBP[c] (mice)		2800					
Di-(1-hydroperoxycyclohexyl) peroxide	2699-12-9						
100% (mice)		900		Irritating	Irritating		
60% in DBP[c] (mice)		850					
1-Hydroperoxy-1-hydroxy-dicyclohexyl peroxide	78-18-2						
100% (mice)		880		Irritating	Irritating		
60% in DBP[c] (mice)		740					

[a]All LC$_{50}$ tests lasted 4 hr unless noted otherwise. [b]DMP dimethyl phthalate. [c]DBP dibutyl phthalate. [d]LD$_{50}$ reported in ml/kg. *All studies used rats unless otherwise specified.

2.4.3.1 Determination in the Atmosphere. No air collection method or analytic method was located for this compound.

2.4.3.2 Physical/Chemical Properties and Uses. No information was located for this compound.

2.4.3.3 Physiological Response. The embryo toxicity of this peroxide was studied using 3-day chicken embryos. The peroxide was delivered directly onto the inner membrane overlaying the embryo. The median effective dose (ED_{50})/egg was 0.34 μmol producing 23% malformed embryos (36). Because of the obvious difference in delivery of the dose in this study versus humans, the relevance of this information to humans is unknown. All other located toxicity data are presented in Table 9.4.

2.4.4 Diacetone Alcohol Peroxide (2,4-Dihydroxy-2-methyl-4-hydroperoxypentane, 4-Hydroxy-4-methyl-2-pentanone Peroxide)

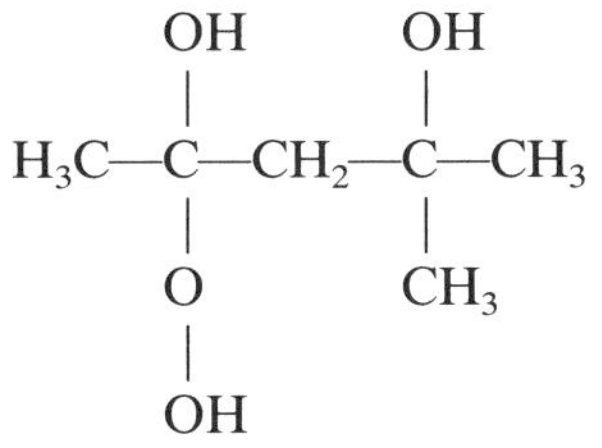

$$
\begin{array}{ccc}
& OH & OH \\
& | & | \\
H_3C-C-CH_2- & C-CH_3 \\
& | & | \\
& O & CH_3 \\
& | & \\
& OH &
\end{array}
$$

Mol. Wt. 150.17
CASRN 54693-46-8

2.4.4.1 Determination in the Atmosphere. No air collection method or analytic method was located for this compound.

2.4.4.2 Physical/Chemical Properties and Uses. No information was located for this compound.

2.4.4.3 Physiological Response. The data elements provided in Table 9.4 were the only toxicity information located on this chemical.

The acute inhalation study used 51 to 52 percent diacetone alcohol peroxide, with the balance being diacetone alcohol, water, and hydrogen peroxide at 0.75, 0.51, 0.37, and 0.27 mg/l (g/m^3). The particle size was determined using a Cascade impactor. Ninety percent of the mist consisted of droplets with a diameter of 1.7 to 2.4 μm. The maximum droplet size was 3.3 μm. During the whole exposure period the animals kept their eyes closed. Mouth breathing and labored respiration occurred during the second half of the exposure period. The exposed animals lost weight during the first 3 days post-exposure, but regained the weight the following week. Based on these data the LC_{50} of this material was determined to be 0.54

mg/l (g/m³), with 0.46 and 0.64 mg/l (g/m³) representing the 95 percent confidence limits (54h).

2.4.5   1,1-Dihydroperoxycyclohexane

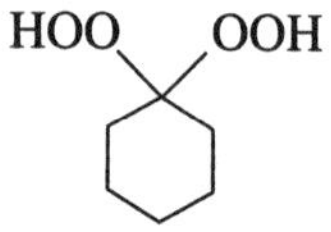

HOO   OOH

Mol. Wt. 148.16
CASRN 2699-11-9

2.4.5.1 Determination in the Atmosphere. No air collection method or analytic method was located for this compound.

2.4.5.2 Physical/Chemical Properties and Uses. No information was located for this compound.

2.4.5.3 Physiological Response. The data elements provided in Table 9.4 were the only toxicity information located on this chemical.

2.4.6   Cyclohexanone Peroxide

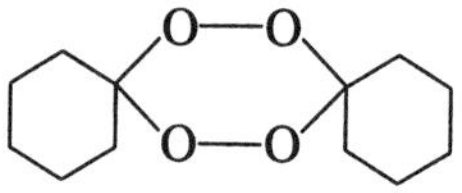

O—O
O—O

Mol. Wt. 228.28
CASRN 12262-58-7

2.4.6.1 Determination in the Atmosphere. No air collection method or analytic method was located for this compound.

2.4.6.2 Physical/Chemical Properties and Uses. No information was located for this compound.

2.4.6.3 Physiological Response. The embryo toxicity of this peroxide was studied using 3-day chicken embryos. The peroxide was delivered directly onto the inner membrane overlaying the embryo. The median effective dose (ED_{50})/egg was 0.13 μmol producing 27 percent malformed embryos (36). Because of the obvious difference in delivery of the dose in this study versus humans, the relevance of this information to humans is unknown. All other toxicity data located are presented in Table 9.4.

2.4.7 Di(1-hydroxycyclohexyl) Peroxide [Dihydroxydicyclohexyl Peroxide, Bis(1-hydroxycyclohexyl) Peroxide, 1,1'-Dioxybiscyclohexanol, 1,1'-Peroxydicyclohexanol]

Mol. Wt. 230.34
CASRN 2407-94-5

2.4.7.1 Determination in the Atmosphere. No air collection method or analytic method was located for this compound.

2.4.7.2 Physical/Chemical Properties and Uses. No information was located for this compound.

2.4.7.3 Physiological Response. A Russian article (English abstract) reports in mice a single oral dose of 65 mg/kg as the threshold dose for nerve-muscle irritation, perturbation in hemoglobin/methemoglobin content, and catalase and peroxidase activity in blood (68a). All other available toxicity data are presented in Table 9.4.

2.4.8 Di-(1-hydroperoxycyclohexyl) Peroxide (Cyclohexyl Peroxide Dihydroperoxide, Dioxydicyclohexylidene Bis-hydroperoxide)

Mol. Wt. 262.34
CASRN 2699-12-9

2.4.8.1 Determination in the Atmosphere. The photocolorimetric determination has been used to analyze for this peroxide in water (69).

2.4.8.2 Physical/Chemical Properties and Uses. No information was located for this compound.

2.4.8.3 Physiological Response. A Russian article (English abstract) reports in mice a single oral dose of 18 mg/kg as the threshold dose for nerve-muscle irritation, xnperturbation in hemoglobin/methemoglobin content, and catalase and peroxidase activity in blood (68b). All other available toxicity data are presented in Table 9.4.

2.4.9 1-Hydroperoxy-1'-hydroxydicyclohexyl Peroxide
[1-(1-Hydroperoxycyclohexyl)dioxycyclohexanol, Cyclohexanone Peroxide,
1-Hydroperoxycyclohexyl-1-hydroxycyclohexyl Peroxide, Cyclohexyl
Peroxide Dihydroperoxide]

Mol. Wt. 246.34
CASRN 78-18-2

2.4.9.1 Determination in the Atmosphere. The photocolorimetric determination has been used to analyze for this peroxide in water (69).

2.4.9.2 Physical/Chemical Properties and Uses. This peroxide is an off-white thick paste (45) used as a catalyst for hardening certain fiber glass resins.

2.4.9.3 Physiological Response. All available acute toxicity data are presented in Table 9.4.

In the workplace the peroxide has caused eye irritation and skin sensitization (70). Prolonged inhalation of vapors result in headache and throat irritation. Prolonged skin contact with clothing contaminated with peroxides may cause irritation and blistering (6).

A Russian article (English abstract) reports in mice a single oral dose of 1.028 g/kg as the threshold dose for nerve-muscle irritation, perturbation in hemoglobin/methemoglobin content, and catalase and peroxidase activity in blood (68b).

When this peroxide was administered to rats orally at 72 mg/kg (10% of the LD_{50}) on a daily basis for 30 days it decreased the mobility of spermatozoids. Peroxide treatment of the males increased embryonic death and decreased the average weight of the first generation offspring. It did not affect the percentage of females impregnated nor the number of embryos per female (68b).

2.5 Dialkyl Peroxides

This group of peroxides is of the general chemical structure:

$$R—O—O—R'$$

These peroxides are practically nontoxic if swallowed. Very few have been tested for dermal absorption and subsequent toxicity; however, the few that have been tested express low toxicity by this route of exposure. Table 9.5 should be referred to for the irritancy of these individual chemicals. Generally, they are not eye or skin irritants.

Table 9.5. Toxic Properties of Dialkyl Peroxides (7)*

Dialkyl Peroxide	CASRN	Oral LD$_{50}$ (mg/kg)	Dermal LD$_{50}$ (mg/kg)	Primary Skin Irritation	Eye Irritation	LC$_{50}$^{a}	Salmonella typhimurium Assay
Di-t-butyl peroxide Mice Mice	110-05-4	>25,000 >20.0^b >50.0^b	>10,000	Not an irritant	Not an irritant	>4103 ppm	Negative
2,5-Dimethyl-2,5-di-(t-butylperoxy)hexane	78-63-7	>32,000	4100				
2,5-Dimethyl-2,5-di-(t-butylperoxy)hexyne-3 In dodecane 90%	1068-27-5	>7680		Moderate irritant Not an irritant	Nontoxic^c		
Dicumyl peroxide 96% min	80-43-3	4100	4100	Mild irritation, no sensitizer			Negative
Dust from 40% on filter							
20% in corn oil 50% in corn oil		~4000			Mild (unwashed) conjunctivitis	2.24 mg/l no effect (6 h)	
α,α'-Bis(t-butylperoxy)diisopropylbenzenes 96% min	25155-25-3 2781-00-2	>23,100		Slight irritation, not a sensitizer	Minimal irritation	>6000 ppm—vapor >180 mg/m³—dust	Negative
t-Butyl cumyl peroxide 92% Mice	3457-61-2	>4500					
4-(t-Butylperoxy)-4-methyl-2-pentanone	26394-04-7	5.18^b 3949	>20,000	Severe irritant	Not an irritant Slight irritation (unwashed)	>140 ppm (1.2 mg/l) >2.3 mg/l (1 hr)	Negative

^{a}All LC$_{50}$ tests lasted 4 hr unless noted otherwise.
^{b}LD$_{50}$ reported in ml/kg.
^{c}According to the Federal Hazardous Substances Act.
*All studies used rats unless otherwise specified.

2.5.1 Di-t-Butyl Peroxide [DTBP, t-Butyl Peroxide, Bis(t-butyl) Peroxide]

$$
\begin{array}{ccc}
CH_3 & & CH_3 \\
| & & | \\
CH_3-C-O-O-C-CH_3 \\
| & & | \\
CH_3 & & CH_3
\end{array}
$$

Mol. Wt. 146.26
CASRN 110-05-4

2.5.1.1 Determination in the Atmosphere. No air collection method exists for sampling this peroxide in the breathing zone of the worker. One method reported for analyzing dialkyl peroxides is high-pressure liquid chromatography (71).

2.5.1.2 Physical/Chemical Properties and Uses. This peroxide is a clear to yellow liquid. It is produced by reacting sulfated t-butyl alcohol or isobutylene with hydrogen peroxide. It is an important initiator for high-temperature, high-pressure polymerizations of ethylene and halogenated ethylenes. It is used in curing resins or allyl acetate and allyl phthalate types and is also used in the synthesis of polyketones from carbon monoxide and ethylene (6).

Di-t-butyl peroxide is classified a deflagration hazard by Noller (2); however, it has low impact sensitivity (1). In the revised time-pressure test this peroxide gave a relatively slow rise in pressure when 3 g of igniter was used. Therefore Tamura (3) did not classify it as a deflagration hazard.

Di-t-butyl peroxide has been reported to increase the degree of purification in the combustion gas of power plants by removing toxic impurities (72).

2.5.1.3 Physiological Response. Acute data elements are provided in Table 9.5.

Di-t-butyl peroxide caused no eye effects when instilled at 0.1 g, 0.5 g, or an unstated amount into rabbit's eyes (9, 73), and only mild irritation was reported in two additional studies (74).

A 4-hr exposure to 24.5 g/m^3 of di-t-butyl peroxide caused excitability and breathing difficulties in 10 mice, but there were no deaths. Similar exposure of six rats caused tremors of the head and neck, weakness of the limbs, and prostration. One animal that became hyperactive died after 7 days. Autopsies of rats and mice that died during this treatment indicated lung damage was the cause of death (9).

A single exposure (route unspecified, but probably subcutaneous) of 14.6 mg (~365 mg/kg) produced unconvincing evidence for carcinogenicity owing to the lack of controls in 50 mice observed for more than 80 weeks. Of 35 survivors, seven (20 percent) had malignant blood tumors (lymphomas) and one had a benign lung tumor (pulmonary adenoma) (75). Owing to its poor design, this study should be judged inadequate to determine carcinogenicity.

Di-t-butyl peroxide was not mutagenic in the bacterial *Salmonella typhimurium* test with or without a liver metabolic activation fraction (54b, 76). No DNA damage was seen in tests with the bacterium *Pneumococcus* (77), but an equivocal result was obtained using *Escherichia coli* bacteria (SOS chromotest) (78). Mutagenicity tests performed in the fungus *Neurospora* have given both negative (79) and reportedly positive results (80), although no results were reported to support the positive findings.

2.5.2 2,5-Dimethyl-2,5-di-(t-butylperoxy)hexane

$$CH_3-\underset{\underset{CH_3}{|}}{\overset{\overset{CH_3}{|}}{C}}-O-O-\underset{\underset{CH_3}{|}}{\overset{\overset{CH_3}{|}}{C}}-CH_2-CH_2-\underset{\underset{CH_3}{|}}{\overset{\overset{CH_3}{|}}{C}}-O-O-\underset{\underset{CH_3}{|}}{\overset{\overset{CH_3}{|}}{C}}-CH_3$$

Mol. Wt. 290.4
CASRN 78-63-7

2.5.2.1 Determination in the Atmosphere. No air collection method exists for sampling this peroxide in the breathing zone of the worker. One method reported for analyzing dialkyl peroxides is high-pressure liquid chromatography (71).

2.5.2.2 Physical/Chemical Properties and Uses. This peroxide is a colorless to slightly yellow liquid used as a melt flow modifier for polyolefins. Its half-life in benzene is 10 hr at 248°F (120°C) (53d).

The deflagration properties of this material was tested by Tamura (3). There was no rise in pressure using 5 g of igniter; therefore, it is not a deflagration hazard.

The thermal decomposition products of this peroxide heated at 125°C in several different organic solvents were basically acetone, *t*-butyl alcohol, *t*-amyl alcohol, methane, ethane, ethylene, *t*-amyl-*t*-butyl peroxide, and some residual parent compound (81).

2.5.2.3 Physiological Response. The data elements provided in Table 9.5 were the only toxicity information located on this chemical.

2.5.3 2,5-Dimethyl-2,5-di-(t-butylperoxy)hexyne-3

$$H_3C-\underset{\underset{CH_3}{|}}{\overset{\overset{CH_3}{|}}{C}}-O-O-\underset{\underset{CH_3}{|}}{\overset{\overset{CH_3}{|}}{C}}-C\equiv C-\underset{\underset{CH_3}{|}}{\overset{\overset{CH_3}{|}}{C}}-O-O-\underset{\underset{CH_3}{|}}{\overset{\overset{CH_3}{|}}{C}}-CH_3$$

Mol. Wt. 286.46
CASRN 1068-27-5

2.5.3.1 Determination in the Atmosphere. No air collection method exists for sampling this peroxide in the breathing zone of the worker. One method reported for analyzing dialkyl peroxides is high-pressure liquid chromatography (71).

2.5.3.2 Physical/Chemical Properties and Uses. This peroxide is a yellow liquid used in the cross-linking of various polyolefins. Its half-life in benzene is 10 hr at 267°F (130°C) (53e).

2.5.3.3 Physiological Response. The data elements provided in Table 9.5 were the only toxicity information located on this chemical. The information reviewed on the acute inhalation study stated that it was a nontoxic compound. The concentration tested was not specified (56d).

2.5.4 Dicumyl Peroxide [Cumyl Peroxide, Bis(α,α-dimethylbenzyl) Peroxide, Diisopropylbenzene Peroxide, Isopropylbenzene Peroxide]

Mol. Wt. 270.37
CASRN 80-43-3

2.5.4.1 Determination in the Atmosphere. No air collection method exists for sampling this peroxide in the breathing zone of the worker. An analytic method exists that consists of using a liquid chromatograph equipped with a 254-nm ultraviolet detector (82a). Howver, the most recent method uses high-pressure liquid chromatography (83).

2.5.4.2 Physical/Chemical Properties and Uses. Dicumyl peroxide is a crystalline solid that melts at 42°C. It is insoluble in water and soluble in vegetable oil and organic solvents (84). It is used as a high-temperature catalyst in production of polystyrene plastics.

The deflagration hazard potential of this peroxide was tested using 5 g of igniter in the revised time-pressure test, but no pressure rise was produced (3). Noller (2) found it to be an intermediate fire hazard.

2.5.4.3 Physiological Response. Mild conjunctivitis was produced in the eyes of rabbits after instillation of 0.1 ml of 50 percent dicumyl peroxide in corn oil. Guinea pigs receiving intradermal injections of 0.1 percent in saline produced no sensitization (82a).

Dicumyl peroxide has been reported to cause slight skin irritation, with no skin sensitization in 200 human volunteers patch tested with a technical grade of 90 percent of this material (82a). However, there has been one report of one female worker allegedly contracting occupational asthma after inhaling the vapors from a heated polyethylene repair tape (85). It was reported that the vapors most likely consisted of dicumyl peroxide and its breakdown products (82b).

A group of 18 workers exposed to dicumyl peroxide in the atmosphere (concentration unspecified) exhibited nasal "crusting" and visible blood vessels (86).

Rabbits' nostrils instilled with 50 μl of 0.001 or 0.0025 percent in saline suffered slight inflammation of the nasal mucosa within 1 hr. Repeated applications of 0.001 percent, three times a day, 5 days/week for up to 4 weeks caused increased vascularization in the nose (87). No overt signs of toxicity were seen in groups of 10 rats and four rabbits exposed for 6 hr to atmospheres containing 21 to 224 mg dust/m^3 (8 to 90 mg dicumyl peroxide/m^3) (82d).

The embryo toxicity of this peroxide was studied using 3-day chicken embryos. The peroxide was delivered directly onto the inner membrane overlaying the embryo (36). The median effective dose (ED_{50})/egg was 1.24 μmol producing 57 percent malformed embryos. Because of the obvious difference in delivery of the dose in this study versus that in humans, the relevance of this information to humans is unknown. All other available toxicity data are presented in Table 9.5.

No evidence of mutagenicity was seen in the *Salmonella typhimurium* assay with or without metabolic activation (54, 76, 88, 89). However, this material was reported to be mutagenic in *Escherichia coli* in the absence of a liver metabolic activation fraction (90).

2.5.5   α,α'-Bis(t-butylperoxy)diisopropylbenzene

$$(CH_3)_3C{-}O{-}O{-}\underset{\underset{CH_3}{|}}{\overset{\overset{CH_3}{|}}{C}}{-}\bigcirc{-}\underset{\underset{CH_3}{|}}{\overset{\overset{CH_3}{|}}{C}}{-}O{-}O{-}C(CH_3)_3$$

Mol. Wt. 338.47
CASRN 25155-25-3, 2781-00-2

2.5.5.1 Determination in the Atmosphere. No air collection method exists for sampling this peroxide in the breathing zone of the worker. One method reported for analyzing dialkyl peroxides is high-pressure liquid chromatography (71).

2.5.5.2 Physical/Chemical Properties and Uses. No information was located for this compound.

2.5.5.3 Physiological Response. The data elements provided in Table 9.5 were the only toxicity information located on this chemical. An acute dust inhalation study was conducted using rats, guinea pigs, and mice. The animals were exposed to 180 mg/m^3 for 6 hr. No deaths or untoward behavioral reactions or effects on body weight were noted during the exposure or during the 5-day post-exposure observation period (91).

An acute vapor inhalation study was carried out on two rats, two guinea pigs, and two mice. The animals were exposed to an average nominal vapor concentration of 0.1 mg/l (100 mg/m^3) for 4 hr. No adverse effects were reported. The LC_{50} was considered to be greater than 0.1 mg/l (91).

2.5.6 t-Butyl Cumyl Peroxide (t-Butyl Cumyl Peroxide, t-Butyl α,α-Dimethylbenzyl Peroxide, Cumyl t-Butyl Peroxide)

$$
\begin{array}{ccc}
CH_3 & & CH_3 \\
| & & | \\
H_3C-C-O-O-C-\bigcirc \\
| & & | \\
CH_3 & & CH_3
\end{array}
$$

Mol. Wt. 208.33
CASRN 3457-61-2

2.5.6.1 Determination in the Atmosphere. Gas chromatography has been used to determine the concentration of this peroxide (54d). An "Intersmat" gas chromatograph with a flame ionization detector with a stainless steel column (06 × 4 mm; 0.30 m) packed with QF-1 (14.8 percent diglycerol, 0.13 percent on chromosorb GAW-DMCS). Column, injector, and detector temperatures were 80, 95, and 100°C, respectively. The gas chromatograph was calibrated by injecting samples of a known concentration of the peroxide in ethanol.

2.5.6.2 Physical/Chemical Properties and Uses. *t*-Butyl cumyl peroxide is a pale yellow liquid.

2.5.6.3 Physiological Response. The data elements provided in Table 9.5 were the only toxicity information located on this chemical.

An acute vapor inhalation study in rats showed the LC_{50} to be greater than 140 ppm (1.2 mg/l) (54e). A later acute vapor inhalation study was conducted using 13 rats. The 4-hr exposure at a nominal concentration of 152 mg/l resulted in no deaths. After the first 15 min of exposure the animals showed reduced activity and partially closed eyes. During the exposure lacrimation and salivation were observed. In addition, coarse intermittent body tremors were observed the last 2 hr of exposure. Slight lacrimation continued through day 9. The LC_{50} was judged greater than 152 mg/l (92).

2.5.7 4-(t-Butylperoxy)-4-methyl-2-pentanone

$$
\begin{array}{cc}
CH_3 & O \\
| & || \\
H_3C-C-CH_2-C-CH_3 \\
| \\
O \\
| \\
O \\
| \\
H_3C-C-CH_3 \\
| \\
CH_3
\end{array}
$$

Mol. Wt. 188.26
CASRN 26394-04-7

2.5.7.1 Determination in the Atmosphere. No air collection method exists for sampling this peroxide in the breathing zone of the worker. One method reported for analyzing dialkyl peroxides is high-pressure liquid chromatography (71).

2.5.7.2 Physical/Chemical Properties and Uses. No information was located for this compound.

2.5.7.3 Physiological Response. The data elements provided in Table 9.5 were the only toxicity information located on this chemical.

2.6 Peroxyketals

This group of peroxides is of the general chemical structure:

$$
\begin{array}{c}
R \\
| \\
O \\
| \\
O \\
| \\
R-O-O-\overset{\displaystyle |}{\underset{\displaystyle |}{C}}-R_2 \\
R_1
\end{array}
$$

These peroxides are practically nontoxic if swallowed. Only one has been tested for any adverse effects following dermal absorption. They are not eye or skin irritants. Generally they are not toxic by inhalation.

2.6.1 1,1-Di-(t-butylperoxy)-3,3,5-trimethylcyclohexane
[(3,3,5-Trimethylcyclohexylidene)bis(t-butyl) Peroxide,
(3,3,5-Trimethylcyclohexylidene)bis(1,1-dimethylethyl) Peroxide]

Mol. Wt. 302.5
CASRN 6731-36-8

2.6.1.1 Determination in the Atmosphere. The extraction and photometric analysis of this peroxide in water was reported by Sevast'yanova and Smirnova (69).

2.6.1.2 Physical/Chemical Properties and Uses. This chemical is a clear liquid with a half-life in benzene of 10 hr at 202°F (94°C).

It is used as an effective source of free radicals in the cross-linking of elastomers. It resists the decomposition effects of most fillers and pigments. This commercially available product is a deflagration hazard.

2.6.1.3 Physiological Response. A foreign study reports that this peroxide is of low toxicity. However, when used at concentrations exceeding maximum permissible levels (unspecified), it caused local skin and eye irritation, as well as irritation of the respiratory tract. It did not show any cumulative effects (time unspecified) (93).

2.6.2 1,1-Di-(t-butylperoxy)cyclohexane

Mol. Wt. 260.36
CASRN 3006-86-8

2.6.2.1 Determination in the Atmosphere. No air collection method or analytic method was located for this compound.

2.6.2.2 Physical/Chemical Properties and Uses. No information was located for this compound.

2.6.2.3 Physiological Response. The data elements provided in Table 9.6 were the only toxicity information located on this chemical.

An acute inhalation study using rats exposed them to 207.2 or 20.8 mg/l for 4 hr. Salivation, nasal discharge, and dyspnea were noted during the study. Slight dyspnea persisted at the lower dose for 1 to 2 days and at the higher dose for up

Table 9.6. Toxic Properties of Peroxyketals (7)*

Peroxyketal	CASRN	Oral LD$_{50}$ (mg/kg)	Dermal LD$_{50}$ (mg/kg)	Primary Skin Irritation	Eye Irritation	LC$_{50}$ (mg/l)[a]	Salmonella typhimurium Assay
1,1-Di-(t-butylperoxy)-3,3,5-Trimethylcyclohexane	6731-36-8						
75% in DBP[b]		>12,918	>8000		Not an irritant	~800	
1,1-Di((t-butylperoxy)cyclohexane	3006-86-8						
65% in DBP[b]		16,653		Not an irritant	Not an irritant	>207.2	
2,2-Di(t-butylperoxy)butane	2167-23-9						
50% in DBP[b]		23.2[c]		Mod. irritating	Slight irritation	>2.42	Negative
50% in mineral oil		>30.0[c]			Slight irritation		
n-Butyl 4,4-Di-(t-butylperoxy)valerate	995-33-5						
40% in chalk		>5000		Not an irritant	Slight irritation		
2,2-Di-(cumylperoxy)propane	4202-02-2						
50% in odorless mineral spirits		11.5[c]		Severe irritant	Not an irritant		Negative

[a] All LC$_{50}$ tests lasted 4 hr unless noted otherwise.
[b] Dibutyl phthalate.
[c] LD$_{50}$ reported in ml/kg.
* All studies used rats.

577

to 7 days. No deaths occurred and no morphological changes were noted at necropsy (56g). The LC_{50} is considered to be greater than 207.2 mg/l.

2.6.3   2,2-Di-(t-butylperoxy)butane

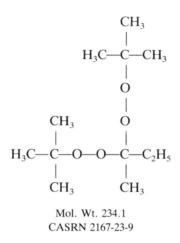

Mol. Wt. 234.1
CASRN 2167-23-9

2.6.3.1 Determination in the Atmosphere. No air collection method or analytic method was located for this compound.

2.6.3.2 Physical/Chemical Properties and Uses. This peroxide dissolved in dibutyl phthalate is a clear, colorless liquid.

2.6.3.3 Physiological Response. The data elements provided in Table 9.6 were the only toxicity information located on this chemical.

An acute inhalation study used the maximum attainable concentration of 2.42 g/m^3 of air. The particle size was determined using a Cascade impactor. Ninety percent of the mist consisted of droplets with a diameter of 1.2 to 3.3 μm. The maximum droplet size was 6.7 μm. No information on the behavior of the rats during the exposure is available because the animals were completely invisible as a consequence of the dense mist prevailing in the inhalation chamber. No mortality occurred. The LC_{50} is considered to be greater than 2.42 g/m^3 (54h).

2.6.4   n-Butyl 4,4-Di(t-butylperoxy)valerate [4,4-Di-(t-butylperoxy) n-Butyl Valerate]

$$\begin{array}{c}
CH_3 \\
| \\
H_3C-C-CH_3 \\
| \\
O \\
| \\
O \\
| \\
H_3C-C-CH_2-CH_2-\overset{\overset{\displaystyle O}{\|}}{C}-O-(CH_2)_3-CH_3 \\
| \\
O \\
| \\
O \\
| \\
H_3C-C-CH_3 \\
| \\
CH_3
\end{array}$$

Mol. Wt. 334.4
CASRN 995-33-5

2.6.4.1 Determination in the Atmosphere. No air collection method or analytic method was located for this compound.

2.6.4.2 Physical/Chemical Properties and Uses. No information was located for this compound.

2.6.4.3 Physiological Response. The data elements provided in Table 9.6 were the only toxicity information located on this chemical.

2.6.5   2,2-Di-(Cumylperoxy)propane (Biscumylperoxypropane)

$$\begin{array}{c}
\text{[phenyl]} \\
| \\
H_3C-C-CH_3 \\
| \\
O \\
| \\
O \qquad\qquad CH_3 \\
\| \qquad\qquad | \\
H_3C-C-O-O-C-\text{[phenyl]} \\
| \qquad\qquad | \\
CH_3 \qquad\qquad CH_3
\end{array}$$

Mol. Wt. 344.43
CASRN 4202-02-2

2.6.5.1 Determination in the Atmosphere. No air collection method or analytic method was located for this compound.

2.6.5.2 Physical/Chemical Properties and Uses. No information was located for this compound.

2.6.5.3 Physiological Response. The data elements provided in Table 9.6 were the only toxicity information located on this chemical.

2.7 Hydroperoxides

This group of peroxides is of the general chemical structure:

$$R—OOH$$

The most striking toxicologic feature of this group is that they are extremely irritating to the skin and eyes. Several are classified by DOT as corrosive. Hence eye and skin contact as well as inhalation of these materials should be carefully monitored.

2.7.1 t-Butyl Hydroperoxide (1,1-Dimethyl Ethyl Hydroperoxide, TBHP)

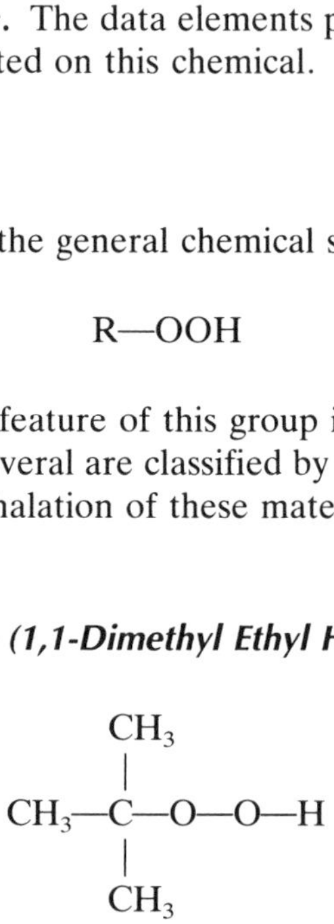

$$CH_3$$
$$|$$
$$CH_3—C—O—O—H$$
$$|$$
$$CH_3$$

Mol. Wt. 90.12
CASRN 75-91-2

2.7.1.1 Determination in the Atmosphere. The odor threshold for this peroxide is reported to be 0.17 mg/m^3. The warning threshold concentration is 1 ppm (94).

No air collection method was located for sampling this peroxide in the breathing zone of the worker. Several analytic methods have been used over the years such as high-pressure liquid chromatography (95, 96), chromatography using a flame ionization detector (97, 98), and colorimetric detection (99).

2.7.1.2 Physical/Chemical Properties and Uses. TBHP is produced by the liquid-phase reaction of isobutane and molecular oxygen or by mixing equimolar amounts of t-butyl alcohol and 30 to 50 percent hydrogen peroxide.

TBHP vapor can burn in the absence of air and may be flammable at either elevated temperature or reduced pressure. Fine mist/spray may be combustible at temperatures below the normal flash point. When evaporated the residual liquid will concentrate the TBHP content and may reach an explosive concentration (>90 percent). Closed containers may generate internal pressure through the degradation of TBHP to oxygen (100).

This chemical has many industrial uses such as an intermediate in the production of propylene oxide and *t*-butyl alcohol from isobutane and propylene (101). TBHP, however, is primarily used as an initiator and finishing catalyst in the solution and emulsion polymerization methods for polystyrene and polyacrylates. It is used in the polymerization of vinyl chloride and vinyl acetate. It is used as an oxidation and sulfonation catalyst in bleaching and deodorizing operations. Two other possible industrial uses are as an anti-slime agent in cooling systems and as a settling agent in aqueous slurries of various mineral tailings (102, 103).

2.7.1.3 Physiological Responses. Oral studies, as shown in Table 9.7, have shown TBHP is moderately toxic if ingested with an LD_{50} of 560 mg/kg. It should be noted that even though only 1 of 10 rats died when given 0.6 ml/kg, the animals in this dose group exhibited signs of depression and lacrimation. As the dose levels increased, 0.8, 1.0, 1.2, 1.4, and 1.6 ml/kg, the animals showed signs of loss of righting, hypothermia, and hematuria.

In a routine primary skin irritation assay, three out of the six rabbits died. The peroxide is a severe dermal irritant, by DOT standards a corrosive. It can also be absorbed through the skin in toxic amounts to cause cyanosis, depression, loss of righting, blanching of the treated skin, convulsions, and death.

Several mutagenicity studies have been carried out with *t*-butyl hydroperoxide (34, 76, 92b). The first two articles cited a positive result with TBHP in test strains of bacteria. The latter study was an in vivo experiment which showed no mutagenesis in the bone marrow cells after rats inhaled 100 ppm for 6 hr/day for 5 days. ARCO Chemical Company (100), concerned about sorting out this information, conducted other tests presented below:

Mutagenicity of TBHP

Test	Result	Comment	Dose
Salmonella typhimurium assay	Positive	TA-98, -100, -1537	Liver S-9 fraction necessary
Mouse lymphoma	Positive	See below	0.018–0.0013 μl/ml
Cell transformation	Negative	C3H/10T-1/2	0.0049–0.0003 μl/ml
Rat bone marrow	Negative	Whole animal	5-day exposure to 100 ppm for 6 hr/day

Of special interest in the Mouse lymphoma test was the indication that TBHP is several times less active when the enzyme preparation is added than without the microsomal preparation. This may indicate that TBHP can be readily metabolized in a whole animal to a much less genotoxic material. To support this theory, the whole animal experiments showed no indication of genotoxic potential. Also, an

Table 9.7. Toxic Properties of Hydroperoxides (7)*

Hydroperoxide	CASRN	Oral LD$_{50}$ (mg/kg)	Dermal LD$_{50}$ (mg/kg)	Primary Skin Irritation	Eye Irritation	LC$_{50}$, ppm (mg/l)^a	Salmonella typhimurium Assay
t-Butyl Hydroperoxide 70%	75-91-2	560	0.5^b	Extreme irritant	Extreme irritant	502 (1.85)	Positive (see text)
80%; 20% di-t-butyl peroxide		406	>10,000	Irritating	Irritant	500	
a-Cumyl hydroperoxide 80–83% in corn oil	80-15-9	800–1600	>200	Severe irritation corrosive (DOT)	Irritant	700 (4.3) (6 hr)	
73%		382		Irritating	Irritating	220	Inconclusive
1-Phenylethyl hydroperoxide 30% in ethylbenzene	3071-32-7	800	1700	Severe irritant	Severe irritant	20–33 mg/l	
1,1,3,3-Tetramethylbutyl hydroperoxide	5809-08-5	0.92^b		Very severe irritant	Very severe irritant	>480 (2.85)	Negative
1,2,3,4-Tetrahydro-1-naphthyl hydroperoxide	771-29-9	250 (unk. route in mice)					
1-Vinyl-3-cyclohexen-1-yl hydroperoxide	3736-26-3		1440				
Diisopropylbenzene hydroperoxide 53%	26762-93-6	6200		Severe irritation (immediate) corrosive (DOT)	Severe irritant	4.5 mg/l (6 hr)	
p-Menthyl Hydroperoxide 55%	26762-92-5	3700		Severe irritation (immediate) corrosive (DOT)	Severe irritation	9.2 mg/l (6 hr)	Positive

^{a}All LC$_{50}$ tests lasted 4 hr unless noted otherwise. ^{b}LD$_{50}$ reported in ml/kg.   *All studies used rats.

in vitro assay looking at cell transformation was negative. This assay has been interpreted to be an in vitro estimation of carcinogenic potential. In conclusion, TBHP is mutagenic in bacterial and mammalian cells in culture, but is non-mutagenic in whole animals.

In cells, TBHP is reduced by glutathione peroxidase to *t*-butyl alcohol and water. TBHP is not destroyed by catalase, an enzyme that destroys other peroxides (75). TBHP has been reported to cause lipid peroxidation and to produce other evidence of oxidative stress, resulting in adverse effects in cells in culture and other in vitro systems (104–107).

A study performed to evaluate the carcinogenicity of TBHP found it was not carcinogenic when applied to the skin of mice at 16.6 percent of the peroxide six times a week for 45 weeks. However, if its application was preceded by 0.05 mg of 4-nitroquinoline-1-oxide as a 0.25 percent solution in benzene applied 20 times over 7 weeks followed by TBHP (16.6 percent in benzene) then malignant skin tumors appeared between days 390 and 405 of the experiment (108). This supports the theory that peroxides are not complete carcinogens, but may act as promoters (109). The effects of TBHP on promotable and nonpromotable mouse epidermal cell culture lines were reported by Muehlematter (110).

The teratogenicity of TBHP has been investigated. Female rats inhaled 226 mg/m^3 for 4 hr during day 19 of gestation. This dose caused nonspecific impairment in the development of their fetuses in addition to general maternal toxicity (111).

A single inhalation study reported a respiratory irritation threshold for humans of 45 mg/m^3. TBHP poisoning in humans causes severe depression, incoordination, cyanosis, and death due to respiratory arrest. However, levels producing such effects would not normally be expected to occur (45).

It appears that TBHP would not be subject to biomagnification in an aquatic environment owing to its low partition coefficient (octanol/water, -1.30) (112) and high reactivity (113).

2.7.2 α-Cumyl Hydroperoxide (Cumene Hydroperoxide, α,α-Dimethylbenzyl Hydroperoxide, Isopropylbenzene Hydroperoxide, CHP)

$$\text{C}_6\text{H}_5-\underset{\underset{\text{CH}_3}{|}}{\overset{\overset{\text{CH}_3}{|}}{\text{C}}}-\text{OOH}$$

Mol. Wt. 152.18
CASRN 80-15-9

2.7.2.1 Determination in the Atmosphere. No air collection method was located for sampling this peroxide in the breathing zone of the worker. Several analytic methods have been used over the years such as high-pressure liquid chromatography (95, 96), chromatography using a flame ionization detector (97, 98), and colorimetric detection (99).

2.7.2.2 Physical/Chemical Properties and Uses. This 90 percent pure, colorless liquid has a flash point of 175°F and is relatively insensitive to shock as indicated by impact and detonation tests (82).

2.7.2.3 Physiological Response. Based on data presented in Table 9.7, this peroxide is moderately toxic when ingested or inhaled. Prolonged inhalation of vapors of CHP causes headache and throat irritation. Necropsy of rats used in the inhalation LC_{50} test showed severe inflammation of the trachea and lungs.

This peroxide causes burning and throbbing when in contact with the skin. Edema, erythema, and vesiculation may take 2 to 3 days to appear. The maximal concentration that produced no irritation to rabbit skin was 7 percent. High concentrations of CHP applied directly to the eyes of rabbits affected the cornea, iris, and conjunctiva extensively. Washing the eyes with water for 4 sec after application prevented any adverse reactions. The maximal nonirritating concentration to rabbits' eyes was 1 percent (82).

Repeated sublethal doses (one-fifth the LD_{50}) of CHP given rats three times a week for 7 weeks either orally or intraperitoneally resulted in cumulative effects. Specific effects were not reported (9).

Inhalation experiments conducted in six female rats exposed to 16 ppm 12 times, each exposure lasting 4 to 5 hr, produced salivation and nose irritation. All organs were normal at necropsy. When the concentration was increased to 31.5 ppm for six rats for seven exposures, 5 hr each, respiratory difficulty was noted along with salivation, tremor, weight loss, and hyperemia of eyes and tail. The lungs were the target organ showing emphysema and thickening of the alveolar walls. Two rats were exposed to 50 ppm; one died of congested lungs and kidneys (114).

2.7.3   1-Phenylethyl Hydroperoxide (Ethylbenzene Hydroperoxide)

$$CH_3-\underset{\underset{\text{(phenyl)}}{|}}{\overset{\overset{\text{H}}{|}}{C}}-OOH$$

Mol. Wt. 138.16
CASRN 3071-32-7

2.7.3.1 Determination in the Atmosphere. No air collection method was located for sampling this peroxide in the breathing zone of the worker. Several analytic methods have been used over the years such as high-pressure liquid chromatography (95, 96), chromatography using a flame ionization detector (97, 98), and colorimetric detector (99).

2.7.3.2 Physical/Chemical Properties and Uses. No information was located for this compound.

2.7.3.3 Physiological Response. The data elements provided in Table 9.7 were the only toxicity information located on this chemical. Two acute inhalation studies have been performed using this material; however, neither study critically evaluated the actual composition of the chamber atmosphere. Therefore the LC_{50} ranges from 20 to 33 mg/l.

2.7.4 1,1,3,3-Tetramethylbutyl Hydroperoxide (2,4,4-Trimethylpentyl-2-hydroperoxide)

$$CH_3-\underset{\underset{CH_3}{|}}{\overset{\overset{CH_3}{|}}{C}}-CH_2-\underset{\underset{CH_3}{|}}{\overset{\overset{CH_3}{|}}{C}}-OOH$$

Mol. Wt. 146.22
CASRN 5809-08-5

2.7.4.1 Determination in the Atmosphere. No air collection method was located for sampling this peroxide in the breathing zone of the worker. Several analytic methods have been used over the years such as high-pressure liquid chromatography (95, 96), chromatography using a flame ionization detector (97, 98), and colorimetric detector (99).

2.7.4.2 Physical/Chemical Properties and Uses. No information was located for this compound.

2.7.4.3 Physiological Response. The data elements provided in Table 9.7 were the only toxicity information located on this chemical.

An acute inhalation study using rats indicated that the LC_{50} was greater than 480 ppm. However, the severity of effects expressed during exposure indicates the LC_{50} will not be exceedingly greater than this value. The effects were mouth breathing and incoordination; severity increased as the exposure progressed (54i).

2.7.5 1,2,3,4-Tetrahydro-1-naphthyl Hydroperoxide (Tetraline Hydroperoxide)

O—OH

Mol. Wt. 164.22
CASRN 771-29-9

2.7.5.1 Determination in the Atmosphere. No air collection method was located for sampling this peroxide in the breathing zone of the worker. Several analytic methods have been used over the years such as high-pressure liquid chromatography (95,

96), chromatography using a flame ionization detector (97, 98), and colorimetric detector (99).

2.7.5.2 Physical/Chemical Properties and Uses. No information was located for this compound.

2.7.5.3 Physiological Response. The data elements provided in Table 9.7 were the only toxicity information located on this chemical.

2.7.6 1-Vinyl-3-cyclohexen-1-yl Hydroperoxide (1-Hydroperoxy-1-vinylcyclohex-3-ene)

OH
|
O
|
-CH=CH₂

Mol. Wt. 140.20
CASRN 3736-26-3

2.7.6.1 Determination in the Atmosphere. No air collection method was located for sampling this peroxide in the breathing zone of the worker. Several analytic methods have been used over the years such as high-pressure liquid chromatography (95, 96), chromatography using a flame ionization detector (97, 98), and colorimetric detector (99).

2.7.6.2 Physical/Chemical Properties and Uses. No information was located for this compound.

2.7.6.3 Physiological Response. When mice were exposed to this material (route unknown) at 1440 mg/kg there was an equivocal increase in incidences of tumors (115).

The data elements provided in Table 9.7 were the only acute toxicity information located on this chemical.

2.7.7 Diisopropylbenzene Hydroperoxide (Isopropyl Cumyl Hydroperoxide, DIBHP)

CH₃   CH₃
| |
H₃C—C   C—OOH
| |
H   CH₃

Mol. Wt. 194.26
CASRN 26762-93-6

2.7.7.1 Determination in the Atmosphere. No air collection method was located for sampling this peroxide in the breathing zone of the worker. Several analytic methods have been used over the years such as high-pressure liquid chromatography (95, 96), chromatography using a flame ionization detector (97, 98), and colorimetric detector (99).

2.7.7.2 Physical/Chemical Properties and Uses. Diisopropylbenzene hydroperoxide is a clear, yellow liquid.

2.7.7.3 Physiological Response. The data elements provided in Table 9.7 were the only toxicity information located on this chemical.

A 4-day subacute skin irritation test in rabbits was performed with this hydroperoxide in propylene glycol. After four applications of one a day, 1.75 percent wt./vol. of this chemical caused mild to moderate dermatitis; 3.5, 7.0, and 14.0 percent caused severe dermatitis at the end of 4 days. In fact, for the two highest concentrations the fourth application was discontinued due to necrosis (91b).

An acute inhalation study using rats exposed to the nominal concentrations of 2.18, 4.26, and 5.59 mg/l caused red nasal discharge at even the lowest dose. The animals also showed signs of shortness of breath, reduced activity, and vasodilitation. Fifty percent of the animals exposed to the 4.26 mg/l level died showing similar symptoms (91c).

2.7.8   p-Menthyl Hydroperoxide (p-Menthane Hydroperoxide)

Mol. Wt. 172.30
CASRN 26762-92-5

*This compound is a mixture with the peroxy group sometimes appearing in this position.

2.7.8.1 Determination in the Atmosphere. No air collection method was located for sampling this peroxide in the breathing zone of the worker. Several analytic methods have been used over the years such as high-pressure liquid chromatography (95, 96), chromatography using a flame ionization detector (97, 98), and colorimetric detector (99).

2.7.8.2 Physical/Chemical Properties and Uses. No information was located for this compound.

2.7.8.3 Physiological Response. The data elements provided in Table 9.7 were the only acute toxicity information located on this chemical. It should be noted that corneal opacity was still present 14 days after instilling this material in the eyes of rabbits.

A 4-day subacute skin irritation test in rabbits was performed with this hydroperoxide in propylene glycol. All concentrations, 1.75, 3.5, 7.0, and 14.0 percent wt./vol., caused severe dermatitis after two applications. Therefore the last two applications were not made for the three highest concentrations. The 1.75 percent was applied for 4 days, resulting in second-degree burns and focal third-degree burns (91b).

An acute inhalation study using 18 rats exposed to the nominal concentrations of 5.37, 8.99, and 10.64 mg/l found red nasal discharge at even the lowest dose. The animals also showed signs of shortness of breath, reduced activity, and vasodilation. Fifty percent of the animals exposed to the 8.99 mg/l level died showing similar symptoms (91c).

This peroxide was tested in five strains of bacteria in the *Salmonella typhimurium* assay. It was negative in four, but weakly positive in TA97 (a frameshift detector) with and without metabolic activation (89).

2.8   Peroxyacids

This group of peroxides is of the general chemical structure:

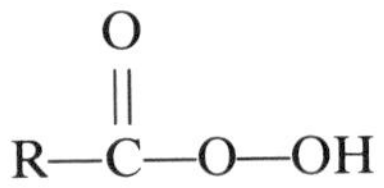

$$
\begin{array}{c}
O \\
\parallel \\
R\!-\!C\!-\!O\!-\!OH
\end{array}
$$

2.8.1   Peroxyacetic Acid (Acetyl Hydroperoxide, Peracetic Acid, Peracetic Acid Solution)

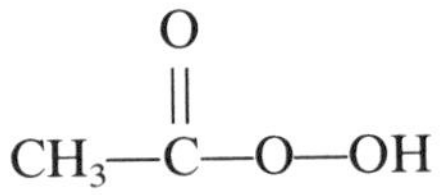

$$
\begin{array}{c}
O \\
\parallel \\
CH_3\!-\!C\!-\!O\!-\!OH
\end{array}
$$

Mol. Wt. 76.06
CASRN 79-21-0

2.8.1.1 Determination in the Atmosphere. No air collection method was located for sampling this peroxide in the breathing zone of the worker. Several analytic methods have been used over the years, such as a gas chromatograph equipped with a flame ionization detector (98).

2.8.1.2 Physical/Chemical Properties and Uses. Peracetic acid has several uses. It is often used as a cold-temperature sterilant because it is an efficient bactericide, fungicide, and viricide (116). Dilute solutions have been recommended for preoperative sterilization of the hands before surgery. However, when this peroxide escapes into the air, it is intensely irritating to human nasal passages (117).

2.8.1.3 Physiological Responses. There were several reports of the oral toxicity of this peroxide being quite low; for example, in guinea pigs, the oral LD_{50} was 10 mg/kg (7) and in an unspecified species it was 17.5 mg/kg, 40 percent peracetic acid in acetic acid (118). The differences from those reported in Table 9.8 could be due to species differences.

Bock (117) has shown that this peroxide in acetone is highly toxic owing to dermal absorption at 2 percent when painted onto the backs of mice. The study was intended to study skin tumors and cocarcinogenicity. Subsequently, the peroxide was applied in water at 3 and 1 percent and was tolerated for the duration of the experiment, which was five times a week for 66 weeks. The peroxide was a potent promoter when its application was preceded with one skin application of dimethylbenz[a]anthracene (DMBA) of 125 μg in 0.125 ml of acetone.

Peracetic acid is an oxidized by-product of ethanol. Experiments have shown it may play a role in ingested ethanol toxicity by interfering with arachidonic acid incorporation into the membranes of erythrocytes. It was suggested it directly inhibits the transferase enzyme (119).

2.8.2 p-*Nitroperoxybenzoic Acid (4-Nitroperoxybenzoic Acid)*

$$O_2N-\!\!\!\left\langle \bigcirc \right\rangle\!\!\!-\!\overset{\displaystyle O}{\overset{\|}{C}}-O-OH$$

Mol. Wt. 183.1
CASRN 943-39-5

2.8.2.1 Determination in the Atmosphere. No air collection method was located for sampling this peroxide in the breathing zone of the worker. Several analytic methods have been used over the years, such as a gas chromatograph equipped with a flame ionization detector (98).

2.8.2.2 Physical/Chemical Properties and Uses. This peroxide can be prepared from *p*-nitrobenzoic acid and 90 percent hydrogen peroxide in methanesulfonic acid medium (120).

2.8.2.3 Physiological Responses. Subcutaneous injections given mice three times a week for 4 weeks caused subcutaneous sarcomas after a total dose of 114 mg of peroxide. However, this is not a relevant route of exposure and stearic acid, used as a control, caused subcutaneous sarcomas with this dosing regimen. No other toxicity data were located for this chemical.

Table 9.8. Toxic Properties of Peroxyacids (7)*

Peroxyacids	CASRN	Oral LD$_{50}$ (mg/kg)	Dermal LD$_{50}$	Primary Skin Irritation	Eye Irritation	LC$_{50}$ (ppm)	*Salmonella typhimurium* Assay (Unless Specified)
Peroxyacetic acid 100%	79-21-0						
40% in acetic acid		1540	1410 mg/kg	Severe irritant	Severe irritant	>500 < 1,000 (4 hr)	Negative
p-Nitroperoxybenzoic acid	943-39-5	1230	0.71 ml/kg				Subcutaneous sarcomas

*All studies used rats.

2.9   Sulfonyl Peroxides

This group of peroxides is of the general chemical structure:

$$R—\overset{\displaystyle O}{\underset{\displaystyle O}{\overset{\|}{\underset{\|}{S}}}}—O—OR'$$

2.9.1   Acetyl Cyclohexanesulfonyl Peroxide

Mol. Wt. 222.2
CASRN 3179-56-4

2.9.1.1 Determination in the Atmosphere. No air collection method or analytic method was located for this compound.

2.9.1.2 Physical/Chemical Properties and Uses. No information was located for this compound.

2.9.1.3 Physiological Responses. The only toxicity data found for acetyl cyclohexanesulfonyl peroxide were as follows (7): it had an oral LD_{50} of >4640 mg/kg, a dermal LD_{50} of >2000 mg/kg, and was classified as an eye irritant. At a concentration of 29% in dimethyl phthalate, its oral LD_{50} was 1710 mg/kg.

An acute inhalation study was performed exposing 40 rats to 25, 50, 100, and 200 mg/l of this chemical. At the lowest concentration, 25 mg/l, bloody nasal discharge and congested lungs were noted. One death was noted. The LC_{50} was judged to be 58.3 mg/l with 95 percent confidence limits of 46 to 74 mg/l (56e).

2.10   Silyl Peroxides

This group of peroxides is of the general chemical structure:

$$(ROO)_n Si(R')_{4-n}$$

2.10.1   Vinyltri-(t-butylperoxy)silane (VTBS)

$$\left(\begin{array}{c} CH_3 \\ | \\ CH_3-C-O-O- \\ | \\ CH_3 \end{array}\right)_3 \begin{array}{c} H \\ | \\ Si-C=CH_2 \end{array}$$

Mol. Wt. 322.47
CASRN 15188-09-7

2.10.1.1 Determination in the Atmosphere. No air collection method or analytic method was located for this compound.

2.10.1.2 Physical/Chemical Properties and Uses. No information was located for this compound. It is a colorless liquid used to promote adhesion in polymer-to-metal bonding applications (54j).

2.10.1.3 Physiological Responses. The data elements provided in Table 9.9 were the only toxicity information located on this chemical. These data indicate that this peroxide is highly toxic by inhalation and toxic by ingestion. The acute inhalation study used rats and was performed using 4, 6, 9, and 18 mg/m^3 of this material diluted 50 percent in Shellsol T. The exposure concentrations were calculated for pure undiluted VTBS. A control experiment exposing the test animals to saturated Shellsol T vapor produced no effects. However, the peroxide exposures caused salivation and nasal discharge. Mouth breathing was observed in the latter part of the experiment. Deaths occurred within 48 hr after the exposure. The LC$_{50}$ was calculated to be 9 mg/m^3 (0.68 ppm) with 11 (0.85 ppm) and 7 mg/m^3 (0.55 ppm) as the 95 percent confidence limits (54j).

2.10.2   Cumylperoxytrimethylsilane

$$\text{Ph}-\begin{array}{c} CH_3 \\ | \\ C \\ | \\ CH_3 \end{array}-O-O-Si(CH_3)_3$$

Mol. Wt. 224.37
CASRN 18057-16-4

2.10.2.1 Determination in the Atmosphere. No air collection method or analytic method was located for this compound.

2.10.2.2 Physical/Chemical Properties and Uses. No information was located for this compound.

Table 9.9. Toxic Properties of Silyl Peroxides (7)*

Silyl Peroxide	CASRN	Oral LD_{50} (mg/kg)	Dermal LD_{50} (mg/kg)	Primary Skin Irritation	Eye Irritation	LC_{50} (ppm)[a]	Salmonella typhimurium Assay
Vinyltri-(t-butylperoxy)-silane	15188-09-7						
100%		>100 <250	>10,000	No toxic signs	Inflammation	0.68 (9 mg/l)	Negative
40% in n-hexane		~2.5	>10,000	No toxic signs	Slight irritation	0.006–0.009 mg/l	
5% in odorless mineral spirits							
Cumylperoxytrimethyl-silane	18057-16-4					>22.3	

[a] All LC_{50} tests lasted 4 hr unless noted otherwise.
*All studies used rats.

2.10.2.3 Physiological Responses. The data elements provided in Table 9.9 were the only toxicity information located on this chemical.

REFERENCES

1. D. Walrod et al., *Plastics Compounding*, 52 (1979, Jan/Feb).
2. D. Noller et al., *Ind. Eng. Chem.*, **56**(12) 18–27 (1964).
3. M. Tamura et al., *J. Hazardous Mater.*, **17**, 89–98 (1987).
4. J. Cywinski, *Reinforced Plastics* (May, 1962).
5. J. Martin, *Ind. Eng. Chem.*, **52**(4), 65, (1960).
6. *Encyclopedia of Occupational Health and Safety*, Vols. I & II, International Labour Office, McGraw-Hill, New York, 1983, p. 1612.
7. SPI Bulletin, *Commercial Organic Peroxide Toxicological Data*, Organic Peroxide Producers Safety Division of the Society of Plastics Industry, Publication #19-B, 9/82.
8. NIOSH Method 5009.
9. E. Floyd and H. Stokinger, *Am. Ind. Hyg. Assoc. J.*, **19**, 205–212 (1958).
10. A. G. McFarland, Elf ATOCHEM North America, Inc. personal communication, May 6, 1992.
11. M. O. Funk and W. J. Baker, *J. Liq. Chromatogr.*, **8**(4), 663–675 (1985).
12. NFPA, *Fire Protection Guide Hazardous Materials*, 49–135 (1978).
13. W. Strong, *Ind. Eng. Chem.*, **33**, 38 (1964).
14. *Code of Federal Regulations* 49:173.225, Department of Transportation, 1992.
15. J. C. Gage, *Brit. J. Ind. Med.* **27**, 1–18 (1970).
16. Patent Specification L040826 No. 20504/65 Application made in Japan (No. 27097), May 14, 1964.
17. W. Eaglstein, *Arch. Dermatol.*, **97**, 527 (1968).
18. American Conference of Governmental Industrial Hygienists, TLV Documentation, 1991.
19. J. R. Tkach, *Cutis*, **29**(2), 187 (1982).
20. F. N. Marzulli and H. L. Maibach, *Food Cosmet. Toxicol.*, **12**(2), 219 (1974).
21. W. M. Grant, *Toxicology of the Eye*, 3rd ed., Charles C Thomas, Springfield, IL, 1986.
22. R. L. Roudabush, Toxicity and Health Hazard Summary, Laboratory of Industrial Medicine, Eastman Kodak Co., Rochester, NY, 1964.
23. M. Sharrat et al., *Food Cosmet. Toxicol.*, **2**, 527 (1964).
24. B. VanDuuren et al., *J. Natl. Cancer Inst.*, **31**, 41 (July 1963).
25. W. C. Hueper, *NCI Monograph No. 10*, 1963, pp. 349–359.
26. J. H. Epstein, *J. Photochem. Photobiol.*, **37**, S-38 (1983).
27. Litton Bionetics, FDA OTC Vol. 070294, June, 1975.
28. T. Slaga et al., *Science*, 213–228 (Aug. 1981).
29. Y. Kurokawa et al., *Cancer Lett.*, **24**, 299 (1984).
30. U. Saffiotti and P. Shubik, *NCI Monograph No. 10*, 1963, p. 489.

31. T. Slaga, *Mechanisms of Tumor Promotion*, Vol. II, *Tumor Promotion and Skin Carcinogenesis*, CRC Press, Boca Raton, FL., 1984.

32. O. Iversen, *J. Invest. Dermatol.*, **86**, 442 (1986).

33. O. Odukoya and G. Shklar, *Oral Surg., Oral Med., Oral Pathol.*, **58**, 315 (1984).

34. National Toxicology Program, unpublished reports on Ames testing of (*a*) BPO (1985) and multistrain comparison of DMBA/MNNG/BPO promotion (1986); (*b*) *t*-butyl hydroperoxide.

35. *IARC Monographs on the Evaluation of the Carcinogen Risk of Chemicals to Humans* (*a*) *No. 36*, 1985, p. 267; (*b*) *No. 36*, 1985, p. 315.

36. A. Korhonen et al., *Environ. Res.*, **33**, 54 (1984).

37. National Research Council, Summary Tables of Biological Tests (*a*) 2:241, 1950; (*b*) 4:110, 1952; (*c*) 2:302, 1950.

38. *Int. Polymer Sci. Technol.*, **3**, 93 (1976).

39. P. R. Dugan, *Anal. Chem.*, **33**(6), 696 (1961).

40. G. G. Hawley, *The Condensed Chemical Dictionary*, 10th ed., Van Nostrand Reinhold, New York, 1981.

41. A. Klein-Szanto and T. Slaga, *J. Invest. Dermatol.*, **79**(1), 30 (1982).

42. F. Orlova, *Sb. Nauch. Tr. Kuibyshev Nauch-Issled Inst. Gig.*, **6**, 101 (1971).

43. National Research Council, *Prudent Practices for Handling Hazardous Chemicals in Laboratories*, National Academy Press, Washington, DC, 1981, p. 106.

44. (The) *Merck Index*, 10th ed., Merck Co., Rahway, NJ., 1983.

45. N. I. Sax, *Dangerous Properties of Industrial Materials*, 5th ed., Van Nostrand Rheinhold, New York, 1979.

46. U. S. Patent No. 4402853, Sept. 6, 1983 (to Sterling Drug, Inc.) CA/099/181508J.

47. V. J. Horgan et al., *Biochem. J.*, **67**, 551 (1957).

48. V. J. Horgan et al., *Organic Peroxides in Radiobiology*, London, 1958, p. 50.

49. Department of Transportation, *Emergency Response Guidebook*, DOT P 5800.4, U.S. Government Printing Office G-49, Washington, DC, 1987.

50. E. Altenburg and L. S. Altenburg, *Genetics*, **42**, 357–358 (1957).

51. E. B. Freese et al., *Mutat. Res.*, **4**(5), 517–531 (1967).

52. M. King, Catalyst Resources, Inc., Elyria, OH, personal communication, 1992.

53. Catalyst Resources Incorporated, Technical Bulletin (*a*) No. 4.1. July, 1983; (*b*) No. 1.1, July, 1983; (*c*) No. 5.1, June, 1985; (*d*) No. 6.1, June, 1985; (*e*) No. 52.1, March, 1987; (*f*) No. 7.1, July, 1983; (*g*) No. 12.1, April, 1990.

54. Central Instituut voor Voedingsonderzoek TNO (CIVO), Zeist, Holland—Reports for Akzo Chemie (*a*) No. R4934, February 1976; (*b*) No. R6141, 1979; (*c*) No. R4707, 1975; (*d*) No. B81/1424, 1981; (*e*) No. V81. 337/211424, 1981; (*f*) No. R5624; 1978; (*g*) No. R5619, 1978; (*h*) No. R6143, 1979; (*i*) No. R5519, 1977; (*j*) No. R5170, 1976.

55. *Code of Federal Regulations* 29 CFR 1910.1200 Appendix A 6(c), Occupational Safety and Health Administration, July 1, 1988.

56. International Research and Development Corp. (*a*) Unpublished data sponsored by Aztec Chemicals. 1979; (*b*) No. 376-007, 1977; (*c*) No. 376-008, 1978; (*d*) No. 164-016, 1970; (*e*) No. 378-005, 1976; (*f*) No. 164-076, 1978; (*g*) No. 378-011, 1978.

57. M. Athar et al., *Carcinogenesis*, **10**(8), 1499–1503 (1989).

58. I. V. Sanotskii et al., *Toksikol. Nov. Prom. Khim. Veshchestv.*, **10**, 55–63 (1968).

59. J. Vennerstrom et al., *Drug Design Delivery*, **4**, 45 (1989).

60. NIOSH Method 3508.

61. R. Hall, "Adhesives," National Clearinghouse for Poison Control Centers Bulletin, Jan-Feb. 1969.

62. F. Fraunfelder et al., *Am. J. Ophthalmol.*, **110**(6), 635–640 (1990).

63. L. A. Tiunov, "Toxicology of Organic Peroxides," *Gig. Sanit.*, **29**(10), 82–87 (1964).

64. M. Sittig, *Handbook of Toxic and Hazardous Chemicals and Carcinogens*, Noyes Publications, Park Ridge, NJ, 1985.

65. National Institute of Environmental Health Sciences, private communication to Lucidol Div., Pennwalt Corp., Dec. 1981. Results also published Table 3, Issue No. 6, Jan. 1982, NTP Tech. Bulletin of NTP Program.

66. R. Mittleman et al., *J. Forensic Sci.*, **31**(1), 312–320 (1986).

67. G. Luft et al., *Angew. Makromol. Chem.*, **141**, 207 (1986).

68. A. N. Klimina, *Sb. Nauch. Tr. Keibyshev Nauch.-Issled. Inst. Epidemiol. Gig.*, (*a*)**5**, 98–100, (1968); (*b*) **6**, 98–100 (1971).

69. E. M. Sevast'yanova and Z. S. Smirnova, (*a*) *Gig. Sanit.*, **0**(2), 48–49 (1987); (*b*) Aug. 8, 90–91, 1990.

70. R. E. Gosselin et al., *Clinical Toxicology of Commercial Products*. 4th ed., Williams & Wilkins, Baltimore, 1976.

71. L. A. Cornish et al., *J. Chromatogr.*, **19**, 85–87 (1981).

72. K. I. Ivanov et al., U.S.S.R. Patent No. 841659, June 30, 1981 (All-Union), Scientific-Research Thermotechnical Institute.

73. Nippon Kayaka K. K., unpublished study sponsored by Kayaka Nory Corp., 1971.

74. H. J. Kuchle, *Zentralbl. ArbMed. ArbSchutz.* (cited in Grant, 1974, ref. 21) **8**, 25 (1958).

75. P. Kotin and H. Falk, *Radiation Res. Suppl.* 3, 193–211 (1963).

76. T. Yamaguchi and Y. Yamashita, *Agric. Biol. Chem.*, **44**(7), 1675 (1980).

77. R. Latarjet et al., in *Organic Peroxides in Radiobiology*, M. Haissinsky, Ed., Pergamon Press, London, 1958.

78. E. Eder et al., *Toxicol. Lett.*, **48**, 225 (1989).

79. K. A. Jensen et al., *Cold Spring Harbor Symp. Quant. Biol.*, **16**, 245 (1951).

80. L. Fishbein. "Pesticidal, Industrial, Food Additive and Drug Mutagens," in *Mutagenic Effects of Environmental Contaminants*, H. E. Sutton and M. I. Harris, Eds., Academic Press, New York, 1972.

81. F. Tang and E. S. Huyser, *J. Org. Chem.* **42**(12), 2160–2163 (1976).

82. Hercules Rubber Chemicals, Hercules Powder Company, Wilmington, DE, 1964 (*a*) Bulletin T-104, Revision I; (*b*) Bulletin ORC-207C.

83. ASTM Standard Test Method E 755-90, American Society for Testing and Materials, Philadelphia, 1990.

84. BIBRA Toxicology International, Toxicity Profile for Dicumyl Peroxide, 1990.

85. S. C. Stenton et al., *J. Soc. Occup. Med.*, **39**, 33 (1989).

86. B. Petruson and B. Järvholm, *Acta Oto-Lar.*, **95**, 333 (1983).

87. H. A. Hansson and B. Petruson, *Acta Oto-Lar.*, **101**, 102 (1986).

88. A. Rannug et al., *Industrial Hazards of Plastics and Synthetic Elastomers*, Alan R. Liss Inc., New York, 1984, p. 407.

89. E. Zeiger et al., *Environ. Mol. Mutagen.* **11**, Suppl. 12, 1–158 (1988).

90. M. R. Chevallier and D. Luzzati, *Compt. Rend.*, **250**, 1572 (1960).

91. Industrial Bio-Test Laboratories, Inc., unpublished data sponsored by Hercules, Inc. (*a*) No. N8328, 1970; (*b*) No. 76-13, 1976; (*c*) No. 8562-09495, 1977.

92. Biodynamics, Inc., (*a*) unpublished data sponsored by Hercules, Inc. No. 81-7514, 1981; (*b*) unpublished data supported by ARCO Chemical Co. No. 81-7532, 1981.

93. Anonymous, "Flammability and Toxic Properties of the Diperoxides DIGIF-40 and BPIB-40," *Kauch. Rezina*, **12**, 28–29 (1985).

94. G. Leonardos, A. D. Little, Inc., memorandum report to Oxirane Corp., dated Dec. 10, 1979.

95. W. J. M. VanTilborg, *J. Chromatogr.*, **115**, 616–620 (1975).

96. B. B. Jones et al., *J. Chromatogr.*, **202**, 127–130 (1980).

97. L. Cerveny et al., *J. Chromatogr.*, **74**, 118–120 (1972).

98. G. T. Cairns et al., *J. Chromatogr.*, **103**, 381–384 (1975).

99. R. S. Deelder and M. G. F. Kroll, *J. Chromatogr.*, **125**, 307–314 (1976).

100. ARCO Chemical Company, Summaries of Acute Toxicity and Mutagenicity Tests, June, 1982.

101. S.R.I. International, (*a*) 1978; (*b*) 1979; Stanford Research Institute International, Inc., *Chemical Economics Handbook*, Menlo Park, CA.

102. R. Brink (to Pentz Laboratories, Inc.), U.S. Patent No. 514278, 1968.

103. M. Hamer and O. Petzen (to International Minerals and Chemicals Corporation), U.S. Patent No. 4017392.

104. R. J. Trotta et al., *Biochim. Biophys. Acta*, **679**, 230–237 (1981).

105. S. A. Jewell et al., *Science*, **217**, 1257–1259 (1982).

106. G. Rush and D. Alberts, *Toxicol. Appl. Pharmacol.*, **85**, 324–331 (1986).

107. D. P. Jones et al., *Adv. Biosci.*, **76**, 13–19 (1989).

108. H. Hoshino et al., *Gann*, **61**(2), 121–124 (1970).

109. P. Cerrutti, paper presented at the 14th Annual Cancer Research Workshop, University of New Orleans, Feb. 12, 1981.

110. D. Muehlematter et al., *Chem.-Biol. Interact.*, **71**, 339–362 (1989).

111. G. A. Shevelava, *Gig. Tr. Prof. Zabol.*, **12**, 40–48 (1976).

112. EPA, "Review of Environmental Fate of Selected Chemicals," EPA 560/5-77-003, Office of Toxic Substances, May, 1977.

113. H. Seis et al., *Fed. Fur. Biochem. Soc.*, **27**, 171–175 (1972).

114. *Brit. J. Ind. Med.*, **27**, 11–12 (1970).

115. "Mycotoxins in Foodstuffs," Proceedings of the symposium held at Massachusetts Institute of Technology, March, 1964.

116. L. B. Kline and R. N. Hull, *Am. J. Clin. Pathol.*, **33**, 30–33 (1960).

117. F. G. Bock et al., *J. Natl. Cancer Inst.*, **55**(6), 1359–1361 (1975).

118. FMC Corporation, data supplied by letter, dated April 1972 from H. M. Castrantas, FMC, to Organic Peroxide Producers Safety Division of the Society of Plastics Industry.

119. D. W. Allen et al., *Biochem. Biophys. Acta.*, **1081**(3), 267–273 (1991).

120. L. Silbert et al., *J. Org. Chem.*, **27**, 133–142 (1962).

CHAPTER TEN

Aliphatic Nitro, Nitrate, and Nitrite Compounds

Richard A. Davis, Ph.D., D.A.B.T.

1 ALIPHATIC NITRO COMPOUNDS

1.1 Introduction

Nitroalkanes, or nitroparaffins, are derivatives of alkanes with the general formula C_nH_{2n+1}, in which one or more hydrogen atoms are replaced by the electronegative nitro group ($-NO_2$). Nitroalkanes are classed as primary, RCH_2NO_2, secondary, R_2CHNO_2, and tertiary, R_3CNO_2, using the same convention as for alcohols. At this time, only primary nitroalkanes are available in commercial quantities with the exception of 2-nitropropane (1). Others available are nitromethane, nitroethane, 1-nitropropane, and tetranitromethane.

1.2 Source and Production

The nitroalkanes are produced in large commercial quantities by direct vapor-phase nitration of propane with nitric acid or nitrogen peroxide. The reaction product is a mixture of nitromethane, nitroethane, and 1- and 2-nitropropane. The individual compounds are obtained by fractional distillation. No recent production figures appear to be available (2, 3), but estimates run as high as 25 to 30 million lb annually.

Patty's Industrial Hygiene and Toxicology, Fourth Edition, Volume 2, Part A, Edited by George D. Clayton and Florence E. Clayton.
ISBN 0-471-54724-7 © 1993 John Wiley & Sons, Inc.

1.3   Uses and Industrial Hazards

1.3.1   Uses

The uses of nitroalkanes depend on their strong solvent power for a wide variety of substances including many coating materials, waxes, gums, resins, dyes, and numerous organic chemicals. They are used in products such as inks, paints, varnishes, and adhesives. Another important use is the production of derivatives, such as nitroalcohols, alkanolamines, and polynitro compounds. In some cases, they provide better methods of manufacturing well-known chemicals, such as chloropicrin and hydroxylamine. They are also used as special fuel additives, rocket propellants, and explosives.

1.3.2   Industrial Hazards

The chief industrial hazard from exposure to nitromethane, nitroethane, or 1- or 2-nitropropane is inhalation of their vapors, because their vapor pressures are sufficient to produce high vapor levels in the workplace unless controlled. Although no injuries from inhalation of nitromethane have been reported, mild dermal irritation has occurred as a result of its solvent action. The more highly toxic 2-nitropropane has produced headache, dizziness, nausea, vomiting, and diarrhea, with complaints of respiratory tract irritation. These signs and symptoms resulted from exposures at concentrations ranging from 30 to 300 ppm. 1-Nitropropane is rated somewhat toxic.

Fire and explosion hazards of nitroalkanes are considered low for they have relatively high flash points. Although their flash points are relatively high, under certain conditions of temperature, chemical reaction, and confinement, shock explosion can result. Nitroethane has a lower explosive limit of 3.4 percent by volume in air. Flammability limits of 2-nitropropane are 2.6 percent to 11 percent by volume in air. Like nitromethane, the nitropropanes react with inorganic bases to form salts that are explosive when dry.

1.4   Physical and Chemical Properties

1.4.1   Physical and Chemical Constants

The constants of six mononitroalkanes and tetranitromethane are given in Table 10.1. Of the seven nitroalkanes listed, only five, nitromethane, nitroethane, 1- and 2-nitropropane, and tetranitromethane, are of commercial interest (1). The nitrobutanes are included because toxicity data generated in the past anticipated their eventual use.

Nitroparaffins are colorless, oily liquids with relatively high vapor pressures. Their solubility in water decreases with increasing hydrocarbon chain length and number of nitro groups, tetranitromethane being completely insoluble. As expected, their boiling and flash points are higher than their corresponding hydrocarbons.

Table 10.1. Properties of the Mononitroalkanes and Tetranitromethane

Name	Mol. Wt.	B.P. (°C)	Specific Gravity	Solubility in H_2O at 20°C (% by Vol.)	Vapor Pressure (mm Hg)(°C)	Vapor Density (Air = 1)	(Closed/ Open) Flash Point (°F)	Conversion Units 1 mg/l (ppm)	1 ppm (mg/m³)	Oral Lethal Dose, Rabbits (g/kg)
Nitromethane	61.04	101.2	1.139 (20/20°C)	9.5	27.8 (20)	2.11	95/110	400.7	2.495	0.75–1.0
Nitroethane	75.07	114.8	1.052 (20/20°C)	4.5	15.6 (20)	2.58	82/106	325.7	3.07	0.50–0.75
1-Nitropropane	89.09	131.6	1.003 (20/20°C)	1.4	7.5 (20)	3.06	96/120	274.7	3.04	0.25–0.50
2-Nitropropane	89.09	120.3	0.992 (20/20°C)	1.7	12.9 (20)	3.06	—/102	274.7	3.64	0.50–0.75
1-Nitrobutane	103.12	151	0.9774 (15.6/15.6°C)	0.5	5 (25)	3.6	—	237.1	4.21	0.50–0.75
2-Nitrobutane	103.12	139	0.9728 (15.6/15.6°C)	0.9	8 (25)	3.6	—	237.1	4.21	0.50–0.75
Tetranitromethane	196.04	125.7	1.6229 (25°C)	Insoluble	8.4 (20)	0.8	—	124.7	8.02	—

1.4.2 Chemical Properties

The nitroalkanes are acidic substances. Polynitro compounds are even stronger acids than the corresponding mononitroalkanes. Thus trinitromethane is a typical strong acid with an ionization constant of the range 10^{-2} to 10^{-3}, as for strong inorganic acids. They are rapidly neutralized with strong bases and readily titrated.

Tautomerism, a general property of primary and secondary mononitroalkanes, gives rise to a more acidic "aci" form, or nitronic acid. In organic solvents they exist as the neutral nitroalkanes. But in aqueous solutions they exist in a state of equilibrium between the protonated neutral nitroalkane, the nonprotonated nitronic acid and its anion, or nitronate.

$$\underset{\text{Nitroalkane}}{R-N\overset{\displaystyle O}{\underset{\displaystyle O}{\big\langle}}} \quad \underset{+H^+}{\overset{-H^+}{\rightleftarrows}} \quad \underset{\text{Nitronate}}{R=N\overset{\displaystyle O^-}{\underset{\displaystyle O}{\big\langle}}}$$

$$\underset{-H^+}{\overset{+H^+}{\rightleftarrows}} \qquad \underset{\text{Nitronic Acid}}{R=N\overset{\displaystyle OH}{\underset{\displaystyle O}{\big\langle}}}$$

Tautomerism is important for understanding the biologic effects of nitroalkanes and also forms the basis of a number of important chemical reactions through the formation of nitroalkane salts. Mercury fulminate, $Hg(ON=C)_2$, is one of the better-known compounds that is derived from the mercury salt of nitromethane, $(CH_2=NO_2)_2Hg$. The production of chloronitroparaffins makes use of this tautomerism, as does the important intermediate methazonic acid,

$$HON=CHCH=N\overset{\displaystyle O}{\underset{\displaystyle OH}{\big\langle}}$$

which serves as a starting product for a number of well-known compounds, for example, nitroacetic acid and glycine. Chloropicrin, trichloronitromethane (CCl_3NO_2), however, is an exception. This compound in turn, like other nitro derivatives, is a source of guanidine [$H_2NC(=NH)NH_2$] by reaction with ammonia.

A number of other useful products such as primary amines, nitrohydroxy compounds, aromatic amines, and β-dioximes, which in turn can yield isoxazoles on hydrolysis, are also among the armamentarium of the nitroalkane reaction possibilities.

Most organic compounds, including aromatic hydrocarbons, alcohols, esters, ketones, ethers, and carboxylic acids, are miscible with the nitroalkanes. This property is the basis for much of their industrial use (Section 1.3.1).

1.5 Analytic Determination

Early methods (1940–1959) of determining nitroalkanes used colorimetric procedures. Since 1970 these have now been replaced by instrumental methods. References to the colorimetric procedures are included here because these procedures were used for monitoring the animal exposures which provide the toxicity data in Table 10.2. Instrumental methods of mass spectroscopy and gas chromatography are now used routinely and infrared has been used in animal exposure studies (7). NIOSH has evaluated exposure monitoring methods for nitromethane, nitroethane, and 2-nitropropane (4).

1.5.1 Determination in Air—Colorimetric Methods

The primary mononitroalkanes on which toxicity data were determined, nitroethane, 1-nitropropane, and 1-nitrobutane, were analyzed colorimetrically by measuring the color developed from an HCl-acidified alkaline solution containing $FeCl_3$. Reproducible results were obtained down to 0.5 mg/25 ml (5). Analytic data on the secondary nitroalkanes were determined by measuring, in a Beckman spectrophotometer, ultraviolet radiation at wavelength 2775 Å through an alcoholic solution (6).

A more sensitive spectrophotometric determination of primary nitroalkanes utilizes the coupling reaction with p-diazobenzenesulfonic acid (7). Simple, secondary nitroalkanes do not interfere as do some complex secondary nitro alcohols. Nitromethane can be determined at 440 mμ up to 50 μg/ml, and nitroethane and 1-nitropropane at 395 mμ up to 80 and 100 μg/ml, respectively.

Tetranitromethane has been measured by collection in reagent-grade methanol followed by reading at 240 mμ on a Beckman spectrophotometer and comparison to reference calibration curves (8).

Chloronitroparaffins have also been determined by colorimetric procedures. Using alkaline resorcinol for 1,1-dichloro-1-nitroethane, as for chloropicrin (9), color density is read at 480 mμ in a spectrophotometer. Color density is linear between 60 and 650 μg/25 ml. For 1-chloro-1-nitropropane, phenylenediamine in concentrated sulfuric acid is used. The absorption curve is linear between 75 and 250 μg/1 at readings at 540 mμ. A colorimetric method for 2-nitro-2-methyl-1-propanol using chromotropic acid is given in Reference 10 (p. 444). It has an accuracy of 2 percent and a precision of 1 percent.

1.5.2 Determination in Air—Instrumental Methods (10)

Mass spectrography, gas chromatography, and infrared spectroscopy are the current methods of choice for the determination of the nitroalkanes.

Mass spectra have been determined on eight C_1 to C_4 mononitroalkanes (10, p. 418). Nitromethane uniquely is the only member to have a major peak at its mass weight, 61. As expected, the two isomers 1- and 2-nitropropane have similar spectra, but can be separated with good precision and accuracy in C_1 to C_3 nitroalkane mixtures by the M/e 42–43 ratio. Conditions for analytic determination are given on pages 424 and 425 of Reference 10.

Table 10.2 Results of Inhalation Experiments (13)[a]

Nitromethane

Concn. (%)	Time (hr)	Concn. × Time	No. Rabbits Killed	No. Guinea Pigs Killed
1.0	6	6	2	2
3.0	2	6	2	2
5.0	1	5	2	2
3.0	1	3	0	2
1.0	3	3	0	2
0.5	6	3	1	1
0.25	12	3	2	2
0.10	30	3	0	2
2.25	1	2.25	0	1
3.0	0.5	1.5	0	1
0.5	3	1.5	0	1
1.0	1	1	0	0
3.0	0.25	0.75	0	0
0.05	140	7.0	0	0[c]
0.1	48	4.8	1[c]	1[c]

Nitroethane

Concn. (%)	Time (hr)	Concn. × Time	No. Rabbits Killed	No. Guinea Pigs Killed
2.5	2	5	2	0
3.0	1.25	3.75	2	2
3.0	1	3	1	1
1.0	3	3	2	1
3.0	0.5	1.5	1	0
0.5	3	1.5	2	0
0.1	12	1.2	1	0
0.5	2	1	1	0
1.0	1	1	1	0
0.25	3	0.75	0	0
0.1	6	0.6	0	0
0.05	30	1.5	0	[b]
0.05	140	7.0	0	0[c]

1-Nitropropane

Concn. (%)	Time (hr)	Concn. × Time	No. Rabbits Killed	No. Guinea Pigs Killed
1.0	3	3	2	2
0.5	3	1.5	2	2
1.0	1	1	0	1

[a] Two rabbits and two guinea pigs in each experiment.
[b] No animals exposed.
[c] One monkey exposed.

The different conditions recommended for analysis of nitroalkane mixtures including chloronitroparaffins by gas chromatography, along with their chromatograms, are given in Reference 10 on pp. 425–429 and 430–433, and for nitroalcohols, on pp. 441–443.

Infrared absorption spectroscopy has been used to monitor animal inhalation chamber concentrations of nitromethane and 2-nitropropane (11). Conditions of use and calibration of a Wilks MIRAN are described. MIRAN was connected to an automatic sampler for hourly analysis of chamber air. Over the 21 weeks exposure concentrations for nitromethane averaged 97.6 ± 4.6 ppm and 745 ± 34 ppm for the two exposure groups, and for 2-nitropropane, 27.2 ± 3.1 ppm and 207 ± 15 ppm, respectively.

1.6 Physiological Response

It should be noted that the inhalation toxicity data on the nitroalkanes gathered in the late 1930s and summarized below lacked some of the refinements of late work with these compounds. Exposure "chambers" consisting of steel drums of 233-liter capacity lacked the space to expose what is now considered an adequate size complement of animal species (12). The exposure concentrations at levels of 5000 to 50,000 ppm must be considered "nominal," because of measurement by interferometer, the chamber airflow characteristics, and fan circulation of air. Lower exposure levels, although measured colorimetrically (Section 1.5.1), lacked the precision of later instrumental methods (Section 1.5.2).

1.6.1 Comparative Animal Acute Toxicities (Table 10.2)

The nitroalkanes act chiefly as moderate irritants when inhaled (13). 2-Nitropropane appears to be more irritating than nitroethane (6), which is more irritating than nitromethane. Animals exposed at levels greater than 10,000 ppm give evidence of restlessness, discomfort, and signs of respiratory tract irritation, followed by eye irritation, salivation, and later central nervous system symptoms consisting of abnormal movements with occasional convulsions. Anesthetic symptoms are generally mild, and appear late. Most animals that manifest anesthesia die eventually. This is more marked with nitroethane and nitropropane than with nitromethane inhalation.

Incomplete information on the polynitroalkanes indicates that an increased number of nitro groups results in increased irritant properties. The chlorinated nitroparaffins are more irritating than the unchlorinated compounds (9). This reaches a severe degree with trichloronitromethane (chloropicrin). Unsaturation of the hydrocarbon chain in the nitroolefins also results in an increase in the irritant effects (14, 15).

The primary nitroalkanes fail to show significant pharmacological effects on blood pressure or respiration (16). Oral doses result in symptoms similar to those produced by inhalation except for the additional evidence of gastrointestinal tract irritation. They are less potent methemoglobin formers than the aromatic nitro compounds, but 2-nitropropane at 80 ppm induced Heinz bodies in 5 to 15 percent of erythrocytes of cats, the most susceptible of species. During prolonged exposure at 80 ppm, 30 to 35 percent of the erythrocytes contained these abnormal structures (6).

The nitroalkanes show no significant percutaneous absorption as judged by lack of any systemic effects or weight loss after application of the pure compounds in five daily treatments to the clipped abdominal skin of rabbits (13).

Animals dying following brief inhalation of the nitroalkanes show general visceral and cerebral congestion. After exposure at high concentrations there is pulmonary irritation and edema, the latter inadequate to be the sole cause of death. Inhalation of 2-nitropropane results in general vascular endothelial damage in all tissues as well as pulmonary edema, hemorrhage, and brain and liver damage (6). Toxic damage to the kidneys and heart is less prominent. Microscopically the liver shows severe parenchymal degeneration and focal necrosis. Sublethal concentrations of nitromethane produce severe liver changes in dogs consisting of infiltration with chronic inflammatory cells, fatty changes, congestion, and some hemorrhage and necrosis (17).

Inhalation of trichloronitromethane produces severe injury of the respiratory tract consisting of inflammation and necrosis of the bronchi, edema, and congestion in the alveoli. The chlorinated nitroparaffins and nitroolefins produce gastrointestinal tract irritation and damage when given by mouth. The monochloronitroparaffins are not markedly irritating to the skin or eyes but the dichloro compounds (9) and particularly the nitroolefins are strong skin and eye irritants (14, 15, 22).

Acute toxicologic and pharmacological studies by Deichmann et al. (14) of 12 nitroolefins from 2-nitro-2-butene to 2-nitro-3-nonene by oral, intraperitoneal, and dermal routes showed those of the series from C_4 to C_8 to be highly toxic to rats orally, and those from C_4 through C_9 intraperitoneally, with toxicities tending to decrease with increasing carbon chain length. Approximate oral lethal doses for rats ranged from 280 to 620 mg/kg, and corresponding intraperitoneal doses from 80 to 280 mg/kg. Corresponding percutaneous doses for the rabbit showed no such regularity.

Absorption of nitroolefins from the respiratory or gastroenteric tract, peritoneal cavity, or skin is very rapid. Signs of systemic intoxication appear promptly, including hyperexcitability, tremors, clonic convulsions, tachycardia, and increased rate and magnitude of respiration, followed by a generalized depression, ataxia, cyanosis, and dyspnea. Death is initiated by respiratory failure and associated with asphyxial convulsions. Pathological changes were most marked in the lungs, regardless of the mode of administration of a compound.

Altered function in animals inhaling nitroolefins was reported by Murphy et al. (15) as increased total pulmonary flow resistance, increased tidal volumes, and decreased respiratory rate of guinea pigs. Decreased voluntary activity of mice occurred during inhalation of nitroolefins at concentrations near or below the threshold for human sensory detection (0.1 to 0.5 ppm). Increasing concentrations increased the magnitude of the effects. Comparison of the effects of 2-nitro-2-butene, 3-nitro-3-hexene, and 4-nitro-4-nonene indicated that the effect on pulmonary function was inversely related to the carbon chain length. However, 4-nitro-4 nonene was slightly more active than the butene and hexene in producing depression of mouse activity. At the low concentrations tested, the effects of nitroolefins were reversible when the animals were returned to clean air. Injection of atropine sulfate overcame the increased pulmonary flow resistance induced by 4-nitro-4-nonene.

1.6.2 Human Experience

Anorexia, nausea, vomiting, and intermittent diarrhea and headaches in men exposed at 20 to 45 ppm of nitropropane during a dipping process ceased to appear when methyl ethyl ketone was substituted (22). Brief exposure to 1-nitropropane concentrations exceeding 100 ppm caused eye irritation (24). Nasal irritation, burning eyes, dyspnea, cough, chest oppression, and dizziness in men handling crude TNT have been attributed to tetranitromethane exposure (22). Headache, methemoglobinemia, and a few deaths have also been attributed to similar exposures.

Ten fatalities have been reported (17, 26–29) among workmen overexposed to solvent mixtures. All cases involved the application of coatings in poorly ventilated confined spaces. The one agent common to all was 2-nitropropane (2-NP), which occurred in the solvent mixtures at concentrations of 11 to 28 percent. All patients showed typical signs of 2-NP overexposure of headache, nausea, vomiting, diarrhea, and chest and abdominal pains, but the prodromal signs were somewhat nonspecific and similar to those from overexposure to any variety of solvents. The characteristic lesion in the fatal cases was destruction of hepatocytes. In all cases, liver failure was the primary cause of death. This was well-documented by antemortem findings of elevations in serum enzymes and postmortem findings of microscopic evidence of liver changes.

Survival time was from 6 to 10 days after acute exposures. A 2-NP serum concentration of 13 mg/l was measured in a patient who died whereas no other solvents were detected (29). In none of the cases were Heinz bodies observed or methemoglobin detected. 2-NP exposure concentrations associated with mortalities or hepatotoxicity are unknown. A study of 49 workers exposed to 2-NP approaching 25 ppm found no liver function abnormalities (30).

There has been more human experience with trichloronitromethane (chloropicrin) because it was used as a war gas, often in mixtures with chlorine or phosgene. Chloropicrin is a potent lacrimator which produces a peculiar frontal headache, coughing, nausea, vomiting, and severe injury of the respiratory tract resulting in pulmonary edema. It has been noted that individuals who have been injured with chloropicrin appear to become more susceptible so that concentrations not producing symptoms in others, cause them distress (31). Chloropicrin is also used as a fumigant for cereals and grains and as a soil insecticide. Misuse of these products has produced respiratory effects in animals and humans (32, 33) and elevated methemoglobin levels in humans (33).

The 2-nitro-2-olefins 2-nitro-2-butene, 2-nitro-2-hexene, and 2-nitro-2-nonene produced distinct eye irritation at low concentrations (22). The butene and hexene derivatives produced irritation with 3 min at concentrations between 0.1 and 0.5 ppm. For the corresponding nonene, irritation occurred only at concentrations above 1.0 ppm. As in the case of chloropicrin, individuals who had been repeatedly exposed to nitroolefins over a sustained period became increasingly sensitive to the eye irritation effects of these compounds. Temperature, within the limits tested of 65 to 90°F, appeared to have no influence on the sensitivity of the eye to a nitroolefin, but brief UV irradiation rapidly destroys the lacrimator.

1.6.3 Pharmacokinetics and Metabolism

The nitroalkanes are absorbed through the lungs and from the gastrointestinal tract. Applications to the skin give no evidence of sufficient absorption to result in systemic injury.

Distribution studies of nitroethane in animals showed rapid disappearance from the body. Within 3 hr only 14 percent of the dose was recovered in rats, and by 30 hr, essentially all the dose had been cleared from the tissues, the blood, lungs, liver, and muscle (34). Partial excretion of nitroethane was via the lungs. By either inhalation or oral administration, nitroethane was shown to be metabolized to aldehyde and nitrite, with the end product the eventual oxidation to nitrate (35). 2-Nitropropane is metabolized to acetone and nitrite. Nitrite can also be found in blood and urine, following the administration of nitrobutanes, but not after nitromethane or 2-nitro-2-methylpropane, a tertiary nitroalkane.

The pharmacokinetics of [^{14}C]-2-nitropropane was studied in male rats exposed for 6 hr to 20 or 154 ppm (36). At least 40 percent of the inhaled compound was absorbed. Blood concentrations of 2-NP measured by gas chromatography (GC) analysis decreased in an apparent first-order manner ($t_{1/2}$ = 48 min). Half-lifes for the biphasic elimination of radioactivity from blood were two to four times longer, indicating that metabolites have a greater potential to accumulate. The major route of excretion was the expired air, with about half of the administered radioactivity recovered as $^{14}CO_2$. Numerous differences in kinetic parameters observed at the two exposure concentrations indicated nonlinear kinetics at 154 ppm and higher. Gas uptake studies in male and female rats have indicated two metabolic processes (19). There were no differences in uptake or exhalation processes between males and females, but metabolic processes differed. A first-order, low-affinity but high-capacity metabolic pathway was similar for both sexes. However, a saturable high-affinity, low-capacity pathway was more than twice as active in females. The saturable pathway was predominant in females at exposure concentrations up to 180 ppm, but only up to 60 ppm in males. Because males are more sensitive to hepatoxicity, genotoxicity, and carcinogenicity produced by inhalation of 2-NP, the authors concluded that these effects are produced by the metabolite resulting from the first-order pathway.

Oxidative denitrification of nitroalkanes has been shown to occur by two mechanisms. The microsomal cytochrome P450 monooxygenase system of rat and mouse liver has been shown to metabolize nitroparaffins in vitro (37, 38). Specific activities are greatest for 2-nitropropane, followed by 1-nitropropane, nitromethane, and tetranitromethane. One study found up to 25 percent residual denitrifying activity with mouse liver microsomes under anaerobic conditions (39), suggesting an oxygen-independent mechanism may exist. Dayal et al. have found that 2-nitro-2-methylpropane was not denitrified by the monooxygenase system (40). This indicates that a hydrogen atom is required in the position alpha to the nitro group for oxidative cleavage of the neutral tautomeric form. They also showed that the nitronate anion of 2-nitropropane was denitrified 5 to 10 times faster than the neutral form. Because for primary nitroalkanes, such as 1-nitropropane, the tautomeric equilibrium lies far to the neutral side of physiological pH, the authors suggest this may explain the slower reaction rate with these compounds relative to 2-nitropropane.

A second mechanism of oxidative denitrification has been demonstrated for various flavoenzyme oxidases (41–43). The relative reactivity rates of the nitroalkanes are similar to that of the microsomal systems except tetranitromethane was inert (42). The interesting aspect of this pathway is that superoxide radical is produced either as an intermediate (42) or as an initiator/propagator of the reaction (43). Superoxide radical and other active oxygen species produced from it (oxygen free radicals, hydrogen peroxide, hydroxyl radical) have been associated with toxicity and mutagenicity.

The chloronitroparaffins appear not to show appreciable percutaneous absorption, as judged by lack of apparent systemic effects (9), but 1,1-dichloro-1-nitroethane induced swelling and irritation after only two applications. The monochloro derivative of 1-nitropropane, however, produced only slight erythema after 10 applications.

Absorption of nitrooleffins from the respiratory or gastrointestinal tract, peritoneal cavity, or skin is very rapid, giving signs of prompt systemic toxicity (14).

1.6.4   Genetic Toxicity

There have been numerous studies on the mutagenic activity of the nitroalkanes in *Salmonella typhimurium* (Ames test). Nitromethane (44, 45), nitroethane (45–47), 1-nitropropane (45, 46, 48, 49), 1-nitrobutane (45, 47), and 1-nitropentane (45) are not mutagenic in multiple strains of *S. typhimurium* with and without the addition of various microsomal metabolic activating systems. However, 2-nitropropane was shown to be active when concurrently tested in all of the above cited studies. It was active in all strains tested both with and without metabolic activation. Generally metabolic activation had no effect or slightly reduced the number of revertants. Löfroth et al. found other secondary nitroalkanes (2-nitrobutane, 2-nitropentane) were also active (45). They attributed the mutagenicity of the secondary nitroalkanes to the fact that their tautomeric equilibrium constants result in relatively higher nitronate and nitronic acid forms at cellular pH.

The genotoxic activity of 2-nitropropane has been evaluated in other test systems and usually compared to 1-nitropropane. Nitroethane and 2-nitropropane did not induce micronuclei in polychromic erythrocytes of mice given two daily oral doses up to 1.0 ml/kg and 0.4 ml/kg, respectively (46). Identical results were obtained in similar studies for 1-nitropropane and 2-nitropropane at doses up to 300 mg/kg in mice (50) and rats (51). In human lymphocytes exposed in culture, 2-nitropropane induced chromosomal aberrations and sister chromatid exchanges (SCE) with metabolic activation (52, 53), whereas 1-nitropropane was not active (53). In rat liver exposed both in vitro and in vivo (20 to 100 mg/kg), 2-nitropropane, but not 1-nitropropane, induced DNA repair synthesis (54, 51). This effect was greater in male rats than in females. Effects of DNA repair synthesis were also studied in numerous mammalian cell lines derived from humans, mice, hamsters, and rats but neither compound was active. The authors concluded the effect was due to a specific liver metabolite of 2-nitropropane. This conclusion was extended in a study with liver cell lines that possess cytochrome P450 metabolic capability versus V79 cells that lack this capability (55). 2-Nitropropane induced DNA repair synthesis, micronuclei, and mutations in the liver cell lines. DNA repair and micronuclei were not induced in V79 cells but mutations (HGPRT assay) were induced. 1-Nitropropane was completely inactive in the liver cell lines but did induce mutations and micronuclei in V79 cells (55).

Tetranitromethane has not been extensively tested for genotoxic activity. It was mutagenic in four *Salmonella* strains with and without metabolic activation (56, 57). In CHO cells exposed in culture, tetranitromethane induced SCEs with and without metabolic activation and chromosomal aberrations only with activation (57).

Chloropicrin (trichloronitromethane) was mutagenic in the Ames test (49) and induced SCEs but not chromosomal aberrations in human lymphocytes exposed in vitro (58). However, chloropicrin did not induce mutations in the *Drosophila* sex-linked recessive lethal test (59).

1.6.5 Animal Carcinogenicity

Inhalation studies in rats on the primary nitroalkanes nitroethane (60) and 1-nitropropane (61) have shown these compounds did not induce cancer. However, 2-nitropropane, a secondary nitroalkane, caused hepatocarcinoma in a 6-month inhalation study on rats (11). Likewise, oral administration of 2-nitropropane caused liver tumors in rats whereas 1-nitropropane administration did not cause tumors (62). A 2-year inhalation study of tetranitropropane in rats and mice conducted by the NTP produced "clear evidence of carcinogenic activity" based on lung cancer produced in both species (57). Details of the results of these studies are discussed in the following sections on the individual compounds.

The International Agency for Research on Cancer (IARC) reviewed the available data on 2-nitropropane in 1982 and concluded there was "sufficient evidence" for carcinogenicity in rats but no adequate epidemiologic data (3). The National Toxicology Program (NTP) classifies 2-nitropropane as a substance that may reasonably be anticipated to be a carcinogen (63) and the American Conference of Governmental Industrial Hygienists (ACGIH) classified it as a suspected human carcinogen. The aliphatic nitro group ($-C-NO_2$) has been specified a structural alert for DNA reactivity (64). These classifications are based on an evaluation of the qualitative carcinogenic information but do not quantify the risk of cancer to humans.

1.6.6 Mechanism of Action

As discussed in the preceding sections, there are some intriguing differences between the chemical and toxicologic properties of primary and secondary nitroalkanes. These differences have provided clues for investigating the mechanism of action by 2-nitropropane.

Differences in tautomeric equilibrium constants are of key importance. The nitronates of primary nitroalkanes have extremely short half-lives relative to secondary nitroalkanes at physiological pH (65). Therefore primary nitroalkanes exist mostly in their neutral protonated form, whereas the secondary alkanes exist mostly in their anionic nitronate form under biologic conditions (65, 66). Comparative studies of 2-nitropropane and propane-2-nitronate have shown the nitronate is a more potent mutagen (67, 18) in the Ames test and is more rapidly metabolized by liver microsomes (40). The two tautomers did not show significant or different degrees of cytotoxicity to mouse hepatocytes in vitro (40) but nitromethane and nitroethane were also not very toxic in this test system.

The difference between male and female rats in their response to 2-nitropropane is another important clue. The compound is more potent in males for liver toxicity, carcinogenicity, and induction of DNA repair synthesis. Differences in saturable metabolism between the two sexes point to a key role for metabolites produced by a first-order pathway (19).

Although it is well known that 2-nitropropane is active in the Ames test without microsomal metabolic activation, Fiala et al. have shown that this activity was inhibited by dimethyl sulfoxide, a hydroxyl radical scavenger (67). Others have shown the oxidation of 2-nitropropane and its nitronate produces superoxide (see Section 1.6.3). Hydroxyl radicals are commonly produced in biologic systems by the Haber–Weiss reaction whereby superoxide radicals reduce Fe^{3+} to Fe^{2+}, which reduces H_2O_2 to OH and OH^-. Hydrogen peroxide is also produced from superoxide by spontaneous or enzymatic dismutation. With this in mind, Fiala et al. suggested the activity of 2-nitropropane was due to the production of DNA-damaging radicals, of either oxygen or 2-nitropropane itself. They demonstrated the in vitro oxidation of the nitronate produced a condensation product of 2-nitropropane free radicals and reaction products of thymidine and active oxygen species (67). They went on to show that 2-nitropropane injection in rats (100 mg/kg, ip) caused substantial increases in the amounts of 8-hydroxydexoyguanosine in liver DNA and 8-hydroxyguanosine in liver RNA. The modified nucleosides are probably the reaction products of active oxygen species with DNA and RNA. Treatment with 1-nitropropane did not cause significant increases in these nucleic acid reaction products. Further studies showed that many secondary nitroalkanes produced oxidative nucleic acid damage but primary nitroalkanes and 2-methyl-2-nitropropane (a tertiary not denitrified by monoxygenases) were not active (65). These results may suggest the mechanism of hepatotoxicity and carcinogenicity of

Table 10.3 Occupational Exposure Limits for Nitroalkanes

Compound	TLV[a]		OSHA Standard	
	ppm	mg/m³	ppm	mg/m³
Nitromethane	20[b]	50	100	250
Nitroethane	100	307	100	310
1-Nitropropane	25	91	25	90
2-Nitropropane	10, A2[c]	36, A2	10	35
Tetranitromethane	1	8	1	8
Choropicrin	0.1	0.67	0.1	0.7
1-Chloro-1-nitropropane	2	10	2	10

[a]From 1991 list.
[b]In Notice of Intended Changes.
[c]Suspected human carcinogen.

2-nitropropane in rats is due to damage of nucleic acids from the intracellular generation of active oxygen species and/or 2-nitropropane radicals.

Another important mechanism has been recently studied by Cunningham and Mathews (68). They administered gavage doses up to 2 mmol/kg of 1-nitropropane or 2-nitropropane to rats for 10 days. Cell proliferation in the liver was measured by the percent of nuclei synthesizing new DNA (determined by incorporation of bromodeoxyuridine). Doses of 2-nitropropane, which were effective in the oral cancer bioassay, increased cell proliferation in a dose-related manner. 2-Nitropropane at 0.5 mmol/kg and 1-nitropropane up to 2 mmol/kg did not affect cell proliferation. This study has been cited as further evidence that the induction of cell proliferation can lead to the fixation of DNA damage (initiation) and/or clonal expansion of preneoplastic cells and that this effect may be critical in the multistage process of carcinogenesis. The significance of cell proliferation (discussed as regeneration subsequent to heptatotoxicity) was recognized earlier by Griffin and co-workers (69). They noted that tumors developed only at concentrations and durations of exposure to 2-NP that produced hepatotoxicity.

1.7 Hygienic Standards of Permissible Exposure

Threshold limit values (TLVs) and official Occupational Safety and Health Administration (OSHA) standards have been adopted for seven nitroalkanes as listed in Table 10.3.

No TLVs have been established for any nitroolefin or nitroalcohol because of little or no industrial interest. For the basis of TLVs, see *Documentation of TLVs* published by the ACGIH.

1.8 Specific Compounds

1.8.1 Nitromethane, CH₃NO₂

The physical and chemical properties of nitromethane (NM) are given in Table 10.1, its analytic determination in Section 1.5, and its toxicity relative to that of other nitroalkanes in Section 1.6.

Table 10.4. Acute Toxicity of Nitromethane (13, 70)

Route	Animal	Dose	Mortality
Oral	Dog	0.125 g/kg	0/2
		0.25–1.5 g/kg	12/12
	Rabbit	0.75–1.0 g/kg	Lethal dose
	Mouse	1.2 g/kg	1/5
		1.5 g/kg	6/10
Subcutaneous	Dog	0.5–1.0 ml/kg	Minimum lethal dose
Intravenous	Rabbit	0.8 g/kg	2/6
		1.0 g/kg	2/6
		1.25–2.0 g/kg	9/9
Inhalation	Rabbit	30,000 ppm, <2 hr	0/6
		2 hr	2/2
		10,000 ppm, 6 hr	3/2
		1–3 hr	0/4
		5,000 ppm, 6 hr	1/2
		3 hr	0/2
		500 ppm, 140 hr	0/2
	Guinea pig	30,000 ppm, 1–2 hr	4/4
		30 min	1/2
		15 min	0/2
		10,000 ppm, 3–6 hr	4/4
		1 hr	0/2
		1,000 ppm, 30 hr	2/3
		500 ppm, 140 hr	0/3
	Monkey	1,000 ppm, 48 hr	1/1
		500 ppm, 140 hr	0/1

1.8.1.1 Acute Toxicity. Animal toxicity has been studied by Machle and co-workers (13) and by Weatherby (70). This information is summarized in Table 10.4. Animals exposed at 30,000 ppm in air for longer than 1 hr developed pronounced nervous system symptoms. At 10,000 ppm nervous system symptoms did not appear until after 5 hr. During exposure at lower concentrations, there was slight irritation of the respiratory tract without evidence of eye irritation. This was followed by mild narcosis, weakness, and salivation. Rabbits, guinea pigs, and monkeys all survived repeated exposures for a total of 140 hr at concentrations of 500 ppm. A single monkey exposed at 1000 ppm for eight 6-hr exposures died. No remarkable changes in either blood pressure of respiration followed intravenous injection of NM in anesthetized dogs.

The histopathological changes observed following acute poisoning by all routes were chiefly confined to the liver and kidneys, with the liver showing the most prominent injury. Subcapsular damage, focal necrosis, both periportal and mid-zonal fatty infiltration, congestion, and edema were observed. Methemoglobinemia has not been observed in rabbits or rats. Nitromethane is apparently metabolized by a mechanism different from nitroethane and nitropropane in that negligible amounts of nitrites are found in the blood following intravenous injection of 1 mM in rabbits. Skin application of NM does not produce irritation or death in animals.

1.8.1.2 Subchronic Toxicity. Three of 10 rats given 0.25 percent NM in their drinking water for 15 weeks and four of 10 rats given 0.1 percent died during the course of the experiment. The surviving animals failed to gain weight normally. Histopathological examination showed mild but definite liver abnormalities.

Six-month inhalation studies of NM in rats or rabbits showed widely differing responses in the two species (11). For exposure at 98 or 745 ppm of NM, only mild to moderate symptoms of toxicity were observed in rats and rabbits. A reduction in body weight gain was observed in rats exposed at 745 ppm of NM. Hematocrit and hemoglobin levels in rats were slightly depressed from 10 days through 6 months of exposure to 745 ppm. Rabbits also provided a suggestion of depression in hemoglobin levels. Other hematologic parameters such as prothrombin time and methemoglobin concentration were unaffected in both rats and rabbits. Ornithine carbamyl transferase in rabbits was elevated after 1 and 3 months, but not 6 months of exposure at 745 ppm. No apparent effects of glutamic–pyruvic transaminase in both species or serum T_4 activity in rats were observed. Serum T_4, however, was statistically significantly depressed in rabbits exposed at either 98 or 745 ppm NM at the 6-month testing period as well as at the 1-month sacrifice for rabbits exposed at 745 ppm. Weights of all organs evaluated were comparable to controls in rats and rabbits except for thyroid weights in rabbits after 6 months of exposure at 745 ppm. The increased thyroid weights after 6 months and decreased thyroxin levels at all testing intervals, for both concentrations, indicated an effect on the thyroid in rabbits by NM. Histopathological evaluation indicated no exposure-related abnormalities in rats due to exposure to NM at 98 or 745 ppm for up to 6 months. Some evidence of pulmonary edema and other pulmonary abnormalities was observed in rabbits exposed to both levels of NM for 1 month.

The most important observations are that inhalation of NM produces mild irritation and toxicity before narcosis occurs, and that liver damage can result from repeated administration at levels in excess of 1000 ppm.

1.8.1.3 Odor and Warning Properties. The odors of nitroalkanes are easily detectable, and concentrations below 200 ppm are disagreeable to most observers (13). An odor threshold of 3.5 ppm has been reported for NM (71). The odor and sensory symptoms are not dependable warning properties; however, no injuries in humans have been reported.

1.8.2 Tetranitromethane, C(NO₂)₄

Tetranitromethane (TNM) is a colorless oily fluid with a distinct, pungent odor. Its physical and chemical properties are given in Table 10.1. It is explosive and is more easily detonated than trinitrotoluene (TNT). Its explosive power is less than that of TNT except when mixed with hydrocarbons, the mixtures being more powerful explosives and very sensitive to shock. Accidental explosions have oc-

Table 10.5. Acute Toxicity of TMN to Rats and Mice (57)

Test	Rats	Mice
Oral LD_{50} (95% C.L.)	130 (83–205) mg/kg	375 (262–511) mg/kg
Intravenous LD_{50} (95% C.L.)	12.6 (10.0–15.9) mg/kg	63.1 (45.0–88.7) mg/kg
4-Hr Inhalation (95% C.L.)	17.5 (16.4–18.7) ppm	54.5 (48.0–61.7) ppm

curred in handling and manufacture. It occurs as a contaminant of crude TNT. It is of interest for use as a propellant and as a fuel additive.

1.8.2.1 Animal Toxicity—Acute. A summary of the data on the response of animals to inhalation of various concentrations of TNM appears in Tables 10.5 and 10.6. In all experiments, exposed animals have exhibited similar symptoms, chiefly those of respiratory tract irritation. The first signs are increased preening, change in the respiratory pattern, and evidences of eye irritation followed by rhinorrhea, gasping, and salivation. The symptoms progress to cyanosis, excitement, and death at higher concentrations. Methemoglobinemia occurred in exposed cats. It should be noted that Sievers et al. (25) exposed their animals to TNM from crude trinitrotoluene, and although the concentrations recorded for TNM were determined by sampling and analysis, other unknown contaminants could have been present. These investigators found that animals exposed at 3 to 9 ppm for 1 to 3 days developed pulmonary edema. Lower concentrations (0.1 to 0.4 ppm) produced only mild irritation.

The results of pathological examinations on animals dying from acute exposures were all similar. There was marked lung irritation with destruction of epithelial cells, vascular congestion, pulmonary edema, and emphysema with tracheitis and bronchopneumonia. Nonspecific changes in the liver and kidney were observed in some animals.

1.8.2.2 Animal Toxicity—Subchronic. Horn (8) exposed two dogs and 19 rats to 6.35 ppm for 6 hr/day, 5 days/week for 6 months. Eleven of the 19 rats died in the

Table 10.6. Effect of Various Concentrations of Tetranitromethane (TNM)

Animal	Concentration (ppm)	Duration of Exposure	Effect (Ref.)
1 cat	100	20 min	Death in 1 hr (72)
1 cat	10	20 min	Death in 10 days (72)
5 cats	7–25	2½–5 hr	Death in 1–5½ hr (25)
2 cats	3–9	6 hr × 3	Severe irritation (25)
2 cats	0.1–0.4	6 hr × 2	Mild irritation (25)
20 rats	1230	1 hr	All died in 25–50 min (8)
20 rats	300	1½ hr	All died in 40–90 min (8)
20 rats	33	10 hr	All died in 3–10 hr (8)
19 rats	6.35	6 months	11 deaths (8)
2 dogs	6.35	6 months	Mild symptoms (8)

course of the exposure with evidence of pulmonary irritation, edema, and pneumonia. Some initial anorexia was observed in the dogs. Repeated examinations did not reveal anemia, Heinz bodies, methemoglobinemia, or biochemical disturbances. Rats surviving 6.35 ppm for 6 months developed pneumonitis and bronchitis of a moderate degree, whereas those dying developed more severe pneumonia. Histopathological examination of two dogs surviving the same concentration for 6 months revealed no evidence of injury (8).

NTP conducted 2-week and 13-week inhalation studies in rats and mice prior to the 2-year bioassay (57). In the 2-week study, rats and mice (five of each sex per group) were exposed to 0, 2, 5, 10, or 25 ppm TNM for 6 hr/day, 5 days/week. A group of mice were also exposed to 50 ppm TNM. All rats exposed to 25 ppm died within 1 day and one male rat in the 10 ppm group died on day 8. Death was probably due to pulmonary edema. Males exposed to 5 ppm and all rats exposed to 10 ppm lost body weight. All mice exposed to 50 ppm and 8/10 mice at 25 ppm died on day 2 and days 3 or 4, respectively. Pulmonary effects were probably the cause of death. Weight loss was observed in mice exposed to 5 ppm and above.

Exposure concentrations for the 13-week study by NTP were 0, 0.2, 0.7, 2, 5, and 10 ppm for both rats and mice 6 hr/day, 5 days/week. There was no treatment-related mortality in rats and body weights for the 10-ppm group were slightly lower than controls. Liver to body weight ratios were elevated at all exposure concentrations but no microscopic changes were observed in the liver. Focal squamous metaplasia of the respiratory epithelium of the nasal passages and minimal to moderate chronic inflammation of the lung was seen in many animals in the 10-ppm group. Effects in mice were very similar except that respiratory tract pathology was also seen at 5 ppm.

1.8.2.3 Animal Carcinogenicity. Based on the above results, NTP selected exposure concentrations for the 2-year inhalation study of 0, 2, and 5 ppm for rats and 0, 0.5, and 2 ppm for mice. Fifty animals of each sex were exposed at each concentration for 6 hr/day, 5 days/week.

Mean body weights of rats at 5 ppm were lower than controls and survival at 5 ppm was reduced due to neoplasia. Significant pathology was limited to the respiratory tract. In the nasal passages, mucosal chronic inflammation and squamous metaplasia and hyperplasia of the respiratory epithelium were observed at elevated incidences at 5 ppm. Alveolar and bronchiolar hyperplasia was observed at both 2 and 5 ppm. Alveolar/bronchiolar adenomas and carcinomas were also seen at both concentrations while squamous cell carcinomas in the lung were significantly elevated at 5 ppm.

Mice at both exposure concentrations had mean body weights lower than controls. Survival was lower in males at 2 ppm owing to neoplasia. Nasal passage lesions such as that seen in rats was seen only in female mice at 2 ppm. Alveolar and bronchiolar hyperplasia was elevated at 0.5 and 2 ppm. Alveolar/bronchiolar adenomas and carcinomas were also elevated at both exposure concentrations.

Based on these results (57) the NTP classified the level of evidence of carcinogenic activity for both sexes of rats and mice as "clear evidence." This classification

Table 10.7. Response to Inhalation of Nitroethane (13)[a]

Concentration (ppm)	Time (hr)	Mortality	
		Rabbit	Guinea Pig
30,000	1.25	2	2
	1	1	1
	0.5	1	None
10,000	3	2	1
	1	1	None
5,000	3	2	None
	2	1	None
2,500	3	None	None
1,000	2	1	None
	6	None	None
500	30	None	None
	140[b]	None	None

[a]Two rabbits and two guinea pigs in each experiment.
[b]One monkey exposed, not fatal.

refers to the strength or amount of experimental evidence for carcinogenicity and not to the potency or quantitative risk of cancer.

1.8.2.4 Human Toxicity. The lowest lethal dose, LD_{LO}, for humans by inhalation is given as 500 mg/kg (~35 g) (73). A few deaths and intoxications that have occurred during handling of heated contaminated TNT have been attributed to TNM (13). Symptoms experienced in the laboratory production of TNM were irritation of eyes, nose, and throat from acute exposures and, after more prolonged inhalation, headache and respiratory distress (74). Skin irritation does not result from repeated contact in humans or animals.

1.8.2.5 Odor and Warning Properties. TNM can be recognized by its characteristic acrid biting odor. It can be measured in air by the methods used by Horn (8) or Vouk and Weber (75).

1.8.3 Nitroethane, $CH_3CH_2NO_2$

Nitroethane (NE) is an oily, colorless liquid with a somewhat pleasant odor. It represents a lesser explosive hazard than nitromethane and tetranitromethane. Unconfined quantities are not exploded by heat or shock. Because under appropriate conditions of confinement or contamination with other materials explosions could result, safe handling procedures have been recommended in detail.

1.8.3.1 Animal Toxicity. The response of rabbits and guinea pigs to inhalation of NE as determined by Machle and co-workers (13) appears in Table 10.7. NE is a moderate respiratory tract irritant. There was more respiratory tract irritation and less narcosis with NE than was observed with NM. Except for this, the symptom-

Table 10.8. Response to Inhalation of 2-Nitropropane (6)

	Highest Tolerable Concentration (ppm)			Lowest Lethal Concentration (ppm)		
Animal	1 hr	2.25 hr	4.5 hr	1 hr	2.25 hr	4.5 hr
Rat	2353	1372	714[a]	3865	2633	1513
Guinea pig	9523	4313	2381		9607	4622[b]
Rabbit	3865	2633	1401	9523	4313	2381
Cat	787	734	328	2353	1148	714

[a]Time 7 hr.
[b]Time 5.5 hr.

atology and pathological findings were similar to NM. Machle et al. found no evidence of skin irritation or skin absorption. Scott found increasing nitrite concentrations in the blood of rabbits during inhalation of NE (35). Nitroethane is excreted via the lungs, is rapidly metabolized, and is completely eliminated in 30 hr (34).

1.8.3.2 Animal Toxicity—Chronic. Griffin and co-workers conducted a 2-year inhalation study of NE in Long-Evans rats (60). Eighty animals per group (40/sex) were exposed to 0, 100, or 200 ppm NE for 7 hr/day, 5 days/week. No clinical signs or body weight effect were observed during the 2-year exposure. No effects on hematology, clinical chemistry, and organ weights were seen at termination. Microscopic examination of tissues showed no neoplastic or other pathology. In particular, no evidence of hepatoxicity was found.

1.8.3.3 Odor and Warning Properties. An odor threshold of 2.1 ppm has been reported for NE (71). This gives an odor safety factor (TLV ÷ odor threshold) of 46. NE was assigned to odor safety class B whereby it is anticipated that 50 to 90 percent of distracted individuals exposed to the TLV would perceive a warning due to odor.

1.8.4   2-Nitropropane, $CH_3CHNO_2CH_3$

2-Nitropropane (2-NP) is a colorless, oily liquid. Its physical and chemical properties are given in Table 10.1. The nitropropanes are less of an explosive hazard than the nitromethanes.

1.8.4.1 Animal Toxicity—Acute. As seen in Table 10.8, considerable differences in species response were observed. Cats were the most susceptible and guinea pigs the least. High concentrations of 2-NP produced dyspnea, cyanosis, prostration, some convulsions, lethargy, and weakness, proceeding to coma and death. Some animals surviving the acute exposure died 1 to 4 days later. These high concentrations of 2-NP caused pulmonary edema and hemorrhage, selective disintegration of brain cells, and hepatocellular damage, with general vascular endothelial injury in all tissues.

1.8.4.2 Animal Toxicity—Subchronic. Cats, rabbits, rats, guinea pigs, and monkeys were exposed repeatedly at 328 and 83 ppm for 7 hr/day. Cats died following several days exposure at 328 ppm but rabbits, rats, and guinea pigs survived 130 exposures and a monkey survived 100. No signs or symptoms were observed in any animals during 130 exposures at 83 ppm (6).

Cats that died following several exposures at 328 ppm had microscopic evidence of focal necrosis and parenchymal degeneration in the liver and slight to moderate degeneration of the heart and kidneys. The lungs showed pulmonary edema, intraalveolar hemorrhage, and interstitial pneumonitis. The other species exposed at this concentration did not exhibit these findings. Except for one cat, no microscopic tissue changes were observed in the animals exposed at 83 ppm. Inhalation of 2-NP induced methemoglobin formation in cats and to a lesser extent in rabbits. Cats developed 25 to 35 percent methemoglobin when exposed at 750 ppm for 4.5 hr and about 15 to 25 percent methemoglobin during repeated, daily, 7-hr exposures at 280 ppm. Heinz bodies appeared in the erythrocytes of cats and rabbits at even lower concentrations.

A later exposure of rats and rabbits for 6 months inhalation to 2-NP at 207 and 27 ppm showed very few classical signs of nitroparaffin toxicity. Ornithine carbamyl transferase was elevated in rabbits after 1 and 3 months of exposure to 207 ppm, but not in rats. No effects on body weight gain or hematology were observed. Liver weights from rats exposed at 27 ppm of 2-NP for up to 6 months were comparable to those of controls. However, severe neoplastic changes were observed in the livers of male rats exposed to 207 ppm 2-NP for 6 months (as discussed in the following section below) (11).

1.8.4.3 Animal Carcinogenicity. Information relevant to the carcinogenicity of 2-NP is presented above in Sections 1.6.4 to 1.6.6. Details of animal bioassays conducted with 2-NP are discussed in this section.

2-NP has been found to be a hepatic carcinogen for male, but not female rats of the Sprague–Dawley derived strain following a 6-month daily 7-hr exposure at 207 ppm. A concurrently exposed group of male New Zealand strain of white rabbits showed no such response (11). Multiple hepatic carcinomas and numerous neoplastic nodules were present in the livers of all 10 rats exposed at 207 ppm of 2-NP for 6 months, but not at 27 ppm. Blood-filled cysts were occasionally seen in the neoplasm, and mitotic figures were frequently present. The hepatocellular carcinomas appeared to be rapidly growing and severely compressing the surrounding parenchyma. No metastatic hepatocellular carcinomas were seen in any of the other tissues examined.

More extensive inhalation studies were conducted by Griffin and co-workers. Initial studies exposing rats to 100 ppm for 18 months and 200 ppm for 6 months were not fully reported (76). After 6 months exposure to 200 ppm, they found slightly suppressed growth and elevated serum glutamic–pyruvic transaminase (GPT) in males with elevated relative liver weights, in both males and females. Microscopic examination of livers showed hyperplastic nodules, cellular necrosis, and multivacuolated fatty metamorphosis. Animals held for six months longer without 2-NP

Table 10.9. Response to Inhalation of 1-Nitropropane

Concentration (ppm)	Time (hr)	Mortality[a]	
		Rabbit	Guinea Pig
10,000	3	2	2
10,000	1	None	1
5,000	3	2	2

[a]Two rabbits and two guinea pigs in each experiment.

exposure developed tumors with metastasis. Similar though less severe pathology was observed with 18 month exposure at 100 ppm.

Results for 25-ppm exposures were reported more completely (69, 77). Sprague–Dawley rats (125 of each sex per group) were exposed to 0 or 25 ppm 2-NP for 7 hr/day, 5 days/week for 22 months. Interim sacrifices were conducted (10 of each sex) after 1, 3, 6, and 12 months and recovery groups were initiated (10 of each sex) after 3 and 12 months of exposure. No effects on body weight, clinical signs, clinical chemistries, or hematology and no methemoglobinemia were observed. There were no tumors in any organ or tissue including the liver. Relative liver weights were slightly elevated in males and focal areas of hepatocellular nodules were seen with greater incidence in males (2/125 for controls, 10/125 for 2-NP at 25 ppm). No other indication of hepatoxicity was observed.

2-NP has also been shown to cause hepatocarcinogenicity in rats given oral exposure (67). Sprague–Dawley rats (29 controls, 22 treated) were given gavage doses of vehicle or 2-NP (1 mmol/kg) three times per week for 16 weeks. Animals were sacrificed in the 77th week after treatment initiation. 2-NP decreased body weight gain and caused massive hepatocellular carcinomas. Metastases were also seen in the lungs of four animals.

1.8.4.4 Human Experience. The first report of worker response from exposure to 2-NP solvent mixtures was made by Skinner (23). Since then there have been several reports of hepatotoxicity and deaths associated with exposure to 2-NP (17, 26–28). These were discussed in Section 1.6.2. Exposure to 25 ppm was reported to cause no liver function abnormalities (30).

1.8.4.5 Odor and Warning Properties. An early report indicated 2-NP odor to be detectable at 294 ppm, but not at 83 ppm (6). A more recent study using controlled exposures at two laboratories reported odor detection at 3.1 ppm and 5 ppm (78).

1.8.5   1-Nitropropane, $CH_3CH_2CH_2NO_2$

1.8.5.1 Animal Toxicity. The acute inhalation toxicity of 1-nitropropane (1-NP) is not greatly different from that of 2-NP (see Tables 10.2 and 10.9). Exposures at 5000 ppm of 1-NP for 3 hr killed rabbits and guinea pigs, whereas the lowest lethal concentrations of 2-NP for these animals after a 2.25-hr exposure were 4313 and

9607 ppm, respectively. The symptoms and gross pathological changes observed in the exposed animals were similar to those exposed to nitroethane.

1.8.5.2 Animal Toxicity—Subchronic and Chronic. Groups of Long–Evans rats (125/sex) were exposed to 0 or 100 ppm 1-NP for 7 hr/day, 5 days/week for 21.5 months (61). Animals (10 of each sex) were sacrificed after 1, 3, 12, and 18 months and the same number were held for recovery without exposure after 3 and 12 months. Exposure had no effect on body weights, organ weights, clinical chemistries, hematology, or microscopic pathology observations. In particular there was no evidence of methemoglobinemia, hepatotoxicity, or hepatocarcinogenicity. These findings are consistent with the information regarding genotoxicity and carcinogenicity discussed in Sections 1.6.4 to 1.6.6.

1.8.5.3 Odor and Warning Properties. In a limited organoleptic test of 1-NP, human volunteer subjects found concentrations exceeding 100 ppm irritating after brief periods of exposure (79). The odor threshold of 1-NP based on the geometric mean of two determinations is reported to be 11 ± 4.2 ppm (S.E.M.) (71).

1.8.6   Nitrobutanes

The formulas and their physical constants are given in Table 10.1. The toxicology of 1-nitrobutane and 2-nitrobutane has not been studied beyond that reported by Machle et al. (13). The effects following oral administration in rabbits were similar to those produced by the other nitroalkanes and the lethal dose range was the same as for 2-NP and NE. As with these materials, no skin irritation or systemic symptoms were observed after five daily open applications to rabbit skin. Less nitrite can be recovered from rabbit blood following intravenous injection of the nitrobutanes than after an injection of equivalent doses of nitropropanes or nitroethanes. It would be expected that the nitrobutanes would present acute hazards qualitatively similar to the nitropropanes. 1-Nitrobutane (1-NB) is not mutagenic in the Ames test where 2-NB is mutagenic (45). Also 1-NB does not produce oxidative damage to DNA whereas 2-NB does after ip administration to rats (65). Therefore the genotoxicity and possibility of carcinogenicity for the nitrobutanes would likely parallel the activity of nitropropanes as discussed in Sections 1.6.4 to 1.6.6.

1.8.7   Chlorinated Mononitroparaffins

The formulas of five of these substances, their physical constants, and the oral lethal doses of four in rabbits appear in Table 10.10. The chlorinated nitroparaffins are of particular interest in the manufacture of highly accelerated rubber cements and insecticides and in chemical synthesis.

1.8.7.1 Animal Toxicity. Comparison of the acute oral lethal doses for rabbits shows that, with the exception of 2-chloro-2-nitropropane, the chlorinated mononitroparaffins were five times more toxic than the unchlorinated compounds (Table 10.10 versus Table 10.1). The same toxicity difference holds for the respiratory

Table 10.10. Physical and Chemical Properties of the Chlorinated Mononitroparaffins

Name	Mol. Wt.	B.p. (°C)	Specific Gravity	H₂O Solubility at 20°C (% by Vol.)	Vapor Pressure (mm Hg) (25°C)	Vapor Density (Air = 1)	Flash Point (°F)	Conversion Units 1 mg/l (ppm)	1 ppm (mg/m³)	Oral Lethal Dose, Rabbits[a] (g/kg)
Trichloronitromethane (chloropicrin)	164.38	111.84	1.656 (20/4°C)	Insoluble	16.9 (20°C)	5.7	—	148.8	6.72	—
1-Chloro-1-nitroethane	109.51	127.5	1.2860 (20/20°C)	0.4	11.9	3.6	133	237	4.21	0.10–0.15
1,1-Dichloro-1-nitroethane	143.9	124	1.4271 (20/20°C)	0.25	16	5.0	168	169.9	5.89	0.15–0.20
1-Chloro-1-nitropropane	123.5	139.5–143.3	1.209 (20/20°C)	0.5	5.8	4.3	144	198	5.05	0.05–0.10
2-Chloro-2-nitropropane	123.5	133.6	1.197 (20/20°C)	0.5	8.5	4.3	135	198	5.05	0.50–0.75

[a]Reference 9.

Table 10.11. Response to Inhalation of 1,1-Dichloronitroethane (9)

Average Concentration (ppm)	Duration of Exposure	Mortality[a]	
		Rabbit	Guinea Pig
4910	30 min	2	2
985	$3\frac{1}{2}$ hr	2	1
594	$2\frac{1}{2}$ hr	1	None
254	1 hr	None	None
169	2 hr	1	1
100	6 hr	2	2
60	2 hr	None	None
52	18 hr, 40 min	2	None
34	4 hr	None	None
25	204 hr	None	None

[a]Two rabbits and two guinea pigs in each experiment.

route. The 4-hr inhalation LC_{50} in rats ranges from 14.4 to 6.6 ppm (80, 81). 1,1-Dichloronitroethane is considerably more irritating to skin and mucous membranes than 1-chloro-1-nitropropane and exhibits greater toxicity by the inhalation route (see Tables 10.11 and 10.12). All three materials are lung irritants. They cause pulmonary edema and death within 24 hr following exposure at high concentrations. The chief site of injury is the lungs, but damage is also observed in the heart, muscle, liver, and kidneys after lethal exposures. Although 1-chloro-1-nitroethane and 2-chloro-2-nitropropane have not been studied in detail, it would be expected that their inhalation toxicity would be qualitatively and quantitatively similar to that of 1-chloro-1-nitropropane.

Very similar findings were made by Soviet scientists for the acute toxicities of 1-chloro-1-nitroethane and propane and for 2-chloro-2-nitropropane and butane (82).

1.8.7.2 Hygienic Standards. OSHA standards and TLVs have been established for 1-chloro-1-nitropropane and chloropicrin as shown in Table 10.3.

Table 10.12. Response to Inhalation of 1-Chloro-1-nitropropane (10)

Average Concentration (ppm)	Duration of Exposure	Mortality[a]	
		Rabbit	Guinea Pig
4950	60 min	2	1
2574	2 hr	2	None
2178	1 hr	None	1
1069	1 hr	None	None
693	2 hr	None	None
393	6 hr	1	None

[a]Two rabbits and two guinea pigs in each experiment.

Table 10.13. Effects of Various Concentrations of Trichloronitromethane—
Animals

Animal	Concentration mg/l	Concentration ppm	Duration of Exposure (min)	Effects
Dog	1.05	155	12	Became ill
	0.08–0.95	117–140	30	Death of 43% of the animals
Mouse	0.85	125	15	Death in 3 hr to 1 day
Cat	0.51	76	25	Death usually in 1 day
Mouse	0.34	50	15	Death after 10 days
Dog	0.32	48	15	Tolerated
Cat	0.32	48	20	Death after 8 to 12 days
	0.26	38	21	Survived 7 days
Mouse	0.17	25	15	Tolerated

1.8.8   Trichloronitromethane, CCl_3NO_2 (Chloropicrin)

The physical constants of chloropicrin are given in Table 10.10. Its uses have
included dyestuffs (crystal violet), organic syntheses, fumigants, fungicides, insecticides, rat exterminator, and poison war gas.

1.8.8.1 Animal Toxicity—Acute. The results of early inhalation studies are sum-
marized in Table 10.13 (72, 80, 81). The acute inhalation LC_{50} of chloropicrin in
Fischer 344 rats using standard whole-body exposure has been determined to be
11.9 and 14.4 ppm (82, 83). Shorter exposures for 30 min resulted in no mortality
at 22 ppm and 100 percent mortality at 46 ppm. Major effects included dyspnea,
cyanosis, pulmonary edema, and increased lung weight. Cause of death was respiratory failure. Nose-only exposure for 4 hr produced an LC_{50} of 6.6 ppm, significantly lower than for whole-body exposure (83). Dermal-only exposure to 25
ppm vapor resulted in no mortality. The RD_{50} (concentration which reduces respiratory rate by 50 percent) in mice, which is a measure of sensory irritation potency,
is 8 ppm (84).

1.8.8.2 Animal Toxicity—Repeated Exposure. Inhalation exposure of mice to 8 ppm
(the RD_{50}), 6 hr/day for 5 days caused no mortality and moderate damage in the
nasal passages and lung (84). However, rats exposed daily to 5 ppm for 6 hr died
after 7 to 10 days. Rats survived 10 exposures to 2.5 ppm but showed increased
lung weights (85). In a 13-week inhalation study, male Fischer 344 rats were exposed
to 0, 0.4, 0.7, 1.6, or 2.9 ppm chloropicrin for 6 hr/day, 5 days/week (85). There
were no deaths, but mean body weights were reduced at the two higher exposure
concentrations. RBC, hematocrit, and hemoglobin concentrations were reduced
at 2.9 ppm. Lung weights were increased and bronchial/bronchiolar lesions were
seen at 1.6 and 2.9 ppm. The authors considered 0.7 ppm a no-observed-adverse-
effect-level (NOAEL).

  The genotoxicity of chloropicrin is discussed in Section 1.6.4. An oral carcino-

Table 10.14. Effects of Various Concentrations of Trichloronitromethane in Humans

Concentration		Duration of Exposure (min)	Effect
mg/l	ppm		
2.0	297.6	10	Lethal concentration
0.8	119.0	30	Lethal concentration
0.1	15.0	1	Intolerable
0.050	7.5	10	Intolerable
0.009	1.3		Lowest irritant concentration
0.0073	1.1		Odor detectable
0.002–0.025	0.3–3.7	3–30 sec	Closing of eyelids according to individual sensitivity

genicity bioassay on chloropicrin sponsored by the National Cancer Institute (NCI) was inconclusive (86).

1.8.8.3 Human Toxicity. Data on exposures of humans to various concentrations of chloropicrin, largely obtained during World War I, are summarized in Table 10.14 (72, 80, 81).

Chloropicrin is both a lacrimator and a lung irritant. Flury and Zernik state that exposure to 4 ppm for a few seconds renders a man unfit for combat, and 15 ppm for approximately the same period of time results in respiratory tract injury (72). Chloropicrin produces more injury to medium and small bronchi than the trachea and large bronchi. Pulmonary edema occurs and is the most frequent cause of early deaths.

Late deaths may occur from secondary infections, bronchopneumonia, or bronchiolitis obliterans. During World War I chloropicrin was noted for its tendency to cause nausea and vomiting. It has been stated that individuals injured by inhalation of chloropicrin become more susceptible, so that concentrations of the gas not producing symptoms in others cause them distress. Based on the information compiled by Vedder (31), Fries and West (81), and quoted by Flury and Zernik (72), concentrations of 0.3 to 0.37 ppm resulted in eye irritation in 3 to 30 sec, depending on individual susceptibility. A concentration of 15 ppm could not be tolerated longer than 1 min even by individuals accustomed to chloropicrin. Chloropicrin is also a potent skin irritant.

1.8.8.4 Odor and Warning Properties. The odor threshold for chloropicrin is reported as 0.78 ppm which does not provide warning of exposure at the current OSHA standard of 0.1 ppm (71).

1.8.9 Nitroolefins

Toxicologic interest in the nitro derivatives of straight-chain olefins has stemmed from human health concerns (eye irritation and carcinogenicity) caused by their presence in urban air pollution derived from automobile exhausts. Deichmann et al. (14) have shown that nitroolefins are indeed emitted in the exhausts from gasoline engines. For this reason, Deichmann et al. studied the toxicologic and pharmacological actions of a series of 21 straight-chain olefins in experimental animals and human volunteers (14). Unfortunately, no physical or chemical constants of any member of the series appear to have been published. Unfortunately also, for the great amount of effort expended, it was learned too late that, because of their high reactivity, the nitroolefins "decompose readily in the presence of sunlight," (14) so that only on sunless days would significant amounts be present temporarily in smog to contribute to effects on urban populations. As industrial chemical entities also, no nitroolefin appears to have any commercial interest (1). Accordingly, the vast amount of data accumulated by Deichmann and associates over a 10-year period, 1955 to 1965, is only summarized here. Detailed tabular data can be found in Reference 14 and in previous references. Table 10.15 gives the comparative toxicities of 12 nitroolefins by four routes of exposure.

1.8.9.1 Human Eye Irritation. A major quest of toxicologic and pharmacological studies of the conjugated, linear nitroolefins was the determination of eye irritation to humans, in an effort to identify the causative agent(s) in Los Angeles smog. This phase of the work was confined to the three nitroolefins 2-nitro-2-butene, 2-nitro-2-hexene, and 2-nitro-2-nonene, and was conducted jointly by members of the Los Angeles County Air Pollution Control District and the staff at the University of Miami. Both groups found the eye irritation produced by 2-nitro-2-butene and 2-nitro-2-hexene to be of the same order of magnitude. At the University of Miami, the threshold for eye irritation was found to be 0.2 to 0.4 ppm for the nitrohexene and 0.1 ppm for nitrobutene. Both groups found the threshold for 2-nitro-2-nonene to be considerably higher. After repeated exposures, subjects became increasingly sensitive.

1.8.9.2 Acute Toxicity—Animals. Acute toxicity of the compounds was investigated by inhalation, oral, intraperitoneal, and cutaneous routes, using rabbits, guinea pigs, rats, mice, chicks, and dogs. The subacute inhalation toxicity of four nitroolefins representative of the series was studied, using rabbits, guinea pigs, rats, and mice.

All nitroolefin compounds were found to be highly toxic as well as irritant. Absorption from the respiratory and gastroenteric tract, peritoneal cavity, or skin was rapid. Signs of systemic intoxication appeared promptly, including hyperexcitability, tremors, clonic convulsions, tachycardia, and increased rate and amplitude of respiration, followed by a generalized depression, ataxia, cyanosis, and dyspnea. Death was initiated by respiratory failure and associated with asphyxial conclusions. Pathological changes were most marked in the lungs, regardless of the mode of administration.

Inhalation toxicity showed no definite relationship to chain length, but the acute oral and intraperitoneal toxicities decreased with increasing length of the carbon chain.

Table 10.15. Acute Effects of Nitroolefins (14)

Name	Vapor Exposure (5 hr) Concn. (ppm)	Survival Times, Rats, 47% Humidity	Oral Toxicity, Rats, Undiluted (Approx. Lethal Dose) g/kg	mmol/kg	Intraperitoneal Toxicity, Rats. Undiluted (Approx. Lethal Dose) g/kg	mmol/kg	Dermal Toxicity, Rabbits, Open, 5-hr (Approx. Lethal Dose) g/kg	mmol/kg
2-Nitro-2-butene	1400	100 min	0.28	2.8	0.08	0.8	0.62	6.1
2-Nitro-2-pentene	240	240 min	0.28	2.4	0.08	0.7	0.94	5.4
	55	Survived						
3-Nitro-2-pentene	268	280 min	0.42	3.7	0.05	0.4	0.62	8.2
2-Nitro-2-hexene	515	50–85 min	0.42	3.3	0.12	0.9	1.40	7.3
	152	Survived						
3-Nitro-3-hexene	557	30–70 min	0.42	3.3	0.08	0.6	0.94	10.9
	50	Survived						
2-Nitro-2-heptene	308	3–18 hr	0.94	6.6	0.28	2.0	0.94	6.6
	135	Survived						
3-Nitro-3-heptene	54	24 hr	0.62	4.3	0.28	2.0	1.40	9.8
2-Nitro-2-octene	47	Survived	1.4	9.0	0.28	1.8	0.62	4.0
3-Nitro-3-octene	142	18–24 hr	0.62	4.0	0.18	1.2	0.94	6.0
	72	Survived						
3-Nitro-2-octene	141	18–72 hr	0.62	4.0	0.18	1.2	0.62	4.0
	44	Survived						
2-Nitro-2-nonene	64	Survived	2.1	12.3	0.28	1.6	0.62	3.6
3-Nitro-3-nonene	59	24 hr	2.1	12.3	0.42	2.5	0.42	2.5
	10	Survived						

1.8.9.3 Chronic Toxicity and Carcinogenicity. The most significant finding in an 18-month chronic inhalation study using dogs, goats, rats, and mice, was five instances of primary adenocarcinoma of the lung in a group of 27 Swiss mice exposed at 0.2 ppm 2-nitro-3-hexene. No such changes were observed in the 21 control mice.

A second chronic inhalation study was performed at 1.0 and 2.0 ppm 3-nitro-3 hexene, in which rats were exposed for 36 months and dogs for 42 months. The histopathological examination of tissues from the rats revealed primary malignant lesions (undifferentiated carcinoma) in the lungs of 6 of 100 rats exposed to 1.0 ppm of 3-nitro-3 hexene, and in the lungs of 11 of 100 rats exposed at 2.0 ppm. The male rat appeared somewhat more susceptible than the female. There were no primary malignant lesions in the lungs of the 100 control rats.

1.8.9.4 Effects on Skin and Eye. Open application to the skin of the rabbit (for 5 hr) resulted in intense local irritation, erythema, edema, and later necrosis. One drop in the eye produced marked irritation and corneal damage.

2 ALIPHATIC NITRATES

2.1 General Considerations

The aliphatic nitrates are nitric acid esters of mono- and polyhydric aliphatic alcohols. The nitrate group has the structure —C—O—NO_2, where the N is linked to C through O, as contrasted to the nitroalkanes in which N is linked directly to C.

The nitric acid esters of the lower mono-, di-, and trihydric alcohols are liquids (methyl nitrate, ethyleneglycol dinitrate, trinitroglycerin), whereas those of the tetrahydric alcohols (erythritol tetranitrate, pentaerythritol tetranitrate) and hexahydric alcohol (mannitol hexanitrate) are solids. They are generally insoluble, or only very slightly soluble in water, but are more soluble in alcohol or other organic solvents. Some physical and chemical properties for this group of compounds are shown in Table 10.16.

Trimethylenetrinitramine (cyclonite, RDX) and cyclotetramethylenetetranitramine (HMX) have been included in this section because they are also used as high explosives. The nitramines contain the grouping —$NHNO_2$.

2.2 Production and Use

Uses of aliphatic nitrates are chiefly as explosives and blasting powders, particularly nitroglycerin (NG) and pentaerythritol tetranitrate (PETN). A small but important amount of NG is used as a vasodilating agent. Amyl nitrate has been used to increase the cetane number of diesel fuels. The lower aliphatic nitrates, methyl, ethyl, propyl, and isopropyl have also been used as rocket propellants and special jet fuels.

Table 10.16. Physical and Chemical Properties of Aliphatic Nitrates and Related Explosive Compounds

Name	Mol. Wt.	Physical State	B.P. (°C)	Vapor Density (air = 1)	H_2O Solubility
Methyl nitrate	77.04	Volatile liquid	66 (explodes)	2.66	Slight
Ethyl nitrate	91.07	Colorless liquid	87.6	3.14	1.3% (55°C)
Propyl nitrate	105.09	Pale yellow liquid	110.5	3.62	Very slight
Amyl nitrate	133.15	Slightly yellow liquid	150 (unstable)	—	0.3%
Ethylene glycol dinitrate	152.07	Colorless liquid	114 (explodes)	5.24	0.52%
Glyceryl trinitrate (nitroglycerin)	227.10	Colorless oily liquid	260 (explodes)	7.80	Slight
Propylene glycol-1,2-dinitrate	166.09	Red-orange liquid	121 (decomp.)	—	—
Pentaerythritol tetranitrate	316.15	White crystalline solid	180 (50 mm Hg)	—	Very slight
Cyclonite (RDX)	222.26	White crystalline solid	276–280 (M.P.)	—	Insoluble
HMX	296.16	Colorless crystalline solid	204 (M.P.)	—	Insoluble

2.3 Pharmacology, Symptomatology, and Mode of Action

After more than 100 years of use for relief of angina pectoris, nitroglycerine is finding wider application in congestive heart failure, limiting myocardial "infarct size," and long-term angina prophylaxis, and as a diagnostic test for the presence of myocardial ischemia. Not only is its clinical potential now better recognized, but its mode of action in normal hearts and in myocardial ischemia is better understood (87, 88).

Because one of the shortcomings of NG is its brief action, long-sought nitrate preparations with more sustained effect have now been found in related nitrate esters, erythritol tetranitrate, pentaerythritol tetranitrate, and isosorbide nitrate. The latter produces significant reduction in arterial and capillary pressures for the first hour after ingestion of a small dose, and decreases cardiac output for up to 4 hr, as well as providing effective prophylaxis in angina.

The chief effects of the aliphatic nitrates are dilation of blood vessels and methemoglobin formation. The vascular dilation accounts for the characteristic lowering of blood pressure and headache. The individual members in the series differ in intensity and duration of these effects. Animals given effective doses by mouth or parenterally exhibit such signs as marked depression in blood pressure, tremors, ataxia, lethargy, alteration in respiration (usually hyperpnea), cyanosis, prostration, and convulsions. Death, when it occurs, is either from respiratory or cardiac arrest. Animals surviving the acute exposure recover promptly.

Nitroglycerin (NG) and erythritol tetranitrate (ETN) are capable of producing approximately the same degree of hypotension in humans but the effect of ETN is more prolonged and requires a larger dose. The maximum blood pressure depression from NG occurs at approximately 4 min, whereas that from ETN occurs at approximately 20 min. Pentaerythritol tetranitrate (PETN) is less effective as a hypotensive agent than ETN. Methyl nitrate causes little depression in blood pressure (89). The outstanding symptom produced in humans is headache. This is usually described as very severe and throbbing, and is often associated with flushing, palpitation, nausea, and less frequently, vomiting and abdominal discomfort. Temporary tolerance developed from continued or repeated daily exposures (88). Doses resulting in headaches are NG, 18 mg (skin); ethylene glycol dinitrate (EGDN), 35 mg (skin); and ETN, 45 mg (oral) (89). PETN in doses of 64 mg orally does not produce headache (90). Thus PETN is the least effective and NG the most potent for headache induction. Other pharmacological consequences of vasodilation are increased pulse rate, an increase in cardiac stroke volume, variable cardiac dilation and cardiac output, and a shift in blood distribution with increased stasis and pressure in pulmonary arteries (89).

For some members of the series, the ease of hydrolysis to the alcohol and nitrate and the degree of blood pressure lowering are parallel. Early studies suggested there is little evidence that hydrolysis to nitrate is necessary for hypotensive action (91–93). It appears that this effect of the nitrate esters does not depend exclusively on the liberation of nitrite groups. Dilatation can occur without measurable nitrite in the blood, or when the amount measured is not sufficient to account for the effect observed. Direct effects of nitrates on smooth muscle cells include stimulation of guanylate cyclase producing increased cyclic guanosine monophosphate (cGMP) levels. cGMP in turn lowers intracellular calcium concentrations, thus relaxing contractile protein causing vasodilation (88).

The in vivo formation of nitrite is commonly assumed to be the explanation for the methemoglobin-forming properties of the aliphatic nitrates (94). The mechanism of formation of nitrite is not clear (95). It is possible that reduction to nitrite occurs before hydrolysis as follows (96):

$$RONO_2 \xrightarrow{+2H} RONO \xrightarrow{+H_2O} ROH + NO_2^-$$

Ethyl nitrate, EGDN, NG, propyl nitrate, and amyl nitrates are known to cause methemoglobin formation in experimental animals. Ethyl nitrate is a weak methemoglobin former. Nitroglycerin is a moderately active methemoglobin former but EGDN is considerably more effective (approximately four times). Ethylene glycol mononitrate, on the other hand, is not very active in this respect (87).

Ethyl, propyl, and amyl nitrate, as well as EGDN and NG, induce Heinz body formation in animals. Although ethyl nitrate induces Heinz body formation, ethyl nitrite does not. EGDN is more effective than NG, which is more effective than ethyl nitrate. The precise nature of these small, rounded inclusion bodies in the red blood cells, described by Heinz in 1890, is not clear. They have been observed in humans and animals after absorption of a variety of chemical compounds, the most prominent of which are the aromatic nitrogen compounds, inorganic nitrites, and the aliphatic nitrates. Their appearance is commonly associated with anemia and the production of methemoglobin. Some evidence indicates that they are proteins in nature, possibly hemoglobin degradation products. Red blood cells containing the inclusion bodies have a shorter life-span and are removed from the circulation by the spleen. Special stains are required to satisfactorily demonstrate their presence. In the case of the aliphatic nitrates, erythrocytes containing Heinz bodies disappear from the circulating blood more slowly than methemoglobin (96–98).

The alkyl nitrates are all absorbed from the gastrointestinal tract (89). PETN is relatively slowly absorbed by this route. EGDN and NG are absorbed through the skin but the absorption of ETN and PETN by this route is slow or absent. The nitric acid esters of the monovalent alcohols are rapidly absorbed from the lung. The absorption of ETN through the lungs is slower than with EGDN. NG is more slowly absorbed from the lungs than EGDN and PETN is more slowly absorbed than ETN.

2.4 Pathology in Animals

Pathological examinations of animals dying following acute intoxication either have been negative or have revealed only slight nonspecific pathological changes consisting of congestion of internal organs. Hueper and Landsberg (99) have described degenerative vascular and parenchymatous lesions in the heart, kidneys, lungs, brain, and testes following several months' administration of large doses of ETN to young rats. They felt that these changes were induced by inadequate nutrition of the tissues following hypoxia and stagnation of the organs' blood supply associated with vasodilatation. On the other hand, von Oettingen et al. were unable to confirm these findings after chronic administration of ETN and PETN at lower doses (100). Evidence of injury to these organ systems in humans chronically exposed to NG, ETN, and PETN is lacking.

2.5 Industrial Experience

No injuries to workers from exposure to any of the lower monohydric alcohol esters of nitric acid (methyl, ethyl, n-propyl, amyl isomers) have appeared in the published literature. However, for polynitrate esters such as NG or EGDN, the occurrence of characteristic and severe headaches in workers was so frequent that such acquired names as "dynamite head" and "powder headache" were common. Similar effects were reported for exposure to ethylhexyl nitrate (101). Since the 1950s hypotension and peripheral vascular collapse have been described as associated with these headaches (101). It is now clear that NG and especially EGDN increase the risk of cardiovascular disease through attacks due to nitrate withdrawal, 1 to 3 days after last exposure, and through a long-term risk which persists long after cessation of exposure (102, 103). The short-term risk was noted first and termed "Monday morning angina." This includes findings of angina, myocardial infarction, arrhythmia, and sudden death. Generally symptoms are not induced by exercise or psychic arousal and no vascular lesions have been found at autopsy. Similar findings have been reported for PGDN (104). PGDN is also known to produce transient decrements in central nervous system performance as discussed in Section 2.8.8. Occupational or other exposures to NG involving repeated dermal contact have been shown to cause irritant contact dermatitis (105).

2.6 Analytic Determination

Early methods used for determining aliphatic nitrate esters consisted of various colorimetric or spectrophotometric procedures. These have now been largely replaced by instrumental methods (since mid 1960). References to the colorimetric procedures are included here (17, 106–113) because they were used for monitoring animal or worker exposures in work summarized later in the sections on specific compounds.

2.6.1 Colorimetric Methods

Colorimetric procedures used for the determination of "traces" of polynitrate esters, NG and EGDN, were reported in the mid 1930s (108), and for PETN and NG a few years later (109). These procedures appear relatively crude by present-day standards, for they were based on the nitration of reagents by the aliphatic nitrate under determination. More precise methods were developed later (110), and used to monitor animal exposures. Colorimetric methods were still being used as late as 1966 for the determination of NG and EGDN in workplace air (111). As little as 0.3 mg NG orEGDN was claimed to be measured in a 10-liter air sample (112). A colorimetric method for determining NG and EGDN in blood and urine was reported by Zurlo et al. (113). The degree of accuracy claimed is ±10 percent; the sensitivity, up to 5 μg.

2.6.2 Instrumental Methods

Infrared spectrography is generally satisfactory for the identification of aliphatic nitrate esters (106, 107). Spectral correlations have been compiled by Pristera et al. using band assignments at 6.0, 7.8, and 12.0 μm. Gas chromatographic procedures have been used for determining isopropyl nitrate, EGDN, and NG in blood and urine (114–116). A comparison of solid sorbents in air sampling using a chromatographic method has been published (117). These methods have the advantage of greater precision and ease of manipulation of samples, and have thus largely replaced the colorimetric methods of the past. More advanced methods of liquid chromatography to measure RDX in biologic fluids (118) and ion mobility spectrometry to measure EGDN in air (119) have been reported more recently.

2.7 Mechanism of Tolerance

The mechanism for tolerance development, a common response to organic nitrates, generally has been elucidated (120). This work was performed because the treatment of angina is hindered by the development of tolerance to the vasodilator action of these esters. It involved the interaction of NG on aortic tissue sulfhydryl in vitro and in the intact animal made tolerant to NG, on the hypothesis that NG oxidizes a critical sulfhydryl group in the NG "receptor." It was found that in tolerant aorta, sensitivity to NG decreased whereas nitrite formation increased, converting the nitrate receptor to the disulfide form, which has a lower affinity for NG. Confirmation of the hypothesis was obtained by reversing the tolerance with dithiothreitol, a disulfide reducing agent, thus demonstrating that tolerance involved chemical alteration of the receptor by NG. Cross tolerance was also demonstrated to other organic nitrates, but lack of it to isoproterenol, inorganic nitrite, and papaverine. Other mechanisms may also be involved. Evidence has been developed showing arterial smooth muscle had altered sensitivity to norepinephrine (121). It has also been suggested that cardiovascular control mechanisms such as sympathetic or renin-angiotensin stimulation may counter the nitrate vasodilation effects in vivo (88).

2.8 Specific Compounds

2.8.1 Methyl Nitrate, CH_3ONO_2

Methyl nitrate (MN) has been used by the military as a munition (122). It appears to be of little commercial or industrial interest (1). Some physical and chemical properties of MN are shown in Table 10.16.

2.8.1.1 Physiological Response. The 4-hr inhalation LC_{50} for MN in the rat was found to be 1275 ppm, whereas for the mouse, the LC_{50} was 5942 ppm (122). Orally the rat was also more than five times more susceptible than the mouse (rat LD_{50}, 344 mg/kg; mouse, 1820 mg/kg), with guinea pig showing an intermediate response (LD_{50}, 548 mg/kg). These acute animal data thus provide no good indication of the acute toxicity for humans. Responses of rats and mice to single, lethal doses of MN by inhalation followed a general pattern of lethargy, decreased respiratory rate, and cyanosis. Similar responses followed oral administration. Death was seldom delayed, with most deaths occurring during the 12-hr period following dosing. Gross examination of the guinea pigs that died following the single oral dose revealed chocolate-brown discoloration of the blood and lungs indicating severe methemoglobinemia. Except for the livers appearing slightly pale, no other lesions were observed. Gross examinations of the animals surviving the 14-day observation period revealed no treatment-related lesions (122).

Doses of 12.5 mg/kg have practically no effect on the blood pressure and pulse rate of rabbits. Doses of 52 mg/kg have a slight transient effect. MN is considerably less effective in these respects than NG. The minimal dose causing headache in man is between 117 and 470 mg. As has been observed with the other nitric acid esters, fractional doses produce tolerance that lasts for several days (89). No injuries from exposure to MN by handlers appear to have been reported in the published literature, and no TLV has been established.

2.8.2 Ethyl Nitrate, $CH_3CH_2ONO_2$

Ethyl nitrate (EN) has a pleasant odor and sweet taste. Some physical and chemical properties of EN are shown in Table 10.16. EN has a flash point (CC) of 50°F and explodes at 185°F. It has been used in organic synthesis of drugs, perfumes, and dyes, and as rocket propellant.

2.8.2.1 Physiological Response.

In cats, 400 mg/kg in olive oil given by intraperitoneal injection produces unconsciousness, increased respiratory rate, dilatation, and fixation of pupils, followed by death in 90 min. At 300 mg/kg, similar effects were followed by recovery. Moderate methemoglobinemia and Heinz body formation are observed after doses of 125 to 250 mg/kg (89).

Thus the effects of EN resemble those of the other aliphatic nitrates that have been studied. No industrial intoxications have been recorded from EN. It is said that EN has anesthetic properties and on inhalation causes headache, narcosis, and vomiting (89). No TLV has been established for EN.

2.8.3 Propyl Nitrate, $C_3H_7ONO_2$

n-Propyl nitrate (PN) has a sweet, sickening odor. It has been tested as a fuel ignition promoter and as a liquid rocket monopropellant, but appears to be no longer used for these purposes (1). Propyl nitrate is a strong oxidizing agent,

Table 10.17. Acute Effects of *n*-Propyl Nitrate-Animals (111, 124)

Animal	Dose (g/kg)	Route	Effect
Rat	7.5	Oral	Approximate lethal dose (sample I)
Rat	5.0	Oral	Approximate lethal dose (sample II)
Rat	1.0	Oral	Weakness, incoordination, cyanosis
Rat	1.5×10	Oral	Weakness, cyanosis, weight loss (first week)
Rabbit	11, 17	Skin	Essentially none
Rabbit	0.2–0.25	IV	Approximate LD_{50}
Dog	0.005	IV	Slight fall in blood pressure
	0.050	IV	Hypotension, cyanosis
	0.2–0.25	IV	Death in respiratory arrest
Cat	0.1–0.25	IV	6/7 died in 1 min
	0.025–0.075	IV	Hypotension, methemoglobinemia, survived

flammable, and a dangerous fire and explosion risk in that it has a flash point of 68°F, autoignition temperature of 350°F, and explosive limits in air of 2 to 100 percent. Other physical and chemical properties of PN are shown in Table 10.16.

2.8.3.1 Physiological Response. The acute toxicity of *n*-propyl nitrate (PN) vapor is relatively low for rat, mouse, and guinea pig, but moderate for the dog (123). The 4-hr LC_{50} values are estimated to be between 9000 and 10,000 ppm for the rat, 6000 and 7000 ppm for the mouse, and 2000 and 2500 ppm for the dog.

For 8-week, daily, 6-hr exposures also, the dog proved to be the most susceptible of the species tested. Approximately half the dogs died at 560 ppm, whereas all guinea pigs exposed for the same period survived 3235 ppm. Rats had intermediate, short-term susceptibility between the dog and guinea pig.

Compared to MN, it should be noted that the species susceptibility ranking represents a reversal of that for MN, where the rat was almost five times more susceptible than the mouse. This difference possibly indicates a widely different rate of metabolism of PN from that of MN in the two species.

Table 10.17 shows the acute response of four laboratory species when PN is administered orally, dermally, and intravenously. It may be seen that percutaneous toxicity is essentially nil, and oral toxicity is very low compared with intravenously administered doses, in which mg/kg doses were lethal compared with g/kg doses orally.

Twenty-six-week inhalation exposures of the dog, guinea pig, and rat showed again that the dog was the most susceptible. The highest dose tested in which all dogs survived was 260 ppm versus 2110 ppm for guinea pigs. Approximately half the rats exposed to 2110 ppm died (123). Thus the same order of susceptibility holds for these species chronically and acutely.

Signs and symptoms resulting from PN exposure differed in kind and degree according to species susceptibility (123). The main effect in rodents was anoxia, resulting from methemoglobin production. Dogs developed hemoglobinuria and hemolytic anemia, together with methemoglobin production and resultant much lower oxygen-carrying-capacity than rodents. The fact that blood levels in dogs returned to normal or near normal on continued exposure, and did not show appreciable development of methemoglobin from continuous daily exposures at levels below 900 ppm, points clearly to development of tolerance to chronic, low level effects.

Examination of tissues from repeatedly exposed dogs and rodents showed no pathological damage except an increase in pigment in spleen and liver, presumably from hemolysis and increased hematopoietic activity. Toxic signs following acute exposures at high levels are obviously more severe, consisting of cyanosis, methemoglobinemia, and uria, hemolytic anemia, vomiting, convulsions, and death in the dogs, and cyanosis, lethargy, convulsions, and death in the rodents.

2.8.3.2 Hygienic Standards of Permissible Exposure.

The current (1991) TLV and OSHA standard for PN is an 8-hr time-weighted average (TWA) of 25 ppm (105 mg/m^3) and a short-term exposure limit (STEL) of 40 ppm (170 mg/m^3). The odor of PN is "presumably detectable at concentration levels of 50 ppm and above." Although "resulting in discomfort in the form of irritation, headache, or nausea," its odor would offer less than the usually desirable warning. But in the absence of on-the-spot monitoring device, these responses should be persuasive.

2.8.4 Isopropyl Nitrate, (CH$_3$)$_2$CHONO$_2$

In contrast to *n*-propyl nitrate (PN) only preliminary acute toxicity determinations have been made on isopropyl nitrate (isoPN), but these have demonstrated that isoPN is qualitatively like PN as far as their acute and subacute toxic effects are concerned (125). Both nitrates show low toxicity orally or absorbed through the skin, and do not produce eye injury on single administrations in small laboratory animals. On the other hand, repeated contact with the skin causes irritation, and inhalation of the vapor produces cyanosis, methemoglobinemia, and even death.

Comparative approximate oral lethal doses for male albino rat were 3.4 g/kg versus 5.0 g/kg for PN. Treatment with 17 g/kg on rabbit skin caused only inflammation for both nitrates, but no overt systemic effects. The 6-hr exposure to isoPN vapor at 8500 ppm was an approximate LC$_{50}$ for the rat, compared with 9000 to 10,000 ppm as an estimated 4-hr LC$_{50}$ of PN.

Comparable subacute oral dosing schedules for isoPN and PN of five times a week for 2 weeks, at approximately one-fifth the LD$_{50}$ (680 mg/kg isoPN, 1500 mg/kg PN) resulted in temporary weakness from PN, cyanosis, weight loss, methemoglobinemia, and congested spleens. Symptoms became less severe on continued treatment, and 10 days after treatment, weight gains occurred. Rats dosed with isoPN showed no overt signs of toxicity. Subacute rabbit skin tests, both at the same dosing schedule (7.5 g/kg, five times a week for 2 weeks), showed no systemic effects from either nitrate while both showed inflammation, staining, and thickening of the skin.

In a study to develop a biologic monitoring method for isoPN, human subjects were exposed under controlled conditions to a mean concentration of 45.8 mg/m^3 (10.5 ppm) for 60 min without apparent adverse effect (116).

Table 10.18. Response to Inhalation of Various Concentrations of Amyl Nitrates (110)

Concentration (ppm)	Duration (hr)	Mortality[a]			
		Guinea Pigs	Rabbits	Rats	Mice
3730	7	2/2	2/2	3/4	5/5
3593	3.5	0/2	2/2	1/4	4/5
3227	1	0/2	0/2	0/4	0/5
3072	3 × 1	2/2	2/2	4/4	5/5
2774	3.5	0/2	0/2	0/4	2/4
2549	0.33	0/2	0/2	0/4	0/5
2380	2 × 7	2/2	2/2	0/4	5/5
2305	1	0/2	0/2	0/4	5/5
1807	7	0/2	1/2	0/4	4/4
1703	3 × 7	2/2	1/2	2/4	5/5
1612	7	0/2	0/2	0/4	0/5
599	9 × 7 + 6.25	0/2	0/2	0/4	0/2
262	20 × 7	0/2	0/2	0/3	0/5

[a]No cats died following any of these exposures.

2.8.5 Amyl Nitrate, $C_5H_{110}NO_2$

As used in the military and as tested toxicologically, amyl nitrate (AN) is a mixture of several primary, normal, and branched-chain amyl nitrates containing only a trace of amyl alcohol. It has been used to increase the cetane number of diesel fuels, and is a component of Otto fuel II. As studied by Treon et al. (110), it was a clear, slightly yellow liquid with a sickening sweet odor. It has a flash point of 42°C (closed cup), and other physical and chemical properties are listed in Table 10.16.

2.8.5.1 Physiological Response. Treon et al. (110) exposed cats, guinea pigs, rabbits, rats, and mice to measured concentrations of amyl nitrates in air. Selected data from their acute and subacute animal exposures are given in Table 10.18.

As far as can be determined from the noncongruent level and duration of exposures, toxicity of the AN isomers is greater than that of *n*-propyl nitrate (PN) by a factor of two- or threefold.

There was also no similarity to PN in species susceptibility. For AN the order of decreasing susceptibility is mouse > rat and rabbit > guinea pig > cat. However, except for the cat, which survived all exposures including the highest, 3730 ppm or 7 hr, the differential susceptibility among species was small. All species survived exposures of 600 ppm for 10 days at 7 hr/day, or 262 ppm for 20 days at 7 hr/day.

At high lethal levels signs and symptoms were characterized by tremors, ataxia, alterations in respiration, lethargy, cyanosis, convulsions, coma, and deaths. All exposures except 262 ppm produced signs in cats and guinea pigs, and some alterations in respiration were observed in rabbits and mice at the lowest concentration, 262 ppm. A cat exposed at 599 ppm developed methemoglobin levels up to 59.5 percent after the seventh exposure. Cats exposed at concentrations ranging from 1700 to 3700 ppm of the amyl nitrates showed Heinz body formation.

Animals dying during exposure had diffuse degenerative changes in the liver, kidneys, and brain with hyperemia and edema of the lungs. Those sacrificed at varying intervals after exposures had normal findings on pathological examination.

Thus the effects of mixed amyl nitrates in animals are qualitatively similar to those of the other alkyl mononitrates. The acute and subacute inhalation toxicity for guinea pigs, rats, and mice is greater than that of PN, but the higher boiling point and lower vapor pressure of the amyl nitrates tend to reduce this differential as far as the industrial hazard is concerned.

Persons exposed in the laboratory during these studies developed nausea and headache. No other illness was observed (110). No observations on people exposed in industry have been recorded and no maximal allowable concentrations have been proposed.

2.8.6 Ethylene Glycol Dinitrate, $C_2H_4(ONO_2)_2$

Ethylene glycol dinitrate (EGDN) is used in conjunction with NG in the manufacture of low-freezing dynamites. EGDN has a vapor pressure of 0.045 mm Hg which is much higher than NG. Therefore, exposure to dynamite primarily involves EGDN. Other physical and chemical properties of EGDN are shown in Table 10.16.

2.8.6.1 Physiological Response. Early acute toxicity tests in animals by Gross et al. (126) were confined to subcutaneous administration, 400 mg/kg being a fatal dose for the rabbit and 100 mg/kg for the cat, a species more susceptible to methemoglobin (MHb) formation. A subcutaneous dose of 60 mg/kg in this species produced 45 percent MHb as well as Heinz bodies. Administration of 0.6 mg/kg caused a rapid but transient drop in pulse pressure. These hypotensive effects of EGDN were found repeatedly to be more marked than those from NG. In Heinz body production, Wilhelmi (127) found EGDN to be four times more effective than NG, and 20 times more than EN in the cat.

Chronic exposure to the vapors of EGDN, 8 hr daily at 2 ppm for 1000 days, caused moderate, temporary blood changes without any clinical after effects in cats. Exposures at 10 times the level resulted in marked blood changes, but otherwise no adverse effect (126).

In humans, acute exposures to EGDN resulted in headache, nausea, vomiting, lowering of blood pressure, increase in pulse rate, and cyanosis. The minimal dose causing headache when applied to the skin is 1.8 to 3.5 ml of a 1 percent alcoholic solution, according to Leake et al. (128). When it was applied in fractional doses totaling 170 mg, tolerance developed in 24 to 36 hr and lasted for 10 to 13 days.

EGDN readily penetrates the skin and is absorbed through the lungs and gastrointestinal tract. Thus exposures from direct skin contact and inhalation of its vapors give rise to symptoms (129).

2.8.6.2 Metabolism. In a study of the metabolism of EGDN and its influence on blood pressure of the rat, Clark and Litchfield (130) found that the breakdown of EGDN in blood results in the liberation of inorganic nitrite and nitrate, and ethylene glycol mononitrate (EGMN). Free EGDN in blood reached a peak in 30 min and fell to zero 8 hr later. Inorganic nitrite was maximal in 1 to 2 hr, falling to zero at 12 hr, whereas nitrate rose more slowly to its maximum in 3 to 5 hr, reaching preinjection levels (approximately 1.0 µg/ml) 12 hr after injection. Nitrite is released from the reduced EGDN ester, later oxidized to nitrate, and excreted in the urine, accounting for 57 percent of the injected dose. A marked fall in blood pressure occurred directly after the injection, reaching its lowest value in 30 min. This was followed at 2 to 3 hr by a significant secondary fall, followed by a steady rise to preinjection levels at 12 hr. Ethylene glycol dinitrate was more effective in this respect that EGMN.

2.8.6.3 Industrial Experience. A virtual worldwide literature on worker exposure to EGDN dates back to the late 1950s, when up to 80 percent EGDN was added to NG to form a lower-freezing dynamite.

Acute effects from exposure to EGDN by either inhalation or by skin contact consist of fall in blood pressure and headache. Four of five volunteer workers exposed to the then TLV of 2.0 mg/m³ (approximately 0.2 ppm, measured as NG) experienced a fall in blood pressure of from 30/20 to 10/8 mm Hg, with severe headaches developing in 1 to 3 min. At 0.7 mg/m³, blood pressure drop was from 30/20 to 0/0 in 10 volunteers, who experienced slight headache or merely slight dullness in the head. At 0.5 mg/m³, when seven volunteers were tested, only three experienced slight or transitory headache, although blood pressure depressions were similar to those from higher exposures. Thus a dose response from short-term inhalation exposures to EGDN has been obtained (111). The investigators noted that, being far more volatile than NG, EGDN was responsible for essentially all the observed effects, and they quoted Rabinowitz (131) as stressing that alcohol accentuates the severity of headaches. In addition to headache, "Pains in the chest, abdomen, and extremities, and symptoms of general fatigue may appear as a result of the so-called acute effect" (132).

A quite different pattern of response develops from repeated, long-term, chronic exposures involving periods of years. Of particular concern are the oft-quoted "Monday morning fatalities" and angina. Carmichael and Lieben (131) have assembled a table of at least 38 sudden deaths in dynamite workers that occurred 30 to 48 hr after absence from work over a period from 1926 to 1961. Fatal heart attacks occurred during the weekend or on a Monday morning with the clinical diagnosis of acute infarction but little evidence of definite coronary occlusion. Narrowed coronary lumen and thickened sclerotic arterial walls were found at postmortem examination. These cases develop the symptomatology of cardiac ischemia, which may be preceded by changes in blood pressure and pulse rate (see also Section 2.5).

The importance of measuring not only exposures from the air but those from skin contact as well has been emphasized in a report by Einert et al. (112). They cited past reports that headache could result from merely "shaking hands with persons who handled dynamite" (133); the minimal effective dose of EGDN when applied to the skin was between 1.8 and 3.5 ml of a 1 percent alcoholic solution (134); and EGDN is more readily absorbed through the skin than NG (135). Einert et al. estimated skin exposures at several work sites in an explosives plant by extracting EGDN from the gloves worn by operators. Values obtained varied from less than 0.1 to 1 mg. These skin values were found comparable to the measured exposures from air inhaled on an 8-hr shift. They indicated 20 percent is the amount of EGDN retained after inhalation exposure (136). The authors recommended the use of lining gloves as a collection method for estimating skin exposures when combined with the simple clinical methods of pulse and blood pressure measurements before and after work.

2.8.6.4 Mechanism. Ever since "sudden death" became generally recognized as a disturbing sequel from exposure to EGDN, investigators have made repeated attempts to determine the biologic mechanisms involved. One such attempt made by Phipps (137) was influenced by an old report that thyroidectomy gives dramatic relief from angina pectoris. Thyroxine sensitizes the myocardium to epinephrine; therefore removal of the thyroid gland lowers the level of hormone and the resultant desensitization reduces the risk of anginal attacks. Using this as background, Phipps found that rats pretreated with thyroid hormone showed increased sensitivity to dynamite mix (85 percent EGDN, 17 percent NG) to the point that an LD_{50} became an LD_{95}. Conversely, thyroidectomy followed by a depletion period made rats resistant in that there was no mortality at the LD_{50}.

An alternative mechanism of tolerance has been studied by Clark (138). Rats dosed with EGDN (65 mg/kg) showed marked increase in plasma corticosterone, rising to a maximum in 15 min and persisting for more than 2 hr. It has been known that systemic hypotension is a potent stimulator of 17-hydroxycorticosteroid secretion, a response dependent on the pituitary. Repeated injections of EGDN lead to a decrease in corticosterone response, as did an injection of EGDN given 24 to 72 hr after the last series of injections. The author suggested, without investigating the possibilities, that the reduced corticosterone response to EGDN is due partly to tolerance to the induced hypotension, and partly to some deficiency in the hypothalamopituitary–adrenal axis (138).

2.8.6.5 Hygienic Standards of Permissible Exposure. The current limits of permissible exposure to EGDN are a 15-min TWA STEL of 0.1 mg/m³ for the OSHA standard and an 8-hr TWA of 0.05 ppm (0.31 mg/m³) for the ACGIH TLV. Both OSHA and ACGIH also include a skin notation for EGDN.

2.8.7 Nitroglycerin, $C_3H_2(ONO_2)_3$

Nitroglycerin (NG), trinitroglycerol or glycerol trinitrate, explodes violently from shock or when heated to about 260°C and thus is a severe explosion risk. Its major use is in explosives and blasting gels. In medicine, it has expanding applications in congestive heart failure, in the reduction of infarct size, in myocardial infarction, and in the long-term prophylaxis of angina (88).

Although NG shares with EGDN most of its toxicologic and pharmacological properties, it should be noted that when used in its usual dynamite mix of 20 percent NG, 80 percent EGDN, NG exposure hazard is essentially nil. This can be determined from its vapor pressure relative to that of EGDN, 0.00025 versus 0.045 mm Hg, yielding a ratio of 1 part in 720 (112). Other physical and chemical properties of NG are shown in Table 10.16.

2.8.7.1 Physiological Response. As with other compounds that have long histories of commercial use, animal toxicity studies were conducted after effects in humans were well known. Most animal toxicity studies on NG are not readily available in the open literature but are reviewed in a U.S. Environmental Protection Agency Health Advisory (139), from which the following information is taken unless otherwise indicated.

2.8.7.2 Animal Toxicity—Acute. The acute oral LD_{50} for male rats and mice are 822 and 1188 mg/kg, respectively. The LD_{50}s for females of each species were not significantly different from male LD_{50}s. All animals became cyanotic and ataxic within 1 hr of dosing. Deaths occurred within 5 to 6 hr, whereas survivors recovered within 24 hr. A paste containing 7.29 percent NG, peanut oil, and lactose caused very mild skin irritation, but no eye irritation in rabbits. NG caused dermal sensitization (40 percent response) in the guinea pig maximization test.

2.8.7.3 Genetic and Reproductive Toxicity. NG has produced mixed results in the Ames test but was not mutagenic in a Chinese hamster ovary cell assay without metabolic activation. Chromosomal aberrations were not found in dogs given up to 5 mg/kg/day for 9 weeks or in rats given up to 234 mg/kg/day for 8 weeks. Also a dominant lethal test in rats given 0, 3, 32, or 363 mg/kg/day in the diet for 13 weeks showed no effect on male fertility and no genotoxic activity. A three-generation reproductive toxicity study in rats showed no effects on fertility or viability, growth, and development of offspring at doses up to 38 mg/kg/day. Adverse fertility effects were seen at doses above 363 mg/kg/day which were secondary to malnutrition and testicular tumors. Developmental toxicity studies have been conducted in rats and rabbits given intravenous injection doses of up to 20 and 4 mg/kg/day, respectively, during typical gestation administration periods. No teratogenic, embryo toxic, or fetotoxic effects were observed.

2.8.7.4 Animal Toxicity—Repeated Exposure. A series of subchronic and chronic studies in rats, mice, and dogs were reported by Ellis et al. (140) showing qualitative and quantitative difference in effects among these species. In dogs, no serious effects were produced other than methemoglobinemia which was mild in animals given 25 mg/kg/day for 12 months. Doses up to 100 mg/kg produced transient increases but 200 mg/kg caused life threatening increases. Ninety-day studies in rats showed only body weight effects up to 234 mg/kg/day but also anemia and testicular degeneration at doses up to 1416 mg/kg. Two-year feeding studies were conducted in both rats and mice. Rats fed 1 percent (363 and 434 mg/kg/day for males and females, respectively) had decreased body weight gain, methemoglobinemia and associated effects on erythropoiesis, liver pathology including cholangiofibrosis and hepatocellular carcinoma, and interstitial cell tumors of the testes. Only a lower incidence of liver lesions was seen in rats given 0.1 percent NG in the diet (31.5 and 38.1 mg/kg/day for males and females, respectively) and no effects were observed at doses about an order of magnitude lower. Mice were less sensitive to chronic NG treatment. The high dose of approximately 1000 mg/kg/day produced only body weight changes and methemoglobinemia with its associated hematologic effects.

2.8.7.5 Human Experience. Nitroglycerin is a potent vasodilator of both arterial and venous vascular smooth muscle. Indeed, the therapeutic dose is 0.2 to 10 mg/day. It acts in a matter of minutes whether exposure is via the lungs, skin, or mucus membranes. The role of skin absorption is particularly significant in view of the small air concentrations resulting from its low vapor pressure (112). However, it takes relatively very small amounts to produce an intense throbbing headache, often associated with nausea, and occasionally with vomiting and abdominal pain. Rabinowitz (131) in discussing tolerance and habituation to NG, reports that as little as 0.001 ml NG is capable of producing a severe headache. Tolerance is developed if the exposure to nitroglycerin is maintained. In most cases this is transient and the headache may reappear after a weekend or holiday. Considerable variability has been pointed out, however. The headache has been described as preceded by a sensation of warmth and fullness in the head which starts at the forehead and moves upward toward the occiput. It may remain for hours or several days and may extend to the back of the neck. The headache is presumably due to cerebral vasodilatation and clinically resembles that produced by histamine. Temporary relief can be obtained from adrenalin or from the administration of ergotamine tartrate.

Larger amounts may result in hypotension, depression, confusion, occasionally delirium, methemoglobinemia, and cyanosis. Aggravation of these symptoms and the occurrence of maniacal manifestations after alcohol ingestion have been repeatedly observed. Fatalities from industrial intoxication are uncommon.

Medical studies of explosives workers with combined NG and EGDN exposures have not given evidence of chronic intoxication or injury despite the occurrence of transient symptoms. An extensive study of 276 workers with long exposure NG and EGDN in three Swedish explosives factories gave no evidence of permanent deterioration in health (141). The average air concentrations of NG–EGDN for most operations were below 5 mg/m^3, usually 2 to 4 mg/m^3. In the group with exposures at concentrations generally below 3 mg/m^3, symptoms such as fatigue and alcohol intolerance were less frequent, but there was little difference in the frequency of headaches.

Nitroglycerin appears to induce a shift of blood flow from relatively well-perfused myocardium to less-adequately nourished endocardium. It also has hypotensive effects largely due to reductions in diastolic pressure distending the relaxed ventricular wall. It appears that NG relieves angina by favorably altering the imbalance between myocardial oxygen supply and demand (87, 88).

2.8.7.6 Hygienic Standards of Permissible Exposure. The current OSHA standard for NG is 0.2 ppm (1 mg/m^3) as a ceiling or 0.1 ppm as a 15-min TWA STEL both with a skin notation. The ACGIH TLV–TWA is 0.05 ppm (0.46 mg/m^3) with a skin notation.

2.8.8 Propylene Glycol 1,2-Dinitrate, $CH_3CHONO_2CH_2ONO_2$

Propylene glycol 1,2-dinitrate (PGDN) is a constituent of Otto fuel II, a torpedo propellant, used by the U.S. Navy. PGDN is volatile, having a vapor pressure of 0.887 mm Hg with a disagreeable odor. Under ordinary conditions it is unstable. Accordingly, it is stabilized by small additions of 2-nitrodiphenylamine and di-n-butyl sebacate, substances that have been shown to have no overt toxic effects at 50 times the maximal animal test dose. Some physical and chemical properties of PGDN are shown in Table 10.16.

2.8.8.1 Physiological Response. Considerable information in this area is available ranging from acute, subacute, and chronic toxicity in animals, to experimental human studies and highly sophisticated neurophysiological measurements in workers performing routine maintenance procedures for prolonged periods. Forman reviewed the toxicology and epidemiology of PGDN in 1988 (142).

2.8.8.2 Animal Toxicity—Acute. In a comparative toxicity study with triethyleneglycol dinitrate (TEGDN), PGDN was 4.0 times more toxic orally for the rat than TEGDN (LD$_{50}$ 250 mg/kg); 4.8 times, subcutaneously (LD$_{50}$ 530 mg/kg); and approximately 1.7 times, intraperitoneally in the guinea pig and rat (LD$_{50}$ 402 and 479 mg/kg, respectively). Only by the oral route in the mouse was PGDN, slightly less toxic than TEGDN (LD$_{50}$ 1047 mg/kg) (143). Both PGDN and TEGDN produced methemoglobin in the rat, with PGDN producing it at a far faster rate, while causing ataxia, lethargy, and respiratory depression. Rats given TEGDN, on the other hand, were hyperactive to auditory and tactile stimulation. Methemoglobin as a contributing cause of death was shown by pretreating rats with methylene blue, in which case the time to death for PGDN was extended 224 min beyond the 197 for the average time to death without methylene blue (143).

A 4-hr inhalation exposure of rats to PGDN mist at 1350 mg/m^3 (approximately 200 ppm) resulted in no deaths and no overt signs of toxicity after 14 days, but methemoglobin values reached 23.5 percent (144).

In ocular irritation tests in rabbits, no immediate reaction occurred after instillation of 0.1 ml PGDN, but redness of the conjunctiva was noted after 5 min. The iris and cornea were not involved, and the redness gradually abated and disappeared within 24 hr (144), indicating PGDN has low eye irritation potential.

2.8.8.3 Human Toxicity—Acute. Because early signs of toxicity from PGDN involve organoleptic sensations which small laboratory animals are incapable of registering, investigators turned to human subjects for more definitive studies (145). Twenty human volunteers served as subjects for inhalation exposure to PGDN vapor at concentrations ranging from 0.01 ppm (approximately 0.075 mg/m^3) to 1.5 ppm for 1 to 8 hr.

The lowest concentration definitely to produce a frontal headache was 0.1 ppm after a 6-hr exposure. This occurred in one of three subjects and persisted for several hours. At 0.2 ppm, 10 of 12 subjects developed headache. Repeated daily exposures at this level resulted in a dramatic decrease in headache intensity. Odor of mild intensity was detected immediately, but after 5 min, the odor was no longer detected. At 0.5 ppm neurological changes as shown by abnormal modified Romberg and heel-to-toe tests, and an elevation of diastolic pressure of 12 mm were seen. Headache, initially mild, became progressively worse and throbbing in nature, with dizziness and nausea after 6-hr exposure. At 1.5 ppm (approximately 11.25 mg/m^3), the highest level that could be tolerated by the subjects for short periods (1.2 and 3.2 hr), headache pain was almost incapacitating, causing termination of the exposure after 3 hr. Coffee ameliorated the pain, which persisted for 1 to 7.5 hr post-exposure. The Flanagan coordinates test was abnormal and eye irritation occurred after 40 min.

Exhaled breath concentrations of PGDN after 1-hr exposure at 1.5 ppm measured 20 to 35 ppb, and remained at this level for the rest of the exposure. At 5 min post-exposure only a trace (5 ppb), the limit of sensitivity of the method, was detected. Only trace amounts of PGDN were found in the blood of subjects exposed at the higher levels, and no levels of exposure were sufficient to elevate blood nitrate or methemoglobin above control values.

Monitored central nervous system effects showed that levels of 0.2 ppm and above produced disruption of the visual-evoked response (VER). Subjects repeatedly exposed at 0.2 ppm for 8 hr on a daily basis developed tolerance to the induction of headaches, but the alteration in VER morphology appeared cumulative. Marked impairment in balance was observed after exposure at 0.5 ppm for 6.5 hr.

2.8.8.4 Animal Toxicity—Repeated Exposure. Monkeys, dogs, rats, and guinea pigs were exposed continuously for 90 days at concentrations of approximately 0, 9, 14, and 31 ppm (144). Except for one monkey that died on day 31 from exposure at 31 ppm (the death probably complicated by a parasitic infection), there were no other deaths or visible signs of toxicity in any of the other animals in the three exposure groups. Post-exposure hematologic values were all within normal limits for all species except dogs exposed at 31 ppm. These animals showed decreases of 63 and 37 percent in their hemoglobin and hematocrit values. Methemoglobin values increased in all exposed species, being most marked in dogs with 23.4 percent and monkeys with 17 percent. Serum inorganic nitrate determined in monkeys rose as high as 375 μg/ml above controls, and as high as 172 μg/ml in dogs, the only two species examined. Heavy iron-positive deposits were also present in the liver, spleen, and kidney sections of dogs and monkeys exposed to 31 ppm. Hepatic iron-positive deposits were commonly associated with vacuolar change, mononuclear cell infiltrates, and focal necrosis. Female rats showed focal necrosis of the liver and acute tubular necrosis of the kidney that appeared to be related to the test material, whereas male rats appeared normal. Vacuolar changes noted in the liver of all guinea pigs and in four of nine monkeys were also attributed to the exposures. No changes were noted in any of the other tissue sections examined from the three exposures. Similar but less severe liver and kidney pathology was observed in the 14-ppm exposed animals.

Squirrel monkeys exposed to 14 and 31 ppm PGDN had elevated serum urea nitrogen and decreased serum alkaline phosphatase levels, indicating the possibility of kidney change in this species.

Behavioral studies were conducted on rhesus monkeys trained to perform in a visual discrimination test (VDT) and in a visual acuity threshold test (VATT). Animals were exposed continuously for 90 days at 31 ppm PGDN, but no changes were seen in the avoidance behavioral pattern as indicated by the VDT and VATT tests (144).

Subacute dermal studies of PGDN in rabbits showed high absorption by this route (144), and hence raise a potential toxic hazard to workers handling PGDN. Doses of 1, 2, and 4 g/kg were applied daily for 90 days to the backs of rabbits. Thirteen of 14 rabbit died after the fifth application at the highest dose. Internal organs took on a dark, blue-gray appearance. At 2 g/kg weakness and slight cyanosis was seen at the start, with one rabbit dead after the sixth application. This was followed by steady physical improvement, except for slight wrinkling and scaling of the skin in the area of application. The animals appeared normal and showed a weight gain of 15 percent on day 20. In the lowest dose group, only minor irritation and roughening of the skin was noted. This cleared by the fifth day. These findings are in agreement with a previous 3-week dermal study by Andersen and Mehl (143). Six of 11 rabbits given topical application of 3.5 g/kg died with a mean time to death of 16 days.

2.8.8.5 Chronic Neurophysiological Effects.

2.8.8.5 *Chronic Neurophysiological Effects*. Horvath et al. studied (146) a group of 87 Navy personnel who were designated as "chronically exposed" to PGDN. Of these, 29 were tested before and immediately after PGDN exposure during routine torpedo maintenance procedures called "turnarounds." Twenty-one non-exposed controls were similar in sex distribution, race, smoking habits, and caffeine intake; however, the exposed group consumed more than twice as much alcohol as did controls. Alcohol is a substance regarded as aggravating the toxicity of aliphatic nitrates. The duration of exposure of the entire group averaged 47.4 months, with a range of 1 to 132 months. Air samples were taken during each turnaround procedure. Concentrations ranged from 0.00 to 0.22 ppm with a mean concentration of 0.03 ppm. Only one sample exceeded the then-current TLV of 0.2 ppm, and 87.5 percent of all peak concentrations were equal to or less than one-half the TLV.

Neurological tests performed on the chronically exposed Otto fuel workers showed no statistically significant differences from controls. These findings held also for a subgroup of workers with a longer mean exposure duration of almost 8 years (range of 5 to 11 years). However, quantitative eye tracking tests conducted for 29 turn-arounds, showed a significant decrease ($p = .03$) in velocity of eye movement and in latency ($p = .04$) when tested before and directly after exposure. Apparent alterations in standing behavior (on one leg for 30 sec) were of questionable significance.

The oculomotor function tests are apparently far more sensitive than the conditioned avoidance behavior test in monkeys. Continuous, 23-hr/day exposures of rhesus monkeys for periods up to 125 days at levels far in excess of those of the workers (0.3 to 4.2 ppm) had no discernible effects on avoidance behavior (147).

2.8.8.6 Metabolism. The metabolism of PGDN as determined in vitro in blood, and in vivo in rats, showed that 50 percent was broken down in 1 hr, and 50 percent of the remainder in the following hour (148). Small concentrations of inorganic nitrite were produced during the incubation in blood, whereas inorganic nitrate accumulated. At the end of 3 hr, the first time it was measured, there were large amounts of propylene glycol 2-mononitrate or PGMN-2, together with small amounts of PGMN-1. The summed quantities of mononitrates, inorganic nitrate, and nitrite represented 95 percent of the initial amount added to blood. This metabolism occurred in the erythrocytes.

In the intact rat, in contrast to in vitro in the blood, the mononitrates undergo further degradation to nitrogen compounds other than the mononitrates and inorganic nitrate. Only 56 percent of the administered PGDN appeared in the urine as inorganic nitrate. Thus there is qualitatively little to distinguish the in vitro and in vivo metabolism of PGDN from that of EGDN (130). The only point of difference is that PGDN gives rise to two mononitrates with the 2-isomer predominant, whereas EGDN gives rise to only EGMN. Quantitatively, there is less dinitrate and inorganic nitrite in the bloodstream after subcutaneous injection from PGDN than from a comparable injection of EGDN. Excretion was complete in 24 hr following a 65 mg/kg PGDN subcutaneous injection in rats (148).

It had previously been noted by Andersen and Mehl (143) that PGDN produces more methemoglobin in vivo than equivalent doses of TEGDN. Accordingly, an effort was made to understand better this oxidative process and explain the role of hemoglobin in detoxifying the nitrate esters (149). The reaction was found to be nonenzymatic and first-order for dinitrate and O_2Hb. The rate of oxidation proceeds linearly with dinitrate concentration, and does not approach a limit as would be the case if enzymatically driven. The rate of oxidation is related complexly to the oxygen concentration. No oxidation occurs at zero oxygen concentration and none at very high concentrations. The stoichiometry was thought to be 1.5 hemes oxidized per ester bond broken in hemolysates, and 1.9 to 2.3 per mole reacted ester in whole cells. From these studies, it was reasoned that hemoglobin would fulfill an important role in detoxifying the effects caused by the dinitrates. Hemoglobin in vivo, with the methemoglobin reductase system, acts catalytically to metabolize dinitrates to nitrite and nitrate, and the mononitrates are further degraded by the denitrifying tissue enzymes.

2.8.8.7 Hygienic Standards of Permissible Exposure. The current OSHA standard for PGDN is an 8-hr TWA of 0.05 ppm (0.3 mg/m³). ACGIH has adopted the same value as the TLV with a skin notation.

2.8.9 Pentaerythritol Tetranitrate, $C(CH_2ONO_2)_4$

Pentaerythritol tetranitrate (PETN) is used as a water-wet product of 40 percent water content, the only state in which it can be shipped. It is used in fuses and detonators. In admixture with TNT it is used for loading small-caliber projectiles and grenades as well as booster charges. PETN has an explosion temperature of 225°C, near that of NG, but is less sensitive to impact and friction. It is extremely sensitive to explosion by lead azide and other initiating agents, much more so than TNT or tetryl, and its explosive strength is at least 50 percent greater than TNT. Although PETN safely withstands storage for 18 months at 65°C, continued storage has marked effects of instability. The presence of as little as 0.01 percent free acid or alkali in PETN markedly accelerates its deterioration. It is the least stable of the standard military bursting-charge explosives. Other physical and chemical properties of PETN are shown in Table 10.16.

2.8.9.1 Physiological Response. Pentaerythritol tetranitrate is absorbed slowly from the gastrointestinal tract and lung, but not appreciably through the skin. Its physiological effects are similar to those of the other aliphatic nitrates, although it is considerably less potent as a vasodilator than NG. Doses of 5 mg/kg by mouth in dogs result in a fall in blood pressure but no effect is observed in humans after 64 mg orally (90). The daily oral administration of 2 mg/kg for 1 year caused no effects on growth, hematology, or pathology in rats. The U.S. National Toxicology Program has conducted dietary toxicity and carcinogenicity studies on PETN in both rats and mice (150). Other than slight body weight effects, PETN caused no adverse effects at feed concentrations up to 10,000 ppm in 14-day and 13-week studies. In the 2-year studies, mice and male rats fed 5000 and 10,000 ppm and female rats fed 1240 and 2500 ppm PETN showed no neoplastic or non-neoplastic lesions clearly related to PETN exposure. Patch tests in 20 persons gave no evidence of skin irritation or sensitization. Although some cases of mild illness and dermatitis have been attributed to contact with PETN in ordinance plants (151), it is apparent that PETN is relatively nontoxic. The controls and good housekeeping necessary to prevent explosions from this shock-sensitive material should be adequate to prevent injurious effects in workers. No TLV has been established for PETN.

2.8.10 Cyclonite, RDX, Hexahydro-1,3,5-trinitro-1,3,5-triazine

Cyclonite is a cyclic nitramine (which has the basic structure -N-NO₂) in which the amino nitrogen is incorporated into the six membered hetrocyclic triazine ring. The British used cyclonite under the name RDX (Royal demolition explosive) which is its military name in the United States. It has been widely used as a base charge for detonators and as an ingredient of bursting-charge and plastic explosives by the military. RDX may be considered to be at least the equal of, it not superior to, any of the solid bursting-charge explosives available in quantity. The stability of cyclonite is considerably superior to that of PETN and nearly equal to that of TNT. It withstands storage of 85°C for 10 months or at 100°C for 100 hr without measurable deterioration. Physical and chemical properties of RDX are shown in Table 10.16.

2.8.10.1 Physiological Response. Information on health effects caused by RDX has been reviewed and used to establish water quality criteria (152) and lifetime health advisory levels (153) by the E.P.A.

2.8.10.2 Animal Toxicity. Oral LD_{50} values for RDX range from 71 to 300 mg/kg for rats and 59 to 97 mg/kg for mice (154). Acute oral toxicity varies with the degree of granulation of the material and the dosing vehicle (155). The LD_{50} of a fine powder in solution or slurry was 100 mg/kg whereas a coarse granular RDX produced an LD_{50} threefold higher.

Hyperactivity, irritability, and generalized convulsions have been seen in rats (100, 155, 156) and dogs (100) given RDX. Burdette et al. conducted more detailed studies on the neurotoxic effects of RDX in rats (157). Spontaneous and audiogenic seizures were induced at acute doses as low as 10 mg/kg. The incidence of spontaneous seizures peaked 2 hr post-dose and reversed by 6 hr. However, audiogenic seizures could be induced only 8 to 16 hr post-dose.

Subchronic toxicity of RDX in rats has been evaluated by feeding doses from 1 to 600 mg/kg/day for 13 weeks (156). Mortality occurred at 100 mg/kg or greater. Hyperactivity was also seen at these doses but convulsions occurred only at 600 mg/kg. The apparent NOAEL was 10 mg/kg/day. Subchronic toxicity was studied in dogs given 50 mg/day, 6 days/week for 6 weeks. A few hours after the first dose they became excited and irritable. As dosing continued, reflexes became hyperactive and within the first week the animals had generalized convulsions characterized by hyperexcitability and increased activity, followed by clonic movements and salivation, then tonic convulsions and collapse. There was weight loss in all animals and death in one. No microscopic pathology was observed (100).

2.8.10.3 Human Experience. RDX does not exhibit pharmacological effects similar to the nitrites or nitrates. Chronic intoxication is characterized by the occurrence of repeated convulsions. It is slowly absorbed from the stomach and apparently from the lungs, but there is no evidence of skin absorption. Although McConnell (151) attributed some dermatitis to RDX manufacture, this probably was due to intermediates because patch testing with the moistened solid did not produce irritation (100).

Epileptiform seizures have occurred in workers manufacturing RDX in Italy (158). The convulsions occurred either without warning or after 1 or 2 days of insomnia, restlessness, and irritability. There were generalized tonic–clonic convulsions resembling in all clinical respects the seizures seen in epilepsy but occurring in individuals without a previous history of seizures. They were most frequent in persons doing the drying, sieving, and packing of RDX where the dust could be inhaled. The seizures were followed by temporary postconvulsive amnesia, malaise, fatigue, and asthenia, but there was eventually complete recovery.

Five cases of convulsions and/or unconsciousness occurred among about 26 workers engaged in pelletizing RDX as late as 1962 in the United States (159). The typical symptoms of RDX intoxication occurred either at work or several hours later at home, with few prodromal signs of headache, nausea, and vomiting. Unconsciousness lasted several minutes to 24 hr with varying periods of stupor, nausea, vomiting, and weakness. Recovery was complete with no sequelae, but two men reexposed to RDX had recurrences of illness. When control measures were installed, illnesses disappeared. A review of additional cases of human intoxication was published by Woody et al. (160).

2.8.10.4 Hygienic Standards of Permissible Exposure. The OSHA standard and the ACGIH TLV for RDX are 1.5 mg/m^3 with a skin notation.

2.8.11 HMX, Octahydro-1,3,5,7-Tetranitro-1,3,5,7-Tetrazocine

HMX (high-melting explosive) is a completely *N*-nitrated eight-member heterocyclic ring compound analogous to RDX with similar uses. Table 10.16 shows some physical and chemical properties of HMX. Virtually none of the toxicity information on HMX has been published, but has been reviewed by the EPA (161). HMX is poorly absorbed with oral administration as reflected by high LD$_{50}$ values of 6.3 g/kg and 2.3 g/kg for mice and rats, respectively. Central nervous system toxicity was observed at the higher doses. Fourteen-day studies in mice showed increased activity and excitability at 100 mg/kg/day. A 13-week study in rats showed histological changes in the liver and/or kidney at doses above 270 mg/kg with a NOAEL of 50 mg/kg/day. No adverse effects have been reported for HMX munitions plant workers.

3 ALKYL NITRITES

3.1 Physical and Chemical Properties and Use

The alkyl nitrites are aliphatic esters of nitrous acid. The nitrite group has the structure –CONO. Except for methyl nitrite, which is a gas, the lower members of the series are volatile liquids. In general they are insoluble or only very slightly soluble in water but are soluble or miscible with alcohol and ether in most proportions. They tend to decompose to oxides of nitrogen with exposure to light or heat. Violent decomposition can occur. As a group, they tend to be flammable and potentially explosive. They are oxidizing materials that present the possibility of violent reactions from contact with readily oxidized compounds. The physical and chemical properties of the alkyl nitrites are given in Table 10.19.

Table 10.19. Physical and Chemical Properties of the Alkyl Nitrites

Name	Molecular Weight	Physical State	Boiling Point (°C)	Specific Gravity
Methyl nitrite	61.04	Gas	−12	0.991 (15°C)
Ethyl nitrite	75.07	Colorless liquid	17	0.900 (15.5°C)
n-Propyl nitrite	89.09	Liquid	57	0.935
Isopropyl nitrite	89.09	Pale yellow oil	45	0.844 (25/4°C)
n-Butyl nitrite	103.12	Oily liquid	78.2	0.9114 (0/4°C)
Isobutyl nitrite	103.12	Colorless liquid	67	0.8702 (20/20°C)
sec-Butyl nitrite	103.12	Liquid	68	0.8981 (0/4°C)
tert-Butyl nitrite	103.12	Yellow liquid	63	0.8941 (0/4°C)
n-Amyl nitrite	117.15	Pale yellow liquid	104	0.8528 (20/4°C)
Isoamyl nitrite	117.15	Transparent liquid	97–99	0.872
n-Hexyl nitrite	131.17	Liquid	129–130	0.8851 (20/4°C)
n-Heptyl nitrite	145.20	Yellow liquid	155	0.8939 (0/4°C)
n-Octyl nitrite	159.23	Greenish liquid	174–175	0.862 (17°C)

The aliphatic nitrites have been of interest mainly because of their pharmacological properties and therapeutic use. They have been used in the treatment of angina. More recently they have been sold as "room odorizers" without prescription and abused for their euphoric effects by adolescents (162) and their sexual stimulatory effects by homosexual men (163). Amyl nitrite has important use as an antidote in the clinical management of cyanide poisoning (164). They are used also to a limited extent as intermediates in chemical syntheses. n-Propyl nitrite, isopropyl nitrite, and tert-butyl nitrite have been used as jet propellants and for the preparation of fuels.

3.2 Physiological Response

The pharmacological and toxicologic effects of the aliphatic nitrites are chiefly characterized by vasodilation resulting in a fall in blood pressure and tachycardia. Methemoglobin is produced by larger doses. In these respects the alkyl nitrites resemble closely the inorganic nitrites (sodium nitrite) and the aliphatic nitrates. Inhalation by animals and humans results in smooth muscle relaxation, vasodilation, increased pulse rate, and decreased blood pressure progressing to unconsciousness with shock and cyanosis. Headache is often a prominent symptom and may be due to meningeal congestion and vascular dilation. The development of tolerance has been observed with the therapeutic use of amyl nitrite for angina pectoris. This disappears after a week or so after discontinuation of use. Methods for the determination of nitrites in air and in biologic fluids have been described (89).

The branched-chain compounds are more effective than the straight chains in lowering blood pressure. Isopropyl nitrite is considerably more effective than n-propyl nitrite and isobutyl nitrite more than n-butyl. The secondary and tertiary butyl compounds also have a more pronounced hypotensive effect than normal butyl nitrite. Methyl nitrite is more effective than are ethyl and propyl nitrites, and amyl nitrite is more effective than ethyl nitrite. As far as the duration of the hypotensive effect is concerned, methyl and ethyl nitrites are more persistent, *n*-propyl is the least persistent of the lower alkyl nitrites, and the iso derivatives of propyl and butyl nitrite are more persistent than the normal compounds (89). However, hypotensive effects of these compounds are relatively transient. Amyl nitrite, for example, produces a rapid fall in blood pressure, which lasts only a few minutes after inhalation.

Krantz et al. have conducted extensive studies on the pharmacology of the alkyl nitrites (91–93). They found that when dogs were exposed by administering 0.3 ml through an aspirating bottle into the trachea, the degree of hypotension produced decreased from *n*-hexyl (58 percent fall) to *n*-heptyl (47 percent), *n*-octyl (30 percent), and *n*-decyl (16 percent). Alkyl nitrites with 11 to 18 carbon atoms in their chain showed slight or no effect on blood pressure under these conditions. If injected, however, they produced hypotension. With chains longer then 2-ethyl-*n*-hexyl-1-nitrite, the duration of action became shorter. Cyclohexyl nitrite produced a fall in blood pressure equivalent to ethyl nitrite or amyl nitrite but the duration was longer. In humans it produced severe headache. Krantz et al. believe that the major effects are related to the relaxing action of the nitrites on smooth muscle.

Methemoglobin formation has been repeatedly observed following administration to humans and animals. The aliphatic nitrites act as direct oxidants of hemoglobin. One molecule of nitrite and two molecules of hemoglobin can react to form two methemoglobin molecules under appropriate conditions. Side reactions to form nitrosohemoglobin and nitrosomethemoglobin may occur. The amount of methemoglobin formed in cats is directly proportional to the intravenous dose (94). The longer-chain compounds induce more methemoglobin formation relative to their hypotensive effect (91).

The therapeutic usefulness of methylene blue in acute intoxications accompanied by methemoglobinemia remains controversial even though support for its effectiveness in severe methemoglobinemia continues to appear (165, 166). Although methemoglobinemia is a prominent effect of nitrite absorption, the action of the alkyl nitrites on the vascular system is also a major determinant in their toxicity.

Reports of industrial intoxications appear to be limited (167–169), but reports of poisoning resulting from "recreational use" have been numerous over the past 15 years and have included cases of fatalities (170). See Section 3.2.2 for further discussion on this topic.

3.2.1 Genotoxicity and Immunotoxicity

Because of the widespread use of alkyl nitrites by homosexual men, it was postulated that alkyl nitrites might be involved in the development of AIDS, prior to the discovery of the HIV virus as the causative agent (163). This led to extensive research (171), particularly in the areas of genetic toxicity and immunotoxicity.

Methyl (172), ethyl (173), propyl (174), butyl (174, 175), isobutyl, *sec*-butyl, amyl, and isoamyl (174) nitrite are mutagenic in the Ames test. Of the six nitrites

Table 10.20. Acute Inhalation Toxicity of Alkyl Nitrites

Nitrite Compound	4-Hr Rat LC$_{50}$ (180) (ppm)	1-Hr Mouse LC$_{50}$ (181) (ppm)	½-Hr Mouse LC$_{50}$ (182) (ppm)
Methyl	176	—	—
Ethyl	160	—	—
Propyl	300	—	—
Butyl	420	567	949
Isobutyl	777	1033	1346
sec-Butyl	—	1753	—
tert-Butyl	—	10,852	—
Isoamyl	716	—	1430

tested (propyl to isoamyl in the above list), only amyl nitrite was not mutagenic in the mouse lymphoma assay (174). Ethyl nitrite also induced sex-linked recessive lethal mutations in *Drosophila* but did not induce micronuclei in mouse bone marrow cells (173). Isobutyl nitrite did not cause mutations in the drosophila test (176).

Studies of effects on immune parameters have produced contradictory results. Amyl nitrite caused functional deficits and structural alterations (seen by electron microscopy) in human mononuclear lymphocytes exposed in vitro (177). However, in a study of mice exposed to 50 or 300 ppm isobutyl nitrite 6.5 hr/day, 5 day/week for up to 18 weeks, no adverse effect on B-cell or T-cell function was detected (178). Another study in mice was done at higher exposure concentrations based on the expectation that abusers use higher concentrations for shorter durations (179). Mice were exposed to increasing isobutyl nitrite concentrations, 100 ppm for 1 day, 600 ppm for 3 days, then 900 ppm for 10 days. Specific decrements in T-cell responsiveness to mitogenic stimulants were seen but B-cell responsiveness was unchanged.

3.2.2 Comparative Animal Toxicity

The acute inhalation LC$_{50}$ values for various alkyl nitrites are shown in Table 10.20. In general, potency decreases as alkyl chain length and branching increases. Very steep dose–response curves were seen for all compounds, indicating that relatively small increases of exposure concentration would produce large increases of mortality. Klonne et al. stated the concentration range corresponding to 0 and 100 percent mortality with four exposures of rats was less than 100 ppm for most of the nitrites (180). The LC$_{50}$/EC$_{50}$ (for decreased motor performance) ratio for a 30-min exposure of mice ranged from 2.0 to 2.4 (182). Therefore, acute inhalation exposure hazard for the alkyl nitrites is greater than would be indicated by their LC$_{50}$s alone.

McFadden et al. measured methemoglobin levels in vivo and in vitro for the butyl nitrite isomers (181). *tert*-Butyl was significantly less toxic than the other isomers and was the least potent methemoglobin inducer. Also pretreatment with methylene blue prior to nitrite exposure greatly increased the mean time to death for butyl, isobutyl, and *sec*-butyl, but only doubled the value for *tert*-butyl. From this the authors concluded methemoglobin formation is the cause of death from the butyl nitrite isomers except for *tert*-butyl. However, Klonne et al. point out that deaths occurred rapidly and always during exposure (180). This was a consistent finding of the other acute inhalation studies (181, 182). Animals recovered rapidly from all signs of exposure except for those associated with methemoglobinemia (cyanosis, bluish coloration of ears and feet). Also exposure concentration rather than cumulative inhaled concentration was the primary determination for mortality (180). Severe hypotension and cardiovascular collapse are more consistent with these findings as a cause of death. Most likely both vasodilation and methemoglobinemia play a role in alkyl nitrite-induced mortality.

The lower aliphatic nitrites are promptly absorbed from the lung. Amyl nitrite is ineffective by mouth because it is destroyed in the gut. It is less effective by injection than by inhalation. Octyl nitrite (2-ethyl-*n*-hexyl-1-nitrite) is not absorbed through the mucous membranes and is ineffective sublingually. It appears that the nitrites are hydrolyzed in vivo to nitrite and the corresponding alcohol, which is then partly oxidized and partly exhaled unchanged.

3.3 Specific Compounds

The pharmacological properties as determined in animals are so uniform within this group that information on the nitrites below can be taken as illustrative of the effects and potential hazards of the other members of the series (see Tables 10.19 and 10.21).

3.3.1 Methyl Nitrite, CH_3ONO

Methyl nitrite is a gas with a severe explosion risk when shocked or heated. It has uses in the synthesis of nitrite and nitroso esters. It is formed as a by-product in the synthesis of a rubber antioxidant (183). Table 10.19 includes some physical and chemical properties of methyl nitrite.

Methyl nitrite is a relatively toxic compound by inhalation for both animals and humans. Rats survived 13 6-hr exposures at 100 ppm with methemoglobin levels of 30 to 40 percent. The 25 ppm level for 15 6-hr exposures was apparently a gross "no-effect" level for rats because no toxic signs were observed and organs were normal. However, 35 ppm produced 6 percent methemoglobin in a cat after a 6-hr exposure (184).

Methyl nitrite was found to be a potent cyanosing agent for workers synthesizing a rubber antioxidant (167). Six cases of methyl nitrite intoxication are described, consisting initially of dizziness and later headache and palpitation, the last more pronounced in two workers who consumed alcohol after exposure at work. All men responded satisfactorily to bed rest for 12 hr and the inhalation of oxygen for about 2 hr. Atmospheric concentrations in the plants where the men had been affected, simulating the conditions at the time of the intoxications, indicated that

Table 10.21. Comparative Toxicity Data on Aliphatic Nitro, Nitrate, and Nitrite Compounds

Chemical Group	Skin Absorption	Irritation	Vascular Dilatation	Methemoglobin Formation	Industrial Experience
Aliphatic nitro compounds (R_3CNO_2)					
Nitroalkanes	None	Moderate	None	Positive	Irritation, systemic symptoms
Chlorinated nitroparaffins	None	Marked	None	Unknown	Lung injury
Nitroolefins	Positive	Marked	Unknown	Not observed	None
Aliphatic nitrates (R_3CONO_2)	Positive	None	Marked	Positive	Systemic symptoms, possible deaths
Aliphatic nitrites (R_3CONO)	Unknown	None	Marked	Positive	Systemic symptoms, fatalities
Nitramines (R_3CNHNO_2)	None	None	None	None	Convulsions

"50 ppm is the uppermost limit of safety." (167) Another incident, when workers were estimated to be exposed to 50 to 100 ppm, caused cyanosis, dizziness, and nausea but complete recovery after exposure ceased (169).

3.3.2 Ethyl Nitrite, CH_3CH_2ONO

Ethyl nitrite is a volatile, flammable, colorless liquid. Its flash point is $-31°F$ and the explosive limits in percent by volume in air are 3.01 to 50. The autoignition temperature of the liquid is 195°F. Thus it has a high potential fire and explosion hazard. It decomposes readily to form oxides of nitrogen. Other physical and chemical properties of ethyl nitrite are shown in Table 10.19.

Inhalation of ethyl nitrite by dogs results in as much as 70 mm Hg drop in blood pressure. This lasts approximately 2 min after a single inhalation. Methemoglobin is formed but Heinz bodies have not been found (89). Mice and cats exposed for 15 min to 15 ppm did not show recognizable effects. Industrial intoxications characterized by headache, tachycardia, and methemoglobinemia have occurred. A fatality has been described following the inhalation of ethyl nitrite after accidental breakage of a 4-liter bottle containing 24 percent ethyl nitrite in alcohol (72).

Three cases of ethyl nitrite intoxication have been reported from Czechoslovakia (168) during a synthesis of hydantoin (glycolylurea). Symptoms of both "nitrite" effect and methemoglobinemia were noted. In addition, there was a vasodilator effect upon the blood vessels of the sclerae, producing a peculiar redness of the eyes, a hitherto undescribed manifestation of the nitrite effect. No Heinz bodies were seen, as is to be expected.

3.3.3 Butyl Nitrites

Butyl nitrite ($CH_3(CH_2)_2CH_2ONO$) and its isomers isobutyl (($CH_3)_2CHCH_2ONO$), sec-butyl ($CH_3CH_2CH(CH_3)ONO$), and tert-butyl nitrite (($CH_2)_3CONO$) have been studied more recently because they are commercially available to drug abusers. Table 10.19 shows some physical and chemical properties of these compounds.

The acute oral LD_{50} value for butyl nitrite in rats is 83 mg/kg (185). The acute inhalation toxicity of butyl nitrites is discussed in Section 3.2 (Table 10.20). A mixture of butyl nitrites (mostly isobutyl) administered by gavage caused hearing impairment in rats at 500 ml/kg (186). Acute LD_{50} values in mice for butyl, isobutyl, sec-butyl, and tert-butyl by ip injection are 158, 184, 592, and 613 mg/kg, respectively, and by oral gavage are 180, 279, 428, and 336 mg/kg, respectively (187).

Subchronic inhalation toxicity of isobutyl nitrite has been studied in mice exposed to 0, 20, 50, or 300 ppm for 6.5 hr/day, 5 days/week for up to 18 weeks (188). Hyperplasia and vacuolization of the bronchial epithelium with some organ weight changes were seen at 300 ppm. Methemoglobinemia was seen at 300 ppm (5 to 10 percent) and in some animals (<5 percent) at 50 ppm. No effects were seen at 20 ppm. In another subchronic inhalation study, mice were exposed to a single concentration of each isomer, which caused less than 20% mortality (i.e., butyl—300 ppm, isobutyl—400 ppm, sec-butyl—500 ppm, tert-butyl—1000 ppm) (189). Animals were exposed 7 hr/day for 60 days. Body weights, organ weights, and methemoglobinemia were the major parameters affected under these exposure conditions.

3.3.4 Isoamyl Nitrite, (CH₃)₂CHCH₂CH₂ONO (Amyl Nitrite)

Isoamyl nitrite is a light yellow, transparent liquid with a pleasant, fragrant, fruity odor. It decomposes upon exposure to air and sunlight. It is flammable and explosive. Table 10.19 shows some other physical and chemical properties of this agent.

Amyl nitrite was introduced to medicine in 1859 and has received considerable pharmacological investigation since that time. Its major use was for the treatment of angina pectoris through its vasodilative effect on the coronary arteries. However, this effect is transient and it has largely been replaced by nitroglycerin and longer acting nitrates. Amyl nitrite has been most helpful in clarifying the differential diagnosis of murmurs due to left ventricular outflow obstruction (which increases following amyl nitrite) from those of mitral regurgitation (which decreases); the apical diastolic rumble of mitral stenosis (which increases) from the Austin-Flint rumble (which decreases); ventricular septal defect (which decreases) from pulmonic stenosis (which increases); and acyanotic tetralogy of Fallot (which decreases) from isolated valvular pulmonary stenosis (which increases) (184).

The symptoms following inhalation of large doses by humans are flushing of the face, pulsatile headache, disturbing tachycardia, cyanosis (methemoglobinemia), weakness, confusion, restlessness, faintness, and collapse, particularly if the individual is standing. The symptoms are usually of short duration. Industrial intoxications have not been reported (89).

4 SUMMARY

Although the aliphatic nitro compounds, the aliphatic nitrates, and aliphatic nitrites have several features in common (nitrogen-oxygen grouping, explosiveness, methemoglobin formation), there are significant differences in their toxic effects. Some of their attributes are summarized in Table 10.21. The esters of nitric and nitrous acid, with the nitrogen linked to the carbon through oxygen, are very similar in their pharmacological effects. Both produce methemoglobinemia and vascular dilatation, with hypotension and headache. These effects are transient. None of the series has appreciable irritant properties. Pathological changes occur in animals only after high levels of exposure and are generally nonspecific and reversible. The nitric acid esters of the monofunctional and lower polyfunctional alcohols are absorbed through the skin. Information is not available on the skin absorption of the alkyl nitrites. Members of both groups are well absorbed from the mucous membranes and lungs. Heinz body formation has been observed with the nitrates but not with the nitrites.

The nitro compounds, like the nitrates and nitrites, cause methemoglobinemia in animals. Heinz body formation parallels this activity within the series. Although some members are metabolized to nitrate and nitrite, there is no significant effect on blood pressure or respiration. As with the lower nitrates and nitrites, anesthetic symptoms are observed in animals during acute exposures, but these occur late. The prominent effect is irritation of the skin, mucous membranes, and respiratory tract. This is most marked with the chlorinated nitroparaffins and the nitroolefins. In addition to respiratory tract injury, cellular damage may be observed in the liver and kidneys. Except for the nitroolefins, skin absorption in negligible.

The nitramines have entirely different activity. RDX is a convulsant for humans and animals. Skin absorption, irritation, vasodilatation, methemoglobin formation, and permanent pathological damage after repeated doses are either insignificant or absent.

Transient illness has been associated with the industrial use or manufacture of these materials but fatalities and chronic intoxication have been uncommon. Some members of each group present extremely high fire and explosion hazards.

ACKNOWLEDGMENTS

The author wishes to acknowledge and thank those who assisted with the preparation of this chapter. Sharon L. Gillen-Davis was very patient and an invaluable assistant in the proofreading of this document. Anita Jones and Allison Price conducted computer database literature searches to identify and obtain studies conducted since publication of the previous edition. Sheila Lund-Pearson provided word processing assistance in preparation of the manuscript.

REFERENCES

1. *Chemical Buyer's Directory*, 1980, 1981, Schnell Publishing Co., New York.
2. *Chemical Economic Handbook*, Stanford Research Institute, September 15, 1977.
3. *IARC* Monographs, **29**, 331 (1982).
4. NIOSH, *Manual of Analytic Methods*, 3rd ed., 2nd Suppl. (1988).
5. W. Scott and J. F. Treon, *Ind. Eng. Chem., Anal. Ed.*, **12**, 189 (1940).
6. J. F. Treon and F.R. Dutra, *Arch. Ind. Hyg. Occup. Med.*, **5**, 52 (1952).
7. R. Cohen and P. Alshuller, *Annl. Chem.*, **31**, 1638 (1959).
8. H. J. Horn, *Arch. Ind. Hyg. Toxicol.*, **27**, 213 (1954).
9. W. Machle et al., *J. Ind. Hyg. Toxicol.*, **27**, 95 (1945).
10. *Encyclopedia of Industrial Chemical Analysis*, F. D. Snell and L. S. Ettre, Eds., Vol. 16, Wiley, New York, pp. 417–448.
11. T. R. Lewis et al., *J. Environ. Pathol. Toxicol.*, **2**, 233 (1979).
12. W. Machle et al., *J. Ind. Hyg. Toxicol.*, **21**, 72 (1939).
13. W. Machle et al., *J. Ind. Hyg. Toxicol.*, **22**, 315 (1940).
14. W. B. Deichmann et al., *Ind. Med. Surg.*, **34**, 800 (1965).
15. S. D. Murphy et al., *Toxicol. Appl. Pharmacol.*, **5**, 319 (1963).

16. W. Machle and E. W. Scott, *Proc. Soc. Exp. Biol. Med.*, **53**, 42 (1943).

17. C. H. Hine et al., *J. Occup. Med.*, **20**, 333 (1978).

18. R. Dayal et al., *Fundam. Appl. Toxicol.*, **13**, 341 (1989).

19. B. Denk et al., *Arch. Toxicol.*, Suppl. **13**, 330 (1989).

20. A. Zitting et al., *Toxicol. Lett.*, **9**, 237 (1981).

21. A. Zitting et al., *Toxicol. Lett.*, **13**, 189 (1982).

22. K. F. Lamper et al., *Ind. Med. Surg.*, **27**, 375 (1958).

23. J. B. Skinner, *Ind. Med.*, **16**, 441 (1947).

24. *ACGIH TLV Documentation*, 1986.

25. R. F. Sievers et al., *U.S. Pub. Health Rep.*, **62**, 1048 (1947).

26. M. Gaultier et al., *Arch. Mal. Prof.*, **25**, 425 (1964).

27. D. Rodia, *Vet. Human Toxicol.*, **21**, 183 (1979).

28. Centers for Disease Control, *MMWR*, **34**, 659 (1985).

29. R. Harrison et al., *Ann. Int. Med.*, **107**, 466 (1987).

30. G. N. Crawford et al., *Am. Ind. Hyg. Assoc. J.*, **46**, 45 (1985).

31. E. B. Vedder, *The Medical Aspects of Chemical Warfare*, Williams and Wilkins, Baltimore, 1925.

32. G. Teslaa et al., *Vet. Human Toxicol.*, **28**, 323 (1986).

33. M. I. Selala et al., *Bull. Environ. Contam. Toxicol.*, **42**, 202 (1989).

34. W. Machle et al., *J. Ind. Hyg. Toxicol.*, **24**, 5 (1942).

35. E. W. Scott, *J. Ind. Hyg. Toxicol.*, **25**, 20 (1943).

36. R. J. Nolan et al., *Ecotoxicol. Environ. Safety*, **6**, 388 (1982).

37. V. Ulrich et al., *Biochem. Pharmacol.*, **27**, 2301 (1978).

38. H. Sakurai et al., *Biochem. Pharmacol.*, **29**, 341 (1980).

39. E. K. Marker and A. P. Kulkarni, *J. Biochem. Toxicol.*, **1**, 71 (1986).

40. R. Dayal et al., *Chem-Biol. Interact.*, **79**, 103 (1991).

41. T. Kido and K. Soda, *Arch. Biochem. Biophys.*, **234**, 468 (1984).

42. T. Kido et al., *Agric. Biol. Chem.*, **48**, 2549 (1984).

43. C. F. Kuo and I. Fridovich, *Biochem. J.*, **237**, 505 (1986).

44. Chung Wai Chiu et al., *Mutat. Res.*, **58**, 11 (1978).

45. G. Löfroth et al., *Prog. Clin. Biol. Res.*, **209B**, 149 (1986).

46. M. Hite and H. Skeggs, *Environ. Mutagen.*, **1**, 383 (1979).

47. J. R. Warner et al., *Environ. Mol. Mutagen*, **11**, Suppl. 11, 111 (1988).

48. W. T. Speck et al., *Mutat. Res.*, **104**, 49 (1982).

49. S. Haworth et al., *Environ. Mutagen*, **8**, Suppl. 7, 1 (1986).

50. U. Kliesch and I-D. Adler, *Mutat. Res.*, **192**, 181 (1987).

51. E. George et al., *Carcinogenesis*, **10**(12), 2329 (1989).

52. M. Bauchinger et al., *Mutat. Res.*, **190**, 217 (1987).

53. W. Göggelmann et al., *Mutagenesis*, **3**(2), 137 (1988).

54. U. Andrae et al., *Carcinogenesis*, **9**(5), 811 (1988).

55. E. Roscher et al., *Mutagenesis*, **5**(4), 375 (1990).

56. F. E. Würgler et al., *Mutat. Res.*, **244**, 7 (1990).

57. National Toxicology Program, Technical Report Series No. 386, March 1990.
58. V. F. Garry et al., *Teratogen. Carcinogen. Mutagen.*, **10**, 21 (1990).
59. R. Valencia et al., *Environ. Mutagen.*, **7**, 325 (1985).
60. T. B. Griffin et al., *Ecotoxicol. Environ. Safety*, **16**, 11 (1988).
61. T. B. Griffin et al., *Ecotoxicol. Environ. Safety*, **6**, 268 (1982).
62. E. S. Fiala et al., *Carcinogenesis*, **8**(12), 1947 (1987).
63. U.S. Depart. of Health and Human Services, Sixth Annual Report on Carcinogens, Summary 1991.
64. R. W. Tennant and J. Ashby, *Mutat. Res.*, **257**, 209 (1991).
65. C. C. Conaway et al., *Cancer Res.*, **51**, 3143 (1991).
66. I. Linbart et al., *Chem.-Biol. Interact.*, **80**, 187 (1991).
67. E. Fiala et al., *Mutat. Res.*, **179**, 15 (1987).
68. M. L. Cunningham and H. B. Mathews, *Toxicol. Appl. Pharmacol.*, **110**(3), 505 (1991).
69. T. B. Griffin et al., *Ecotoxicol. Environ. Safety*, **4**, 267 (1980).
70. J. H. Weatherby, *Arch. Ind. Health*, **11**, 103 (1955).
71. J. E. Amoore and E. Hautala, *J. Appl. Toxicol.*, **3**(6), 272 (1982).
72. F. Flury and F. Zernik, *Schädliche Gase*, Springer, Berlin, 1931.
73. *NIOSH Registry of Toxic Effects of Chemical Substances*, U.S.D.H.E.W., Rockville, MD, 1976.
74. K. F. Hager, *Ind. Eng. Chem.*, **41**, 2168 (1949).
75. V. B. Vouk and O. A. Weber, *Brit. J. Ind. Med.*, **9**, 32 (1952).
76. T. B. Griffin et al., *Pharmacologist*, **20**(3), 145 (1978).
77. T. B. Griffin et al., *Ecotoxicol. Environ. Safety*, **5**, 194 (1981).
78. G. N. Crawford et al., *Am. Ind. Hyg. Assoc. J.*, **45**(2), B-7 (1984).
79. L. Silverman et al., *J. Ind. Hyg. Toxicol.*, **28**, 262 (1946).
80. A. M. Prentiss, *Chemicals in War*, McGraw-Hill, New York, 1937.
81. A. A. Fries and C. J. West, *Chemical Warfare*, McGraw-Hill, New York, 1921, p. 143.
82. M. Yoshida et al., *J. Pesticide Sci.*, **12**, 237 (1987).
83. M. Yoshida et al., *J. Pesticide Sci.*, **16**, 63 (1991).
84. L. A. Buckley et al., *Toxicol. Appl. Pharmacol.*, **74**, 417 (1984).
85. M. Yoshida et al., *J. Pesticide Sci.*, **12**, 673 (1987).
86. National Cancer Institute, Carcinogenesis Technical Report Series, No. 65, 1978.
87. S. E. Warren and G. S. Francis, *Am. J. Med.*, **65**, 53 (1978).
88. P. A. Todd et al., *Drugs*, **40**(6), 880 (1990).
89. W. F. von Oettingen, "The Effects of Aliphatic Nitrous and Nitric Acid Esters on the Physiological Functions with Special Reference to Their Chemical Constitution," *Natl. Inst. Health Bull. No. 186*, 1946.
90. W. F. von Oettingen et al., "Toxicity and Potential Dangers of Pentaerythritol-Tetranitrate (PETN)," *U.S. Pub. Health Bull. No. 282*, 1944.
91. J. C. Krantz et al., *Proc. Soc. Exp. Biol. Med.*, **42**, 472 (1939).
92. J. C. Krantz et al., *J. Pharmacol. Exp. Therap.*, **70**, 323 (1940).
93. M. Rath and J. C. Krantz, *J. Pharmacol. Exp. Therap.*, **76**, 33 (1942).

94. O. Bodansky, *Pharmacol. Revs.*, **3**, 144 (1951).

95. R. T. Williams, *Detoxication Mechanisms*, 2nd ed., Wiley, New York, 1959.

96. P. Rofe, *Brit. J. Ind. Med.*, **16**, 15 (1959).

97. J. B. Hughes and J. F. Treon, *Arch. Ind. Hyg. Occup. Med.*, **10**, 192 (1954).

98. J. F. Treon et al., *Arch. Ind. Health*, **11**, 290 (1955).

99. W. O. Hueper and J. W. Landsberg, *Arch. Pathol.*, **29**, 633 (1940).

100. W. F. von Oettingen et al., *J. Ind. Hyg. Toxicol.*, **31**, 21 (1949).

101. S. Someroja and H. Savolainen, *Toxicol. Lett.*, **19**, 189 (1983).

102. W. E. Morton, *J. Occup. Med.*, **19**, 197 (1977).

103. T. S. Kristensen, *Scand. J. Work Environ. Health*, **15**, 245 (1989).

104. S. A. Forman et al., *J. Occup. Med.*, **29**(5), 445 (1987).

105. L. Kanerva et al., *Contact Dermatitis*, **24**, 356 (1991).

106. F. Pristera et al., *Anal. Chem.*, **32**, 495 (1960).

107. F. Pristera et al., "Compilation of Infrared Spectra of Ingredients of Propellants and Explosives," Tech. Memo. Rept. 1889, AMCMS Code H 30.11.1161.1, Picatiny Arsenal, Dover, NJ (no date given).

108. J. H. Foulger, *J. Ind. Hyg. Toxicol.*, **18**, 127 (1936).

109. H. Yagoda and F. H. Goldman, *J. Ind. Hyg. Toxicol.*, **25**, 440 (1943).

110. J. F. Treon et al., *Arch. Ind. Health*, **11**, 290 (1955).

111. D. C. Trainor and R. C. Jones, *Arch. Environ. Health*, **12**, 231 (1966).

112. C. Einert et al., *Am. Ind. Hyg. Assoc. J.*, **24**, 435 (1963).

113. N. Zurlo et al., *Med. Lav.*, **54**, 166 (1963).

114. A. F. Williams et al., *Nature*, **210**, 816 (1966).

115. Y. Fukuchi, *Int. Arch. Occup. Environ. Health*, **48**, 339 (1981).

116. I. Ahonen et al., *Toxicol. Lett.*, **47**, 205 (1989).

117. K. Andersson et al., *Chemosphere*, **12**(6), 821 (1983).

118. C. P. Turley and M. A. Brewster, *J. Chromatogr. Biomed. Appl.*, **421**, 430 (1987).

119. A. H. Lawrence and P. Neudorfl, *Anal. Chem.*, **60**, 104 (1988).

120. P. Needleman and E. M. Johnson, Jr., *J. Pharm. Exp. Ther.*, **184**, 709 (1973).

121. P. Johansson et al., *Pharmacol. Toxicol.*, **61**, 172 (1987).

122. E. R. Kinkead et al., *Toxic Hazard Evaluation of Five Atmospheric Pollutant Effluents from Ammunition Plants*, AMRL Rept. TR-76-XX, October 1976.

123. W. E. Rinehart et al., *Am. Ind. Hyg. Assoc. Quart.*, **19**, 80, (1958).

124. E. F. Murtha et al., *J. Pharmacol. Exp. Ther.*, **118**, 77 (1956).

125. D. B. Hood, Haskell Laboratory for Toxicology and Industrial Medicine, E.I. du Pont de Nemours & Co., unpublished data, Rept. No. 21-53, 1953.

126. E. Gross et al., *Arch. Exp. Pathol. Pharmakol.*, **200**, 271 (1942).

127. H. Wilhelmi, *Arch. Exp. Pathol. Pharmakol.*, **200**, 305 (1942).

128. C. D. Leake et al., *J. Pharm. Exp. Ther.*, **35**, 143 (1931).

129. R. A. Lange et al., *Circulation*, **46**, 666 (1972).

130. D. G. Clark and M. H. Litchfield, *Brit. J. Ind. Med.*, **24**, 320 (1967).

131. I. M. Rabinowitz, *Can. Med. Assoc. J.*, **50**, 199 (1944).

132. P. Carmichael and J. Lieben, *Arch. Environ. Health*, **7**, 424 (1963).

133. G. E. Ebright, *J. Am. Med. Assoc.*, **62**, 201 (1914).

134. L. A. Crandell et al., *J. Pharm. Exp. Ther.*, **41**, 103 (1931).

135. E. Gross et al., *Arch. Exp. Pathol. Pharmakol.*, **200**, 271 (1942).

136. E. Gross et al., *Arch. Toxicol.*, **18**, 200 (1960).

137. F. C. Phipps, *Proc. Soc. Exp. Biol. Med.*, **139**, 323 (1972).

138. D. G. Clark, *Toxicol. Appl. Pharmacol.*, **21**, 355 (1972).

139. U.S. Environmental Protection Agency, Office of Drinking Water, *Trinitroglycerol Health Advisory*, 1987.

140. H. V. Ellis III et al., *Fundam. Appl. Toxicol.*, **4**, 248 (1984).

141. S. Forssman, *Arch. Gewerbepathol. Gewerbehyg.*, **16**, 157 (1958).

142. S. A. Forman, *Toxicol. Lett.*, **43**, 51 (1988).

143. M. E. Andersen and R. G. Mehl, *Am. Ind. Hyg. Assoc. J.*, **34**, 526 (1973).

144. R. A. Jones et al., *Toxicol. Appl. Pharmacol.*, **22**, 128 (1972).

145. R. D. Steward et al., *Toxicol. Appl. Pharmacol.*, **30**, 377 (1974).

146. E. P. Horvath et al., *Am. J. Ind. Med.*, **2**, 365 (1981).

147. J. L. Mattsson et al., *Aviat. Space Environ. Med.*, **52**(6), 340 (1981).

148. D. G. Clark and M. H. Litchfield, *Toxicol. Appl. Pharmacol.*, **15**, 175 (1969).

149. M. E. Andersen and R. A. Smith, *Biochem. Pharmacol.*, **22**, 3247 (1973).

150. J. R. Bucher et al., *J. Appl. Toxicol.*, **10**(5), 353 (1940).

151. W. J. McConnell et al., *Occup. Med.*, **1**, 551 (1946).

152. E. L. Etnier, *Reg. Toxicol. Pharmacol.*, **9**, 147 (1989).

153. E. L. Etnier and W. R. Hartley, *Reg. Toxicol. Pharmacol.*, **11**, 118 (1990).

154. U.S. Environmental Protection Agency, Criteria and Standards Division, Office of Drinking Water, *Health Advisory for Hexahydro-1,3,5-trinitro-1,3,5-triazine (RDX)*, 1988.

155. N. R. Schneider et al., *Toxicol. Appl. Pharmacol.*, **39**, 531 (1977).

156. B. S. Levine et al., *Toxicol. Lett.*, **8**, 241 (1981).

157. L. J. Burdette et al., *Toxicol. Appl. Pharmacol.*, **92**, 436 (1988).

158. M. Barsotti and G. Crotti, *Med. Lav.*, **40**, 107 (1949).

159. A. S. Kaplan et al., *Arch. Environ. Health*, **10**, 877 (1965).

160. R. C. Woody et al., *Clin. Toxicol.*, **24**(4), 305 (1986).

161. U.S. Environmental Protection Agency, Criteria and Standards Division, Office of Drinking Water, *Health Advisory for Octahydro-1,3,5,7-tetranitro-1,3,5,7-tetrazocine (HMX)*, 1988.

162. R. H. Schwartz and P. Peary, *Clin. Pediatrics*, **25**(6), 308 (1986).

163. G. R. Newell et al., *Am. J. Med.*, **78**, 811 (1985).

164. M. A. Holland and L. M. Kozlowski, *Clin. Pharmacy*, **5**, 737 (1986).

165. R. Shesser et al., *Ann. Emerg. Med.*, **10**(5), 262 (1981).

166. D.A. Guss et al., *Am. J. Emerg. Med.*, **3**(1), 46, (1985).

167. W. G. F. Adams, *Trans. Assoc. Ind. Med. Officers*, **14**, 24 (1964).

168. T. Beritic, *Arch. Hig. Rada*, **8**, 333 (1957); *Ind. Hyg. Digest Abstr.*, **55** (Jan. 1959).

169. A. J. M. Slovak and R. N. Hill, *J. Occup. Med.*, **23**(12), 857 (1981).

170. J. B. O'Toole III et al., *J. Forensic Sci*, **32**(6), 1811 (1987).

171. H. W. Haverkos and J. Dougherty, *Am. J. Med.*, **84**, 479 (1988).

172. M. Törnqvist et al., *Mutat. Res.*, **117**, 47 (1983).

173. D. Wild et al., *Food Chem. Toxicol.*, **21**(6), 707 (1983).

174. V. C. Dunkel et al., *Environ. Mol. Mutat.*, **14**, 115 (1989).

175. J. Osterlok and D. Goldfield, *J. Anal. Toxicol.*, **8**, 164 (1984).

176. R. C. Woodruff et al., *Environ. Mutagen.*, **7**, 677 (1985).

177. R. F. Jacobs et al., *J. Toxicol.—Clin. Toxicol.*, **20**(5), 421 (1983).

178. D. M. Lewis et al., *J. Toxicol. Environ. Health*, **15**, 835 (1985).

179. L. S. Soderberg et al., *Adv. Exp. Med. Biol.*, **288**, 265 (1991).

180. D. R. Klonne et al., *Fundam. Appl. Toxicol.*, **8**, 101, (1987).

181. D. P. McFadden et al., *Fundam. Appl. Toxicol.*, **1**, 448 (1981).

182. D. C. Rees et al., *Neurobehav. Toxicol. Teratol.*, **8**, 139 (1986).

183. J. C. Gage, *Brit. J. Ind. Med.*, **27**, 1 (1970).

184. P. T. Cochran, *Am. Heart J.*, **98**, 141 (1979).

185. R. W. Wood and C. Cox, *J. Appl. Toxicol.*, **1**(1), 30 (1981).

186. L. D. Fechter et al., *Fundam. Appl. Toxicol.*, **12**, 56 (1989).

187. R. P. Maickel, *NIDA Res. Gr.*, **83**, 15 (1988).

188. D. W. Lynch et al., *J. Toxicol. Environ. Health*, **15**, 823 (1985).

189. D. P. McFadden and R. P. Maickel, *J. Appl. Toxicol.*, **5**(3), 134 (1985).

CHAPTER ELEVEN

N-Nitrosamines

Mark A. Thomson, Ph.D., Charles R. Green, Ph.D., Rasma B. Balodis, J.D., Joseph L. Holtshouser, C.I.H., C.S.P., Robert K. Hinderer, Ph.D., Raymond J. Papciak, Hon-Wing Leung, Ph.D., D.A.B.T., C.I.H., Susan R. Howe, and T. Dee Kuhn*

1 GENERAL CONSIDERATIONS

1.1 Nomenclature, Chemistry, and Industrial Occurrence

N-Nitroso compounds are characterized by a nitroso group ($-N=O$) bonded to a nitrogen atom. These compounds may be divided into two types, N-nitrosamines and N-nitrosamides. The N-nitrosamides are produced by nitrosation of ureas, guanidines, carbamates, urethanes, and other aliphatic and aromatic amides. N-Nitrosamides are relatively unstable under physiological conditions and their toxic actions are clearly different from those of N-nitrosamines (1). This chapter primarily addresses N-nitrosamines.

N-Nitrosamines are formed by nitrosation of aliphatic and aromatic amines with nitrosating agents related to nitrous acid. A common source of nitrosating agents is ambient nitrogen oxides or nitrates, but nitrosation may be also effected by nitrosyl halides (2–4), nitrosyl tetrafluoroborate (5, 6), and nitrosyl sulfuric acid (1). N-Nitrosamines may be also formed from N,N-disubstituted hydrazines (7).

*This chapter was prepared by members of the Chemical Manufacturers Association Nitrosamines Panel.

Patty's Industrial Hygiene and Toxicology, Fourth Edition, Volume 2, Part A, Edited by George D. Clayton and Florence E. Clayton.
ISBN 0-471-54724-7 © 1993 John Wiley & Sons, Inc.

In a classical reaction, N-nitrosamines are formed by the reaction of secondary amines and nitrous acid under acidic conditions (8):

$$R_2NH + HONO \xrightarrow{H^+} R_2NNO + H_2O$$

More recent studies indicate that N-nitrosamines may be formed by other reaction schemes. Depending on the precursors and the presence of catalysts such as formaldehyde, ozone and metal ions, N-nitrosamines can be formed under neutral or alkaline conditions. Mixtures containing N-nitrosamines are produced by the rapid reaction of amines with the hydroxy radical (9). Besides secondary amines, both primary and tertiary amines may form N-nitrosamines by different routes in the presence of catalysts and inhibitors (10, 11). For instance, if the reaction of a tertiary aliphatic amine goes beyond the normal salt formation, nitroso derivatives of secondary amines may be formed (8).

Exposure of humans to N-nitrosamines can occur through the food chain, in the environment, and in the industrial setting. N-Nitrosamines can form in the body from nitrosating agents, particularly nitrites and amine precursors.

The major human intake of nitrates comes from vegetables, water supplies, or additives in the meat and fish curing processes. Nitrates are converted to nitrites in the upper part of the gastrointestinal tract by the nitroreductase-containing bacteria normally present in the lower bowel (12). High concentrations of nitrates are present in spinach, beets, lettuce, radishes, eggplant, celery, and turnips. The average nitrate intake is about 52 mg/day (13). Amine precursors are present in vegetables, wine, spirits, beer, tea, fish, tobacco, tobacco smoke, food flavoring agents, and some drugs (14). Many sources also present the potential for direct exposure from N-nitrosamines that are already formed. Cosmetics (15) and food products, alcoholic beverages, beer, and tobacco have been reported to contain trace amounts of these chemicals (16). Water may contain N-nitrosamines derived from herbicides and pesticides or ground water contamination. Common sources of human exposure to various N-nitrosamines are tabulated in Table 11.1 (17).

In industry, very few N-nitrosamines are used in manufacturing processes. Dimethylnitrosamine is used, however, as an intermediate in the production of 1,1-dimethylhydrazine for rocket fuel applications. Its use as a solvent in the automotive industry has been discontinued. In the past, N-nitroso compounds have been used as vulcanizing retarders and blowing agents in the rubber industry, but they have been replaced with other materials. Various types of N-nitrosamines have been also described for use in the polymerization of polyacrylate-type sealants (18).

Industrial exposure to N-nitrosamines occurs as the result of by-products formed via the reaction of a constituent amine with a nitrosating agent or a compound capable of producing nitrite ion or nitrosonium cation during the reaction process. This unintentional formation occurs in dye manufacture, metalworking, rubber and tire industry, leather tanneries, foundries, and fish meal processing. Table 11.2 identifies the N-nitrosamines commonly found in these industries.

In the rubber industry, the presence of N-nitrosamines in the workplace is

Table 11.1. Common Sources of Human Exposure to N-Nitrosamines[a]

Sources	N-Nitrosamines[b]	Concentration	Detection Methods[b]	Ref.
MEAT PRODUCTS				
Smoked sausages, bacon, ham	Unidentified	0.6–6.5 ppb	TLC	26
Bacon, dry-cured	NTHZC	7–6487		
	NTHZ	<1–22.3 ppb	CG–TEA	27
Bacon, fried	DMN, DEN, NPYR, NPIP	1–40 ppb	GC, MS	28
Bacon, netted	DBN	1—123 ppb		29
Bacon, uncooked and fried	DMN, NPYR	2–30 ppb	TLC, GC, MS	30
Cured meat (Kasserel)	DEN	40 ppb	TLC	31
Salami, dry sausages	DMN	10–80 ppb	TLC, GC, MS	32
Luncheon meat, salami, pork chopped Danish	DMN, DEN	1–4 ppb	GC, MA	28
Pressed ham, hamburger	DMN	15–25 ppb	TLC	33
Mettwurst sausages	NPYR, NPIP	13–105 ppb	TLC, GC, MS	34
Various meat products	NPRO, NSAR, NMA	0.4–440 ppb	TLC, HPLC, GC, MS	35–47
FISH				
Smoked herring, haddock, mackerel, kipper	Unidentified	0.5–40 ppb	TLC	26
Chinese marine salted fish	DMN	0.05–0.3 ppm	GC, MS	48
	DEN	1.2–21 ppm		
Raw smoked and smoked nitrite–nitrate treated: salmon, shad, sable	DMN	0–26 ppb	GC, MS	49
Fresh, salted, fried cod: fresh, fried hake	DMN	1–9 ppb	GC, MS	28
Dried mackerel, pike, salted salmon roe	DMN, DEN	Trace	TLC	33
OTHER FOODS				
Cheese	DMN	1–4 ppb	GC, MS	28
Flour, wheat	DEN	0–10 ppb	TLC, GC	50–56

Table 11.1. (Continued)

Sources	N-Nitrosamines[b]	Concentration	Detection Methods[b]	Ref.
	MISCELLANEOUS			
Tobacco smoke condensate	DEN, NPIP	0.2 ppm/1200 cig.	TLC	53
	DMN, DEN, DETN, MEN, MBN, DPN, NPYR, NPIP, DBN, NNN	0–180 ng/cig.	GC, MS	57–62
Soya bean oil	DMN	300 ppb	TLC, GC, MS	63, 64
Beer, Polish	DMA	<0.3 µg/kg	GC–TEA	65
Beer, U.S. and Canadian	DMA	0–0.58 µg/kg	LC	66
Beer, German	DMA	0–1.7 ng/kg		67
	COMMERCIAL PRODUCTS			
Cosmetics	DETN	1 ppb–48 ppm	HPLC–TEA	68
Latex gloves	DMN	37–329 ppb	GC–MS, W–SV	69, 21
	NPIP	115–1.879 ppb	GC–MS, W–SV	69, 21
	NMOR, NPYR	Trace	GC–MS	69
	DEN	<10 µg/kg	W–SV	21
Metalworking fluids	DETN	0.02–3%	HPLC–TEA	70
		0.4–4.2 µg/g		71
	NOZ	60 ppm		72
Pesticide formulations	DMN, DPN	300 ppb–640 ppm	GC–TEA, HPLC–TEA	73
Rubber nipples	DMN	<200 µg/kg	W–SV	21
	DEN, NPIP	<100 µg/kg	W–SV	21
	DBN	<70 µg/kg	W–SV	21
Rubber nipples, FDA	Total volatile N-Nitrosamines	<60 ppb		74
Rubber toys	DMN	<25 µg/kg	W–SV	21
	DBN	<10 µg/kg	W–SV	21
Rubber balloons	DMN	<150 µg/kg	W–SV	21
	DEN	>30 µg/kg	W–SV	21

Diesel crankcase emission	DMA		
	DMA, DEA, DPA	4.4–136 mμ >6.0 ng/m³	75, 76 77

[a] Adapted from *ACS Monograph 173* (1).
[b] Abbreviations:

N-Nitrosamines

DBN (NDBA)	N-Nitrosodibutylamine
DEN (NDEA)	N-Nitrosodiethylamine
DETN (NDELA)	N-Nitrosodiethanolamine
DMN (NDMA)	N-Nitrosodimethylamine
DPN	N-Nitrosodipropylamine
EBN	N-Nitrosoethylbutylamine
MBN	N-Nitrosomethylbutylamine
MEN	N-Nitrosomethylethylamine
NDELA	N-Nitrosodiethanolamine
NDCHA	N-Nitrosodicyclohexylamine
NDPhA	N-Nitrosodiphenylamine
NEMA	N-Nitrosoethylmethylamine
NEU	N-Nitroso-N-ethylurea
NMA	N-Nitroso-N-methylaniline
NMAB	N-Nitroso-4-methylaminobenzaldehyde
NMNG	N-Nitroso-N-methyl-N'-nitroguanidine
NMOR	N-Nitrosomorpholine
NMOZ	N-Nitroso-5-methyl-1,3-oxazolidine
NMU	N-Nitroso-N-methylurea
NMUA	N-Nitrosomethylurethane

NMVA	N-Nitrosomethylvinylamine
NNN	N-Nitrosonornicotine
NOZ	N-Nitroso-1,3-oxazolidine
NPIP	N-Nitrosopiperidine
NPRO	N-Nitrosoproline
NPYR	N-Nitrosopyrrolidine
NSAR	N-Nitrososarcosine
NTHZ	N-Nitrosothiazolidine
NTHZC	N-Nitrosothiazolidine-4-carboxylic acid

Analytical Methods and Terms

GC	Gas chromatography
HPLC	High-pressure liquid chromatography
IR	Infrared spectroscopy
LC	Liquid chromatography
MS	Mass spectrometry
NMR	Nuclear magnetic resonance spectroscopy
PC	Photoconductivity detector
PL	Polarography
TEA™	Thermal energy analyzer
W–SV	Water-steam vacuum method
ND	Not determined

Table 11.2. Industrial Exposure to *N*-Nitrosamines[a]

Industry	*N*-Nitrosamine[b]	Amount Detected ($\mu g/m^3$)	Test Method[b]	Ref.
Dye industry	Not identified	<40	GC, HPLC, TEA	78
Fish meal processing	NDMA	0.06	GC, HPLC, TEA	78
Foundries, U.S.	NDMA	0.062–2.4	GC–TEA	78
	NEMA	0.021	GC–TEA	78
Foundries, France	NDMA	av. 062	GC–TEA	25
	NEMA	av. 0.020		25
Foundries, Germany	Total	0.1–2.1		79
Leather tanneries	NDMA	0.1–47	GC, HPLC, TEA, GC–MS	78, 80
	Total (DEN, DMN, NMOR)			79
	Total	400 μg/(day)/ (person)		81, 82
Metalworking industry	NDELA	0.08 ng/m³	GC, HPLC, TEA	78
	Total	0.2–2.3		79
		5–50 μg/(8 hr shift)/(person)		81, 82
Rubber chemical manufacture	NMOR	0.07–4.6	GC, HPLC, TEA	78
	NDMA	<0.3	GC, HPLC, TEA	78
	NDPhA, NDEA, NPYR	<47	GC, HPLC, TEA	78
Rubber curing	NMOR	4.6	TEA, GC–MS	
	NDPhA	5–150 ng/m³		
	DMNA	9.0–136 ng/m³		21, 84
Rubber curing, salt bath	NDMA	1–40		
	NDEA	0.1–5		
	NMOR	0.1–4700		21, 84
Rubber curing, injection molding	NDMA	40–1060		
	NMOR	120–380		

[a]Adapted from *Carcinogenesis*, **4** (84) and IARC Sci. Publ. No. 41 (21).
[b]For list of abbreviations used, see Table 11.1.

directly related to the use of the corresponding vulcanization accelerators and production processes that release nitrogen oxides (19). Amine precursors, accelerators, and stabilizers of the dithiocarbamate, sulfenamide, thiuram sulfide, and thiourea type are known to yield nitrosatable fragments during curing (20). The role of these ingredients is evidenced by the fact that a decrease in the amount of nitrosatable chemicals in the formulation results in a decreased *N*-nitrosamine formation.

The knowledge of the mechanism of *N*-nitrosamine formation has provided a basis for the development of various approaches to minimize generation of these chemicals. Some investigators have noted that the substitution of a non-nitrosatable accelerator will eliminate or decrease *N*-nitrosamine formation (21). Others have suggested the use of accelerators that generate nonvolatile *N*-nitrosamines and noncarcinogenic *N*-nitrosamines.

Cutting fluids are a potential source of *N*-nitrosamine formation because they contain sodium nitrate together with a triethanolamine salt of a tricarboxylic acid complex or a similar amine (22). This formation can also occur in lubricating oils and greases that contain sodium nitrite (23). Elimination of nitrites from these formulations has not entirely solved the problem. Lubricants, cutting fluids, and other functional fluids, such as coolants, can absorb nitrogen dioxide from air and also form *N*-nitrosamines from other components in the formulation (24).

In the leather tanning industry, the dimethylamine sulfate salt used in the de-hairing process is the prime source of *N*-nitrosamines (24). In foundries, the main source of *N*-nitrosamines appears to be from dimethylethylamine (25).

In summary, human exposure to *N*-nitrosamines occurs via four main routes: In vivo nitrosation, ingestion, inhalation, and/or dermal contact (11). Occupational exposure is only one source. The total exposure that one receives is greatly influenced by an individual's life-style.

1.2 Environmental Occurrence and Fate

N-Nitrosamines have been detected in the environment, generally in low concentrations. The mechanism by which they are generated and degraded depends on the medium, physiochemical characteristics of the molecule, and site-specific chemical and biological activity.

1.2.1 Atmosphere

NDMA has been detected in the air following contamination from industrial sources involved in the production of asymmetrical dimethylhydrazine, where average on-site concentrations were found to be 11,600 ng/m^3 (11.6 mg/m^3) (87); at factory locations involved with the production of dimethylamine (88); and in various urban locations (88). NDMA, NDPhA, NMOR, and NDELA (see Table 11.1 for names of compounds) have been identified and in some cases quantified in confined environments within the rubber, leather tanning, and metalworking industries (89).

There is limited information on the atmospheric degradation of *N*-nitrosamines. Because *N*-nitrosamines absorb light at differing rates, they are subject to varying rates of photolytic degradation. For example, the photolytic half-life of NDMA and NDPhA have been determined to be 5 to 30 min (90) and 7 hr, respectively (91).

1.2.2 Water

The occurrence and disposition of N-nitrosamines in water have received considerably more attention recently as a result of their detection in groundwater supplies used ultimately as a source of public drinking water. Although studies have shown that N-nitrosamines can be generated as a result of bacterial metabolism (92), particularly in water containing precursors such as secondary amines and nitrites, or compounds that can be converted to the precursors such as sewage, N-nitrosamines are more likely to be found in water as a result of industrial contamination (93). In the case of NDMA, the occurrence and concentration has been limited (87). NDMA was detected in 0.9 percent of 2308 groundwater samples between 1980 to 1988. In these samples, the average concentration was determined to be 12.4 μg/l (93). Other studies have revealed the presence of NDMA in water samples from hazardous waste sites (94), tap water (95), and deionized water (96). In these cases, the levels were in the low μg/l range. It is unclear whether the N-nitrosamines were introduced as contaminants or were generated in situ. NDPhA has been detected at unspecified concentrations in river samples (97).

Biodegradation represents the major environmental fate process for those N-nitrosamines that have been detected in water, although photolysis may play some role. Limited data show NDMA to be slowly degraded by sunlight in water (98) with ultimate fate related to biodegradation, although the rate under natural conditions is not known. Additional studies on NDPhA show that it is degradable with rapid microbial adaptation at concentrations of 5 ppm, and more gradual adaptation at concentrations of 10 ppm (99). Additional factors contributing to the degradation of NDMA and possibly N-nitrosamines in general include changes in pH and the concurrent presence of nitrite ions (100).

1.2.3 Soil

N-Nitrosamines have been detected in soil. Their presence may occur as a result of (1) inadvertent generation by the presence of amines in contact with nitrogen oxides, nitrous acid, nitric salts, or by trans-nitrosation via nitro or other nitroso compounds (101); (2) bacterial metabolism through the nitrification–denitrification process; and (3) chemical generation, where the presence of secondary amines and high local concentrations of nitrites can under certain conditions lead to the formation of N-nitrosamines, particularly NDMA (102, 103). Although the microbial synthesis of N-nitrosamines has been shown to occur under experimental conditions (92, 104), other studies have found little if any evidence for the formation of N-nitrosamines in soils or wastewater systems (105).

The extent to which N-nitrosamines are likely to migrate in soil and groundwater depends upon several physiochemical factors. For example, although NDMA exhibits a relatively high vapor pressure (2.7 mm Hg at 20°C) compared to that of NDPhA (0.1 mmHg at 25°C), suggesting that both would exist in the vapor phase, NDMA is relatively less likely to attrition from the vapor phase to organic particulates in the atmosphere (106). NDMA has a soil sorption coefficient (K_c^o) of 12, indicating that it will be highly mobile in soils, whereas NDPhA, with a K_c^o of 830-

1, is expected to have low mobility. As such, significant leaching would not be expected to occur in most types of soil for this *N*-nitrosamine (107, 108).

1.3 Methods of Detection

Numerous methods have been developed for the detection and quantification of *N*-nitrosamines. Thin layer chromatography (TLC) (109–112), polarography (113–116), and spectrometry (117–120) were some of the first methods used to detect *N*-nitrosamines. Gas chromatography (GC) with the flame ionization detector (FID) (121–124) became the most used method for *N*-nitrosamines in the 1960s. However, the need to detect *N*-nitrosamines in the parts per billion (ppb) range brought about the development of new detectors with greater sensitivity. The FID detector was replaced with the nitrogen-phosphorus detector (NPD) which is still used today. By coupling a mass spectrometer to a gas chromatograph (GC/MS) (125–129) the detection limits for *N*-nitrosamines dropped to the parts per trillion (ppt) range. The advent of the thermal energy analyzer (TEA) for *N*-nitrosamines coupled with a gas chromatograph of high-performance liquid chromatograph (HPLC) (130, 131) further facilitated low level analyses. The TEA detector enabled analytical chemists to develop methods for the separation and detection of *N*-nitrosamines that previously had not been reported.

The development of the second-generation TEA detector and the advent of the high-resolution MS detector provides the greatest analytical sensitivity. These two detectors can detect certain *N*-nitrosamines with a sensitivity of 10^{-12}.

The most applicable analytic methods of detection are the GC/MS, GC/TEA, HPLC/TEA, and GC/NPD. All these instruments require methods of specific sample preparation to utilize these detectors to their designed potential.

Analytic techniques have been modified for nonoccupational exposures to *N*-nitrosamines. These have allowed the detection of *N*-nitrosamines in mainstream cigarette smoke (132) at levels of 0.21 μg/cigarette. The total *N*-nitroso content of foods also has been measured by the addition of chemiluminescent detection to the HPLC–TEA technique (133).

The use of microbore chromatographic columns has led to nanogram detection capability in baby bottle nipples. Very good reproducibility was demonstrated following replicate injections of 1.0 ng *N*-nitrosodiethylamine (134). Although the use of selective fluorescence detection allows similarly low detection levels, the sample preparation steps may be more cumbersome (135). Detection limits of 5 μg/kg has also been demonstrated for nonvolatile *N*-nitrosamines as well (136).

Table 11.3 provides specific analytic conditions and the resolution possible using microbore chromatography.

1.4 Regulations

Regulations aimed at reducing health or environmental effects associated with exposure to *N*-nitrosamines currently exist at the national and local levels in several industrialized countries.

Table 11.3. Analytical Conditions For Volatile *N*-Nitrosamines by Gas Chromatography

Gas chromatograph: Hewlett Parkard 5890
Column: 105 m by 0.53 mm carbowax 20M (Restek Corp.)
Column flow: 15 ml/min He
Temperature program: 24 min.
                               Ramp A 125/2/5/200
                               Ramp B 35/240/5.86
Detector: TEA model 543 analyzer
Interface temperature 250°C
Pyrolysis temperature 500°C

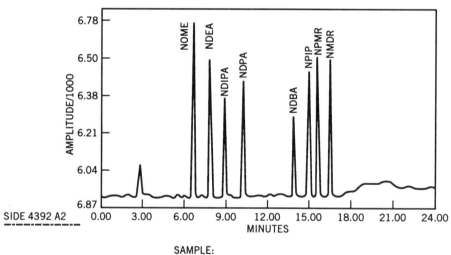

SIDE 4392 A2

SAMPLE:
ANALYZED: Fri Apr 3, 1992 8:15:32 pm

RESULT: /DATA/LOOP/RESULT/STD2_4092_A2.RES   METHOD: I IIX_TIL4392_AR

1.4.1 Occupational

1.4.1.1 United States. Workplace standards for *N*-nitrosamines are limited to NDMA, which is categorized by the Occupational Safety and Health Administration (OSHA) as a cancer suspect agent. Under this regulation, exposures are to be minimized (137). Field directives, issued by the Department of Labor in 1990 to alert regional and area OSHA offices of occupational hazards, have advised its personnel that process-generated *N*-nitrosamines pose a health risk. *N*-Nitroso compounds identified as carcinogens by the International Agency for Research on Cancer (IARC) and the National Toxicology Program (NTP) are subject to the OSHA Hazard Communication Standard (HCS) (138), as shown in Table 11.4.

1.4.1.2 Germany. In 1987 under authority of the Ausschuss für Gefahrenstoffe

Table 11.4. *N*-Nitroso Compounds Regulated in the United States

Compound	CASRN[a]	IARC[b]	NTP[c]
N-Nitrosomethylvinylamine	4549-40-0	2B	L
N-Nitroso-*N*-ethylurea	59-73-9	2A	L
N-Nitroso-*N*-methyl-*N*′-nitroguanidine	70-25-7	2A	
N-Nitroso-*N*-methylurea	684-93-5	2A	L
N-Nitrosomethylurethane	615-53-2	2B	
N-Nitrosoanabasine	1133-64-8	3, 4	
N-Nitrosotetrahydrobipyridine	71267-22-6	3, 4	
N-Nitrosodi-*n*-butylamine	924-16-3	2B	L
N-Nitrosodi-*n*-propylamine	621-64-7	2B	L
N-Nitrosodicyclohexylamine	947-92-2	NL	
N-Nitrosodiethanolamine	1116-54-7	2B	L
N-Nitrosodiethylamine	55-18-5	2A	L
N-Nitrosodiisopropanolamine	53609-64-6	NL	
N-Nitrosodimethylamine[d,e]	62-75-9	2A	L
p-Nitrosodimethylaniline	138-89-6	NL	
N-Nitrosodiphenylamine	86-30-6	3, 4	L
p-Nitrosodiphenylamine	156-10-5	3, 4	L
4-(*N*-Nitrosomethylamino)-1-(3-pyridyl)-1-butanone	64091-91-4	2B	
3-methylnitrosaminoproprionitrile	60153-49-3	2B	
N-Nitrosomethylethylamine	10595-95-6	2B	
N-Nitrosomorpholine	59-89-2	2B	L
N-Nitrosonornicotine	16543-55-8	2B	L
N-Nitrosopiperidine	100-75-4	2B	L
N-Nitrosoproline	7519-36-0	3, 4	
N-Nitrosopyrrolidine	930-55-2	2B	L

[a]CASRN—Chemical Abstracts Service Registry Number.
[b]IARC—International Agency for Research on Cancer: 1, sufficient evidence for cancer in humans; 2A, probable human carcinogen; 2B, possible human carcinogen; 3, inadequate evidence to determine carcinogenic status; 4, not likely to be carcinogenic to humans.
[c]NTP—National Toxicology Program, 6th ed.: L—listed in the Annual Report on Carcinogens.
[d]OSHA Regulated Cancer Suspect Agent.
[e]American Conference of Governmental and Industrial Hygienists (ACGIH) reported carcinogen (1991–1992).

(AGS), guidelines were issued for occupational exposure to atmospheric *N*-nitrosamines in the workplace. Unlike the U.S. OSHA guideline for NDMA, the AGS guidelines were intended to apply to *N*-nitrosamines as a class. The technical orientation values (TRG) of 2.5 and 1.0 μg/m³ (measured as 8-hr time-weighted averages) are intended to be applied to designated sections of production facilities in industries known to generate *N*-nitrosamines (79).

1.4.1.3 Canada. Canada's Workplace Hazardous Materials Information System (WHMIS) incorporates therein the Hazardous Products Act's Ingredient Disclosure

Table 11.5. United States Federal Regulations for N-Nitroso Compounds

Federal Water Pollution Control Act (Public Law 92-500, 1972) as amended by the Clean
Water Act of 1977 (Public Law 95-217), commonly called Clean Water Act (CWA)
and amendments
    Section 304(a)(1) Ambient Water Quality Criteria
    Section 307(2)(1) Toxic Priority Pollutants (40 CFR 401.15)

Comprehensive Environmental Response, Compensation, and Liability Act of 1980
(CERCLA or Superfund, Pub. L. 96-510, 1980) and Emergency Planning and
Community Right-To-Know Act or Title III of the Superfund Amendments and
Reauthorization Act of 1986 (SARA)
    Release Reporting of Hazardous Substances (40 CFR 302.4)
    Section 313, Toxic Chemical Release Reporting (40 CFR 372)

Resource Conservation and Recovery Act (RCRA, Public Law 94-580, 1976) and
amendments. Hazardous Waste [40 CFR 261.33(e)]
        Regulated Substance (40 CFR 261)
        Groundwater Monitoring (40 CFR 261)
        Land Disposal Restriction (40 CFR 268.1)

Occupational Safety and Health Act (OSHA, Public Law 91-596, (1970)
    29 CFR 1910.1016 N-Nitrosodimethylamine

List. A product containing ≥ 1 percent of a listed hazardous substance or ≥ 0.1
percent of a listed highly hazardous substance is subject to WHMIS requirements.

The Ingredient Disclosure List classifies as highly hazardous, among others,
N-nitrosodibutylamine, N-nitrosodiethanolamine, N-nitrosodiethylamine, N-nitroso-
dimethylamine, N-nitrosodiphenylamine, N-nitrosodipropylamine, N-nitroso-N-ethyl-
urea, N-nitroso-N-methylurea, N-nitrosomethylvinylamine, N-nitrosomorpholine, N-
nitrosonornicotine, N-nitrosopiperidine, N-nitrosopyrrolidine, and N-nitrososarcocine.

1.4.2 Environmental

1.4.2.1 United States. Over 20 N-nitrosamines are regulated in the Unites States.
One or more of these chemicals are covered by the regulations listed in Table 11.5.
At the state or local levels, many standards are derived from existing federal criteria
for environmental containment. Predominant among state legislation for N-nitro-
samines are regulations aimed at disclosing hazards to individuals, which include
right-to-know laws and the reporting of environmental spills, as shown in Table
11.6.

A number of states also have enacted standards and advisories for N-nitrosa-
mines in water and air. These include Kansas, Minnesota, Indiana, California, and
Wisconsin. Each has issued quality guidelines and standards relating to maximal
concentrations in drinking water and water intended for recreation (139–141).
Maryland, Wisconsin, Kansas, North Carolina, Pennsylvania, Virginia, and Ken-
tucky have issued Acceptable Ambient Air Concentrations (AAAC) (142, 140).

The City of Philadelphia also has a N-nitrosamine standard. An AAAC for
NDMA is set at 0.0004 ppb (1-year average) (142).

Table 11.6. United States Local Regulations for N-Nitroso Compounds

California: Toxic Chemicals and Safe Drinking Water Act (Proposition 65)
Connecticut Carcinogen Reporting
Connecticut Hazardous Materials Survey
Florida Toxic Substances Right-To-Know Reporting
Illinois Toxic Substances Disclosure to Employees Act
Illinois Chemical Safety Act
Louisiana Hazardous Materials Information Development and Prep. Act
Louisiana Hazardous Materials Information Development and Prep. Act Release Reporting
Louisiana Spill Reporting (Chapter 39)
Massachusetts Right-To-Know Reporting
Massachusetts Spill Reporting
New Jersey Right-To-Know Reporting
New Jersey Special Hazard
New Jersey Spill Tax
New York Bulk Storage Registration/Release Reporting
Pennsylvania Right-To-Know Reporting
Rhode Island Right-To-Know Reporting

1.4.2.2 Canada. Concerns about environmental contamination by N-nitrosamines also have surfaced in Canada. Recently, the Department of the Environment of Ontario has issued an interim maximum acceptable concentration (IMAC) of 7 ppt for NDMA in municipal water supplies (143). This temporary standard will be reissued as the maximum acceptable concentration (MAC) pending review by an Advisory Committee for Environmental Standards.

2 PHYSICAL PROPERTIES

N-Nitrosamines are typically liquids, oils, or volatile solids. Absorption of visible light by the N-NO chromophore generally results in colors ranging from light yellow to red-brown. Many N-nitrosamine compounds undergo relatively rapid photolytic degradation in the presence of ultraviolet or visible light (144).

IARC (144) summarized some of the biologic, physical, and chemical properties of selected N-nitrosamines. Table 11.7 lists physical and chemical properties of selected N-nitrosamines.

2.1 Acute Toxicity

The acute lethality for selected N-nitrosamines is presented in Table 11.8. In general, these compounds appear to exhibit a low order of acute toxicity. Structure and molecular weight have been noted to play a role in determining the acute lethal toxicity (148). For example, the lower alkyl amines exhibited the highest order of acute lethality, whereas cyclic N-nitrosamines appeared to be somewhat lower in acute lethality.

Table 11.7. Physical Properties of N-Nitrosoamines

Name, N-Nitroso	Formula	Mol. Wt.	Physical Form	Color	Specific Gravity	M.P. (°C)	B.P. (°C, mm Hg)	Ref.
-diphenylamine (N)	$(C_6H_5)_2N(NO)$	198.224	Crystal	Orange	1.23	64.66	—	
-di-n-propylamine	$(C_3H_7)_2N(NO)$	130.19	Liquid	Yellow	0.9163	6.6 (est.)	206 (760)	145
-diethanolamine	$(HOCH_2CH_2)_2N(NO)$	134.13	Oil	Light Yellow	1.4849	—	125	145
-diethylamine	$(C_2H_5)_2N(NO)$	102.14	Liquid	Light Yellow	1.4388	—	175–177	145
-dimethylamine	$(CH_3)_2N(NO)$	74.08	Liquid	Yellow	1.4368	—	151–153	145
-morpholine	$OC_4H_8N(NO)$	116.11	Crystals	Yellow	—	29	139–140	145
-pyrrolidine	$C_4H_8N(NO)$	100.11	Liquid	Yellow	—	—	104–106	145
-ethylaniline	$C_6H_5N(NO)—C_2H_5$	154.88	Oil	Yellow	1.037	—	119–120 (25)	146
-N-ethylurethane	$C_2H_5N(NO)CO_2—C_2H_5$	146.15	Oil	Red	1.071	86 (36)	128 (19)	146
-methylaniline (N)	$C_6H_5N(NO)—CH_3$	136.15	Oil	Yellow	1.124	14.15	65 (13)	146
-N-methylurethane	$CH_3(NO)—CO_2—C_2H_5$	132.12	Liquid	Yellow-red	1.122	-24	217–222 (760)	146
-piperidine(N)	$C_5H_{11}N(NO)$	184.15	Oil	Light Yellow	1.063	—	Sublimes	146
-triacetonamine	$C_9H_{16}N(NO)$	184.24	Needles	—	1.340	72–73		146
-dioctylamine	$(C_8H_{17})_2N(NO)$	270.46	Liquid	Yellow	0.8930	5.5	202–203 (13)	147
-propyl-isopropylamine	$(C_3H_7)_2N(NO)$	130.19	Liquid	Yellow	—	—	91 (17)	147
-dipentylamine	$(C_5H_{11})_2N(NO)$	186.30	Liquid	Yellow	0.8930	—	146 (12)	147
-dipropylamine	$(C_3H_7)_2N(NO)$	130.19	Liquid	Yellow	0.9153	—	104–105 (22)	147

Table 11.8. Acute Lethality of Selected N-Nitrosamines

Chemical	CASRN[a]	Species	Route	LD$_{50}$ (mg/kg)
N-Nitrosodi-n-butylamine	924-16-3	Rat	Oral	1200 (148)
		Hamster	Subcutaneous	561 (150)
N-Nitrosodiethylamine	1116-54-7	Rat	Oral	7500 (149)
		Hamster	Subcutaneous	11000[b]
N-Nitrosodiethylamine	55-18-5	Rat	IV	280 (149)
			IP	216 (149)
			Oral	280 (149)
			Subcutaneous	204 (149)
		Mouse	Oral	190–220 (149)
		Guinea pig	Oral	250 (149)
		Hamster	Subcutaneous	246 (149)
N-Nitrosodimethylamine	62-75-9	Rat	Oral	27–41 (149)
			IP	26.5 (149)
			Inhalation	78[c] (149)
			Subcutaneous	45 (149)
			IV	40 (149)
		Mouse	Inhalation	57[c] (149)
		Hamster	Oral	21 (149)
			Subcutaneous	28 (149)
N-Nitrosodi-n-propylamine	621-64-7	Rat	Oral	480 (149)
N-Nitroso-N-ethylurea	759-73-9	(None found)		
N-Nitroso-N-methylurea	684-93-5	Rat	Oral	180 (149)
N-Nitrosomethylvinylamine	4549-40-0	Rat	Oral	24 (149)
			Inhalation	24 (149)
N-Nitrosomorpholine	59-89-2	Rat	Oral	320 (149)
			IV	98 (149)
		Hamster	Oral	956 (149)
			Subcutaneous	491 (149)
N-Nitrosopiperidine	100-75-4	Rat	Oral	200 (149)
			Subcutaneous	100 (149)
			IV	60 (149)
		Hamster	Oral	617 (149)
			Subcutaneous	324 (149)
N-Nitrosopyrrolidine	930-55-2	Rat	Oral	900 (149)
		Hamster	Oral	1023 (149)
N-Nitrososarcosine	13256-22-9	Rat	Oral	>4000 (149)

[a]Chemical Abstracts Service Registry Number.
[b]LD$_{LO}$ (151).
[c]4-hr LC$_{50}$, ppm.

The liver appears to be the target organ for toxicity and liver injury a common result of acute toxicity for a number of N-nitrosamines, including N-nitrosodimethylamine, N-nitrosodiethylamine, N-nitroso-n-butylmethylamine, N-nitrosovinylethylamine, N-nitrosopiperidine, and N-nitrosomorpholine (148).

Case reports have indicated liver injury in humans from exposure to N-nitrosodimethylamine. Freud (149) reported liver necrosis and regenerative cellular proliferation in one of two chemists accidentally exposed to N-nitrosodimethylamine. Liver damage also was noted among workers involved in the production of 1,1-dimethylhydrazine, a liquid rocket propellant in which N-nitrosodimethyl-amine was used as an intermediate (150).

2.2 Pharmacokinetics and Metabolism

2.2.1 Absorption

N-Nitrosamines are well absorbed after oral administration. Approximately 30 to 87 percent of an oral dose of N-nitrosodiethanolamine could be absorbed within 24 hr (152–155). Less than 2 percent of the labeled compound was recovered from the gastrointestinal tract 15 min after oral administration of ^{14}C-N-nitrosodime-thylamine (156). Similarly, N-nitrosamines can be readily absorbed following inhalation exposure. About 70 to 90 percent of the dose was reported to be absorbed in animals inhaling N-nitrosodimethylamine and N-nitrosodiethanolamine vapors (157, 158). N-Nitrosamines can also be absorbed after application to the intact skin. However, the rate of absorption depends on a variety of factors including physicochemical properties of the N-nitrosamine, thickness of skin, area of application, and vehicle effects. Owing to the complexities involved, exact skin penetration rates of N-nitrosamines are unknown. For N-nitrosodiethanolamine about 20 to 30 percent of an applied dermal dose could be absorbed in rats (152–153, 159–160) and about 2 to 5 percent was estimated to be absorbed by humans (161). In vitro studies with excised human abdominal skin indicated that N-nitrosodiethanolamine penetrated slowly. The permeability constant ranged from 3.2 × 10^{-6} to 5.5 × 10^{-6} cm/hr (162).

2.2.2 Distribution

The distribution of radio-labeled N-nitrosodiethanolamine has been evaluated in rats given a single 0.5 or 50 mg/kg dose via either the oral or dermal route (152). The radio label was distributed throughout the organs and tissues from 4 to 168 hr after dosing. There were no apparent differences in the distribution of the radio label in animals dosed orally and those dosed dermally. In general, the tissue content was low, and there was no evidence for retention of radio label by any organ. However, elevated levels of radio label were noted in the stomachs and livers of animals dosed orally during the first 24 hr. Airoldi et al. (160) showed that a three-compartment model could adequately describe the kinetics of N-nitrosodiethanol-amine following intravenous and cutaneous administration to rats. Daugherty and Klapp (163) reported that after oral administration of ^{14}C-N-nitrosodimethylamine to mice the relative amounts of radioactivity in the heart, forestomach, esophagus, liver, and lungs were 1, 2, 3, 10, and 70, respectively. Unmetabolized N-nitroso-dimethylamine was found to be evenly distributed among the main organs of mice and rats shortly after parenteral injection to animals in which the metabolism of N-nitrosodimethylamine had been inhibited (164–166). Johnson et al. (167) reported that 1 hr after a dose of 6 mg ^{14}C-N-nitrosodimethylamine per kilogram was administered by intraperitoneal injection to mice, the liver contained two times as much radioactivity as the kidney, spleen, and thymus. Measurable amounts of

N-nitrosodimethylamine were reported in the blood, liver, kidney, lungs, and brain of mice exposed to 5 mg/kg/day in drinking water for up to 4 weeks (168). After a single 100 mg/kg subcutaneous injection to pregnant hamsters, N-nitrosodi-n-propylamine was detected in the maternal blood, placenta, fetus, and amniotic fluid (169–170). The concentration of the chemical in maternal blood reached a maximum at 45 and 90 min after the injection, whereas a single peak at 90 min was observed in the fetus. About 1.6 percent of the compound was found in the placenta and 1.3 percent in the fetus at day 14 of gestation. After intravenous injection of ^{14}C-N-nitrosodi-n-butylamine to rats the highest concentrations of the radio label occurred in the nasal mucosa, liver, and preputial gland (171). N-Nitrosomorpholine, N-nitrosodiethylamine, and N-nitrosopyridine were reported to distribute evenly in most tissues in animals following intravenous injection (172–175).

2.2.3 Metabolism

Although N-nitrosamines can be metabolized via oxidation at the α-, β-, and γ-carbon positions, most dialkylamines are primarily oxidized at the α-carbon to the nitroso group by tissue-specific microsomal mixed-function oxidases. In general, this oxidation results in an unstable product that collapses to a diazohydroxide, and ultimately to an electrophile and a carbonyl compound (Figure 11.1). Thus N-nitrosodi-n-propylamine is oxidized at the α-carbon to produce propionaldehyde and propyldiazohydroxide (176), N-nitrosodimethylamine to formaldehyde (177), and N-nitrosodiethylamine to acetaldehyde (178–179).

N-Nitrosodi-n-propylamine is metabolized via α-oxidation to propionaldehyde, 1-propanol and 2-propanol (176, 180, 181). The latter two are formed via propyldiazohydroxide and a propyl cation (carbonium ion). It is generally believed that carbonium ions can also react with nucleic acids to form adducts. This is the proposed mechanism for the carcinogenic properties of the N-nitrosamines. β-Oxidation yields N-nitroso-2-hydroxypropylpropylamine, which is excreted as the glucuronide or further oxidized to a small extent to N-nitroso-2-oxopropylpropyl-amine (180, 182–184). Methylated hepatic nucleic acids have been recovered from rats treated with N-nitrosodi-n-propylamine (185–189). Putative methylating intermediates formed from N-nitroso-2-oxo-n-propylamine are N-nitrosomethylpropylamine and diazomethane. γ-Oxidation of N-nitrosodi-n-propylamine yields N-nitroso-3-hydroxypropylpropylamine and N-propyl-N-(2-carboxyethyl)nitrosamine (182, 184, 190).

N-Nitrosodimethylamine is metabolized by α-oxidation to formaldehyde, molecular nitrogen, and methylating agent. Recent evidence suggests that a significant proportion of N-nitrosodimethylamine is metabolized via a denitrosation mechanism (191, 192). Metabolism of N-nitrosodimethylamine varies among species (193, 194), dose (195, 196), age, and routes of administration (197).

N-Nitrosomorpholine is oxidized via α-oxidation to 2-hydroxyethoxyacetic acid, N-nitroso-(2-hydroxyethyl)glycine, and N-nitrosodiethanolamine (198, 199). Cytochrome P450 dependent metabolism of N-nitrosomorpholine has been shown to occur in the nasal and esophageal mucosa, as well as in the liver (172).

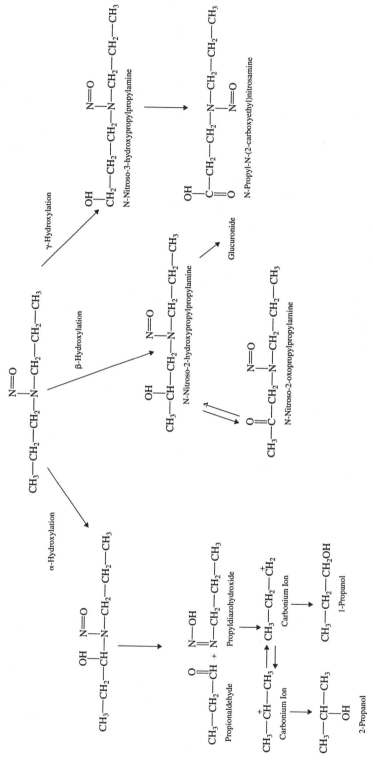

Figure 11.1.

N-Nitrosodiethanolamine is apparently not metabolized at the α-carbon but at the β-carbon by non-microsomal enzymes to N-(2-hydroxyethyl)-N-carboxymethyl-nitrosamine and N-(2-hydroxyethyl)-N-(formylmethyl)nitrosamine (152, 176, 200). An alternative two-step mechanism has been proposed in which N-nitrosodiethanolamine is first transformed by alcohol dehydrogenase into N-nitroso-2-hydroxymorpholine, the ring-closed hemiacetal form of N-(2-hydroxyethyl)-N-(formylmethyl)nitrosamine. This cyclic β-hydroxynitrosamine appears to be a substrate for sulfotransferase. The resulting sulfate conjugate is suggested to be the ultimate genotoxic metabolite of N-nitrosodiethanolamine (201).

Unlike the dialkylnitrosamines, the diarylnitrosamine N-nitrosodiphenylamine does not appear to undergo hydroxylation of its side chain. In experiments with animals, N-nitrosodiphenylamine seemed to undergo denitrosation reaction to form diphenylamine and nitric oxide (202). Nitrate was identified as the major metabolite, whereas nitrite, diphenylamine, and a monohydroxydiphenylamine were found in smaller amounts.

2.2.4 Excretion

Labeled CO_2 can be detected in the exhaled air 1 hr after intraperitoneal administration of ^{14}C-N-nitrosodimethylamine to rats (197). Three fractions identified as amino acids, allantoin and a metabolite of thiazolidine-4-carboxylic acid, and 7-methylguanine were recovered in the urine over a period of 5 days (203). Klein and Schmezer (157) reported that 10 to 30 percent of N-nitrosodimethylamine was excreted by exhalation 10 min following endotracheal intubation exposure in rats. In beagle dogs 23 percent of the administered dose was exhaled in 30 min after a 3-hr inhalation exposure to N-nitrosodimethylamine (204). Spiegelhalder et al. (155) reported that humans excreted 0.5 to 2.4 percent of an ingested dose of N-nitrosodimethylamine in the urine over a 24-hr period. Unchanged N-nitrosodimethylamine was recovered in the urine and feces of rats up to 24 hr after a single oral dose of 50 mg (164). Swann et al. (205), however, did not detect labeled N-nitrosodimethylamine in the urine of rats after oral administration of 30 μg/kg of ^{14}C-N-nitrosodimethylamine in water. Phillips et al. (197) determined that after a single oral dose of 5 mg of ^{14}C-N-nitrosodimethylamine to female rats the maximum rate of [14]CO_2 production was 12.4 percent of the dose per hour, and that 48 percent of the dose could be recovered as [14]CO_2 in the exhaled air in 7 hr and 5.7 percent as [14]C (total label) in a 24-hr urine sample.

Rats excreted about 5 percent of the administered dose as metabolites, but not unchanged N-nitrosodi-n-propylamine in the urine following oral dosing (184, 190).

After oral administration of a single 1000 mg/kg dose of N-nitrosodiphenylamine to female Wistar rats, about 25 and 30 percent of the administered dose were excreted in the urine as nitrate and nitrite in 36 and 96 hr, respectively (202). Excretion was found to be twice as fast when the N-nitrosodiphenylamine was administered by intraperitoneal injection as compared to oral administration. There appears to be a wide variation in the rates of N-nitrosodiphenylamine excretion into the bile from various animal species: half-life of 95 min for rabbits, 240 min for guinea pigs, and 510 min for rats (206).

Approximately 20 percent of the radio label was recovered as unchanged *N*-nitrosodiethanolamine in the urine of rats collected over 96 hr after dosing, and about 50 percent was recovered after 1 week (152). In contrast to these results, it was reported that rats excreted 20 to 30 percent of the dose in the urine 24 hr after topical administration (153, 159, 160). Preussmann et al. (158) reported even higher rate of urinary excretion (70 to 80 percent) within 24 hr after rats were given a cutaneous dose of *N*-nitrosodiethanolamine using either water or acetone as a vehicle.

2.3 Reproductive/Developmental Toxicity

Much of the information on the reproductive and teratogenic effects of *N*-nitrosamines is limited to transplacental carcinogenesis. Pregnant rats, at or near term, appear to be more sensitive to the acute effects of *N*-nitrosamines (NDMA) than when they were administered earlier in the pregnancy (207). Teratogenic effects include the induction of kidney tumors in the offspring of pregnant rats exposed during the last week of pregnancy (208). Single or repeated infections of up to 75 mg/kg body weight of NDMA during the last days of pregnancy resulted in lung adenomas and hepatomas in the offspring (209).

NDEA has been shown to act as a transplacental carcinogen when pregnant mice were treated intraperitoneally on days 16 or 19 with up to 51 mg/kg. As was noted with NDMA, NDEA had no effect when mice were exposed earlier in the pregnancy, for example, at 16 days, but induced a significant increase in liver and lung tumors when exposed at day 19 (210). In other experiments, NDEA, when administered to pregnant rats intraplacentally at single doses of up to 180 mg, did not elicit teratogenic effects. Repeated intraplacental doses of 0.1 to 0.3 mg/kg in contrast resulted in increased fetal mortality (144). These observations are in agreement with studies showing that animal and human fetal tissues, though capable of metabolizing low-molecular-weight chemicals including *N*-nitrosamines, appear to do so substantially more near the end of gestation in rodents (210).

Experimental attempts to induce effects in the fetuses of species other than rodents following exposure to *N*-nitrosamines including NDMA and NDEA have resulted in equivocal results with predominant effects being liver tumors. In fish, no tumors were induced in trout 9 months after intraplacental exposure to 100 μg NDMA per egg (211). Equivocal results were noted in Japanese medakas, suggesting that certain stages of development are more susceptible to the teratogenic effects of NDMA than others (212). Although no neoplastic changes were noted in the offspring of patas monkeys following administration of NDMA during pregnancy, subsequent exposures of the offspring to phenobarbital beginning at 4 years of age resulted in the rapid appearance of multiple hepatocellular tumors suggesting that in the case of *N*-nitrosamines such as NDMA postnatal exposure to promoting agents may greatly accelerate the carcinogenic process (213).

2.4 Mutagenicity

The general progression of mutagenicity testing of the N-nitrosamines and N-nitrosamides have been outlined by Magee and Barnes (214) and further refined by Montesano and Bartsch (215). Their combined historical perspective indicates that documentation of the mutagenic activity of the N-nitrosamides preceded documentation of the mutagenicity of the N-nitrosamines, owing to the nonenzymatic formation of DNA alkylating agents in biologic systems by the N-nitrosamides. This mechanism of action resulted in the detection of the mutagenicity of the N-nitrosamides in a wide range of assay systems. In contrast, the N-nitrosamines, generally requiring metabolic activation to the bioactive form, were active in only a limited number of mutagenicity assays, notably in assays using "whole animal" exposure to the N-nitrosamines, such as *Drosophila*- or mouse-based mutagenicity assays (214, 215).

Around 1970 several major advances in genetic toxicology testing methodology resulted in significantly enhanced detection of the mutagenic activity of the N-nitrosamines. These advances were (1) the use of the Udenfriend hydroxylation activation system to establish the mutagenicity of DMN and DEN in *Neurospora crassa* (216); (2) the development of the host-mediated assay (HMA) (217), which provided a host-based mammalian metabolic activation system for *Salmonella typhimurium*; and (3) the successful development of the in vitro *S. typhimurium* mutagenicity assay (218) using rat liver microsomal extracts, generally referred to as S9, for metabolic activation of mutagens to their bioactive form.

The importance of these major advances cannot be underestimated because they have, either singly or in combination, greatly enhanced our ability to detect mutagenic agents, especially N-nitrosamines. Malling (216) demonstrated the necessity of an activation system to activate (hydroxylate) N-nitrosamines to a mutagenic form in *Neurospora crassa*. Although the Udenfriend system was a nonenzymatic activation system, these studies were important from the standpoint that they combined a mutagenesis assay (*N. crassa*) that was nonresponsive to N-nitrosamines with an activation system, to result in a mutagenesis assay that was able to detect mutagenic N-nitrosamines. Gabridge and Legator (217) combined the *S. typhimurium* G-46 mutation assay, previously shown to be inactive with N-nitrosamines, with an intact mammalian activation system in the host-mediated assay. The recovered *S. typhimurium* demonstrated mutagenicity of the N-nitrosamines, activated by the intact host metabolic activation system. Although many short-term assays employ the liver microsome preparation described below, other assays have used primary hepatocytes as metabolic activation systems, or have used these cells as the target cells in in vitro mutagenicity assays (219). Finally, the work of Ames and colleagues (218) combined the *S. typhimurium* mutagenicity assay with a mammalian liver-derived metabolic activation system, resulting in a completely in vitro assay (Ames assay) that was sensitive to the mutagenicity of a wide variety of chemicals, including N-nitrosamines and N-nitrosamides. The advantages of low cost, high degree of concordance to the results of carcinogenicity assays, and limited resources necessary to perform this assay have resulted in the Ames assay as a de facto standardized assay for mutagenicity testing. Although this work was significant in adapting the *S. typhimurium* mutation assay to detect the mutagenicity of *N*-nitrosamines, the S9 activation system has been successfully used in a variety of other short-term in vitro tests to activate metabolically chemicals to their mutagenic form, resulting in a variety of short-term mutagenicity assays investigating different genotoxicity end points (220).

Experience has shown, however, that the standard Ames assay and other short-term assays relying upon rat liver S9 fraction for metabolic activation have limited sensitivity to detect the mutagenicity of *N*-nitrosamines. When the results of these assays are compared to the results of carcinogenicity assays, the concordance of the assays, either singly or as a group, is low. To enhance the predictive value of these short-term assays, they have been modified to include the use of hamster or human liver as the source of S9 microsomal enzymes, and have utilized S9 prepared from the tissues of the known target organs of *N*-nitrosamine carcinogenicity (221, 222). Additional modifications have included increases in the amount of S9 used in the short-term tests, consideration of the age of the animals used for the S9 preparation, and modification of some of the Ames assay procedures, especially those that result in enhanced exposure of the bacteria to the *N*-nitrosamine (223–227).

The data presented in Table 11.9 are a summary of mutagenicity data for certain *N*-nitrosamines and *N*-nitrosamides. In this table, tests have been selected to provide information on a variety of mutagenicity end points, and test optimization strategies, but are not meant to be an exhaustive anthology of all mutagenicity testing conducted on *N*-nitrosamines or *N*-nitrosamides. In the Results column, for negative (non-mutagenic) results, the concentration indicated is the highest dose level tested. For positive (mutagenic) results, the concentration indicated is the lowest value to give a positive result under the criteria of the test.

The results of mutagenicity assays investigating the mutagenicity of *N*-nitrosamines and *N*-nitrosamides, taken singly, or as in a "battery approach," have indicated a high degree of concordance with the results of carcinogenicity assays for *N*-nitrosamines (228, 229, 215). These short-term mutagenicity assays have been demonstrated to be an accurate, repeatable, and standardized tool for routine screening of the mutagenic/carcinogenic potential of *N*-nitrosamines and *N*-nitrosamides.

2.5 Chronic Effects

The *N*-nitroso compounds as a class are potent animal carcinogens. Of 300 *N*-nitroso compounds tested for carcinogenicity, about 90 percent of these have shown carcinogenic activity in laboratory animals (312) in more than 30 species. A summary of tumor sites and routes of exposure are provided for selected *N*-nitrosamines in Table 11.10.

All those compounds listed are considered "reasonably anticipated" to pose a risk of cancer in humans according to the NTP (313). Many are recognized by IARC as having sufficient evidence to classify them as carcinogens in experimental animal studies (144).

Table 11.9. Mutagenicity of Certain *N*-Nitrosamines

N-Nitrosamine[a]	Assays[b]	Test System	Activation[c]	Dosage[d]	Result	Ref.
N-Nitrosodimethylamine CASRN 62-75-9	Ames assay	Multiple tester strains	Rat S9	1.35 μmol/plate	+	221
	Ames assay	TA1535	Rat S9	1000 μ/plate	+	230
	Ames assay	TA1535	Hamster S9	250 μg/plate	+	230
	Ames assay	TA100	Rat S9	Graphed data	+	231
	Ames assay	TA100 with rat or rabbit S9s	Nasal, lung, and liver S9s	100 μmol/plate	–	222
	Ames assay	TA100	Rat or hamster S9	Graphed data	+	232
	E. coli mutation assay	*E. coli* WP2 uvrA/pKM101	Rat or hamster S9	Graphed data	+	232
	E. coli reverse mutation assay	*E. coli* Sd-B(TC)	Rat S9	0.2 mmol/ml	+	233
	Yeast mutation assay	*S. cerevisiae*	Udenfriend system	100 μM/ml	+	234
	Fungal mutation assay	*N. crassa*	Udenfriend system	0.1 mM	+	216
	Mutation and SCE assays (mult. end pt.)	V79 cells and Ames strains	Rat/hamster hepatocytes	Assay dependent	+	235
	CHO/HGPRT assay	CHO-K1-BH4	Rat S9	150 μM	+	236
	CHO/HGPRT assay	CHO-K1-BH4	Rat S9	400 μM	+	237
	Ouabain-resistance mutation assay	Chinese hamster V79	Rat hepatocytes	1 μg/ml	+	238
	Ouabain-resistance mutation assay	Chinese hamster V79	Hamster hepatocytes	Graphed data	+	239
	8-Azaguanine mutation assay	Chinese hamster V79	Rat S15	10 mM	+	240
	Chromosomal aberrations (in vitro)	CHO-K1-BH4	Rat S9	1110 μM	+	236
	Chromosomal aberrations (in vivo/in vitro)	Rat lymphocyte	N/A	30 mg/kg (ip)	+	241
	Chromosomal aberrations (in vivo)	Chinese hamster liver	N/A	5 g/kg (ip)	+	242
	Sister chromatic exchange (in vitro)	CHO	Rat S9	2.7 mM	+	243
	Sister chromatic exchange (in vitro)	CHO-K1-BH4	Rat S9	33.8 μM	+	236
	Sister chromatic exchange (in vitro)	Chinese hamster V79	Rat hepatocytes	1 μM	+	244
	Sister chromatic exchange (in vitro)	Chinese hamster V79 or CHO	Human hepatocytes	0.01–0.1 mM	+	245
	Sister chromatid exchange (in vivo/in vitro)	B6D2F1 mice/lymphocytes	N/A	0.25 mM/kg	+	246
	Sister chromatid exchange (in vivo)	B6D2F1 mice/multicellular end pt.	N/A	0.03–0.27 mM/kg	+	246
	Unscheduled DNA synthesis assay	Rat hepatocyte	N/A	0.01 M	+	219
	Unscheduled DNA synthesis assay	Human fibroblast	Mouse S9	0.05 M	+	247
	Spermatocyte UDS assay	ddY Mouse	N/A	20 mg/kg (ip)	–	248
	Cell transformation assay (in vitro)	Fischer 344 rat	N/A	100 μg/ml	+	249
	Cell transformation assay (in vitro)	Rat liver epithelium	Feeder cell layer	0.1 μg/ml	+	250
	Cell transformation assay (in vivo/in vitro)	Syrian hamster embryo	N/A	60 mg/rat	+	251
	Mouse lymphomas assay	Wistar rat kidney	Rat S9	0.5 μg/ml	+	252
	Mouse lymphoma assay	L5178Y/TK +/−	Rat/hamster hepatocytes	Graphed data	+	253
	Micronucleus test	L5178Y/TK +/−	N/A	12.5 mg/kg (gav)	+	254
	Micronucleus test	Rat/hamster hepatocytes	N/A	17–40 mg/kg	+	255
	X-linked recessive lethal assay	Rat liver or esophagus	N/A	0.0005%	+	256
	Host-mediated mutation assay in rats	*Drosophila*	N/A	10 mM	+	257
	Host-mediated mutation assay in mice	*S. cerevisiae* D4	N/A	0.1% in 0.1 ml	+	217
	Host-mediated mutation assay in mice	*S. typhimurium* G-46	N/A	500 mg/kg (im)	+	258
	Host-mediated mutation assay in mice	*S. typhimurium* G-46	N/A	25 mg/kg	+	259

Table 11.9. (Continued)

N-Nitrosamine[a]	Assays[b]	Test System	Activation[c]	Dosage[d]	Result	Ref.
	Host-mediated mutation assay in mice	E. coli K12/343/113	N/A	0.81 mM/kg	+	260
	Host-mediated mutation assay in Drosophila	E. coli K12/343/113	N/A	Graphed data	+	261
	X-linked recessive lethal assay	Drosophila–larval stage	N/A	1–2.5 ppm	+	262
	X-linked recessive lethal assay	Drosophila melanogaster	N/A	500 ppm	+	263
	Drosophila wing spot test	Drosophila melanogaster	N/A	0.5 μM/vial	+	227
	Reciprocal translocation assay	Drosophila—larval stage	N/A	2 ppm	+	262
	Mammalian (mouse) spot test	(C57BL/6HanxT)F1	N/A	7.5 mg/kg (ip)	+	264
	Mouse dominant lethal assay	ICR/Ha Swiss	N/A	9 mg/kg (ip)	–	265
	Mouse dominant lethal assay	(101xC3H)F1	N/A	4.4 mg/kg (sc)	+	266
	Mouse specific locus test (Dlb-1 locus)	(C57BL/6JxSWR)F	N/A	0.001% po (8 wks)	+	267
N-Nitrosodiethylamine CASRN 55-18-5	Ames assay	Multiple tester strains	Rat S9	250 μg/plate	+	268
	Ames assay	Multiple tester strains	Rat S9	10 μmol/plate	+	221
	Ames assay	TA100	Rat S9	Graphed data	+	231
	Ames assay	TA1535	Rat S9	To 1000 μg/plate	–	230
	Ames assay	TA1535	Hamster S9	100 μg/plate	+	230
	Ames assay	TA100 with rat or rabbit S9s	Nasal, lung, and liver S9s	100 μmol/plate	–	222
	E. coli mutation assay	E. coli WP2 uvrA/pKM101	Rat or hamster S9	Graphed data	+	232
	E. coli reverse mutation assay	E. coli Sd-B(TC)	Rat or hamster S9	Graphed data	+	232
	Mitotic recombination—yeast	S. cerevisiae D3	Rat S9	0.1 mmol/ml	+	233
	Yeast mutation assay	S. cerevisiae	Udenfriend system	3%	+	269
	Fungal mutation assay	N. crassa	Udenfriend system	100 mM/ml	+	234
	Ouabain-resistance mutation assay	Chinese hamster V79	Rat hepatocytes	0.1 mM	+	216
	8-Azaguanine mutation assay	Chinese hamster V79	Rat S15	10 μg/ml	+	238
	CHO/HGPRT assay	CHO-K1-BH4	Rat S9	10 mM	+	240
	Chromosomal aberrations (in vitro)	CHO	Rat S9	290 μM	+	237
	Sister chromatid exchange (in vitro)	CHO	Rat S9	4 mM	+	243
	Cell transformation assay (in vitro)	Syrian hamster embryo	Feeder cell layer	4 mM	+	243
	Cell transformation assay (in vitro)	Syrian hamster embryo	Feeder cell layer	To 0.1%	–	250
	Cell transformation assay (in vivo/in vitro)	Syrian hamster embryo (in utero)	N/A	100 μg/ml	–	270
	Unscheduled DNA synthesis assay	Rat hepatocyte	N/A	0.08 g/kg ip	+	270
	Spermatocyte UDS assay	Fischer 344 Rat	N/A	1 mM	+	271
	Mouse lymphoma assay	L5178Y/TK +/–	Rat hepatocytes or S9	450 mg/kg (ip)	–	248
	Mouse lymphoma assay	L5178Y/TK +/–	Rat/hamster hepatocytes	50 μg/ml	+	252
	Micronucleus test	Rat liver or esophagus	N/A	Graphed data	+	253
	Host-mediated mutation assay in Drosophila	E. coli K12/343/113	N/A	33.5 mg/kg	+	255
	Host-mediated mutation assay in mice	S. typhimurium G-46	N/A	Graphed data	+	261
	Host-mediated mutation assay in mice	S. typhimurium (TA1530, 1535)	N/A	To 400 mg/kg	–	259
	Host-mediated mutation assay in mice	E. coli K12/343/113	N/A	To 3.77 g/kg	–	269
	Host-mediated mutation assay in mice	E. coli K12/343/113	N/A	0.68 mM/kg	+	260
	Host-mediated mutation assay in rats	S. cerevisiae (ade2 & trp5 loci)	N/A	6.3 mM (sc)	+	272

Compound	Assay	Test system	Metabolic activation	Dose	Result	Ref.
N-Nitrosodipropylamine CASRN 621-64-7						
	X-linked recessive lethal assay	Drosophila melanogaster	N/A	0.01%	+	256
	Drosophila wing spot test	Drosophila melanogaster	N/A	1.0 μM/vial	+	227
	Drosophila wing spot test (SMART)	Drosophila-ORR crosses	N/A	1.0 mM	+	273
	X-linked recessive lethal assay	Drosophila melanogaster	N/A	0.10 mM	+	274
	Mammalian (mouse) spot test	(TxHT)F1	N/A	5 mg/kg (ip)	+	264
	Dominant lethal assay	(101xC3H)F1	N/A	13.5 mg/kg (ip)	−	266
	Mouse specific-locus assay (visible)	Mouse	N/A	119 mg/kg (ip)	+	275
	Ames assay	Multiple tester strains	Varied	Graphed data	+	231
	Ames assay	TA1530	Rat or human S9	0.1 μmol/plate	+	221
	Ames assay	TA1535	Rat S9	500 μg/plate	+	230
	Ames assay	TA100	Rat S9	Graphed data	+	231
	Ames assay	TA100 and rat/rabbit S9s	Rat nasal only	100 μmol/plate	+	222
	Ames assay	TA100	Rat or hamster S9	Graphed data	+	232
	E. coli mutation assay	E. coli WP2 uvrA/pKM101	Rat or hamster S9	Graphed data	+	232
	E. coli reverse mutation assay	E. coli Sd-B(TC)	Rat S9	0.03 mmol/ml	+	233
	Ouabain-resistance mutation assay	Chinese hamster V79	Hamster hepatocytes	Graphed data	+	235
	8-Azaguanine mutation assay	Chinese hamster V79	Rat S15	20 mM	+	240
	Unscheduled DNA synthesis assay	ACI rats	Rat hepatocytes	1.0 μM	+	276
	Mouse lymphoma assay	L5178Y/TK +/−	Rat/hamster hepatocytes	Graphed data	+	253
	Mouse lymphoma assay	L5178Y/TK +/−	Rat hepatocytes or S9	100 μg/ml	+	252
	Host-mediated mutation assay in Drosophila	E. coli K12/343/113	N/A	Graphed data	+	261
N-Nitrosodibutylamine CASRN 924-16-3						
	Ames assay	TA100, TA1530, TA1535	Rat S9	0.5 μM/plate	+	277
	Ames assay	TA1530	Rat or human S9	0.5 μmol/plate	+	221
	Ames assay	TA1535	Rat S9	10 μg/plate	+	230
	Ames assay	TA1535	Hamster S9	50 μg/plate	+	230
	Ames assay	TA100	Rat or hamster S9	Graphed data	+	232
	E. coli mutation assay	E. coli WP2 uvrA/pKM101	Rat or hamster S9	Graphed data	+	232
	E. coli reverse mutation assay	E. coli SD-B(TC)	Rat S9	0.003 mmol/ml	+	233
	Ouabain-resistance mutation assay	Chinese hamster V79	Hamster hepatocytes	Graphed data	+	239
	8-Azaguanine mutation assay	Chinese hamster V79	Rat S15	0.5 mM	+	240
	Host-mediated mutation assay in rats	S. cerevisiae (ade2 & trp5 loci)	N/A	4.6 mM (ip)	+	272
	Drosophila wing spot test	Drosophila melanogaster	N/A	10.0 μM/vial	+	227
N-Nitrosopyrrolidine CASRN 930-55-2						
	Ames assay	TA1530	Rat S9	1 μmol/plate	+	221
	Ames assay	TA1535	Rat S9	1 mg/plate	+	230
	Ames assay	TA100 and rat/rabbit S9s	Nasal, lung, and liver S9s	100 μmol/plate	+	222
	E. coli reverse mutation assay	E. coli WU3610	Rat S9	16.7 mM	+	278
	8-Azaguanine mutation assay	Chinese hamster V79	Rat S9	10 mM	+	240
	CHO/HGPRT assay	CHO-K1-BH4	Rat S9	5 mM	+	237
	Unscheduled DNA synthesis assay	Rat hepatocyte	N/A	1 mM	+	271
	Host-mediated mutation assay in Drosophila	E. coli K12/343/113	N/A	Graphed data	+	261
	Drosophila wing spot test	Drosophila melanogaster	N/A	10 μM/vial	+	227
N-Nitrosoproline CASRN 7519-36-0						
	Ames assay	TA1535	Rat S9	Graphed data	−	279
	Wing spot test	Drosophila	N/A	To 100 μM/vial	−	227

Table 11.9. (Continued)

N-Nitrosamine[a]	Assays[b]	Test System	Activation[c]	Dosage[d]	Result	Ref.
N-Nitrosopiperidine CASRN 100-75-4	Ames assay	TA1530	Rat S9	1 μmol/plate	+	221
	Ames assay	TA1535	Rat S9	50 μg/plate	+	280
	Ames assay	TA1535	Rat S9	100 μg/plate	+	230
	Ames assay	TA100 and rat/rabbit S9s	Nasal, lung, and liver S9s	100 μM/plate	+	222
	Ames assay	TA100	Rat or hamster S9	Graphed data	+	232
	E. coli mutation assay	E. coli WP2 uvrA/pKM101	Rat or hamster S9	Graphed data	+	232
	E. coli reverse mutation assay	E. coli Sd-B(TC)	Rat S9	0.67 mmol/ml	+	233
	E. coli reverse mutation assay	E. coli WU3610	Rat S9	16.7 mM	+	278
	Unscheduled DNA synthesis assay	Syrian hamster embryo	Rat PMS	1 μg/ml	+	281
	Cell transformation assay (in vitro)	Syrian hamster embryo	Rat PMS	1 μg/ml	+	281
	Cell transformation assay (in vitro)	Syrian hamster embryo	Rat PMS	0.1 μg/ml	+	250
	Sister chromatid exchange (in vitro)	Human lymphocyte	Rat S9	0.01 M	+	237
	Host-mediated mutation assay in rats	S. cerevisiae (ade2 & trp5 loci)	N/A	5.7 mM (sc)	+	272
	X-linked recessive lethal assay	Drosophila melanogaster	N/A	Graphed data	+	256
	Drosophila wing spot test	Drosophila melanogaster	N/A	10.0 μM/vial	+	227
N-Nitrosomorpholine CASRN 59-89-2	Ames assay	TA1530	Rat S9	5 μmol/plate	+	221
	Ames assay	TA1535	Rat S9	500 μg/plate	+	230
	Ames assay	TA1535	Hamster S9	10 μg/plate	+	230
	Ames assay	TA100	Rat or hamster S9	Graphed data	+	232
	E. coli mutation assay	E. coli WP2 uvrA/pKM101	Rat or hamster S9	Graphed data	+	232
	E. coli reverse mutation assay	E. coli Sd-B(TC)	Rat S9	0.1 mmol/ml	+	233
	E. coli mutation assay	E. coli	Rat S9	16.7 mM	+	278
	8-Azaguanine mutation assay	Chinese hamster V79	Rat S15	10 mM	+	240
	CHO/HGPRT assay	CHO-K1-BH4	Rat S9	500 μM	+	237
	Sister chromatid exchange (in vitro)	Chinese hamster V79	Rat S9	10 mM	+	244
	Sister chromatid exchange (in vitro)	B6C3F1 mice/lymphocytes	Rat hepatocytes	75 mg/kg	+	256
	Unscheduled DNA synthesis assay	Rat hepatocyte	N/A	0.001 M	+	271
	Host-mediated mutation assay (mouse)	S. typhimurium G-46	N/A	100 mg/kg (im/po)	+	282
	Host-mediated mutation assay (Drosophila)	E. coli K12/343/113	N/A	Graphed data	+	261
	Drosophila wing spot test	Drosophila melanogaster	N/A	5.0 μM/vial	+	227
	Heritable translocation assay	Drosophila melanogaster	N/A	1% in 1 ml (ip)	+	283
	X-linked recessive lethal assay	Drosophila melanogaster	N/A	1% in 1 ml (ip)	+	283, 284
	X-linked recessive lethal assay	Drosophila melanogaster	N/A	4 mM	+	285
	Mouse dominant lethal assay	(C57BLxBALB/C)F1	N/A	To 100 mg/kg (ip)	−	286
N-Nitrosodiethanolamine CASRN 1116-54-7	Ames assay	TA100	Rat or hamster S9	1 to 5 mg/plate	+	289
	Ames assay	TA1535	Rat or hamster S9	To 1 mg/plate	−	230
	Ames assay	Various strains	Rat S9	10 mg/plate	−	290
	Unscheduled DNA synthesis assay	Rat hepatocyte	N/A	Graphed data	+	291
	Sister chromatid exchange (in vitro)	Human lymphocyte	N/A	26.6 mM	+	292
	Chromosomal aberrations (in vitro)	Human lymphocyte	N/A	103.1 mM	+	292

Compound	Assay	Strain/Cell	Activation	Dose	Result	Reference
	Chromosomal aberrations (in vivo)	BALB/c mouse—bone marrow	N/A	10 g/kg (ip)	−	290
	Micronucleus test (in vitro)	Human lymphocyte	N/A	65.4 mM	+	292
	Micronucleus test (in vivo)	BALB/c Mouse—bone marrow	N/A	10 g/kg (ip)	−	290
	Host-mediated mutation assay in mice	*E. coli* K12/343/113	N/A	0.75 g/kg	+	293
	Host-mediated mutation assay in mice	*E. coli* K12/343/113	N/A	0.45 mM/kg	+	260
	Host-mediated mutation assay in *Drosophila*	*E. coli* K12/343/113	N/A	Graphed data	+	261
N-Nitrosodiphenylamine CASRN 86-30-6	Ames assay	Multiple tester strains	Rat S9	To 250 µg/plae	−	268
	Ames assay	TA1530	Rat S9	To 1 µmol/plate	−	221
	Ames assay	TA1535	Rat S9	To 1 mg/plate	−	230
	Ames assay	TA1535	Hamster S9	To 1 mg/plate	−	230
	Ames assay	TA100	Rat or hamster S9	Graphed data	+	232
	Ames assay	TA104	Rat or hamster S9	1 mM	+	294
	E. coli mutation assay	*E. coli* WP2 uvrA/pKM101	Rat or hamster S9	Graphed data	+	232
	E. coli mutation assay	*E. coli* (pol locus)	Rat S9	250 µg/ml	+	295
	E. coli reverse mutation assay	*E. coli* Sd-B(TC)	Rat S9	0.016 mmol/ml	−	233
	Mitotic recombination—yeast	*S. cereviae* D3	Rat S9	5%	−	269
	Fungal mutation assay	*N. crassa*	None indicated	1 mM	−	296
	8-Azaguanine mutation assay	Chinese hamster V79	Rat S15	0.5 mM	−	240
	Sister chromatid exchange (in vitro)	Chinese hamster—Don	None indicated	10 mM	+	297
	Unscheduled DNA synthesis assay	Rat hepatocyte	N/A	1 mM	−	271
	Cell transformation assay	Fischer rat embryo	Rat S9	To 1 µg/ml	+	298
	Cell transformation assay (in vitro)	Syrian hamster embryo	Feeder cell layer	To 10 µg/ml	+	250
	Cell transformation assay (in vitro)	BHK-21/Cl 13	Rat S9	To 250 µg/ml	+	299
	Mouse lymphoma assay	L5178Y/TK +/−	Rat S9	100 µg/ml	+	252
	Host-mediated mutation assay in *Drosophila*	*E. coli* K12/343/113	N/A	Graphed data	+	261
	Host-mediated mutation assay in mice	*S. typhimurium* (TA1530,1535)	N/A	To 5 g/kg	+	269
	Drosophila wing spot test	*Drosophila*	N/A	To 100 µM/vial	+	227
	X-linked recessive lethal assay	*Drosophila*	N/A	4 mM	−	285
N-Nitroso-*N*-dibenzylamine CASRN 5336-53-8	Ames assay	TA100	Hamster S9	1000 µg/plate	+	300
	Ames assay	TA98 and TA100	Rat S9	100–200 µg/plate	+	301
	Thioguanine-resistance mutation assay	Chinese hamster V79	Rat hepatocytes	10 µg/ml	+	300
	Sister chromatid exchange (in vitro)	Chinese hamster V79	Hamster hepatocytes	25 µg/ml	+	300
	Unscheduled DNA synthesis assay	Rat hepatocyte	N/A	To 750 mg/kg (gav)	−	300
	Micronucleus test	Rat bone marrow	N/A	500 mg/kg (gav)	+	300
N-Nitrososarcosine CASRN 13256-22-9	Host-mediated mutation assay in mice	*S. typhimurium* G-46	N/A	1 g/kg	−	258
N-Nitrosonornicotine CASRN 53759-22-1	Ames assay	TA100	Rat S9	2.2 µm/plate	+	287
	Ames assay	TA1530	Rat S9	5.7 µm/plate	+	287
	Ames assay	TA1535	Rat S9	500 µg/plate	+	230
	Ames assay	TA100	Rat S9	250 µg/plate	+	288
	Unscheduled DNA synthesis assay	Rat hepatocyte	N/A	1 mM	+	271
	Chromosomal aberrations (in vitro)	Human lymphocytes	N/A	100 µg/ml	−	288
	Sister chromatid exchange (in vitro)	Human lymphocytes	N/A	100 µg/ml	−	288
	Micronucleus test	Swiss mouse—bone marrow	N/A	250 mg/kg	+	288

Table 11.9. (Continued)

N-Nitrosamine[a]	Assays[b]	Test System	Activation[c]	Dosage[d]	Result	Ref.
4-(Methylnitrosamino)-1-(3-pyridyl)-1-butanone CASRN 64091-91-4	Ames assay	TA100	Rat S9	250 µg/plate	+	288
	Chromosomal aberrations (in vitro)	Human lymphocytes	N/A	100 µg/ml	+	288
	Sister chromatid exchange (in vitro)	Human lymphocytes	N/A	100 µg/ml	+	288
	Unscheduled DNA synthesis assay	Rat hepatocyte	N/A	1 mM	−	271
	Micronucleus test	Swiss mouse–bone marrow	N/A	250 mg/kg	+	288
	Micronucleus test—Syrian hamsters	Fetal liver cells	N/A	50 mg/kg	+	311
	Micronucleus test—Syrian hamsters	Fetal liver cells	N/A	50 mg/kg	+	311
	Micronucleus test—Syrian hamsters	Maternal bone marrow	N/A	25 mg/kg	−	311
	Micronucleus test—Syrian hamsters	Maternal bone marrow	N/A	25 mg/kg	−	311
N-Nitroso-N-ethylurea CASRN 759-73-9	Ames assay	Multiple tester strains	Rat S9	250 µg/plate	+	268
	Ames assay	TA1535	Rat S9	10 µg/plate	+	230
	Ames assay	TA1535	No activation	10 µg/plate	+	230
	Mitotic recombination—yeast	S. cereviae D3		0.4%	+	269
	E. coli reverse mutation assay	E. coli	Rat S9	0.5 mM	+	302
	Chromosomal aberrations (in vitro)	Human lymphocyte	N/A	1 mM	+	303
	Chromosomal aberrations (in vitro)	Human lymphocyte	N/A	25 µg/ml	+	304
	Chromosomal aberrations (in vivo)	Mouse/rat bone marrow	N/A	100 mg/kg (ip)	+	304
	Cell transformation assay	Syrian hamster embryo	Feeder cell layer	0.5 µg/ml	+	250
	Spermatocyte UDS assay	Fischer 344 rat	N/A	150 mg/kg (ip)	+	248
	Host-mediated mutation assay in mice	S. typhimurium (TA1530, 1535)	N/A	12 mg/kg	+	269
	X-linked recessive lethal assay	Drosophila melanogaster	N/A	4–5 mg	+	305
N-Nitroso-N-methylurea CASRN 684-93-5	Ames assay	Multiple strains	Rat S9	Graphed data	+	215
	Ames assay	TA1535	Rat S9	250 µg/plate	+	230
	Ames assay	TA1535	No activation	10 µg/plate	+	230
	E. coli reverse mutation assay	E. coli	Rat S9	0.4 mM	+	302
	Sister chromatid exchange (in vitro)	Chinese hamster—Don	None indicated	1 mM	+	297
	Sister chromatid exchange (in vivo/in vitro)	B6D2F1 mice/lymphocytes	N/A	0.19 mM/kg	+	246
	Sister chromatid exchange (in vivo)	B6D2F1 mice/multicellular end pt.	N/A	Graphed data	+	246
	Chromosomal aberrations (in vitro)	Mouse bone marrow	N/A	0.08 mM (ip)	+	306
	Chromosomal aberrations (in vitro)	Human lymphocyte	N/A	1 mM	+	303
	Unscheduled DNA synthesis assay	Multiple cell lines	N/A	125 µg/ml	+	307

Assay	Test system	Activation	Concentration[c]	Result[d]	Reference
Spermatocyte UDS assay	Fischer 344 rat	N/A	75 mg/kg (ip)	+	248
Cell transformation assay (in vitro)	Chinese hamster fibroblasts	None	10 µg/ml	+	308
Cell transformation assay	Rat embryos	None	60 µg/ml	+	309
Host-mediated mutation assay in rats	S. cerevisiae (ade2 & trp5 loci)	N/A	2.0 mM (sc)	+	272
Drosophila wing spot test	Drosophila melanogaster	N/A	0.25 µM/vial	+	227
Mouse dominant lethal assay	(C57BLxBALB/C)F1	N/A	50 mg/kg (ip)	−	286
Mouse dominant lethal assay	NMRI mice	N/A	5 mg/kg (sc)	+	266
Mouse dominant lethal assay	(102/ElxC3H/El)F1	N/A	70 mg/kg (ip)	+	310
Mouse specific-locus assay (visible)	(102/ElxC3H/El)F1	N/A	70 mg/kg (ip)	+	310

N-methyl-N'-nitroso-N-nitrosoguanidine
CASRN 70-25-7[a]

Assay	Test system	Activation	Concentration[c]	Result[d]	Reference
Ames assay	Multiple tester strains	Rat S9	2 µg/plate	+	268
Mitotic recombination—yeast	S. cerevisiae D3	Rat S9	5%	−	269
CHO/HGPRT assay	CHO-K1-BH4	Rat S9	0.06 µM	+	236
Chromosomal aberrations (in vitro)	CHO-K1-BH4	Rat S9	4.96 µM	+	236
Sister chromatid exchange (in vitro)	CHO-K1-BH4	Rat S9	0.18 µM	+	236
Mouse lymphoma assay	L5178Y/TK +/−	Rat S9	5 mg/ml	+	252
Cell transformation assay (in vitro)	Hamster embryo	None	0.5 µg/ml	+	270
Cell transformation assay (in vitro)	Syrian hamster embryo	None	0.5 µg/ml	+	250
Unscheduled DNA synthesis assay	Rat hepatocyte	N/A	50 µM	−	219
Spermatocyte UDS assay	Fischer 344 rat	N/A	150 mg/kg (ip)	+	248
Host-mediated mutation assay in mice	S. typhimurium TA1530, 1535)	N/A	5 mg/kg	+	269
Host-mediated mutation assay in rats	S. cerevisiae (ade2 & trp5 loci)	N/A	0.7 mM (sc)	+	272
Micronucleus test	ddY mouse	N/A	50 mg/kg (gav)	+	254
Drosophila wing spot test	Drosophila melanogaster	N/A	50 µM/vial	+	227
Mammalian (mouse) spot test	(C57BL/6Han x T)F1	N/A	35 mg/kg (ip)	+	264
Mouse dominant lethal assay	(C57BLxBALB/C)F1	N/A	50 mg/kg (ip)	+	286
Specific-locus assay (visible)	Mouse	N/A	250 mg/kg (ip)	+	275

[a] CASRN = Chemical Abstracts Service Registry Number.

[b] CHO = Chinese hamster ovary cells; HGPRT = hypoxanthine–guanine phosphoribosyl transferase (assay); UDS = unscheduled DNA synthesis (assay); SCE = sister chromatid exchange (assay).

[c] PMS = post-mitochondrial supernatant; N/A = not applicable to this test system.

[d] For negative (−) results, this number is the highest concentration tested; for positive (+) results, this number is the lowest concentration producing a positive response. Graphed data = study data was not tabulated, actual concentrations difficult to observe from graph.

ip = intraperitoneal
im = intramuscular
sc = subcutaneous
gav = gavage
po = peroral

Table 11.10. Summary of Tumorigenicity Data for Selected *N*-Nitrosamines[a]

Chemical	CASRN	Species	Route(s) of Administration	Tumor sites
N-Nitrosodi-*n*-butylamine	924-16-3	Rat, mouse, hamster, guinea pig, rabbit	Diet, drinking water, gavage, subcutaneous	Urinary bladder, forestomach, intestine, esophagus, pharynx, tongue, soft palate, trachea, lung, respiratory tract, liver, nasal cavity, pulmonary system, mammary gland, leukemia, fibrosarcomas
N-Nitrosodiethanolamine	1116-54-7	Rat, hamster	Oral, subcutaneous	Liver, kidney, nasal cavity, trachea, fibrosarcomas
N-Nitrosodiethylamine	55-18-5	Mouse, rat, hamster, guinea pig, rabbit, dog, pig, monkey, gerbil, hedgehog, parakeet	Oral, intragastric, skin, inhalation, intratracheal, intramuscular, intraperitoneal	Liver, esophagus, forestomach, lung, nasal cavity, trachea, kidney, brain, larynx, bile duct, leukemia, ovary
N-Nitrosodimethylamine	62-75-9	Mouse, rat, hamster, guinea pig, rabbit, mastomy, newts	Diet, drinking water, gavage, inhalation, intratracheal, subcutaneous, intramuscular, intraperitoneal	Liver, lung, kidney, glandular stomach, nasal cavity, abdominal tissue, forestomach
p-Nitrosodiphenylamine	156-10-5	Mouse, rat	Diet	Liver
N-Nitrosodi-*n*-propylamine	621-64-7	Rat, hamster	Drinking water, subcutaneous injection	Liver, tongue, esophagus, nasal, and/or paranasal, cavity, lung, kidney, laryngo-bronchial tract
N-Nitroso-*N*-ethylurea	759-73-9	Rat, opposum, mouse, monkey	Oral, drinking water, subcutaneous, intraperitoneal, intravenously	Brain, spinal cord, peripheral nervous system, eye, liver, kidney, muscle, jaw, stomach, large intestine, mammary gland, leukemia, lung, ovary, harderian gland, lymphoreticular system, thymus, small intestine, uterus, vagina, bone, bone marrow, vascular endothelium, skin

Compound	CAS No.	Species	Route	Target organs/sites
N-Nitroso-*N*-methylurea	684-93-5	Rat, guinea pig, hamster, pig, monkey, mouse, gerbil	Diet, drinking water, intragastric, oral, dermal, intratracheal, subcutaneous, intraperitoneal, intravenous, urethrae catheterization, intrarectal, intracerebral injection	Forestomach, brain, nervous system, stomach pancreas, ear duct, leukemia, kidney, small and large intestine, skin, jaw, oral cavity, colon, mesenteric lymph nodes, oropharynx, esophagus, skin, leukemia, nasopharyngeal tube, pharynx, larynx, trachea, bronchi, thymus, myocardium, lung, spleen, lymph nodes, liver, bone marrow, pulmonary, peritoneal cavity, mammary gland, mesentery, peripheral nervous system, heart, oral cavity, sebaceous gland, spleen, bladder, distal colon and rectum, anal canal, large bowel
N-Nitrosomethylvinylamine	4549-40-0	Rat	Drinking water, inhalation	Esophagus, tongue, pharynx, nasal cavity, esophagus, ethmoturbinals
N-Nitrosomorpholine	59-89-2	Rat, mouse, hamster	Drinking water, oral subcutaneous, intravenous, intragastric, intraperitoneal	Liver, lung, kidney, ovary, nasal cavity, olfactory, analplastic carcinomas, adeno carcinomas, ethmoturbinals, neoplastic nodules
N-Nitrosonornicotine	1654-55-8	Rat, hamster, mice	Drinking water, diet, subcutaneous, intraperitoneal	Nasal cavity, esophagus, trachea, forestomach, liver, sensory tissue, lung, salivary gland
N-Nitrosopiperidine	100-75-4	Mouse, monkey, rat	Diet, drinking water, subcutaneous, intraperitoneal, intravenous	Forestomach, esophagus, liver, leukemia, lung, nasal cavity, respiratory system, tongue, palate, pharynx
N-Nitrosopyrrolidine	930-55-2	Mouse, rat	Drinking water	Lung, liver, leukemia, bile duct, olfactory
N-Nitrososarcosine	13256-22-9	Mouse, rat	Diet, drinking water, intraperitoneal	Nasal, small intestine, vagina, testis, kidney, skin, thymus, bladder, pancreas, esophagus, liver

*a*Adapted from National Toxicology Program, "Fifth Annual Report on Carcinogens" (313).

Organotropic effects, as organ specificity for the carcinogenic effect, are noted for these compounds (312). Tissues affected appear to depend upon the structure of the compound, the dosage, the route of administration, and the animal species. Changes in the alkyl chain length and the chains containing even or odd numbers of carbons have elicited different tumor types (314). Woo and Arcos (150) have noted that all symmetrically substituted dialkyl nitrosamines tested produced liver tumors. Di-*n*-butyl and di-*n*-amyl derivatives also produced bladder tumors and pulmonary tumors, respectively. Asymmetric dialkyl nitrosamines appeared to affect the esophagus. Of the cyclics, *N*-nitrosopyrrolidine induced liver tumors and *N*-nitrosopiperidine demonstrated organ specificity toward the esophagus of the rat (150).

In general, the predominant sites of tumor induction include the liver, esophagus, kidney, urinary bladder, nasal cavities, brain and nervous system, oral cavity, stomach, gut, pancreas, hematopoietic system, lungs, heart, and skin (312).

N-Nitrosodimethylamine has been shown to be carcinogenic in all animals species in which it has been tested, producing tumors in the liver, kidney, and lung by the oral route (315). Magee and Barnes (316) first demonstrated that *N*-nitrosodimethylamine was not only highly toxic, but a potent carcinogen, producing liver tumors in mice and rats following oral administration. Malignant primary hepatic tumors were produced in the rat by feeding DMNA (316, 150, 148).

The carcinogenic response is dose related. *N*-Nitrosodimethylamine administered to rats in low doses over a long period of time leads to liver cancer, whereas a single or few large doses have resulted in renal carcinomas (317). The lowest effective daily dose noted from chronic feeding studies has been determined to be 2 mg/kg for *N*-nitrosodimethylamine, 0.75 mg/kg for *N*-nitrosodiethylamine, and 10 mg/kg for *N*-nitrosopyrrolidine (318).

Several studies have also shown that a single dose may be effective. Subcutaneous injections of single doses of *N*-nitrosodiethylamine in rats at 1.25 mg/kg body weight produced one kidney tumor (319). *N*-nitrosamines induce malignant tumors at a number of anatomic sites, the principal neoplastic effects of long-term exposure of rodents appear to be those arising in the liver as seen with NDMA and of the esophagus with NDELA (320).

NYPR, unlike NDMA, appears to exert clear effects only on the rodent liver. By contrast, NPIP has been shown in rodents to exert significant effects on the liver, esophagus, and lower jaw which is similar to those for NDEA. Tumors of the liver, esophagus, and stomach have been observed in mice and hamsters exposed to NDEA. NDEA-induced tumors of the trachea also have been noted in hamsters.

The route of exposure, the dose, and the need for metabolic activation does appear to play a significant role in the type of cancer induced by certain *N*-nitrosamines. NDMA administered to rats via the oral route at low exposures can result in liver tumors (321). However, higher oral doses, and exposures via direct injection, can produce a higher incidence of renal malignancies or lung adenoma and adenocarcinoma (144). In contrast, NDMA via inhalation exposure induces malignant nasal cavity tumors in rats (322).

Studies aimed at elucidating the relative carcinogenic potency of *N*-nitrosamines demonstrate a spectrum of activities. NDEA, for example, induced esophageal cancer in 85 to 95 percent of rats at doses from 0.56 to 3.6 mg/(kg)/(day). The average time to death in these experiments ranged from 26 to 50 weeks (323). In contrast, NDPhA induced transitional cell carcinomas of the urinary bladder in 38 percent of the males and 86 percent of the females when receiving the highest dose or 200 mg/(kg)/(day) (323). Potency ranking of *N*-nitrosamines (NDEA = NDMA > NMOR > NPYR > NNN > NDELA > NDPhA), was established on the basis of structure–activity relationships, target organ effects, and tumor incidence data.

The relationship between dose and the induction of cancer was reviewed in a large study where rats, mice, and hamsters were exposed to NDMA, NDEA, NPYR, and NPIP at 16 different concentrations (325). The results of this study confirmed previous observations made by an earlier study showing the predominant induction of esophageal and liver tumors by NDMA and NDEA (321). In addition, a linear dose relationship for cancer of the liver occurred at extremely low doses of NDMA (325).

Comparison of the effect of age and a fixed duration by treatment on NDEA carcinogenicity show differences in the cumulative incidences among older animals. The data indicate that older animals are less likely to get esophageal and liver tumors than the younger animals. The higher incidence of lesions of the liver in this study also suggests that the liver is extremely susceptible to the carcinogenic effects of NDEA at a young age.

2.6 Epidemiology

No published studies, to date, have provided conclusive evidence of a possible causal relationship between *N*-nitrosamine exposures and cancer in humans. Evidence of cancer excesses in industrial populations where *N*-nitrosamines are known to occur, continues to fuel concerns about the possible involvement of these chemicals. However, because of the probable confounding effect of numerous exposures to other chemical agents in these populations, these studies do not provide adequate evidence of a causal relationship between *N*-nitrosamine exposure and cancer in humans.

Several authors have suggested a role for *N*-nitrosamines in excess cancer of the bladder, lung, stomach, or other sites noted among studies of workers in the rubber industry. Again, the multiple exposures present in that occupational setting do not permit conclusions regarding the role of any single etiologic agent (148). The exposure associated with rubber industry has been identified by IARC as posing a carcinogenic risk to man (326).

Occupational studies of workers with exposure to metalworking fluids have not demonstrated excess cancer risks. Spiegelhalder et al. have monitored urinary excretion of *N*-nitrosodiethanolamine in metal grinders (327).

Tobacco habits in humans such as chewing, smoking, and dipping snuff have been shown to cause similar cancers in humans, such as gum, buccal mucosa, oral cavity, and respiratory tract. Although the presence of *N*-nitrosamines in tobacco products may be an important consideration in the etiology of these human cancers

(328), the complex nature of tobacco and tobacco smoke, and the presence of known human carcinogens, makes it difficult to draw any conclusions about N-nitrosamines.

Based on the assumption that high nitrate and nitrite levels are indicative of high N-nitrosamine exposure, dietary surveys were conducted using a population in Lin-xian county in Northern China possessing a high incidence of esophageal cancer. A correlation between dietary nitroso compounds consumption (food source and drinking water) and mortality roles for esophageal cancer was observed. Levels of nitrate and nitrite in drinking water, food and saliva were higher in Lin-xian county than in Fan-xian county, where the mortality rates for esophageal cancer are seven times lower (328). Again, the potential contribution of other confounding factors was not ruled out.

Other studies also have examined possible correlations of nitrates in the diet and cancer (329). Variable results have been observed between dietary nitrate and nitrite and gastric cancer. However, the protective role of fresh fruits and vegetables, Vitamin C, carotenoids, and tocopherols cannot be ruled out.

2.7 Occupational Exposure

The industries with the most data available concerning occupational exposures were the tanning industry and rubber manufacturing.

NIOSH first reported occupational exposures to nitrosamines in 1979 (330). This was followed by a more comprehensive NIOSH survey of 40 different facilities that was published four years later (331). A survey of nitrosamine exposures at 19 different German facilities was also published that year (332). The studies examined a variety of conditions that might lead to nitrosamine formation.

In tanneries, the only substance that has been linked with nitrosamine formation is dimethylamine sulfate. It is used as an accelerator in the hide unhairing process. NIOSH measured area airborne exposure concentrations of NDMA in the range of 0.1 to 4.7 $\mu g/m^3$ produced in tanneries (331).

In general, the highest airborne exposure levels found in rubber manufacturing were associated with jobs involving processes that heated rubber stock. These included Banbury mixing, milling, extrusion, calendering, tire curing, and injection molding. Exposures in other areas such as administration, lunchrooms, tire-building, and shipping and warehouse operations were reported in some cases, indicating that residual nitrosamines from freshly cured stock could contribute to exposure. Exposures were also found in showrooms, installation areas and stock-rooms of retail tire stores (333–337).

The most comprehensive study of rubber worker exposure to nitrosamines was performed by NIOSH in a Health Hazard Evaluation (HHE) of a Kelly Springfield Tire Company plant located in Cumberland, Maryland (338). The study was conducted during four surveys of the plant in September 1979, November 1979, February 1980, and May 1980. Interim reports were distributed after each survey and a summary report was issued in December, 1984.

In the first and second surveys, airborne levels of NMOR were measured in the calendering area (over the rolls of a ventilated rubber mill) as high as 248 μg/m³. Air monitoring in selected plant areas detected various levels of NDMA (0.05–2.93 μg/m³), NPYR (ND–2.61 μg/m³), and NMOR (ND–109 μg/m³). Personal samples worn by workers detected levels of NDMA (ND–0.11 μg/m³) and NMOR (ND–1.61 μg/m³).

In the third survey, NIOSH conducted a pilot study to determine if nitrosamines could be detected in workers with potential exposure to nitrosamines. Earlier work by Fine et al. described *in vivo* testing of six students who ingested 170 gm of cooked bacon each. Detectable levels of NDMA and NDEA were found in their blood. Selected workers in high exposure (hot process) areas wore personal sampling pumps to measure nitrosamine exposure over the workshift. These workers provided blood, urine, and stool samples at the end of the workshift for nitrosamine analysis. Air sampling results detailed 8-hour exposures for these workers to NDMA (ND–0.18 μg/m³), NPYR (ND–0.23 μg/m³), and NMOR (0.44–18.2 μg/m³).

The 28 biological samples provided by the workers were analyzed for nitrosamines. No levels of nitrosamines could be detected. Detection limit for the biological samples was one part per billion. The urine specimens from two highly exposed workers were also examined for the presence of glucuronide-bound nitrosamines. Results were again negative. Also, nitrosodiethanolamine, a known metabolite of NMOR, was not detected in either urine sample of the highly exposed workers, as determined by personal sampling over the workshift.

In the fourth survey, four urine samples collected from production workers were analyzed for volatile nitrosamines and NDPhA. In no case was any nitrosamine found. Mutagenic testing of the urine specimens collected from nine nonsmoking workers showed no evidence of mutagenicity when subjected to the Ames assay (Salmonella reverse mutation test).

NIOSH also conducted an analysis of the causes of death for 503 employees known to the Kelly Springfield Tire Company Cumberland Plant to have died between 1968 and 1980. Results showed no significant difference in the proportion of deaths because of cancer among the 236 former employees who had ever worked in the factory departments where elevated levels of nitrosamines were measured compared to 267 individuals who had worked in the remaining factory departments or offices.

3 RISK ASSESSMENT

Quantitative risk assessment (QRA) of carcinogens is commonly used by governmental agencies to identify the range of risks posed by various concentrations or doses and to identify diminimus risk levels. Though debate continues over which specific methodologies and assumptions are appropriate, certain approaches have become standard for various regulatory agencies. In the United States, the Linearized Multistage Model (LMM) is employed by the Environmental Protection Agency (EPA) and other agencies, although other models such as the One-Hit, Weibull, Probit, and Multihit may be considered. However, interest in new physiologically based pharmacokinetic models (PB-PK) has grown as our knowledge that the differences between animals and humans in physiology, metabolism, and pharmacokinetics are other important considerations in estimating the cancer potency and exposure risks has matured. These new models provide a vehicle to reduce our reliance on worst-case assumptions and to increase the accuracy of our QRAs.

A number of risk assessments have been developed for N-nitrosamines under different conditions of exposure. Using the Weibull model, the EPA has calculated a level of 7 ppt as representing a 10^{-6} risk of cancer for NDMA in water used as a source for public drinking, fishing, or recreation (339). On the basis of intake, the state of California also has determined that 0.04 and 0.004 µg NDMA/day are the equivalent of 10^{-5} and 10^{-6} risks of cancer, respectively (340).

However, a recent risk assessment using the LMM has shown that substantial differences in the estimate can occur based on the data base and assumptions used to generate the value. Nestmann et al. (341) reported that exposure to 205.6 ppt NMDA in drinking water per day over a lifetime would lead to a 10^{-6} excess cancer risk.

QRAs also have been conducted for NDELA. The cancer risk from exposure to NDELA in metalworking fluids has been determined by the EPA. By employing the LMM, the EPA has calculated that routine occupational exposure to 10^{-5} mg/(kg)/(day) of NDELA results in an approximate 10^{-6} cancer risk (342). In the personal care products area, where products such as facial creams contain quantifiable amounts of NDELA, conservative estimates of cancer risk have been developed. Based on conservative use assumptions, an exposure level of approximately 0.0002 µg/(kg)/(day) NDELA is expected. At this level the risk of developing cancer is calculated to be 0.16×10^{-6} (148).

REFERENCES

1. P. N. Magee, R. Montesano, and R. Preussmann, "N-Nitroso Compounds and Related Carcinogens," in *Chemical Carcinogens*, ACS Monograph 173, C. E. Searle, Ed., American Chemical Society, Washington, DC, 1976, p. 491.

2. F. Klages and H. Sitz, *Chem. Ber.*, **96**, 2394 (1963).

3. M. Weissler, *Angew. Chem. Int. Ed.*, **13**, 743 (1974).

4. M. Weissler, *Tetrahedron Lett.*, 2575 (1975).

5. R. E. Lyle, J. E. Saaverdra, and G. G. Lyle, *Synthesis*, **7**, 462.

6. G. A. Olah, J. A. Olah, and N. A. Overchuk, *J. Org. Chem.*, **30**, 3373 (1965).

7. J. H. Boyer et al., *J. Chem. Soc. Chem. Commun.*, **11**, 715–716 (1988).

8. C. R. Noller, *Chemistry of Organic Compounds*, 3rd ed., Saunders, Philadelphia, 1966, p. 226.

9. D. Grosjean, *J. Air Waste Manag. Assoc.*, **41**(3), 306–311 (1991).

10. L. E. Ember, *Chem. Eng. News*, 20 (March 31, 1980).

11. D. H. Fine, *Adv. Environ. Sci. Technol.*, **10**, 39–123 (1980).

12. W. E. J. Phillips, *Food Cosmet. Toxicol.*, **9**, 219 (1971).

13. G. Ellen et al., *Food Additives Contaminants*, **7**(2), 207–221 (Mar/April 1990).

14. Ref. 1, p. 587.

15. E. Fine, ACS Meeting, March 22, 1977.

16. Ref. 1, pp. 591–595.

17. Ref. 1, pp. 592–593.

18. P. S. Patel and D. J. McDowell, U.S. Pat. No. 4,166,169 (1979).

19. C. W. Frank and C. M. Berry, "*N*-Nitrosamines," in Patty's *Industrial Hygiene and Toxicology*, 3d ed., G. D. Clayton and F. E. Clayton, Eds., Wiley, New York, 1981, pp. 3117–3133.

20. F. W. Yeager, N. N. Van Gulick, and G. A. Lasoski, *Am. Ind. Hyg. Assoc. J.*, **41**, 148–150 (Feb. 1980).

21. B. Spiegelhalder and R. Preussmann, in "N-Nitroso Compounds: Occurrence and Biological Effects," IARC Sci. Publ. No. 41, Lyon, 1982, pp. 231–243.

22. W. J. Nicholson, *Environ. Occup. Med.*, 609–619 (1983).

23. R. L. Frye, U.S. Pat. No. 3,488,721 (1970).

24. B. Spiegelhalder, in "*N*-Nitroso Compounds: Occurrence, Biological Effects and Relevance to Human Cancer," IARC Sci. Publ. No. 57, Lyon, 1984, pp. 937–942.

25. P. Ducos et al., *Environ. Res.*, **41**(1), 72–78 (1988).

26. F. Ender and L. Ceh., "Alkylierend wirkende Verbindungen, Wissenschaftliche Forschungsstelle im Vergand der Cigarettenindustrie," Hamburg, 1968, p. 83.

27. W. Fiddler, J. W. Pensabene, J. M. Foster, and R. A. Gates, *J. Food Safety*, **9**(4), 225–233 (1989).

28. N. Crosby, J. K. Foreman, J. F. Palframan, and R. Sawyer, *Nature*, **238**, 342 (1972).

29. Anonymous, *Food Chem. News*, 15 (March 1991).

30. N. P. Sen, B. Donaldson, J. R. Iyengar, and T. Panalaks, *Nature*, **241**, 473 (1973).

31. U. Freimuth and E. Glaser, *Nahrung*, **14**, 357 (1970).

32. N. P. Sen, *Food Cosmet. Toxicol.*, **10**, 219 (1972).

33. M. Ishidate, A. Tanimura, Y. Ito, A Sakay, H. Sakuta, T. Kawamura, K. Sakai, and F. Miyazana, in *Topics in Chemical Carcinogenesis*, W. Nakahara et al., Eds., 1972, p. 313.

34. N. P. Sen, W. F. Miles, B. Donaldson, T. Panalaks, and J. R. Iyengar, *Nature*, **245**, 104 (1973).

35. P. J. Groenen et al., in "Environmental *N*-Nitroso Compounds: Analysis and Formation," IARC Sci. Publ. No. 14, Lyon, 1976, p. 321.

36. N. P. Sen, J. R. Iyengar, W. F. Miles, and T. Panalaks, in "Environmental *N*-Nitroso Compounds: Analysis and Formation," IARC Sci. Publ. No. 19, Lyon, 1976, p. 133.

37. R. W. Stephany, J. Freudenthal, and P. L. Schuller, in "Environmental *N*-Nitroso Compounds: Analysis and Formation," IARC Sci. Publ. No. 19, Lyon, 1976, p. 343.

38. J. H. Dhont and C. Van Ingen, in "Environmental *N*-Nitroso Compounds: Analysis and Formation," IARC Sci. Publ. No. 19, Lyon, 1976, p. 355.

39. T. A. Gough and C. L. Walters, in "Environmental *N*-Nitroso Compounds: Analysis and Formation," IARC Sci. Publ. No. 19, Lyon, 1976, p. 195.

40. J. Kann, O. Tauts, K. Raja, and R. Kalve, in "Environmental *N*-Nitroso Compounds: Analysis and Formation," IARC Sci. Publ. No. 19, Lyon, 1976, p. 385.

41. D. C. Harvey et al., *J. Assoc. Offic. Anal. Chem.*, **59**, 540 (1976).

42. N. P. Sen, S. Seaman, and W. F. Miles, *Food Cosmet. Toxicol.*, **14**, 167 (1976).

43. J. J. Wartheson, D. D. Bills, R. A. Scanlan, and L. M. Libbey, *J. Agric. Food Chem.* **24**, 892 (1976).

44. A. Mirna et al., *Fleischwirtschaft*, **56**, 1014 (1976).

45. Kotter, A. Fischer, and H. Schmidt, *Fleischwirtschaft*, **56**, 997 (1976).

46. P. Cooper, *Food Cosmet. Toxicol.*, **14**, 205 (1976).

47. N. P. Sen et al., *J. Agric. Food Chem.*, **38**(4), 1007–1011 (Apr. 1990).

48. Y. Y. Fong, W. C. Chan, *Food Cosmet. Toxicol.*, **11**, 841 (1973).

49. T. Fazio, J. N. Damico, J. W. Howard, R. H. White, and J. O. Watts, *J. Agric. Food Chem.*, **19**, 250 (1971).

50. L. Hedler and P. Marquardt, *Food Cosmet. Toxicol.*, **6**, 341 (1968).

51. H. J. Petrowitz, *Arzneim. Forsch.*, **18**, 1486 (1968).

52. P. Marquardt, L. Hedler, *Arzneim. Forsch.*, **16**, 778 (1966).

53. E. Kroller, *Deut. Lebensm. Rundsch.*, **63**, 303 (1967).

54. B. H. Thewlis, *Food Cosmet. Toxicol.*, **5**, 333 (1967).

55. B. H. Thewlis, *Food Cosmet. Toxicol.*, **6**, 822 (1968).

56. K. Mohler and O. L. Mayrhofer, *Z. Lebensm. Unters. Forsch.*, **135**, 313 (1968).

57. D. Hoffman, G. Rathkamp, and Y. Y. Liu, in "*N*-Nitroso Compounds in the Environment," IARC Sci. Publ. No. 9, Lyon, 1974, p. 159.

58. D. Hoffman, S. S. Hecht, R. M. Ornaf, and E. L. Wynder, T. Tso, in "*N*-Nitroso Compounds in the Environment," IARC Sci. Publ. No. 9, Lyon, 1974, p. 307.

59. I. Schmeltz, S. Abidi, and D. Hoffman, *Cancer Lett.*, **2**, 125 (1976).

60. S. S. Hecht, R. M. Ornaf, and D. Hoffman, *J. Natl. Cancer Inst.*, **54**, 1237 (1975).

61. G. Neurath, B. Pirmann, and H. Wichern, *Beitr. Tabakforsch.*, **2**, 311 (1964).

62. G. Neurath, B. Pirmann, H. Wichern, and W. Luttich, *Beitr. Tabakforsch.*, **3**, 251 (1965).

63. L. Hedler et al., in "*N*-Nitroso Compounds: Analysis and Formation," IARC Sci. Publ. No. 3, Lyon, 1972, p. 71.

64. L. Hedler and P. Marquardt, in "*N*-Nitroso Compounds in the Environment," IARC Sci. Publ. No. 9, Lyon, 1974, p. 183.

65. S. J. Kubacki, D. C. Havery, and T. Fazio, *Food Additives Contam.*, **6**(1), 29–34 (Jan./March 1989).

66. R. A. Scanlan, J. F. Barbour, and C. I. Chappel, *J. Agric. Food Chem.*, **38**(2), 442–443 (1990).

67. R. Frommberger, *Food. Chem. Toxicol.*, **27**(1), 27–29 (1989).

68. Anonymous, *Chem. Eng. News*, 7 (March 28, 1977).

69. W. Fiddler et al., *Am. Ind. Hyg. Assoc.*, **46**(8), 463–465 (1985).

70. T. Y. Fan et al., *Science*, **196**, 70 (1977).

71. D. T. Williams et al., *Bull. Environ. Contam. Toxicol.*, **20**, 206–211 (1978).

72. E. G. Freudenthal, *J. R. Neth. Chem. Soc.*, **101**, 357–359 (1982).

73. D. H. Fine et al., paper presented at the 172nd ACS National Meeting, San Francisco, September 2, 1976.

74. D. C. Harvey, G. A. Perpetti, B. J. Canas, and T. Fazio, *Food Chem. Toxicol.*, **23**(11), 991–993 (1985).

75. E. Goff et al., SAE Prepr. No. 80, 374 (meeting of Oct. 1980).

76. R. Ziskind, *Los Angeles Counc. Eng. Sci. Proc. Ser.*, **4** (May 1978).

77. G. Choudhary, G. D. Foley, NIOSH, USG. 8713, PB87-163515, 1983.

78. J. M. Fajen, First NCI/EPA/NIOSH Collaborative Workshop, NIOSH 00123099, May 1980.

79. German Technical Regulations for Hazardous Materials (TRGS) *N*-Nitrosamines, TRGS No. 552, 1986.

80. D. P. Rounbehler, J. W. Reisch, J. R. Combs, D. H. Fine, and J. Fajen, ACS Symposium Series No. 149, 1981, pp. 343–356.

81. E. O. Benett and D. L. Bennett, *Tribology Int.*, **17**(6), 341–346 (Dec. 1984).

82. D. H. Fine, *Oncology*, **37**, 199–202 (1980).

83. C. Evans, *Tribology*, **11**, 47 (1977).

84. B. Spiegelhalder and R. Preussman, *Carcinogenesis*, **4**(a), 1147–1152 (1982).

85. B. A. Hollett et al., NIOSH Rep. No. HETA 81-045B-1216, Nov. 1982.

86. J. D. McGlothlin, J. M. Fajen, G. S. Edwards, ACS Symposium Series No. 149, 1981, pp. 283–299.

87. D. H. Fine, D. P. Rounbehler, A. Rounbehler, et al., *Environ. Sci. Technol.*, **11**, 581–584 (1977).

88. D. H. Fine, D. P. Rounbehler, E. D. Pellizari, et al., *Bull. Environ. Contam. Toxicol.*, **15**, 739–746 (1976).

89. D. Dropkin, Sampling of Automobile Interiors for Organic Emissions Report EPA 600/3-85-008 NITS PB85-172567/GARm, 1985, p. 29.

90. E. C. Tuazon, W. P. L. Carter, and R. Atkinson, *Environ. Sci. Technol.*, **18**, 49–54. (1984).

91. HSDB, Hazardous Substances Data Base, "Nitrosodiphenylamine," Toxicology Data Network, National Library of Medicine, Bethesda, MD, File 2875, 1990.

92. A. Ayanaba and M. Alexander, *Appl. Microbiol.*, **25**, 862–868 (1973).

93. EPA Storage and Retrieval Water Quality Data Base (STORET) Online: September 6, 1988, EPA, Washington, DC, 1988.

94. VIEW Database Agency for Toxic Substances and Disease Registry (ATSDR), Vol. 6, Office of External Affairs, Exposure and Disease Registry Branch, Atlanta, GA, 1989.

95. W. I. Kimoto, C. J. Dooley, and J. Carre, *Water Res.*, **15**, 1099–1106 (1981).

96. T. A. Gough, K. S. Webb, and M. F. McPhail, *Food Cosmet. Toxicol.*, **21**, 151–156 (1977).

97. Great Lakes Water Quality Board, "An Inventory of Chemical Substances Identified in the Great Lakes Ecosystem," Vol. 1, *Summary Report to the Great Lakes Water Quality Board*, 1983, Windsor, Ontario, Canada, 1983, pp. 1–195.

98. J. Polo and U. L. Chow, *J. Natl. Cancer Inst.*, **56**, 997–1001 (1976).

99. H. H. Tabak, S. A. Quave, and C. I. Maschni, *J. Water Pollut. Cont. Fed.*, **53**, 1503–1518 (1981).

100. M. G. MacNaughton and T. B. Stauffe, Natl. Technical Info. Services AD A020922, Vol. 3, 1975.

101. J. M. Fajen, "Industrial Hygiene Study of Workers Exposed to Nitrosamines," in Proceedings of the First NCI/EPA/NIOSH collaboration Workshop: Progress on Joint Environmental and Occupational Cancer Studies, Rockville, MD, 1980.

102. M. Kobayashi and Y. T. Tohan, *Water Res.*, **17**, 199–201 (1978).

103. M. A. B. Mallik and K. Tesfa, *Bull. Environ. Contam. Toxicol.*, **27**, 115–121 (1981).

104. A. L. Mills and M. Alexander, *J. Environ. Qual.*, **5**, 437–440 (1976).

105. M. A. B. Voets and K. Tesfai, *Bull. Environ. Contam. Toxicol.*, **27**, 115–121 (1981).

106. S. J. Eisenreich, B. B. Looney, and D. J. Thornton, *Environ. Science Technol.*, **15**, 30–31 (1981).

107. W. J. Lyman, W. F. Reehl, and D. H. Rosenblatt, in *Handbook of Property Estimation Methods*, McGraw Hill, New York, 1982, pp. 4–9, 15–16, 15–29.

108. R. L. Swann, D. A. Laskowski, and P. J. McCall, *Residue Rev.*, **85**, 17–28 (1983).

109. R. Preussmann, D. Daiber, and H. Hengy, *Nature*, **201**, 502 (1964).

110. R. Preussmann et al., *Z. Anal. Chem.*, **202**, 187 (1964).

111. C. L. Walters, E. M. Johnson, N. Ray, and G. Woolford, in "*N*-Nitroso Compounds: Analysis and Formation," IARC Sci. Publ. No. 3, Lyon, 1972, p. 79.

112. A. A. L. Gunatilake, *J. Chromatogr.*, **120**, 229 (1976).

113. F. L. English, *Anal. Chem.*, **23**, 344 (1951).

114. C. L. Walters, E. M. Johnson, and N. Ray, *Analyst*, **95**, 485 (1970).

115. C. L. Walters, *Lab. Prac.*, **20**, 574 (1971).

116. W. F. Smyth et al., *Anal. Chim. Acta*, **78**, 81 (1975).

117. A. A. Forist, *Anal. Chem.*, **36**, 1338 (1964).

118. G. Eisenbrand and R. Preussmann, *Arzneim-Forsch.*, **20**, 1513 (1970).

119. E. M. Johnson and C. L. Walters, *Analy. Lett.*, **4**, 383 (1973).

120. M. J. Downers, M. W. Edwards, T. S. Elsey, and C. L. Walters, *Analyst*, **101**, 742 (1976).

121. T. F. Kelly and J. R. Nunn, in "*N*-Nitroso Compounds in the Environment," IARC Sci. Publ. No. 9, Lyon, 1974, p. 26.

122. J. F. Palframan, J. McNab, and N. T. Crosby, *J. Chromatogr.*, **76**, 307 (1973).

123. E. Von Rappardt, G. Eisenbrand, and R. Preussmann, *J. Chromatogr.*, **124**, 247 (1976).

124. N. P. Sen, *J. Chromatogr.*, **51**, 301 (1970).

125. T. A. Gough and K. S. Webb, *J. Chromatogr.*, **64**, 201 (1972).

126. T. A. Gough and K. S. Webb, *J. Chromatogr.*, **95**, 59 (1974).

127. T. A. Gough and K. Sugden, *J. Chromatogr.*, **109**, 265 (1975).

128. R. W. Stephany et al., *J. Agric. Food Chem.*, **24**, 536 (1976).

129. E. D. Pellizzari et al., *Analy. Lett.*, **9**, 579 (1976).

130. G. B. Cox, *J. Chromatogr.*, **83**, 471 (1973).

131. H. J. Klimisch and D. Ambrosius, *J. Chromatogr.*, **121**, 93 (1976).

132. S. S. Hecht, J. D. Adams, and D. Hoffman, in "Environmental Carcinogens. Selected Methods of Analysis, Vol. 6, "*N*-Nitroso Compounds," IARC Sci. Publ. No. 45, Lyon, 1983, pp. 429–436.

133. R. C. Massey, P. E. Key, D. J. McWeeny, and M. E. Knowles, in "*N*-Nitroso Compounds: Occurrence, Biological Effects and Relevance to Human Cancer," IARC Sci. Publ. No. 57, Lyon, 1984, pp. 131–136.

134. C. Ruehl and J. Reusch, *J. Chromatogr.*, **326**, 362–355 (1985).

135. S. H. Lee and L. R. Field, *J. Chromatogr.*, **386**, 137–148 (1987).

136. N. P. Sen, S. W. Seamen, and S. C. Rushwaha, *J. Chromatogr.*, **463**(2), 419–428 (1989).

137. CFR Title 29, Part 1910.1016.

138. CFR Title 29, Part 1910.1200.

139. FSTRAC, Federal-State Toxicology and Regulatory Alliance Committee, "Summary of State and Federal Drinking Water Standards and Guidelines." Chemical Communication Subcommittee, March 10, 1988, p. 17.

140. CELDS, Computer Environmental Legislative Data System, University of Illinois, Urbana, IL, Nov. 28, 1990.

141. NATICH, National Air Toxics Information Clearinghouse NATICH Data Base Report on State, Local and EPA Air Toxics Activities, Washington, DC, 1987.

142. NATICH, National Air Toxics Information Clearinghouse NATICH Data Base Report on State, Local and EPA Air Toxics Activities, Washington, DC, 1991.

143. "Rationale Document for the Interim Maximum Acceptable Concentration for *N*-Nitrosodimethylamine (NDMA) in Drinking Water," Ontario, Canada, April, 1991.

144. *IARC Monograph*, Vol. 17, WHO, 1978.

145. S. Budavari, Ed., *Merck Index*, Rahway, NJ, 1989, pp. 49–1050.

146. J. A. Dean, Ed., *Lange's Handbook of Chemistry*, 12th ed., Cleveland, OH, 1989.

147. E. Behrle, Ed., *Beilstein Handbook of Organic Chemistry*, Berlin, 1942.

148. ECETOC (1990) Evaluation of Risk from Exposure to N-Nitrosodiethanolamine in Personal Care Products, Draft Document ISSN 07773-8072-##, Brussels, Belgium.

149. Freud (1937).

150. Y. T. Woo and J. C. Arcos, "Environmental Chemicals," In *Carcinogens in Industry and the Environment*, J. M. Sontag, Ed., 1981, pp. 200–204.

151. J. Hilfrich, I. Schmeltz, and D. Hoffmann, *Cancer Lett.*, **4**, 55 (1977).

152. E. J. Lethco, W. C. Wallace, and E. Brower, *Food Chem. Toxicol.*, **20**, 401 (1982).

153. E. B. Sansone, A. M. Losikoff, and W. Lijinsky, in "*N*-Nitroso Compounds: Analysis, Formation, and Occurrence," IARC Sci. Publ. No. 31, Lyon, 1980. p. 705.

154. R. Preussmann et al., *Cancer Lett.*, **4**, 207 (1978).

155. B. Spiegelhalder, G. Eisenbrand, and R. Preussmann, "*N*-Nitroso Compounds: Occurrence and Biological Effects," IARC Sci. Publ. No. 41, Lyon, 1982, p. 443.

156. M. I. Diaz-Gomez, P. F. Swann, and P. N. Magee, *Biochem. J.*, **164**, 497 (1977).

157. R. G. Klein and P. Schmezer, in "*N*-Nitroso Compounds: Occurrence, Biological Effects and Relevance to Human Cancer," IARC Sci. Publ. No. 57, Lyon, 1984, p. 513.

158. R. Preussmann et al., *Cancer Lett.*, **13**, 227 (1981).

159. W. Lijinsky, A. M. Losikoff, and E. B. Sansone, *J. Natl. Cancer Inst.*, **66**, 125 (1981).

160. L. Airoldi et al., *Food Chem. Toxicol.*, **22**, 133 (1984).

161. G. S. Edwards et al., *Toxicol. Lett.*, **4**, 217 (1979).

162. R. L. Bronaugh, E. R. Congdon, and R. J. Scheuplein, *J. Invest. Dermatol.*, **76**, 94 (1981).

163. J. P. Daugherty and N. K. Klapp, *Life Sci.*, **19**, 265 (1976).

164. P. N. Magee, *Biochem. J.*, **64**, 676 (1956).

165. E. B. Johansson and H. Tjalve, *Toxicol. Appl. Pharmacol.*, **45**, 565 (1978).

166. J. S. Wishnok et al., *Toxicol. Appl. Pharmacol.*, **43**, 391 (1978).

167. K. W. Johnson, A. E. Munson, and M. P. Holsapple, *Immunopharmacology*, **13**, 47 (1987).

168. L. M. Anderson et al., *Drug Metab. Dispos.*, **14**, 733 (1986).

169. J. Althoff et al., *Z. Krebsforsch. Klin. Onkol.*, **90**, 79 (1977).

170. J. Althoff and C. Grandjean, *Natl. Cancer Inst. Monogr.*, **51**, 251 (1979).

171. E. B. Brittebo and H. Tjalve, *Chem-Biol. Interact.*, **38**, 231 (1982).

172. B. Loefberg and H. Tjalve, *Food Chem. Toxicol.*, **23**, 647 (1985).

173. E. B. Brittebo, B. Loefberg, and H. Tjalve, *Chem-Biol. Interact.*, **34**, 209 (1981).

174. E. B. Brittebo, B. Loefberg, and H. Tjalve, *Xenobiotica*, **11**, 619 (1981).

175. B. Loefberg and H. Tjalve, *Acta Pharmacol. Toxicol.*, **54**, 104 (1984).

176. J. G. Farrelly, M. L. Stewart, and W. Lijinsky, *Carcinogenesis*, **5**, 1015 (1984).

177. J. A. J. Brouwers and P. Emmelot, *Exp. Cell Res.*, **19**, 467 (1960).

178. S. Magour and J. G. Nievel, *Biochem. J.*, **123**, 89 (1971).

179. J. C. Arcos et al., *Z. Krebsforsch. Klin. Onkol.*, **86**, 171 (1976).

180. K. K. Park and M. C. Archer, *Chem-Biol. Interact.*, **22**, 83 (1978).

181. K. K. Park, J. S. Wishnok, and M. C. Archer, *Chem-Biol. Interact.*, **18**, 349 (1977).

182. P. A. Bauman, J. J. Hotchkiss, and R. S. Parker, *Cancer Lett.*, **28**, 229 (1985).

183. K. H. Leung and M. C. Archer, *Carcinogenesis*, **2**, 859 (1981).

184. E. Suzuki and M. Okada, *Gann*, **72**, 552 (1981).

185. J. Althoff et al., *J. Natl. Cancer Inst.*, **58**, 439 (1977).

186. F. W. Kruger, *Z. Krebsforsch. Klin. Onkol.*, **76**, 145 (1971).

187. F. W. Kruger, *Z. Krebsforsch. Klin. Onkol.*, **79**, 90 (1973).

188. F. W. Kruger and B. Bertram, *Z. Krebsforsch. Klin. Onkol.*, **80**, 189 (1973).

189. K. H. Leung and M. C. Archer, *Chem-Biol. Interact.*, **48**, 169 (1984).

190. L. Blattmann and R. Preussman, *Z. Krebsforsch. Klin. Onkol.*, **79**, 3 (1973).

191. L. K. Keefer et al., *J. Natl. Cancer Inst.*, **51**, 299 (1987).

192. J. S. H. Yoo, F. P. Guengerich, and C. S. Yang, *Cancer Res.*, **88**, 1499 (1988).

193. H. R. Prassana et al., *Cancer Lett.*, **26**, 25 (1985).

194. R. Montessano, H. Bresil, and A. E. Pegg, "Metabolism of Dimethylnitrosamine and Repair of O-6-Methylguanine in DNA by Human Liver," in *Nitrosamines and Human Cancer*, Banbury Report, Vol. 12, P. N. Magee, Ed., Cold Spring Harbor Laboratory, Cold Spring Harbor, NY, 1982, pp. 141–152.

195. M. B. Kroeger-Koepke and C. J. Michejda, *Cancer Res.*, **39**, 1587 (1979).

196. P. D. Lotlikar, Y. S. Hong, and W. S. Baldy, Jr., *Cancer Lett.*, **4**, 355 (1978).

197. J. C. Phillips et al., *Food Cosmet. Toxicol.*, **13**, 203 (1975).

198. S. D. Hecht and R. Young, *Cancer Res.*, **41**, 5039 (1981).

199. K. D. Brunnemann, S. S. Hecht, and D. Hoffmann, *J. Clin. Toxicol.*, **19**, 661 (1983).

200. L. Airoldi et al., *Chem-Biol. Interact.*, **51**, 103 (1984).

201. W. Sterzel and G. Eisenbrand, *J. Cancer Res. Clin. Oncol.*, **11**, 20 (1986).

202. K. E. Appel et al., *Toxicol. Lett.*, **23**, 353 (1984).

203. K. Hemminki, *Chem-Biol. Interact.*, **39**, 139 (1982).

204. O. G. Rabbe, "Inhalation Uptake of Selected Chemical Vapors at Trace Levels," Report to California State Air Resources Board, Sacramento, NTIS PB86-209863, 1986.

205. P. F. Swann, A. M. Coe, and R. Mace, *Carcinogenesis*, **5**, 1337 (1984).

206. S. E. Atawodi and E. N. Maduagwu, *Eur. J. Drug Metab. Pharmacokinet.*, **15**, 27 (1990).

207. K. Nichia, *Food Chem. Toxicol.*, **21**(4), 453–462 (1983).

208. V. A. Alexander, *Vopr. Onkol.*, **36**(4), 387–395 (1990).

209. E. E. Smetanin, *Vopr. Onkol.*, **17**, 75–81 (1971).

210. L. M. Anderson, A. B. Jones, M. S. Miller, and D. P. Chauhan, in "Perinatal and Multigeneration Carcinogenesis," IARC Sci. Publ. No. 96, Lyon, 1989, pp. 155–188.

211. J. J. Black, A. E. Maccubbin, and M. Schiffert, NTIS Technical Report-NTIS/PB90-146267 (1985).

212. G.D. Marty, J. M. Nunez, D. J. Lauren, and D. E. Hinton, *Aquatic Toxicol.*, **17**(1), 45–62 (1990).

213. J. M. Rice, S. Rehm, P. J. Donovan, and A. O. Peratoni, in "Perinatal and Multigeneration Carcinogenesis," IARC Sci. Publ. No. 96, Lyon, 1989, pp. 17–34.

214. P. N. Magee and J. M. Barnes, *Adv. Cancer Res.*, **10**, 163–246 (1967).

215. R. Montesano and H. Bartsch, *Mutat. Res.*, **32**, 179–228 (1976).

216. H. V. Malling, *Mutat. Res.*, **3**, 537–540 (1966).

217. M. G. Gabridge and M. S. Legator, *Proc. Soc. Exp. Biol.*, **130**, 831–834 (1969).

218. B. N. Ames, J. McCann, and E. Yamasaki, *Mutat. Res.*, **31**, 347–364 (1975).

219. G. M. Williams, *Cancer Res.*, **37**, 1845–1851 (1977).

220. D. Brusick, *Environ. Mol. Mut.*, **14** (Suppl. 16), 60–65 (1989).

221. H. Bartsch, C. Malaveille, A. Camus, G. Martel-Planche, G. Brun, A. Hautefeuille, N. Sabadie, A. Barbin, T. Kuroki, C. Drevon, C. Piccoli, and R. Montesano, *Mutat. Res.*, **76**, 1–50 (1980).

222. A. R. Dahl, *Mutat. Res.*, **158**, 141–147 (1985).

223. J. B. Guttenplan, *Mutat. Res.*, **64**, 91–94 (1979).

224. T. K. Rao, D. W. Ramey, W. Lijinsky, and J. L. Epler, *Mutat. Res.*, **67**, 21–26 (1979).

225. M. Zielenska and J. B. Guttenplan, *Mutat. Res.*, **180**, 11–20 (1987).

226. T. Yahagi, M. Nagao, Y. Seino, T. Matsushima, T. Sugimura, and M. Okado, *Mutat. Res.*, **48**, 121–130 (1977).

227. T. Negishi, T. Shiotani, K. Fujikawa, and H. Hayatsu, *Mutat. Res.*, **252**, 119–128 (1991).

228. J. B. Guttenplan, *Mutat. Res.*, **186**, 81–134 (1987).

229. W. Lijinsky, *Mol. Toxicol.*, **1**, 107–199 (1987).

230. A. W. Andrews and W. Lijinsky, "*N*-Nitrosamine Mutagenicity Using the Salmonella/Mammalian-Microsome Mutagenicity Assay," in *Topics in Chemical Mutagenesis*, Vol. 1, *Genotoxicity of N-Nitroso Compounds*, T. K. Rao, W. Lijinsky, and J. L. Epler, Eds., 1984, pp. 13–43.

231. T. Sugimura, T. Yahagi, M. Nagoa, M. Takeuchi, T. Kawachi, K. Hara, E. Yamasaki,

T. Matsushima, Y. Hashimoto, and M. Okada, in "Screening Tests in Chemical Carcinogenesis," IARC Sci. Publ. No. 12, Lyon, 1976, pp. 81–101.

232. A. Araki, M. Muramatsu, and T. Matsushima, *Gann*, **75**, 8–16 (1984).

233. T. Nakajima, A. Tanaka, and K.-I. Tojyo, *Mutat. Res.*, **26**, 361–366 (1974).

234. V. W. Mayer, *Mol. Gen. Genet.*, **112**, 289–294 (1971).

235. R. Langenbach, S. Leavitt, C. Hix, Y. Sharief, and J. W. Allen, *Mutat. Res.*, **161**, 29–37 (1986).

236. A. W. Hsie, J. R. San Sebastian, S. W. Perdue, R. L. Schenley, and M. D. Waters, *Mol. Toxicol.*, **1**, 217–234 (1987).

237. T. Hoe, J. R. San Sebastian, and A. W. Hsie, "Mutagenic Activity of Nitrosamines in Mammalian Cells," in, *Topics in Chemical Mutagenesis*, Vol. 1, *Genotoxicity of N-Nitroso Compounds*, T. K. Rao, W. Lijinsky, and J. L. Epler, Eds., 1984, pp. 13–43.

238. Y. Katoh, M. Tanaka, and S. Takayama, *Mutat. Res.*, **105**, 265–269 (1982).

239. R. Langenbach, *Mutat. Res.*, **163**, 303–311 (1986).

240. T. Kuroki, C. Drevon, and R. Montesano, *Cancer Res.*, **37**, 1044–1050 (1977).

241. L. J. Lilly, B. Bahner, and P. N. Magee, *Nature*, **258**, 611–612 (1975).

242. A. L. Brooks and V. Cregger, *Mutat. Res.*, **21**, 214 (1973).

243. A. T. Natarajan, A. D. Tates, P. P. W. Van Buul, M. Meijers, and N. De Vogel, *Mutat. Res.*, **37**, 83–90 (1976).

244. P. Kasper, S. Madle, and E. George, *Mutagenesis*, **3**, 521–525 (1988).

245. W. Blazak, B. Stewart, D. DiBlasio-Erwin, K. Allen, and C. Green, *Environ. Mutagen.*, **7** (Suppl. 3), 32 (1985).

246. R. E. Neft and M. K. Conner, *Teratogen. Mutagen. Carcinogen.*, **9**, 219–237 (1989).

247. B. A. Laishes and H. F. Stich, *Biochem. Biophys. Res. Commun.*, **52**, 827–833 (1973).

248. K. S. Bentley and P. K. Working, *Mutat. Res.*, **203**, 135–142 (1988).

249. R. Montesano, L. Saint Vincent, and L. Tomatis, *Brit. J. Cancer*, **28**, 215–220 (1973).

250. R. J. Pienta, J. A. Poiley, and W. B. Lebherz, III, *Int. J. Cancer*, **19**, 642–655 (1977).

251. R. Borland and G. C. Hard, *Eur. J. Cancer*, **10**, 177–184 (1974).

252. D. Clive, K. O. Johnson, J. F. S. Spector, A. G. Batson, and M. M. M. Brown, *Mutat. Res.*, **59**, 61–108 (1979).

253. D. E. Amacher and S. C. Paillet, *Mutat. Res.*, **106**, 305–316 (1982).

254. M. Watanabe, S. Honda, M. Hayashi, and T. Matssuda, *Mutat. Res.*, **97**, 43–48 (1982).

255. R. Mehta, K. C. Silinskas, P. F. Zucker, A. Ronen, J. A. Heddle, and M. C. Archer, *Cancer Lett.*, **35**, 313–320 (1987).

256. L. Pasternak, "Mutagene Wirkung von Dimethylnitrosamin bei Drosophila melanogaster," *Naturwissenschaften*, **49**, 381 (1962).

257. R. Fahrig, *Mutat. Res.*, **31**, 381–394 (1975).

258. D. B. Couch and M. A. Friedman, *Mutat. Res.*, **38**, 89–96 (1976).

259. H. V. Malling, *Mutat. Res.*, **26**, 465–472 (1974).

260. P. Kerklaan, G. Mohn, and S. Bouter, *Carcinogenesis*, **2**, 909–914 (1981).

261. S. Knasmüller, A. Szakmary, and M. Kehrer, *Chem.-Biol. Interact.*, **75**, 17–29 (1990).

262. R. Valencia, J. M. Mason, and S. Zimmering, *Environ. Mol. Mutat.*, **14**, 238–244 (1989).

263. R. K. Brodberg, M. J. Mitchell, S. L. Smith, and R. C. Woodruff, *Environ. Mol. Mutat.*, **10**, 425–432 (1987).

264. R. Fahrig, G. W. P. Dawson and L. B. Russell, *Environ. Sci. Res.*, **24**, 709–727 (1981).

265. S. S. Epstein, E. Arnold, J. Andrea, W. Bass, and Y. Bishop, *Toxicol. Appl. Pharmacol.*, **23**, 288–325 (1972).

266. P. Propping, G. Röhrborn, and W. Buselmaier, *Mol. Gen. Genet.*, **117**, 197–209 (1972).

267. D. J. Winton, N. J. Gooderham, A. R. Boobis, D. S. Davies, and B. Ponder, *Cancer Res.*, **50**, 7992–7996 (1990).

268. V. F. Simmon, *J. Natl. Cancer Inst.*, **62**, 893–899 (1979a).

269. V. F. Simmon, H. S. Rosenkranz, E. Zieger, and L. A. Poirer, *J. Natl. Cancer Inst.*, **62**, 911–918 (1979).

270. J. A. DiPaolo, R. L. Nelson, and P. J. Donovan, *Nature*, **235**, 278–280 (1972).

271. G. M. Williams and M. F. Laspia, *Cancer Lett.*, **6**, 199–206 (1979).

272. R. Fahrig and H. Remmer, *Teratogen. Carcinogen. Mutagen.*, **3**, 41–49 (1983).

273. A. Frölich and F. E. Würgler, *Mutat. Res.*, **216**, 179–187 (1989).

274. E. Vogel and B. Leigh, *Mutat. Res.*, **29**, 383–396 (1975).

275. W. L. Russell, "Positive Genetic Hazard Predictions From Short-Term Tests Have Proved False for Results in Mammalian Spermatogonia With All Environmental Chemicals So Far Tested," in *Genetic Toxicology of Environmental Chemicals*, Part B; *Genetic Effects and Applied Mutagenesis*, C. Ramel, B. Lambert, and J. Magnusson, Eds., 1986, pp. 67–74.

276. H. Yamazaki, Y. Mori, K. Toyoshi, H. Mori, S. Sugie, N. Yoshimi, and Y. Konishi, *Mutat. Res.*, **144**, 197–202 (1985).

277. M. Nagoa, E. Suzuki, K. Yasuo, T. Yahagi, Y. Seino, T. Sugimura, and M. Okada, *Cancer Res.*, **37**, 399–407 (1977).

278. R. K. Elespuru and W. Lijinsky, *Cancer Res.*, **36**, 4099–4101 (1976).

279. D. R. Stoltz and N. P. Sen, *J. Natl. Cancer Inst.*, **58**, 393–394 (1977).

280. T. K. Rao, A. A. Hardigree, J. A. Young, W. Lijinsky, and J. L. Epler, *Mutat. Res.*, **56**, 131–145 (1977).

281. T. Tsutsui, N. Suzuki, H. Maizumi, and J. C. Barrett, *Mutat. Res.*, **129**, 111–117 (1984).

282. Zeiger and M. S. Legator, *Mutat. Res.*, **12**, 469–471 (1971).

283. H. Henke, G. Höhne, H. A. Künkel, and A. Trams, "Zur Frage der differentiellen Mutationswirkung einiger neuer organotroper Cancerogene," *Arch. Gynäkol.*, **202**, 475–479 (1965).

284. H. Henke, G. Höhne, and H. A. Künkel, "Über die mutagene Wirkung von Röntgenstralen, *N*-nitroso, *N*-methyl-urethan und *N*-nitroso-morpholin bei Drosphila melanogaster," *Biophysik*, **1**, 418–421 (1964).

285. E. Vogel, "Mutagenicity of Carcinogens in Drosophila as Function of Genotype-controlled Metabolism," in *In Vitro Metabolic Activation in Mutagenesis Testing*, F. J. de Serres, J. R. Fouts, J. R. Bend, and R. M. Philpot, Eds., 1976, pp. 63–79.

286. R. Parkin, H. B. Waynforth, and P. N. Magee, *Mutat. Res.*, **21**, 155–161 (1973).

287. *IARC Monograph*, Vol. 37, WHO, 1985, pp. 241–261.

288. P. R. Padma, A. J. Amonkar, and S. V. Bhide, *Cancer Lett.*, **46**, 173–180 (1989).

289. Y. Mori, H. Yamazaki, and Y. Konsihi, *Mutat. Res.*, **192**, 91–94 (1987).

290. P. Gilbert, L. Fabry, B. Rollmann, P. Lombart, J. Rondelet, F. Poncelet, A. Leonard, and M. Mercier, *Mutat. Res.*, **89**, 217–228 (1981).

291. E. Denkel, B. L. Pool, I. R. Schlehofer, and G. Eisenbrand, *J. Cancer Res. Clin. Oncol.*, **111**, 149–153 (1986).

292. U. Dittberner, G. Eisenbrand, and H. Zankl, *J. Cancer Res. Clin. Oncol.*, **114**, 575–578 (1988).

293. S. Knasmüller, G. Stehlik, and G. Mohn, *J. Cancer Clin. Oncol.*, **112**, 266–271 (1986).

294. M. Zielenska and J. B. Guttenplan, *Mutat. Res.*, **202**, 269–276 (1988).

295. H. S. Rosenkranz and L. A. Poirer, *J. Natl. Cancer Inst.*, **62**, 873–892 (1979).

296. H. Marquardt, R. Schwaier, and F. Zimmerman, *Naturwissenschaften*, **50**, 135–136 (1963).

297. S. Abe and M. Sasaki, *J. Natl. Cancer Inst.*, **58**, 1635–1641 (1977).

298. N. K. Mishra, C. M. Wilson, K. J. Pant, and F. O. Thomas, *J. Toxicol. Environ. Health*, **4**, 79–91 (1978).

299. J. A. Styles, *Brit. J. Cancer*, **36**, 558–563 (1977).

300. B. G. Boyes, G. C. Rogers, T. I. Matula, R. Stapley, and N. P. Sen, *Mutat. Res.*, **241**, 379–385 (1990).

301. P. Schmezer, B. L. Pool, P. A. Lefevre, R. D. Callander, F. Ratpan, and H. Tinwell, *Environ. Mol. Mutagen.*, **15**, 190–197 (1990).

302. S. Neale, *Mutat. Res.*, **32**, 229–266 (1976).

303. W. G. Sanger and J. D. Eisen, *Mutat. Res.*, **34**, 415–426 (1976).

304. S. W. Soukoup and W. Au, *Humangenetik*, **29**, 319–328 (1975).

305. I. A. Rapoport, *Dokl. Biol. Sci.*, **146**, 1044–1046 (1962).

306. J. V. Frei and S. Venitt, *Mutat. Res.*, **30**, 89–96 (1975).

307. L. Zardi, L. Saint Vincent, A. Barbin, R. Montesano, and G. P. Margison, *Cancer Lett.*, **3**, 183–188 (1977).

308. F. K. Sanders and B. O. Burford, *Nature*, **213**, 1171–1173 (1967).

309. D. J. Kirkland, C. Armstrong, and R. J. C. Harris, *Brit. J. Cancer*, **31**, 329–337 (1975).

310. U. H. Ehling and A. Neuhäuser-Klaus, *Mutat. Res.*, **250**, 447–456 (1991).

311. M. A. Alaoui-Jamali, G. Rossignol, H. M. Schuller, and A. Castonguay, *Mutat. Res.*, **223**, 65–72 (1989).

312. R. Preussmann, in "Environmental Carcinogens. Selected Methods of Analysis," Vol. 6, "*N*-Nitroso Compounds," IARC Sci. Publ. No. 45, Lyon, 1983, p. 4.

313. National Toxicology Program, "Fifth Annual Report on Carcinogens," DHHS, U.S. Public Health Service, NTP 89-239, 1989.

314. W. Lijinsky, J. E. Saavedra, and M. D. Reuber, *Cancer Res.*, **41**, 1288–1292 (1988).

315. ACGIH Documentation of the Threshold Limit Values and Biological Exposure Indices, 5th Ed., 1986.

316. R. N. Magee and J. M. Barnes, *Brit. J. Cancer*, **10**, 114–122 (1956).

317. J. M. Doull, C. S. Klaassen, and M. O. Amdur, Eds. Casarett and Doull's *Toxicology: The Basic Science of Poisons*, 2nd ed., Macmillan, New York, 1980.

318. L. Griciute, in "*N*-Nitroso Compounds: Analysis, Formation and Occurrence," IARC Sci. Publ. No. 31, Lyon, 1980.

319. Mohr and Hilfrich, *J. Natl. Cancer Inst.*, **49**, 1729–1731 (1972).

320. Preussmann and Weissler, 1987.

321. H. Druckrey, R. Preussmann, S. Ivankovic, and S. Schmael, *Z. Krebsforsch.*, **69**, 103–201 (1967).

322. Druckrey, 1967.

323. W. Lijinsky and M. D. Reuber, *Cancer Lett.*, **14**, 297–309 (1981).

324. R. H. Cardy, W. Lijinsky, and P. K. Hildebrandt, *Ecotoxicol. Environ. Safety*, **3**, 29–35 (1979).

325. R. Peto, R. Gray, P. Brantom, and P. Grasso, in "*N*-Nitroso Compounds: Occurrence, Biological Effects and Relevance to Human Cancer," IARC Sci. Publ. No. 57, 1984, pp. 627–666.

326. *IARC Monograph*, Supplement 7, WHO, 1987.

327. B. Spiegelhalder, J. Müller, H. Drasche, and R. Preussmann, in "The Relevance of *N*-Nitroso Compounds to Human Cancer: Exposure and Mechanisms," IARC Sci. Publ. No. 84, Lyon, 1987.

328. S. Preson-Martin, in "The Relevance of *N*-Nitroso Compounds to Human Cancer: Exposure and Mechanisms," IARC Sci. Publ. No. 84, Lyon, 1987.

329. P. Correa, in "The Relevance of *N*-Nitroso Compounds to Human Cancer: Exposure and Mechanisms," IARC Sci. Publ. No. 84, Lyon, 1987.

330. J. M. Fajen et al., in "N-Nitrosamines in the Rubber and Tire Industry," *Science*, **205**, 21 (1979).

331. D. B. Rounbehler and J. M. Fajen, in "N-Nitroso Compounds in the Factory Environment," NIOSH Technical Report 83-114, US HHS, PHS, CDC, NIOSH, 1983.

332. B. Spiegelhalder and R. Preussman, in "Occupational Nitrosamine Exposure, 1. Rubber and Tyre Industry," *Carcinogenesis*, **4**, (9), 1147–1152, 1983.

333. NIOSH, Health Hazard Evaluation Report HETA 79-126-951, US, HHS, CDC, NIOSH, St. Clair Rubber Company, Marysville, Ohio, 1979.

334. NIOSH, Health Hazard Evaluation Report HETA 85-003-1834, US, HHS, CDC, NIOSH, B. F. Goodrich Company, Woodburn, Indiana, 1979.

335. NIOSH, Health Hazard Evaluation Report HETA 80-121-919, US, HHS, CDC, NIOSH, Kelly-Springfield Tire Company, Freeport, Illinois, 1980.

336. NIOSH, Health Hazard Evaluation Report HETA 81-045B-1216, US, HHS, CDC, NIOSH, Uniroyal, Inc., Mishakawa, Indiana, 1981.

337. NIOSH, Health Hazard Evaluation Report HETA 81-107-1331, US, HHS, CDC, NIOSH, Geauga Company, Middlefield, Ohio, 1981.

338. NIOSH, Health Hazard Evaluation Report HETA 79-109-1538, US, HHS, CDC, NIOSH, Kelly-Springfield Tire Company, Cumberland, Maryland, 1984.

339. Environmental Protection Agency Clean Water Act—Water Quality Standards Part 131, 40 CFR Chapter 1, 1991, pp. 272–276.

340. Risk Specific Intake Levels for the Proposition 675 Carcinogen: *N*-Nitrosodimethylamine, Reproductive and Cancer Hazard Assessment Section. Office of Environmental Health Hazard Assessment, California Department of Health Services, 1988.

341. E. R. Nestmann, K. M. Gowdy, B. Lynch, and A. Feniak, Biological Risk Assessment of *N*-Nitrosodimethylamine (NDMA), CanTox Inc., Oakville, Canada L6J 5Z7, 1990.

342. Environmental Protection Agency, Assessment of Cancer Risks to Machinists from Exposure to *N*-Nitrosodiethanolamine (NDELA) in Metalworking Fluids, Office of Toxic Substances, U.S. EPA Document, Feb. 1, 1985.

CHAPTER TWELVE

Organic Phosphates

Robert J. Weir, Ph.D.

1 GENERAL CONSIDERATIONS

The organic phosphates, generally as esters, are most widely recognized in their usage as insecticides, and the bulk of this chapter is devoted to discussion of chemicals intended for this use. Their biocidal properties, which were responsible for their early synthesis as war gases, present appreciable toxicologic problems from the standpoint of manufacture and use. Organic phosphates are also used as gasoline additives, hydraulic fluids, cotton defoliants, fire retardants, plastic components, growth regulators, and industrial intermediates, where their highly toxic effect is neither desirable nor always apparent. Some members of the class are still used as war gases. In dealing with this group of chemicals, it is clear that the toxicity of the organic phosphate pesticide is not a universal characteristic of the organic phosphates as a class. One need consider only those that occur naturally in the body, such as the phospholipids, phosphonucleotides, and phosphoproteins, to observe that some of the class members are relatively nontoxic.

The importance of this class of chemicals as insecticides rests in their high biocidal activity and their short life as residues. The loss from the market of DDT (1) and other chlorinated hydrocarbon insecticides such as aldrin, dieldrin, endrin, heptachlor, and chlordane because of their persistence and toxicity has enhanced the use of the organic phosphate insecticides.

The nomenclature of the organic phosphates frequently is confusing. Trade names, generic names, and manufacturers' experimental designations add to the confusion. For the purpose of this chapter, class names follow those outlined by

Patty's Industrial Hygiene and Toxicology, Fourth Edition, Volume 2, Part A, Edited by George D. Clayton and Florence E. Clayton.
ISBN 0-471-54724-7   © 1993 John Wiley & Sons, Inc.

Negherbon (2). The individual examples are listed as trade names or generic names depending on common usage, but in all cases both are given if used. The chemical name is also given in the listing of individual compounds.

1.1 Symptoms in Animals

The universal signs of intoxication of the organic phosphates that are insecticidal appear to result from the inhibition of the cholinesterases (che) (these esterases hydrolyze acetylcholine, butyrylcholine, benzoylcholine, acetyl-β-methylcholine, etc., depending on the species). It is important to mention that this is not an exclusive function of the organic phosphate insecticides, for it is shared with the class of insecticides known as carbamates and typified by Sevin®.

Following excessive exposure, the signs of toxicity reflect stimulation of the autonomic and central nervous systems, resulting from inhibition of acetylcholinesterase and consequent accumulation of acetylcholine. Prolongation and intensification of the acetylcholine action results in two degrees of response, depending on dosage and specific action of the inhibitor. The initial action is on smooth muscles, cardiac muscle, and exocrine glands and, in general, is comparable to stimulation of the postganglionic parasympathetic nerves. This phase of action results in the early signs of toxicity resembling those of muscarine and hence is referred to as the muscarinic action of acetylcholine. The action, and hence the signs, can be counteracted by atropine. The most common early signs in humans are intestinal cramps, tightness in the chest, blurred vision, headache, diarrhea, decrease in blood pressure, and salivation.

The second stage of intoxication results from stimulation of the peripheral motor system and of all autonomic ganglia. Experimentally, these actions can be counteracted by curare and ganglionic blocking agents and in other respects resemble the classical action of nicotine; hence they are referred to as the nicotinic action of acetylcholine. The complexity of toxic action during the second stage includes neuromuscular and ganglionic blockade; thus curare therapy would be contraindicated. Ultimately the toxic manifestations of poisoning are referable to stimulation and/or paralysis of the somatic, autonomic, and central nervous system. A more detailed understanding of this complex action and the influence of adequate atropinization may be obtained from Goodman and Gilman (3). A basic review in terms of health problems has been published by Hazleton (4). The role of cholinesterase activity is reviewed by Wills (5).

The signs of toxicity outlined above do not apply to animals that receive small doses over a long period of time. In this case, the correlation of signs of toxicity and inhibition of the cholinesterase(s) is not clear-cut. With many organic phosphates, inhibition in rats may be so complete in plasma and red blood cells that the cholinesterase activity is immeasurable with present techniques, and the brain activity may be markedly inhibited, although the animals appear normal in all respects. With other members of the organic phosphate class, chronic exposure in rats produces marked inhibition of plasma and red cell cholinesterase and moderate inhibition of brain activity, and results in diarrhea and tremors as the only toxic signs. These examples point out the variability of signs of toxicity, which depend on the toxicant, vehicle, route of administration, dosage, species studied, specific enzyme system on which the toxicant acts (true or pseudocholinesterase), metabolic conversion products, reversibility of the inhibition, and so on. These variables and their ultimate discussion are beyond the scope of this chapter and only those factors that have bearing on industrial hygiene are covered here.

1.2 Gross and Microscopic Pathology in Animals

The greatest bulk of the organic phosphate insecticides do not produce morphological alterations in animals. Some of the more recent additions to the group are chlorinated and are suspected of being capable of producing liver and kidney damage similar to that produced by the halogenated hydrocarbons. There is almost no literature to support the supposition. Even the organic phosphates that have been fed to rodents at maximum tolerated doses according to National Cancer Institute (NCI) carcinogenesis protocol are not carcinogens (6–12). Generally speaking the organic phosphates as a class are not mutagenic. Exceptions are Dematon, dichlorvos, and thio TEPA (13, 14).

It has long been known that certain organic phosphates are capable of producing a delayed paralysis in animals and humans; this effect is the result of a syndrome resembling "jake leg" or "ginger jake" paralysis, which has been described for Jamaica ginger poisoning. Early investigators described the associated histopathology as demyelination of the peripheral nerves; this terminology has been widely used, as is evidenced by the following reports. Paralytic effects of TOCP (tri-*o*-cresyl phosphate) have been described by Smith et al. (15, 16). More recent reports on this phenomenon after administration of TOCP have been made by Durham et al. (17), Barnes and Denz (18), Hine et al. (19), and Frawley et al. (20). Durham et al. (17) also studied Cholorthion® (*O,O*-dimethyl-*O*-3-choloro-4-nitrophenyl thionophosphate), DDVP (2-2-dichlorovinyl dimethyl phosphate), Systox® [*O,O*-diethyl *O*-2-(ethylmercapto)ethyl thionophosphate], Diazinon® [*O-O*-diethyl *O*-(2-isopropyl-6-methyl-4-pyrimidyl) phosphorothioate], OMPA (octamethyl pyrophosphoramide), EPN (ethyl-*p*-nitrophenyl benzenethionophosphonate), malathion [*O-O*-dimethyl *S*-(1,2-dicarboethoxyethyl) dithiophosphate], and Isopestox® [bis(monoisopropylamino)fluorophosphine oxide]. Of these, only Isopestox® was found to produce the demyelination syndrome. Later, Frawley et al. (20) showed that EPN also produced demyelination, contrary to the findings of Durham et al. (17); Barnes and Denz (18) and Austin and Davies (21) demonstrated that DFP (diisopropyl fluorophosphate) also produced demyelination in animals. The symptomatic observations have been confirmed with histological evidence in all cases (DFP, EPN, TOCP, and Isopestox®).

Later investigations revealed that the causative lesion for the delayed paralysis syndrome was not demyelination, but rather a degeneration of the axons in the spinal cord and peripheral nerves. A complete review of this subject, with references, is available as the proceedings of a conference sponsored by the U.S. Environmental Protection Agency (EPA) (22). An even newer review of this subject can be found in the extensive work on neurotoxicity by Spencer and Schaumberg (23). EPA is requiring that greater emphasis be placed on neurotoxic studies. There are several important aspects to be considered in regard to this syndrome as a feature of organic phosphate poisoning: it is always delayed, never acute; it is not due to cholinesterase inhibition; it is structure specific and not caused by all organic phosphates; there is considerable species susceptibility variation; it may follow a single massive dose (if cholinergic effects are antidoted) or may result from the cumulative action of repeated small exposures; there is no known effective treatment; mild cases may slowly return to normal whereas in severe cases the paralysis is usually permanent; and finally, there are well-documented cases of delayed neuropathy due to organic phosphates in humans. In Japan it was discovered that there was a high degree of correlation between eye defects reported in children in the 1950s and 1960s and chronic exposure to fields sprayed with organophosphates. There have also been reports of degeneration of the optic nerve in some animal feeding studies using organophosphates. (See Section 2.23.4 on Systox® and Disystox®). Special ocular studies where the use of fundic photography and electroretinography studies are helpful in determining electroretinographic activity as a precursor to optic nerve degeneration are recommended in the study of these materials in animals. It may be useful to employ this type of ocular study to severely exposed individuals or following repeated industrial exposure.

1.3 Exposure in Humans

The most common, and most important, route of industrial exposure to the organic phosphates is by accidental spillage on the skin. Most of the materials later discussed in detail are rapidly absorbed through the skin. Percutaneous absorption frequently is unnoticed because dermal irritation rarely occurs, unless the solvent systems of the formulated materials possess this irritative property.

The second most frequent exposure route is through the respiratory tract. Intoxication may occur with some of the more highly toxic members of the group, such as TEPP (tetraethyl pyrophosphate) and Phosdrin® (alpha isomer of 2-carbomethoxy-1-methylvinyl dimethyl phosphate), but in general it is agreed that exposure is due to particulate matter rather than to vapor (4). The organic phosphates as a group have extremely low vapor pressures; despite this, Kay and coworkers (24) give analytic values for parathion (vapor pressure 3.78×10^{-5} mm Hg) in the air of treated orchards for up to 3 weeks after application. Summerford et al. (25) correlated blood cholinesterase level with symptomatology in plant personnel, orchard workers, and other groups during and after a spray season. Inhalation exposure was not distinguished from dermal exposure. Fatal or near-fatal illnesses have resulted from brief, massive exposure to parathion due to gross carelessness rather than repeated exposure.

Oral exposure is rarely a problem, except for accidental ingestion by children and in the case of suicide. The more toxic members of the organic phosphate group may be an ingestion problem in manufacturing and spraying operations if good personal hygiene practices are not followed.

The organic phosphate insecticides share the biologic action of inhibiting cholinesterase(s). Although this is generally recognized not to be the sole toxic action, it does provide the toxicologist and industrial hygienist with an excellent tool for the measurement of exposure of animals or workers to the toxicant. This measurement of exposure serves as a warning of impending toxicity and is useful in prophylactic programs. Beyond this its reliability, either diagnostic or prognostic, is of little value.

As noted above, the delayed neuropathy caused by organic phosphates may not be due to inhibition of cholinesterase, and hence the assay of that enzyme level is not an indicator of risk. A "neurotoxic esterase" (NTE) has been described (26); this enzyme is believed to be responsible for the delayed neurotoxicity or paralytic effect displayed by some organophosphates.

In past years, the most widely used method of determination of cholinesterse inhibition appears to be that of Michel (27) or some modification of it. The manometric method of Ammon (28) is also reliable. The colorimetric method of Metcalf (29) and the modified method for whole blood described by Fleisher and Pope (30) are also useful, and the field kits such as described by Edson (31) are based on this method. More recently, many investigators have been using the colorimetric method of Ellman (32) because it can be used in enzyme–substrate complexes where the action is rapidly reversible. This is more important with carbamates but can be true with organic phosphates. When the cause of intoxication is unknown, this is an extremely important consideration. A kit has been developed by Boehringer Mannheim for this method. The radiometric method of Johnson and Russell (33) has received much consideration but radiological waste disposal gives concern.

In a recent workshop on cholinesterase methodologies (34) held by EPA, concern about the above methods' accuracy was expressed in terms of risk assessment. Where a poor method may allow an unfair advantage for one chemical by allowing a higher dose to be assessed to be the "no observed effect level" (NOEL) in animal studies, another chemical may have a lower NOEL simply because the method was more accurate. The need for standard operating procedures, inter-laboratory method evaluation studies, and standardization of methodology was emphasized at the workshop. The Ellman method appears to be the recommended method until further evaluation can be made.

Using the electrometric technique, Wolfsie and Winter (35, 36) have evaluated the plasma and red blood cell cholinesterase levels of men and women who were not exposed to organic phosphate insecticides (ingestion of residues from treated crops could not be eliminated). Mean values were as follows:

| Red blood cell | 0.67 to 0.86 ΔpH units/hr |
| Plasma | 0.70 to 0.97 ΔpH units/hr |

In a prophylactic program, both red blood cell and plasma values should be obtained. Measurement should be made frequently and with regularity. The workers' previous values should be available for inspection and comparison. In addition to knowledge of the specific action of the compound, judgment and experience are essential to adequate interpretations of results. Marked or severe depression of either the plasma or red blood cell value may be considered strong evidence of exposure, whether accompanied by gross signs or not. The red cell level is more significant because it represents the true neurohormone esterase level for humans, acetylcholinesterase. Plasma enzyme inhibition is less specific but may be important as a diagnostic aid in acute exposure because it usually, but not always, responds more rapidly and at lower dosage. Whole blood analysis is least reliable because it may give a composite effect and may mask the individual level of either the plasma or cells.

In the event exposure occurs, the symptoms in humans are qualitatively similar to those described for animals and include, but are not limited to, headache, vertigo, blurred vision, dilated pupils, lacrimation, salivation, sweating, muscular weakness and ataxia, dyspnea, nausea, diarrhea, abdominal cramps, vomiting, coma, pulmonary edema, and death.

Although the acute signs in humans may at times resemble paralysis, this is not to be confused with the possibility of delayed neuropathy and paralysis, which would occur only after 8 or more days following acute exposure. Commercial organic phosphate pesticides and many other organic phosphate chemicals have generally been screened for neurotoxic action and do not present a hazard under ordinary use conditions. In cases of severe exposure, which require intensive control of the cholinergic signs and symptoms, it would be well to follow the patient carefully for some time after the acute phase.

1.4 Treatment

The onset of symptoms is rapid and maximum effects may develop within a few hours. It is thus important that medical care be obtained without delay. Because the early symptoms of headache, malaise, and so on, are easily confused with other diseases, it is important that workers exposed to the organic phosphates be instructed to report any such indications.

Adequate atropinization is essential to relieve the muscarinic effects and to provide central respiratory stimulant action. An average adult may require 12 to 24 mg total dose of atropine intravenously during the first 24 hr. Because this is far in excess of the usual therapeutic dose, the physician unacquainted with the mutually antagonistic action of this drug and the organic phosphate may be hesitant to employ such large doses. A general rule is that atropine should be administered until visible effects of atropinization are observed. Because, as pointed out above, the muscarinic effects are only a part of the action produced by heavy exposure, it is essential that the patient be treated symptomatically with artificial respiration, postural drainage, warmth, and so on. Prognosis depends largely upon the exposure, type of compound, and adequacy of treatment. Care should be exercised until the patient is obviously free of any signs of toxicity. A detailed survey of this subject is presented by Gordon and Frye (37) and has been more recently updated by Clyne and Shaffer (38).

Because the toxicity of the organic phosphates is due to the inhibition of cholinesterase, the reactivation of these enzymes would offer great promise as a therapeutic measure. This action is apparently achieved by pyridine-2-aldoxime methiodide (PAM, PAM-2, 2-PAM, P-2-AM), diacetyl monoxime (DAM), and other oximes through a reaction called nucleophilic displacement. Namba and Hiraki (39) report both experimental and clinical investigations of PAM. In cases of parathion poisoning, they recommended intravenous doses of 1 g of PAM or more, if indicated.

Grob and Johns (40) give a step-by-step outline of combined therapy for organic phosphate intoxication. After removal from exposure, a patent airway and artificial respiration should be established. The therapeutic regimen includes atropine, 2 to 4 mg intravenously, repeated frequently until muscarinic symptoms disappear. PAM or DAM is recommended in doses of 2000 mg intravenously. The speed of the treatment is critical for survival.

The industrial hygiene physician should contact the local poison control center if there is a likelihood of organic phosphate intoxication. The most recent advice on the use of oximes can thus be obtained. They are, however, in no way a substitute for atropine.

2 SPECIFIC COMPOUNDS

These compounds are representative of the organic phosphates used to the greatest degree, but this list is not all-inclusive. Information on methods for determination in the atmosphere has come from methods for crop residues which emphasize the need for cleanup. Because air samples do not usually have pigments and other interfering substances, the cleanup steps in these methods can be eliminated. If the method employs gas or high-pressure liquid chromatography, or similar analytic technique, there is all the more reason to eliminate cleanup. Air samples are easily collected by drawing a known volume through chemosorb 102, XAD-2 sorbent, or a glass fiber filter and desorbing into a suitable solvent.

2.1 Abate®(*O,O,O′,O′*-Tetramethyl *O,O′*-Thiodi-*p*-phenylene Phosphorothioate, Temephos, Biothion®)

$$\left[\begin{matrix} CH_3O \\ CH_3O \end{matrix} \right\rangle \overset{\overset{S}{\|}}{P}{-}O{-}\langle\bigcirc\rangle{-}\Big]_2 S$$

2.1.1 Source, Uses, and Industrial Exposure

The major use of temephos is in the public health field for control of mosquito larvae and a number of adult flies and biting midges. The product's low toxicity to birds, fish, and other beneficial species is an advantage.

2.1.2 Physical and Chemical Properties (41)

Physical state	Yellow to brown, viscous liquid (technical)
Purity	90% (technical)
Boiling point	Decomposes at 120 to 125°C
Melting point	10 to 15°C
Refractive index, $n_D^{25°C}$	1.586 to 1.588
Vapor pressure	7.17×10^{-8} mm Hg at 25°C
Solubility	Soluble in acetonitrite, ethylene dichloride, toluene; insoluble in hexane, methylcyclohexane, water

2.1.3 Determination in the Atmosphere

A method is provided by Pasarela and Orloski (42) that may be suitable for determination in the atmosphere.

2.1.4 Physiological Response

2.1.4.1 Acute. The acute oral LD_{50} of Abate® (technical) is 2030 and 2300 mg/kg in male and female rats, respectively (41). The acute dermal LD_{50} is 1930 and 970 mg/kg, respectively, for male and female rabbits. It is not an irritant either in the eye or on the skin of rabbits. Rats showed no toxic signs when exposed to 1 hr of saturated vapor in an inhalation (whole-body) exposure.

2.1.4.2 Subchronic and Chronic. Male rats fed dietary levels of 250, 500, and 1000 ppm Abate® for 30 days resulted in significant deaths at 1000 ppm; growth and food consumption were reduced at 500 ppm. Tremors were apparent at all levels. Cholinesterase was depressed at all levels. There was no gross pathology at necropsy.

In 90-day dietary feeding studies male and female rats survived dosages of 350 ppm. A dietery level of 2 ppm produced no effects in either sex. At 6 and 18 ppm red cell cholinesterase inhibition was observed (only males were affected at the lower dose).

In a second 3-month study in rats at dietary levels of 0, 6, 18, or 54 ppm there was no cholinesterase depression at 6 ppm. Although 18 ppm produced slight red cell cholinesterase depression, the red cell effect returned to normal levels in a group of rats fed 18 ppm and then placed on a control diet for 2 weeks. At 54 ppm, brain, red cell, and plasma cholinesterase activity were moderately depressed.

Dogs were fed dietary levels of Abate® at 2, 6, and 18 ppm and a fourth level of 700 ppm to assure no pathological tissue changes would occur. No deaths occurred at any level. There were no changes in weight or food consumption. Signs of cholinergic stimulation occurred at 70 ppm but lower doses were without signs. The high dose was accordingly reduced to 500 ppm where red cell and plasma levels were almost completely inhibited, without toxic signs. The lower dose groups showed no anticholinesterase activity, and no histopathology was observed at any level.

In a 21-day repreated dermal study in rats at 12 and 60 mg/kg, there were no observed effects.

Neurotoxicity studies in hens fed 920 ppm Abate® (approximately one-fourth the single dose LD_{50}) as a daily dose for 30 days resulted in no neurological pathology. This test is designed to determine potential for peripheral neuropathy.

Three-generation reproduction studies at 25 or 125 ppm in albino rats resulted in no effect on fertility, gestation, reproduction, or lactation. No terata associated with Abate® were produced in either dosage group.

2.1.5 Hygienic Standard of Permissible Exposure

The American Conference of Governmental Industrial Hygienists (ACGIH) (43) has adopted a time-weighted average (TWA) of 10 mg/m^3 for Abate® as the threshold limit value (TLV) for an 8-hr workday.

2.2 Acephate (O,S-Dimethyl Acetylphosphoramidothioate, Orthene®)

$$
\begin{array}{c}
\text{CH}_3\text{S} \quad \text{O} \qquad \text{O} \\
\diagdown \; \| \qquad\quad \| \\
\text{P—NHC—CH}_3 \\
\diagup \\
\text{CH}_3\text{O}
\end{array}
$$

2.2.1 Source, Uses, and Industrial Exposure (44, 45)

Acephate has moderate persistence, with residual systemic activity of 10 to 15 days, and is used to control insects and aphids in ornamentals, where it has a reasonably broad spectrum. It is also used on beans, cotton, head lettuce, celery, soybeans, and bell peppers. It controls parasites of cattle, goats, hogs, horses, poultry, and sheep, where tolerances have been set for milk, eggs, fat, and meat.

2.2.2 Physical and Chemical Properties (44, 46)

Physical state	White powder with rotten cabbage odor
Specific gravity	1.35
Vapor pressure	1.7×10^{-6} mm Hg at 24°C
Molecular weight	183
Melting point	86.9 to 91.0°C
Stability	Relatively stable; store in cool place
Solubility	Moderate to good in water, alcohol, or acetone; relatively low in organic solvents (less than 1% kerosene and less than 10% in xylene)

2.2.3   Determination in the Atmosphere

The method used (47) is for multiple pesticides, but this should not be a problem with air analysis because no cleanup is necessary; the method can be applied as is to the trapped acephate (48).

2.2.4   Physiological Response (40, 46)

2.2.4.1 Acute. The technical material has an acute oral LD_{50} in male and female rats of 900 and 700 mg/kg, respectively. The MLD (minimum lethal dose) in the dog is 681 mg/kg. Eye exposure in the rabbit results in slight conjunctival irritation that clears in less than 24 hr. The acute dermal LD_{50} in the rabbit is greater than 10 g/kg body weight. Dermal irritation studies in the rabbit and sensitization studies in the guinea pig are without effect. Four-hour acute inhalation studies with the rat indicated an $LC_{50} > 61$ mg/l.

2.2.4.2 Subchronic and Chronic. In 90-day subchronic toxicity studies using the rat, 300 ppm revealed no change in weight gain, food consumption, survival, blood and urine values, organ weights and ratios, or gross and microscopic pathology.

In 90-day rat and dog subchronic studies, feeding at doses of 10 ppm, no reduction of cholinesterase activity was observed, and this dose is considered to be the no-effect dose (NOEL) in both species.

In teratological studies in rats and mice, no effect was noted. Mutagenic effects were absent in a dominant lethal study.

Chronic 2-year dog and rat studies conducted at 100 ppm showed no gross or histological changes. In the dog, cholinesterase was slightly depressed at 100 ppm, but in the rat 30 ppm caused slight to moderate cholinesterase inhibition. Growth was slightly depressed in the rat at 100 ppm.

Atropine sulfate is antidotal.

2.2.5   Hygienic Standard of Permissible Exposure (43, 46)

The ACGIH has not yet set a TLV for acephate. Chevron Chemical Co. has set an internal TLV at 1.0 mg/m³.

2.3   Chlorpyrifos [O,O-Diethyl O-(3,5,6-Trichloro-2-pyridyl) Phosphorothioate, Dursban®]

2.3.1   Source, Uses, and Industrial Exposure (52)

Chlorpyrifos is used for the control of insects on turf, ornamental plants, and stored products and of fire ants, mosquitos, and household insects. It is used for foliar application for controlling borers, aphids, and scale in trees and agronomic crops (51).

2.3.2 Physical and Chemical Properties (51, 53)

Physical state	White granular crystalline material
Melting point	45.5 to 46.5°C
Molecular weight	350.6
Vapor pressure	1.9×10^{-5} mm Hg

2.3.3 Determination in the Atmosphere

The method is to absorb on Gn-4 membrane filters and chromosorb 102 tubes at flow rates of 0.125 to 2 l/min for 0.25- to 8-hr time intervals, then desorb into high-pressure liquid chromatography (HPLC) grade hexane and use gas chromatography with electron capture detection (53).

2.3.4 Physiological Response (54)

2.3.4.1 Acute. The acute oral LD_{50} for chlorpyrifos in the rat ranges from 118 to 270 mg/kg for males and 96 to 174 mg/kg for females based on a number of studies. The oral LD_{50} for other species follow: mice, 64 to 102; rabbits, 1000 to 2000; guinea pigs, 500; chickens, 32 to 64 mg/kg. The dermal LD_{50} for rabbits ranged from 1580 to 1801 mg/kg. All rats survived a 4-hr inhalation exposure to vapor concentrations of 200 mg/m³.

2.3.4.2 Subchronic and Chronic (54). Feeding studies conducted in rats, mice, dogs, and monkeys have shown in general plasma cholinesterase (ChE) depression as the most sensitive end point, followed by erythrocyte ChE depression; higher doses resulted in brain ChE depression. ChE activity can be depressed considerably in these three tissues before clinical signs are apparent. Rats fed a dietary concentration of 1.0 percent [~100 mg/(kg)(day)] showed weight loss, tremors, other signs of ChE depression, and high mortality. The no-observable effect level (NOEL) in rat studies was generally concluded to be 0.1 mg/(kg)(day). A 3-week repeated dermal study at 5.0 mg/(kg)(day) (highest dose tested) resulted in no observable effects whereas at 10 mg/(kg)(day) in a 4-day repeated dermal exposure, plasma and erythrocyte ChE depression resulted without other evidence of toxicity. In a "nose-only" 13-week inhalation study at up to 21 ppb chlorpyrifos in air, no adverse effects were noted.

In two chronic rat feeding studies the results were essentially similar. In one study with Sherman rats, cholinesterase activity was depressed in plasma and erythrocytes of rats receiving 1.0 or 3.0 mg/(kg)(day) (highest dose treated), whereas brain ChE activity was depressed at the highest dose only. No effects of any kind were seen at 0.1 mg/(kg)(day). In Fischer 344 rats at 10 mg/(kg)(day) males showed decreased body weight gain, and depression of ChE in plasma, erythrocytes, and brain ChE activity and an increase in adrenal size accompanied by adrenal micro-scopic alteration. Similar but less pronounced effects were seen in the females. At 1.0 mg/(kg)(day), effects were limited to depressed plasma (males and females) and erythrocyte (males only) cholinesterase. There were no increases in tumor incidences in either study at any dose. In a chronic mouse feeding study, there were no increases in tumor incidence at the maximum tolerated dose, 1.5 mg/(kg)(day).

In acute and subchronic chicken studies at high doses of 100 mg/kg as a single dose, or 10 mg/(kg)(day) as a repeated dose, chlorpyrifos produced ataxia or muscle weakness, respectively, without histopathological changes associated with organo-phosphate-induced delayed neurotoxicity.

In an attempted suicide case where an individual ingested approximately 300 mg/kg chlorpyrifos, the patient was in a coma for 17 days, during which time he was on a respirator and received atropine and 2-PAM as antidotes. About 6 weeks after the attempt, he complained about weakness and paresthesia of the legs. There is no evidence that chlorpyrifos causes organophosphate-induced delayed neuro-toxicity at doses less than those expected to be lethal without medical intervention.

Male human volunteers receiving a single oral dose of 0.5 mg/kg chlorpyrifos showed 85 percent depression of plasma ChE activity with complete recovery in 30 days. Erythrocyte activity was not affected and no other signs or symptoms of toxicity were observed. In a 28-day repeated oral exposure with human subjects, ChE activity was not affected at 0.014 mg/(kg)(day). Slight plasma ChE depression was observed in the volunteers at 0.03 mg/(kg)(day). In volunteers receiving 0.1 mg/(kg)(day), plasma ChE was depressed after 6 days.

2.3.5 Hygienic Standard of Permissible Exposure

The ACGIH TLV for chlorpyrifos is 0.2 mg/m^{-3} (43).

2.4   DDVP (2,2-Dichlorovinyl Dimethyl Phosphate, Vapona®, Dichlorvos)

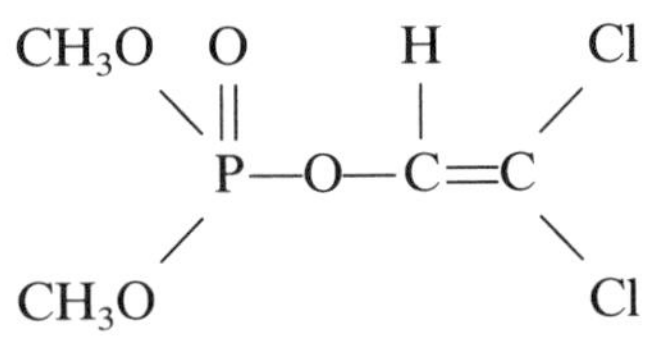

$$
\begin{array}{ccccc}
CH_3O & O & & H & Cl \\
\diagdown & \| & & | & \diagup \\
 & P & -O-C & = & C \\
\diagup & & & & \diagdown \\
CH_3O & & & & Cl
\end{array}
$$

2.4.1 Sources, Uses, and Industrial Exposures

DDVP was used almost exclusively as a quick knockdown agent for the control of houseflies. Its action depended on its fumigant action, for it was a poor stomach and contact poison. It is no longer available for this use.

2.4.2 Physical and Chemical Properties (55)

Physical state	Oily liquid
Molecular weight	221.0
Density	1.415 (25°C)
Boiling point	84°C (1 mm Hg)
Vapor pressure	0.01 mm Hg (30°C)
Refractive index, $n_D^{25°C}$	1.451
Solubility	Miscible with alcohol and most nonpolar solvents; 1.0 percent in water and 0.5 percent in glycerin at room temperature

2.4.3 Determination in the Atmosphere (56)

The method is to absorb on XAD-2, desorb with toluene, and analyze using gas chromatography with a flame photometric detector and a phosphorus filter.

2.4.4 Physiological Response

2.4.4.1 Acute. The acute oral LD_{50} is 80 mg/kg in male rats and 55 mg/kg in female rats (57). The acute dermal LD_{50} is 107 mg/kg in male rats and 75 mg/kg in female rats.

2.4.4.2 Subchronic and Chronic (58). In a 2-year gavage study in B6C3F1 mice, dichlorvos was fed at 10 and 20 mg/(kg)(day) (males) and 20 and 40 mg/(kg)(day) (females) for 5 days/week for 104 weeks. Corn oil was the solvent. Squamous papilloma incidence was increased in the 40 mg/kg females as compared to the controls. The incidence of carcinomas/papillomas in the forestomach showed a significant trend in the high-dose females and a significant pairwise comparison. In a 2-year gavage study using male and female Fischer 344 rats, dichlorvos was administered at 4 and 8 mg/(kg)(day). The males showed increased incidence of pancreatic acinar adenoma at both doses by pairwise comparison and an increased dose-related trend. There was also an increased incidence of leukemia in the males at both doses and a dose-related trend. Based on these studies and mutagenicity for both dichlorvos and its metabolite dichloroacetaldehyde, EPA has classified DDVP (dichlorvos) as B2, a probable human carcinogen. The Q_1^* is 2.9×10^{-1} mg/(kg)(day) using the linearized multistage procedure, extra risk.

2.4.5 Hygienic Standard of Permissible Exposure (43)

A TLV has been adopted for DDVP by ACGIH at 0.1 ppm or 0.9 mg/m³.

2.5 Delnav® [70% of 2,3-*p*-Dioxanedithiol *S,S*-Bis (*O,O*-diethyl Phosphorodithioate) and 30% Related Compound, Dioxanthion, Hercules AC528, Navadel®]

$$
\begin{array}{ccc}
& O & S \\
& \diagup \ \diagdown & \| \\
H_2C & HC\!-\!S\!-\!P\!-\![OC_2H_5]_2 \\
| & | \\
H_2C & HC\!-\!S\!-\!P \\
& \diagdown \ \diagup & \|\!-\![OC_2H_5]_2 \\
& O & S
\end{array}
$$

2.5.1 Source, Uses, and Industrial Exposure (51)

Delnav® is an insecticide–miticide used on citrus, stone fruit, grapes, and walnuts. It is also used for control of ticks, lice, and horn fly on cattle, goats, hogs, horses, and sheep when sprayed or dipped.

2.5.2 Physical and Chemical Properties (51)

Physical state	Nonvolatile tan liquid
Melting point	$-20°C$
Solubility	Insoluble in water; soluble in aromatic hydrocarbons, ethers, esters, ketones
Refractive index, $n_D^{20°C}$	1.5420

2.5.3 Determination in the Atmosphere

A method is provided in the compendium by Thompson and Watts (47) that is suitable for determination in the atmosphere.

2.5.4 Physiological Response

2.5.4.1 Acute. The acute oral LD_{50} of Delnav® has been investigated for a number of species including rats, dogs, and mice; the oral LD_{50} ranged between 43 mg/kg and 176 mg/kg (59). The acute oral toxicity of 15 other pharmacologically related pesticides has also been studied for evidence of potentiation. Delnav® does not potentiate the acute toxicity of the other pesticides. The percutaneous LD_{50} in rats and rabbits ranges between 63 and 235 mg/kg. The 1-hr LC_{50} for mice and rats by inhalation is 340 and 398 μg/l, respectively. Instillation in the eye of rabbits of 0.1 ml concentrated delnav, as well as 5 and 25 percent solutions in corn oil, produced no corneal damage and only mild transient conjunctivitis.

2.5.4.2 Subchronic and Chronic. Subchronic exposure of chickens to Delnav® did not produce neurotoxic effects under conditions that are positive for known neurotoxic organic phosphates.

Morphological changes could not be produced in rats at 90-day dietary levels as high as 100 ppm Delnav®. Rats fed Delnav® in the diet at various doses for 90 days showed no plasma, brain, or erythrocyte inhibition at 1.0 and 3.0 ppm. Dogs administered delnav orally for 90 days showed no cholinesterase depression at doses of 0.075 mg/kg and below. Human volunteers have tolerated daily oral doses of Delnav® for 5 weeks at 0.075 mg/kg without plasma, erythrocyte cholinesterase depression, or any other toxic effect.

2.5.5 Hygienic Standard of Permissible Exposure

A TLV has been established for Delnav® by the ACGIH (43). The TWA for occupational exposure is 0.2 mg/m^3.

2.6 Diazinon® [O,O-Diethyl O-(2-Isopropyl-6-methyl-4-pyrimidyl) Phosphorothioate, G-24480]

2.6.1 Source, Uses, and Industrial Exposure (51)

Diazinon® is a broad-spectrum insecticide and acaricide. It has also received much use in the control of cockroaches, particularly those resistant to chlorinated hydrocarbon pesticides.

2.6.2 Physical and Chemical Properties (55)

Physical state	The pure material is a colorless liquid; the technical material is a pale to dark brown liquid
Molecular weight	340.4
Density	1.116 to 1.118 (25°C)
Boiling point	83 to 84°C (0.002 mm Hg)
Vapor pressure	1.4 × 10^{-4} mm Hg (20°C)
Refractive index, $n_D^{20°C}$	1.4978 to 1.4981
Solubility	0.004 percent in water at room temperature; miscible with alcohol, xylene, acetone, petroleum oils

2.6.3 Determination in the Atmosphere

A method of analysis for Diazinon® has been described by Harris (60) and can be adapted to air analysis.

2.6.4 Physiological Response

2.6.4.1 Acute. The acute oral LD_{50} of 95 percent technical Diazinon® in rats is 100 to 150 mg/kg. The LD_{50} of a 23 percent wettable powder is 264.5 mg/kg on the basis of active ingredient (61, 62). This discrepancy was later explained by Gysin and Margot (63) as resulting from the formation of a number of possible isomerization and decomposition products resulting from the technical form. Presumably the responsible form is monothio-TEPP (64). Formulation prevented the degradation.

2.6.4.2 Subchronic and Chronic. Male and female rats receiving 100 and 1000 ppm (weight) technical Diazinon® in the diet for 4 weeks showed no gross signs of intoxication, alteration of growth, or gross pathology at autopsy. Red blood cells and brain cholinesterase levels were significantly inhibited at both dosages. Plasma activity at both dosages was comparable to the control.

Rats received 10, 100, and 1000 ppm (weight) active Diazinon® as a wettable powder in the diet for 72 weeks with no apparent gross signs of toxicity. Dogs received orally various doses of active Diazinon® as a wettable powder for 46 weeks. No pathology, gross or microscopic, was observed at the lowest dosage [4.6 mg/(kg)(day)] in 2 weeks. After 12 weeks cholinesterase inhibition was complete at the lowest dosage. At a dosage of 9.3 mg/(kg)(day) for 5 weeks signs of toxicity and complete cholinesterase inhibition were observed. Withdrawal of Diazinon® at the highest dosage resulted in reversal of signs and regeneration of cholinesterase activity to normal limits after 2 weeks.

Fischer 344 rats fed Diazinon® in the diet at 400 or 800 ppm for 2 years and chronic dietary studies in B6C3F1 mice at dose levels of 100 or 200 ppm showed no evidence of carcinogenesis in this bioassay (65). When dogs were fed Diazinon® at 0.25, 0.75, and 75 ppm in the diet for 90 days, erythrocyte cholinesterase was depressed in the highest dose only. Plasma cholinesterase activity was depressed at the two highest doses. No other adverse effect was observed (66).

Diazinon® was not teratogenic in rabbits receiving oral doses of 7 or 30 mg/kg during organogenesis. In hamsters, Diazinon® was not teratogenic at dosage levels of 0.125 to 0.25 mg/kg. In the rabbit teratogenic effects were not seen at 30 mg/(kg)(day), but cholinergic effects were noted (67).

2.6.5 Hygienic Standard of Permissible Exposure

A TLV has been adopted for Diazinon® by the ACGIH (43). The TWA is 0.1 mg/m^3.

2.7 Dimethoate [*O,O*-Dimethyl *S*-(*N*-Methylcarbamoylmethyl) Phosphorodithioate, Cygon®, Rogor®, Folimate®]

$$[CH_3O]_2 \overset{\overset{\displaystyle S}{\|}}{P}-S-CH_2CONHCH_3$$

2.7.1 Source, Uses, and Industrial Exposures

Dimethoate is a very low toxicity pesticide that can be used on fruits, vegetables, cereals, coffee, cotton, olives, rice, tea, tobacco, and so on (68). The EPA has set tolerances of 2.0 ppm in many cases.

2.7.2 Physical and Chemical Properties (68)

Molecular weight	229.3
Color and state	White crystalline solid
Flash point	130 to 132°C
Melting point	43.5 to 45.8°C
Refractive index	1.5377 at 50°C
Specific gravity	1.28 at 20°C
Solubility	Soluble in most organic solvents; soluble in water (7 percent at 80°C); very slightly soluble in aliphatic hydrocarbons
Stability	Unstable upon heating and at low pH; decomposes above 170°C

2.7.3 Determination in the Atmosphere

The residues in the atmosphere may be captured by drawing a known quantity of air through several scrubbers containing a useful solvent. The entrapped dimethoate may be analyzed by the method of Chilwell and Beecham (69) without cleanup.

2.7.4 Physiological Response (38)

2.7.4.1 Acute. The acute oral LD_{50} of technical dimethoate was 280 mg/kg for the male and 240 mg/kg for the female rat. In the mouse these figures are 510 and 220 mg/kg for the male and female, respectively. The dermal LD_{50} for the male rat and guinea pig is >800 mg/kg. Dimethoate is not an irritant in the eyes or on the skin of rabbits.

2.7.4.2 Subchronic and Chronic. Dietary levels of 2, 8, and 32 ppm were tolerated by rats for 90 days without toxic signs or effects on cholinesterase activity. Diets containing 50 ppm dimethoate for 33 days produced decreased plasma, red cell, and brain cholinesterase activity.

Repeated feeding of dimethoate to dogs for 13 weeks at 2, 10, and 50 ppm produced no effect at 2 ppm, slight red cell cholinesterase depression at 10 ppm and further suppression at 50 ppm. Dogs fed 1500 ppm or more for 13 weeks were without effect with regard to gross or microscopic pathology.

Hens fed 130 ppm dimethoate for 4 weeks showed no effect on the neurological system.

Three-generation reproduction studies in mice at 5, 15, or 50 ppm were carried out. There were no alterations of fertility, gestation, viability, lactation, tissue pathology, or fetus morphology that could be related to the dimethoate treatment.

2.7.5 Hygienic Standard of Permissible Exposure

No TLV has been established for dimethoate by the ACGIH (43).

2.8 Dipterex® (O,O-Dimethyl 2,2,2-Trichloro-1-hydroxyethylphosphonate, Bayer L 13/59, trichlorfon)

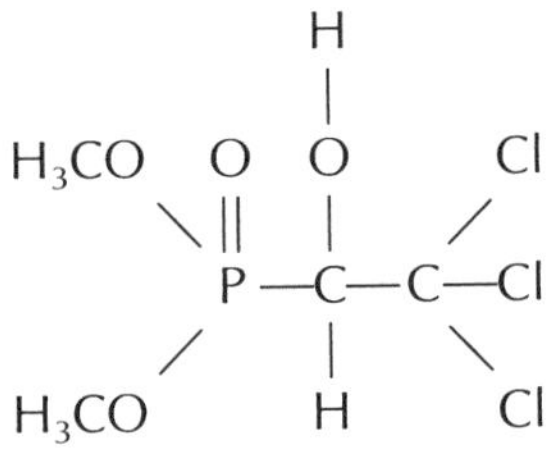

$$
\begin{array}{ccccccc}
& & & & \text{H} & & \\
& & & & | & & \\
\text{H}_3\text{CO} & & \text{O} & \text{O} & & \text{Cl} & \\
& \diagdown & \| & | & & \diagup & \\
& & \text{P} & \text{C} & \text{C} & \text{Cl} & \\
& \diagup & & | & & \diagdown & \\
\text{H}_3\text{CO} & & & \text{H} & & \text{Cl} &
\end{array}
$$

2.8.1 Source, Uses, and Industrial Exposure (51)

Dipterex has been used to control houseflies. It has shown promise in agriculture against lepidopterous and dipterous insects and mites.

2.8.2 Physical and Chemical Properties (52)

Physical state	White to pale yellow crystalline solid
Molecular weight	275.5
Density	1.73 (20°C)
Melting point	78 to 80°C
Boiling point	120°C (0.4 mm Hg)
Vapor pressure	Volatile
Refractive index, $n_\text{D}^{20°\text{C}}$	1.3439 (10 percent aqueous solution)
Solubility	Soluble in water to 13 to 15 percent at 25°C; soluble in alcohols, benzene, toluene, chloroform; slowly unstable in water; decomposition speeded by heat or alkali; decomposition product DDVP

2.8.3 Determination in the Atmosphere

A method suitable for air analysis of Dipterex® has been published by Giang et al. (70).

2.8.4 Physiological Response

2.8.4.1 Acute. The acute oral and intraperitoneal LD_{50} in rats has been reported by DuBois and Cotter (71) as 450 and 225 mg/kg, respectively. The dermal LD_{50} is 72000 mg/kg.

2.8.4.2 Subchronic and Chronic. Dipterex® was fed to rats at 500 mg/(kg)(day) for 1 year without histopathological change (72). Mice received Dipterex® by gavage twice weekly for 75 weeks without evidence of increased tumor incidence. Rats received Dipterex® by gavage at approximately 155 mg/kg twice weekly at 22 mg/kg for 90 weeks; there was no difference in tumor incidence between treated and control animals (73). Olajos et al. (74) reported hens exposed to oral or subcutaneous doses of 50 to 300 mg Dipterex® showed signs of neurological dysfunction at 12 to 28 days. Multifocal neuropathy affected the CNS and the peripheral nervous system about 1 month after dosing. Dipteryx® was teratogenic in rats when administered by gavage at a dose level of 480 mg/(kg)(day) or days 6 to 15 of gestation, but not teratogenic when administered on days 8 or 10 only. It was also teratogenic in hamsters at 400 mg/(kg)(day) after administration on days 7 to 11 of gestation. No effect occurred at 200 mg/(kg)(day). When administered on day 8 of gestation, it was only embryo toxic but not teratogenic. The mouse was less susceptible to Dipterex® than were the rat or hamster but a significant increase in incidence of cleft palates resulted from exposure on days 10 to 14 or 12 to 14 of gestation (75). Berge et al. (76) administered Dipterex® to pregnant guinea pigs at doses of 0 or 100 mg/kg on days 36, 37, 38, 51, 52, and 53 of gestation. The number of abortions and stillborn fetuses and the mean fetal weights were lower in the treated animals than in the controls. Pups from some treated litters showed marked trembling and locomotor disturbances, as well as skeletal muscular atrophy and cardiovascular functional abnormalities. There were decreases in weights of the brain, cerebellum, and medulla oblongata. The choline-acetyl transferase and glutamate decarboxylase enzymes were both significantly reduced in the test animals as compared to the controls.

2.8.5 Hygienic Standard of Permissible Exposure

No TLV has been set for Dipterex® (43).

2.9   EPN (Ethyl *p*-Nitrophenyl Thionobenzenephosphonate)

2.9.1   Source, Uses, and Industrial Exposure

EPN has been used as an insecticide and acaricide. It has shown a broad spectrum of activity against mites and insects. Present use in agriculture is limited.

2.9.2   Physical and Chemical Properties (52)

Molecular weight	323.3
Density	1.268 (25°C)
Melting point	36°C
Vapor pressure	3.0×10^{-4} mm Hg (100°C)
Refractive index, $n_D^{30°C}$	1.5978
Solubility	Practically insoluble in water; soluble in most of the common organic solvents; stable at ordinary temperatures and in neutral and acid media

2.9.3   Determination in the Atmosphere (77)

A known volume of air is drawn through a glass fiber filter. The filter is transferred to a screw-cap bottle within 1 hr of sampling. The analyte is extracted from the filter with isooctane. An aliquot of the extract is analyzed with GC using flame photometric detection with a phosphorus filter.

2.9.4   Physiological Response

2.9.4.1 Acute. The acute oral LD_{50} for pure EPN is 42 and 14 mg/kg for male and female rats, respectively; for the technical materials, values of 28 to 33 and 7 to 13 mg/kg are given by Hodge et al. (78).

2.9.4.2 Subchronic and Chronic. In 2-year chronic feeding studies in rats (78), doses of 150 ppm for males and 75 ppm for females produced no effect. Doses of 2.0 mg/(kg)(day) were administered to male and female dogs for 1 year without effect. EPN has been found to potentiate the effects of malathion (details under malathion, Section 2.14).

EPN produced neurotoxic effects (79) in chickens when administered subcutaneously to atropine-protected hens at dosages of 40 mg/kg or more. A single oral dose as low as 25 mg/kg was neurotoxic. Repeated doses of 0.1 to 10 mg/(kg)(day) for 90 days produced delayed neurotoxicity. Histological examination revealed marked axon and myelin degeneration in sciatic nerve and spinal cord of some hens receiving a single oral dose (67). Abou-Donia (79) reported rodents metabolize and excrete the neurotoxic organophosphorus esters with great efficiency. Adult chickens carry out these processes with great difficulty. The cat is intermediate. Findings from chickens, the classical model, may overestimate the risk for humans. This may explain why no human case of EPN-induced delayed neurotoxicity has been reported despite the fact that EPN has been in use for more than 35 years.

2.9.5 Hygienic Standard of Permissible Exposure

The TLV for EPN as a dust, fume, or mist is 0.5 mg/m^3 (43).

2.10 Ethion [O,O,O,O-Tetraethyl S,S-Methylene Bis(phosphorodithioate)]

2.10.1 Source, Uses, and Industrial Exposure (45, 51)

Ethion is used as an insecticide/acaricide for nuts, citrus, apples, other fruits, and cotton.

2.10.2 Physical and Chemical Properties (51)

Physical state	Colorless to amber-colored liquid
Molecular weight	384.48
Density	1.215 to 1.230 20°C
Vapor pressure	1.5×10^{-6} 25°C
Solubility	Slightly soluble in water; soluble in acetone or alcohol

2.10.3 Determination in the Atmosphere

The method for analysis of ethion residues in vegetables can easily be modified for atmospheric determination: Collect the vapors in any suitable system. Desorb into a suitable solvent; analyze with gas chromotography using a KCL thermionic or flame photometric detector (80).

2.10.4 Physiological Response

2.10.4.1 Acute. The acute oral LD$_{50}$ in rats is 208 mg/kg. The acute percutaneous LD$_{50}$ in rabbits is 915 mg/kg (81).

Groups of beagle dogs (four to six per group) were administered ethion orally at dietary concentrations of 0, 0.5, 2.5, 25, and 300 ppm for 90 days. At the 300-ppm level, cholinergic signs occurred togther with decreased body weight and food consumption. Plasma cholinesterase depression was observed at 2.5 ppm and above. Erythrocyte and brain cholinesterase activity was inhibited at 25 and 300 ppm. No histopathological effects were observed (82).

In a 2-year rat feeding study, plasma cholinesterase was depressed at 40 ppm (highest dose tested, HDT) as the only effect. No carcinogenicity was evident (83). In a chronic mouse study, plasma cholinesterase was depressed at 8 ppm (HDT) as the only effect. No carcinogenicity was evident. In a 2-year dog study, plasma cholinesterase was decreased at 20 ppm as the only effect (84).

A delayed neurotoxic study in hens at antidoted dose of 2792 mg/(kg)(day) did not produce evidence of neurotoxicity (85).

2.10.5 Hygienic Standard of Permissible Exposure (43)

The ACGIH has set a TLV of 0.4 mg/m^3 for ethion.

2.11 Fonofos (O-Ethyl S-Phenyl Ethylphosphonodithioate, Dyfonate®)

2.11.1 Source, Uses, and Industrial Exposure (45)

Fonofos is an organic phosphate insecticide useful on sorghum, soybean, spearmint, peanuts, peas, peppermint, corn, and beans (as both the crop and its hay), as well as for asparagus, sugar beets, strawberries, and vegetables (fruit, leaf, root, seed, and pod).

2.11.2 Physical and Chemical Properties (51)

Physical state	Liquid
Flash point	>200°F (Tagliabue closed cup)
Specific gravity	1.154 at 20/20°C
Solubility	13 ppm in water at 20°C
Vapor pressure	0.21 μm Hg at 25°C
Boiling point	100°C at 0.3 mm Hg

2.11.3 Determination in the Atmosphere

A known volume of air can be trapped in a series of scrubbers containing a suitable solvent. The method for fonofos (86) may then be applied without further cleanup.

2.11.4 Physiological Responses (87)

2.11.4.1 Acute. The acute oral LD$_{50}$ of fonofos in rats ranges between 4 and 43 mg/kg with typical signs of toxicity of organic phosphate esters. The acute dermal LD$_{50}$ in rabbits is 32 to 261 mg/kg. The acute dermal LD$_{50}$ in rats is 147 mg/kg and in the guinea pig 278 mg/kg. Fonofos is not irritating, but in one of three Draize tests fonofos was lethal. Application to the eyes of rabbits also produced death.

The acute LC$_{50}$ in rats was 0.46 and 0.9 mg/l for 4- and 1-hr inhalation exposures, respectively.

2.11.4.2 Subchronic and Chronic.

2.11.4.2 Subchronic and Chronic. In a 90-day dog study, daily ingestion of 0.4 to 6.0 mg/kg fonofos resulted in plasma, red cell, and brain cholinesterase inhibition and moderate toxic signs. In 2-year dog studies with daily ingestion of 0.4 to 6.0 mg/kg, in addition to cholinesterase inhibition there was increased liver weight, congestion of the small intestine, decreased weight gain, soft stools, alopecia, increased nasal, salivary, and lacrimal secretions, nervous behavior, tremors, increased serum alkaline phosphatase, and altered liver morphology. No effects were seen at a dosage of 0.2 mg/(kg)(day).

In a 90-day rat study, daily ingestion of 100 ppm resulted in moderate plasma, red cell, and brain cholinesterase inhibition. No compound-related effects were seen at 10 and 31.6 ppm. In a 2-year chronic dietary rat study, 31.6 and 100 ppm technical fonofos resulted in inhibition of plasma and red cell cholinesterase, nervous behavior, and tremors. No effects were noted at 10 ppm.

In a three-generation reproduction study in rats, doses of 1 to 3 mg/(kg)(day) technical fonofos resulted in no adverse effects.

In a 46-day ingestion study in chickens, doses of 2 to 20 mg/kg daily produced no suggestion of delayed neurotoxicity.

2.11.5 Hygienic Standard of Permissible Exposure (43)

The ACGIH has established a TWA of 0.1 mg/m^3.

2.12 Imidan® [N-(Mercaptomethyl)phthalimide-S-(O,O-dimethyl Phosphorodithioate), Prolate® (Animal Health), phosmet]

2.12.1 Source, Uses, and Industrial Exposure (45)

Imidan® is useful as an insecticide on grapes, kiwi and citrus fruit, nuts, peaches, plums, and potatoes, as well as for parasites of goats, hogs, horses, and sheep.

2.12.2 Physical and Chemical Properties (51)

Physical state	White or off-white crystalline solid
Water solubility	25 ppm at 20°C
Vapor pressure	6 × 10^{-2} mm Hg at 25°C
Melting point	67 to 70°C

2.12.3 Determinations in the Atmosphere

A known volume of air is collected in a series of scrubbers containing a suitable solvent. Imidan® is the oxygen analogue of Phosmet® and can be analyzed by the Phosmet® method (88). In order to be sure all of the Phosmet® is converted to Imidan®, it may be oxidized with dilute bromine water. No cleanup is required.

2.12.4 Physiological Response (89)

2.12.4.1 Acute. The acute oral LD_{50} in rats is in the range of 147 to 316 mg/kg following ingestion of technical Imidan®. Signs of toxicity include tremors, ataxia, salivation, cyanosis, diarrhea, excessive urination, and death. The acute oral LD_{50} in mice is in the range of 23 to 43 mg/kg with similar signs. The acute dermal LD_{50} in rabbits is greater than 4640 mg/kg and no signs of toxicity were observed at this dose. In dermal and eye irritation studies following the Draize technique using the rabbit, Imidan® was mildly irritating to rabbit skin and eye.

2.12.4.2 Subchronic and Chronic. Daily ingestion of 20, 40, and 400 ppm technical Imidan® by dogs for 2 years resulted in decreased red cell and plasma cholinesterase activity and lacrimation. One dog in six dosed at 400 ppm also showed hyperactivity, salivation, hyperemia of mouth, mucoid feces, and mortality. The same doses were used in a 2-year rat feeding study and no effect was observed at 20 and 40 ppm. Plasma and red cell cholinesterase activity was depressed at 400 ppm; a reduction in weight gain and liver cell vacuolation also occurred.

Daily ingestion of 40 ppm Imidan® by rats in a three-generation reproduction study or 80 ppm in a two-generation reproduction study resulted in no changes in reproduction, body weight gain, general observations, or survival.

Repeated dermal application of 10, 30, or 60 mg/kg 5 days a week for 3 weeks prior to mating and for 3 weeks after mating in rabbits resulted in reduced plasma and red blood cell cholinesterase activity but reproduction and teratogenic parameters were unaffected.

Daily ingestion of 100, 316, and 1000 ppm Imidan® technical by chickens for 6 weeks did not result in neurotoxicity.

2.12.5 Hygienic Standard of Permissible Exposure

The ACGIH has not established a TLV for Imidan® (43).

2.13 Isopestox® [Bis(monoisopropylamino)fluorophosphine Oxide, Mipafox®]

$$
\begin{array}{ccccccc}
CH_3 & H & H & O & H & H & CH_3 \\
\diagdown & | & | & \| & | & | & \diagup \\
& C & N & P & N & C & \\
\diagup & & & | & & & \diagdown \\
CH_3 & & & F & & & CH_3
\end{array}
$$

2.13.1 Source, Uses, and Industrial Exposure

Isopestox® was introduced as an effective systemic insecticide and acaricide. After being placed on the market, it was found responsible for near fatalities and paralysis of several workers in England, which resulted in its being withdrawn from commerce. It is included here for its historical value as one of the first organic phosphates that caused organic phosphate neurotoxicity in animals and humans.

2.13.2 Physical and Chemical Properties (52)

Physical state	Crystalline solid
Molecular weight	182.2
Density	1.2 (25°C)
Melting point	60 to 65°C
Boiling point	125 to 126°C (2 mm Hg)
Vapor pressure	0.001 mm Hg (5°C)
Solubility	Soluble in water and polar organic solvents; slightly soluble in petroleum oils

2.13.3 Determination in the Atmosphere

Presumably the general method of Giang and Hall (90) for enzymatic determination of organic phosphorus insecticides could be applied to measure Isopestox® air contamination.

2.13.4 Physiological Response

2.13.4.1 Acute. The acute oral LD_{50} in various species has been reported to range from 25 to 100 mg/kg (52, 91, 92). Near-lethal doses produce severe neurotoxic signs.

2.13.4.2 Subchronic and Chronic. In chickens, rabbits, and humans, irreversible flaccid paralysis has been demonstrated and appears to result from "demyelinization" of nerves, as described for TOCP (tri-*o*-cresyl phosphate) (92–94). In chickens, a single dose of 1.0 mg/kg orally produced the response in 10 to 14 days. Like other organic phosphates, Isopestox® reduces the activity of cholinesterase(s), both in vitro and in vivo. The signs of toxicity, except for the deymelination syndrome, are typical of the class, as indicated in the discussion of neurotoxicity, above.

2.13.5 Hygienic Standard of Permissible Exposure

No TLV or pesticide tolerance has been set for Isopestox® (43).

2.14 Malathion [*O,O*-Dimethyl *S*-(1,2-Dicarboethoxyethyl) Dithiophosphate, Experimental Insecticide No. 4049 (American Cyanamid Co.)]

$$
\begin{array}{c}
\mathrm{H_3CO\ \ S\qquad\quad H\ \ O} \\
\diagdown\ \| \qquad\quad |\ \ \| \\
\mathrm{P{-}S{-}\ C{-}C{-}O{-}C_2H_5} \\
\diagup \qquad\qquad | \\
\mathrm{H_3CO\qquad\ \ H{-}C{-}C{-}O{-}C_2H_5} \\
|\ \ \| \\
\mathrm{H\ \ O}
\end{array}
$$

2.14.1 Source, Uses, and Industrial Exposures

Malathion is regarded as the least toxic of the class and is generally considered as a wide spectrum insecticide for fruits, vegetables, and ornamental plants.

2.14.2 Physical and Chemical Properties (52, 95)

Physical state	Clear amber liquid, dependent on purity
Molecular weight	330
Density	1.2315 at 25°C
Melting point	156 to 157°C (0.7 mm Hg)
Vapor pressure	4.0×10^{-5} mm Hg (30°C)
Refractive index, $n_D^{25°C}$	1.4985
Solubility	145 ppm (weight in water); miscible with alcohols, ethers, vegetable oils; decomposes above the boiling point

2.14.3 Determination in the Atmosphere (96)

A known volume of air is drawn through a glass fiber filter. The filter is transferred to a screw-cap bottle within 1 hr of sampling. The analyte is extracted from the filter with isooctane. An aliquot of the extract is analyzed with gas chromatography using a flame photometric detection with a phosphorus filter.

2.14.4 Physiological Response

2.14.4.1 Acute. The acute LD_{50} for various species and routes has been complied by Golz and Shaffer (97). The acute oral LD_{50} has been reported to vary widely, but it is generally agreed to range around 1400 mg/kg for female rats when administered in vegetable oils. Acute vapor inhalation exposure does not appear to be too great a problem owing to the low vapor pressure and low inherent toxicity of malathion. Single dermal application of 2460 to 6150 mg/kg (90 percent grade) to rabbits produces cholinergic signs.

2.14.4.2 Subchronic and Chronic. Subchronic and chronic exposures of rats (97, 98) indicate that an oral dosage of 100 ppm (weight) can be tolerated without effect on cholinesterase(s) activity. Although 5000 ppm (weight) has a slight effect on survival, food consumption, and growth, some rats have survived 20,000 ppm (weight) in the diet for 2 years. Frawley et al. (99, 100) demonstrated that the simultaneous administration of malathion and EPN resulted in mortality greater than expected from the administration of the components alone (potentiation). Slight potentiation was also found in subchronic studies of these materials when cholinesterase activity was studied as the criterion of evaluation. DuBois and Coon (101) found that EPN interfered with the enzymatic hydrolysis of malathion and thus produced the potentiating effect by disrupting the detoxication mechanism of malathion.

In a 2-year feeding/oncogenic study in rats the NOEL was 5 mg/(kg)(day); 50 mg/(kg)(day) produced inhibition of brain cholinesterase and reduced body weight. No evidence of carcinogenicity was observed (102).

In a human study malathion was administered by gelatin capsules to groups of five male volunteers, 23 to 63 years of age, at doses of 8 mg/day for 32 days, 16 mg/day for 47 days, or 24 mg/day for 56 days. No observed effects were noted at the low and mid doses (NOEL = 16 mg/day), but at the high dose plasma and erythrocyte activity were depressed with no clinical signs (102).

In a reproductive study in rats there was a reduced number of live pups and reduced pup body weight at 240 mg/(kg)(day), the only dose tested. In a rat teratology study, 900 (only dose tested) mg/(kg)(day) (ip) produced no reproductive effects and no terata (102).

2.14.5 Hygienic Standard of Permissible Exposure

The TLV for malathion as established by the ACGIH is 10 mg/m^3 (43).

2.15 Methamidophos (*O,S*-Dimethyl Phosphoramidothioate, Acuphate-met, Monitor®)

$$
\begin{array}{c}
\text{CH}_3\text{S} \quad \text{O} \\
\diagdown \; \| \\
\text{P—NH}_2 \\
\diagup \\
\text{CH}_3\text{O}
\end{array}
$$

2.15.1 Source, Use, and Industrial Exposure (51)

Used for the control of cutworms, loopers, aphids, mites, and white flies in a number of crops.

2.15.2 Physical and Chemical Properties (51, 103)

Physical state	Clear liquid with rotten cabbage odor.
Solubility	Soluble in water or alcohol, 1% soluble in kerosene, 10% soluble in benzene or xylene

Boiling point	340 to 350°C
Molecular weight	141
Melting point	39 to 41°C
Specific gravity	1.31 at 20/4°C
Vapor pressure	1×10^{-4} mm Hg at 20°C

2.15.3 Determination in the Atmosphere

Product analysis is by infrared spectrometry or by gas–liquid chromatography. Details of these methods are available from Chevron Chemical Co., Richmond, CA or from Bayer, Leverkusen, Germany (104).

2.15.4 Physiological Response (103)

2.15.4.1 Acute. The oral LD_{50} in male and female rats is 21 and 16 mg/kg, respectively. The dermal LD_{50} in rabbits is 118 mg/kg. The 4-hr inhalation LC_{50} in male and female rats is 63 and 77 mg/kg, respectively. Methamidophos is irritating to the cornea, with clearing in 8 to 21 days.

2.15.4.2 Subchronic and Chronic. Dogs received daily oral doses of 0.75 mg/kg for 2 years without effect. Rats fed diets containing 10 mg/(kg)(day) for 2 years showed no adverse effect (81).

Mutagenic activity was negative in the microbial reverse mutation assay (Ames), chromosomal aberration assay (in vivo) in mouse bone marrow cells, and mouse dominant lethal assay (103). Methamidophos produced delayed neurotoxicity in chickens at antidoted doses 5 to 16 times the dose expected to kill 50 percent of the treated birds. This effect is not expected to occur at workplace exposure levels when used according to recommended precautions.

Methamidophos did not cause birth defects or impair reproductive performance when tested in experimental animals (103).

2.15.5 Hygienic Standard of Permissible Exposure

The ACGIH has not set a TLV for methamidophos (43).

2.16 Methyl Parathion [O,O-Dimethyl O-(p-Nitrophenyl) Phosphorothioate, Metacide®]

2.16.1 Source, Uses, and Industrial Exposure (51)

Methyl parathion is closely related to parathion in its chemistry and toxicology. It controls aphids, boll weevils, and mites especially well, although its spectrum for control of insects is nearly as broad as parathion.

2.16.2 Physical and Chemical Properties (52)

Physical state   White crystalline solid in pure form; brown liquid crystallizing at 29°C as the technical material

Molecular weight   263.3

Density   1.358 (20°C)

Melting point   35 to 36°C (pure)

Vapor pressure   0.5 mm Hg (109°C)

Refractive index, $n_D^{35°C}$ 1.5515

Solubility   50 ppm (weight) in water at 25°C; soluble in most aromatic solvents; slightly soluble in paraffin hydrocarbons

2.16.3 Determination in the Atmosphere (96)

A known volume of air is drawn through a glass fiber filter. The filter is transferred to a screw-cap bottle within 1 hr of sampling. The analyte is extracted from the filter with isooctane. An aliquot of the extract is analyzed with gas chromatography using a flame photometric detection with a phosphorus filter.

2.16.4 Physiological Response

2.16.4.1 Acute. The acute oral LD_{50} for methyl parathion in the rat ranges between 9 and 25 mg/kg (52), depending on the purity of the material studied. Methyl parathion is toxic by all routes but has been described as especially hazardous via the eye.

2.16.4.2 Subchronic and Chronic. DuBois and Coon (101) state the methyl parathion is approximately as toxic as parathion to rats but is a much less potent cholinesterase inhibitor. This material, like parathion, is a poor cholinesterase inhibitor in vitro but is converted in the mammalian liver to the toxic form, which is the oxygen analogue.

2.16.5 Hygienic Standard of Permissible Exposure

A threshold limit in air has been established by the ACGIH at 0.2 mg/m³ (43).

2.17 Naled (1,2-Dibromo-2-dichloroethyl Dimethyl Phosphate, Dibrom®)

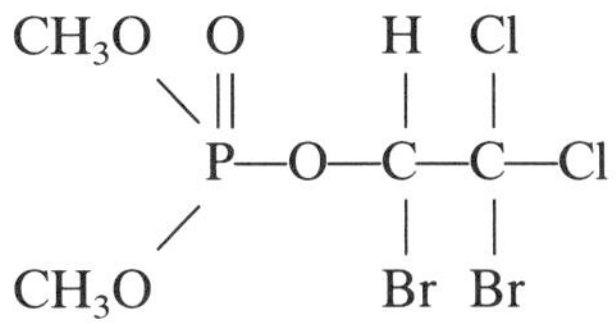

2.17.1 Sources, Uses, and Industrial Exposure (45)

Naled is useful as an insecticide and a miticide on nuts, seed, root, and leaf vegetables; cotton; peppers, eggplant, tomatoes; citrus fruit, grapes, hops, squash, melons, mushrooms, stone fruit, pumpkin, rice, strawberries, and grasses; it has therapeutic use in cattle, goats, hogs, horses, poultry, and sheep for the control of parasites. In this use it combines fast knockdown and broad spectrum with short residual life of deposits, plus a low order of mammalian toxicity.

2.17.2 Physical and Chemical Properties (52, 105)

Physical state	Pure—white solid; technical—light straw-colored liquid
Melting point	27°C (pure)
Molecular weight	381
Specific gravity	1.97 at 20°C/20°C (technical)
Volatility	Low
Vapor pressure	0.0002 mm Hg at 20°C
Boiling point	110°C at 0.5 mm Hg
Solubility	Hydrolyzes in water or aliphatic solvents, highly soluble in aromatic solvents

2.17.3 Determination in the Atmosphere

A known volume of air can be collected in a suitable solvent in a series of scrubbers. The method of McKinley (106) can be applied without cleanup.

2.17.4 Physiological Response (105)

2.17.4.1 Acute. The acute oral LD_{50} in rats is 430 mm/kg for purified naled. The acute dermal LD_{50} in rabbits is 1100 mg/kg for purified naled.

2.17.4.2 Subchronic and Chronic. Technical naled was fed in the diets of rats at 100 ppm for 12 weeks without effect. In chronic 2-year studies, technical naled was fed in the diet of rats at 100 ppm. No toxic effects were observed. In a 2-year chronic dog study, naled was administered at 7.5 mg/(kg)(day) without effect.

Rats and guinea pigs were exposed to the vapor of naled at 0.5 mg/ft^3 on a 6 hr/day, 5 day/week basis for 5 weeks. No toxic effects were observed.

2.17.5 Hygienic Standard of Permissible Exposure (43)

The ACGIH has published a TLV for Dibrom® in air as 3 mg/kg.

2.18 Parathion (O,O-Diethyl O-p-Nitrophenyl Phosphorothioate, E-605, Compound 3422)

$$O_2N-\!\!\!\!\bigcirc\!\!\!\!-O-\underset{\underset{\displaystyle OC_2H_5}{|}}{\overset{\overset{\displaystyle S}{\|}}{P}}\!\!\diagup\!\!\!OC_2H_5$$

2.18.1 Source, Uses, and Industrial Exposure (45)

Parathion is one of the best known of the class and, despite relatively high toxicity, it has received extensive use in agriculture as a broad spectrum insecticide. It is useful for use on fruits, vegetables, grasses, and nuts.

2.18.2 Physical and Chemical Properties (52)

Physical state	Technical-grade clear, medium to dark brown liquid
Molecular weight	291.27
Density	1.265 (25°C)
Melting point	6.1°C
Boiling point	157 to 162°C (0.6 mm Hg); 375°C (760 mm Hg)
Vapor pressure	0.00003, 0.00066, and 0.0028 mm Hg at 24.0, 54.5, and 70.7°C, respectively
Refractive index, $n_D^{20°C}$	1.53668
Solubility	24 ppm (weight) in water at 25°C; miscible with most organic solvents; decomposes at the boiling point

2.18.3 Determination in the Atmosphere (77)

A known volume of air is drawn through a glass fiber filter. The filter is transferred to a screw-cap bottle within 1 hr of sampling. The analyte is extracted from the filter with isooctane. An aliquot of the extract is analyzed with gas chromatography using flame photometric detection with a phosphorus filter.

2.18.4 Physiological Response

2.18.4.1 Acute. According to Hazleton and Holland (107), the oral LD_{50} for parathion is 3.5 and 12.5 mg/kg for female and male rats, respectively. Essential aspects of pharmacology and acute toxicology and antidotes for parathion appeared in 1948 in reports by DuBois et al. (108), Hagan and Woodard (109), and Hazleton and Godfrey (110).

2.18.4.2 Subchronic and Chronic. Parathion was not found to be stored in tissues of rats in subacute feeding studies (107). No effect was noted in growth, food consumption, gross or microscopic morphology, or survival when 100 ppm (weight) parathion was fed to male rats for a period of 2 years. Cholinesterase inhitation was observed at lower levels. Parathion was found by Gardocki and Hazleton (111) to be excreted in the urine as p-nitrophenol. This is a useful method, together with cholinesterase determinations, for measurement of exposure to parathion. The toxicity to humans can be evaluated from the studies of Grob et al. (112) and acute human intoxication incidences; a picture similar to that observed in laboratory animals is apparent. According to the International Agency for Research on Cancer Committee (113), parathion is not carcinogenic. EPA has placed parathion in category C, possible human carcinogen, based on increased adrenal cortical tumors in Osborne–Mendel rats and positive trends in thyroid follicle adenomas and pancreatic carcinomas in one study (114). Parathion was negative for mutational changes in a large number of tests including the microbial tests in *E. coli*, *S. typhimurium*, *Serratia marcesens*, *Saccharomyces cerevisiae*, or *Schizosaccharomyces pombe*, with and without metabolic activation; mouse dominant lethal assay (oral and ip); and *Drosphila* sex-linked recessive lethal assay (113). An EPA work group is reviewing the risk of parathion as a carcinogen. This panel has previously classified parathion as class C, possible human carcinogen, as indicated above.

2.18.5 Hygienic Standard of Permissible Exposure

The TLV for parathion as a dust, fume, or mist is 0.1 mg/m^3 (43).

2.19 Phorate (*O,O*-Diethyl *S*-(Ethylthiomethyl) Phosphorodithioate, Thimet®, Rampart®]

$$\begin{array}{c} C_2H_5O \quad \ S \\ \diagdown \ \| \\ P\!-\!S\!-\!CH_2SC_2H_5 \\ \diagup \\ C_2H_5O \end{array}$$

2.19.1 Source, Uses, and Industrial Exposure (51)

Phorate is a soil and systemic insecticide used to control a wide range of insects on various crops, primarily corn and grains.

2.19.2 Physical and Chemical Properties (51)

Phorate is a clear liquid with low solubility in water; it is miscible in xylene, vegetable oils, carbon tetrachloride, alcohols, ethers, and esters. At room temperature, phorate is stable but will hydrolyze under alkaline conditions.

Boiling point	125 to 127°C
Molecular weight	260.40 at 25°C
Vapor pressure	8.4×10^{-4} mm Hg at 20°C

2.19.3 Determination in the Atmosphere

Phorate can be analyzed with a gas chromatography method using a glass column, 1.8 m $\times$ 2 mm inner diameter, packed with Supelcoport (100/120 mesh) coated with 5 percent SP-240, and flame photometric detection (115).

2.19.4 Physiological Response (116)

2.19.4.1 Acute. The LD$_{50}$ in the male rat is 2 to 4 mg/kg. The percutaneous LD$_{50}$ in the male rat is 98 to 137 mg.

2.19.4.2 Subchronic and Chronic. In a 13 to 15 week subchronic study in rats, the no-effect level was defined as 0.01/mg/(kg)(day). Cholinesterase activity was severely depressed at higher dose [0.05, 0.25, and 1.24 mg/(kg)(day)] and mortality occurred at the two highest doses.

A three-generation reproduction study performed in Carworth Farms mice indicated there were no adverse effects on reproductive performance nor during lactation at doses as high as 1.5 ppm.

2.19.5 Hygienic Standard of Permissible Exposure

A TLV of 0.05 mg/m³ has been set by the ACGIH for phorate.

2.20 Phosdrin® (Alpha Isomer of 2-Carbomethoxy-1-methylvinyl Dimethyl Phosphate, Mevinphos®)

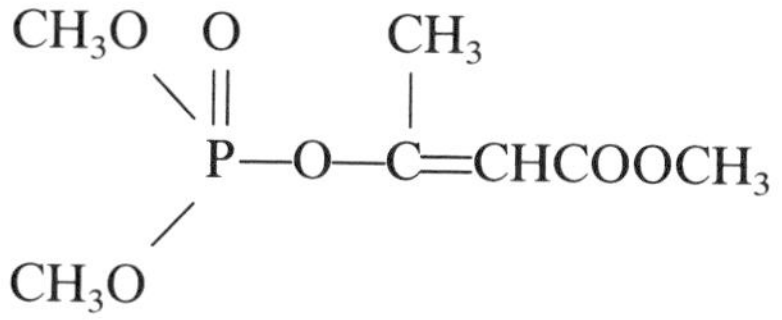

$$
\begin{array}{ccc}
CH_3O & O & CH_3 \\
\diagdown \, \| & & | \\
& P\!\!-\!\!O\!\!-\!\!C\!\!=\!\!CHCOOCH_3 \\
\diagup & & \\
CH_3O &
\end{array}
$$

2.20.1 Sources, Uses, and Industrial Exposure

Phosdrin® is useful in the control of aphids, mites, thrips, and lepidopterous larvae on a wide variety of crops (117).

2.20.2 Physical and Chemical Properties (118)

Physical state	Light yellow to orange liquid
Molecular weight	224.1
Density	1.23 (20°C)
Boiling point	106 to 107.5°C (1.0 mm Hg)
Vapor pressure	0.0029 mm Hg (21°C)
Refractive index, $n_D^{25°C}$	1.4493
Solubility	Miscible with water, alcohols, ketones, chlorinated and aromatic hydrocarbons; slightly soluble in aliphatic hydrocarbons

2.20.3 Determination in the Atmosphere (119)

Phosdrin® can be determined by drawing air through chromosorb 102 to trap, transferring the chromosorb 102 to vials and desorbing with toluene, and analyzing with gas chromatography equipped with a flame photometric detector.

2.20.4 Physiological Response

2.20.4.1 Acute. The acute oral LD_{50} for rats is 6.0 to 7.0 mg/kg. The dermal LD_{50} for rabbits is approximately 34 mg/kg. The LC_{50} in inhalation studies of 1-hr duration is approximately 14 ppm for female rats.

2.20.4.2 Subchronic and Chronic. In chronic studies the minimal lethal dose for rats has been established as between 100 and 200 ppm in the diet (weight). Lower doses affect tissue cholinesterase levels. Dosages below 5.0 ppm have no effect on cholinesterase activity (120).

2.20.5 Hygienic Standard of Permissible Exposure

A TLV has been set for Phosdrin® insecticide at 0.01 ppm or 0.092 mg/m³ (43).

2.21 Phosphamidon [2-Chloro-2-(diethylcarbamoyl)-1-methylvinyl Dimethyl Phosphate, Dimecron®]

2.21.1 Source, Uses, and Industrial Exposure (45)

Phosphamidon is useful as an insecticide on pome and citrus fruit, cole crops, melons, cucurbits, tomatoes and bell peppers, potatoes, sugarcane, walnuts, and cotton.

2.21.2 Physical and Chemical Properties (121)

Physical state	Colorless and odorless liquid: 89% phosphamidon; 3% related compounds
Specific gravity	1.2 at 20°C
Molecular weight	299.5
Volatility	Low
Boiling point	160°C at 1.5 mm Hg
Stability	Stable under ordinary storage conditions
Solubility	Miscible in all proportions with water, alcohol, and many other organic solvents

2.21.3 Determination in the Atmosphere (122)

Phosphamidon may be analyzed in the atmosphere by first collecting a sample in a suitable solvent in a series of scrubbers. Phosphamidon can be analyzed by gas–liquid chromatography with internal standard.

2.21.4 *Physiological Response (121)*

2.21.4.1 Acute. The acute oral toxicity of phosphamidon technical (89% phosphamidon and 3% related compounds) in the albino rat is 28.3 mg/kg. The acute dermal LD_{50} in rabbits is 267 mg/kg.

2.21.4.2 Subchronic and Chronic. In subchronic feeding studies in dogs, phosphamidon was tolerated at a dose of 5 mg/kg for 90 days without effect. In subacute inhalation studies, dogs, guinea pigs, and rats were exposed to an atmosphere containing 0.125 mg/l phosphamidon for 6 hr/day, 5 days/week for 90 days without any marked toxic effects. In a 62 to 80 week study (123) there was no evidence to indicate carcinogenesis when phosphamidon was fed to rats or mice at doses of 80 or 160 ppm.

2.21.5 *Hygienic Standard of Permissible Exposure*

No TLV has been established for phosphamidon by the ACGIH (43).

2.22 Schradan [Octamethyl Pyrophosphoramide, OMPA, Tetrakis(dimethylamino) Phosphonous Anhydride]

$$(CH_3)_2N \quad O \qquad O \quad N(CH_3)_2$$
$$\diagdown \| \qquad \| \diagup$$
$$P-O-P$$
$$\diagup \qquad \diagdown$$
$$(CH_3)_2N \qquad \qquad N(CH_3)_2$$

2.22.1 *Source, Uses, and Industrial Exposure*

Schradan is a systemic insecticide that is included here for historical value. It was developed in Germany by Schrader during World War II as a war gas. It is the first member of the class used as an insecticide. It is no longer used.

2.22.2 *Physical and Chemical Properties (52)*

Physical state	Viscous, dark brown liquid
Molecular weight	286.3
Density	1.1343 (25°C)
Melting point	Below −10°C
Boiling point	135 to 137°C (1 mm Hg)
Vapor pressure	0.0003 mm Hg (25°C)
Refractive index, $n_D^{25°C}$	1.4612
Solubility	Miscible with water; soluble in ethanol, acetone, chloroform, benzene; insoluble in heptane and petroleum ether

2.22.3 Determination in the Atmosphere

A method suitable for the analysis of schradan in air has been reported by Hartley et al. (124).

2.22.4 Physiological Response

2.22.4.1 Acute. In acute studies, the LD_{50} has been reported by Lehman (125) to be 13.5 mg/kg orally in rats. DuBois et al. (126) reported the oral LD_{50} to be 8.0 mg/kg. Frawley et al. (127) listed the oral LD_{50} as 35.5 mg/kg for female rats and 13.5 mg/kg for male rats, thus indicating some sex variation. Reports from these and other investigators indicate that there is very little, if any, species variation to the acute response.

2.22.4.2 Subchronic and Chronic. The chronic toxicity in rats has been studied by Barnes and Denz (128). At 50 ppm (weight) in the diet for 1 year, male rats showed toxic signs, growth suppression, and marked cholinesterase depression without producing tissue pathology. At dosages of 10 and 50 ppm (weight) whole blood cholinesterase activity was reduced but brain activity was unaffected. Further studies in male rats at 1.0 ppm (weight) showed that the acetylcholinesterase (true cholinesterase) in the red cell was unaffected. A dosage of 0.3 ppm produced no effect. This was confirmed by Edson et al. (129), who fed 0.25 ppm (weight) to rats without effect. Studies conducted on humans indicate that a level of 0.6 ppm (weight) was without effect in six male and six female subjects. One subject was administered 2.4 ppm (weight) in the total diet, which produced marked plasma and red cell cholinesterase depression (129).

2.22.5 Hygienic Standard of Permissible Exposure

No threshold limit has been set for schradan (43).

2.23 Systox® (O,O-Diethyl O((and S))-[2-(Ethylthio)ethyl] Phosphorothioate, Dematon®) Mixture of the Thiono- and Thiol-isomers

$$
\begin{array}{c}
C_2H_5\!-\!O \qquad\qquad \overset{\displaystyle H}{|}\;\; \overset{\displaystyle H}{|} \\[2pt]
\diagdown \\
P\!-\!S\!-\!C\!-\!C\!-\!S\!-\!C_2H_5 \\
\diagup \qquad\qquad |\;\; | \\
C_2H_5\!-\!O \qquad\qquad H\;\; H
\end{array}
$$

(Thiono Isomer)

2.23.1 Source, Uses, and Industrial Exposures (52)

Systox® is a systemic insecticide, which also has contact action. It is particularly useful for the control of sucking insects such as mites and aphids.

2.23.2 Physical and Chemical Properties (52)

Systox® is a mixture of thiol and thiono isomers.

Physical state	Pale yellow to light brown liquid
Molecular weight	258
Density	1.1183 (20°C)
Boiling point	134°C (2 mm Hg)
Vapor pressure	0.001 mm Hg (33°C)
Refractive index, $n_D^{20°C}$	1.4875
Solubility	0.01 percent in water; miscible with most organic solvents

2.23.3 Determination in the Atmosphere (130)

Demeton-O and Demeton-S are the two isomers found in Systox®. Air volume is drawn through a liter cassette containing a 37-mm mixed cellulose ester filter (MCEF) followed by a glass tube containing XAD-2 sorbent to trap the two isomers. The filter and sorbent are combined and are desorbed with tolune. The sample is analyzed by gas chromatography utilizing a phosphorus-sensitive flame photometric detector. Each isomer elutes separately and is quantified individually.

2.23.4 Physiological Response

2.23.4.1 Acute.
The acute oral LD_{50} in rats ranges from 2.5 to 12 mg/kg. The dermal LD_{50} is 8.2 to 14 mg/kg (51).

The isomeric confusion surrounding Systox® in 1954 has since been resolved. Two isomers, the thiol and thiono isomers, have been identified. Furthermore, the thiol isomer is the same as the oxon metabolite of disulfoton, and the thiono isomer is structurally similar to diisulfoton. On this basis, the chronic data submitted for disulfoton were fully accepted by EPA to fulfill the chronic requirements for demeton (Systox®) (131).

Disulfoton (disystox) (131) was fed to male and female Fischer 344 rats at 0.8, 3.3, or 13 ppm in the diet for 105 weeks (intended doses were 1, 4, and 16 ppm). Dose-related plasma, erythrocyte, and brain cholinesterase inhibition was observed in both sexes at all doses. Optic nerve degeneration was observed in a dose-dependent fashion at all doses but was statistically significant only at the two highest dose levels. In a 2-year feeding study in dogs a no observed effect level (NOEL) of 1.0 ppm was established based on cholinesterase depression and the lowest effect level (LEL) was 2 ppm. In a rat 2-year feeding study, the NOEL was 1.0 ppm based on cholinesterase depression and the LEL was 2.0 ppm. No evidence of carcinogenicity was observed in any chronic study regardless of species. In teratology studies in the mouse, the developmental NOEL was 7 mg/(kg)(day) with the LEL at 10 mg/(kg)(day) (ip). There were possible increases in abnormalities in the digestive tract and evidence of cleft palate.

2.23.5 Hygienic Standard of Permissible Exposure (43)

The TLV adopted by the ACGIH for Systox® is 0.01 ppm or 0.11 mg/m³.

2.24 Trithion® [O,O-Diethyl S-(p-Chlorophenyl)thiomethyl Phosphorodithioate, R-1303, Carbophenothion]

2.24.1 Source, Uses, and Industrial Exposure (52)

Trithion® is a nonsystemic insecticide, miticide, and ovicide with a relatively long residual action.

2.24.2 Physical and Chemical Properties (132)

Physical state	Light amber liquid
Molecular weight	342.9
Density	1.265 to 1.285 (25°C)
Vapor pressure	Very low
Refractive index, $n_D^{25°C}$	1.590 to 1.597
Solubility	Not appreciably soluble in water; miscible with vegetable oils and most organic solvents

2.24.3 Determination in the Atmosphere

Air may be analyzed for Trithion® by the method described by Patchett (133).

2.24.4 Physiological Response

2.24.4.1 Acute. The acute oral LD_{50} of Trithion® for male albino rats is 17.2 to 28 mg/kg (134).

2.24.4.2 Subchronic and Chronic. Exposure of rats and dogs to a near saturated vapor (in air) of Trithion® for 4 weeks resulted in plasma cholinesterase depression in dogs, but there was no reduction in enzyme activity in rats. The animals appeared normal throughout the exposure. It would appear that Trithion® is not volatile enough to be a hazard by the inhalation route. Subchronic feeding studies in rats and dogs have been performed by Weir and Fogleman (134). Dosages of 100 ppm by weight in the diet of rats produced tremors and reduced body weight gains. Cholinesterase activity was markedly depressed, but no gross or microscopic morphological changes occurred. Dosages of 5.0 ppm (weight) produced no effect. Dosages of 1.0 ppm (weight) produced plasma and red cell cholinesterase depression without overt signs or pathology. No neurotoxicity was apparent when Trithion® was fed to hens at concentrations up to 100 ppm, although this dose affected egg production (67).

2.24.5   Hygienic Standard of Permissible Exposure

No threshold limit has been established for Trithion® (39).

REFERENCES

1. *Fed. Reg.* **37**, 13369 (July 1972).
2. W. O Negherbon, *Handbook of Toxicology*, Vol. III, Saunders, Philadelphia, 1959.
3. L. S. Goodman and A. Gilman, *The Pharmacological Basis of Therapeutics*, 8th ed., Pergamon Press, New York, 1990.
4. L. W. Hazleton, *J. Agric. Food Chem.*, **3**, 312 (1955).
5. J. H. Wills, *Crit. Rev. Toxicol.*, **1**, 153 (1972).
6. Anon., *National Cancer Institute Carcinogenesis Technical Report Series*, NCI-CG-TR-10, 1977.
7. NCI-CG-TR-24, 1978.
8. NCI-CG-TR-33, 1978.
9. NCI-CG-TR-69, 1978.
10. NCI-CG-TR-70, 1979.
11. NCI-CG-TR-192, 1979.
12. NCI-CG-TR-16, 1979.
13. D. J. Brunsick, V. F. Simmon, H. S. Rosendranz, V. A. Ray, and R. S. Stafford, *Mutat. Res.*, **76**, 169–190 (1980).
14. J. McCann, E. Choi, E. Yamasaki, and B. M. Ames, *Proc. Natl. Acad. Sci., U.S.*, **72**(12), 5135–5139 (1975).
15. M. I. Smith, E. Ehrve, and W. H. Frazier, *Public Health Rep. U.S.*, **45**, 2509 (1930).
16. M. I. Smith and R. D. Lillie, *Am. Med. Assoc. Arch. Neurol. Psychiatr.*, **26**, 976 (1931).
17. W. F. Durham, T. B. Gaines, and W. J. Hayes, Jr., *Am. Med. Assoc. Arch. Ind. Health*, **13**, 326 (1956).
18. J. M. Barnes and F. A. Denz, *J. Pathol. Bacteriol.*, **65**, 587 (1953).
19. C. H. Hine, E. F. Gutenburg, M. M. Coursey, K. Seligman, and R. M. Gross, *J. Pharmacol. Exp. Ther.*, **113**, 28 (1955).
20. J. P. Frawley, R. E. Zwickey, and H. N. Fugat, *Fed. Proc.*, **15**, 424 (1956).
21. L. Austin and D. R. Davies, *Brit. J. Pharmacol.*, **9**, 145 (1954).
22. R. Baron, Ed., *Pesticide Induced Delayed Neurotoxicity*, Proceedings of a Conference Sponsored by the Environmental Protection Agency, 600/1-76-025, July, 1975.
23. P. S. Spencer and H. H. Schaumberg, Eds., *Experimental and Clinical Neurotoxicology*, Williams and Wilkins, Baltimore/London, 1980.
24. K. Kay, L. Monkman, J. P. Windish, T. Doherty, J. Park, and C. Racicat, *Arch. Ind. Hyg. Occup. Med.*, **6**, 252 (1952).

25. W. T. Summerford, W. J. Hayes, Jr., J. M. Johnson, K. Walker, and J. Spillane, *Arch. Ind. Hyg. Occup. Med.*, **7**, 383 (1953).

26. M. K. Johnson, "Mechanism of Action of Neurotoxic Organophosphorous Esters," in R. Baron, Ed., *Pesticide Induced Delayed Neurotoxicity*, Proceedings of a Conference Sponsored by EPA, 600/1-76-025, July 1975, p. 5.

27. H. O. Michel, *J. Lab. Clin. Med.*, **34**, 1564 (1949).

28. R. Ammon, *Arch. Ges. Physiol. Pflüger's*, **233**, 57 (1933).

29. R. L. Metcalf, *J. Econ. Entomol.*, **44**, 883 (1951).

30. J. H. Fleisher and E. J. Pope, *Arch. Ind. Hyg. Occup. Med.*, **9**, 323 (1954).

31. E. F. Edson, *World Crops*, **1**, (August 1958).

32. G. L. Ellman et al., *Biochem. Pharmacol.*, **7**, 88 (1961).

33. C. D. Johnson and R. L. Russell, *Anal. Biochem.*, **64**, 229–238 (1987).

34. *Proceedings of the U.S. EPA Workshop on Cholinesterase Methodologies*, March, 1992.

35. J. H. Wolfsie and G. D. Winter, *Arch. Ind. Hyg. Occup. Med.*, **6**, 43 (1952).

36. J. H. Wolfsie and G. D. Winter, *Arch. Ind. Hyg. Occup. Med.*, **9**, 396 (1954).

37. A. S. Gordon and C. W. Frye, *J. Am. Med. Assoc.*, **159**, 1181 (1955).

38. R. M. Clyne and C. B. Shaffer, *Toxicological Information, Cyanamid Organic Pesticides*, 3rd ed., American Cyanamid Company, Princeton, NJ, 1975.

39. T. Namba and K. Hiraki, *J. Am. Med. Assoc.*, **166**, 1834 (1955).

40. D. Grob and R. J. Johns, *J. Am. Med. Assoc.*, **166**, 1855 (1955).

41. *Abate®, Product Bulletin, American Cyanamid Company*, Princeton, NJ, 1980.

42. N. R. Pasarela and E. J. Orloski, "Abate® Insecticide," in *Analytical Methods for Pesticides and Plant Growth Regulators*, Vol. 8, *Thin-layer and Liquid Chromatography, Pesticides of International Importance*, J. Sherma and G. Zweig, Eds., Academic Press, New York, 1973.

43. *Threshold Limit Values for Chemical Substances and Physical Agents in the Workroom Environment*, American Conference of Governmental Industrial Hygienists, Cincinnati, OH, 1990–1991.

44. "Orthene® Insecticide," Technical Information, Exp. Data Sheet, Chevron Chemical Co., Research Labs, Richmond, CA, 1976.

45. *Tolerances for Pesticides In or On Agricultural Commodities*, EPA, Washington, DC, July 9, 1980.

46. *MSDS No. 1354, Chevron Chemical Co.*, Richmond, CA, 1991.

47. J. F. Thompson and R. R. Watts, *Analytical Reference Standards and Supplementary Data for Pesticides and other Organic Compounds*, EPA-600/9-78-0012, U.S. EPA, Health Effects Research Lab, ORD, Research Triangle Park, NC, 1978.

48. M. A. Luke, J. E. Froberg, and H. T. Masumoto, *J. Offic. Anal. Chem.*, **58**, 1020 (1975).

49. E. Y. Spencer, *Guide to Chemicals in Crop Protection*, Queens Printer, Ottawa, Canada, 1973.

50. *Merck Index*, 10th ed., 1983.

51. Charlotte Sine, *Farm Chemicals Handbook*, Meister Publishing Co., Willough, OH, 1989.

52. American Conference of Governmental Industrial Hygienists, TLVs, 4th edition and Suppl., 1980.

53. Dow Chemical Co., Industrial Hygiene Monitoring Method, DOWM 100863-HE90A HEH-IHM-90-02, Midland, MI, 1990.

54. Dow Chemical Co., *Toxicological Properties of Chlorpyrifos*, Midland, MI, 1990.

55. *Pesticide Official Publication and Condensed Data on Pesticide Chemicals*, Association of American Pest Control Officials, College Park, MD, 1957.

56. U.S. Dept. Health, Education and Welfare, *NIOSH Manual of Analytical Methods*, 2nd ed., 1979.

57. A. M. Mattson, J. T. Spillane, and G. W. Pearce, *J. Agric. Food Chem.*, **3**, 319 (1955).

58. U.S. EPA, Memorandum Third Peer Review of Dichlornos-Reevaluation following April 18, 1988 meeting of the NTP Panel of Experts, 1988.

59. J. P. Frawley, R. Weir, T. Tusing, K. P. DuBois, and J. C. Calandra, *Toxicol. Appl. Pharmacol*, **2**(5), 605–624 (1963).

60. H. J. Harris, *A Tentative Ultraviolet Method for Analysis of Diazinon® in Spray Residues*, Geigy Chemical Corp., Yonkers, NY, 1953.

61. R. B. Bruce, *J. Agric. Food Chem.*, **3**, 1017 (1955).

62. R. B. Bruce, *Fed. Proc.*, **13**, (1954).

63. H. Gysin and A. Margot, *J. Agric. Food Chem.*, **6**, 900 (1958).

64. H. Gysin, personal communication.

65. *National Cancer Institute Carcinogenesis Technical Report Series*, NCI-CG-TR-137, 1979.

66. National Research Council, *Drinking Water and Health*, 1977, p. 610.

67. W. J. Hayes, *Pesticides Studied in Man*, Williams and Wilkins, Baltimore, MD, 1982.

68. *Cygon®, Dimethoate Systemic Insecticide*, Cyanamid International Technical Information, American Cyanamid Co., Wayne, NJ, 1967.

69. E. D. Chilwell and P. I. Beecham, *J. Agric. Food Chem.*, **13**, 178 (1964).

70. P. A. Giang, W. F. Barthel, and S. A. Hall, *J. Agric. Food Chem.*, **2**, 1281, (1954).

71. K. P. DuBois and C. J. Cotter, *Arch. Ind. Hyg. Occup. Med.*, **11**, 53 (1955).

72. *IRAC Monogr.*, **30**, 215 (1983).

73. *IARC Monogr.*, **30**, 212 (1983).

74. E. J. Olajos, et al., *Ecotoxicol. Environ. Safety*, **3**(3), 245 (1979).

75. R. E. Staples and E. H. Goulding, *Environ. Health Perspect.*, **30**, 105 (1979).

76. Berge et al., *Arch. Toxicol.* **59**, 30–35 (1986).

77. U.S. Dept. Health, Education and Welfare, *NIOSH Manual of Analytical Methods*, 2nd ed., Vol. 3, Part II, 1979, p. S285.

78. H. C. Hodge, E. A. Maynard, L. Hurwitz, V. DiStefano, W. L. Downs, C. K. Jones, and H. J. Blanchet, Jr., *J. Pharmacol. Exp. Ther.*, **112**, 29 (1954).

79. M. B. Abou-Donia, *Neurotoxicology* (Park Forest, IL), **4**(1), 113–129 (1983).

80. AOAC, 10th ed., (1965).

81. Charles E. Worthing, *Pesticide Manual*, 7th ed., British Crop Protection Council, Croydon, England, 1983.

82. FMC, unpublished study, available from EPA under Freedom of Information Act, 1988.

83. FMC, unpublished study, available from EPA under Freedom of Information Act, 1985.

84. FMC, unpublished study, available from EPA under Freedom of Information Act, 1972.
85. FMC, unpublished study, available from EPA under Freedom of Information Act, 1986.
86. M. C. Bowman and M. Beroza, *J. Assoc. Offic. Anal. Chem.*, **54**, 1086 (1971).
87. *Prod. Safety Info. Dyfonate® Technical*, Stauffer Chemical Co., Agricultural Chem. Div., Westport, CT, 1978.
88. B. D. Ripley, R. J. Wilkinson, and A. S. Y. Chau, *J. Assoc. Offic. Anal. Chem.*, **57**, 1033 (1974).
89. *Prod. Safety Info., Imidan®*, Stauffer Chemical Co., Agricultural Chem. Div., Westport, CT, 1978.
90. P. A. Giang and S. A. Hall, *Anal. Chem.*, **23**, 1830 (1951).
91. H. Martin, *Guide to the Chemicals Used in Crop Protection*, 2nd ed., University of Western Ontario, 1955.
92. W. E. Ripper, *Ct. Rd. III Congr. Int. Phytopharm*, Paris, 1952.
93. D. R. Davies, *J. Pharm. Pharmacol.*, **6**, 1 (1954).
94. D. R. Davies, *Proc. Roy. Soc. Med.*, **45**, 570 (1952).
95. *Malathion Concentrate (Technical Bull.)*, Cyanamid Intl. (American Cyanamid Co.), Wayne, NJ, 1966.
96. U.S. Health, Education and Welfare, *NIOSH Manual of Analytical Methods*, 2nd ed., Vol. 3, Part II, 1979, p. S370.
97. H. H. Golz and C. B. Shaffer, *Malathion, Summary of Pharmacology and Toxicology*, Central Medical Dept., American Cyanamid Co., New York, revised January, 1955.
98. L. W. Hazleton and E. G. Holland, *Arch. Ind. Hyg. Occup. Med.*, **8**, 339 (1953).
99. J. P. Frawley, E. C. Hagan, O. G. Fitzhugh, H. N. Fuyat, and W. I. Jones, *J. Pharmacol. Exp. Ther.*, **119**, 147 (1957).
100. J. P. Frawley, H. N. Fuyat, E. C. Hagan, J. R. Blake, and O. G. Fitzhugh, *J. Pharmacol. Exp. Ther.*, **121**, 96 (1957).
101. K. P. DuBois and J. M. Coon, *Arch. Ind. Hyg. Occup. Med.*, **6**, 9 (1952).
102. American Cyanamid, unpublished studies, EPA Oral RFD documents, available from EPA under Freedom of Information Act, 1987.
103. *Material Safety Data Sheet, No. 1338*, Chevron Chemical Co., Richmond, CA, 1991.
104. E. Mollhoff, *Pflschutz-Nach. Bayer*, **24**, 254 (1971).
105. *Technical Info., Ortho Dibrom®*, Chevron Chemical Co., Richmond, CA, 1970.
106. W. P. McKinley, *J. Assoc. Offic. Anal. Chem.*, **48**, 748 (1965).
107. L. W. Hazleton and E. G. Holland, *Adv. Chem. Ser.*, **1**, 31 (1950).
108. K. P. DuBois, J. Doull, and J. M. Coon, *Fed. Proc.*, **7**, 216 (1948).
109. E. C. Hagan and G. Woodard, *Fed. Proc.*, **7**, 216 (1948).
110. L. W. Hazleton and E. Godfrey, *Fed. Proc.*, **7**, 226 (1948).
111. J. F. Gardocki and L. W. Hazleton, *J. Am. Pharm. Assoc. Sci. Ed.*, **40**, 491 (1951).
112. D. Grob, W. L. Garlick, and A. M. Harvey, *Bull. Johns Hopkins Hosp.*, **87**, 106 (1950).
113. *IARC Monographs 1972*, Vol. 30, 1983, p. 168.

114. EPA memorandum, Toxicology Branch Peer Review Committee, Office of Pesticides, July, 1986.

115. H. J. Stan and D. J. Mrowetz, *J. High Resolution Chromatogr. Commun.*, **6**(15), 255–263 (1983).

116. National Research Council, *Drinking Water and Health*, 1977, p. 617.

117. *Pesticide Official Publication, Supplement*, Association of American Pest Control Officials, College Park, MD, 1957.

118. *Bull. SC: 59-37, Summary of Basic Data for Phosdrin® Insecticide*, Shell Chemical Corp., New York, 1959.

119. U.S. Dept. of Health, Education and Welfare, *NIOSH Manual of Analytical Methods*, 2nd ed., Vol. 6, 1979, p. 296.

120. J. K. Kodama, C. H. Hine, and M. S. Morse, *Arch. Ind. Hyg. Occup. Med.*, **9**, 54 (1954).

121. *Tech. Info. Bulletin: Phosphamidon*, Chevron Chemical Co., Richmond, CA, 1970.

122. J. R. Markus and B. Puma, *Phosphamidon, Pesticide Analysis Manual*, Vol. 2, *Pesticide Regulation*, Sec. 120.239, Food & Drug Administration, August, 1977.

123. National Cancer Institute, *NCI Technical Report Series*, No. 16, 1976.

124. G. S. Hartley, D. F. Health, J. M. Hulme, D. W. Pound, and M. Whittaker, *J. Sci. Food. Agric.*, **2**, 303 (1951).

125. A. J. Lehman, *Assoc. Food Drug. Ofic. U.S. Q. Bull*, **15**, 122 (1951).

126. K. P. DuBois, J. Doull, and J. M. Coon, *J. Pharmacol. Exp. Ther.*, **99**, 376 (1950).

127. J. P. Frawley, E. C. Hagan, and O. G. Fitzhugh, *J. Pharmacol. Exp. Ther.*, **105**, 156 (1952).

128. J. M. Barnes and F. A. Denz, *Brit. J. Ind. Med.*, **11**, 11 (1954).

129. E. P. Edson, K. P. Fellowes, and F. MacL. Carey, *Med. Dept. Rep.*, Fisons Pest Control Ltd., England, 1954.

130. U.S. Dept. of Health, Education and Welfare, *NIOSH Manual of Analytical Methods*, Vol. 6, 1979, p. S280.

131. Unpublished reports considered by EPA to establish RFD for Syntox®, (1988). Mobay Chemical MRID No. 00129456, 00146873, 41115401, and other studies from Mobay Chemical, (1975, 1985); Pensalt Chemicals (1954); and Ciba-Geigy (1957). Reports available from EPA under Freedom of Information Act.

132. *Prod. Safety Info. Trithion® Technical*, Stauffer Chemical Co., Westport, CT, 1979.

133. G. G. Patchett, *Determination of R-1303 Spray Residues in Oranges, Lemons and Alfalfa*, Stauffer Chemical Co., Richmond, CA, 1956.

134. R. J. Weir and R. W. Fogleman, unpublished data.

CHAPTER THIRTEEN

Alkaline Materials

James O. Pierce, Sc.D.

The primary health hazards arising out of undue exposures to the alkaline chemicals are mainly those of irritation and corrosion of sensitive body tissues that come into direct contact with the agent. The most commonly encountered alkaline materials include the alkaline salts of ammonia, calcium, potassium, and sodium. The tissues most susceptible to biologic damage are the eyes, which, if not treated rapidly after exposure, can be severely damaged. Most exposures come about as a result of accidental spills by way of splashes of solid and liquid materials through carelessness or mishandling. Employee training and the use of appropriate personal protective equipment are the most effective way to reduce or minimize exposures.

In addition to appropriate training and proper use and care of personal protective equipment, all facilities using or handling alkaline materials should have emergency eye washes available in readily located areas for immediate and prolonged washing of the eyes with water in the case of exposure. It is extremely important that these emergency washing devices be periodically checked for proper working order: many instances of serious injuries resulting from accidental splashes have been reported that could have been avoided had the emergency wash devices been in working order. Direct contact of the alkaline materials with the skin or respiratory tract may result in irritation and corrosion, because these chemical agents react directly with tissue proteins to form aluminates. Long-term respiratory exposure, even for low airborne concentrations, can result in permanent damage to the respiratory track.

Patty's Industrial Hygiene and Toxicology, Fourth Edition, Volume 2, Part A, Edited by George D. Clayton and Florence E. Clayton.
ISBN 0-471-54724-7   © 1993 John Wiley & Sons, Inc.

1 AMMONIA (CASRN 7664-41-7), AMMONIUM HYDROXIDE (CASRN 1336-21-6), AND AMMONIUM SALTS

1.1 Sources, Uses, and Occupational Exposures

Ammonia (NH_3) is a naturally occurring substance that plays a vital role in protein metabolism in almost all species including humans. In humans it also is an important component of the balanced acid–base, electrolyte system (1). The mean blood level in the normal human is 0.08 mg percent measured as N (2). Excess ammonia is detoxified in the liver by conversion to urea (3). In the event an individual's liver function is greatly reduced, any source of ammonia, such as protein catabolism, ingestion of ammonium salts, or inhalation of ammonia, can lead to hepatic coma with increased circulating ammonia.

Ammonia is manufactured primarily by a modified Haber reduction process using atmospheric nitrogen and a hydrogen source, for example, methane, ethylene, or naphtha, at high temperatures (400 to 6500°C) and pressures (100 to 900 atm) in the presence of an iron catalyst (4, 5). It is a by-product of coal carbonization. Calcium cyanamide, formed by reacting nitrogen with calcium carbide, can be reacted with water to produce ammonia. Most of the ammonia produced worldwide is used for fertilizer. It is used extensively in Japan and Russia (6). Much of it is applied either by injection of the anhydrous gas directly into the soil or as an aqueous solution. Some is applied as various salts, such as the nitrate, sulfate, or diphosphate. Ammonia is used extensively as a refrigerant gas in commercial installations (5, 7). It is also used in the manufacture of chemicals such as plastics, explosives, nitric acid, urea, hydrazine, pesticides, and detergents (4, 5).

When the National Institute for Occupational Safety and Health (NIOSH) first published its Criteria Document for Ammonia in 1974, it was estimated that 500,000 workers were potentially exposed to ammonia (8). Although NIOSH has not updated these figures, private communication with industry representatives has indicated that this number has risen significantly by 1991, with estimates ranging from 750,000 to over a million. These exposures may occur at points of manufacture, transfer, use, or disposal. When ammonia is used as a developer in photocopying processes, for example, blueprint and diazo, it may be released into the workplace (8). Ammonia is transported by way of pipelines, barges, trucks, and cylinders, and may be stored under high pressure, refrigerated at low pressure, or as aqueous ammonia in low pressure tanks (4).

In addition to being handled as a compressed gas, ammonia is commonly encountered as aqueous solutions of 28 percent (aquammonia), called ammonium hydroxide (NH_4OH), and 10 percent, called household ammonia.

1.2 Physical and Chemical Properties

Ammonia is a colorless gas at ambient temperatures and pressures with a strong, irritating odor. It has the following characteristics:

Molecular weight	17.03
Melting point	−77.7°C

Boiling point	$-33°C$
Specific gravity (liquid)	$0.682(-33.35°C/4°C)$
Specific gravity (28 percent aqueous)	0.90 (25°C/25°C)
Vapor density	0.59 (air = 1) at 25°C, 760 mm Hg
Solubility	90 g in 100 ml water at 0°C
	13.2 g in 100 ml ethanol at 20°C
Alkalinity	pH of 1 percent aqueous solution is about 11.7
Odor threshold	3.5 to 37 mg/m³ (5 to 53 ppm)
Autoignition temperature	651°C (1204°F)
Explosive limits	16 to 25 percent by volume in air
Critical temperature	132.9°C
Pressure at critical temperature	111.5 atm
Ionization constants	$K_b 1.774 \times 10^{-5}$, $K_a 5.637 \times 10^{-10}$ at 25°C

1 mg/l = 1438 ppm and 1 ppm = 0.7 mg/m³ at 25°C and 760 mm Hg

In addition to the foregoing, extensive thermodynamic data have been tabulated (4, 9, 10).

Ammonia is classified as a flammable gas by the National Fire Protection Association. The fire hazard of high concentrations of ammonia is increased when other combustibles, such as oil, are also present. The critical temperature of 133°C is easily exceeded in fires so that containers of liquefied ammonia may explode unless their rupture strength is safely in excess of 112 atm (11). Concentrations of 28,800 mg/m³ or greater are considered by the Occupational Safety and Health Administration (OSHA) to be a potential fire or explosion hazard. Dry chemical or carbon dioxide are recommended extinguishing media (8).

The strong, pungent, penetrating odor of low levels of ammonia (± 35 mg/m³) becomes increasingly irritating as concentrations exceed 70 mg/m³ (12, 13).

1.3 Atmospheric Analysis

Analytic procedures for ammonia in a variety of media have been reviewed and extensively discussed (4). Absorption of the ammonia in a standard acid by means of an impinger followed by back-titration with standard base is generally applicable but is subject to interference from acids and bases (14). The traditional method using Nessler's reagent is sensitive to 3 ppm in a 50-liter air sample (14). Alternatives to the Nessler procedure with fewer interferences are the indophenol (15) and the pyridine–pyrazolone (16) techniques. Infrared spectrophotometry may be used for direct air analysis at 10.34 nm (17). Specific ion electrodes are now available commercially for the ammonium ion in aqueous systems. Gas detector tubes for use in conjunction with handheld pumps are commercially available. These are certified by NIOSH to have an accuracy of ± 35 percent at one-half the exposure limit, and ± 25 percent at one to five times the limit (18). They are useful and convenient for exploratory surveys but more precise techniques are essential for assuring compliance with standards. Numerous techniques have been critically reviewed (19). The newer, more sophisticated analytic methods include the use of passive dosimeters and ion chromatography (NIOSH Vol. III, No. 6701).

1.4 Biologic Effects

1.4.1 Animal Exposures

Static exposures of cats and rabbits for 1 hr to ammonia at 7000 mg/m^3 resulted in the death of approximately 50 percent of the animals. Postmortem examination showed severe effects on the upper respiratory tract, indicating high absorption by these tissues. Less severe effects in the lower respiratory tract included damage to the bronchioles and alveolar congestion, edema, atelectasis, hemorrhage, emphysema, and fluid (20). Alpatov (21) and Mikhailov (22) found the LC$_{50}$ for 2-hr exposures of rats and mice to ammonia to be 7.6 mg/l and 3.31 mg/l, respectively. Coon et al. (17) exposed 15 guinea pigs, 15 rats, three rabbits, two squirrel monkeys, and two beagle dogs in 30 repeated exposures to ammonia of 8 hr/day, 5 days/week at concentrations of 155 and 770 mg/m^3. There were no signs of toxicity, no hematologic changes, and no gross or histopathological changes at necropsy from the lower concentration. The higher concentration caused moderate lacrimation in the dogs and rabbits initially, but this was not observed after the first five exposures. Continuous exposure of rats for 114 days at 4 mg/m^3 ammonia resulted in no signs of toxicity, and at necropsy the only finding was clinically insignificant lipid-filled macrophages in the lungs of both dogs, one monkey, and one rat. Rats were also exposed continuously for 90 days at 127 and 262 mg/m^3 and for 65 days at 455 mg/m^3. The 127 mg/m^3 ammonia exposure induced no changes for the 48 rats in gross or microscopic pathology, hematology, or liver histochemistry for NADH, NADPH, or dehydrogenases for succinate, isocitrate, lactate, and β-hydroxybutyrate. The exposure at 262 mg/m^3 also was without specific effect other than mild nasal discharge in about 25 percent of the 49 rats. All the 51 rats exposed at 455 mg/m^3 showed mild dyspnea and nasal irritation. There were 32 deaths by day 25 and 50 by day 65. These authors also exposed rats, dogs, guinea pigs, rabbits, and monkeys to 470 mg/m^3 ammonia continuously for 90 days. Mortalities were 13/15 rats, 4/15 guinea pigs, 0/3 rabbits, 0/2 beagles, and 0/3 squirrel monkeys. The dogs had heavy lacrimation and nasal discharge. There were erythema, discharge, and corneal opacity in the rabbits. There were no hematologic effects. The gross necropsies revealed moderate lung congestion in two of three rabbits and one of two dogs. Histopathological examinations found focal or diffuse interstitial pneumonitis in all animals with epithelial calcification in the renal tubules and bronchi, epithelial proliferation of the renal tubules, myocardial fibrosis, and fatty liver plate cell changes in several of the exposed animals of each species. Control animals showed less severe similar changes (17). Stolpe and Sedleg (23) exposed rats continuously for 50 days at concentrations of 20, 35, and 63 mg/m^3 of ammonia. There were no effects on weight gained or hematologic parameters at 35 mg/m^3 at normal temperature, 22°C. There was slight weight reduction compared to controls at 10°C.

Swine exposed for 2 to 6 weeks at 100 ppm developed conjunctival irritation and a thickening of the nasal and tracheal epithelium without injury to the bronchi or alveoli (24).

Mayan and Merilan (25) reviewed the literature on the effect of ammonia on respiratory rates in connection with their studies of the biologic effect in rabbits. Exposures of 2.5 and 3.0 hr at 35 and 70 mg/m³ decreased the respiratory rate by 33 percent. Blood pH was not affected but blood urea was elevated by about 25 percent, and blood CO_2 increased by 32 percent. These latter determinations are consistent with the known rapid conversion of absorbed ammonia to ammonium carbonate and urea in most mammals. Postmortem examination revealed no adverse effects of the exposure on the lungs, liver, spleen, or kidneys.

In its review of the toxicity of ammonia to aquatic organisms, the U.S. Environmental Protection Agency (EPA) emphasized that the adverse effects are due to the un-ionized ammonia (26).

Intraperitoneal or intravenous injections of ammonium salts produce neurotoxicity, evidenced by increased respiration, tremors, convulsions, and coma in proportion to the ammonia content in the blood and brain. Death is apparently due to cardiotoxicity (27).

1.4.2 Human Exposures

Accidental exposures of humans to ammonia may arise from failure of equipment containing either liquid or gaseous ammonia. The chemical injuries are the same, however; liquid ammonia exposures may be complicated by freezing of tissues and by injection of a liquid stream under high pressure. The biologic effects of ammonia in humans clearly depend on concentration. Six volunteers inhaled ammonia at 21 and 35 mg/m³ for 10 min. Five reported faint to moderate irritation and one reported no irritation at 35 mg/m³. The exposure was described as penetrating but not discomforting or painful (28). Another group of volunteers was exposed for 5 min to 22, 35, 50, and 94 mg/m³. The 35 mg/m³ was not irritating to the eyes, nose, throat, or chest, whereas the 94 mg/m³ exposure caused eye irritation with lacrimation, nose and throat irritation, and in one volunteer, chest irritation (29). Six subjects were exposed to ammonia at 17, 35, and 70 mg/m³ for 2, 4, and 6 hr. They reported no discomfort during any of the exposures. Medical examination revealed mild nasal irritation in all at 35 and 70 mg/m³. Two had nasal irritation at 17 mg/m³. The examining physician could not detect any difference in the degree of irritation at any of the levels. Three individuals showed no eye irritation from any exposure and two had no throat irritation (3).

Upon 30-min exposure to 350 mg/m³ by means of a nose-only mask, all of seven volunteers reported upper respiratory irritation; two described it as "severe." There was lacrimation in two persons. There were no changes in blood urea nitrogen and nonprotein nitrogen, nor any changes in urinary urea or ammonia (31).

At the other end of the exposure spectrum, severe injuries and death have resulted from accidental exposures to anhydrous ammonia or solutions of ammonia at high concentrations. A bank teller received an ammonia solution directly into the eyes, nose, and throat during the course of a robbery. Severe damage to one eye consisted of gross chemosis, corneal staining, loss of pupillary reaction, lens pigmentation, and uveitis, resulting in greatly decreased vision. The nasopharynx and glottis were extremely swollen so as to prevent swallowing. The trachea and lungs had chemical pneumonitis. There were extensive burns of the face and mouth. All lesions except the eye healed slowly following treatment (32). Zygladowski (33) has described 44 cases of exposure to ammonia vapors and Walton (34) has reported on seven patients. When first seen, such patients are near collapse, in great pain, or unconscious. The exposed surfaces, including the eyes, nose, mouth, and throat show chemical burns, blisters that rupture and bleed, severe local edema, coughing, dyspnea, and progressive cyanosis. Examination reveals characteristic ecchymoses of the soft palate and swelling of the larynx and glottis to the point of respiratory obstruction. Shock and chemical pneumonitis may ensue. If death occurs, it is usually due to suffocation or pulmonary edema. Depending on the severity of exposure and the promptness of treatment, there may be full or partial recovery. Residual effects may include visual impairment, decreased respiratory function, or hoarseness. In contradistinction to the animal studies by Ballantyne et al. (35), White reports a case of near-fatal poisoning and ocular damage with no increase in ocular tension (36). Sugiyama et al. report a patient developing acquired bronchiectasis 1 year after ammonia inhalation (37). Verberk (38) exposed 16 volunteers to ammonia vapor for 2 hr at concentrations of 50, 80, 110, and 140 ppm. Subjective responses and pulmonary function parameters were reported. There were no effects on ventilatory capacity or 1-sec forced expiratory and inspiratory volumes. Eight of the volunteers found the irritation from 140 ppm so severe that they terminated the exposure prematurely. Other subjective responses such as smell and irritation of the eyes and nose increased with exposure concentration. Frosch and Kligman (39) describe an experimental dermatology technique wherein a 1:1 aqueous solution of ammonia is applied to human skin under a plastic closure. In 13 min an intra-epidermal blister results, which is virtually painless and which heals rapidly without scarring. The technique provides a means of comparing the irritancy of chemicals contacting the skin.

1.4.3 Absorption, Metabolism, Excretion

Ammonia is absorbed by inhalation, by ingestion, and probably percutaneously at concentrations high enough to cause skin injury. Data are not available on absorption of low concentrations through the skin. Once absorbed, ammonia is converted to the ammonium ion as the hydroxide and as salts, especially as carbonates. The ammonium salts are rapidly converted to urea, thus maintaining an isotonic system. Ammonia is also formed and consumed endogenously by the metabolism and synthesis of amino acids. Excretion is primarily by way of the kidneys, but a not insignificant amount is passed though the sweat glands.

1.5 Threshold Limit Values and Permissible Exposure Levels

The data cited above generally support the effects in the classic tabulation by Henderson and Haggard (40) (Table 13.1), taken from earlier editions of this book.

Table 13.1. Acute Toxicity: Physiological Response

Response	Concentration (ppm)
"Immediately dangerous to life and health"	500
Minimal irritation	5
Moderate irritation	9 to 50
Definite irritation	125 to 137
Cyclic hyperpnea/upper respiratory irritation, persistent	500 (30 min)
Immediate irritation	700
Dyspnea, convulsive coughing, chest pain, pulmonary edema, may be fatal	1500 to 10,000

The NIOSH-recommended time-weighted average (TWA) for anhydrous ammonia is 25 ppm (18 mg/m^3) and the short-term exposure limit (STEL) is the same. The 1991–1992 threshold limit value (TLV) from the American Conference of Governmental Industrial Hygienists (ACGIH) is also in agreement with the above recommended limits. The Japanese (1971) standard is the same as OSHA's (44). Winell (45) reports the following standards: Federal Republic of Germany (1974), 35 mg/m^3; German Democratic Republic (1973), 25 mg/m^3; Sweden (1975), 18 mg/m^3; Czechoslovakia (1969), 40 mg/m^3; and the Soviet Union (1972), 20 mg/m^3.

1.6 Personal Protective Equipment

The most obvious protective measure to be employed in reducing exposures is to provide local exhaust ventilation and/or dilution ventilation to meet published exposure limits. In the case of excessive exposures, respirators are the method of choice. The U.S. Department of Health and Human Services publishes a NIOSH Guide to chemical hazards which suggests specific respirator selection based upon contamination levels found to be present. In addition, the respirator must be one jointly approved by NIOSH and the Mine Safety and Health Administration (MSHA).

1.6.1 Respirators

The following depicts the respirator to be selected:

250 ppm

Any chemical cartridge respirator with cartridges providing protection against ammonia

Any supplied-air respirator

Any self-contained breathing apparatus

500 ppm

Any powered, air-purifying respirator with appropriate cartridge

Any self-contained breathing apparatus with a full face piece

Any chemical cartridge respirator with a full facepiece and appropriate cartridges

Any supplied-air respirator operated in a continuous flow mode

For firefighting and other immediately dangerous to life or health conditions:

Any self-contained breathing apparatus that has a full facepiece and is operated in a pressure demand or other positive-pressure mode

1.6.2 Clothing

Appropriate protective clothing that is impervious to ammonia should be worn to lessen the probability of skin contact. Appropriate, impervious gloves should also be worn. Splash-proof safety goggles and a face shield should be worn to pevent possible eye injuries.

1.7 Fire and Explosion Data

The upper explosive limit for anhydrous ammonia is 28 percent and the lower explosive limit is 15 percent. There is a moderate explosion hazard when exposed to heat or flame. The autoignition temperature is 1205°F (651°C). Dry chemicals or carbon dioxide is the fire-fighting medium of choice for small fires; for larger fires, water spray, fog, or regular foam should be used (*1990 Emergency Response Guidebook*, DOT P S800.5).

The following are additional general references on anhydrous ammonia:

J. W. Lloyd and R. J. Young, *Hazardous Materials Management Journal*, 1980, pp. 20–24

I. D. Gadaskina, *Encyclopedia of Occupational Health*, Vol. 1, 1983, pp. 148–150

M. Malanchuk, N. P. Barkley and G. L. Contner, *Journal of Environmental Pathology and Toxicology*, Vol. 4, Number 1, 1980, pp. 265–276

2 CALCIUM HYDROXIDE (CASRN 1305-62-0)

2.1 Sources, Uses, and Occupational Exposures

Calcium hydroxide, $Ca(OH)_2$ (slaked lime, hydrated lime), is formed by adding calcium oxide to water in an exothermic reaction. Calcium oxide itself may pose a problem, especially if inhaled. It is also very irritant to mucous membranes and moist skin.

Calcium hydroxide has many industrial applications. Its major uses are in mortar, plaster, cement, and building and paving materials; it is also used in lubricants, drilling fluids, and pesticides. Other uses of calcium hydroxide include the neutralization of acids and dehairing of animal hides. Commercially available calcium hydroxide usually contains 95 percent or more of calcium hydroxide.

2.2 Physical and Chemical Properties

Calcium hydroxide is an inorganic base with a molecular weight of 74.09. Trade names or synonyms are hydrated lime, slaked lime, caustic lime, calcium dihydroxide, and calcite. It is highly alkaline, with a pH of 12.8 (saturated solution at 25°C).

Calcium hydroxide is an odorless, white crystalline powder with a bitter taste. Its melting point is 1076°F (580°C) and it is soluble in water and insoluble in alcohol. There is a negligible fire hazard when exposed to heat.

2.3 Biologic Effects

2.3.1 Animal Exposures

The single oral intubation LD_{50} for rats is between 4.83 and 11.14 g/kg (46). Male rats were given tap water containing 50 and 350 mg/l (47). At 2 months they were restless and aggressive and had a reduced food intake. At 3 months there were a loss in body weight, decreased counts for erythrocytes and phagocytes, and decreased hemoglobin. At sacrifice the gross necropsy showed inflammation of the small intestine and dystrophic changes in the stomach, kidneys, and liver. Rabbits were exposed for 1 min to a paste of $Ca(OH)_2$ in the eyes followed by cleaning and rinsing with a physiological salt solution. This resulted in a gradual decrease in mucopolysaccharides of the cornea, reaching a maximum at 24 hr, which did not return to normal levels in 3 months (48).

2.3.2 Human Effects

Direct contact with calcium hydroxide can result in skin and eye irritation. Because it is a relatively strong base it can cause irritation to all exposed surfaces of the body including the respiratory tract.

Acute exposures may cause such irritation, along with coughing, pain, and possibly burns of the mucous membranes with, in severe acute exposures, pulmonary edema and hypotension with weak and rapid pulse. Chronic exposures may cause inflammatory and ulcerative changes in the mouth and gastrointestinal problems. Persons with preexisting skin problems should avoid any exposure to alkaline corrosives such as calcium hydroxide.

First aid for acutely exposed persons includes immediate removal from the exposure and the immediate and vigorous irrigation of the eyes and affected skin with running water. Medical attention should then be sought at once.

2.3.3 Personal Protective Equipment

Local exhaust ventilation should be employed to reduce workplace exposures to acceptable limits. Personal protection should include appropriate impervious clothing and gloves to prevent the possibility of skin contact, and eye protection should be in the form of splash-proof or dust-resistant safety goggles and a face shield to prevent contact. Emergency wash facilities should be readily available and include both an eye wash and a drench shower.

Properly selected respirators should be used when workplace limits are exceeded. Respirators must be selected based upon the levels of calcium hydroxide found to be present and must be jointly approved by NIOSH and MSHA.

The recommendations of these federal agencies as to the selection of respirators to be used should be followed. In the case where respirators are used, federal laws (OSHA) require a respirator training program to be in place.

2.4   Permissible Exposure Levels and Recommended Levels

The OSHA time-weighted average (TWA) exposure in 1991 is 5 mg/m^3 (respirable fraction) and 15 mg/m^3 for total dust. The ACGIH-recommended TWA is also 5 mg/m^3, as is the NIOSH recommendation.

Additional references for calcium hydroxide follow:

L. Remeo, L. Perbelline, P. Apostoli, and F. Malesani. *Proceedings of the VIIth International Pneumoconioses Conference*, Part II, Pittsburgh, PA, August 23–26, 1988
NIOSH, U.S. Department of Health and Human Services, DHHS (NIOSH) Publication No. 90–108, 1990, pp. 1304–1305

3   CALCIUM OXIDE, CaO (CASRN 1305-78-8)

3.1   Sources, Uses, and Occupational Exposures

Calcium oxide, CaO (lime, burnt lime, quicklime), is produced by the kiln-roasting of limestone, CaCO$_3$, to drive off carbon dioxide. It is used in construction materials, steel manufacturing, aluminum production, and glass and paper making processes, as well as in plaster, mortar, bricks, and construction materials. It also finds use in fungicides, insecticides, and some lubricants. An interesting application for calcium hydroxide is its use in the beet sugar processing industry. Commercial grades usually contain 95 percent or more of CaO.

Other synonyms for calcium oxide are quicklime, lime, and burnt lime.

3.2   Physical and Chemical Properties

Calcium oxide comes from the kiln as white or grayish white, porous lumps which may be crushed or powdered for distribution.

Molecular weight	56.08
Specific gravity	3.35

Solubility	1 g in 835 ml water at 25°C
	1 g in 1670 ml water at 100°C
Alkalinity	ph of 12.8 for a saturated solution at 25°C

Calcium oxide reacts exothermically with water to form the hydroxide (Section 2). It also reacts readily with carbon dioxide from the air to form calcium carbonate at the matrix in mortar.

3.3 Biologic Effects

There is a scarcity of published data on the toxicity of calcium oxide. Calcium oxide dust irritates the eyes and upper respiratory tract primarily because of its alkalinity. Inflammation of the respiratory passages, ulceration and perforation of the nasal septum, and pneumonia have been attributed to inhalation of calcium oxide dust. Particles of calcium oxide have caused severe burns of the eyes. Some contact dermatitis cases have also been observed (50).

3.4 Personal Protective Equipment

Employees should be provided with clothing and gloves that are impervious to prevent skin contact. Eye protection in the form of dust-resistant goggles and face shields should be used. Where exposures do occur, facilities for quick drenching of the body should be provided, as should eye washing stations. Control of exposures in the workplace can easily be achieved by the application of good industrial hygiene application. In the event of workplace exposures exceeding permissible levels, respirators approved and recommended by NIOSH/MSHA should be provided. Selection depends upon levels encountered in the work area. As is usual, if respirators are employed, the federally mandated respiratory training program should be in place.

Sampling and analysis for calcium oxide should be conducted using the standard NIOSH recommended methods.

3.5 Hygienic Standards

The ACGIH recommends a time-weighted average TLV of 2 mg/m^3 for calcium oxide (43). The Italian standard is 5 mg/m^3 (51). The OSHA TWA for calcium oxide is 5 mg/m^3.

4 POTASSIUM (CASRN 7440-06-4)

4.1 Source, Uses, Occupational Exposures

Potassium metal (K) is made by thermal reduction of the chloride with sodium or by electrolysis of molten salts. It is used in the production of NaK and of KO_2, the superoxide, used as an oxygen source in self-contained breathing apparatus.

4.2 Physical and Chemical Properties

When freshly cut, this ductile metal is soft and waxy with a metallic sheen that is soon lost by reaction with atmospheric oxygen, carbon dioxide, or moisture.

Molecular weight	39.10
Specific gravity	0.86 (20°C)
Melting point	62.3°C
Boiling point	758°C
Vapor pressure	8 mm HG at 432°C

Potassium reacts readily with most gases and liquids. It does not react with the noble gases such as helium or argon or with hydrocarbons such as hexane or high alkanes. It reacts violently with water to release hydrogen, which may ignite from the heat of the reaction. The resulting solution of potassium hydroxide (Section 5) may have a pH of 13 or greater.

4.3 Biologic Effects

Potassium's high reactivity makes it strongly caustic and corrosive in contact with tissues as described at the beginning of this chaper. The products of potassium combustion include oxides (see sodium peroxide, Section 9). It is an essential element, commonly found in most foods as a salt, which regulates osmotic pressure within cells, maintains the acid – base balance, and is necessary for many enzymatic reactions, especially those involving energy transfer.

5 POTASSIUM HYDROXIDE (CASRN 1310-58-3)

5.1 Source, Uses, and Occupational Exposures

Potassium hydroxide, KOH (caustic potash), is produced by electrolysis of potassium chloride solution. Its principal use is in the manufacture of soft and liquid soaps. It is also used to make high-purity potassium carbonate, K_2CO_3, for use in the manufacture of glass. Synonyms for potassium hydroxide are potassium hydrate, caustic potash, and potassa.

5.2 Physical and Chemical Properties

A white deliquescent solid, potassium hydroxide has a molecular weight of 56.10 and a specific gravity of 2.044. It may be in the form of pellets, sticks, lumps, or flakes. The melting point is about 360°C. Potassium hydroxide is soluble in water, alcohol, and glycerin and slightly soluble in ether. An aqueous solution may have a pH of 13 or higher.

5.3 Atmospheric Analysis

Dusts, mists, or vapors of potassium hydroxide or similar alkaline materials in the atmosphere may be determined by passing a measured volume of air through an impinger containing a measured volume of a standard solution of sulfuric acid. The excess acid is titrated with a standard alkaline solution using methyl red or other means to indicate the point. The NIOSH-recommended method of sampling and analysis is the procedure of choice.

5.4 Biologic Effects

Potassium hydroxide, when inhaled in any form, is strongly irritating to the upper respiratory tract. Severe injury is usually avoided by the self-limiting sneezing, coughing, and discomfort. Contact with eyes or other tissues can produce serious injury, as described at the beginning of this chapter. Rubber gloves should be worn when handling potassium hydroxide in order to prevent irritation, burns, or contact dermatitis. When exposed to air, potassium hydroxide forms the bicarbonate and carbonate. Very little is known of their biologic effects. Because they are less alkaline in aqueous solutions they may be expected to be less irritant or corrosive to skin and eyes. Bailey and Morgareidge (52), in a study for the Food and Drug Administration, found potassium carbonate to be nonteratogenic in mice when they were given daily oral intubations of up to 290 mg/kg on days 6 through 15 of gestation. Accidental ingestion of a solution of potassium hydroxide may be expected to produce rapid corrosion and perforation of the esophagus and stomach (53). Frequent applications of aqueous solutions (3 to 6 percent) of potassium hydroxide to the skin of mice for 46 weeks produced tumors identical to those from coal tar (54). Acute exposures involving the inhalation of dust or mist may cause symptoms in the respiratory tract including severe coughing and pain. Additionally, lesions may develop along with burning of the mucous membranes. Pulmonary edema can develop within a latency period of 5 to 72 hr.

Chronic exposures may cause inflammatory and ulcerative changes in the mouth and possibly bronchial and gastrointestinal disorders.

5.5 Permissible Exposure Levels and Recommended Levels

The OSHA ceiling limit for occupational exposures is 2 mg/m^3, as are the ACGIH and NIOSH recommended levels for the ceiling limit.

5.6 Personal Protective Equipment

Local exhaust ventilation should be employed to reduce workplace exposures to within acceptable levels. Respirators as recommended and approved by NIOSH/MSHA should be used when levels exceed established standards. If respirators are used, a respirator training program as mandated by OSHA regulations should be followed.

Employees should wear impervious clothing and gloves to prevent skin contact. Eye protection in the form of splash-proof or dust-resistant goggles and a face shield should be required. Emergency wash facilities, including drench showers and eye wash fountain, should be present within the immediate work area.

6 SODIUM (CASRN 7440-23-5)

6.1 Source, Uses, and Occupational Exposures

Sodium (Na) is manufactured by electrolysis of a molten mixture of sodium and calcium chlorides. It is a soft, waxy material having a silvery sheen on freshly cut surfaces. These surfaces quickly change to a coating of the white peroxide, Na_2O_2, by reaction with oxygen in the air. One-pound bricks and smaller amounts may be encountered in chemistry laboratories, where it is usually protected from the air by immersion in an aliphatic oil. Larger amounts are shipped in drums or tank cars. The manufacture of organometallic compounds, such as tetraethyllead, consumes the bulk of the production. Significant amounts are also used as a heat-exchange medium, frequently as the alloy with potassium known as NaK. Metal descaling baths also use a sodium–sodium hydride mixture.

6.2 Physical and Chemical Properties

Metallic sodium is a light, ductile material having a high electrical conductivity.

Molecular weight	23.00
Specific gravity	0.9684 (20°C)
Melting point	97.83°C
Boiling point	886°C
Vapor pressure	1 mm Hg at 432°C

Sodium reacts violently with carbon dioxide, water, and most oxygenated and halogenated organic compounds. It may ignite spontaneously in air at temperatures above 115°C, producing a sodium peroxide fume (Section 9), which is strongly alkaline and is thus a serious hazard for inhalation or skin and eye contact. Procedures for handling sodium and similar materials are described in the Atomic Energy Commission *Liquid Metals Handbook* (55). Fire-fighting and waste disposal methods are included. The behavior of the aerosol from a sodium fire has been described by Clough and Garland (56).

6.3 Biologic Effects

Vapors and fumes arising from sodium are strongly alkaline and are highly irritating and corrosive to the respiratory tract, eyes, and skin. Physiologically, sodium is an essential element encountered as a salt in most foodstuffs. Its ion is the principal electrolyte in extracellular fluid, which is excreted in the urine. Prolonged dietary excess may lead to renal hypertension. Sodium metal may react with moisture to form sodium hydroxide. Acute exposures may cause irritation and burning of the respiratory tract with severe coughing and pain. Pulmonary edema may develop within 72 hr following exposure. Severe cases may be fatal. Chronic exposures may cause inflammation, ulcerate changes in the mouth, and bronchial and gastrointestinal disturbances.

6.4   Atmospheric Analysis

The analysis is similar to that of potassium hydroxide (Section 5).

6.5   Permissible Exposure Level and Recommended Level of Exposure

No occupational exposure levels for metallic sodium per se have been established by OSHA, NIOSH, or ACGIH. The OSHA permissible exposure level (PEL) of sodium as sodium hydroxide is 2 mg/m^3 as a ceiling value.

7   SODIUM CARBONATE (CASRN 497-19-8)

This compound, Na_2CO_3 (soda ash), is usually encountered as the decahydrate, $Na_2CO_3.10H_2O$, commonly called sal soda or washing soda.

7.1   Sources, Uses, and Occupational Exposures

Sodium carbonate occurs naturally in large deposits in Africa and the United States as either the carbonate or trona, a mixed ore of equal molar amounts of the carbonate and bicarbonate. Soda ash is manufactured primarily by the Solvay process, whereby ammonia is added to a solution of sodium chloride and carbon dioxide is then bubbled through to precipitate the bicarbonate, $NaHCO_3$. Calcination of the bicarbonate produces sodium carbonate. It may also be produced by injecting carbon dioxide into the cathode compartment, containing sodium hydroxide, of the diaphragm electrolysis of sodium chloride.

The glass industry consumes about one-third of the total production of sodium carbonate. About one-fourth is used to make sodium hydroxide by the double decomposition reaction with slaked lime, $Ca(OH)_2$. Large amounts are also used in soaps and strong cleansing agents, water softeners, pulp and paper manufacture, textile treatments, and various chemical processes. Synonyms are carbonic acid, bisodium carbonate, sod, soda ash, and calcined soda.

7.2   Physical and Chemical Properties

This hygroscopic, white powder is strongly caustic.

Molecular weight	106.0
Specific gravity	2.53 (20°C)

Melting point	851°C
Boiling point	Decomposes
Solubility	7.1g/100 ml water at 0°C
	45.5 g/100 ml water at 100°C
Alkalinity	pH of 11.5 for a 1 percent aqueous solution

7.3 Atmospheric Analysis

Sodium carbonate and similar strongly alkaline materials when airborne as mists or dusts may be sampled with an impinger containing standard sulfuric acid. Quantitative determination is by forward titration of the excess acid using methyl red as an indicator.

7.4 Biologic Effects

Male rats were exposed to an aerosol of a 2 percent aqueous solution of sodium carbonate, 4 hr/day, 5 days/week, for $3\frac{1}{2}$ months. The particle size of the aerosol was less than 5 μm in diameter. A concentration of 10 to 20 mg/m³ did not cause any pronounced effect. In observations from exposure at 70 ± 2.9 mg/m³, the weight gain of the exposed group was 24 percent less than that of controls. There were no differences in hematologic parameters. Histological examination of the lungs showed thickening of the intra-alveolar walls, hyperemia, lymphoid infiltration, and desquamation (65).

An aqueous solution, 50 percent (w/v), of sodium carbonate was applied to the intact and abraded skins of rabbits, guinea pigs, and human volunteers. The sites were examined at 4, 24, and 48 hr and scored for erythema, edema, and corrosion. The solution produced no erythema and edema. In humans, the rabbit and human skins showed tissue destruction at the abraded sites (66).

Pregnant mice were dosed daily by oral intubation with aqueous solutions of sodium carbonate at levels of 3.4 to 340 mg/kg on days 6 through 15 of gestation. There were no effects on nidation or survival of the dams of fetuses. The number of abnormalities in soft and skeletal tissues in the experimental group did not differ from sham-treated controls. Positive controls gave the expected results (67). Similar studies in rats at doses up to 245 mg/kg and in rabbits at doses up to 179 mg/kg produced similar negative results.

Sodium bicarbonate, $NaHCO_3$, was evaluated for teratologic effects by the same procedures as for sodium carbonate. Maximum dose levels were as follows: mice, 580 mg/kg; rats, 340 mg/kg; and rabbits, 330 mg/kg. No effects were found in any of these species (68).

Twenty-seven Army inductees assigned to dishwashing immersed their bare hands for 4 to 8 hr in hot water containing a strong detergent blend of sodium carbonate, sodium metasilicate, and sodium tripolyphosphate. All developed irritation of the exposed surfaces. Six developed vesicles and giant bullae within 10 to 12 hr after exposure. Three also had subungual purpura. Secondary infections were noted in several individuals (69). Acute exposures of dusts or vapors of sodium carbonate may cause irritation of mucous membranes with subsequent coughing and shortness of breath.

Chronic exposures may lead to perforation of the nasal septum; skin exposure may cause irritation and redness with concentrated solutions causing erythema. Chronic skin exposures may cause dermatitis and ulceration.

7.5 Permissible Exposure Levels and Recommended Levels of Exposure

No occupational exposure levels have been established in the United States, and there are no ACGIH- or NIOSH-recommended levels. The level of 5 mg/m^3 has been tentatively recommended in the (former) USSR for sodium carbonate (66).

7.6 Personal Protective Equipment

Local exhaust ventilation or process enclosure systems should be employed to prevent workplace exposures. If respirators are used, they should be ones recommended and approved by NIOSH/MSHA, and a mandated respirator training program should be in place.

Appropriate impervious clothing and gloves should be required along with splash-proof or dust-resistant goggles to prevent skin contact.

Eye wash fountains and drench showers should be immediately available in the work area.

8 SODIUM HYDROXIDE (CASRN 1310-73-2)

8.1 Sources, Uses, and Occupational Exposures

The primary source of sodium hydroxide, NaOH (caustic soda, caustic flake, lye, liquid caustic), is the electrolysis of sodium chloride solutions, which also yields chlorine. In this process the anode may be surrounded by an asbestos diaphragm to isolate the chlorine. The caustic soda so produced may contain a significant amount of asbestos fibers. As noted above, sodium hydroxide may also be produced from sodium carbonate.

The millions of tons of sodium hydroxide produced annually in the United States are used in the manufacture of chemicals, rayon, soap and other cleansers, pulp and paper, petroleum products, textiles, and explosives. Caustic soda is also used in metal descaling and processing and in batteries.

As indicated in the synonyms, sodium hydroxide may be encountered as solids in various forms (pellets, flakes, sticks, cakes) and as solutions, usually 45 to 75 percent in water. Mists are frequently formed when dissolving sodium hydroxide in water, which is an exothermic process.

8.2 Physical and Chemical Properties

Molecular weight	40.01
Specific gravity	2.13 (20°C)

Melting point	318.4°C
Boiling point	1390°C
Vapor pressure	1 mm Hg at 739°C
Solubility	42 g in 100 ml H_2O at 0°C
	347 g in 100 ml H_2O at 0°C
	Soluble in aliphatic alcohols
Refractive index	1.3576
Alkalinity	The pH of a 1 percent aqueous solution is about 13

8.3 Atmospheric Analysis

See the procedures for potassium hydroxide.

8.4 Biologic Effects

This strong alkali is irritating to all tissues and requires extensive washing to remove it. Eye splashes are especially serious hazards. See the opening discussion in this chapter and Section 5 on potassium hydroxide. Protective equipment essential and treatment must be prompt. A 5 percent aqueous solution of sodium hydroxide produced severe necrosis when applied to the skin of rabbits for 4 hr (70). Rats were exposed to an aerosol of 40 percent aqueous sodium hydroxide whose particles were less that 1 μm in diameter. Exposures were for 30 min, twice a week. The experiment was terminated after 3 weeks when two of the 10 rats died. Histopathological examination showed mostly normal lung tissue with foci of enlarged alveolar septa, emphysema, bronchial ulceration, and enlarged lymph adenoidal tissues (71). Nagao and co-workers (72) examined skin biopsies from volunteers having 1 N sodium hydroxide applied to their arms for 15 to 180 min. There were progressive changes beginning with dissolution of the cells in the horny layer and progressing through edema to total destruction of the epidermis in 60 min.

Sodium hydroxide concentrations of 250 mg/m³ are considered immediately dangerous to life or health.

Acute exposures involving inhalation of dusts or mist may cause mucous membrane irritation with subsequent cough and dyspnea. Intense exposure may result in pulmonary edema and shock may result.

Prolonged or chronic exposures to high concentrations of sodium hydroxide may lead to ulceration of the nasal passages. All skin contact with this corrosive materials should be avoided. Acute skin exposures may cause cutaneous burns and skin fissures. Chronic skin exposures can lead to dermatitis.

8.5 Permissible Exposure Levels and Recommended Levels

A ceiling level for exposure to sodium hydroxide has been adopted by OSHA as 2 mg/m³. This level is also recommended by NIOSH and ACGIH.

8.6 Personal Protective Equipment

Local exhaust ventilation should be provided to reduce exposure levels to acceptable levels. If respirators are to be used when permissible exposure levels are exceeded, they should be selected as recommended by NIOSH/MSHA. An OSHA-mandated respirator training program should be established if respirators are worn.

Appropriate, impervious clothing and gloves should be worn to prevent skin contact.

Eye protection in the form of splash-proof or dust-resistant safety goggles should be worn.

Emergency wash facilities in the form of drench showers and eye wash fountains should be available in the immediate work area.

9 SODIUM PEROXIDE (CASRN 1313-60-6)

9.1 Sources, Uses, and Occupational Exposures

Metallic sodium reacts in dry air to form sodium monoxide and sodium peroxide, Na_2O_2 (sodium dioxide, sodium superoxide). It may be encountered as an oxidant in chemical processes or as a bleaching agent, for example, of textiles. Its reactivity with carbon dioxide finds utility in self-contained breathing apparatus. It may also be encountered in the aerosol from sodium fires.

9.2 Physical and Chemical Properties

This white powder is a very strong oxidizing agent. A vigorous exothermic reaction takes place with water to form sodium hydroxide and oxygen.

Molecular weight	77.99
Specific gravity	2.805 (20°C)
Melting point	460°C with decomposition

9.3 Atmospheric Analysis

See the procedure for potassium hydroxide and observe personal protection precautions when collecting samples.

9.4 Biologic Effects

In order to simulate the products of an alkali metal fire, sodium vapor was used to form a fresh aerosol composed mainly of sodium peroxide with some sodium monoxide. It was passed into an aging chamber to reach equilibrium with atmospheric carbon dioxide and water and stabilize its particle size. The resulting aerosol was thought to represent the product of an accidental sodium metal fire in a nuclear reactor. Juvenile and adult rats were exposed for up to 2 hr to various dilutions of the aerosol. The final particle size was 2.5 μm or less and was predominantly sodium carbonate with some sodium hydroxide. At necropsy following sacrifice, the only lesion observed was necrosis of the surface of the larynx; the area of affected epithelium and the depth of penetration were related to increased concentrations of aerosol. The ED_{50}, based on the number of animal affected, was about 510 μg/l for adults and 489 μg/l for juveniles. The severity of the injury was significantly greater in the juvenile rats. Animals sacrificed 4 to 7 days post-exposure had no exposure-related lesions, suggesting a healing process (73). The characteristics of the experimental aerosol are in accord with those projected by Clough and Garland (56). Hughes and Anderson (74) measured the decreased visibility encountered in a sodium fire. They collected data on volunteers exposed for a short time to the fumes. They concluded that, "Short term exposures of unprotected workers up to 40 mg/m³ NaOH in air is unlikely to result in any serious discomfort. A concentration of 100 mg/m³ produced serious discomfort promptly and precluded continuing work."

Acute exposures to sodium peroxide via inhalation may cause respiratory irritation and difficulty in breathing.

Chronic airborne exposures result in severe irritation of the eye, mucous membranes, and skin. Skin contact, either acute or chronic, may result in severe irritation, redness, and pain. High concentrations can burn the skin.

9.5   Permissible Exposure Levels and Recommended Levels

No recommendation has been made but an acceptable level probably should not exceed the ceiling value of 2 mg/m³ of sodium hydroxide. There are no OSHA, NIOSH, or ACGIH recommendations for sodium peroxide.

9.6   Personal Protective Equipment

As is usually true for reducing workplace exposures to within an acceptable level, process enclosure and/or local exhaust ventilation should be employed. Thermal decomposition of sodium peroxide leads to oxygen evolution, which increases fire hazards.

If respirators are used as a method of avoiding excessive exposures, they must be NIOSH/MSHA approved and selection must be based upon specific contamination levels found at the work site. The OSHA mandated respirator training program must be in place.

Appropriate impervious clothing and gloves should be worn.

Eye protection in the form of splash-proof or dust-resistant safety goggles should be worn. Emergency drench showers and eye wash fountains should be available in the immediate work area.

10   TRISODIUM PHOSPHATE (CASRN 7601-54-9)

10.1   Sources, Uses, and Occupational Exposures

Trisodium phosphate, $Na_3PO_4 \cdot 12H_2O$ (TSP, sodium orthophosphate) is produced by neutralization of disodium phosphate with sodium hydroxide. The disodium salt is produced from phosphoric acid and sodium carbonate. Trisodium phosphate is an important ingredient in soap powders, detergents, and cleaning agents. It is also used as a water softener to remove polyvalent metals and in the manufacture of paper and leather. Products for removing or preventing boiler scale often contain trisodium phosphate, as do those for removing insecticide residues from fruit and inhibiting mold.

10.2 Physical and Chemical Properties

This strongly basic, tertiary salt is usually seen as colorless crystals.

Molecular weight	380.21
Specific gravity	1.62 (20°C)
Melting point	75°C with decomposition
Solubility	25.8 g/100 g water at 20°C
	157 g/100 g water at 70°C
Alkalinity	The pH of a 1 percent aqueous solution is 11.6

10.3 Atmospheric Analysis

It is suggested that dusts of trisodium phosphate be collected in an impinger containing distilled water which may then be titrated with a standard solution of sulfuric acid. They may also be analyzed for P_2O_3 content by the precipitation of ammonium phosphomolybdate followed by redissolving in standard alkali and back-titrating with acid. Gravimetric procedures are also available, as are nuclear magnetic resonance spectroscopy and ion-exchange chromatography (75).

10.4 Biologic Properties

The toxicity of trisodium phosphate has not been investigated but it may be expected to be related only to its alkalinity since its ions are normal constituents of all living matter. Its alkalinity is close to that of sodium carbonate (Section 7).

10.5 Permissible Exposure Levels and Recommended Levels

There is no OSHA-mandated PEL for trisodium phosphate. There are also no ACGIH or NIOSH recommended levels; however, the American Industrial Hygiene Association (AIHA) recommends a TWA of 5 mg/m^3 for trisodium phosphate. Acute exposures to trisodium phosphate may cause irritation of the respiratory tract with subsequent coughing and pain. In severe exposure pulmonary edema may develop.

Chronic exposures can result in inflammatory or ulcerative changes in the mouth and possible gastrointestinal disorders.

10.6   Personal Protective Equipment

Appropriate workplace controls, such as local exhaust ventilation, should be employed to reduce exposures to an acceptable level. Respirators should be worn in areas of unacceptable airborne concentration; they should be NIOSH/MSHA approved and, if employed, the OSHA-mandated respirator training program should be in place.

Appropriate impervious clothing and gloves should be worn to reduce skin contact.

Eye protection in the form of splash-proof or dust-resistant safety goggles should be worn.

Emergency drench showers and eye wash fountains should be available for emergency use.

11   SODIUM METASILICATE (CASRN 6834-92-0)

11.1   Sources, Uses, and Occupational Exposures

The various hydrates of sodium metasilicate, $Na_2SiO_3.nH_2O$, range from the anhydrous to the nonahydrate, with the anhydrous and the penta- and nonahydrates forms being the most common. The sodium metasilicates are differentiated from other sodium salts of silicic acid by the molar ratio of the Na_2O and SiO_2 components. In the metasilicates this ratio is 1:1. A continuum of sodium silicates of other ratios may also be commonly encountered. Their chemical and biologic properties are essentially similar to those of the metasilicates. Fusing silica (sand) with sodium carbonate at 1400°C produces sodium metasilicate. A major use is as a builder in soaps and detergents. It is also used extensively as an anti-corrosion agent in boiler-feed water. The metasilicates should not be confused with other, less alkaline silicates used as adhesives of corrugated paper and as an additive to alfalfa cattle feed. Annual U.S. production of the metasilicates exceeds 400 million lb.

11.2   Physical and Chemical Properties

The metasilicates are highly water soluble. The anhydrous and the pentahydrate are produced as amorphous beads, whereas the nonahydrate appears as efflorescent sticky crystals (76).

Molecular weight	122.07 anhydrous
	212.15 pentahydrate
	284.21 nonahydrate
Specific gravity	2.614 anhydrous
	1.749 pentahydrate
	1.646 nonahydrate

Melting point	1089°C anhydrous
	72°C pentahydrate
	47.8°C nonahydrate
Refractive index	1.49 anhydrous
	1.447 pentahydrate
	1.451 nonahydrate
Alkalinity	pH of a 1 percent aqueous solution is about 13

Solutions of sodium metasilicate, when heated or acidified, are hydrolyzed to free sodium ions and silicic acid. The latter polymerizes through oxygen bridges to form amorphous silica. At high initial concentrations a gel is formed, whereas silica sols arise from dilute solutions. When dried, these colloidal forms of silica have great absorbing power.

11.3 Atmospheric Analysis

Neat material collected in an impinger may be titrated for total alkalinity to a methyl orange end point. Silicon dioxide in the titrate is determined by evaporating to dryness by ignition, then treating the weighed residue with $HF-H_2SO_4$ and calculating the weight loss as SiF_4 as above (77). The theoretical ratio of percent Na_2O:percent SiO_2 is 1.032.

Sodium metasilicate in soaps and detergents is determined by ashing to remove organic matter, dissolving the residue in HCl, igniting and redissolving twice, and measuring SiO_2 and above (78).

For microgram quantities in air, as from escaping steam of treated boiler water, a colorimetric method using an ammonium molybdate–sulfuric acid reagent has been described (79).

11.4 Biologic Effects

The acute oral toxicity (LD_{50}) of sodium metasilicate to rats is 1280 mg/kg as a 10 percent aqueous solution, and for mice the LD_{50} is 2400 mg/kg (76). The intraperitoneal injection in rats of the nonahydrate solution in amounts of 300 mg on day 1 and 200 mg on days 2 and 3 produced lesions in the spleen and lymph nodes and caused mitotic changes in the nuclei of cells resembling those from ionizing radiation or hypoxia (80). Weaver and Griffith (81), in their studies of detergentemesis in 11 dogs by gastric intubation, found that 8 mg/kg as a 10.5 percent aqueous solution of a sodium silicate (SiO_2:Na_2O, 3.2:1) produced emesis in 6 min which continued for up to 33 min. Dogs were fed sodium silicate in their diet at a dose of 2.4 g/kg per day for 4 weeks (82). Polydipsia and polyuria were observed in some animals. Damage to renal tubules was observed in 15/16 dogs.

Radio-labeled (^{31}Si) sodium metasilicate, partially neutralized, was given orally was rapidly absorbed and excreted in the urine but a significant amount was retained in the tissues (82). These findings are consistent with the recognition that silicon is an essential trace element for bone formation in animals (83).

As might be expected, detergents containing sodium metasilicate and other alkaline materials are strong irritants to the skin, eyes, and respiratory tract (84–86). Seabaugh (84) has also shown that, of 134 detergents containing alkaline silicates, 81 percent were irritant or corrosive. Automatic dishwashing detergents were the most frequently corrosive.

"Soluble silica" in the drinking water of rats at 1200 ppm as silicon dioxide from time of weaning through reproduction reduced the numbers of offspring by 80 percent and decreased the number of pups surviving to weaning by 24 percent (87).

Acute exposures involving the inhalation of dusts of sodium metasilicate may result in irritation of the respiratory tract and corrosive damage may result from contact with mucous membranes.

Prolonged exposures can lead to inflammatory changes and ulcerative problems in the mouth. Possible bronchial and gastrointestinal problems can exist, depending upon concentration and duration of exposure.

Sodium metasilicates have not been evaluated in humans. Experience has shown that skin contact with solutions of strong detergents containing this builder produces severe skin irritation (69). However, other components of these detergents undoubtedly contribute to the irritancy.

Inhalation of dusts from soluble silicate powders is irritating to the upper respiratory tract (88). Exposure to such dusts is not related to the development of silicosis because their solubility permits them to be readily eliminated. Confirmation of this is found in the work of Svinkina (89), who was unable to produce immunologic sensitization in rabbits with sodium silicate bound to protein.

11.5 Permissible Exposure Levels and Recommended Levels

There is no OSHA-mandated PEL for sodium metasilicate and there are no NIOSH- or ACGIH-recommended levels.

11.6 Personal Protective Equipment

Standard industrial hygiene procedures should be employed to keep workplace exposure levels to an acceptable amount.

If respirators are to be used, they should be NIOSH/MSHA approved and selected on the basis of specific contamination levels found at the worksite. If respirators are used, the OSHA-mandated respirator training program should be in place.

Protective clothing and gloves should be worn to prevent the possibility of skin contact.

Eye protection should be provided in the form of splash-proof or dust-resistant safety goggles with a face shield to prevent eye contact.

Emergency wash facilities including an eye wash fountain should be available in case of emergencies.

REFERENCES

1. J. M. Lowenstein, *Physiol. Rev.*, **52**, 382 (1972).

2. K. Diem, Ed., *Documenta Geigy. Scientific Tables*, 6th ed., Geigy Pharmaceuticals, Ardsley, NY, 1962.

3. W. J. Visek, *Fed. Proc.*, **31**, 1178 (1972).

4. *Ammonia*, National Academy of Sciences, Washington, DC, 1977.

5. U.S. Dept. Labor, *Fed. Reg.*, **40**, 54864 (1975).

6. Anonymous, *Jap. J. Ind. Health*, **14**, 45 (1972).

7. E. S. White, *J. Occup. Med.*, **13**, 549 (1971).

8. *Criteria Document for Occupational Exposure to Ammonia*, Pub. No. HEW (NIOSH) 74–136, U.S. Dept. Health, Education and Welfare (NIOSH), Cincinnati, OH, 1974.

9. N. A. Lange, Ed., *Handbook of Chemistry*, 8th ed., Handbook Publishers, Sandusky, OH, 1952.

10. R. C. Weast, Ed., CRC *Handbook of Chemistry and Physics*, 59th ed., CRC Press, West Palm Beach, FL, 1978.

11. *Fire Protection Guide on Hazardous Materials*, 6th ed., National Fire Protection Association, Boston, MA, 1975.

12. *Guides for Short-Term Exposures of the Public to Air Pollutants. IV. Guide for Ammonia*, Committee on Toxicology, National Academy of Sciences, Washington, DC, 1972.

13. *Hygienic Guide Series*, "Anhydrous Ammonia," *Am. Ind. Hyg. Assoc. J.* **32**, 139 (1971).

14. H. B. Elkins, *The Chemistry of Industrial Toxicology*, Wiley, New York, 1959.

15. M. C. Rand, A. E. Greenberg, and M. J. Taras, Eds., *Standard Methods for the Examination of Water and Wastewater*, 14th ed., American Public Health Association, Washington, DC, 1976.

16. T. Okita and S. Kanamori, *Atm. Environ.*, **5**, 621 (1971).

17. R. A. Coon, R. A. Jones, L. J. Jenkins, Jr., and J. Siegel, *Toxicol. Appl. Pharmacol.*, **16**, 646 (1970).

18. CFR 42 Chap. 1, Subchapter G, Part 84, *Certification of Gas Detector Tube Units*, 1973.

19. D. N. Kramer and J. M. Sech, *Anal. Chem.*, **44**, 395 (1972).

20. E. M. Boyd, M. L. Maclachlan, and W. F. Perry, *J. Ind. Hyg. Toxicol.*, **26**, 29 (1944).

21. I. M. Alpatov, *Gig. Tru. Prof. Zabol.*, **8**,(2), 14 (1964).

22. V. I. Mikhailov, *Probl. Kosm. Biol. Akad. Nauk. SSR*, **4**, 531 (1965).

23. J. Stolpe and R. Sedleg, *Arch. Exp. Vet. Med. Leipzig*, **30**(4), 533 (1976).

24. P. A. Doig and R. A. Willoughby, *J. Am. Vet. Med. Assoc.*, **159**, 1353 (1971).

25. M. H. Mayan and C. P. Merilan, *J. Animal Sci.*, **34**, 448 (1972).

26. W. T. Willingham, EPA-908/3-76-001 National Technical Information Service, Springfield, VA, Doc. No. PB-256447, 1976.

27. National Research Council, Committee on Medical and Biological Effects of Environmental Pollutants, *Ammonia*, 1977.

28. J. D. MacEwen, J. Theodore, and E. H. Vernot, Eds., *Proceedings First Annual Conference on Environmental Toxicology*, AMRL-TR-70-102, Wright Patterson Air Force Base, Ohio, 1970.

29. Industrial Bio-Test Laboratories, Inc., IBT 663-0 3161, *Report to the International Institute of Ammonia Refrigeration of Irritational Threshold Evaluation Study*, 1973.

30. Allied Chemical Corp., *Preliminary Report to OSHA of Ammonia Test Program*, 1975.

31. L. Silverman, J. L. Whittenberger, and J. Muller, *J. Ind. Hyg. Toxicol.*, **31**, 74 (1949).

32. A. H. Osmund and C. J. Tallents, *Brit. Med. J.*, **3**, 740 (1968).

33. J. Zygladowski, *Otolaryngol. Polska*, **22**, 773 (1968).

34. M. Walton, *Brit. J. Ind. Med.*, **30**, 78 (1973).

35. B.Ballantyne, J. F. Gazzard, and P. W. Swanston, *J. Physiol.*, **226**, 12P (1972).

36. E. S. White, *J. Occup. Med.*, **13**, 549 (1971).

37. K. Sugiyama, M. Yoshida, and M. Koyamada, *Jap. J. Chest. Dis.*, **27**, 797 (1968).

38. M. M. Verberk, *Int. Arch. Occup. Environ. Health*, **39**, 73 (1977).

39. P. J. Frosch and A. M. Kligman, *Brit. J. Dermatol.*, **96**, 461 (1977).

40. Y. Henderson and H. W. Haggard, *Noxious Gases*, Reinhold, New York, 1943.

41. National Institute for Occupational Safety and Health, *NIOSH Recommended Standard for Occupational Exposure to Ammonia*, U.S. Department of Health, Education, and Welfare, Rockville, MD, 1974.

42. Occupational Safety and Health Administration, U.S. Department of Labor, *Fed. Reg.*, **40**(228), 54692 (1975).

43. American Conference of Governmental Industrial Hygienists, *Threshold Limit Values for Chemical Substances in Workroom Air*. Adopted by ACGIH, Cincinnati, OH, 1991–1992.

44. Japanese Association of Industrial Health, Subcommittee on Permissible Concentrations, *Sangyo Igaku*, **14**, 45 (1972).

45. M. Winell, *Ambio*, **4**(1), 34(1975).

46. H. F. Smyth, Jr., C. P. Carpenter, C. S. Weil, U. C. Pozzani, J. A. Striegel, and J. S. Nycum, *Am. Ind. Hyg. Assoc. J.*, **30**, 470(1969).

47. L. I. El'piner and A. M. Voitenko, *Tr. Nauch.Issled.Inst. Gig. Vod. Transp.*, **1**, 304 (1968).

48. J. Teterwak, *Klin. Oczna*, **39**(4), 543 (1969).

49. *Fed. Reg.* (Sect. 311 Clean Water Act), **44**, 65400 (November 13, 1979).

50. Pennsylvania Department of Health, personal communication to the Division of Occupational Health, P.O. Box 90, Harrisburg, PA, by N. E. Whitman.

51. Italian Association of Industrial Hygienists, *Med. Lavoro*, **66**, 361 (1975).

52. D. E. Bailey and K. Morgareidge, *Teratologic Examination of FED 73–76 (K_2CO_3) in Mice*, National Technical Information Service PB-245522, 1975.

53. National Information Service PB-265507, 1976.

54. J. K. Narat, *J. Cancer Res.*, **9**, 135 (1925).

55. C. B. Jackson, *Liquid Metals Handbook*, Sodium, NaK Supplement, Atomic Energy Commission, July 1955.

56. W. S. Clough and J. A. Garland, *J. Nucl. Energy*, **25**, 425 (1971).

57. Kirk-Othmer, *Encyclopedia of Chemical Technology*, 3rd ed., Vol. 4, Wiley, New York, 1978, p. 86.

58. B. W. Bailley and F. C. Lo, *J. Environ. Anal. Chem.*, **1**, 267 (1972).

59. W. J. Underwood, *Trace Elements in Human and Animal Nutrition*, Academic Press, New York, 1971.

60. A. A. Kasparov, *Encyclopedia of Occupational Health and Safety*, Vol. 1, McGraw-Hill, New York, 1971, p. 203.

61. R. J. Weir, Jr., and R. S. Fisher, *Toxicol. Appl. Pharmacol.*, **23**, 351 (1972).

62. R. E. Gosselin, H. C. Hodge, R. P. Smith, and M. N. Gleason, *Clinical Toxicology of Commercial Products*, 4th ed., Williams and Wilkins, Baltimore, 1976.

63. D. J. Birmingham, in *Cutaneous Toxicity*, V. A. Drill and P. Lazar, Eds., Academic Press, New York, 1977, p. 58.

64. L. Kato and B. Gozsy, *Can. Med. Assoc. J.*, **73**, 31 (1955).

65. A. L. Reshetyuk and L. S. Shevchenko, *Hyg. Sanit.* **33**(1–3), 129 (1968).

66. G. A. Nixon, C. A. Tyson, and W. C. Wertz, *Toxicol. Appl. Pharmacol.*, **31**, 481 (1975).

67. K. Morgareidge, *Teratologic Evaluation of Sodium Carbonate in Mice, Rats, and Rabbits*, PB-234 868, National Technical Information Service, Springfield, VA, 1974.

68. K. Morgareidge, *Teratologic Evaluation of Sodium Bicarbonate in Mice, Rats and Rabbits*, PB-234 871, National Technical Information Service, Springfield, VA, 1974.

69. N. Goldstein, *J. Occup. Med.*, **10**, 423 (1968).

70. E. Horton, Jr. and R. R. Rawl, *Toxicological and Skin Corrosion Testing of Selected Hazardous Materials*, PB-264 975, National Technical Information Service, Springfield, VA, 1976.

71. M. Dluhos, B. Sklensky, and J. Vyskocil, *Vnitr. Lek.*, **15**(1), 38(1969).

72. S. Nagao, J. D. Stroud, T. Hamada, H. Pinkus, and D. J. Birmingham, *Acta Dermatovener* (Stockholm), **52**, 11(1972).

73. G. M. Zwicker, M. D. Allen, and D. L. Stevens, *J. Environ. Pathol. Toxicol.*, **2**, 1139 (1979).

74. G. W. Hughes and N. R. Anderson, "Visibility in Sodium Fume," presented at International Atomic Energy Agency International Working Group on Fast Reactor Meeting, March 17–19, 1971, IAE-NPR-12.

75. Kirk-Othmer *Encyclopedia of Chemical Technology*, 2nd ed., Vol. 15, Wiley, New York, 1968, p. 267.

76. A. Weissler, "Monograph on Sodium Metasilicate," PB287766, National Technical Information Service, Springfield, VA, 1978.

77. Philadelphia Quartz Company, *Technical Bulletin on Sodium Metasilicate and Industrial Alkalines and Detergents*, Philadelphia Quartz Company, Valley Forge, PA, 1978.

78. *1976 Annual Book of ASTM Standards*, "Part 30, Soaps and Detergents," American Society for Testing and Materials, Philadelphia, PA, 1976, pp. 14, 27–28, 70–72.

79. *Standard Methods for the Examination of Water and Wastewater*, 14th ed., American Public Health Association, Washington, DC, 1975, pp. 484–492.

80. L. Nanetti, *Zacchia*, **9**(1), 96 (1973).

81. J. E. Weaver and J. F. Griffith, *Toxicol. Appl. Pharmacol.*, **14**, 214 (1969).

82. F. Sauer, D. H. Laughland, and W. M. Davidson, *J. Biochem. Physiol.*, **37**, 1173 (1959).

83. E. M. Carlisle, *Science*, **178**, 619 (1972).

84. V. M. Seabaugh, *Detergent Survey Toxicity Testing*, PB264698/AS, National Technical Information Service, Springfield, VA, 1977.

85. G. E. Morris, *Arch. Ind. Hyg. Occup. Med.*, **7**, 411 (1953).

86. L. G. Scharpf, Jr., I. D. Hill, and R. E. Kelly, *Food Cosmet. Toxicol.*, **10**, 829 (1972).

87. G. S. Smith, A. L. Newmann, A. B. Nelson, and E. E. Ray, *J. Animal Sci.*, **36**, 271 (1973).

88. PhiladelphiaQuartz Company, *Soluble Silicates Bulletin T-17-65*, Philadelphia Quartz Company, Valley Forge, PA, 1965.

89. N. V. Svinkina, *Labor Hyg. Occup. Diseases* (USSR), **10**, 20 (1966).

CHAPTER FOURTEEN

Phosphorus, Selenium, Tellurium, and Sulfur

Robert P. Beliles, Ph.D., D.A.B.T., and Eloise M. Beliles

Phosphorus and sulfur are elements 15 and 16 on the periodic chart. Selenium and tellurium are in the same group as sulfur. Sulfur was not covered in the preceding edition. Because of the importance of the sulfur compounds, they have been added in this edition.

1 PHOSPHORUS

1.1 Sources, Uses, and Industrial Exposures

Phosphorus (P) is not found free in nature. The name is from the Greek for light-bring. Phosphorus is found in all life forms as phosphate. Phosphates are essential for energy transfer reactions. Of major importance is adenosine triphosphate (ATP), involved in nearly all metabolic reactions. Phosphates are an important ingredient of bone. The human skeleton contains about 1.4 kg as calcium phosphate. Phosphates also form part of a number of coenzymes and nucleic acids. Most of the available forms are as phosphates. The elemental material is produced as a by-product or intermediate in the production of phosphate fertilizer. Phosphorus itself is not found in the elemental state in nature. The environmental contamination with phosphorus results from its manufacture into phosphorus compounds and during the transport and use of these compounds. In the manufacturing process

Patty's Industrial Hygiene and Toxicology, Fourth Edition, Volume 2, Part A, Edited by George D. Clayton and Florence E. Clayton.
ISBN 0-471-54724-7 © 1993 John Wiley & Sons, Inc.

phosphate rock containing the mineral apatite (tricalcium phosphate) is heated, and elementary phosphorus is liberated as a vapor. Phosphorus is used in the manufacture of explosives, incendiaries, smoke bombs, chemicals, rodenticides, phosphor bronze, and fertilizer. The use of phosphate fertilizers, resulting in increased nutrients in fresh water, is a major environmental pollution problem.

Phosphorus exists in several allotropic forms: white (or yellow), red, and black (or violet). The last is of no industrial importance. Elemental yellow phosphorus extracted from bone was used to make "strike-anywhere" matches. In 1845, the occupational disease "phossy jaw," a jaw bone necrosis, was recognized in workers who manufactured such matches. A prohibitive tax (1912) on matches made from yellow phosphorus led to the use of less toxic materials, red phosphorus and phosphorus sesquisulfide. The United States appears to have lagged behind European countries in that the Berne Convention of 1906 agreed not to manufacture or import matches made with yellow phosphorus. Occasional injuries continued to result from the use of yellow phosphorus in the manufacture of fireworks until 1926, when an agreement was reached to discontinue the use of yellow phosphorus for this purpose.

The world production of elemental phosphorus exceeds 1,000,000 metric tons. It is manufactured either in the electric or the blast furnace. Both depend on silica as a flux for the calcium present in the phosphate rock. Nearly all of the phosphorus is converted into phosphoric acid or other phosphorus compounds (1).

1.2 Physical and Chemical Properties

The properties of phosphorus (CASRN* 7664-14-0) and some selected phosphorus compounds are presented in Table 14.1.

Yellow or white phosphorus ignites spontaneously in air at 34°C. It should be stored under water. Under this condition, however, it may form phosphoric acid. Stainless steel containers should be used to hold the corrosive material. Fires of white phosphorus can be controlled by the use of water or sand or the exclusion of air.

Red phosphorus does not ignite spontaneously but may be ignited by friction, static electricity, or heating or oxidizing agents. Handling it in an aqueous solution helps prevent fires.

1.3 Determination in the Atmosphere

The separation of free phosphorus from many phosphorus compounds in the air is a major problem. Air samples may be obtained using an impinger containing xylene, depending on the sampling rate. The exit side of the impinger should be equipped with a filter to catch fumes that may pass through the xylene (2). Analysis of the collected samples can be accomplished using gas (flame photometric detector) chromatography (3).

*Chemical Abstracts Service Registry Number.

Table 14.1. Physical and Chemical Properties of Selected Phosphorus Compounds

Phosphorus Compound	Physical State	Mol. Wt.	Specific Gravity	M.P. (°C)	B.P. (°C)	Solubility
Phosphorus, P	Yellow (white) wax-like solid or red (reddish) brown powder	123.92	1.82 (20°C)	44	280	0.0003 g in 100 ml of water; very soluble in carbon disulfide, ether, chloroform, and benzene
Phosphorus pentoxide (phosphoric anhydride), P_2O_5	White fluffy powder	142		347		Very soluble in water; reacts violently with evolution of heat to form phosphoric acid, H_3PO_4
Phosphoric acid (orthophosphoric acid), H_3PO_4	Clear, thick liquid or deliquescent crystals	98	1.83	42.3	213	Soluble in water and ethanol
Tetraphosphorus trisulfide (phosphorus sesquisulfide), P_4S_3	Yellow crystals	220.12	2.03	172.5	407.5	Insoluble in cold water, decomposes in hot water
Phosphorus pentachloride, PCl_5	Pale yellow, fuming solid	208.31		148	160 sublimes at 100	Decomposes in water to form phosphorus oxychloride, phosphoric acid, and HCl
Phosphorus oxychloride, $POCl_3$	Clear colorless fuming liquid	153.35	1.645	1.25	105.8	Decomposes in water to form phosphoric acid and hydrogen sulfide
Phosphorus pentasulfide, P_2S_5 (generally exists as P_4S_{10})	Light yellow crystals	222.29	2.09	286	513	Decomposes in water to form phosphoric acid and hydrogen sulfide
Phosphorus trichloride, PCl_3	Colorless liquid	137.39	1.574 (21°C)	−91	75.5	Reacts with water to produce HCl and H_3PO_4
Phosphine, PH_3	Colorless gas	34.04	1.17 (air = 1)	−133.5	87.4	26 ml of gas dissolves in 100 ml of water at 17°C

Alternative methods of collection using filtration procedures have been suggested. Likewise, other methods of analysis are available, based mostly on the conversion of phosphorus to phosphate, which is then treated with ammonium molybdate to form a colored complex (4).

1.4 Physiological Response

Phosphorus (white-yellow) can be absorbed through the skin, respiratory tract, and gastrointestinal (GI) tract. Experimental investigations in rats show the highest retention 5 days after oral administration in the liver, skeletal muscle, GI tract, blood, and kidney. In the body, phosphorus is converted to phosphates. Urinary excretion, the chief mode of elimination, is largely as organic and inorganic phosphates (5).

The human lethal oral dose of phosphorus (white) is about 1 mg/kg body weight, and as little as 0.2 mg/kg may produce adverse effects (6). Acute oral phosphorus intoxication generally has two stages. In the initial phase GI effects predominate and may include nausea, vomiting, and belching. The onset may be within 30 min after ingestion. Death from cardiovascular collapse can occur in about 12 hr. A period of regression and apparent recovery lasting about 2 days may occur. The second stage is characterized by the return of the GI distress plus signs of hepatic, renal, and cardiovascular problems, for example, jaundice, pitting edema, oliguria, high pulse rate, and low blood pressure. In either phase, smoking, luminescence, and a garlic odor of the vomitus and feces are characteristic but not diagnostic. The most common pathological finding in deaths has been fatty degeneration of the liver and kidneys (7).

Acute yellow phosphorus poisoning from rat poison consists of garlic odor, mucosal burns, and phosphorescent vomitus or feces occurring from a few minutes to 24 hr after ingestion and are variable in occurrence. These initial symptoms are related to GI or central nervous system (CNS) effects. Mortality rates of 23 percent were associated with the early symptoms of vomiting or abdominal pain, whereas the mortality rate was 73 percent for victims with the initial manifestations related to CNS effects consisting of restlessness, irritability, drowsiness, stupor, or coma. The mortality rate for patients having both GI- and CNS-related symptoms was 47 percent (8). These findings differ somewhat from the classical three stages of phosphorus poisoning. McCarron et al. (8) recommend gastric lavage with 1:5000 solution of potassium permanganate. They did not recommend treating with copper sulfate, which reacts to form insoluble copper phosphide because the dose of copper sulfate is near the lethal dose (15 g) for that compound.

Skin contact with white phosphorus results in severe and painful burns. Usually the affected area turns grayish white and infection ensues. Treatment should be prompt to minimize deeper penetration. Ocular damage may result from direct contact. Animal studies have shown that liver and kidney damage may result from dermal application (9).

Acute inhalation of white phosphorus vapors or smoke, usually containing phosphorus pentoxide at concentrations of about 0.035 mg/l, has resulted in respiratory tract irritation. In at least one case hepatic involvement was reported (10). In humans short-term exposure, 10 to 15 min, at concentrations above 400 mg/m^3 has resulted in signs of respiratory tract irritation (11).

Chronic toxicity following exposure to phosphorus vapors has occurred in factory workers. Generally, the onset of signs is only after many years of exposure. Chronic poisoning following ingestion is rare. Cases of occupational exposure in which the onset of signs has occurred after the personnel have ceased working in a phosphorus-contaminated area have been reported. Early symptoms of chronic intoxication include gastrointestinal distress and sometimes a phosphorus odor (garlic-like) of the breath. A slight jaundice is not uncommon. The classical effect of chronic phosphorus intoxication is on the bone. Most typically this involves the jaw and leads to the so-called "phossy-jaw" or necrosis of the jaw (12).

In chronic phosphorus intoxication, lowered potassium blood levels or increased chloride concentrations along with leukopenia and anemia have been reported. Hepatic and renal involvement are generally not characteristic (13). The hematologic changes have been produced in experimental animals as have the classical effects on bone. In addition, hepatic degeneration leading to cirrhosis from chronic oral administration of phosphorus has been produced in experimental animals (14, 15).

In general, red phosphorus is considered much less toxic than the white. Experimental animals subjected to red phosphorus have shown liver and kidney effects similar to those reported for white phosphorus (16). However, exposure data are nonexistent for humans. The cellular basis for the toxic effects of phosphorus is probably its reducing properties, which may cause disturbance of intracellular oxidative processes.

1.5 Health Standards

The American Conference of Governmental Industrial Hygienists (ACGIH) and Occupational Safety and Health Administration (OSHA) time-weighted average (TWA) for phosphorus (yellow) is 0.1 mg/m^3 (17). No standards for red phosphorus have been established.

Suitable skin and eye protection should be required. Pre-employment medical examination should include X-ray studies of the teeth and jaw, and good dentition should be required for placement. Routine dental examination should be made monthly if exposure is high or prolonged. Dental work, fillings, and extractions should be followed by exclusion from exposure for several months, and any suspicion of jaw injury should result in permanent removal from phosphorus exposure.

2 PHOSPHORUS COMPOUNDS

2.1 Phosphorus Pentoxide

2.1.1 Sources, Uses, and Industrial Exposures

Phosphorus pentoxide or phosphoric anhydride, P_2O_5, is formed by burning yellow phosphorus in dry air or oxygen. It has a great affinity for water and is used as a desiccating agent.

2.1.2 Physical and Chemical Properties

The properties of phosphorus pentoxide (CASRN 1314-56-3) are listed in Table 14.1.

2.1.3 Physiological Response

Phosphorus pentoxide is corrosive and irritating to mucosal surfaces, eyes, and skin. The resulting phosphoric acid is less harmful than sulfuric acid. See Section 2.2 on phosphoric acid.

2.1.4 Health Standards

The hygienic standard recommended in the American Industrial Hygiene Association (AIHA) Hygienic Guide is 1 mg P_2O_5/m^3 of air.

2.2 Phosphoric Acid

2.2.1 Sources, Uses, and Industrial Exposures

Phosphoric acid (orthophosphoric acid), H_3PO_4, is a weak acid. It is an intermediate in the manufacture of superphosphate fertilizers. In addition, it may be used in cleaning metals. With regard to this use, impurities in the metals may lead to the formation of phosphine.

2.2.2 Physical and Chemical Properties

The properties of phosphoric acid (CASRN 7664-14-0) are given in Table 14.1.

2.2.3 Physiologic Response

Irritation of the eyes, respiratory tract, and mucous membranes may occur as with any weak acid. There is no evidence that phosphorus poisoning results from contact with phosphoric acid.

2.2.4 Health Standards

The ACGIH/OSHA TWA is 1 mg H_3PO_4/m^3, and the short-term exposure limit (STEL) is 3 mg/m^3 (17). Eye and skin protection should be provided for personnel likely to be involved with accidental spills (18).

2.3 Tetraphosphorus Trisulfide

2.3.1 Sources, Uses, and Industrial Exposures

Tetraphosphorus trisulfide, or phosphorus sesquisulfide, P_4S_3, is used in making matches and friction strips for safety-match boxes.

2.3.2 Physical and Chemical Properties

Some of the properties of tetraphosphorus trisulfide are given in Table 14.1.
The ignition temperature is about 100°C.

2.3.3 Physiological Response

Phosphorus sesquisulfide is not sufficiently volatile to present a vapor hazard at the temperatures of occupied spaces. The dust or fume is an irritant to the eyes, the respiratory tract, and the skin. Its toxicity, however, is minor when compared with that of yellow phosphorus; and serious ill effects other than eczema have not been reported (19).

2.3.4 Health Standards

No hygienic standard has been suggested.

2.4 Phosphorus Pentachloride

2.4.1 Sources, Uses, and Industrial Exposures

Phosphorus pentachloride, PCl_5, is made by reacting yellow phosphorus with chlorine; it is used in chemical manufacturing. It produces phosphorus trichloride and chlorine when heated, and phosphorus oxychloride, phosphoric acid, and hydrochloric acid when decomposed in water.

2.4.2 Physical and Chemical Properties

Properties of phosphorus pentachloride (CASRN 10026-13-8) are given in Table 14.1.

2.4.3 Physiological Response

Phosphorus pentachloride has a pungent, unpleasant odor; and its vapor or fume is very irritating to all mucous surfaces including the lungs.

2.4.2 Health Standards

The threshold limit value (TLV) as a TWA has been set by ACGIH at 0.85 mg of PCl_5/m^3 (0.1 ppm). OSHA has not established a permissible exposure limit (PEL) at this time (17). It would seem prudent to control for the possible breakdown products: phosphorus trichloride, chlorine, phosphorus oxychloride, phosphoric acid, and hydrochloric acid.

2.5 Phosphorus Oxychloride

2.5.1 Sources, Uses, and Industrial Exposures

Phosphorus oxychloride, $POCl_3$, is used in the manufacture of plasticizer, hydraulic fluids, and gasoline additives (20).

2.5.2   Physical and Chemical Properties

Properties of phosphorus oxychloride (CASRN 10025-87-3) are given in Table 14.1.
The reaction of phosphorus oxychloride with water and alcohol generates heat.

2.5.3   Physiological Response

The main effects of phosphorus oxychloride inhalation are on the mucous membranes, and effects as severe as emphysema have been reported. The ACGIH (20) reports that cases of both chronic and acute intoxication have been observed. Some instances of nephritis have occurred.

2.5.4   Health Standards

The ACGIH and the OSHA TWAs for phosphorus oxychloride (POCl$_3$) (CASRN 10025-87-3) are 0.1 ppm. The ACGIH limit is based largely on analogy to phosphorus trichloride. A STEL of 0.5 ppm was deleted (20).

2.6   Phosphorus Pentasulfide

2.6.1   Sources, Uses, and Industrial Exposures

Also known as phosphoric sulfide, thiophosphoric anhydride, or phosphorus persulfide, this sulfide, P$_2$S$_5$, generally exists as P$_4$S$_{10}$ and is prepared by fusing red phosphorus with sulfur. It is used in the manufacture of safety matches and ignition chemicals and in chemical processes for introducing sulfur into organic compounds.

2.6.2   Physical and Chemical Properties

The properties of phosphorus pentasulfide (CASRN 134-80-3) are included in Table 14.1.

2.6.3   Physiological Response

Irritation due to the decomposition products of phosphorus pentasulfide is the major toxicity.

2.6.4   Health Standards

The ACGIH TWA–TLV is 1 mg/m^3, with a STEL of 3 mg/m^3. OSHA has adopted the same standards (17).

2.7   Phosphorus Trichloride

2.7.1   Sources, Uses, and Industrial Exposures

Phosphorus trichloride, PCl$_3$, is made by reacting yellow phosphorus with chlorine and is used in chemical manufacturing. It hydrolyzes to phosphoric acid and hydrochloric acid.

2.7.2 Physical and Chemical Properties

The properties of phosphorus trichloride (CASRN 7719-12-2) are given in Table 14.1.

The conversion factors are as follows:

$$1 \text{ ppm } = 5.6 \text{ mg/m}^3$$
$$1 \text{ mg/m}^3 = 0.18 \text{ ppm at } 25°C, 760 \text{ mm Hg}$$

2.7.3 Physiological Response

Phosphorus trichloride reacts with water exothermically, giving off hydrochloric acid and phosphoric acid gases. Phosphorus trichloride itself volatilizes at room temperature. As a vapor it is irritating to the skin, eye, and mucous membranes. Severe acid burns can occur. The onset of acute pulmonary edema as a consequence of inhalation exposure may be delayed 2 to 6 hr or 12 to 24 hr in minor and moderate to severe exposures, respectively.

2.7.4 Health Standards

The ACGIH recommends 0.2 ppm (1.1 mg of PCl_3/m^3) for the TWA and 0.5 ppm for the STEL. The OSHA standards are the same (17).

2.8 Phosphine

2.8.1 Sources, Uses, and Industrial Exposures

Phosphine, or phosphorated hydrogen or hydrogen phosphide, PH_3, has no direct commercial use; however, it may be generated from aluminum or zinc phosphide and water for grain fumigation. It may be present in phosphorus as a polymer or generated at low rates under alkaline conditions and at a temperature of 85°C. The generation of acetylene from calcium carbide containing calcium phosphide as an impurity and metal processing procedures in which phosphides are formed are the most frequent sources of industrial hygiene problems with phosphine.

2.8.2 Physical and Chemical Properties

Properties of phosphine (CASRN 7803-51-2) are included in Table 14.1.

The conversion factors are as follows:

$$1 \text{ ppm } = 1.39 \text{ mg/m}^3$$
$$1 \text{ mg/m}^3 = 0.72 \text{ ppm}$$

2.8.3 Determination in the Atmosphere

Airborne concentrations of phosphine can be collected in a fritted glass bubbler containing silver diethyldithiocarbamate. A complex is formed and analyzed spectrographically (21). Also gas chromatography using a flame photometric (phosphorus mode) detector is applicable (22).

2.8.4 Physiological Response

Phosphine differs from arsine in that red blood cell hemolysis does not occur. When low concentrations are inhaled, headaches, dizziness, tremors, general fatigue, GI distress, and burning substernal pain may result. Toxic exposures to phosphine have been documented as a result of grain fumigation, attempted suicide, and ferro-silicon decomposition. A productive cough with fluorescent-green sputum, acute dyspnea, and pulmonary edema may develop. Death may be preceded by tonic convulsions, which may ensue after apparent recovery. At higher concentrations death may occur after $\frac{1}{2}$ to 1 hr of exposure at concentrations of 400 to 600 ppm. Serious effects may be produced by exposure to 5 to 10 ppm for several hours. Phosphine's characteristic decayed fish odor is barely detectable at concentrations of 1.5 to 3 ppm. The chronic effects produced by phosphine are essentially the same as those produced by phosphorus (23, 24).

Aluminum phosphide is used worldwide as a rodenticide to protect stored grain. It reacts with moisture to release the highly toxic gas phosphine. It normally dissipates into the air, leaving little residue. Thus the grain is fit for human consumption. Fifteen cases of aluminum phosphide oral intake poisoning have been reported (25). The clinical and pathological changes were in the GI tract and in the respiratory, cardiovascular, and central nervous systems. These changes have been reported after inhalation exposure to phosphine. Hypotension and hepatic toxicity were a usual feature of these poisonings. The time between ingestion and hospital admission averaged 5.3 hr (0.5 to 16 hr). The average time between ingestion and death in patients was 31 hr (1 to 106 hr). The average dose was 1.5 to 90 g (average 4.7 g). Assuming 65 kg of body weight, this amounts to 72 mg/kg and is greater than the 40.5 mg/kg LD_{50} in rats for zinc phosphide.

Of the 1,000,000 U.S. workers possibly exposed to phosphine, 10,000 of them may be exposed aboard ships as a result of generation of phosphine to fumigate grain. Unlike industrial workers, crew members and others living aboard ships may be exposed constantly and for long durations. Aboard-ship exposures may be intensified by the confined nature of the shipboard working conditions and life.

The wife of the captain and two children, plus 29 of 31 crew members on a grain carrier, became ill after inhaling phosphine generated from the use of aluminum phosphine. The first cases occurred 2 days after the fumigation started. The predominant symptoms were fatigue (86 percent), nausea (72 percent), headache (66 percent), vomiting, cough, and shortness of breath. Abnormal physical findings included paresthesia (59 percent), jaundice (52 percent), tremor (31 percent), ataxia, and diplopia. The younger of the two children, a 2- and a 4-year-old, died. Necropsy findings included focal myocardial infiltration with necrosis, pulmonary edema, and widespread small vessel injury. The older surviving daughter had ECG changes and echo-cardiographic changes indicating heart muscle damage accompanied by an increase of skeletal muscle serum creatinine phosphokinase. She recovered in 24 hr. Except for nausea, vomiting, and paresthesia, values of parameters from physical, neurological, and clinical pathological evaluation of the mother were normal; she recovered 24 hr after removal of exposure. Concentrations ranged from 30 ppm in void space near the midship ventilator intake to 0.5 ppm in the living quarters (26). The severity of the effects in the younger child and recovery time course of the surviving child as compared with that of the mother suggest that children may be especially susceptible to the toxicity of phosphine at low levels. This may be related to the greater intake rate in the young because of a higher rate of metabolism.

2.8.5 Health Standards

The TWA–TLV (ACGIH) is 0.42 mg/m^3 (0.3 ppm). The STEL (15 min) is 1.4 mg/m^3 (1 ppm). OSHA has adopted the same standards (17). The TLV of 0.3 ppm is below the odor threshold of 1.5 to 3 ppm.

Because of the widespread generation of phosphine aboard ships, a maritime standard should consider the possibility of 24-hr exposure and the exposure of non-crew members, who may be more susceptible.

3 SELENIUM

3.1 Sources, Uses, and Industrial Exposures

Selenium (Se), a nonmetallic element of the sulfur group, is widely distributed in nature. It is obtained along with tellurium as a by-product of metal ore refining, chiefly copper. About 1600 tons are mined per year on a global basis (27).

Selenium exists in three allotropic forms. When heated, the red amorphous powder cools to the vitreous form. A gray form with hexagonal structure is photoconducting and semiconducting. The third form is that of monoclinic red crystals. Selenium burns in air with a bright blue flame and emits a horseradish-like odor.

Up to 90 percent of the selenium content in ambient air is emitted during the burning of fossil fuels. Selenium dioxide is formed during combustion of elemental selenium present in the fossil fuels. Air pollution concentrations averaged from 13 ng/mg^3 in urban areas to 0.38 in remote areas. The mass medium diameter was 0.92 μm. The worldwide emissions of 10,000 tons/year from natural sources exceeds the atmospheric emissions from anthropogenic sources (5100 tons). However, 41,000 tons is emitted into aquatic ecosystems. The largest contributor is electric power generating plants that produce 18,000 tons; manufacturing processes account for 12,000 tons (27).

Selenium is an essential trace metal. Because of data suggesting that it may inhibit chemical carcinogenesis, it has been widely promoted as a dietary supplement (28). Selenium may replace sulfur and forms selenoproteins in plants and animal systems.

Most of the world's selenium today is provided by its recovery from anode muds of electrolytic copper refineries. Selenium is recovered by roasting muds with soda or sulfuric acid, or by smelting them with a soda and niter.

One of the important uses of selenium is in photoelectric cells. In addition, it is used in the manufacture of rectifiers to convert alternating current to direct current. Its conductivity increases up to 1000 times on exposure to light. The greatest use is as a decolorizer for glass and ceramic. Cadmium selenide is the red pigment used for the ruby glass of automobile taillights. It is also used in various alloys and in rubber manufacture. Some compounds have been used as insecticides. Selenium dioxide is the most widely used selenium compound. It is produced by burning selenium in oxygen and is employed in the production of other selenium compounds (1). Approximately 1,000,000 lb of selenium are used annually in the United States. Use in electronic and photocopier components accounts for 35 percent; glass manufacturing for 30 percent; and chemicals and pigments for 25 percent (29). For radionuclide imaging of the pancreas ^{75}Se-selnomethionine is used.

3.2 Physical and Chemical Properties

Properties of selenium (CASRN 7782-49-2) and some of its compounds are listed in Table 14.2.

Toxic gases and vapors may be released in a fire involving selenium. Selenium can react violently with chromic oxide (CrO_3), lithium silicon (Li_6Si_2), nitric acid, nitrogen trichloride, oxygen, potassium bromate, silver bromate, and fluorine.

3.3 Determination in the Atmosphere

Particulates containing selenium can be collected on filters treated with nitric acid to extract the selenium. Gases or vapors may be scrubbed through 40 to 48 percent hydrobromic acid containing 5 to 10 percent free bromine. Soda lime has been used to collect hydrogen selenide. Selenium dioxide can be collected in 10 ml of water with a midget impinger.

The use of an atomic absorption spectrophotometer with argon hydrogen flame and an electrodeless discharge lamp for selenium to provide a characteristic selenium line at 196.0 nm is recommended (30). Gravimetric, volumetric, colorimetric, or spectrophotometric methods have been used for analysis.

3.4 Physiological Response

Selenium is an essential element. It interacts with a wide variety of vitamins, xenobiotics, and sulfur-containing amino acids. Selenium reduces the toxicity of many metals such as mercury, cadmium, lead, silver, copper, and arsenic (31).

Selenium and most of its compounds are readily absorbed as a result of oral intake or by breathing. Dermal exposure generally does not result in elevated selenium blood concentration. After absorption, high concentrations are found in the liver and kidney. In humans, dimethylselenide is formed and may account for

Table 14.2. Physical and Chemical Properties of Selenium and Selected Compounds

Form of Selenium	Physical State	Atomic or Mol. Wt.	Sp. Gr.	M.P. (°C)	Solubility
Selenium (Se)	Red amorphous powder turning black on standing and vitreous on heating	78.96	4.3–4.8 (20°C)	217 (b.p. 688)	Soluble in H_2SO_4, ether, chloroform
Selenium dioxide (SeO_2)	White crystalline powder	110.96	3.95	340 under pressure (subl. 315)	Readily sol. in hot or cold water to form selenious acid, H_2SO_3
Selenium trioxide (SeO_3)	Yellowish white hygroscopic powder	126.96	3.6		Readily sol. in water to form selenic acid similar to H_2SO_4
Sodium selenite ($Na_2SEO_3 \cdot 5H_2O$)	White powder	263.04			Freely sol. in water to form a slightly alkaline solution
Sodium selenate	Colorless crystals, with or without 10 mol H_2O	188.95			Very sol. in water

the garlic odor of the breath. Selenium crosses the placenta in rats, dogs, mice, and humans. Trimethylselenium is a predominant urinary metabolite, at least in rats. Urinary and fecal excretion of selenium accounts for about 50 percent of the total selenium output. At higher oral doses excretion in the exhaled air becomes more important. In humans as in rats, the elimination of selenium is triphasic. The half-life of the first phase in humans is about 1 day; the second phase has a $t_{1/2}$ of 8 to 9 days; and the third phase has a $t_{1/2}$ of 115 to 116 days. The first two elimination phases represent the fecal excretion of unabsorbed selenium and the urinary excretion of absorbed but unmetabolized material (31). In rats, selenium is highly concentrated in bile relative to plasma. The liver and kidney contain the highest concentrations of selenium. Selenium is removed predominantly by the urinary tract (32). In the liver, many selenium compounds are biotransformed to excretable metabolites. Identified metabolites are trimethylselenide in urine and dimethylselenide in breath (33).

The organic selanoamino acids are among the most toxic forms of ingested selenium and are more toxic to primates than rodents. In a 30-day oral study the maximum tolerated dose of L-selenomethionine to female macaques (*Macaca fascicularis*) was 188 µg/(kg)(day). Signs of adverse effects included hypothermia, weight loss, dermatitis, and disturbance in menstruation. The addition of dietary supplements and fruits was necessary to prevent death at this dose (34).

Acute selenium toxicity is primarily limited to inhalation of selenium dust or fumes, which are irritants. Oral doses of 1 to 2 mg can rapidly be fatal in animals. Two deaths caused by the ingestion of gun bluing solutions containing selenious acid have been reported. In one case the estimated dose was 10 to 20 mg/kg of selenious acid. Hypertension and garlicky breath (from dimethylselenide) were the major signs on hospital admission 1 hr after ingestion. Autopsy findings on day 8 indicated death from respiratory problems, chiefly edema and hemorrhage. The patient's serum concentration and urinary excretion of selenium were 20 times normal. However, 24-hr selenium levels of 100 µg have been associated with vomiting and other gastrointestinal problems. Other clinical findings of acute inorganic selenium toxicity included vomiting, diarrhea, labored breathing, weakness, unsteady gait, and coma. The histological evaluation may include pulmonary edema, hepatic necrosis, skeletal muscle degeneration, renal tubular hydropic degeneration, and swelling and disruption of myocardial mitochondria (27).

Selenium dioxide is the primary problem involved with most industrial exposure to the element because the oxide is formed when selenium is heated. The dioxide itself forms selenious acid with water or sweat, and the acid is an irritant. Selenium dioxide splashed in the eye or entering the eye because of airborne concentrations causes the development of a pink allergic-type reaction of the eyelids, which may become puffy. There is usually conjunctivitis of the palpebral conjunctiva (35). Two hours after a fatal suicidal ingestion of selenium dioxide, a 17-year-old male was admitted to a hospital with asystolia and apnea. Autopsy findings included congestion of the lungs and kidneys with diffuse swelling of the heart and brain edema. The most striking finding was an orange-brown discoloration of the skin and all viscera, probably due to hemolysis. The selenium blood and tissues levels were 100 to 1000 times normal. The concentration in the pancreas was remarkably high. Tissue concentrations were predominantly elemental selenium or selenium disulfide, although the authors hypothesized that the discoloration was due to hemolysis caused by hydrogen selenide formed in the metabolism of selenium (36).

Most of the concern has related to the possible effects resulting from chronic oral intake and arises from results of long-term oral feeding studies in rodents. The U.S. governmental authorities have now accepted that the use of selenical feed additives does not pose a carcinogenic hazard for humans. The industrial evidence is a little clearer with regard to this question. A series of oral rodent bioassays was performed on selenium compounds in the period 1943 to 1977. The results were mixed, with some indicating a carcinogenic potential (most frequently in the liver), and some showing a reduction of cancer. In 1975 the International Agency for Research on Cancer (IARC) considered data on sodium selenate, sodium selenite, and organic forms of selenium contained in plants, and judged that the available data provided no suggestion that selenium was carcinogenic in humans (37). Since then, the National Toxicological Program (NTP) has shown that selenium sulfide, a material used in shampoo but not absorbed through the skin, causes increased hepatocellular cancer in both rats and mice, and alveolar/bronchial carcinomas and adenomas in mice when administered orally (38). Dermal administration to mice did not produce significantly more tumors in the treated than in the controls (39), although in the males the incidence of lung tumors was increased when a prescription dandruff shampoo containing 2.5 percent selenium sulfide was tested (40). The latter could be the result of grooming by the test animals.

In a 26-year period of selenium rectifier processes in Britain, the cancer deaths were comparable to the national average (35). In addition, epidemiologic studies in geographic areas of low and high selenium (forage and blood concentrations) levels suggested an inverse relationship between selenium and cancer in humans (41, 42).

Symptoms and signs of occupational disease resulting from long-term industrial exposure to selenium were recognized early. These were bronchial irritation, gastrointestinal distress, nasopharyngeal irritation, and a persistent garlic odor on the breath (43). This odor, however, is not reliable and may be present when urinary excretion is as high as 0.1 mg/l. The normal value is about 0.05 mg/l, but can vary somewhat because of large differences in dietary content. A metallic taste has also been reported frequently in instances of chronic industrial exposure. Other signs include pallor, irritability, and excessive fatigue (35). There have been only occasional associations made to suggest that either industrial or environmental exposure of humans may result in the renal or hepatic lesions that have been produced experimentally by selenium (44, 45).

In farm animals (cattle, sheep, hogs, and horses) toxicity from intake of feed containing excessive selenium has resulted in blind staggers and/or alkali disease, characterized by anorexia, emaciation, and collapse. At lower concentrations, loss of hair and hooves has occurred. In addition, the reproductive capacity was reduced. Pathology includes hepatic necrosis, nephritis, hyperemia, and ulceration in the upper GI tract. In experimental animals liver injury, disturbance of the endocrine system, anemia, nephritis, myocarditis, and pancreatitis have been produced (46).

An accidental exposure of female laboratory workers to selenite resulted in termination of several pregnancies by miscarriages and one bilateral club foot infant (47). Although results of investigations in the chick embryo and experience in farm animals would tend to support this possibility (48), experimental investigations in the hamster failed to show a teratogenic potential (49). Epidemiologic investigations have actually suggested that neonatal deaths declined as the environmental test selenium concentration rose (50). Positive mutagenic effects in various experimental test systems are supported by reports of excess deaths before weaning, failure to breed, and when selenium is incorporated into the drinking water of rats and mice for three generations, an unusual sex ratio (48, 51–53).

Although environmental levels of selenium have been associated with embryonic mortality and teratogenesis in birds, especially in the western United States, few investigators other than Robertson (47) have shown it to be a problem in humans or animals. Tarantal et al. (54) tested L-selenomethionine in cynomolgus monkeys. Embryonic toxicity occurred in three monkey fetuses at 300 μg/kg. All the dams were compromised at this dose, and one of the three dams died.

Human whole blood cultures were exposed to selenium dioxide and selenium in other valence states. The ability of the selenium compound to induce sister chromatid exchanges in decreasing order of their effectiveness were selenium, selenium dioxide, sodium selenide, sodium selenite, and sodium selenate (55).

Both experimental evidence and epidemiologic studies support the contention that selenium increases dental caries. This aspect of selenium industrial toxicity has not been investigated.

3.5 Health Standards

Most of the toxicologic information related to selenium itself is applicable to selenium salts. The analytic procedures will be useful in determining the airborne concentration of most materials. The TLV–TWA (ACGIH) for selenium and its compounds is 0.2 mg Se/m^3. The 1989 OSHA standard is 0.2 mg/m^3. IARC currently classifies selenium and its compounds as a Group 2B carcinogen (probably carcinogenic in humans, but having [usually] no human evidence), whereas they classify selenium sulfide as a Group 3 carcinogen (sufficient evidence in experimental animals). The NTP indicates that selenium sulfide is "reasonably anticipated to be a carcinogen with limited evidence in humans or sufficient evidence in animals" (17).

Dermal and oral exposure to some selenium compounds may present a risk of cancer. Medical examination should be made available annually to each employee who is exposed to selenium and its inorganic compounds at potentially hazardous levels. Persons with a history of asthma, allergies, or known sensitization to selenium, or with a history of other chronic respiratory disease, GI disturbances, disorders of liver or kidneys, or recurrent dermatitis would normally be at increased risk from exposure. Special consideration should be given to women of childbearing age, since the possibility that selenium may be teratogenic could place these women in a high risk group. Because selenium causes liver damage and tumors in animals, a profile of liver function should be obtained. Whole blood selenium levels in individuals range from 1.33 to 7.5 µg/ml. Normal levels in the United States are 0.1 µg/ml, but no biologic exposure indexes are established.

The maximum contaminant level (MCL) of selenium in drinking water is 0.01 mg/l (40 CFR 141.11).

4 SELENIUM COMPOUNDS

4.1 Hydrogen Selenide

4.1.1 Sources, Uses, and Industrial Exposures

Hydrogen selenide, H_2Se, may be formed by the reaction of acids or water with metal selenides or wherever nascent hydrogen is in contact with soluble selenide compounds. It has no commercial use.

4.1.2 Physical and Chemical Properties

The properties of hydrogen selenide (CASRN 7783-07-5) are as follows:

Physical state	Colorless gas
Molecular weight	80.98
Specific gravity	2.79 (air = 1)
Melting point	$-64°C$
Boiling point	$-42°C$
Solubility	270 ml in 100 ml of water at 22.5°C, more soluble in alkaline water

1 mg/m^3 = 0.3 ppm; 1 ppm = 3.3 mg/m^3 at 25°C, 760 mm Hg

4.1.3 Physiological Response

The results of acute hydrogen selenide inhalation intoxication in humans include irritation of the respiratory tract mucous membranes, pulmonary edema, severe bronchitis, and bronchial pneumonia. Accidental inhalation of hydrogen selenide has produced upper respiratory tract irritation and wheezing, followed by progressive dyspnea and reduced expiratory flow rates in 18 hr. The patients improved over 5 days, but some had pulmonary function changes remaining 3 years later (56). Similar signs have been observed in rats at concentrations of 1 to 4 µg/l (57). Five cases of subacute intoxication from less than 0.2 ppm of hydrogen selenite, probably generated from the use of selenious acid in an etching and printing operation, have been reported (58). Gastrointestinal distress, dizziness, increased fatigue, and a metallic taste in the mouth were reported. A single case has been reported in which exposure to a high concentration caused severe hyperglycemia, controllable by increasing doses of insulin (59).

4.1.4   Health Standards

The ACGIH and OSHA (17) TWAs are both 0.05 ppm (0.16 mg/m^3). Hydrogen selenide has an offensive odor somewhat resembling that of decayed horseradish. In the lower toxic range this odor is not a dependable warning. Eye and nasal irritation is moderate.

4.2   Selenium Hexafluoride

4.2.1   Sources, Uses, and Industrial Exposures

Selenium hexafluoride, F_8Se, is used as a gaseous electric insulator. It may be prepared by passing gaseous fluorine over finely divided selenium in a copper vessel.

4.2.2   Physical and Chemical Properties

The properties of selenium hexafluoride (CASRN 7783-79-1) follow:

Physical state	Gas
Molecular weight	192.96
Density	3.25
Melting point	$-50.8°C$
Solubility	Insoluble in water

4.2.3   Physiological Response

Short-term exposures of experimental rodents to airborne concentrations of selenium hexafluoride produced signs of pulmonary edema. Exposure to 5 ppm for 4 hr was the lowest effect level, and 1 ppm showed no effect. In a 5-day, 1-hr exposure regimen, 5 and 1 ppm produced similar results (60).

4.2.4   Health Standards

The ACGIH TWA–TLV is 0.05 ppm measured as selenium. This is equivalent to the OSHA limit for selenium hexafluoride (18). The National Institute for Occupational Safety and Health (NIOSH) (18) indicated that 5 ppm is immediately dangerous to life or health (IDLH).

4.3   Selenium Oxychloride

4.3.1   Sources, Uses, and Industrial Exposures

Selenium oxychloride, $SeOCl_2$, is a powerful solvent, chlorinating agent, and resin plasticizer used in the chemical industry.

4.3.2 Physical and Chemical Properties

The properties of selenium oxychloride are listed below:

Physical state	Clear pale yellow liquid
Molecular weight	165.87
Specific gravity	2.42 (22°C)
Melting point	8.5°C
Boiling point	176.4°C
Vapor pressure	~0.05 mm Hg, 20°C
Solubility	Hydrolyzes in water to HCl and selenious acids

4.3.3 Physiological Response

Selenium oxychloride is strongly vesicant and will rapidly destroy the skin upon contact unless immediately removed by washing. Less than 0.1 ml on the skin of rabbits has proved fatal within 24 hr, and selenium was found in the blood and liver. The minimum lethal dose, when applied to the skin of a rabbit, is 7 mg/kg. This would be equivalent to approximately 0.2 ml applied to a man of average size. The application of less than 0.005 ml to the arm of a man caused a painful burn with swelling, and its healing required a month. The vapors of selenium oxychloride are toxic, but their irritant and corrosive actions on the respiratory tract are not so great as might be supposed because the vapor readily decomposes in air. Also, its low vapor pressure limits the concentration possible in air (61).

4.3.4 Health Standards

No health standard has been published.

5 TELLURIUM

5.1 Sources, Uses, and Industrial Exposures

The element tellurium (Te) has some metallic properties, although it is classed as a nonmetal or metalloid. The name is derived from the Latin word for earth. Tellurium is occasionally found native, more often as telluride of gold, calaverite. The elemental form has a bright luster, is brittle, readily powders, and will burn slowly in air. Tellurium exists in two allotropic forms, as a powder and in the hexagonal crystalline form (isomorphous) with gray selenium. The concentration in the earth's crust is about 0.002 ppm. It is recovered from anode muds during the refining of blister copper. It is also found in various sulfide ores along with selenium and is produced as a by-product of metal refineries. The United States, Canada, Peru, and Japan are the largest producers (1).

Tellurium's industrial applications include its use as a metallurgical additive to improve the characteristics of alloys of copper, steel, lead, and bronze. Increased ductility results from its use in steel and copper alloys. Addition of tellurium to cast iron is used for chill control, and it is a basic part of blast caps (1). It is used in some chemical processes as a catalyst. Tellurium vapor is used in "daylight lamps." The use of tellurium, along with selenium, in semiconductors is expanding. Tellurium's use in pottery glazes is limited. Its major use is in the vulcanization of rubber.

5.2 Physical and Chemical Properties

The properties of tellurium (CASRN 13494-80-9) follow:

Physical state	Hexagonal, brittle, metallic, silvery semiconductor or an amorphous, brownish black powder
Atomic weight	127.61
Specific gravity	Metallic, 6.24 at 20°C and amorphous powder, 6.00 at 20°C
Melting point	450°C
Boiling point	1390°C
Solubility	Insoluble in water, benzene, and carbon disulfide; soluble in oxidizing acids and alkalies

Elemental tellurium will burn slowly in air; a finely divided suspension in air can be exploded. Reactions with zinc, chlorine, fluorine, and solid sodium are vigorous and have a potential to cause fires (62).

5.3 Determination in the Atmosphere

Dusts and fumes can be collected on a cellulose membrane filter. The samples are then washed using perchloric and nitric acid to destroy other organic materials and the filter. A diluted solution of the tellurium in nitric acid is prepared and aspirated into the oxidizing air–acetylene flame of an atomic absorption spectrophotometer. This is good in the range of 0.0495 to 0.240 mg/m^3 (30).

Alternative methods of sampling for tellurium hydride or methyl telluride include absorption in concentrated hydrochloric acid containing 5 to 10 percent free bromine. The hydride can be collected in dilute sodium hydroxide. When it is collected in alkali (63) in the presence of air, polytellurides are formed. Alternative analytical methods include determinations as the iodotellurite complex or precipitation and determination as a hydrosol (64).

5.4 Physiological Response

Elemental tellurium is poorly absorbed, but its more soluble compounds may undergo some oral absorption. Soluble tellurium can be absorbed through the skin, although ingestion or inhalation of fumes presents the greatest industrial hazard. A metallic taste in the mouth may result from excessive absorption. The characteristic sign of tellurium absorption is the garlic-like odor attributed to dimethyl-telluride in the breath and sweat. This may persist for many days after exposure. Urinary, fecal, and biliary excretion also occur. The urinary excretion is probably more important than the respiratory excretion in the elimination of absorbed tellurium (65, 66). The normal concentration of tellurium in the urine is 0.2 to 1.0 µg/ml. Tellurium is complexed to the plasma proteins and little is found in the red blood cells (67). In the nervous system tellurium accumulates in the gray matter, not the white matter, when injected intracerebrally. The metal is found in phagocytic and ependymal cells and in the lysosomes as fine needles (68). The whole-body retention model assumes a long half-life based on tellurium dioxide (69).

Acute toxicity of tellurium has not been well explored, but it does not seem to be an industrial problem. In animals, repeated administration by the oral route has produced kidney and nerve damage in several species (70, 71). The appearance of endoneurial edema early in the evolution of tellurium neuropathy raises the possibility that a breakdown of the blood–nerve barrier (BNB) plays a role in the pathogenesis of the tellurium-induced demyelination (72). Inclusion of 1.1 percent tellurium in the diet of developing rats causes a highly synchronous primary demyelination of peripheral nerves, which is followed closely by a period of rapid remyelination. The demyelination is related to the inhibition of squalene epoxidase activity, which results in a block in cholesterol synthesis and accumulation of squalene (73).

Teratogenesis characterized by hydrocephalus has also been reported following the administration of tellurium to pregnant rats (74). Neonatal rats were exposed to tellurium, via the mother's milk, from the day of birth until sacrifice at 7, 14, 21, and 28 days of age. Light and electron microscopy revealed Schwann cell and myelin degeneration in the sciatic nerves at each age studied. These changes were similar to those described in weanling rats as a result of tellurium intoxication. In the CNS, hypomyelination of the optic nerves was demonstrated at 14, 21, and 28 days of age, accompanied by some evidence of myelin degeneration. These changes were also seen in the ventral columns of the cervical spinal cords, although less markedly, and were confirmed by quantitative methods (75).

Exposure of iron foundry workers to concentrations of 0.01 and 0.1 mg Te/m³ for 22 months produced mild GI distress, the characteristic garlic odor, dryness of the mouth, metallic taste, and somnolence (65). A tellurium-containing catalyst from an industrial plant in the Netherlands caused a serious odor problem during biologic treatment of waste water from about 35 plants producing organic and inorganic chemicals. Odor problems, not noticeable during the first years of waste-water plant operation, developed on the skin of the operators and at a distance of 0.5 to 1 km from the plant. It was determined that the odor-producing organic tellurium compound was produced under anoxic conditions in the denitrification step. Faint odor was detected with the addition of as little as 0.01 mg/l of tellurium tetrachloride. This odor was apparent earlier under anoxic conditions than under aerobic conditions. Strong odor was produced under anoxic conditions after 20 hr at concentrations of 0.02, 0.05, and 0.10 mg/l. Under aerobic conditions the higher concentrations produced faint odor except for one trial. To avoid odor problems a process was developed to reduce the tellurium content of the waste water to 5 to 10 percent of the usual values before allowing it to leave the plant of origin (76).

5.5   Health Standards

The ACGIH/OSHA (17) TWA limit is 0.1 mg/m^3. This is about 10 times greater than the concentration that will cause the garlic odor on the breath. Protective equipment (hands and respiratory) should be used in dust operations such as grinding. Local ventilation is recommended (77). Women of childbearing age should be informed of a possible adverse effect from tellurium on offspring when exposed in utero or by nursing.

6   TELLURIUM COMPOUNDS

Tellurium hexafluoride, F_6Te, is the only tellurium compound for which a separate permissible exposure level has been developed. Tellurium dioxide (TeO_2), potassium tellurite (K_2TeO_3), and hydrogen telluride (H_2Te) have a significant potential for industrial exposure. The oxide is found only at temperatures over 450°C and is almost insoluble in water and body fluids, which limits the hazard from this compound. The only cases of poisoning with potassium tellurite have been associated with accidental poisoning.

6.1   Tellurium Hexafluoride

6.1.1   Sources, Uses, and Industrial Exposures

Tellurium hexafluoride, F_6Te, is prepared by the direct fluorination of tellurium metal.

6.1.2   Physical and Chemical Properties

The properties of tellurium hexafluoride are as follows:

Physical state	Colorless gas
Molecular weight	241.61
Melting point	$-37.6°C$
Solubility	Slowly absorbed by water with hydrolysis to telluric acid

6.1.3   Physiological Response

Short-term exposure of experimental rodents to airborne concentrations of tellurium hexafluoride produced evidence of pulmonary edema. The animals exposed to 1 ppm for 4 hr were adversely affected. Those exposed for 1 hr to 1 ppm had an increased respiratory rate; however, repeated exposure to 1 ppm for 5 days produced no effect (78). Tellurium hexafluoride is considered toxic by inhalation and may produce pulmonary edema and death.

Two cases of excessive occupational exposure to tellurium hexafluoride have been reported. Because both workers were also handling volatile liquid esters, some increased absorption and deposition of elemental tellurium in the skin may have occurred. The signs included garlic breath. An unusual feature was bluish black discoloration of the webs of the fingers and streaks on the face and neck. No permanent damage was noted (79).

6.1.4   Hygienic Standard of Permissible Exposure

The ACGIH/OSHA (17) TWA exposure limit is 0.2 mg/m^3 (0.02 ppm). This limit is consistent with that for tellurium and its other compounds. If there has been a significant exposure to tellurium hexafluoride, a chest X-ray may be necessary.

6.2   Hydrogen Telluride

6.2.1   Sources, Uses, and Industrial Exposures

Hydrogen telluride, H_2Te, is a flammable, colorless gas with an offensive odor similar to hydrogen sulfide. It can be synthesized by the action of acid on aluminum telluride. It has no industrial uses.

6.2.2   Physical and Chemical Properties

The properties of hydrogen telluride are as follows:

Physical state	Colorless gas
Molecular weight	129.63
Melting point	$-51°C$
Boiling point	$-4°C$
Vapor density	4.5 (air = 1)
Solubility	Dissolves and decomposes in water, precipitating elemental Te

1 mg/m^3 = 0.2 ppm; 1 ppm = 5.3 mg/m^3 at 25°C, 760 mm Hg

6.2.3   Physiological Response

Hydrogen telluride is an irritant at relatively low concentrations. Headaches may follow an exposure to hydrogen telluride. Dry mouth and throat have also occurred after exposure to this compound. Hydrogen telluride causes pulmonary irritation and edema in animals. It produces symptoms similar to those of hydrogen selenide or arsine, including headaches, malaise, weakness, dizziness, and respiratory and cardiac symptoms. It is probably less toxic than arsine or hydrogen selenide, owing to its ready decomposition; however, exposure of guinea pigs to airborne concentrations caused pronounced hemolysis (80).

7 SULFUR

7.1 Sources, Uses, and Exposures

Sulfur (S) occurs in nature as a yellow, water-insoluble solid. The name is from the Latin *sulphur*. The early Greek physicians mention sulfur and the fumes from burning sulfur in religious ceremonies. Sulfur constitutes about 0.053 percent of the earth's crust and occurs in two allotropic crystalline forms, rhombic and monoclinic. Below 96°C only the rhombic form is stable. Large sedimentary deposits of almost pure element are mined in Texas and Louisiana. Sulfur can be extracted from crude oil in the refining process, as well as from stack gases resulting from coal burning. Sulfur occurs in fossil fuels and in metal (Fe, Pb) ores. Exposure may occur in numerous operations related to the mining and recovery of sulfur. The recovery of sulfur as a by-product accounts for a larger portion of the world's production than the mined mineral. Sulfur is one of the most important raw materials, particularly in the fertilizer industry (1).

Organic sulfur compounds occur in garlic, mustard, onions, and cabbage and are responsible for the odor of skunks. Sulfur occurs in living tissue and is part of some amino acids. Unlike many of the other inorganic elements, sulfur itself is relatively nontoxic. Sulfur and some of its salts have been used medicinally. The consumption of sulfur is a measure of national industrial development and economic activity. Sulfur is most often used as a chemical reagent, rather than as part of a finished product (1).

7.2 Physical and Chemical Properties

The properties of sulfur are as follows:

Physical state	Yellow, solid
Molecular weight	32.064
Density	1.96(20°C)
Melting point	119°C
Boiling point	444.6°C
Solubility	Very soluble in carbon disulfide

Sulfur and its dust suspensions are flammable; the lower flammable limit for sulfur dust in air is 35 mg/l. The ignition temperature is 190°C.

7.3 Determination in the Atmosphere

Normally, airborne sulfur would be collected and analyzed based on the individual sulfur compound.

7.4 Physiological Response

Various sulfur preparations have been used therapeutically. For example, sodium thiosulfate has been used as a cathartic, in the treatment of parasitic diseases, and in the treatment of some types of poisoning. Up to 12 g/day has been consumed without illness except for emesis or vomiting. Orally, colloid preparation of sulfur leads to urinary excretion of the sulfur load as sulfate in 24 hr (81). Exposure to sulfur particulates produces tracheobronchitis, characterized by cough, sore throat, chest pain, and lightheadedness (82).

7.5 Health Standards

There are no airborne standards for elemental sulfur.

8 SULFUR OXIDES

Sulfur oxides include major pollutants, and whereas other sulfur compounds are highly toxic, either type of compound can have a major impact on health and the quality of life.

8.1 Sources, Uses, and Exposures

Sulfur dioxide (SO_2, CASRN 7446-09-5) is formed whenever sulfur is burned in the air. It is perhaps the most widely encountered and best known irritant gas, not only because of its wide usage, but also because of its frequent occurrence as an undesired by-product in the smelting of sulfide ores, in paper manufacture, in the combustion of sulfur-bearing coals and petroleum fuels, and in the action of sulfuric acid on reducing agents. The predominant use of sulfur dioxide is in the production of sulfuric acid. Sulfur dioxide is also used as a refrigerant (1).

Sulfur dioxide is one of the most prominent gases contributing to atmospheric pollution in large cities and in areas surrounding smelters. Mobile sources (automobiles) were one of the main contributors, but catalytic convertors have reduced emissions. The sulfur oxides, along with those of nitrogen, are the principal cause of acid rain, which adversely affects the ecology. The residence time for sulfur dioxide in the ambient air is between 2.5 and 5 days (83).

Sweetening plants for petroleum products sometimes dispose of sulfide gases by burning them to sulfur dioxide and discharging the resulting gas from high stacks. The terrain, height of the stack, rate of gas discharge, and atmospheric conditions present variable factors that have made the success of their dilution method unpredictable and frequently disappointing. The gas rises vertically for some distance above the stack, then spreads out laterally. The important factors in its dispersion are fog, wind direction and velocity, inversion, and turbulence. Sulfur dioxide in moist air or fog combines with the water to form sulfurous acid and is slowly oxidized to sulfuric acid. In the United States, coal-fired electric power plants generate about 70 percent of the sulfur dioxide. The Clean Air Act of 1990 calls on utilities to reduce their annual emissions of SO_2 from about 17.9 million tons to 8.9 million tons by the year 2000.

Sulfur dioxide is an intermediate in the manufacture of sulfuric acid. It is also used in the manufacture of sodium sulfite and in other chemical processes. Large quantities are used in refrigeration, in bleaching, fumigating, and preserving. It is used as an antioxidant in melting and pouring magnesium, where it is applied as the gas or generated by adding powdered sulfur to the surface of the molten metal in the ladle and to the surface of the poured casting. Sulfur dioxide up to 0.5 percent is also used for the prevention of oxidation in controlled-atmosphere heat-treat ovens for magnesium (1).

Prior to or during inhalation, sulfur dioxide (SO_2) may react with water to form sulfurous acid (H_2SO_3) and be oxidized to form sulfur trioxide. The latter reacts rapidly to form sulfuric acid (H_2SO_4), which in the presence of ammonia forms ammonium. Sulfurous acid dissociates to sulfite and bisulfite ions.

Sulfuric acid (H_2SO_4, CASRN 7664-93-9) is made by oxidation of sulfur dioxide in the lead chamber and the contact processes. The acid is used in great quantities in fertilizer manufacture, chemical manufacture, petroleum refining, the production of rayon and film, iron and steel, explosives, textiles, and in pickling and anodizing. The fume of SO_3, or mist of H_2SO_4, may result also whenever sulfuric acid is heated in the open air or gas bubbles are released from a liquid surface containing sulfuric acid. Batteries may contain a 25 to 30 percent solution, and toilet bowl cleaners contain 8 to 10 percent (1).

8.2 Physical and Chemical Properties

The physical and chemical properties of the oxides are listed in Table 14.3.

8.3 Determination in the Atmosphere

Workplace sulfur dioxide may be collected on a charcoal tube. The maximum flow rate of 0.2 l/min is used until 24 liters are collected. Analysis is by ion chromatography (84). As an alternate, a maximum flow rate of 1.5 l/min is used and 400 liters are collected. The sulfur dioxide is desorbed using sodium bicarbonate solution and the analysis is again by ion chromatography (30).

Automatic recorders for SO_2 are available and are widely used for area or atmospheric sampling. Passive dosimeters, detector tubes, and portable direct-reading meters, some pocket sized, can be used to augment the approved workplace monitoring methods.

Airborne sulfuric acid in the workplace is collected on a mixed cellulose ester filter (0.8 μm, 37 mm) in a two- or three-piece cassette and supported by a cellulose backup pad. The samples should be collected at a maximum flow rate of 2 l/min until 480 liters are collected. Analysis is by ion chromatography (84).

Sampling of the minute particles in sulfuric acid "fumes" or fog requires either very efficient scrubbing, electrostatic collectors employing acid-resistant linings or glass tubes, or collection on a suitable filter. The SO_4 may also be precipitated by barium chloride.

Table 14.3. Physical Characteristics of Sulfur Oxides

Characteristic	Sulfur Dioxide (SO$_2$)	Sulfuric Acid (H$_2$SO$_4$)	Sulfuric Trioxide (SO$_3$)
Molecular weight	64.07	98.06	80.07
Specific gravity	1.434 (liquid) at 0°C	1.834 (20°C)	1.857
Melting point	−72.7°C	10.5°C	16.83°C
Boiling point	−10°C	330°C (98.3%)	44.8°C
Vapor density	2.3 (air = 1)		2.8 (air = 1)
Physical state	Colorless irritant gas having a characteristic odor and taste	Colorless liquid	Colorless liquid or crystals
Solubility	0.6 g in 100 g of water at 90°C, 8.5 g in 100 g at 25°C, and 22.8 g in 100 g at 0°C; more soluble in methyl and ethyl alcohol; soluble in acetic and sulfuric acids, chloroform, and ethyl ether	Miscible with water. Pure anhydrous acid decomposes into sulfur trioxide and water at 340°C. It is a strong dehydrating agent and reacts with water or alcohol. The vapor pressure is neglible; thus it is unlikely to exist as such in the workplace	Forms 100% H$_2$SO$_4$ in water
Conversion factors	1 ppm = 2.62 mg/m^3; 1 mg/m^3 = 0.38 ppm at 25°C, 760 mm Hg	1.2 mg of H$_2$SO$_4$ contains 1 mg of SO$_3$	1 mg SO$_3$/m^3 = 0.3 ppm; 1 ppm = 3.2 mg/m^3 at 25°C, 760 mm Hg

8.4 Physiological Response

Only a small portion of inhaled sulfur dioxide penetrates the lower respiratory tract because it is water soluble. However, lower respiratory tract penetration may occur during exercise. The major effects of sulfur dioxide are on the upper respiratory tract. It may cause edema of the lungs or glottis and can produce respiratory paralysis. On contact with the moist mucous membranes, sulfur dioxide produces sulfurous acid. This is a direct irritant and inhibits mucociliary transportation. Most of the inhaled sulfur dioxide is detoxified in the liver by a molybdenum-dependent sulfite oxidase pathway. Concentration rather than duration of the exposure is the more important determinant of histopathological damage (85).

The acute effects of sulfur dioxide and causal concentrations are listed below:

400 to 500 ppm	Immediate danger to life
100 ppm	NIOSH IDLH
	Maximum tolerated exposures for 30 to 60 min

20 ppm	Chronic respiratory symptoms
6 to 12 ppm	Nasal and throat irritation
3 ppm	Odor detected
0.3 to 1 ppm	Threshold for smell or taste

Individuals with hyperactive airway disease, including asthma, may be particularly sensitive to the bronchospastic properties of sulfur dioxide. Mild asthmatics selected for methacholine sensitivity have as a group significant bronchoconstriction in response to short-term moderate exercise when breathing in 1.0 and 0.5 ppm sulfur dioxide (86). In addition, the induced bronchoconstriction is decreased after short-term repeated exercise in the presence of elevated sulfur dioxide concentrations. Particulates and water droplets can take sulfur dioxide into the lower respiratory tract and thereby worsen cardiopulmonary diseases.

Exposures to sulfur dioxide (SO_2) have been associated with progressive, dose-dependent bronchoconstriction in sensitive individuals. The clinical significance of such changes remains poorly characterized. Witek et al. (87) studied subjective responses following exposure to low-level concentrations of SO_2 (less than 1 ppm) in a group of 10 healthy and 10 asthmatic subjects. The number and severity of complaints associated with SO_2 increased with concentrations in both healthy and asthmatic subjects. Asthmatics had progressive lower respiratory complaints, such as wheezing, chest tightness, dyspnea, and cough, with increasing levels of SO_2, whereas healthy subjects complained more frequently of upper airway symptoms such as taste and odor with increasing levels of SO_2. Although exercise increased the frequency of lower airway symptoms in asthmatics, there was no increase in symptoms for healthy subjects.

Acute exposure to sulfur dioxide has caused severe obstructive pulmonary disease, unresponsive to bronchodilators, lasting for 3 months after exposure. Only on occasion have short-term acute exposures resulted in moderate to severe obstructive defects accompanied by persistent and productive cough (82). A chronic cough and sputum production are associated with exposure to sulfur dioxide in ambient concentrations in women and nonsmokers (88).

Signs and symptoms of respiratory distress were reported among 100 refrigeration workers exposed to 20 to 30 ppm. In this study the exposure was almost purely sulfur dioxide (89). More recently, a high incidence of respiratory symptoms was reported in pulp mill workers exposed to 10 to 20 ppm of sulfur dioxide (90). However, a 10-year follow-up in workers exposed to concentrations ranging from 4 to 33 ppm did not reveal an increased prevalence of respiratory disease or deteriorating pulmonary function as compared with the control group (91). Smelter workers exposed to sulfur dioxide at 2 ppm or lower developed pulmonary disease, and workers exposed to 1 ppm or more had accelerated loss of pulmonary function (92).

Sulfur dioxide may act as a cancer promoter. The mortality of arsenic smelter workers was higher when they had also been exposed to sulfur dioxide (93). Additionally, rats exposed to 3.5 or 10 ppm of sulfur dioxide developed squamous cell carcinomas from inhalation of benzo[a]pyrene, but neither compound alone produced carcinomas under the conditions of this experiment (94).

Amdur (95) exposed guinea pigs to concentrations of 2 to 200 mg of sulfuric acid mist/m^3 of atmosphere for 1-hr periods. Using particle sizes of 0.8, 2.5, and 7 μm, Amdur (95) found the smallest size was the most effective at the lowest level (2 mg/m^3). The 2.5-μm droplets caused the greatest response in the higher concentrations and at 200 mg/m^3 caused death within 1 hr to all four animals exposed to this concentration. The 7-μm particles in concentrations up to 30 mg/m^3 produced only a slight response because they did not penetrate beyond the upper respiratory tract.

Sulfur trioxide and sulfuric acid mists are strongly irritant, and inhaling concentrations of approximately 3 mg/m^3 causes a choking sensation. Persons accustomed to the exposure are unable to notice concentrations at this order of magnitude. Sulfur trioxide is irritating and corrosive to all mucous surfaces, causing inflammation of the upper respiratory tract, and possible lung injury. Sulfuric acid also attacks the enamel of the teeth.

8.5   Health Standards

The 1989–1990 ACGIH TLV for sulfur dioxide is 5.2 mg/m^3 (2 ppm). A STEL of 5 ppm is recommended (13 mg/m^3). OSHA's PELs are similar, with a slight variation based on conversion from ppm to mg/m^3, resulting in a PEL of 5 mg/m^3 for the 8-hr TWA and 10 for the STEL (17). The irritating effects of this concentration are not sufficient to provide ample warning. Those with reactive respiratory tracts and asthmatics are sensitive to concentrations below these levels. In contrast, NIOSH recommends a TWA of 0.5 ppm (1.3 mg/m^3). The NIOSH IDLH estimation is 100 ppm (18).

The 1989–1990 ACGIH TLV for sulfuric acid is 1 mg/m^3. The STEL is 3 mg/m^3. OSHA only has an 8-hr TWA of 1 mg/m^3. The NIOSH recommended exposure limit (REL) is also 1 mg/m^3. The NIOSH IDLH estimation is 80 mg/m^3 (18).

9   HYDROGEN SULFIDE

9.1   Sources, Uses, and Industrial Exposures

Excessive exposure to hydrogen sulfide, H_2S, results occasionally from its use as an industrial chemical and, frequently, from its occurrence as a by-product in industrial or natural processes wherever proteins decompose. It is encountered in mining, especially where sulfide ores are found; in excavating swampy or filled ground, and hence sometimes in wells, caissons, and tunnels; in natural gas; in the production and refining of petroleum; in the waters of certain natural springs; in volcanic gases; in the low-temperature carbonization of coal; in the manufacture of chemicals, dyes, and pigments; in the rayon industry; in the rubber industry; in tanneries; in the manufacture of glue; in the washings from sugar beets; and in sewer gases. Because hydrogen sulfide is soluble in water and oil, it may flow for a considerable distance from its place of origin to escape at unexpected areas. Many budding chemists have developed a casual disregard for the toxicity of hydrogen sulfide because of its general, and sometimes careless, use in the teaching of qualitative and quantitative analysis; and it is with great surprise that they later learn that the gas they used so consistently has a toxicity comparable to that of hydrogen cyanide. It is detectable by odor at about 1/400th of the lowest amount that can cause injurious effects. Hydrogen sulfide is used in the manufacturing of chemicals; in metallurgy; as an analytical reagent; as an agricultural disinfectant; as an intermediate for sulfuric acid, elemental sulfur, sodium sulfide, and other inorganic sulfides; as an additive in extreme-pressure lubricants and cutting oils; and as an intermediate for organic sulfur compounds (1). It is not registered as a pesticide in the United States (96).

A hygienic survey for hydrogen sulfide, methyl mercaptan and its derivatives, and sulfur dioxide in kraft mills and in sulfite mills revealed concentrations varying from 0 to 20 ppm hydrogen sulfide, 0 to 15 ppm methyl mercaptan, and comparable amounts of dimethyl sulfide, with dimethyl disulfide up to 1.5 ppm. The greatest emissions were detected at chip chutes and evaporation vacuum pumps. Batch operations yielded clearly higher sulfur dioxide concentrations (up to 20 ppm) as compared with a continuous ammonia-base digester. Furthermore, there was a strong correlation with the season in the sulfite mills, where higher concentrations were found in the winter when natural ventilation was poorer. As to the health effects, the exposed workers complained of headaches and a decrease in concentration capability more often than did matched controls. The number of sick days was greater for the exposed workers than among the controls (97).

Hydrogen sulfide is a by-product of organic decomposition, the petroleum industry, tanning, rubber vulcanizing, and heavy water production. Deaths have been reported as a result of agitation of underground manure tanks, the addition of sulfuric acid to drains, HCl to wells, cleaning of propane tanks, and entry of both victims and rescuers into sewers and shipholds containing fish meal (82). The 1969 air pollution levels ranged from an average of 1 to 6 $\mu g/m^3$, with concentrations as high as 300 $\mu g/m^3$ near industrial sources, largely from burning sulfur-containing materials. Nonindustrial sources of H_2S include volcanoes, bacteria, and plants (98).

9.2 Physical and Chemical Properties

The physical and chemical properties of hydrogen sulfide (CASRN 7783-06-4) are listed below:

Physical state	Colorless gas with rotten egg odor
Molecular weight	34.08
Vapor density (air = 1)	1.19
Specific gravity (H_2O = 1)	0.916 at $-60°C$ (liquid) 1.54 g/l at 0°C
Melting point	$-82.9°C$

Boiling point	$-61.8°C$
Vapor pressure (mm Hg)	20 atm at 25.5°C
Solubility	1 g dissolves in 242 ml water at 20°C; also soluble in alcohol, petroleum solvents, and crude petroleum
Conversion factor	1 ppm = 1.4 mg/m^3

Hydrogen sulfide is flammable within the range of 4.30 to 45.5 percent by volume. The ignition temperature is 558°F. When heated to decomposition, hydrogen sulfide emits highly toxic fumes of oxides of sulfur (99).

9.3. Determination in the Atmosphere

Passive dosimeters, detector tubes, and portable direct-reading meters, some pocket size, can be used to augment the approved workplace monitoring methods for hydrogen sulfide. An automatic detection and control system has been described.

The OSHA method (84) calls for collection on a silver nitrate-impregnated cellulose filter, 17 mm in diameter. The maximum flow rate is 0.2/(l)(min). The samples are analyzed by a polarographic analyzer system using differential pulse polarography. The limit of detection is 0.4 ppm (0.5 mg/m^3) for a 2-liter volume.

9.4 Physiological Response

The absorption of hydrogen sulfide is almost exclusively through the respiratory tract. Absorption through the skin has been demonstrated and discoloration of the skin reported. This is not a significant source of systemic poisoning, as indicated by the fact that the routine for gas mask approval testing by the U.S. Bureau of Mines has included the wearing of gas masks for 30-min periods in atmospheres containing 2 percent (20,000 ppm) hydrogen sulfide. During these tests, which include strenuous exercise, the subjects have noted slight skin irritation, but no systemic effects indicative of hydrogen sulfide absorption and no discoloration of the skin. When free sulfide exists in the circulating blood, a certain amount of hydrogen sulfide is excreted in the exhaled breath. This is sufficient to be detected by odor. Although the characteristic odor of gas is detectable in concentrations as low as 0.025 ppm, is distinct at 0.3 ppm, is offensive and moderately intense at 3 to 5 ppm, and is strong and marked, but not intolerable, at 20 to 30 ppm, the odor of higher concentrations does not become more intense; above about 200 ppm the disagreeable odor appears less intense. These perceptions are based upon initial inhalations, and with continuous inhalation the olfactory sense fatigues rapidly. The initial event in the absorption of H$_2$S is its disassociation into HS$^-$ anion (pK_a = 7.08 at 18°C). The reverse reaction does not occur at normal blood pH or in histotoxic acidosis caused by H$_2$S. The greater portion of hydrogen sulfide, however, is excreted in the urine, chiefly as sulfate (particularly thiosulfate); but some is also excreted as sulfide. The blood sulfide has a short half-life of about 60 min, thus any sampling for this possible marker would need to be accomplished shortly after exposures cease (100).

After inhalation for 1 min to 10 hr at high concentrations, changes were noted in the brain, pancreas, liver, kidney, and small intestine of rats and guinea pigs. The formation of methemoglobinemia protects against H_2S toxicity experimentally, although oxygen therapy is recommended in humans (98). Detoxification of hydrogen sulfide occurs rapidly (85 percent of a lethal dose per hour in animals). The red blood cells and the liver mitochondria are the main sites. Hydrogen sulfide intracellular cytochrome oxidase by altering electron transport is the most likely mechanism of action (82).

By far the greatest danger from the inhalation of hydrogen sulfide is its acute effects. Whether the effects are to be acute or subacute and chronic depends on the concentration of the gas in the atmosphere. Also, death or permanent injury may occur after very short exposure to small quantities of hydrogen sulfide (99). It acts directly upon the nervous system, resulting in paralysis of respiratory centers. Contact with eyes causes painful conjunctivitis, sensitivity to light, tearing, and clouding of vision. Inhalation of low concentrations causes a runny nose with a sense of smell loss, labored breathing, and shortness of breath. Direct contact with skin causes pain and redness. Other symptoms of exposure include profuse salivation, nausea, vomiting, diarrhea, giddiness, headache, dizziness, confusion, rapid breathing, rapid heart rate, sweating, weakness, sudden collapse, unconsciousness, and death due to respiratory paralysis (101). Deaths occur rapidly on the site. Patients who have vital signs at hospital arrival usually survive, if severe hypoxic encephalopathy is absent (84). Histopathological evaluation most often reveals changes in the lung, brain, and heart. The thyroid and heart may also be target organs (98).

Hydrogen sulfide is acutely toxic to humans as evidenced by the numerous reports of individuals fatally poisoned by accidental exposure (102–108). According to NIOSH, hydrogen sulfide is a leading cause of sudden death in the workplace (109). The odor threshold is reported to be 25 ppb (0.035 mg/m^3). Levels in the 3 to 5 ppm range cause an offensive odor. At levels around 100 ppm, no odor is detectable, owing to loss of the olfactory sensation, which results in loss of warning properties at lethal levels. In reports of acute poisoning, systemic intoxication can result from a single (one to two breaths), massive exposure to concentrations usually greater than 1000 ppm (105, 107). Inhalation of high levels of hydrogen sulfide act directly on the respiratory center, causing respiratory paralysis with consequent asphyxia and subsequent death (102, 103, 110). At levels between 500 and 1000 ppm, acute effects include symptoms of sudden fatigue, headache, dizziness, intense anxiety, loss of olfactory function, nausea, abrupt loss of consciousness, disturbances of the optic nerves, hypertension, insomnia, mental disturbances, pulmonary edema, coma, and convulsions and respiratory arrest, followed by cardiac failure and, often, death (111–113). Levels estimated at 250 ppm resulted in unconsciousness in three workers after several minutes of exposure (106). Cardiac effects in acute hydrogen sulfide intoxication have been reported in humans (114) and animals (115). If exposure is terminated promptly, recovery occurs quickly. However,

Table 14.4. Exposure and Blood Sulfide Concentrations and Clinical Effects in Acute Hydrogen Sulfide Exposure (98, 100)

Concentration (mg/m³)	Duration (min)	Blood Concn. (mmol/l)[a]	Effect
14	480		TLV–TWA
21	15		STEL
16–32	360–420		Eye irritation
		1.3	Clear smell, worker immediately escaped, minor headache
150–300	2–30		Loss of smell
1350	<30		Marked systemic symptoms
		5.0	No smell, worker collapsed
		19	Worker found unconscious
1350	>60		Death
		53–117	Five fatal cases from oil refineries, tanneries, and sewage work
2250	15–30		

[a]In the nonfatal cases, the blood samples were taken within 30 min of the event. Maximum concentrations were possibly higher.

neurological effects have been reported to persist in survivors of high-level exposure (116).

Two case studies noted neuropsychological dysfunction characterized by cognitive impairment, deficits of verbal fluency, disorders of written language, and impairment of various memory, psychomotor, and perceptual abilities in individuals acutely exposed to hydrogen sulfide (117, 118). The damage that has been observed to persist after hydrogen sulfide exposure is not distinguishable from the effects of systemic anoxia or ischemia of the brain or heart, and no specific hydrogen sulfide chronic systemic toxicity has been defined (119).

Hydrogen sulfide is also a potent eye and mucous membrane irritant, even at low concentrations (50 to 200 ppm). Pulmonary edema is often a clinical finding in persons who have been rendered unconscious by hydrogen sulfide exposure (108, 111, 113, 114). In several of the reported fatalities, the individuals apparently died of acute respiratory distress syndrome due to pulmonary edema. Irritation of the eye results in initial lacrimation, loss of coronary reflex, and changes in visual acuity and perception (usually at concentrations in excess of 50 ppm), which may progress to inflammation and ulceration, with the possibility of permanent scarring of the cornea in severe cases. Inflammation of the cornea of the eye has been reported in workers exposed to as little as 10 ppm hydrogen sulfide for 6 to 7 hr (103, 112). The blood sulfide concentrations, air concentrations, and effects are tabulated in Table 14.4.

Subchronic (90-day) vapor inhalation studies of hydrogen sulfide were conducted using Sprague–Dawley rats and Fischer-344 rats (120, 121). These animals were exposed to 0, 10.1, 30.5, and 80 ppm for 6 hr/day, 5 days/week for at least 90 days, simultaneously and in the same chambers as the mice in the principal study. For each dose, three groups of 15 males and 15 females of each strain of rats were employed. In addition, a control group consisting of 15 of each sex was exposed to clean air only. At termination of exposure, the animals were examined as described for the critical study (122). A significant reduction in body weight gain was noted in all animals exposed to 80 ppm. In F-344 and male Sprague–Dawley rats, the effect on body weight was statistically significant at some times in all exposed groups, but mean body weights were never less than 93 percent of control. In female Sprague–Dawley rats, mean body weight in the 80 ppm group was <90 percent of the control groups during most of the study. Brain weight was significantly reduced in the male Sprague–Dawley rats in the high-dose group and slightly, but not significantly, reduced in females (121), indicating a lowest observed adverse effect level (LOAEL) of 80 ppm. No clinical signs in rats were observed to be exposure related. Neurological function examinations yielded negative results. Blood volume, appearance, occult blood, specific gravity, protein, pH, ketone, and glucose values were all normal. Ophthalmoscopic examination, hematology, serum chemistry parameters, and urinalysis were also normal. Histopathological examination, which included four sections of the nasal turbinates, revealed no abnormalities in comparison with controls.

Male F-344 rats were exposed to 0, 10, 200, or 400 ppm nominal concentration (14, 279, and 557 mg/m^3) of hydrogen sulfide for 4 hr (123); and samples were collected by broncho-alveolar lavage and nasal lavage at 1, 20, and 44 hr after exposure (four rats/exposure level). Increased number of cells in nasal lavage and increased protein and lactate dehydrogenase in both broncho-alveolar and nasal lavage fluid from rats exposed to 400 ppm were observed. Male F-344 rats (four rats/exposure level) were exposed to 0, 116, or 615 mg/m^3 for 4 hr (124). Rats exposed to 615 mg/m^3 had marked perivascular and alveolar edema and bronchioles containing granular leukocytes, proteinaceous fluid, fibrin, and exfoliated cells. Necrosis of bronchiolar ciliated cells and hyperplasia of alveolar type II cells were observed in rats exposed to 615 mg/m^3. In rats exposed to 116 mg/m^3, only mild perivascular edema was observed. Nasal structures were not examined. Male F-344 rats were exposed to 0, 14, 280, or 560 mg/m^3 hydrogen sulfide for 4 hr; and four levels of the nasal cavity were examined histologically at 1, 18, and 44 hours after exposure (125). Necrosis and exfoliation of respiratory and olfactory mucosal cells, but not squamous epithelial cells, were observed in rats exposed to 560 mg/m^3. No nasal lesions were seen in the controls or the two lower exposure levels.

Subchronic (90-day) inhalation studies were conducted using B6C3F1 mice (122). Three groups of 10 males and 12 female mice each were exposed to 0, 10.1, 30.5, and 80 ppm six hr/day, five days/week for 90 days. A control group of mice was exposed to clean air only. Animals were examined using neurological function tests of posture, gait, facial muscle tone, and reflexes. Ophthalmological examination using a slit lamp scope was performed. The only exposure-related histopathological lesion was inflammation of the nasal mucosa in the anterior segments of the nose, which was observed in eight of nine male mice and in seven of nine female mice in the group exposed to 80 ppm. This lesion was also present in two high-dose mice that died during the course of the study. The lesion was generally minimal to mild in severity and was located in the anterior portion of the nasal structures, primarily in the squamous portion of the nasal mucosa, but extending to areas covered by respiratory epithelium. This lesion was not observed in any animals in the other exposure groups. Thus for mice, 80 ppm is considered a LOAEL for nasal inflammation, and 30.5 ppm is a no observed adverse effect level (NOAEL).

A minimal submucosal lymphocytic cellular infiltrate was observed in the posterior section of the nasal structures. This lesion occurred with approximately equal frequency in control and exposed animals and was not considered to be exposure related. Histopathological evaluation revealed no other abnormalities in comparison with controls. Significant reductions in body weight gain were noted in all exposure groups at various times during the study. Decreased weight gain in animals exposed to 80 ppm occurred consistently in both male (approximately 90 percent of control during last 7 weeks of study) and female (<90 percent of control during last 3 weeks of study) mice, and this level is considered a LOAEL. Statistically significant changes in absolute kidney, liver, and spleen weight were also observed in the male rats exposed to 80 ppm, but no differences were apparent when organ weights were normalized to body weight. Neurological function examinations yielded negative results. Ophthalmoscopic examinations and clinical pathological values were also normal.

No data on human developmental effects of inhaled hydrogen sulfide were found; but based on the limited information available in animals, hydrogen sulfide does not appear to induce developmental effects. In a preliminary study, Saillenfait et al. (126) administered 0, 50, 100, or 150 ppm hydrogen sulfide 6 hr/day to pregnant rats during gestational days 6 to 20. Maternal body weight gain was significantly reduced at 150 ppm, and fetal body weight was slightly (4 to 7 percent) reduced in all exposed groups. In dams exposed to 100 or 150 ppm, reduced absolute weight gain, increased implantations, and increased live fetuses were observed. In a follow-up experiment, 20 pregnant females were exposed to 100 ppm for 6 hr/day on days 6 to 20 of gestation. Fetal weights, number of live and dead fetuses, number of implantation sites and resorptions, and external malformations were recorded. Viable fetuses were then prepared for soft tissue and skeletal examination. Neither maternal toxicity nor adverse effects on the developing embryo or fetus was observed. The preliminary study identifies a LOAEL of 50 ppm for maternal weight gain. Because of the larger number of animals in the main study, a NOAEL of 100 ppm for maternal effects and developmental effects is identified.

Concentrations of 700 ppm and above may result in acute poisoning, and although the gas is an irritant, the systemic effects from absorption of hydrogen sulfide into the bloodstream overshadow the irritant effects. These acute systemic effects result from the action of free hydrogen sulfide in the blood stream and occur whenever the gas is absorbed faster than it can be oxidized to pharmacologically inert compounds, such as thiosulfate or sulfate. Such oxidation occurs rapidly in humans or animals, and even following inhalation exposure to concentrations up to 700 ppm hydrogen sulfide in the atmosphere, hydrogen sulfide does not appear in the exhaled breath. Relatively massive doses are required to overcome this protective activity of the body. Sodium sulfhydrate, NaHS, solution injected intravenously into dogs rapidly disappears from the circulating blood when a rate equivalent to 0.1 to 0.2 mg of hydrogen sulfide per kilogram of body weight per minute is not exceeded. When the amount absorbed into the bloodstream exceeds that which is readily oxidized, systemic poisoning results. There is a general action on the nervous system; hyperpnea occurs shortly; and respiratory paralysis may follow immediately. This condition may be reached almost without warning, as the originally detected odor of hydrogen sulfide may have disappeared as a result of olfactory fatigue. Unless the victim is removed to fresh air within a very few minutes and breathing stimulated or induced by artificial respiration, death occurs. Unconsciousness and collapse occur within seconds in high concentrations; for that reason many persons have lost their lives attempting to save a victim who has collapsed from exposure. In such a case, holding the breath will permit a brief stay in the atmosphere, whereas inhaling would cause almost immediate collapse.

Hydrogen sulfide is an irritant gas and exposure to concentrations between 70 and 700 ppm may irritate the mucous membranes of the eyes and of the respiratory tract. Pulmonary edema or bronchial pneumonia is likely to follow prolonged exposure to concentrations on the order of 250 to 600 ppm. These levels of exposure may cause such symptoms as headache, dizziness, excitement, nausea or GI disturbances, dryness and sensation of pain in the nose, throat, and chest, and coughing. Table 14.4 indicates responses to various concentrations of hydrogen sulfide in the atmosphere.

Among the subacute and chronic effects of exposure to hydrogen sulfide, eye irritation resulting in conjunctivitis or "gas eyes" is the most common and, ranging from mild to severe with extent and intensity of exposure, may include itching and smarting, a feeling of sand in the eyes, marked inflammation and swelling, cloudy cornea, and destruction of the epithelial layer with scaling resulting in blurring of vision. Exposure to light may increase the painful effect. Atmospheric concentrations above 50 ppm and up to 300 ppm are conducive to this condition.

9.5   Health Standards

Based on epidemiologic data (127) the ACGIH (20) has recommended a TLV–TWA of 10 ppm (13.9 mg/m^3) for hydrogen sulfide. However, citing evidence of eye injury, headaches, nausea, and insomnia after exposure to H_2S at low concentrations for several hours, NIOSH (109) adopted a ceiling occupational exposure limit of 10 ppm (15 mg/m^3) with a 10-min maximum exposure to this concentration. The OSHA PEL is 10 ppm as an 8-hr TWA with a STEL of 15 ppm (17).

10   CARBON DISULFIDE

10.1   Sources, Uses, and Industrial Exposures

Carbon disulfide, CS_2, or carbon bisulfide, is used in the xanthation of cellulose in the preparation of viscose; exposures exist not only in the xanthating process, but also during spinning and washing of the viscose. In the rubber industry carbon disulfide has been used as a solvent for sulfur or as a diluent for sulfur chloride in vulcanizing and as a solvent for rubber cement. It has been used as an insecticide and in the chemical industry as a solvent for phosphorus, fats, oils, resins, and waxes. It is used in the manufacture of optical glass. Carbon disulfide is also encountered in the destructive distillation of coal. Carbon disulfide is used in the manufacture of soil disinfectants and vacuum tubes and is used as a solvent for cleaning and extractions, especially in metal treatment and plating. It is a fumigant for grain, a corrosion inhibitor, and a polymerization inhibitor for vinyl chloride (1).

10.2 Physical and Chemical Properties

The properties of carbon disulfide (CASRN 75-15-0) are as follows:

Physical state	Colorless liquid
Molecular weight	76.13
Specific gravity	1.2626 (20°C)
Melting point	− 108.6°C
Boiling point	46.3°C
Vapor density	2.63 (air = 1)
Solubility	0.22 g in 100 ml of water at 22°C; miscible with alcohol, ether, and benzene
Flash point	− 22°F (closed cup)

$1 \text{ mg/m}^3 = 0.32$ ppm; 1 ppm $= 3.12 \text{ mg/m}^3$ at 25°C, 760 mm Hg

Carbon disulfide has a range of flammability of 1.25 to 50.0 percent by volume in air. The ignition temperature is 248°F, a temperature commonly encountered in steam pipes, electric light bulbs, and elsewhere. The flash point by the closed cup method is − 22°F. Carbon disulfide has a foul, slightly ethereal (cabbage-like) odor that nevertheless does not offer adequate warning in the lower harmful concentrations. It is very fat soluble. When heated to decomposition, this highly flammable material emits highly toxic fumes of sulfur oxides and can react vigorously with oxidizing materials. The density of the vapor is 2.5 greater than air.

10.3 Determination in the Atmosphere

Carbon disulfide can be collected using a sodium sulfide drying tube (270 mg) connected by a 20-mm section of tubing to the front section of a coconut shell charcoal tube (100/50 mg sections, 20/40 mesh. The maximum flow rate is 0.2 l/min for maximum air volume of 25 l (TWA) or 2 l (peak). The samples may be shipped or stored under refrigeration until analysis can be performed. The samples are desorbed with toluene, and analysis is by gas chromatography (flame photometric detection). The detection limit is 0.02 mg per sample (30).

Table 14.5. Effects of Various Concentrations of Carbon Disulfide on Humans

	Concentration	
Effects	mg/l	ppm
Slight or no effect	0.5–0.7	160–230
Slight symptoms after a few hr	1.0–1.2	320–390
Symptoms after $\frac{1}{2}$ hr	1.5–1.6	420–510
Serious symptoms after $\frac{1}{2}$ hr	3.6	1150
Dangerous to life after $\frac{1}{2}$ hr	10.0–12.0	3210–3850
Fatal in $\frac{1}{2}$ hr	15.0	4815

10.4 Physiological Response

Carbon disulfide is absorbed by the lungs; some dermal absorption can also occur. Inhaled carbon disulfide reaches equilibrium in 1 to 2 hr and 40 to 50 percent is retained. Once absorbed, more carbon disulfide is carried in the red blood cells than in the plasma. About 70 to 90 percent is metabolized. Carbon dioxide and carbon monoxide are metabolites.

Although the total sulfur in the urine of exposed persons is considerably elevated, many other factors cause that same result. Carbon disulfide in the blood or urine, however, is indicative of exposure; and its concentration can give some indication of the severity of the exposure. Only about 1 percent of the inhaled carbon disulfide is eliminated in the urine. Other urinary metabolites are thiocarbamide (thiourea) and mercaptothioaxolinone (128, 129).

Peak plasma concentration occurs about 2 hr after inhalation. The plasma half-life is approximately 1 hr. Carbon disulfide may be stored in large amounts in the fat and liver, and it binds to microsomal enzymes, reducing their activity (82).

Carbon disulfide affects the central nervous system, cardiovascular system, eyes, kidneys, liver, and skin. It may be absorbed through the skin as a vapor or liquid, inhaled, or ingested. The probable oral lethal dose for a human is between 0.5 and 5 g/kg or between 1 oz and 1 pt (or 1 lb) for a 70-kg person (101). The lowest lethal dose for humans has been reported at 14 mg/kg.

Most industrial poisonings from carbon disulfide occur in the viscose rayon manufacturing industry. The neuropsychiatric signs include mania, hallucinations, memory loss, increased suicide rates, tremors, and optical and peripheral neuropathy, as well as accelerated atherogenic changes. Underlying changes in mineral balance may be the cause for some of these changes (82).

Table 14.5 lists representative levels of effect upon humans, with corresponding ranges in concentration of inhaled carbon disulfide. The predominant effect of high concentrations of carbon disulfide is narcosis, and death may result from respiratory failure. Less severe exposures may result in headache, giddiness, respiratory disturbances, precordial distress, and gastrointestinal disturbances. The possibility of injury to the CNS from a single severe acute exposure has been reported.

Repeated brief exposures to high concentrations or prolonged exposures to low concentrations are of much greater industrial importance than are the single acute exposures. Among the subjective complaints that characterize chronic carbon disulfide poisoning are fatigue, loss of memory, insomnia, listlessness, headache, excessive irritability, melancholia, vertigo, weakness, loss of appetite, gastrointestinal disturbances, and impairment of sexual functions. Visual disturbances, loss of reflexes, hallucinations, mania, or chronic dementia may occur. Lung irritation has been reported. Degenerative changes in the blood and blood-forming organs are reported to occur, sometimes after poisoning has progressed. Dermatitis, and even blistering, may result from contact with vapor or liquid on the skin or mucous surfaces. One hundred workers were exposed for 4 years to average concentrations of 1.0 to 5.5 ppm hydrogen sulfide and 1.9 to 26.4 ppm carbon sulfide, or a combined sulfide gas and vapor concentration of 2.9 to 31.9 ppm. There was no indication of intoxication. Improvement or complete recovery can be expected if the exposure is discontinued before severe damage results. Barthelemy (130), reporting on 10 years of experience in the manufacture of viscose rayon, cites three cases of poisoning due to excessive exposure to carbon disulfide: one of mental derangement, and two with impaired motor nerves adversely affecting the leg muscles. All three workmen recovered completely within a few months after termination of exposure. He further states that when the carbon disulfide in the air was kept below 30 ppm and the hydrogen sulfide below 20 ppm, no trouble whatsoever was experienced.

Recent epidemiologic studies of female workers in China suggested that there were increased teratogenic effects. Congenital heart defects in offspring were most frequently observed. When the workers were divided on the basis of exposure above or below 10 mg/m^3, there was no difference between the groups, suggesting that the threshold for these effects was below the encountered exposures or unrelated to dose (131).

Rats and rabbits were exposed to 20 ppm or 62.3 mg/m^3 (recommended occupational exposure limit) and 40 ppm or 124.6 mg/m^3 of carbon disulfide during the entire length of the pregnancy period and also 34 weeks before breeding to simulate occupational exposure (132). Hardin et al. (132) observed no effects on fetal development in rats or rabbits following inhalation exposure to 62.3 or 124.6 mg/m^3, which corresponds to estimated equivalent oral dosages of 5 and 10 mg/kg for rats and 11 and 22 mg/kg for rabbits. In an oral study in which rabbits received 25 mg/(kg)(day), fetal resorption occurred (133, 134). Fetotoxicity and fetal malformations in this study were not observed in rats at the lowest level (100 mg/(kg)(day) of carbon disulfide exposure. The data from this study also suggest that the rabbit fetus is more sensitive than the rat fetus to carbon disulfide-induced toxicity. Johnson et al. (135) reported an epidemiologic study that employed a wide range of exposure with carbon disulfide, such as 0.04 to 5 ppm (mean 1.2 ppm, low exposure), 0.04 to 33.9 ppm (mean 5.1 ppm, medium exposure) and 0.04 to 216 ppm (mean 12.6 ppm, high exposure). In this study the entire population was exposed to a combined exposure of 7.3 ppm over a period of 12 or more years. Of the several clinical findings, the exposed population showed significant alterations in sensory conduction velocity and peroneal motor conduction velocity. However, in the opinion of the authors, the data indicated that minimal neurotoxicity was evident because the reduction in nerve conduction velocity was still within a range of clinically normal values and thus not associated with specific health consequences. Additionally, the exposed population had blood lead levels <40 mg/dl and the exposed air alone contained H_2S, H_2SO_4, and tin oxide. Therefore 7.3 ppm CS_2 can be considered as a NOAEL for neurotoxicity. Bulgarian investigators (136) reported significant fetal malformations in rats exposed to a low carbon disulfide dose of 0.3 mg/m^3 over three generations. However, the details of the procedures used in the study were not sufficiently presented to allow one to fully validate the findings.

10.6 Health Standards

The 1989–1990 ACGIH TLV for carbon disulfide is 10 ppm (31 mg/m^3). OSHA's PEL is 4 ppm, but the agency recommends a STEL of 12 ppm. Both ACGIH and OSHA have a skin notation. In contrast, NIOSH recomends a TWA of 1 ppm (3 mg/m^3) and a STEL of 10 ppm (17). NIOSH indicates that 500 ppm is immediately dangerous to life or health (17). The ACGIH TLV was lowered from 20 ppm to 10 ppm on the basis of cardiovascular effects. The 20 ppm limit was based on CNS effects (20).

The current PEL has little margin of safety for developmental effects. Alcoholics and those suffering from neuropsychic trouble are at special risk.

11 SULFUR MONOCHLORIDE

11.1 Sources, Uses, and Industrial Exposures

Sulfur monochloride, S_2Cl_2, is used in vulcanizing and in curing rubber. In the manufacture of rubber-coated fabrics, sulfur chloride has been used in oven "curing" atmospheres, and in some such operations, has been poured into open containers and placed on steam coils on the floor of the curing oven with little or no ventilation. The leakage into the room in such instances causes pronounced irritation to the eyes and nose of anyone working there. The distribution and collection of open containers involve brief exposures to relatively high concentrations. Men who do this work sometimes inadvisedly rely upon holding their breath during the period of exposure to high vapor concentration, rather than wearing gas masks. Sulfur monochloride is also used in the manufacture of organic chemicals, printers' inks, varnishes, and cements, in hardening soft woods, and as an agricultural insecticide (1).

11.2 Physical and Chemical Properties

The properties of sulfur monochloride (CASRN 10025-67-9) are as follows:

Physical state   Yellowish-red viscous liquid
Molecular weight   135.05

Specific gravity   1.678 (20°C)
Melting point $-80°C$
Boiling point   135.6°C
Solubility   Soluble in carbon disulfide, benzene, and ether; the liquid and its vapor decompose in water (or humid air) to form sulfur, HCl, and sulfur dioxide

$1 \text{ mg/m}^3 = 0.2 \text{ ppm}$; $1 \text{ ppm} = 5.5 \text{ mg/m}^3$ at 25°, 760 mm Hg

11.3 Determination in the Atmosphere

Sulfur monochloride vapor in the air may be determined by scrubbing through two scrubbers containing a measured quantity of 0.1 N silver nitrate acidified with nitric acid. When sampling has been completed, add 0.1 N sodium chloride solution equivalent to the silver nitrate used; and titrate the excess chloride with additional 0.1 N silver nitrate. The reaction is as follows:

$$2 \text{ S}_2\text{Cl}_2 + 2 \text{ H}_2\text{O} + 4 \text{ AgNo}_3^{\rightarrow} 4 \text{ AgCl} + 3 \text{ S} + \text{SO}_2 + 4 \text{ HNO}_3$$

11.4 Physiological Response

Sulfur monochloride has a suffocating odor and is strongly irritant to the eyes, nose, and throat. A concentration of 150 ppm has been stated as fatal to mice after an exposure of 1 min, but the degree of toxicity has not yet been well established. The irritant effects are due to the sulfur dioxide and hydrochloric acid liberated by hydrolysis. Because this occurs rather readily, most of the irritant action likely is expended upon the upper respiratory tract. However, if the hydrolysis should not be completed in the upper respiratory tract, injury to the bronchioles and alveoli would result.

11.5 Health Standards

The 1989–1990 ACGIH ceiling suggested for sulfur monochloride is 1 ppm (5.5 mg/m³). OSHA has established a similar ceiling (17). NIOSH indicates that 10 ppm is immediately dangerous to life or health (18).

12 THIONYL CHLORIDE

12.1 Sources, Uses, and Industrial Exposures

Thionyl chloride, $SOCl_2$, is used as a chlorinating agent in the chemical manufacture of organic compounds. It is also used as a solvent in high-energy density lithium batteries (1).

12.2 Physical and Chemical Properties

The properties of thionyl chloride (CASRN 7719-09-7) are as follows:

Physical state	Colorless liquid
Molecular weight	118.97
Specific gravity	1.655 (10°C)
Melting point	−105°C
Boiling point	75.5°C
Vapor density	4.1 (air = 1)
Density	1.5 (air = 1) at 26°C

Decomposes in water or moist atmosphere, yielding sulfur dioxide, sulfur, chlorine, sulfur monochloride, and hydrochloric acid

$1 \text{ mg/m}^3 = 0.2 \text{ ppm}$; $1 \text{ ppm} = 4.87 \text{ mg/m}^3$ at 25°C., 760 mm HG

12.3 Determination in the Atmosphere

Thionyl chloride may be determined by the same method as sulfur monochloride.

12.4 Physiological Response

Thionyl chloride is an irritant to the eyes, mucous membrane, and skin. The information on this compound in the open literature is very limited. Hooker Chemical Company recommended 1.0 ppm as an industrial standard (20). A 20-min exposure to 17.5 ppm is said to have proved fatal to cats (99).

12.5 Health Standards

The 1989–1990 ACGIH ceiling suggested for thionyl chloride ($SOCl_2$) is 1 ppm (4.9 mg/m^3). OSHA has established a similar ceiling (17). NIOSH indicates that 500 ppm is immediately dangerous to life or health (18). The production of 3 ppm of chlorine from 1 ppm of thionyl chloride is given as the rationale for the ACGIH ceiling. Thionyl chloride also produces sulfur dioxide, another irritant. Thus it is assumed that the 1-ppm ceiling would prevent the irritant effects of thionyl chloride's reaction products (20).

REFERENCES

1. Kirk-Othmer *Concise Encyclopedia of Chemical Technology*, 3rd ed., Wiley-Interscience, New York, 1985.

2. D. E. Rushing, *Am. Ind. Hyg. Assoc. J.*, **23**, 383 (1962).

3. M. J. Prager and W. R. Seitz, *Anal. Chem.*, **47**, 148 (1975).

4. N. A. Tavilie, E. Perez, and D. P. Illustre, *Anal. Chem.*, **34**, 866 (1962).

5. J. M. Cameron and R. S. Patrick, *Med. Sci. Law*, **8**, 209 (1966).

6. R. S. Diaz-Rivera, P. J. Collazo, E. R. Pons, and M. V. Torregrosa, *Medicine*, **29**, 269 (1950).

7. F. A. Simon and L. K. Pickering, *J. Am. Med. Assoc.*, **235**, 1343 (1976).

8. M. M. McCarron, G. P. Gaddus, and A. T. Trotter, *Clin. Toxicol.*, **18**, 693 (1981).

9. T. J. Orcult and B. A. Pruitt, *Major Probl. Clin. Surg.*, **19**, 84 (1976).

10. V. S. Aizenshtadt, S. M. Neruabai, and I. I. Voronin, *Gig. Tr. Prof. Zavol.*, **15**, 48 (1971).

11. S. A. White and C. C. Armstrong, Project A5.2-1 Chemical Warfare Service, Edgewood Arsenal, MD, 1935.

12. J. P. W. Hughes, R. Baron, D. H. Buckland, M. A. Cooke, J. D. Craig, D. P. Duffield, A. W. Grosart, P. W. J. Parkes, and A. Porter, *Brit. J. Ind. Med.*, **19**, 83 (1962).

13. H. Heimann, *J. Ind. Hyg. Toxicol.*, **28**, 142 (1946).

14. C. O. Adams and B. G. Sarnat, *Arch. Pathol.*, **30**, 1192 (1940).

15. L. L. Ashburn, A. J. McQueeney, and R. R. Faulkner, *Proc. Soc. Exp. Biol. Med.*, **67**, 351 (1948).

16. T. Dalhamm and B. Holma, *Arch. Pathol.*, **20**, 429 (1050).

17. American Conference of Governmental Industrial Hygienists, *Guide to Occupational Exposure Values*, 1990.

18. National Institute for Occupational Safety and Health, *Pocket Guide to Chemical Hazards*, 2nd Printing, DHHS (NIOSH) Publ. No. 85-114, U.S. Dept. of Health and Human Services, Washington, DC, 1990.

19. J. F. Burgess, *Can. Med. Assoc. J.*, **65**, 567 (1951).

20. American Conference of Governmental Industrial Hygienists, *Documentation of the Threshold Limit Values and Biological Exposure Indices*, 5th ed., ACGIH, Cincinnati, OH, 1986.

21. R. Dechant, G. Sanders, and R. Graul, *Am. Ind. Hyg. Assoc. J.*, **27**, 75 (1966).

22. B. Berk, W. E. Westlake, and F. A. Gunther, *J. Agric. Food Chem.*, **18**, 143 (1970).

23. A. Glass, *J. R. Navy Med. Serv.*, **42**, 184 (1956).

24. R. N. Harger and L. W. Spoylar, *Arch. Ind. Health*, **18**, 497 (1958).

25. S. Singh, J. B. Dilawari, R. Vashist, H. S. Malhotra, and B. K. Sharma, *Brit. Med. J.*, **290**, 1111 (1985).

26. R. Wilson, F. H. Lovejoy, R. J. Jaeger, and P. L. Landrigan, *J. Am. Med. Assoc.*, **244**, 148 (1980).

27. J. O. Nriagu, *Environment*, **32**, 7 (1990).

28. P. Pentel, D. Fletcher, and J. Jentzen, *J. Forensic Sci.*, **30**, 556 (1985).

29. U.S. Bureau of Mines, Mineral Commodity Summaries, 1986.

30. National Institute for Occupational Safety and Health, *Manual Of Analytical Methods*, 3rd ed., Cincinnati, OH, 1984.

31. Agency for Toxic Substances and Disease Registry (ATSDR), Toxicological Profile for Selenium, 1989.

32. Z. Gregus and O. Klaassen, *Toxicol. Appl. Phar.*, **85**(1), 24–38 (1986).
33. L. Friberg, G. F. Nordberg, E. Kessler, and V. B. Vouk, Eds., *Handbook of the Toxicology of Metals*, 2nd ed., Vols. I, II, Elsevier, Amsterdam, 1986, V2 482.
34. M. J. Cukierski, C. C. Willhite, B. Lasley, T. A. Hendrie, S. A. Book, D. N. Cox, and A. G. Hendrickx, *Fundam. Appl. Toxicol.*, **13**, 26 (1989).
35. J. R. Glover, *Ind. Med. Surg.*, **39**, 50 (1970).
36. S. E. Koppel, C. Koppel, H. Baudisg, and I. Kloppel, *Clin. Toxicol.*, **24**, 21 (1986).
37. International Agency for Research on Cancer (IARC), *IARC Monographs on the Evaluation of Carcinogenic Risk of Chemicals to Man: Some Aziridine, N-, S-, and O-Mustards and Selenium*, Vol. 9, IARC, Lyon, 1975.
38. National Toxicology Program, Bioassay of Selenium Sulfide (Dermal Study) for Possible Carcinogenicity, Bethesda, MD, National Toxicology Program, National Cancer Institute, National Institutes of Health, NCI Technical Report Series No. 197, NTP No. 80–18, 1980.
39. National Toxicology Program, Bioassay of Selsun for Possible Carcinogenicity, Bethesda, MD, National Toxicology Program, National Cancer Institute, National Institutes of Health, NCI Technical Report Series No. 199, NTP No. 80–19, 1980.
40. National Toxicology Program, Bioassay of Selenium Sulfide (Gavage) for Possible Carcinogenicity, Bethesda, MD, National Toxicology Program, National Cancer Institute, National Institutes of Health, NCI Technical Report Series No. 194, NTP No. 80–17, 1980.
41. R. J. Shamberger and D. V. Frost, *Can. Med. Assoc. J.*, **100**, 682 (1969).
42. R. J. Shamberger and C. E. Willis, *Crit. Rev. Clin. Lab. Sci.*, **2**, 211 (1971).
43. A. Hamilton, *Industrial Toxicology*, Harper, New York, 1934.
44. A. J. Natal, M. Brown, and P. Dery, *Vet. Human Toxicol.*, **27**, 531 (1985).
45. E. Holstein, *b. Arbeitsmed.*, **1**, 102 (1951).
46. J. R. Harr, "Biological Effects of Selenium," in *Toxicity of Heavy Metals in the Environment*, F. W. Oehme, Ed., Dekker, New York, 1978, p. 393.
47. D. S. F. Robertson, Lancet, **1970-I**, 518.
48. I. S. Palmer, R. C. Arnold, and C. W. Carlson, *Poult. Sci.*, **52**, 1984 (1973).
49. V. H. Ferm, *Adv. Teratol.*, **6**, 51 (1972).
50. R. J. Shamberger, Lancet, **1971-II**, 1316.
51. G. W. R. Walker and A. M. Bradley, *Can. J. Genet. Cytol.*, **11**, 677 (1969).
52. P. Sentein, *Chromasoma*, **23**, 95 (1967).
53. H. A. Schroeder and M. Mitchner, *Arch. Environ. Health*, **23**, 102 (1971).
54. A. F. Tarantal, C. C. Willhite, B. L. Lasley, C. J. Murphy, C. J. Miller, M. J. Cukierski, S. A. Book, and A. G. Hendrickx, *Fundam. Appl. Toxicol.*, **16**, 147 (1991).
55. J. H. Ray and L. C. Altenburg, *Mutat. Res.*, **78**, 261 (1980).
56. A. Schecter, W. Shanske, and A. Stenzier, *Chest*, **77**, 554 (1980).
57. R. F. Buchan, *Occup. Med.*, **3**, 439 (1947).
58. I. Rosenfeld and O. A. Beath, *Selenium—Geobotany, Toxicity, and Nutrition*, Academic, New York, 1964.
59. R. Rohmer, E. Carrot, and J. Gouffault, *Bull. Soc. Chim. Fr.*, **275** (1950).
60. G. Kimmerle, *Arch. F. Toxikol.*, **18**, 140 (1960).

61. H. C. Dudley, *Publ. Health Rep.*, **53**, 94 (1938).
62. National Fire Protection Association, Fire Protection Guide on Hazardous Materials, 7th ed., National Fire Protection Association, Boston, MA, 1978.
63. R. A. Johnson and F. P. Kwan, *Anal. Chem.*, **25**, 11017 (1953).
64. R. A. Johnson, F. P. Kwan, and D. Westlake, *Anal. Chem.*, **25**, 1017 (1953).
65. H. H. Steimberg, S. C. Massari, A. C. Miner, and R. Rink, *J. Ind. Hyg. Toxicol.*, **24**, 183 (1942).
66. R. H. DeMeio and F. C. Henriques, Jr., *J. Biol. Chem.*, **169**, 609 (1947).
67. B. Venugopal and T. D. Luckey, *Metal Toxicity in Mammals*, Vol. 2, Plenum Press, New York, 1978, p. 246.
68. C. Thienes and T. J. Haley, *Clinical Toxicology*, 5th ed., Lea and Febiger, Philadelphia, 1972, p. 199.
69. L. Friberg, G. R. Nordberg, and V. B. Vouk, *Handbook on the Toxicology of Metals*, Elsevier North Holland, New York, 1979, p. 591.
70. M. I. Amdur, *Arch. Ind. Health*, **17**, 665 (1958).
71. P. Lampert, F. Garro, and A. Pentschew, *Acta Neuropathol.*, **15**, 308 (1970).
72. A. D. Toews, S. Y. Lee, B. Popko, and P. Morell, *J. Neurosci. Res.*, **26**, 501 (1990).
73. T. W. Bouldin, T. S. Earnhardt, N. D. Goines, and J. Goodrum, *Neurotoxicol.*, **10**, 79 (1989).
74. S. Duckett, *Experientia*, **26**, 1239 (1970).
75. K. F. Jackson, J. P. Hammang, S. F. Worth, and I. D. Duncan, *Acta Neuropathol. (Berl.)*, **78**, 301 (1989).
76. F. Dijkstra, *Water Sci. Technol.*, **20**, 83 (1988).
77. International Labour Office, *Encyclopedia of Occupational Health and Safety*, Vols. I and II, McGraw-Hill, New York, 1971, p. 1391.
78. G. Kimmerele, *Arch. F. Toxikol.*, **18**, 40 (1960).
79. E. S. Blackadder and W. G. Manderson, *Brit. J. Ind. Med.*, **32**, 59 (1975).
80. E. A. Cerwenka and W. C. Cooper, *Arch. Environ. Health*, **3**, 189 (1961).
81. T. Sollmann, *A Manual of Pharmacology*, W. B. Sanders, Philadelphia, 1957, p. 150.
82. M. J. Ellenhorn and D. G. Barceloux, *Medical Toxicology*, Elsevier, 1988.
83. National Research Council of Canada, *Sulphur and its Inorganic Derivatives in the Canadian Environment*, 1977, p. 56.
84. Occupational Safety and Health Administration, *Analytical Methods Manual*, U.S. Department of Labor, OSHA Analytical Laboratory, Salt Lake City, UT, 1985.
85. World Health Organization, *WHO Tech. Rep. Ser.*, **707**, 1115 (1984).
86. L. J. Roger, H. R. Kehrl, M. Hazucha, and D. H. Horstman, *J. Appl. Physiol.*, **59**, 784 (1985).
87. T. J. Witek, E. N. Schachter, G. J. Beck, and W. S. Cain, *Int. Arch. Occup. Environ. Health*, **55**, 179 (1985).
88. R. S. Chapman, D. C. Calafiorne, and V. Hasslelblad, *Ann. Rev. Respir. Dis.*, **132**, 261 (1985).
89. R. A. Kehoe, W. F. Machhle, K. Kitzmiller, and T. J. Leblanc, *J. Ind. Hyg.*, **14**, 159 (1932).
90. I. O. Skaple, *Brit. J. Ind. Med.*, **21**, 69 (1964).

91. B. G. Ferris, S. Puleo, and H. Y. Chen, *Brit. J. Ind. Med.*, **36**, 127 (1979).

92. T. J. Smith, J. M. Peters, J. C. Reading, and C. H. Castle, *Am. Rev. Respir. Dis.*, **37**, 149 (1987).

93. A. M. Lee and J. F. Fraumeni, *J. Nat. Cancer Inst.*, **42**, 1045 (1969).

94. S. Laskin, M. Kuschner, A. Sllakumae, and G. Katz, "Combined Carcinogen-irritant Animal Inhalation Studies," presented at OHOLO Biological Conference, New Zion, Israel, 1975.

95. M. O. Amdur, *A.M.A. Arch. Ind. Health*, **18**, 407 (1958).

96. U.S. Environmental Protection Agency, *Pesticide Index*, 1985.

97. J. Kangas, P. Jappinen, and H. Savolainen, *Am. Ind. Hyg. Assoc. J.*, **45**, 787 (1984).

98. R. O. Beauchamp, J. S. Bus, J. A. Popp, C. J. Boreiko, and D. A. Andjelkovich, *CRC Crit. Rev. Toxicol.*, **13**, 25 (1984).

99. N. I. Sax, *Dangerous Properties of Industrial Materials*, 5th ed., Van Nostrand, New York, 1979.

100. H. Savolainen, *Biol. Monitoring*, **1**, 27 (1991).

101. R. E. Gosselin, *Clinical Toxicology of Commercial Products*, Williams and Wilkins, Baltimore, MD, 1984.

102. L. Adelson and I. Sunshine, *Arch. Pathol.*, **81**, 375 (1966).

103. T. H. Milby, *J. Occup. Med.*, **4**, 431 (1962).

104. I. H. Ohya, H. Komoriya, and Y. Bunai, *Res. Pract. Forensic Med.*, **28**, 119 (1985).

105. L. W. Spolyar, *Ind. Health Monthly*, **11**, 116 (1951).

106. J. M. McDonald and A. P. McIntosh, *Arch. Ind. Hyg. Occup. Med.*, **3**, 445 (1951).

107. J. Deng and S. Chang, *Am. J. Ind. Med.*, **11**, 447 (1987).

108. M. Campanya, P. Sanz, and R. Reig, *Med. Lav.*, **80**, 251 (1989).

109. National Institute for Occupational Safety and Health, Criteria for a Recommended Standard, Occupational Exposure to Hydrogen Sulfide, 1977.

110. H. W. Haggard, *J. Ind. Hyg. Toxicol.*, **7**, 113 (1925).

111. W. Burnett, E. King, M. Grace, and W. Hall, *Am. J. Ind. Med.*, **117**, 1277 (1977).

112. R. Frank, "Acute and Chronic Respiratory Effects of Exposure to Inhaled Toxic Agents," in *Occupational Respiratory Diseases*, J. A. Merchant, Ed., Division of Respiratory Disease Studies, Appalachian Laboratory for Occupational Safety and Health, NIOSH, DHHS (NIOSH) Publication No. 86-102, 1986, pp. 571–605.

113. M. Thoman, *Clin. Toxicol.*, **2**, 383 (1969).

114. M. Arnold, R. Dufresne, B. Alleyne, and P. Stuart, *J. Occup. Med.*, **27**, 373 (1985).

115. S. Kosmider, E. Rogala, and A. Pacholek, *Arch. Immunol. Ther. Exp.*, **15**, 731 (1967).

116. G. Ahlborg, *Arch. Ind. Hyg. Occup. Med.*, **3**, 247 (1951).

117. M. S. Hua and C. C. Huang, *J. Clin. Exp. Neuropsychol.*, **10**, 328 (1988).

118. H. H. Wasch, W. J. Estrin, P. Yip, R. Bowler, and J. E. Cone, *Arch. Neurol.*, **46**, 902 (1989).

119. U.S. Environmental Protection Agency, Health Assessment Document for Hydrogen Sulfide, prepared by the Office of Health and Environmental Assessment, Environmental Criteria and Assessment Office, Research Triangle Park, NC (External Review Draft), EPA/600/8-86/026A, 1990.

120. Chemical Industry Institute of Toxicology, 90-Day Vapor Inhalation Toxicity Study of

Hydrogen Sulfide in Fischer-344 Rats, U.S. EPA, Office of Toxic Substances Public Files, Fiche Number 0000255-0, Document Number FYI-OTS-0883-0255, 1983.

121. Chemical Industry Institute of Toxicology, 90-Day Vapor Inhalation Toxicity Study of Hydrogen Sulfide in Sprague-Dawley Rats, U.S. EPA, Office of Toxic Substances Public Files, Fiche Number 0000255-0, Document Number FYI-OTS-0883-0255, 1983.

122. Chemical Industry Institute of Toxicology, 90-Day Vapor Inhalation Toxicity Study of Hydrogen Sulfide in B6C3F1 Mice, U.S. EPA, Office of Toxic Substances Public Files, Fiche Number 0000255-0, Document Number FYI-OTS-0883-0255, 1983.

123. A. Lopez, M. Prior, S. Yong, M. Albassam, and L. Lillie, *Fundam. Appl. Toxicol.*, **9**, 753 (1987).

124. A. Lopez, M. Prior, L. Lillie, C. Gulayets, and O. Atwal, *Vet. Pathol.*, **25**, 376 (1988).

125. A. Lopez, M. Prior, S. Yong, L. Lillie, and M. Lefebvre, *Am. J. Vet. Res.*, **49**, 1107 (1988).

126. A. P. Saillenfait, P. Bonnet, and J. deCeaurriz, *Toxicol. Lett.*, **48**, 57 (1989).

127. G. A. Poda, *Arch. Environ. Health*, **12**, 795 (1966).

128. M. Pergal, N. Vukojevic, D. Djuric, and T. Bojovic, *Arch. Environ. Health*, **25**, 38 (1972).

129. M. Pergal, N. Vukojevic, and D. Djuric, *Arch. Environ. Health*, **25**, 42 (1972).

130. H. L. Barthelmey, *J. Ind. Hyg. Toxicol.*, **21**, 141 (1939).

131. Y. S. Bao, S. Cai, S. F. Zha, X. C. Xhang, M. Y. Huang, O. Zheng, and H. Jiang, *Teratology*, **43**, 451 (1991).

132. B. D. Hardin, G. P. Bond, M. R. Sikor, F. D. Andrew, R. P. Beliles, and R. W. Niemeir, *Scand. J. Work Environ. Health*, **7** (Suppl. 4), 66 (1981).

133. C. Jones-Price, R. W. Tyl, M. C. Marr, and C. A. Kimmel, "Teratologic Evaluation of Carbon Disulfide (CAS No. 75-15-0) Administered to CD Rats on Gestational Days 6 through 15," National Center for Toxicological Research, Jefferson, AR, *Govt. Reports Announcements and Index*, Issue 15, NTIS PB 84-192343, 1984.

134. C. Jones-Price, R. W. Tyl, M. C. Marr, and C. A. Kimmel, "Teratologic Evaluation of Carbon Disulfide (CAS No. 75-15-0) Administered to New Zealand White Rabbits on Gestational Days 6 through 15," National Center for Toxicological Research, Jefferson, AR, *Govt. Reports Announcements and Index*, Issue 15, NTIS PB 84-192350, 1984.

135. B. L. Johnson, J. Boyd, J. R. Burg, S. T. Lee, C. Xintaras, and B. E. Albright, *Neurotoxicology*, **4**, 53 (1983).

136. S. Tabacova, B. Nikiforov, and L. Balabaeva, *J. Appl. Toxicol.*, **3**, 223 (1983).

CHAPTER FIFTEEN

Silicon and Silicates, Including Asbestos

Carl O. Schulz, Ph.D., D.A.B.T.

1 INTRODUCTION

Just as carbon serves as the elemental building block of the living world, silicon is the elemental building block of the mineral world. As a result it is second only to oxygen in its abundance on earth. Silica and naturally occurring silicate minerals make up about 90 percent of the earth's crust. There are interesting physical and chemical parallels between silicon and carbon. Both belong to Group 4A and primarily form compounds in the $+4$ oxidation state. Both elements form an almost infinite number of compounds, both natural and synthetic. In the vast majority of these compounds the carbon or silicon atom occupies the center of a tetrahedron of four covalent bonds. Both elements form large homologous series of compounds in which they are bound to oxygen, hydrogen, and halogens, for example, alkanes and silanes. There are important differences between these elements, however. Whereas carbon–carbon bonds are the rule in organic compounds, silicon forms relatively few compounds containing silicon–silicon bonds. Instead, the backbones of silicates and silicones are chains of alternating silicon and oxygen atoms. Furthermore, the presence of a full complement of $3d$ electrons in silicon results in this element having more metallic properties than carbon. Finally, whereas elemental carbon is found naturally in coal, graphite, and diamond, elemental silicon does not occur naturally.

Because of the abundance of silicon-containing compounds in the earth's crust,

Patty's Industrial Hygiene and Toxicology, Fourth Edition, Volume 2, Part A, Edited by George D. Clayton and Florence E. Clayton.
ISBN 0-471-54724-7 © 1993 John Wiley & Sons, Inc.

general environmental and occupational exposure to silicon by inhalation and ingestion is unavoidable and has occurred throughout human history. By and large,
naturally occurring silicon compounds are poorly absorbed, and if absorption does
occur they tend not to react with biomolecules or undergo biochemical transformation to more reactive species. For these reasons, few naturally occurring silicon
compounds are "toxic" in the classical biochemical sense of this term. However,
epidemiologic evidence amassed over the past half century has clearly demonstrated
that occupational and environmental exposure to several naturally occurring silicon-
containing compounds, for example, crystalline silica, and certain fibrous silicates,
for examples, asbestos, has resulted in extensive morbidity and mortality from
benign and malignant respiratory diseases. Although the mechanism by which these
compounds cause these diseases is not fully understood, it is abundantly clear that
the physical characteristics of the silicon material is the predominant, if not the
only, determinant of biologic activity.

The ability of certain forms of silicon-containing compounds to cause respiratory
diseases as a consequence of inhalation exposure is the primary and almost the
only occupational health concern associated with this entire family of compounds.
Although many of the industrial hygiene aspects are compound-specific, there are
certain generic issues applicable to all the different types of silicon-containing
compounds. These include the deposition of particulate matter in the respiratory
tract and the sampling and analysis of atmospheres for evaluation and control of
occupational exposures.

The fourth revision of this book marks the first time that the toxicologic aspects
of silicon and its compounds have been discussed in a separate chapter. This
emphasis is not driven so much by the toxicity of these materials as it is by the
increasing diversity of silicon-containing compounds, both natural and synthetic,
to which workers are exposed and the unique public health issues that accompany
a number of these exposures. A number of significant changes have taken place
in the decade since this volume was updated. Primary among these is the increased
recognition of the long-term health consequences of occupational exposure to asbestos. Because of this recognition most of the traditional uses of asbestos have
been strictly curtailed, if not eliminated altogether. This, however, has not eliminated occupational exposure nor does it put an end to the ongoing investigation
into the mechanism by which asbestos causes respiratory diseases or into the quantitative relationship between exposure and incidence of those diseases. Occupational exposure has shifted from those involved in the production and use of asbestos
and asbestos-containing products to those who might be exposed during construction, demolition, or maintenance of buildings or equipment that contain asbestos.
A better understanding of the nature of asbestos-induced disease is of critical
importance to protecting the long-term health of this population of workers.

As a consequence of the diminished use of asbestos for a number of applications,
there has been an increase in the use of alternate materials for the traditional
applications of asbestos and new substitutes continue to be developed. A number
of these materials are natural or synthetic silicates or silicon-containing compounds.
Thus there have been dramatic increases in the past decade in the production and

use of man-made mineral fibers (MMMF) and certain natural silicate materials. The long-term health effects of exposure to most of these have been less thoroughly investigated than the effects of asbestos or crystalline silica. Because our knowledge of the pathogenesis of respiratory disease caused by fibrogenic dusts is incomplete, there is a great deal of uncertainty about the long-term health risks of exposure to these substitute materials.

Another important aspect of the past decade has been the development of space-age composite materials designed for applications where chemical and thermal resistance, strength, and low electrical conductivity are essential. More than any other type of material, ceramics, of which silicon dioxide (SiO_2) is the basic building block, have filled this need. Of additional commercial importance are fiber-reinforced plastics. Many of the fibers used for these composites are MMMF. This increased use of silica and silicate-based materials is expected to increase in the foreseeable future.

Even an entire chapter is insufficient for an in-depth discussion of the toxicologic and occupational health aspects of silicon and silicates. Rather, this chapter is intended to provide the reader with an overview of these subjects. Wherever possible, the reader will be referred to more in-depth treatments of the topics discussed in specific sections. This chapter begins with a general discussion of some of the toxicologic aspects of exposure to respiratory particulates and fibers. Subsequent sections discuss occupational exposures to, and the toxicology of, silicon, silicon dioxide (silica), naturally occurring silicates (including asbestos), MMMF, synthetic silicates, and other industrially important silicon compounds.

2 TOXICOLOGIC ASPECTS OF INHALED DUSTS AND FIBERS

As indicated above, few silicon-containing compounds act as classical systemic toxicants. In other words, they are not absorbed into the circulation and transported to specific target tissues, where they may be biotransformed to more or less active compounds and interact with specific receptors or target molecules to initiate the biochemical changes that result in toxicity. Instead, the biologic effects of the silicon-containing compounds of primary interest in this chapter, that is, crystalline silica, asbestos, natural and synthetic silicate fibers, and silicon carbide, are the result of inhalation of dusts or fibers of these compounds and are, for the most part, confined to the respiratory system. The mechanism by which this entire process occurs is not fully understood and significant questions remain. For example, as discussed below, there is considerable controversy regarding the relative potency of different asbestos fiber types for causing cancer and mesothelioma. Although fiber geometry and durability are important determinants of their potential to cause disease, other unidentified factors apparently also play a role. In a similar manner, it is not clear why different forms of silica have different biologic activity. Recognizing that these questions remain, there are some principles of inhalation toxicology that are generally accepted and that are common to the toxicology of all of these different forms of silicon-containing compounds.

Although the pathogenesis of malignant and nonmalignant respiratory diseases

Table 15.1. Relationship Between Particle Size and Respirability (3)

Particle Aerodynamic Diameter (μm)	Respirable Particulate Mass (%)
0	100
1	97
2	91
3	74
4	50
5	30
6	17
7	9
8	5
10	1

caused by inhalation exposure to silica and the silicates is not fully understood, it is clear that an essential step in the process is the deposition and retention of specific particulate forms of the material in the gas-exchange portion of the lung, that is, the alveoli. The ability of particles to penetrate to the alveoli and deposit there depends upon their size and shape.

A detailed discussion of the inhalation toxicology of particulate matter is beyond the scope of this chapter. For more detailed treatments of this topic the reader is referred to other resources (1, 2). Typically, when workers are exposed to airborne particulate matter, the matter is a mixture of different sizes and shapes of particles and often different chemical compounds. It is well established that the biologic effects of materials such as silica, asbestos, and other fibrous minerals depend in part upon the amount of the material that reaches key regions of the respiratory tract and remain there. Three key components determine the ultimate fate of particulate matter in the respiratory system, namely, respiration, deposition, and retention.

The respirability of particles and fibers, that is, the ability to enter the respiratory system, depends entirely upon their geometry. For particles that are approximately spherical in shape, the relationship between mean particle size and the proportion of particles that enter the respiratory system is fairly well understood. The fraction of total airborne dust that enters the respiratory system is called the "respirable fraction" and increases with decreasing average diameter of the particles in the dust. The relationship between particle size and respirability has been tabulated by the American Conference of Governmental Industrial Hygienists (ACGIH) and is shown in Table 15.1.

For irregularly shaped particles, including fibers, respirability is determined by their "aerodynamic diameter," which is an empirically determined parameter. In general, fibers tend to align with the direction of airflow when they enter the inhalation airstream. Therefore the aerodynamic diameter is generally only slightly greater than the diameter of the fiber itself, and fibers of diameters of 5 μm or less are generally considered to be within the respirable range.

Once particles enter the respiratory tract they may deposit or remain suspended in the airstream and be exhaled with the respired air. The fraction of respired particles that deposit and the location within the respiratory tract where that deposition takes place are not fully understood. Deposition is determined by a number of factors, the most important of which are the physical characteristics of the particle (size, shape, density, and electrical charge), but they also include ventilation characteristics (respiratory rate, tidal volume, mode of breathing, etc.) (4).

Larger particles deposit primarily by impaction upon the surface of the respiratory system at those locations where the direction of the airflow changes. The largest particles tend to deposit in the nasopharyngeal region and may be expelled by sneezing or coughing or may enter the gastrointestinal system by swallowing. Smaller particles may reach the tracheobronchial tree where they tend to deposit by impaction at the bifurcations of the bronchi. Particles that deposit in this region of the respiratory tract may be cleared by coughing or dissolution, but for insoluble particulates the most important mechanism is by mucociliary transport upward through the respiratory tract and into the gastrointestinal system.

As particle sizes decrease sedimentation becomes an increasingly important additional mechanism by which they deposit on the respiratory epithelium. Particles larger than approximately 0.5 μm deposit by sedimentation. In other words, they fall downward because of gravity and settle on the respiratory epithelium. This may occur at any location within the respiratory tract.

Particles less than 0.5 μm in diameter tend not to deposit by impaction or sedimentation but move by diffusion from areas of higher to lower concentration and, in doing so, may deposit on the respiratory epithelium. Diffusional deposition is most likely in that region of the respiratory system where air velocity is lowest, that is, the alveoli.

The available research suggests that for the materials discussed in this chapter, that is, silica, fibrous silicates, and silicon carbide, only that material that deposits in the nonciliated, gas-exchange portion of the lung (alveoli and terminal bronchioles) is significant in the pathogenesis of respiratory disease. The deposition of essentially spherical particles in this pulmonary region of the lung has been studied (4). When inhalation is via the nose, particles greater than 5μm in diameter are deposited in the upper airways. With decreasing diameter pulmonary fractional deposition increases from 0 to 20 percent for particles having an aerodynamic diameter of 1 μm. Fractional deposition then decreases, with decreasing particle size reaching a minimum of 15 percent for particles approximately 0.5 μm in diameter. Deposition, primarily by diffusion, then increases as particles become increasingly smaller. This general pattern of deposition can be altered dramatically by changes in ventilatory rate, respiratory disease, or concurrent exposure to irritants. For example, breathing through the mouth rather than through the nose, as occurs during physical exertion, changes the deposition pattern considerably, with as much as 50 percent of particles 2 to 6 μm in diameter depositing in the deep lung.

The deposition of fibers in the deep lung is less well understood than that of approximately spherical particles. Clearly the elongated form of fibers increases the likelihood that they will contact the walls of the respiratory system by impaction or sedimentation more frequently than spherical particles of similar diameter. This tendency will be greater for fibers that are curly or relatively inflexible, compared to straight and flexible fibers. Because they tend to be oriented along the direction of airflow, fibers that are quite long have the potential to reach the alveoli. Fibers with a diameter of 1 μm or less have the potential to deposit in the alveolar portion of the lung. Fibers of sufficiently small diameter to reach the alveoli may have very different diffusional behavior than roughly spherical particles of the same diameter. Because of their greater mass relative to diameter, these fibers would be less likely to deposit by diffusion.

The ability of particulates in the lung to cause fibrosis and other long-lasting changes appears to be related, among other factors, to their retention in the area of the lung where they deposit. There are many mechanisms that may operate to remove particulate matter from the lung. Those particles that deposit in the alveolar portion of the lung may be cleared by dissolution and absorption. Insoluble materials must be cleared by other mechanisms, however, and these are mediated primarily by alveolar macrophages. These macrophages engulf or otherwise surround deposited particulate material and transport it away from the site of deposition. Movement may involve intraluminal movement to the ciliated bronchial epithelium where the macrophage and ingested material are cleared by mucociliary action. Macrophages may also pass through the alveolar epithelium into the interstitial space and be cleared by lymphatic drainage to the lymph nodes. They may also pass through the pulmonary capillary epithelium and enter the general circulation.

Theories regarding the pathogenesis of pulmonary diseases induced by insoluble particulates frequently involve impaired clearance as a key feature. These include the hypothesis that fibrogenic particulates such as asbestos are cytotoxic to alveolar macrophages and thus interfere with their own clearance, resulting in increased lung retention. Similar theoretical explanations have been used to explain the observation of fibrosis in response to high concentrations of inert dusts, the so-called dust overloading condition (5). These hypotheses are not entirely successful, however, in explaining all the observed differences between fibrogenic and non-fibrogenic dusts and in accounting for differences in potency among disease-causing particulates.

Uncertainties regarding the pathogenesis of respiratory diseases by dusts and fibers have important implications for the control of exposures that lead to these diseases. Of primary concern is the uncertainty regarding the characteristics of those particles or fibers that are most likely to be pathogenic and the difficulty in accurately monitoring exposure to them. As will be seen in specific sections below, the methods that have historically been used, and continue to be the methods of choice, to evaluate exposure to these materials do not necessarily detect the particulate matter that is most likely to deposit and be retained in the alveolar region of the lung. Furthermore, the studies that purport to relate disease incidence and severity to lung burdens of dust are based on total lung dust burden rather than the amount of particulate matter retained in the alveolar regions. These limitations raise the possibility that those exposures that determine future adverse health consequences are not being adequately evaluated and controlled.

3 SILICON, Si

3.1 Sources, Production, Use, and Industrial Exposures

Elemental silicon (CASRN* 7440-21-3) does not occur in nature. Silicon for industrial applications is prepared from naturally occurring silicon compounds, almost exclusively quartz sand or gravel (SiO_2). The primary industrial applications of silicon are in metallurgy, where it is alloyed with other metals to impart desirable physical properties. It also serves as an oxidizer in steelmaking (6). Another industrial application of silicon is in the manufacture of electronic equipment. Pure silicon acts as a semiconductor and its electrical resistivity can be precisely manipulated by the inclusion of minute quantities of elemental impurities into the crystal lattice (doping). These properties are responsible for the use of very pure single crystals of silicon in the manufacture of transistors and integrated circuits for computers and other electronic equipment. Whereas metallurgical-grade silicon need only have a purity of 95 to 99 percent, impurities in semiconductor-grade silicon, if present, are usually measured in parts per billion (ppb) quantities.

The consumption of silicon for metallurgical purposes dwarfs that for electronics manufacturing but consumption of silicon for the latter use has grown exponentially over the past few decades. In 1980, the total consumption of silicon for all uses was estimated to be 5 million tons with semiconductor-grade silicon accounting for about 2750 tons, or less than 0.1 percent (6). However, consumption of semiconductor-grade silicon had increased by nearly 100 percent during the five-year period from 1975 to 1980.

Metallurgical-grade silicon is prepared by heating quartz sand or gravel with coke in an electric oven (1300 to 1400°C), thereby reducing silicon dioxide to silicon. Semiconductor-grade silicon is prepared from metallurgical-grade silicon by converting it to silicon halide or trihalosilane, which is then reduced back to silicon in the absence of contaminating impurities. This may be accomplished by heating the silicon halide compound in a pure hydrogen atmosphere with deposition of silicon on a pure silicon filament heated to about 1150°C (6).

Semiconductor-grade silicon may also be prepared from metallurgical-grade silicon or from chemically purified silicon by zone refining. In this process a long rod of silicon is heated by a moving heater in such a way that a very thin zone of the rod is melted and that molten zone is moved along the rod from one end to the other. Impurities in the silicon are moved along through the rod ahead of the molten zone. By repeating sweeps of melted zones along the rod in the same direction, impurities are swept to one end of the rod, which is discarded.

Most modern electronic uses of silicon require single-crystal material. Crystals are grown from high-purity silicon that has been melted and to which a precisely

*Chemical Abstracts Service Registry Number.

measured amount of the dopant material has been added. In one process, molten silicon freezes slowly onto a properly oriented seed crystal of silicon that is rotated and withdrawn from the melt. A long crystal is produced that can then be thinly sliced to provide the basic semiconductor material that may then be etched to produce integrated circuits.

There is little opportunity for occupational exposure to silicon in its elemental form. Heating of silicon in an oxygen atmosphere results in oxidation to silicon dioxide (silica) before there is any potential for volatilization of the metal. Therefore no exposure to silicon per se is anticipated from its metallurgical uses. Occupational exposure by inhalation and ingestion of mixed dusts containing silicon and silica may potentially occur during the manufacture of semiconductor-grade silicon and the manufacture of electronic equipment. The importance of keeping silicon semiconductors and electronic parts free of impurities necessitates that these processes be carried out in very clean environments and, as a result, exposures are minimized.

3.2 Physical and Chemical Properties

The physical and chemical properties of silicon and some of the compounds that are discussed in the remainder of this chapter are summarized in Table 15.2. Silicon itself is a gray material with a metallic luster. It is very hard and brittle. There are three stable isotopes of silicon, ^{28}Si, ^{29}Si, and ^{30}Si, having a relative natural abundance of 92.2, 4.7, and 3.1 percent. Silicon undergoes spontaneous oxidation in air to form a thin layer of silicon dioxide (silica) on the surface. The rate of spontaneous oxidation increases with temperature. Silicon semiconductors are usually deliberately oxidized to form a protective layer of silica. At temperatures below 1200°C the silica layer is amorphous and at temperatures from 1200 to 1600°C it is crystalline.

As stated in the introduction, silicon forms an almost infinite number of individual compounds. The compounds shown in Table 15.2 are selected to show the variety of these compounds. A number of the compounds of silicon, especially those in which silicon is bound to oxygen and/or carbon, are exceptionally stable and resistant to chemical attack. Ceramics and silicone elastomers, for example, exhibit thermal stability and are quite resistant to degradation by acids and bases. Many of the uses of these materials are based on these properties. Compounds in which silicon is bound to oxygen may be either insoluble like silica and most of the natural silicates or soluble when the silicate or meta- or orthosilicate ions are present, as in the soluble silicates. Silicon forms a number of compounds in which it is bound to hydrogen, halogens, or both. These families of compounds, the silicon hydrides (silanes) and silicone halides, tend to be much more reactive than the compounds in which silicon is bound to oxygen or carbon.

3.3 Analytical Determination

As indicated above, occupational exposure to silicon is primarily through inhalation of airborne mixed dust containing silicon. Thus sampling of the occupational en-

Table 15.2. Physical and Chemical Properties of Silicon and Some of Its Compounds

Element or Compound	CASRN[a]	Atomic or Mol. Wt.	Physical Properties	M.P. (°C)	B.P. (°C)	Solubility
Silicon, Si	7440–21–3	28.086	Gray cubic crystals; hard, brittle	1410	2355	Insol. in water; sol. in HF + HNO_3
Silicon dioxide, SiO_2 (quartz)	14808–60–7	60.08	Colorless hexagonal crystals	1610	2230	Insol. in water; sol. in HF; sl. sol. in alkali
Silicon dioxide (cristobalite)	14464–46–1	60.08	Colorless cubic or tetragonal crystals	1713	2230	Insol. in water; sol. in HF; sl. sol. in alkali
Silicon dioxide (tridymite)	15468–32–3	60.08	Colorless rhombic crystals	1703	2230	Insol. in water; sol. in HF; sl. sol. in alkali
Silicon dioxide (amorphous silica)	7631–86–9	60.08	Transparent to gray powder, noncrystalline	1711	2230	Insol. in water; sol. in HF; sl. sol. in alkali
Sodium metasilicate, Na_2SiO_3	6834–92–0	122.08	A glass or othorhombic crystals	1089	—	Sol. in cold water; decomp. in hot water
Silicon carbide, SiC	409–21–2	40.07	Green to black, sharp crystals; very hard	~2600	—	Insol. in water, acids, or bases
Silane, SiH_4 (Silicon tetrahydride)	7803–62–5	32.09	Colorless gas with an offensive odor	–185	–112	Slowly decomp. in water; insol. in organic solvents
Trichlorosilane, $HSiCl_3$	10025–78–2	135.47	Colorless, fuming liquid	–126.5	31.8	Decomp. in water; sol. in benzene, CS_2, $CHCl_3$, and CCl_4
Silicon tetrachloride, $SiCl_4$	10026–04–7	169.89	Colorless, fuming liquid	–70	59	Decomp. in water; miscible with benzene, ether, $CHCl_3$
Ethyl silicate, $C_2H_5SiO_4$	78–10–4	208.3	Colorless liquid	–117	110	Slowly decomp. in water; miscible with alcohol

[a]Chemical Abstracts Service Registry Number.

vironment is accomplished as for any respirable dust. Typically, air in the breathing zone of employees is drawn through a device, such as a cyclone, that removes particles other than those of respirable dimensions, and then through a filter at a known rate for a known period of time. The filter is dried and weighed just prior to and just after sampling, and the weight difference is the total mass of respirable particulate matter in the air. The percentage of the respirable dust that is silicon can be determined by recovering and analyzing the collected dust either by wet chemical or instrumental methods.

For the wet chemical methods of analysis, silicon is first converted to silica or silicic acid. thus these methods will not distinguish between silicon and silica in the sample. Silicic acid, obtained either directly or by dissolution of silica in hydrogen fluoride solution, is then treated with ammonium molybdate to form a yellow silicomolybdate complex, the concentration of which can be determined colorimetrically.

Instrumental methods for the determination of elemental silicon are X-ray powder diffraction, X-ray fluorescence, energy-dispersive X-ray analysis, or neutron activation analysis.

3.4 Physiological Response and Hygienic Standards of Permissible Exposure

Almost no information is available on the possible adverse effects of long-term exposure to silicon. Elemental silicon dust has historically been considered to be essentially inert. The lack of demonstrated toxicity together with limited opportunities for occupational exposure to elemental silicon are reflected by the existing guidelines and standards for occupational exposure limits. The Occupational Safety and Health Administration (OSHA) permissible exposure limit (PEL) for silicon is an air concentration of 10 mg/m^3 of total dust or 5 mg/m^3 of respirable dust as a time-weighted average (TWA) over a 6 hr work shift (7). The threshold limit value (TLV) for occupational exposure to silicon dust as recommended by the ACGIH is 10 mg/m^3 of dust containing less than 1 percent respirable quartz and no asbestos (3). The National Institute for Occupational Safety and Health (NIOSH) has not recommended an exposure limit for silicon (8).

It is important to recognize that silicon will undergo spontaneous oxidation to silicon dioxide in air. Thus silicon dust particles typically have a surface layer of silicon dioxide. As temperatures increase above 1200°C there is a theoretical possibility that silicon may be converted to crystalline silica which is a substantially more hazardous material. Thus occupational settings in which there is potential exposure to silicon dust generated at high temperatures should be evaluated for potential exposure of workers to crystalline silica-like moieties.

4   SILICON DIOXIDE, SILICA, SiO$_2$

Silica (silicon dioxide, SiO$_2$) is the most common naturally occurring compound of silicon and oxygen and constitutes about 60 percent of the earth's crust, occurring either alone or combined with other oxides in the silicate minerals (9). Silica occurs naturally, and is used commercially, in both its crystalline and amorphous forms.

4.1 Crystalline Silica

4.1.1 Sources and Production

By far the most common naturally occurring form of crystalline silica is quartz (CASRN 14808-60-7). Other forms of crystalline silica include cristobalite (CASRN 14464-46-1), tridymite (CASRN 15468-32-3), tripoli (CASRN 1317-95-9), coesite, and stishovite. Crystalline silica is also a major component of sandstone. Quartz, cristobalite, and tridymite all occur in slightly different forms depending on the specific orientation of tetrahedral silicon units within the crystal lattice. These forms are interchangeable at elevated temperatures and at very high temperatures (>1710°C) the crystalline structure is lost entirely and crystalline silica is converted to amorphous silica (10).

The estimated U.S. production of sand and gravel for construction use was approximately 0.9 million tons in each of the four years from 1986 to 1989 (11). The production of silica sand and gravel of sufficient purity for use as a raw material for the production of glass, ceramics, and the synthesis of silicon and silicon compounds remained stable over a 10 to 15 year period prior to 1983, and total world production in that year was estimated to be 182 million tons (10). This does not include sand and gravel produced for construction uses, most of which is used in the country in which it is produced. Sand used in making glass and ceramics contains greater than 98.5 percent silica and has very low iron content. U.S. production of silica sand and gravel for glass making was estimated to be 23.7 million tons in 1983.

The production of high-quality quartz crystals (lasca) in the United States was reported to be 363 tons in 1983 (10) and 300 tons in 1989 (11). Single quartz crystals are used extensively in the electronics industry because of their dielectric and piezoelectric properties. As such, they are used in radio oscillator circuits to control electromagnetic wave frequencies as well as in watches. Historically, natural quartz crystals were used for these applications but in recent years man-made crystals grown from lasca have surpassed natural crystals for these applications. There is also an increased use of quartz crystals (synthetic) for fiber optics.

4.1.2 Uses and Industrial Exposures

Silica sand and gravel are used, as is, in a variety of applications, with the largest being as a constituent of road construction and building materials, that is, concrete, mortar, and bricks. Silica (usually quartz) is also the raw material used for the production of metallurgical-grade and semiconductor-grade silicon, soluble silicates, silicon carbide, silicones, and a number of glass, ceramic, and refractory materials (10). Silica sand is used to make foundry castings for iron, aluminum, and copper alloys. An estimated 850,000 tons of silica sand were used in the United States in 1983 for hydraulic fracturing to increase oil and gas recovery. The breakdown of silica consumption for industrial uses is as follows:

Glass sand	37.4%
Foundry sand	26.7%
Abrasives	8.0%
Hydraulic fracturing	4.0%
Other	23.9%

Occupational exposure to silica is widespread and is not confined to its production and use. Exposure to crystalline quartz occurs in mining, manufacturing, construction, and agriculture. In 1983, NIOSH estimated that 3.2 million workers in 238,000 plants were exposed to crystalline silica (10). Key occupations in which there is significant potential for occupational exposure to crystalline silica are as follows:

Metal mining
Coal mining
Nonmetallic mineral mining
Quarrying
Sand, gravel, and clay mining
Stone, clay, and glass products
Iron and steel foundries
Nonferrous foundries

Not included are workers who may be exposed to crystalline silica in agriculture, construction, and in the manufacture of chemicals and allied products from silica.

4.1.3 Analytic Determination

Prior to 1970, exposure to crystalline silica was determined by counting of particles in an air sample by optical microscopy. The quartz content of the particulate was inferred from knowledge of the composition of the material that was being worked with. At present the preferred analytic method for the quantitative determination of crystalline silica in the occupational environment is X-ray diffraction spectrometry (12). This method involves collection of respirable dust on a 37-mm low-ash PVC membrane filter. The filter is then dissolved and the dust is analyzed by X-ray diffraction analysis using external or internal standards of known crystal composition. The detection limit for this method in respirable dust samples is approximately 5 μg for quartz and 10 μg for cristobalite, which would translate to an air concentration of approximately 0.01 mg/m^3 of quartz for a sample size of 500 l.

4.1.4 Physiological Response

Respiratory impairment and disease among workers who were exposed to mineral dusts were recognized in ancient times and have continued to be identified throughout history. The disease now recognized as silica-induced pneumoconiosis (silicosis) has been known for centuries by a number of names usually derived from the occupations in which it was recognized, such as grinders' asthma, masons' disease, miners' asthma, and potters' rot (13). The first pathological account of silicosis in stonecutters was provided by van Diemerbroeck in 1672. Visconti was the first to use the term "silicosis" in 1870 and by the first decade of the twentieth-century silicosis was well recognized as an occupational disease among tin miners, granite workers, sandstone quarry workers, and gold miners. In the period from 1950 to 1964, 27,000 worker compensation claims were settled in 18 states for a total of $132 million for silicosis/pneumoconiosis.

Silicosis is a nodular pulmonary fibrosis that results from the deposition of particles of crystalline silica in the lung tissue. This interstitial lung disease is characterized by the accumulation of activated inflammatory cells, granuloma formation, and a proliferation of connective tissue (14). A clear exposure–response relationship has been shown between the development of radiological and functional evidence of fibrosis and the cumulative respirable exposure to respirable crystalline silica (10). At some stage, fibrotic changes in the lung become progressive even if exposure to silica ceases. Typically silicosis is not diagnosed until at least 20 years after initial exposure, although rapidly progressive, fatal silicosis has developed within months of massive exposures (14).

Radiographically, silicosis appears as small rounded opacities distributed primarily in the upper lobes of the lungs. In its early stages the opacities are of moderate radio density and average from 1 to 3 mm in diameter. With longer and more severe exposure large (>1 cm) dense coalesced radiographic opacities may be seen. Pathologically, silicosis is characterized by the presence of multiple fibrohyaline nodules having a characteristic whorled pattern. Nodules (granulomas) may be present in the lungs and in the hilar lymph nodes. In severe cases or after fibrosis becomes progressive, massive fibrotic lesions are seen. Fibrosis may be accompanied by emphysema.

Clinically, the earliest symptoms of silicosis are usually cough and expectoration (14). This is often accompanied by shortness of breath upon exertion. These symptoms do not ordinarily become noticeable until there is radiographic evidence of complicated silicosis. As the condition progresses this condition worsens and respiratory failure is the most important consequence of complicated silicosis. The effect of silicosis on respiratory function is not consistent among individuals. Patients with symptomatic silicosis will display ventilatory function patterns characteristic of restrictive or obstructive lung disease or a mixture of the two. Pulmonary function tests are frequently normal among people with simple silicosis although appropriate testing and statistical analysis may provide a means for the differential diagnosis of simple silicosis (15).

Silicosis, once it becomes progressive, is frequently fatal. Severe silicosis may also result in secondary coronary insufficiency (cor pulmonale). The presence of silica-induced fibrosis apparently increases susceptibility of the lung to infection, and the increased risk of tuberculosis among silicotics has long been recognized in occupational medicine. The risk of tuberculosis is also increased among workers who are exposed to silica but show no radiographic evidence of silicosis (10).

Significantly increased mortality from tuberculosis has been observed in the pottery and granite industries. Animal studies also show enhanced susceptibility to tuberculosis and other lung infections following exposure to silica. There does not appear to be an interaction between silica dust and tobacco smoke in the induction of pulmonary fibrosis. Epidemiologic evidence suggests that the risk of coal-workers' pneumoconiosis (black lung) is increased by concomitant exposure to crystalline silica (quartz dust).

The pathogenesis of the fibrogenic response induced in tissues by crystalline silica is not known. Much of the activity must be attributable to the physical rather than the chemical characteristics of the material, in that amorphous silica is far less fibrogenic than crystalline silica. In a recent study in which rats were exposed by inhalation to amorphous silica and two different samples of quartz, the authors found that one sample of quartz was significantly more active than the other, even though these materials were identical by X-ray diffraction analysis (16). A similar finding was reported by Stober and Brieger (14) nearly 20 years earlier. The difference in activity was not attributable to differences in particle size, deposition, or clearance. The authors concluded that undetermined surface properties must play a role. The relatively greater fibrogenic activity of cristobalite as compared to quartz has been noted in epidemiologic studies as well, and is the basis for a separate and lower OSHA PEL for cristobalite (7).

The predominant mechanism for the removal of relatively insoluble particulate matter from the alveoli is phagocytosis by alveolar macrophages followed by clearance of the macrophages from the respiratory system either by mucociliary action or via the lymphatic system. Ziskind et al. suggested that the relative cytotoxicity of the particulate matter to alveolar macrophages was important in determining the fibrogenicity of inhaled dust (14). Thus when crystalline silica is phagocytized by macrophages, the macrophages die, whereas other materials such as silicon carbide and diamond fail to demonstrate macrophage toxicity. These authors claimed that there was good correlation between macrophage toxicity and fibrogenicity for a number of mineral dusts. More recently Morrow (17) reaffirmed the hypothesis that fibrogenic materials, such as crystalline silica, are cytotoxic to macrophages and thus interfere with the normal clearance mechanisms of the lung. They extended the hypothesis to explain the observation that relatively inactive dusts can generate a fibrogenic response at sufficiently high exposures. In a condition of "dust overloading" even relatively inactive "nuisance particulates" cannot be effectively cleared because the sheer volume of macrophages with engulfed particulate matter leads to aggregation and immobilization of macrophages. Thus residence time in the lung increases and pathological changes including fibrosis may develop.

The role of differential deposition and/or clearance of particles from the lung in the pathogenesis of silicosis is not clear. Warheit et al. (18) studied the deposition of amorphous silica in the lungs of rats exposed by nose-only inhalation for 6 hr/day, 5 days/week for 4 weeks. The relationship between silica in the lung and its concentration in the air was linear up to the highest concentration of 150 mg/m^3. In a similar, but not identical, study, Hemenway et al. (16) studied the deposition of crystalline silica (cristobalite) in the lungs of rats exposed for 6 hr/day, 4 days/

week for 2 weeks. The rates of deposition of silica were comparable in the two studies. However, Hemenway et al. found that the proportion of cristobalite that remained in the lung decreased with increasing concentration of cristobalite in the air. There was a significant departure from linearity in the relationship between deposition and air concentration even though the highest concentration of cristobalite (81 mg/m³) was much lower than the highest concentration of amorphous silica in the Warheit et al. study (18). The linear relationship between deposition and exposure concentration in the latter study argues against the hypothesis that the fibrogenicity of crystalline silica is due to greater retention in the lung than nonfibrogenic dusts. Additional evidence that differences in clearance do not account for differences in pathogenicity was provided by the study of Hemenway et al. (16), in which two different forms of quartz were cleared at comparable rates, yet one form was much more biologically active than the other.

Additional evidence that the fibrogenic potency of silica may not be a simple function of increased retention/decreased clearance in the lung as compared to other particulate matter is provided by studies indicating that tracheobronchial clearance is faster in persons with silicosis than those without (19). Furthermore, unlike many dusts including relatively inactive ones, the retention of crystalline silica by the lungs appears to be limited, possibly by the clearance mechanisms described above. Thus the maximum amount of silica that is found in the lungs of heavily exposed individuals is on the order of 5.0 g (10). This is in contrast to coal dust, in which lung burdens of 100 g have been determined.

Although removal of silica from the lung by macrophages may be the predominant clearance mechanism, other mechanisms may play a role. Crystalline silica is slightly soluble in biological fluids. Thus it can be dissolved and absorbed into the circulation. This mechanism is of sufficient significance that it leads to increased levels of silica in the blood and urine of exposed individuals (10).

The evidence showing that occupational exposure to crystalline silica results in increased incidences of, and mortality from, nonmalignant respiratory and cardiovascular diseases is conclusive, but the evidence for an association with malignant diseases is less than definitive. For this reason, the ACGIH does not identify crystalline silica as either a confirmed or suspected human carcinogen nor is it regulated as a carcinogen by OSHA. Crystalline silica *is* classified as an occupational carcinogen by NIOSH, however (8). The International Agency for Research on Cancer (IARC) reviewed the available evidence regarding the carcinogenic potential of silica in 1987 and concluded that there was "sufficient" evidence that crystalline silica was carcinogenic in experimental animals but the evidence that it caused cancer in humans was "limited" (10).

The epidemiologic data base relevant to a possible association between occupational exposure to silica and cancer is equivocal and difficult to interpret. The occupational cohorts that have been studied are ore miners; coal miners; stone quarriers, cutters, and sandblasters; ceramic and glass industry workers; and foundry and metallurgical industry workers. In many of these industries workers are exposed to other agents that are known or suspected carcinogens. Thus, for example, miners may be exposed to asbestos, arsenic, chromium, nickel, or radon gas and its decay products. Foundry and metallurgical workers are frequently exposed to polycyclic aromatic hydrocarbons, heavy metals, and formaldehyde. In addition to possible confounding due to concurrent exposures, many studies suffer from lack of information on smoking behavior and data on the amount of dust to which workers were exposed and the silica content of that dust.

One industry in which there is less potential for confounding exposure is the glass and ceramics manufacturing industry. The IARC working group reviewed a proportionate mortality study, a case-control study, and a cohort study among workers in this industry (10). All three studies showed a significant increase in the risk of lung cancer among the exposed individuals. The case-control study, in particular, provides strong support for a causal association in that smoking behavior was accounted for in the analysis of the data and the risk of lung cancer was shown to increase both with duration of exposure and with medical diagnosis of silicosis (20).

The IARC working group also reviewed a number of epidemiologic studies of lung cancer among persons who were diagnosed with silicosis as a result of occupational exposure (10). Most of these studies showed a statistically significant increase in the risk of lung cancer that was associated with a diagnosis of silicosis. The working group cautioned, however, that all these studies may have been biased toward an association by noncomparability of the reference populations. These studies were not controlled for such confounding variables as smoking behavior and other occupational exposures.

In a study published after the IARC review, Amandus et al. (21) described the results of a mortality study among individuals with silicosis in North Carolina. These investigators determined the cause of death for 546 individuals who had been diagnosed with silicosis on the basis of radiography between 1940 and 1983 as part of the North Carolina's pneumoconiosis surveillance program for dusty trades workers. Age-adjusted mortality rates within the cohort were compared to those for the U.S. population at large as well as to referent groups comprising coal miners with coal workers' pneumoconiosis and metal miners without radiographic evidence of silicosis. Mortality from lung cancer was significantly higher than in the U.S. population and in metal miners, even when adjusted for smoking behavior. Significant differences persisted even when workers with possible exposure to other carcinogens in the workplace were eliminated from the cohort. Lung cancer rates among persons with silicosis were elevated compared to that in coal miners with pneumoconiosis, but the difference was not statistically significant. This study, in which confounding by age, smoking behavior, and exposure to other occupational carcinogens was controlled for, provides strong support for a hypothetical association between silicosis and the later development of lung cancer.

A possible causal association between exposure to crystalline silica and the development of autoimmune diseases is supported by a variety of observational evidence. As summarized by Ziskind et al. (14) these include the findings that autoantibodies are frequently present in individuals with silicosis; silicotic nodules contain plasma cells and immunoglobulins; individuals with silicosis have elevated plasma levels of gamma-globulin; and several autoimmune diseases, for example, scleroderma, rheumatoid arthritis, and systemic lupus erythematosus, appear to be more prevalent among individuals with silicosis.

4.1.5 Hygienic Standards of Permissible Exposure

The occupational exposure standard for crystalline silica depends on the crystal type. The PELs for quartz and tripoli are both 0.1 mg/m³ of respirable dust as a 6-hr TWA (7). The PELs for cristobalite and tridymite are half this value (0.05 mg/m³). These standards are the same as the ACGIH TLVs for these materials (3). NIOSH has recommended an exposure level of 0.5 mg/m³ of respirable dust for any form of crystalline silica and considers this material to be a human carcinogen (8).

4.2 Amorphous Silica

4.2.1 Sources and Production

Amorphous silica (CASRN 7631-86-9) occurs naturally in the form of opal and opaline materials, diatomite, chert, and flint. A major portion of commercial amorphous silica is produced synthetically either by vapor-phase hydrolysis of silicon halides (fumed silica, CASRN 69012-64-2), by the reaction of sodium silicate with acid in solution (silica gel, CASRN 112926-00-8), or by the precipitation of fine silica particles from solution (precipitated silica, CASRN 112926-00-8). Silica gel is a coherent three-dimensional network of spherical particles of colloidal silica. A hydrogel is one in which the pores are filled with water, whereas a xerogel is one in which the water has been removed and the gel structure has collapsed.

Diatomite (diatomaceous earth, CASRN 61790-53-2) is a form of amorphous silica that occurs in natural deposits resulting from the accumulation of microscopic forms of animal life (diatoms) on ancient seabeds. It is estimated that 623,000 tons of diatomite were produced in the United States in 1989, continuing a growth trend that began in 1982 (22).

4.2.2 Uses and Industrial Exposures

Amorphous silica is used as a desiccant, adsorbent, reinforcing agent, filler, and catalyst component. Half the world's production is used in filtering applications (dry-cleaning solvents, pharmaceuticals, beer, whiskey, wine, municipal and industrial water, fruit and vegetable juice, etc.). Other important uses include its use as a filler in paint, paper, and scouring powders (10). The use pattern for amorphous silica in 1985 was 67 percent for filter aids, 22 percent for fillers, and 11 percent for other uses (23).

Natural diatomite (diatomaceous earth) typically is relatively free of crystalline silica. The cristobalite content was less than 1 percent in one U.S. processing plant. However, when diatomite is heated to high temperatures during calcining, cristobalite is formed and the cristobalite content of calcined diatomite may be from 10 to 60 percent. Occupational exposures to amorphous silica occur during its mining and processing, and airborne concentrations well in excess of 1 mg/m^3 have been observed in such operations in Sweden and Germany (10). Workers in foundries and metallurgical industries are also exposed to amorphous silica dust. This dust is frequently a mixed dust and may contain high percentages of crystalline silica.

4.2.3 Physiological Response

Silicosis has rarely been observed in workers exposed to relatively pure amorphous silica. Recent studies in experimental animals suggest that the inhalation of amorphous (colloidal) silica produces dose-related pulmonary effects qualitatively similar to the initial inflammatory changes induced by crystalline silica but that much higher concentrations are required to produce the effect and the lesion is rapidly reversible when exposure ceases (18, 24). Studies of rats exposed to similar concentrations of amorphous silica, two different samples of quartz, and cristobalite indicated that cristobalite produced a more severe cellular response than did quartz, which was in turn more active than amorphous silica (25, 16).

The IARC working group reviewed the available information on the possible carcinogenic activity of amorphous silica and concluded that it was inadequate (10). They were unable to identify epidemiologic studies relevant to the human carcinogenesis of amorphous silica. They reviewed a number of studies in experimental animals in which poorly characterized samples of "amorphous" silica were administered by the oral, inhalation, intraperitoneal, subcutaneous, and intrapleural routes. Most of these studies were negative. A study in which mice were exposed to precipitated silica by inhalation resulted in an increased incidence of lung tumors (adenomas and adenocarcinomas) in the exposed mice (21 percent) compared to controls (9 percent). Another study in which amorphous silica was administered to mice by intraperitoneal injection showed an increased incidence of lymphosarcomas of the abdominal cavity. The working group considered these results to be of questionable relevance in that the physical characteristics of the administered silica were not given. If the cancer that results from exposure to silica is secondary to the local tissue response and subsequent fibrosis that develops at the site of deposition, then one might expect that amorphous silica would be much less carcinogenic than crystalline forms.

4.2.4 Hygienic Standards of Permissible Exposure

The OSHA standard for exposure to amorphous silica is 6 mg/m^3 as a TWA (7). The ACGIH TLVs for three forms of amorphous silica, diatomaceous earth, precipitated silica, and silica gel, were all 10 mg/m^3 for dust containing no asbestos and < 1 percent crystalline silica (3). In 1991 the ACGIH issued a notice of intended change to add TLVs for two additional forms of amorphous silica. The proposed TLVs are 2 mg/m^3 TWA respirable dust for silica fume and 0.1 mg/m^3 TWA respirable dust for fused silica. At present NIOSH has not recommended an exposure limit for amorphous silica and is engaged in a review of the potential health effects of this material in the workplace (8).

Table 15.3. Commercially Important Asbestos Materials

Mineral Name	Commercial Mineral Name for asbestos	Mineral Group	Chemical formula
Chrysotile	Chrysotile	Serpentine	$(Mg)_6(OH)_8Si_4O_{10}$ (Fe)
Riebeckite	Crocidolite	Amphibole	$Na_2(Fe)_5(OH)_2Si_8O_{22}$
Anthophyllite	Anthophyllite	Amphibole	$(MG,Fe)_7(OH)_2Si_8O_{22}$
Grunerite	Amosite	Amphibole	$Fe_7(OH)_2Si_8O_{22}$
Actinolite	Actinolite	Amphibole	$Ca_2Fe_5(OH)_2Si_8O_{22}$
Tremolite	Tremolite	Amphibole	$Ca_2Mg_5(OH)_2Si_8O_{22}$

5 NATURALLY OCCURRING SILICATES

5.1 Asbestos

Some silicate minerals occur naturally in the physical form of asbestos. The term "asbestos" comes from the Greek *a-sbestos* meaning inextinguishable (26). This term has been used for several thousand years to describe a wide variety of natural fibrous minerals found in concentrated aggregates or veins amenable to mining and having the properties of high tensile strength, flexibility, and relative thermal and chemical resistance. Although all forms of asbestos are fibrous, not all fibrous minerals are asbestos. Traditionally, the term "asbestos" has been reserved for those fibrous minerals that have the proper combination of physical characteristics and properties that allow for commercial exploitation in the manufacture of thermal-, friction-, and fire-resistant textiles and materials. In 1990 the American Society for Testing Materials (ASTM) (27) adopted the following official definition: "Asbestiform mineral fiber populations generally have the following characteristics when viewed by light microscopy: (1) many particles with aspect ratios ranging from 20:1 to 100:1 or higher (greater than 5 μm in length); (2) very thin fibrils generally less than 0.5 μm in width; and (3) in addition to the mandatory fibrillar crystal growth, two or more of the following attributes: (a) parallel fibers occurring in bundles, (b) fibers displaying splayed ends, (c) matted masses of individual fibers, and (d) fibers showing curvature." There are many minerals that occur naturally in a fibrous form that are not asbestos because they lack the requisite physical properties. For example, nemalite, the fibrous variety of brucite, occurs naturally in fibrous aggregates but does not exhibit either the strength or resilience of asbestos minerals and therefore is not properly classified as asbestos (26).

The commercially important forms of asbestos are all silicate minerals from the serpentine and amphibole mineral groups. These forms are listed in Table 15.3.

The two major mineral groups to which asbestos belongs, serpentine and amphibole, include minerals that usually do not occur in fibrous form. Thus, for example, chrysotile is commonly fibrous whereas two other common serpentines, lizardite and antigorite, are rarely found in fibrous form (26). Furthermore, all the minerals listed in Table 15.3 exhibit polymorphism and occur naturally in both asbestos and nonasbestos forms. The commercial names chrysotile, crocidolite, and amosite are reserved for the asbestos forms of those minerals but the names anthophyllite, actinolite, and tremolite are used for both the asbestos and nonasbestos forms. Although the asbestos forms of anthophyllite and actinolite are fairly common, the asbestos form of tremolite is unusual. Confusion regarding the mineralogical and chemical identity of asbestos materials is increased by the use of commercial terms such as amosite and crocidolite to describe specific mineral types for which mineralogists have a different name. Readers with an interest in the mineralogy and crystal structure of asbestos and other fibrous minerals are referred to the excellent book by Skinner et al. (26).

5.1.1 History

The useful characteristics of asbestos minerals were recognized in ancient times. Deposits of "mineral wool" that could be spun and woven into fabrics were exploited in the Alps by the Romans at the time of Caesar, and Marco Polo observed the weaving of cloth from asbestos in Mongolia (28). In 1829, Chevalier Jean Aldini demonstrated a fire-resistant suit made from asbestos and felt at the Royal Institution in London. A marked increase in the industrial use of asbestos paralleled the development of the steam engine resulting from its use as packing and insulation. By 1892, more than 100 commercial uses for asbestos had been identified (28). Most of the industrial uses of asbestos take advantage of its fibrous character coupled with its thermal, chemical, and friction resistance. In its pure form it has been used as insulation and packing. It has been combined with other fibers for weaving into fire-resistant or heat-resistant textiles. It is incorporated into cement which is then used to manufacture fire- and heat-resistant construction materials. Composite materials containing asbestos have been used to manufacture gaskets and friction products such as brake linings and clutch plates. Asbestos has also been added to roofing, floor tiles, and other building materials to impart strength and fire retardancy. It has been estimated that 30 million tons of asbestos were used in the United States between 1900 and 1980, mostly for insulation and fireproofing (29). Chrysotile is estimated to account for 95 percent of this, but recent surveys indicate that the use of amosite and some crocidolite in buildings was more common than this figure would indicate (30).

With the increased commercial use of asbestos in the last decades of the nineteenth century, adverse health effects began to be noticed in exposed workers in the early decades of the twentieth century. The first report of asbestos-induced respiratory disease is generally attributed to H. Montague Murray in 1906 (28, 31). This report elicited little attention because of the similarity of the described condition to tuberculosis, which was very prevalent among industrial workers at that time. Pulmonary fibrosis resulting from asbestos exposure was not reported again until the 1920s when W. E. Cooke described several cases of the disease and coined the term "asbestosis" (28). In 1931 a set of comprehensive regulations designed to limit occupational exposure to asbestos was adopted in Great Britain as a result of a systematic survey by Cooke and a colleague.

The issue as to when a causal association between asbestos and malignant tumors was identified is controversial. Cases of lung cancer among asbestos-exposed workers were reported in the 1930s but definitive epidemiologic evidence of a cause and effect relationship was not forthcoming until 1955 (28). A possible association between asbestos exposure and mesothelioma was first noted in an article by Wagner et al. (as cited in Reference 28) in 1960.

5.1.2   Uses and Industrial Exposures

The nature of occupational exposures to asbestos has evolved and changed over the past 30 years with increasing recognition of the adverse health consequences of such exposures. Historically, large populations of workers were exposed to asbestos during its mining and milling, during primary manufacture of asbestos-containing materials (ACM), during secondary manufacture of products containing ACM, and from the installation, maintenance, and removal of products containing asbestos. Typical occupational exposures are listed in Table 15.4.

In recent years there has been a dramatic decrease in the use of asbestos in commerce and the nature of occupational exposures has changed considerably. The Environmental Protection Agency (EPA) estimated that the total annual consumption of asbestos in the United States was 240,000 metric tons in 1984, and this dropped to 85,000 metric tons in 1987 (32). Even so, EPA considered this reduction in use as inadequate to prevent adverse health effects of asbestos from the release of asbestos into the ambient environment, and in 1989, issued a final order under Section 6 of the Toxic Substances Control Act (TSCA) to prohibit almost all remaining commercial uses of asbestos in the United States (32). Although this action was taken to reduce environmental exposures to asbestos, it will have the ancillary effect of reducing occupational exposures as well.

The EPA rule provided for a phased discontinuation of the manufacture, importation, and processing of asbestos-containing products. The types of products that are being banned and the effective dates of these prohibitions are shown in Table 15.5.

In effect this ban will greatly reduce, if not totally eliminate, many of the historical occupational exposures to asbestos in the United States, that is, mining, milling, manufacturing, and installation of asbestos and asbestos products. However, significant occupational exposures will continue to occur in the foreseeable future. These exposures are those of construction and maintenance workers involved in demolition, remodeling, or maintenance of buildings in which ACM are present and those workers engaged in asbestos abatement and removal activities. In addition, workers maintaining or dismantling mechanical equipment that contains asbestos materials will also continue to be exposed in the near future.

5.1.3   Analytic Determination

Sampling and analysis of occupational environments for airborne asbestos concentration has been and continues to be problematic. As described below, the adverse health effects of inhaled asbestos appear to be very dependent upon fiber shape and dimension but the exact relationships have not been determined. In most

Table 15.4. Occupational Exposures to Asbestos

Asbestos Mining and Milling

Primary Manufacturing

    Asbestos cement pipe
    Asbestos cement sheet
    Asbestos textiles
    Vinyl/asbestos floor tiles
    Coatings
    Friction products
    Insulation
    Paper
    Gaskets
    Plastics

Secondary Manufacturing

    Asbestos cement sheet
    Asbestos textiles
    Friction products
    Gaskets
    Plastics
    Automobiles

Service and Repair

    Ship repair
    Automobile repair

Construction and Maintenance

    Shipbuilding
    New construction
    Building demolition
    Building renovation
    Insulation installation
    Building maintenance
    Asbestos abatement

---

occupational settings, the distribution of fiber size and shape is heterogeneous and may be highly dependent on the mineral type of the asbestos and the mechanical procedures to which it has been subjected. Furthermore, workplace air is likely to contain nonasbestos particulate matter in both fibrous and nonfibrous form, depending on the nature of other materials handled in the workplace. For these reasons, the proportion of respirable particulate matter that is in the form of biologically active asbestos fibers is variable within a given workplace and especially between different occupational settings. The proportion of respirable dust that is

Table 15.5. Types of Products and the Effective Dates of the Order Removing them from Commerce

Stage 1: Asbestos-containing products whose manufacture, importation, and processing must cease by August 27, 1990 and whose distribution in commerce must cease by August 25, 1992

Flooring felt	Asbestos cement corrugated sheet
Roofing felt	Vinyl/asbestos floor tile
Pipeline wrap	Asbestos clothing
Asbestos cement flat sheet	

Stage 2: Asbestos-containing products whose manufacture, importation, and processing must cease by August 25, 1993 and whose distribution in commerce must cease by August 25, 1994

Beater-add gaskets (except specialty industrial gaskets)
Sheet gaskets (except specialty industrial gaskets)
Clutch facings
Automatic transmission components
Commercial and industrial friction products
Drum brake linings (original equipment market))
Disk brake pads for light- and medium-weight vehicles (original equipment market)

Stage 3: Asbestos-containing products whose manufacture, importation, and processing must cease by August 26, 1996 and whose distribution in commerce must cease by August 25, 1997

Asbestos cement pipe	Roof coatings
Commercial paper	Non-roof coatings
Corrugated paper	Brake blocks
Rollboard	Drum brake linings (aftermarket)
Millboard	Disk brake pads (after market)
Asbestos cement shingle	
Specialty paper	

asbestos may, in fact, be quite small in those occupational settings of greatest current concern, such as construction, demolition, and maintenance work.

The only available analytic method that allows the direct determination and positive identification of asbestos particles in the size range most likely to be responsible for chronic adverse health effects is transmission electron microscopy (TEM). In this procedure, a measured quantity of air is drawn through a membrane filter. The collected particulate matter may then be washed from the filter and redeposited for microscopic examination (indirect method) or the sample may be examined as collected on the original membrane following several manipulations (direct method). Typically, small areas of the total surface area of the sample are selected at random and examined under the electron microscope for the purpose of identifying and counting asbestos fibers. Asbestos fibers can be distinguished from nonasbestos fibers using energy-dispersive X-ray analysis, and different mineral types of asbestos can be distinguished from one another by fiber shape and the use of selected area electron diffraction.

A similar method involves the examination of samples collected on the membrane filter using scanning electron microscopy (SEM). Like the TEM method, SEM allows the visualization of most fibers likely to be of biological significance. SEM has some slight practical advantages in terms of sample preparation and the ability to scan a larger area of the sample more rapidly. However, SEM does not allow the positive identification of fiber and mineral type and is comparable to the TEM method in terms of time and expense.

The equipment and techniques necessary for TEM and SEM analysis have been developed only relatively recently and were considered by OSHA to be too expensive and time consuming for the routine monitoring of occupational environments. The regulatory method adopted by OSHA involves the examination of fibers on the membrane filter using phase contrast optical microscopy (PCOM) (33). This method cannot detect fibers that are less than about 0.25 μm in diameter and cannot distinguish between asbestos and nonasbestos fibers. Basically, all particles meeting the designated size and shape criteria, that is, greater than 5.0 μm in length with an aspect ratio greater than 3 and roughly parallel sides, are counted as asbestos fibers.

Inherent in all the analytic methods for asbestos are numerous sources of both random and systematic error. The relative magnitude of these potential errors increases with decreasing concentration of asbestos and with increasing proportions of nonasbestos particulate matter in the sample. Common to all procedures are problems inherent in sample collection and preparation. The collection efficiency of asbestos fibers from the air on the membrane filter may be influenced by a number of factors including sample volume, flow rate, cross-sectional area of the membrane filter, total particulate concentration in the air, relative humidity, temperature, and surface charge of the particulate matter. Once the sample is deposited on the membrane the membrane filter is removed and manipulated in ways that depend on the ultimate method of visualization. Fibers can be lost or contamination can occur at several steps in this process.

No satisfactory automated method for counting asbestos fibers has been developed for any of the microscopic techniques. Therefore all methods depend on manual counting and are subject to human error. Furthermore, time considerations prohibit the visualization and counting of the entire surface area of filters. All the existing methods involve sampling of a portion of the filter and the counting of fibers in randomly sampled grid squares on resulting microscopic specimens. In those techniques involving electron microscopy, only a very small percentage of the actual membrane filter surface area is evaluated. If this sample is not representative of the concentration of fibers across the entire filter, error is introduced.

Recognizing the potential for error inherent in these analytic methods, both EPA and OSHA provided extremely explicit directions for sample collection, handling, treatment, and analysis in the analytical methods that accompanied their standards (33, 34). These methods specify the size and type of membrane filters to use; the range of acceptable airflow through the filter; the range of total air volume to be sampled; detailed procedures to be followed to collect, transport, handle, and treat membrane filters for analysis; specific methods for selecting and preparing portions of the filter for microscopic analysis; procedures for calibration of the microscope and measuring devices; guidance for selecting the number and location of grid squares to be observed under the microscope; detailed counting rules including size and shape criteria; data analysis and transformation guidance; and specific quality assurance procedures to check accuracy and precision.

5.1.4 Physiological Response

The health effects of occupational exposure to asbestos have been studied throughout this century and this investigation has been quite intense during the past two decades. Despite this, scientific knowledge of the mechanism(s) by which asbestos causes disease is incomplete, and many controversies exist regarding the exact nature and extent of the health risks associated with exposure to asbestos. A detailed description and discussion of all the issues surrounding the possible health effects of asbestos are beyond the scope of this chapter. However, several recent publications have provided extensive overviews of the state of current knowledge of asbestos. These include the preamble to OSHA's final rule for occupational exposure to asbestos, tremolite, anthophyllite, and actinolite (35); a report entitled, "Asbestos in Public and Commercial Buildings: A Literature Review and Synthesis of Current Knowledge" prepared by the Health Effects Institute–Asbestos Research Literature Review Panel and staff (30); the published proceedings of a conference entitled, "The Third Wave of Asbestos Disease: Exposure to Asbestos in Place—Public Health Control" held in New York in June, 1990 (36); and the proceedings of a NATO Advanced Research Workshop on Mechanisms in Fiber Carcinogenesis held in Albuquerque, New Mexico in October, 1990 (37).

There is an enormous body of scientific and medical literature relevant to the physiological effects of asbestos in humans and in experimental animals. The extensive investigation of this class of materials dating back to the 1920s has resulted in a reasonably complete *qualitative* understanding of the effects of inhaled asbestos, although questions remain about the relationship of certain nonrespiratory cancers and alterations in immune function or status to asbestos exposure. Unfortunately, extensive research has failed to reveal the mechanism by which asbestos materials cause benign and malignant respiratory diseases. In vitro experiments have shown that asbestos is not mutagenic and does not initiate the carcinogenic process via the production of reactive chemical species that covalently bind to macromolecules. It has been suggested that asbestos acts as a complete carcinogen in mesothelial cells but only as a promoter in bronchial epithelial cells and that promotion requires that fibers be trapped in the fibrotic lung tissue of asbestosis where it then interacts with classical initiators (38). However, the basis for this hypothesis is speculative and not testable by classical epidemiologic methodology.

The well-characterized physiological effects of asbestos occur in those tissues where it deposits and appear to be highly dependent on the physical form of the material, although chemical composition may also be quite important. There is no evidence of primary systemic toxicity associated with asbestos exposure and little indication that exposure to asbestos by routes other than inhalation results in adverse effects. All the identified effects of respiratory exposure to asbestos are subacute or chronic and exhibit a latent period. Nonmalignant respiratory diseases or conditions that are attributable to asbestos exposure include chronic pulmonary fibrosis (asbestosis) and three pleural diseases: fibrotic pleural plaques, pleuritis, and diffuse pleural thickening (30).

Deposition of asbestos fibers in the alveoli results in an inflammatory reaction (alveolitis). Under conditions of sustained exposure the inflammation may be chronic and result in the development of fibrosis. Asbestosis is diffuse, multifocal interstitial fibrosis resulting from the inhalation of asbestos fibers. Clinically, asbestosis is nonspecific. The most frequent symptom is shortness of breath, usually upon exertion but occasionally at rest. This is often accompanied by a persistent cough, either dry or productive. Fine crackles may be heard posteriorly in the lung bases and in the axillae. Rarely, finger clubbing may be observed.

Radiologically, asbestosis is characterized by irregular linear opacities, predominantly in the lower lobes. Reduced lung function may be present in the absence of clear radiological evidence of asbestosis. Iron- and protein-coated asbestos fibers, known as "asbestos bodies" are usually present in the sputum or in bronchoalveolar lavage fluid from individuals exposed to asbestos and can also be observed in lung tissues at autopsy. Their presence has often been used as the basis for the differential diagnosis of asbestosis. However, analysis of the cores of "asbestos bodies" indicates that such bodies form around other minerals including non-asbestos fibers, and therefore the presence of "asbestos bodies" is not specific for asbestos exposure.

Fibrotic pleural plaques are focal areas of hyaline fibrous thickening. These appear most commonly in the parietal and less commonly in the visceral pleura. The prevalence of these plaques depends on dose and latency, and the latency appears to be in excess of 15 years. Radiological evidence of pleural plaques is usually correlated with impaired lung function and respiratory symptoms. Calcification may be present on occasion. Plaques are not thought to have malignant potential.

Asbestos-related pleuritis occurs in about 5 percent of occupationally exposed workers about 7 to 15 years after first exposure (30). This subacute exudative lesion is reversible but can result in adhesive fibrothorax, causing severe impairment of lung function.

Diffuse pleural thickening (fibrosis) results from three conditions: confluent pleural plaques, extension of lung fibrosis into the subvisceral pleural interstitium, or adhesive fibrothorax from pleuritis. Diffuse pleural thickening is accompanied by reductions in respiratory function similar to those seen in asbestosis. Diffuse pleural thickening may have malignant potential, and the presence of this condition may be a useful surrogate for mesothelioma risk.

Epidemiologic evidence has clearly indicated that occupational exposure to airborne asbestos is causally associated with two neoplastic diseases, lung cancer and mesothelioma. Cohort mortality studies showing an association between mesothelioma and/or lung cancer and asbestos exposure are summarized in Table 15.6.

Taken together, these studies provide compelling evidence that there is an association between exposure to asbestos and both lung cancer and mesothelioma, regardless of fiber type or exposure scenario.

In addition to lung cancer and mesothelioma, there are inconsistent reports of excess incidences or mortality from cancers at other sites among workers exposed to asbestos. These include cancers of the gastrointestinal system (esophagus, stomach, colon, and rectum), laryngeal cancer, kidney cancer, and ovarian cancer (32). Both EPA and OSHA considered the evidence for an association between asbestos exposure and gastrointestinal cancers to be rather compelling (35, 32). For these and cancers at other sites, however, the data were insufficient to derive dose–response relationships. It does appear, however, that the risk of these cancers is significantly lower than those for lung cancer and mesothelioma in similarly exposed cohorts.

When the health effects of asbestos exposure are ranked according to primary causes of death by mortality rate, the malignant diseases are most important with a probable ranking of mesothelioma > lung cancer > asbestosis > diffuse pleural thickening > pleural plaques (75). However, when these diseases are ranked according to frequency of occurrence (prevalence) among workers with 20 or more years of exposure, the order is exactly reversed. Thus the estimated mortality due to mesothelioma among this population of workers is 1200 cases per 100,000, and due to lung cancer, 2500 cases per 100,000. The estimated prevalences of asbestosis and pleural plaques, on the other hand, are 20,000 and 40,000, respectively, per 100,000.

It is well accepted that the risks of both nonmalignant and malignant respiratory diseases are directly proportional to the intensity and duration of asbestos exposure. In addition, it is clear that the potency of asbestos dust for causing these conditions depends on the geometry of the fibers present in the dust, because fiber size and shape play a role in the deposition, retention, and elimination of these materials from the respiratory system. In 1972, Merle Stanton advanced the hypothesis that the ability of inorganic fibers to cause cancer was related to their fibrous form and that the longer and thinner the fiber, the more likely it was to cause mesothelioma (76). More recently, Lippman (77) reviewed the available evidence and concluded that short fibers are relatively free of biologic activity and serve primarily as nuisance dusts. He further postulated that the risk of *chronic fibrosis* is most closely related to the *total surface area* of fibers that are longer than 2 μm and thicker than 0.15 μm. The risk of *lung cancer* is most closely related to the *number* of fibers longer than 10 μm and of similar diameter. The risk of *mesothelioma* is most closely related to the *number* of very thin (< 0.1 μm diameter) fibers greater than 5 μm in length. Experimental evidence in support of this hypothesis has been provided by a number of animal studies. This evidence has most recently been reviewed by Davis (78).

Lippman (77), Davis (78), and others have hypothesized that the pathogenicity of asbestos fibers in the respiratory system also depends on their durability in the lung or pleural tissues. This durability is determined, in part, by the chemical composition and the crystal structure of the asbestos mineral. These investigators attribute the apparently greater potency of amphibole forms of asbestos in causing

Table 15.6. Epidemiology of Lung Cancer and Mesothelioma Mortality Among Asbestos-Exposed Workers[a]

Nature of Exposure	Size of Exposed Cohort[b]	Mortality				Ref.
		Lung Cancer		Mesothelioma		
		Observed/ Expected	Excess Deaths	Pleural	Peritoneal	
CHRYSOTILE						
Manufacture of gas masks, Blackburn, U.K.	570 F	6/4.5	1.5	1	0	39
Asbestos workers	1,261	35/11.1	23.9	0	1	40, 41
Textile plant in South Carolina	2,543	59/29.6	29.4	0	1	42
Asbestos mining and milling in Quebec	11,379	230/184	46.0	10	0	43
Asbestos mining and milling in Quebec	544	25/11.1	13.9	1	0	44
Asbestos friction products plant in Connecticut	3,641	73/49.1	23.9	0	0	45
Asbestos mining in Balangero, Italy	1,058	22/19.9	2.1	2	0	46
Asbestos products manufacturing plant	264	4/4.3	-0.3	0	0	47
PREDOMINANTLY CHRYSOTILE						
Textile plant in Pennsylvania	4,137	53/50.5	2.5	10	4	48
Asbestos friction and packing products plant	c	49/36.1	12.9	4	5	49
Same as above	c F	14/1.7	12.3	1	1	49
Asbestos workers	1,493	33/14.8	18.2	1	8	50
Asbestos textile plant in Rochdale, U.K.	3,356	152/106	46.0	18	0	51
Asbestos cement factory in Cardiff, Wales, U.K.	1,970	22/25.8	-3.8	2	0	52
Asbestos cement factory in Sweden	1,176	9/5.7	3.3	0	0	53
Asbestos cement factory in Tamworth, U.K.	2,167	41/42.5	-1.4	1	0	54
AMOSITE						
Insulation board manufacturing plant, Uxbridge, U.K.	5,969	57/29.1	27.9	4	1	55
Asbestos factory in Paterson, NJ	820	93/21.9	71.1	7	7	56

PREDOMINANTLY CROCIDOLITE					
Manufacture of gas masks, Leyland, U.K.	757 F	13/6.6	6.4	2	0	39
Asbestos mining and milling in western Australia	4,760	91/34.5	56.5	32	1	57
Manufacture of gas masks, Nottingham, U.K.	500 F	10/3.7	6.3	9	3	58
Manufacture of gas masks in Canada	199	7/2.4	4.6	2	7	59
ANTHOPHYLLITE					
Asbestos mining in North Savo, Finland	1,092	21/12.6	8.4	0	0	60
TALC (TREMOLITE)					
Talc mining and milling in New York State	260	12/3.7	8.3	0	1	61
Talc mining and milling in New York State	741	12/5.0	7.0	1	0	62
MIXED FIBER TYPES					
Asbestos cement factory in Paray-Le-Monial, France	1,506	12/5.5	6.5	3	1	63
Asbestos friction products plant	12,571	241/242.5	-1.5	13	0	64
Insulation workers	162	27/5.0	22.0	8	5	65
Asbestos cement factory	241	20/3.3	16.7	6	5	66
Asbestos products	1,074	77/28.4	48.6	5	2	67
Asbestos insulation in Canada	17,800	486/106	380	63	112	68
Asbestos insulation workers in New York and New Jersey	632	93/13.3	79.7	11	27	68
Asbestos insulation workers	152	10/1.4	8.6	1	2	69
Shipyard workers in Hawaii	4,779	35/32.5	2.5	0	0	70
Asbestos products manufacturing	4,255	196/73.9	122.1	38	29	71
Asbestos products manufacturing	694 F	37/5.0	32.0	14	11	71
Asbestos cement factory in Aalborg, Denmark	7,996	162/89.8	72.2	10	1	72
Shipyard workers in Genoa, Italy	2,190	123/54.9	68.1	?	?	73
Two asbestos cement factories in New Orleans	6,931	154/115.5	38.5	4	0	74

[a]Adapted from Reference 29.

[b]F = females; all other cohorts are either all male or predominantly male.

[c]Information is not available.

mesothelioma to the fact that they are much more durable in lung tissue than are chrysotile fibers. These authors also attribute the apparent lack of fibrogenic and carcinogenic activity associated with inhalation of similarly sized fibers of other natural and man-made minerals, such as fiber glass and rock wool, to their relative solubility in lung tissues. In addition to fiber size and durability in tissues, the physicochemical characteristics of the fiber surface, that is, surface charge, crystal lattice defects, and catalytic properties, may play a role in the relative biologic activity of different asbestos and nonasbestos fibers (79).

Another important factor in determining the risks of the health effects of asbestos is concurrent exposure to tobacco smoke. Epidemiologic evidence has clearly shown that the risks of chronic pulmonary fibrosis (asbestosis) and of lung cancer are significantly higher in individuals who smoke than in nonsmoking individuals with similar asbestos exposures (30, 80, 81). The data on possible interactions between cigarette smoke and mesothelioma or benign pleural diseases is not as clear, but it appears that smoking does not increase the risk of these conditions (30, 80). The mechanism by which cigarette smoke increases the risk of fibrosis and lung cancer is not clear but studies in experimental animals have shown that tobacco smoke enhances the penetration of asbestos fibers into the respiratory epithelium, increases epithelial proliferation, and decreases the rate of clearance of asbestos fibers from the respiratory system (82, 83). It has also been hypothesized that embedded asbestos fibers provide a matrix for the deposition of the carcinogenic factors present in tobacco smoke and thus enhance the initiating and promoting activity of these factors (37).

There has been some controversy on the quantitative characteristics of the interaction between asbestos and tobacco smoke in the induction of lung cancer. In the 1970s some investigators held that asbestos did not induce lung cancer in the absence of smoking, but more recent epidemiologic evidence has indicated that there may be an increased risk for lung cancer among nonsmokers who are heavily exposed to asbestos (30). Whether or not the combined risks of lung cancer from asbestos exposure and smoking are additive or multiplicative has also been controversial. A review by Berry et al. (84) indicated that the available epidemiologic evidence gave a wide range of estimates for the relative interaction of asbestos with cigarette smoke and that an additive relationship could not be ruled out. Both the Health Effects Institute–Asbestos Research (HEI–AR) (30) and OSHA (35) consider the weight of the available evidence as supporting a multiplicative model for the interaction of these factors.

The available epidemiology studies of the association of asbestos exposure with various respiratory diseases do not provide a definitive basis for estimating quantitative dose–response relationships for these conditions in humans nor do they confirm the relationship of pathogenic potency to fiber type and fiber dimensions. Many of the studies summarized in Table 15.4 contain no information on intensity and duration of exposure. Various authors and groups have reviewed these studies and identified those studies with sufficient exposure information to allow the estimation of quantitative exposure–response relationships. Thus, for example, Nicholson (85) identified 14 studies of the association of lung cancer with asbestos exposure with sufficient exposure information to estimate a dose–response relationship. Calculation of the unit risk for lung cancer resulted in a very broad range of estimates (0.01 to 6.7 percent increase per fiber-year per milliliter). Liddell and Hanley (86) considered only eight epidemiologic studies to be adequate for the purpose of estimating lung cancer risks whereas Doll and Peto (1985, as cited in Reference 30) found only three studies meeting their criteria. Studies useful for estimating the risks of mesothelioma from asbestos exposure are even more rare. Nicholson (85) identified only four studies that were suitable for this purpose. Estimates of the unit risk for mesothelioma derived from these studies varied by more than an order of magnitude.

Some of the variability in estimated unit risks of lung cancer or mesothelioma derived from these studies is attributable to uncertainties in the exposure data in the studies themselves. Often such exposure information consists only of the dates and duration of employment of subjects in areas where they had exposure to asbestos. Asbestos concentrations to which workers were exposed were estimated from measurements made long after the highest and most critical exposures were experienced. In those few studies where historical air monitoring data are available, the concentrations of airborne asbestos were determined by different methods over the years and conversion of these measurements to standard units of exposure is problematic. Prior to about 1965, almost all data on airborne asbestos concentrations were generated by methods in which particulate matter was collected on a filter that was then incinerated (ashed). Total remaining particulates were determined either by weighing or by counting particles with a light microscope. Such methods did not distinguish between fibers and other particles nor between asbestos and other relatively refractory particulate matter. Particle counting using phase contrast light microscopy was developed in the 1960s, and airborne asbestos concentrations began to be reported as fibers per unit volume (usually cm^3), with a fiber being defined as any particle with roughly parallel sides more than 5 μm long having a length to diameter ratio of 3:1 or larger. It was not until the 1970s that sample collection and handling procedures were essentially standardized. The 1970s also saw the development of new fiber counting methods involving transmission electron microscopy or scanning electron microscopy that allowed better visualization of small-diameter fibers (less than 0.25 μm in diameter) and definitive identification of asbestos materials, but cost and time considerations have precluded the adoption of these methods for the routine analysis of asbestos in air samples.

It is difficult, if not impossible, to convert airborne asbestos concentrations determined by one method to those determined by another. This is particularly true for those early measurements in which total particulate matter was determined. The proportion of respirable, fibrous asbestos to total particulate matter in a sample was certainly highly variable and depended to a great degree on the source of the raw asbestos material, the mechanical operations to which it was subjected, and the presence of other materials or operations that might serve as source for fibrous and nonfibrous particulate matter.

As indicated above, it is widely held that there are inherent differences among various types of asbestos in their ability to cause respiratory diseases including lung cancer and mesothelioma. Liddell (80) reviewed the epidemiologic evidence and concluded that the proportional mortality from mesothelioma in 33 male cohorts varied more between fiber types than within fiber types, with amphiboles being much more active than serpentine, and that this difference could not be due to chance. In fact, the highest proportional mortality for mesothelioma among workers exposed to chrysotile was lower than the lowest proportional mortality among workers exposed to amosite. This difference is so great that one investigator has argued that pure chrysotile does not cause mesothelioma at all, the few cases of mesothelioma seen in chrysotile-exposed workers being due to the presence of very small quantities of tremolite fibers in the chrysotile (87).

The role of fiber type in determining the potency of asbestos for inducing lung cancer is less clear and more controversial. It appears that the mortality from lung cancer is much lower among chrysotile miners and millers than among workers exposed to amphibole forms of asbestos (80). However, lung cancer mortality among cohorts predominantly exposed to chrysotile in the manufacture of textiles and in insulation work is comparable to that among cohorts exposed to amphiboles. In fact, lung cancer mortality among workers at a textile plant in Charleston, South Carolina who were exposed to chrysotile asbestos was 35 times greater than lung cancer mortality among workers with comparable exposures at the mine in Canada where the asbestos was produced (80). These results suggest that the same fiber type of asbestos may exhibit different intrinsic activity depending on how it is handled and treated. Because existing analytic methods measure the concentration only of asbestos fibers greater than 0.2 μm in diameter, it is possible that asbestos dust in textile mills and from insulation materials contains a different proportion of very thin fibers than does asbestos dust at the mine.

In its review of the literature on the risks of cancer associated with occupational exposure to asbestos, OSHA concluded that the evidence that different fiber types of asbestos had different carcinogenic potency was less than definitive and established a single PEL for all forms of asbestos (35). The effective regulatory definition of asbestos as adopted by OSHA in 1986 is fibers of chrysotile, crocidolite, amosite, tremolite, anthophyllite, and actinolite having an aspect ratio of at least 3:1 and a minimum length of 5 μm (35).

The lack of definitive information regarding the exact mechanism of action of asbestos, together with limitations inherent in the epidemiologic investigations of asbestos-related diseases, has resulted in a great deal of uncertainty and controversy regarding *quantitative* relationships between asbestos exposure and the risks of various adverse health effects. There is general agreement among scientists that the risk of lung cancer is linearly related to cumulative asbestos exposure within the range of epidemiologic observation. However, there is uncertainly about the shape of the dose–response curve at exposures below the observed range. If asbestosis is a necessary precondition for the later development of asbestos-induced lung cancer, and there is a threshold for the development of asbestosis, then the dose–response curve for asbestos-induced lung cancer would also be expected to indicate a threshold (37). Both OSHA and the HEI-AR panel rejected this hypothesis for lack of evidence and adopted similar models for estimating the risks of cancer at exposure levels below the observed range. The model for lung cancer is based on the available epidemiologic data indicating that the increase in the relative risk of lung cancer is proportional to cumulative asbestos exposure. Mathematically,

$$R_L = R_E[1 + (K_L \times f \times d)]$$

where R_L = lung cancer mortality
R_E = expected mortality in the absence of exposure
f = intensity of exposure in fibers/cm^3
d = duration of the exposure in years
K_L = the proportionality constant (slope of the dose–response curve)

Actual and expected mortality from lung cancer can be replaced by incidences if such data are available. In the OSHA rule making the cumulative exposure ($f \times d$) was calculated using a delay of 10 years from initial exposure to account for a minimum latency of 10 years for the onset of cancer.

OSHA relied on data from eight studies of lung cancer mortality to calculate values for K_L (67, 88, 66, 51, 40, 41, 89, 56, 68). These studies all involved cohorts of workers exposed to asbestos in settings other than mining and milling. The values of K_L ranged from 0.0006 (89) to 0.048 (66), with an arithmetic mean of 0.019 and a geometric mean of 0.01. OSHA accepted the geometric mean as the best estimate of K_L.

The model used by OSHA to calculate the dose–response relationship for mesothelioma (AR$_M$) is based on the assumption that the absolute risk of mesothelioma (AR$_M$) is proportional to the amount of the exposure to the third power of time since it occurred (35). Mathematically,

$$AR_M = f \times K_M[(t - 10)^3 - (t - 10 - d)^3] \quad \text{for } t > 10 + d$$

$$AR_M = f \times K_M(t - 10)^3 \quad \text{for } 10 + d > t > 10$$

$$AR_M = 0 \quad \text{for } 10 > t$$

This is a more specific case of the same model adopted by the HEI-AR panel (30). OSHA calculated values of K_M from six studies (40, 51, 56, 66, 68, 88). These values ranged from 7.0×10^{-10} to 1.2×10^{-7}. OSHA derived its "best estimate" of the value of K_M (1.0×10^{-8}) from these values.

The interrelationship of asbestosis and lung cancer is not clear. As indicated by the Lippman hypothesis, fibers of the same diameter appear to be involved in the etiology of both conditions. Epidemiologic evidence indicates that the risk of lung cancer is considerably higher in those workers with radiographic evidence of asbestosis, but there is no proof that asbestosis is a necessary precursor for lung cancer or that asbestosis progresses to lung cancer (80). Furthermore, the proportion of individuals with asbestosis who eventually develop lung cancer is small. The available quantitative epidemiologic information on lung cancer and the experimental dose–response information on asbestos-induced pulmonary fibrosis suggests that the former relationship is linear even at low exposures, whereas the latter suggests a threshold below which there is little or no risk of asbestosis. If these conclusions are valid, there may be some risk of lung cancer at exposures that are unlikely to cause asbestosis.

5.1.5   Hygienic Standards of Permissible Exposure

The recent history of the TLV for asbestos as recommended by the ACGIH reflects the controversy regarding the "toxicity" of various mineralogical forms of the material. In 1991 the ACGIH had adopted three different exposure limits for four different types of asbestos (3). While all forms of asbestos were designated as "confirmed human carcinogens," the 8-hr TWA TLVs were 0.2 fibers/cm^3 for crocidolite, 0.5 fibers/cm^3 for amosite, and 2 fibers/cm^3 for chrysotile and all other forms. However, in the 1991 edition of the TLVs, ACGIH published a notice of intended change to a TLV of 0.2 fibers/cm^3 for all forms of asbestos. The current OSHA PEL for asbestos is 0.2 fibers/cm^3 as a TWA, with asbestos being defined as chrysotile, amosite, crocidolite, tremolite asbestos, anthophyllite asbestos, and actinolite asbestos (7). NIOSH has recommended a limit of 0.1 fibers/cm^3 as determined by collection of a sample of at least 400 liters and analyzed using NIOSH method 7400 (8).

5.2   Other Naturally Occurring Silicate Minerals

A number of silicate or aluminosilicate minerals are commercially important in their own right and/or as starting materials for manufacturing other compounds or materials. These minerals encompass a wide range of chemical compositions and display a variety of physical properties. Thus, for example, the materials discussed below range from minerals that occur naturally as loose fibrous networks, for example, erionite, to dense cohesive clays, for example, kaolin and attapulgite. Occupational health interest in these minerals stems from two concerns. First, they may be toxic in their own right. Second, some of them may occur naturally in association with other silicon-containing minerals that are toxic, such as crystalline silica or asbestos. For the second reason, many epidemiologic studies of workers who are occupationally exposed to these minerals are difficult to interpret because observed health effects may be attributable to any or all of the components of the mixed exposure. For example, the observation of increased incidences of malignant and nonmalignant respiratory diseases among workers exposed to talc may be attributable to asbestos present in the talc or to talc itself.

5.2.1   Erionite

Erionite is a natural zeolite consisting of aluminosilicate tetrahedra occurring in finely fibrous or wool-like form. The name derives from the Greek word for wool. When ground, erionite particles resemble amphibole asbestos fibers morphologically.
   Worldwide production of natural zeolites increased dramatically in 1989 and was

estimated to be 250,000 tons (90), of which 12,000 tons was mined in the United States. Up until 1990, erionite was mined at two locations in the United States but these operations have been stopped. No known production of erionite takes place in the United States at present.

Natural zeolites have a number of commercial uses based on their ability to adsorb molecules selectively from air or liquids. They are used in waste-water treatment, in odor control products, and in cat litter. Historically, erionite was used as a metal-impregnated catalyst in a hydrocarbon cracking process. At present, however, because of health concerns described below, there are no known commercial uses of erionite in the United States.

Erionite is of particular occupational health interest because of experimental and epidemiologic evidence that it is similar to asbestos in its ability to induce mesothelioma and, perhaps, lung cancer. Groups of 40 (20 per sex) Fischer 344 rats were exposed to either fibrous erionite, a synthetic nonfibrous zeolite having the same composition as erionite, or crocidolite asbestos by inhalation for 7 hr/day, 5 days/week for 1 year (91). Pleural mesotheliomas were found in 27 of 28 rats exposed to fibrous erionite that survived for at least 12 months. One pulmonary adenocarcinoma and one mesothelioma were observed in rats exposed to the non-fibrous synthetic erionite and one pulmonary squamous-cell carcinoma was observed in the rats exposed to crocidolite asbestos. In other studies in which erionite was administered to rats and mice by intrapleural or intraperitoneal administration, high incidences of mesotheliomas were produced (10).

Pulmonary and pleural fibrosis have been observed in humans living in areas where there are natural deposits of fibrous erionite (10). High mortality due to malignant pleural mesothelioma was noted in three Turkish villages where there was natural contamination of the environment by erionite. Erionite had been used in construction. Descriptive studies conducted in these villages strongly suggest that the incidences of mesothelioma and nonmalignant radiographic changes in the lung and pleura among residents are correlated with exposure to erionite. The IARC working group concluded that there was sufficient animal and human evidence of the carcinogenicity of erionite (10).

Erionite is a striking illustration that the ability to cause mesothelioma and nonmalignant pulmonary and pleural fibroses is not confined to minerals of the serpentine and amphibole groups, that is, asbestos. Clearly the physical form of the fibrous material is as important in determining its biologic activity as its chemical and mineralogical characteristics.

5.2.2 Wollastonite

Wollastonite is a naturally occurring calcium silicate that typically occurs in deposits with other silicate minerals. When crushed it tends to cleave into particles that have length to diameter ratios of 7 or 8 to 1. Fibrous forms of wollastonite are not uncommon.

The largest commercially exploited natural deposits of wollastonite are in the United States and Finland. Significant commercial production began in the United

States in about 1950. The United States is currently the largest producer. It is estimated that 83,000 tons was produced in the United States in 1983 (10). There has been a dramatic increase in use (>10 percent per year) in recent years, with production estimated to be 180,000 tons in 1988 (92).

Wollastonite was first mined for the production of mineral wool. The most important use at present is in ceramics, accounting for more than half of the consumption. Ceramic materials may include up to 70 percent wollastonite. It is also used as an extender in paints and coatings and as a filler in plastics. Some of the recent increase in use is attributable to its increasing importance as a replacement for asbestos. It is combined with binders, fillers, and organic fibers to make heat containment panels, ceiling and floor tiles, brake linings, and high-temperature appliances.

Occupational exposures to wollastonite involve significant exposure to fibers (10). Fiber concentrations ranging from 1 to 45 fibers/cm^3 have been measured in air at a Finnish quarrying operation and concentrations between 8 and 37 fibers/cm^3 were measured in the flotation and bagging plant. Fiber concentrations in the air in a mill in the United States ranged from 0.8 to 48 fibers/cm^3.

Very little relevant information is available on the potential health effects of wollastonite. Intrapleural administration of wollastonite to rats resulted in a significant increase in pleural sarcomas when the implanted material contained fibers >4 μm in length and <0.5 μm in diameter (76). Mild changes characteristic of pneumoconiosis and pleural thickening have been seen in some workers exposed to wollastonite at facilities in Finland and the United States (10). In one small cohort mortality study of workers at a Finnish quarry there was no indication of increased cancer mortality. In view of the apparent increase in the use of wollastonite as an asbestos replacement, it would appear that much more research is needed regarding its potential health effects.

5.2.3 Talc, Soapstone, and Pyrophyllite

Talc is a soft, hydrous magnesium silicate with the molecular formula, $Mg_3Si_4O_{10}(OH)_2$. The basic crystal structure of talc consists of stacks of double sheets of tetrahedral SiO_4 units bonded together by magnesium atoms. There are no covalent bonds holding the layers together, so that the individual sheets slip easily along cleavage planes; this accounts for many of the macroscopic physical characteristics of this material, including the ease with which it is pulverized and the "slippery" character of the resulting powder (10). Talc is a metamorphic material that occurs naturally in deposits of varying purity. It occurs naturally in different physical forms in which individual particles may be fibrous, platy, acicular, laminar, or granular. Talc is often associated with other mineral types including other natural silicates. Of most interest from a toxicologic perspective is the possibility that natural talc may contain crystalline silica (quartz) and fibrous forms of tremolite, anthophyllite, or chrysotile (asbestos). The term "soapstone" is used to designate a rocklike form of talc deposit usually containing up to 50 percent impurities. Another form of natural talc deposit is steatite, a mixture of talc, clay,

Table 15.7. Uses of Talc as a Percentage of Total Consumption

Use	Percent
Ceramics	34
Paint and coatings	14
Paper	13
Roofing materials	11
Plastics	10
Cosmetics	5
Other	13

and alkaline earth oxides. Pyrophyllite is an anhydrous aluminum silicate that is similar to talc in appearance and physical characteristics.

An estimated 1.3 million tons of talc was produced in the United States during 1989, with 86 percent being mined in Montana, New York, Texas, and Vermont (93). The use pattern for this material is shown in Table 15.7.

It is estimated that 128,000 tons of pyrophyllite were used in the United States in 1989 (94). The use pattern for this material is shown in Table 15.8.

Occupational exposure to talc and pyrophyllite occurs by inhalation of dust during mining, milling, and most uses. Because most commercial forms are rarely pure minerals, the composition of the inhaled material is generally a mixture of mineral dusts. Geometric mean respirable dust concentrations were 0.14, 0.45, and 0.66 mg/m^3 at talc mines in North Carolina, Texas, and Montana, respectively (10). Mean respirable dust concentrations in the milling operations at these three locations were 0.26, 1.1, and 1.56 mg/m^3. Examination of bulk samples from these three mines revealed no fibrous material in the talc from Montana. The talc from Texas contained fibrous tremolite and antigorite whereas that from North Carolina contained cleavage fragments with high aspect ratios and diameters as small as 0.1 μm or less, but these were not asbestos.

The potential adverse health effects of talc and similar minerals depend on the properties of the talc mineral. The presence of crystalline silica or asbestos increases the potential hazard of exposure to these materials. Pneumoconiosis has been described as being associated with the intensity and duration of exposure to talc dust in mining and milling operations in the United States and Italy (10). The nature of the pneumoconiosis is highly variable, ranging from mild asymptomatic types to disabling conglomerate pneumoconiosis. The severity of the disease is

Table 15.8. Uses of Pyrophyllite as a Percentage of Total Consumption

Use	Percent
Ceramics	55
Refractories	14
Insecticides	7
Plastics	3
Other	21

related to the content of other dusts in the mixed dust exposure, and silicosis and asbestosis are frequently seen in talc miners as well. In one study, in which workers were exposed to talc that was relatively free of silica and asbestos, no radiographic abnormalities were seen although exposed workers did report an increased number of respiratory symptoms and there was a decrement in ventilatory function among the exposed workers. It appears, therefore, that long-term high-level exposure to relatively pure talc may be associated with respiratory symptoms and the development of simple pneumoconiosis but that severe progressive pneumoconiosis occurs only when there is concomitant exposure to other fibrogenic dusts such as silica and/or asbestos.

Two epidemiologic studies of workers who were exposed to talc containing fibrous tremolite and anthophyllite at a production facility in New York State revealed a significant excess mortality from lung and pleural cancers (10). One death from peritoneal mesothelioma was also observed in the cohort. On the other hand, several mortality studies conducted among workers exposed to talc that contained no more than trace amounts of asbestos indicated no excess risk of cancer. The IARC working group reviewed the evidence available and concluded that there was inadequate evidence both from animal studies and from studies of exposed humans for the carcinogenicity of talc that did not contain asbestos. However, the working group did consider that the evidence was sufficient that asbestos-containing talc was carcinogenic to humans.

The OSHA PEL for talc containing <1% quartz and no asbestos is 2 mg/m^3 TWA of the respirable dust (7). This is the same as the NIOSH recommended exposure limit (REL) (8). For soapstone the OSHA PEL and NIOSH REL is 6 mg/m^3 TWA for total dust or 3 mg/m^3 for the respirable dust. The ACGIH TLVs for these materials are the same as the OSHA PELs and NIOSH RELs (3). No separate standards or guidelines exist for pyrophyllite.

5.2.4 Mica

Mica is a nonfibrous, naturally occurring silicate, found in plate form in nine different species. These materials are hydrous silicates with the predominant minerals of commerce being muscovite, a hydrous aluminosilicate, and phlogopite, a magnesium silicate (95). Sheet forms of mica in the form of muscovite have historically been mined by hand from pegmatites. However, in the past decade demand for sheet forms of mica have dropped dramatically as the electronics industry has switched to synthetic quartz to meet most of the traditional uses of this material. Most modern uses of mica involve ground material and almost all of this is derived as a by-product from the mining of lithium, feldspar, or kaolin. The state of North Carolina accounted for 75 percent of the domestic production of mica in 1988.

Major uses of mica in 1988 were wallboard joint cement, paint, rubber, and oil well drilling fluids. These uses totaled 95,000 tons (95).

The OSHA PEL for mica dust in air is 3 mg/m^3 TWA of respirable dust containing less than 1 percent quartz (7). This is the same as the NIOSH REL and the ACGIH TLV for this material (3, 8).

5.2.5 Attapulgite and Sepiolite

These two materials are closely related and are included in the category of hormitic clays. Sepiolite is a magnesium silicate whereas attapulgite is a magnesium aluminum silicate. Both have structures that are similar to those of minerals of the amphibole group. This structure results in long, thin crystals that are similar to chrysotile asbestos fibrils (10). Attapulgite occurs in large deposits in the southeastern United States. The term "fuller's earth" has been used to describe commercially mined absorbent clays in the United States and most of this material is attapulgite. A particularly pure form of sepiolite mined in Europe and the Middle East is known as "meerschaum" and has been used historically for the carving of pipes and cigarette holders.

Worldwide attapulgite production in 1983 was estimated to be about 1.1 million tons, of which 84 percent came from the United States (10). Sepiolite production was less than half of this. Spain accounts for more than 90 percent of worldwide sepiolite production. The primary use of both attapulgite and sepiolite is as animal waste absorbents (cat litter). Other important uses of attapulgite in the United States are as a component of drilling muds, as oil and grease absorbents, and in fertilizer and pesticide formulations.

The results of long-term surveillance of workers at two sites in the United States where attapulgite was mined and milled indicated that there was an increased prevalence of pneumoconiosis and that the incidence increased with age and with duration of exposure (10). A decrease in pulmonary function was associated with total cumulative exposure to respiratory dust in the workers at one of these facilities.

The evidence relevant to the possible carcinogenic effects of attapulgite was reviewed by the IARC working group (10). Studies in which attapulgite was administered to rats by either intraperitoneal or intrapleural administration indicated that attapulgite containing a significant number of fibers >5 μm in length produced mesotheliomas and sarcomas. A single epidemiologic study of miners and millers exposed to high concentrations of attapulgite dust for long duration indicated that there was increased mortality from lung cancer, but no information on smoking behavior was determined. The working group concluded that there was limited evidence that attapulgite was carcinogenic in experimental animals but that the human evidence was inadequate to support a conclusion.

Little information is available regarding the potential effects of sepiolite. A limited study of workers and residents in a village in Turkey who were exposed to sepiolite during mining and trimming indicated that exposed individuals did have clinical and radiological evidence of pulmonary fibrosis but no cases of mesothelioma or other pleural diseases were observed. The IARC working group concluded that the animal evidence was inadequate and that there was no human evidence available to evaluate the potential carcinogenicity of sepiolite (10).

No exposure standards or guidelines have been developed for attapulgite or sepiolite by OSHA, NIOSH, or ACGIH (3, 7, 8).

5.2.6 Kaolin

Kaolin is a hydrous aluminosilicate mineral that is found in large natural deposits of kaolinite in Georgia, South Carolina, and Texas (96). A typical kaolin contains 38.5 percent by weight aluminum oxide, 45.5 percent silicon dioxide, 13.9 percent water, and 1.5 percent titanium dioxide, with small amounts of calcium oxide, magnesium oxide, and iron oxide. A single crystal consists of a layer of silicon dioxide that is covalently bonded to a layer of aluminum oxide. When the clay is processed by centrifugal classification it can be separated into fractions consisting of thin hexagonal plates (< 2 μm in diameter) and particles greater than 2 μm in diameter consisting of stacks of the hexagonal plates.

Kaolin, as mined, contains other minerals including quartz, muscovite, and altered feldspars (97). The purification process removes much of the crystalline silica so that commercial products typically contain less than 3 percent crystalline silica and the respirable dust contains less than 1 percent. On the other hand, if kaolin is calcined some of it may be converted into cristobalite.

Domestic production of kaolin was estimated to be 8.6 million tons in 1988 (98). More than 80 percent of this was produced in Georgia. A major use of kaolin is as a filler and pigment in the manufacture of coated paper. Kaolin is also used as an extender and pigment in paints. It is also used in ceramics, rubber, thermosetting resins, and adhesives.

The health effects of exposure to kaolin dust by inhalation have not been adequately studied. Historically, reports of respiratory diseases among workers exposed to kaolin were attributed to possible contamination of the kaolin by crystalline silica. Prior to 1991 the ACGIH TLV for kaolin was that for a nuisance dust, namely, 10 mg/m^3 TWA. However, in 1991 ACGIH reviewed the available information and issued a notice of intended change (97). The ACGIH cited a number of case reports and epidemiologic studies of workers who were exposed to kaolin during its mining and processing in Georgia. For the most part, Georgia kaolinite contains little or no crystalline silica. In workers who were exposed to kaolin dust during the milling and bagging of kaolin, there was an increased prevalence of pneumoconiosis. The prevalence of pneumoconiosis was correlated with both the intensity and duration of exposure. Pneumoconiosis incidence was not increased in open pit miners, who were exposed to significantly lower dust concentrations than workers involved in milling, bagging, and loading. As with many epidemiologic studies of the effects of respiratory particulate matter, quantitative data on past exposures for these workers were of poor quality but exposure levels in the past were unquestionably very high.

There are no reports that suggest that workers exposed to kaolin free of silica have an excess risk of malignant respiratory diseases. The carcinogenic potential of kaolin has not been systematically studied in either experimental animals or exposed workers, however. Based on the available evidence that kaolin induces pneumoconiosis, ACGIH has issued a notice that it intends to revise the TLV for kaolin downward to 2 mg/m^3 of respirable dust (97).

5.2.7 Perlite

Perlite is a natural glass formed by volcanic action. It is a sodium potassium aluminosilicate having an amorphous structure. It possesses an unusual physical characteristic in that it expands to about 20 times its original volume when heated to
temperatures within the softening range, that is, somewhere between 1400 and
2000°C. Expanded perlite is either a fluffy, highly porous substance or a glassy-
white particulate with low porosity, depending on how the material is heated. The
bulk material has a density of between 3 and 20 lb/ft^3.

The crystalline silica content of 16 samples of perlite ore collected from 19
deposits in 16 western states was quite low, that is, < 2 percent in 15 and 3 percent
in one sample (99). The crystalline silica content of typical expanded perlite ranged
from 0 to 2 percent (100).

An estimated 517,000 tons of perlite was produced in the United States during
1988 (101). Most uses of perlite involve the expanded form. It is used in abrasives,
acoustical plaster and tile, charcoal barbecue base, cleanser base, concrete aggregates, filter aids, fertilizer, metal foundries, insulation, and refractory products. It
is also used as a filler in numerous materials. Incorporation into construction
materials accounts for about 70 percent of the domestic use of perlite.

The ACGIH TLV for perlite is 10 mg/m^3 for dust containing less than 1 percent
crystalline silica and no asbestos (3).

5.2.8 Pumice and Pumicite

Domestic production of pumice and pumicite was estimated to be 400,000 tons in
1988 (102). About 70 percent of this came from mines in New Mexico and Idaho.
Concrete admixtures and building blocks are the major uses for this material accounting for 86 percent of the total production. A very finely ground form of
pumice known as micronized pumice is used as an abrasive in toothpaste, polishes,
and soaps. There is very little published information on potential occupational
exposures to, and adverse health effects of, pumice. There are no hygienic standards
or guidelines for controlling occupational exposure to these compounds. By default,
pumice and pumicite containing no asbestos and less than 1 percent crystalline
silica are considered to be inert or nuisance dusts for which the OSHA PEL is 15
mg/m^3 total dust or 5 mg/m^3 respirable dust TWA (7).

5.2.9 Kyanite

Kyanite, andalusite, and sillimanite are anhydrous aluminosilicate minerals that
are closely related to several other aluminum silicate minerals such as topaz. Calcination of this mineral produces a refractory material that can be used in the
manufacturing of high-performance, high-alumina refractories. It was estimated
that in 1987, 90 percent of the kyanite produced in the United States was used in
refractories, 55 percent of this for smelting and processing iron, 20 percent for
smelting and processing nonferrous metals, and 15 percent for refractories in glass
making and ceramics (103). As with pumice and pumicite little is known about the

occupational health and industrial hygiene aspects of this class of materials. There are no standards or guidelines for limiting occupational exposure to this material. By default air concentrations are limited to those for inert or nuisance dusts according to OSHA policy (7).

6 PORTLAND CEMENT

Although not a naturally occurring silicate mineral, portland cement resembles this family of minerals in its physical and chemical characteristics. Portland cement is the most common form of cement used throughout the world and was named because of its resemblance to a well-known English building stone from the Isle of Portland. It is manufactured by blending lime, alumina, silica, and iron oxide as tetracalcium aluminoferrate (with the theoretical formula $4CaO \cdot Al_2O_3 \cdot Fe_2O_3$), tricalcium aluminate ($3CaO \cdot Al_2O_3$), tricalcium silicate ($3CaO \cdot SiO_2$), and dicalcium silicate ($2CaO \cdot SiO_2$). Small amounts of magnesia (MgO), Na, K, and S are also present. Sand is added to make concrete. Modern cements may be augmented with a variety of natural or synthetic additives to impart specific physical properties. Natural or synthetic fibers may be added to improve strength or thermal resistance. Polymeric materials such as epoxy resins may be added to impart strength, flexibility, improved curing properties, or moisture resistance.

Portland cement is produced by grinding and mixing the starting materials and calcining this mixture in rotary kilns at about 1400°C. The cooled clinker that is formed is ground and mixed with additives such as gypsum to form the final cement. The cement may then be mixed with sand or gravel to make concrete. The quartz content of cement is usually less than 1 percent.

Because most portland cement is used in the construction of buildings, consumption parallels building trends. The total consumption of cement in the United States in 1988 was 84 million tons (104). This was similar to the consumption in the previous 2 years and considerably higher than the approximately 65 million tons consumed in 1982, a recession year. Occupational exposure to portland cement occurs during its manufacture and use and is limited primarily to inhalation of, or dermal contact with, the dry material. The majority of cement used in construction is mixed with sand or gravel and water at a central plant and trucked in wet form to the construction site. Thus there is little opportunity for exposure of workers at the site to cement dust. Nevertheless, many construction workers may be exposed to cement dust when small amounts are mixed on site or when cement or concrete materials are cut or ground.

The potential adverse health effects of portland cement have not been extensively studied. The available evidence suggests that it has a low degree of toxic hazard. There does not appear to be any evidence that pneumoconiosis is associated with exposure to portland cement dust if that exposure is not accompanied by exposure to other fibrogenic dusts such as crystalline silica. Deposition of portland cement in the eye can result in alkali burns due to the presence of calcium oxide (CaO) in the cement if the material is not washed out (105). The OSHA PEL for exposure to Portland cement dust is 10 mg/m^3 TWA for total dust or 5 mg/m^3 for respirable dust (7). The NIOSH REL is the same (8). The ACGIH TLV is 10 mg/m^3 TWA for dust containing < 1 percent quartz and no asbestos (3).

7 SYNTHETIC SOLUBLE SILICATES

The synthetic, soluble silicates of commercial importance have the general formula

$$M_2O \cdot mSiO_2 \cdot nH_2O$$

where M is sodium, potassium, or lithium and m and n are the number of moles of SiO_2 and H_2O relative to one mole of the alkali metal oxide. Because of the variability of m and n, this class of material includes a broad range of glasses in which the m value ranges from 0.5 to 4.0. The most common form of soluble silicate, sometimes called water glass, has an m value of 3.3. In the anhydrous state these materials are vitreous and crystalline. In water they form viscous, alkaline solutions.

Soluble silicates are typically manufactured in conventional open-hearth glass furnaces. A mixture of soda ash (sodium carbonate) and quartz sand (silica) of relatively high purity is heated to the melting point. The molten glass is then drawn or formed into solid lumps or it may be dissolved at the manufacturing facility. The resulting solutions may be filtered to remove suspended unreacted sand particles. The m value of the resulting silicate is controlled by the relative proportions of soda ash and sand.

Worldwide production of soluble silicates was estimated to be about 3 million tons in 1981 (106). These materials were produced at about 30 locations in the United States. The largest single use for soluble silicates in the United States is as a functional component of soaps and detergents. They are useful for this purpose because of their alkalinity and buffering ability. They also enhance the effectiveness of surfactants, maintain soil-particle dispersion, and form complexes with metal ions. Some of the same characteristics account for the use of these materials in water treatment. Silicates are used in solution to beneficiate minerals by froth flotation. Adhesives are an important application of these materials. Silicates are used to bind sand molds in foundries, and in a number of applications in binding paper such as corrugated paper and laminating foil to paper. Solutions of silicates can be pumped into soil to bind the soil and stabilize it, a technology that has found increasing use as a means of limiting migration of hazardous wastes from contaminated sites. Soluble silicates are also used as binders for weather-resistant roofing materials. An increasingly important use of soluble silicates is in the recovery of oil from depleted wells by flooding the reservoir with water. The presence of silicates in the flooding solution enhances the effectiveness of surfactants and helps prevent mixing of oil and water. Finally, soluble silicates are used as starting materials for the manufacture of silica-based products such as silica gels and pigment grade silica.

Apart from the high alkalinity of some soluble silicates that could result in irritation and burns from direct contact, these materials have a low degree of health hazard. Commercial liquids have an oral LD_{50} of approximately 2500 mg/kg in rodents. Small amounts have been incorporated in feed for dairy cattle without reports of noticeable adverse consequences. Sodium silicate is generally recognized as safe (GRAS) by the Food and Drug Administration for some food uses and in the treatment of potable water. No TLVs or OSHA standards have been developed for these materials.

8 MAN-MADE MINERAL FIBERS

"Man-made mineral fibers" (MMMF) is an operational term encompassing a number of fibrous amorphous silicate materials made from rock, clay, slag, or glass. The unifying characteristic of the group is that they are produced by drawing, centrifuging, or blowing a molten preparation of the mineral material into fibers or filaments. When the materials are drawn through a nozzle (extruded) the result is usually referred to as a filament or a continuous fiber. Materials produced by drawing without a nozzle or by blowing or centrifugation are referred to as fibers or wools.

The category of MMMF contains a number of different materials including glass filaments, glass fibers, glass wool, rock wool, slag wool, and ceramic fibers. The term man-made vitreous fiber (MMVF) is also used to describe some materials in this category, that is, glass fibers and ceramic fibers, but is usually reserved for the latter. Continuous glass filaments are manufactured by drawing molten glass through a nozzle. Glass filaments may be coated with metallic or polymeric materials to impart desirable properties. Because they are extruded through a nozzle, glass filaments usually are relatively uniform and usually range from 6 to 15 μm in diameter. Glass filaments are usually cut to produce fibers of the desired length for specific applications. Insulation wools include glass wool, rock wool, and slag wool, and are produced by centrifuging or blowing molten starting material into an airstream. This results in an interwoven mass of fibrous material in which the fibers are quite variable with respect to length and diameter. Typically, these fibers are thinner than glass filaments with diameters ranging from 2 to 9 μm. Insulation wools may be chemically treated to impart moisture resistance. The term "mineral wool" is commonly used in the United States to describe rock wool and slag wool collectively. In Europe, glass wool is also included in this category.

The composition of MMMF is highly variable. Even within the category of glass fibers the silicon dioxide content of the glass ranges from 34 to 73 percent by weight (107). Other constituents include aluminum oxide (Al_2O_3), lime (CaO), magnesium oxide (MgO), borax, alkaline oxides (Na_2O, K_2O), and lead oxide. Physical and chemical properties such as tensile strength, chemical resistance, thermal conductivity, dielectric strength, and flame resistance are influenced by the composition of the glass. Rock wool and slag wool are prepared by drawing or centrifuging molten rock or slag and the composition of these materials depends on the source

Table 15.9. Man-made Mineral Fibers (MMMF)

Type of Fiber	Typical Uses
Continuous glass filaments	Reinforcement of cement, plastics, paper, and elastomers
	Electrical insulation
Glass, rock, and slag wool	Thermal and acoustic insulation
	Acoustic tile and panels
	Ventilation and air-conditioning ducts
Refractory fibers	High-temperature insulation and packing
Special-purpose fibers	High-performance insulation (aircraft)
	High-efficiency filtration

material. Rock wool is typically made from igneous rocks such as diabase and basalt. These typically contain less than 50 percent SiO_2 and relatively high percentages of lime (CaO). The composition of slag wools is characteristic of the smelting process from which the slag came and is quite variable. These materials also tend to have lower SiO_2 content than do glass fibers and may contain significant amounts of iron, copper, and magnesium.

Refractory fibers are often referred to as ceramic fibers or refractory ceramic fibers (RCF). They are a special class of MMMF that are able to withstand temperatures of 1000°C or more without distortion or softening. Most refractory fibers are made from kaolin or from an approximately equal mixture of alumina and silica but there are many variations. Zirconia, boria, and other metal oxides may be added to impart desired properties for specific high-temperature applications. The fibers are produced by blowing or spinning the molten mineral mixture. The term "refractory fiber" has also been used to describe a number of synthetic fibers other than silicates that are used for high temperature applications. Included, for example, in some descriptions of refractory fibers are silicon carbide whiskers, described in Section 9, below.

Table 15.9 shows the general classes of man-made fibers and their major uses.

8.1 Production

The manufacture of rock wool and slag wool began in the United States at the turn of the century (107). It did not develop significantly, however, until after World War I. Production of glass fibers developed in the 1930s and after World War II the use of glass fiber for thermal insulation became more common; at the same time the use of rock and slag wools for this application decreased. The use of glass fiber insulation has increased in proportion to the decrease in the use of asbestos.

The World Health Organization (WHO) (108) estimated that global production of MMMF was 6 million tons in 1985 and 80 percent of that production was fibrous glass. Approximately 700,000 tons of glass fiber was produced in the United States in 1984 (107). Worldwide production of continuous glass filaments was estimated to be 1384 million kg (approximately 1.5 million tons) in 1984. Worldwide production of ceramic fibers was estimated to be 70 to 90 million kg (0.08 to 0.1 million tons) in the early 1980s and the United States accounted for approximately half of this. The production of ceramic fibers has been increasing in recent years in response to demand for these components in a number of high-temperature applications.

8.2 Uses and Industrial Exposures

About 80 percent of the glass wool, rock wool, and slag wool produced in the United States is used for acoustic and thermal insulation, either as loose material or in batts, blankets, or rolls (108). The remainder is used in the manufacture of reinforced textiles and fabrics.

Occupational exposure to MMMF occurs during manufacture and use. When MMMF are released into the environment, the longer and thicker fibers tend to settle, whereas shorter thinner fibers tend to stay suspended in the air column. Thus the fiber size distribution to which workers are exposed is different from that in the material that is being produced or used. Average air concentrations of glass fibers in plants where glass fiber insulation is manufactured range from 0.01 to 0.05 fibers/cm³ (108). In plants where continuous glass filaments are manufactured concentrations were about 10-fold lower. Fiber concentrations in mineral wool insulation plants were found to be considerably higher, on the order of 0.032 to 0.72 fibers/cm³. A wide range of concentrations (0.0082 to 7.6 fibers/cm³) has been observed in ceramic fiber manufacturing facilities. Airborne fiber concentrations during the installation of insulation tend to be equal to or lower than those encountered in manufacturing except when insulation is blown in. Concentrations of 1.8 and 4.2 fibers/cm³ were measuring during the blowing of fibrous glass and mineral wool insulations, respectively.

The respirability of MMMF is very dependent upon their dimensions and especially their diameter. The upper limit of fiber diameter for respirability in humans appears to be about 3.5 μm. This is slightly higher (5 μm) by mouth breathing. Fibers that are < 5 μm in length are efficiently cleared by alveolar macrophages. Fibers longer than 10 μm are not cleared by alveolar macrophages. Little is known about the durability of MMMF in lung tissue. There appears to be a wide range of solubility.

Exposure to refractory ceramic fibers (RCF) occurs during their manufacture and use. Exposures range from 0.01 to 6.4 fibers/cm³ in manufacturing plants (109). Exposures during processing and end use are in the ranges 0.02 to 56 fibers/cm³ and 0.01 to 25 fibers/cm³, respectively. The EPA estimated that 800 people are exposed to RCF in manufacturing and 31,500 during installation of RCF-containing materials.

8.3 Analytic Determination

The monitoring of workplace atmospheres for MMMF is accomplished in the same manner as for asbestos and is subject to the same limitations and potential errors. Optical and electron microscopy are used to determine the number of fibers per unit volume of sampled air, whereas gravimetry is used to determine the mass of total or respirable particulate matter in the air sample. In all cases the sample is collected by drawing a measured volume of air through a filter mounted in a cassette, preferably in a worker's breathing zone. Gravimetric determination is accomplished by weighing the filter prior to and after sampling, with the difference being the mass of particulate matter in the known volume of sampled air. Fiber counting is usually accomplished using phase contrast optical microscopy (PCOM). Either scanning or transmission electron microscopy allows the detection and counting of thinner fibers than does PCOM. The use of energy-dispersive X-ray analysis in conjunction with transmission electron microscopy allows the analyst to determine the elemental composition of individual fibers.

An important limitation of PCOM for assessing exposure to asbestos is that fibers less than about 0.25 μm in diameter are not detectable. This is not as critical a limitation for assessing concentrations of MMMF fibers because only a very small proportion of commercial MMMF is thinner than this limit. Only certain specialty fibers and RCFs have a significant proportion of particles thinner than 0.25 μm. On the other hand, the use of PCOM has a limitation not shared by asbestos, in that some MMMF have refractive indexes close to that of the filter medium and may be difficult to visualize under the optical microscope.

8.4 Physiological Response

Studies among humans exposed to various MMMF indicate that fibers larger than 4.5 to 5 μm in diameter cause mechanical irritation of the skin as a result of direct contact. These particles penetrate the skin and cause a local erythematous response that can be characterized by itching and burning pain. The reaction often disappears upon continuing exposure. Eye irritation has also been reported. A recent survey of 2654 Danish construction workers who were exposed to MMMF insulation materials indicated a greater frequency of eye, skin, and upper respiratory tract irritation than in construction workers who were not exposed to these materials (110).

Studies in experimental animals *suggest* that MMMF are far less fibrogenic than are asbestos fibers. Furthermore, the fibrosis that is observed is reversible after exposure ceases. However, because there is little information from animal studies on lung burdens, this apparent difference may be due to fewer fibers actually reaching the lung. In a review of the total body of experimental evidence available at the time, WHO (108) found little evidence that glass fibers caused fibrosis in a wide range of animal species when exposed by inhalation to concentrations as high as 100 mg/m^3 for anywhere from 2 days to 24 months. In some of these studies parallel groups were exposed to similar mass concentrations of asbestos fibers. The reaction to glass fibers was consistently less than that to asbestos.

In long-term bioassays in which animals were exposed to glass fibers by inhalation, no statistically significant increase in lung tumors has been observed. In some of these studies a statistically nonsignificant increase in the incidence of lung tumors was observed. In all of these studies exposure of animals to similar mass concentrations of chrysotile asbestos clearly caused lung cancer but exposure to crocidolite asbestos did not. In three studies in which glass wool was administered to rats or hamsters by intratracheal installation, lung tumors were observed in one rat study and lung tumors and mesotheliomas were seen in hamsters (107). The glass wool in these studies had a mean fiber diameter of less than 0.3 μm, which is at the low end of the range of typical fiber diameters in commercial glass wools.

In two studies in which rats were exposed to rock wool by inhalation there was no evidence of a significant increase in lung tumors. Control groups exposed to similar concentrations of chrysotile asbestos had high incidences of lung tumors. Administration of rock wool by intraperitoneal injection led to a high incidence of tumors. In one study where rats were exposed to slag wool by inhalation there was no increase in the incidence of respiratory-tract tumors.

Mesotheliomas have been observed in a number of animal studies in which MMMF, other than glass filaments, were administered by intrapleural or intra-peritoneal installation. The development of mesotheliomas in these studies appears to be correlated with the number of fibers less than 0.25 μm in diameter and more than 8 μm in length.

In their review of all the available data from studies in experimental animals, the IARC working group concluded that there was "sufficient evidence" that glass wool and ceramic fibers caused cancer in animals (107). There was "limited evidence" that rock wool caused cancer and "inadequate evidence" for assessing the carcinogenicity of slag wool and glass filaments in animals.

WHO considered two large historical prospective (cohort) epidemiologic studies to be the most definitive for detecting health effects among workers exposed to MMMF. One of these was a study of 16,661 male workers in 11 fibrous glass and six mineral wool manufacturing plants in the United States (111). The vast majority of the cohort members were employed in the manufacture of fibrous glass. The study subjects had been employed for at least 6 months beginning between 1945 and 1963 and were followed through 1981. Most exposures were to mean concentrations of less than 0.5 fibers/cm^3. There were 4986 deaths in the cohort and cause-specific mortality rates within the cohort were compared with age- and calendar-rate-adjusted mortality for males in the United States or in the county where the plant was located. An associated case-control study was conducted of all workers from these plants who died of nonmalignant respiratory diseases or of respiratory cancer between January 1950 and December 1982. The controls were a random sample of workers from the same plants, stratified by plant and year of birth.

The other large study was conducted within a cohort of 21,967 employees of 13 MMMF (seven rock wool, four glass fiber, and two continuous filament) plants in seven countries in Europe (112). These workers, first employed between 1900 and 1955, were followed through 1981. There were 2719 deaths in the cohort and mortality rates were compared to national rates and adjusted in some cases for local conditions.

In neither of these studies was there evidence of excess mortality attributable to nonmalignant respiratory diseases among the workers who were employed in the manufacture of glass fibers. There was a slight increase in nonmalignant respiratory disease among workers in the mineral wool plants in the United States but this was not statistically significant. There were no trends toward increased mortality from respiratory diseases with time from first exposure, duration of exposure, or estimated cumulative fiber exposure.

The only statistically significant findings with regard to mortality from lung cancer in either of the two studies was increased mortality due to lung cancer among workers exposed to rock or slag wool in the European study. There was a significant increasing trend with time since first exposure, and the risk of lung cancer was greatest among those exposed when there was the least control of respirable fibers in the workplace. However, there was no consistent relationship between lung cancer mortality and duration of exposure or cumulative exposure in this cohort. Neither the U.S. nor the European study showed a significant excess of mortality from lung cancer among workers exposed to glass wool or glass filaments. In the case-control study of workers with lung cancer in the United States, there was no association between the incidence of cancer and cumulative exposure to MMMF.

These two studies were reviewed by the IARC working group, which noted that in the United States study there was a slightly raised mortality due to lung cancer among workers exposed to glass wool when compared to the local control cohort (107). Mortality increased with time from first exposure but not with duration of exposure or with cumulative exposure. None of these findings was statistically significant. In the European study, there was no excess mortality from lung cancer among workers exposed to glass wool when compared to regional rates. Based on their evaluation, the IARC working group concluded that there was inadequate evidence that glass wool and glass filaments were carcinogenic in humans, and there was limited evidence that rock wool and slag wool were carcinogenic in humans.

The potential health effects of refractory ceramic fibers (RCF) have become the focus of significant regulatory concern. In November, 1991 the EPA published notice that it would undertake a priority review of the potential health effects of RCF under section 4(f) of the Toxic Substance Control Act (109). This action was precipitated by an interim report of preliminary results of an ongoing multiple-dose lifetime inhalation toxicology study in rats and hamsters. According to EPA, this report indicated that inhalation of RCF caused dose-related nonmalignant pleural and pulmonary changes in rats. In addition, rats exposed to RCF developed more lung tumors than did control animals, and 42 percent of the hamsters developed mesotheliomas (113). An earlier study in which rats and hamsters were exposed to a commercial RCF by inhalation, or by intratracheal or intraperitoneal administration, indicated that the RCF was comparable to crocidolite asbestos in inducing fibrosis and mesotheliomas of the abdominal cavity following intraperitoneal administration (114). However, this material was not active in inducing fibrotic changes or lung tumors when administered by inhalation or intratracheal instillation. The important qualification to this was the appearance of a mesothelioma in one of 70 hamsters exposed via inhalation. Though not statistically significant, this finding is biologically significant in view of the relationship of this tumor to fiber toxicology. No systematic studies have been found of workers exposed to RCF but such studies are reported to be underway (109).

8.5 Hygienic Standards of Permissible Exposure

OSHA has not promulgated any standards for occupational exposure to MMMF. By default, exposure is limited to that for particulates not otherwise regulated, which is a TWA of 15 mg/m^3 for total dust or 5 mg/m^3 for the respirable fraction thereof (7). The ACGIH has established TLVs for fibrous glass and mineral wool fiber of 10 mg/m^3 TWA (3).

9 SILICON CARBIDE

9.1 Production, Uses, and Industrial Exposures

Silicon carbide, also known by the trade name Carborundum, has been manufactured and used as an abrasive material for over a century. It combines desirable properties of hardness and thermal resistance. It is produced by heating high-grade silica sand with finely ground carbon at 2400°C in an electric furnace (115). In its powdered or granular form, it has been used as the abrasive material in "emery" paper and wheels. It is also used as an abrasive in sandblasting and engraving. It has been incorporated into ceramics and glass and especially into refractory ceramic materials.

In recent years a crystalline form of silicon carbide, known as silicon carbide whiskers, has become an important industrial material. A silicon carbide whisker is a single crystal of silicon carbide having a cylindrical shape and an aspect ratio of greater than 3 and a diameter of less than 5 μm (116). Silicon carbide whiskers are used to impart strength and increased thermal resistance to structural materials that are used at high temperatures. Composite ceramics containing silicon carbide whiskers have been used in the manufacture of sandblasting nozzles, rocket motor nozzles, heat shields for reentry vehicles, and parts for nuclear reactor fuel assemblies.

9.2 Physiological Response

Silicon carbide dust has been considered to be relatively inert when inhaled. However, in recent years a number of publications have appeared, suggesting that inhalation of silicon carbide during its manufacture or use as an abrasive may result in pneumoconiosis. Individual cases were described in reports by Funahashi et al. (117), De Vuyst et al. (118), and Hayashi and Kajita (119). Peters et al. (120) found radiographic abnormalities and altered pulmonary function in 171 men employed in the manufacture of silicon carbide. Osterman et al. (121, 122) and Gauthier et al. (123) studied workers from this same plant and also reported decrements in pulmonary function, increased respiratory symptoms, and radiographic changes related to duration of exposure. Edling et al. (124), on the other hand, found no increases in total mortality, cancer mortality, or mortality from nonmalignant respiratory diseases among 521 men who manufactured abrasive materials using silicon carbide. Interpretation of the results from the studies by Peters et al. (120) and Osterman et al. (121, 122) is complicated by the fact that workers involved in the manufacture of silicon carbide are also exposed to sulfur dioxide and polycyclic aromatic hydrocarbons. Furthermore, the particulate matter to which they were exposed contained small quantities of quartz and cristobalite (125).

Durand et al. (126) have examined chest radiographs of 200 workers at a Quebec silicon carbide manufacturing plant. Twenty-eight had abnormal radiographs providing clinical evidence of pneumoconiosis, half of which were typical of pure silicosis. Examination of these workers over a 7-year period indicated that the condition did not appear to progress. These same authors examined the particulate materials to which these workers were exposed and conducted experiments in sheep designed to identify the agent that might cause the conditions that were seen (115). They discovered that the particulate matter in the air of the plant contained silicon carbide in both particulate and fibrous form. In the sheep model, silicon carbide particles were no more active than inert materials such as latex beads and graphite powder. The silicon carbide fibers, on the other hand, had fibrogenic activity that was comparable to that of crocidolite and chrysotile asbestos fibers. They concluded that workers in silicon carbide manufacturing plants may be exposed to silicon carbide fibers, which can contribute to the induction of interstitial lung disease.

There is little information available on the potential health effects of silicon carbide whiskers as distinguished from silicon carbide in its granular form. In order to begin to fill this gap, Lapin et al. (116) exposed 50 Sprague–Dawley rats per sex per group to an aerosol of silicon carbide whiskers in inhalation chambers for 6 hr/day, 5 days/week for 13 weeks. The mean mass concentrations of silicon carbide in the chamber air for the three treatment groups were 3.93, 10.7, and 60.5 mg/m^3 and the mean concentrations of whiskers were 630, 1746, and 7276 fibers/cm^3. The average dimensions of the fibers to which the treated rats were exposed were 4.41 μm long and 0.545 μm in diameter. The control group was exposed to filtered air under the same conditions as the treated groups. Half of the rats in each group were sacrificed at the end of the treatment period and the other half were maintained for a 26-week recovery period.

Silicon carbide whiskers accumulated in a dose-related manner in the lung tissue during the exposure period and resulted in a significant increase in lung weight for the rats in the highest exposure group compared to controls. The whiskers were most concentrated at the bifurcations of the alveolar ducts and respiratory bronchioles. Most whiskers were either engulfed by alveolar macrophages or located intracellularly in the interstitial tissues. Whiskers also accumulated in the bronchial and mediastinal lymph nodes. Whiskers were present in the interstitial lung tissues and lymph nodes after the 26-week recovery period. Histopathologically, there was evidence of inflammation in both the alveoli and in the lymph nodes. Bronchiolar, alveolar, and pleural thickening, focal pleural fibrosis, and reactive lymphoid hyperplasia were observed in treated rats, and the incidence and severity were dose-related. At the end of the 26-week recovery period, inflammation and lymph node hyperplasia had regressed but there was an increased incidence of alveolar and pleural thickening accompanied by a dose-related incidence of adenomatous hyperplasia of the lungs. Despite the relatively short exposure period in this study, the changes seen are consistent with early pulmonary responses to fibrogenic and carcinogenic mineral fibers, such as asbestos, and some of these changes did not appear to be reversible. Lapin et al. also noted that no "no-effect level" had been demonstrated in their study and that lower exposure levels would need to be examined to detect such a level.

In vitro and in vivo studies of silicon carbide whiskers deposited on the ciliated epithelium of the respiratory system indicated that the whiskers were swept to the nonciliated regions by ciliary action (127). There they penetrated the epithelial layers and caused cell damage and death. The observed cytotoxicity was similar to that of asbestos.

Intrapleural injections of 20 mg of silicon carbide whiskers once a month for 3 months caused a 47.7 percent incidence of pleural mesotheliomas in rats compared to a 34.1 percent incidence in rats treated with UICC chrysotile asbestos (128).

9.3 Hygienic Standards of Permissible Exposure

The OSHA standard for occupational exposure to silicon carbide dust is 10 mg/m^3 TWA for total dust containing less than 1 percent crystalline silica or 5 mg/m^3 TWA for respirable dust (7). This standard is the same as the ACGIH-recommended TLV for total dust (3). These standards and guidelines reflect the traditional view of this material as being relatively innocuous. It would appear that silicon carbide whiskers in particular, and perhaps silicon carbide dust also, may be more biologically active than is reflected by these numbers. Occupational exposure should be minimized while additional information is being developed.

10 OTHER SILICON COMPOUNDS

10.1 Silicon Halides

Of the four silicon tetrahalides (also known as tetrahalosilanes), only three, the fluoride, chloride, and bromide, are of commercial significance, with silicon tetrachloride being of greatest importance. The tetrafluoride is a gas at room temperature whereas the tetrachloride and tetrabromide are fuming liquids. They are prepared by the direct halogenation of pure quartz or silicon carbide at elevated temperature and pressure.

The primary use of silicon tetrachloride is as the starting material for the manufacture of high-purity silicon, amorphous fumed silica, and ethyl silicate. Silicon tetrafluoride is used as a starting material for fluosilic acid (H_2SiF_6) for water fluoridation.

The silicon tetrahalides are readily hydrolyzed to their corresponding hydrogen halides and silica upon contact with moisture. For this reason all three silicon tetrahalides are highly toxic by either inhalation or ingestion and can cause severe irritation of the skin and mucous membranes. In this respect their occupational hazards are qualitatively similar to those associated with the hydrogen halides. No TLVs or OSHA standards have been developed for these compounds.

10.2 Silanes

Silicon forms a large homologous series of silicon hydride compounds analogous to the alkane series of hydrocarbons. The simplest member of this series is silicon tetrahydride, better known as silane. Subsequent members of the series are disilane (Si_2H_6), trisilane (Si_3H_8), and so on. As with the hydrocarbons, individual hydrogen atoms may be replaced by certain functional groups such as halogens, aliphatic hydrocarbons, and hydroxyl groups to form parallel series of compounds.

The silicon–hydrogen bond in the silanes is much weaker than the corresponding carbon–hydrogen bond in aliphatic hydrocarbons. Therefore these compounds are much more reactive and undergo spontaneous oxidation in air. They also are readily hydrolyzed to silicic acid and silica in the presence of water. They decompose upon heating to liberate hydrogen and free silicon.

Because of their reactivity and specialized usage there is relatively little opportunity for significant occupational exposure to the silanes and their derivatives. Perhaps the most extensively used compounds are trichlorosilane and hexachlorodisilane. These compounds decompose upon contact with moisture to form hydrogen chloride. The primary hazard associated with silanes and chlorosilanes is their extreme flammability. Silane ignites spontaneously in air and trichlorosilane has a flash point of 7°F. Separate from their flammability, these compounds have a relatively low order of acute toxicity relative to other volatile inorganic hydrides. The LC_{50} of trichlorosilane for rats is 1000 ppm and the oral LD_{50} is 1030 mg/kg. Silane is less toxic than trichlorosilane.

The OSHA PEL for silane (silicon tetrahydride) is 5 ppm (7 mg/m³) TWA, which is identical to the ACGIH TLV for this compound (7, 3).

10.3 Silicones (Polydimethylsiloxanes)

The silicones are a large family of siloxane polymers having the basic structural unit:

$$\left[\begin{array}{c} R \\ | \\ -O-Si- \\ | \\ R \end{array} \right]_n$$

where R is most typically the methyl group ($-CH_3$). These materials are made by reducing purified quartz with carbon to form elemental silicon. The silicon is then heated in the presence of methyl chloride (CH_3Cl) vapor with a copper catalyst to form dichlorodimethylsilane. This can then be polymerized by carefully controlled hydrolysis. The resulting polymeric product is a mixture of polymers of different chain length normally distributed about a mean chain length. The length of the polymer chain can be controlled by adding specific proportions of end-blocking monomer units and carefully controlling the polymerization conditions. Linear polydimethylsiloxanes are liquids whose viscosities increase with increasing average molecular weight.

If some of the methyl side chains are replaced with a more reactive functional group such as hydride and vinyl groups, the linear chains may be cross-linked in the presence of a suitable catalyst. Cross-linking linear polysiloxane chains results in the formation of an increasingly solid material. Slightly cross-lined polymers have a gel-like consistency, whereas more heavily cross-linked materials are increasingly hard. Cross-linked silicone polymers are elastomers; that is, they tend to return to their original shape following deformation. Highly cross-linked dimethylsiloxane polymers are often referred to as silicone rubbers. Fillers are often incorporated into silicone elastomers to increase strength. The most common filler is finely divided amorphous silica, that is, fumed silica. The silica is incorporated between polymer chains through electrostatic interaction. The physical properties of both silicone fluids and elastomers can be altered by the substitution of different functional groups, such as phenyl or trifluoropropyl groups, for some of the methyl side chains. These may alter the permeability, strength, or oxidative and chemical resistance of the basic polymeric material.

Silicones have a very large number of applications. Relatively low-molecular-weight fluids of low viscosity (silicone oils) are used as pump fluids, lubricants, defoamers, and heat-exchange liquids. Silicone fluids are incorporated into adhesives and serve as release agents in other polymeric materials. Silicone elastomers find a wide range of uses in the manufacture of chemically resistant and biologically stable materials. Silicone is one of the least reactive synthetic materials that can be implanted in the human body so it is the material of choice for a number of implantable devices, including cardiovascular prostheses, shunts, and joint replacements. Significant attention has been focused in recent years on the use of silicone materials in mammary prostheses. Silicone elastomers have also found extensive use in the manufacture of tubing, O-rings, wire and cable, and linings of reactors and storage vessels.

The physical properties of silicones limit the potential for occupational exposure. Except for very low-molecular-weight polymers, that is, four monomer units or less, these materials have negligible vapor pressures at ambient temperatures. Thus there is little or no potential for inhalation exposure to vapors. The high viscosity of liquid silicones and the elastomeric properties of cross-linked silicone polymers minimize the likelihood of aerosol or dust generation when these materials are manipulated during manufacture or use, so there is little opportunity for inhalation exposure to respirable particulate forms of these materials. Silicones are resistant to chemical attack and are therefore neither absorbed nor metabolized when ingested. Dermal contact can occur.

Extensive toxicologic testing of silicone fluids and elastomers by injection or implantation has shown that these materials possess a low degree of toxicity. In a series of teratology tests performed in rats and rabbits, the administration of relatively low-molecular-weight polydimethylsiloxane fluids by dermal application or subcutaneous implantation resulted in inconsistent findings of increased fetal resorptions and skeletal defects at daily doses of 200 mg/kg or higher (129). Subsequent teratology studies of higher-molecular-weight silicones have failed to reveal any adverse effects on reproductive parameters.

Inhalation exposure of rats, mice, hamsters, and monkeys to relatively high concentrations (>100 ppm) of very low-molecular-weight silicone polymers, primarily octamethylcyclotetrasiloxane for 14 to 90 days resulted in increased liver weights in some studies (127). As indicated above, however, volatile compounds are unlikely to be present in significant quantities in commercially important silicone materials. In view of the limited potential for occupational exposure and the low degree of toxicity of silicones, it is not surprising that there are no recommended occupational exposure limits of hygienic standards for these materials.

REFERENCES

1. P. F. Holt, *Inhaled Dust and Disease*, Wiley, New York, 1987.

2. R. O. McClellan and R. F. Henderson, *Concepts in Inhalation Toxicology*, Hemisphere Publishing Corporation, New York, 1989.

3. *1991–1992 Threshold Limit Values for Chemical Substances and Physical Agents and Biological Exposure Indices*, American Conference of Governmental Industrial Hygienists, 1991.

4. R. B. Schlesinger, "Deposition and Clearance of Inhaled Particles," in *Concepts in Inhalation Toxicology*, R.O. McClellan and R. F. Henderson, Eds., Hemisphere Publishing Company, New York, 1989, p. 163.

5. P. E. Morrow, *Fundam. Appl. Toxicol.*, **10**, 369 (1988).

6. W. Runyan, "Pure Silicon," in M. Grayson and D. Eckroth, Eds., Kirk-Othmer *Encyclopedia of Chemical Technology*, 3rd ed., Wiley, 1982, Vol. 20, p. 826.

7. Limits for Air Contaminants, U.S. Occupational Safety and Health Administration, 29 CFR 1910.000, Table Z-1-A.

8. *NIOSH Pocket Guide to Chemical Hazards*, National Institute for Occupational Safety and Health, 1990.

9. T. D. Coyle. "Silica," in Kirk-Othmer Encyclopedia of Chemical Technology, 3rd ed., Vol. 20, Wiley, New York, 1982, p. 748.

10. *Silica and Some Silicates*, International Agency for Research on Cancer (IARC) Monographs on the Evaluation of the Carcinogenic Risk of Chemicals to Humans, Vol. 42, 1987.

11. *Sixth Annual Report on Carcinogens 1991 Summary*, National Toxicology Program, National Institute of Environmental Health Sciences, 1991.

12. *Quartz and Cristobalite in Workplace Atmospheres, Method Number ID-142*, U.S. Occupational Safety and Health Administration, Salt Lake City Analytical Laboratory, 1981.

13. *Criteria for a Recommended Standard . . . Occupational Exposure to Crystalline Silica*, National Institute for Occupational Safety and Health, 1974.

14. M. Ziskind, R. N. Jones, and H. Weill, *Am. Rev. Resp. Dis.*, **113**, 643 (1976).

15. M. Nakamura, K. Chiyotani, and T. Takishima, *Pulmonary Dysfunction in Pneumoconiosis*, MARUZEN Planning Network Co., Ltd., Japan, 1991.

16. D. R. Hemenway, M. P. Absher, L. Trombley, and P. M. Vacek, *Am. Ind. Hyg. Assoc. J.*, **51**, 363 (1990).

17. P. E. Morrow, *Toxicol. Appl. Pharmacol.*, **113**, 1 (1992).

18. D. B. Warheit, M. C. Carakostas, D. P. Kelly, and M. A. Hartsky, *Fundam. App. Toxicol.*, **16**, 590 (1991).

19. A. Sardi and A. Tomcsanyi, *Analyst*, **92**, 529 (1967).

20. F. Forastiere, S. Lagorio, P. Michelozzi, F. Cavariani, M. Arca, P. Borgia, C. Perucci, and O. Axelson, *Am. J. Ind. Med.*, **10**, 363 (1986).

21. H. E. Amandus, C. Shy, S. Wing, A. Blair, and E. F. Heineman, *Am. J. Ind. Med.*, **20**, 57 (1991).

22. C. Coombs, *Mining Eng.*, **42**, 560 (1990).

23. *Hazardous Substances Data Bank* (HSDB), An On-line Data File in the National Library of Medicine's Toxicology Data Network (TOXNET), 1991.

24. D. P. Kelly and D. P. Lee, *Toxicologist*, **10**, 202A (1990).

25. D. R. Hemenway, M. Absher, M. Landesman, L. Trombley, and R. Emerson "Differential Lung Response Following Silicon Dioxide Polymorph Aerosol Exposure," in *Silica, Silicosis and Cancer*, D. F. Goldsmith, D. M. Winn, and C. M. Shy, Eds., Praeger, New York, 1986, p. 105.

26. H. C. W. Skinner, M. Ross, and C. Frondel, *Asbestos and Other Fibrous Materials*, Oxford University Press, New York-Oxford, 1988.

27. *Standard Method for Testing for Asbestos Containing Materials by Polarized Light Microscopy*, Draft Report, D22.05 Committee, American Society for Testing Materials, Philadelphia, 1990.

28. R. Murray, *Brit. J. Ind.Med.*, **47**, 361 (1990).

29. M. Reisch, *Chem. Eng. News*, **68**, 10 (1990).

30. *Asbestos in Public and Commercial Buildings: A Literature Review and Synthesis of Current Knowledge*, Health Effects Institute–Asbestos Research, Cambridge, MA, 1991.

31. I. J. Selikoff, *Environ. Health Perspect.*, **88**, 269, (1990).

32. *Asbestos: Manufacture, Importation, Processing, and Distribution in Commerce Prohibitions; Final Rule*, U.S. Environmental Protection Agency, 54 FR 29460, July 12, 1989.

33. *OSHA Reference Method for Asbestos*, U. S. Occupational Safety and Health Administration, 29 CFR 1926.58, Appendix A.

34. *Mandatory Transmission Electron Microscopy Method*, U.S. Environmental Protection Agency, 40 CFR 763, Subpart E, Appendix A.

35. *Occupational Exposure to Asbestos, Tremolite, Anthophyllite, and Actinolite; Final Rules*, U.S. Occupational Safety and Health Administration, 51 FR 22612, June 20, 1986.

36. P. J. Landrigan and H. Kazemi, Ed., *Ann. N.Y. Acad. Sci.*, **643**, 1 (1991).

37. R. C. Browne, J. A. Hoskins, and N. F. Johnson, Eds., *Mechanisms in Fiber Carcinogenesis*, Plenum Press, New York, 1991.

38. K. Browne, *Brit. J. Ind. Med.*, **43**, 145 (1986).

39. E. D. Acheson, M. J. Gardner, E. C. Pippard, and L. P.Grime, *Brit. J. Ind. Med.*, **39**, 344 (1982).

40. J. M. Dement, R. L. Harris, M. J. Symons, and C. M. Shy, *Am. J. Ind. Med.*, **4**, 399 (1983).

41. J. M. Dement, R. L. Harris, M. J. Symons, and C. M. Shy, *Am. J. Ind. Med.*, **4**, 421 (1983).

42. A. D. McDonalds, J. S. Fry, A. J. Wolley, and J. McDonald, *Brit. J. Ind. Med.*, **40**, 361 (1983).

43. J. C. McDonald, F. D. K. Liddell, G. W. Gibbs, G. E. Eyssen, and A. D. McDonald, *Brit. J. Ind. Med.*, **37**, 11 (1980).

44. W. J. Nicholson, I. J. Selikoff, H. Seidman, R. Lilis, and P. Formby, *Ann. N.Y. Acad. Sci.*, **330**, 11 (1979).

45. A. D. McDonald, J. S. Fry, A. J. Woolley, and J. C. McDonald, *Brit. J. Ind. Med.*, **41**, 151 (1984).

46. G. Piolatto, E. Negri, C. LaVecchia, E. Pira, A. Decarli, and J. Peto, *Brit. J. Ind. Med.*, **47**, 810 (1990).

47. W. Weiss, *J. Occup. Med.*, **19**, 737 (1977).

48. A. D. McDonald, J. S. Fry, A. J. Woolley, and J. C. McDonald, *Brit. J. Ind. Med.*, **39**, 368 (1982).

49. C. Robinson, R. Lemen, and J. K. Wagner, "Mortality Patterns, 1940–75, Among Workers Employed in an Asbestos Textile Friction and Packing Products Manufacturing Facility," in *Dust and Disease: Proceedings of the Conference on Occupational Exposures to Fibrous and Particulate Dust and Their Extention into the Environment*, R. Lemen and J. M Dement, Eds., Pathotox Publishers, New York, 1979, p. 131.

50. T. R. Mancuso and A. A. El-Attar, *J. Occup. Med.*, **9**, 147 (1967).

51. J. Peto, R. Doll, C. Hermon, W. Binns, R. Clayton, and T. Goffe, *Ann. Occup. Hyg.*, **29**, 305 (1985).

52. H. F. Thomas, I. T. Benjamin, P. C. Elwood, and P. M. Sweetnam, *Brit. J. Ind. Med.*, **39** 273 (1982).

53. C. G. Ohlson and C. Hogstedt, *Brit. J. Ind. Med.*, **42**, 397 (1985).

54. M. J. Gardner, P. D. Winter, B. Pannett, and C. A. Powell, *Brit. J. Ind. Med.*, **43**, 726 (1986).

55. E. D. Acheson, M. J. Gardner, P. D. Winter, and C. Bennett, *Int. J. Epidemiol.*, **13**, 3 (1984).

56. H. Seidman, I. J. Selikoff, and E. C. Hammond, *Ann. N.Y. Acad. Sci.*, **330**, 61 (1979).

57. B. K. Armstrong, N. H. de Klerk, A. W. Musk, and M. S. T. Hobbs, *Brit. J. Ind. Med.*, **45**, 5 (1988).

58. B. K. Wignall and A. J. Fox, *Brit. J. Ind. Med.* **39**, 34 (1982).

59. A. D. McDonald and J. C. McDonald, *Environ. Res.*, **17**, 340 (1978).

60. L. O. Meurman, R. Kiviloto, and M. Hakama, *Brit. J. Ind. Med.*, **31**, 105 (1974).

61. M. Kleinfield, J. Messite, and M. G. Zaki, *J. Occup. Med.*, **16**, 345 (1974).

62. S. H. Lamm, M. S. Levine, J. A. Starr, and S. L. Tirey, *Am. J. Epidemiol.*, **127**, 1202 (1988).
63. A. M. Alies-Patin and A. J. Vallerou, *Brit. J. Ind. Med.*, **42**, 219 (1985).
64. M. L. Newhouse and K. R. Sullivan, *Brit. J. Ind. Med.*, **46**, 176 (1989).
65. P. C. Elmes and M. J. C. Simpson, *Brit. J. Ind. Med.* **34**, 174 (1977).
66. M. M. Finkelstein, *Brit. J. Ind. Med.*, **40**, 138 (1983).
67. P. Enterline and V. Henderson, *J. Natl. Cancer Inst.*, **79**, 31 (1987).
68. I. J. Selikoff, E. C. Hammond, and H. Seidman, *Ann. N.Y. Acad. Sci.*, **330**, 91 (1979).
69. M. Kleinfeld, J. Messite, and O. Kooyman, *Arch. Environ. Health*, **15**, 177 (1967).
70. L. N. Kolonel, T. Hirohata, B. V. Chappell, F. V. Viola, and D. E. Harris, *J. Natl. Cancer Inst.*, **64**, 739 (1980).
71. M. L. Newhouse, G. Berry, and J. C. Wagner, *Brit. J. Ind. Med.*, **42**, 4 (1985).
72. E. Raffn, E. Lynge, K. Juel, and B. Korsgaard, *Brit. J. Ind. Med.*, **46**, 90 (1989).
73. R. Puntoni, M. Vercelli, F. Merlo, F. Valerio, and L. Santi, *Ann. N.Y. Acad. Sci.*, **330**, 353 (1979).
74. J. M. Hughes, H. Weill, and Y. Y. Hammad, *Brit. J. Ind. Med.*, **44**, 161 (1987).
75. J. A. Merchant, *Environ. Health Perspect.*, **88**, 287 (1990).
76. M. F. Stanton and C. Wrench, *J. Natl. Cancer Inst.*, **48**, 797 (1972).
77. M. Lippmann, *Environ. Health Perspect.*, **88**, 311 (1990).
78. J. M. G. Davis, "Experimental Studies on Mineral Fibre Carcinogenesis: An Overview," in *Mechanisms in Fibre Carcinogenesis*, R. C. Brown, J. A. Hoskins, and N. F. Johnson, Eds., Plenum Press, New York, 1991, p. 51.
79. K. R. Spurny, "Carcinogenic Effect Related to the Fiber Physics and Chemistry," in *Mechanisms in Fibre Carcinogenesis*, R. C. Brown, J. A. Hoskins, and N. F. Johnson, Eds., Plenum Press, New York, 1991, p. 103.
80. D. Liddel, "Gaps in Knowledge of Fibre Carcinogenesis: An Epidemiologist's View," in *Mechanisms in Fibre Carcinogenesis*, R. C. Brown, J. A. Hoskins, and N. F. Johnson, Eds., Plenum Press, New York, 1991. p. 3.
81. K. H. Kilburn, R. Lilis, H. A. Anderson, A. Miller, and R. H. Warshaw, *Am. J. Med.*, **80**, 377 (1986).
82. D. McFadden, J. L. Wright, B. Wiggs, and A. Churg. *Am. Rev. Respir. Dis.*, **133**, 372 (1986).
83. J. Hobson, B. Gilks, J. Wright, and A. Churg, *J. Natl. Cancer Inst.*, **80**, 518 (1988).
84. G. Berry, M. L. Newhouse, and P. Antonis, *Brit. J. Ind. Med.*, **42**, 12, (1985).
85. W. J. Nicholson, *Airborne Asbestos Health Assessment Update*, U.S. Environmental Protection Agency, Report Number EPA-600/8-84/003F, 1986.
86. F. D. K. Liddell and J. A. Handley, *Brit. J. Ind. Med.*, **42**, 389 (1985).
87. J. C. Wagner, "Mesotheliomas in Man and Experimental Animals," in *Mechanisms in Fibre Carcinogenesis*, R. C. Brown, J. A. Hoskins, and N. F. Johnson, Eds., Plenum Press, New York, 1991, p. 45.
88. H. Weill, J. Hughes, and C. Waggenspack, *Am. Rev. Respir. Dis.*, **20**, 345 (1979).
89. G. Berry and M. L. Newhouse, *Brit. J. Ind. Med.*, **40**, 1 (1983).
90. T. H. Eyde, *Mining Eng.*, **42**, 582 (1990).

91. J. C. Wagner, J. W. Skidmore, R. J. Hill, and D. M. Griffiths, *Br. J. Cancer*, **51**, 727 (1985).

92. L. W. Choate, *Mining Eng.*, **41**, 424 (1989).

93. F. W. Pereira, *Mining Eng.*, **42**, 578 (1990).

94. K. C. Rieger, *Mining Eng.*, **42**, 571 (1990).

95. W. H. Stewart, *Mining Eng.*, **41**, 413 (1989).

96. R. C. Schiek, "Pigments (Inorganic)," in Kirk-Othmer *Encyclopedia of Chemical Technology*, 3rd ed., Vol. 17, M. Grayson and D. Eckroth, Eds., Wiley, New York, 1982, p. 807.

97. American Conference of Governmental Industrial Hygienists, *Appl. Occup. Environ. Hyg.*, **6**, 1044 (1991).

98. D. W. L. Spry, *Mining Eng.*, **41**, 410 (1989).

99. F. G. Anderson, *U.S. Bur. Mines, Rep. Invest. No. 5199*, 1956.

100. *Technical Data Sheet No. 1-1*, Perlite Institute, New York, 1962.

101. J. R. Larson, *Mining Eng.*, **41**, 415 (1989).

102. G. C. Presley, *Mining Eng.*, **41**, 417 (1989).

103. M. J. Potter, *Mining Eng.*, **40**, 425 (1988).

104. W. D. Toal, *Mining Eng.*, **41**, 403 (1989).

105. A. M. Potts, "Toxic Responses of the Eye," in *Casarett and Doull's Toxicology: The Basic Science of Poisons*, 4th ed., M. O. Amdur, J. Doull, and C. D. Klaassen, Eds., Pergamon Press, New York, 1991, p. 524.

106. J. S. Falcone, Jr., "Synthetic Inorganic Silicates," in Kirth-Othmer *Encyclopedia of Chemical Technology*, 3rd ed., Vol. 20, M. Grayson and D. Eckroth, Eds., Wiley, New York, 1982, p. 855.

107. *Man-Made Mineral Fibers and Radon*, International Agency for Research on Cancer Monographs on the Evaluation of the Carcinogenic Risk of Chemicals to Humans, Vol. 43, 1988.

108. *Environmental Health Criteria 77: Man-made Mineral Fibres*, World Health Organization, Geneva, 1988.

109. *Refractory Ceramic Fibers; Initiation of Priority Review: Notice*, U.S. Environmental Protection Agency, 56 FR 58693, November 21, 1991.

110. R. Petersen and S. Sabroe, *Am. J. Ind. Med.*, **20**, 113 (1991).

111. P. E. Enterline, G. M. March, V. Henderson, and C. Callahan, *Ann. Occup. Hyg.*, **31**, 625 (1987).

112. L. Simonato, A. C. Fletcher, J. W. Cherrie, A. Andersen, P. Bertazzi, N. Charnay, J. Claude, J. Dodgson, J. Esteve, R. Frentzel-Beyme, M. J. Gardner, O. Jensen, J. Olsen, L. Teppo, R. Winkelmann, P. Westerholm, P. D. Winter, C. Zocchetti, and R. Saracci, *Ann. Occup. Hyg.*, **31**, 603 (1987).

113. Anonymous, "Fiber Flap," *Science*, **255**, 1356 (1992).

114. D. M. Smith, L. W. Ortiz, R. F. Archuleta, and N. F. Johnson, *Ann. Occup. Hyg.*, **31**, 731 (1987).

115. R. Begin, A. Dufresne, A. Cantin, S. Masse, P. Sebastien, and G. Perrault, *Chest*, **95**, 842 (1989).

116. C. A. Lapin, D. K. Craig, M. G. Valerio, J. B. McCandless, and R. Bogoroch, *Fundam. Appl. Toxicol.*, **16**, 128 (1991).

117. A. Funahashi, D. P. Schlueterk, K. Pintar, K. A. Siegesmund, G. S. Mandel, and N. S. Mandel, *Am. Rev. Respir. Dis.*, **129**, 635 (1984).

118. P De Vuyst, R. V. Weyer, A. DeCoster, F. X. Marchandise, P. Dumortier, P. Ketelbant, J. Jedwab, and J. C. Yernault, *Am. Rev. Respir. Dis.*, **133**, 316 (1986).

119. H. Hayashi and A. Kajita, *Am. J. Ind. Med.*, **14**, 145 (1988).

120. J. M. Peters, T. J. Smith, L. Bernstein, W. E. Wright, and S. K. Hammond, *Brit. J. Ind. Med.*, **41**, 109 (1984).

121. J. W. Osterman, I. A. Greaves, T. J. Smith, S. K. Hammond, J. M. Robins, and G. Theriault, *Brit. J. Ind. Med.*, **46**, 629 (1989).

122. J. W. Osterman, I. A. Greaves, T. J. Smith, S. K. Hammond, J. M. Robins, and G. Theriault, *Brit. J. Ind. Med.*, **46**, 708 (1989).

123. J. J. Gauthier, H. Ghezzo, and R. R. Martin, *Am. Rev. Respir. Dis.*, **131**, A191 (1985).

124. C. Edling, B. Jarvholm, L. Andersson, and O. Axelson, *Brit. J. Ind. Med.*, **44**, 57 (1987).

125. T. J. Smith, S. K. Hammond, F. Laidlaw, and S. Fine, *Brit. J. Ind. Med.*, **41**, 100 (1984).

126. P. Durand, R. Begin, L. Samson, A. Cantin, S. Masse, A. Dufresne, G. Perreault, and J. Laflamme, *Am. J. Ind. Med.*, **20**, 37 (1991).

127. G. L. Vaughn, J. R. Kennedy, and S. A. Trently, *Environ. Res.*, **56**, 178 (1991).

128. L. A. Vasileva, L. N. Pylev, V. V. Kiianenko, and A. A. Nikolaishville, *Eksp. Onkol.*, **11**, 13 (1989); through MEDLINE, a bibliographic data base of the National Library of Medicine's MEDLARS On-line System.

129. "Biocompatibility Studies of Materials of Composition," in *Dow Corning Wright Silastic Gel Saline Mammary Implant H.P. and Silastic MSI Gel Saline Mammary Implant H.P. Premarket Approval Application*, 1991, unpublished.

Subject Index

Chemical Index